ACCIDENT PREVENTION MANUAL FOR BUSINESS & INDUSTRY

Administration & Programs
14TH EDITION

EDITORS:

Philip E. Hagan, JD, MBA, MPH, ARM, CIH, CET, CHMM, CHCM, CHSP, CEM

John F. Montgomery, PHD, CSP, CHMM

James T. O'Reilly, JD

National Safety Council
Itasca, IL

NSC Press Editor: Deborah Meyer
Cover Design, Interior Design, and Composition: Jennifer Villarreal
Executive Director Publications: Suzanne Powills
Cover Photo: Ingram Publishing/Thinkstock

Copyright, Waiver of First Sale Doctrine

The National Safety Council's materials are fully protected by the United States copyright laws and are solely for the noncommercial, internal use of the purchaser. Without the prior written consent of the National Safety Council, purchaser agrees that such materials shall not be rented, leased, loaned, sold, transferred, assigned, broadcast in any media form, publicly exhibited or used outside the organization of the purchaser, reproduced, stored in a retrieval system or transmitted in any form or by any means, electronic, mechanical, photocopying, recording, or otherwise. Use of these materials for training for which compensation is received is prohibited, unless authorized by the National Safety Council in writing.

Disclaimer

Although the information and recommendations contained in this publication have been compiled from sources believed to be reliable, the National Safety Council makes no guarantee as to, and assumes no responsibility for, the correctness, sufficiency, or completeness of such information or recommendations. Other or additional safety measures may be required under particular circumstances.

Copyright © 1946, 1951, 1955, 1959, 1964, 1969, 1974, 1981, 1988, 1992, 1997, 2001, 2009, 2015
by the National Safety Council
All Rights Reserved
Printed in the United States of America
28 27 26 25   10 9 8 7 6 5 4 3

Library of Congress Cataloging-in-Publication Data
Accident prevention manual for business & industry. Administration & programs/Philip E. Hagan, John F. Montgomery, James T. O'Reilly. — 14th edition.
 pages cm
 Includes bibliographical references and index.
 ISBN 978-0-87912-321-5
 1. Industrial safety—United States—Handbooks, manuals, etc. 2. Accidents—United States—Prevention—Handbooks, manuals, etc. I. Hagan, Philip (Philip E.), editor. II. Montgomery, John F. (Johnny Franklin), 1944– editor. III. O'Reilly, James T., 1947– editor. IV. Title: Administration & programs. V. Title: Accident prevention manual for business and industry. Administration and programs.
  T55.A33325 2015
  658.3'82—dc23
   2014045957
Product Number: 121580000

CONTENTS

Preface v
New and Revised Material v
Definitions of Terms vi
Acknowledgments vi
Contributors vi

PART 1 INTRODUCTION TO SAFETY AND HEALTH 1

1 Historical Perspectives 3
2 The Safety, Health, and Environmental Professional 25
3 Safety Culture 33
4 Regulatory History 47
5 Legal and Regulatory Issues for the Safety Manager 97

PART 2 LOSS CONTROL INFORMATION AND ANALYSIS 129

6 Loss Control Programs 131
7 Safety, Health, and Environmental Auditing 167
8 Workers' Compensation 183
9 Identifying Hazards 199
10 Incident Investigation, Analysis, and Costs 237
11 Injury and Illness Record Keeping, Incidence Rates, and Analysis 265

PART 3 SAFETY/HEALTH/ENVIRONMENT PROGRAM ORGANIZATION 293

12 Occupational Health Programs 295
13 Industrial Hygiene Program 317
14 Environmental Management 329
15 Indoor Air Quality 353

16	Ergonomics Yesterday, Today, and Tomorrow	391
17	Employee Assistance Programs	427
18	Emergency Preparedness	439
19	Workplace Violence	467
20	Product Safety Management	489
21	Industrial Sanitation and Personnel Facilities	495
22	Occupational Medical Surveillance	515
23	Workers with Disabilities	525
24	Retail/Service Facilities Logistics	547
25	Transportation Safety Programs	565
26	Office Safety	597
27	Laboratory Safety	617
28	Contractor and Customer Safety	649

PART 4 PROGRAM IMPLEMENTATION AND MAINTENANCE 673

29	Homeland Security Compliance in the Workplace	675
30	Motivation	681
31	Safety and Health Training	705
32	Media	731
33	Safety Awareness Programs	755

APPENDIX 1 SOURCES OF HELP 791

APPENDIX 2 BIBLIOGRAPHY 797

Index 803

PREFACE

The 14th edition of the *Accident Prevention Manual for Business & Industry: Administration & Programs* continues a tradition begun in 1946 with the publication of the first *Accident Prevention Manual*. This Manual brings to the safety/health/environmental professional the broad spectrum of topics, specific hazards, best practices, control procedures, resources, and sources of help known in the field today. To accommodate the expansion of knowledge and topics, the Manuals are now printed in five volumes: *Administration & Programs* (14th edition), *Engineering & Technology* (14th edition), *Environmental Management* (3rd edition), *Security Management* (2nd edition), and *Accident Prevention Manual Essentials* (1st edition).

This 14th edition builds on the excellent work of previous contributors to the National Safety Council's flagship series. Volunteer experts from many different subject areas have come together to make this book an important resource to be used in support of safety programs and related education. In addition to the expertise of National Safety Council volunteers and staff, we have received expert assistance in developing, writing, and reviewing from contributors representing various disciplines and from the editors, Philip E. Hagan, John F. Montgomery, and James T. O'Reilly. If you have different ideas and want them to be considered for the 15th edition, your suggestions are welcome and can be sent to the National Safety Council, 1121 Spring Lake Drive, Itasca, IL 60143, attn. Deborah Meyer or deborah.meyer@nsc.org.

The audience served by this textbook is widespread. Safety professionals with years of experience, individuals new to the field, managers tasked with safety responsibilities, and educators preparing students for careers in the field of safety will find that these volumes are a valuable source of information. Those who work in the fields of risk management and loss control, human resources, and engineering will also find programs and information that can be incorporated successfully into working goals and objectives that will add value to any organization's safety program.

NEW AND REVISED MATERIAL

The Accident Prevention Manuals are intended for a wide range of users: for students using them as textbooks, for corporate or company managers searching for solutions to safety and health problems, for new safety specialists who must plan and organize a safety and health program within a company, or for experienced safety professionals seeking to improve an operating program and to learn more about advances in the field of safety and health. To increase their usefulness, the 14th editions of the *Administration & Programs* volume and the *Engineering & Technology* volume contain new chapters as well as completely revised material in all chapters.

All chapters in both volumes were reviewed, revised, and updated by safety professionals with expertise in the specific subject area. In addition to a new layout and two-column design of the textbook, major changes include:

Administration & Programs volume:
- Chapter 3: Safety Culture—updated and expanded
- Chapter 5: Legal and Regulatory Issues for the Safety Manager—extensively rewritten
- Chapter 11: Injury and Illness Record Keeping, Incidence Rates, and Analysis—new data and analysis added
- Chapter 14: Environmental Management—extensively rewritten
- Chapter 19: Workplace Violence—updated and expanded
- Chapter 21: Industrial Sanitation and Personnel Facilities—updated and moved from *Engineering & Technology* volume
- Chapter 22: Occupational Medical Surveillance—updated and moved from *Engineering & Technology* volume
- Chapter 23: Workers with Disabilities—updated and moved from *Engineering & Technology* volume
- Chapter 25: Transportation Safety—extensively rewritten
- Chapter 29: Homeland Security Compliance in the Workplace—updated and expanded
- Chapter 30: Motivation—extensively rewritten
- Chapter 31: Safety and Health Training—extensively rewritten
- Chapter 32: Media—extensively rewritten

Engineering & Technology volume:
- Chapter 6: Safeguarding—updated and expanded
- Chapter 11: Nanomaterials in the Workplace—updated and expanded
- Chapter 19: Welding and Cutting—extensively rewritten
- Chapter 21: Working with Hot and Cold Metals—updated and combines two metals chapters into one
- Chapter 24: Process Safety Management—updated and moved from *Administration & Programs* volume

- Chapter 25: Aviation Safety—new chapter
- Chapter 26: Oil and Gas Safety—new chapter
- Chapter 27: Waste and Recycling Safety—new chapter

DEFINITIONS OF TERMS

As the concerns and responsibilities of safety/health/environmental professionals expand, so must their ability to communicate and educate. Technical terms are defined in the text where they are used and also in Appendix 3, Glossary, in the *Engineering & Technology* volume. However, the terms *incident* and *accident* deserve a special note. In the years since the original publication of this manual, many theories of accident causation and definitions of the term *accident* have been advanced. The National Safety Council continues to work to increase awareness that an *incident* is a near-accident and that so-called *accidents* are not random events but rather preventable events. To that end, the term *incident* is used in its broadest sense to include incidents that may lead to property damage, work injuries, or both. The following definitions are generally used in this manual:

- **Accident**: That occurrence in a sequence of events that produces unintended injury, death, or property damage. Accident refers to the event, not the result of the event (see unintentional injury).
- **Incident**: An unintentional event that may cause personal harm or other damage. In the United States, OSHA specifies that incidents of a certain severity be recorded.
- **Near-miss incident**: For purposes of internal reporting, some employers choose to classify as "incidents" the near-miss incident; an injury requiring first aid; the newly discovered unsafe condition; fires of any size; or nontrivial incidents of damage to equipment, building, property, or product.
- **Unintentional injury**: The preferred term for accidental injury in the public health community. It refers to the result of an accident.

With proper hazard identification and evaluation, management commitment and support, preventive and corrective procedures, monitoring, evaluation, and training, unwanted events can be prevented.

ACKNOWLEDGMENTS

General Editor Phil Hagan thanks his many friends who still remain friends even when he disappears for long periods of time to work on book projects and such—Bil and Sharon, Paul and Shenfen, Fannie and Michael, Enrique and Michael, Dale and Miyoko, Gloria, Aster, Larry, Don, and Randy. He also thanks Deborah Meyer for keeping the team on the straight and narrow—always moving forward and keeping the team focused on getting to press.

General Editor John Montgomery thanks his wife, Karen, and the memory of Christopher for their support and encouragement during the editing and rewriting process. He also wishes to thank all chapter contributors and reviewers for their diligent review and manual updates, with special thanks to Teddy Gil, Curt Lewis, and JB Gregory for their willingness to take on the task of writing new chapters. He also wishes to thank Air Serv Corp/ABM for its support and encouragement during the review of the manuals.

General Editor Prof. O'Reilly thanks Richard Hackman, Jack Hulon, Charles Geraci, and Jack McAneney for their insights on these complex issues. His research assistant, Marina Schemmel, University of Cincinnati Law Class of 2015, provided exceptional aid in locating and digesting source materials. He also thanks his family—Carol, Jessie, and CB—for their support and encouragement.

Thank you for taking the time to consider and utilize some of these ideas. Of course, none of the book's comments take the place of legal advice, medical advice, or professional advice. Please be certain to discuss the contents of this text with the appropriate professional advisers. We do not offer this as a substitute for the timely, prudent expertise of your organization's regular advisers.

CONTRIBUTORS

The following safety, health, and environmental professionals have contributed to the 14th editions as editors, writers, and/or reviewers of chapters or sections. The National Safety Council very much appreciates the dedication and professional expertise they have contributed to the cause of safety, health, and environmental education.

Alan Barr, co-founder of ErgoMek, LLC, has been a Development Engineer for the UCB/UCSF Ergonomics Program since 2001. He has extensive knowledge in design, CAD modeling, and fabrication of mechanical interventions designed to reduce injury in the workplace. He received a BS in Biomechanics and an MS in Biomedical Engineering from the University of California at Davis in 2000.

Stephen Bennett, ARM, is co-founder of a web-based membership cooperative that brings together a network of independent safety and health consultants, providing a wide range of safety, loss prevention, regulatory management, industrial hygiene, workers' compensation, and return-to-work solutions to public and private employers. Previously, Bennett has held regional to global consulting leadership positions at Sedgwick CMS; Marsh, Johnson & Higgins; and The Travelers. E-mail: smbennett54@sbcglobal.net.

Jairo Betancourt, has more than 25 years of experience in biomedical research and laboratory safety. His experiences include designing, implementing and managing laboratory safety programs for universities and research institutions both domestically and internationally.

He is an active member of the American Biological Safety Association (ABSA). Currently he is involved in the International Working Biosafety Group (IBWG, www.internationalbiosafety.org), the ABSA Philanthropy task force, and co-editor of the Biosafety Compendium. He frequently conducts education and training programs on biosafety and biosecurity in Spain and Latin American countries throughout Central and South America (Colombia, Mexico, Dominican Republic, Venezuela, and Argentina).

Stephen Blackwell, MS, REHS, is an environmental health consultant retired from the U.S. Public Health Service Commissioned Corps. He has extensive experience in the environmental health field with the cruise ship industry, Centers for Disease Control, the Indian Health Service, the U.S. Coast Guard, and the Agency for Toxic Substances and Disease Registry, a federal public health agency of the U.S. Department of Health and Human Services. He has a BS from East Carolina University and an MPH from Florida International University. E-mail: srb0@comcast.net.

Janice Comer Bradley, MS, CSP, is Senior Vice President, National Waste & Recycling Association, where she manages the program group addressing safety, ANSI standards, technical issues, statistics, and education for the waste and recycling industry. She is also the past Vice President of the International Safety Equipment Association, where she managed the activities of Fortune 500 companies involved in supplying safety and health equipment and services. Prior to her work at ISEA, she was the director of environmental health and safety for the Rockefeller University in New York City, the university health and safety officer for Brown University, and the safety specialist for the Department of Veterans Affairs Medical Center in Dayton, Ohio. Bradley is a member of the National Academy of Science, Institute of Medicine, and Research Programs Board on Health Sciences Policy, and she lectures at Georgetown University. She earned a BS from the University of Dayton and a master's degree in environmental studies from Brown University. E-mail: jcbmarch@yahoo.com.

Thomas Bush is a safety professional with close to 30 years of extensive experience in the transportation, manufacturing, insurance, training, aviation, ground handling, and construction industries. He has worked as an Incident Commander on numerous emergency response projects and as a Project Manager for large environmental remediation projects in North America. Bush has been a Professional Member of the American Society of Safety Engineers for the past 28 years. Tom is an Instructor for the Texas A&M University Engineering Extension Service.

Salvatore Caccavale, CHMM, CPEA, is Corporate Senior Manager Environmental, Health and Safety at A.M. Castle & Co., 3400 North Wolf Road, Franklin Park, Illinois 60131, (847) 349-2601.

Denis Clark, MS, PE, is a consulting Welding Engineer. He holds a BS from Cornell University and an MS from The Ohio State University, has a PE license in Metallurgical and Materials Engineering from the State of Idaho, and is an AWS Certified Welding Inspector. He has worked in the welding area for 40 years—most of that time at the Idaho National Laboratory. His areas of research have included welding process sensing and control, spray forming, cupola furnace control, the weldability of new alloys, and the application of standard AWS and ASME codes to the welding of nuclear components. Clark is an adjunct professor at Montana Tech University in Butte and has also taught as an adjunct professor at the University of Idaho. He is currently Chair of the AWS Safety and Health Committee. E-mail: denis.clark.51@gmail.com or www.declark-engineering.com.

Robert Wayne Clifton, CSP, PE, ALCM, CPCU, CIE, is presently working as the Global AVP for ESIS Inc. In his current role, he manages a staff of safety professionals in the United States, United Kingdom, Singapore, China, and Malaysia. Clifton has more than 37 years of experience as a safety consultant. He holds a master's degree in safety and health. E-mail: wayne.clifton@esis.com.

Patrick J. Conroy, OHST, CHST, is Special Assistant to the Board of Directors Council on Certification of Health, Environmental and Safety Technologists, 208 Burwash Avenue, Savoy, IL 61874-9571, Office (239) 599-8907. E-mail: pat@cchest.org.

John DeLaHunt, MBA, ARM, has managed environmental, health, safety, and risk issues in higher education since 1989. At Colorado College, he launched a comprehensive EHS program. At the University of Texas at San Antonio, he manages property insurance, workers' compensation, property conservation, and fire protection while serving as the university's Fire Marshal and Risk Manager.

DeLaHunt writes a bimonthly column for the *Journal of Chemical Health & Safety*, edited both editions of the *Environmental Compliance Assistance Guide for Colleges & Universities*, and has presented on diverse topics at dozens of conferences. He has been published in the *Journal of Chemical Health and Safety* and the Association of Higher Education Facilities Officer's Body of Knowledge. He was the charter President of the College and University Hazardous Waste Conference. He serves on the University Risk and Insurance Management Association's government relations and affairs committee and the editorial board for the *Journal of Chemical Health and Safety*.

DeLaHunt holds a bachelor's degree in Chemistry from Colorado College and an MBA in Finance and Management from the University of Colorado–Colorado Springs. E-mail: john_delahunt@msn.com.

Jane Dolezal is the Safety and Compliance Manager for Homewood Disposal Service. She has more than 20 years of combined experience in both the regulatory and waste and the recycling sectors involving the implementation and management of DOT, EPA, and OSHA compliance programs. She is a member of several ANSI committees and has a BS in Environmental and Natural Resources Policy and Studies from Michigan State University.

J. Nigel Ellis, PhD, PE, CSP, CPE, can be contacted at Ellis Fall Safety Solutions, 306 Country Club Drive, Wilmington, DE 19803. E-mail: www.Fall-Safety.com or efss@fallsafety.com.

Michael J. Fagel, PhD, CEM, has more than three decades of public service. He has been in fire service, emergency medical service, public health, law enforcement, and emergency management, as well as corporate safety, security, and threat risk management. Fagel is currently an instructor at the University of Chicago in its new Masters of Threat Risk Management program. He is also an instructor for Benedictine University's Masters in Public Health Program, as well as an instructor at Eastern Kentucky University in its Loss Prevention Masters Program. He was a team leader at the Louisiana State University's National Center for Bio Medical Research and training in its Response to Agricultural Terrorism Training program, as well as its Public Health Programs in Response to High Consequence Events. Fagel has delivered more than 350 lectures and has written more than 100 articles on safety and disaster planning. He has published two textbooks on safety and disaster management and was on the National Domestic Preparedness Office SLAG team at the FBI.

Fagel spent 10 years with the Federal Emergency Management Association in its Occupational Safety & Health Cadre, responding to incidents and disasters such as the Oklahoma City bombing, where he worked as a safety officer and CISD Debriefer. He has also consulted on terrorism-related issues at home and abroad. Fagel can be reached at PO Box 211, Sugar Grove, IL 60554, (630) 907-2020. E-mail: mjfagel@aol.com.

Cristine Z. Fargo has held the position of Manager, Standards Programs for the International Safety Equipment Association for close to 20 years. She manages the voluntary standards setting activities of 13 product groups representing manufacturers and suppliers of safety and health equipment. She represents the ISEA on numerous industry standards committees and works with federal regulatory agencies and outside standards bodies to influence activities that affect the manufacture, use, and distribution of safety equipment. In addition, Fargo speaks at various industry functions about ISEA and its role in the safety equipment business. She holds a degree in Political Science from West Virginia University and can be reached at the ISEA, (703) 525-1695. E-mail: cfargo@safetyequipment.org.

Dave Felinski, Safety Director, Association of Manufacturing Technology, 7901 Westpark Drive, McLean, VA 22102, (703) 893-2900. E-mail: www.amtonline.org.

Anne M. Germain, PE, BCEE, has been the Director of Waste & Recycling Technology for the National Waste & Recycling Association since September 2013. Prior to that, she was the Chief of Engineering and Technology for the Delaware Solid Waste Authority. In addition, she is a Past President of the Solid Waste Association of North America. She has written more than 20 papers and has presented nationally and internationally on solid waste matters. Germain is a Professional Engineer and a Board Certified Environmental Engineer. She has been active in ABET, evaluating environmental engineering programs for accreditation. She graduated from Virginia Tech with a BS in Civil Engineering and received her master's, also in Civil Engineering, from the University of Delaware. E-mail: amgermain@wasterecycling.org.

Teddy Gil, BME, MBA, ASNT Level III ET, FAA A&P License, has held varying positions throughout his career. Most recently, he has served as Process Improvement & Engineering Consultant for Delta Air Lines, implementing executive leadership for short fuse interior configuration upgrades and fabrication; Vice President of Business Development & Quality for Global Integrated Security Services, providing risk assessment consulting and training; and Vice President of People Development and Training for Air Serv Corporation. He also led the Delta Air Lines Technical Operations Employee Council responsible for implementing positive improvement and positive cultural changes throughout the organization for the more than 12,000 Technical Operations employees. Gil was also the presiding officer of the Conflict Resolution Program from 2002 to 2004 for Delta Air Lines Technical Operations. In addition, Gil has published extensively, including several book chapters for the National Safety Council.

Allen Gilley, CIH, CSP, ARM, ALCM, is a Managing Consultant in the Marsh Insurance Company's Atlanta, Georgia, office. He is responsible for the coordination and delivery of professional loss-control services to his clients, which include a wide variety of manufacturers, broadcast and print media, public entities, health care providers, contractors, restaurants, distributors, property management firms, and service organizations. Gilley is manager of the Atlanta Workforce Strategies Group, practice leader of the Southern Regional Workforce Strategies Group, and national Safety and Health practice leader. He also serves as the chairman of the Marsh Atlanta Restaurant Roundtable and chairman of the Marsh Newspaper/Media Risk Control Roundtable. He is also the MRC Practice Leader for Hospitality/Restaurants. E-mail: allen.m.gilley@marsh.com.

JB Gregory, MEd, is an experienced trainer of OSHA regulations for both general industry and construction. He has more than 28 years of experience working with federal, state, and local agencies as well as private employers in the field of regulatory compliance. Gregory has worked with both national and international clients in companies ranging from 2 to more than 800,000 employees in both union and nonunion environments. Services provided to past clients/employers include EHS project management and oversight; EHS program evaluation, development, and implementation; EHS audits; and EHS training.

Gregory has worked with various oil and gas companies in both upstream and downstream EHS operations and has conducted Hazardous Waste Operations Training for federal, state, and local agencies, as well as private employers. He has been on staff at the Texas A&M Engineering Extension Service, OSHA Southwest Education Center, for 15 years. He received a BA and a MEd with an emphasis in Training and Development from the University of Oklahoma.

Von M. Griggs-Laws has 28 years of experience as a safety professional with practical, hands-on occupational safety and health experience in such areas as aircraft maintenance, medical facilities, transportation, civil engineering, communications, supply/warehousing, petroleum distribution, food services, retail, and other industrial and administrative functions. Griggs-Laws is certified as an OSHA Instructor—General Industry & Construction Safety, Oil & Gas, and is a Certified Safety & Health Official/CSHO Texas A&M; a Certified Government Environmental Specialist/World Safety Organization; a Safety Trained Supervisor/SCHEST; a Drug & Alcohol Defensive Driving Instructor/National Safety Council; a Construction Engineering Concepts/Turner Construction Instructor; and a Work Zone Trainer/MUTCD and TxDOT. She is an Adjunct Instructor in Health & Safety for Texas A&M University, Texas Engineer Extension Service (TEEX).

William Grimes, CSP, is currently the Vice President of Safety and Security for CitationShares, a division of Cessna Aircraft Company that provides fractional jet service. Grimes is a CSP with more than 25 years of safety and health experience and also holds an FAA Airline Transport Pilot rating and type ratings in the Citation 550 and Citation 560 Excel aircraft. Prior to CitationShares, Grimes worked for TAG Aviation and Marsh Risk Consulting as a managing director. He was recognized by the Flight Safety Foundation with the President's Safety Citation award in 2007. E-mail: wgrimes@citationshares.com.

Richard J. Hackman, CIH, QEP, is Associate Director at The Procter & Gamble Co. Hackman has been with P&G for 28 years. He directs the P&G North America Regulatory and Technical Relations organization, which is responsible for regulatory compliance, regulatory/legislative influence, and technical external relations activities across all products and operations in the region. Hackman has a BS in Biology from the University of Cincinnati, an MS in Environmental Sciences from the University of Cincinnati, and an MBA in Finance from Xavier University. He is certified by the American Board of Industrial Hygiene in Comprehensive Practice, as well as by the Institute of Professional Environmental Practice as a Qualified Environmental Professional. E-mail: hackman.rj@pg.com.

Philip E. Hagan, JD, MBA, MPH, ARM, CIH, CHMM, CHCM, CET, CHSP, CEM, is Assistant Professor in the Department of Human Sciences at Georgetown University, Washington DC, and National Practice Leader for Partner Engineering and Science. In addition, he is a practicing attorney specializing in tort, safety, environmental, and business-related law issues. He is the former Director of Safety & Environmental Management at Georgetown University. He has consulted internationally on risk, safety, environmental, and emergency management issues in Italy, Qatar, India, Turkey, and China. Hagan has co-authored texts on environmental and workplace safety, legal liability, training, and indoor environmental quality; and he has been a general editor for the *Accident Prevention Manual for Business & Industry* for three editions. He was also the lead editor on the new *Accident Prevention Manual Essentials* (2014). In addition, he has presented to a diverse group of audiences on subjects ranging from hazardous waste disposal to business continuity and emergency management. He is a member of several American Bar Association committees dealing with environmental issues in business transactions, toxic torts, workers' compensation, and international environmental law. He has been a peer reviewer for various safety-related ANSI consensus standards and guidelines. He holds a BS from East Carolina University in Environmental Health, an MPH from George Washington University, an MBA from Georgetown University, and a JD from the George Mason University School of Law. E-mail: haganp@georgetown.edu.

Valienti Antonio Henry, MBA, is board certified in safety and a Certified Lean-Six Sigma Black Belt. Currently, he is a senior manager of Loss Prevention and Reduction at the University of Miami. In addition, Henry also maintains a consulting practice in both the private and public sectors, which includes the development of loss control and risk management programs using specialized Lean-Six Sigma methodologies to maximize efficiency and decrease costs for diverse clients in the health care and financial sectors. He has an MBA in International Business and Finance from the University of Miami. E-mail: vhenry@miami.edu.

David Hibbard, MPH, CIH, is the Director of Environmental Health & Safety at the University of Kentucky. He has more than 25 years of combined experience in the military, regulatory, chemical manufacturing, and higher-education sectors involving the implementation and management of health and safety and industrial hygiene programs. He holds a BS from East Carolina University and a master's in Public Health from Eastern Kentucky University. E-mail: dwhibb0@uky.edu.

Gary A. Higbee, EMBA, CSP, is President/CEO of Higbee & Associates Inc., a full-service international consulting firm. Higbee has more than 40 years of experience in the safety field and is an international expert in industrial and construction safety. He can be reached at (515) 270-6623. E-mail: g.higbee@mchsi.com.

Richard Hislop, PE, CSP, is a Chicago resident and Management Consultant focusing on safety program development and implementation of work planning and control processes for construction projects and operating facilities. E-mail: richard.hislop@gmail.com.

Dwight Hyche, Vice President Senior Boiler, Machinery & Equipment, of the Marsh Risk Consulting Practice Group, serves as a Boiler, Machinery & Equipment Risk Consultant for the Southeast Region of Marsh USA. In this capacity, Hyche specializes in servicing pulp and paper clients, electric utilities, food processors, and other clients with large investments in machinery and equipment assets. He also reviews and coordinates insurance carrier inspection services, provides guidance on recommended risk improvements and plant ratings, and assists in determining marketing strategies. In addition, Hyche performs boiler and machinery risk assessments, loss control inspections, and marketing reports where these services are needed. E-mail: dwight.hyche@marsh.com.

Wendy R. Keys, BS, MS, CS, is certified by the Board of Certification in Professional Ergonomics. Fifteen years ago, she established Ergonomics Engineering Consultants, providing ergonomic and safety services to the chemical, food processing, and manufacturing industries. Keys has been an Instructor for the Texas A&M University Engineering Extension Service since 2007, specializing in Ergonomics and Process Safety Management. Prior to EEC, Keys was employed by Neutral Posture Ergonomics, providing ergonomic consultation to NPE customers nationwide. She also spent 5 years at the corporate headquarters of El Paso Energy as a Senior Safety Engineer. Her primary responsibility with El Paso Energy included office and field ergonomics, process safety management, hearing conservation, asbestos and lead management, OSHA injury reporting, case management, incident statistics, and corporate safety auditing for construction and operational sites. She earned a BS in Industrial Engineering in 1992 and an MS in Industrial Engineering in 1994 at Texas A&M University.

Ken Kolosh directs the National Safety Council statistical reporting and statistical estimating systems. Kolosh leads the development of *Injury Facts*, an annual NSC statistical report on unintentional injuries, their characteristics, and costs. Prior to joining NSC, Kolosh worked in the corporate e-learning industry. He served as advanced strategy and systems consultant with Element K and managing consultant with NETg's strategic services team. He was also senior researcher for NETg's research and development group.

Kolosh holds a BA from DePauw University and an MA from Western Kentucky University. He is on the editorial board of the NSC's *Journal of Safety Research*. E-mail: ken.kolosh@nsc.org.

John Kurtz, International Staple, Nail and Tool Association, 512 West Burlington Avenue, Suite 203, La Grange, IL 60525-2245. E-mail: isanta@ameritech.net.

Jim E. Lapping, MS, PE, has more than 30 years of construction industry experience with 20 years as an administrator of the National Safety and Health Education and Training Program. His responsibilities included representation before congressional committees and regulatory agencies on safety, health, and environmental issues. Other accomplishments include development, implementation, and evaluation of more than 40 cooperative safety programs for major construction projects. Lapping has also held positions as Senior Advisor to Assistant Secretary of OSHA, Washington DC, and vice president and safety director of a major construction company.

James Larson, JD, MBA, PE, has a long history of providing consulting services to both internal and external customers. As a student of the Theory of Constraints, Larson has used root cause problem solving to tackle many substantial and pressing issues in the business, construction, manufacturing, educational, and energy industries. He has worked for such notable names as Booz Allen Hamilton, Intel Corporation, Georgetown University, and the U.S. Department of Veterans Affairs. Larson is a licensed professional engineer and attorney, and holds an MBA from Georgetown University.

Curt L. Lewis, PE, CSP, is the President/Owner of Curt Lewis & Associates LLC, a consulting firm specializing in aviation/airline safety, accident investigation and reconstruction, industrial safety, forensic investigation, product safety, system safety, automotive crash worthiness, railroad crossing collision investigations and reconstruction, and airport and aircraft security. He worked for American Airlines/AMR Corporation for 17 years as the Corporate Manager System Safety and previously as the Corporate Manager of Flight Safety and Flight Operational Quality Assurance for American Airlines. Currently, he is an Adjunct Assistant Professor of Occupational Safety & Health at Southeastern Oklahoma State University and an Adjunct Assistant Professor with Embry-Riddle Aeronautical University, teaching aviation, system, human factors, and industrial safety courses (Outstanding Faculty Award—2003). E-mail: curt@curt-lewis.com.

Bob LoMastro is a former Army Green Beret, Navy Hospital Corpsman, and Supervisor of the National Safety Council's Safety Training Institute. He holds a master's degree in Safety Management & Engineering and several bachelor's degrees. LoMastro developed his unique, interactive teaching style as an instructor at the Naval School of Health Sciences. As President of LoMastro & Associates Inc., he draws on a career spanning more than 30 years of teaching health- and safety-related topics worldwide for the military, general industry, and construction companies.

Patrick Lorimer, BS, MPH, has more than 20 years of experience in the Health and Safety industry and is currently the Director of Operations for the EH&S practice Partner Engineering & Science, a large, national environmental engineering firm. Lorimer has worked as a Corporate Risk Manager for a large regional health system and has presented many professional seminars in the field of health safety and environmental hygiene on topics ranging from indoor air quality and risk management to environmental management systems. He holds a BS from East Carolina University and an MPH from George Washington University. E-mail: plorimer@partneresi.com.

Steven L. Lubetkin, APR, Fellow, PRSA, is managing partner of Lubetkin Communications LLC, a strategic communications consulting firm in Cherry Hill, New Jersey. His background includes nearly 30 years in senior corporate communications positions with Fortune 100 companies in transportation, technology, and financial services. E-mail: steve@lubetkin.net.

Brian Maitland, CSP, CET, is a Technical Director in Partner Engineering and Science Inc.'s Health, Safety & Environmental Hygiene discipline. He currently performs regulatory compliance services for public and private entities. His areas of expertise include loss prevention safety audits, accident investigations, job safety observations, and education and training on regulatory compliance issues standards. He has a BS in Environmental Science and Public Health from East Carolina University.

David E. Marquette, CSP, is the owner of SafeMarq Risk Advisors LLC. He has 40 years of consulting experience in a broad spectrum of risk control topics, including occupational safety, ergonomics, product safety and liability prevention, business continuity plans, and property loss prevention. He has consulted with industrial manufacturers of all types, major airlines, railroads, wholesalers and retailers, and heavy-highway and commercial building contractors. He specializes in aviation ground safety, behavior-based safety, ergonomics, and safety management systems. Marquette is a respected lecturer and regularly presents papers at national and international safety conferences. He was co-editor and contributor to the *Aviation Ground Operations Safety Handbook*, 6th edition, published by the National Safety Council in 2007.

James D. Mayers II, JD, is a Georgia- and DC-barred attorney who earned his JD from Mercer University. His undergraduate concentration was in biology while at Harvard. He has worked as a legal editor as well as clerked for the Macon Circuit Public Defenders' Office.

John J. McAneny, CIH, The Procter & Gamble Company, 2 Procter & Gamble Plaza, Cincinnati, OH 45202.

Bradley A. McPherson, MS, CSP, is a Certified Safety Professional and has worked for Allegheny Energy as a Safety and Health Consultant for 9 years. He has a master's degree in Occupational Hygiene and Safety from West Virginia University and a bachelor's degree in Safety Engineering Technology from Fairmont State University. He can be reached at Route 4 Box 695C, Fairmont, WV 26554.

Bill Montante, CSP, is Vice President and Senior Consultant in casualty hazard control, serving Marsh US and global clients. He has more than 25 years of manufacturing, safety management, ergonomics, and consulting experience. He has written numerous articles and received awards for technical writing excellence. He served as Ergonomics Practice Leader and continues active involvement with this practice group as well as other specialty technical task groups within and external to Marsh Risk Consulting. He has a BSE in Industrial Engineering and Human Factors from the State University of New York and holds several certifications.

John F. Montgomery, PhD, CSP, CHMM, is the Senior Vice President of ES&H for Air Serv Corp, with responsibility for 54 domestic and international stations, and is an Instructor for the Texas A&M University Engineering Extension Program. Prior to joining Air Serv, he was with American Airlines for 18 years, where he served as the Corporate Manager of Ground Safety, Corporate Manager/Acting Managing Director of the Environmental Department, and the Manager of the Noise and Emissions Regulatory Program. Prior to joining American, he was the Corporate Manager of Safety and Lost Time at Sky Chefs, spent time in the industrial sector, and was an Assistant Professor/Lecturer at several universities, including Texas A&M, University of Central Missouri, Central Oklahoma University, and Lamar University.

Montgomery holds three advanced degrees, including a PhD in Philosophy from Texas A&M University, and has two Professional Certifications: Certified Safety Professional (Safety) and Certified Hazardous Material Manager (Environmental). He is a frequent speaker at industry meetings and seminars, was an editorial advisor to *Safety and Health Magazine*, and was the General Editor for the 1997, 2001, and 2009 two-volume *Accident Prevention Manual for Business & Industry*, published by the National Safety Council. He was also a contributor to the *Accident Prevention Manual for Environmental Management* (1995 and 2000) and the first edition of *Accident Prevention Manual Essentials* (2014). He was named Rapporteur of the International Air Transportation Association Emissions Sub Group and served as a delegate to the United Nations' International Civil Aviation Organization. He was a contributor/reviewer for the United Nations' Intergovernmental Panel on Climate Change's review of the effect of aircraft emissions on the environment. E-mail: drjfmonty@sbcglobal.net.

Patrick Moylan, MBA, CSP, is the Senior Director of Safety and Security at BBA Aviation's flight support businesses, Signature Flight Support and Aircraft Service International Group—two leading ground service providers to the global commercial airline and business aviation sectors. In this capacity, he is responsible for the design and implementation of the company's Safety Management System and for nurturing a positive safety culture in its 200+ airport locations around the world.

Prior to his current position, Moylan served as a safety management consultant at the Federal Aviation Administration's Office of System Safety and the Global Aviation Information Network, an international aviation safety program promoting the collection, analysis, and sharing of aviation safety information. His professional experience also includes positions in airline flight operations, in general aviation operations management, and as an air traffic controller in the U.S. Marine Corps. He has served as an adjunct faculty member at the University of Maryland Eastern Shore and the Community College of Baltimore County, teaching courses in aviation safety, airport management, and air traffic control. Moylan earned his BA and MBA from Michigan State. E-mail: patrick. moylan@bbaviation.com.

Michael O'Berry, MEd, has 35 years of industrial, fire, and oil field safety experience and is currently the Program Chair of Occupational Safety Engineering and Environmental Management at Eastern New Mexico University. He also is a Senior Training Specialist at the OSHA Training Institute Southwest Education Center and is on staff at the Texas Engineering Extension Service within the Texas A&M University System and is an Adjunct Instructor in the New Mexico Junior College System. His expertise includes oil field and natural gas training and emerging alternative energy field sources.

O'Berry hold several degrees, including an AAS from Trinidad State Junior College, Colorado; a BS in Safety and Occupational Education from Wayland Baptist University, Plainview, Texas; and an MEd, in Curriculum Development and Instruction: Technology Integration, from Grand Canyon University, Phoenix, Arizona.

James T. O'Reilly, JD, College of Law and College of Medicine, University of Cincinnati, Cincinnati, Ohio, has authored 45 texts and 200 articles. His scholarly work was acknowledged in a March 2000 decision of the U.S. Supreme Court, quoting one of his textbooks as the "expert" in its field. He was formerly Associate General Counsel of The Procter & Gamble Company and Chair of the Local Emergency Planning Committee for Cincinnati and has served as chair of the American Bar Association's Section of Administrative Law, 1996–1997. He also served as vice mayor of an Ohio city and as a member of the regional council of governments. He has acted as a general editor for both volumes of the *Accident Prevention Manual for Business & Industry* for three editions in addition to the new *Essentials* text produced in 2014. E-mail: joreilly@fuse.net.

Richard Payant, BS, MA, CFM, CPE, CHS, has more than 20 years of experience as Director of Facilities Management at Georgetown University and more than 23 years of experience with the Army Corps of Engineers. He is a Certified Facility Manager and Plant Engineer and holds a certification in Homeland Security. Payant is also an Adjunct Professor teaching Facilities Management at George Mason University. He is the author of several professional publications and co-authored the *Facility Inspection Field Manual*, the *Facility Manager's Emergency Preparedness Handbook*, and the *Facility Manager's Maintenance Handbook*, 3rd and 4th editions. He holds a BS from Norwich University, an MA from Central Michigan University, and a PhD in Business Administration from Northcentral University. E-mail: rich.payant@gmail.com.

Richard Pifer, BS, MS, is currently a consultant in facilities management. He is the past Associate Vice President for Facilities and Services at the University of Rochester and was in that position from 1999 to 2014. Prior to assuming his role at the University of Rochester, he worked for more than 9 years in the Georgetown University Facilities Department, holding a variety of positions in Facilities Management Administration. He was a career military officer, has a graduate degree in management, and is active on a variety of university and community boards and committees.

Antonello Pileggi, MD, PhD, is a Research Professor at the Division of Cellular Transplantation of the DeWitt-Daughtry Family Department of Surgery and at the Departments of Microbiology and Immunology and Biomedical Engineering at the University of Miami. Since 2003, he directs the Preclinical Cell Processing and Translational Models Program at the Cell Transplant Center of the Diabetes Research Institute. His research has been funded through the National Institutes of Health, Juvenile Diabetes Research Foundation, the Diabetes Research Institute Foundation, and the University of Miami, as well as by other industries. He has lectured at national and international institutions and professional meetings.

Pileggi has served as ad hoc Reviewer and/or Study Section Member for the National Institutes of Health; the Italian Republic's Ministry of Health; American Diabetes Association; Czech Science Foundation; Regenerative Medicine Research Committee, Medical Research Council, United Kingdom Diabetes UK; and the Biomedical Research Council & National Medical Research Council, Singapore; among other national and international agencies. Pileggi has authored 20 scientific book chapters and more than 150 peer-reviewed publications in the fields of organ and cellular transplantation, immunobiology, and regenerative medicine.

Joy Prescott, MS, is Manager of Training at Texas A&M Engineering Extension Service, Infrastructure Training and Safety Institute and OSHA Training Institute Education Center. She has more than 30 years of experience in the Safety and Health profession. Prescott has served several large corporations directing both national and global safety, health, environmental, and medical programs over her tenure as well as actively participating in many professional organizations. For the past 2 years, her efforts have centered on giving back her industry knowledge through her work at the OSHA Training Institute Education Center.

Cynthia Roth, RN, has been a professional in the ergonomics, safety, and health industry since 1987. In 1993, she co-founded Ergonomic Technologies Corp. Prior to ETC, Roth was Executive Vice President of Biomechanics Corporation of America and the Senior Vice President and Business Manager for the Langer Biomechanics Group. She has lectured on Ergonomics/Biomechanics/Safety and Health to Fortune 500 companies, to Fortune 200 international companies, and at universities and colleges around the world. She is a Trustee and past Chairperson of the American Society of Safety Engineers Foundation Board. With extensive international experience, she has also been appointed a permanent member of New York State's Commission on International Trade and is on the Advisory Boards of the NYC Department of Mental Health and Hygiene, the Ergonomics Exposition, and the publications of *Occupational Hazards* and *CTDNews*. Roth serves on many national committees and is very well published, with articles appearing worldwide. She received a degree from the University of Pittsburgh as a professional registered nurse with specialties in Occupational Nursing and Biomechanics. E-mail: Croth@ergoworld.com.

Steven G. Schoolcraft, PE, CSP, MBA, is the Examination Director for the Board of Certified Safety Professionals and has served in this role since 2002. Prior to joining the board staff, he worked for the American Bureau of Shipping, where he worked closely with the Coast Guard and other components of the U.S. government and government contractors in areas such as risk management and safety engineering. His undergraduate degree is in chemical engineering from Texas A&M University, and he joined NASA's Goddard Space Flight Center after college. Schoolcraft left NASA in 1997 and joined the American Bureau of Shipping. He is a Certified Safety Professional and a licensed professional engineer in several states. He has also earned an MBA and has been in a management role in the safety profession for more than 15 years.

Bonnie Martin Steward is the founder and owner of Martin Safety Consulting, a 10-year-old ES&H consultant company. She has been an Environmental Health and Safety Professional for 30 years and has worked in a number of companies: Adjunct Instructor in Health & Safety faculty for Texas A&M University, Texas Engineer Extension Service; External Reviewer for ISNetworld; Environmental Health & Safety Engineer, Reckitt Benckiser Pharmaceutical Manufacturer, Fort Worth, Texas; Campus Safety Trainer and Training Coordinator, The University of Texas at Arlington; Safety/Environmental Manager, Paragon Trade Brands; Instructor—Occupational Safety & Health Program, Texas State Technical College; Safety Engineer, LTV Aerospace and Defense; Safety Technician, United Technologies; and Safety Technician, M&M Mars Candy. She holds certifications as an OSHA 500 Trainer; Medic First Aid Certified—Train the Trainer; and First Aid/AED/CPR for Adult, Youth, and Infants; and is a Certified NIOSH Defensive Driving Instructor.

Ralph Stuart, MS, CIH, CCHO, is Chemical Hygiene Officer at Keene State College in Keene, NH. He has an MS from the University of Vermont in Environmental Engineering and has been active at the national level in laboratory safety innovations since 1989. These innovations include development of professional health and safety Internet information resources, as well as the EPA Project XL regulatory reinvention project for laboratory chemical waste management. He is currently secretary of the Division of Chemical Health and Safety of the American Chemical Society and chair of the ACS Safety Advisory Panel. E-mail: rstuartcih@me.com.

Patricia L. Thomas, CSP, CHMM, CET, is a Certified Safety Professional, Certified Hazmat Manager, Certified Environmental Safety and Health Trainer, and the founder and CEO of a small specialty consulting firm which provides safety, industrial hygiene, fire protection, and emergency response services, training, and consulting to a variety of industrial, municipal and construction clients. Thomas has responded as the Site Safety Officer to large hazardous material spills on the waterways including the BP Gulf Oil Spill. She has developed and presented a number of specialty courses on loss control including regulatory industrial training classes and safety and health manuals for clients.

Thomas has served at the corporate level as a consultant to five refining facilities in addition to working in several refineries and chemical plants, has provided technical expertise and consultation on refinery/chemical plant safety issues, writing safety procedures, conducting training, and emergency response. She has provided liaison services to OSHA and insurance inspectors/auditors, conducted design reviews, and has consulted on new construction projects. She also has experience directing activities for implementation of the Process Safety Management Standard.

Thomas has dual Bachelor of Science degrees from Oklahoma State University in Fire Protection and Safety Engineering Technology and Business Administration. She has spent 30 years as a master instructor at Texas A&M Industrial Fire School and has been a part-time instructor for the Southwest OSHA Education Training Center for 14 years. She is a member of NFPA, a Professional Member of ASSE, and an Associate Member of the Ft. Worth and Dallas County IEC chapters. She is also an advisory board member to the Environmental Safety and Health Program at Tarrant County Community College, and a member of the National Safety Committee for Independent Electrical Contractors.

Treasa Turnbeaugh, PhD, MBA, CSP, CET, is the Chief Executive Officer for the Board of Certified Safety Professionals. She is responsible for the overall operations of the BCSP as well as its contribution to the safety, health, and environmental profession. Turnbeaugh is experienced in the safety, health, and environmental field and in the field of professional certification. Additionally, she brings experience and leadership in the business arena of both for-profit and not-for-profit organizations.

Turnbeaugh has more than 25 years of experience in the safety profession, with experience in workers' compensation cost reduction, ergonomics, industrial hygiene, indoor air quality, behavior-based safety, cultural assessments, diagnostics and metrics, injury management, and safety process improvement. She is experienced in servicing a variety of industries, including manufacturing, health care, gaming, higher education, agribusiness, and municipalities.

Turnbeaugh holds a PhD in Health Services Research, with a minor in Epidemiology, and an MPH from Saint Louis University; an MBA from Lindenwood University; and both an MS and BS in Occupational Safety and Health, with a specialization in Industrial Hygiene, from Murray State University. She is a member of the American Society of Safety Engineers, the American Industrial Hygiene Association, and the American Society of Association Executives. She has held her CSP certification more than 20 years and is a Certified Environmental, Safety & Health Trainer.

Nicholas Valter, JD, MBA, has more than 20 years of international experience in monitoring safety in work environments. He is currently a practicing attorney and an in-flight safety coordinator for a major airline. He has experience as a union negotiator for airline safety. He has worked internationally with his own companies in both the retail and materials distribution/warehouse export sectors. His undergraduate degree is from Chaminade University, his MBA is from Georgetown University, and he holds a JD from the University of Hawaii. E-mail: nvalter@gmail.com.

Sherrie Wilson was the first female fire fighter–paramedic with the Dallas Fire Rescue Department. She currently serves as the Founder, President, and CEO of Emergency Management Resources LLC and FireHouseCommunications.com. She served on editorial and advisory boards for *Industrial Fire World Magazine*, *Emergency Medical Services Magazine*, and *Texas EMS Magazine* and as a reviewer of EMS text for both Mosby Lifeline and Brady Publishing. She has more than 50 published fire-, EMS-, and emergency-incident-related articles, publications, and conference presentations. She has authored two manuals, *Rescue Team Training* and *Medical Terminology*.

Wilson received a BS in Public Administration from Hawthorne University and completed Paramedic training at the University of Texas Health Science Center, Dallas, Texas. She serves as adjunct Health & Safety faculty for Texas A&M University, Texas Engineer Extension Service.

James A. Wolf, CFPS, CXLT, CMGT, retired in 2007 after serving the last 20 years of his career as a risk control consultant with ESIS Global Risk Control Services. He dedicated many years of service to risk management and safety engineering, most particularly in the manufacturing and aerospace industries. E-mail: wolfhamm@comcast.net.

Lynne Zarate, MSE, CIH, CHMM, is Director of the Division of Maintenance and formerly an Environmental Safety Coordinator with the state of Maryland, Montgomery County Public Schools System, which is the 17th largest public school system in the United States. She has been with the school system since 2003, managing a variety of systemwide facilities-related programs. Prior to joining the school system, she worked as the Safety Manager at Georgetown University. Other experience includes biomechanics research in automobile safety, research and operations in the paper manufacturing industry, and space shuttle development and testing at Kennedy Space Center. Zarate holds a master's degree in Environmental Engineering and a bachelor's in Chemical Engineering. E-mail: zaratelm@hotmail.com.

PART 1

Introduction to Safety and Health

Growth in personal and professional competence requires flexibility and awareness of change. Part 1 explores the roots of current safety functions and discusses the culture of safety in organizations today. The first two chapters offer a brief background for the safety function and its history. Chapter 3 discusses how the profession has adapted to change and offers ideas for the promotion of the safety culture within companies and organizations. Chapter 4 explains the regulatory agencies and regulations that have such a great impact on the modern workplace.

Historical Perspectives

James T. O'Reilly, JD

Philosophy of Incident Prevention

The Industrial Revolution

History of the U.S. Safety and Health Movement
Birth of the National Safety Council ▸ American Standards Association Beginnings ▸ Incident-Prevention Discoveries

Acceleration of the Drive for Safety and Health

Evaluation of Accomplishments
The Dollar Values ▸ Industry and Nonwork Injuries

Resources for Safety
Knowledge and Experience ▸ The Heritage of Cooperation ▸ Goodwill ▸ Professionalism ▸ Advancement of Knowledge

Achievements of the Safety Movement
Small Establishments ▸ Labor's Participation in Safety and Health ▸ Statistics, Standards, and Research ▸ Safety and the Law ▸ International Standards ▸ Safety and Occupational Health ▸ Psychology and "Accident Proneness"

Current Issues
Technology and Public Interest ▸ Political Problems ▸ Organizational Problems ▸ A Look to the Future

Summary

General Safety Books

References

Review Questions

Protecting life and promoting health is the basic mission of the National Safety Council. The council's hallmarks over the years have been flexibility and the ability to create new programs to cope with ever-changing challenges. Many of the major safety and health problems that faced safety professionals at the beginning of the 20th century are virtually nonexistent today. However, occupational and environmental health, bloodborne diseases, and drug and alcohol abuse present new challenges to today's safety professionals.

In an imperfect world, there will always be risks, and the National Safety Council will continually strive to reduce the number and severity of those risks—no matter what the cause. In working to protect people from unintentional-injury death or injury, the council seeks ways to ensure that everyone enjoys a safe and healthful environment.

In this, the 14th edition of the *Accident Prevention Manual for Business & Industry*, the National Safety Council presents a compilation of knowledge and experience that forms part of the safety movement's general heritage. For specific information regarding occupational health and industrial hygiene, see two other National Safety Council books: *Occupational Health and Safety* and the *Fundamentals of Industrial Hygiene*, respectively.

PHILOSOPHY OF INCIDENT PREVENTION

In medieval times, the master craftsman instructed apprentices and journeymen to work skillfully and safely because he knew the value of high-quality, uninterrupted production. However, the industrial revolution, which began in England during the 18th century, shifted the emphasis to faster and greater production and created the conditions that inspired the development of incident prevention as a specialized field.

The industrial safety philosophy developed because the hazardous work environment of early factories and other production and distribution sites produced an appalling rate of worker injuries and deaths. If these conditions had not been corrected to stop the waste of personnel and resources, the growing number of accidents and injuries would have staggered the imagination. In the beginning, one way to encourage management to accept responsibility for preventing unintentional injuries was to pass workers' compensation laws. This "new" line of thinking held the employer responsible for a share of the economic loss suffered by an employee involved in an incident.

It was a rather short step from this approach to the realization that most unintentional injuries could be prevented, and that the same industrial knowledge used to develop mass production methods also could be applied to incident prevention. Managers soon discovered that efficient production and safety were closely related. From this beginning grew the safety movement as it is known today.

The progress in reducing the number of incidents and injuries in the relatively short time since this movement began has exceeded the most optimistic expectations of early safety pioneers. The unintentional-injury death rate per 100,000 population in the United States decreased 57% from 1912 to 2006 (National Safety Council, *Injury Facts*, 2008).

Experience has shown that virtually any hazard can be overcome by practical safety measures. To further that belief, the National Safety Council continues its concerted efforts to prevent unintentional injuries and occupational illnesses.

In summary, here are six reasons for working hard to prevent accidents and occupational illnesses:

1. Needless destruction of life and health is morally unjustified.
2. Failure to take necessary precautions against predictable accidents and occupational illnesses makes management and workers morally responsible for those accidents and occupational illnesses.
3. Accidents and occupational illnesses severely limit efficiency and productivity.
4. Accidents and occupational illnesses produce far-reaching social harm.
5. The safety movement has demonstrated that its techniques are effective in reducing accident rates and promoting efficiency.
6. Recent state and federal legislation mandates management responsibility to provide a safe, healthful workplace.

THE INDUSTRIAL REVOLUTION

Until the 1700s, production methods were labor intensive, with work being done by hand in cottages. Three developments were to change this way of life: (1) in England, inventors developed the spinning jenny in 1764; (2) the power loom was perfected in 1784; and (3) in America, Eli Whitney developed the cotton gin in 1792. These and other innovations ushered in what would later be called the industrial revolution. What began in Britain in the 18th century and spread to the European continent and the United States transformed the life of Western culture and society and had a dramatic impact on traditional relationships between groups of people.

Specifically, the innovations in the processes and organization of production included the following:

- substitution of mechanical energy for animal sources of power, particularly steam power through the combustion of coal
- substitution of machines for human skills and strength
- invention of new methods for transforming raw materials into finished goods, particularly in iron and steel production and industrial chemicals
- organization of work into large units, such as factories, forges, and mills. This made possible direct supervision of the manufacturing process and an efficient division of labor. Paralleling these production changes were the altered technologies employed in agriculture and transportation.

Initially, this new way of organizing work was termed the *factory system*. Later, however, when it reached larger and more complex scales, it was designated the *industrial revolution* by A. Toynbee (Toynbee 1884). His nephew, Arnold J. Toynbee, is described as "the first economic historian to think of, and to set out to describe, the Industrial Revolution as a single great historical event, in which all the details come together to make an intelligent and significant picture."

Unfortunately, these changes in production methods with their need for masses of workers also created hazards never before encountered. These conditions greatly affected the history of occupational safety and health. Many health workers and industrial experts recognized the increasing need for hazard control.

The effects of the industrial revolution were first felt in the United States about a century after the revolution began in Great Britain. Before the 19th century, most families in the United States lived and worked on farms. Some industries had developed in the new country—namely, printing, shipbuilding, quarrying, cabinetmaking, bookbinding, clock making, and the production of paper, chocolate, and cottonseed oil. However, it was the textile industry that introduced the new factory system into the United States, especially in New England, where hundreds of spinning mills were built. As the industrial revolution continued its rapid growth, unsafe production methods exacted a heavy toll on the work force in terms of job-related injuries and deaths (Felton 1994).

HISTORY OF THE U.S. SAFETY AND HEALTH MOVEMENT

During the last half of the 19th century, American factories were expanding their product lines and producing at previously unimagined rates. Although the factories were far superior in terms of production to the preceding small handicraft shops, they were often vastly inferior in terms of human values, health, and safety.

In the area of human values alone, the facts make a grim picture. The 1900 census showed 1,750,178 working children between ages 10 and 15. Some 25,000 were employed in mines and quarries; 12,000 in making chewing tobacco and cigars; 5,000 in sawmills; 5,000 at or near steam-driven planers and lathes; 7,000 in laundries; 2,000 in bakeries; and 138,000 as servants and waiters in hotels and restaurants. These children often worked 12 to 14 hours a day; no health or safety guidelines were in effect, even for children under age 10 (Figure 1–1).

The lag between the emergence of new working methods and the creation of health and safety standards was probably inevitable. The tools of mass production had to be invented and applied before anyone could begin to imagine the problems they might create. In turn, the problems had to be known before corrective measures could be considered, tested, and proved. Thus, for some time, deaths and injuries were accepted as part of "industrial progress"—one of the costs of doing business.

While this revolution in the work environment was taking place, the thinking of the public, management, and the law still reflected the past, when the worker was an independent craftsman or a member of the family-owned shop. Common law provided the employer with a defense that gave the injured worker little chance for compensation. The three doctrines of common law that favored the employer were as follows:

- *Fellow servant rule*—The employer was not liable for injury to an employee that resulted from negligence of a fellow employee.
- *Contributory negligence*—The employer was not liable if the employee was injured because of his or her own negligence.

Figure 1–1. When the safety movement began (c. 1900–1910), children commonly worked amid heavy machinery—even without shoes or any protective equipment.

- *Assumption of risk*—The employer was not liable because the employee took the job with full knowledge of the risks and hazards involved.

In 1906, the Pittsburgh Survey, sponsored by the Russell Sage Foundation, marked the first attempt to pinpoint the serious nature of occupational accidents and deaths. The survey team realized they did not have the resources to survey the problem throughout the United States and instead concentrated on Allegheny County, Pennsylvania. They constructed a "death calendar" of the county, showing that industrial accidents accounted for an average of nearly two deaths per day throughout the year. The number of crippling injuries was far higher. If this was the case in only one county, people asked, what must the situation be for the entire United States? The Pittsburgh Survey made it clear that the accident and death rate was serious and gave the safety movement a much-needed boost.

In large industrial centers, the ugly results of industrial unintentional injuries and poor occupational health conditions became more and more obvious. Individuals and public and private organizations raised their voices to protest these conditions. Though some employers denied that the problem existed, wiser managers began to take action to improve the work environment.

As early as 1867, Massachusetts had begun to use factory inspectors. Ten years later, the state passed a law requiring employers to safeguard hazardous machinery. During 1877, Massachusetts also passed the Employer's Liability Law, which made employers liable for damages when a worker was injured. However, court decisions based on common law often let the employer escape liability.

From 1898 on, there were additional efforts to make the employer financially liable for accidents. In his presidential message of 1908, Theodore Roosevelt stated: "The number of accidents which result in the death or crippling of wage earners is simply appalling. In a very few years it runs up a total far in excess of the aggregate of the dead in any major war." Roosevelt's message acquired force when his social legislation passed that year in Congress. Although this first workers' compensation law covered only federal employees, it set a precedent for state laws to follow.

The first bill for workers' compensation (the Wainwright Law) was passed in New York in 1910, but it was declared unconstitutional by the New York Court of Appeals. The court ruled that the law violated both the federal and New York State constitutions, "because it took property from the employer and gave it to his employee without due process of law." On the same day the 1910 act was declared unconstitutional, March 25, 1911, a devastating fire in New York City's Triangle clothing factory killed 146 employees. This disaster, called the Triangle Fire, outraged the public and spurred demand for factory legislation and health and safety reform. After an amendment to the state constitution was approved in 1913 at the general election, a compulsory Workmen's Compensation Act finally became effective in mid-1914.

In 1911, Wisconsin passed its first effective Workers' Compensation Act, but the act was declared unconstitutional by the Wisconsin Supreme Court within a few months. New Jersey and Washington also passed laws that year.

At first, the courts continued to declare such laws invalid because they conflicted with the due process of law provisions of the 14th Amendment. However, after the U.S. Supreme Court in 1916 declared workers' compensation to be constitutional in *New York Central Railroad Co. v White*, 243 US 188, many states passed compulsory workers' compensation laws.

By the late 1800s and early 1900s, the railroads had crisscrossed the East and West but exacted a heavy toll among employees. It was said that a man was killed for each mile of track laid. By 1907, annual railroad employee deaths had reached 4,353.

Industry experts made some progress on the technical side of the problem. The railroads adopted air brakes and the automatic coupler well before the beginning of the 20th century. They also worked on guarding and fire prevention. They came to realize, however, that guarding was not the total answer. People's actions were equally important factors in creating accident situations.

At the same time, insurance companies started to relate the cost of premiums for workers' compensation insurance to the cost of accidents. Management began to understand the close relationship between successful production and safe production.

During the first decade of the 20th century, two giant industries, railroads and steel, began the first large-scale organized safety programs (Figure 1–2). From this period comes one of the historic documents of safety. In 1906, Judge Elbert Gary, president of the United States Steel Corporation, wrote:

> The United States Steel Corporation expects its subsidiary companies to make every effort practicable to prevent injury to its employees. Expenditures necessary for such purposes will be authorized. Nothing which will add to the protection of the workmen should be neglected.

The Association of Iron and Steel Electrical Engineers, organized soon after this announcement, devoted considerable attention to safety problems.

Figure 1–2. Because of hazards encountered in railroad operations, the railroad industry was one of the first industries to develop organized safety programs. This photograph shows some of the hazards associated with building a bridge at Rockford, Illinois, in 1869. *(Courtesy Chicago & North Western Railroad.)*

Birth of the National Safety Council

The year 1912 proved to be a landmark for accident prevention. In the previous year, the Association of Iron and Steel Electrical Engineers (which had been formed in 1907) had called for a general industrial safety conference on a national scale. The result was the First Cooperative Safety Congress, which met in 1912 in Milwaukee. The following year, at a second national meeting in New York, the delegates formed the National Council for Industrial Safety. Shortly afterward, members changed the organization's name to the National Safety Council and broadened its program to include all aspects of accident prevention. The program also included occupational health. Yet it must be remembered that the council was the creation of industry and that its activities have always been heavily concentrated on industrial safety.

The group that met in Milwaukee and New York was composed of a few safety professionals, some management leaders, public officials, and insurance specialists. Their one point in common was a desire to attack a problem that most people considered either unimportant or insoluble. The determination of these safety pioneers helped create the safety movement as we know it today.

Actually, the members' underlying objective in forming the National Safety Council in 1913 was standardization. Thus, the primary purpose of the council was to provide an avenue of communication, an exchange of views, and various solutions to common problems in accident prevention.

In 1918, the council conducted the first national survey of state, federal, and municipal regulations, together with a study of insurance recommendations, technical association recommendations, and the practices of industry. The survey revealed utter chaos in industrial safety and a clear need for industrywide methods and practices.

Realizing its own limitations, the council consulted the National Bureau of Standards, which agreed to call a conference to discuss establishing procedures to standardize safety methods and practices. Meeting in Washington DC in 1919, the attendees expressed the belief that uniform industry standards were not only desirable but essential to promote effective worker health and safety. The conference voted to formulate safety standards under the auspices and procedures of the American Engineering Standards Committee (AESC), which had been formed in 1918 by five engineering societies and three governmental departments.

American Standards Association Beginnings

In 1920, the National Safety Code Program was brought into the AESC. This resulted in the first reorganization of the committee and marked the beginning of what later became the American Standards Association (ASA). A national code committee was organized to suggest the initial safety code projects. This later became the Safety Codes Correlating Committee, the first of ASA's 18 standards boards. Bringing manufacturing companies and trade associations into AESC membership also initiated a broader program of engineering standards. These steps launched an enlarged national standardization program.

In 1928, recognizing that the extensive activities called for a more formal type of organization, the member groups reorganized the AESC as the American Standards Association, now known as the American National Standards Institute (ANSI). ASA continued to be an important partner in the safety movement. This group handled the "materials" aspect of safety while the National Safety Council focused on the "people" portion of accident and occupational illness prevention.

Incident-Prevention Discoveries

As industry developed some experience in safety, it discovered that engineering could prevent unintentional injuries, that employees could be reached through education, and that safety rules could be established and enforced. Thus the "Three Es of Safety"—engineering, education, and enforcement—were developed.

Two of the many breakthroughs that safety groups made during the 1900–1990 era were identification of occupational diseases such as mercury and lead poisoning and efforts to control these hazards. Other discoveries have had equally profound implications.

For example, asbestos was found to be a carcinogen that causes lung cancer and another type of cancer, mesothelioma. Health professionals also studied the effects of chromium compounds and beryllium on industrial workers.

These and other discoveries led safety professionals to argue that savings in compensation costs and medical expenses would repay safety expenditures many times over. Thoughtful business leaders soon learned that these savings were only a fraction of the financial benefits to be derived from accident prevention work. Newer, more effective techniques have been discovered and are described elsewhere in this volume.

ACCELERATION OF THE DRIVE FOR SAFETY AND HEALTH

Industrial safety received wide acceptance in the years between World Wars I and II. During World War II, the growth of safety procedures and policies intensified, particularly as the federal government began encouraging its contractors to adopt safe work practices. As industry expanded to meet the needs of the war effort, additional safety personnel were hastily trained in an effort to keep pace with the new risks and hazards in the workplace. The acceptance of safety activities as part of the industrial picture did not diminish with the end of the war. By then, the importance of safety to quality production was well established. The small handful of people dedicated to safety in 1912 had grown to tens of thousands of trained personnel. In 1948, for example, Admiral Ben Moreell, then president of Jones and Laughlin Steel Corporation, wrote:

> Although safe and healthful working conditions can be justified on a cold dollars-and-cents basis, I prefer to justify them on the basic principle that it is the right thing to do. In discussing safety in industrial operations, I have often heard it stated that the cost of adequate health and safety measures would be prohibitive and that "we can't afford it."
>
> My answer to that is quite simple and quite direct. It is this: "If we can't afford safety, we can't afford to be in business."

A discussion of current U.S. safety legislation follows later in this chapter under "Safety and the Law" and also in Chapter 4, Regulatory History.

Figure 1–3. These women working during World War I are shown risking their lives on unsecured scaffolding. Note also the version of protective footwear of that era. *(Courtesy Women's Bureau, National Archives.)*

and injuries. In 1939, the American Industrial Hygiene Association was established to promote the recognition, evaluation, and control of environmental stresses arising in or from the workplace (Figures 1–3 and 1–4).

EVALUATION OF ACCOMPLISHMENTS

Because safety factors are complex, no simple rating scale can yield all the answers to the question, "What has the safety movement accomplished?" Instead, an attempt to answer the question must be made by assembling several kinds of data.

First, the following question must be asked: "Has the safety movement, in fact, really helped prevent unintentional injuries?" The answer is a clear "Yes!" If the annual unintentional-injury death rate per 100,000 population recorded in 1912 had continued, more than 3,700,000 additional deaths would have occurred since that time. Between 1912 and 2006, unintentional work deaths per 100,000 population were reduced 92%, from 21 to 1.7. In 1912, an estimated 18,000 to 21,000 workers' lives were lost. In 2005, in a work force nearly quadrupled in size and producing 10 times the goods and services, there were only 4,988 accidental work deaths (National Safety Council, *Injury Facts*, 2008, p. 48). Medical progress accounts for some of this gain, but the larger part is certainly the product of organized safety work.

Since World War II, the number of work-related deaths per 100,000 population, standardized to the age distribution of the population in 1940, also has decreased steadily. This fact indicates that the risk of on-the-job death has declined for the population as a whole. Part of the progress made in lowering the overall death rate, however, can be attributed to the rapid growth in recent years of the economy's service sector with its lower death rate, and the decline of some fairly high-risk segments of the manufacturing sector. In 1945, 43% of the nonagricultural work force was in production-related industries (mining, construction, and manufacturing). By 1994, that proportion had declined to 21%.

Long-term trends in nonfatal occupational injury rates cannot be examined because of a break in continuity of the historical statistical series. Until the early 1970s, injury rates were based on the voluntary American National Standards Method of Recording and Measuring Work Injury Experience, ANSI Z16.1. With the passage of the Occupational Safety and Health Act of 1970, it became mandatory for most private-sector employers in the United States to keep occupational injury and illness records in accordance with OSHA record-keeping requirements.

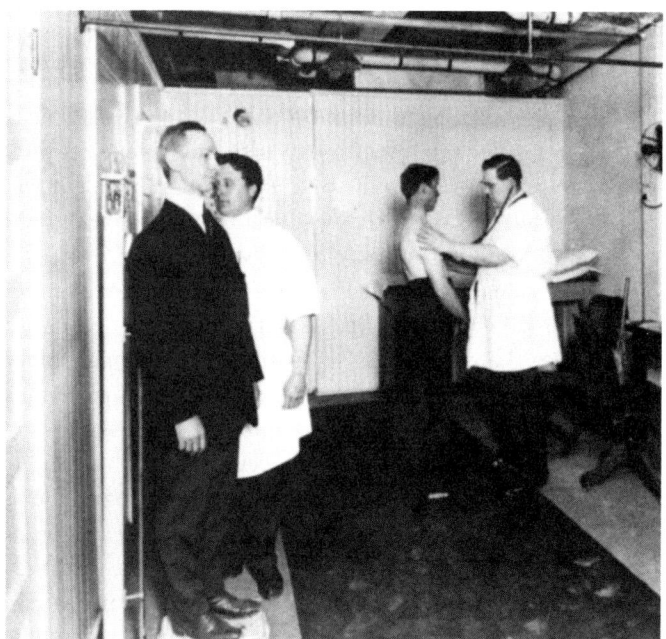

Figure 1–4. As the relationships between health and employment hazards became recognized, employers began to provide medical examinations.

One by-product of organized safety activities has been a growing interest in safety engineering on the part of colleges and universities. Many schools offer degrees and advanced courses on this subject and are contributing to a higher standard of knowledge among professionals in the field.

In addition, the World War II labor shortage dramatically brought home to management the magnitude and seriousness of the problem of off-the-job accidents to industrial employees. The wartime theme of the National Safety Council, "Save Manpower for Warpower," focused attention on the need to reduce off-the-job accidents in order to maintain efficient, safe production on the job.

An increasing number of employers are including off-the-job safety in their overall safety programs. Companies realize that their operating costs and production schedules are affected almost as much when employees are injured away from work as when they are injured on the job. Off-the-job safety generally is an extension of a company's on-the-job safety program and is intended to educate the employee to follow, in outside activities, the safe practices used on the job. Companies have found that on-the-job and off-the-job programs complement each other.

From the earliest days of industrial safety, it has been difficult to make a clear distinction between illness and injury (accident hazard). Is dermatitis an injury or an illness? What about hernias, hearing loss, and heart trouble? Inevitably, safety professionals have become interested in many health problems on the borderline between illnesses

A clear trend has not yet emerged in the occupational injury and illness incidence rates published by the Bureau of Labor Statistics since 1972. Business cycles and changes in the distribution of the labor force among industries can mask any short-term changes in rates caused by more effective or more intensive safety efforts.

The Dollar Values

It has been estimated that the annual cost of occupational unintentional injuries in the United States exceeds $120 billion. If the 1912 unintentional-injury rate had continued unchanged and if there had been no organized safety movement, this annual cost could easily be two to three times as great, even in constant dollars.

Against such dollar savings, the relatively small expenditures for safety throughout the United States provide a striking contrast. Each dollar industry spends for safety may be returning a clear profit of up to several hundred percent.

Industry and Nonwork Injuries

Directly and indirectly, industry bears a substantial part of the cost related to nonwork unintentional injuries and their prevention. Although the National Safety Council is the creation of industry and largely supported by it, the council, along with state and local safety organizations, plays a major role in the fight to prevent such incidents. Industry is a major supporter of the efforts to inform the general public on safety issues through the press, radio, and television.

This nonwork incident-prevention campaign is having a definite impact on public safety. From the time records on nonwork unintentional injuries were first kept in 1921, both home and public unintentional-injury death rates have substantially declined.

Although industry has been a large contributor to this successful work, it has also been a major beneficiary. The reduction in nonwork injuries, illnesses, and deaths has lessened disruption of the labor force and has reduced hardship among employees, consumers' loss of purchasing power, and tax burdens required to support hospitals and relief agencies.

RESOURCES FOR SAFETY

Safety statistics measure what has been accomplished in this area. Safety resources describe the tools, methods, and knowledge developed for safety professionals in their efforts to overcome future accident and occupational illness problems.

Knowledge and Experience

This *Accident Prevention Manual for Business & Industry*, for example, is an accumulation of facts and experiences that are a part of the safety and health movement. Its purpose is to present key points of knowledge to people interested in safety, whether they are students new to the field or advanced and experienced practitioners.

An individual using this manual can find better answers to a wider range of industrial safety problems than were available to the wisest and best-trained professional safety practitioner several decades ago. Yet even this manual cannot contain all of the knowledge available to fight the never-ending war against accidents and occupational illnesses.

Other material may be found in numerous pamphlets, books, and periodicals published by safety and health organizations, government agencies, and insurance companies and in the studies and directives of individual industrial concerns. The literature of various trades and professions is likewise rich in safety information. A list of handbooks is presented in Safety Tables, an appendix of the *Engineering & Technology* volume. At the end of this chapter is a list of the general safety books used as sources of questions for the Certified Safety Professional examination.

The National Safety Council offers a series of training courses, at both the beginning and advanced levels, for professionals. The council also offers extensive consulting services, books, and software.

Finally, through conferences, technical seminars, newsletters, and other publications, professional safety engineers, executives, supervisors, and rank-and-file employees regularly exchange safety information. The annual National Safety Congress and Exhibition is an excellent means of enhancing professional development.

The Heritage of Cooperation

The safety movement would be far less effective if its members had concealed their discoveries from their colleagues who worked in competing companies. It was teamwork that created the safety activities of the Association of Iron and Steel Electrical Engineers. It was broadened teamwork that organized the first Milwaukee Conference, which led to the formation of the National Safety Council and other safety-related organizations.

Effective incident prevention requires cooperation at all levels of industry and government. Through the council and other safety organizations, safety professionals meet to exchange ideas, develop safety publications, and stimulate one another in friendly competition. The tradition that there should be "no secrets in safety," no denial of help even to a competitor when it involves saving lives, is one of the great strengths in the safety movement.

Goodwill

In its early days, safety did not rank highly among management concerns. Today, a significant part of the safety professional's capital is the prestige and goodwill associated with this area that have accumulated over the years, making management more receptive to safety proposals and expenditures. Whereas yesterday's safety pioneers had to battle management every step of the budgetary way, today's safety professionals usually obtain a far more sympathetic hearing regarding safety and health issues.

Professionalism

Dedicated safety professionals continue to be the most valuable asset of accident and occupational illness prevention. Their ranks have grown. In 2008, membership in the American Society of Safety Engineers (ASSE) was more than 32,000. This organization, dedicated to these professionals' interests and development, has 151 chapters in the United States and Canada. Individual membership is worldwide. Other professional societies include the National Safety Management Society, the Board of Certified Hazard Control Management, and the System Safety Society.

In 1968, the ASSE was instrumental in forming the Board of Certified Safety Professionals (BCSP). Its purpose is to certify qualified people as safety professionals once they meet strict educational and experience requirements and pass an examination. Similarly, professional certification of industrial hygienists (CIH) was sponsored by the American Industrial Hygiene Association (AIHA). Both the ASSE and the AIHA are described in Appendix 1, Sources of Help.

Figure 1–5a. In 1919, women making small parts for telephones sat in uncomfortable chairs without back support, working at unguarded, inadequately lit machinery.

Advancement of Knowledge

The tremendous increase in scientific knowledge and technological advancement since the 1950s has added to the complexities of safety work (Figures 1–5a and 1–5b). Prevention and control measures have oscillated between those that emphasize environmental control or engineering and those that emphasize human factors. From this, several important trends in safety work and the safety professional's development have emerged. All are discussed in subsequent chapters of this volume.

- First, more emphasis is being placed on analyzing the loss potential of any organization or projected activity. Such analysis requires the ability (1) to predict where and how loss and injury-producing events will occur and (2) to find ways of preventing such events.
- Second, industry is developing more factual, unbiased, and objective information about loss-producing problems and incident causation to help those who are ultimately responsible for worker health and safety make sound decisions.

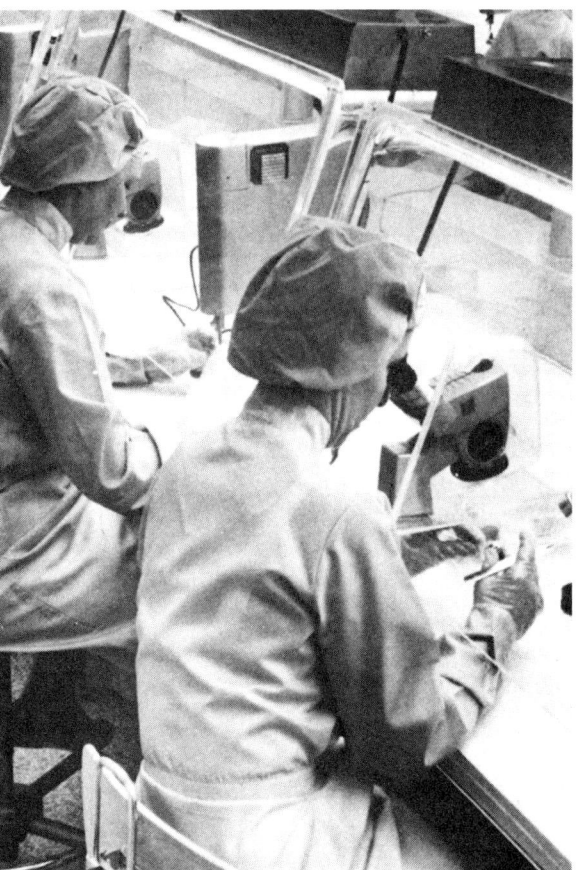

Figure 1–5b. In 1969, workstations were better lit and more comfortable. Improvements continue today. *(Courtesy Western Electric Company.)*

- Third, management is making greater use of the safety professional's knowledge and assistance in developing safe products. The application of the principles of incident causation and control to product manufacturing is assuming more importance because of the rise in product liability cases, the recent legal emphasis on the concept of negligent design, and the potential impact of the life-cycle analysis of a product on the environment.

To identify and evaluate the magnitude of the safety problem, safety professionals must be concerned with all facets of the problem—personal and environmental, transient and permanent. This perspective will help determine the causes of unintentional injuries or identify loss-producing conditions, practices, or materials. On the basis of this collected and analyzed information, safety professionals propose alternate solutions, together with recommendations founded on their specialized knowledge and experience, to those who have decision-making responsibilities.

Therefore, application of this knowledge—whether in industry, transportation, the home, or recreation—makes it imperative that those in the safety field be trained to use scientific principles and methods to achieve adequate results. Most important, they need to have the knowledge, skill, and ability to integrate machines, equipment, and environments with people and their capabilities.

In performing these functions, safety professionals draw on specialized knowledge in both the physical and social sciences. They should know how to apply the principles of measurement and analysis to evaluate safety performance and must have a fundamental knowledge of statistics, mathematics, physics, chemistry, and engineering. Students in safety and health degree programs should be certain that these subjects are included. Safety professionals must know and understand management systems so they can properly advise line managers concerning safety and health management.

They also require training in the fields of behavior, motivation, communications, and management principles along with the theory of business and government organization. Their specialized knowledge must include a thorough understanding of the causative factors contributing to accidents and occupational illnesses as well as methods and procedures designed to control such events.

Safety professionals also need a good general education if they are to meet future challenges. The population explosion, problems of urban areas and future transportation systems, weakening of the family structure, decline of respect for authority, and an uncertain economy, coupled with the increasing complexities of everyday life, will create many problems and stretch safety professionals' creativity to its maximum. They will need all their skills if they are to provide knowledge and leadership to conserve life, health, and property.

Training for the safety professional of the future can no longer be solely "on-the-job" or one-on-one education. It must include specialized undergraduate courses that lead to a bachelor's degree or higher. Training courses, such as those offered by the National Safety Council, will continue to educate a large number of people who began performing safety functions as part of their jobs and who need initial or advanced training in certain specialized areas.

A large number of four-year U.S. colleges and universities offer courses in safety and health, and several dozen offer a bachelor's degree or higher in safety. Two-year community colleges offer associate degrees or certificates for courses designed for the safety technician or part-time administrator.

ACHIEVEMENTS OF THE SAFETY MOVEMENT

The safety movement has helped save more than 3 million lives. It is saving industry and its employees billions of dollars every year. It faces the future with numerous resources for preventing and controlling unintentional incidents and occupational illnesses—resources in know-how, teamwork, goodwill, and education programs that produce trained and dedicated safety workers.

The safety movement has, therefore, done much to meet the double challenge presented to it. The movement has dealt with current unwanted incidents and occupational illnesses and built a solid foundation for the long-range attack on these problems in the future.

To answer the question, "What has the safety movement accomplished?" we look at continued growth in safety awareness and incident reduction. To answer the question, "Where does the safety movement stand?" we must look at what is wrong, as well as what is right, with the present situation. The answer can be found by comparing where the safety movement now stands to where it ought to be. The first point to be considered is simple and grim.

In 2006, unintentional injuries from all sources were estimated to cost this country more than 120,000 lives, cause about 26.2 million disabling injuries, and account for a total financial loss of more than $652.1 billion (National Safety Council, *Injury Facts*, 2008). Occupational unintentional-injury deaths take 4,988 lives a year. Disabling work injuries affect about 3.7 million persons annually and cost more than $164.7 billion (2006 figures).

In recent years, the ratio of off-the-job deaths to on-the-job deaths has been about 7.2 to 1; more than half of employee injuries occurred off the job. In terms of time lost, all injuries to workers, both on and off the job, resulted in a loss of about 100 million staff days of work per year.

Unintentional-injury rates vary widely from industry to industry and from company to company. The wholesale and retail trade, services, finance, insurance, and real estate industries all have occupational injury and illness incidence rates below the private-sector average. On the other hand, rates are above average in construction, agriculture, manufacturing, transportation and public utilities, and mining industries.

Injury incidence rates by size of establishment are lowest for businesses with 1 to 19 employees but rise steadily until they reach a maximum in establishments of 100 to 249 employees. The rates then decline steadily as establishment size continues to increase (U.S. Department of Labor 1995).

Small Establishments

As has been stated, businesses with 100 to 249 employees have proportionately more work injuries than large corporations or companies with 1 to 19 employees. As a general rule, companies—large or small—that ignore safety and health efforts have more than their share of unintentional injuries and occupational illnesses.

The serious safety and health problems of small enterprises are widely recognized, and the National Safety Council has devoted considerable effort to reducing incident and illness rates in this sector. One way has been to establish a liaison between the National Safety Council and the trade associations representing many small companies.

Certain aspects of the small-company problem can be stated with assurance:

1. The small establishment may not be able to employ specialized safety and health personnel to deal with its incident and occupational health problems.
2. The number of unintentional injuries or the financial position of many small companies makes it difficult to convince them that spending money for proper equipment, layout, guarding, and other elements is important.
3. Managers of small operations deal with a host of problems and seldom have the expertise or time for the proper study of incidents and occupational illnesses and their causes.
4. In small units, statistical measures of performance are unreliable. As a result, it is difficult to produce clearcut evidence of the cost of incidents versus the effectiveness of incident-prevention work. In other words, a small operation may have, by luck, a good or bad incident record over a few years, whether or not its safety program is sound.

These aspects present serious obstacles to any progress in worker safety and health. They are not, of course, excuses for failure to prevent unintentional injuries and occupational illnesses. The trade association approach offers the best hope for improvement in this sector.

Labor's Participation in Safety and Health

One of the primary goals of organized labor has always been to protect the safety and health of its members. Many of today's international unions were originally organized to deal with and improve severely hazardous situations in the workplace. They have a sincere desire to work with management on methods to prevent occupational injury and illness to all workers.

In 1949, the National Safety Council issued a policy statement declaring the common interest of labor and management in incident prevention. Even before this date, representatives of leading labor organizations served as members of the council's governing boards. In 1955, a Council Labor Department and a Labor Conference (now known as the Labor Division) were formally established. The Labor Division serves as a vital link between industry management and the nation's labor unions.

Labor Division representatives review products, training materials, and policy statements. The division shares information with labor leaders and nearly 500 volunteers from international and local labor unions. These groups have combined their educational efforts to help organized labor's millions of members improve the quality of their lives, both on and off the job. Some unions have done extensive safety work, published printed matter, and released films or videos that promote the safety and health movement.

The National Safety Council's Labor Division and its Industrial Division, along with other affected divisions, often prepare council position statements on such matters as standards action, oversight testimony, publicity releases, and other areas bearing on occupational safety and health. As a result of these efforts, council positions on health and safety issues are recognized as representing all elements of society, which gives them even greater impact on administrative agencies and legislative bodies.

As a result of a program started in January 1978 with a symposium of leaders from government, industry, and organized labor, the National Safety Council has launched an extensive inquiry to determine causal factors of occupational injuries. Because most data in the past have cataloged only types of injuries, the focus of this program is twofold: (1) to change investigatory and reporting methods and (2) to provide an information exchange bank that lists the factors actually causing injury or occupational illness.

Scope of Safety and Health Committees

The purpose of an occupational safety and health committee is to make the workplace a safer, healthier environment. Some people may want the committee to address matters not directly connected with safety and health. The committee must be careful to stay clear of these side issues, which can detract from its ability to handle its own work objectively. Other issues, such as employee relations problems, should be kept separate from safety- and health-related items or concerns.

The committee also must be careful not to assume the employer's ultimate responsibility for providing a safe and healthful workplace. The committee should be active and encourage management to develop a system for responding to committee recommendations, but this should not be interpreted as reducing the employer's liability. Many unions have included language that specifies the committee structure and duties in labor agreements with the employer in an attempt to better define some of these critical points. A similar system occurs in many European countries—for example, France, where there are legally mandated health and safety committees.

Experience has shown that certain elements are necessary for a joint safety and health activity to be successful in any workplace:
- Both labor and management leaders in the workplace must display a sincere commitment to the safety and health effort.
- Specific roles and responsibilities should be defined for all committee members.
- An effective communication link must be established between the committee members and all employees in the workplace. This includes feedback to workers from committee members.
- Measurable, realistic goals and objectives should be established for committee activity.

The committee must be prepared to evaluate its progress and effectiveness on a regular basis and to modify or change its activities as required. Because of changes in the culture of many workplaces, such as the introduction of total quality management (TQM) or similar programs, those involved with safety and health activities are becoming more well known as teams, rather than as committees. Regardless of the terminology used, the basic concept involves people working together to find and reduce or eliminate workplace hazards that affect employees' health and safety and cost employers in lost time and money every day. Joint safety and health programs can provide a win–win atmosphere for labor and management in most workplaces.

Unions and Safety Committees

Because assigning total responsibility for a union's safety program may be too much for one person, many unions have organized committees to seek out and provide corrections for job hazards. Workers concerned about hazards in the workplace often do not know where to turn for help and information. The union steward or president may not know much about correcting these problems, and the contract may make no provisions for filing grievances about these matters. By setting up a safety and health committee, the union establishes a regular procedure for dealing with problems. Members know where to go first to get matters resolved. Many problems can be resolved immediately without relying on government agencies.

Companies have structured their safety and health committees in as many different ways as there are different workplaces. The only constant in developing these groups is whatever works in that particular worksite. Some committees may work with management; others may be totally separate. In fact, many unions have discovered that in some workplaces, forming two occupational safety and health committees—one joint committee for coordination with management and one limited to union members—works even better.

Besides protecting workers from hazards, a union occupational safety and health committee can also encourage interested members to get involved in union activities. The union safety and health committee shows employees that the union is concerned with their safety and health. Workers will realize that by participating actively in the union's functions, they are helping to improve their own work environments.

Independent Union Safety and Health Committees

A union occupational safety and health committee is totally independent of management and is not to be confused with a joint (union and management) safety and health committee. A union committee can be created without special contract language. The scope, function, and constitution are entirely matters for the union to decide.

A union safety and health committee should be elected or appointed according to the union constitution. To be effective, the committee should be fully integrated into the local union organization. It should be involved at contract negotiation time and should meet regularly with the grievance committee. If a joint union/management committee is also established, some workers should be members of both committees, which makes it easier to coordinate activities. An independent committee can be important for several reasons:

- It allows representatives of the union to meet and consider the safety of the workplace, without any interference from management.
- It gives the union its own forum to discuss and set priorities and strategies for dealing with workplace hazards.
- It allows the committee members to gain appropriate expertise in researching hazards and seeking effective solutions. (Note that there is an important difference between gaining expertise and becoming an expert.)
- It can be used to monitor the performance of a joint union/management committee, if one exists.
- If a joint committee runs into roadblocks or becomes ineffective and the union side withdraws, the union will already have a structure for handling safety and health concerns.

A union and its members should not hesitate to form a safety and health committee. The OSH Act not only encourages worker participation in workplace safety and health efforts, but makes it illegal for an employer to punish or discriminate against a worker for job safety and health activities, such as participating in a workplace safety and health committee.

Joint Union/Management Safety and Health Committees

In addition to a local union committee, the union and management at a workplace may choose to form a joint committee to handle occupational safety and health issues. It makes sense to set up a formalized joint committee because safety and health in the workplace depend on communication with the employer. It is important, however, to be sure that the committee is set up in such a way that both parties can freely express their views and positions. This independent posture is often established through specific contract provisions.

Because every work situation is different, joint committees should be organized to meet the specific needs of their particular workplace. Both labor and management should consider several guidelines when establishing a joint committee at a workplace.

- Union and management should have an equal number of members serving on the committee.
- The union should elect or select the members who will serve as its representatives on the committee.
- The role of committee chairperson should be rotated between labor and management, or the committee may choose to have co-chairs.
- Management should consider appointing committee representatives who have enough authority to make real decisions about projects or spending money to avoid creating delays or unnecessary interference.
- The committee should be able to recommend corrections or request assistance with any occupational safety and health concern or issue.
- The union and management should have an equal voice in the decision-making process and in planning committee actions and agendas.
- A method or procedure should be established to monitor and evaluate the effectiveness of joint committee activities.

Merely establishing a joint safety and health committee does not guarantee that it will be effective. During the planning and early development phases, certain commitments or guarantees must be obtained from the union leadership and from management if the committee is to be successful. Many times these guarantees are stated in a labor agreement and may include some of the following:

- Management should issue a policy statement directly related to safety and health. Usually, these statements cover issues such as implementing committee decisions or responding quickly to employee concerns or complaints.
- The committee should be funded by the employer, with committee members being compensated at their normal rate of pay for time spent on committee activities or projects.
- The committee should have the right to request assistance, such as environmental monitoring, from qualified specialized personnel.
- The committee should have access to useful information in company files, such as monitoring or exposure records, accident and injury reports, and records or lists of chemicals used in the workplace.

Statistics, Standards, and Research

Statistical data on occupational injuries and illnesses have been compiled by the National Safety Council since the 1930s. Analyses are prepared annually and published in *Injury Facts* (formerly *Accident Facts*).

Some industries, through their trade associations, have recorded injury rates for almost 60 years. In most instances, even the divisions of an industry can record specific numbers and types of injuries and compare their experiences with national averages.

A large number of standards enacted by ANSI and other professional organizations relate to safety. Continuing research over the years has kept these standards in line with current safety and health developments and the creation of new products, processes, and materials.

The council's Occupational Safety and Health Services group has developed several survey instruments that measure employees' perceptions of an organization's safety and health operations and management practices. Survey results are maintained in a database that is used to benchmark the adequacy of the organization's safety and health programs against those of a nationwide sample.

The Research and Statistical Services group conducts consulting and research projects for business and industry. Recently, the group completed a series of studies on the risk of occupational fatalities associated with hazardous waste site remediation. The findings showed that in many situations, the fatality risks to workers engaged in remediation were greater than the cancer risks to nearby residents. The group also conducts special studies focused on critical issues, such as a nationwide survey of employers to investigate the acceptance, makeup, activities, and perceived effectiveness of safety and health committees in industry.

Safety and the Law

Early legal action in industrial safety took the form of laws to regulate and investigate industry working conditions and death and injury rates. The next phase was largely concerned with workers' compensation payments.

In subsequent years, all governments gradually expanded their roles in regulating industry on safety matters. The Walsh-Healey Act, which mandates safety measures in companies having supply contracts with the federal government, is an example of such regulation.

In certain industries—notably mining and transportation—U.S. government regulation and inspection have been extensive. The Construction Safety Act, which Congress passed in 1969, addresses the particular health and safety problems of that industry.

In 1970, the Williams-Steiger Occupational Safety and Health Act (OSH Act) was passed. For the first time, the United States had a comprehensive national safety law. The legislation covers every business affected by interstate commerce and employing one or more persons. Safety took on a new direction and meaning as a result of the OSH Act.

But change was also happening worldwide. For example, in the United Kingdom the Health and Safety at Work Act was passed in 1974. The act permits ministers to make regulations to replace existing piecemeal legislation with regulations and codes of practice requiring improved standards of safety, health, and welfare. It provides for a coordinating enforcement authority and gives inspectors powers to initiate actions. Coverage is extended to anyone employed. The act addresses discussions on the atmosphere and also deals with certain building codes and standards. It was a major step involving legislation.

In Australia, the Victoria Occupational Health and Safety Act of 1985 provides for mandatory regulation and is supplemented by codes of practice. The act requires union, industry, and government input, and it intends that most safety and health issues be resolved at the workplace.

In Canada, the Ontario Occupational Health and Safety Act of 1978 for Industrial Establishments, Construction, Mines, and Mining Plants requires the recording and posting of accident records, penalties, and inspection of workplaces. The act's provisions were coupled with the workers' compensation program.

Another Canadian act was developed from provisions of the U.S. OSH Act. Working with all Canadian provinces, the Workplace Hazardous Management Information System (WHMIS) was created. WHMIS has three parts: (1) labels, (2) Material Safety Data Sheets (MSDSs), and (3) a worker education program.

A third Canadian act, the Workers' Compensation Act, became law on January 1, 1990. However, it covers only the province of Ontario. The law deals with a number of subjects: injury, employee retention, accommodation for disabled employees, employee benefit continuation after employee injury, and employee disability benefits.

Besides the OSH Act, several other laws have affected both industry and the safety professional, particularly those laws dealing with environmental issues. These laws include the Clean Air and Clean Water Acts (CAA, CWA), the Toxic Substance Control Act (TSCA), the Resource Conservation and Recovery Act (RCRA), the Comprehensive Environmental Response, Compensation, and Liability Act (CERCLA, "Superfund"), and the Superfund Amendments and Reauthorization Act (SARA, or Community Right-to-Know). These acts are presented in greater detail in Chapter 14, Environmental Management, in this volume and in the *Accident Prevention Manual for Business & Industry: Environmental Management*.

Concern with the safety and health of workers is a major priority for management. This concern goes beyond the obvious benefits of less downtime; reduced costs for workers' compensation insurance; and lower medical and administrative expenses resulting from disability, death, and impaired productivity. Management also knows that there are penalties for not adhering to the law. For example, failure to comply with health and safety requirements can mean citations, which (at the least) create administrative costs but could also lead to serious monetary penalties. The federal government can also institute criminal sanctions against employers and even against individual managers who ignore or disregard the law. Such criminal action has not only come from federal and state job safety and health agencies, however. Local prosecutors have successfully convicted individual managers for murder and aggravated assault in the deaths and injuries of workers on the job.

Management must address serious emerging issues in worker health and safety law. These issues include off-the-job safety and ways to deal with the special problem of employees who are at risk in the work environment because of physical conditions, language problems, or a particular susceptibility to injury or disease. Another issue is the burgeoning paperwork required to comply with OSHA and other agency record-keeping regulations, with the Medical Access Standard, and with the Hazard Communication Standard. Industry accepts almost without question the concept of financial responsibility for work injuries. Not all of industry, however, is convinced of the cost-effectiveness of government regulation of safety procedures.

In addition, some states have gone further and established laws requiring compulsory health and accident insurance to cover employee disabilities from diseases or unintentional injuries occurring off the job. This compulsory insurance might be considered either a drastic extension of the principle of workers' compensation or an extension of Social Security legislation. It differs from workers' compensation in that it puts a financial burden on management for diseases and unintentional injuries brought about by conditions beyond its control.

Whatever the theory, the result of these laws is to give the employer a direct financial stake in dealing with the off-the-job-incident problem. (See Chapter 7, Safety, Health, and Environmental Auditing.)

International Standards

Several factors have spurred the drive for international standardization of health and safety regulations. These include the following:
- Emerging global markets have intensified the need for international standardization.
- Worldwide technological innovations result in changes to industrial methods and organizations that threaten worker and consumer safety.
- The rapid pace of change in science and technology is outstripping standards development in most countries.
- Developing countries' efforts to industrialize mean that they may downplay safety and health regulations in favor of rapid economic growth.

Regulatory change has occurred not only in the United States, but also in other nations. For example, the formation of the European Union (EU; formerly the European Economic Community) has caused changes affecting worker safety, products, and other goods and services. The nations involved are Austria, Belgium, Denmark, Finland, France, Germany, Greece, Iceland, Italy, Luxembourg, the Netherlands, Portugal, Spain, Sweden, and the United Kingdom. Three of the more important regulatory guidelines are the Framework Directive (89/391/EEC) of June 12, 1989; the International Organization for Standardization (ISO) 9000 series; and the ISO 14000 series, issued in 1996. Although compliance with ISOs 9000 and 14000 is voluntary, companies doing business overseas, and their domestic suppliers, have found compliance necessary if their overseas ventures are to succeed.

The Framework Directive

The Framework Directive covers the introduction of measures to encourage improvements in on-the-job safety and health of workers. The range of application for the Framework Directive is broad indeed. It applies to all sectors of activity, both private and public, with the exception of specific working forces such as the armed forces and the police.

The Framework Directive lays down the obligations of employers relating to the safety and health of workers:
- taking the measures necessary for their safety and health, bearing in mind technical progress
- evaluating hazards and instructing workers accordingly
- setting up protection and prevention services—for example, the precision of safety and health practitioners—within the workplace, possibly by enlisting competent external services or people
- organizing first aid
- evacuating workers in the event of serious danger.

The directive also lists workers' obligations in relation to their own health and safety:
- correct use of machinery
- correct use of protective equipment
- the need to report defects in equipment, defects in procedures, and potentially dangerous situations such as near-miss accidents.

To those who export goods and products to the EU countries, the materials must conform to only one standard rather than 15 different ones. Although exporting may be easier, conforming to EU product standards may be more difficult. Under current regulations, testing and certification is required before exporting to EU countries. Companies that are not in EU countries will not be permitted to self-certify. Companies in countries that are EU members can self-certify.

ISO 9000 Series

The ISO 9000 series differs from the Framework Directive in that it is not a set of product standards. Instead, these standards describe a process for establishing quality management and quality assurance. It consists of five quality management standards:

- ISO 9000: provides general guidelines for applying ISO 9001–9003
- ISO 9001: provides a quality systems model for quality assurance in design, development, production, installation, and servicing
- ISO 9002: provides a quality systems model for quality assurance in production and installation
- ISO 9003: offers a quality systems model for quality assurance in final inspection and testing
- ISO 9004: provides guidelines for quality management and quality system elements.

The basic goal of the 9000 series is to give companies guidelines for achieving consistency and uniformity of products or services through the supply chain from the primary supplier to the final customers. Companies can use ISO 9000 as an international benchmark against which they can measure their own performance.

Companies do not need to follow all the ISO 9000 standards, however. For example, organizations involved in manufacturing, shipping, research and development, and service contracts would choose ISO 9001 as their benchmark. ISO 9001 consists of 20 sections (4.1–4.20). The following list indicates how this quality standard would apply to health and safety or environmental management in a company's operations.

- 4.1 Management Responsibility: responsibility and commitment required of management to establish, disseminate, monitor, and refine an environmental health and safety program in the company
- 4.2 Management Systems: description of various systems that organizations can use to define and implement health and safety activities
- 4.3 Contract Review: description of the company's responsibilities to safeguard the health and welfare of all outside contractors and the general public affected by the company's operations
- 4.4 Design Control: defines how the company controls and implements change in terms of the effect on quality and safety
- 4.5 Document Control: describes procedures for writing instructions, operational procedures, and other controlled documents
- 4.6 Purchasing: relates to supply selection and assurance of product quality
- 4.7 Purchaser Supplied Product: describes quality standards for products given to a manufacturer by another supplier to complete
- 4.8 Product Identification and Traceability: establishes methods to trace problem materials throughout the manufacturing process
- 4.9 Process Control: ensures consistency of quality and performance in manufacturing processes
- 4.10 Inspection and Testing: establishes a quality assurance system to confirm quality, quantity, and uniformity of products against predetermined standards
- 4.11 Inspection Measuring and Testing Equipment: refers to calibration, maintenance, and operation of test equipment
- 4.12 Inspection and Test Status: establishes a system that ensures reliability and test status of all test equipment
- 4.13 Control of Nonconforming Product: describes emergency procedures and training to handle incidents relating to nonconforming products
- 4.14 Corrective Action: establishes a system to identify problems and correct them on the job
- 4.15 Handling, Storage, Packaging, and Delivery: establishes systems to ensure quality of product after manufacture
- 4.16 Environmental Health and Safety Records: refers to records needed to demonstrate compliance with standards and policies
- 4.17 Internal Audits: describes a verification system to ensure compliance with policies, procedures, and standards for the system being controlled
- 4.18 Training: ensures that workers are competent to perform their tasks and are continually trained in health and safety matters on the job
- 4.19 Servicing: refers to product stewardship and responsibility for safe, environmentally sound operations and waste disposal
- 4.20 Statistical Techniques: refers to techniques used to monitor performance or progress in safety and health areas.

ISO 14000 Series

The ISO 14000 standards are also related to quality assurance but are primarily directed toward environmental management systems. Companies have a major incentive to comply with these standards because of possible regulatory relief. For instance, the Environmental Protection Agency (EPA) has stated that it might consider up to a 100% decrease in National Pollutant Discharge Elimination System reporting requirements for companies that comply with the ISO standards. Also, penalties for violations might be reduced for certified companies. Although the first ISO 14000 standards were issued in 1996, momentum has been building since then to create another set of standards for occupational health and safety performance.

The ISO 14000 shares many management system principles with ISO 9000 quality system standards. Organizations may choose to use their management systems developed in conformity with the ISO 9000 series as a basis for environmental management. Quality management systems deal with customer needs, while environmental management systems handle issues related to environment protection. Under the ISO 14000, companies are audited by third-party registrars to earn certification from appropriate registration agencies. Companies may also declare themselves to be in conformity with the standard. In Europe, nations have adopted the ISO standard as meeting the requirements of EU environmental standards. Public pressure in Europe may force U.S. companies to do the same.

The time frame for implementing the ISO 14000 standard as follows:
- 1996: environmental management systems specification ("specification" indicates that the company will be audited by a third party to confirm progress); guidelines for auditing environmental management systems (guidelines are voluntary)
- 1997: guidelines for environmental labeling (product claims)
- 1998: environmental performance evaluation guidelines
- 1999: life-cycle assessment guidelines.

Safety and Occupational Health

Cooperation between medical and safety personnel in incident-prevention activities began during the earliest days of the safety movement. Nevertheless, interest in safety on the part of the medical profession and, conversely, interest in employee and public health on the part of the safety professional is increasing. This trend has increased since the early 1980s and assumed increasing importance in the 1990s.

Part of the interest in workers' health results from concern with occupational disease, noise, radiation, and other problems that extend beyond the former concepts of occupational incident prevention. Interest in protecting the health of citizens in nearby communities from industrial unintentional injuries or hazardous materials releases comes partly from such infamous and well-publicized events as the Three Mile Island incident; the Chernobyl, Ukraine, meltdown; and the Bhopal, India, disaster.

Work Environment

Over time, safety professionals became aware of the relationship between physical illness and working conditions. They found that workers in certain industries such as mining, chemical manufacturing, and steel exhibit a higher-than-normal incidence of such problems as dermatitis, musculoskeletal problems, pulmonary disease, mental illness, and cancer. The improved safety and health of today's workers is the result of concerted efforts by a safety, industrial hygiene, and occupational health team working with a management that realizes that an organization's primary asset is a safe and healthy work force (Figure 1–6).

For more details, refer to the council's *Fundamentals of Industrial Hygiene* and *Occupational Health and Safety*, part of the Occupational Safety and Health Series.

Figure 1–6. Only time, experience, and a concerted effort by all involved disciplines revealed the many relationships between physical illness and work environment. *(Courtesy Library of Congress.)*

Community Environments

Increasingly, the public has demanded a larger role in the management of community environmental risks. Both public and private risk managers realize that providing avenues for public participation is a necessary part of their decision-making process. The problem is how to ensure public involvement and at the same time improve the quality of safety and health decisions.

To help fill the gap in credible risk communication on environmental health and safety issues, the National Safety Council established the Environmental Health Center. This special-purpose organization is led by a board of governors and operates mostly through philanthropic funding from concerned corporations, foundations, labor unions, and individuals. Its goals include development of accurate and objective information on environmental and public health risks, improvement of public knowledge about these risks, and dissemination of this information to the public.

Workers with Disabilities

The employment of workers with disabilities by progressive companies and the impact of federal and state equal opportunity laws have modified the practice of preemployment examinations to screen out unfit or undesirable prospects. Medical personnel perform a preplacement examination simply to determine what physical or mental restrictions are appropriate to the prospective employee; they do not determine fitness for a specific job. The job description must specify realistic physical and mental requirements that the human resources department can match to medical restrictions. The Americans with Disabilities Act (ADA) requires that the job be modified to accommodate the disabled worker. (See the discussion in Chapter 23, Workers with Disabilities, in this volume.)

Psychology and "Accident Proneness"

Safety professionals who are thoughtfully looking for ways to improve their work encounter a great deal of useful information in modern psychological writing—and a great deal of careless and misleading generalizations. For example, concern about the so-called accident-prone individual in industry is as old as the safety movement itself. Statistical information suggests that such individuals exist, though clear and sharp data demonstrating this suggestion are remarkably hard to obtain. Too many alleged proofs turn out to be statistically deceptive, based on inadequate samples, or the result of highly subjective diagnoses. Justification for use of "proofs" of accident proneness may reflect the common-law doctrines in which employers tried to justify accidents as being caused by the employee's action.

Statistical support for the existence of accident-prone individuals is elusive. Some safety professionals believe in accident proneness and feel it may be a passing phase in the individual and not a permanent characteristic. At most, it may be a problem encountered only by an insignificant minority. Realistically, objective analysis might reveal a supervisory deficiency or procedural weakness that can increase the risks of certain operations or interfere with the performance of individuals or groups of workers.

The same observation applies to psychological tests used as screening devices for new employees. So-called experts may make spectacular claims for the ability of these tests to predict accident proneness, but so far no test has proven itself to the general satisfaction of the safety profession. In the past, the work of psychologists such as Dunbar and the Menningers aroused great interest among safety professionals. However, the best weapons in the practical, day-to-day fight against unintentional injuries tend to come from strong line management safety awareness and from the disciplines of engineering and behavioral psychology, such as human factors engineering, system safety, and risk management or assessment. (Refer to the discussions in Chapter 30, Motivation, and Chapter 31, Safety and Health Training.)

The field of industrial safety is one of progress and improvement, largely through the continued application of techniques and knowledge slowly and systematically acquired through the years. There appears to be no limit to the progress possible through the application of the universally accepted safety techniques of education, engineering, and enforcement.

Yet serious problems remain unsolved. A number of industries still have high unintentional-injury rates. In far too many instances, management and labor either fail to work together or have different goals for the safety program. The efforts to form effective joint safety and health committees are helping labor and management meet common goals in safety and health.

The resources of the safety movement to tackle these problems are impressive: a growing body of knowledge, a corps of able safety professionals, a high level of prestige, and strong organizations for cooperation and exchange of information.

CURRENT ISSUES

Some problems of the safety movement are directly related to the movement's traditional strengths and weaknesses. A few of these problems are social and political, while others are essentially organizational.

Technology and Public Interest

There is no reason for the safety professional to be alarmed at the public's interest in product safety, a better environment, and general technological trends. An informed, educated public can be a powerful ally in safety work. Emphasis on automation and more refined instrumentation will probably continue. New problems will arise, but they will generally be those that well-established methods of safety engineering can solve.

The use of new materials and techniques—particularly radioactive materials, automation, and lasers—is likely to present more serious difficulties to the safety professional. However, even here, the professional can draw on the movement's considerable experience to handle these problems.

Finally, safety professionals need to keep up with the rapid developments in communications and information technology as they affect the safety field. (See Chapter 23, The Computer as a Safety Information Tool, in the *Engineering & Technology* volume.)

Political Problems

On the political side remains the age-old problem of industry-union-government relations. The key issue here is what regulatory role government should play in protecting worker health and safety and what aspects of life national government should regulate. This issue is likely to remain a source of controversy for years to come.

Organizational Problems

On the worldwide scale, a wide variety of organizations are attacking specific aspects of occupational safety and health problems. The National Safety Council is a strong, constructive, nonpolitical, nonprofit leader among these organizations. It has repeatedly sought and often achieved a cooperative division of labor between itself and other organizations in the safety and health field. One of the guiding principles of the council has been that there is work enough and credit enough for all involved.

It remains to be seen whether the best organizational forms have been found for participation by all businesses in safety and health work. Safety professionals should be ready to consider new ideas and new solutions.

A Look to the Future

The greatest reasons for intensifying the safety effort are humane and moral. The worth of neighbors, friends, and family cannot be measured in dollars or coded into computer records. In the coming years, the U.S. population will continue to grow, although more slowly than in the past. By 2010 it was about 308 million. The average life expectancy of these people will increase, partly because of health care advances.

The shift from extractive and manufacturing to service areas of employment will continue in the next several decades. This fact will hopefully contribute to a decline in occupational death rates, as fewer people will be employed in high-risk industries.

The work force continues to undergo major changes as minorities and women move into more industries and management levels. Safety issues associated with the growing numbers of female workers are complicated, and many difficult choices regarding working conditions will have to be made.

As social changes continue to take place—including a high divorce rate, more single-parent families, and more two-income households—the effect on the structure and values of family life will be felt more strongly. Many perceive a reduced respect for either parental or social authority, a factor they believe creates considerable conflict and safety problems in the workplace.

One current trend is the evolution of a world market. The establishment of a European Union market in the early 1990s has had a major impact on world economics due to the EU regulations on products and goods. The complete restructuring of the economies of the former Soviet Union and eastern Europe continues to impact the world markets in the early 21st century. As a result, there will be a great demand for standardizing injury and illness record keeping so that comparisons can be made between various segments of the global market. Labeling of hazardous materials must also be standardized so that all personnel handling chemicals manufactured in one country and shipped to another will understand the warnings on the container.

Incident-prevention information will become a much more global commodity than it has been. Although to some extent incident-prevention information is shared with worldwide affiliates of large multinational corporations, most information developed in one country rarely finds its way to others. Excellent behavioral techniques are in use in Sweden today that are hardly known in the United States. As the world becomes a global market, these national barriers will come down and information sharing will also become global.

The United States, western Europe, and Japan are moving into a post–industrial revolution era or information age. Developing countries, however, are just coming out of the early industrial revolution as new technologies and manufacturing industries spread to these nations. This trend should create a strong demand for U.S. safety and health expertise in the developing countries.

What will the government's regulatory agencies be doing in the next few years? What kind of standards can be expected from OSHA, EPA, and the Mine Safety and Health Administration (MSHA) in the next few years? Probably the most far reaching will be the "generic standards" regulating exposure to chemicals. In 1989, OSHA promulgated its permissible exposure limits (PELs) standard. Because of rules laid down in the 1970 OSH Act, OSHA was able to adopt at that time an existing federal standard (the 1968 Walsh-Healey Act) to establish permissible exposure levels for about 400 hazardous chemicals.

The days of the facility that operated incognito are gone forever. In the past, industry operators and owners felt that the less anyone knew about their businesses, the better off they were. That philosophy became obsolete as the public and industry members alike realized the need to control accidents, injuries, and work-related illnesses. The trend toward openness and cooperation with the public, the government, and the media has spread to all industries. In fact, the leading companies will be those that are most successful in convincing the public and its local governments that they are good neighbors. Safety professionals will have to understand the legal and political ramifications of their companies' safety practices in light of the public's changing expectations.

These are issues of major importance and will need the expertise and guidance of everyone in the safety and health field. The future is, as it always has been, most uncertain. By working together, everyone in the safety and health community can reduce some of that uncertainty by helping to make workplaces and off-the-job environments safer and healthier.

SUMMARY

- According to the philosophy of incident prevention, the employer is primarily responsible for ensuring a safe, healthy work environment. Employees are held accountable for following prescribed safety standards and guidelines.
- During the industrial revolution, industry, government, and the public began to realize that productivity and safety are closely related. In 1913, the National Council for Industrial Safety, later renamed the National Safety Council, was founded. Other groups soon followed, and these groups began to keep statistics on accident and injury rates in industry.
- The safety movement began to identify health and safety issues and to develop methods to help protect workers on and off the job. Since the end of World War II, the safety movement has become an accepted and valued part of industry.
- Nations have started to develop international safety and health regulations because of the emergence of global markets, the rapid pace of industrialization, and advances in science and technology that put workers and consumers at risk.
- Over the past eight decades, safety workers have gradually become safety professionals. Today's safety and health problems require a highly trained, experienced, and dedicated team of professionals to find solutions.
- The National Safety Council seeks to improve investigative methods and to establish an information exchange bank regarding work-related hazards. The council also supports special research projects to study the effectiveness of safety measures and legislation.
- The safety movement will face serious challenges in the future, including new technologies, major changes in the work force, formidable environmental hazards, bloodborne diseases such as AIDS, and widespread drug and alcohol abuse.

GENERAL SAFETY BOOKS

In addition to this manual, the following books cover the basics of occupational safety and health. Note: These resources serve as the basis for the questions on management in the Board of Certified Safety Professional examinations.

Browning, R. L. *The Loss Rate Concept in Safety Engineering.* New York: Marcel Dekker, 1980.

DeReamer, R. *Modern Safety and Health Technology.* New York: Wiley, 1981.

Ferry, T. S. *Modern Accident Investigation and Analysis.* 2nd ed. New York: Wiley, 1988.

Firenze, R. J. *The Process of Hazard Control.* Dubuque, IA: Kendall-Hunt, 1978.

Gilmore, C. L. *Accident Prevention and Loss Control.* New York: American Management Association, 1970.

Grimaldi, J. V. *Safety Management.* 5th ed. Burr Ridge, IL: Irwin, 1988.

Heinrich, H. W. *Industrial Accident Prevention*, 5th ed. New York: McGraw-Hill, 1980.

Manuele, F. *On the Practice of Safety.* 2nd ed. New York: Wiley, 1997.

Petersen, D. C. *Analyzing Safety Performance.* Reprint, Goshen, NY: Aloray, 1984.

———. *Safety by Objectives.* Goshen, NY: Aloray, 1978.

———. *Techniques of Safety Management.* 3rd ed. Goshen, NY: Aloray, 1989.

Tarrents, W. E. *The Measurement of Safety Performance.* New York: Garland STPM Press, 1980.

REFERENCES

American Engineering Council. *Safety and Production*. New York: Harper & Brothers, 1928.

American National Standards Institute, 11 West 42nd Street, New York, NY 10036.

Andrews, E. W. "The Pioneers of 1912." *National Safety News* 66 (1952): 24–25, 64–65.

Beyer, D. S. *Industrial Accident Prevention*, 3rd ed. Boston: Houghton Mifflin, 1928.

Campbell, R. W. "The National Safety Movement." Proceedings of the Second Safety Congress of the National Council for Industrial Safety, 1913, 188–92.

DeBlois, L. A. *Industrial Safety Organization for Executive and Engineer*. New York: McGraw-Hill, 1926.

Eastman, C. *Work Accidents and the Law*. New York: Charities Publication Committee, 1910. Reprint, New York: Arno Press, 1969.

Felton, J. S. "History of Occupational Health and Safety." In Occupational Health and Safety, 2nd ed., edited by J. LaDou. Itasca, IL: National Safety Council, 1994.

Heinrich, H. W. *Industrial Accident Prevention*, 5th ed. New York: McGraw-Hill, 1980.

Holbrook, S. H. *Let Them Live*. New York: Macmillan, 1939.

International Organization for Standardization, 1 rue de Varembe, Case Postale 56, CH-1211, Geneva, Switzerland.

Menninger, K. A. *Man against Himself*. New York: Harcourt, Brace & World, 1956.

Meyer, R. L. "Series Commemorating the Diamond Anniversary of the Council." *Safety and Health* (January–October 1987).

Mock, H. E. *Industrial Medicine and Surgery*. Philadelphia: Saunders, 1920.

National Safety Council, 1121 Spring Lake Drive, Itasca, IL 60143.

"Golden Anniversary Issue." *National Safety News* 87 (1963): 5.

Injury Facts (formerly *Accident Facts*). Issued annually.

Proceedings of the First Co-Operative Safety Congress, 1912.

Proceedings of the National Safety Congress. Issued annually, 1914–1925.

Proceedings of the Second Safety Congress of the National Council for Industrial Safety, 1913.

Safety and Health. Issued monthly.

Schaefer, V. G. *Safety Supervision*. New York: McGraw-Hill, 1941.

Schulzinger, M. S. "Accident Syndrome—A Clinical Approach." *Archives of Industrial Health* 11 (1955): 66–71.

Schwedtman, F. A. *Accident Prevention and Relief*. New York: National Association of Manufacturers in the United States of America, 1911.

Toynbee, A. *The Industrial Revolution*. Boston: Beacon Press, 1956. First published 1884.

U.S. Department of Labor, Bureau of Labor Statistics. *Occupational Injuries and Illnesses: Counts, Rates, and Characteristics*, 1992. Bulletin 2455. Washington DC: Government Printing Office, April 1995.

REVIEW QUESTIONS

1. List the six reasons for preventing injuries and occupational illnesses as given in the text.
   a.
   b.
   c.
   d.
   e.
   f.
2. With the industrial revolution came innovations in the processes and organization of production known as the factory system. List four components of the factory system that emerged at that time.
   a.
   b.
   c.
   d.
3. What industry introduced the factory system in the United States?
   a. shoe manufacturing
   b. farm implements
   c. textiles
   d. food processing
4. Define the fellow servant rule.
5. Define contributory negligence.
6. Define assumption of risk.
7. Which president's legislation set up the first workers' compensation laws covering only federal employees?
   a. Theodore Roosevelt
   b. Dwight Eisenhower
   c. Harry Truman
   d. Franklin D. Roosevelt

8. Which state passed the first effective workers' compensation act?
   a. New Jersey
   b. Wisconsin
   c. Washington
   d. New York
9. What industry was the first to realize that the actions of people are important in creating unintentional-injury situations?
   a. textiles
   b. railroads
   c. steel
   d. agriculture
10. When was the National Safety Council started?
    a. 1910
    b. 1905
    c. 1912
    d. 1915
11. What is the "Three Es of Safety" concept?
12. List the three trends in safety work and the safety profession that have emerged from the prevention and control measure concepts of the 1950s.
    a.
    b.
    c.
13. List the necessary elements for a joint safety and health activity to be successful in any workplace.
    a.
    b.
    c.
    d.
14. Regarding safety matters in the United States, which act regulates companies with government contracts?
    a. Walsh-Healey Act
    b. Williams-Steiger Occupational Safety and Health Act
    c. Construction Safety Act
    d. 1974 Health and Safety at Work Act
15. Which items were included in the national safety law brought about by the Williams-Steiger Occupational Safety and Health Act?
    a. one or more employees
    b. businesses affected by interstate commerce
    c. manufacturers
    d. businesses selling to the general public
16. List four factors that have spurred the drive for international standardization of safety and health regulations.
    a.
    b.
    c.
    d.
17. List five employer obligations relating to the safety and health of workers stated in the Framework Directive (89/391/EEC) of ISO 9000 and ISO 14000.
    a.
    b.
    c.
    d.
    e.
18. List three employee obligations relating to the safety and health of workers stated in the Framework Directive (89/391/EEC) of ISO 9000 and ISO 14000.
    a.
    b.
    c.
19. What were the purpose and significance of the death calendar used by the Russell Sage Foundation in Allegheny County, Pennsylvania?
20. What contributed to the lowering of the overall occupational death rate in the United States since World War II?
21. Explain why an independent union safety and health committee can be important.
22. What guidelines should both labor and management consider when establishing a joint safety and health committee at a workplace?
23. Discuss the major safety and health issues management must address.

The Safety, Health, and Environmental Professional

2

James T. O'Reilly, JD

Richard J. Hackman, CIH, QEP

Defining the Role

Economic Issues

Ethical Issues

The Effects of Globalization

Presenting SH&E Benefits to Managers

Training the SH&E Professional for New Roles

Consulting and Expert Witness Roles

Personal Liability Questions

Professional Liability of Consultants

Future Opportunities for the SH&E Professional
Technology Changes ▸ Better Control of Risks ▸ Product Stewardship Roles ▸ Indoor Air Quality ▸ "The Shift to the Subtle" ▸ Regulatory Agency Influences ▸ Nongovernmental Organizations

Summary

References

Review Questions

DEFINING THE ROLE

The safety, health, and environmental (SH&E) professional plays several critically important roles in the business and the governmental settings: to save lives, to prevent harm to workers, to maintain productivity, and to encourage retention of productive workers by enhancing the perception that working at the facility is consistent with the individual's desire for a safe and healthy work environment. This chapter addresses some of the current topics that challenge the SH&E professional.

Some professions have well-defined roles as well as laws and licensing structures to shield them, such as the physician's right to prescribe and the lawyer's right to appear in court. SH&E professionals can benefit their careers and their status on the job by exhibiting attitudes and actions that demonstrate professional objectivity, moral and ethical standards, and a concern for accurate delivery of technically sound advice. To be an effective corporate "team player" or a valued member of a public-sector health program, the SH&E professional should understand where his or her role fits within the set of complex challenges that face modern private-sector and public-sector organizations. "Seek first to understand" may be the best opening advice for the new entrant into this field.

ECONOMIC ISSUES

Corporate or government employers have productivity goals to achieve, and they often have a leadership team whose vision of the future includes improvements in profitability (or constituent service, for government entities). Safety, health, and environmental topics are usually not a focus of new initiatives. New SH&E professionals should not expect business managers to arrive in management with worker safety as a high initiative or goal. Instead, the culture of the facility should make safety, health protection, and environmental compliance a foundation that underlies the methods by which the employer does what it is expected to do. This should be a core expectation—similar to quality, codes of conduct, and so on. Managers who recognize the costs of lost productivity and adverse external criticism will increase the attention given to SH&E issues, as Alcoa did in the 1990s with great success.

So, the SH&E role is not at the forefront of the goals of the chief executive officer (CEO), mayor or governor, or other top managers in the organization. For the professional, this is good and bad news—bad in the sense that management does not intuitively think about SH&E roles as part of productivity and worker retention, but good in the sense that as a culture, the healthy workplace is expected and encouraged within most companies. The professional's task is to discern attitudes, educate workers, enhance attention to workplace health and safety, and keep a vibrant culture of safety awareness. Money matters and budgets are important, and a good connection can be documented between a safe working environment and strong productivity. In a labor market that allows workers many choices, the best workers are going to leave a company where health or safety is endangered. In a social structure that awards positive media coverage to socially responsible managers, the CEO and management team recognize that positive SH&E performance can be a key component in an overall strong corporate governance program.

ETHICAL ISSUES

The ethical behavior of professionals involves making decisions that reflect sound choices that conform to the peer group's expectations of prudent conduct. Ethical behavior is newsworthy. The news media focus on medical ethics in the care of dying patients and on lawyers' ethical obligation to client confidentiality. But the media rarely broadcast the message that SH&E professionals have ethical norms of behavior as well. As a preventive profession, the successful SH&E program prides itself on what does not occur—and the professional's choices should be sound, prudent, and practical even when management applies pressures that would make the work environment more hazardous.

A sense of ethics and "doing the right thing" should be at the core of the professional's dedication to the work. SH&E professionals who belong to the American Industrial Hygiene Association (AIHA) subscribe to six canons of ethical behavior, including the obligation to apply recognized scientific principles to protect the health and well-being of the employees. Professionals should maintain confidentiality of the business and personal information shared with them, with a few exceptions. The potential health risks and precautions needed to prevent adverse health effects should be disclosed through fact-based worker counseling.

Beyond a code approach, a prudent SH&E professional understands that this profession touches people's lives and health, making his or her responsibilities more significant than those of someone who deals only with assets or equipment. When worker health issues are compromised by corporate cost demands, the strain on the professional's loyalties raises tough ethical choices. "First, do no harm" is a medical norm of ethics for physicians. "First, ensure reasonable protection for the health of individuals" may be the SH&E professional's guide to follow.

Ethical choices become difficult when the employer asserts a legitimate business need to delay disclosures to workers about the risks of particular job assignments. What is the duty to disclose which pieces of information? False statements about health risks are unwise, unethical, and possibly subject to liability for damages or prosecution. But less clear-cut examples of delays in disclosure or partial release of information carry some ethical implications for the professional.

THE EFFECTS OF GLOBALIZATION

Several converging trends affect the work of the SH&E professional, and globalization is certainly a prime consideration. Increasingly, heavy manufacturing has moved to developing countries in the Far East, such as China. Research and development of new products is more global, and marketing opportunities for U.S. facilities are increasingly found in exports to developing nations. Labels are becoming multilingual to a degree never before seen in U.S. retail stores. The World Trade Organization (WTO) and North American Free Trade Agreement (NAFTA) treaty provisions that promote environmental and labor standards have received broad public attention; WTO's role in the shifting of industrial production has been criticized, and NAFTA's impetus to export markets has been widely reported. Whether makers of widgets in Wisconsin or tractors in Texas will benefit from any particular economic trade innovation is a topic beyond our scope. What we see in modern facilities is an increasing impact of global standards and public concern over whether workers in developing countries who make products for the U.S. market will be accorded safe working conditions. This has become the focus of several prominent nongovernmental organizations (NGOs) that have led sophisticated media campaigns highlighting improper working conditions in foreign manufacturing sites.

SH&E professionals with an interest in traveling can be found all over the globe, advising local and multinational facilities about workplace safety methods and goals. Developing nations are found at various stages of safety and environmental regulation and law enforcement. SH&E consultants can help make worker protections a reality in those countries with the leadership of progressive managers who foresee the benefits of being leaders in safe workplace environments. With today's rapid communication systems, a local issue can become publicized globally almost instantaneously—especially stories of poor working conditions and abusive behavior toward employees.

PRESENTING SH&E BENEFITS TO MANAGERS

In a corporate setting, the internal competition for budget resources can get intense. A fixed number of dollars must be allocated among manufacturing equipment, advertising, new packaging, salaries, new safety measures, and so on. In order to compete for these scarce resources, SH&E professionals must learn to effectively articulate that capital spent on safety, health, and environmental measures is a good investment. They must be able to demonstrate that the return on investment for these areas exceeds that of the other opportunities. Although this may be difficult to prove in certain situations, the SH&E professional must be able to provide compelling information that shows the business value created by SH&E expenditures and activities. Points that can be made to support these arguments include the following:

Public trust—The public trust that companies are expected to maintain with respect to employees, investors, news media, consumer organizations, and so on, must be satisfied by providing a safe and healthy workplace and environmental compliance. Corporations and public agencies are vulnerable to criticism of their safety, health, and environmental records, and investors and voters are attentive to such criticisms. Loss of corporate or organizational reputation can have significant financial ramifications—loss of market share, diminished ability to generate revenue, product boycotts, negative media coverage, loss of consumer trust, and so on.

Cost reduction—Comparative benchmarking of costs can be an effective method of assessing SH&E performance. When managers of a consumer paper company observed that their safety and health programs had kept workers' compensation costs at about $8 million while competing firms' costs were about $60 million, managers of the "better" company validated their return on investment. Reducing costs further while making prudent choices is a model easily understood by business leaders or elected officials in public agencies.

Worker retention—Retaining good workers increasingly means that the employer must provide a work environment where employees believe that their personal safety and health are of primary importance. Dreary and dangerous worksites with no concern for the individual's health are the first ones to lose qualified workers when the economy offers a variety of job opportunities.

Increased productivity—The increase in productivity at a safe workplace can be effectively measured over time. Enlisting employees in the process of making the work environment more safe and productive can be an energizing activity that further elevates morale and creates a positive culture at the site. This translates directly to increased performance and productivity.

TRAINING THE SH&E PROFESSIONAL FOR NEW ROLES

Education does not end with a diploma. SH&E professionals face three competing needs for training: specialized training in a particular discipline, generalist training to understand multiple disciplines or areas of expertise, and business training to gain a deeper understanding of business measures and practices. A rich abundance of training opportunities is available—including online coursework. Deciding what the SH&E professional must learn is like choosing desserts at a buffet line. Within a reasonable budget and time allowed for training, the professional has options.

In general, the prudent SH&E professional will get a good grounding in the general topics first, move to specialized role training in depth, and then pursue training in business communication skills. But no career can or should be static and rigid; flexibility and continuous improvement of professional knowledge must be parts of the professional's work life. Physical safety issues, emergency planning, measurement and abatement of health risks, regulatory responsibilities and inspection responses, emerging health risk topics relevant to one's industry, and the like, should be parts of the basic menu of continuing education.

CONSULTING AND EXPERT WITNESS ROLES

SH&E professional roles have been affected by a significant shift in employment relationships. According to the AIHA, approximately one-fourth of the membership of the AIHA are consultants and companies' use of nonemployee consultants for safety and health tasks has increased. Several associations whose members are active in the field have noted similar shifts to outsourcing of health-related service functions.

Expert witness testimony roles in occupational injury claims such as chemical exposure liability cases and indoor air quality lawsuits have presented a new set of problems for some SH&E professionals. Since the mid-1990s, federal courts and most states have applied standards that check potential expert witnesses for accuracy before they are allowed to testify. Such an examination, often called the *Daubert test* because of a Supreme Court decision on expert testimony, is an important such device.

The pretrial hearing for the judge's questioning of the expert acts as a screen to prevent the trial jury from considering evidence or opinions that are not well supported by rigorous testing or empirically verifiable findings. As a response to concerns that juries would be misled or that "you can find somebody to testify to anything," the courts use screening techniques that emphasize pretrial evaluation of the accuracy and technical validity of the expert testimony. A court applying the Daubert test looks at the scientific principles being offered in the expert's report; the experimental basis, if any, for the opinion; the potential rate of error for the findings; the existence of a uniform standard that is being applied; the degree of peer review and general acceptance of the claimed findings; and other factors that would make the expert's opinion reliable.

PERSONAL LIABILITY QUESTIONS

The SH&E professional is probably not going to be sued or held legally liable in the event of injury to a worker or others. Workers' compensation laws usually prevent lawsuits against fellow employees. Lawsuits against corporations do not usually include claims against the individual employees who acted for the company; it is usually simpler and faster to sue the corporate defendant alone. Most corporate systems include provisions that indemnify or pay the costs of defense for employees who are sued when performing their assigned duties. In the event that an individual SH&E manager is sued, the corporate employer will probably pay the legal expenses to have the individual suit dropped.

When the SH&E professional is acting as a consultant, vulnerability to lawsuits is a constant cost of doing business. Such professionals should carry adequate *errors and omissions* insurance so that allegations of harm that ripen into suits can be competently defended by lawyers paid by the insurance carrier. This is the equivalent of malpractice insurance for health care professionals.

Government penalties are usually imposed against the employer when a safety penalty or environmental enforcement case is brought. The rare few cases that have named individuals have been criminal prosecutions of managers whose actions seriously jeopardized human safety or the environment. In some cases, the consultant who advised the company to take some action may be charged along with the company.

PROFESSIONAL LIABILITY OF CONSULTANTS

The SH&E consultant develops important sets of data about facility conditions that may affect worker health and the environment. Some of these data are evidence in court cases, and some SH&E professionals spend much of their careers preventing or responding to liability challenges. An expert consultant on indoor air quality complaints, for example, may advise a client to settle a workers' compensation claim rather than be exposed to extensive disputes among a wider class of workers. Or the professional may assure a client that an OSHA regulation does not apply and should not be complied with.

Under these scenarios, the employer that issued the contract for expert services may be surprised when a court or administrative body punishes the employer. In turn, the consultant on whose advice the company relied has a possible vulnerability to professional negligence if the advice did not meet the normal standards of care of similar advisers within that area or subject category of expertise.

FUTURE OPPORTUNITIES FOR THE SH&E PROFESSIONAL

Technology Changes

New chemicals, new technologies, and new biologically active materials are contributing to a dramatic growth in "high-tech" industries. Nanotechnology is a prime example of an emerging area that contains significant potential benefits as well as potential safety, health, and environmental risks. The SH&E professional must enable these cutting-edge technologies to realize their potential benefits to society, yet safeguard employees who work with these materials. These new technologies will challenge the existing SH&E body of science and will require professionals to expand their skillsets. Understanding the toxicological and environmental effects of these new materials will be paramount, as will the ability to communicate effectively the risks as they become recognized. Renewal of SH&E education, tied to evolving awareness of the types of technologies presented, must be constant. Educating oneself about the new materials and then dealing effectively with their control, containment, and so on, will demand special attention.

Better Control of Risks

The "state of the art" of worker protection is moving rapidly, and there is a premium for quickly resolving workplace and environmental hazards. Speed of response to newly identified risk issues is important. Not only is optimum worker and environmental protection the right thing to do, but with the rapid communication mechanisms of today, tolerance for unacceptable conditions has also diminished. Issues that previously remained inside factory walls or within a community now quickly become news items on the Internet or in the media. Professionals who seek to help the organization stay free of criticism, penalties, and lawsuits and to maximize workplace performance need to respond quickly to newly identified areas of risk. Staying aware of the latest control technologies is also critical.

Product Stewardship Roles

Although SH&E professionals are not usually true product development experts, they must play a key stewardship role in bringing new technologies from the laboratory to the marketplace. The occupational setting is often where some of the highest concentrations and durations of exposures occur, and this is frequently where health issues first arise. This brings the SH&E professional into the world of "product stewardship"—and well beyond the walls of the traditional workplace. Here, the professional's experience with exposure assessment and risk communication can be invaluable to the product development process. Some of the same basic tools of exposure assessment can be applied to institutional and consumer uses of a product. This type of involvement by the SH&E professional will ensure the safety of the product throughout its life cycle—from the laboratory bench top, through manufacturing, to its use by customers and consumers.

Indoor Air Quality

With today's increased concern with health and well-being, the complex mixture of perceptions, fears, and measurable contaminants that is collectively known as indoor air quality will likely increase. Medical and psychological literature on assertions of illness from indoor air has blossomed in recent years, and more work needs to be done with more combinations of ventilation, chemical exposure, and pre-existing worker health status.

Much can be learned from the guidance of those who have studied this phenomenon. Doing the initial investigation objectively, quickly, and correctly is vital. The reader should examine some of the reference materials cited at the end of this chapter for more information about this area of concern. We predict that the SH&E professional will have a few interesting opportunities to be diplomatic, creative, and data focused when indoor air disputes arise.

"The Shift to the Subtle"

In developed countries, the good news is that the modern workplace is likely to have few acute hazards or major physical risks. However, more subtle hazards, , such as chemical exposures, ergonomic risks, and physical agents, can still persist. Safeguards for this set of hazards can be more elusive and often more expensive than the traditional engineering controls that mitigate explosion risks, prevent falls, or guard machinery. The job of the SH&E professional may shift over time to protecting employees in specialized industries from potent compounds or to safeguarding vulnerable populations from other, more recognized hazards. More research is being dedicated to more subtle health effects such as endocrine disruption and reproductive problems. The synergistic interactions between chemicals are also areas of ongoing research.

Regulatory Agency Influences

Where you stand depends on where you sit; in other words, a person's position in an SH&E organization affects how he or she responds to government enforcement of occupational safety, health, and environmental standards. To the extent that an organization's management views government enforcement as a threat and views government agencies as adversaries to be fought, the in-house SH&E professional will be on the firing line whenever government inspectors arrive. It does not have to be that way. The optimal approach may be for the professional to take advantage of state and federal agency programs for cooperative compliance evaluation and an agreed method of dealing with areas of potential concern. Surely it is better to have a cooperative relationship with government experts helping to find solutions to a safety, health, or environmental concern than to have an onerous relationship in which the employer is seen as a stubborn offender with persistent problems.

OSHA's decline in budget power and in perceived stature in recent decades may affect the ability of the professional to gain certain managers' attention. Sometimes, a company in the private sector disdains OSHA and discounts the value of compliance. Or the company may perceive OSHA as weak, complex, or too distracted by other battles to become an active regulator of the company's operations. Experience teaches that underestimating a regulator is a recipe for disaster. The regulatory agency is capable of harsh punishment or even criminal prosecution, so it would be folly to take regulators too lightly. The "culture of compliance" should accept that OSHA and its sister agencies have an important public role that deserves high respect. The progressive manager will view regulatory agencies with respect and acknowledge their missions. That person will develop a relationship with these agencies based on trust and collaboration.

Nongovernmental Organizations

With the rapid advancement of today's communication technologies, one group in particular has grown substantially in importance and influence: nongovernmental organizations (NGOs). The SH&E professional would be wise to understand and recognize the importance of these groups. Many of the major NGOs are increasingly well funded, globally coordinated, and sophisticated users of the media. Their credibility is greater and their social networking is greater than in their formative decades. NGOs are often organized to support a particular cause or area of interest, and they vary in the ways they carry out their advocacy. NGOs have played a critical role in highlighting worker safety issues in developing countries, significant environmental issues, and newly discovered product safety concerns. Because of their ability to react to an issue quickly, they can often call attention to an emerging problem and obtain widespread media coverage. This media coverage often results in swift remediation by the targeted organization. These NGOs have helped create a global set of public expectations when it comes to safety, health, and environmental topics. Building a relationship with NGOs pertinent to the SH&E professional's organization is a proactive and positive step to take.

SUMMARY

- Traditional duties and responsibilities of the safety professional are expanding from loss control to health and environmental concerns.
- Safety, health, and environmental professionals need to understand the economic, ethical, and international issues affecting their roles.
- Training for new roles must be a process of continuous improvement on a good foundation of scientific, mathematical, and engineering knowledge.

REFERENCES

For a general list of readings for the SH&E professional, see Appendix 2, Bibliography, at the end of this volume. In addition, topic-specific resources are listed at the end of each chapter.

REVIEW QUESTIONS

1. List four major roles of the safety, health, and environmental professional.
   a.
   b.
   c.
   d.
2. What ethical code or guide should the SH&E professional adopt?
3. List two ways in which globalization affects the SH&E professional.
   a.
   b.
4. What four benefits can the SH&E professional use to sell safety to management?
   a.
   b.
   c.
   d.
5. What is the SH&E professional's personal and professional liability for errors and omissions?
6. Discuss five future opportunities a SH&E professional can expect.
   a.
   b.
   c.
   d.
   e.

Safety Culture

Treasa M. Turnbeaugh, PhD, CSP, CET
John F. Montgomery, PhD, CSP, CHMM
Stephen M. Bennett, ARM
Allen M. Gilley, CIH, CSP, ARM
William H. Grimes, CSP

Defining Culture
Artifacts ▶ Espoused Values ▶ Basic Assumptions

Climate or Culture

Organizational Culture
Safety Culture ▶ Levels of Positive Safety Culture

Measuring Safety Culture and Climate
Job Satisfaction Impact ▶ Trust Impact ▶ Planning the Safety Climate Survey

Group Behavior

Changing Culture
Levels, Goals, and Strategies ▶ Measurement Systems ▶ Basic Options ▶ Implementation and Organizational Change

Behavior-Based Safety as a Change Catalyst
Conclusion

Summary

References

Review Questions

DEFINING CULTURE

There are many ways to describe what comprises safety culture. The Safety Culture Maturity model espoused by Fleming (2000) provides one example and includes the following 10 elements:
- management commitment and visibility
- communication
- productivity versus safety
- learning organization
- safety resources
- participation
- shared perceptions about safety
- trust
- industrial relations and job satisfaction
- training.

To better understand *safety culture*, the safety and health professional needs to understand how *culture* is defined. Culture originated as a key concept in anthropology, but as noted in the earlier discussion of industrial safety history, management and organizational researchers have shown an interest in culture in an attempt to explain change within organizations.

Schein (1992) implies that culture can be changed and used to improve safety-related matters. He holds strong positions on organizational culture as well as on safety culture. His literature is the basis for other researchers who have built their analytical approaches on his works. This chapter is based, in large part, on Schein's views on these areas of culture.

There are many definitions of culture and many include terms such as *shared perceptions*, *norms*, and *values*. The best-known and most succinct definition of organizational culture is that it is "the way we do things around here" (Schein 1992). There are many academic definitions as well, but it is more important to understand the concepts behind those definitions than to agree on an elaborate definition of the term when it has been so nicely summed up by Schein.

In 1990, Schein theorized that the problem with culture is that a set of people must have had enough stability and common history to have allowed a culture to form. This means that some organizations have no overarching culture because they have no common or long-term history. Other organizations can be presumed to have a strong culture because of a long, shared history or because they have shared important, intense experiences. It should be noted that "strong culture" does not imply "good culture" but is an indicator of the strength and intensity of the culture.

Culture is what a group learns over a period of time as that group solves problems of the external environment and of internal integration. Such learning is simultaneously a behavioral, a cognitive, and an emotional process (Schein 1990).

Culture can now be further defined as a pattern of basic assumptions that are invented, discovered, or developed by a given group.

Once a group has learned to hold common assumptions, the resulting automatic patterns of perceiving, thinking, feeling, and behaving provide meaning, stability, and comfort. The anxiety that results from the inability to understand or predict events is reduced by shared learning. The strength of a culture is derived in part from this anxiety-reducing factor. In a way, some aspects of culture are to the group what defense mechanisms are to the individual (Hirschhorn 1987; Menzies 1960; Schein 1985).

There are three fundamental levels at which culture manifests itself: (1) observable artifacts, (2) espoused values, and (3) basic underlying assumptions (see Figure 3–1).

Artifacts
Artifacts can be easily observed and include how people address each other, dress codes, physical layout of the environment, emotional intensity, company records, and policies. These are the most superficial manifestations of culture. They may be apparent to the outside observer, but the values and assumptions behind these artifacts are much more difficult to identify and are, therefore, potentially misleading if used alone to draw conclusions about a culture.

Espoused Values
Espoused values are basically people's expressed beliefs. These values typically relate to basic assumptions (discussed next) but may be in conflict with them if social pressure or motivation is present. Management may say that safety is important, but only time and deeper probing will indicate whether this is true (Hale 2006).

Basic Assumptions
Basic assumptions are the deep-rooted, subconscious perceptions, thought processes, feelings, and behaviors of a group. Deeply held assumptions often start out as values, but as they stand the test of time, they are gradually taken for granted and then take on the character of assumptions, at

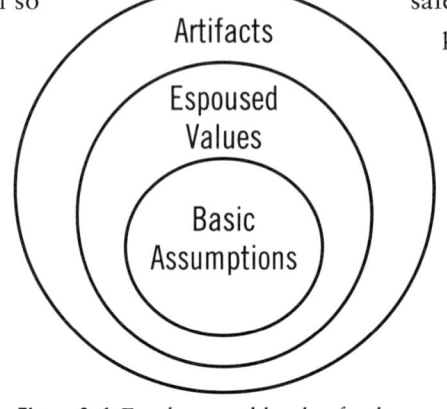

Figure 3–1. Fundamental levels of culture.

which point they are no longer questioned and become less open for discussion. Such avoidance behavior occurs particularly if the learning was based on traumatic experiences in the organization's history. This often leads to the group counterpart of what would be repression in the individual and can help to explain why culture is so difficult to change (Schein 1990).

Once the assumptions are understood, it is easier to determine the meanings of various behavioral and artifactual observations and understand how cultures can seem to be ambiguous or even self-contradictory (Martin and Meyerson 1988). It is possible for a group to hold conflicting values that manifest themselves in inconsistent behavior while having complete consensus on underlying assumptions. It is equally possible for a group to reach consensus on the levels of values and behavior and still develop serious conflict later because there was no consensus on critical basic assumptions. This is frequently observed in mergers or acquisitions, where initial synergy is gradually replaced by conflict, leading to dissension and ultimately divestiture.

If we combine insider knowledge with outsider questions, assumptions can be brought to the surface, but the process of inquiry has to be interactive—with the outsider continuing to probe until assumptions have been discovered and validated. This will lead to a feeling of greater understanding on the part of both the outsider and the insiders.

The culture of a group can be defined as:

> A pattern of shared basic assumptions that the group learned as it solved its problems of external adaptation and internal integration that has worked well enough to be considered valid and, therefore, to be taught to new members as the correct way to perceive, think, and feel in relation to those problems. (Schein 1992)

This definition presents three key concepts: how new members are socialized into the group, group behavior, and the existence of multiple cultures within a single organization. Schein submits that how new members learn in an organization is more revealing than what they learn because the former exposes deep assumptions of the group. Schein's definition does not overtly mention behavior, but behavior reflects basic cultural assumptions. Further, he suggests that behavior is determined by how we perceive, think, and feel in relation to situational factors that arise from the immediate external environment. Finally, Schein does not mention size of the cultural group but claims that organizations can have variations among the subgroups. Therefore, multiple cultures can coexist within the larger group. Some of these subcultures will typically be in conflict with other subcultures.

CLIMATE OR CULTURE

The terms *safety culture* and *safety climate* are often used interchangeably. However, these terms should be viewed as different concepts that are subsets of an organization's overall culture (Mearns and Flin 1999). Understanding the differences between these terms will allow the safety and health professional to differentiate between what is observable and measurable and what is highly subjective.

Generally speaking, culture is more complex than climate and includes the underlying assumptions, values, norms, and expectations of an organization. It has its roots in sociology and social anthropology, with an emphasis on symbols, myths, collective values, norms, and the interactions of groups (Mead 1934). Researchers agree that culture is a valid measure for assessing organizations and making improvements (Cooke and Rousseau 1988; Schein 1984; Schneider 1990). However, culture cannot be easily measured nor easily interpreted.

Climate is a reflection of culture, and is often assessed by gathering information through questionnaires or surveys that provide a snapshot of individual perceptions, attitudes, and beliefs (Sarkus 2001). Climate has its roots in social psychology as a reflection of culture and as the interaction between an individual and a situation (Ashforth 1985; Cooke and Rousseau 1988; Killman et al. 1985; Lewin 1951; Schneider and Gunnarson 1996). Climate focuses on individual perceptions related to the work environment. This concept is what most safety culture surveys—or, more properly, safety climate surveys—are based upon.

According to Denison and Mishra (1996), a way to differentiate between culture and climate is to view climate as a thermometer, reading the current temperature of an organization—a surface-level indicator. It leads to quantitative measures and standardized instruments with established reliability and validity. Climate is predicted to differ to a greater extent among groups than it does within a group.

Culture is analogous to a barometer, reading aspects of the weather system and measuring a deeper level of conditions, which can aid in the prediction of changes in organizational patterns. Culture uses qualitative observations and in-depth interviews of organizational members. Culture is unique to a given organization, making comparisons across organizations and the use of standardized tools of limited value (Schein 1985, 1990).

ORGANIZATIONAL CULTURE

The concept that organizational cultures contribute to the performance effectiveness of a company is not new. The well-known Hawthorne study from the late 1920s and early 1930s observed and documented the social, technical, and ideological relations among employees (Roethlisberger and Dickson 1946). Organizational culture has remained a topic of research and management interest in terms of exploring performance effectiveness, and degreed programs have been created to formalize this pursuit.

Figure 3–2 illustrates the relationship between performance effectiveness (productivity) and losses/safety. The relationship suggests that there is an organizational culture connection that ties low productivity to high losses and higher productivity to lower losses. This relationship holds until the upper one-third of high productivity is reached, where it appears that a culture of productivity takes precedence over a culture of safety.

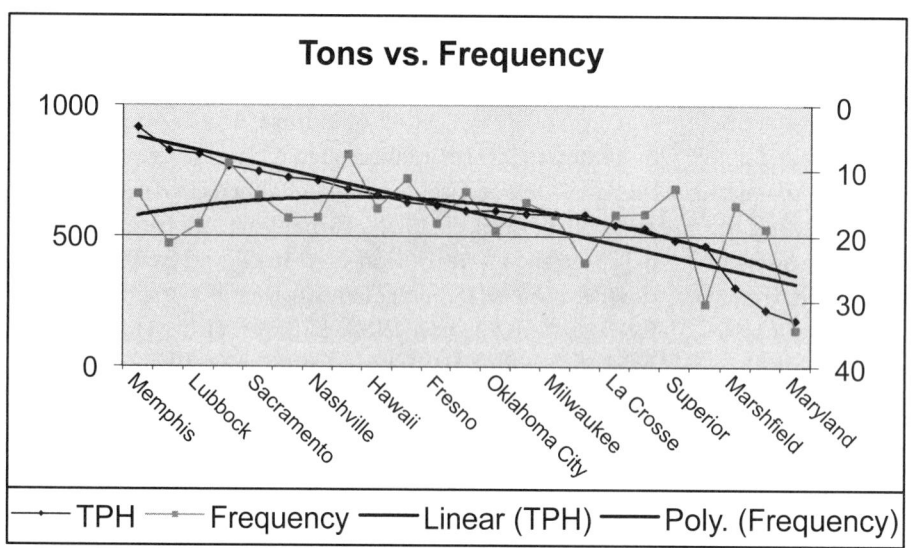

Figure 3–2. Cultural relationship of productivity to safety.

Safety Culture

The term *safety culture* first appeared in the 1987 Organization for Economic Cooperation and Development (OECD) Nuclear Energy Agency report on the 1986 Chernobyl disaster (INSAG 1988). Subsequently, researchers have come to believe that safety culture is a subcomponent of corporate culture (Hudson and van der Graaf 2002; Westrum 1993) that affects the safety and health of the group members and of others outside of the group as well. Safety culture affects and is affected by other operational processes and systems—it is inherently coupled to the overall corporate culture. Thus, any dominant subcomponent such as productivity, turnover, or quality will influence the safety processes and vice versa (Williams 1991).

Creek (1995) suggests that a safety program will be successful only if it is an integrated part of the overall organizational culture; it has to "fit" the culture of the organization. Hansen (1993) agrees, stating that safety activities in most companies are treated as separate from and unrelated to other organizational activities, which limits the success of many safety efforts.

Hansen (1993) also identifies the safety culture of an organization as crucial to safety performance success. He cites the National Institute for Occupational Safety and Health (NIOSH) studies from the 1970s that found "best practices" in safety—such as safety rules, safety committees, accident investigations, and safety promotions—are not factors of good safety performance companies versus poor ones. These studies did, however, show a significant correlation between safety performance and core management competencies such as management commitment and the communication and feedback process with employees.

Hansen (1993) characterizes a successful safety culture as one with humanistic employee relations policies and informal communications between management and employees in which feedback is encouraged and in which methods to produce the company's product safely are built into the standard operating procedures. Organizations without a positive safety culture and with negative employee attitudes typically have poor management practices, which cause such attitudes (Hansen 1995). The responsibility for changing that culture is ultimately management's (Arden 1993). Management must take an active and visible role in the safety program and in encouraging employee participation, input, and feedback. An example of active and visible management participation is the plant manager beginning each management meeting with departmental accountability for safety (i.e., having managers report on safety performance in their departments).

Geller (1994) concurs that safety must be integrated into the overall corporate culture for sustained improvement in safety performance to occur and thereby create a "total safety culture." He further defines a *total safety culture* as an organization that creates an environment where everyone takes responsibility for his or her individual safety as well as for that of others. He asserts that building a total safety culture requires continuous improvement in three areas: (1) environmental conditions, (2) personal factors (attitudes and beliefs), and (3) behavioral factors.

PATHOLOGICAL	BUREAUCRATIC	GENERATIVE
Do not want to know.	May not find out.	Actively seek information.
Messengers are shot.	Listened to if they arrive.	Messengers are trained.
Responsibility is shirked.	Responsibility is compartmentalized.	Responsibility is shared.
Bridging is discouraged.	Bridging is allowed but neglected.	Bridging is rewarded.
Failure is punished or covered up.	Organization is just and merciful.	There is inquiry and redirection.
New ideas are actively crushed.	New ideas present problems.	New ideas are welcomed.

Figure 3–3. How organizations treat information. *(Westrum 1993)*

Levels of Positive Safety Culture

There are different levels of cultural maturity that organizations must navigate. Organizations with a mature culture address issues quickly, have open discussions as issues arise, and welcome new ideas as to how to address the issues. These organizations encourage active thinking and involvement from all levels of the organization. By contrast, those organizations with a less mature culture address issues only when absolutely necessary and may address only the symptoms of the issue as opposed to the root causes.

Westrum (1993) differentiates an effective organization from an ineffective one by an organization's ability to handle issues that arise. An organization must have a culture of conscious inquiry in order to ensure its safety. This indicates that individuals and groups are empowered to observe, inquire, and make their conclusions known to higher management. Individuals will participate in offering observations, conclusions, and suggestions when they believe that their thoughts will be considered and used in a constructive manner. Westrum also proposes that the "license to think" is one of the key features of an effective organization and links the license to think with what an organization empowers individuals to do.

The strength of the safety culture depends on how well information flows inside the organization, and Westrum (1993) divides organizations into three categories based on their flow of information (Figure 3–3). Westrum's ideas were expanded by Hudson (2001) and by Hudson and van der Graaf (2002), who present an evolutionary model of safety culture and suggest that organizations can work their way from one level of maturity to the next (Figure 3–4).

Hudson and van der Graaf (2002) present stair steps that begin with the *pathological culture*. This culture is ruled by the desire for status quo; such organizations do not care to understand why accidents happen or how to prevent them. This aligns with Westrum's (1993) definition of a pathological culture. Next, Hudson and van der Graaf include the *reactive culture*, in which much attention is given to safety—but only after an accident happens.

Hudson and van der Graaf's (2002) *calculative culture* is synonymous with Westrum's (1993) bureaucratic class, in which the organization holds fast and hard to its rules and ignores signals of needed change. This organization is inspection- and audit-driven, and individuals assume all systems are in place for compliance purposes; individual participation will not make much difference. The next level of safety maturity as described by Hudson and van der Graaf is the *proactive culture*, in which the organization believes all safety systems are in place and adhered to but is still looking for ways to make continuous improvement (Hudson 2001).

The final step of the safety culture maturity model (Hudson and van der Graaf 2002) is the *generative culture*, where safety is no longer a separate issue but is entirely

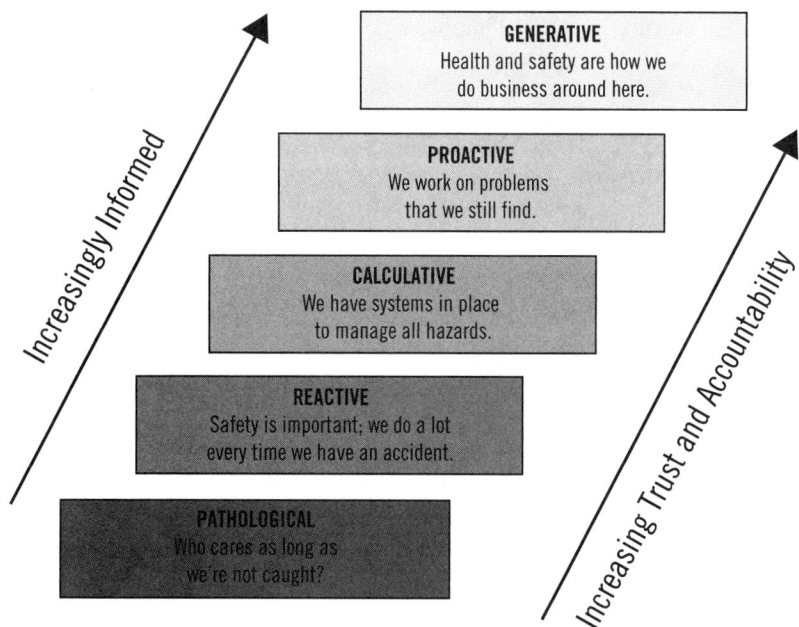

Figure 3–4. Adapted from the safety culture maturity model. *(Hudson and van der Graaf 2002)*

integrated into the business and is a part of everyday business processes within the organization. Organizations at this level are learning organizations with an effective feedback system in place. This aligns directly with Westrum's (1993) generative level, where organizations welcome signals for change, look at change as an opportunity for overall system improvement, and are enthusiastic about making such changes for continuous improvement of the overarching business systems.

Reason (1997) identifies characteristics of a successful safety culture that align well with the generative culture:
- *reporting culture*—one in which people are willing to report errors and near misses
- *just culture*—one of "no blame," where trust is present and individuals are encouraged or rewarded for providing safety information but also where there is a clear distinction between acceptable and unacceptable behavior
- *flexible culture*—one that can take on different forms but that shows evidence of moving from a conventional, hierarchical mode to a flatter structure
- *learning culture*—one that has the ability and the desire to draw correct conclusions from its safety information and seeks to implement major improvements when the need is evident.

Combined, these subcultures or characteristics create a well-informed organizational culture. Reason (1997) concludes that in most respects, an informed culture is a successful culture. Hudson (2001) interprets that Reason's four characteristics form a culture of trust. Thus, as organizations ascend the safety maturity model, they attempt to create a more informed culture—one that implies a culture rich in trust.

MEASURING SAFETY CULTURE AND CLIMATE

The most popular way of measuring safety culture today is by using *safety culture perception surveys*. More correctly, they should be called *safety climate perception surveys*, as discussed earlier in this chapter. For the purposes of this section, all surveys will be called perception surveys. Petersen (1989) espouses that safety culture surveys (or safety climate surveys) are important tools that can be used to measure safety program effectiveness. While perception surveys cannot measure attitudes and beliefs directly, they are the best means available to extrapolate information about attitudes and beliefs, which, behaviorists propose, affect behavior or actions. Conversely, behaviorists also believe that behaviors or actions performed repeatedly can affect a person's attitudes and beliefs. Hellriegel and Slocum (2004) state that general attitudes best predict general behaviors and that specific attitudes best predict specific behaviors.

One issue with perception surveys is the lack of psychometrically sound measures for this purpose (McLain, 1995). There is confusion among safety practitioners about what a perception survey should be. Many people believe that asking a multitude of questions constitutes a sound, valid survey. Nothing could be further from the truth. Doing so will simply garner answers to the questions asked and the way in which they were asked. On the other hand, a psychometrically valid survey has been based on theory, has been tested many times, and has an acceptable level of validity and reliability. However, this chapter is not on survey methods, so only the cursory elements of these factors are discussed.

A psychometric model is based on a theory formed from an in-depth literature search. The theory helps establish constructs of measures and items. Figure 3–5 depicts a theoretical model of a safety culture. This model was developed based on a literature review and on many cumulative years of consulting experience. "Measure" is the category of interest for researchers, and the "items" are the questions that collectively assist in making determinations about the measure (see Figure 3–6). Through advanced statistical calculations, a reliability score, or alpha score, can be developed for the measure. This alpha score indicates the certainty to which the researcher believes that he or she has asked the correct questions and in the correct manner to explain what the measure indicates. An alpha score of 0.60 (60%) or better is considered a strong measure. Asking more questions does not necessarily make the measure construct stronger and could actually make it weaker. However, questions can be systematically removed and alpha scores recalculated to see whether fewer questions affect the measure.

The advanced statistical procedure to develop the measure constructs and their reliability scores is called *factor analysis*. This is also a good technique for refining a survey so that it is more succinct as well as more reliable.

Job Satisfaction Impact

The research on safety-specific climate gave way to considerations that underlying workplace issues may contribute to employee perceptions of safety. For example, McLain (1995) proposes that perceptions of risk might contribute to the safety climate—that is, that the perceived likelihood of being harmed in the workplace can affect psychological interpretations and perceptions, which thereby influence work attitudes and behaviors. More specifically, McLain researched the impacts of job satisfaction, satisfaction with physical working conditions, stress, and distraction from task performance. He does not speculate as to the direction of the relationship between perception of risk and job

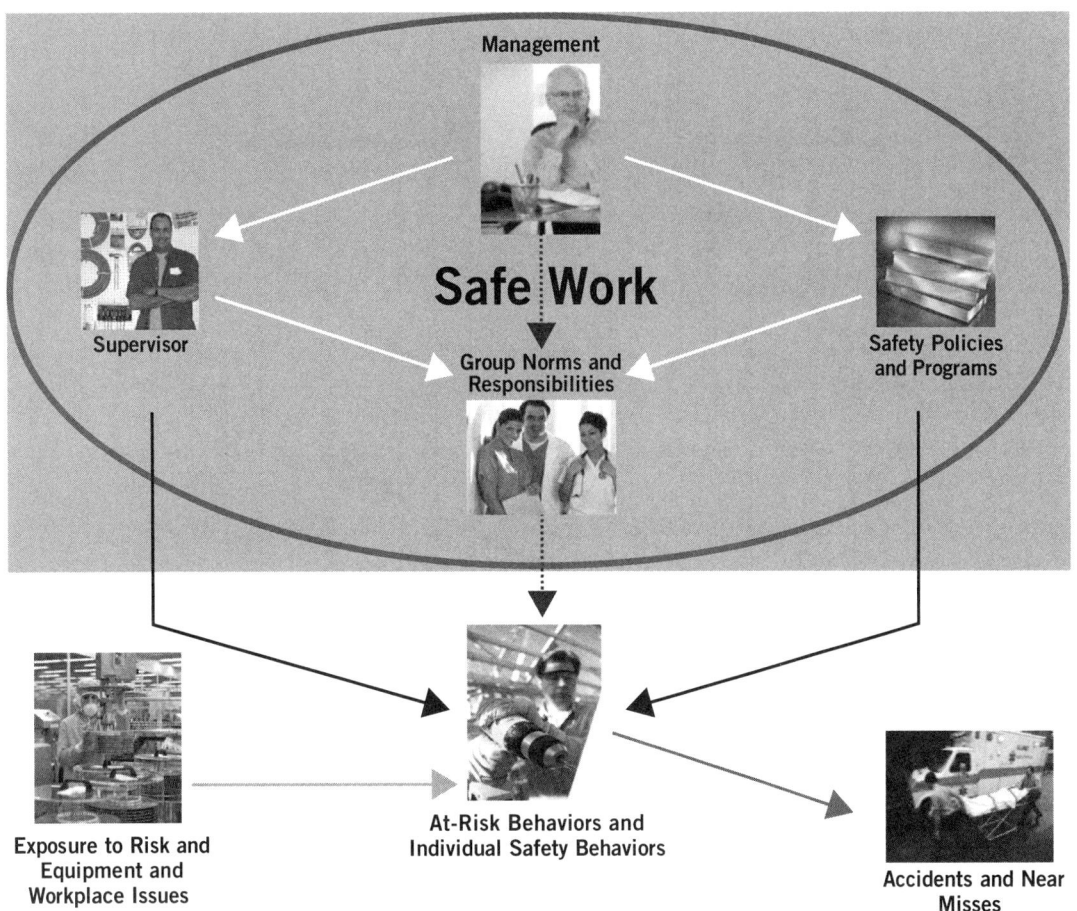

Figure 3–5. Example of a theoretical model of a safety culture.

Measure

Management Trust	Strongly Disagree	Disagree	Slightly Disagree	Slightly Agree	Agree	Strongly Agree
Management can be trusted.	6%	16%	14%	24%	33%	7%
I trust management to tell me if I am in danger at work.	4%	12%	6%	18%	43%	18%
If I were hurt on the job, I trust management to treat me fairly.	4%	8%	11%	19%	42%	16%
I trust management to do what they say they are going to do.	8%	15%	13%	29%	32%	4%
I feel free to discuss safety concerns with management without the fear of having it used against me later.	2%	4%	9%	25%	47%	13%

Items

Figure 3–6. Example of measure and items construct—alpha (0.79).

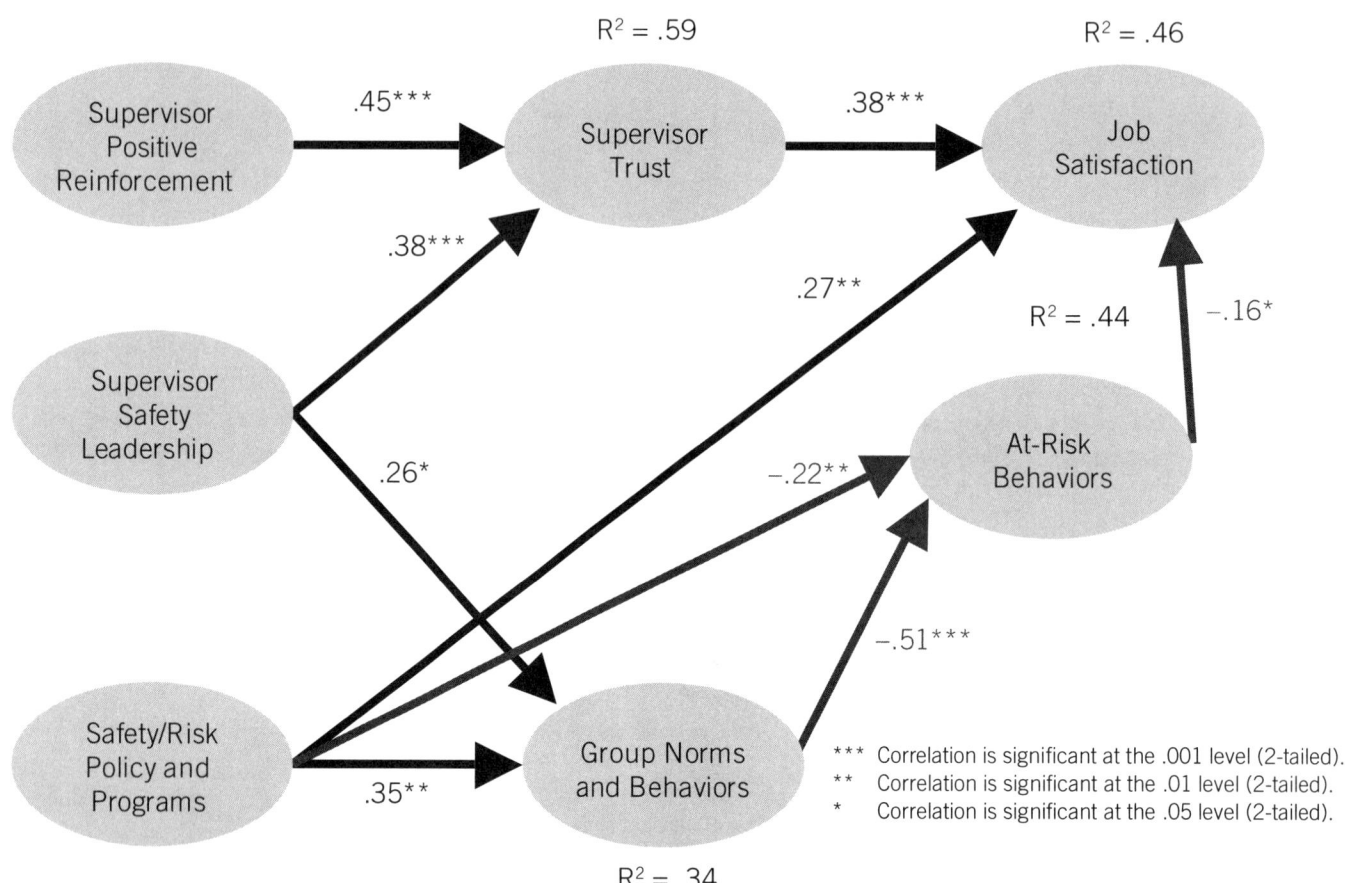

Figure 3–7. Example of path analysis results for safety climate measures.

satisfaction, but this relationship could be presumed to be an inverse one.

Extrapolating McLain's (1995) work, it might be assumed that the relationship of job satisfaction to safety climate is a positive one. Other researchers have found a similar relationship as it relates to overall organizational climate and job satisfaction (Downey et al. 1974; James and James 1992). Research by Bigoes (1986) and Greenwood and Wolf (1987) shows a significant relationship between employee attitudes and accident rates; they suggest that increasing job satisfaction is just as important as eliminating physical hazards in the workplace. Thus, a review of the literature reveals that many factors play significant roles in creating a safety climate. Unfortunately, the relationship among these factors is not a simple, linear one.

Another tool that can be used to explain how these factors interrelate is *path analysis*. Path analysis can show the relationship of safety climate measures to one another as well as the direction and relative strength of those relationships. Figure 3–7 indicates that 46% of the respondents' perceptions about job satisfaction are directly related to their trust in their supervisor and safety policies and procedures and are inversely related to their perceptions of their own at-risk behaviors. This example also shows an interesting relationship between supervisor trust and job satisfaction.

Trust Impact

Many researchers have proposed that *trust* plays a central role in safety culture (Hudson 2003; Reason 1997; Westrum 1995). Helmreich and Merritt (1998) have stated that safety cultures are built on trust and that trust needs to exist at multiple levels: among workers, in supervisors, and in management. Mayer, Davis, and Schoorman (1995) studied the relationship between trust and perceived risk in the work environment and found that if trust is greater than perceived risk, one employee may try to stop another from acting unsafely. If risk is greater than trust, the employee may not try to stop the other for fear of some sort of reprisal. However, the employee must trust both his or her co-worker and his or her management's commitment to safety in order to proceed with promoting safe behaviors. The willingness of employees to assume the role of "my brother's keeper" is often cited as an indicator of a safe work culture, and based upon the preceding research, this willingness appears to be trust-dependent.

Burns (2004) also found a relationship between trust and self-reported safety behaviors. Specifically, he found a positive relationship between an employee's trust in co-workers and whether the employee would then challenge those co-workers regarding unsafe acts. He also found a positive relationship between an employee's trust in a supervisor and whether the employee would report an incident or safety concern. Mayer and Davis (1999) suggest that ability, benevolence, and integrity are distinct factors of trustworthiness—and provide a solid framework on which to build trust.

Burns (2004) also suggests that in organizations with a hierarchical structure, hierarchy is an antecedent of trust. Turner (1987) believes that employees trust their co-workers more than their supervisors and trust their supervisors more than senior managers. However, employees typically do not have much regular interaction with senior management. This may help to explain why safety climate surveys usually show a discrepancy in perceptions based on the level of the employee answering the perception survey.

Planning the Safety Climate Survey
Before embarking on a safety survey, several factors should be considered. Management should realize that one-third of the survey results will indicate what they already know, one-third will indicate what they suspect, and one-third will indicate what they do not know. Management must also realize that if they survey their employees, there will be an implied promise to provide feedback and make changes based on the findings. If there is no feedback and no change, employees will lose trust in management and in the organization.

There are several key items to keep in mind when surveying:
- Keep the data anonymous, and report back only at the aggregated level.
- Decide before surveying how to analyze the data so that the proper demographic data for stratification can be included.
- Survey response rates will be highest—about 80% to 90%—if the survey is proctored by an outside party and taken on company time; survey rates via the Web usually yield about a 50% return rate; surveys mailed to homes usually yield a 20% to 30% return rate.
- Involve influential employees in the survey effort.
- Have management communicate the purpose and anonymity of the survey.
- Use survey tools that are psychometrically sound and have good reliability and validity.
- Never survey without subsequently providing feedback and modifications based on the results.

GROUP BEHAVIOR

As alluded to, culture and subcultures lead to and are exhibited by group behavior. Interestingly, the 19th-century British anthropologist Edward Tylor defined culture as "socially patterned human thought and behavior" (Bodley 1994). With this in mind, an individual either is integrated into the group's behavior or must choose to behave independently of the group. A driving factor in the individual's decision to comply with group cultural behavior is the individual's sense of belonging.

In order to change, people must first feel secure in their belonging to the group—this includes leaders having a sense of belonging to the group as well. If there is evidence that the group is not changing in the desired fashion, it will be very difficult for individuals to change due to the need for acceptance. If the evidence indicates that desired change is occurring, individuals will change with the group for continued acceptance.

CHANGING CULTURE

This chapter has so far established the importance of organizational culture and a safe work culture. How to measure the safety climate to get a sense of how safety processes are perceived at various levels of the organization has been discussed. The need for possible improvement in several areas, including trust and overall job satisfaction, has also been described, and the importance of groups, group behavior, and leadership in terms of understanding the current culture and in making necessary changes has been determined. Now it is time to discuss how to actually make changes in the organization—which can be a daunting task.

First, the objectives of change must be established. It would be very difficult to make positive changes in an organization if the company's present circumstances and its ultimate goal are not known. Chaudron (2003) suggests that organizations need to go through a formal decision-making process that has four major components:
- levels, goals, and strategies
- measurement systems
- basic options
- implementation and organizational change.

Levels, Goals, and Strategies
Deciding the level at which to begin is one of the most difficult decisions to make. Chaudron (2003) suggests that there are four levels of organizational change:
- shaping and anticipating the future (level 1)
- defining what business(es) to be in and the core competencies of each (level 2)

- structurally changing or reengineering processes (level 3)
- incrementally improving the processes (level 4).

Based on experience, most organizations are eager to jump to level 3 or level 4 without giving careful consideration to the entire process. However, management should begin with an open mind and few assumptions about the organization. Level 1 helps management to reassess its future opportunities and examine its strengths and weakness to determine the need for change in its mission and measurement system of other systems changes (Chaudron, 2003a).

Chaudron believes that many attempts at strategic planning skip Level 1 and go straight to Level 2 with preconceived notions such as: (1) the future is predictable and will be like the past; (2) the future is determined by the CEO's "vision for the future"; (3) management doesn't know where to begin the change process; (4) management is afraid to start at Level 1 because of the magnitude of the changes that are really needed; or (5) the only instructions they have is to refine the current processes.

Level 3 focuses on altering the fundamentals of how work is completed. This takes away from the idea of modest improvements and considers major reengineering processes with the goal of substantial improvement in productivity, efficiency, quality, etc. (Chaudron, 2003a). Conversely Level 4 does focus on making many small, incremental changes to existing work processes and can often prove successful when everyone in the organization gets involved. However, Level 4 does not take into account that there may be a need to abandon some existing methods, in favor of a different approach (Chaudron, 2003a).

Levels 1–3 focus on "high level" elements and tend to be used by companies with a significant need for change—those industries whose environments require rapid adaptation to fast-moving events (e.g., electronics, information systems, and telecommunication) (Chaudron, 2003a). Level 4 is typically used by companies that have only a modest need for change or are risk-avoidant; they focus on clarifying what they already do rather than on new opportunities. Examples include such industries (but not all members of these industries) as the military, aerospace, and health care organizations (Chaudron, 2003a). Different organizations can be successful entering the change cycle from different levels; one is not better than the other. However, the most successful organizations are those that take a step back and consider how much change is needed and openly communicates about those needs to determine where to begin. This is the juncture where many companies choose to consult with outside firms for assistance.

Chaudron continues explaining the formal change process by discussing goals, which need to have management resources and commitment to see success. Chaudron, like others, recommends using the SMART system as the guidance for establishing goals:

- Specific—concrete action, step-by-step actions needed to make the goal succeed
- Measurable—observable results from the goals' accomplishment
- Attainable—goals are possible and done at the right time with sufficient attention and resources
- Realistic—the probability of success is good, given the resources and attention given it
- Time-bound—the goal is achieved within a specified time period in a way that takes advantage of the opportunity before it passes by

Measurement Systems

All organizations need measurement to assess the progress of change and to continue to improve. The measurement system needs to be implemented both before a change (to help in directing the change) and after a change (to assess the success of the change made). Measurement of culture/climate change can be in the form of perception surveys; daily observations; or downstream measures such as frequency of losses, severity of losses, productivity numbers, or quality measurements.

Basic Options

Chaudron (2003) suggests that there are four basic options for implementation: (1) the whole organization is involved from the start and is intensively and simultaneously working on making the change; (2) divisions or business units go at their own pace and generally use an incremental approach; (3) all business units implement the same things according to the same time schedule; and (4) there is a pilot project in one division or business unit so the organization can learn from its mistakes and successes and then apply those lessons to the rest of the organization.

Implementation and Organizational Change

Some organizations involve employees from the beginning and allow them to influence the strategic plans. Involving employees (and union representatives) from the start tends to reduce resistance to change, and employee buy-in is critical to the success of culture change. The problem some organizations have with involving employees is the threat this creates to traditional hierarchies and power structures; such organizations tend to involve employees only at the implementation phase (Chaudron 2003).

Simon and Leik (1999) conclude that communication and feedback are the keys to a successful safety culture and to successful culture change initiatives that lead to a safe work culture. Schein (1992) agrees that communication is a powerful mechanism for change and says that managers should systematically pay attention to how and what they communicate. Schein indicates that managers can change the culture by:

- what they pay attention to, measure, and control on a regular basis
- how they react to critical incidents
- how they allocate scarce resources
- how they role-model, teach, and coach
- how they allocate rewards and plentiful resources
- how they recruit and select new employees.

Many behaviorists would agree with these tenants, and as discussed in the next section, on behavior-based safety, many of these factors are addressed within this process. What there is very little debate over, however, is the role that management must take in change initiatives. Management must understand the need for change, lead the change initiative, and model desired behaviors to achieve the desired outcomes.

BEHAVIOR-BASED SAFETY AS A CHANGE CATALYST

One approach to bringing about a safety culture is the use of behavior-based safety (BBS). While there are a number of BBS models that have gained recognition over the past decade, most are based on the same behavioral foundation, which includes two commonly accepted principles:

1. It is desirable to focus on "upstream" or leading indicators.
2. Effective use of consequences can modify behaviors.

Traditionally, safety performance has been measured based on the end results or the by-products of accidents and injuries—in other words, *lagging indicators*. These measures include Occupational Safety and Health Administration injury rates, workers' compensation loss data, and negative effects on production characteristics such as quality and productivity. These types of measures do not provide a reliable predictor of the upstream factors that can lead to employee injuries or human error potentials. Measuring only outcomes does not provide guidance on how to incrementally change the work environment. In addition, traditional safety measures do not provide an effective feedback mechanism to change the actions a person can exhibit that may lead to on-the-job errors. As discussed earlier in this chapter, increased feedback is essential to a positive safety culture.

The primary objectives of a behavior-based safety system are to provide methods to analyze an organization's safety climate and to catalog critical behaviors associated with specific tasks. The combination of an organizational climate assessment and the implementation of a systematic process for shaping individual behaviors can provide an organization with a powerful tool to change safety values and culture at both the individual and the group levels.

BBS is grounded in the psychology of B. F. Skinner (1953), who scientifically demonstrated a process of pinpointing, observing, measuring, and providing feedback to change human behaviors. The well-researched benefits of a BBS process include the following:

1. Providing a system to measure upstream indicators of safe behaviors and a platform to give individual feedback on safety performance
2. Influencing the development of a safe work culture in which everyone accepts responsibility for safety and does something about it on a daily basis
3. Developing a methodology for safe work practices that is supported by reinforcing feedback from peers, managers, and members of the safety team
4. Establishing a culture where employees go beyond the call of duty and identify unsafe conditions and employee at-risk behaviors and intervene to correct them
5. Developing a foundation of positive recognition of safe behaviors
6. Developing a system for identifying the root causes of unsafe behavior and the barriers to safe behavior in the existing management system
7. Developing an employee involvement mechanism that encourages employee input and participation and contributes to the overall safety process
8. Integrating seamlessly with the current systems for problem identification, problem solving, and continuous improvement
9. Identifying and addressing the common pathway for injuries occurring at the facility
10. Identifying and correcting system-related issues that can compel at-risk behaviors to occur

The underlying philosophy of the behavioral approach is to give ownership of safety to at-risk people through involvement and leadership at all levels of the organization. Traditionally, people not at risk manage the safety of those who are at risk. It is a considerable change for organizations to acknowledge the roles of the individuals in the workplace and to instill in each person a sense of control over and ownership of his or her actions. At least initially, some organizations are not ready to make this change, and it must be stressed that an environment that allows open communication between employees and supervisors/managers needs to be developed. Supervisors also must accept a more proactive and positive role. Rather than being in total control, they should become lead figures—facilitators—in a team context. Some managers and their supervisors may be threatened by this perceived erosion of control and the existence of an empowered workforce. It is thus essential that they are involved, not isolated.

Conclusion

The processes of observation and feedback continue until the employees reach 90% to 100% of a target goal for safe behaviors on a given objective and maintain that level of safe behavior for longer than a 30-day period. At that time, the safe behavior is said to be at "habit strength." Then another objective is chosen, and the processes begin again. This systematic approach to altering the work environment helps to implicitly and incrementally change the work culture. The processes embed the essential cultural elements of communication and feedback into the organization, thus giving the entire work force the feeling of being able to bring about change with their efforts.

The use of a BBS process can be a powerful tool to supplement existing safety management processes. Its concepts stem from identifying the safe behaviors an organization wants its employees to engage in, observing and tracking the behaviors on a daily basis, and implementing a positive reinforcement system that recognizes teams when they achieve desired behavior changes. Some have summarized the BBS method as "catching the employee doing it right." While this view might be somewhat simplistic, it does reflect the idea that the proper use of consequences—especially positive reinforcement—can be effective in modifying behavior and, ultimately, changing culture.

SUMMARY

Organizational culture and safety culture are undeniably important in creating a successful and safe work environment. Many researchers agree on the relevance and importance of culture in creating this success. However, what is not as understood is which specific elements of an organization help to create or reinforce a positive safety culture. The 10 Safety Culture Maturity Model elements espoused by Fleming (2000) include the following:
- management commitment and visibility
- communication
- productivity versus safety
- learning organization
- safety resources
- participation
- shared perceptions about safety
- trust
- industrial relations and job satisfaction
- training.

It is generally recognized that organizations with effective safety cultures are successful in convincing individuals at every level of the organization that safety is at least as important as other business characteristics such as productivity and quality. One of the most important elements of a culture is shared values, and if each individual does not share the organization's belief and vision, the elevation of safety as a top priority will not occur, and by definition, a world-class safety culture will not exist. Merely declaring within a safety policy statement that safety occupies the position of top priority will not, by itself, convince anyone of that assertion.

At a minimum, visible demonstrations of management's commitment, support, and involvement; genuine employee involvement; effective training and communication; appropriate budgeting and allocation of resources; and credible behavioral management are necessary. Employees are very perceptive and can readily differentiate between lip service and genuine commitment. If safety is promoted in a fashion different from that used to promote other operational aspects, an organization should not be surprised when the supposedly shared perceptions of safety do not correlate well with the company's safety policy.

There are many ideas on what a safety culture and what a positive safety process look like and how to arrive at that safe work culture. No two organizations need develop a safe work culture in the same fashion. This is a custom process and must be worked on by those internal to the organization. Many consulting firms and cultural assessment tools can assist in the change initiatives, but ultimately, the answers must be discovered and implemented by those within the organization—those who are part of the culture.

REFERENCES

Arden, P. "Create a Corporate Safety Culture." *Safety and Health* 147, no. 2 (1993).

Ashforth, B. E. "Climate Formation: Issues and Extensions." *Academy of Management Review* 10 (1985): 837–47.

Bigoes, S. "Back Injuries in Industry: A Retrospective Study; Employee Related Factors." *Spine* 11, no. 1 (1986).

Bodley, J. H. *Cultural Anthropology: Tribes, States and the Global System.* New York: McGraw-Hill, 1994.

Burns, C. "The Role of Trust in Safety Culture." PhD thesis. University of Aberdeen, 2004.

Chaudron, D. "Building a Framework: Begin at the

Beginning in Organizational Change." (2003). www.organizedchange.com/decide.htm.

Cooke, R. A., and D. M. Rousseau. "Behavioral Norms and Expectations: A Quantitative Approach to the Assessment of Organizational Culture." *Group and Organizational Studies* 13 (1988).

Creek, R. N. "Organizational Behavior and Safety Management: Evolution of Safety." *Professional Safety* 40 (1995).

Denison, D., and A. Mishra. "Toward a Theory of Organizational Culture and Effectiveness." *Organizational Science* 6, no. 2 (1996): 204–33.

Downey H. K., D. Hellriegel, M. Phelps, and J. W. Slocum. "Organizational Climate and Job Satisfaction: A Comparative Analysis." *Journal of Business Research* 2 (1974).

Fleming, M. "Safety Culture Maturity Model." The Keil Centre for the Health and Safety Executive (HSE). (2000). www.hse.gov.uk/research/otopdf/2000/oto00049.pdf.

Geller, E. S. "Ten Principles for Achieving a Total Safety Culture." *Professional Safety* 39, no. 9 (1994).

Greenwood, J., and H. Wolf. "Job Satisfaction Affects Rates of Occupational Accidents." *West Virginia News Digest,* November 1987.

Hale, A. "Organizational Culture and Safety." In *International Encyclopedia of Ergonomics and Human Factors* 2nd ed. Vol. 3. Boca Raton, FL: Taylor & Francis Group, 2006.

Hansen, L. "Re-braining' Corporate Safety and Health." *Professional Safety* 40, no. 10 (1995).

———. "Safety Management: A Call for (R)evolution." *Professional Safety* 38 (1993).

Hellriegel, D., and J. W. Slocum. *Organizational Behavior,* 10th ed. Mason, OH: Southwestern College Publishing, 2004.

Helmreich, R., and A. Merritt. *Culture at Work in Aviation and Medicine.* Aldershot, U.K.: Ashgate, 1998.

Hirschhorn, L. *The Workplace Within.* Cambridge, MA: MIT Press, 1987.

Hudson, P. "Applying the Lessons of High Risk Industries to Health Care," supplement 1, *Quality and Safety in Health Care* 12 (2003)

———. "Safety Management and Safety Culture—The Long, Hard and Winding Road." Proceedings of the First National Conference, Occupational Health and Safety Management Systems, Crown Content, Work Cover NSW, 2001.

Hudson, P., and G. C. van der Graaf. "Hearts and Minds: The Status After 15 Years Research." Paper presented at the International Conference on HSE in Oil and Gas Exploration and Production, Society of Petroleum Engineers (SPE 73941), Kuala Lumpur, 2002.

INSAG. "Basic Safety Principles for Nuclear Power Plants." Safety Culture Safety Series No. 76-INSAG-3. Vienna, Austria: International Atomic Energy Agency, 1988.

James, L. R., and L. A. James. "Psychological Climate and Affect: Test of a Hierarchical Dynamic Model." In *Job Satisfaction: How People Feel about Their Jobs, and How It Affects Their Performance,* edited by C. J. Cranny, P. C. Smith, and E. F. Stone. New York: Lexington Books, 1992.

Kilmann, R. H., M. J. Saxton, R. Serpa, and Associates. *Gaining Control of the Corporate Culture.* San Francisco: Jossey-Bass, 1985.

Lewin, K. *Field Theory in Social Science; Selected Theoretical Papers.* Edited by D. Cartwright. New York: Harper & Row, 1951.

Martin, J., and D. Meyerson. "Organizational Cultures and the Denial, Channeling, and Acknowledgement of Ambiguity." In *Managing Ambiguity and Change,* edited by L. R. Pondy, R. J. Boland, and H. Thomas. New York: Wiley, 1988.

Mayer, R. C., and J. H. Davis. "The Effect of the Performance Appraisal System on Trust of Management: A Field Quasi-Experiment." *Journal of Applied Psychology* 84 (1999).

Mayer, R. C., J. H. Davis, and F. D. Schoorman. "An Integrative Model of Organizational Trust." *Academy of Management Review* 20 (1995).

McLain, D. L. "Responses to Health and Safety Risk in the Work Environment." *Academy of Management Journal* 38 (1995).

Mead, M. *Mind, Self and Society.* Chicago: University of Chicago Press, 1934.

Mearns, K. J., and R. Flin. "Assessing the State of Organizational Safety: Culture or Climate?" *Current Psychology* 18 (1999).

Menzies, I. E. P. "A Case Study in the Functioning of Social Systems as a Defense against Anxiety." *Human Relations* 13 (1960).

Petersen, D. *Techniques of Safety Management: A Systems Approach.* New York: Aloray Inc., 1989.

Reason, J. *Managing the Risks of Organizational Accidents.* Aldershot, U.K.: Ashgate, 1997.

Roethlisberger, F. J., and W. J. Dickson. *Management and the Worker.* Cambridge, MA: Harvard University Press, 1946.

Sarkus, D. J. "Safety and Psychology." *Professional Safety* January 2001.

Schein, E. H. "Coming to a New Awareness of Organizational Culture." *Sloan Management Review,* 25 (1984).

———. *Organizational Culture and Leadership.* 2nd ed. San Francisco: Jossey-Bass, 1992.

———. *Organizational Culture and Leadership: A Dynamic View.* 2nd ed. San Francisco: Jossey-Bass, 1990.

———. "Organizational Culture: Skill, Defense Mechanism or Addiction?" In *Affect Conditioning, and Cognition*, edited by F. R. Brush and J. B. Overmier. Hillsdale, NJ: Erlbaum, 1985.

Schneider, B. *Organizational Climate and Culture.* Oxford, Jossey-Bass, 1990.

Schneider, B., and S. Gunnarson. "Organisational climate and culture: The psychology of the workplace." In *Applying Psychology in Business*, edited by J. James, B. Steffy, and D. Bray. Mass: Lexington, 1996.

Simon, S. I., and M. Leik. "Breaking the Safety Barrier: Implementing Culture Change." *Professional Safety*, (March 1999).

Skinner, B. F. *Science and Human Behavior.* New York: Macmillan, 1953.

Turner, J. C., "A Self Categorization Theory." In *The Blackwell Encyclopedia of Social Psychology*, 1987

Westrum, R. "Cultures with Requisite Imagination." In *Verification and Validation of Complex Systems: Human Factors Issues*, edited by J. A. Wise, V. D. Hopkin, and P. Stager. NATO ASI Series. New York: Springer-Verlag, 1993.

———. "Organizational Dynamics and Safety." In *Applications of Psychology to the Aviation System*, edited by N. McDonald, N. Johnston, and R. Fuller. Aldershot, U.K.: Avebury, 1995.

Williams, J. C. "Safety Cultures: Their Impact on Quality, Reliability, Competitiveness and Profitability." In *Reliability '91*, edited by R. H. Matthews. London: Elsevier Applied Science, 1991.

REVIEW QUESTIONS

1. List the four basic characteristics of a successful safety culture that would align well with the generative culture.
   a.
   b.
   c.
   d.
2. List the four levels of an organization's cultural change.
   a.
   b.
   c.
   d.
3. When planning a safety survey, what are five key items to keep in mind?
   a.
   b.
   c.
   d.
   e.
4. Chaudron (2003) suggests that organizations need to go through a formal decision-making process that has what four major components?
   a.
   b.
   c.
   d.
5. What are four of the six ways that managers can change the culture?
   a.
   b.
   c.
   d.
6. The well-researched benefits of a behavior-based safety process (BBS) include:
   a.
   b.
   c.
   d.
   e.
   f.
   g.
   h.
   i.
   j.
7. The 10 Safety Culture Maturity Model elements espoused by Fleming include:
   a.
   b.
   c.
   d.
   e.
   f.
   g.
   h.
   i.
   j.
8. Define *safety culture*.
9. Define *safety climate*.

Regulatory History

James T. O'Reilly, JD

Part 1—Understanding the Occupational Safety and Health Act

Legislative History of OSHA

Administration
Occupational Safety and Health Administration
▶ Occupational Safety and Health Review Commission
▶ National Institute for Occupational Safety and Health
▶ Injury and Illness Data ▶ Advisory Committee

Major Provisions of the OSH Act
Coverage • Employer and Employee Duties ▶ Employer Rights ▶ Onsite Consultation ▶ Voluntary Protection Programs ▶ Employee Rights ▶ The OSHA Poster ▶ Occupational Safety and Health Standards ▶ Input from the Private Sector ▶ Variances from Standards ▶ Record-Keeping Requirements ▶ Reporting Requirements ▶ Workplace Inspection ▶ Violations ▶ Citations ▶ Penalties ▶ Egregious Policy

Contested Cases

Small-Business Loans

Federal–State Relationships

What Does It All Mean?

Part 2—Key Standards and Regulations

Access to Employee Exposure and Medical Records
Records ▶ Toxic Substances ▶ Trade Secrets

Hazard Communication Standard
Elements of a Hazard Communication Program
▶ Written Hazard Communication Program ▶ Hazard Evaluation ▶ Employee Training ▶ Safety Data Sheets (SDSs) ▶ Labeling

Occupational Exposure to Hazardous Chemicals in Laboratories
Training and Information ▶ Hazard Evaluation and Medical Care

Workplace Emergencies
Emergency-Action and Fire-Prevention Plans ▶ Training ▶ Medical Care

Confined-Spaces Standard
Elements of a Confined-Space Program ▶ Training and Education

Control of Hazardous Energy (Lockout/Tagout)
Elements of an Energy Control Program ▶ Employee Training ▶ Inspections

Personal Protective Equipment Standard

Process Safety Management Standard
Compliance Elements for Process Safety Management ▸ New Facilities and Modified Worksites

Respiratory Protection Standard
Elements of a Respiratory Protection Program ▸ Major Changes ▸ Immediately Dangerous to Life or Health

Bloodborne Pathogens Standard
Exposure Plan ▸ Medical Care

Occupational Exposure to Asbestos Standards
Exposure Assessment ▸ Hazard Control ▸ Medical Care

Occupational Exposure to Formaldehyde
Exposure Assessment and Management ▸ Controls and Training

Lead Exposure in Construction

State Plan Programs

Part 3—Mine Safety and Health Act and Other Programs

Legislative History of MSHA

Administration
Mine Safety and Health Administration ▸ Mine Safety and Health Review Commission

Major Provisions of the Mine Act
Coverage ▸ Advisory Committees ▸ Miners' Rights ▸ Duties ▸ Miner Training

Mine Safety and Health Standards
Judicial Review ▸ Input from the Private Sector ▸ Emergency Temporary Standards ▸ Variances ▸ Granting of Petition ▸ Accident, Injury, and Illness Reporting ▸ Inspection and Investigation Procedures ▸ Withdrawal Orders ▸ Citations ▸ Penalties

Contested Cases

Part 4—Requirements beyond U.S. Borders
European Union ▸ ISO 9000 Series ▸ ISO 14000 Series ▸ Total Quality Management

Summary

Directory of Federal Agencies
The Occupational Safety and Health Administration ▸ OSHA Regional Offices ▸ The Mine Safety and Health Administration

Compilations of Regulations and Laws
The *Federal Register* ▸ The *Code of Federal Regulations*

References

Review Questions

Since 1970, Congress has enacted two major pieces of federal legislation and a number of key standards and regulations affecting the field of occupational safety and health. The following laws and standards will be covered in this chapter:
- Occupational Safety and Health Act (OSH Act)
- Access to Employee Exposure and Medical Records
- Hazard Communication Standard
- Occupational Exposure to Hazardous Chemicals in Laboratories
- Workplace Emergencies
- Confined-Spaces Standard
- Control of Hazardous Energy (Lockout/Tagout)
- Personal Protective Equipment (PPE) Standard
- Process Safety Management (PSM) Standard
- Respiratory Protection Standard
- Bloodborne Pathogens Standard
- Occupational Exposure to Asbestos Standards
- Occupational Exposure to Formaldehyde
- Lead Exposure in Construction
- State Plan Programs
- U.S. Mine Safety and Health Act.

PART 1—UNDERSTANDING THE OCCUPATIONAL SAFETY AND HEALTH ACT

A new national policy was established on December 29, 1970, when President Richard M. Nixon signed into law the Occupational Safety and Health Act (Public Law 91-596, found in 29 United States Code [USC] §§651–678; see gpo.gov/fdsys/pkg/USCODE-2012-title29/pdf/USCODE-2012-title29.pdf).

Congress declared that the purpose of this piece of legislation was "to assure so far as possible every working man and woman in the Nation safe and healthful working conditions and to preserve our human resources."

The OSH Act took effect on April 28, 1971. Coauthored by Senator Harrison A. Williams (D-NJ) and Congressman William Steiger (R-WI), the act is sometimes referred to as the Williams-Steiger Act. It is regarded by many as landmark legislation because it goes beyond the existing workplace and considers long-term health hazards in the working environment of the future.

The information provided in Part 1 of this chapter focuses on federal OSHA programs. State OSHA programs may differ from the federal programs in certain areas, but they must be equal to the federal requirements. However, unless specifically stated to the contrary, the recommendations in this chapter can be followed whether jurisdiction rests at the federal or the state level.

LEGISLATIVE HISTORY OF OSHA

Historically, the enactment of safety and health laws had been left to the states. Before the 1960s, only a few federal laws (such as the Walsh-Healey Public Contracts Act and the Longshoremen's and Harbor Workers' Compensation Act) directed any attention to occupational safety and health. Several pieces of legislation passed by Congress during the 1960s, including the Service Contract Act of 1965, the National Foundation on Arts and Humanities Act, the Federal Metal and Nonmetallic Mine Safety Act, the Federal Coal Mine Safety and Health Act, and the Contract Workers and Safety Standards Act (Construction Safety Act), focused industry attention on occupational safety and health.

Each of these federal laws was applicable only to a limited number of employers. The laws were either directed at those who had obtained federal contracts or targeted a specific industry. Even collectively, all the federal safety legislation passed before 1970 was not applicable to most employers or employees. Up to that time, congressional action on occupational safety and health issues was, at best, sporadic, covering only specific sets of employers and employees. There was little attempt to establish the omnibus coverage that is a central feature of the OSH Act.

Proponents of a more significant federal role in occupational safety and health, mostly represented by organized labor, based their stance primarily on the following:
1. With rare exceptions, the states failed to meet their obligations in regard to occupational safety and health. A few had reasonable or adequate safety and health legislation, but most states legislated safety and health only in specific industries. In general, states had inadequate safety and health standards, inadequate enforcement procedures, inadequate staff with respect to quality and quantity, and inadequate budgets.
2. In the late 1960s, approximately 14,300 employees were killed annually on or in connection with their jobs, and more than 2.2 million employees incurred a disabling injury each year as a result of work-related incidents. The injury/death toll was considered by most to be unacceptably high.
3. The nation's work-injury rates in most industries increased throughout the 1960s. Because the trend was moving in the wrong direction, proponents of federal intervention felt that national legislation would help reverse this trend.

The act evolved amid stormy controversy in both houses of Congress as the legislators debated state versus federal roles, industry versus government control, and the like. Such issues were responsible for sharply drawn lines between political parties and between the business community and organized labor. After 3 years of political tug-of-war, numerous compromises were made to allow passage of the OSH Act by both houses of Congress.

ADMINISTRATION

Administration and enforcement of the OSH Act are vested primarily with the Secretary of Labor, the Assistant Secretary of Labor for OSHA, and the Occupational Safety and Health Review Commission (OSHRC) as an appellate agency, which is discussed later. With respect to the enforcement process, the Secretary of Labor, through the Assistant Secretary, performs the investigation and prosecution aspects, and OSHRC performs the administrative adjudication portion, with possible appeal through the courts.

Research and related functions and certain educational activities are vested in the Secretary of Health and Human Services. These responsibilities, for the most part, are carried out by the National Institute for Occupational Safety and Health (NIOSH) established within the Department of Health and Human Services (DHHS). Compiling injury and illness statistical data is handled by the Bureau of Labor Statistics (BLS), part of the U.S. Department of Labor (DOL).

To assist the Secretary of Labor, the act authorizes the appointment of an Assistant Secretary of Labor for Occupational Safety and Health. This position is filled by presidential appointment with the advice and consent of the Senate. The Assistant Secretary is the chief of the Occupational Safety and Health Administration (OSHA) established within the DOL. The Assistant Secretary acts on behalf of the Secretary of Labor. For the purposes of this chapter, *OSHA* is synonymous with the term *Secretary of Labor* or *Assistant Secretary of Labor*.

The primary functions of the four major governmental units assigned to carry out the provisions of the act are described in this section.

Occupational Safety and Health Administration

The Occupational Safety and Health Administration (OSHA) came into existence officially on April 28, 1971, the date the OSH Act became law. This agency was created by the DOL to discharge the department's responsibilities assigned by the act.

Major Areas of Authority

The OSH Act grants OSHA the authority, among other things, (1) to promulgate, modify, and revoke safety and health standards; (2) to conduct inspections and investigations and to issue citations, including proposed penalties; (3) to require employers to keep records of safety and health data; (4) to petition the courts to restrain imminent-danger situations; and (5) to approve or reject state plans for programs under the act.

The act also authorizes OSHA (1) to provide training and education to employers and employees; (2) to consult with employers, employees, and organizations regarding prevention of injuries and illnesses; (3) to grant funds to the states for identification of program needs and for plan development, experiments, demonstrations, administration, and operation of programs; and (4) to develop and maintain a statistics program for occupational safety and health.

Major Duties Delegated

In establishing OSHA, the Secretary of Labor delegated to the Assistant Secretary for Occupational Safety and Health the authority and responsibility for safety and health programs and activities of the DOL, including responsibilities derived from the following:
1. Occupational Safety and Health Act of 1970
2. Walsh-Healey Public Contracts Act of 1936, as amended
3. Service Contract Act of 1965
4. Public Law 91-54 of 1969 (construction safety amendments)
5. Public Law 85-742 of 1958 (maritime safety amendments)
6. National Foundation on the Arts and Humanities Act of 1965
7. Longshoremen's and Harbor Workers' Compensation Act (33 USC §§901, 904)
8. Federal safety program under 5 USC §7902.

Similarly, the Commissioner of the BLS was delegated the authority and given the responsibility for developing and maintaining an effective program for collection, compilation, and analysis of occupational safety and health statistics; for providing grants to the states to assist in developing and administering programs for compiling such statistics; and for coordinating functions with the Assistant Secretary for Occupational Safety and Health.

The Solicitor of Labor is assigned responsibility for providing legal advice and assistance to the Secretary and all officers of the department in the administration of statutes and Executive Orders relating to occupational safety and health. In enforcing the act's requirements, the Solicitor of Labor represents the Secretary in litigation before OSHRC and, subject to the control and direction of the Attorney General, before the federal courts.

To assist in carrying out its responsibilities, OSHA has scientists, administrators, and inspectors. Its organization is explained on its website (osha.gov). OSHA has established regional offices in Boston, New York, Philadelphia, Atlanta, Chicago, Dallas, Kansas City, Denver, San Francisco, and Seattle. (See Directory of Federal Agencies at the end of this chapter.) The primary mission of the regional office chief, known as the Regional Administrator, is to supervise, coordinate, evaluate, and execute all OSHA programs in the region. Assisting the Regional Administrator are Assistant Regional Administrators for (1) training, education, consultation, and federal agency programs; (2) technical support; and (3) state and federal operations. (Some functions are combined in certain regions.)

Area offices have been established within each region, each office headed by an Area Director. The mission of the Area Director is to carry out the compliance program of OSHA within the designated geographic area. The area office staff carries out its activities under the general supervision of the Area Director and with guidance of the Regional Administrator, who uses policy instructions received from the national headquarters. Federal enforcement of the OSH Act is carried out by the area offices in states that do not have an approved state plan. In states with an approved plan, the area office monitors state activities. (See Federal–State Relationships later in this chapter.)

Occupational Safety and Health Review Commission

The Occupational Safety and Health Review Commission (OSHRC) is a quasi-judicial board of three members appointed by the president and confirmed by the Senate. OSHRC is an independent agency of the executive branch of the U.S. government and is not a part of the DOL. The principal function of the commission is to adjudicate cases when an enforcement action taken by OSHA against an employer is contested by the employer, the employees, or their representatives.

OSHRC's actions are limited to contested cases. In such instances, OSHA first notifies the commission of the contested case. The commission then hears all appeals on actions taken by OSHA concerning citations, proposed penalties, and abatement periods and determines the appropriateness of such actions. When necessary, the commission may conduct its own investigation and may affirm, modify, or vacate OSHA's findings.

There are two levels of adjudication within the commission: (1) the administrative law judge (ALJ) and (2) the three-member commission. All cases not resolved in OSHA informal proceedings are heard and decided by one of the commission's ALJs. The judge's decision can be changed by a majority vote of the commission if one of the members, within 30 days of the judge's decision, directs that the decision be reviewed by the commission members. The commission is the final administrative authority to rule on a particular case, but its findings and orders can be subject to further review by the courts. (For further information, see Contested Cases later in this chapter.)

National Institute for Occupational Safety and Health

The National Institute for Occupational Safety and Health (NIOSH) was established within the Department of Health and Human Services under the provisions of the OSH Act. Administrative headquarters for NIOSH is within the Centers for Disease Control and Prevention (CDC) in Atlanta, Georgia. NIOSH is the principal federal agency engaged in research, education, and long-term training related to occupational safety and health.

The primary functions of NIOSH are (1) to develop and establish recommended occupational safety and health standards, (2) to conduct research experiments and demonstrations related to occupational safety and health, and (3) to develop educational programs to provide an adequate supply of qualified personnel to carry out the purposes of the OSH Act.

Research and Related Functions

Under the OSH Act, NIOSH is responsible for conducting research to develop new occupational safety and health standards. NIOSH develops criteria for establishing these standards and transmits the criteria to OSHA. OSHA is then responsible for the final establishment, promulgation, and enforcement of the standards.

Education and Training

NIOSH also has the responsibility to develop (1) education and training programs aimed at providing an adequate supply of qualified personnel to carry out the purposes of the OSH Act and (2) informational programs on the importance and proper use of adequate safety and health equipment. The long-term approach to having an adequate supply of trained personnel in occupational safety and health is found in colleges, universities, and other institutions in the private sector. NIOSH encourages such institutions, via contracts and grants, to expand their curricula in occupational medicine, occupational health nursing, industrial hygiene, and occupational safety engineering.

Employer and Employee Services

Of principal interest to individual employers and employees are the technical services offered by NIOSH. The five main services are provided on request to NIOSH's Division of Technical Services, 4676 Columbia Parkway, Cincinnati, OH 45226 (800-232-4636; cdc.gov/niosh). These services are as follows:

1. *Hazard evaluation*—onsite evaluations of potentially toxic substances used or found on the job
2. *Technical information*—detailed technical information concerning health or safety conditions at workplaces, such as the possible hazards of working with specific solvents, and guidelines for use of protective equipment
3. *Incident prevention*—technical assistance for preventing on-the-job injuries, including the evaluation of particular problems and recommendations for corrective action
4. *Industrial hygiene*—technical assistance in the areas of engineering and industrial hygiene, including the evaluation of special health-related problems in the workplace and recommendations for control measures
5. *Medical service*—assistance in eliminating medical and nursing problems in the workplace, including assessment of existing medically related needs and development of recommended means for meeting such needs.

NIOSH and the Mine Safety and Health Administration (MSHA) test and approve personal sampler units for coal-mine dust and respiratory protective devices, including self-contained breathing apparatus; gas masks; supplied-air respirators; chemical-cartridge respirators; and dust, fume, and mist respirators.

NIOSH representatives, although not authorized to enforce the OSH Act, are authorized to make inspections and to question employers and employees in carrying out the duties assigned to the DHHS under the act. NIOSH has both warrant and subpoena power, if necessary, to obtain the information needed for its investigations. It may also request access to employee records. However, it must obtain the consent of employees or use methods that maintain the employee's right to privacy concerning information in the records.

Injury and Illness Data
The responsibility for conducting statistical surveys and establishing methods to acquire injury and illness data assigned to in the Bureau of Labor Statistics. Questions regarding record-keeping requirements and reporting procedures can be directed to any of the OSHA regional or area offices. (See the directory at the end of this chapter.)

Advisory Committee
The OSH Act established a 12-member National Advisory Committee on Occupational Safety and Health (NACOSH) to advise, consult with, and make recommendations to the Secretaries of Labor and Health and Human Services with respect to the administration of the act. Eight members are designated by the Secretary of Labor and four by the Secretary of Health and Human Services. Members include representatives from management, labor, occupational safety and health professions, and the public.

MAJOR PROVISIONS OF THE OSH ACT

This section discusses the major regulations and provisions of the OSH Act with which employers are expected to comply. The act represents one of the most far-reaching efforts in U.S. history to provide safer, healthier, and cleaner conditions, not only for employees but also for communities living near office and manufacturing facilities and for consumers who use the products and services of these firms.

Coverage
Except for specific exclusions, the act applies to every employer that has one or more employees and that is engaged in a business affecting interstate commerce. The law applies to all 50 states, the District of Columbia, Puerto Rico, and all U.S. possessions.

Specifically excluded from coverage are all federal government employees. However, the act includes special provisions for federal employees. The act requires each federal agency head to establish and maintain an occupational safety and health program consistent with the standards promulgated by the Secretary of Labor. Executive Orders setting requirements for federal programs have been issued over the years. OSHA regulations implementing Executive Orders are found in Title 29 of the *Code of Federal Regulations* (CFR), Part 1960 (1983).

All employees of states and political subdivisions of the states are also excluded from the federal OSH Act. However, states with approved state plans are required to provide coverage for these public employees. Public employees in states without approved plans are not covered by the OSH Act in any manner.

However, two states—Connecticut and New York—have state plans covering only public employees. The District of Columbia and the states of Mississippi, New Hampshire, New Jersey, Rhode Island, and Wisconsin have laws specifically providing job safety and health protection to public employees.

The OSH Act also does not apply to operations in which a federal agency (and state agencies acting under the Atomic Energy Act of 1954), other than the DOL, already has authority to prescribe or enforce standards or regulations affecting occupational safety or health and is performing that function. An example of this exclusion is specific issues covered by Department of Transportation regulations in the railroad industry.

Additionally excluded from the OSH Act are mine operators and miners covered by the U.S. Mine Safety and Health Act of 1977. This act applies to mines of all types: coal and noncoal, surface and underground.

In its yearly appropriations bills since 1977, Congress has restricted OSHA enforcement. Previously, for example, items exempt from inspection included the following:
- farmers with 10 or fewer employees on the day of inspection and the 12 months preceding the day of inspection
- any work activity in any recreational, hunting, fishing, or shooting area
- employers with 10 or fewer employees in selected Standard Industrial Classification (SIC) industries with rates below the national average in lost-workdays case rates (the 1989 rate was 3.9 per 100).

However, the exemptions do not apply to situations involving such issues as employee complaints, referrals, health hazards, incidents resulting in a fatality or inpatient hospitalization of three or more employees, or discrimination complaints.

OSHA clarified its interpretation of coverage with respect to certain employees by issuing a regulation. (This policy, regarding "Coverage of Employees under the Williams-Steiger Occupational Safety and Health Act of 1970," is contained in 29 CFR 1975.) OSHA has also stated that churches and other religious organizations are not regarded as employers when performing ecclesiastical activities, nor are persons who, in their own residences, employ others to perform domestic household tasks. Further, any person engaged in agriculture who is a member of the farmer's immediate family is not regarded as an employee and hence is not covered by the act.

Employer and Employee Duties

OSHA clearly delineates employer responsibilities:
1. Each employer covered by the act has the *general* duty to furnish each employee with employment and places of employment free from recognized hazards causing or likely to cause death or serious physical harm (this is commonly known as the *general-duty clause*).
2. Each employer covered by the act has the *specific* duty to comply with safety and health standards promulgated under the act.

Each employee, in turn, has the duty to comply with the safety and health standards and with all rules, regulations, and orders that apply to employee actions and conduct on the job.

For employers, the general-duty provision is used only when no specific standards are applicable to a particular hazard. A hazard is "recognized" if it is a condition generally regarded as a hazard in the particular industry in which it occurs and is detectable (1) by means of human senses or (2) by accepted tests known in the industry to reveal its presence to an employer. An example of a recognized hazard in the second category is an excessive concentration of a toxic substance in the work area atmosphere, a concentration that could be detected only through the use of measuring devices. The general-duty clause is applicable to a specific condition only when there is no OSHA standard addressing the condition.

During the course of an inspection, a compliance safety and health officer (CSHO) is concerned primarily with determining whether the employer is complying with the promulgated safety and health standards. However, the officer also tries to discover whether the employer is complying with the general-duty clause.

The law provides for sanctions against the employer in the form of citations and civil and criminal penalties if the employer fails to comply with the general and the specific duties. Citations increased more than 9% in 2000–2007 (oshrc.gov). However, there is no provision for government sanctions against an employee for failure to comply with the employee's duty. Although some may view this policy as unjust, such employee failures were not controversial issues in the formative stages of the act.

Both management and organized labor have long agreed that safety and health on the job are management responsibilities. The business community generally did not want the law structured to provide for government sanctions against an erring employee. This is because management can invoke its own measures against an employee who obstructs the employer's efforts to provide a safe workplace.

Although the law expressly gives each employee the obligation to comply with the act's standards, final responsibility for compliance rests with the employer. As a result, employers should take all necessary actions to ensure that employees follow the promulgated standards. Employers should also establish within their safety systems a means to detect when employees are not complying with applicable standards.

The duty of an employer to protect employees against health hazards and safety hazards continues to gain emphasis. The growing awareness of the hazard of chemical exposure has caused OSHA to focus on the employer's obligation to monitor the work environment, to provide periodic medical examinations, and to make available a range of protective measures to guard against the hazards of the work environment.

Employer Rights

An employer has the following rights:
1. to seek advice and offsite consultation as needed by writing, calling, or visiting the nearest OSHA office
2. to request and receive proper identification of the OSHA CSHO before inspection
3. to be advised by the CSHO of the reason for the inspection
4. to have an opening and a closing conference with the CSHO
5. to file a notice of contest with the OSHA Area Director within 15 working days after receiving a citation notice and proposed penalty
6. to apply to OSHA for a temporary variance from a standard if unable to comply because the needed materials, equipment, or personnel are not available to make the necessary changes within the required time
7. to take an active role in developing safety and health standards through participating in OSHA Standards Advisory Committees, or nationally recognized standards-setting organizations and through presenting evidence and views in writing or at hearings
8. if a small-business employer, to apply for long-term loans through the Small Business Administration (SBA) to help bring the establishment into compliance, either before or after an OSHA inspection
9. to be assured of the confidentiality of any trade secrets observed by an OSHA compliance officer.

Onsite Consultation

Congress has authorized, and OSHA now provides through a state agency or private contractors, free onsite consultation services for employers in every state. The onsite consultants help employers identify hazardous conditions and determine corrective measures.

The service is available on employer request. Priority is given to businesses with fewer than 150 employees. These firms are generally less able to afford private-sector consultation. OSHA assigns higher priority to companies whose employees have highly hazardous jobs.

The consultative visit consists of an opening conference, a walk-through of the company's facility, a closing conference, and a written summary of findings.

During the walk-through, the consultant tells the employer which OSHA standards apply to company operations and what they mean. The employer is informed of any apparent violations of those standards and, where possible, is given suggestions on how to reduce or eliminate the hazards.

Because employers, not employees, are subject to legal sanctions for violating OSHA standards, the employer determines to what extent employees or their representatives will participate in the visit. However, the consultant must be allowed to confer with individual employees during the walk-through in order to identify and judge the nature and extent of hazards. No citations or penalties are issued, and OSHA is not notified of the results except when an employer refuses to correct a significant hazard. Follow-up actions are taken to ensure that appropriate corrections were made.

Voluntary Protection Programs

OSHA developed several Voluntary Protection Programs (VPPs), including Star and Merit status, which offer recognition to facilities that satisfy detailed criteria for quality of systems and performance. These programs were held illegal by a federal court but remain in place as a voluntary option for firms wishing to participate. Their purpose is to emphasize the importance of, encourage the improvement of, and recognize excellence in employer-provided, site-specific occupational safety and health programs. These programs must not only meet, but exceed, the standards. When employers apply and are accepted, they are removed from the regular inspection list (except for valid formal employee complaints, fatalities, or other catastrophic events).

The VPP consists of three programs: a Star program for the most fully compliant sites, a Merit program for sites that aspire to achieve Star status, and a Demonstration Project program for sites that demonstrate alternate means of compliance. OSHA's 32-page application booklet cautions the applicant that this is a "major undertaking," with a comprehensive examination of the site. The benefit that OSHA offers, in recognition of the additional paperwork and meetings undertaken for the voluntary programs, is that OSHA general inspections (other than "for cause" incident investigations) are scheduled once every 3 years for Star program participants. Star recognition indicates special efforts and by commitment of the company which has undergone the inspection and demonstrated its qualifications.

Detailed data on submitting and being reviewed for these programs are available on the OSHA website (osha.gov). About 100 companies are participants in the three VPP programs, according to the website.

The federal courts struck down OSHA's separate Cooperative Compliance Program in April 1999 because it had not been subject to the public rulemaking procedures required for new rules. The industry objections, asserting that OSHA was imposing new, extra workplace reporting and compliance obligations on companies, were accepted by the appeals court. OSHA did not choose to appeal further; the agency stated that it would continue to target high-hazard workplaces (based on their records of injuries and illnesses) for inspections.

Employee Rights

Although the employee has the legal duty to comply with all the standards and regulations issued under the OSH Act, many employee rights are also incorporated into the act. Because these rights may affect labor relations as well as labor negotiations, employers should also be aware of the employee rights contained in the act. These rights fall into three main areas related to (1) standards, (2) access to information, and (3) enforcement. With respect to standards:

1. Employees may ask OSHA to begin proceedings for adoption of a new standard or to amend or revoke an existing standard.
2. Employees may submit written data or comments on proposed standards and may appear as interested parties at any hearing held by OSHA.
3. Employees may file written objections to a proposed federal standard and/or appeal the final decision of OSHA.
4. Employees must be informed when an employer applies for a variance of a promulgated standard.
5. Employees must be given the opportunity to participate in a variance hearing as interested parties and have the right to appeal OSHA's final decision.

With respect to access to information:

1. Employees have the right to information from the employer regarding employee protection and obligations under the act and to review appropriate OSHA standards, rules, regulations, and requirements, all of which should be available at the workplace.
2. Employees whose jobs may expose them to chemicals, radiation, or other hazardous substances have an OSHA-protected right to information about the risks,

precautions, and safe use of the substances. Three methods are prescribed in OSHA's Hazard Communication Standard: training must be given and records kept; Material Safety Data Sheets (MSDSs), or SDSs, must be available to workers concerning chemicals present in the workplace; and labels must be adequate to advise workers about container contents and safe use.
3. If employees are exposed to harmful materials in excess of the levels set by the standards, the affected employees must be so informed by the employer, who must also tell them what corrective action is being taken.
4. If an OSHA inspector finds an imminent hazard during an inspection, the workers and the employer must be told. If the employer refuses to act, OSHA may seek a court order to protect workers against the danger.
5. On request, employees must be given access to their medical records and history of exposure to toxic materials or harmful physical agents that must be monitored or measured and recorded. SDSs must be immediately available or accessible.
6. If a standard requires monitoring or measuring hazardous materials or harmful physical agents, employees must be given the opportunity to observe such monitoring or measuring.
7. Employees have the right of access to (1) the list of toxic materials published by NIOSH, (2) criteria developed by NIOSH describing the effects of toxic materials or harmful physical agents, and (3) industrywide studies conducted by NIOSH regarding the effects of chronic, low-level exposures to hazardous materials.
8. On written request to NIOSH, employees have the right to obtain the determination of whether a substance found or used in the establishment is harmful.
9. On request, employees must be allowed to review the Log and Summary of Work-Related Injuries and Illnesses (OSHA Forms 300 and 300A) at a reasonable time and in a reasonable manner.

With respect to enforcement:
1. Employees have the right to confer in private with the CSHO and to respond to questions from the CSHO during an inspection of an establishment.
2. An authorized employee representative must be given the opportunity to accompany the compliance officer during the inspection. (This is commonly known as the "walkaround" provision.) Also, an authorized employee has the right to participate in the opening and the closing conferences during the inspection.
3. An employee has the right to make a written request to OSHA for a complaint inspection if the employee believes that a violation of a standard presents physical harm; the employee also has the right to request that OSHA keep his or her identity confidential.
4. An employee who believes that a violation of the act has occurred has the right to notify OSHA or a compliance officer in writing of the alleged violation, either before or during an inspection of the establishment.
5. If OSHA denies an employee's request for a special inspection, the agency must notify the employee in writing that the complaint was not valid and explain the reasons for this decision. The employee has the right to object to such a decision and may request a hearing by OSHA.
6. If a written complaint concerning an alleged violation is submitted to OSHA and the compliance officer responding to the complaint fails to cite the employer for the alleged violation, OSHA must furnish the employee or an authorized employee representative with a written statement explaining the reasons for its final disposition.
7. If OSHA cites an employer for a violation, employees have the right to review a copy of the citation, which must be posted by the employer at or near the place where the violation occurred. If "systemwide" agreements are made, all locations must be notified.
8. An employee has the right to appear as an interested party or to be called as a witness in a contested enforcement matter before the OSHRC.
9. If OSHA arbitrarily or capriciously fails to seek relief to counteract an imminent danger and an employee is injured as a result, that employee has the right to bring action against OSHA for relief as may be appropriate.
10. An employee has the right to file a complaint with OSHA within 30 days if the employee believes that he or she has been discriminated against as a result of asserting employee rights under the act.
11. An employee has the right to contest the abatement period established in the citation issued to the employer. This can be done within 15 working days of the issuance of the citation by notifying the OSHA Area Director who issued the citation.

The OSHA Poster
The OSHA poster (OSHA 3165) must be prominently displayed in the work environment where notices to employees are customarily posted. The poster informs employees of their rights and responsibilities under the act.

Occupational Safety and Health Standards
The act authorizes OSHA to promulgate, modify, or revoke occupational safety and health standards. The rules of pro-

cedure for promulgating, modifying, or revoking standards are spelled out in 29 CFR 1911. The current requirements are available at all OSHA area and regional offices and are printed annually in the *Code of Federal Regulations*. OSHA is responsible for promulgating legally enforceable standards that may require stipulations or the adoption of practices, means, methods, or processes that are reasonably necessary and appropriate to protect employees on the job. Employers are responsible for becoming familiar with the standards applicable to their firms and ensuring that employees have and use the personal protective equipment required for safety. In addition, employers are responsible for complying with the act's general-duty clause. The *Code of Federal Regulations*, Title 29, is available online at gpoaccess.gov. Standards contained in Part 1910 apply to general industry, while those contained in Part 1926 apply to construction. Standards that apply to ship repairing, shipbuilding, shipbreaking, and longshoring are contained in Parts 1915 through 1918, respectively. Agricultural standards are contained in Part 1928. As new equipment, methods, and materials are developed, these standards are updated via modification.

OSHA standards incorporate by reference certain other standards adopted by industry organizations. Standards incorporated by reference in whole or in part include those adopted by the following organizations:
- American Conference of Governmental Industrial Hygienists
- American National Standards Institute
- American Petroleum Institute
- American Society of Agricultural Engineers
- American Society of Mechanical Engineers
- American Society for Testing and Materials
- American Welding Society
- Compressed Gas Association
- Crane Manufacturers Association of America
- The Fertilizer Institute
- Institute of Makers of Explosives
- National Electrical Manufacturers Association
- National Fire Protection Association
- National Institute for Occupational Safety and Health
- National Plant Food Institute
- Society of Automotive Engineers
- Underwriters Laboratories Inc.
- U.S. Department of Commerce
- U.S. Public Health Service.

OSHA has the authority to promulgate emergency temporary standards in situations in which employees are exposed to grave danger. Emergency temporary standards take effect immediately on publication in the *Federal Register*. They remain in effect until superseded by a standard promulgated under procedures described in the act. The law requires OSHA to develop a permanent standard no later than 6 months after the publication of the emergency temporary standard. Any person adversely affected by any standard issued by OSHA has the right to challenge its validity by petitioning the U.S. Court of Appeals within 60 days of the standard's promulgation.

Input from the Private Sector

Occupational safety and health standards promulgated by OSHA will never cover every conceivable hazardous condition that could exist in any workplace. Nevertheless, new standards and modifications of existing standards are of significant interest to employers and employees alike. Industry organizations along with individuals and employee organizations need to express their views in two ways: (1) by responding to OSHA's advance notice of proposed rulemaking, which usually calls for information on which to base proposed standards, and (2) by responding to OSHA's proposed standards (most of the expertise and the technical competence lie within the private sector). To do less means that industry and employees are willing to let the standards-development process rest in the hands of OSHA.

Additional sources that OSHA uses to revise existing occupational safety and health standards or develop new ones are NIOSH criteria documents and standards advisory committees. These committees are appointed by the Secretary of Labor.

In order to promulgate, revise, or modify a standard, OSHA must first publish in the *Federal Register* a notice of any proposed rule that will adopt, modify, or revoke any standard and invite interested persons to submit their views on the proposed rule. The notice must include the terms of the new standard and provide an interval of at least 30 days (usually 60 days or more) from the date of publication for interested persons to respond. These persons may file objections to the rule and are entitled to a hearing on their objections if they request that one be held. However, they must specify the parts of the proposed rule to which they object and the grounds for such objection. If a hearing is requested, OSHA must hold one. Based on (1) the need for control of an exposure to an occupational injury or illness and (2) the reasonableness, effectiveness, and feasibility of the control measures required, OSHA may either issue a rule promulgating an additional standard or modify or revoke an existing standard.

Variances from Standards

On some occasions, and for various reasons, standards cannot be met. In other cases, the protection already afforded by an employer to employees is equal or superior to the protection that would be provided if the standard were strictly followed. The OSH Act provides an avenue of relief from these situations by empowering OSHA to grant variances from the standards, provided that doing so would not degrade the purpose of the act. The detailed "Rules of Practice for Variances, Limitations, Variations, Tolerances, and Exemptions" are codified in 29 CFR 1905.11(b).

OSHA can grant two types of variances—temporary and permanent. Temporary variances are generally concerned with compliance with new standards. Employers may apply for an order granting a temporary variance if they can establish that (1) they cannot comply with the applicable standard because they do not have the personnel, equipment, or time to construct or alter facilities; (2) they are taking all available steps covered by the standard to protect employees against exposure; and (3) their own programs will effect compliance with the standard as soon as possible.

Employer applications for an order for a temporary variance must contain at least the following:
1. the name and address of the applicant
2. the address(es) of the place(s) of employment involved
3. identification of the standard from which the applicant seeks a variance
4. representation by the applicant that he or she is unable to comply with the standard and a detailed statement of the reasons
5. a statement of the steps the applicant has taken and will take, with dates, to protect employees against the hazard covered by the standard
6. a statement of when the applicant expects to be able to comply with the standard and what steps have been taken, with dates, to come into compliance with the standard
7. certification that the employer has informed employees of the application and of their right to petition OSHA for a hearing. A description of how employees have been informed is to be included in the certification.

Employers may also apply for a permanent variance from a standard. Variance orders can be granted if OSHA finds that the employer has demonstrated, by a preponderance of evidence, that it will provide a place of employment as safe and healthful as the one that would exist if it complied with the standards.

Employer applications for a permanent variance order must contain at least the following:
1. the name and address of the applicant
2. the address(es) of the place(s) of employment involved
3. a description of the countermeasures used or proposed to be used by the applicant
4. a statement showing how such countermeasures would provide a place of employment that is as safe and healthful as that required by the standard for which the variance is sought
5. certification that the employer has informed employees of the application
6. any request for a hearing
7. a description of how employees were informed of the application and of their right to petition for a hearing.

An employer may request an interim order permitting either kind of variance until the formal application can be processed. Again, the request for an interim order must contain statements of fact or arguments why such interim order should be granted. If the request is denied, the applicant will be notified promptly and informed of the reasons for the decision. If the order is granted, all concerned parties will be informed and the terms of the order will be published in the *Federal Register*. In such cases, the employer must inform the affected employees about the interim order in the same manner used to inform them of the variance application.

Upon the filing of an employer's application for a variance, OSHA will publish a notice of such filing in the *Federal Register* and invite written data, views, and arguments regarding the application. Those affected by the petition may request a hearing. After review of all the facts, including those presented during the hearing, OSHA will publish its decision regarding the application in the *Federal Register*.

Beginning in the early 1980s, OSHA authorized its Regional Administrators to "interpret" standards in a way that essentially became variances for individual employers. Such interpretations had no effect on other employers and reflected specific conditions at a particular workplace. OSHA has also been issuing "clarifications" of standards for employers asking for deviations from standards. While granting less than 10% of employers' requests for variances since enforcement began, OSHA has issued about eight times as many clarifications.

Record-Keeping Requirements

Most employers covered by the OSH Act are required to maintain company records of all occupational injuries and illnesses. Regulations describing how to properly record and report injuries and illnesses are codified in 29 CFR 1904. Such records consist of the following:

- a log and summary of occupational injuries and illnesses, OSHA Form 300
- a supplementary record of each occupational injury or illness, OSHA Form 101 (or a state form)
- an annual summary of the total number of occupational injuries and illnesses. This must be posted by February 1 of the following year and remain posted until March 1 (Figure 4–1).

Figure 4-1. OSHA poster, "Job Safety and Health Protection," must be posted conspicuously at every plant, job site, or other facility.

For details concerning recording and reporting occupational injuries and illnesses, see Chapter 11, Injury and Illness Record Keeping, Incidence Rates, and Analysis, in this volume. Current OSHA forms for record keeping are available from OSHA area and regional offices. In states with an OSHA-approved plan, employers should check for any additional record-keeping requirements.

OSHA has made an effort to relieve small businesses from many burdensome record-keeping requirements. As a result, most employers that had no more than 10 employees at any time during the calendar year immediately preceding the current calendar year need not comply with the record-keeping requirements. However, any employer, regardless of size, can be notified in writing by OSHA or BLS that the organization has been selected to participate in a statistical survey of occupational injuries and illnesses. The employer will then be required to maintain a log and summary and to make reports for the time specified in the notice. Further, no employer is relieved of the obligation to report any fatalities or multiple-hospitalization incidents to the nearest OSHA area office.

In 1996, most employers with 10 or fewer employees that engaged in retail trade, finance, insurance, real estate, and services were also exempted from most of the record-keeping requirements (Standard Industrial Classification 52–89, except 52–50, 70, 75, 76, 79, and 80).

Reporting Requirements

Within 8 hours after an unintentional injury occurs that is fatal to one or more employees or that results in the inpatient hospitalization of three or more employees, the employer must report the incident either orally or in writing to the nearest Area Director of OSHA. In states with approved state plans, the report must be made to the state agency that has enforcement responsibilities for occupational safety and health. If an oral report is made, it should always be followed with a confirming letter written the same day. The report must relate the circumstances of the incident, the number of fatalities, and the extent of any injuries.

Workplace Inspection

Before the U.S. Supreme Court decision on the controversial *Barlow* case—*Marshall v Barlow's Inc.*, 436 U.S. 307 (1978)—the DOL's compliance safety and health officers could enter, at any reasonable time and without delay, any establishment covered by the OSH Act to inspect the premises and all its facilities. (See 29 CFR 1903.) However, since the *Barlow* decision, the OSHA compliance officer must obtain an inspection warrant and present it if the employer requires a warrant to permit an inspection.

Most employers readily consent to inspection and do not require an OSHA inspector to obtain a search warrant. OSHA's entitlement to a warrant does not depend on demonstrating probable cause to believe that conditions on the premises violate the OSHA regulations. Rather, the agency merely has to show that reasonable legislative or administrative standards for conducting an inspection have been satisfied. An organization needs to determine, well in advance, whether an inspection warrant will be requested. As a general rule, it is not advisable for the employer to refuse entry to a CSHO who has no inspection warrant. Such action only delays an inspection and increases the officer's suspicion about working conditions. In many of these cases, OSHA can obtain an inspection warrant within 48 hours; however, generally a longer period is involved.

The OSH Act authorizes an employer representative as well as an authorized employee representative, if one is designated, to accompany the CSHO during the official inspection of the premises and all its facilities. Employee representatives also have the right to participate in both the opening and the closing conferences.

Usually, the authorized employee representative is the union steward or the chairman of the employee safety committee. Occasionally, there may be no authorized employee representative, especially in non-union establishments. In this instance, the CSHO will select employees at random and confer with them on matters of safety and health and work conditions.

An employer should not refuse to compensate employees for the time spent participating in an inspection tour and for related activities such as attending the opening and the closing conferences.

Inspection Priorities
OSHA has established priorities for assignment of staff and resources. The priorities are as follows:
1. *Investigation of imminent dangers*—Allegations of an imminent-danger situation ordinarily trigger an inspection within 24 hours of notification.
2. *Catastrophic and fatal*—Accidents will be investigated if they include any one of the following:
   - one or more fatalities
   - three or more employees hospitalized as inpatients
   - significant publicity and property damage
   - issuance of specific instructions for investigations in connection with a national office special program.
3. *Investigations of employee complaints*—Highest priority is given to complaints that allege an imminent-danger situation. Complaints reporting a "serious" situation are given high priority. If time and resources allow, the CSHO may attempt to inspect the entire workplace and not just the condition reported in the complaint, particularly if a high-hazard industry is involved.
4. *Programmed high-hazard inspections*—Industries are selected for inspection based on their death, injury, and illness incidence rates; employee exposure to toxic substances; and national and local inspection scheduling programs.
5. *Reinspections*—Establishments cited for alleged serious violations may be reinspected to determine whether the hazards have been abated, particularly if the employer does not provide adequate abatement information to OSHA.

General Inspection Procedures
The primary responsibility of the 1,100 OSHA inspectors and their state counterparts is to conduct an effective inspection to determine whether employers and employees are in compliance with the requirements of the standards, rules, and regulations promulgated under the OSH Act. OSHA inspections are almost always conducted without prior notice.

To enter an establishment, the CSHO presents proper

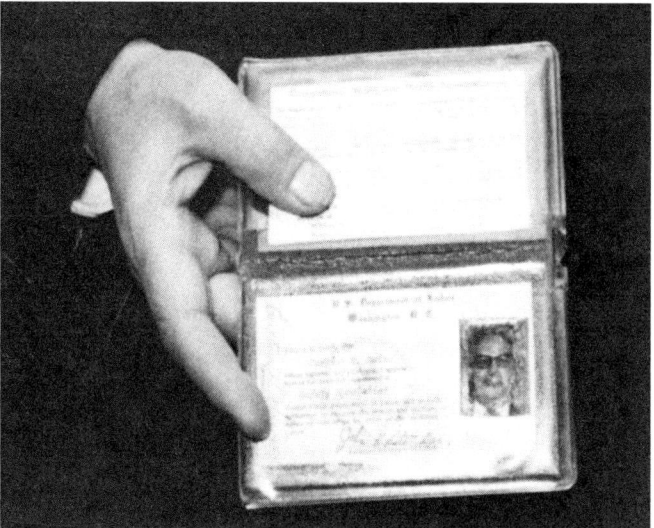

Figure 4–2. Bona fide OSHA compliance officers are equipped with official identification as shown here. The credentials are signed by the current or former Assistant Secretary of Labor for occupational safety and health. If in doubt about the validity of the credentials, the employer should contact the nearest OSHA area office, determine whether the area office has scheduled an inspection at the establishment in question, and verify the serial number on the credentials.

credentials to a guard, receptionist, or other person acting in such a capacity. Employers should always insist on seeing and checking the CSHO's credentials carefully before allowing the individual to enter their establishment for the purpose of an inspection (Figure 4–2). Anyone who tries to collect a penalty or promotes the sale of a product or service is not a CSHO.

The CSHO will usually ask to meet with an appropriate employer representative. It is recommended that employers furnish written instructions to security, the receptionist, and other affected personnel regarding the CSHO's right of entry and initial treatment, whom should be notified, and to whom and where the CSHO should be directed to avoid undue delay.

Opening Conference
The CSHO will conduct a joint opening conference with employer and employee representatives. Where it is not practical to hold a joint conference, separate conferences are to be held for employer representatives. If there is no employee representative, then a joint conference is not necessary. When separate conferences are held, a written summary of each conference should be made and the summary provided on request to employer and employee representatives.

Because the CSHO will want to talk with the firm's safety personnel, these employees should participate in the opening conference. The employer representative who accompanies the CSHO during the inspection should also participate in the opening conference.

At the opening conference, the CSHO will do the following:
1. inform the employer that the purpose of the officer's visit is to investigate whether the establishment, procedures, operations, and equipment are in compliance with OSH Act requirements
2. give the employer copies of the act, standards, regulations, and promotional materials, as necessary
3. outline in general terms:
   - the scope of the inspection
   - the records the officer wants to review
   - the officer's obligation to confer with employees
   - the physical inspection of the workplace
   - the closing conference
4. if applicable, furnish the employer with a copy of any complaint(s) that is the basis for the inspection
5. answer questions from those attending the conference.

In the opening conference, the employer representative should find out which areas of the establishment the CSHO wishes to inspect. In some cases the inspection may include areas of the facility in which trade secrets are maintained. If this is the case, the employer representative should orally request confidential treatment of all information obtained from such areas. The employer should follow up with a trade-secret letter to the CSHO requesting the officer to keep information identified in the letter strictly confidential. The CSHO should not discuss any part of this information or provide copies to any person not authorized by law to receive the data without prior written consent of the employer.

During the course of the opening conference, the CSHO may ask to review company records. The CSHO is authorized to review only the records required to be maintained by the OSH Act, regulations, and standards. In general, these records include the "OSHA Injury and Illness Log and Summary" (OSHA Form 300, available at osha.gov/recordkeeping/RKforms.html) and the "Supplemental Record of Occupational Injuries and Illnesses" (OSHA Form 301). Such records should be made readily available to the OSHA inspector. Prompt, thorough, and complete cooperation in the inspection will always make a good impression.

The CSHO may request information about the employer's current safety and health program in order to evaluate it. Naturally, a comprehensive safety and health program that shows evidence of effective incident prevention will be impressive to all concerned.

The CSHO will also ask the employer whether workers of another employer (e.g., a maintenance or remodeling contractor) are working in or on the establishment. If so, the CSHO will give the authorized representative of those employees a reasonable opportunity to participate in the inspection of their work areas.

During the conference, the CSHO will meet with the employee's authorized representative and explain the person's rights. Generally, the representative will be an employee of the establishment inspected. However, the CSHO may judge that good cause has been shown to require a third party (such as an industrial hygienist or safety consultant) who is not an employee but is still an authorized employee representative to accompany the CSHO on the inspection to ensure an effective and thorough job. The final decision on this matter will rest with the CSHO.

The employer is not permitted to designate the employee representative. Employee representatives may change as the inspection process moves from department to department. The CSHO may refuse to allow any person to accompany him or her whose conduct interferes with a full, orderly inspection. If there is no authorized employee representative, the CSHO will consult with a reasonable number of employees concerning matters of safety and health in the workplace during the course of the inspection.

Inspection of Facilities

The CSHO will have the necessary instruments for checking items such as noise levels, certain air contaminants and toxic substances, and electrical grounding. During the course of inspection, the CSHO will note and usually record any apparent violation of the standards, including its location, and any comments regarding the violation. The officer will do the same for any apparent violation of the general-duty clause. These notes will serve as a basis for the Area Director's citations or proposed penalties. For these reasons, the employer representative should find out what apparent violations the CSHO has detected during the actual inspection of the facilities. The employer representative also should take the same notes as the CSHO during the inspection, including names of employees interviewed, so that the employer will have the same information as the CSHO.

The CSHO is required only to record apparent violations and is not required to present a solution or method of correcting, minimizing, or eliminating the violation. OSHA, however, will respond to requests for technical information regarding compliance with given standards. In such cases, the employer is urged to contact the regional or area office.

During the course of an inspection, the CSHO may receive a complaint from an employee regarding a condition alleged to be in violation of an applicable standard. The CSHO, even though the complaint is brought via an informal process, will normally inspect for the alleged violation.

In the course of a normal inspection, the CSHO may make some preliminary judgments regarding environmental conditions affecting occupational health. In such cases, the officer will generally use direct-reading instruments. Should this occur, and if proper instrumentation is available, it would be prudent for the employer to have qualified personnel at the establishment make duplicate tests in the same area at the same time and under the same conditions. In addition, the employer representative should again take careful notes on the CSHO's methods as well as the results. If the inspection indicates a need for further investigation by an industrial hygienist, the CSHO will notify the Area Director, who may assign a qualified industrial hygienist to investigate further. If a laboratory analysis is required, samples will be sent to OSHA's laboratory in Salt Lake City and the results will be reported back to the Area Director. Initial inspections may also be originated by industrial hygienists, with subsequent referrals to a CSHO. Photos and videos may be taken by the CSHO to document apparent violations.

Closing Conference

Upon completion of the inspection, the CSHO will hold a joint closing conference with employee representatives and representatives of the employer. If a joint conference is not possible, separate conferences will be held. The employer's safety personnel should be present at the closing conference. At this time, the CSHO will advise the employer and employee representatives of all conditions and practices that may constitute an apparent safety or health violation. The officer should also indicate the applicable section or sections of the standards that may have been violated.

The CSHO will normally advise that citations may be issued for alleged violations and that penalties may be proposed for each violation. The authority for issuing citations and proposed penalties, however, rests with the Area Director or the director's representative.

The employer will also be informed that the citations will establish a reasonable time for abatement of the violations alleged. The CSHO will attempt to obtain from the employer a reasonable estimate of the time required to control or eliminate the alleged violation. The officer will take such estimates into consideration when recommending a time for abatement. Although the employer is not required to do so, it may be advantageous to give the officer copies of any correspondence or orders concerning equipment to achieve compliance. This act of good faith may help establish a reasonable abatement period and may reduce the proposed penalty. The CSHO should also explain the appeal procedures with respect to any citation or any notice of a proposed penalty.

Informal Postinspection Conferences

Issues raised by inspections, citations, proposed penalties, or notice of intent to contest may be discussed at the request of an affected employer, employee, or employee representative at an informal conference held by the Area Director or his or her representative. Whenever the employer or employee representatives request an informal conference, both parties should be afforded the opportunity to participate fully.

Follow-Up Inspections

Follow-up inspections are always conducted for situations involving imminent danger and may be conducted where citations have been issued for serious, repeated, or willful violations. Follow-up inspections are ordered at the discretion of the Area Director.

The follow-up inspection should be limited to verifying compliance of the conditions alleged to be in violation. The follow-up inspection is conducted with all of the usual formality of the original inspection, including the opening and the closing conferences and the walkaround rights of the employer and employee representatives.

Violations

In addition to the general-duty clause, OSH Act occupational safety and health standards are used to determine alleged violations. There are five categories of violations: willful, serious, repeated, other than serious, and *de minimis* (very minor).

Willful Violations

The following definitions and procedures apply whenever the CSHO suspects that a willful violation may exist:
- A willful violation exists under the act when the evidence shows either an intentional violation of the act or obvious indifference to its requirements.
- The violation need not be committed with a hurtful purpose or an evil intent to be deemed "willful." It is sufficient that the violation was deliberate, calculated, or intentional as distinguished from inadvertent, accidental, or ordinarily negligent.
- The determination of whether to issue a citation for a willful or repeated violation frequently raises difficult issues of law and policy and requires the evaluation of complex factual situations. Accordingly, a citation for a willful violation should not be issued without consulting the Regional Administrator, who should, as appropriate, discuss the matter with the Regional Solicitor. A repeat violation is a subsequent violation of the same or a similar standard.

Serious Violations

A serious violation involves hazardous conditions that could cause death or serious physical harm to employees and conditions that the employer knew, or should have known, existed.

OSHA's *Field Operations Manual* (specifically Chapter 4, Violations) sets forth four steps for the CSHO to follow to determine whether a violation is serious or other than serious. Section 17(k) of the act provides that a serious violation should be deemed to exist in a place of employment when there is a substantial probability that death or serious physical harm could result from a condition that exists—or from one or more practices, means, methods, operations, or processes that have been adopted or are in use—in that place of employment unless the employer did not, and could not with the exercise of reasonable diligence, know of the presence of the violation.

The CSHO should take four steps to determine that a violation is serious. The first three steps determine whether there is a substantial probability that death or serious physical harm could result from an incident or exposure relating to the violative condition. (The probability of an incident or illness occurring should not be considered when determining whether a violation is serious.) The fourth step determines whether the employer knew or could have known of the violation.

Apparent violations of the general-duty clause should also be evaluated on the basis of these steps to ensure that these violations are serious. The four elements the CSHO should consider are as follows:

Step 1—the type of accident or health hazard exposure that the violated standard or the general-duty clause is designed to prevent

Step 2—the type of injury or illness that could reasonably be expected to result from the type of accident or health hazard exposure identified in Step 1
- In making this determination, the CSHO should consider all factors that would affect the severity of the injury or illness that could reasonably be predicted to result from an accident or health hazard exposure. (At this point, the CSHO should not consider factors that relate to the probability that an injury or illness will occur.) The following are examples of the types of injuries that could reasonably be predicted to result from an accident:
  - If an employee falls from the edge of an open-sided floor 30 ft to the ground below, that employee could break bones, incur a concussion, or experience other, more serious injuries.
  - If an employee trips on debris, that employee could sustain abrasions or bruises, but it is only marginally predictable that the employee could sustain substantial impairment of a bodily function.
  - If an employee is exposed regularly and continually to beryllium at 0.004 mg/m^3, it is reasonable to predict that berylliosis or cancer could result.
  - If an employee is exposed regularly and continually to acetic acid at 20 ppm, it is reasonable to predict that the illness that could result (irritation to nose, eyes, or throat) would not involve serious physical harm.

Step 3—whether the types of injury or illness identified in Step 2 could include death or a form of serious physical harm, such as impairment of the body in which part of the body is made functionally useless or is substantially reduced in efficiency on or off the job. Such impairment may be permanent or temporary, chronic or acute. Injuries involving such impairment usually require treatment by a medical doctor.

Examples include the following:
- amputation (loss of all or part of a bodily appendage, including the loss of bone)
- concussion
- crushing (internal, even if the skin surface is intact).

Examples of illnesses that constitute serious physical harm include the following:
- cancer
- poisoning (resulting from inhalation, ingestion, or skin absorption of a substance that adversely affects a bodily system)
- lung diseases, such as asbestosis, silicosis, and anthracosis
- hearing loss.

Step 4—whether the employer knew, or with the exercise of reasonable diligence could have known, of the presence of the hazardous condition.

The knowledge requirement is met if it is determined that the employer actually knew of the hazardous condition that constituted the apparent violation.

As a general rule, if the CSHO was able to discover a hazardous condition, it can be presumed that the employer could have discovered the same condition through the exercise of reasonable diligence.

Repeated Violations

An employer may be cited for a repeated violation if that employer has been cited previously for a substantially similar condition and the citation has become a final order.

Identical Standard. Generally, similar conditions can be demonstrated by showing that in both situations, the identical standard was violated.

Different Standards. In some circumstances, similar conditions can be demonstrated when different standards were violated.

Other-than-Serious Violations

This type of violation should be cited in situations in which the injury or illness that would be most likely to result from a hazardous condition would probably not cause death or serious physical harm but would have a direct and immediate effect on the safety and health of employees.

De Minimis *Violations*

De minimis violations are conditions that represent no immediate or direct threat to safety or health. *De minimis* is short for the legal maxim *De minimis non curat lex*, "The law does not concern itself with trifles." No written document is issued for such violations.

Citations

An investigation or inspection may reveal a condition that is alleged to be in violation of the standards or the general-duty clause. In such instances, the employer may be issued a written citation that describes the specific nature of the alleged violation, cites the standard allegedly violated, and establishes a time for abatement. The employer must prominently post each citation, or a copy thereof, at or near the place where the alleged violation occurred. All citations are issued by the Area Director or a designee and are sent to the employer by certified mail.

A "Citation for Serious Violation" is prepared to cover violations that fall into the "serious" category. This type of violation must be assessed a monetary penalty.

A citation used for other-than-serious violations may or may not carry a monetary penalty. A citation may be issued to the employer for employee actions that violate the safety and health standards (either serious or other than serious).

A verbal notice, in lieu of a citation, is issued for *de minimis* violations, which have no direct relationship to safety and health.

If an inspection has been initiated in response to an employee complaint, the employee or authorized employee representative may request an informal review of any decision not to issue a citation. However, employees may not contest citations, amendments to citations, penalties, or lack of penalties. Employees may contest the time for abatement of a hazardous condition specified in a citation. They also may contest an employer's Petition for Modification of Abatement (PMA), which requests an extension of the abatement period. Employees must contest the PMA within 10 working days of its posting or within 10 working days after an authorized employee representative has received a copy.

Within 15 working days after an employer receives the citation, the employer may submit a written objection to the citation to OSHA. The OSHA Area Director then forwards the objection to OSHRC. Employees may request an informal conference with OSHA to discuss any issues raised by an inspection, citation, notice of proposed penalty, or employer's notice of intention to contest.

Petition for Modification of Abatement (PMA)

Upon receiving a citation, the employer must correct the cited hazard by the prescribed date. However, factors beyond the employer's reasonable control may prevent the work from being completed on time. In such a situation, an employer who has made a good-faith effort to comply may file for a PMA date.

The written petition should specify (1) all steps the employer took to achieve compliance, (2) the additional time needed to complete the work, (3) reasons why additional time is needed, (4) all temporary steps being taken to safeguard employees against the cited hazard during the intervening period, (5) that a copy of the PMA was posted prominently at or at least near each place where a violation occurred, and (6) the employee representative (if there is one) who received a copy of the petition.

Penalties

Penalties may range up to $7,000 for serious, other-than-serious, failure to abate and posting violations and up to $70,000 for repeated or willful violations. Willful violation penalties are no less than $5,000. Penalties are based on the gravity of the violation, good faith, size, and history of the employer. A "failure to abate" penalty may be a daily penalty for each day past the stated abatement date the employer fails to correct a violation.

Egregious Policy

If OSHA considers the apparent violations flagrant, the agency, instead of grouping similar violations, may propose a separate penalty for each violation or employee exposed.

The factors used to determine the penalty include the following:
- the number of worker fatalities, a worksite catastrophe, or a large number of similar injuries or illnesses
- a violation that results in high rates of injuries or illnesses
- whether the organization has a considerable history of workplace violations
- whether the employer flagrantly disregarded workplace safety and health responsibilities
- a large number of violations found at the worksite.

States can adopt their own penalty structures, which may exceed the one set up under OSHA.

CONTESTED CASES

An employer has the right to contest any OSHA action. The employer may contest one or more of the following: a citation, a proposed penalty, a notice of failure to correct a violation, or the time allotted for abatement of an alleged violation. OSH Act regulations that cover procedures for contesting cases are codified in 29 CFR 2200. On the other hand, an employee or authorized employee representative may contest only the time allotted for an abatement of an alleged violation.

Subsequent to initiating a formal contest of a citation, employers should request an informal conference with the Area Director or the Area Director's representative. Many times, such informal sessions resolve questions and issues, thus eliminating the need for formal contested case proceedings. The informal conference should occur within 15 working days of receipt of the violation and does not extend the Notice of Intent to Contest period. If the 15-day time period is exceeded, the citation becomes a final order.

The informal conference should be attended by representatives from OSHA, the company, and the company's union representatives (if applicable). Conference members may be asked to obtain a more complete explanation of the violation cited by OSHA, to obtain an understanding of the specific standards cited in the violation, or to negotiate and enter into an informal settlement agreement. Other areas that may be discussed include the company's difficulty in meeting the abatement dates, employer defenses, problems concerning employee safety practices, and procedures for obtaining answers to any questions that the employer brings to the table. The informal conference provides an opportunity for both sides to resolve the disputed citation and associated penalties in a relatively cordial atmosphere and is specifically intended to eliminate the need for a more formal contest of the citation. The choice to contest the citation sends the process up to the OSHRC.

However, the informal conference may fail to resolve the dispute between OSHA and the employer. If the employer elects to contest the case, affected employees or the authorized employee representatives are automatically deemed to be parties to the proceeding. In contesting an OSHA action, the employer must comply with the following rules, which apply to each specific case:

1. The employer must notify the Area Office that initiated the action that the employer is contesting the case. This must be done within 15 working days after receiving OSHA's notice of proposed penalty and should be sent by certified mail. If the employer does not contest within the required 15 working days, the citation and proposed assessment of penalties are deemed to be a final order of OSHRC and are not subject to review by any court or agency. As a result, the alleged violation must be corrected within the abatement period specified in the citation.
2. If any of the employees working at the site where the alleged violation exists are union members, a copy of the notice of contest must be served upon their union.
3. If employees who work on the site are not represented by a union, a copy of the notice of contest must either be posted at a place where the employees will see it or be served upon them personally.
4. The notice of contest must also list the names and addresses of parties who have been personally served a notice, or, if such notice is posted, it must contain the address of the posted location.
5. In some cases, the employees at the site of the alleged violation are not represented by a union and have not been personally served with a copy of the notice to contest. If so, posted copies must specifically advise the unrepresented employees that they may not be able to assert their status as parties to the case if they fail to properly identify themselves to OSHRC or to the Hearing Examiner before the hearing begins or when it first opens.
6. There is no specific form for the notice of contest. However, such notice should clearly identify what is being contested—the citation, the proposed penalty, the notice of failure to correct a violation, or the time allowed for abatement—for each alleged violation or combination of alleged violations.

If the employer contests an alleged violation in good faith and not solely for delay or variance of penalties, the abatement period does not begin until OSHRC enters the final order. When a notice of contest is received by an Area Director from an employer, an employee, or an authorized employee representative, the Director will file with OSHRC the notice of contest and all contested citations, notice of proposed penalties, or notice of failure to abate.

Upon receiving the notice of contest from the Area Director, the commission assigns the case a docket number. Ultimately, an administrative law judge (ALJ) will be assigned to the case and will conduct a hearing at a location reasonably convenient to those concerned. At the hearing, OSHA presents its case and is subject to a cross-examination by other parties. The party contesting then presents its case and is also subject to a cross-examination by other parties. Affected employees or an authorized employee representative may participate in the hearings. The decision by the ALJ will be based only on what is in the record. Therefore, if statements go unchallenged, they will be assumed to be fact.

After the hearings are completed, the ALJ will submit the record and a report with the decisions to OSHRC. If no commissioner orders a review of an ALJ's decisions, they will stand as OSHRC's decision. If any commissioner orders a review of the case, the commission itself must render a decision to affirm, modify, or vacate the judge's decision. The commission's orders become final 15 days after issuance, unless stayed by a court order.

Any person adversely affected or aggrieved by an order of the commission may obtain a review of the order in the U.S. Court of Appeals. However, the person must seek a review within 60 days of the order's issuance.

SMALL-BUSINESS LOANS

The act enables small businesses to obtain economic assistance for health- and safety-related issues. It amends the Small Business Act to provide for financial assistance to small firms that must make changes to comply with the standards promulgated under the OSH Act or by a state under a state plan. Before approving any assistance, the Small Business Administration (SBA) must first determine that the small firm is likely to suffer substantial economic hardship without financial help.

An employer can apply for a loan using one of two procedures: (1) before federal or state inspection in order to come into compliance or (2) after federal or state inspection to correct alleged violations.

When an employer has not been inspected and requests a loan to bring the establishment into compliance before an inspection, the employer must submit the following to the SBA:
- a statement of the conditions to be corrected
- a reference to the OSHA standards that require the employer to make corrections
- a statement of the firm's financial condition showing that a loan is needed.

The employer should submit this information to the nearest SBA field office along with any background material. The SBA will then send the application to the appropriate OSHA Regional Office, Office of Technical Support. The OSHA Regional Office will review the application and advise SBA whether the employer is required to correct the described conditions in order to come into compliance and whether the proposed use of funds will accomplish the needed corrections. OSHA will initiate direct contact with the applicant only after clearance from the SBA.

If the employer makes an application after an inspection to correct alleged violations, the procedure is the same as before inspection, except that the applicant also must furnish the SBA with a copy of the OSHA citation(s). The SBA then sends the application to the OSHA Area Office that conducted the inspection. That office notifies the SBA whether the proposed use of loan funds will adequately correct cited violations.

Forms for loan applications may be obtained from any SBA field office. In some instances, private lending institutions will be able to provide the form for SBA/bank participation loans.

FEDERAL–STATE RELATIONSHIPS

The OSH Act encourages states to assume the fullest responsibility for administering and enforcing their own occupational safety and health laws. However, in order to assume this responsibility, states must submit a state plan to OSHA for approval. If such a plan satisfies designated conditions and criteria, OSHA will approve the plan. The regulations pertaining to state plans for the development and enforcement of state standards are codified in 29 CFR 1902. The states and possessions listed in Figure 4–3 have approved plans.

The basic criterion for approval of a state plan is that the plan is "at least as effective as" the federal program. It was not Congress's intent to require that state programs be a mirror image of the federal program. Congress believed that rules for developing state plans should be flexible to allow consideration of local problems, conditions, and resources. The act provides for funding up to half the costs of the implementation of the state program.

A state plan must include any occupational safety and health issue (industrial, occupational, or hazard group) for which a corresponding federal standard has been promulgated. A state plan cannot be less stringent, but it may include subjects not covered in the federal standards. However, state plans that do not include issues covered by the federal program effectively surrender such issues to OSHA. For example, a state plan may cover all industries except construction. When such is the case, the state surrenders its jurisdiction for safety and health programs in construction operations to OSHA, which is then obligated to enforce the federal standards for operations not covered by the state plan.

Following approval of a state plan, OSHA will continue to exercise its enforcement authority until it determines on the basis of actual operations that the state plan is indeed being satisfactorily carried out. If the implementation of the state plan is satisfactory during the first 3 years after the plan's approval, then the relevant federal standards and

Alaska Department of Labor
PO Box 111149
Juneau, AK 99811
907-465-2700

Industrial Commission of Arizona
800 West Washington
Phoenix, AZ 85007
602-542-4661

California Department of Industrial Relations
455 Golden Gate Ave.
San Francisco, CA 94102
415-972-8846

Connecticut Department of Labor
200 Folly Brook Boulevard
Wethersfield, CT 06109
860-263-6000

Hawaii Department of Labor and Industrial Relations
830 Punchbowl Street
Honolulu, HI 96813
808-586-8842

Indiana Department of Labor
402 West Washington, Room W-195
Indianapolis, IN 46204
317-232-2655

Iowa Division of Labor Services
1000 East Grand Avenue
Des Moines, IA 50319
515-242-5870

Kentucky Labor Cabinet
1047 U.S. Highway 127 South
Frankfort, KY 40601
502-564-3289

Maryland Division of Labor and Industry
Department of Labor, Licensing, and Regulation
1100 N. Eutaw St.
Baltimore, MD 21201
410-767-2241

Michigan Department of Labor
PO Box 30643
Lansing, MI 48079
517-322-1814

Minnesota Department of Labor and Industry
443 Lafayette Road
St. Paul, MN 55155
651-284-5050

Nevada Department of Industrial Relations
Division of Occupational Safety and Health
1309 Green Valley Parkway, Suite 200
Henderson, NV 89074
702-486-9044

New Mexico Environment Department
525 Camino de los Marquez
Santa Fe, NM 87505
505-476-8700

New York State Department of Labor
522 State Office Campus
Albany, NY 12240
518-457-3518

North Carolina Department of Labor
1101 Mail Service Center
Raleigh, NC 27699-1101
919-779-8560

Oregon OSHA
PO Box 14480
Salem, OR 97309
503-378-3272

Puerto Rico Department of Labor and Human Resources
Prudencio Rivera Martinez Building
505 Munoz Rivera Avenue
Hato Rey, PR 00918
787-754-2119

South Carolina Department of Labor
PO Box 11329
Columbia, SC 29211-1329
803-896-7665

Tennessee Department of Labor
220 French Landing Drive
Nashville, TN 37243
615-741-2793

Utah Occupational Safety and Health
PO Box 146650
Salt Lake City, UT 84114-6650
801-530-6010

Vermont Department of Labor and Industry
PO Box 488
Montpelier, VT 05601
802-828-2138

Virginia Department of Labor and Industry
Powers-Taylor Building
13 South 13th Street
Richmond, VA 23219
804-371-2327

Virgin Islands Department of Labor
53A &54B Kronprindense Gade,
St. Thomas, USVI 00803
340-776-3700 x 2617

Washington Department of Labor and Industries
PO Box 44600
Olympia, WA 98504-4600
360-902-5495

Wyoming Department of Occupational Health and Safety
1510 E. Pershing Blvd. – West Wing
Cheyenne, WY 82002
307-777-7786

Figure 4–3. States with approved plans.

OSH Act enforcement of such standards no longer apply to issues covered under the state plan. This means that for the interim period when dual jurisdiction exists, employers must comply with both state and federal standards.

Although the state agencies administering the state plan are vitally concerned with its success, members of the state legislature do not always share their enthusiasm. The legislature not only must appropriate an adequate budget, but in many cases must also pass legislation enabling the state agency to carry out all the functions incorporated in the state plan. Because of the legislature's involvement, sometimes the state agency responsible may not be able to fully implement the state plan, resulting in the state's performance falling short of being "at least as effective as" the federal program. In such instances, OSHA has the right and the obligation to withdraw its approval of the state plan and once again assume full jurisdiction in that state.

WHAT DOES IT ALL MEAN?

Congressional action in creating the OSH Act is only one step toward achieving the full purpose underlying the act. Getting it to work with reasonable efficiency is the second and more difficult task. Achieving this goal of providing a safe, healthy environment on and off the job depends on the willingness and cooperation of all concerned—employees and organized labor as well as business and industry.

The act has given new visibility to the whole realm of occupational safety and health. Because many employee rights are incorporated into the OSH Act, it has given employees a significant role in occupational safety and health matters. It has moved the laggards from "little or no safety" to "some safety," but not to "optimum safety." It has elevated the priority of occupational safety and health issues in business management. It has given new status and responsibilities to professionals working in the occupational safety and health field. Management is now relying more heavily on these safety professionals for advice. And the act has bestowed a new status on nationally recognized organizations that develop industry standards.

The OSH Act has also given new impetus to the field of occupational health, a much more demanding discipline compared to occupational safety. In occupational health, much more needs to be done to determine what kinds of exposures are indeed hazardous to humans and under what conditions. Further, a great deal more needs to be done to determine what countermeasures are not only adequate but also reasonable and feasible to eliminate or minimize exposures to occupational health hazards. Far more research and data about occupational health will be required to achieve the best occupational safety and health programming.

The OSH Act has encouraged greater training for professionals in occupational safety and health. Several universities have developed new curricula and programs leading to various degrees in this field—and more are yet to come.

The OSH Act also gave new emphasis to the product safety discipline. Until the passage of the U.S. Consumer Product Safety Act, the OSH Act was the most significant piece of legislation affecting product safety ever passed by Congress. Designers and manufacturers of equipment now used by industry have a moral (although not legal) obligation to design, deliver, and install such equipment in accordance with the applicable standards.

The OSH Act is not without limitations, however. Mere compliance with the requirements of the act will not achieve optimum safety and health in terms of cost, benefits, and human values. All those concerned must recognize that occupational safety and health cannot be handed to the employer or the employee by legislative enactment or administrative decree. At best, state or federal occupational safety and health standards are minimum standards and can cover only areas that are enforceable—namely, control over physical conditions and the environment.

As a matter of hard reality, enforcement standards simply do not adequately relate to the human in the human–machine–environment system. Important elements of a complete safety program—such as (1) establishment of work procedures to limit risk, (2) supervisory training, (3) job instruction training for employees, (4) job safety analysis, and (5) human factors engineering—by and large have not been included in the standards promulgated under the OSH Act. Neither do the standards address such issues as employee attitudes, morale, and teamwork.

For the most part, the occupational safety and health standards developed under the OSH Act are minimal criteria and represent a foundation rather than goals to achieve. Thus, to rely on mere compliance with these standards is to invite disaster because the risk remaining after compliance often remains unacceptable. Effective unintentional-injury prevention and control of occupational health hazards must go beyond the OSH Act.

Generally, a violation of a standard is only a symptom of something wrong with the management safety system as a whole. Only complete occupational safety and health programming as described elsewhere in this manual can achieve a level of risk acceptable to employers and employees. The real objective of the OSH Act is improved occupational safety and health performance and not merely compliance with a set of standards.

PART 2—KEY STANDARDS AND REGULATIONS

In the 1980s, two important standards came into effect: (1) final rules for access to exposure and medical records and (2) hazard communication. Both were designed to provide employees with information about the hazardous conditions to which they are or have been exposed and to give them access to their own medical and exposure records and documented information about chemicals used in the workplace. In the years following, a number of other key standards and regulations came into effect to extend worker protections.

ACCESS TO EMPLOYEE EXPOSURE AND MEDICAL RECORDS

Employers in general industry and the maritime and construction industries must provide records access (the right to examine and copy records) to all employees exposed to toxic substances and harmful physical agents, to their employee representatives, to health professionals, and to OSHA. The rule does not require creation of any records, only their preservation. Employee medical records should be retained by the company for 30 years from the termination date of the employee.

Employers must provide records promptly (within 15 working days), including a date for release of the information and an explanation of the delay should it take longer to process the request. Employers should inform their workers initially and at least annually of their rights to access to medical and exposure records. OSHA may obtain personal medical records promptly without the written consent of the particular employees, but must adhere to rules of agency practice and procedure governing OSHA access to employee medical records contained in 29 CFR 1913.10.

Requests for records need not be in writing except when trade secrets are involved. Union and health professionals must have specific written consent to gain access to employees' personal medical records. Exposure records may be examined without such consent. However, these professionals must state the specific record needed and the occupational health need for gaining access to the information. Health professionals include physicians, occupational health nurses, industrial hygienists, toxicologists, and epidemiologists who provide medical or other occupational health services to exposed employees.

Unless a physician representing an employer believes that direct employee access to certain sensitive information in the employee's medical record could be detrimental to the employee, employees should have prompt access to their own exposure and medical records without any unreasonable barriers. Employees should have access to exposure records of others when these exposures are representative of their past or present exposures.

Records
The rule covers the following types of records:
- exposure to toxic substances and harmful physical agents
- employee personal medical records.

Exposure records must be maintained for 30 years and medical records for the duration of employment plus 30 years. First-aid records and experimental toxicological research records are excluded from the 30-year retention requirements.

If a company maintains a chemical inventory or set of SDSs, it need not retain production records, shipping records, invoices, batch cards, or other similar documents. Biological monitoring results, except those pertaining to alcohol or drugs, are to be retained. Employers need not copy x-rays and may require viewing on site or at some other suitable location. All x-rays, except chest x-rays, may be microfilmed for records storage. Records created in anticipation of litigation (e.g., workers' compensation examinations) do not need to be retained. If provided on termination to the employee, personal medical records for short-term employees (less than 1 year) do not have to be retained.

Toxic Substances
Toxic substances and harmful agents include the following:
- materials listed in the National Institute for Occupational Safety and Health (NIOSH) Registry of Toxic Effects of Chemical Hazards (RTECS) (see cdc.gov/niosh/rtecs/RTECSaccess.html)
- substances that have evidenced an acute chronic health hazard in testing conducted by or known to the employer
- substances named in an SDS kept by or known to the employer indicating that the material may pose a health hazard.

Except for trade secrets, employers must disclose the specific chemical identity (chemical name and Chemical Abstract Service [CAS] number) of materials for which exposure records are requested.

Trade Secrets
Although employers may withhold the specific chemical identity of a toxic substance if the information represents a stated trade secret, all other information concerning the toxic substance must be disclosed as required. The chemical name should be made available to health professionals,

employees, and designated representatives under certain specified conditions.

In a medical emergency, an employer must immediately disclose the specific chemical identity of a toxic substance to a treating physician or nurse when needed for emergency or first-aid treatment. The employer may obtain a statement of need and a confidentiality agreement as soon as circumstances permit.

When there is no emergency, requestors seeking trade-secret identity must put their request in writing, describing the medical or health need for the request and explaining why other information (health risks of the chemical, proper protective measures, etc.) is insufficient. Requestors must also describe the procedures they will take to protect confidentiality, agree not to use the information except for health purposes, and agree not to disclose the information to anyone except OSHA. Confidentiality agreements must be signed and may include a liquidated damages provision, but no penalty bond.

Employers' denials of request for specific chemical identities must be in writing within 30 days of the request. Denials must provide evidence that the information is a trade secret and explain how alternate information will suffice. The requestor can appeal the denial to OSHA. If the agency finds the denial improper, the employer can be cited and penalties proposed.

HAZARD COMMUNICATION STANDARD

In 1983, OSHA promulgated the Hazard Communication Standard (HCS; 29 CFR 1910.1200). This standard presently covers all workers exposed to hazardous chemicals in all industrial sectors. Under the HCS, federal workers are covered by Executive Order.

The HCS governs exposure to physical hazards (e.g., flammability, corrosivity, and reactivity) and both acute and chronic effects due to health hazards (e.g., irritation, sensitization, toxicity, and carcinogenicity). The HCS requires information about these hazards to be made available to employers and employees, including recommended precautions for safe use.

The HCS accomplishes this by establishing uniform requirements to ensure that the hazards of all chemicals imported into, produced, or used in U.S. workplaces are evaluated and that this hazard information is transmitted to affected employers and exposed employees.

Chemical manufacturers and importers must convey the hazard information they learn from their evaluations to downstream employers by means of labels on containers and SDSs. In addition, all covered employers must have a hazard communication program to convey this information to their employees through labels on containers, SDSs, and training.

In 1994, a modified final rule included a number of minor changes and technical amendments to further clarify the requirements of the HCS to help ensure full compliance and achieve protection for employees. These changes included the following:

- adding and clarifying certain exemptions from labeling and other requirements modifying and clarifying aspects of the written hazard communication program and labeling requirements (e.g., the purpose of the label is to provide an immediate visual warning of the hazards)
- explaining and slightly modifying the duties of distributors, manufacturers, and importers to provide SDSs to employees (e.g., retail distributors selling hazardous chemicals to commercial customers should provide an SDS on request and should post a sign or otherwise inform that one is available)
- clarifying certain provisions regarding SDSs (e.g., electronic access, microfiche, and other alternatives to maintaining paper copies of the SDSs are permitted as long as no barriers to ready employee access in each workplace are created by such options).

Elements of a Hazard Communication Program

The basic goal of a hazard communication program is to ensure that employers and employees know about work hazards due to potential chemical exposure and how to protect themselves. The three information components—labels, SDSs, and worker training—are all essential to the effective functioning of the program. The program should ensure that all employers receive the information they need to inform and train their employees properly and to design and implement employee protection programs.

- SDSs provide comprehensive technical information and serve as a reference document for exposed workers as well as health professionals providing services to those workers.
- Labels provide a brief summary of the hazards of the chemicals used in the work area.
- Training ensures that workers understand the information on both SDSs and labels, know how to access this information when needed, and are aware of the proper protective procedures to follow. Employers should provide employees with information and training on hazardous chemicals in their work area at the time of their initial assignment and whenever a new hazard is introduced into their work area. Retraining is required when a new hazard, but not a new chemical, is brought into the workplace. It should also provide necessary hazard information to employees

so that they can participate in, and support, the protective measures in place at their workplaces.

All employers are responsible for informing and training workers about the hazards in their workplaces, retaining warning labels, and making available SDSs with hazardous chemicals.

Workplaces where employees are exposed to hazardous chemicals must have a written plan describing how the standard will be implemented in that facility. The written program must reflect what employees are doing in a particular workplace. For example, the written plan must list the chemicals present at the site, indicate who is responsible for the various aspects of the program in that facility, and state where written materials will be made available to employees. The written program must describe how the requirements for labels and other forms of warning, SDSs, and employee information and training will be met in the facility.

The only work operations that do not have to comply with the written plan requirements are laboratories and work operations where employees handle chemicals only in sealed containers. Laboratories are covered in the Occupational Exposure to Hazardous Chemicals in Laboratories section of this chapter.

Employers with workplaces where employees deal only with chemicals in sealed containers under normal conditions of use must comply with the following provisions:
- ensure that labels affixed to incoming containers of hazardous chemicals are kept in place
- maintain SDSs received in accessible locations
- obtain SDSs when requested by an employee
- train workers on what to do in the event of a spill or leak.

In 1983, after 6 years of effort, OSHA adopted its complex and comprehensive rules on labels, training, and information flow about workplace chemicals. The premise of the HCS was that exposure to chemicals causes adverse health effects that can be lessened or avoided if workers know the specific risks of the specific chemicals with which they are dealing.

The HCS applies to all employers covered by OSHA in both manufacturing and nonmanufacturing sectors, including construction. It also covers workers who may be exposed to hazardous materials under normal conditions or in a foreseeable emergency.

As stated earlier, the basic purpose of the HCS is to establish uniform requirements to ensure that the hazards of all chemicals produced, imported, or used within the United States are evaluated. This hazard information must be transmitted to affected employers and employees. This is accomplished through the following:

- a written hazard communication program
- hazard evaluation
- employee training
- SDSs
- container labeling.

Written Hazard Communication Program

All employers using materials that may pose a hazard to employees must establish a written, comprehensive hazard communication program that includes provisions for container labeling, SDS availability, and employee training. The program must also include a list of hazardous chemicals in each work area, how the employer will inform employees of the hazards of nonroutine tasks and unlabeled pipes, and how the employer will inform contractors in manufacturing facilities of the hazards to which their employees may be exposed. The program need not be lengthy or complicated but must be available to employees.

Hazard Evaluation

The quality of a hazard communication program depends on the accuracy of the initial hazard assessment. The primary responsibility for hazard evaluation lies with the chemical manufacturer or importer. If a company uses a process that produces a chemical to which employees are exposed, the employer/owner is considered a "chemical manufacturer" and must evaluate the chemical's hazards. This is true for chemical intermediates or for decomposition products such as welding fumes.

A producer or user of a material that may be hazardous should develop a hazard evaluation, starting with an inventory of chemicals used or produced. The inventory can be developed by reviewing purchase orders and performing a physical inventory of all containers of chemicals. After developing an inventory, the next step is to determine which substances are hazardous. The products in the inventory and their components should be compared to the following lists:
- 29 CFR 1910.1000–1047, Toxic and Hazardous Substances, OSHA
- American Conference of Governmental Industrial Hygienists (ACGIH), *TLVs: Threshold Limit Values for Chemical Substances in the Work Environment* (latest edition)
- National Toxicology Program (NTP), *Annual Report on Carcinogens* (latest edition)
- International Agency for Research on Cancer (IARC), *Monograph* (latest edition).

The next step is to determine whether any of the remaining chemicals in the inventory possess physical or health hazards. Consult the SDS provided by the manufacturer, importer, or distributor, or discuss questions with a knowledgeable industrial hygienist or safety professional (call the National Safety Council).

Employee Training

An employer must provide training for employees exposed to hazardous chemicals. Training must be done when employees are first assigned to an operation and whenever a new hazard is introduced into the work area. Training must cover the following topics:
- existence and requirements of the HCS
- operations in the work area where hazardous chemicals are present and the hazards of these chemicals
- how the hazard communication program is implemented in the workplace, how to read and interpret information on labels and SDSs, and how employees can obtain and use available hazard information
- measures employees can take to protect themselves from hazards
- specific procedures adopted to provide protection, such as work practices and the use of engineering controls or personal protective equipment.

Safety Data Sheets (SDSs)

Chemical manufacturers and importers must develop an SDS for each hazardous chemical they produce or import. Employers must obtain SDSs for every hazardous chemical in the workplace, and copies must be readily available to employees. The HCS requires that specific information be on the SDS:
- *Section I*—manufacturer's name, address, and phone number, and the date the sheet was prepared
- *Section II—Hazardous Ingredients/Identity Information*—chemical identity of components, exposure limits (OSHA, PEL, ACGIH, TLV, and other recommended limits)
- *Section III—Physical/Chemical Characteristics*—boiling point, vapor pressure and density, specific gravity, melting point, and so on, and the physical and chemical data that indicate the potential for vaporization
- *Section IV—Fire and Explosion Hazard Data*—flash point, flammable limits, extinguishing media, unusual fire and explosion hazards, special fire-fighting procedures
- *Section V—Reactivity Data*—stability of product, potential for polymerization and decomposition, materials and conditions to avoid
- *Section VI—Health Hazards*—acute and chronic hazards, carcinogenicity, signs and symptoms of exposure, emergency and first-aid procedures
- *Section VII—Precautions for Safe Handling and Use*—procedures to be used for spills, waste disposal, handling, storage
- *Section VIII—Control Measures*—personal protective equipment, ventilation, special worker or hygienic practices.

The producer of the SDS is responsible for the information on it and must ensure that all sheets are up to date.

Labeling

Chemical manufacturers, importers, and distributors must be sure that containers of hazardous chemicals leaving the workplace are labeled with the following:
- identity of the product
- written hazard warnings (in English)
- name and address of the manufacturer or other responsible party.

In the workplace, each container of hazardous chemicals must be labeled, tagged, or marked with the following:
- identity of the product (such that the name can be referenced to an SDS)
- written hazard warnings (in English), including target organ(s), if applicable
- graphic symbols. However, OSHA cites studies indicating that graphic symbols are not as quickly recognized as word statements. The warnings may be printed in other languages besides English.

There are several exemptions for onsite labeling of containers. A sign or placard can be used for a number of stationary containers within a work area with similar contents. Operating procedures, process sheets, batch tickets, blend tickets, and similar written materials can be substituted for container labels if they contain the same information and are readily available in the work area to the employees.

Portable containers intended for immediate use by the employee who makes the transfer are also exempted. The employer is not required to label pipes or piping systems. However, the means that the employer will use to train employees on contents of piping systems must be described in the written hazard communication program.

OCCUPATIONAL EXPOSURE TO HAZARDOUS CHEMICALS IN LABORATORIES

The Occupational Exposure to Hazardous Chemicals in Laboratories standard (29 CFR 1910.1450) covers all laboratories engaged in the laboratory use of chemicals. The standard does not apply to uses of hazardous chemicals that do not meet the definition of laboratory use or that provide no potential for employee exposure. The contents of this standard and its appendices must be made available to all affected employees.

Where hazardous chemicals are used in a laboratory covered by the laboratory standard, the employer must develop and carry out the provisions of a written chemical hygiene plan (CHP). The CHP must be made available to all affected parties and include the necessary work practices, procedures, and policies to ensure that employees are protected from all potentially hazardous chemicals in use in their work area.

Training and Information

The employer must provide employees with the following information and training to ensure that they are aware of the hazards of the chemicals present in their work area. This should be done at the time of initial assignment to a work area where hazardous chemicals are present and before assignments involving new exposure situations.
- the location, availability, and details of the employer's chemical hygiene plan (CHP)
- the permissible exposure limits (PELs) for OSHA
- signs and symptoms associated with exposures to hazardous chemicals used in the laboratory
- the location and availability of known reference material on the hazards, safe handling, storage, and disposal of hazardous chemicals found in the laboratory, including but not limited to SDSs received from chemical suppliers
- methods and observations that may be used to detect the presence or release of a hazardous chemical
- the physical and health hazards of chemicals in the work area
- measures employees can take to protect themselves from these hazards, including specific procedures implemented by the employer to protect employees from exposure to hazardous chemicals, such as appropriate work practices, emergency procedures, and the personal protective equipment to be used.

Labels on incoming containers of hazardous chemicals must not be removed or defaced. SDSs on incoming hazardous chemicals must be retained and made available to lab employees.

Hazard Evaluation and Medical Care

When required by an applicable OSHA standard, workplace exposures should be monitored in cases in which exposure levels may routinely exceed the action level (or in the absence of an action level, the PEL). The employer must notify the employee of the monitoring results within 15 working days after receipt of the results. Where the use of respirators is necessary to maintain exposure below permissible exposure limits, the employer must provide, at no cost to the employee, the proper respirator equipment in accordance with the Respiratory Protection standard (29 CFR 1910.134).

All employees who work with hazardous chemicals must be given the opportunity to receive medical attention, including any follow-up examinations determined necessary under certain circumstances by the examining licensed physician. Medical examinations and consultants must be provided without cost to the employee, without loss of pay, and at a reasonable time and place. The employer must provide certain information to the physician, including the identity of the hazardous chemicals, a description of the conditions under which the exposure occurred, and a description of the signs and symptoms of exposure that the employee is experiencing.

The employer must establish and maintain an accurate record of any measurements taken to monitor employee exposure and any medical consultation and examination, including tests or written opinions.

WORKPLACE EMERGENCIES

When an OSHA standard requires either emergency-action or fire-prevention plans in accordance with 29 CFR 1910.38, firms with more than 10 employees must have a written emergency-action plan; smaller companies may communicate their plans orally (29 CFR 1910.38[a]).

Emergency-Action and Fire-Prevention Plans

Management should review plans with employees initially and whenever the plan itself or employee responsibilities under it change. Plans should be reevaluated and updated periodically. An emergency-response coordinator and a backup coordinator must be designated. The coordinator may be responsible for facilitywide operations, public information, and ensuring that outside aid is called in. A backup coordinator ensures that a trained person is always available. Procedures should be in place for the emergency-response coordinator to operate from an alternate communications center if necessary. A current list of key personnel and off-duty telephone numbers should be maintained. Additional duties of the coordinator include the following:
- determining what emergencies may occur and ensuring that emergency procedures are developed to address them
- directing all emergency activities, including evacuation and accounting for personnel
- ensuring that outside emergency services are called when necessary
- directing the shutdown of facility operations when necessary.

Emergency procedures, including the handling of any toxic chemicals, should include the following:
- escape procedures and escape route assignments
- special procedures for employees who perform or shut down critical facility operations
- a system to account for all employees after evacuation
- rescue and medical duties for affected employees
- a process for reporting fires and other emergencies
- contacts for information about the plan.

A fire action plan should include a list of the major workplace fire hazards and their proper handling and storage procedures, potential ignition sources (welding, smoking, etc.), control procedures, and the type of fire protection equipment or systems that can be used to control a fire.

Training
Training must be conducted initially, when new employees are hired, and at least annually. Every employee needs to know details of the emergency-action plan, including evacuation plans, alarm systems, reporting procedures for personnel, shutdown procedures, and types of potential emergencies. Drills should be held at random intervals, at least annually, and include outside police and fire authorities if possible. Additional training should be provided when new equipment, materials, or processes are introduced; when procedures have been updated or revised; or when exercises show that employee performance is inadequate. Members of emergency-response teams should be trained for potential emergencies and physically capable of carrying out their duties. Training should ensure that response team members are familiar with the toxic hazards in the workplace and can judge when to evacuate personnel or rely on outside help.

Medical Care
Employers not near an infirmary, clinic, or hospital should have someone on site trained in first aid, have medical personnel readily available for advice and consultation, and develop written emergency medical procedures. It is essential that first-aid supplies are available to the trained medical personnel, emergency phone numbers are placed in conspicuous places near or on telephones, and prearranged ambulance services for any emergency are available.

CONFINED-SPACES STANDARD

The Permit-Required Confined-Spaces standard (29 CFR 1910.146) applies to all of general industry. However, note that 29 CFR 1910.146 does not cover the agriculture (1928), construction (1926), and shipyard employment (1915) industries. The focus of the standard is on an employer's whole program as a primary safeguard for employees and on the capacity of that program to detect confined-space hazards and to respond to them appropriately. Minimum safety and health program management practices are described for a permit-required confined space (permit space).

The standard identifies an employer's general obligations to identify and evaluate confined spaces in the workplace and to take protective action because of existing or potential hazards. A confined space has the following characteristics:
- It has limited or restricted means of entry or exit.
- It is large enough to enter and perform assigned work.
- It is not designed for continuous occupancy.

Confined spaces may include underground vaults, sewers, tanks, storage bins, diked areas, vessels, and silos.

Hazards specific to a confined space are dictated by the following:
- the material stored or used in the confined space
- the activity performed by the worker
- the external environment.

The most hazardous kind of confined space combines limited access and mechanical devices. Such confined spaces may also contain physical hazards that further complicate the work environment and the entry and exit process.

In general, employers must evaluate the workplace to determine whether spaces are permit-required confined spaces. Although the evaluation need not be documented, the employer must be able to explain how the evaluation was conducted and describe the results. If there are permit spaces in the workplace, the employer must inform applicable employees of the existence of, location of, and danger posed by the spaces. This can be accomplished by posting danger signs or other equally effective means.

Elements of a Confined-Space Program
Many workplaces contain spaces that meet the definition of "confined" because their configurations impede the activities of employees who must enter into, work in, and exit from them. In many instances, employees who work in confined spaces also face increased risk of exposure to serious physical injury from hazards such as entrapment, engulfment, and hazardous atmospheric conditions.

Hazardous atmospheres encountered in confined spaces can be divided into four distinct categories: flammable, toxic, irritant and/or corrosive, and asphyxiating. Physical hazards encountered while working in a confined space include thermal effects (heat and cold), noise, vibration, radiation, and fatigue. Confinement itself may pose entrapment hazards, and work in confined spaces may keep employees closer to hazards, such as an asphyxiating atmosphere, than they would be otherwise. For example, confinement, limited access, and restricted airflow can result in hazardous conditions that would not arise in an open work area.

The term "permit-required confined space" (i.e., permit space) refers to spaces that meet the definition of a "confined space" and pose health or safety hazards, thereby requiring a permit for entry.

A permit-required confined space has one or more of the following characteristics:
- contains or has the potential to contain a hazardous atmosphere
- contains a material that has the potential for engulfing an entrant
- has an internal configuration that might cause an entrant to be trapped or asphyxiated by inwardly converging walls or by a floor that slopes downward and tapers to a smaller cross-section
- contains any other recognized serious safety or health hazards.

Pertinent employees should be informed by posted danger signs or by any other equally effective means of the existence of, location of, and danger posed by the permit spaces.

If employees are not to enter and work in permit spaces, employers must take effective measures to prevent employees from entering those spaces.

An employer that allows employee entry must develop and implement a written program for permit-required confined spaces. The program should identify employee job duties and establish and implement a system for the preparation, issuance, use, and cancellation of entry permits. Appropriate procedures for rescue and emergency services should be detailed.

A permit, signed by the entry supervisor and verifying that preentry preparations have been completed and that the space is safe to enter, must be posted at entrances or otherwise made available to entrants before they enter a permit space.

The duration of entry permits must not exceed the time required to complete an assignment. Also, the entry supervisor must terminate entry and cancel permits when an assignment has been completed or when new conditions exist. New conditions must be noted on the canceled permit and used in revising the permit space program. The standard also requires the employer to keep all canceled entry permits for at least 1 year.

If hazardous conditions are detected during entry, employees must immediately leave the space, and the employer must evaluate the space to determine the cause of the hazardous atmospheres. At least one attendant should be stationed outside the permit space for the duration of employee presence in the confined space.

When testing and inspection data prove that a permit-required confined space no longer poses hazards, that space may be reclassified as a nonpermit confined space. If entry is required to eliminate hazards and to obtain the data, the employer must follow special procedures as set forth by the standard. A certificate documenting pertinent data—such as date, location of the space, and the signature of the person making the certification—must be made available to those entering the space.

Contractors also must be informed of permit spaces and permit-space entry requirements, any identified hazards, any known hazardous conditions, and precautions or procedures to be followed when in or near permit spaces.

Training and Education

Before the initial work assignment begins, the employer must provide proper training that covers the duties of authorized entrants, attendants, entry supervisors, and emergency rescue personnel who are required to work in or in support of entry into permit spaces. Upon completing this training, employers must ensure that employees have acquired the understanding, knowledge, and skills necessary for the safe performance of their duties. Additional training is required when the job duties change, the permit-space program changes or the permit-space operation presents a new hazard, or when an employee's job performance shows deficiencies. Training required for rescue team members includes cardiopulmonary resuscitation (CPR) and first-aid training. Upon completion of training, employees must receive a certificate of training that includes the employee's name, signature or initials of the trainer(s), and dates of training. The certificate must be made available for inspection by employees and their authorized representatives.

CONTROL OF HAZARDOUS ENERGY (LOCKOUT/TAGOUT)

The final rule on Control of Hazardous Energy (Lockout/Tagout) (29 CFR 1910.147) helps safeguard employees from the unexpected start-up of machines or equipment or release of hazardous energy during servicing or maintenance. The standard specifies the practices and procedures necessary to shut down and lock out or tag out machines and equipment through the development and use of written procedures, employee training, and periodic inspections conducted to maintain or enhance the energy control program.

The lockout/tagout standard applies to general industry employment and covers the servicing and maintenance of machines and equipment in which unexpected start-up or release of stored energy could cause injury to employees. Any source of mechanical, hydraulic, pneumatic, chemical, thermal, or other energy except electrical hazards is covered. Subpart S of 29 CFR 1910 covers electrical hazards, and 29 CFR 1910.333 contains specific lockout/tagout provisions for electrical hazards. If employees are performing service or maintenance tasks that will not expose them to the unexpected start-up of machines or equipment, energization, or release of hazardous energy, the standard does not apply.

The standard also does not apply in the following situations:
- normal process operations, including repetitive, routine, and minor adjustment that would be covered under OSHA's machine guarding standards
- work on cord- and plug-connected electric equipment when the equipment is unplugged and the employee working on the equipment has complete control over the plug
- hot tap operations involving transmission and distribution systems for gas, steam, water, or petroleum products when the employer shows that continuity of service is essential, shutdown is impractical, and documented procedures are followed to provide proven effective protection for employees
- generation, transmission, and distribution of electric power by utilities and work on electric conductors and equipment.

Some servicing and troubleshooting operations must be performed with the power on; effective protection must be provided for employees performing such operations.

Employees performing minor tool changes and adjustments or other minor servicing activities that are routine, repetitive, and integral to the use of the production equipment and that occur during normal production operations are not covered by the lockout/tagout standard. However, the work must be performed using alternative measures that provide effective employee protection.

Elements of an Energy Control Program

The standard requires that, in general, before service or maintenance is performed on machines or equipment, the machines or equipment must be turned off and disconnected from the energy source, and the applicable energy-isolating device(s) must be either locked out or tagged out.

Employers must establish a program with procedures for isolating machines or equipment from their source of energy and affixing appropriate locks or tags to energy-isolating devices to prevent any unexpected energization, start-up, or release of stored energy that could injure workers. When tags are used on energy-isolating devices not capable of being locked out, the employer must provide additional controls to ensure a level of protection equivalent to that of locks.

An energy control program should include the following:
- documented energy control procedures
- an employee training program
- periodic inspections of the use of the procedures.

Employers have the flexibility to develop programs and procedures that meet the needs of their workplace and the particular types of machines and equipment being maintained or serviced.

The written energy control procedures must outline the scope, purpose, authorization, rules, and techniques that will be used to control hazardous energy sources as well as the means that will be used to enforce compliance. At a minimum, they should include, but not be limited to, the following elements:
- a statement on how the procedures will be used
- the procedural steps needed to shut down, isolate, block, and secure machines or equipment
- the steps designating safe placement, removal, and transfer of lockout/tagout devices and who is responsible for them
- specific requirements for testing machines or equipment to determine and verify the effectiveness of locks, tags, and other energy control measures.

The employer or an authorized employee must notify affected employees before lockout or tagout devices are applied and after they are removed from the machine or equipment.

In addition, before lockout or tagout devices are removed and energy is restored to the machine or equipment, certain steps must be taken to reenergize equipment after servicing is completed, including the following:
- ensuring that machines or equipment components are operationally intact
- ensuring that all employees are safely positioned or removed from equipment and that lockout or tagout devices are removed from each energy-isolating device by the employee who applied the device.

The primary tool for providing protection is the energy-isolating device that prevents the transmission or release of energy and to which locks or tags are attached.

This mechanism guards against unintended start-up or unexpected reenergization of machines or equipment during servicing or maintenance.

There are two types of energy-isolating devices: those that can be locked and those that cannot. A lockout device provides protection by preventing the machine or equipment from becoming energized. A tagout device does so by identifying the energy-isolating device as a source of potential danger and indicating that the energy-isolating device and the equipment being controlled may not be operated while the tagout device is in place.

If the energy-isolating device is lockable, the employer must use locks unless he or she can demonstrate that the use of tags would provide protection at least as effective as locks and would ensure "full employee protection."

When the energy-isolating device cannot be locked out, the employer must either modify or replace the device or use tagout. When using tagout, the employer must comply with all tagout-related provisions of the standard and any additional safety measures necessary to ensure that the level of safety is equivalent to that obtained by using lockout. In addition to the normal training required for employees, tagout-related provisions should be implemented, with training detailing the limitations inherent in the use of tags. All newly purchased equipment must be lockable.

When attached to an energy-isolating device, both lockout and tagout devices used in accordance with the requirements of the standard help protect employees from hazardous energy. Whichever devices are used, they must be singularly identified; must be the only devices used for controlling hazardous energy; and must be durable, standardized, substantial, and identifiable. Locks and tags must clearly identify the employee who applies them. Tags also must warn against hazardous conditions if the machine or equipment is energized.

Employee Training
The training program must ensure that all employees understand the purpose, function, and restrictions of the energy control program.

The standard requires different levels of training for three types of employees: "authorized," "affected," and "other." Authorized employees must possess the knowledge and skills necessary for the safe application, use, and removal of energy controls. The training should deal with the equipment, type(s) of energy, and hazard(s) specific to the workplace being covered.

Because an "affected" or "other" employee is not performing the servicing or maintenance, training goals and objectives under the energy control program are simple: whenever a lockout or tagout device is in place on an energy-isolating device, the "affected" or "other" employee must be taught to leave it alone and not attempt to energize or operate the equipment.

Retraining must be provided whenever there is a change in job assignments; a change in machines, equipment, or processes that present a new hazard; or a change in energy control procedures. Additional retraining must be conducted whenever an employer has reason to believe that there are deviations from or inadequacies in an employee's knowledge or use of the energy control procedure.

Inspections
A periodic inspection of each procedure should be conducted, and any deviations or inadequacies observed should be corrected. This inspection must be performed at least annually to ensure that energy control procedures continue to be implemented properly and that employees are familiar with their responsibilities under those procedures. An authorized employee other than the one(s) using the energy control procedure(s) must perform the periodic inspections.

PERSONAL PROTECTIVE EQUIPMENT STANDARD

At times, engineering, work practice, or administrative controls cannot be used to feasibly eliminate employee exposure or potential exposure to a workplace hazard. In those cases, OSHA's personal protective equipment (PPE) standard (29 CFR 1910.132) requires employers to establish general procedures, called a PPE program, to give employees necessary protective equipment and to train them to use it properly. Respirators and insulating devices are not included in this standard because OSHA requires employers to develop separate programs specifically addressing the issues associated with those types of protective devices (29 CFR 1910.134 and 29 CFR 1910.137, respectively). The construction industry PPE requirements are addressed in 29 CFR 1926.95; requirements for the maritime industry are listed in 29 CFR 1915.152.

Criteria used to determine whether PPE is needed to protect employees are as follows:
- The work environment presents a hazard or is likely to present a hazard.
- Work processes present a hazard or are likely to present a hazard.
- The work involves contact with hazardous chemicals, radiation, or mechanical irritants.

Many factors must be considered when selecting PPE to protect employees from workplace hazards. OSHA requires that a PPE program be developed that can be used to systematically assess the hazards in the workplace and select appropriate PPE that will protect workers from those hazards. The program should set out procedures for selecting, providing, and using PPE as part of a routine operation. Although not required, a written PPE program is easier to establish and maintain as company policy and easier to evaluate than an unwritten one. The use of checklists can be an effective aid to complying with this standard.

The PPE program should include the following constituents:

- assessing the workplace to identify equipment, operations, chemicals, and other workplace components that could harm employees (areas that should be covered include use and care of eye and face, head, foot and leg, hand and arm, body, and hearing protection)
- implementing engineering controls and work practices to control or eliminate identified or potential hazards to the extent feasible
- selecting appropriate types of PPE to protect employees from hazards that cannot be eliminated or controlled through engineering controls and work practices (PPE includes such items as goggles, face shields, safety glasses, hard hats, safety shoes, gloves, vests, earplugs, and earmuffs; respirators and rubber insulating equipment [gloves, sleeves, blankets] are also considered PPE, but because OSHA has specific requirements for those kinds of PPE, the standard does not address such equipment)
- informing employees why the PPE is necessary and when it must be worn
- training employees on how to use and care for the selected PPE and how to recognize PPE deterioration and failure
- requiring employees to wear the selected PPE in the workplace.

PROCESS SAFETY MANAGEMENT STANDARD

The Process Safety Management (PSM) of Highly Hazardous Chemicals (HHCs) standard (29 CFR 1910.119) is intended to prevent or minimize the consequences of a catastrophic release of toxic, reactive, flammable, or explosive HHCs from a process and was fully implemented by May 26, 1997. A process is any activity or combination of activities including any use, storage, manufacturing, handling, or onsite movement of HHCs. A process includes any group of vessels that are interconnected and separate vessels that are located such that an HHC could be involved in a potential release. Process hazard analyses (PHAs) should be conducted for each identified process, updated and revalidated at least every 5 years, and retained for the life of the process.

The standard applies to processes that contain a threshold quantity or greater amount of a toxic or reactive HHC specified in the OSHA standard. It also applies to amounts of flammable liquids and gases that are 10,000 lb or greater and to the process activity of manufacturing explosives and pyrotechnics.

The standard does not apply to the following:
- retail facilities
- normally unoccupied remote facilities
- oil or gas well drilling or servicing activities
- hydrocarbon fuels used solely for workplace consumption as a fuel, if such fuels are not part of a process containing another HHC covered by the standard
- atmospheric tank storage and associated transfer of flammable liquids that are kept below their normal boiling point without benefit of chilling or refrigeration, unless the atmospheric tank is connected to a process or is sited in close proximity to a covered process such that an incident in a covered process could involve the atmospheric tank.

Compliance Elements for Process Safety Management

The standard describes the following requirements and covers worksite and contractor employees:

- Written process safety information (PSI) must be compiled, including hazard information on HHCs, technology information, and equipment information on covered processes.
- There must be a written plan of action regarding employee participation.
- There must be consultation with employees and their representatives on the conduct and development of process hazard analyses and on the development of other elements of process safety management required under the rule.
- Provide employees and their representatives with access to process hazard analyses and all other information required to be developed under the rule.
- Employees operating a covered process must be trained in the overview of the process and in the operating procedures. Written operating procedures provide clear instructions for safely conducting activities involving a covered process, including steps for each operating phase, operating limits, safety and health considerations, and safety systems and their functions. The operating procedures must be readily accessible to employees and be reviewed as often as necessary to ensure that they reflect current operating practice, including special circumstances such as lockout/tagout and confined-space entry. Training must emphasize

specific safety and health hazards, emergency operations, and safe work practices. Documented refresher training is required at least every 3 years.
- The onsite employer must establish and implement written procedures for the ongoing mechanical integrity of process equipment.
- Hot-work permits must be issued for hot-work operations conducted on or near a covered process.
- Employers, along with an investigation team that includes at least one person knowledgeable in the process involved, must investigate and analyze as soon as possible (within 48 hours) incidents that did result or could reasonably have resulted in catastrophic releases of covered chemicals and develop a written report on the incident that must be retained for 5 years.
- An emergency-action plan (including procedures for handling small releases) must be developed.
- Compliance audits are required to certify that compliance with process safety requirements has been evaluated at least every 3 years. Prompt response to audit findings and documentation that deficiencies were corrected is required. Employers must retain the two most recent audit reports.

New Facilities and Modified Worksites

New facilities and significantly modified worksites are required to conduct a pre-start-up safety review. For new facilities, the PHA must be performed and recommendations resolved and implemented before start-up. Modified facilities must meet the following management-of-change requirements:
- The construction and equipment of a process must be in accordance with design specifications.
- Adequate safety, operating, maintenance, and emergency procedures must be in place.
- Process operator training must be completed.

RESPIRATORY PROTECTION STANDARD

OSHA's revised Respiratory Protection standard (29 CFR 1910.134 and 29 CFR 1926.103) reflects current respirator technology and better ways to ensure fit and is applicable to general industry, construction, shipyard, longshoring, and marine terminal workplaces. The entire previous respirator standard, 29 CFR 1910.134, has been redesignated as 29 CFR 1910.139 (respiratory protection for *M. tuberculosis*) and will continue to apply to respirator use for protection against exposure to TB until OSHA finalizes its TB standard.

The revised standard clarifies responsibility for administering a respirator program and its provisions; adds definitions; and provides specific guidance on respirator selection, use, hazard evaluation, medical evaluations, fit testing, and training.

Elements of a Respiratory Protection Program

The revised standard requires employers to establish or maintain a respiratory protection program to protect their respirator-wearing employees. The general requirements are as follows:
- a written plan with worksite-specific procedures that tailor the program to each worksite
- a hazard evaluation that characterizes respiratory hazards and conditions of work to ensure that appropriate respirators are selected for use
- medical evaluation to determine the ability of workers to wear the respirators selected
- fit testing of tight-fitting respirators to reduce face-seal leakage and ensure that respirators provide adequate protection
- employee training to ensure that respirators are used safely
- periodic program evaluation to ensure that respirator use continues to be effective.

Major Changes

Other changes to the revised standard will have a positive effect on program implementation:
- It simplifies respirator requirements for employers by deleting respiratory provisions in other OSHA health standards that duplicate those in the final standard and by revising other respirator-related provisions to make them consistent.
- It supersedes existing standards that require semiannual fit testing and requires only annual fit testing.
- Use of portable quantitative fit-testing devices is permitted.
- The employer can simply provide enough respirator choices to obtain an acceptable fit among employees (instead of being required to have at least three different sizes of facepieces from two different manufacturers).
- Disposable respirators can be reused if they will continue to protect employees.
- The requirement for an annual review of the employee's medical status is eliminated.
- A medical questionnaire rather than a hands-on, physical examination can be used to evaluate an employee's ability to wear a respirator.
- The revised standard accepts previous training in lieu of full initial training requirements.

Immediately Dangerous to Life or Health

The OSHA standard requires at least one standby person when work is conducted in atmospheres that are most immediately dangerous to life or health (IDLH). IDLH

atmospheres resulting from interior structural fires trigger additional provisions:
- At least two fire fighters must enter the burning building and remain in visual and voice contact with each other at all times.
- At least two standby persons are required when two persons are engaged in interior structural fire fighting in a burning building (this protective practice is known as "two-in/two-out").

These changes are applicable to state and local government fire fighters in states that operate OSHA-approved state plans through the adoption of an identical or "at least as effective" standard.

BLOODBORNE PATHOGENS STANDARD

On March 6, 1992, the bloodborne pathogens standard (29 CFR 1030) was promulgated with the intent of limiting occupational exposure to blood and other potentially infectious materials (PIM). By June 4, 1992, all provisions of the standard were in place for affected workplaces.

The standard covers all employees who could be "reasonably anticipated" to come in contact with blood and other potentially infectious materials as the result of performing their job duties.

Infectious materials include semen, vaginal secretions, cerebrospinal fluid, synovial fluid, pleural fluid, pericardial fluid, peritoneal fluid, amniotic fluid, saliva in dental procedures, any body fluid visibly contaminated with blood, and all body fluids in situations in which it is difficult or impossible to differentiate among the body fluids. They also include any unfixed tissue or organ other than intact skin from a human (living or dead); human immunodeficiency virus (HIV)–containing cell or tissue cultures, organ cultures, and HIV or hepatitis B (HBV)–containing culture media or other solutions; and blood, organs, or other tissues from experimental animals infected with HIV or HBV.

Exposure Plan
The employer is required to develop a written exposure control plan that identifies, in writing, tasks and procedures as well as job classifications in which potential occupational exposure to blood occurs. The plan must be reviewed and updated at least annually, or more often if necessary to accommodate workplace changes. The following items should be addressed by the plan:
- Universal precautions (treating body fluids/materials as if infectious) that emphasize engineering and work practice controls should be used to comply with the standard.
- Employers must provide, at no cost, and require employees to use appropriate personal protective equipment such as gloves, gowns, masks, mouthpieces, and resuscitation bags.
- A written schedule for cleaning must be provided that identifies the method of decontamination to be used when cleaning following contact with blood or other potentially infectious materials.
- Methods for disposing of contaminated sharps must be specified.
- Standards for containers of contaminated sharps and other regulated waste must be provided.
- Provisions must be made for handling contaminated laundry to minimize exposure.
- Warning labels including the orange or orange-red biohazard symbol must be affixed to containers of regulated waste, refrigerators and freezers, and other containers that are used to store or transport blood or other potentially infectious materials. Red bags or containers may be used instead of labeling.
- Training must be provided, initially on assignment and annually (employees who have received appropriate training within the past year need only receive additional training in items not previously covered).

Medical Care
Medical vaccinations must be made available—within 10 working days of assignment, at no cost, and at a reasonable time and place—to all employees who have occupational exposure to blood. Employees must sign a declination form if they choose not to be vaccinated, but they may later opt to receive the vaccine at no cost to the employee.

Follow-up to an exposure incident must include a confidential medical evaluation documenting the circumstances of exposure, identifying and testing the source individual if feasible, testing the exposed employee's blood if he or she consents, administering postexposure prophylaxis, providing counseling, and evaluating reported illnesses. The need for hepatitis B vaccination should be determined by a qualified health care professional, and all diagnoses must remain confidential. Medical records must be kept for each employee with occupational exposure for the duration of employment plus 30 years.

OCCUPATIONAL EXPOSURE TO ASBESTOS STANDARDS

Revised occupational exposure to asbestos standards (29 CFR 1910.1001, 1926.1101, and 1915.1001) became effective in October 1994. These final standards for occupational exposure to asbestos in general industry and the construction industry (29 CFR 1926.1101, previously 1926.58) amended

OSHA's asbestos standards issued in June 1986 (51 FR 22612; 29 CFR 1910.1001). In addition, a separate standard covering occupational exposure to asbestos in the shipyard industry (29 CFR 1915.1001) was issued. Major revisions in the standards include the following:
- a reduced 8-hour time-weighted average (TWA) permissible exposure limit (PEL) of 0.1 fiber per cubic centimeter (f/cc) for all asbestos work in all industries
- a new classification scheme for asbestos construction and shipyard industry work that links mandatory work practices to work classification
- a presumptive asbestos identification requirement for certain asbestos-containing building materials
- limited notification requirements for employers who use unlisted compliance methods in high-risk asbestos abatement work
- mandatory methods of control for brake and clutch repair.

All standards set a maximum exposure limit and include provisions for engineering controls, respirators, protective clothing, exposure monitoring, hygiene facilities and practices, warning signs, labeling, record keeping, and medical exams.

Housekeeping practices should be implemented whenever asbestos-containing material (ACM) and presumed asbestos-containing material (PACM) is present. When flooring materials contain or are presumed to contain asbestos:
- Sanding is prohibited.
- Stripping of these materials may be conducted using only low abrasion pads of speeds below 300 rpm and wet methods.
- Burnishing or dry buffing may be performed on these materials only when the finish on these materials is sufficient (e.g., covered by three layers of wax) to prevent the pad from contacting them.
- Dust or debris in an area containing certain types of ACM/PACM may not be removed by a dry method other than vacuuming using a HEPA filter.

Hazards must be communicated to employees and contractors whenever ACM/PACM is present. Employers must identify and label ACM/PACM or their containers, place signs to mark regulated areas, and provide contractors with information on the presence and location of ACM/PACM in areas where they work.

Appropriate employee information and training must be provided at least initially and with annual retraining. In addition, an expanded training curriculum includes a 2-hour asbestos awareness training course, which must be provided to all employees who perform housekeeping operations in a facility that contains ACM/PACM.

Nonasbestiform tremolite, anthophyllite, and actinolite were excluded from coverage under the asbestos standard in May 1992.

Exposure Assessment

In addition to the PEL, an excursion or short-term limit of 1 fiber per cubic centimeter of air (1.0 f/cc) averaged over a sampling period of 30 minutes is permitted for both general industry and construction.

The employer must keep an accurate record of all measurements taken to monitor employee exposure, including the following information:
- date of measurement
- operation involving exposure
- sampling and analytical methods used and evidence of their accuracy
- number, duration, and results of samples taken
- type of respiratory protective devices worn
- employee name and Social Security number and the results of all employee exposure measurements.

These records must be kept for 30 years.

Hazard Control

In both general industry and construction, employers must reduce exposures using engineering controls to the extent feasible. Where engineering controls do not reduce exposures to below the exposure limit, they must be supplemented by the use of respiratory protection. In general industry and construction, the level of exposure determines what type of respirator is required; the standards specify the respirator to be used.

In general industry and construction, regulated areas must be established where the 8-hour TWA or 30-minute excursion values for airborne asbestos exceed the prescribed PEL. Only authorized persons wearing appropriate respirators may enter a regulated area. In regulated areas, eating, smoking, drinking, chewing tobacco or gum, and applying cosmetics are prohibited.

Warning signs must be displayed at each regulated area and posted at all approaches to regulated areas. CAUTION labels must be placed on all raw materials, mixtures, scrap, waste, debris, and other ACM and PACM.

For any employee exposed to asbestos airborne concentrations exceeding the PEL, protective clothing must be provided and used. Wherever the possibility of eye irritation exists, appropriate eye protection must be provided and worn (see Chapter 7, Personal Protective Equipment, in the *Engineering & Technology* volume).

In construction, there are special regulated-area requirements for asbestos removal, renovation, and demolition operations:
- a negative-pressure area
- decontamination procedures for workers
- a "competent person" with the authority to identify and control asbestos hazards.

Clean change rooms, separate lockers or storage facilities, showers, and lunchroom facilities must be designed and constructed to ensure that contamination does not occur. Employees must enter and exit the regulated area through a decontamination area. The equipment room must be supplied with impermeable, labeled bags and containers for the containment and disposal of contaminated protective clothing and equipment. Sanitary and work practices should preclude the spread of contamination.

Medical Care

In general industry, employees who will be exposed to asbestos airborne concentrations at or above the PEL or the excursion level must have a preplacement physical examination before being assigned to work. The physical examination must include the following:
- chest x-ray
- medical and work histories
- pulmonary function tests.

Subsequent exams must be given annually and on termination of employment, although chest x-rays are required annually only for older workers whose first asbestos exposure occurred more than 10 years ago.

In construction, examinations must be made available annually for workers exposed above the action level or excursion limit for 30 or more days per year or who are required to wear negative-pressure respirators. Chest x-rays are at the discretion of the physician.

OCCUPATIONAL EXPOSURE TO FORMALDEHYDE

The OSHA standard in 29 CFR 1910.1048 protects workers exposed to formaldehyde. The standard covers formaldehyde gas; its solutions; and a variety of materials such as trioxane, paraformaldehyde, resin formulations, and solids and mixtures containing formaldehyde that are potential sources of the substance. The estimated number of potentially exposed workers is more than 2 million, with a large number found in the apparel, furniture, foundry, textile-finishing, laboratory, paper mill, and plastic-molding industries.

The standard sets permissible exposure levels (PELs), exposure monitoring and training, requirements for medical surveillance and medical removal, record keeping, regulated areas, hazard communication, emergency procedures, primary reliance on engineering and work practices to control exposure, and maintenance and selection of personal protective equipment.

Exposure Assessment and Management

The PEL for formaldehyde in all workplaces (including general industry, construction, and the maritime industry, but not agriculture) covered by the OSH Act is 0.75 ppm measured as an 8-hour time-weighted average (TWA). The standard includes a 2-ppm short-term exposure limit (STEL) (the maximum exposure allowed during a 15-minute period). The "action level" is 0.5 ppm measured over 8 hours.

The employer should conduct initial monitoring to identify all employees who are exposed to formaldehyde at or above the action level or STEL. If the exposure level is maintained below the STEL and the action level, employers may discontinue exposure monitoring. Monitoring should be conducted whenever the potential for an exposure changes or on receiving reports of formaldehyde-related signs and symptoms.

Employees who experience significant adverse effects from formaldehyde exposure must be reassigned to jobs with less exposure for up to 6 months—until their conditions improve or a physician determines that the employees will not ever be able to return to a job involving formaldehyde exposure, whichever occurs first.

Controls and Training

Engineering and work practice controls should be instituted to reduce and maintain employee exposure to formaldehyde at or below the TWA and the STEL. If engineering and work practice controls cannot feasibly reduce employee exposure to or below the PEL, respirators must be used.

Specific hazard-labeling requirements are needed for all forms of formaldehyde, including mixtures and solutions, either composed of 0.1% formaldehyde or capable of releasing at least 0.1 ppm of formaldehyde. Hazard labeling, including a warning that formaldehyde presents a potential cancer hazard, is required where formaldehyde levels, under reasonably foreseeable conditions of use, may potentially exceed 0.5 ppm.

Training is required at least annually for all employees exposed to formaldehyde concentrations of 0.1 ppm or greater and should increase employees' awareness of specific hazards in their workplace and of the control measures used.

LEAD EXPOSURE IN CONSTRUCTION

Employers of construction workers are responsible for the development and implementation of a worker protection program in accordance with 29 CFR 1926.20 and 29 CFR 1926.62(e). OSHA's interim final standard for lead in construction limits worker exposures to 50 micrograms of lead per cubic meter of air averaged over an 8-hour workday. Because construction projects vary in their scope and potential for exposing workers to lead, the most effective way to protect workers is to minimize their exposure through the use of engineering controls and good work practices. OSHA policy states that respirators are not to be used in lieu of engineering and work practices to reduce employee exposures to below the PEL. Respirators may be used only in combination with engineering controls and work practices. The employer should, as needed, consult a qualified safety and health professional to develop and implement an effective worker protection program.

At the minimum, the following elements should be included in the employer's worker protection program for employees exposed to lead:
- hazard determination, including exposure assessment
- engineering and work practice controls
- respiratory protection
- protective clothing and equipment
- housekeeping
- hygiene facilities and practices
- medical surveillance and provisions for medical removal
- training
- signs
- record keeping.

A competent person should be designated to implement the worker protection program. This individual should be able to identify existing and predictable working conditions that are hazardous or dangerous to employees, in accordance with the general safety and health provisions of OSHA's construction standards. This person must be authorized to take prompt corrective measures to eliminate such problems. Qualified medical personnel must be available to advise the employer and employees on the health effects of employee lead exposure and supervise the medical surveillance program.

STATE PLAN PROGRAMS

States and territories with their own occupational safety and health plans have adopted comparable standards; they include Alaska, Arizona, California, Hawaii, Indiana, Iowa, Kentucky, Maryland, Michigan, Minnesota, Nevada, New Jersey, New Mexico, New York, North Carolina, Oregon, Puerto Rico, South Carolina, Tennessee, Utah, Vermont, Virginia, Virgin Islands, Washington, and Wyoming.

OSHA Program Documentation
- Fact Sheet 93-29: Access to Employee Exposure and Medical Records, 01/01/1993
- Fact Sheet 93-26: Hazard Communication Standard, 01/01/1993
- Fact Sheet 95-33: Occupational Exposure to Hazardous Chemicals in Laboratories, 01/01/1995
- Fact Sheet 92-19: Responding to Workplace Emergencies, 01/01/1992
- 29 CFR 1910.0146: Permit-Required Confined Space:
  - Appendix A: Permit-Required Confined Space Decision Flow Chart
  - Appendix B: Procedures for Atmospheric Testing
  - Appendix C: Examples of Permit-Required Confined Space Program
  - Appendix D: Confined Space Pre-Entry Check List
  - Appendix E: Sewer System Entry
- Fact Sheet 93-32: Control of Hazardous Energy (Lockout/Tagout), 01/01/1993
- OSHA 3151, "Assessing the Need for Personal Protective Equipment: A Guide for Small Business Employers" (1997)
- Fact Sheet 93-45: Process Safety Management of Highly Hazardous Chemicals, 01/01/1993
- News Release USDL 98-04, "OSHA Improves Respirator Protection for Five Million Workers in 1.3 Million Worksites," 01/08/1998
- "Asbestos Standard for General Industry" and "Asbestos Standard for Construction Industry"; 29 CFR 1926.1101, osha.gov/OshDoc/data_AsbestosFacts/asbestos-factsheet.pdf
- Fact Sheet 93-06: Better Protection against Asbestos in the Workplace, 01/01/1993
- Fact Sheet 95-27: Occupational Exposure to Formaldehyde, 01/01/1995
- Fact Sheet 93-47: Lead Exposure in Construction (1 of 6)—Worker Protection Programs, 01/01/1993

PART 3—MINE SAFETY AND HEALTH ACT AND OTHER PROGRAMS

On November 9, 1977, President Jimmy Carter signed into law the U.S. Mine Safety and Health Act of 1977 (subsequently referred to as the Mine Act), Public Law 95-164. The act became effective March 9, 1978.

The Mine Act is intended to ensure, as far as possible, safe and healthful working conditions for miners. It applies to operators of all types of mines, both coal and metal/nonmetal and both surface and underground. The Mine Act states that mine operators are responsible for preventing unsafe, unhealthful conditions or practices in mines that could endanger the lives and health of miners.

Mine operators are required to comply with the safety and health standards promulgated and enforced by the Mine Safety and Health Administration (MSHA), an agency within the DOL. Like OSHA, MSHA may issue citations and propose penalties for violations. Unlike employees under the OSH Act, miners (employees) are subject to government sanctions for violating safety standards relating to smoking in or near mines and mining machinery. Similarly, employers and other supervisory personnel may be held personally liable for civil penalties and may be prosecuted criminally for violations of Mine Act standards.

LEGISLATIVE HISTORY OF MSHA

Historically, the Bureau of Mines within the Department of the Interior administered mine safety and health laws. However, the bureau was eliminated at the end of 1995, and some of its functions were transferred to other sectors within the Department of the Interior. Before Congress passed the Mine Act, mine operators were governed by two separate laws: the Federal Coal Mine Safety and Health Act of 1969 and the Federal Metal and Nonmetallic Mine Safety Act of 1966. Under the U.S. Department of Labor, MSHA now administers mining safety laws.

Because the Bureau of Mines was also responsible for promoting mine production, critics charged that this responsibility produced an inherent conflict of interest with respect to enforcement of safety and health laws. The establishment of the Mine Enforcement Safety Administration (MESA) in 1973 within the Department of the Interior failed to answer the criticism. Congress looked for alternative solutions, including transferring mine safety and health to the OSH Act. Finally, Congress resolved the issue by adopting the Mine Act, which repealed the Federal Coal Mine Safety and Health Act of 1969 and the Federal Metal and Nonmetallic Mine Safety Act of 1966.

ADMINISTRATION

The administration and enforcement of the U.S. Mine Safety and Health Act are vested primarily with the Secretary of Labor and the Mine Safety and Health Review Commission. MSHA administers the investigation and prosecution aspects of the enforcement process. The Mine Safety and Health Review Commission, an independent agency created by the Mine Act, reviews contested MSHA enforcement actions.

The Mine Act distinguishes between health research and safety research. Miner health research and standards development are the responsibility of NIOSH, in cooperation with MSHA.

Mine Safety and Health Administration

The Mine Safety and Health Administration, located within the DOL, administers and enforces the Mine Act. MSHA is headed by the Assistant Secretary of Labor for Mine Safety and Health, who is appointed by the president with the advice and consent of the Senate. The Assistant Secretary acts on behalf of the Secretary of Labor. For the purposes of this chapter, *MSHA* is synonymous with the term *Secretary of Labor* or *Assistant Secretary of Labor*.

MSHA is authorized to adopt procedural rules and regulations to carry out the provisions of the Mine Act. The agency also has the responsibility and authority to perform the following:
- promulgate, revoke, or modify safety and health standards
- conduct mine safety and health inspections
- issue citations and propose penalties for violations
- issue orders for miners to be withdrawn from all or part of the mine
- investigate mine accidents (as defined later under Accident, Injury, and Illness Reporting)
- grant variances
- seek judicial enforcement of its orders.

Aiding the Assistant Secretary in carrying out the provisions of the Mine Act are, among others, (1) an Administrator for Coal Mine Safety and Health and (2) an Administrator for Metal and Nonmetal Mine Safety and Health. Each administrator is responsible for a Division of Safety and a Division of Health.

Mine Safety and Health Review Commission

The five-member Mine Safety and Health Review Commission serves as the administrative adjudication body. The Review Commission is completely independent of the DOL and has authority to assess all civil penalties provided in the Mine Act. It reviews contested citations, notices of proposed penalties, withdrawal orders, and employee discrimination complaints. Commission members are appointed by the president for 6-year terms with the advice and consent of the Senate. The first commissioners took office for staggered terms of 2, 4, and 6 years.

The commission appoints administrative law judges (ALJs) to conduct hearings on behalf of the commission. The decision of an ALJ becomes a final decision of the commission 40 days after its issuance unless the commission directs a review.

MAJOR PROVISIONS OF THE MINE ACT

This section describes the general scope of the regulations and guidelines that apply to the mining industry. Since passage of the act, industry management has worked with union representatives and employees to improve the health and safety of workers in this hazardous occupation.

Coverage

The Mine Act covers all mines that affect commerce. The act defines "mines" as all underground or surface areas from which minerals are extracted and all surface facilities used in preparing or processing the minerals. Structures, equipment, and facilities including roads, dams, impoundments, and tailing ponds used in connection with mining and milling activities are also included. The Mine Act provides that MSHA will develop and implement regulations, provide for state grants, and offer training at the National Mine Academy.

Because some facilities have operations under OSHA and others under MSHA, OSHA and MSHA have established an interagency agreement that, among other things, delineates certain areas of authority and provides for coordination between OSHA and MSHA in all areas of mutual interest. (This was published in 44 CFR 22827 on April 17, 1979.) In case of jurisdictional disputes between OSHA and MSHA, the Secretary of Labor is authorized to assign enforcement responsibilities to one of the agencies.

Advisory Committees

The act requires the Secretary of the Interior to appoint an Advisory Committee on Mine Safety Research. The Secretary of Health and Human Services is required to appoint an Advisory Committee on Mine Health Research. The Secretary of Labor may appoint other advisory committees as needed to aid in carrying out the provisions of the act.

Miners' Rights

The act affords miners a number of rights, including the following:
- Miners may request an inspection in writing if they believe that a violation of a standard or an imminent-danger situation exists in the mine. Similarly, written notification of alleged violations or imminent-danger situations may be given to an inspector before or during an inspection.
- A representative designated by miners must be given the opportunity to accompany the inspector during the inspection process. Also, representatives have the right to participate in postinspection conferences held by the mine inspector on the premises.
- Miners are entitled to observe monitoring and to examine monitoring records when the standards require tracking exposure to toxic materials or harmful physical agents.
- Miners, including former miners, must be given access to medical and other records documenting their own exposures.
- Operators must notify miners if they are exposed to toxic substances in concentrations that exceed prescribed limits of exposure. Further, those miners must be informed of the corrective action being taken.
- Miners given new work assignments for medical reasons because of their exposure to hazardous substances must be paid at their regular rate if the related standard so provides.
- Miners who are not working because of a withdrawal order are entitled to be compensated subject to certain limits.
- Miners or their authorized representatives may contest the issuance, modification, or termination of any MSHA order or the time period set for abatement.
- Miners adversely affected or aggrieved by an order of the Review Commission may obtain judicial review.
- Miners may file a complaint with the Review Commission concerning their compensation for not working as the result of a withdrawal order issued by MSHA or, when notified by MSHA, for acts of employer discrimination.
- Miners, through their authorized representative, may petition for a variance from mine safety standards.
- MSHA is required to send to the miners' authorized representative copies of proposed safety or health standards. In addition, the mine operator must post a copy of such standards on the office bulletin board.
- To keep miners informed, mine operators are required to post copies of orders, citations, notices, and decisions issued by MSHA or the Review Commission. Posting of these items should be on the mine bulletin board (§ 109a).
- Miners are entitled to receive training for their specific jobs and must be given refresher training annually. They are entitled to their customary compensation while being trained. Miners who leave the operator's employ are entitled to copies of their training certificates.
- Operators may not discriminate against miners or representatives of miners for the exercise of miners' rights under this act.
- Miners who have black-lung disease are entitled to extensive black-lung benefits.

Duties

Mine operators are required to comply with the safety and health standards and other rules promulgated under the act and are subject to sanctions for failing to comply. Similarly, every miner is required to comply with the safety and health standards promulgated under the act. Any miner, whether management or shiftworker, may be cited for smoking or carrying smoking materials, matches, or lighters in certain situations. For knowingly and willfully violating mandatory safety and health standards, any miner (management or shiftworker) may be fined or sentenced to jail.

Miner Training

Mine operators are required to have a safety and health training program approved by MSHA that provides the following:

- New underground miners must receive at least 40 hours of instruction. The training must include the statutory rights of miners and their representatives under the act, use of the self-rescue device and respiratory devices, hazard recognition, escapeways, walkaround training, emergency procedures, basic ventilation, basic roof control, electrical hazards, first aid, and the safety and health aspects of the task assignment.
- New surface miners must receive 24 hours of instruction. The training must include all of the items listed for underground miners except escapeways, basic ventilation, and basic roof control, none of which is essential to surface mining.
- All miners must receive at least 8 hours of annual refresher training.

The Mine Act requires that training be conducted during normal working hours and that the miners be paid at their normal rates during the training period. Regulations concerning training and retraining of miners are codified in 30 CFR 48. (See Chapter 31, Safety and Health Training, in this volume.)

MINE SAFETY AND HEALTH STANDARDS

The Mine Act authorizes MSHA to promulgate, modify, or revoke mine safety and health standards. To implement the initial set of standards without delay, the safety and health standards in the Coal Mine Safety and Health Act of 1969 were adopted in the Mine Act. These standards are codified in 30 CFR 70, 71, 74, 75, 77, and 90.

Similarly, the Mine Act adopted the mandatory standards that prevailed in the Metal and Nonmetallic Mine Safety Act of 1966. Later, many of the advisory standards were adopted as mandatory standards in the Mine Act. All of the metal/nonmetal standards are codified in 30 CFR 56, 57, and 58. Part 55 (Metal/Nonmetal) was eliminated on April 15, 1985, and recodified as Part 56. (A list of all of the standards promulgated under authority of the Mine Act is provided in the References at the end of this chapter.)

If MSHA determines that a standard is needed, it may propose a standard or seek assistance from an advisory committee. MSHA must publish the proposed standard in the *Federal Register* and establish a time period of at least 30 days for public comment. MSHA may hold public hearings if objections are raised about a proposed standard. Upon adoption by MSHA, the standard must be published in the *Federal Register*. The new standard becomes effective on publication or on a date specified.

Judicial Review

Any person adversely affected by any standard issued by MSHA has the right to challenge its validity by petitioning the U.S. Court of Appeals within 60 days after promulgation of the standard. Although filing such a petition does not stay enforcement of the standard, the court may order a stay before conducting a hearing on the petition. Objections that were not raised during rulemaking will not be considered by the court, unless good cause is shown for why an objection was not raised.

Input from the Private Sector

Mine safety and health standards promulgated by MSHA can never cover every conceivable hazardous condition that might exist in mines. Nevertheless, new standards and modification or revocation of existing standards are important to mine operators and miners alike. Mining operator organizations, miner organizations, and individuals should express their views during the rulemaking process by responding to MSHA's proposed standards. Their participation is important because most expertise and technical competence lie in the private sector. To do otherwise means that mining operators and miners are willing to let the standards-development process be controlled solely by MSHA.

Emergency Temporary Standards

MSHA has the authority to publish emergency temporary standards if it deems that immediate action must be taken to protect miners from toxic substances or physically harmful agents. The emergency temporary standard is effective immediately upon publication in the *Federal Register* and remains in effect until superseded by a permanent standard developed under normal rulemaking procedures. MSHA is required to establish a permanent standard within 9 months after publication of an emergency temporary standard.

Variances
When a petition for modification is filed by an operator or a representative of miners, MSHA may modify the application of any mandatory safety standard. The act does not allow for variances of health standards.

A petition for modification may be granted under two conditions. First, MSHA must find that an alternative method of compliance will achieve the same measure of protection for miners as the standard would provide. Second, MSHA must determine that the standard in question will provide less safety to miners.

A petition for modification should be filed with the Assistant Secretary of Labor for Mine Safety and Health. If the mining operator submits a petition, a copy must be served on the miners' representative. Similarly, if the miners' representative petitions for modification, a copy must be served on the mine operator. The petition must include the name and address of the petitioner and the mailing address, identification, and name or number of the affected mine. It must also identify the standard, describe the desired modification, and state the basis for the request.

MSHA will publish a notice of the petition in the *Federal Register*. The notice will summarize information contained in the petition. Interested parties have 30 days to comment. MSHA then will conduct an investigation on the merits of the petition, and the appropriate Administrator will issue a proposed decision. The proposed decision becomes final 30 days after service, unless a hearing request is filed within that time.

Granting of Petition
If MSHA agrees, a petition can be granted if (1) the alternative method will achieve the same measure of protection for miners as compliance with the standard would provide or (2) application of the standard will result in a diminution of miners' safety.

A petition request is filed with the Director of the Office of Standards, Regulations, and Variance. A copy of the request must be posted on the mine bulletin board. Until a petition is granted, there can be no discussion of mine operators requesting temporary relief from enforcement of a mandatory safety standard (30 CFR 44.16).

Accident, Injury, and Illness Reporting
For the purpose of reporting accidents, injuries, and illnesses under the Mine Act, the term "accident" includes the following:
- an unplanned inundation of a mine by a liquid or gas
- a fatality at a mine
- an injury to an individual at a mine that has a reasonable potential to result in the worker's death
- an injury that may result in death
- entrapment for more than 30 minutes
- an unplanned ignition or explosion of gas or dust
- an unplanned fire not extinguished within 30 minutes of its discovery
- an unplanned ignition or explosion of a blasting agent or an explosive
- an unplanned roof collapse in active work areas where roof bolts are in use or a roof collapse that impairs ventilation or impedes passage
- coal or rock outbursts that cause withdrawal of miners or that disrupt mining activity for more than 1 hour
- an unstable condition at an impoundment, refuse pile, or culm bank requiring emergency action or for which individuals must evacuate an area because of failure of an impoundment, refuse pile, or culm bank
- damage to hoisting equipment in a shaft or slope that endangers an individual or interferes with the use of equipment for more than 30 minutes
- "an event at a mine which causes death or bodily injury to an individual not at the mine at the time the event occurs."

"Occupational injury" means an injury that results in death, loss of consciousness, medical treatment, temporary assignment to other duties, transfer to another job, or inability to perform all duties on any day after the injury. "Occupational illness" is an illness or disease that may have resulted from work at a mine or for which a compensation award is made.

All mine operators are required to immediately report accidents (as defined earlier) to the nearest MSHA district or subdistrict office. Similarly, operators must investigate and submit to MSHA, on request, an investigation report on accidents and occupational injuries. The investigation report must include the following:
- the date and hour of the occurrence
- the date the investigation began
- the names of the individuals participating in the investigation
- a description of the site
- an explanation of the accident or injury
- the name(s), occupation(s), and experience(s) of any miner(s) involved
- if appropriate, a sketch of the accident site, including dimensions
- a description of actions taken to prevent a similar occurrence
- identification of the accident report submitted.

All mine operators must submit to MSHA within 10 days of the incident a report of each accident, occupational injury, or illness on Form 7000-1. A separate form is to be prepared for each miner affected.

Accident investigation reports and the injury/illness reports filed by means of Form 7000-1 must be kept for 5 years at the mine office closest to the mine in which the accident, injury, or illness occurred.

Inspection and Investigation Procedures

Inspections of a mine are conducted by MSHA to determine whether an imminent danger exists in the mine and whether the mine operator is complying with the safety and health standards and with any citations, orders, or decisions issued. Mine inspectors from MSHA, called "authorized representatives" of the Secretary of Labor, or representatives of NIOSH have the right to enter any mine to inspect the site or to conduct an investigation. However, NIOSH representatives have no enforcement authority. The Mine Act's provision for conducting inspections without securing a search warrant has been held valid. MSHA inspectors, on certain occasions, can give advance notice of an impending inspection.

As in OSHA, MSHA inspectors cannot give employers advance warning of an inspection conducted to determine compliance. However, NIOSH may give advance notice of inspections carried out for research or other purposes.

Frequency
MSHA must inspect underground mines in their entirety at least four times a year. Surface mines are to be inspected at least two times a year. Normally, these inspections require multiple visits. MSHA must also conduct spot inspections based on the number of cubic feet of methane or other explosive gases liberated in mining operations during a 24-hour period. The act authorizes MSHA to develop guidelines for additional inspections based on other criteria.

Miner Complaints
A miner's authorized representative, or any individual miner if there is no authorized representative, may request in writing an immediate inspection by MSHA if he or she has reasonable grounds to believe that a violation of a standard or an imminent-danger situation exists. MSHA will normally conduct a special inspection soon after receiving the complaint. If MSHA determines that a violation does not exist, it must notify the complainant in writing. Similarly, before or during an inspection, the miners' authorized representative, or an individual miner if no representative exists, may notify the inspector in writing of any alleged violation or imminent-danger situation believed to exist in the mine.

Health Hazard Evaluations
Upon written request of an operator or authorized representative of miners, NIOSH is authorized to enter a mine to determine whether any toxic substance, physical agent, or equipment found or used in the mine is potentially hazardous. A copy of the evaluation will be submitted to both the operator and the miners' representative.

The Inspection Procedure
A MSHA inspector will normally begin the inspection at the mine office. The inspector will inform the mine operator of the reason for the inspection and request all needed records. The inspector's review of the records will likely focus on the preshift or on-shift examination records. Such records help the inspector determine where to concentrate his or her attention during the inspection.

An operator's representative and a representative authorized by the miners must be given the opportunity to accompany the MSHA inspector during the inspection. Similarly, each representative must be given the opportunity to participate in the postinspection conference. The miner representative (who is an employee of the operator) must be paid the regular wage for the time spent accompanying the inspector.

Whenever the inspector observes a condition that appears to be a violation of the standards, he or she must issue a citation. If, in the opinion of the mine inspector, an imminent-danger condition exists, then the inspector must issue a withdrawal order.

After completing the inspection, the inspector will hold a closing conference with the representatives of the mine operator and the miners to discuss all findings. The closing conference is mandatory. Occasionally, in the interests of those concerned, one closing conference may be held with the mine operator and another closing conference held with the miners' representative (see Program Policy Letter P94-III-1 [3/31/94], page 9, in 30 CFR 100.6 for additional information).

Withdrawal Orders
MSHA has the authority, under specified conditions, to order an operator to withdraw the miners from all or part of a mine. Miners idled by such an order are entitled to receive compensation at their regular rates of pay for specified periods of time. Everyone in the affected area must be withdrawn except those miners necessary to eliminate the hazard, public officials whose duty requires their presence in the area, representatives of the miners qualified to make mine examinations, and consultants.

If an imminent danger is found to exist during an inspection, MSHA is required to order the withdrawal of all persons from the affected area, except those referred to in Section 104 of the act, until the danger no longer exists.

The order must describe the conditions or practices both causing and constituting the imminent danger and the area affected. The withdrawal order does not preclude issuance of a citation and proposed penalty.

Other situations for which MSHA may issue a withdrawal order include the following:

1. If, during a follow-up inspection, MSHA finds that a mine operator has failed to abate a cited violation and there is no valid reason to extend the abatement period, MSHA must issue a withdrawal order until the violation is abated.
2. If a mine operator fails to abate a respirable dust violation for which a citation has been issued and the abatement period has expired, MSHA must either extend the abatement period or issue a withdrawal order.
3. If two violations constituting "unwarrantable failures" to comply with the standards are found during the same inspection, or if the second unwarrantable violation is found within 90 days of the first, a withdrawal order must be issued. An unwarrantable failure violation refers to a situation in which the operator knew or should have known that a violation existed and yet failed to take corrective action.
4. Miners may be ordered withdrawn from a mine if they have not received the safety training required by the act. Miners withdrawn for this reason are protected by the Mine Act from discharge or loss of pay.

Except for withdrawal orders issued for respirable dust violations and imminent-danger situations, an operator or a miner may file a written request for a temporary stay of the order with the Review Commission. Also, they may request temporary relief from any modification or termination of a withdrawal order. The Review Commission may grant a stay of a withdrawal order provided that granting such relief would not endanger the safety and health of the miners.

Both operators and miners, or their representatives, may contest an imminent-danger withdrawal order or any modification or termination of such an order. They must file with the Review Commission an application for review of the order within 30 days after receiving it or after receiving any modification or termination of such an order.

Citations

If a MSHA inspector or the inspector's supervisors believe that the mine operator is in violation of any standard, rule, order, or regulation promulgated under the Mine Act, they must issue a citation to the operator with "reasonable promptness." A citation may be issued immediately at the site of the alleged violation. In any case, the inspector must provide a citation for each alleged violation before leaving the mine property, unless mitigating circumstances exist.

Citations must be in writing, must describe the nature of the violation, and must include a reference to the provision of the Mine Act, standard, rule, regulation, or order allegedly violated. The citation, based on the inspector's opinion, will establish a reasonable time for the employer to correct the violation. Termination dates of citations are established after consultation with the mine operator and on consideration of the hazard that is cited.

The act requires the operator to post all citations on the mine's bulletin board. Copies are sent to the miners' representative, to the state agency charged with administering mine safety and health laws, and to those designated by the operator as having responsibility for safety and health in the mine.

Within 10 days after an operator receives a citation or an order for an alleged violation, the operator and/or miners' representative has the right to request a safety and health conference with MSHA management. The conferencing process is designed to allow parties an opportunity to present additional evidence or mitigating facts or circumstances concerning the citation or order. The conferencing officer considers information from these sources and has the authority to modify, vacate, or affirm the citation.

This procedure enables mine operators and/or representatives of miners to resolve some issues before the penalty stage.

Penalties

MSHA must assess a civil penalty of not more than $50,000 for each violation of the act. The agency may assess penalties up to $5,000 (maximum) per day for each day the operator fails to correct a cited violation. If an operator is convicted of willfully violating a standard, a federal judge may assess a fine of $25,000 and/or 1-year imprisonment. Miners who willfully violate a standard that prohibits smoking or carrying smoking materials, matches, or lighters into or near a mine or mining equipment may be assessed a penalty of up to $250 per occurrence.

After the alleged violation has been corrected and any safety and health conference has been conducted, MSHA will issue the proposed penalty. Nonsignificant and substantial (commonly termed "non-S&S") violations that are abated in a timely fashion usually result in a minimal penalty. In determining the amount of the penalty for all violations, MSHA considers six criteria:

1. the operator's history of previous violations
2. size of the operator's business
3. evidence of operator negligence
4. impact on the operator's ability to remain in business

5. gravity of the violation
6. demonstrated good faith to achieve rapid compliance after notification of the alleged violation.

Most of the proposed penalty assessments are assigned a range of penalty points based on each of the six criteria. The total points are then converted into a dollar penalty. In addition, serious violations or those involving negligence are usually given an additional penalty assessment.

CONTESTED CASES

Operators and miners (or miners' representatives) have 30 calendar days after receiving notice to contest a citation, a withdrawal order, or a proposed penalty. The notice of contest must be sent by registered or certified mail to the Assistant Secretary of Labor for Mine Safety and Health at the MSHA headquarters, 1100 Wilson Boulevard, Arlington, VA 22209. The miners' representative must also receive a copy of the notice.

The notice of contest may be settled informally by MSHA, whose nonlawyer employees are trained to settle cases for which fines are under $1,000. If the case is not settled, the charges will be scheduled for a hearing before an administrative law judge (ALJ) in the Falls Church, Virginia, or Denver, Colorado, office of the Mine Safety and Health Review Commission. If the ALJ's decision is appealed by either the mine operator or the DOL, it goes to the five-member Mine Safety and Health Review Commission at 601 New Jersey Avenue, NW, Washington, DC 20001. A small number of cases reach the federal courts of appeals after a final Review Commission decision is issued.

If a mine operator fails to notify MSHA within the 30-day period and no notice is filed by any miner or miners' representative, the citation and/or the proposed penalty is deemed a final order of the Mine Safety and Health Review Commission and is not subject to review by any court or agency. However, it should be understood that the citation and the penalty have separate 30-day periods within which each may be contested. For instance, a mine operator that fails to contest the citation is not without options. When the proposed penalty is received at some later date, the mining operator has another 30 days to contest the penalty. In addition, if the penalty is contested, the citation may be reopened for negotiation at the same time.

Even if the alleged violation has been abated, a citation may be contested before the operator receives a notice of proposed penalty. A notice of contest states what is being contested and the relief sought. A copy of the order or citation being contested must accompany the notice of contest.

Upon receiving the notice of contest, MSHA immediately notifies the Review Commission. A docket number and an ALJ are then assigned to the case. The Review Commission will provide an opportunity for a hearing via the ALJ. The ALJ may hold an informal conference with all parties involved to clarify and settle the issues. If the issues are not settled, then a formal hearing conducted by the ALJ will take place.

Mine operators, miners or representatives of miners, and applicants for employment may be parties to the Review Commission proceedings. Miners or their representatives may become parties by filing a written notice with the Executive Director of the Review Commission before the hearing.

The Review Commission's ALJs are authorized to, among other things, administer oaths, issue subpoenas, receive evidence, take depositions, conduct hearings, hold settlement conferences, and render decisions. The decision will include findings of facts, conclusions of law, and an order. A copy of the decision will be issued to each of the parties involved and to each of the commissioners. Any person aggrieved by the decision of the ALJ may, within 30 days after an order or decision is issued, file a petition for a discretionary review by the Review Commission.

The Review Commission on its own motion and with the affirmative vote of two members may direct review of an ALJ's decision within 30 days of issuance only under two conditions: (1) when the decision may be contrary to law or to Commission policy and (2) when the petition raises a novel question of policy.

Any entity adversely affected by a decision of the Review Commission, including MSHA, may appeal to the U.S. Court of Appeals within 30 days of issuance. The court may affirm, modify, or set aside the Commission's decision in whole or in part.

PART 4—REQUIREMENTS BEYOND U.S. BORDERS

The rapid emergence of worldwide markets, the pace of technological and scientific changes, and the sweeping social changes occurring around the world have all created a need for a uniform set of international safety and health standards. So far, Europe and the United States have led the way, although other nations are beginning to realize the importance of protecting the worker and the environment as they seek to industrialize their economies.

The goals of international standardization of safety and health regulations include the following:
- setting a common ground for market agreements and technological applications

- improving the quality and reliability of products and services
- protecting the user and/or the environment
- providing compatibility of goods and services.

Some of the more important regulations include those developed by the European Union and the International Organization for Standardization.

European Union
The European Union (EU, formerly the EEC) has sought to create safety and health regulations that afford the greatest protection for workers and consumers in member nations while providing uniform standards for all nations doing business either in western Europe itself or with western European nations. The countries involved are Austria, Belgium, Denmark, Finland, France, Germany, Greece, Ireland, Italy, Luxembourg, the Netherlands, Portugal, Spain, Sweden, and the United Kingdom. One standard, the Framework Directive (89/391/EEC) of June 12, 1989, encourages improvements in the on-the-job safety and health of workers and describes the obligations of employers to ensure the safety and health of their employees. This standard has been adopted by EU countries and most nations doing business with these countries. (See Chapter 1, Historical Perspectives, in this volume for a more detailed discussion of the Framework Directive; see also Chapter 4, International Legal and Legislative Framework, in the *Environmental Management* volume.)

ISO 9000 Series
The ISO 9000 series applies to companies that export goods and products to the EU countries. These standards describe a process for establishing quality management and quality assurance in companies. The basic goal of the ISO 9000 series is to provide guidelines to help companies achieve consistency and uniformity in products or services throughout the entire chain of supply. (See Chapter 1, Historical Perspectives, and Chapter 14, Environmental Management, both in this volume.)

ISO 14000 Series
The ISO 14000 series, which is also related to establishing standards for quality assurance, is aimed primarily at protecting the environment from the by-products of industrial processes. Companies can use their management systems developed under ISO 9000 regulations as a basis for establishing environmental management. ISO 14000 was phased in throughout the 1990s. The European Union has adopted the ISO 14000 standards as meeting the requirements of EU environmental standards. (See Chapter 1, Historical Perspectives, and Chapter 14, Environmental Management, both in this volume.)

Total Quality Management
Total quality management (TQM) principles, which emphasize continuous improvement, have been applied to safety and health issues on both the domestic and the international fronts. TQM should not be thought of as a technical program, however; like any quality assurance program, it is a management system. As applied to health and safety procedures, TQM requires the following:
- commitment by top management to quality in all areas of organizational life, particularly health and safety
- training and education in safety matters at all levels of the organization
- continuous improvement programs implemented throughout the organization
- measurement systems that identify areas in which improvements can be made
- communication fostered between management and employees that encourages employee involvement and participation in quality improvement.

SUMMARY

- OSHA's primary responsibilities are (1) to promulgate, modify, and revoke safety and health standards; (2) to conduct inspections and investigations and to issue citations, including proposed penalties; (3) to require employers to keep records of safety and health data; (4) to petition the courts to restrain imminent-danger situations; and (5) to approve or reject state plans for programs under the OSH Act.
- The Occupational Safety and Health Review Commission (OSHRC) is a quasi-judicial, three-member board that hears cases when OSHA actions are contested by employers or employees.
- The primary functions of NIOSH are (1) to develop and establish recommended occupational safety and health standards, (2) to conduct research experiments and demonstrations, and (3) to develop educational programs to provide qualified safety and health personnel.
- With some exceptions, the OSH Act applies to every employer in all 50 states and U.S. possessions that has one or more employees and that is engaged in a business affecting commerce. Under OSHA, the employer has a general and a specific duty to provide safe, healthy work environments and comply with all applicable standards. Employees, in turn, must comply with all standards that apply to their actions and conduct on the job.
- Sanctions take the form of citations for violating standards and civil and criminal penalties if the employer failed to comply with duties under the act. OSHA can grant tem-

- porary and permanent variances from standards.
- Employers and employees are granted certain rights under OSHA regulations to seek advice and consultation with OSHA staff, participate in inspections, take an active role in developing safety and health standards, apply for financial assistance, and appeal or contest OSHA findings and decisions.
- OSHA requires employers to keep records of all occupational injuries and illnesses.
- General inspection procedures include an opening conference, a walk-through of the establishment, documenting alleged violations, interviewing workers, and a closing conference. Employers or employees can request informal postinspection conferences to discuss problems and solutions. Follow-up inspections are conducted to ensure correction of alleged violations.
- All citations except those for *de minimis* violations must be posted near the place where the violation occurred. The employer can object to or contest a citation and petition for a modification of the time for correcting the violation. If a citation or penalty is not contested, it takes immediate effect.
- An informal conference may be held before the citation is formally contested. If the parties do not agree with the results of the informal conference, the employer may formally contest the citation to the Occupational Safety and Health Review Commission (OSHRC).
- The OSH Act encourages states to assume the fullest responsibility for administering and enforcing their own occupational safety and health laws.
- Employers must inform employees of any hazardous materials used in the workplace and train them in methods of handling these materials and self-protection.
- The U.S. Mine Safety and Health Act of 1977 is intended to ensure safe, healthful working conditions for miners in all types of mines. Its primary duties are (1) to develop, revoke, or modify safety and health standards; (2) to conduct inspections; (3) to issue citations and propose penalties for violations; (4) to order miners withdrawn from parts or all of mines; (5) to grant standard variances; and (6) to seek judicial enforcement of its orders. MSHA has the authority to propose new standards, emergency temporary standards, and variances.
- Miners (employees) have the right to request inspections, accompany inspectors during their facility tour, observe monitoring procedures, access their own medical and employment records, be informed of their exposure to toxic substances, contest any alteration in MSHA orders, and receive training for their jobs. Mine operators must comply with safety and health standards and are subject to sanctions if they fail to do so.
- Within 10 days of an accident, mine operators must file a report on the incident with the nearest MSHA district or subdistrict office. They must follow up with a report on their investigation of the accident's causes.
- Mine inspections are usually carried out without prior warning to the operator. The inspector will conduct an opening conference and a walk-through of the premises, document all alleged violations, and hold a closing conference.
- Inspectors or their supervisors can issue citations for violations. Penalties are assessed for each violation and for each day the violation remains uncorrected. Operators, miners, or others affected have 30 days in which to contest the citations either at an informal conference or through formal procedures.
- International and multinational regulations seek to create standards and guidelines to ensure the health and safety of workers, consumers, and the environment wherever companies operate across national boundaries.

DIRECTORY OF FEDERAL AGENCIES

The Occupational Safety and Health Administration

National Headquarters

Occupational Safety and Health Administration, U.S. Department of Labor, Department of Labor Building, 200 Constitution Avenue NW, Washington DC 20210; 202-523-8017

National Institute for Occupational Safety and Health, U.S. Department of Health and Human Services, 1600 Clifton Road NE, Atlanta, GA 30333; 404-639-3061

Bureau of Labor Statistics, U.S. Department of Labor, 200 Constitution Avenue NW, Washington DC 20210; 202-523-7943

Occupational Safety and Health Review Commission, 1120 20th Street NW, Washington DC 20036; 202-606-5380

OSHA Regional Offices

Region I (Connecticut, Maine, Massachusetts, New Hampshire, Rhode Island, Vermont): JFK Fed. Bldg. E-340, Boston, MA 02203; 617-565-9860

Region II (New Jersey, New York, Puerto Rico, Virgin Islands): 201 Varick Street, Room 670, New York, NY 10014; 212-337-2378

Region III (Delaware, District of Columbia, Maryland, Pennsylvania, Virginia, West Virginia): Suite 740 West, 170 S. Independence Mall West, Philadelphia, PA 19106-3309; 215-861-4900

Region IV (Alabama, Florida, Georgia, Kentucky, Mississippi, North Carolina, South Carolina, Tennessee): 61 Forsyth Street SW, Atlanta, GA 30303; 404-562-2300

Region V (Illinois, Indiana, Michigan, Minnesota, Ohio, Wisconsin): J. C. Kluczynski Federal Building, 230 South Dearborn Street, Chicago, IL 60604; 312-353-2220

Region VI (Arkansas, Louisiana, New Mexico, Oklahoma, Texas): 555 Griffin Square Building, Griffin & Young Streets, Dallas, TX 75202; 972-850-4145

Region VII (Iowa, Kansas, Missouri, Nebraska): Two Pershing Square, 2300 Main Street, Suite 1010, Kansas City, MO, 64108; 816-283-8745

Region VIII (Colorado, Montana, North Dakota, South Dakota, Utah, Wyoming): Federal Building, 1999 Broadway, Suite 1690, Denver, CO 80202; 720-264-6550

Region IX (American Samoa, Arizona, California, Guam, Hawaii, Nevada, Trust Territory of the Pacific Islands): 90 7th Street, Suite 18100, San Francisco, CA 94103; 415-625-2547

Region X (Alaska, Idaho, Oregon, Washington): 1111 Third Avenue, Suite 715, Seattle, WA 98101-3212; 206-553-5930

The Mine Safety and Health Administration

Mine Safety and Health Administration, U.S. Department of Labor, Room 601, 1100 Wilson Boulevard, Arlington, VA 22209; 202-693-9400

National Institute for Occupational Safety and Health, U.S. Department of Health and Human Services, 1600 Clifton Road NE, Atlanta, GA 30333; 404-329-3061

National Mine Safety and Health Academy, 1301 Airport Rd., Beaver, WV 25813

Mine Safety and Health Review Commission, 601 New Jersey Avenue NW, Washington DC 20001; 202-434-9906

COMPILATIONS OF REGULATIONS AND LAWS

The safety and health professional and industrial hygienist should be familiar with two U.S. government publications:
- *Federal Register* (FR)
- *Code of Federal Regulations* (CFR)

They are published by the Office of the *Federal Register*, National Archives and Records Service, General Services Administration, and are available from the Superintendent of Documents, U.S. Government Printing Office, Washington DC 20402. Every safety office should obtain them.

The *Federal Register*

The *Federal Register*, published daily Monday through Friday, is a mechanism for making regulations and legal notices issued by all federal agencies available to the public. In general, an agency issues a regulation as a proposal in the *Federal Register*, which is followed by a comment period, and then the agency promulgates or adopts the regulation in the publication. References to material published in the *Federal Register* are usually in the format A *FR* B, in which A is the volume number, *FR* indicates *Federal Register*, and B is the page number. For example, 43 *FR* 58946 refers to volume 43, page 58,946.

The *Code of Federal Regulations*

The *Code of Federal Regulations*, published annually in paperback volumes, is a compilation of the general and permanent rules and regulations that were previously published in the *Federal Register*.

The CFR is divided into 50 different titles, each of which represents a broad subject area of federal regulations—for example, Title 29: Labor; Title 40: Protection of Environment; and Title 49: Transportation. Each title is divided into chapters (usually bearing the name of the issuing agency), which are then further divided into parts and subparts covering specific regulatory areas. References are usually in the format 40 CFR 250.XX, meaning Title 40 CFR, Part 250 (Hazardous Waste Guidelines and Regulations), or 49 CFR 172.XX (Hazardous Materials Table and Hazardous Materials Communications Regulations). The "XX" refers to the number of the specific regulatory paragraph.

The *Code of Federal Regulations* is kept up to date because it contains the individual issues of the *Federal Register*. These two publications must be used together to determine the latest version of any given rule or regulation.

REFERENCES

American Conference of Governmental Industrial Hygienists (ACGIH). *TLVs: Threshold Limit Values for Chemical Substances in the Work Environment.* Latest ed.

Balge, M. Z., and G. R. Krieger, eds. *Occupational Health and Safety.* 3rd ed. Itasca, IL: National Safety Council, 2000.

Bureau of National Affairs, 1801 South Bell Street,

Alexandria, VA 22202.
> *Occupational Safety and Health Reporter.*

International Agency for Research on Cancer (IARC). *Monograph.* Latest ed.

National Institute for Occupational Safety and Health, 5600 Fisher Lane, Rockville, MD 20857.
> *Occupational Safety and Health Directory.*
> *OSHA Compliance Advisor.* Issued twice monthly.

National Safety Council, 1121 Spring Lake Drive, Itasca, IL 60143.
> *Fundamentals of Industrial Hygiene*, 6th ed., 2012.
> *OSHA Up-to-Date.* Issued monthly.
> *Safety and Health.* Issued monthly.

National Toxicology Program (NTP). *Annual Report on Carcinogens.* Latest ed.

Price, M. O., and H. Bitner. *Effective Legal Research.* 4th ed. Boston: Little, Brown, 1979.

Rothstein, M. *Occupational Safety and Health Law.* Updated annually.

Superintendent of Documents, U.S. Government Printing Office, Washington DC 20402.
> Annual List of Toxic Substances.
> Directory of Federal Agencies.
> *Federal Register.*
> Field Operations Manual.
> Industrial Hygiene Technical Manual.
> Mine Safety and Health Act of 1977 (Public Law 95-164).
> Mine Safety and Health Regulations and Standards.
>> 29 CFR Part 2700—Mine Safety and Health Review Commission, Rules of Procedure.
>> 30 CFR Part 11—Respiratory Protective Devices; Tests for Permissibility; Fees.
>> Part 40—Representative of Miners.
>> Part 41—Notification of Legal Identity.
>> Part 43—Procedures for Processing Hazardous Condition Complaints.
>> Part 44—Rules of Practice for Petitions for Modification of Mandatory Safety Standards.
>> Part 45—Independent Contractors.
>> Part 46—State Grants for Advancement of Safety and Health in Coal and Other Mines.
>> Part 47—National Mine Health and Safety Academy.
>> Part 48—Training and Retraining of Miners.
>> Part 49—Mine Rescue Teams.
>> Part 50—Notification, Investigation, Reports and Records of Accidents, Injuries, Illnesses, Employment, and Coal Production in Mines.
>> Part 56—Safety and Health Standards: Surface Metal and Nonmetal Mines.
>> Part 57—Safety and Health Standards: Underground Metal and Nonmetal Mines.
>> Part 58—Health Standards for Metal and Nonmetal Mines.
>> Part 70—Mandatory Health Standards: Underground Coal Mines.
>> Part 71—Mandatory Health Standards: Surface Coal Mines and Surface Work Areas of Underground Coal Mines.
>> Part 74—Coal Mine Dust Personal Sampler Units.
>> Part 75—Mandatory Safety Standards: Underground Coal Mines.
>> Part 77—Mandatory Safety Standards: Surface Coal Mines and Surface Work Areas of Underground Coal Mines.
>> Part 90—Mandatory Health Standards: Coal Miners Who Have Evidence of the Development of Pneumoconiosis.
>> Part 100—Criteria and Procedures for Proposed Assessment of Civil Penalties.
> 42 CFR Part 37—Specifications for Medical Examinations of Underground Coal Miners.
>> Part 85—Requests for Health Hazard Evaluations.
>> Part 85a—NIOSH Policy on Workplace Investigations.
> Occupational Safety and Health Act of 1970 (Public Law 91-596).
> Occupational Safety and Health Regulations Title 29, *Code of Federal Regulations* (CFR):
>> Part 11—Department of Labor, National Environmental Policy Act (NEPA) Compliance Procedures.
>> Part 1901—Procedures for State Agreements.
>> Part 1902—State Plans for the Development and Enforcement of State Standards.
>> Part 1903—Inspections, Citations and Proposed Penalties.
>> Part 1904—Recording and Reporting Occupational Injuries and Illnesses.
>> Part 1905—Rules of Practice for Variances, Limitations, Variations, Tolerances, and Exemptions.
>> Part 1906—Administration Witnesses and Documents in Private Litigation.
>> Part 1907—Accreditation of Testing Laboratories.
>> Part 1908—Consultation Agreements.
>> Part 1910—Occupational Safety and Health Standards.
>> Part 1911—Rules of Procedure for Promulgating, Modifying, or Revoking Occupational Safety or Health Standards.

Part 1912—Advisory Committees on Standards.
Part 1912a—National Advisory Committee on Occupational Safety and Health.
Part 1913—Rules of Agency Practice and Procedure Concerning OSHA Access to Employee Medical Records.
Part 1915—Occupational Safety and Health Standards for Shipyard Employment.
Part 1917—Marine Terminals.
Part 1918—Safety and Health Regulations for Longshoring.
Part 1919—Gear Certification.
Part 1920—Procedure for Variations from Safety and Health Regulations under the Longshoremen's and Harbor Workers' Compensation Act.
Part 1921—Rules of Practice in Enforcement Proceedings under Section 41 of the Longshoremen's and Harbor Workers' Compensation Act.
Part 1922—Investigational Hearings under Section 41 of the Longshoremen's and Harbor Workers' Compensation Act.
Part 1924—Safety Standards Applicable to Workshops and Rehabilitation Facilities Assisted by Grants.
Part 1925—Safety and Health Standards for Federal Service Contracts.
Part 1926—Safety and Health Regulations for Construction.
Part 1928—Occupational Safety and Health Standards for Agriculture.
Part 1949—Office of Training and Education, Occupational Safety and Health Administration.
Part 1950—Development and Planning Grants for Occupational Safety and Health.
Part 1951—Grants for Implementing Approved State Plans.
Part 1952—Approved State Plans for Enforcement of State Standards.
Part 1953—Changes to State Plans for the Development and Enforcement of State Standards.
Part 1954—Procedures for the Evaluation and Monitoring of Approved State Plans.
Part 1955—Procedures for Withdrawal of Approval of State Plans.
Part 1956—State Plans for the Development and Enforcement of State Standards Applicable to State and Local Government Employees in States without Approved Private Employee Plans.
Part 1960—Basic Program Elements for Federal Employee Occupational Safety and Health Programs and Related Matters.
Part 1975—Coverage of Employees under the Williams-Steiger Occupational Safety and Health Act of 1970.
Part 1977—Discrimination against Employees Exercising Rights under the Williams-Steiger Occupational Safety and Health Act of 1970.
Part 1990—Identification, Classification, and Regulation of Potential Occupational Carcinogens.
Part 2200—Review Commission Rules of Procedure.
Part 2201—Regulations Implementing the Freedom of Information Act.
Part 2202—Standards of Ethics and Conduct of Occupational Safety and Health Review Commission Employees.
Part 2203—Regulations Implementing the Government in the Sunshine Act.
Part 2204—Implementation of the Equal Access to Justice Act.
Part 2205—Enforcement of Nondiscrimination on the Basis of Handicap.

REVIEW QUESTIONS

1. What is the purpose of the Williams-Steiger Act (OSH Act)?
2. Before the passage of the OSH Act, other pieces of national safety legislation had been passed in the United States; what was the main drawback of these pieces of legislation?
3. Labor was a strong proponent of giving the federal government a more significant role in occupational safety and health. On what is this position based?
4. List the two organizations that are vested with the administration and enforcement of the OSH Act.
   a.
   b.
5. Name the quasi-judicial board of three members appointed by the president and confirmed by the Senate to adjudicate cases.

6. What are the primary functions of NIOSH?
   a.
   b.
   c.
   d.
7. List the five main technical services provided by NIOSH.
   a.
   b.
   c.
   d.
   e.
8. The final responsibility for compliance rests with the _____.
9. What are the three voluntary safety programs established by OSHA for companies that have exceptional safety programs?
   a.
   b.
   c.
10. In what publication are possible federal safety standards published before they are brought up for formal consideration to be enacted into law?
11. What two types of variances can OSHA grant?
    a.
    b.
12. Employers must report to OSHA within _____ after an unintentional injury occurs that is fatal to one or more employees or that results in the inpatient hospitalization of three or more employees.
13. Are compliance safety and health officers (CSHOs) required to present solutions or methods of correcting, minimizing, or eliminating a violation?
14. List the five categories of violations.
    a.
    b.
    c.
    d.
    e.
15. What four steps must a CSHO take to determine that a violation is serious?
    a.
    b.
    c.
    d.
16. How may an employer contest a violation?
17. How long should employee medical records be retained by employers?
18. What is the basic purpose of the hazard communication standard?
19. What does the quality of a hazard communication program depend on?
20. Who is responsible for the information on MSDSs?
21. Whom does the OSH Act cover, and who is excluded from its coverage?
22. What are the general duties of employers and employees under the OSH Act?
23. What is the purpose of the OSHA Voluntary Protection Programs?
24. What concepts must an organization have present to qualify as having an exemplary safety program?
25. What does the informal conference provide regarding a citation?
26. How does a state go about having a "state OSHA plan" approved?
27. How has the OSH Act given new visibility to occupational safety and health?

Legal and Regulatory Issues for the Safety Manager

Philip Hagan, JD, MBA, MPH, ARM, CIH, CET, CHMM, CHCM, CHSP, CEM

Overview

Scope of the Occupational Safety and Health Act
General-Duty Clause ▸ Required Written Plans and Other Documentation ▸ Most Frequently Cited OSHA Standards

Overview of OSHA Regulations Most Often Cited
Fall-Protection Standard ▸ Hazard Communication Standard ▸ Control of Hazardous Energy (Lockout/Tagout) Standard ▸ Respiratory Protection Standard ▸ Machine-Guarding Standard ▸ Electrical Safety Standards ▸ Powered Industrial Trucks Standard ▸ Ladder Standards ▸ Scaffolding Standard ▸ Excavations Standard ▸ Walking/Working Surfaces Standard ▸ Process Safety Management of Highly Hazardous Chemicals Standard ▸ Confined-Space Entry Standard

OSHA Log 300: Recording and Reporting Occupational Injuries and Illness Standard
OSHA Record Keeping ▸ Exemptions for Recording Injuries and Illnesses ▸ Exemption for Low-Hazard Industries ▸ OSHA Record Keeping Forms ▸ Determining Work-Relatedness and OSHA Recordability ▸ Definitions of Medical Treatment and First Aid ▸ Restricted Work Activity ▸ Classifying Injuries and Illnesses

Summary

Review Questions

OVERVIEW

The safety environment is shaped by legal requirements from potential liability issues that arise from court proceedings associated with workplace torts and by the government regulatory arena. The court system addresses issues related to workers' compensation, negligence, and other related torts. Legislatures at both the federal and state levels (and sometimes municipal level) promulgate regulations that govern safety practices in the workplace. The executive branches (i.e., president and governors) of government entities enforce these regulations through agencies like the Occupational Safety and Health Administration (OSHA) using the Occupational Safety and Health Act (OSH Act) or a similar regulation. The courts are used to interpret differences of opinion when two or more sides disagree on either the meaning or enforcement of a regulation.

Legal problems and adversarial trial systems are familiar to anyone who watches television or movie dramas. Judges and juries exist to determine compensation for losses associated with disputes that arise from claims that a company or its product caused death, injury, or economic harm. Regulatory systems inside the government are less visible to the general public but are even more important to the plant operator.

The modern safety manager has six significant roles in dealing with a company's legal obligations.

First, the safety manager uses knowledge based on experience, education, training, and good instincts to facilitate compliance with corporate safety programs while addressing legal liabilities associated with safety in the workplace. The work of the safety manager is an important part of protecting workers and avoiding liabilities associated with injuries, fatalities, and loss of customer confidence in the company.

Second, the role of the safety manager is usually to track and maintain relevant data in records that must be kept to satisfy regulatory obligations for the company with regard to safety and environmental reporting.

Third, the safety manager is a key member of the compliance team, which reduces liability of the company by preventing work-related injuries and fatalities, preventing potentially bad outcomes from government safety inspections, and implementing the requirements of new regulations and standards.

Fourth, the safety manager acts as a valued participant in addressing workers' compensation, arbitration, or litigation actions involving the company and its operations.

On the regulatory side, the safety manager keeps up with proposed regulation changes from the comment stage to implementation in order to ensure that resulting regulations are reasonable, effective in addressing workplace hazards, and implemented in a timely manner.

Lastly, the safety manager assists at the time of the inspections conducted by third-party auditors or government regulators to ensure that inspections are conducted in a transparent fashion and include all affected stakeholders.

SCOPE OF THE OCCUPATIONAL SAFETY AND HEALTH ACT

In general, the Occupational Safety and Health Act extends to all employers and their employees in the 50 states, the District of Columbia, Puerto Rico, and all other territories under federal government jurisdiction. Coverage is provided either directly by the federal Occupational Safety and Health Administration (OSHA) or, in states that have approved programs, through an OSHA-approved state occupational safety and health program.

Each supervisor should know which set of regulations governs the workplace. Twenty-five states and the District of Columbia are covered by federal OSHA regulations. The other 25 states and Puerto Rico and the Virgin Islands (see Table 5–A) have their own regulatory plans that have been approved by federal OSHA and cover the private and government sectors.

The Connecticut, New Jersey, New York, and Virgin Islands plans cover public-sector (state and local government) employment only. This means that in these states, government workers are covered by the state plan and private-sector employees are covered by federal OSHA regulations.

In some cases, there are regulatory requirements from other sources that need to be followed. The Environmental Protection Agency (EPA), the Department of Transportation (DOT), and the Nuclear Regulatory Commission (NRC) are examples of sources of regulatory requirements besides OSHA that may govern a workplace. The safety supervisor should ask his or her employer for advice in determining which regulations are applicable if the safety supervisor has not already done so. This chapter deals primarily with OSHA regulations.

OSHA regulations can be found in Title 29 of the *Code of Federal Regulations* (CFR), Parts 1902 through 1990. These regulations, which cover general industry, maritime, construction, and agricultural workplaces, can be found at the osha.gov website.

Table 5–A is provided as a source of OSHA regulatory information for states with their own plans (some plans do not cover the whole working sector).

TABLE 5–A. List of State OSHA Regulations

State	Description
Alaska	Alaska Division of Labor Standards and Safety labor.state.ak.us/lss/home.htm
Arizona	Arizona Secretary of State Website www.ica.state.az.us/Divisions/osha/index.html
California	California DOSH www.dir.ca.gov/dosh/
Connecticut	CONN-OSHA adopts federal OSHA regulations and standards identically.
Hawaii	Hawaii Standards hawaii.gov/labor/hiosh/
Illinois	Illinois Public Employee Only (PEO) State Plan was approved as a developmental plan. www.illinois.gov/idol/Pages/default.aspx
Indiana	Indiana OSHA adopts federal OSHA regulations and standards identically.
Iowa	Iowa OSH Administrative Rules www.iowaworkforce.org/labor/iosh/index.html
Kentucky	OSH Regulations www.labor.ky.gov/osh/
Maryland	Maryland Standards www.dllr.state.md.us/labor/mosh.html
Michigan	Standards and Legislation www.michigan.gov/miosha
Minnesota	Minnesota Standards and Regulations www.dli.mn.gov/MnOsha.asp
Nevada	Nevada Standards dirweb.state.nv.us/OSHA/osha.htm
New Jersey	See Regulations and Standards section. lwd.dol.state.nj.us/labor/lsse/employer/Public_Employees_OSH.html
New Mexico	New Mexico Regulations www.nmenv.state.nm.us/Ohsb_Website/index.htm
New York	See Regulations and Standards section. www.labor.state.ny.us/workerprotection/safetyhealth/DOSH_PESH.shtm
North Carolina	NCDOL Standards www.nclabor.com/osha/osh.htm
Oregon	Oregon OSHA Rules and Laws www.orosha.org/
Puerto Rico	Spanish translations of safety and health standards/Normas traducidas al español radicadas www.trabajo.pr.gov/prosha/index.asp
South Carolina	South Carolina Standards www.llr.state.sc.us/Labor/Osha/index.asp
Tennessee	Tennessee Department of Labor and Workforce Development www.state.tn.us/labor-wfd/tosha.html
Utah	R614—Labor Commission, Occupational Safety and Health www.laborcommission.utah.gov/UOSH/index.html
Vermont	Vermont Standards www.labor.vermont.gov/Default.aspx?tabid=74
Virginia	Virginia Standards www.doli.virginia.gov
Virgin Islands	VIDOSH adopts federal OSHA regulations and standards identically.
Washington	Washington Standards www.lni.wa.gov/Safety/default.asp
Wyoming	Wyoming OSHA Standards wydoe.state.wy.us/osha

Workplaces covered by these regulations should ensure that the poster "Job Safety and Health: It's the Law" (OSHA 3165) is displayed in a conspicuous place to inform employees and applicants of the protections afforded by the OSH Act. (Federal government agencies must post a Federal Agency Poster.) The poster is available for free from the OSHA Office of Publications. Reproductions or facsimiles of the poster shall be at least 8 ½ × 14 in. (21 ¼ × 35 cm) with 10-point type. Employers do not need to replace previous versions of the poster.

General-Duty Clause

In many cases, there are specific regulations that govern different workplace activities. For those cases in which there are no specific regulations, there is Section 5(a)(1) of the OSH Act, often referred to as the *general-duty clause*. The general-duty clause requires employers to "furnish a place of employment that is free from recognized hazards which are causing or are likely to cause death or serious physical harm to employees." When OSHA inspectors identify hazards that are not covered by other regulations, the inspectors use the general-duty clause to cite employers. However, for inspectors to be able to use the general-duty clause, the identified hazard needs to meet several criteria:

- There is no applicable OSHA standard for the identified hazard.
- The employer failed to keep the workplace free of a hazard to which employees were exposed.
- The hazard was recognized or should have been recognized:
  - The employer knew about the hazard, as shown by written or oral statements made during or before an OSHA inspection.
  - The hazard is recognized by others in the same industry.
  - Common sense indicates that any reasonable person would recognize the hazard.
- The hazard was causing or was likely to cause death or serious physical harm.
- There was a feasible and effective method to correct the hazard.

A general-duty citation must involve both the presence of a serious hazard and exposure of the cited employer's own employees.

Required Written Plans and Other Documentation

Regulations can take different forms, and their requirements can vary widely. Some require written plans or other forms of documentation that will serve as a guide for complying with the regulation. Common examples of regulations requiring written plans are 29 CFR 1910.1030(c)(1)(i) and 1910.1030(c)(1)(iii) (bloodborne pathogens), 29 CFR 1910.1200(e)(1) and 1910.1200(e)(4) (hazard communication), and 29 CFR 1910.146(c)(4) (permit-required confined spaces). Traditionally, these written programs have been kept in separate binders in appropriate work areas so that employees can comply with the standards. Generally, these written plans need to be readily available to employees or their representatives.

These written plans can be general in nature (such as the hazard communication program), or they can have very specific requirements (like those found in the bloodborne pathogens standard). Sometimes, the only requirement is that written records be maintained. Because computers are more common in the workplace now than in years past, OSHA allows a written program to be in either paper or electronic format as long as the program meets all other requirements of the standard in question and is accessible by employees. This in turn allows an employer to obtain significant benefits in consistency, ease of use, and accuracy in maintaining and updating the materials in a timely manner.

A partial listing of OSHA regulations with written components required in many workplaces is the following:

OSHA Log 300, Recording and Reporting Occupational Injuries and Illness, § 1904
Emergency-Action Plan, § 1910.38(a)
Fire-Prevention Plan, § 1910.38(a)
Hearing Conservation, § 1910.95
Process Safety Management, § 1910.119
HAZWOPER, § 1910.120
Respiratory Protection, § 1910.134
Confined Spaces, § 1910.146
Control of Hazardous Energy, § 1910.147
Fire Extinguishers, § 1910.157
Toxic/Hazardous Substances (Subpart Z), § 1910.1000
Bloodborne Pathogens, § 1910.1030
Hazard Communication, § 1910.1200
Laboratory Standard, § 1910.1450

It is important for safety and health professionals to determine whether regulations governing their workplace require written documentation. It is very difficult to comply with a regulation if there are no written components available. When an OSHA inspector visits, quite often the first area that is evaluated involves those regulations with written requirements.

Most Frequently Cited OSHA Standards

This chapter addresses those standards that have resulted in frequent OSHA citations and large monetary penalties.

Recently, the following standards were the top 10 most frequently cited by OSHA inspectors:
1. Fall Protection, § 1926.501
2. Hazard Communication, § 1910.1200
3. Scaffolding, § 1926.451
4. Respiratory Protection, § 1910.134
5. Electrical, Wiring Methods, § 1910.305
6. Powered Industrial Trucks, § 1910.178
7. Ladders, § 1926.1053
8. Lockout/Tagout, § 1910.147
9. Electrical, General Requirements, § 1910.303
10. Machine Guarding, § 1910.212

OVERVIEW OF OSHA REGULATIONS MOST OFTEN CITED

Each supervisor should determine whether any of the following regulations discussed in this section are applicable in his or her workplace; if so, he or she should ensure that a compliance program is in place.

Fall-Protection Standard

Fall protection must be provided at 4 ft (1.2 m) in general industry, at 5 ft (1.5 m) in maritime industries, and at 6 ft (1.8 m) in the construction industry.

There are a number of ways to protect workers from falls, including conventional systems such as guardrail systems, safety net systems, and personal fall-protection systems (fall-arrest systems, positioning systems, and travel restraint systems), as well as through the use of safe work practices and training. The use of warning lines, designated areas, control zones, and similar systems are permitted by OSHA in some situations and can provide protection by limiting the number of workers exposed and instituting safe work methods and procedures. These alternative systems may be more appropriate than conventional fall-protection systems when performing certain activities.

Whether conducting a hazard assessment or developing a comprehensive fall-protection plan, thinking about fall hazards before the work begins will help to manage fall hazards and focus attention on prevention efforts. If personal fall-protection systems are used, particular attention should be given to identifying attachment points and to ensuring that employees know how to properly don and inspect the equipment.

Fall protection in the construction industry is often cited by OSHA inspectors. In general, any employee working in a location that is at least 6 ft (1.8 m) or more above a lower level should be protected by guardrail systems, safety net systems, or personal fall-arrest systems. If using one of these systems is not feasible or would create its own hazards when employees are performing leading-edge work, precast concrete erection work, or residential construction work, then it is permissible to develop and implement a fall-protection plan that meets the following requirements:

- It must be prepared by a qualified person, developed specifically for the worksite, and kept up to date.
- Any changes to the fall protection plan needs to be approved by a qualified person.
- A copy of the fall-protection plan needs to be at the job site.
- The fall-protection plan should be under the supervision of a competent person.
- The fall-protection plan should document the following items:
  - The reasons the use of conventional fall-protection systems (guardrail systems, personal fall-arrest systems, or safety net systems) will not work or why their use would create a greater hazard
  - The other measures that will be taken to reduce or eliminate the fall hazard for workers who cannot be provided with the conventional fall-protection systems. These other measures can include discussing the extent to which scaffolds, ladders, or vehicle-mounted work platforms can be used to provide a safer working surface and thereby reduce the hazard of falling.
- The fall-protection plan should identify each location where conventional fall-protection methods cannot be used.
  - These locations should then be classified as controlled-access zones and defined by control lines or something similar to restrict access.
  - When control lines are used, they should be erected not less than 6 ft (1.8 m) nor more than 25 ft (7.7 m) from the unprotected or leading edge, except when erecting precast concrete members.
  - When erecting precast concrete members, the warning line should be erected not less than 6 ft (1.8 m) nor more than 60 ft (18 m)—or half the length of the member being erected, whichever is less—from the leading edge.
  - The control line should extend along the entire length of the unprotected or leading edge, be approximately parallel to the unprotected or leading edge, and be connected on each side to a guardrail system or wall.
- Where overhand bricklaying and related work will take place, the controlled-access zone should be defined by a control line erected not less than 10 ft (3.1 m) nor more than 15 ft (4.5 m) from the working edge.

Hazard Communication Standard

OSHA's website indicates that more than 30 million American workers are exposed to hazardous chemicals in their workplaces. The OSHA Hazard Communication

Standard (HCS) is intended to ensure that these workers and their employers are informed of the identities of these hazardous chemicals, associated health and safety hazards, and appropriate protective measures. Hazard communication is addressed in standards specific to general industry, shipyard employment, marine terminals, longshoring, and the construction industry. The HCS covers some 650,000 hazardous chemical products found in more than 3 million establishments. As a supervisor, it is important to determine whether employees in the workplace could be exposed to hazardous chemicals. Given the right (or wrong) conditions, most chemicals can be hazardous.

In general, the Hazard Communication Standard states that chemical manufacturers/importers must determine the hazards of each chemical product and communicate the hazard information to customers through labels and Safety Data Sheets (SDSs). The employers (and supervisors) then have the following responsibilities:
- Identify and list hazardous chemicals in their workplaces.
- Obtain SDSs and labels for each hazardous chemical.
- Develop and implement a written hazard communication program, which should include information on container labels, SDSs and their availability, and employee training.

Written Program
The written hazard communication program does not have to be lengthy or complicated but must be available on the request of employees and employees' designated representatives. The written program should contain a list of the hazardous chemicals in each work area, indicate how the employer will inform employees of nonroutine task hazards, and specify the hazards associated with chemicals in unlabeled pipes. If the workplace has multiple employers on site, it is necessary to ensure that information regarding hazards and protective measures be made available to the other employers on site as appropriate.

Labels
OSHA has updated the requirements for labeling of hazardous chemicals under its Hazard Communication Standard (HCS). As of June 1, 2015, all labels are required to have pictograms, a signal word, hazard and precautionary statements, the product identifier, and supplier identification.

All containers of hazardous chemicals must be labeled, tagged, or marked with the identity of the material and must show hazard warnings appropriate for protecting employees. As long as the hazards of the chemicals are conveyed, the hazard warning can use words, pictures, or symbols. Labels must be legible so that they are understandable to anyone who would be using the chemical.

There are some exceptions regarding container labels:

- Signs or placards that convey hazard information can instead be posted if there are stationary containers within a work area that have similar contents and hazards.
- Standard operating procedures, batch tickets, blend tickets, process sheets, and other similar written materials can be used instead of container labels on stationary process equipment if the same information is covered and readily available to employees in the work area.
- Portable containers into which hazardous chemicals are transferred from labeled containers and that are intended only for the immediate (i.e., within the current workshift) use of the employee who makes the transfer are permissible.
- Pipes or piping systems are not required to be labeled.

A sample revised HCS label that identifies the required label elements is shown in Figure 5–1. Supplemental information can also be provided on the label as needed.

Safety Data Sheets
Each SDS must be in English (although they can also be in other languages) and include information regarding the specific chemical identity of the hazardous chemical(s) involved and the common names. In addition, information must be provided on:
- identification and contact information for the organization responsible for preparing the SDS
- names, alternate names, and other identification information for the chemical
- physical and chemical characteristics of the chemical
- fire and exposure data
- health hazard information (including whether the chemical is considered a carcinogen)
- exposure limits
- procedures for safe handling and use
- control methods
- emergency and first-aid procedures.

Copies of the SDSs need to be readily accessible by employees in their work areas. This means that SDSs have to be available during each workshift in case someone wants or needs to review their information.

List of Hazardous Chemicals
The written plan needs to contain a list of all hazardous chemicals in the workplace. There should also be an SDS for each chemical on the list. If there are hazardous chemicals without SDSs, the employer needs to contact the supplier, manufacturer, or importer to obtain the missing SDSs. If the SDS is not received within a reasonable period of time, OSHA should be contacted. It is a good idea to document all activities when trying to obtain an SDS.

Sample Label

Product Identifier
CODE _____
Product Name _____

Supplier Identification
Company Name_____
Street Address _____
City _____ State_____
Postal Code _____ Country _____
Emergency Phone Number _____

Precautionary Statements
Keep container tightly closed. Store in cool, well-ventilated place that is locked.
Keep away from heat/sparks/open flame. No smoking.
Use only nonsparking tools.
Use explosion-proof electrical equipment.
Take precautionary measures against static discharge.
Ground and bond container and receiving equipment.
Do not breathe vapors.
Wear protective gloves.
Do not eat, drink, or smoke when using this product.
Wash hands thoroughly after handling.
Dispose of in accordance with local, regional, national, and international regulations as specified.

In Case of Fire: Use dry chemical (BC) or carbon dioxide (CO_2) fire extinguisher to extinguish.

First Aid
If exposed, call the nearest Poison Center.
If on skin (or hair): Immediately take off any contaminated clothing. Rinse skin with water.

Hazard Pictograms

Signal Word:
Danger

Hazard Statement
Highly flammable liquid and vapor.
May cause liver and kidney damage.

Supplemental Information
Directions for use:
_____
_____
_____

Fill Weight: _____ Lot Number: _____
Gross Weight: _____ Fill Date: _____
Expiration Date: _____

Figure 5–1. Sample Label

Employee Training

Employees exposed to hazardous chemicals in their work areas need to receive training. This training needs to occur at the time of the initial assignment and/or whenever a new hazard is introduced into the work area. The training plan should cover the following elements as of December 1, 2013:
- how the hazard communication program is implemented and how to read and interpret information in SDSs and on labels
- the hazards of the chemicals in the work area—hazards may be discussed on a chemical-by-chemical basis or by hazard categories such as corrositivity or flammability
- measures that can be used to protect themselves from hazards
- procedures used to provide protection, such as engineering controls, work practices, and the use of personal protective equipment (PPE)
- observations workers can make—such as a smell or noticing a spill from a container—to detect the presence of a hazardous chemical.

Trade Secrets—Medical Emergency

Sometimes, chemicals come under the trade secret provision and employees do not know the identities and ingredients. However, employers must immediately disclose the specific chemical identity of a hazardous chemical to a treating physician or nurse when the information is needed for proper emergency or first-aid treatment. If an employer needs the information from the SDS, the chemical manufacturer, importer, or employer may submit a written statement of need and a confidentiality agreement that will protect the manufacturer after having disclosed the trade secret.

Control of Hazardous Energy (Lockout/Tagout) Standard

Approximately 3 million workers service equipment and face the greatest risk of injury if a lockout/tagout program is not properly implemented. "Lockout/tagout (LOTO)" refers to specific practices and procedures safeguard employees from the unexpected energization or start-up of machinery and equipment and from the release of hazardous energy during service or maintenance activities.

Compliance with the lockout/tagout standard (29 CFR 1910.147) prevents an estimated 120 fatalities and 50,000 injuries each year. Workers injured on the job because of exposure to hazardous energy lose an average of 24 workdays to recuperation. In a study conducted by the United Auto Workers (UAW), (83 of 414) 20% of the fatalities that occurred among their members between 1973 and 1995 were attributed to inadequate hazardous energy control procedures—specifically, lockout/tagout procedures. Based on these statistics, it is clear that a supervisor should pay special attention to ensuring that an effective LOTO program is implemented throughout designated work areas.

Lockout/Tagout

Because workers performing service or maintenance on machinery and equipment may be exposed to injuries from the unexpected energization, start-up of the machinery or equipment, or release of stored energy in the equipment. Thus, a LOTO standard requires the adoption and implementation of practices and procedures to shut down equipment, isolate it from its energy source(s), and prevent the release of potentially hazardous energy while maintenance and servicing activities are being performed. As long as minimum performance requirements are met, employers have the flexibility to develop lockout/tagout programs that are suitable for their respective facilities.

Program Elements

The LOTO program should include the following elements: energy control procedures, employee training, and periodic inspections to ensure that before service and maintenance are performed, machines and equipment that could unexpectedly start up, become energized, or release stored energy are isolated from their energy source(s) and rendered safe.

This program should cover servicing and/or maintenance of machines or equipment and workers who are exposed to the unexpected energization, start-up, or release of hazardous energy. "Unexpected" includes situations in which servicing and/or maintenance is performed during ongoing, customary production operations when an employee is required to remove or bypass machine guards or other safety devices or place any part of his or her body into a point of operation, an area on a machine or piece of equipment where work is performed, or the danger zone associated with a machine's operation.

Activities and operations covered by this standard include any servicing and/or maintenance of machines or equipment whose source of energy is electrical, mechanical, hydraulic, pneumatic, chemical, thermal, or another energy source. This also includes activities like constructing, installing, setting up, adjusting, inspecting, and modifying machines or equipment, including lubricating, cleaning or unjamming, and making adjustments or tool changes.

Exceptions

This standard does not cover work on cord- and plug-connected electrical equipment when the equipment is unplugged from the energy source and the authorized employee has exclusive control of the plug.

Another exception to this standard is when continuity of service is essential, shutdown of the system is impractical, documented procedures are followed, and employees are effectively protected by special equipment.

Lockout

Lockout is where placement of a lockout device on an energy-isolating device is used, in accordance with an established procedure, thus ensuring that the energy-isolating device and the equipment being controlled cannot be operated until the lockout device is removed. This would include the use of any device that uses positive means—such as a lock, blank flanges, and bolted slip blinds—to hold an energy-isolating device in a safe position and thereby preventing the energizing of machinery or equipment.

Tagout

Tagout is used where lockout is not feasible and placement of a tagout device on an energy-isolating device is used, in accordance with an established procedure, to indicate that the energy-isolating device and the equipment being controlled should not be operated until the tagout device is removed. This would include the use of any prominent warning device—such as a tag and a means of attachment—that can be securely fastened to an energy-isolating device.

Required Procedures

Employers must develop, document, and use specific procedures to control potentially hazardous energy while employees are servicing equipment or machinery. The procedures must specify the scope, purpose, authorization, rules, and techniques that the employer will use to control hazardous energy. At a minimum, the procedures must include:
- a specific statement of the intended use of the procedures
- specific procedural steps for shutting down, isolating, blocking, and securing machines or equipment to control hazardous situations
- specific procedural steps for the placement, removal, and transfer of lockout devices or tagout devices and a description of who has responsibility for them
- specific requirements for testing a machine or piece of

equipment to determine and verify the effectiveness of lockout devices, tagout devices, and other energy control measures.

Inspections should be performed at least annually to identify any deficiencies or deviations and correct them. When lockout is used, the inspector must review with each authorized employee his or her responsibilities under the procedure (group meetings are acceptable). When tagout is used, the inspector must review with both the authorized and the affected employee their responsibilities in the energy control procedure being inspected and any additional training responsibilities. The inspection report should include the following information:
- identity of the machine that was inspected
- date of inspection
- identity of the employees included in the inspection
- identity of the person who performed the inspection.

Removing Lockout or Tagout Devices

Before lockout or tagout devices are removed and energy is restored, the work area must be inspected to ensure that nonessential items (e.g., tools, spare parts) have been removed and that all of the machine's or equipment's components are operationally intact.

The work area must be checked to ensure that all employees have been safely positioned or have cleared the area. In addition, all affected employees must be notified before the equipment is energized that the lockout or tagout devices have been removed.

Each lockout or tagout device must be removed from the energy-isolating device by the employee who applied the device. When the employee who applied the lockout or tagout device is not available to remove it, the device may be removed under the direction of the employer, provided that specific procedures and training for such removal have been developed, documented, and incorporated into the employer's energy control program.

Employers must ensure the continuity of employee protection by providing for the orderly transfer of lockout or tagout devices between outgoing and incoming employees. This continuity will help to minimize exposure to hazards from the unexpected energization or start-up of the machine or equipment or the release of stored energy.

Training

The standard requires different levels of training for each of the three categories of employees:
- Authorized employees must receive training in the recognition of applicable hazardous energy sources, the type and magnitude of the energy available in the workplace, and the methods and means necessary for energy isolation and control.
- Affected employees must receive training on the purpose and use of the energy control procedure.
- Other employees (those whose work activities are or may be in an area where energy control procedures may be utilized) must be instructed in the procedure and about the prohibition relating to attempts to restart or reenergize machines or equipment that is locked out or tagged out. Employers must train employees in the limitations of tags when used instead of lockout when addressing energy control measures.

If changes occur in a job or procedures or problems are identified, then employees should be retrained. Training should be documented with a certificate that includes each employee's name and the dates of training and/or retraining.

Respiratory Protection Standard

An estimated 5 million workers in 1.3 workplaces throughout the United States are required to wear respirators. Respirators protect workers in insufficient-oxygen environments and from harmful dusts, fogs, smoke, mists, gases, vapors, and sprays. These hazards may cause cancer, lung impairment, other diseases, or death.

Whenever possible, engineering controls should be used to restrict employee exposure to airborne hazards. When such controls are not feasible, a respiratory protection program that complies with OSHA's respirator standard, 29 CFR 1910.134, should be instituted. This standard requires the use of respirators to protect employees from breathing contaminated and/or oxygen-deficient air when effective engineering controls are not feasible or while they are being instituted. Several other OSHA regulations also require the use of respirators.

Respirators should be selected on the basis of hazards to which the worker is exposed (i.e., particulates, vapors, oxygen deficiency, or a combination of these). In addition, only respirators certified by the National Institute for Occupational Safety and Health (NIOSH) should be used. Certified respirators are marked with "NIOSH," the manufacturer's name, the respirator's part number, and an abbreviation that indicates the cartridge or filter type.

Samples of approval labels are shown in Figures 5–2 and 5–3.

When respirators are required in the workplace, a written respiratory protection program with worksite-specific procedures and elements of required respirator use is also required. The provisions of the program should include procedures for fit testing, medical evaluation, training, and use and care of respirators.

PART 84 MATRIX APPROVAL LABEL
FPR P100 FILTER

DEF MANUFACTURING COMPANY
ANYWHERE, USA
1-800-555-1234

THESE RESPIRATORS ARE APPROVED ONLY IN THE FOLLOWING CONFIGURATIONS:

TC-	PROTECTION	RESPIRATOR	CAUTIONS AND LIMITATIONS
84A-00X	P100	HALO 2000	ABCJMNO

1. PROTECTION

P100–Particulate Filter (99.97% filtered efficiency level) is effective against all particulate aerosols.

2. CAUTIONS AND LIMITATIONS
A—Not for use in atmospheres containing less than 19.5% oxygen.
B—Not for use in atmospheres immediately dangerous to life or health.
C—Do not exceed the maximum use concentrations established by regulatory standards.
J—Failure to use and maintain this product properly could result in injury or death.
M—All approved respirators should be selected, fitted, used, and maintained in accordance with MSHA, OSHA, and other applicable regulations.
N—Never substitute, modify, add, or omit parts. Use only exact replacement parts in the configuration specified by the manufacturer.
O—Refer to user instructions and/or maintenance manuals for information about the use and maintenance of these respirators.

Figure 5–2.

PART 11 LABEL FOR HEPA FILTER

PERMISSIBLE

PERMISSIBLE PARTICULATE FILTER RESPIRATOR FOR DUSTS, FUMES, AND MISTS, INCLUDING ASBESTOS-CONTAINING DUSTS AND MISTS, AND RADIONUCLIDES

MINE SAFETY AND HEALTH ADMINISTRATION
NATIONAL INSTITUTE FOR OCCUPATIONAL
SAFETY AND HEALTH

APPROVAL NO. TC-21C-XXX

 ISSUED TO
ABC Company
Anywhere, USA

LIMITATIONS
Approved for respiratory protection against dusts, fumes, and mists having a time-weighted average less than 0.05 milligram per cubic meter, including asbestos-containing dusts and mists, and radionuclides.

Not for use in atmospheres containing toxic gases or vapors.

Not for use in atmospheres immediately dangerous to life or health. Not for use in atmospheres containing less than 19.5% oxygen.

CAUTION
In making renewals or repairs, parts identical to those furnished by the manufacturer under the pertinent approval should be maintained.

Follow the manufacturer's instructions for changing filters.

The respirator should be selected, fitted, used, and maintained in accordance with the regulations of the Mine Safety and Health Administration, the Occupational Safety and Health Administration, and other applicable agencies.

MSHA/NIOSH Approval TC-21C-XXX
Issued to ABC Co. February 28, 1990
The approved assembly consists of the following part numbers:
000-000
000-000
etc.

Figure 5–3.

Fit Testing

Proper respirator size is determined through a *fit test*, which is a method used to select the right size respirator for the user. Employees who will be using negative- or positive-pressure, tight-fitting facepiece respirators must undergo an appropriate fit test using the procedures detailed in OSHA's respirator standard.

Fit testing of all negative- or positive-pressure, tight-fitting facepiece respirators is required prior to initial use, whenever a different respirator facepiece is used, and at least annually thereafter. An additional fit test is required whenever there are changes in the user's physical condition that could affect respirator fit (e.g., facial scarring, dental changes, cosmetic surgery, or an obvious change in body weight). The employee must be fit-tested with the make, model, style, and size of respirator that the employee will use.

Employees using tight-fitting facepiece respirators are required to perform a user seal check each time they put on the respirator. A *user seal check* is a method to verify that the user has correctly put on the respirator and adjusted it to fit properly.

Manufacturers make several different sizes of respirators. Respirators may also vary in size from manufacturer to manufacturer. Users may be able to get a better fit by trying a respirator made by another manufacturer. In some cases, the use of powered air-purifying respirators may be appropriate. Employers must help employees find a suitable respirator.

Tight-fitting facepiece respirators must not be worn by employees who have facial hair that goes between the sealing surface of the facepiece and the wearer's face or that interferes with valve function. Respirators that do not rely on a tight face seal, such as hoods and helmets, may be used by bearded individuals.

Medical Evaluation

Before each employee is fit-tested or required to use a respirator in the workplace, the employer must provide a medical evaluation to determine the employee's suitability to use a respirator. A physician or other licensed health care professional must perform the medical evaluation using the medical questionnaire contained in Appendix C of 29 CFR 1910.134 or a preliminary medical examination that procures the same information.

Training

Training must be provided to all employees who are required to use respirators. The training must be comprehensive, understandable, and recur at least annually and more often if necessary. This training should include at a minimum:

- why the respirator is necessary and how improper fit, use, or maintenance can compromise its protective effect
- limitations and capabilities of the respirator
- effective use in emergency situations
- how to inspect, put on and remove, use, and check the seals
- maintenance and storage
- recognition of medical signs and symptoms that may limit or prevent effective use.

Maintenance, Storage, and Replacement

The employer must provide for the cleaning and disinfecting, storage, inspection, and repair of respirators used by employees.

Some respirators are disposable and cannot be disinfected and are, therefore, assigned to only one person. Disposable respirators must be discarded if they are soiled, physically damaged, or at the end of their service life. Replaceable filter respirators may be shared, but they must be thoroughly cleaned and disinfected after each use and before being worn by a different person.

Respirators with replaceable filters are reusable, and a respirator classified as disposable may be reused by the same worker as long as it functions properly. All filters must be replaced whenever they are damaged, soiled, or causing noticeably increased breathing resistance (e.g., breathing discomfort in the wearer). Before each use, the outside of the filter should be inspected. If the filter material is physically damaged or soiled, the filter should be changed (in the case of respirators with replaceable filters) or the respirator discarded (in the case of disposable respirators). Always follow the respirator filter manufacturer's service-time-limit recommendations. Standard operating procedures for storing, reusing, and disposing of respirators that have been designated as disposable and for disposing of replaceable filter elements need to be developed.

Respirators must be stored in a way that protects them from damage, contamination, dust, sunlight, extreme temperatures, excessive moisture, and damaging chemicals. They must also be packed or stored in a way that prevents the facepiece and exhalation valve from deformation or other damage. A good method is placing respirators in individual storage bins. Keep in mind that respirator facepieces can become deformed and that the straps can lose their elasticity if a respirator is hung on a peg for a long time. Check for these problems before each use.

Storing a respirator in a plastic sealable bag after use is not considered good practice. The respirator may be damp after use, and sealing prevents drying and encourages microbial growth. If plastic bags are used, respirators must be allowed to dry before they are stored in these bags.

When an employee uses a respirator voluntarily, the employer must implement those elements of the written respiratory protection program necessary to ensure that the employee is medically capable of using a respirator and that the respirator is cleaned, stored, and maintained so that its use does not present a health hazard to the user. Also, employers must provide voluntary respirator users with the information contained in Appendix D of 29 CFR 1910.134.

Employers are not required to include in a written respiratory program those employees whose only use of respirators involves the voluntary use of filtering facepieces (dust masks).

Machine-Guarding Standard

Moving machine parts have the potential to cause severe workplace injuries, such as crushed fingers or hands, amputations, burns, or blindness. Safeguards should thus be used for any machine part, function, or process that presents a danger that may result in an injury. When the operation of a machine or the accidental contact with a machine could injure the operator or others in the vicinity, the hazards must be eliminated or controlled.

Machine-guarding hazards are covered in the specific standards for general industry, marine terminals, longshoring, and the construction and agriculture industries. The machine-guarding standard, 29 CFR 1910.212, provides general guarding requirements for machines and a blueprint for machine safety.

All machines consist of three fundamental regions: the point of operation, the power transmission device, and the operating controls. Even though all machines have these same basic components, safeguarding needs often differ due to varying physical characteristics and operator involvement.

The *point of operation* is where work is performed on the material, such as cutting, shaping, boring, or forming stock. When a machine is in use, the point of operation is also where an employee is potentially exposed to injury and needs to be guarded. The guarding device needs to conform with any appropriate standard or, in the absence of applicable specific standards, be designed and constructed so as to prevent the operator from having any body parts in the danger zone during the operating cycle.

The following is a list of machines that usually require point-of-operation guarding:
- guillotine cutters
- shears
- alligator shears
- power presses
- milling machines
- power saws
- jointers
- portable power tools
- forming rolls and calenders
- press brake
- barrels, containers, and drums—revolving barrels, containers, and drums must be guarded by an enclosure interlocked with the drive mechanism so that the barrel, drum, or container cannot revolve unless the guard enclosure is in place
- exposure of blades—when the periphery of the blades of a fan is less than 7 ft (2 m) above the floor or working level, the blades must be guarded. The guard must not have openings larger than ½ inch (1.25 cm).
- anchoring fixed machinery—a machine designed for a fixed location must be securely anchored to prevent its walking or moving.

The *power transmission apparatus* consists of all components of the mechanical system that transmit energy to the part of the machine performing the work. These components include flywheels, pulleys, belts, connecting rods, couplings, cams, spindles, chains, cranks, and gears.

Operating controls are all parts of a machine that move while the machine is working. These can include reciprocating, rotating, and transverse moving parts as well as feed mechanisms and auxiliary parts of the machine.

Operating controls are also input devices that the operator uses to activate or control the mechanical system. They need to be arranged so that a worker is protected from danger while operating the machinery. Often, in addition to required machine guards, safety devices are configured as part of the operating control process in order to increase the safety of an operator.

One or more methods of machine guarding need to be provided to protect the operator and other employees in the machine area from hazards such as those created by the point of operation, ingoing nip points, rotating parts, flying chips, and sparks. Examples of guarding methods include barrier guards, two-hand-tripping devices, electronic safety devices, and so forth.

Machine Guarding

The purpose of machine guarding is to protect the machine operator and other employees in the work area from potential hazards. Guards should be affixed to the machine where possible and secured elsewhere if attachment to the machine is not possible.

Safeguards must meet minimum general requirements:
- The safeguard must prevent hands, arms, and any other part of a worker's body from making contact with moving parts.
- Workers should not be able to easily remove or tamper with the safeguard.

- The safeguard should ensure that no objects can fall into moving parts.
- If feasible, the machine should be able to be lubricated without removing the safeguards.

A fixed guard is a permanent part of a machine and is usually preferable to all other types of machine guarding because of its relative simplicity. It is not dependent upon moving parts to function. It may be constructed of sheet metal, screen, wire cloth, bars, plastic, or any other material that is substantial enough to withstand whatever impacts it may receive and to endure prolonged use.

The safeguard must create no new hazards and create no interference for the worker. The guard should also not be an accident hazard in itself.

Additional Safety Measures

In some cases, additional measures will have to occur so that an operator is protected. Use of safety devices may enhance the safety provided by a machine guard. For example, they may:

- Stop the machine if a hand or any other part of the body is inadvertently placed into the danger area.
- Restrain or withdraw the operator's hands from the danger area during operation.
- Require the operator to use both hands on machine controls, thus keeping both hands and body out of danger.
- Provide a barrier that is synchronized with the operating cycle of the machine in order to prevent entry to the danger area during the hazardous part of the cycle.

Special hand tools for placing and removing material without the operator needing to place a hand into the danger zone are good examples of safety devices. Such tools should not be used in place of other guarding required by this standard but only to supplement the protection provided by a machine guard.

Some commonly used safety devices include the following types:

- photoelectric (optical)—a presence-sensing device that uses a system of light sources and controls that can interrupt the machine's operating cycle
- radiofrequency (capacitance)—another presence-sensing device that uses a radio beam in the machine control circuit to either stop or prevent the machine from activating if the beam is broken
- electromechanical—uses a probe or contact bar that descends a predetermined distance when the machine cycle is initiated, unless there is an obstruction preventing it from descending its full predetermined distance, at which time the machine will not start
- pullback—uses a series of cables attached to the operator's hands, wrists, and/or arms that automatically ensure withdrawal of the hands from the point of operation when a slide/ram begins to cycle and starts its descent
- restraint (holdback)—uses cables or straps that are attached to the operator's hands and a fixed point and requires adjustment to let the operator's hands travel within a predetermined safe area
- safety trip controls—use a pressure-sensitive body bar, safety tripod, safety tripwire, or similar device to quickly deactivate a machine in an emergency situation
- two-hand control—requires constant, concurrent pressure by the operator to activate the machine so the operator's limbs are kept in a safe location from the danger area
- two-hand trip—requires concurrent application of both of the operator's control buttons to activate the machine cycle
- gate—a movable barrier that protects the operator at the point of operation and before the machine cycle can be started
- awareness barrier—serves as a reminder that a person is approaching a danger area and that these barriers are generally not considered adequate for continual exposure to the danger
- interlocks—removal of this type of guard results in the tripping mechanism engaging and/or power automatically shutting off or disengaging; the moving parts of the machine then stop, and the machine cannot cycle or be started until the guard is back in place.

While these aids do not give complete protection from machine hazards, they may provide the operator with an extra margin of safety.

Eye and face protection must be provided to each employee who will be exposed to eye or face hazards from flying particles.

Servicing or Maintenance

The employer must establish an energy control program consisting of energy control procedures, employee training, and periodic inspections to ensure that before any employee performs any servicing or maintenance on a machine or piece of equipment, the machine or equipment is isolated from the energy source and rendered inoperative. The maintenance and repair crew should always replace the guards before considering the job finished and before the machine is released from lockout. In order to prevent hazards while servicing machines, each machine or piece of equipment should be safeguarded by:

- notifying all appropriate employees (usually machine or equipment operators or users) that the machine or

equipment must be shut down before performing maintenance or servicing
- stopping the machine
- isolating the machine or piece of equipment from its energy source
- locking out or tagging out the energy source
- releasing any stored or residual energy
- verifying that the machine or equipment is isolated from the energy source.

When the servicing or maintenance is completed, specific steps must be taken to return the machine or piece of equipment to service. These steps include:
- inspecting the machine or equipment to ensure that all guards and other safety devices are in place and functional
- checking the area to ensure that energization and start-up of the machine or equipment will not endanger employees
- removing the lockout devices
- reenergizing the machine or equipment
- notifying appropriate employees that the machine or equipment may be returned to service.

Training

A worker needs to know how and why to use a safeguarding system. The training should be specific and detailed. Operator training should involve instruction or hands-on training in the following:
- a description and identification of the hazards associated with particular machines
- the safeguards themselves, how they provide protection, and the hazards for which they are intended
- how to use the safeguards and why
- how and under what circumstances safeguards can be removed and by whom (in most cases, repair or maintenance personnel only)
- when a lockout/tagout program is required
- what to do (e.g., contact the supervisor) when a safeguard is damaged, missing, or unable to provide adequate protection.

This kind of safety training is necessary for new operators and maintenance or setup personnel, when any new or altered safeguards are put into service, and when workers are assigned to a new machine or operation.

Electrical Safety Standards

Electricity has long been recognized as a significant workplace hazard. The Occupational Safety and Health Administration of the U.S. Department of Labor (OSHA) considers electrical safety a high priority. The U.S. Bureau of Labor Statistics indicates that over the decade from 1992 to 2002, deaths due to electrical contact averaged almost 300 per year and lost-time injuries numbered more than 4,000 per year. OSHA's electrical standards are designed to protect employees from electric shock, electrocution, fires, and explosions. Electrical hazards are addressed in specific standards for general industry, shipyard employment, and marine terminals.

OSHA added requirements that govern electrical work in the workplace by revising 29 CFR 1910 Subpart S, Electrical Standard, which went into effect on August 15, 2007.

The final rule of this revision is partly based on Part I of NFPA 70E, Standard for Electrical Safety in the Workplace, 2000 edition, and on NFPA 70, National Electrical Code, 2002 edition. This final rule revises OSHA's prior standard for electrical installations and focuses on safety in the design and installation of electric equipment in the workplace. It applies, as the previous standard did, to employees in general industry and in shipyard employment, longshoring, and marine terminals. It is important for supervisors of employees who work with electricity to become familiar with relevant parts of NFPA 70E and NFPA 70.

Frequently Cited Standards Related to Electricity

In recent years, the following standards have resulted in either frequent or costly citations:
- Electrical Systems Design, General Requirements, General Industry, § 1910.303
- Electrical, Wiring Methods, Components and Equipment, General Industry, § 1910.305
- Electrical, Hazardous (Classified) Locations, § 1910.307.

The following standards, in order, were the most frequently cited by federal OSHA plans in a recent 1-year time span for the Electrical Work Industry Group (SIC code 1731):
- Wiring Methods, Components, and Equipment for General Use, § 1926.405
- General Requirements (Electrical), § 1926.403
- Wiring Design and Protection, § 1926.404
- Duty to Have Fall Protection, § 1926.501
- Aerial Lifts, § 1926.453
- Ladders, § 1926.1053
- General Requirements (Electrical), § 1926.416
- Stairways, § 1926.1052
- General Safety and Health Provisions, § 1926.20
- Safety Training and Education, § 1926.21.

In addition, OSHA has extended the ground-fault protection requirement to temporary receptacles used in construction activities performed in general industry. Ground-fault circuit interrupter (GFCI) protection is required for all receptacle outlets in temporary wiring installations used during maintenance, remodeling, or repair of buildings, structures, or equipment or during other construction activities. A GFCI device protects a worker by opening a circuit if the currents going into and coming out of an electric circuit do not equal each other. This prevents another current pathway—such as flow through the worker's body to ground—from injuring the worker.

To summarize, a supervisor should understand that electrical safety is a complex issue:
- There are new requirements based on a revised standard.
- It is important to ensure that workers receive specialized education and training for working with electrical systems.
- Consensus standards can be used to help select the best method to achieve compliance with OSHA regulations.
- Some OSHA state plans are more restrictive than the federal OSHA requirements and, as such, may have adopted or incorporated consensus standards; but this adoption is on a state-by-state basis and should be evaluated for each employer location.
- In most cases, electrical shock is a recognized hazard that could—or likely will—cause death or serious physical harm to employees. This means that electrical shock hazards not covered by existing regulations can be cited by use of the general-duty clause described earlier.

Powered Industrial Trucks Standard
Requirements and Recommended Practices
OSHA requires that all forklifts be examined at least daily before being placed into service. Forklifts used on a round-the-clock basis must be examined after each shift.

The operator should conduct a pre-start-up visual check with the engine off and then perform an operational check with the engine running. The forklift should not be placed into service if the examinations show that the vehicle may not be safe to operate. If a vehicle is in need of repair, defective, or unsafe in any other way, it should not be driven and should be taken out of service immediately. Any problems should be recorded on the appropriate documents and reported to a supervisor.

Preoperation Inspection
Before a vehicle is started, a preoperation (or pre-start-up) inspection that checks the following should take place:
- fluid levels—oil, water, and hydraulic fluid
- visible leaks, cracks, or any other defect, including in hydraulic hoses and mast chains
  NOTE: Operators should not place their hands inside the mast. Use a stick or other device to check chain tension.
- tire condition , including cuts and gouges, and pressure
- condition of the forks and the top clip retaining pin and heel

- load backrest extension
- finger guards
- safety decals and nameplates
  NOTE: Ensure that all warning decals and plates are in place and legible. Check that the information on the nameplate matches the model and serial numbers and attachments.
- operator compartment for grease and debris
- operation of all safety devices, including the seat belt.

In addition to this general inspection, additional items should be checked depending on the forklift type (electric, internal combustion, or liquid propane):

Electric Forklifts
- cables and connectors for frayed or exposed wires
- battery restraints
- electrolyte levels
- hood latch.
  NOTE: Always use personal protective equipment such as a face shield and goggles, rubber apron, and rubber gloves when checking electrolyte levels.

Internal Combustion Forklifts
- engine oil
- brake reservoir
- engine coolant
- air filter
- belts and hoses
- radiator
- hood latch.

Liquid Propane Forklifts
- properly mounted tank
- pressure relief valve pointing up
- hose and connectors
- tank restraint brackets
- tank for dents and cracks
- tank fitting within profile of truck
- leaks; use a soapy solution.
  NOTE: Always use personal protective equipment such as a face shield, long sleeves, and gauntlet gloves when checking liquid propane tanks and fittings.

Operational Inspection
After completing the preoperation inspection, operators should conduct an operational inspection with the engine running. This inspection includes:
- accelerator linkage
- inch accelerator (if equipped)
- brakes

- steering
- drive control: forward and reverse
- tilt control: forward and back
- hoist and lowering control
- attachment control
- horn
- lights
- back up alarm (if equipped)
- hour meter.

  NOTE: Unusual noises or vibrations should be reported immediately.

The OSHA powered industrial truck standard lists a number of conditions under which a forklift must be removed from service. If the operator notes these conditions while driving, the driver must stop, park the vehicle, and get assistance.

- Any powered industrial truck not in safe operating condition should be removed from service. All repairs should be made by authorized personnel.
- Defects when found must be immediately reported and corrected.
- Any vehicle that emits hazardous sparks or flames from the exhaust system should immediately be removed from service and not returned to service until the cause of the emission of such sparks and flames has been eliminated.
- When the temperature of any part of any truck is found to be in excess of its normal operating temperature, thus creating a hazardous condition, the vehicle should be removed from service and not returned to service until the cause of such overheating has been eliminated.
- No truck should be operated until the leak has been corrected.

Scheduled maintenance is critically important to the safe operation of a vehicle. Never operate a forklift requiring maintenance, and always ensure that repair problems are reported to a responsible party. Never operate a vehicle that requires maintenance or is thought to be unsafe in any way.

- Keep industrial trucks clean and free of lint, excess oil, and grease.
- Use noncombustible agents for cleaning trucks.
  - High-flash-point (at or above 100°F [38°C]) solvents may be used. Do not use low-flash-point (below 100°F [38°C]) solvents.
  - Take recommended precautions regarding toxicity, ventilation, and fire hazards.

Training

The standard requires employers to develop and implement a training program based on the general principles of safe truck operation, the types of vehicle(s) being used in the workplace, the workplace hazards created by the use of the vehicle(s), and the general safety requirements of the OSHA standard. Trained operators must know how to do the job properly and safely as demonstrated in workplace evaluations. Formal (lecture, video, etc.) and practical (demonstration and practical exercises) training must be provided.

Employers must also certify that each operator has received the training and evaluate each operator at least once every 3 years. Prior to operating a truck in the workplace, the employer must evaluate each operator's performance and determine the operator to be competent to operate a powered industrial truck safely.

Refresher training is needed whenever an operator demonstrates a deficiency in the safe operation of the truck. Training should consist of a combination of formal instruction (e.g., lecture, discussion, interactive computer learning, videotape, and written material), practical training (demonstrations performed by the trainer and practical exercises performed by the trainee), and evaluation of the operator's performance in the workplace.

Training Program Content

Powered industrial truck operators should receive initial training in the following topics, except for topics that the employer can demonstrate are not applicable to safe operation of the truck in the workplace.

Truck-Related Topics
- operating instructions, warnings, and precautions for the types of trucks the operator will be authorized to operate
- differences between the truck and an automobile
- truck controls and instrumentation: where they are located, what they do, and how they work
- engine or motor operation
- steering and maneuvering
- visibility (including restrictions due to loading)
- fork and attachment adaptation, operation, and use limitations
- vehicle capacity
- vehicle stability
- any vehicle inspection and maintenance that the operator will be required to perform
- refueling and/or charging and recharging of batteries
- operating limitations.

Workplace-Related Topics
- surface conditions where the vehicle will be operated
- compositions of loads to be carried and load stability

- load manipulation, stacking, and unstacking
- pedestrian traffic in areas where the vehicle will be operated
- locations of narrow aisles and other obstructed places where the vehicle will be operated
- hazardous (classified) locations where the vehicle will be operated
- ramps and other sloped surfaces that could affect the vehicle's stability
- enclosed environments and other areas where insufficient ventilation or poor vehicle maintenance could cause a buildup of carbon monoxide or diesel exhaust
- other unusual or potentially hazardous environmental conditions in the workplace that could affect safe operation.

If an operator was previously trained in one of these topics and if the training is appropriate to the truck and working conditions encountered, additional training on that topic is not required as long as the operator has been evaluated and found competent to operate the truck safely.

Trainees may operate a powered industrial truck only:
- under the direct supervision of persons who have the knowledge, training, and experience to train operators and evaluate their competence
- where such operation does not endanger the trainee or other employees.

Ladder Standards

Never use a ladder for any purpose other than the one for which it was designed. A competent person should inspect ladders for visible defects on a periodic basis and after any occurrence that could affect their safe use. Portable or fixed ladders with structural defects—such as, but not limited to, broken or missing rungs, cleats, or steps; broken or split rails; corroded components; or other defective components—should either be immediately marked in a manner that readily identifies them as defective or be tagged with "Do Not Use" or similar language, and they should be withdrawn from service until repaired.

Other requirements for ladder use include the following:
- Position portable ladders so that the side rails extend at least 3 ft (0.9 m) above the landing.
- Secure side rails at the top to a rigid support and use a grabbing device when a 3-ft (0.9-m) extension is not possible.
- Make sure that the weight on the ladder will not cause the ladder to slip off its support.
- Before each use, inspect ladders for cracked or broken parts such as rungs, steps, side rails, feet, and locking components.
- Do not place more weight on the ladder than the ladder is designed to support.
- Ladders should not be moved, shifted, or extended while occupied.
- Use only ladders that comply with the OSHA design standards listed in 29 CFR 1926.1053(a)(1).

Ladder Design Parameters

Loads
Self-supporting (foldout) and non-self-supporting (leaning) portable ladders must be able to support at least four times the maximum intended load, except for extra-heavy-duty metal or plastic ladders, which must be able to support 3.3 times the maximum intended load.

Angle
Non-self-supporting ladders, which must lean against a wall or other support, should be positioned at such an angle that the horizontal distance from the top support to the foot of the ladder is about one-fourth of the working length of the ladder.

In the case of job-made wooden ladders, that angle should equal about one-eighth of the working length. This minimizes the strain of the load on ladder joints, which may not be as strong as the joints on commercially manufactured ladders.

Rungs
Ladder rungs, cleats, or steps must be parallel, level, and uniformly spaced when a ladder is in position for use. Rungs must be spaced between 10 and 14 in. (25 and 35 cm) apart.

For extension trestle ladders, the spacing must be 8–18 in. (20–45 cm) for the base and 6–12 in. (15–30 cm) on the extension section.

Rungs must be shaped so that an employee's foot cannot slide off and must be skid-resistant.

To Minimize Slipping or Falling
- Ladders are to be kept free of oil, grease, wet paint, and other slipping hazards.
- Wood ladders should not be coated with any opaque covering except identification or warning labels on one face only of a side rail.
- Foldout or stepladders must have a metal spreader or locking device to hold the front and back sections in an open position when the ladder is in use.
- When two or more ladders are used to reach a work area, they must be offset with a landing or platform between them.
- The area around the top and bottom of a ladder must be kept clear.

- Ladders should not be tied or fastened together to provide longer sections, unless they the ladders specifically designed for such use.
- The top or top step of a stepladder should not be used as a step.

The regulatory requirements for ladder use are relatively straightforward, but slips and falls from ladders continue to be a major source of injuries in the workplace.

Scaffolding Standard

The scaffolding standard has the following key provisions:
- Designate a "competent person" and his or her responsibilities under the scaffolding standard:
  - "one who is capable of identifying existing and predictable hazards in the surroundings or working conditions, which are unsanitary, hazardous or dangerous to employees, and who has authorization to take prompt corrective measures to eliminate them."
- Designate a "qualified person" and his or her responsibilities under the scaffolding standard:
  - "one who by possession of a recognized degree, certificate, or professional standing, or who by extensive knowledge, training, and experience has successfully demonstrated his/her ability to solve or resolve problems related to the subject matter, the work, or the project."
- Fall protection for employees is required at a 10-ft (3.1-m) height above a lower level.
- Guardrail height:
  - The height of the toprail for scaffolds manufactured and placed into service before January 1, 2000, can be between 36 in. (0.9 m) and 45 in. (1.2 m).
  - The height of the toprail for scaffolds manufactured and placed into service after January 1, 2000, must be between 38 in. (0.97 m) and 45 in. (1.2 m). [When the crosspoint of crossbracing is used as a toprail, it must be between 38 in. (0.97 m) and 48 in. (1.3 m) above the work platform.]
  - Midrails must be installed approximately halfway between the toprail and the platform surface.
  - When the crosspoint of crossbracing is used as a midrail, it must be between 20 in. (0.5 m) and 30 in. (0.8 m) above the work platform.
- Erecting and dismantling—When erecting and dismantling supported scaffolds, a competent person must determine the feasibility of providing a safe means of access and fall protection for these operations.
- Training—Employers must train each employee who will work on a scaffold in the procedures to control or minimize the hazards.
- Inspections—Before each workshift and after any occurrence that could affect a scaffold's structural integrity, a competent person must inspect the scaffold and scaffold components for visible defects.
- Overhand bricklaying—A guardrail or personal fall-arrest system on all sides except the side where the work will be done must protect employees doing overhand bricklaying from supported scaffolds. The standards for aerial lifts have been relocated from § 1926.556 to § 1926.453.

Competent Person

In general, a competent person performs the following duties:
- Selects and directs employees who erect, dismantle, move, or alter scaffolds.
- Determines whether it is safe for employees to work on or from a scaffold during storms or high winds and ensures that a personal fall-arrest system or windscreens protect these employees. (NOTE: Windscreens should not be used unless the scaffold is secured against the anticipated wind forces.)

Qualified Person

When necessary, a qualified person should design and load scaffolds in accordance with §1926.451, §1926.452, §1926.454, and conduct training for employees working on the scaffolds to recognize the associated hazards and understand procedures to control or minimize those hazards.

Engineering Requirements

The standard requires a registered professional engineer to perform the following duties when required by §1926.451 and §1926.452.

Suspension Scaffolds

- Design the direct connections of masons' multipoint adjustable suspension scaffolds.

Design

- Design scaffolds that can be moved while employees are on them
- Design pole scaffolds over 60 ft (18.3 m) in height
- Design tube and coupler scaffolds over 125 ft (38 m) in height
- Design fabricated frame scaffolds over 125 ft (38 m) in height above their base plates
- Design brackets on fabricated frame scaffolds used to support cantilevered loads in addition to workers
- Design outrigger scaffolds and scaffold components.

Each employee on a scaffold more than 10 ft (3.1 m) above a lower level should be provided with fall protection.

In addition, a competent person must determine the feasibility and safety of providing fall protection for employees erecting or dismantling supported scaffolds.

Fall protection includes guardrail systems and personal fall-arrest systems. Guardrails and toeboards should be installed on all open sides and ends of platforms more than 10 ft (3.1 m) above the ground or floor, except needle beam scaffolds. Scaffolds 4 ft (1.2 m) to 10 ft (3.1 m) in height, having a minimum horizontal dimension in either direction of less than 45 in. (112.5 cm), should have standard guardrails installed on all open sides and ends of the platform.

Guardrails should be 2 × 4 in. (5 × 10 cm), or the equivalent, approximately 42 in. (105 cm) high, and with a midrail when required. Supports should be at intervals not to exceed 8 ft (2.4 m). Toeboards should be a minimum of 4 in. (10 cm) in height.

A personal fall-arrest system is a system used to catch an employee falling from a working level. Personal fall-arrest systems include harnesses; components of the harness/belt, such as D-rings; and snap hooks, lifelines, and anchorage points.

Only a few of the requirements of 29 CFR 1926.451 are provided in this overview of the scaffolding standard, and there are many other requirements that a supervisor should be familiar with when working with scaffolding is part of a job's requirements.

At the minimum, a competent person should be involved in all phases of a scaffold job from design, erection, and operation to dismantling.

Excavations Standard

Many construction projects involve excavations. *Excavation* means any manmade cut, cavity, trench, or depression in an earth surface that is formed by earth removal.

Any site with excavation work occurring as part of the project should ensure that a competent person inspects excavations before construction begins, daily before each shift, as needed throughout the shift, and following rainstorms or other hazard-increasing events.

As used here, a competent person is an individual who has training in soil analysis and use of protective systems; is knowledgeable about OSHA requirements; is capable of identifying existing and predictable hazards in the surroundings and working conditions that are unsanitary or dangerous to employees; and is authorized to take prompt corrective measures to eliminate such hazards where identified. Prior to working at an excavation, the following issues must be addressed by a competent person:
- Evaluating soil and selecting appropriate protective systems
  - *Protective system* involves a method of protecting employees from cave-ins, from material that could fall or roll from an excavation face or into an excavation, or from the collapse of adjacent structures.
  - Protective systems include support systems, sloping and benching systems, shield systems, and other systems that provide the necessary protection.
- Constructing protective systems in accordance with the standard's requirements
- Contacting utilities (gas, electric) to learn the locations of underground lines, planning for traffic control if necessary, and determining the lines' proximity to structures, which could affect the choice of protective system
- Testing for low oxygen levels, hazardous fumes, and toxic gases, especially when gasoline engine–powered equipment is running or the dirt has been contaminated by leaking lines or storage tanks. Ensure adequate ventilation or respiratory protection if necessary.
- Providing safe access into and out of the excavation
- Providing appropriate protections where water accumulation is a problem
- Keeping excavations open the minimum amount of time needed to complete the operations.

Trenches and excavations should be inspected daily for evidence of possible cave-ins, hazardous atmospheres, failure of protective systems, and other unsafe conditions.

Excavated materials (spoils) can be hazardous when they are set too close to the edge of a trench/excavation. The weight of the spoils can cause a cave-in, or spoils and equipment can roll back on top of workers, resulting in serious injuries or death. Spoils and equipment should be set at least 2 ft (0.6 m) back from the excavation. Retaining devices, such as a trench box, that will extend above the top of the trench should be used to prevent equipment and spoils from falling back into the excavation. Where the site does not permit a 2-ft (0.6-m) setback, spoils may need to be temporarily hauled to another location.

Stairways, ladders, ramps, or other safe means of egress in all trenches that are 4 ft (1.2 m) deep or more should be provided. These means of egress should be positioned within 25 lateral feet (7.5 m) of workers. Structural ramps that are used solely for access or egress from excavations must be designed by a competent person. When two or more components form a ramp or runway, they must be connected to prevent displacement and be of uniform thickness. Cleats or other means of connecting runway components must be attached in a way that will not cause tripping. Structural ramps used in place of steps must have a nonslip surface. Use earthen ramps as a means of egress only when a worker can ascend them in an upright position and only when they have been evaluated by a competent person.

Walking/Working Surfaces Standard

Slips, trips, and falls constitute the majority of general industry accidents. They cause 15% of all accidental deaths and are second only to motor vehicles as a cause of fatalities. The OSHA standards for walking/working surfaces apply to all permanent places of employment, except where only domestic, mining, or agricultural work is performed.

Walking/working surfaces are addressed in specific standards for general industry, shipyard employment, marine terminals, longshoring, and the construction industry. For general industry, the following sections apply:

Walking/Working Surfaces, § 1910 Subpart D
Definitions, § 1910.21
General requirements, § 1910.22
Guarding floor and wall openings and holes, § 1910.23
Fixed industrial stairs, § 1910.24
Portable wood ladders, § 1910.25
Portable metal ladders, § 1910.26
Fixed ladders, § 1910.27
Safety requirements for scaffolding, § 1910.28
Manually propelled mobile ladder stands and scaffolds (towers), § 1910.29

Housekeeping

Housekeeping is a very important aspect of OSHA's Walking/Working Surfaces standard. OSHA requires that all places of employment, passageways, storerooms, and service rooms be kept clean and orderly and in a sanitary condition and, as far as possible, dry. Where wet processes are used, drainage should be maintained, and false floors, platforms, mats, or other dry places should be provided when practicable. To facilitate cleaning, every floor, working place, and passageway should be kept free of protruding nails, splinters, holes, and loose boards.

Aisles and Passageways

Where mechanical handling equipment is used, sufficient safe clearances should be allowed in aisles, at loading docks, through doorways, and wherever turns or passage must be made. Aisles and passageways should be kept clear and in good repair, with no obstruction across or in aisles that could create a hazard. Permanent aisles and passageways should be appropriately marked.

Covers and Guardrails

Covers and/or guardrails should be provided to protect personnel from the hazards of open pits, tanks, vats, ditches, holes, and so forth.

Floor Loading Protection

In every building, other structure, or part thereof that is used for mercantile, business, industrial, or storage purposes, the loads approved by the building official should be marked on plates of approved design, which should be supplied and securely affixed by the owner of the building, or by his or her duly authorized agent, in a conspicuous place in each location where they are applicable. Such plates should not be removed or defaced, but if they are lost, removed, or defaced, they should be replaced by the owner or his or her agent.

It is unlawful to place, cause, or permit to be placed on any floor or roof of a building or other structure a load greater than that for which such floor or roof is approved by the building official.

Protecting Openings

OSHA's 29 CFR 1910.23 provides extensive guidelines for many situations in which there is an opening that could result in a fall if the opening is left unprotected. Included in OSHA's extensive list are stairway floor openings, ladderway floor openings or platforms, hatchway and chute floor openings, skylight floor openings and holes, and pits and trapdoor floor openings; infrequently used wall openings where there is a drop of more than 4 ft (1.2 m); manhole floor openings, every temporary floor opening, every floor hole into which persons can accidentally walk; places where doors or gates open directly onto a stairway; and those situations in which there is a hazard of materials falling through a wall hole. It would be a good idea for each supervisor to be familiar with both the unsafe conditions described by OSHA and the remedies for addressing those conditions.

Process Safety Management of Highly Hazardous Chemicals Standard

OSHA Fact Sheet 93-45 does a good job of explaining the general requirements for complying with the Process Safety Management (PSM) of Highly Hazardous Chemicals (HHCs) standard, 29 CFR 1910.119. This standard is intended to prevent or minimize a catastrophic release of toxic, reactive, flammable, or explosive HHCs from a process. A *process* is any activity or combination of activities that include any use, storage, manufacturing, handling, or the onsite movement of HHCs. A process includes any group of vessels that are interconnected and separate vessels that are located such that an HHC could be potentially released.

Application

The standard applies to any process that contains a threshold quantity or greater amount of a toxic or reactive HHC as specified in Appendix A of 29 CFR 1910.119. Also, it applies to 10,000 lb or (4,500 kg) greater amounts of flammable liquids and gases and to the process activity of manufacturing explosives and pyrotechnics.

Exceptions

The standard does not apply to retail facilities; normally unoccupied, remote facilities; and oil- or gas-well-drilling or -servicing activities. Hydrocarbon fuels used solely for workplace consumption as a fuel are not covered when such fuels are not part of a process containing another HHC covered by the standard. Atmospheric tank storage and associated transfer of flammable liquids that are kept below their normal boiling points without the benefit of chilling or refrigeration are not covered by the PSM standard unless the atmospheric tank is connected to a process or is sited in close proximity to a covered process such that an incident in a covered process could affect the atmospheric tank.

Process Safety Information

The standard requires compilation of written process safety information (PSI), including hazard information on HHCs, technology information, and equipment information on covered processes.

Employee Involvement

The standard requires developing a written plan of action regarding employee participation, consulting with employees and their representatives on conducting and developing process hazard analyses and on developing the other elements of process safety management required by the rule, and providing to employees and their representatives access to process hazard analyses and to all other information required to be developed under the rule. Employees are both worksite and contract employees.

Process Hazard Analysis

The standard specifies that process hazard analyses (PHAs) must be conducted as soon as possible for each covered process using compiled process safety information in an order based on a set of required considerations. At least 25% of initial process hazard analyses must have been completed by May 26, 1994; 50% by May 26, 1995; 75% by May 26, 1996; and 100% by May 26, 1997. Process hazard analyses must be updated and revalidated at least every 5 years and must be retained for the life of the process.

Operating Procedures

The standard requires that all operating procedures be in writing. The procedures must provide clear instructions for safely conducting activities involving covered processes and be consistent with PSI; include steps for each operating phase, operating limits, safety and health considerations, and safety systems and their functions; be readily accessible to employees who work on or maintain a covered process; be reviewed as often as necessary to ensure that the procedures reflect current operating practice; and implement safe work practices for special circumstances such as lockout/tagout and confined-space entry.

Training

Employees operating a covered process must be trained in the process and in the operating procedures addressed previously. This training must emphasize specific safety and health hazards, emergency operations, and safe work practices. Initial training must occur before assignment, or employers must certify that employees involved in the process as of May 26, 1992, have the required knowledge, skills, and abilities. Documented refresher training is required at least every 3 years.

Contractors

The standard identifies the responsibilities of worksite employers and contract employers with respect to contract employees involved in maintenance, repair, turnaround, major renovation, or specialty work on or near covered processes. Contract employers are required to train their employees to safely perform their jobs, document that employees received and understood training, ensure that contract employees know about potential process hazards and the worksite employer's emergency-action plan, ensure that employees follow the safety rules of the facility, and advise the worksite employer of the hazards that the contract work itself presents or the hazards identified by contract employees.

Pre-Start-Up Safety Review

The standard mandates a safety review of new facilities and significantly modified work-sites to confirm that the construction and equipment of a process are in accordance with design specifications; to ensure that adequate safety, operating, maintenance, and emergency procedures are in place; and to ensure that process operator training has been completed. Also, for new facilities, the PHA must be performed and recommendations resolved and implemented before start-up. Modified facilities must meet the management-of-change requirement.

Mechanical Integrity

The standard requires the onsite employer to establish and implement written procedures for the ongoing integrity of process equipment, particularly those components that contain and control a covered process.

Hot Work
Hot-work permits must be issued for hot-work operations conducted on or near a covered process.

Management of Change
The worksite employer must establish and implement written procedures to manage changes except "replacements in kind" to facilities that affect a covered process. The standard requires that prior to start-up, the worksite employer and contract employers inform and train their affected employees on the changes. Process safety information and operating procedures must be updated as necessary.

Incident Investigation
The standard requires employers to investigate as soon as possible (but no later than 48 hours after) incidents that did result or could reasonably have resulted in catastrophic releases of covered chemicals. The standard calls for an investigation team—including at least one person knowledgeable in the process involved; a contract employee, when the incident involved contract work; and others with the appropriate knowledge and experience—must investigate and analyze the incident and develop a written report on the incident. Reports must be retained for 5 years.

Emergency Planning and Response
The standard requires employers to develop and implement an emergency-action plan. The emergency-action plan must include procedures for handling small releases.

Compliance Audits
The standard calls for employers to certify that they have evaluated compliance with process safety requirements at least every 3 years. Prompt response to audit findings and documentation that deficiencies were corrected are required. Employers must retain the two most recent audit reports.

Trade Secrets
The standard sets requirements similar to the trade secret provisions of the 1910.1200 Hazard Communication standard requiring information required by the PSM standard to be available to employees (and employee representatives). Employers may enter into a confidentiality agreement with employees to prevent disclosure of trade secrets.

Although information on the OSHA Fact Sheet provides a good overview of the regulatory requirements for complying with the Process Safety Management of Highly Hazardous Chemicals standard, compliance can be ascertained only by carefully addressing each of the provisions in the standard.

Confined-Space Entry Standard
Because the Confined Space standard—29 CFR 1910.146—applies to all of general industry, a performance-oriented standard was developed rather than a specification standard. Currently, construction, marine terminal, shipyard employment, and agriculture are not subject to the OSHA General Industry Permit-Required Confined Spaces regulation. However, employers in those industries should be aware that their workers are covered by the standard when they do work that falls under the general industry category. For example, maintenance, repair, and refurbishing work are covered under general industry rules even though done by "construction" contractors.

It is an employer's obligation to have someone evaluate the workplace to determine whether any spaces are permit-required confined spaces. The health and safety professional must first determine whether a space is a confined space. If it is a confined space, then he or she must determine whether it is a permit-required confined space. If it is a permit-required confined space, then it must be determined whether full permit entry rules apply or less restrictive, alternative entry rules apply.

A *confined space* is characterized by restricted means of entry/exit, size sufficient to contain a worker, and not specifically designed for worker occupancy. Many workplaces contain areas that are considered "confined spaces" because, although they were not necessarily designed to accommodate people, they are large enough for workers to enter and perform certain jobs. Confined spaces include but are not limited to tanks, vessels, silos, storage bins, hoppers, vaults, pits, manholes, tunnels, equipment housings, ductwork, pipelines, and so forth.

OSHA uses the term *permit-required confined space* (permit space) to refer to a confined space that has one or more of the following characteristics: contains, or has the potential to contain, a hazardous atmosphere; contains a material that has the potential to engulf an entrant; has walls that converge inward or floors that slope downward and taper into a small area that could trap or asphyxiate an entrant; or contains any other recognized safety or health hazard, such as unguarded machinery, exposed live wires, or heat stress.

In general, the Permit-Required Confined Spaces standard requires the employer to evaluate the workplace and determine whether any spaces are permit-required confined spaces. If permit spaces are present and if workers are ever authorized to enter such spaces, a comprehensive permit space program must be developed, which is an overall plan/

policy for protecting employees from permit space hazards and for regulating employee entry into permit spaces.

Permit spaces must be identified by signs, and entry must be controlled and limited to authorized persons. An important element of the requirements is that entry be regulated by a written entry permit system and that entry permits be recorded and issued for each entry into a permit space. The standard specifies strict procedures for evaluating and atmospheric testing of a space before and during an entry by workers.

The standard requires that entry be monitored by an attendant outside the space and that provisions be made for rescue in the event of an emergency. The standard specifies training requirements and specific duties for authorized entrants, attendants, and supervisors. Rescue service provisions are required, and, where feasible, rescue must be facilitated by a nonentry retrieval system, such as a harness and cable attached to a mechanical hoist. Many would-be rescuers have become trapped by the same hazard requiring the rescue in the first place.

The OSHA Permit-Required Confined Spaces Standard provides for alternative entry procedures (which are less stringent than full permit procedures) in cases where the only hazard in a space is atmospheric and the hazard can be controlled by forced air. The alternative procedure is allowed only when specific requirements for substantiation and notification are met.

Many hazards related to uncontrolled releases of energy may be located in confined spaces, and any program should ensure that measures are available to assess potential hazards from the following:
- high-pressure fluids and gases
- mechanical energy
- oxygen deficiency
- toxics
- flammable materials
- engulfment
- wet/slick surfaces
- electrical energy.

OSHA requires the following elements to be covered in a permit-required confined-space entry program:
- identification and evaluation of confined spaces and permit-required confined spaces
- identification of hazards in those spaces
- procedures for controlling the hazards
- written program
- permit system
- employee training
- record keeping
- rescue provisions.

The next step is to evaluate potential hazards, which include (but are not limited to) the following:
- oxygen deficiency, toxics, flammable/explosive materials
- engulfment, mechanical and electrical energy sources, release of materials, noise, wet/slick surfaces, falling objects, hot and cold temperatures.

Oxygen deficiency, which refers to oxygen levels below 19.5%, is a major potential hazard in confined spaces and is described in Table 5–B.

TABLE 5–B. Oxygen-Deficiency Levels

Oxygen Level (%)	Effects
16 to 12%	Deep breathing, accelerated heartbeat, impaired attention, impaired thinking, impaired coordination
14 to 10%	Very faulty judgment, very poor coordination, rapid fatigue from exertion, which may cause permanent heart damage, intermittent breathing
≤10%	Nausea, vomiting, inability to perform vigorous movement or loss of all movement, unconsciousness followed by death
<6%	Spasmodic breathing, convulsive movements, death in minutes

Confined-space entry requires that specific procedures be developed that cover both pre-entry and entry activities. Pre-entry consists of evaluating potential hazards, organizing the entry, and completing the written permit.

An entry process/system needs to include the following information and actions:
- work to be performed
- identity of individuals performing the work
- time period
- identification and evaluation of potential hazards in the space
- completion of written permit
- equipment procurement
- permit review with team members
- rescue procedure(s)
- communication processes
- ways to contact the rescue team if they are not near the confined space
- completion of LOTO procedures
- cleaning/purging atmosphere of any hazardous constituents
- ventilation system in place based on identified requirements
- documentation of atmospheric conditions
- entry to perform work
- exit with all entrants
- cancellation of permit by appropriate party
- entry equipment maintenance.

Procedures for multiple operations in the same space also need to be developed to cover incompatible jobs, oversight for the operation, notification procedures, scheduling, and an entry permit system. As needed, air monitoring devices, portable ventilation, appropriate personal protective equipment (PPE), communication devices, and retrieval equipment should be obtained prior to entry.

Additionally, other equipment might be specified when indicated, such as:
- explosion-proof lighting
- GFCI electrical protection
- nonsparking tools
- barriers to shield the opening
- entry harnesses
- nonentry retrieval.

Specific duties must be assigned to members of the entry team, including the entry supervisor, the authorized entrant, and the attendant. The *authorized entrant* must be trained to understand hazards, symptoms of possible health problems and their consequences, proper use of the equipment, the importance of maintaining ongoing communication with the attendant, and the need to immediately leave the space when directed to do so by a team member.

The *entry supervisor* must follow the procedures to end an entry, including the following: (1) notifying all team members that the entry has ended, (2) following the appropriate method(s) for notifying team members, (3) ensuring that all equipment and materials have been removed from the space, and (4) securing the space to prevent unauthorized entry.

A review process should be carried out after each entry and periodically for appropriate confined spaces to ensure that procedures are appropriate and circumstances that would require entry procedures to change have not occurred.

Training for all confined-space entry personnel must be completed and documented. Retrain employees when there are changes in the program, workplace, or equipment; employee deviations during an entry; and inadequate knowledge observed.

No program, permit system, entrant, or rescue procedures are required when:
- the only hazard is a hazardous atmosphere *and*
- ventilation will be maintained to provide a safe atmosphere *and*
- ventilation will be documented through monitoring and inspections *and*
- entry will take place only to verify the atmosphere.

Reclassification as a "nonpermit" space is allowed under the following conditions:
- no actual or potential atmospheric hazards exist
- entry space to eliminate hazards follows §§ 1910.146(d)–(k) of standard
- reclassification process is documented.

Contractors

OSHA's Construction Safety and Health Regulations Part 1926 does not currently contain a permit-required confined-space regulation. Subpart C of §1926.21, Safety Training and Education, specifies training for personnel who are required to enter confined spaces and defines a "confined or enclosed space."

A confined or enclosed space is any space having a limited means of egress that is subject to the accumulation of toxic or flammable contaminants or has an oxygen-deficient atmosphere. Confined or enclosed spaces include, but are not limited to, storage tanks, process vessels, bins, boilers, ventilation or exhaust ducts, sewers, underground utility vaults, tunnels, pipelines, and open-top spaces more than 4 ft (1.2 m) in depth such as pits, tubs, vaults, and vessels.

The host employer must provide contractors hired to enter confined spaces with the following information regarding the confined space:
- the nature of the hazards involved
- the necessary precautions that should be taken
- the required protective and emergency equipment.

Dual-entry missions should be coordinated between contractors and the employer. Contractors should be debriefed after entry.

Before allowing any employees to enter a confined space, the contractor should make sure to have the following information from the host company: "dual-entry" coordination, contractor program elements, and a list of hazards confronted or created during entry. This information should be shared with the employees through a briefing and prior to entry.

OSHA's construction regulations also contain requirements dealing with confined-space hazards in underground construction (Subpart S), underground electric transmission and distribution work (§ 1926.956), excavations (Subpart P), and welding and cutting (Subpart J).

Further guidance may be obtained from the American National Standards Institute's Safety Requirements for Confined Spaces, ANSI Z117.1–1989. This standard provides minimum safety requirements that should be followed when entering, exiting, and working in confined spaces at normal atmospheric pressure. This standard does not pertain to underground mining, tunneling, caisson work, or other, similar tasks that have established national consensus standards.

OSHA LOG 300: RECORDING AND REPORTING OCCUPATIONAL INJURIES AND ILLNESS STANDARD

Reporting work-related injuries, illnesses, or deaths does not mean someone was at fault or that an OSHA regulation was violated. When an employee dies or when three or more employees are admitted to a hospital because of a work-related incident, an oral report to OSHA must occur within 8 hours. This can be done in person or over the phone (1-800-321-OSHA/1-800-321-6742).

Illnesses or injuries related to the workplace must be recorded on the OSHA Log 300. A current OSHA Log 300 can be downloaded from www.osha.gov. For an injury or illness to be recorded, it should be new, work related, and serious. New injuries and illnesses are not the same type and do not involve the same part of the body of an injury or illness that was recorded previously, unless the injuries or illnesses had completely recovered.

OSHA Record Keeping

The purpose of this section is to provide the supervisor with information about the uses and benefits of keeping records of occupational injuries and illnesses as well as how to comply with OSHA injury and illness record-keeping requirements. Types of records and reports, record-keeping requirements, and determining recordability will be reviewed here and in the following sections. Although the supervisor may not maintain the OSHA record-keeping log, it is important to fully understand the requirements to be able to assist the individual who makes the final determination for record keeping.

The OSH Act of 1970 requires covered employers to prepare and maintain occupational injury and illness records. The OSH Act and record-keeping regulations in 29 CFR 1904 provide specific recording and reporting requirements.

There are many specific OSHA standards and regulations that require record maintenance and retention of medical surveillance, inspections, exposure monitoring, and other activities not covered here. Companies are responsible for keeping informed of current OSHA regulations at all times.

The purpose of this rule (Part 1904) is to require employers to record and report work-related fatalities, injuries, and illnesses.

NOTE: Recording or reporting a work-related injury, illness, or fatality does not indicate that the employer or employee was at fault, that an OSHA rule was violated, or that the employee is eligible for workers' compensation or other benefits.

All employers covered by the Occupational Safety and Health Act (OSH Act) are covered by these Part 1904 regulations. States that operate their own job safety and health programs have adopted comparable record-keeping regulations, which have been in effect since January 1, 2002. States must have the same requirements as OSHA's regarding which injuries and illnesses are recordable and how they are recorded. Other provisions may be different from the federal requirements as long as they are as stringent. Employers in some state plan states may be subject to more stringent reporting requirements (e.g., California requires that every case of "serious injury or illness" be reported).

However, most employers do not have to keep OSHA injury and illness records unless OSHA or the Bureau of Labor Statistics (BLS) informs them in writing that they must do so. For example, employers with 10 or fewer employees and business establishments in certain industry classifications are partially exempt from keeping OSHA injury and illness records.

NOTE: All employers covered by the OSH Act must report any workplace incident resulting in a fatality or the hospitalization of three or more employees within 8 hours to the nearest office of the OSHA Area Director by telephone or in person (no voice mail messages allowed). If no one answers at the area office, call the OSHA toll-free telephone number: 1-800-321-OSHA (6742).

This rule promotes improved employee awareness and involvement in the record-keeping process, thus providing workers and their representatives with access to information on the record-keeping forms and increasing their awareness of potential workplace hazards. Employee privacy concerns have been addressed—the former rule had no privacy protections included in the log used to record work-related injuries and illnesses.

The rule uses a question-and-answer format written in plain language and uses checklists and flowcharts to provide easier interpretations of record-keeping requirements. Employers are afforded more flexibility in using computers and telecommunications technology to meet record-keeping requirements. The term *lost workdays* was eliminated, and the record keeping focuses on days away or days restricted or transferred. Calendar days instead of workdays are used when tabulating.

The company executive must sign and certify that the 300 Form is accurate.

Exemptions for Recording Injuries and Illnesses

OSHA has identified two categories of *exemptions* that may affect a company's obligation to record injuries and illnesses sustained by its employees: an exemption for size and an exemption for low-hazard industries. However, remember that *all employers* must report to OSHA any workplace incident that results in a fatality or the hospitalization of three or more employees.

If a company had 10 or fewer employees at all times during the last calendar year, the company does not need to keep OSHA injury and illness records *unless OSHA or the BLS informs the company in writing that it must do so.*

- The partial exemption for size is based on the number of employees in the entire company.
- To determine whether a company is exempt because of size, the company's peak employment during the last calendar year must be determined. If the company had no more than 10 employees at any time in the last calendar year, it qualifies for the partial exemption for size.

Exemption for Low-Hazard Industries

If a business establishment is classified as a specific, low-hazard retail, service, finance, insurance, or real estate industry, it does not need to keep OSHA injury and illness records *unless the government asks the company to keep the records.*

- All employers, however, must report to OSHA any workplace incident that results in a fatality or the hospitalization of three or more employees.
- The partial industry classification exemption applies to individual business establishments. If a company has several business establishments engaged in different classes of business activities, some of the company's establishments may be required to keep records whereas others may be exempt.

OSHA Record-Keeping Forms

The injury and illness records required by OSHA's record-keeping rule are an important source of information for OSHA, employers, and employees. OSHA requires completion of the following forms:

- OSHA 300 Log: Log of Work-Related Injuries and Illnesses
- OSHA 300: A Summary Form: Summary of Work-Related Injuries and Illnesses
- OSHA 301: Incident Report: Injury and Illness Incident Report

NOTE: There are penalties for failure to comply with OSHA's record-keeping obligations.

The OSHA 300 Log of Work-Related Injuries and Illnesses (Form 300) is used to classify work-related injuries and illnesses and to note the extent and severity of each case.

- When an incident occurs, use the log to record exact details about what happened and how it happened.

The OSHA 300 Summary Form of Work-Related Injuries and Illnesses is a separate form (Form 300A) that shows the totals for the year in each category.

- Each covered employer must complete the summary at the end of the year and post it for 3 months (February 1 to April 30).
- The employer must review the records at year-end for accuracy before summarizing them.

Additional certification of accuracy by a company executive and additional data on the average employment and hours worked at the establishment are also required.

The OSHA 301 Injury and Illness Incident Report (OSHA Form 301) or equivalent form must be completed for each injury or illness recorded on Form 300.

- This form or an equivalent must be filled out within 7 calendar days after information is received that a recordable work-related injury or illness has occurred. Some state workers' compensation, insurance, or other reports may be acceptable substitutes.
- To be considered an equivalent form, any substitute form must contain all the information asked for on Form 301.
- According to Public Law 91-596 and 29 CFR 1904, OSHA's record-keeping rule, this form must be kept on file for 5 years following the year to which it pertains.
- If the supervisor is unsure whether a case is recordable, he or she should call the local OSHA office for help. When in doubt, record it. It can always be crossed out later.

NOTE: Federal OSHA requires other types of record keeping for many other reasons. For example, some OSHA regulations require training records documenting that training was conducted and understood. It is beyond the scope of this chapter to discuss those various record-keeping requirements.

Each employer that is required to keep records of fatalities, injuries, and illnesses must record each fatality, injury, and illness that:

- is work related
- is a new case
- meets one or more of the general recording criteria (see Figure 5–4).

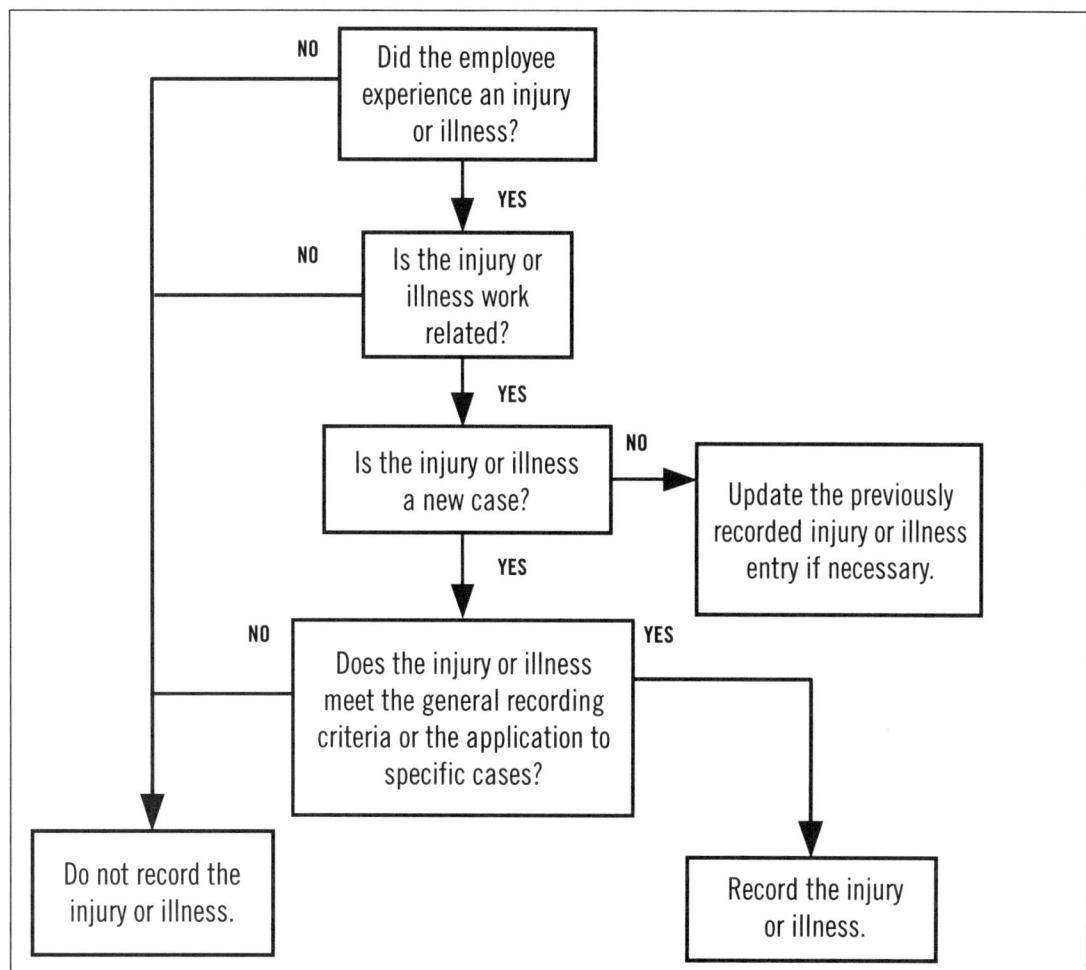

Figure 5-4.

Determining Work-Relatedness and OSHA Recordability

There are a number of points to consider in the process of accurately determining whether an injury or illness is work related.

When an Injury or Illness Is Considered Work Related

An injury or illness is considered *work related* if an event or exposure in the work environment caused or contributed to the condition or significantly aggravated a preexisting condition. Work-relatedness is presumed for injuries and illnesses resulting from events or exposures occurring in the workplace, unless an exception specifically applies.

Definition of the Work Environment

OSHA defines the *work environment* as "the establishment and other locations where one or more employees are working or are present as a condition of their employment. The work environment includes not only physical locations, but also the equipment or materials used by the employee during the course of his or her work."

Situations in Which an Injury or Illness Occurs in the Work Environment and Is Not Considered Work Related

An injury or illness occurring in the work environment that falls under one of the following exceptions is not work related and, therefore, is not recordable.

A company is *not* required to record injuries and illnesses if the injury or illness:
- Occurred when the employee was present in the work environment as a member of the general public rather than as an employee.
- Involves signs or symptoms that surface at work but that result solely from a non-work-related event or exposure that occurred outside the work environment.

An injury or illness that occurs in the work environment is not considered work related when it meets the following criteria:
- Results solely from voluntary participation in a wellness program or in a medical, fitness, or recreational activity such as blood donation, physical examination, flu shot, exercise class, racquetball, or baseball
- Is solely the result of an employee eating, drinking, or preparing food or drink for personal consumption. Example: If the employee is injured by choking on a sandwich while in the employer's establishment, the case would not be considered work related.
  NOTE: If an employee is made ill by ingesting food contaminated by workplace toxins (such as lead) or gets food poisoning from food supplied by the employer, the case would be considered work related.
- Is solely the result of an employee doing personal tasks (unrelated to his or her employment) at the establishment outside of the employee's assigned working hours
- Solely results from personal grooming or self-medication for a non-work-related condition or is intentionally self-inflicted
- Is caused by a motor vehicle accident and occurs on a company parking lot or company access road while the employee is commuting to or from work
- Is the common cold or flu
  NOTE: Contagious diseases such as tuberculosis, brucellosis, hepatitis A, and plague are considered work related if the employee is infected at work.
- Is a mental illness. Mental illness is not considered work related unless the employee voluntarily provides the employer with an opinion from a physician or other licensed health care professional with appropriate training and experience (psychiatrist, psychologist, psychiatric nurse practitioner, etc.) stating that the employee has a mental illness that is work related.

Injuries or Illnesses While Traveling

Injuries and illnesses that occur while an employee is on travel status are work related if, at the time of the injury or illness, the employee was engaged in work activities "in the interest of the employer."
- Examples of such activities include traveling to and from customers; conducting job tasks; and entertaining or being entertained while transacting, discussing, or promoting business (work-related entertainment includes only the entertainment activities engaged in at the direction of the employer).

New Cases

An employer must consider an injury or illness a new case to be evaluated for recordability if the employee:
- Had not previously experienced a recorded injury or illness of the same type that affected the same part of the body.
- Previously experienced a recorded injury or illness of the same type that affected the same part of the body but had recovered completely (all signs/symptoms of the previous injury or illness had disappeared) and an event or exposure in the work environment caused the injury or illness, or its signs or symptoms, to reappear.

Recordable Injuries and Illnesses

Work-related injuries and illnesses that result in the following must be recorded:
- death
- loss of consciousness
- days away from work
- restricted work activity or job transfer
- medical treatment beyond first aid.

Any significant work-related injury or illness that is diagnosed by a physician or other licensed health care professional must be recorded. Any work-related case involving cancer; a chronic, irreversible disease; a fractured or cracked bone; or a punctured eardrum must also be recorded.

Additional Criteria

A company must also record work-related injuries and illnesses that meet any of the following additional criteria:
- any needlestick injury or cut from a sharp object that is contaminated with another person's blood or other potentially infectious material
- any case requiring an employee to be medically removed under the requirements of an OSHA health standard
- a tuberculosis infection evidenced by a positive skin test or diagnosed by a physician/other licensed health care professional after exposure to a known case of active tuberculosis
- an employer's hearing test (audiogram) revealing:
  - that the employee has experienced a standard threshold shift (STS) in hearing in one or both ears (averaged at 2,000, 3,000, and 4,000 Hz)
  - that the employee's total hearing level is 25 dB or more above audiometric zero (also averaged at 2,000, 3,000, and 4,000 Hz) in the same ear(s) as the STS.

Definitions of Medical Treatment and First Aid
The following information from OSHA defines medical treatment and first aid by providing a variety of examples that will assist in classification.

Medical Treatment
Medical treatment means the management and care of a patient to combat disease or disorder. The following are not considered medical treatment and are thus not recordable:
- visits to a doctor or other health care professional solely for observation or counseling
- diagnostic procedures, such as x-rays and blood tests, and the administration of prescription medications used solely for diagnostic purposes (e.g., eyedrops that dilate pupils)
- any procedure that can be considered first aid.

First Aid
If the incident required only the following types of treatment, consider it *first aid*. Do not record the case if it involves only:
- using a nonprescription medication at nonprescription strength
- administering tetanus immunizations (immunizations such as hepatitis B vaccine or rabies vaccine are considered medical treatment)
- cleaning, flushing, or soaking wounds on the skin's surface
- using wound coverings such as bandages, Band-Aids™, gauze pads, or the like, or using butterfly bandages or Steri-Strips™ (other wound coverings such as sutures, staples, etc., are considered medical treatment)
- using hot or cold therapy
- using any nonrigid means of support, such as elastic bandages, wraps, nonrigid back belts, or the like (devices with rigid stays or other systems designed to immobilize parts of the body are considered medical treatment)
- using temporary immobilization devices while transporting an accident victim (splints, slings, neck collars, backboards)
- drilling a fingernail or toenail to relieve pressure or draining fluid from a blister
- using eye patches
- removing foreign bodies from the eye using only irrigation or a cotton swab
- removing splinters/foreign material from areas other than the eye by irrigation, tweezers, cotton swabs, or other simple means
- using finger guards
- using massages (physical therapy or chiropractic treatment are considered medical treatment)
- drinking fluids for relief of heat stress.

Restricted Work Activity
Situations may arise in the workplace that are defined as *restricted work activities*. To handle these situations effectively, it is important to understand the parameters of restricted work, as defined by OSHA, as well as how to count the number of days of restricted work activity or the number of days away from work.

Deciding Whether the Case Involves Restricted Work
An employee's work activity is considered to be restricted when, as the result of a work-related injury or illness, an employer or health care professional prevents, or recommends preventing, an employee from doing the routine functions of his or her job or from working the full workday that the employee had been scheduled to work before the injury or illness occurred.

Routine Job Function
A *routine job function* is defined as a work activity that the employee regularly performs at least once per week.

How to Count the Number of Days of Restricted Work Activity or the Number of Days Away from Work
- Count the number of calendar days the employee was on restricted work activity or was away from work as a result of the recordable injury or illness.
- Do not include in this number the day on which the injury or illness occurred.
- Begin counting days from the day the incident occurs.
- If a single injury or illness resulted in both days away from work and days of restricted work activity, enter the total number of days for each.
- Stop counting days of restricted work activity or days away from work once the total of either or the combination of both reaches 180 days.

Classifying Injuries and Illnesses
The information from OSHA in Table 5–C defines injuries and illnesses by providing a variety of examples that will assist you in your classification.

TABLE 5–C. Injuries and Illnesses

Classifying Injuries	Classifying Illnesses
An *injury* is any wound or damage to the body resulting from an event in the work environment. *Examples:* Cut, puncture, laceration, abrasion, fracture, bruise, amputation, insect bite, electrocution, or a thermal, chemical, electrical, or radiation burn. Sprain and strain injuries to muscles, joints, and connective tissues are classified as injuries when they result from a slip, trip, fall, or other similar accidents.	The following list provides a summary of major types of *illnesses*. • Musculoskeletal disorders (MSD illnesses) are disorders of the muscles, nerves, tendons, ligaments, joints, cartilage, or spinal discs. MSDs do not include disorders caused by a slip, trip, motor vehicle accident, fall, or other similar accident. • Skin diseases or disorders are illnesses involving the worker's skin that are caused by work exposure to chemicals, plants, or other substances. • Respiratory conditions are illnesses associated with breathing hazardous biological agents, chemicals, dust, gases, vapors, or fumes at work. • Poisoning includes disorders evidenced by abnormal concentrations of toxic substances in blood, other tissues, other bodily fluids, or the breath that are caused by the ingestion or absorption of toxic substances into the body. • Noise-induced hearing loss is defined for record-keeping purposes as a change in the hearing threshold relative to the baseline audiogram of an average of 10 dB or more in either ear at 2,000, 3,000, and 4,000 Hz and when the employee's total hearing level is 25 dB or more above audiometric zero (also averaged at 2,000, 3,000, and 4,000 Hz) in the same ear(s). • All other occupational illnesses.

SUMMARY

The following is an at-a-glance overview of how to handle situations that require a supervisor to assess a case involving an employee injury or illness. It should be remembered that records are not only a requirement of OSHA. OSHA's record-keeping rule results in important information for employers and employees as well.

What to do
- Within 7 calendar days after you receive information about an injury or illness, decide whether the case is recordable under the OSHA record-keeping requirements.
- Determine whether the incident is a new case or a recurrence of a previous one.
- Establish whether the incident was work related.
- If the incident is recordable, complete the OSHA 301 form.
  NOTE: Use OSHA's Injury and Illness Incident Report (Form 301) or an equivalent form. Some state workers' compensation, insurance, or other reports may be acceptable substitutes, as long as they provide the same information as OSHA Form 301.

How to work with the 300 Log
- Identify the employee involved unless it is a privacy concern case as discussed later.
- Identify when and where the incident occurred.
- Describe the incident, as specifically as possible.
- Classify the seriousness of the incident by recording the most serious outcome associated with the case, with column J (Other Recordable Cases) being the least serious and column G (Death) being the most serious.
- Identify whether the case is an injury or an illness. If the case is an injury, check the injury category. If the case is an illness, check the appropriate illness category.

Circumstances under which the employee's name should not be entered on OSHA Form 300
Consider the following types of injuries or illnesses to be privacy concern cases. In these cases, do *not* enter the employee's name on the OSHA 300 Form. Instead, enter "privacy case" in the space normally used for the employee's name. Keep a separate, confidential list of the case numbers and employee names for the establishment's privacy concern cases so that the cases can be updated and, if requested, the information provided to the government.
- an injury or illness to an intimate body part or to the reproductive system
- an injury or illness resulting from a sexual assault
- a mental illness
- a case of HIV infection, hepatitis, or tuberculosis
- a needlestick injury or a cut from a sharp object that is contaminated with blood or other potentially infectious material (see 9 CFR Part 1904.8 for a definition)
- other illnesses, if the employee independently and voluntarily requests that his or her name not be entered on the log.
  NOTE: Musculoskeletal disorders (MSDs) are not considered privacy concern cases.

What to do if the outcome changes after the case is recorded
If the outcome or extent of an injury or illness changes after recording the case, simply draw a line through the original entry or, if desired, delete or use correction fluid to cover up the original entry. Then write the new entry where it belongs. Remember, it is important to record the most serious outcome for each case.

When to post the Form 300A Summary
The Summary only—not the Log—must be posted by February 1 of the year following the year covered by the form, and it must remain posted until April 30 of that year.

How long to keep the Log and Summary on file
The Log and Summary must be kept for 5 years following the year to which they pertain.

Sending the forms to OSHA at the end of the year
These completed forms do not need to be sent to OSHA unless OSHA specifically requests them.

Although there are many more regulations that a workplace has to comply with, the regulations given in this summary affect a majority of workplaces. For questions regarding a regulation, sources of help include internal company resources, associations, the Internet, and the regulatory agencies that enforce the regulations.

REFERENCES

Commerce Clearing House, CCH Occupational Safety & Health Guide (periodical).

O'Reilly, J. T., *Administrative Rulemaking*, Thomson/West 2nd ed. 2007.

Occupational Safety & Health Act, 29 United States Code 651 et seq., available on www.gpoaccess.gov/uscode/title29/chapter 15_.html.

OSHA Regulations, http://www.access.gpo.gov/nara/cfr/waisidx_07/29cfrv5_07.html.

OSHA Regulations, http://www.osha.gov/doc/outreachtraining/htmlfiles/introsha.html.

REVIEW QUESTIONS

1. List five of the top-cited OSHA workplace violations.
    a.
    b.
    c.
    d.
    e.
2. For a company to be cited using the general-duty clause, name the five criteria that must exist.
    a.
    b.
    c.
    d.
    e.
3. What is the definition of a confined space?
4. What is the definition of a permit-required confined space?
5. When recording information on the OSHA 300 Log or the 301 Injury and Illness Incident Report forms, how is a routine job function defined?
6. Name three regulations that require written programs.
    a.
    b.
    c.
7. What are the recent changes to labeling under the Hazard Communication Standard, 29 CFR 1910.1200?
8. True or False: OSHA has extended the ground-fault protection requirement to temporary receptacles used in construction activities performed in general industry.

PART 2

Loss Control Information and Analysis

The best response to safety concerns is a diligent preventive effort. Part 2 focuses on the means and methods for evaluating potential risks. In the past, it might have been enough to study incidents and change systems; today, a prudent manager spends more time actively considering and planning for positive improvements that can make the workplace systems more protective of worker health and safety.

Information gathering and analysis is a fundamental basis for the rest of the professional's duties, so Part 2 explores the means by which data are collected, compiled, considered, and coordinated. Perhaps the most visible sign of the changes in the safety profession is the use of computerized data systems in the auditing of safety in workplaces. Safety professionals are encouraged to adapt and adjust the loss control systems and checklists to fit their own facility and organizational needs.

Loss Control Programs

6

Pat Lorimer, MPH
Val Henry, MBA

Need for a Balanced Approach
Management Oversight and Omission ▸ Examining Incident Causation

Incidents and Loss Control
Definition of Hazards ▸ Effects of Hazards on the Work Process ▸ Controlling Hazards: A Team Effort ▸ Management Support ▸ Safety Management and Productivity Improvement ▸ Worker-Equipment-Environment System

Protecting against Liability
Types of Liability ▸ Special Concerns of Government Contractors ▸ Prevention Strategies

Incident Causes and Their Control
Unsafe Practices or Procedures ▸ Situational Factors ▸ Environmental Factors ▸ Sources of Situational and Environmental Hazards

Principles of Loss Control

Processes of Loss Control
Hazard Identification and Evaluation ▸ Ranking Hazards by Risk (Severity, Probability, and Exposure) ▸ Risk Assessment ▸ Management Decision Making ▸ Establishing Preventive and Corrective Measures ▸ Monitoring ▸ Evaluating Program Effectiveness

Organizing an Occupational Safety and Health Program
Establishing Program Objectives ▸ Establishing Organizational Policy ▸ Responsibility for the Hazard Control Program ▸ Professionals in Loss Control

Purchasing
Purchasing-Safety Liaison ▸ Safety Considerations ▸ Price Considerations ▸ Codes and Standards ▸ Specifications

Safety and Health Committees

Off-the-Job Safety Programs
What Is Off-the-Job Safety? ▸ Program Benefits ▸ Promoting Off-the-Job Safety ▸ Off-the-Job Safety Policy ▸ Getting Started ▸ Select Program Details ▸ Employee, Family, and Community Involvement ▸ Programs That Work

Summary

References

Review Questions

NEED FOR A BALANCED APPROACH

Before the concept of loss control was developed, accidents were regarded as either chance occurrences or acts of God—a view still held by some—or as an inherent consequence of production. Such approaches accept accidents as inevitable and, therefore, yield no information about causation and prevention. Control strategies are limited to mitigating the consequences of the occurrence.

In the early days of loss control, accident prevention activities focused on the human element. Findings indicated that a small proportion of workers accounted for a significant percentage of accidents. Control strategies were devised to reduce human error through training, education, motivation, communication, and other forms of behavior modification. During World War II, industrial psychology was aimed at matching employees to particular jobs. Personnel screening and selection were seen as the primary ways to prevent accidents. However, accident proneness and other behavior models have a glaring weakness: although they are useful for understanding human behavior, they do not consider the interaction between the worker and the other parts of the system. (See the discussion in Chapter 30, Motivation, and Chapter 31, Safety and Health Training.)

The 1950s and 1960s saw the emphasis change to engineering and control programs aimed at machines and equipment. Further, the implementation of the Occupational Safety and Health Act of 1970 (OSH Act) emphasized preventing accidents through control of the work environment and the elements of the workplace. This act, along with other legislation, specified compliance by employers with promulgated safety and health standards and other rules and regulations.

In the 1980s and 1990s, there was a realization that even with all of the earlier emphases, "accidents" still occurred. Focus returned to an emphasis on the human element, and the commitment and culture of the organization were important issues. Clearly, emphasis on any one area does not bring about permanent reduction of unintentional injuries and losses. A balance between applying all feasible engineering and control programs aimed at machines and equipment and applying human behavior modification, motivation, and training techniques must be achieved.

In the 1990s, the National Safety Council continued to work to increase awareness that an incident is a near accident and that so-called accidents are not random events but rather preventable events. The Council changed its mission statement to eliminate the word *accident*:

> The mission of the National Safety Council is to educate and influence society to adopt safety, health, and environmental policies, practices, and procedures that prevent and mitigate human suffering and economic losses arising from preventable causes.

To further increase awareness that so-called accidents are preventable, the Council has tried to reduce use of the word *accident* in its publications, substituting more specific terms such as *unintentional injury* or *accidental injury*.

During the late 1990s to current, the Council further clarified its mission statement to read:

> The mission statement of the National Safety Council is to save lives by preventing injuries and deaths at work, in homes, and on the roads through leadership, research, education and advocacy.

Unintentional injury or *accidental injury* is the preferred term in the public health community. It refers to the result of an accident.

With proper hazard identification and evaluation, management commitment and support, preventive and corrective procedures, monitoring, evaluation, and training, most unintentional injuries can be prevented.

Management Oversight and Omission

Over the past few decades, many organizations seeking to reduce hazards have focused on system defects, which result from management's lack of oversight or omission, or on malfunction of the management system. A balanced approach to loss control looks at each component of the system and includes such weaknesses as inadequate training and education, improper assignment of responsibility, unsuitable equipment, and failure to fund hazard control programs. Because managers are responsible for the design, implementation, and maintenance of systems, management errors can result in system defects.

Examining Incident Causation

There are two basic approaches to examining how incidents are caused: after-the-fact and before-the-fact.

After-the-Fact

This approach relies on examining incidents after they have occurred to determine the cause and to develop corrective measures. Evaluation of past performance uses information derived from incident and inspection reports, workers' compensation data, and insurance audits. This approach is too often used only after a serious incident has resulted in injury or damage or system ineffectiveness. Furthermore, incident frequency and severity rates do not answer the crucial questions of how, what, why, and when incidents or near misses occur.

Before-the-Fact
This method relies on inspecting and systematically identifying and evaluating the nature of undesired events in a system. One such method is the critical-incident technique.

Critical-Incident Technique
The critical-incident technique measures safety performance and identifies through direct observations of human behavior practices or conditions that need to be corrected. This technique can identify the cause of an incident before the loss occurs. To obtain a representative sample of workers exposed to hazards, management selects workers from various departments of the facility. An interviewer questions workers who have performed particular jobs within certain environments. They are asked to describe only those existing hazards and unsafe conditions they are aware of. These are called *incidents* (Figure 6–1). Management then classifies incidents into hazard categories and identifies problem areas. The investigative team can also analyze the management systems that should have prevented the occurrence of unsafe practices or the existence of unsafe conditions. The technique can lead to improvements in loss control program management.

Critical-incident technique relies on five major components. The first step is to determine the incident; the second is to then evaluate the facts related to the incident, which involves collecting the details of the incident from the participants. When all of the facts are collected, the next step is to identify the issues contributing to the incident cause. Following this step, a decision can be made on how to resolve the issues based on various possible solutions. The final component is the evaluation, which will determine if the solution selected will solve the root cause of the situation and resolve the problems. This final aspect is most critical.

The procedure needs to be repeated because the worker-equipment-environment system is not static. Repeating the technique with a new sample of workers can reveal new problem areas and measure the effectiveness of the incident-prevention program.

Safety Sampling
Also called *behavior* or *activity sampling*, safety sampling is another technique that uses the expertise of those within the organization to inspect, identify, and evaluate hazards. This method relies on personnel—usually management or safety staff members—who are familiar with operations and well trained in recognizing unsafe practices. While they make rounds of the facility or establishment, they record on a safety sampling sheet both the number and type of safety defects they observe. A code number can be used to designate

LIST OF TYPICAL INCIDENTS

An incident is any observable human activity sufficiently complete in itself to permit references and predictions to be made about the persons performing the act.

1. Adjusting and gaging (calipering) work while the machine is in operation
2. Cleaning a machine or removing a part while the machine is in motion
3. Using an air hose to remove metal chips from table or work (a brush or other tools should be used for this purpose, except on recessed jigs)
4. Using compressed air to blow dust or dirt off clothing or out of hair
5. Using excessive pressure on air hose
6. Operating machine tools (turning machines, knurling and grinding machines, drill presses, milling machines, boring machines) without proper eye protection (including side shields)
7. Not wearing safety glasses in a designated eye-hazard area
8. Failing to use protective clothing or equipment (face shield, face mask, ear plugs, safety hat, cup goggles)
9. Failing to wear proper gloves or other hand protection when handling rough or sharp-edged material
10. Wearing gloves, ties, rings, long sleeves, or loose clothing around machine tools
11. Wearing gloves while grinding, polishing, or buffing
12. Handling hot objects with unprotected hands
13. No work rest or poorly adjusted work rest on grinder ($1/8$ in. maximum clearance)
14. Grinding without the glass eye shield in place
15. Making safety devices inoperative (removing guards, tampering with adjustment of guard, beating or cheating the guard, failing to report defects)
16. Using an ungrounded or uninsulated portable electric hand tool
17. Improperly designed safety guard, for example, a wide opening on a barrier guard, which will allow the fingers to reach the cutting edge.

Figure 6–1. This list represents a sample of work practices that, if not corrected, can result in an unintentional injury or loss.

specific unsafe conditions, such as hands in dies, failure to wear eye protection and protective clothing, failure to lock out sources of power while working on machinery, crossing over belt conveyors, working under suspended loads, improper use of tools, or transporting unbanded steel.

Safety personnel or managers should make observations at different times of the day, on a planned or random basis in the actual work setting, and throughout the various parts of the facility. In a short time, they can easily convert observations to a simple report showing what specific unsafe conditions exist in which areas and which supervisors and foremen need help in enforcing good work practices. The information is unbiased and therefore irrefutable. What has been recorded is what has been observed. This may also determine the need for further or recurrent training of employees.

INCIDENTS AND LOSS CONTROL

This section focuses on the definition of hazards, their effects on the workplace, and efforts by management, safety professionals, and employees to control them. The more that management and workers realize that safety involves day-to-day teamwork and mutual support, the closer a company will come to achieving its safety goals.

Definition of Hazards

A workable definition of *hazard* is any existing or potential condition in the workplace that, by itself or by interacting with other variables, can result in deaths, injuries, property damage, and other losses. This definition carries with it two significant points:
- A condition does not have to exist at the moment to be classified as a hazard. When the total hazard situation is being evaluated, *potentially* hazardous conditions must be considered.
- Hazards may result not only from independent failure of workplace components, but also from one workplace component acting on or influencing another. For instance, if gasoline or another highly flammable substance comes in contact with sulfuric acid, the reaction created by the two substances produces both toxic vapors and sufficient heat for combustion.

Hazards are generally grouped into two broad categories: those dealing with safety and injuries and those dealing with health and illnesses. However, hazards that involve property and environmental damage must also be considered.

Effects of Hazards on the Work Process

In a well-balanced operation, workers, equipment, and materials are brought together in the work environment to produce a product or to perform a service. When operations go smoothly and time is used efficiently and effectively, production is at its highest.

When an incident interrupts an operation, it sets in motion a different chain of events and carries its own price tag. An incident is an unplanned, undesired event, not necessarily resulting in injury, but damaging to property and/or interrupting the activity in process. An incident increases the time needed to complete the job, reduces the efficiency and effectiveness of the operation, and raises production costs. If the incident results in injury, materials waste, equipment damage, or other property loss, there is a further increase in operational and hidden costs and a decrease in effectiveness.

Controlling Hazards: A Team Effort

Traditionally, most managers have relied solely on their safety and operations people to anticipate, identify, evaluate, and control hazardous situations. However, the more that is learned about hazard and loss control, the more evident it becomes that the job is too large for any individual or small group to do alone. Incident and hazard reduction requires a team effort by employees and management.

Here is how several departments and employee teams can work together:
- The engineering departments can design facilities to be free of uncontrolled hazards and provide technical hazard identification and analysis services to other departments. Their designs must comply with federal, state or provincial, and local laws and standards in addition to meeting the overall goals established by the organization.
- Manufacturing departments can reduce hazards through efforts such as effective tool design, changes in processes, job hazard analysis and control, and coordinating and scheduling production.
- Quality control can test and inspect all materials and finished products. It can conduct studies to determine whether alternate design, materials, and methods of manufacture could improve the quality and safety of the product and the safety of the employees making the product.
- Purchasing departments can ensure that materials and equipment entering the workplace meet established safety and health standards and that adequate protective devices are an integral part of equipment. They may also evaluate similarly effective materials that are less hazardous substitutions for existing products used in the workplace. They should disseminate information received from suppliers to line management and workers about safety and health hazards associated with workplace substances and materials.
- Maintenance can provide planned preventive maintenance on electrical systems, machinery, and other equip-

ment to prevent abnormal deterioration, loss of service, or safety and health hazards.
- Human resources often administers programs directly related to health and safety.

Input also can come from the joint safety and health committee (discussed later in this chapter) and from quality circles and safety circles.

Management Support

To coordinate the organizational and departmental efforts, a program of loss control is necessary as part of the management process (Windsor 1979). Such a program provides hazard control with management tools such as programs, procedures, audits, and evaluations. Sometimes, hazard control program teams neglect the basics in their rush to be competitive and innovative, to deal with complex employee relations issues and government involvement, and to address the technical aspects of the programs. A program of hazard control ensures that safety fundamentals also will be addressed. These basics include sound operating and design procedures, operator training, inspection and test programs, and communicating essential information about hazards and their control.

A loss control program establishes facility-wide safety and health standards and coordinates responsibilities among departments. For example, if one department makes a product and another distributes it, they share responsibility for hazard control. The producer knows the nature of the process, its apparent and suspected hazards, and how to control the hazards. The producer and the distributor are responsible for making sure this information does not end with the production department but is available to the purchaser or the next unit in the manufacturing process.

Coordination is also important when manufacturing responsibility is transferred from one department to another (as when a pilot program becomes a complete manufacturing unit). In addition, when a process is phased out, departments need to coordinate efforts to ensure that personnel who know the hazards are retained throughout the phase-out and that appropriate hazard control activities continue until the end.

Safety Management and Productivity Improvement

The process of identifying and eliminating or controlling hazards in the workplace is one way of making the best use of human, financial, technological, and physical resources. Optimizing these resources results in higher productivity. For purposes here, *productivity* is defined as producing more output with a given level of input resources.

Loss control—like productivity, quality, costs, and personal relations—is a strategic process. To be effective, it must be integrated into the day-to-day activities and management systems of the organization and must become institutionalized—an operating norm and a strategic part of the organization's culture.

There are other similarities between efforts aimed at hazard control and productivity improvement. To achieve both objectives, an organization must intelligently manage its financial and human resources and use the most appropriate technology. It must illustrate innovative, enlightened, and efficient use of its facilities, equipment, materials, and work force and have a trained, educated, and skilled work force.

An incident interrupts the production process. It not only increases the time needed to complete a production task, but may also reduce the efficiency and effectiveness of the overall operation and increase production costs. Sometimes, a succession of interruptions, or one long one, prevents the production schedule or desired product quality from being met. Such conditions make it difficult to attract new business.

Production Accomplishment and Control

Control of an operation, by definition, means keeping the system on course and preventing problems from occurring. However, it also implies some allowance for variations within the system, provided they remain within controlled limits. Any production system has built-in control limits, both upper and lower. These limits provide direction and also any acceptable leeway for the system's operation.

There are many aspects to control, including control over the quality of products and services, personnel, capital, energy, materials, and the facility environment. Each of these factors interacts with the other factors to produce the desired effect.

Determining Incident Factors

In order to set realistic goals for its process, the organization should first determine the major factors likely to cause loss of control. It should then identify the location, importance, and potential effects. Control measures can then be instituted to help reduce risk and potential losses. Factors responsible for incident losses may be identified by either inspection or detailed hazard analyses. The control measures may be administrative changes, some type of process innovation or machine safeguarding, personal protective equipment, or training. In addition to the control measures, monitoring systems should be used to continuously assess the effectiveness of these hazard-reducing controls. The results of the monitoring systems should result in changes to processes to abate potential hazards by incorporating changes to systems or production.

Worker-Equipment-Environment System

Professionals involved in establishing effective loss control programs must understand the interrelationships in the worker-equipment-environment system. (Chapter 16, Ergonomics Yesterday, Today, and Tomorrow, examines the system in greater detail.) This chapter explains the elements of the system (Figure 6–2). As shown in Figure 6–3, an incident can intervene between the system and the task to be accomplished.

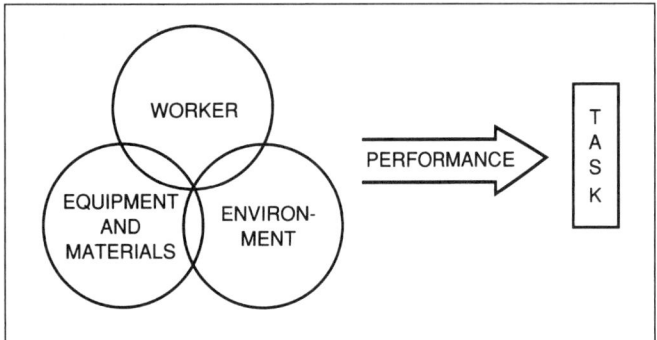

Figure 6–2. A system approach to hazard control recognizes the interaction between worker, equipment, materials, and environment in the performance of work.

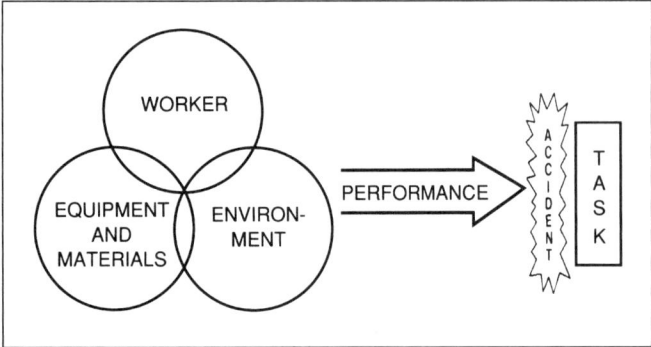

Figure 6–3. An unintentional incident causes the work system to break down. It intervenes between the worker, equipment, and environment and the task to be performed.

Worker

In any worker-equipment-environment system, the worker performs three basic functions: (1) sensing, (2) information processing, and (3) controlling.
- As a sensor, the worker monitors or gathers information.
- As an information processor, the worker uses the information collected to make a decision about the relevance or appropriateness of various courses of action.
- The third function, control, flows from the first two. Once information is collected and processed, the worker keeps the situation within acceptable limits or takes the necessary action to bring the system back into an acceptable or safe range.

Evaluating an incident in the light of these three functions can pinpoint the causes. Did the error occur while the worker was gathering information as a sensor? Was the worker able to gather information accurately—for example, in adequate illumination without glare? Did the error occur as a result of faulty information processing and decision making? Did the error occur because an appropriate control option was not available or because the worker took inappropriate action?

In order for the system to move toward its production objectives, the employee must perform work effectively and avoid taking unnecessary risks. To do this, workers must be educated about the following (Firenze 1978):
- necessary requirements of the task and the steps needed to accomplish it
- personal knowledge, skill, and limitations and how they relate to the task
- what will be gained if the worker attempts the task and succeeds
- what will result if the worker attempts the task and fails
- what will be lost if the worker makes no attempt to accomplish the task.

Equipment

Equipment (materials), the second component in the system, must be properly designed, maintained, and used. Hazard control can be affected by the shape, size, and thickness of tools; the weight of equipment; operator comfort; and the strength required to use or operate tools, equipment, and machinery. These variables influence the interaction between worker and equipment. Other equipment variables important in hazard recognition include speed of operation and mechanical hazards.

Environment

Special consideration must be given to environmental factors that may detract from the comfort, health, and safety of the worker. Emphasis should be placed on factors such as the following:
- layout: the worker should have sufficient room while performing the assigned task
- maintenance and housekeeping
- adequate illumination: poorly lit areas increase eyestrain and also the chance of making an incident-causing mistake
- temperature, humidity, noise, vibration, and control of emission of toxic materials.

Interpersonal relationships are another system factor that plays an important role in operational effectiveness.

The task performed by one worker is related to tasks performed by others. Special consideration must be given to coordinating information, materials, and human effort.

PROTECTING AGAINST LIABILITY

Types of Liability

When an unintentional incident causes injuries, legal liability may result. This discussion of the liability consequences of unintentional incidents is brief and general; consult legal counsel for advice before making important decisions. After a discussion of the types of liability, this section briefly explores the protections available to avoid those risks.

Civil Liability of Organizations and Potential Damages

The challenge for the safety professional within a company or organization is often to "do more with less." When he or she asks management to invest more in order to achieve more protection against hazards, the safety professional often is asked, "What's at stake for us if we don't do this?" The focus of this section is an outline of the legal system's consequences for lapses, omissions, or misconduct in the handling of health and safety protections that results in injuries.

Civil Tort Remedies. The U.S. system of awarding damages against people responsible for causing injury takes its name from the French and English legal systems, which used the term *tort* to describe an action that was wrong or harmful and against which some legal consequence such as a jury verdict for damages may be awarded. The law of torts has evolved in the United States for dual purposes: as a compensation for losses and as a deterrent to misconduct. The tort law system focuses on whether the defendant who caused the harm was at fault. In some cases, the legal system has imposed "strict liability without fault" for a narrow range of hazardous activities. The U.S. system allows more civil cases to be won for larger amounts of monetary damages than other nations, but it remains difficult for the injured individual to prevail if legal liability is not proven.

Often, proving legal liability is a matter of showing that someone was negligent. Negligence is typically the failure to exercise the care toward others that a reasonable person would exercise in similar circumstances or taking action when a reasonable person would not. Most jurisdictions recognize three different levels of negligence.

- *Simple negligence* is the failure to use the degree of care an ordinary person would exercise to avoid hurting another person.
- *Gross negligence*, a higher level of negligence, is action that shows indifference to others and shocks fair-minded people, with the result of neglecting the safety of others.
- *Willful and wanton negligence*, the third and most egregious level of negligence, occurs when one acts consciously with reckless indifference to the consequences and when knowledge of existing circumstances and conditions indicates that the conduct would probably cause injury to another.

The Elements of a Negligence Action. A typical formula for evaluating negligence in a civil context requires that the following five elements be proved by a "preponderance of the evidence"—in which *preponderance of the evidence* simply means that it was more likely than not that the defendant was at fault.

- The defendant owed a duty to the plaintiff (or a duty to the general public, including the injured party).
- The defendant violated that duty.
- A plaintiff must prove that the defendant's actions actually caused the plaintiff's injury. This is often referred to as "but-for" causation. In other words, but for the defendant's actions, the plaintiff's injury would not have occurred.
- As a result of the defendant's violation of that duty, the plaintiff suffered injury.
- The injury was a reasonably foreseeable consequence of the defendant's action or inaction, and the defendant's actions actually caused the injury—but for the defendant's actions, the injury would not have occurred.

In claims of liability, usually the difficult element to prove is *causation*—a showing that the injury was a proximate result of what the defendant did and not the result of some other person's actions. It is essential for a plaintiff, the person suing, to show that the harm was caused by this particular defendant—for example, by its truck hitting the delivery driver or its pile of boxes falling on a warehouse visitor.

This legal concept of *negligence* in the context of the safety professional means that a duty existed for the company, employer, shipper, and so on to take reasonable steps to avoid injury to workers, users, carriers, and so on, but the duty was breached.

The following situations could support a finding of negligence:

- failure to safeguard a machine or process against the foreseeable failure that causes an injury, such as omitting a guard screen on a machine that stamps metal
- removing a safety device on a machine, thus creating a hazardous situation for workers who might have been protected by the previous design of the machine
- maintaining facilities, premises, grounds, and so on in an unsafe condition that is not reasonable so that people

who work or visit there may be harmed by slipping, falling, or otherwise encountering unexpected dangers
- failure to supervise employees, allowing the poorly trained or poorly managed worker to cause harm—for example, by reckless driving of a vehicle on which the employee had no safety training
- negligent selection of employees that places the visitor or customer in a hazardous situation—for example, an employee with a history of erratic, violent behavior assigned as an armed security guard.

In addition, *negligence per se* is a legal doctrine that holds a company or person liable for damages if an applicable law or rule was violated and, as a result, the injury occurred. Not all laws and rules are affected by this doctrine, and not all states follow it in the same way. Often *negligence per se* is used as evidence in evaluating the concept of negligence.

Organizational Liability. The Latin phrase *respondeat superior* is used in modern negligence law to impose on an employer the liability and associated obligation to pay damages if one of its workers causes harm to another person during his or her assigned duties, such as a driver hit by a recklessly operated dump truck. The law's attribution of liability to the company that owns the truck recognizes that people, not anonymous concepts such as a corporation, do the work that earns the money for the company; the company acts through people when it selects, hires, trains, and disciplines employees; and so the company or organization is held accountable for the actions of its employees.

A later subsection of this chapter discusses the individual liability of supervisors and co-workers when injuries occur at the workplace.

Punitive Damages. After an injury occurs, a retrospective look at its causes may show such reckless misconduct by the company or employer that a severe penalty should be used to deter future misconduct. Different states have used different norms for the jury award of these extra punishments, called *punitive damages* because their goal is to punish the perpetrator rather than to compensate the victim. In simplest generalization, a company faces punitive damage liability if its actions are so sharply different from the socially acceptable behavior of similar companies that the misconduct is shocking. The result of such extremely careless or intentionally harmful actions is the severe harm caused to these injured people, whom the jury sees in this particular lawsuit. The response to the act is a jury award of money that the jury hopes will punish the offending company and deter misconduct in the future.

Workers' Compensation. Civil tort liability lawsuits brought against employers in the courts for workplace injuries would be a major problem for both employers and injured workers if no alternative system had been developed to channel these injury claims. The state laws that have created systems of postinjury payments for injured workers, called *workers' compensation systems* (see Chapter 8, Workers' Compensation), are a compromise that balances the rights of workers with those of employers. Employers have won the right to pay some level of compensation instead of being sued for damages by their workers if they are injured on the job. In most workplace injury cases, the law bars workers from suing their employer if the injury fits within the state workers' compensation system. Workers get an assurance of predictable, understandable health care benefits and payments for lost work time—but give up the right to sue employers—in virtually all workplace injury situations. Compensation checks come from the state fund to the worker; employers pay insurance premiums corresponding to a formula that takes into account the risk aspects of the particular employment.

Subrogation of Workers' Compensation. Some of the workers' compensation cases that result in payments to injured workers also reflect a harm that was attributable to a third person, not just the employer or the worker. In these cases, the product supplier, the machine builder, the other vehicle in a collision, and so on may be sued in civil tort liability. The damages awarded against that person must, in many cases, be paid back to the state insurance fund from which compensation payments had been made. This is called *subrogation*, in which the injured person receives payment and the one who contributed to the injury is held liable to pay damages and to reimburse the state fund.

Intentional Torts. The employer responsible for an injury may, in extreme cases, face lawsuits by injured workers or their families, asserting that the danger was so severe and obvious that the company intended that workers be vulnerable to injury. These *intentional tort* cases are an exception to workers' compensation in states where they have been allowed. Because the normal standard of workers' compensation laws bars lawsuits, a worker who claims an intentional tort must bear a heavy burden to prove that the employer was aware that a serious or fatal injury would be likely to result if the worker were forced to undertake the risky task. Some states require a showing that the employer was substantially certain that injury would occur from the dangerous task or hazardous equipment.

Administrative Penalties against Organizations

Managers who need some incentive to invest in safety

programs should consider government's role as well as the court system's potential effects. In a system that is roughly parallel to the "fault"-based tort liability system, government agencies can impose administrative penalties and fines against an organization when an injury has been caused by the actions or omissions of the company. These administrative sanctions are discussed in this section according to the federal agency responsible; parallel systems for imposing penalties exist under state laws, as well, especially where the state enforces occupational safety and health programs under a "state OSHA plan."

These government penalty systems apply to the company or organization and not to the individual supervisors or managers in virtually all cases. For example, if a crane collapses at a job site and workers are injured, the federal Occupational Safety and Health Administration (OSHA) will investigate. The OSHA inspection may result in a recommended penalty of $80,000 against the crane service company. After a hearing before an administrative law judge of the Occupational Safety and Health Review Commission, the penalty may be upheld, reduced, or dismissed entirely. The $80,000 federal penalty acts as a punishment for not following safety requirements imposed by OSHA; OSHA expects that its penalties will be publicized as a deterrent to other crane companies as well.

Statutory Penalties as an Argument for Preventive Efforts

The government imposition of penalties for violations of safety rules has a benefit to society, augmenting the deterrent effect of the civil tort system described earlier. The safety professional who asks for more company effort or resources may want to study the recent history of similar organizations that were hit with large OSHA penalties. Financially oriented company managers who see the fines imposed on other companies may become more receptive to investing in a safety device or a training course for employees.

The OSHA Penalties. The most relevant penalty system is that of the federal OSHA and the state systems that carry out the same congressional mandate. OSHA civil penalties are imposed by area office directors after review of the facts found by OSHA inspectors. These include "willful" violations, the most severely punished forms of violations; "serious" violations, in which a health risk or injury potential may have caused significant harm; and routine, less-than-serious violations, usually of paperwork omissions or record-keeping errors. Penalties can range from a few thousand dollars to millions of dollars, depending on the number of workers affected, the harm created by the violation, and other factors.

The OSHA penalty becomes final and payment is due, unless the employer makes a timely request for a hearing. The hearing request must be made within 15 days or the total fine must be paid—so employers have to react quickly to respond to the OSHA letter announcing the decision to charge a penalty for certain violations.

Other Federal Penalties. Beyond OSHA, other federal agencies also have civil penalty systems. The Mine Safety and Health Administration (MSHA) follows the OSHA model closely and applies to workers in surface and underground mines and related processing facilities.

The Environmental Protection Agency (EPA) can impose large fines for the handling of chemicals that are regulated by several of its programs. For example, the Toxic Substances Control Act (TSCA) regulates the handling of experimental batches of new chemicals, whereas agricultural workers on a farm must be protected against exposures to certain hazardous pesticide sprays that are regulated by the Federal Insecticide, Fungicide, and Rodenticide Act (FIFRA). EPA can impose huge penalties for improper handling of asbestos in building demolition, for the misuse of certain refrigerant gases that may affect stratospheric ozone, or for air pollution or water pollution violations that result from an industrial illness.

The Department of Transportation (DOT) can impose penalties for various types of violations. Rail and highway carriers are subject to penalties for unsafe practices. The Federal Aviation Administration (FAA) closely regulates airline workplace safety conditions, as well as the safe transport of freight and cargo by air. The DOT also levies fines and penalties for mishandling of shipments containing hazardous materials; a poor label or a false statement on shipping papers can draw a fine of thousands of dollars. DOT penalties seek to deter accidental injuries to workers who ship, move, and receive packages of chemicals and other potentially harmful cargoes.

The federal Consumer Product Safety Commission (CPSC) can impose fines of more than $1 million when a product intended to be used by consumers violates one of the CPSC's mandatory product safety standards or when the producer fails to comply with the CPSC's requirement for immediate reporting of substantial product hazards. A lawn tractor made for consumers that causes accidental amputations is an example of a product for which the reporting of adverse effects is encouraged through the use of federal administrative penalties.

The federal Nuclear Regulatory Commission's role in safe handling of radioactive materials and the Food and Drug Administration's control on laser products used in workplaces also include civil administrative penalties. Both agencies use their detailed technical standards as the basis for charging violators. For example, factories whose measurement equipment includes radioisotopes, x-ray devices, or sophisticated laser machines can expect that incidents involving this equipment will be the subject of investigations, and possibly enforcement penalties, by these federal agencies.

Civil penalties are also the means by which the states enforce their reporting, record-keeping, and safety-protection rules. After an incident injures a worker, the state agencies responsible for safe workplaces may investigate and may impose a penalty against the employer.

Citizen Suits/Private Rights of Action. When a facility contractor, visitor, guest, or passerby is injured by actions of the company or organization, a lawsuit may result that includes the negligence arguments discussed in the preceding section. In some circumstances, the injured person can use special clauses in regulatory laws to also charge that the defendant owes damages for violating a government rule. This charge adds to the claim that the company or organization must pay the normal tort law damages. But only a minority of laws have been written to provide an extra basis on which to recover for losses by victims. For example, a neighbor may argue that a factory's fuel tank leak ruined its adjacent stream property and also that damages should be awarded under the state water pollution laws as a "private enforcement" of the statewide water quality rules. Because these vary so widely across programs, states, and government agencies, it is simply enough to remind managers that when a violation occurs, the opponents may include not only the government, but also local residents seeking to collect on the consequences of the violation.

Criminal Liabilities

Criminal law exists to punish and deter offenders whose actions harm society, including others who are victims of the offense. Assault, robbery, and other crimes are well known; most workplace injuries involve no criminal liability at all. Unlike the long history of development of civil damages in tort cases for private remedies, the use of criminal sanctions for workplace injury cases is a recent phenomenon. Criminal law is the public's enforcement mechanism—and not the individual's or company's means of getting revenge against people who caused harm. Using criminal sanctions for the prosecution of harm caused by industrial incidents reflects a modern U.S. trend to deter mistakes by adding to the negative consequences of unintentional injuries and losses. Those who wish to prevent liability should be aware that the extra power of criminal enforcement may serve as a tougher sanction against the company and managers—so an investment in cautious compliance makes sense.

This section will discuss the potential for criminal liability consequences of an industrial incident. Only a tiny percentage of injury cases lead to these prosecutions. Of those, only a small percentage ever get to trial because settlement by *plea bargain* is so widely used in criminal cases. The reader should keep in mind that state or federal prosecutors must show the jury proof that goes *beyond a reasonable doubt* before a conviction can be secured. The standard of *beyond a reasonable doubt* is the highest level of proof that is required in the U.S. judicial system. The particular law sets the standard of what actions are criminal—for example, willfully exposing workers to radiation. To the extent possible, defense attorneys will argue that the action or omission by the defendants was not willful, intentional, or otherwise criminal beyond a reasonable doubt, but was merely a mistake due to some level of negligence.

No generalizations can be made about the severity of criminal penalties. Federal sentencing guidelines impose prison sentences of predetermined length for certain classes of crimes, but additional months of prison time can be included in the sentence if the defendant met certain criteria of concealment or recklessness. State laws give trial judges more leeway to select from a range of possible sentences. In general, a criminal conviction for a violation related to an industrial injury will result in prison terms under the federal system but may receive probation under the state sentencing program.

Criminal Remedies versus a Company or Organization. A corporation is legally treated as a "person" under state and federal laws, and it can be fined or penalized in criminal trials, even though the organization itself cannot be sent to jail. The industrial event that leads to criminal indictments against a company will probably also result in charges against supervisors, facility managers, the chief executive, and others.

Criminal Prosecution versus a Company for Negligence. When a criminal law forbids a company from allowing or requiring certain actions by employees, the company could be criminally charged when its actions were negligent. The jury would be asked by the prosecutor to find, beyond a reasonable doubt, that the company's actions were below the level at which a reasonable company would have undertaken this task. For example, a radio tower construction worker's fatal fall may result in a charge that the company negligently failed to comply with a state safety law governing construction safeguards.

Criminal Prosecution versus an Individual for Willful Actions. The criminal prosecution of a supervisor or manager at an industrial site for a safety-related violation of laws is a rare event. The prosecutor will conduct interviews, gather evidence, and try to reconstruct the scene. If the incident was the result of actions or commands by a supervisor that were more than merely negligent, the individual may be criminally charged. These individual prosecutions are unusual, but they offer some deterrence against future incidents.

Criminal Prosecution versus an Individual for Strict-Liability Offenses. A law that seeks to deter managers from allowing injuries to happen may impose *strict liability*, which means the jury can convict the individual with no proof of his or her fault or knowledge of the specifics of the action. For example, a company CEO who is responsible for operating a factory that uses laser cutting tools could be criminally liable for violating the safety standards for lasers that are set by the Food and Drug Administration (FDA). Strict liability means the prosecutors could win a conviction without needing to prove that the CEO had personal knowledge or involvement with the laser equipment, but only that the violation occurred.

Criminal Prosecution versus an Individual for Fraud or Concealment. If an injury occurs and the employer tries to cover up the causes of the incident, a prosecutor can criminally charge the responsible officials with obstruction of justice, violations of the relevant safety statute, or other crimes. Fraud is a deliberate concealment of a material fact from one who has a right to accurate disclosure, such as the federal safety inspectors who inquire about a trench collapse injury at a construction site. The chances of criminal liability being imposed are greatly increased if the government inspectors believe that the company management has lied about the facts or altered records—for example, to falsely claim that required training had been given to the injured worker.

Individual Civil Liability

Protecting the company or organization against liability will be the principal task of safety managers. But in what circumstances will the individual supervisor, manager, or worker be successfully sued for damages, in addition to cases brought against the company?

When the injury situation arises inside a company's own workplace, and the injured person is also employed by the same company, then state workers' compensation law generally bars the "fellow servant" (co-worker) from winning such a suit.

In the case of a serious injury in which insurance covers the defendant company (or the company is large and self-insured), damages will probably be imposed on the company or organization, and most of the time, the injured person will settle with that company rather than pursue a separate lawsuit against one individual employee. When the injury appears to have been the result of misconduct by the employee, the normal practice is to sue both the company and the employee, in case the company successfully avoids being held liable. It would be rare that a company is held not liable but its supervisory employee or manager is held liable.

Negligence is the primary claim for damages lawsuits against individuals, as discussed earlier. The duty is owed by the individual, and the injury arises from that person's acts or omissions. The proof of causation is even more difficult because the claim asserts that this particular employee of the defendant company caused this particular harm.

Assault is the civil charge that alleges some physical contact by the named individual against the injured plaintiff. A tavern patron's claim of assault against a bouncer may be the simplest analogy. Unlike negligent failures to act, some direct adverse act by the defendant was done that harmed this individual person.

Statutory violations such as the OSHA and EPA civil penalty cases discussed earlier are rarely, if ever, brought against individuals. The exception may be in a discrimination claim under which a partially disabled worker alleges that a particular supervisor violated her rights by refusing to correct an unsafe situation at work.

Special Concerns of Government Contractors

Because this chapter encourages the reader to protect against liability, it is useful to note that sometimes the adverse effects of a particular liability case run well beyond one particular lawsuit or administrative penalty. This is, of course, true with bad publicity, which negatively smears the reputation of the company long after a check is written to pay a fine.

Government contractors must pay special attention to avoiding liability in certain categories. The special risks that government contractors face are debarment and disqualification orders. *Debarment* prevents a person who has violated certain laws, such as a serious environmental violator, from performing any responsible job at a federal contractor's facilities. *Disqualification* means the company is "blacklisted" and cannot be a successful bidder for federal contracts, such as for tire sales or for service to postal vehicles. To avoid these unfortunate results, the affected company should carefully maintain its OSHA compliance and should respond promptly and effectively to claims that it discriminated against disabled workers. The details are beyond the scope of this book, but when the safety professional meets with the organization's legal counsel, the subject of contract disqualification as a risk of noncompliance is likely to be addressed.

Prevention Strategies

Historically, the most successful efforts to prevent liability have been through a proactive program that uses audits, training, monitoring, and continual revisions of the internal compliance efforts of the organization. Incident prevention is discussed throughout this text, and, of course, no liability arises if no incidents occur.

The best alternative means of avoiding criminal liability is an effective internal monitoring and self-auditing system, such as has been suggested by the U.S. Sentencing Commission's guidelines, that will justify reduction of criminal sentences in case a violation is prosecuted. An organization's best efforts to avoid violating criminal laws can include classroom and video training sessions, internal and external audits of the quality and effectiveness of internal compliance systems, and top management's expressed commitment to comply with the law. The company should consult with legal counsel regarding methods by which self-audit systems can provide some level of protection against prosecutions.

The best alternative means for reducing civil tort liability is to invest in safeguards that will minimize losses, such as smoke detectors, and that will mitigate the losses that could occur, such as fire suppression equipment near fuel tank farms. The safeguard investment reduces the risk of punitive damages for flagrant or reckless conduct. If an incident occurs, the immediate postincident response should be appropriate and not adversarial to the injured people in order to reduce their interest in aggressive pursuit of damages.

INCIDENT CAUSES AND THEIR CONTROL

Close examination of each incident shows that it can be attributed, directly or indirectly, to an oversight, omission, or malfunction of the management system regarding one or more of the following three items (refer also to the discussion later in this chapter):

1. unsafe practices or procedures; either the worker or another person
2. situational factors, such as facilities, tools, equipment, and materials
3. environmental factors, such as noise, vibration, temperature extremes, and illumination.

If an adequate line management hazard control system is properly designed for the organization's workers, equipment, and environment, then the likelihood of injuries occurring in the workplace is greatly reduced.

Unsafe Practices or Procedures

An unsafe practice is generally described either as a human action departing from prescribed hazard controls or job procedures or practices, or as an action causing a person unnecessary exposure to a hazard. Both workers and management can cause injuries by commission; for example, a worker may sharpen a wood gouge on a grinder without placing the tool on the grinder's rest. In this case, management should ask questions such as, "Was the worker trained properly or pressured to rush the job?" or "Were procedures enforced?" Conversely, a supervisor contributes to the cause of an incident by omission when failing to have an oil spot on the floor wiped up.

An unsafe practice often is a deviation from the standard job procedures. Examples of such actions include the following:
- using equipment without authority
- operating equipment at an unsafe speed or in another improper way
- removing safety devices, such as guards, or rendering them inoperative
- using defective tools.

Unsafe practices also can be a deviation from safety rules or regulations, instructions, or job safety analyses. Why the deviation occurred is the real issue. Some causes and their countermeasures are given in the following examples. When implementing a hazard control program, emphasis should be placed on the countermeasure. The following examples show how to do this.
- No known standard for a safe job procedure exists. Countermeasure: Perform a job safety analysis (JSA) and develop a good procedure through job instruction training (JIT).
- The employee did not know the standard job procedure. Countermeasure: Train employees in the correct procedure.
- The employee knew, but did not follow, the standard job procedure. Countermeasure: Consider an employee performance evaluation. Test the validity of the procedure and motivation.
- The employee knew and followed the procedure. Countermeasure: Develop a safer job procedure.
- The procedure encouraged risk taking, such as incentive pay for piecework. Countermeasure: Change the unsafe job design, procedure, or incentive program.
- The employee changed the approved job procedure or bypassed safety equipment. Countermeasure: Change the method or safeguards so safety measures cannot be bypassed.
- The employee did not follow the correct procedure because of work pressure or the supervisor's influence. Countermeasure: Counsel employee and supervisor; consider change in work procedures or job requirements.

- Individual characteristics, which may involve a disability, made the employee unable or unwilling to follow the correct procedure. Countermeasure: Counsel employee; consider a change in work procedures, workstation design, or job requirements; also consider in-depth training.

Many incidents are the result of someone deviating from the standard job procedures, doing something prohibited, or failing to do something that should be done. In other situations, however, the worker unfairly becomes the target for criticism when other factors actually caused the mishap. The following example illustrates this point. Suppose a newly hired worker, after receiving what was thought to be sufficient instruction on the use of a table saw guard, is required to make a particular cut that cannot be made with the guard in proper position. In this case, the required task causes the worker to remove the guard temporarily so the cut can be made. While removing the guard, the worker's hand slips off the wrench and is cut on the saw blade. Obviously, the worker was instrumental in the incident situation, and consequently, many people would view the procedure as unsafe. A closer analysis, however, reveals that the primary cause was the failure of training to identify equipment limitations.

In this instance, a failure in the management system contributed to the incident. First, a better guard should have been purchased. Second, the new worker should have been instructed more carefully in the use of the guard, including how to remove it when necessary. Most important, a contingency plan should have provided protection for times when the saw has to be used without adequate safeguarding.

An important first step in loss control is distinguishing between worker error and supervisory error, then addressing what caused the system to break down. Don't just look for a "scapegoat"; look for the cause(s). Human error is reduced when:

- supervisors and workers know the correct methods and procedures to accomplish given tasks
- workers demonstrate a skill proficiency before using the particular piece of equipment
- higher management and supervisors consider the relationship between worker performance and physical characteristics and fitness
- the entire organization gives top priority and continuous regard to potentially dangerous situations and the corrective action necessary to avoid unintentional injuries
- supervisors provide proper direction, training, and surveillance. The supervisor must be aware of the worker's skill level with each piece of equipment and process, and adjust the supervision of each worker accordingly. The supervisor shapes worker attitudes and actions by letting employees know that nothing less than safe work practices and the safest possible workplace will be accepted.

Situational Factors

Situational factors are another major cause of unintentional injuries. These factors are materials that make incidents likely and unsafe operations, tools, equipment, and facilities. Examples are unguarded, poorly maintained, and defective equipment; ungrounded equipment that can cause shock; equipment without adequate warning signals; poorly arranged equipment, buildings, and layouts that create congestion hazards; and equipment located in positions that expose more people to a potential hazard.

The following are some causes of situational problems:
- defects in design—for example, a lightweight, unvented metal container for use with flammable materials or no guard on a power press
- poor, substandard construction—for example, a ladder built with defective lumber or with a variation in the space between its rungs
- improper storage of hazardous materials—for example, oxygen and acetylene cylinders stored in an unstable manner and ready to topple over with the slightest impact
- inadequate planning, layout, and design—for example, a welding station located near combustible materials or placed where many workers without eye protection are exposed to the intense light of the welding arc.

An example of a situational problem occurred in a light industrial manufacturing facility where maintenance workers too often found themselves replacing a bearing on an expensive machine. Something had to be done to save downtime, labor, and the cost of the bearing. The industrial engineering and maintenance departments jointly devised the solution: a system that fed oil to the bearing at set intervals, keeping it well lubricated. It was no longer necessary to replace the bearing so often.

But the solution created new hazards. When oil was fed to the bearing, it dripped onto the floor in the aisle adjacent to the machine. Workers could slip on the oil spot and sustain serious injuries. Forklifts drove over the oil. With oil on the rubber wheels, the driver might not be able to stop the vehicle.

Had the maintenance and industrial engineering organizations been thinking of incident prevention, they could have avoided situational hazards by correcting their design. As an interim step, they might have collected the oil by placing a pan under the motor where the bearing was housed. They would then have time to install a tube that would return the oil to the system, thus saving oil while eliminating the hazard.

Environmental Factors

The third factor in incident causation is environmental—that is, the way in which the workplace directly or indirectly causes or contributes to incident situations. Environmental factors fall into four broad categories: human, chemical, biological, and ergonomic.

Human Factors

Noise, vibration, radiation, illumination, and temperature extremes are examples of factors that can influence or cause injuries and illnesses. Operations on a machine lathe, for example, may produce high noise levels that prevent workers from hearing other sounds and impair communication with others or may damage the workers' hearing over time. Thus, workers may be unable to warn one another of a hazard in time to avoid an incident.

Chemical Factors

Classified under this category are toxic gases, vapors, fumes, mists, smokes, and dusts. In addition to causing illnesses, these often impair a worker's skill, reactions, judgment, or concentration. A worker exposed to the narcotic effect of some solvent vapors, for example, may experience a loss of judgment and fail to follow safe procedures.

Biological Factors

Biological factors refer to items capable of making a person ill through contact with bacteria, viruses, fungi, or parasites. For example, workers may suffer boils and inflammations caused by staphylococci and streptococci or experience groin itch caused by parasites.

Ergonomic Factors

See Chapter 16, Ergonomics Yesterday, Today, and Tomorrow.

Sources of Situational and Environmental Hazards

Actions by purchasing agents; those responsible for tool, equipment, and machinery placement and for providing adequate machine guards; and those responsible for maintaining shop equipment, machinery, and tools may result in situational and environmental hazards.

Employee contributions to situational and environmental hazards include disregarding safety rules and regulations by (1) making safety devices inoperative, (2) using equipment and tools incorrectly, (3) using defective tools rather than obtaining serviceable ones, (4) failing to use engineering controls such as exhaust fans when required, and (5) using toxic substances in unventilated areas or without proper protection.

Purchasing agents can be instrumental in creating situational and environmental hazards if they disregard safety engineering recommendations. These agents may acquire tools, equipment, and machinery without adequate guards and other safety devices, especially if such items are selected with only cost in mind. Sometimes, highly toxic and hazardous materials are purchased when less toxic and hazardous materials could be substituted. Other times, purchasing agents fail to acquire from the vendor the necessary warning and control information that must be given to those in charge of the particular process. In many companies, however, the purchasing agent's choices are controlled by engineers, safety professionals, and government or consensus standards and other regulations. The safety professional, in conjunction with engineers, should provide the necessary criteria, specifications, and so on to assist the purchasing agents.

Those involved in layout, design, and placement of equipment and machinery also must consider adequate safeguarding and safety devices or equipment. Otherwise, they contribute to hazardous situations in the workplace. Examples are as follows:
- placing equipment and machinery with reciprocating parts where workers can be crushed between the equipment and substantial objects
- installing electrical control switches on machinery where the operator will be exposed to the hazards of cutting tools or blades in order to start and stop the equipment
- installing equipment without providing for adequate lockout/tagout
- installing equipment and machinery guards that interfere with work operations
- locating high-hazard workstations where they expose workers unnecessarily; for example, placing a welding station in the middle of a floor area instead of locating it in a corner or along a wall where better control over the welding arc is possible.

Those responsible for maintenance—both management and employees, among others—sometimes cause hazards in the workplace. Examples are as follows:
- improperly identifying high- and low-pressure steamlines, compressed air, and sanitary lines
- failing to detect or replace worn or damaged machine and equipment parts, such as abrasive wheels on power grinders
- failing to adjust and lubricate equipment and machinery on a scheduled basis
- failing to inspect and replace worn hoisting and lifting equipment
- failing to replace worn and frayed belts on equipment
- over-oiling motor bearings, resulting in oil being thrown

onto the insulation of electrical wiring and onto the floor, and possible damage to the bearings
- failing to replace guards
- failing to lock out and tag unsafe equipment.

More details are included later in this chapter under Responsibility for the Hazard Control Program.

PRINCIPLES OF LOSS CONTROL

Loss control is the function directed toward anticipating, recognizing, evaluating, and eliminating, or at least controlling, the potentially negative effects of occupational hazards. These hazards generally result from human errors and from the situational and environmental aspects of the workplace (Firenze 1978). The primary function of a loss control system is to locate, assess, and set effective preventive and corrective measures for elements that are detrimental to operational efficiency and effectiveness.

The process exists on three levels:
1. national—laws, regulations, exposure limits, codes, and standards of governmental, industrial, and trade bodies
2. organizational—management of the hazard control program, safety and health committees, task groups, teams, and so on
3. component—worker-equipment-environment.

Loss control can be thought of as "looking for defects." First, there are fewer defects, or failures, than successes. Second, it is easier to agree on what constitutes failure than on what constitutes success. Failure is the inability of a system or a part of a system to perform as required under specified conditions for a specific length of time. The causes of failures often can be determined by answering a series of questions. What can fail? How can it fail? How frequently can it fail? What are the effects of failure? What is the importance of the effects? The manner in which a system, or portion of a system, can exhibit failure is commonly known as the *mode of failure*.

The opposite of failure is not necessarily total success. After all, totally error-free performance is an ideal state, not a reality. Rather, the opposite of failure is the *minimum acceptable* success. This is the condition in which operations are run with a minimum number of losses and interruptions, keeping efficiency and effectiveness of the operation within acceptable limits of control.

Management builds into each of its systems lower and upper limits of control. Each of these interfacing subsystems—maintenance, quality control, production control, personnel, and purchasing, to name a few—is designed to move the system within acceptable limits toward its objective. This concept of keeping operations within acceptable limits gives substance and credibility to the process of loss control. In addition to familiarizing management with the full consequences of system defects, loss control can pinpoint hazards before failures occur. The anticipatory character of loss control increases productivity.

PROCESSES OF LOSS CONTROL

The processes of an effective loss control program should be directed toward evaluating, eliminating, and preventing workplace hazards. Management and safety officials can implement many preventive measures when designing a loss control program. An effective loss control program consists of several steps or processes:
1. hazard identification and evaluation
2. ranking hazards by risk
3. risk assessment
4. management decision making
5. establishing preventive and corrective measures
6. monitoring
7. evaluating program effectiveness, including employee evaluation and/or corrective action and recurrent training.

Hazard Identification and Evaluation

The first step in a comprehensive loss control program is to identify and evaluate workplace hazards. These hazards are associated with machinery, equipment, tools, operations, materials, and the physical facility.

There are many ways to acquire information about workplace hazards. A good place to begin is with those who are familiar with facility operations and the hazards associated with them. (See Appendix 1, Sources of Help, for a description of many organizations that can be of help.) The critical-incident technique (described earlier) is useful for obtaining information from workers and supervisors. Insurance company loss control representatives know which hazards are most likely to cause damage, injuries, and fatalities. In addition to the National Safety Council (NSC), professional societies such as the American Society of Safety Engineers (ASSE), American Industrial Hygiene Association (AIHA), and the American Conference of Governmental Industrial Hygienists (ACGIH) have information about safety and health experience. Manufacturers of industrial equipment, tools, and machinery offer information about the hazards associated with their products, as do suppliers of materials and substances. Labor representatives and business agents can offer a perspective on hazards overlooked by others. Safety and health personnel in organizations doing similar work can be of inestimable value.

A second place to look would be previously prepared inspection reports, either internal (by a safety and health committee or company management and specialists) or external (by local, state or provincial, or federal enforcement agencies). OSHA can supply information describing violations uncovered in similar operations and outlining compliance regulations. (See Chapter 4, Regulatory History, and Appendix 1, Sources of Help, for descriptions of state agencies and private concerns that give onsite inspection and consultation services under OSHA and the National Institute for Occupational Safety and Health [NIOSH].)

Hazard information also can be obtained from incident reports. Information explaining how a particular injury, illness, or fatality occurred often reveals hazards requiring control. Close review of incident reports filed in the past 3 to 5 years will identify the individuals and specific operations involved, the department or section where the incident occurred, the extent of supervision, and possibly the injured person's deficiencies in knowledge and skill.

OSHA incident rates also are useful. Although they are historical and reflect what has happened, not the current status of safety performance, they provide, from a large sample, data that reflect what actually has occurred in the workplace. Other valuable sources can be found in other chapters of this volume, in the *Occupational Safety and Health Data Sheets* of the NSC, and in the specifications for particular equipment and machines published by the American National Standards Institute (ANSI), Underwriters Laboratories Inc. (UL), the American Society for Testing and Materials (ASTM), and the National Fire Protection Association (NFPA). Information about work activities, facilities, and equipment is distributed by NIOSH.

Hazard analysis is another way to acquire meaningful hazard information and a thorough knowledge of the demands of a particular task. Analysis probes operational and management systems to uncover hazards that (1) may have been overlooked in the layout of the facility or the building and in the design of machinery, equipment, and processes; (2) may have developed after production started; or (3) may exist because original procedures and tasks were modified.

The greatest benefit of hazard analysis is that it forces those conducting the analysis to view each operation as part of a system. In doing so, they assess each step in the operation while keeping in mind the relationship between steps and the interaction between workers and equipment, materials, the environment, and other workers. Other benefits of hazard analysis include (1) identifying hazardous conditions and potential incidents; (2) providing information with which effective control measures can be established; (3) determining the level of knowledge and skill as well as the physical requirements workers need to execute specific shop tasks; and (4) discovering and eliminating unsafe procedures, techniques, motions, positions, and actions.

The topic of hazard analysis—its underlying philosophy, the basic steps to be taken, and its ultimate use as a safety, health, and decision-making tool—will be treated in Chapter 9, Identifying Hazards.

Ranking Hazards by Risk (Severity, Probability, and Exposure)

The second step in the loss control process is to rank hazards by risk. Such ranking takes into account the consequence (the severity), the probability, and the exposure index. The purpose of this second process is to address hazards according to the principle of "worst first." Ranking provides a consistent guide for corrective action, specifying which hazardous conditions warrant immediate action, which have secondary priority, and which can be addressed in the future (Figure 6–4).

Severity: Consider the potential losses or destructive and disruptive consequences that are most likely to occur if the job task is performed improperly. Use the following point values for this activity:

1 **Negligible**—probably no injury or illness; no production loss; no lost workdays
2 **Marginal**—minor injury or illness; minor property damage
3 **Critical**—severe injury or occupational illness with lost time; major property damage; no permanent disability or fatality
4 **Catastrophic**—permanent disability; loss of life; loss of facility or major process

The higher numbers are assigned to the most severe consequences. This example uses a four-point scale.

Figure 6–4. Ranking criteria for a hazard consequence (severity).

Once safety personnel or others have ranked hazards according to their potential destructive consequences, the next step is to estimate the probability of the hazard resulting in an incident situation. Quantitative data for ranking hazard probability are desirable, but, almost certainly, they will not be available for each potential hazard being assessed. Whatever quantitative data exist should be part of the risk-rating formula used to estimate probability. Qualitative data—estimates based on experience—are a necessary supplement to quantitative data. Figure 6–5 shows how probability estimates should be made.

After estimating both consequence and probability, the next and final step is to estimate worker exposure to the hazard. The exposure classification scheme in Figure 6–6 is suggested for rating exposure.

Probability: Consider the probability of loss that occurs each time the job is performed. In this evaluation, the key question is How likely is it that things will go wrong when this job is performed? The probability is influenced by a number of factors, such as the hazards associated with the job, the difficulty of performing the job, and the complexity of the job. Use the following point values for this activity:

1 Low probability of loss occurrence
2 Moderate probobility of loss occurrence
3 High probability of loss occurrence

The higher numbers are assigned to those jobs for which the likelihood of loss is greatest. This example uses a three-point scale.

Figure 6–5. Ranking criteria for evaluating probability of loss.

Exposure: Consider the number of employees that perform the job, the number of times an individual employee performs the job, or both. This is an attempt to evaluate the frequency of exposure to the hazards of the job. Use the following point values for this activity:

1 A few employees perform the task up to a few times a day
2 A few employees perform the task frequently
3 Many employees perform the task frequently

The higher numbers are assigned to the jobs with the highest frequency of performance, whether by number of employees, by number of performances, or both. This example uses a three-point scale.

Figure 6–6. Ranking criteria for evaluating hazard exposure.

Risk Assessment

The next step is to assign a single risk number or risk assessment code (RAC). After evaluating the job task according to criteria that are assigned a numerical ranking—severity, frequency, and probability—the point values are added. This produces a single number that allows for a risk ranking. There are many types of RACs. The RAC described in this paragraph combines the three evaluations into a single scale that provides a ranking for job tasks, making it easier to select specific job tasks to analyze. Management can use these values to select particular jobs for immediate analysis (Figure 6–7). The RAC rating scale is a guideline and is not intended to be used as an absolute measurement system.

Management Decision Making

The fourth step involves providing management with full and accurate information, including all possible alternatives, so managers can make intelligent, informed decisions concerning loss control. Such alternatives include recommendations for training and education, better methods and procedures, equipment repair or replacement, environmental controls, and—in rare cases where modification is not

Job Task	Severity 1–4 points	Exposure 1–3 points	Probability 1–3 points	Total
Conduct grinding mill area preshift inspection	3	2	2	7
Start complete grinding circuit	2	1	2	5
Start particle size monitor	4	2	3	9
Change steel rods	3	2	2	7
Routine rod mill shut-down	2	1	2	5
Charge rod mill with ore	2	2	2	6

Figure 6–7. Sample ranking system.

enough—recommendations for redesign. Information must be presented to management in a way that clearly states the actions required to improve conditions. The person who reports hazard information must do so in a manner that promotes, rather than hinders, action.

After management's decision makers receive hazard reports, they normally have four alternatives:
1. Take no action.
2. Modify the workplace or its components.
3. Redesign the workplace or its components.
4. Discuss the hazard and possible methods of elimination with affected workers.

When management chooses to take no positive steps to correct hazards uncovered in the workplace, it usually is for one of three reasons:
1. Management feels that it cannot take the required action. Immediate constraints—be they financial, crucial production schedules, or limitations of personnel—loom larger than the risks involved in taking no action.
2. Management is presented with limited alternatives. For example, it may receive only the best and most costly solutions with no less-than-totally-successful alternatives to choose from.
3. Management does not agree that a hazard exists. However, the situation can require additional consultation and study to resolve any problem.

When management chooses to modify the system, it does so with the idea that its operation is generally acceptable but, with the reported deficiencies corrected, performance will be improved. Examples of modification alternatives are the acquisition of machine guards, personal protective equipment, or ground-fault circuit interrupters to prevent electrical shock; a change in training or education; a change in preventive maintenance; isolating hazardous materials and processes; replacing hazardous materials and processes with nonhazardous or at least less-hazardous ones; and purchasing new tools.

Although redesign is not a popular alternative, it is sometimes necessary. When redesign is selected, management must be aware of certain problems. Redesign usually involves substantial cash outlay and inconvenience. For example, assume that the air quality in a facility is found to be below acceptable standards. The only way to correct this situation is to completely redesign and install the facility's general ventilation system. The cost and inconvenience can be formidable.

Another problem is the fact that the new designs usually contain hazards of their own. For this reason, whenever redesign is offered as an alternative, those making the recommendation must establish and execute a plan to detect problems in design and the early stages of construction so hazards can be eliminated, reduced, or controlled.

One way to expedite decision making regarding actions for loss control is to present findings clearly so that management understands the nature of the hazards, their location, their importance, the necessary corrective actions, and the estimated cost. Figure 6–8 shows a record of occupational safety and health deficiencies and illustrates one approach for recording and displaying hazard information for decision making. It indicates the hazard ranking, the specific location and nature of the hazard, and what costs are likely to be incurred. It also clearly states the recommended corrective action. At a glance, it shows whether the corrective action has been taken and the final cost.

Establishing Preventive and Corrective Measures

After the safety team or others have identified and evaluated hazards and provided data for informed decisions, the next step involves implementing control measures.

Controls are of several kinds:
1. elimination—physically removing a hazard, the most effective hazard control (e.g., moving a device that produced excessive noise to a location away from employees)
2. substitution—involves replacing something that produces a hazard (similar to elimination) with something that does not produce a hazard (e.g., replacing lead-based paint with a water-based paint)
3. engineering (isolation of source; lockout procedures; design, process, or procedural changes; monitoring and warning equipment)
4. administrative (personnel, management, monitoring, limiting worker exposure, measuring performance, training and education, housekeeping and maintenance, purchasing)
5. personal protective equipment (body protection, fall protection, and so on). See Chapter 7, Personal Protective Equipment, in the *Engineering & Technology* volume.

Before control installation takes place, those involved in safety and health activities must understand how hazards are controlled. Figure 6–9 illustrates the three major areas where hazardous conditions can be either eliminated or controlled.
- The first and perhaps best control alternative is to attack a hazard at its source—elimination or substitution. One method is to eliminate or substitute a less harmful agent for the one causing the problem. For example, if a certain solvent is highly toxic and flammable, the first step

RECORD OF OCCUPATIONAL SAFETY AND HEALTH DEFICIENCIES

Location: Shipping
Inspector: Pete Varga

Deficiency No.	Date Recorded	Description of Hazardous Condition	Specific Location	Identification of Acceptable Standard	Hazard Rating Consequence	Hazard Rating Probability	Corrective Action	Estimated Cost of Correction	Date Deficiency Corrected	Resources Used for Correction
S-1	12/11/9-	Ungrounded Tools and Equipment	Throughout Shop	OSHA; Subpart S National Electrical Code, Article 250; 4S	I	A	Provide receptacles with the 3-prong outlet. Test each to make certain it is grounded. Make sure that all tools (other than double-insulated) have a grounding plug.	$5,000	1/9/9-	$4,900

Figure 6–8. One approach to recording and displaying hazard information for decision making. *(Reprinted with permission from RJF Associates Inc.)*

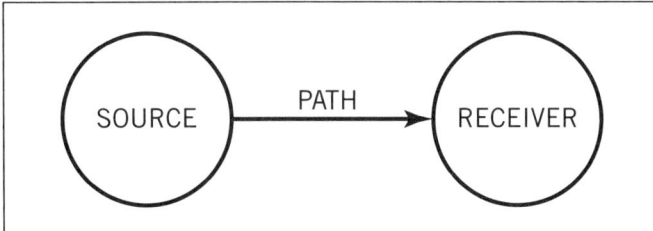

Figure 6–9. Three major areas where hazards can be controlled: the contaminant source, the path it travels, and the employee's work pattern and use of personal protective equipment.

is to determine whether the hazardous substance can be exchanged for one that is nontoxic and nonflammable, yet still capable of doing the job. If a nonhazardous substance meeting these criteria is not available, then a less toxic, less flammable substance can be substituted and additional safeguards employed.

- The second alternative is to control the hazard along its path—engineering controls. This could be done by erecting a barricade between the hazard and the worker. Examples of engineering controls are (1) machine guards, which prevent a worker's hands from making contact with the table saw blade; (2) protective curtains, which prevent eye contact with welding arc flashes; and (3) a local exhaust system, which removes toxic vapors from the breathing zone of the workers.
- The third alternative is to direct control efforts at the receiver—that is, the worker—using administrative controls. Removing the worker from exposure to the hazard can be accomplished by (1) employing automated or remote-control options (e.g., automatic feeding devices on planers, shapers); (2) providing a system of worker rotation or rescheduling some operations to times when there are few workers in the facility; or (3) providing personal protective equipment when all options have been exhausted and the hazard cannot be corrected through substitution or engineering redesign.

Protective equipment may be selected for use in two instances: when there is no immediate way to control the hazard by more effective means and when it is employed as a temporary measure while more effective solutions are being installed. However, major shortcomings are associated with the use of personal protective equipment (PPE):

- Nothing has been done to eliminate or reduce the hazard.
- If the protective equipment (such as gloves or an eye shield) fails for any reason, the worker is exposed to the full destructive effects of the hazard.
- The protective equipment may be cumbersome and interfere with the worker's ability to perform tasks, thus compounding the problem.

Chapter 7, Personal Protective Equipment, in the *Engineering & Technology* volume, discusses these subjects more fully.

Monitoring

The sixth step in the process of hazard control deals with monitoring activities to locate new hazards and assess the effectiveness of existing controls. Monitoring includes inspection, industrial hygiene testing, and medical surveillance. These subjects are covered in Chapter 9, Identifying Hazards.

Monitoring is necessary (1) to ensure that hazard controls are working properly, (2) to ensure that modifications have not so altered the workplace that current hazard controls can no longer function adequately, and (3) to discover new or previously undetected hazards.

Evaluating Program Effectiveness

The final process in hazard control is to evaluate the effectiveness of the safety and health program. Evaluation involves answering the following questions: What is being done to locate and control hazards in the facility? What benefits are being received—for example, reduction of injuries, workers' compensation cases, and damage losses? What impact are the benefits having on improving operational efficiency and effectiveness? What employee evaluation, corrective actions, and recurrent training are occurring?

The evaluation team examines the program to see whether it has accomplished its objectives (effectiveness evaluation) and whether they have been achieved in accordance with the program plan (administrative evaluation, including such factors as schedule and budget). Evaluation must be adapted to (1) the time, money, and kinds of equipment and personnel available for the evaluation; (2) the number and quality of data sources; (3) the particular operation; and (4) the needs of the evaluators.

Effectiveness criteria include the number and severity of injuries to workers compared with work hours, the cost of medical care, material damage costs, facility damage costs, equipment and tool damage or replacement costs, and the number of days lost from injuries or illnesses.

An indicator of the effectiveness of a hazard control program is the experience rating given a company by the insurance carrier responsible for paying workers' compensation. Experience rating is a comparison of the actual losses of an individual (company) risk with the losses that would be expected from a risk of such size and classification. Experience rating determines whether the individual risk is better or worse than the average and to what extent the premium should be modified to reflect this variation. Experience modification is determined in accordance with the experience rating plan (ERP) formula, which has been approved by the insurance commissioners in most states. Loss frequency is penalized more heavily than loss severity because it is assumed that the insured can control small losses more easily than less frequent, severe losses.

ORGANIZING AN OCCUPATIONAL SAFETY AND HEALTH PROGRAM

The purposes of loss control program organization are to help management develop and operate a program designed to protect workers, to prevent and control incidents, and to increase effectiveness of operations. Figure 6–10 illustrates the major organizational components of a safety and loss control program.

Establishing Program Objectives

Critical to the design and organization of a safety and health program is the establishment of objectives and policy to guide the program's development. If the organization has a joint safety and health committee, it could be the body chosen to set the program objectives. It is assumed that those making recommendations to management would be employee representatives, supervisors, middle management, and safety professionals (safety directors, managers, supervisors, and administrators; industrial hygiene technicians and professionals; and fire protection engineers).

Among the program objectives should be the following:
- gaining and maintaining support for the program at all levels of the organization
- motivating, educating, and training the program team to recognize and correct or report hazards located in the workplace
- engineering hazard control into the design of machines, tools, and facilities
- providing a program of inspection and maintenance for machinery, equipment, tools, and facilities
- incorporating hazard control into training and educational techniques and methods
- complying with established safety and health standards
- looking at the overall training the employee is given to perform the work.

Establishing Organizational Policy

Once the objectives have been formulated, the second step is for management to adopt a formal policy. A written policy statement, signed by the chief executive officer or president of the organization, should be made available to all personnel. It should state the purpose of the hazard control program and require the active participation of all those involved in the program's operation. The policy statement also should reflect the following:
- the importance that management places on the health and well-being of employees
- management's commitment to occupational safety and health
- the emphasis the company places on efficient operations, with a minimum of incidents and losses
- the intent to integrate loss control into all operations, including compliance with applicable standards
- the necessity for active leadership, direct participation, and enthusiastic support of the entire organization.

The National Safety Council's data sheet on *Management Safety Policies*, 12304-0585, covers this subject more fully.

After management (in cooperation with labor) has established a safety policy, it should be publicized so that each

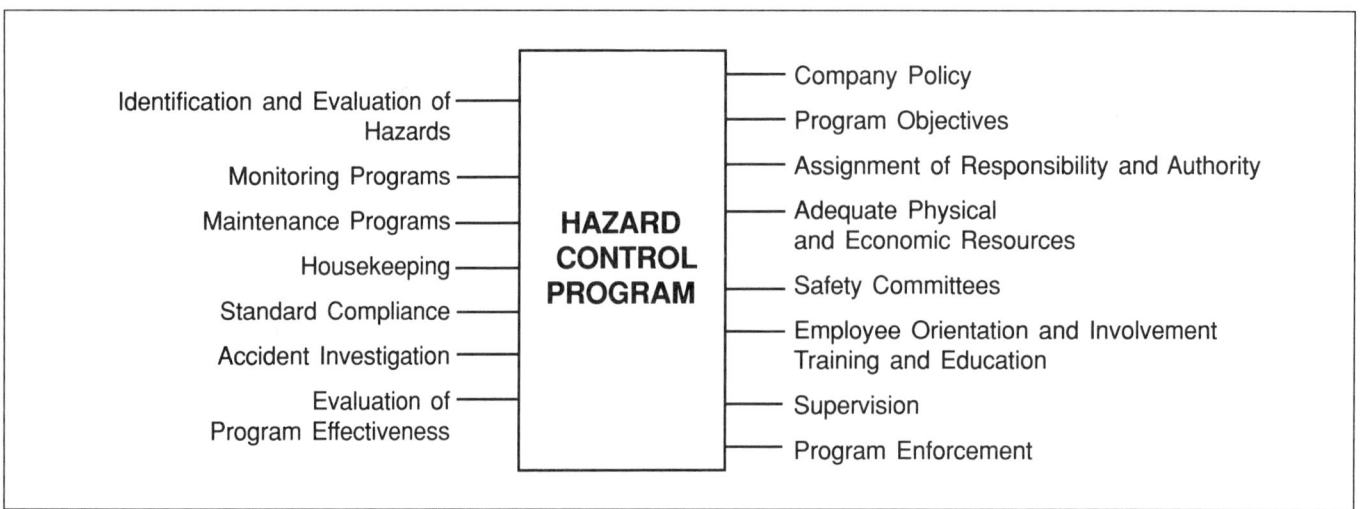

Figure 6–10. Major components of a loss control program. *(Reprinted with permission from RJF Associates Inc.)*

employee becomes familiar with its content, particularly how it applies directly to him or her. Ways to publicize the statement include meetings, letters, pamphlets, and bulletin boards. The policy should also be posted in management offices to serve as a constant reminder of management's commitment and responsibility.

Responsibility for the Hazard Control Program

Responsibility for the safety program can be established at the following levels: board of directors, chief executive officers, managers, and administrators; department heads, supervisors, foremen, and employee representatives; purchasing agents; housekeeping and maintenance personnel; employees; safety personnel; staff medical personnel; and safety and health committees.

Management and Administration

Before any safety program gets under way, it must receive full support and commitment from top management and administration. The president, board members, directors, and other management personnel have primary responsibility for the safety program, which involves the continuing obligation to monitor the program's effectiveness. Management provides the motivation to get the program started and to oversee its operations. Management must initiate discussions with personnel during planning meetings and periodically review the performance of its safety program. Discussions should cover program progress, specific needs, and a review of company procedures and alternatives for handling emergencies in the event an incident occurs.

Specifically, responsibility at this level consists of setting objectives and policy and supporting safety personnel in their requests for necessary information, facilities, tools, and equipment to conduct an effective safety program and to establish a safe, healthy work environment. Management must realize that it is not fulfilling its organization's potential efficiency and effectiveness until it brings its operations at least into compliance with mandatory and voluntary regulatory safety and health standards.

Management and administration must delegate the necessary tasks at all organizational levels to ensure the safety program's success. Although management cannot delegate to others its responsibility for employee safety and health, it can assign responsibility for certain parts of the safety program. However, the authority to act must be delegated along with responsibility. Although authority always starts with those in the highest administrative levels, it eventually must be passed down to middle and line management to achieve the desired results. If safety professionals, safety and health committees, department heads, supervisors, foremen, and employee representatives are to conduct a vigorous and thorough safety program, and if they are to accept and assert the authority delegated to them when circumstances warrant it, they must be fully confident that they have administrative support.

Management must understand that although it can assert authority, it may encounter resistance unless it has enlisted employee support from the earliest stages of the program. If supervisors, employees, and their representatives are not aware of the reasons for and the benefits of an effective safety program, they may resist any changes in their methods of operation and instruction and may do as little as possible to assist the overall program effort.

Management must insist that safety and health information be an integral part of training, methods, materials, and operations. It must guarantee a system in which loss control is considered an important part of equipment purchase and process design, operation, preventive maintenance, and layout and design. It must make sure that effective fire prevention and fire protection controls exist. Management also is responsible for informing subcontractors, at the time of negotiating the contract, of all applicable safety and health standards. Management must see that subcontractors comply fully with company and other safety regulations.

Management is required to safeguard employees' health by ensuring that the work environment is adequately controlled. Managers must be aware that operations producing airborne fumes, mists, smoke, vapors, gases, dust, noise, and vibration have the capacity to cause impaired health or discomfort among their workers. Management must be aware that occupational illnesses beginning in the workplace can take their toll later, even after a worker retires. To protect the future of its employees, management must maintain an effective industrial hygiene monitoring system.

Management should also provide meaningful criteria to measure the success of the safety program and to provide information on which to base future decisions. It must decide the safety program goals for reduced incidents, injuries, and illnesses, and their associated losses.

There are many concrete ways that management can show evidence of its commitment to safety: managing safety and health programs, attending safety and health meetings, conducting periodic walk-through inspections, reviewing and acting on incident reports, investigating incidents, reviewing safety records through conferences with department heads and joint employee–management committees, providing awards, and setting a good example. Management must also audit the program to ensure that it was not "lost" in the management chain.

Department heads, supervisors, foremen, and employee representatives are in strategic positions within the organization to implement safety policies. Their leadership and influence should ensure that safety and health standards are enforced and upheld in each work area and that standards and enforcement are uniform throughout the workplace.

What are some responsibilities of department heads? They ensure that materials, equipment, and machines slated for distribution to their areas are hazard free or that adequate control measures have been provided. They ensure that equipment, tools, and machinery are being used as designed and are properly maintained. They keep abreast of incident and injury trends occurring in their areas and take proper corrective action to reverse these trends. They investigate all incidents occurring within their jurisdiction. They ensure that all safety and health rules, regulations, and procedures are enforced in their departments. They require that a hazard analysis be conducted for certain operations, particularly those that they regard as dangerous, either from past incident history or from their perception of incident potential. They require that hazard recognition and control information be included in instruction, training, and demonstration sessions for both supervisors and employees. They actively participate in and support the safety and health committee and follow up on its recommendations.

Supervisors, foremen, and employee representatives carry great influence. With their support, top management can be assured of an effective safety and health program. Supervisors have a moral and professional responsibility to safeguard, educate, and train those who have been placed under their direction. Thus, they are generally responsible for creating a safe, healthy work setting and for integrating hazard recognition and control into all aspects of work activities. By their careful monitoring, they can prevent incidents.

For all practical purposes supervisors, foremen, and employee representatives are the eyes and ears of the workplace control system. On a day-to-day basis, they must be aware of what is happening in their respective areas, who is doing it, how various tasks are being performed, and under what conditions. As they monitor their areas, they must prevent incidents from occurring. If, despite their controls, they see danger, they must be prepared and have the authority to intervene in the operation and take immediate corrective action. What are the chief safety and health responsibilities of supervisors? They train and educate workers in safe working methods and techniques. They should be certain that employees understand the properties and hazards of the materials they store, handle, and use. They make sure that employees observe necessary precautions, including proper guards and safe work practices. They furnish employees with the proper personal protective equipment, instruct them in its use, and ensure that it is worn.

Supervisors demonstrate an active interest in and comply with loss control policy and safety and health regulations. They actively participate in and support the safety and health committees. They supervise and evaluate worker performance, with consideration given to safe behavior and work methods. They should first try to convince employees of the need for safe performance, and then, unpleasant though it may be, they should administer appropriate corrective action when health and safety rules are violated. Correcting employees requires tact and good judgment. Enforcement should be viewed as education rather than discipline. However, if a supervisor feels that a worker is deliberately disobeying rules or endangering his or her life and the lives of others, then prompt and firm action is called for. Laxity in the enforcement of safety rules undercuts the entire safety and health program and allows incidents to happen. Supervisors monitor their area on a daily basis for human, situational, and environmental factors capable of causing incidents. They should ensure that meticulous housekeeping practices are developed and used at all times. They should correct hazards detected in their monitoring or report such hazards to people who can take corrective action. They should investigate all incidents occurring within their areas to determine causes.

Foremen and employee representatives share much of the responsibility for safety and health with upper management. Their job is to inspect, detect, and correct. What are the specific responsibilities of foremen and representatives? They should encourage other workers to comply with the organization's safety and health regulations. They detect safety violations and hazardous machinery, tools, equipment, and other implements. They take corrective action when possible and report the hazard to the supervisor, along with the corrective action taken or still required. They may participate in incident investigations. They represent the interests of the workers on the safety and health committee.

Practical training aids for employee representatives, foremen, and supervisors do exist. The Council's *Supervisors' Safety Manual* and accompanying "Supervisors' Development Program" have been widely accepted by the industry.

Housekeeping and Preventive Maintenance

These can be regarded as two sides of the same coin. No safety program can succeed if housekeeping and maintenance are not seen as an important element to the overall process.

Good housekeeping reduces incidents, improves morale, and increases efficiency and effectiveness. Most people appreciate a clean and orderly workplace where they can accomplish their tasks without interference and interruption.

An industrial organization, by its very nature, uses tools

that must be kept clean. In some operations, workers use flammable substances and materials requiring special storage and removal. The processes can generate dust, scrap metal filings and chips, waste liquids, scrap lumber, and countless other by-products that must be properly disposed in accordance with applicable regulations.

Housekeeping is a continuous process involving both workers and custodial personnel. A good housekeeping program incorporates the housekeeping function into all processes, operations, and tasks performed in the workplace. The ultimate goal is for each worker to see housekeeping as part of his or her job performance and not as an extra task that someone else should do.

When the workplace is clean and orderly and housekeeping becomes a standard part of operations, less time and effort will be spent keeping it clean, making repairs, and replacing equipment, fixtures, and the like. When the worker can concentrate on required tasks without excess scrap material, tools, and equipment interfering with work, operations will be more efficient and the product will be of higher quality. Time will be used for work, not searching for tools, materials, or parts. When a facility is clean and orderly, employee morale is heightened.

- When everything has an assigned place, there is less chance that materials and tools will be taken from the facility or misplaced. In a few moments before quitting time, the supervisor can determine what is missing with minimal effort.
- Different colors of paint can be applied to tools to identify the department to which they belong. Tool racks or holders could be painted a contrasting color as a reminder to workers to return the tools to their proper places. When stored on a rack, the space directly behind each tool should be painted or outlined in color to call attention to a missing tool. Inadvertent use of materials and supplies is minimized when items are stored in assigned places.

Savings and efficiency are increased when workers minimize spillage and scrap, save leftover pieces of material for use in future projects, and return even small parts to their storage area. When aisle and floor space is uncluttered, movement within the facility is safer, and workers can easily clean and maintain their machinery and equipment. When the facility has adequate work space and when oil, grease, water, and dust are removed from floors and machinery, workers are less likely to slip, trip, or fall or inadvertently come into contact with dangerous parts of machinery.

Chances of fires also are minimized when a workplace is kept free from accumulations of combustible materials that can burn on ignition or, in the case of certain material relationships, spontaneously ignite. Proper storage of chemicals will minimize the existence of hazardous conditions due to the potential incompatibility of these agents. Furthermore, an orderly facility permits easy exit by keeping exits and aisles leading to exits free from obstructions. A neat, well-organized workplace also makes it easier to locate and obtain fire emergency and extinguishing equipment.

Preventive maintenance is typically comprised of orderly, uniform, continuous, scheduled actions to prevent breakdown and prolong the useful life of equipment and buildings. Preventive maintenance is a shared responsibility. Other maintenance duties, such as oiling, tightening guards, adjusting tool rests, and replacing wheels, are routinely performed by workers. Advantages to be gained from preventive maintenance include safer working conditions, decreased downtime of equipment because of breakdown, and increased life of the equipment.

Satisfactory production depends on having buildings, equipment, machinery, portable tools, safety devices, and the like in operating condition and maintained so that production activities will not be interrupted while repairs are being made or equipment replaced. Preventive maintenance prolongs the life of the equipment by ensuring its proper use. When tools are kept dressed or sharpened and in satisfactory condition, the right tool will be used for the job. When safe and properly maintained tools are issued, workers have an added incentive to give the tools better care. When repairs are made quickly, workers do not need to improvise—for example, using a strip of metal as a screwdriver or crowbar. Sound, efficient maintenance management anticipates machine and equipment deterioration and sets up overhaul procedures designed to correct defects as soon as they develop. Such a repair and overhaul system obviously requires close integration of maintenance with inspection.

Preventive maintenance has four main components:
- scheduling and performing periodic maintenance functions
- keeping records of service and repairs
- repairing and replacing equipment and equipment parts
- providing spare parts control.

Maintenance schedules can be set up on several different bases. Factors usually considered include the following:
- manufacturer's recommendations
- age of the machine
- number of hours per day the machine is used
- past experience
- machine changes with use.

The manufacturer's specifications contain standards that need to be maintained for safe and economical use of the machine. These specifications can provide maintenance personnel definite guidelines to follow. Examples of scheduled activities include lubricating each piece of equipment; replacing belts, pulleys, fans, and other parts; and checking and adjusting brakes.

Management should keep two types of maintenance records. The first is a maintenance service schedule for each piece of equipment. This schedule indicates the date the equipment was purchased or placed in operation, its cost (if known), where it is used, each part to be serviced, the kind of service required, the frequency of service, and the person assigned to the servicing. Each piece of equipment also requires a repair record, which includes an itemized list of parts replaced or repaired and the name of the person who did the work.

In addition to scheduled adjustments and replacements, maintenance personnel must repair malfunctioning or broken equipment in accordance with the manufacturer's specifications. Sometimes, equipment must be sent back to the manufacturer or its service representative for repair. Maintenance personnel should be aware of their limitations and recognize that their experience and expertise are not sufficient to do all repairs. Those assigned repair responsibilities require special safety training. Many of the jobs to be performed include testing or working on equipment with guards and safety devices removed. Therefore, a statement of necessary precautions should accompany the repair directive. Maintenance personnel (along with others) have a responsibility to lock out and tag defective equipment, or simply tag if the equipment, such as a ladder, cannot be locked out. (See Chapter 6, Safeguarding, in the *Engineering & Technology* volume for the proper lockout/tagout procedure.)

The second maintenance record itemizes spare parts stock. Management should conduct routine surveys of spare parts requirements. To keep needed repair parts on hand, management should schedule review and reordering of stocked spare parts. When maintenance personnel keep purchasing agents informed of their anticipated stock needs, they help reduce or eliminate downtime that occurs while workers wait for parts.

The difference between a mediocre maintenance program and a superior one is that although the former is aimed at maintaining facilities, the latter is designed to improve them. If conditions are good, a mediocre program keeps them that way but does not make them better; if conditions are not good, a mediocre program does not improve them. Preventive maintenance, on the other hand, is a program of mutual support that creates safe conditions, eliminates costly delays and breakdowns, and prolongs equipment life.

Employees

Employees make the safety and health program succeed. Well-trained and educated employees are the greatest deterrent to damage, injuries, and health problems in the facility or establishment. What are the specific ways a safety program can be rooted in employee involvement and concern? Employees can observe safety and health rules and regulations and work according to standard procedures and practices. They can recognize and report to the foreman or supervisor hazardous conditions or unsafe work practices in the workplace. They can develop and practice good habits of hygiene and housekeeping. They can use protective and safety equipment, tools, and machinery properly. They can report all injuries or hazardous exposure as soon as possible. They can report near misses that may result in future losses. Employees can help develop safe work procedures and make suggestions for improving work procedures. Management should encourage employees to participate on safety committees.

What shapes an employee's attitude toward safety? From the first day an employee goes to work, the employee starts to form attitudes about the job and organization. Substantial if subtle influence is exerted by the attitudes employees observe in management, supervisors, and fellow workers. If these individuals regard safety as vital to an effective operation (a mark of skill and good sense), if they all participate actively and cooperatively in the safety program, and if employees are recognized for having good safety records, then the employees will regard safety as something important, not as window dressing or an empty policy to which others pay lip service.

Introduction to the safety program should come on the employee's first day on the job. A three-pronged approach is suggested (Kane 1979):

1. Coverage of general company policy and rules and discussion of various benefit programs, such as hospitalization, pension plan, holidays, and sick leave. The personnel department usually is responsible for giving this information immediately after the new employee is added to the payroll.
2. Discussion of general safety rules and the safety program. This part of the program should be the responsibility of a safety professional. The company's safety handbook should be given to the new employee, and the company's safety policy statement should be explained. The reasons behind the general safety rules should be explored with the employee, who is more likely to follow rules when they are understood.
3. Explanation of specific safety rules that apply to the new employee's department. At this point the supervisor's role overlaps with that of the safety professional. The supervisor can show how some hazards have been eliminated whereas others that could not be designed out of the operation are guarded against.

This discussion provides an opportunity to talk about safe work practices and emergency procedures as well as to show how engineering controls and personal protective equipment can further reduce the effects of the hazard. The employee's responsibilities for safety and health also must be stressed. These responsibilities include reporting all injuries and incidents, checking equipment and tools before use, operating equipment only with proper authorization and prior instruction, and asking questions when the procedure is unclear.

Figure 6–11 lists the company safety rules developed by the Construction Advancement Foundation SAFE Committee and distributed to construction employees in Indiana.

Purchasing Agents

Those responsible for purchasing items for organizations are in a key position to reduce hazards associated with operations. The purchasing department has much latitude in selecting machinery, tools, equipment, and materials used in the organization. In maintaining standards of quality, efficiency, and price, the purchasing department must ensure that safety has received adequate attention in the design, manufacture, and shipping of items.

Depending on the company organization, other departments—such as safety, engineering, quality control, maintenance, industrial hygiene, and medical—should indicate to the purchasing department what equipment and materials meet with their approval. The purchasing department is responsible for soliciting such guidance and direction. (Purchasing responsibilities are covered in depth later in this chapter under Purchasing.)

First, the purchasing agent must ensure that all items comply with regulatory standards. A statement to this effect must be part of the purchase order. Purchasing agents also will be guided by (1) the standards of ANSI, the Canadian Standards Association (CSA), and other standards and specifications groups; (2) products approved or listed by such agencies as UL and the NFPA; and (3) recommendations by such agencies as the NSC, insurance carriers or associations, the Factory Mutual System in its *Factory Mutual Handbook of Industrial Loss Prevention and Loss Prevention Data*, and trade or industrial organizations. (See Appendix 1, Sources of Help.) Purchasing agents must be aware of any product recalls that may be issued by manufacturers and respond in a timely manner. Each year, federal agencies such as the CPSC and the FDA recall or issue warnings about hundreds of products and devices as a means to prevent harm from those that have been found to be defective. Product hazards may occur because of design flaws, defects, or dangers based on new scientific data about certain materials. Most recalls are carried out voluntarily by manufacturers under the supervision of these federal agencies. Purchasing agents can play an important role in ensuring that this product safety issue is performed as designed. However, this process is not widely understood, and these recalls are often not as effective as they could be.

COMPANY SAFETY RULES

All Employees Will Abide By The Following Rules:
1. Report unsafe conditions to your immediate supervisor.
2. Promptly report all injuries to your immediate supervisor.
3. Wear hard hats on the job site at all times.
4. Use eye and face protection where there is danger from flying objects or particles, such as when grinding, chipping, burning and welding, etc.
5. Dress properly. Wear appropriate work clothes, gloves, and shoes or boots. Loose clothing and jewelry should not be worn.
6. Never operate any machine unless all guards and safety devices are in place and in proper operating condition.
7. Keep all tools in safe working condition. Never use defective tools or equipment. Report any defective tools or equipment to immediate supervisor promptly.
8. Properly care for and be responsible for all personal protective equipment.
9. Be alert and keep out from under overhead loads.
10. Do not operate machinery if you are not authorized to do so.
11. Do not leave materials in aisles, walkways, stairways, roads or other points of egress.
12. Practice good housekeeping at all times.
13. Do not stand or sit on sides of moving equipment.
14. The use of, or being under the influence of, intoxicating beverages or illegal drugs while on the job is prohibited.
15. All posted safety rules must be obeyed and must not be removed except by management's authorization.
16. Comply at all times with all known federal, state and local safety laws as well as employer regulations and policies.
17. Horseplay causes accidents and will not be tolerated.

Violations of any of these rules may be cause for immediate disciplinary action.

Figure 6–11. Company safety rules. *(Reprinted with permission from the Construction Advancement Foundation SAFE Committee.)*

Second, the purchasing agent must ensure that tools, equipment, materials, machinery, and chemicals are purchased with adequate regard for safety. This requirement applies even to ordinary items such as boxes, cleaning rags, paint, and common hand tools. It is essential to compare safety features among the various brands when purchasing personal protective equipment and larger items, especially machines. Sometimes, the cost of an adequately guarded machine seems out of proportion to that of an unguarded machine to which makeshift guards can be added. But experience has repeatedly proven that the best time to eliminate or minimize a hazard is in the design stage. Safeguards that are integral parts of a machine are the most efficient and durable.

Third, the purchasing agent must be cost conscious, realizing that every incident has both direct and indirect costs. The agent should understand that the organization cannot afford bargains that later result in injuries and occupational illnesses.

Professionals in Loss Control

Safety, health, and environmental professional staff can include loss control specialists, industrial hygienists, medical personnel, and support staff, depending on the size of the organization. Some of these functions may be handled by contracted consultants.

PURCHASING

The safety department should have excellent liaison not only with the engineering department but also with the purchasing department, as discussed in the previous section. The safety department should coordinate or develop written safety standards to guide the purchasing department. These standards should help eliminate the hazards associated with a particular kind of equipment or material—for example, by substituting a safe material for a dangerous one, or safeguarding equipment for the protection of worker, machine, and product.

The purchasing agent is not closely concerned with educational and enforcement activities but is vitally concerned with many phases of engineering activities. The agent selects and purchases the various items of machinery, tools, equipment, and materials used in the organization. The agent also—in part or to a considerable degree—sees that safety and health have received adequate attention in the design, manufacture, and shipment of these items.

In one facility, a lead hazard occurred when workers unloaded litharge (lead oxide) that had been shipped in 10-gallon paint pails with covers. Some of the litharge had leaked, coating the pails with a film of litharge on the outside. When the pails were moved, litharge was released into the air, creating a lead concentration 30 to 40 times the permissible limit and contaminating the skin and clothing of workers.

Management and safety workers tried several solutions before finding the best approach. They eliminated the hazard by instructing the purchasing department to specify a rubber gasket under the pail lid as a part of the purchasing requirements. Thus, the leakage, which created a serious health hazard, was easily controlled. The company also required that the shipper label the containers with information meeting the requirements of the OSHA hazard communication standard.

Purchasing-Safety Liaison

Having background knowledge of engineering specifications and codes and standards, the safety professional should be well prepared to advise the purchasing department.

What Purchasing Can Expect

The purchasing agent can reasonably expect that the safety professional will do the following:
- give specific information about process and machine hazards that can be eliminated by changing the design or by installing manufacturer-designed guarding
- supply similar information about other equipment, tools, and materials along with facts about injuries caused
- give specific information about health and fire hazards in the workplace
- provide information on federal and state safety requirements
- supply, on request, additional special information on incident experience with machines, equipment, or materials when such articles are about to be reordered
- request assistance in the investigation of incidents that may have been caused by faulty equipment or material
- request a list of equipment and materials requiring safety approval before going out for bids.

What the Safety Department Can Expect

When there is effective liaison between the safety and purchasing departments, the safety professional can expect that the purchasing agent will do the following:
- become familiar with the departmental and facility process hazards, especially in relation to machinery, equipment, and materials
- ask the safety department for information on hazards and incident costs, for federal and state safety requirements, and for lists of approved devices and appliances before making purchases
- become acquainted with the specific location and

departmental use of machinery or equipment about to be ordered
- participate in incident investigations when injuries may have been caused by failure of machinery, equipment, or materials.

Safety Considerations

The purchasing staff should always review applicable safety specifications and guidelines before purchasing supplies and equipment. For many articles, there is no need to consider safety. Some items, however, have a more important bearing on safety than may be suspected.

The agent must exercise utmost caution when purchasing personal protective equipment, such as eye protection, respirators, and masks; equipment to move suspended loads, such as ropes and chains; equipment to move and store materials; and miscellaneous substances and fluids for cleaning and other purposes that may constitute or aggravate a fire or health hazard. The agent should specify adequate labeling that identifies contents and calls attention to hazards.

Investigation, however, may show that unsuspected hazards arise in the purchase of ordinary items, such as common hand tools, reflectors, tool racks, cleaning rags, paint for shop walls and machinery, and even filing cabinets. Among the factors purchasing agents need to consider include maximum load strength; long life without deterioration; reduction of sharp, rough, or pointed characteristics; less frequent need for adjustment; ease of maintenance; reduction of fatigue-causing characteristics; and minimal hazard to workers' health.

Here are a few examples of hazards created by purchased items that were considered safe. Goggles supplied to one group of workers were found to have imperfections in the lenses that caused eyestrain and headache, which led to fatigue and incidents. The toes of a laborer were crushed because his safety shoe had an inadequate metal cap and collapsed under a weight that would have been easily supported by a shoe meeting the ANSI Z41 standard. In another facility, workers were supplied with wooden carrying boxes, when a proper type of metal box could have eliminated the hazard of splinters and perhaps an infected hand. More impressive examples of purchasing for safety are found in the purchase of larger items, especially machines. Today, machines of many types are manufactured and can be bought with adequate safeguards in place as integral parts of the machine. The enclosed motor drive is an outstanding example of engineering machine construction for safety.

When an order for equipment is about to be placed, the purchasing agent, if possible, should not consider any machine that has been only partly guarded by the manufacturer and needs to be fitted with makeshift safeguards. The agent should consult frequently with the safety department before making any purchase for which safety is a factor. The agent also should ensure that every purchased machine complies fully with the safety regulations of the state in which it is to be operated.

Price Considerations

When considering facility purchases, the purchasing agent must struggle to reconcile quality, work efficiency, and safety with the price of an item. As an example, a purchasing department was offered a distributor's discount if it ordered a larger quantity of a chemical that the facility used routinely as part of the manufacturing process. This agent had some inherent hazardous characteristics and needed to be handled with caution. Without adequate storage space for this larger order, a spill occurred, resulting in employee injury. In addition, additional hazardous waste disposal costs will be incurred if the larger quantity of chemicals is not used up.

At times, the cost of a well-safeguarded machine may seem out of proportion to the cost of an unguarded machine, including the estimated expense of adding homemade safeguards. But experience has proved that the best time to safeguard a machine or process is in the design stage. Safeguards planned and built as integral parts of a machine are the most efficient and durable.

The purchasing agent, through incident information supplied by the safety department, should be familiar with the costs of specific incidents in the facility, especially those in which a purchased item has been found to contribute to an incident. As a result, the purchasing agent should be able to defend the decision to buy a slightly higher-priced component when appropriate. Cooperation among the safety, engineering, manufacturing, and purchasing departments is absolutely necessary if injuries and illnesses are to be eliminated.

These arguments should appeal to all executives responsible for the success of the industrial organization. The executive who is already sold on safety is agreeable to expenditures reasonably justified in the interest of injury and illness prevention. If an executive does not have this attitude, the purchasing agent should find ways to stimulate the executive's interest in the organizational safety program. In this undertaking, the purchasing agent can undoubtedly count on the active cooperation of the safety professional.

In some instances, the purchase of machinery or equipment involves important engineering details. For such purchases, the company undoubtedly has a system whereby engineers first prepare definite specifications, perhaps including drawings. The safety professional then carefully checks these plans and specifications before the purchasing agent solicits bids and cost estimates. The purchasing agent has the plans and specifications at hand when asking for prices. He or she keeps in close touch with the safety professional throughout the negotiations to use the latter's knowledge and experience in injury and illness prevention.

After the purchase order has been made but before it is signed, one other important detail should not be overlooked: a statement, in language that cannot possibly be misinterpreted, that the articles ordered must comply fully with the applicable federal and state safety laws and regulations of the locality in which they are to be used. This statement must be made a part of the purchase order.

Codes and Standards

In purchasing, the safety professional must have a thorough knowledge of the facility's incident history, the costs involved in injuries and illnesses, and the probable benefits of changes suggested. To fulfill this function in cooperation with the purchasing department, the safety professional must be familiar with codes and standards. When a specific item of equipment is recommended, the safety professional should be able to state that it is a type approved by authoritative bodies and that it meets regulatory requirements.

Generally, the safety professional should consult with everyone concerned before setting up company standards to guide the purchasing department. Many guidelines and standards can be used as models. Accordingly, the safety professional (and all others concerned with setting company standards) should be familiar with the following:

- codes and standards approved by ANSI and other standards and specifications groups—see Appendix 1, Sources of Help
- codes and standards adopted or set by federal, state, and local governmental agencies, such as OSHA, the Bureau of Mines, and the National Bureau of Standards
- codes, standards, and lists of approved or tested devices published by agencies such as NIOSH, MSHA, UL, and fire protection organizations; for fire protection, the standards and codes of the NFPA (see Appendix 1, Sources of Help)
- safe-practice recommendations of such agencies as the NSC, insurance carriers or their associations, and trade and industrial organizations.

Specifications

The engineering department, with the help of the safety department, should specify the necessary safeguarding to be built into a machine before it is purchased. People responsible for purchasing in an industrial facility are necessarily cost conscious. Consequently, the safety professional must become aware of incident costs associated with specific machines, materials, and processes. For instance, if the individual recommends spending several thousand dollars for a superior grade of tool, he or she must have evidence to justify the investment.

Because of highly competitive marketing, manufacturers of machine tools and processing equipment often list safety devices as accessories. The safety professional must be familiar with regulatory-required auxiliary equipment and be able to justify its inclusion in the original order.

In some organizations, the safety professional is charged with checking all plans and specifications for machinery and other equipment. In many organizations, particularly where certain items, such as goggles or safety shoes, are to be reordered from time to time, various operating officials cooperate to prepare standard lists, and purchases are selected only from among the types and companies shown on these approved lists. In still other establishments, the responsibility for design, quality, safety, and other features rests with the employees who are to use the articles. In such cases, the purchasing agent is responsible only for price, date of delivery, and similar details.

In many companies where purchases are made in huge quantities and at a great investment of money, important duties are placed in three coordinated departments: (1) the engineering department, whose staff prepare plans and specifications for all machinery and equipment to be purchased; (2) the safety department, where the staff carefully check these plans and specifications for safety and carry out final inspections of articles purchased; and (3) the purchasing department, which still has latitude in making selections as well as in determining standards of quality, efficiency, and price.

Still another variable must be mentioned. Many companies have both a full-time purchasing agent and a full-time safety professional. However, in many other companies, especially smaller ones, these important duties are assumed by executives who devote part of their time to other activities. Nevertheless, the measures that should be taken to prevent unintentional incidents in the small facility are substantially the same as those taken in the large facility. The part the purchasing agent can play in the safety program is similar; the interest is the same and the activities vary only by degree. Success lies in adopting as fully as possible all the suggestions that are presented here and in cooperating closely with others in the company to promote safety for all workers.

Specification of Shipping Methods

When the purchasing agent orders materials, it may be desirable to specify that they be shipped in a particular manner. If safe and efficient shipping methods are worked out and then specified in the orders, the suppliers will be better able to deliver materials on time, in good condition, and in a shape or form that can be easily and safely handled by employees.

The agent should specify that all hazardous materials must be labeled with DOT-authorized shipping labels. Safety Data Sheets (formerly Material Safety Data Sheets) should accompany all chemicals or products that contain chemicals, as well as some substances such as solid metal that are designed to be remelted or equipment such as welding rods.

SAFETY AND HEALTH COMMITTEES

Safety and health committees can be invaluable to the loss control program by providing the active participation and cooperation of many key people in the organization. They also can be unproductive and ineffective. The difference between success and failure lies with the original purpose of the committee, its staffing and structure, and the support it receives while carrying out its responsibilities.

A safety and health committee is a group that aids and advises both management and employees on matters of safety and health pertaining to facility or company operations. In addition, it performs essential monitoring, educational, investigative, and evaluative tasks.

Committees may represent various constituencies or levels within the organization or may be management or workplace committees. The joint safety and health committee (discussed here) is responsible for the following:

- actively participating in safety and health instruction programs and evaluating the effectiveness of these programs
- regularly inspecting the facility to detect unsafe conditions and practices and hazardous materials and environmental factors
- planning improvements to existing safety and health rules, procedures, and regulations
- recommending suitable hazard elimination, reduction, or control measures
- periodically reviewing and updating existing work practices and hazard controls
- assessing the implications of changes in work tasks, operations, and processes
- field-testing personal protective equipment and making recommendations for its use or alteration based on the findings
- monitoring and evaluating the effectiveness of safety and health recommendations and improvements
- compiling and distributing safety and health and hazard communications to the employees
- immediately investigating any workplace incident
- studying and analyzing incident and injury data.

Section 2(b)(13) of the OSH Act clearly allows for the possibility of joint safety and health initiatives as a supplementary approach to accomplishing OSHA's objectives more effectively. Joint committees have considerable potential for reducing injuries and illnesses, thus leaving OSHA free to target enforcement according to the worst-first principle.

The joint committee concept stresses cooperation and a commitment to safety as a shared responsibility between management and workers. Employees can become actively involved in and make positive contributions to the company's safety and health program. The committee serves as a forum for discussing changes in regulations, programs, or processes and potential new hazards. Employees can communicate problems to management openly and face-to-face, allowing information and suggestions to flow both ways. The knowledge and experience of many people combine to accomplish the objectives of creating a safe workplace and reducing incidents. The approach can produce effective solutions to safety problems more easily. Because joint committees facilitate communication and cooperation, they usually raise employee morale as well.

Even though a joint committee represents both employees and management, the committee's analyses and recommendations—whether they pertain to policy or practice—should be reviewed and confirmed by experts when they relate to specialized areas (such as electrical safety or exposure levels).

Labor/management cooperation was discussed in Chapter 1, Historical Perspectives. Committee organization and operation are covered in the NSC publication "You Are the Safety and Health Committee."

OFF-THE-JOB SAFETY PROGRAMS

According to National Safety Council estimates, 9 out of 10 deaths and about 70% of the medically consulted injuries suffered by workers in 2011 occurred off the job. While more than 13 times the number of deaths occurred off the job compared to on the job (13.3 to 1), more than twice as many medically consulted injuries occurred off the job (2.6 to 1). Production time lost due to off-the-job injuries totaled about 235,000,000 days in 2011 compared with 60,000,000 days lost by workers on the job. Off-the-job injuries cost the nation at least $262.2 billion in 2011 compared with $188.9 billion for on-the-job injuries. Clearly, reducing the number and severity of off-the-job injuries should be a major concern for all industries.

There is a certain amount of confusion, however, as to what off-the-job safety really includes. Essentially, *off-the-job safety* is a term used by employers to designate the part of their safety program directed to employees when they are not at work.

The principal aim of off-the-job safety is to get employees to follow the same safe practices used on the job while pursuing outside activities. Therefore, off-the-job safety should not be a separate program but rather an extension of a company's on-the-job safety program. Although companies have a legal responsibility to prevent injuries on the job, they have a moral responsibility to try to prevent injuries away from the job. The other reason for an off-the-job safety program is cost. Operating costs and production schedules are affected as much when employees are injured away from work as when they are injured on the job. (These costs are discussed in Chapter 10, Incident Investigation, Analysis, and Costs.)

What Is Off-the-Job Safety?

Off-the-job (OTJ) safety is a logical extension of the occupational safety program. Injury/illness prevention at work is cost effective, while fulfilling an organization's moral and legal responsibilities. An effective off-the-job safety program meets these same needs: reduction of costly employee absences due to incidents, injuries, or deaths and commitment to employee well-being. Preventing off-the-job incidents that could result in injury or illness can be accomplished by using methods proven successful for increasing safety awareness at work.

Responsible employers should also be concerned about the well-being of their employees' families. Many times, injury to a family member affects an employee's work performance. Family involvement in safety can be a key factor to reducing employee off-the-job injuries. This approach can be accomplished in many ways, but education and peer pressure are extremely important to the success of an OTJ safety program.

Complicating the company's efforts is the fact that current methods of gathering, recording, and measuring employee OTJ injury information vary substantially from company to company. Only when companies with OTJ safety programs treat their injury data uniformly can they compare their experience to other companies and help determine OTJ injury rates nationwide. While compiling OTJ injury data, companies should follow the practices recommended in ANSI Z16.3, Recording and Measuring Employee Off-the-Job Injury Experience.

Program Benefits

The aim of safety education—namely, changing the employee's behaviour—is especially true of off-the-job safety. No asset is more important to a company than its employees. They should not only be protected during working hours, but also be given every incentive to practice safety off the job. A company can realize three benefits from expanding its safety program to include off-the-job safety. The first is a reduction in lost production time and operating costs from both on-the-job and off-the-job injuries. Second, companies have found that efforts in off-the-job safety increase employees' interest in their on-the-job safety program. The third benefit, often overlooked, is that of better public relations.

Many benefits are derived by developing or revitalizing an off-the-job safety program, including the following:
- fewer off-the-job incidents, injuries, and deaths
- fewer employee absences
- reduced operating costs
- safety awareness at home carried over to the workplace
- improved work efficiency and performance
- enhanced employee and employer relationships
- participation of employees and their families in community safety actions
- positive, viable demonstration of the organization's commitment to employee-family well-being and social issues affecting the employee's family.

As with any effective occupational safety program, management involvement and participation at all levels in off-the-job safety must be vocal, visible, and continuous.

Promoting Off-the-Job Safety

Techniques for promoting off-the-job safety are essentially the same principles and techniques used on the job. From a safety standpoint, operating power tools at home involves the same risks as operating the equipment at work; likewise, driving the family car is the same as driving a company vehicle.

The only difference between these two safety programs is that organizations must depend more on education and persuasion to get their message across. This is because once employees leave the office, facility, or job site, they tend to believe that the risks they assume are a private matter and no longer fall under the organization's policies and guidelines.

As with any other program—such as attendance, quality control, or waste reduction—management support and guidance are essential. Once safety personnel show management the seriousness of the problem (through experience and cost records), there should be little difficulty in obtaining support.

Off-the-Job Safety Policy

The organization should communicate management commitment to employees and their families. For example, a written policy statement concerning off-the-job safety or reference to off-the-job safety should be in the occupational safety policy statement signed by the top organizational official.

Getting Started

Various methods and sources of information on off-the-job injuries are available to an organization. For example, management can keep records documenting the causes of employee absences due to unintentional injuries away from the workplace; health and injury insurance claims may record the incident cause on the form for payment; and the NSC's annual publication *Injury Facts* can be used to pinpoint the leading causes of unintentional deaths and injuries nationally (these data help to determine the most common incidents occurring in a particular organization).

It is important to tailor an off-the-job safety program to the special needs of an organization. The location of the facility, its environment, and the special interests of employee groups such as skiing, boating, mountain climbing, cave exploring, hunting, camping, or flying are factors that can help in selection of topics and activities.

The types and number of injuries/deaths people suffer off the job, however, are not usually related to how dangerous the activities appear to be. Many injuries occur when people are doing ordinary, everyday activities that do not appear to be risky, such as lifting objects, working on the car, walking up and down stairs, or taking a shower.

Select Program Details

A good topic breakdown provides a solid structural framework for the development of off-the-job safety programs. For example, content may be based on seasonal hazards; on home, traffic, and public injuries; on health risk assessment, such as exercise and fitness, stress management, alcohol and drugs, and community right-to-know; or according to injury types and causes.

Timely, interesting, and practical topics will attract the attention of employees, create and maintain enthusiasm, and encourage active participation to develop patterns of safe behavior.

A seasonal emphasis outline may include the following:
- spring—good housekeeping, lawn mowers, garden tools, do-it-yourself activities, pruning/planting trees, bicycles/helmets
- summer—sunburn, swimming, camping, boating, hiking, field sports, fishing, insects, vacation hazards
- fall—hunting, home power tools, back-to-school hazards, home-heating equipment, yard cleanup, repair and storage of tools
- winter—winter sports, holiday safety, severe weather, overexertion, winter driving.

Topics can tie into programs of national scope or interest. The following national programs provide radio, TV, newspaper, and other forms of publicity that help promote program content: National Child Passenger Safety Awareness Week in February, National Safe Boating Week in May, National Fire Prevention Week in October, and National Drunk and Drugged Driver Awareness Month in December.

A program on home fire safety may include showing a video on the proper use of fire extinguishers or on general fire prevention or safety practices as well as handout literature on fire safety topics. Combined with this program could be a company discount for employee purchase of smoke alarms, fire extinguishers, and fire escape ladders for home, workshop, and auto use.

Employee, Family, and Community Involvement

The assistance of a special off-the-job safety group may be valuable. Membership in this group can include employees, family members of employees, local civic and school groups, and community people with an interest or role in safety in general. Organizations that use such a committee often find that its members contribute immeasurably to the success of the program by providing special information that represents their background and understanding of off-the-job safety. Another benefit of the group is that members share serious safety convictions with peers, friends, neighbors, and their families.

Every community has special-interest groups and organizations already concerned with various phases of off-the-job safety that will lend their resources and personnel to assist in company activities. Among such groups are the NSC; Chambers of Commerce, including the Jaycees; service clubs; Red Cross chapters; local newspapers; radio and TV stations; health, police, and fire departments; parent–teacher associations; rescue squads; and emergency-service groups. The organizations may adopt home, traffic, or recreation safety as a project for the year. In so doing, they ensure the participation of many stay-at-home parents and family members who are not exposed to a formal safety program. (See also Chapter 33, Safety Awareness Programs.)

A representative from safety and health disciplines, such as the ASSE, can present an off-the-job safety program to an organization's members or employees. He or she can obtain assistance from state safety organizations; medical, visiting nurse, and other associations; local utilities; poison control centers; state police; insurance companies; and public health groups. A few of these groups maintain incident records, some participate in special programs, some publish bulletins on health and safety subjects, and some conduct courses in subjects related to off-the-job safety. Safe behavior booklets and posters on seasonal activities are available from the NSC.

An organization can cooperate with the municipal recreation department in promoting swimming classes or courses in boating safety. Also, it can request that local police supervise an auto inspection clinic. Often, insurance companies will provide the necessary equipment for testing drivers' physical qualifications and skills.

Organization personnel can, in turn, offer leadership and support for community activities. For instance, employees can help form local safety councils, assist schools and churches by making safety inspections on request, or act as volunteer members of local fire departments.

Several small companies in a community may consider the possibility of pooling their resources and talents in a joint off-the-job safety program, as is sometimes done with disaster and rescue programs. Advance publicity ensures employee support by communicating objectives, plans, and activities. An informal letter from top management sent to employees' homes personalizes the organization's concern for the safety of not only the employees, but their families as well.

The initial meeting for employees can be followed by other meetings that include family members. Many organizations hold picnics and other outings featuring various types of entertainment and safety exhibits as a means of reaching the families. An organization's newsletter, magazine, or paper is an effective way to carry the word to employees. Company bulletin boards often reinforce these messages. The community can be kept informed of plans and progress through spot announcements and stories on home, traffic, and recreational safety carried by the local radio and TV stations and the local newspapers. Preparation of such publicity can be financed by several companies together or by the local safety council.

Meetings of local organizations—such as church groups, women's and service clubs, PTAs, and similar groups—are also good resources to educate the community about off-the-job safety programs and activities.

Programs That Work

The following sections discuss promotional methods that have proven successful. They can easily be tailored to the needs of any organization.

Contests

Offering some incentive, as minimal as a savings bond or as generous as a college scholarship, can usually elicit good participation from employees' children. The resulting posters, essays, calendars, or slogans can then become vehicles for promoting safety.

The two most widely used contest ideas involve essays (with a limit on the number of words) or posters (with restrictions on size and materials), with themes such as "What My Dad's/Mom's Safety Means to Me," "Vacation Safety," "Community Safety," and "The Importance of Off-the-Job Safety." Competitors should be divided into age groups, from about age 5–7 for the youngest up to age 16–18 for the oldest. Each entrant should receive a token of participation, such as a keychain, certificate, embroidered patch, or even a model of the organization's product. To ensure impartiality, judges may be from outside the community or the organization.

Traffic Safety

Offering the NSC's Defensive Driving Program, at the organization's expense and possibly with organization facilities and instructors, to all driving members of the employees' families is one of the most positive home, off-the-job, or community safety efforts an organization can make. It can improve the driving habits of those who drive and can also promote car pooling as a safe answer to energy, traffic, and pollution problems. As an incentive to participate, several states offer insurance discounts on personal automobile insurance policies.

Auto safety checks tie in well with vacation and holiday programs and work best on weekends. Have plenty of qualified inspectors available so participants will not be discouraged by long lines.

Company Picnic

Safety picnics should involve the whole family and can be as expensive as a completely catered affair or as simple as a family picnic. A safety theme can be included in drawings for door prizes, activities for all ages, and presentations of awards to or recognition of employees for safety achievements.

Family Night

Quite different from picnics, family night gatherings are built around the presentation of some discussion or audiovisual presentation on safety, accompanied by refreshments. Sometimes, the theme shifts from general safety to a program to make the family aware of and gain support for the safety efforts made at work. Sometimes, an "open house" tour of the facility can be combined with the family night get-together.

Family first-aid programs, including CPR training, can be held at business locations after business hours or on weekends to involve the family. First-aid kits available at a company discount purchase price could be offered for family use in the home, auto, or workshop.

Youth Activities

Sponsorship of activities such as softball and football teams and bicycle rallies usually is not aimed directly at promoting safety. However, a poor safety record among young people taking part in the activity may hurt the sponsor's image. Therefore, financial sponsorship of such activities should be only part of an organization's participation. Employee leaders who are knowledgeable in the activity and trained in safety and first aid (at the organization's expense) should be on hand to teach and help participants. Such actions can produce a strong, positive image of the organization in the community. Involvement of groups such as the Boy and Girl Scouts, 4-H clubs, Camp Fire USA, FFA, and FHA will expand off-the-job safety efforts.

Recreational Programs

In any organization, many employees and their families are sports enthusiasts. Their pastimes may include hunting, boating, camping, swimming, fishing, and skiing. At the season openings of these activities, organizations may sponsor clinics featuring registered/competent instructors to check equipment and to provide instruction on improving skills. Sources of help include local gun clubs, powerboat squadrons, the Coast Guard Auxiliary, the National Recreation and Park Association, the President's Council on Physical Fitness and Sports, the National Red Cross, YMCA, police and fire departments, and health organizations. All of these groups can provide ideas for safety activities.

Vacation/Holiday Program

Some organizations close down for regular summer vacation; others offer year-round vacation periods and 3-day weekends. A good time to offer safe driving tips and safety literature to employees is at vacation times.

Other promotional methods include the following:
- Publicity can be distributed in-house that reinforces safety, both on the job and off the job (pamphlets, press releases, posters, and billboards are ideal vehicles for in-house promotions). The NSC has an excellent selection of this material, aimed at a variety of off-the-job safety topics. Employee-generated, original posters are also effective in personalizing safety efforts.
- Nearly every organization has some sort of in-house journal or newsletter that is either given to employees at work or sent to their homes. No other publication enjoys wider readership within an organization. As a result, it provides an excellent forum for safety education and safety program promotion; both employees and their families see it. Articles can be written on all aspects of a safety program, and the employee and family can be solicited for safety-related story ideas as well.
- NSC's periodical *Safety + Health* is an excellent way to promote off-the-job safety.

SUMMARY

- Hazards are defined as any existing or potential condition in the workplace that, by itself or interacting with other variables, can result in deaths, injuries, property damage, and other losses. The two broad categories of hazards are (1) those dealing with safety and injuries and (2) those dealing with health and illnesses.
- The primary function of loss control is to help management locate, assess, and set effective preventive and corrective measures for hazards. Controlling hazards is a team effort between management and employees. Loss control programs set facility-wide safety and health standards and coordinate responsibility among departments. These programs must take into consideration the worker-equipment-environment system.
- Both the organization and its responsible officials can be punished if a serious health or safety violation results in serious injury. In some cases, a criminal penalty can be imposed against the company as well as against the "responsible official" who could have taken actions to prevent the injury.
- Apart from these government sanctions, the large risk of liability in civil lawsuits concerning worksite injuries has caused many companies to heed the advice that careful documentation, careful planning, and effective training are essential. Lapses in safe work practices may injure several workers or passersby, and a later jury verdict may include punitive damage awards. Winning damages against an employer or the supplier of equipment is easier today than it has ever been for plaintiffs. From the employer's viewpoint, a damages suit costs money, distracts managers, damages the product's or company's reputation, and boosts insurance premiums for that organization for years to come. The classic "ounce of prevention" is far better than the legal system's "pound of cure."
- Safety managers must take time to understand the state rules governing the handling of workers' compensation claims. The workers' compensation system is a form of no-fault payment for industrial injuries, spreading the losses from individual injuries across a pool of insurance premiums so that injured workers are certain that they will receive some compensation and coverage of medical bills if the injury was related to the job. Injured employees in some states have been able to claim that the employer owes additional money to the worker because their injury had been so obvious as to be an "intentional" harming of the injured worker. Preventing the harm is the best solution.
- Incidents can be attributed to the oversight, omission, or malfunction of the management system regarding work factors, human factors, and environment factors.
- Two basic approaches to examining incident causation are after-the-fact and before-the-fact investigations.
- An effective loss control program has seven processes: hazard identification and evaluation, hazard ranking, risk management, management decision making, establishment of preventive and corrective measures, monitoring, and evaluating program effectiveness.

- To organize an effective safety and health program, management must establish program objectives, develop and publish an organizational policy, and delegate responsibility and authority for promoting the program.
- Management and administration, along with housekeeping and preventive maintenance, are a few of an organization's most potent weapons against hazards. A three-pronged approach to employee safety programs involves presenting general company policy and rules, discussing overall safety rules and the safety program, and explaining the safety rules that apply to each employee's department.
- Safety and health professionals, industrial hygienists, and staff medical personnel advise and guide organizations on safety matters, investigate incidents, maintain injury and illness records, ensure line management's cooperation, and perform other duties as required to assist line management in safeguarding employees.
- Purchasing plays a key role in safety by ensuring that all items comply with government regulations and ordinances, that safety is engineered into items, and that safety is weighed into the purchasing price.
- The purpose of organizational safety and health committees is to aid and advise management and employees on safety and health and to perform monitoring, educational, investigative, and evaluative tasks in the safety program.
- Off-the-job safety programs educate employees to apply many of the same safety practices they use at work to activities away from the job.

REFERENCES

American National Standards Institute, ANSI Z16.3–1997, Recording and Measuring Employee Off-the-Job Injury Experience.

Board of Certified Safety Professionals of the Americas, 208 Burwash, Savoy, IL 61874. "Curricula Development and Examination Study Guidelines," Technical Report No. 1.

Boylston, R. B. "Managing Safety and Health Programs." Speech given before the Textile Section, National Safety Congress, October 1989.

Construction Advancement Foundation, 6050 Southport Road, Suite A, Portage, IN. *Safety Manual*.

Dennis, L. E., and M.E. Onion. *Out in Front: Effective Supervision in the Workplace*. Chicago: National Safety Council, 1990.

Factory Mutual Engineering Corp., 500 River Ridge Road, Norwood, MA 02062. "Loss Prevention Data."

Firenze, R. J. *Guide to Occupational Safety and Health Management*. Dubuque, IA: Kendall/Hunt, 1973.

———. *The Process of Hazard Control*. Dubuque, IA: Kendall/Hunt, 1978.

———. *Safety and Health in Industrial/Vocational Education*. Cincinnati, OH: National Institute for Occupational Safety and Health, 1981.

Hogan, R. B. *Occupational Safety and Health Law*. New York: Lexis, 2000 Supplement.

Johnson, W. G. *MORT Safety Assurance Systems*. New York: Dekker, 1980.

Kane, A. "Safety Begins the First Day on the Job." *National Safety News* (January 1979).

Larsen, R. *Workers' Compensation Laws*. New York: Lexis.

Manuele, F. A. "How Effective Is Your Hazard Control Program?" *National Safety News* (February 1980): 53–58.

———. *On the Practice of Safety*. 2nd ed. New York: Wiley, 1997.

National Association of Suggestion Systems, 230 North Michigan Avenue, Chicago, IL 60611.
Performance Magazine. Issued bimonthly.
"Suggestion Newsletter." Issued bimonthly.

National Safety Council, 1121 Spring Lake Drive, Itasca, IL 60143.
Fundamentals of Industrial Hygiene. 4th ed. 1996.
Injury Facts (formerly *Accident Facts*). Issued annually.
Management Safety Policies, Occupational Safety and Health Data Sheets, 12304-0585, 1995.
Supervisors' Safety Manual, 9th ed., 1997.
"You Are the Safety and Health Committee."

O'Reilly, J. T. *Product Warnings, Defects, and Hazards*. 2nd ed. New York: Aspen, 1999.

Peters, G. A. "Systematic Safety." *National Safety News* (September 1975): 83–90.

Rothstein, M. *Occupational Safety and Health Law*. 4th ed. St. Paul, MN: West, 1998. Updated annually.

U.S. Department of Human Resources, National Institute for Occupational Safety and Health, Division of Technical Services, Cincinnati, OH 45226. "Self-Evaluation of Occupational Safety and Health Programs," Publication 78-187, 1978.

U.S. Department of Labor, Occupational Safety and Health Administration. "Organizing a Safety Committee," OSHA 2231, June 1975.

U.S. Department of Transportation, Office of Hazardous Materials, Washington DC 20590.
"Newly Authorized Hazardous Materials Warning Labels," latest edition. (Based on Title 49 CFR, Parts 173.402, 403, and 404.)

Windsor, D. G. "Process Hazards Management." Speech

given before the Chemical Section, National Safety Congress, October 17, 1979.

Websites

Consumer Product Safety Commission Enforcement Manual and descriptive information, cpsc.gov and 16 CFR 1000 ff.

Department of Justice, Civil Division, Office of Consumer Litigation, usdoj.gov.

Department of Transportation, Federal Highway Administration and Federal Aviation Administration, dot.gov and 49 CFR 10200 ff.

Environmental Protection Agency, Office of Enforcement and Compliance Assistance, epa.gov/oeca and Title 40 CFR, various sections.

Mine Safety and Health Administration, msha.gov and Title 30 CFR, Part 1 ff.

Occupational Safety and Health Administration, osha.gov and Title 29 CFR, Part 1000 ff.

U.S. Sentencing Guidelines, ussc.gov and reprinted in *18 United States Code Annotated*, Appendix. St. Paul, MN: West, 2000.

REVIEW QUESTIONS

1. Give three of the five benefits of hazard analysis.
   a.
   b.
   c.
2. What is the purpose of ranking hazards by risk?
3. What types of negligence liability does a company assume when it takes on a contractor role inside a workplace, such as by repairing equipment or providing specialized mechanical service?
4. When can injured workers sue their employers?
5. What federal government programs regulate workplace safety?
6. Is proof of the manager's knowledge of the action necessary when the action violates a law with "strict liability" penalty provisions?
7. What extra risks do government contractors face from violations of federal safety rules?
8. Name the three major areas where hazardous conditions can be either eliminated or controlled, and give an example of each.
   a.
   b.
   c.

Safety, Health, and Environmental Auditing

Brian Maitland, CSP, CET, CPS
Robin Izzo, MS
Philip E. Hagan, JD, MBA, MPH, ARM, CIH, CET, CHMM, CHCM, CHSP, CEM

Development of Safety, Health, and Environmental Auditing
What Is Safety, Health, and Environmental Auditing?
▶ Purposes of Auditing ▶ Audit Scope and Focus

Current Safety, Health, and Environmental Audit Practices
Safety, Health, and Environmental Audit Process
▶ Basic Audit Tools ▶ Designing Audit Programs
▶ Program Organization and Staffing

Government-Related Regulatory Audits
Auditing ▶ EPA Audit Policy ▶ Corporate Audit Agreements

Enterprise-Related Audits
Internal Audits ▶ Consultant Audits ▶ Uses of Audits

Future Directions
Emergence of Safety, Health, and Environmental Auditing Standards ▶ Increased Growth ▶ Broader Scope ▶ Increased Rigor and Depth of Review
▶ Increased Effectiveness of Field Resources
▶ Increased Emphasis on Basic Skills ▶ Expanded Nature of Reporting ▶ Self-Reporting and Immunities

Summary

References

Review Questions

The events of recent years demonstrate that inadequate performance in the safety, health, and environmental (SH&E) arenas can have severe consequences for both an organization and its management. A company responsible for environmental damage is at risk for fines, legal liabilities, and/or economic losses. Moreover, corporate managements continue to come under strong pressure from the public, customers, and regulators to create safe, healthy workplaces and to help preserve the environment. This chapter covers the following topics:

- the nature, purposes, and scope of SH&E auditing
- the basic steps in conducting SH&E audits
- major issues in designing and staffing audit programs
- future directions of SH&E auditing and the emergence of auditing standards.

Many organizations are devoting a sizable and increasing proportion of capital expenditures, operating costs, and managerial resources to programs for pollution control, occupational health and safety, product safety, and loss prevention, among others. In addition to using sophisticated environmental control equipment and other technology, companies are also formalizing many aspects of their safety, health, and environmental management systems. One such area of management that has received considerable attention since the mid-1970s is that of safety, health, and environmental (SH&E) auditing.

This chapter discusses the development of SH&E auditing, describes current practices, and identifies likely future trends.

DEVELOPMENT OF SAFETY, HEALTH, AND ENVIRONMENTAL AUDITING

In this chapter, the term *SH&E auditing* is used to cover the full range of pollution control, occupational safety, process safety, industrial hygiene, occupational health and medicine, and product safety. Many companies have now established programs to monitor and audit the performance of SH&E activities and have come to regard SH&E auditing as a powerful management tool to help determine the compliance status and SH&E performance of their operating facilities. Audits can be performed to minimize liability, comply with regulatory requirements, or enhance existing programs or in response to management concerns over potential problems.

What Is Safety, Health, and Environmental Auditing?
Auditing, in its most common sense, is a methodical examination that involves analyses, tests, and confirmations of an organization's procedures and practices to verify whether they comply with legal requirements and internal policies and evaluates whether they conform with good SH&E practices.

In this context, auditors base their judgments of compliance or deficiency on evidence gathered during the audit and documented in the auditor's working papers. In publicly held companies, auditing has frequently been associated with the review of financial accounting statements by accountants. In addition, such terms as *management audit*, *technical audit*, *operational audit*, *quality assurance audit*, and *energy audit* are commonly used in many other corporate settings.

Purposes of Auditing
Motivations for developing a SH&E audit program include the desire to measure compliance with specific regulations, standards, or policies and the goal of identifying potentially hazardous conditions for which standards may not exist. Thus, although auditing may appear to serve the universal need of evaluating and verifying SH&E compliance, in practice, auditing programs are designed to meet a broad range of objectives, depending on the needs of their various stakeholders.

Companies have established SH&E auditing programs to do the following:

- determine and document compliance status
- improve overall SH&E performance at operating facilities
- assist facility management in identifying potential areas of concern
- increase the overall level of SH&E awareness
- accelerate the overall development of SH&E management and control systems
- improve the SH&E risk management system
- protect the company from potential liabilities
- develop a basis for optimizing SH&E resources
- assess facility management's ability to achieve SH&E goals.

Though these objectives can all be viewed as addressing compliance, they can also produce differences in program scope and focus. For example, some programs focus on determining present and past compliance over a specified time; others focus on determining compliance only at the time of the audit; still others focus on helping the facility manager achieve or maintain compliance.

The extent to which the goals and objectives of audit programs are documented and communicated varies from company to company. Some programs have a written corporate statement describing the audit program, whereas others have written position descriptions for the audit program manager and audit staff.

Audit Scope and Focus
Companies can define the scope and focus of an audit in many different contexts. Some common examples of differ-

ent audit types are organizational, geographical, locational, functional, media specific, and compliance.

Organizational boundaries address which of the company's operations (e.g., manufacturing, research and development, and distribution) are included in the audit program. They generally depend on the organizational structure, business unit reporting relationships, and corporate culture.

Geographical boundaries address how far or wide the program applies (state, province, regional, national, or international). Selection of geographical boundaries generally depends on the location of facilities and offices and the nature of products and services.

Locational boundaries address what "territory" is included in a specific audit. In many cases, the audit focus is mostly on the activities within the facility boundary, although some companies audit beyond this boundary. For example, an audit could also include offsite manufacturing or packaging activities, offsite waste disposal activities, local residences, or a nearby river or lake if there is a potential for environmental damage. Additionally, many companies include the activities of a tenant located on facility property in the scope of their audit.

A number of specific *functional* areas can be included in a SH&E audit program. Although most audit programs cover air and water pollution control, solid and hazardous waste management, employee safety, and industrial hygiene, many now also include occupational medicine, life safety, loss prevention, process safety, and product safety. If all SH&E disciplines are housed in the same organizational unit, the scope of the audit program is more likely to include many, if not all, of these subject areas.

Media audits examine a specific medium, such as water or air. These are often developed in response to comprehensive audits and can be scheduled to occur more often and in a shorter time span than a full audit.

Finally, *compliance* boundaries define the standards against which the facility is measured. These standards can include federal, state, provincial, regional, and local laws and regulations; corporate or division policies, procedures, standards, and guidelines; local facility operating procedures; or standards established by an outside group such as an industry or trade association.

CURRENT SAFETY, HEALTH, AND ENVIRONMENTAL AUDIT PRACTICES

An audit process built on a variety of experiences—such as safety audits and operations audits, and using verification techniques patterned after the financial audit—can be a powerful means of achieving the objectives of SH&E auditing. A generalized description of such a SH&E auditing process includes the following:

- reviewing the company's internal SH&E management and control systems, comprising the following:
  - *policies*: information, directives, guidelines, and standards concerning operations and performance
  - *procedures*: instructions to ensure that operations are carried out as planned
  - *controls*: checks and balances built into SH&E operations, record keeping, and reporting
- developing a complete understanding of the internal control systems that are in place and recording this understanding in a flowchart or narrative form in the auditor's working papers
- evaluating the strengths and weaknesses of the internal control systems
- gathering audit evidence to meet the objectives of the audit program through inquiry, observation, and verification testing
- discussing the issues with management, presenting the exceptions found, and judging their importance
- advising management on improvements needed in the SH&E management systems
- creating a corrective action plan to address deficiencies and discoveries
- following up to maintain corrective actions.

Safety, Health, and Environmental Audit Process

A number of basic activities are common among most audit programs. Some activities are undertaken before the onsite audit (planning), some during the audit field work (understanding systems, assessing controls, gathering and evaluating internal documents), and others after the field audit has been completed (reporting the results and follow-up). Virtually all SH&E audits involve gathering information, analyzing facts, judging the status of the facility, and reporting the results to some level of management. A team approach is commonly used to conduct these activities. Even with these basic similarities, there can be a number of important differences. Figure 7–1 presents a simplified model identifying the key steps in the audit process. Most companies make some provision for including each of these steps in their audit process.

Planning

The SH&E audit process begins with a number of activities before the actual onsite audit takes place. Initial arrangements relating to a facility audit include scheduling the visit, selecting the audit team, and gathering and reviewing background information.

Some companies audit all facilities on a repeating cycle

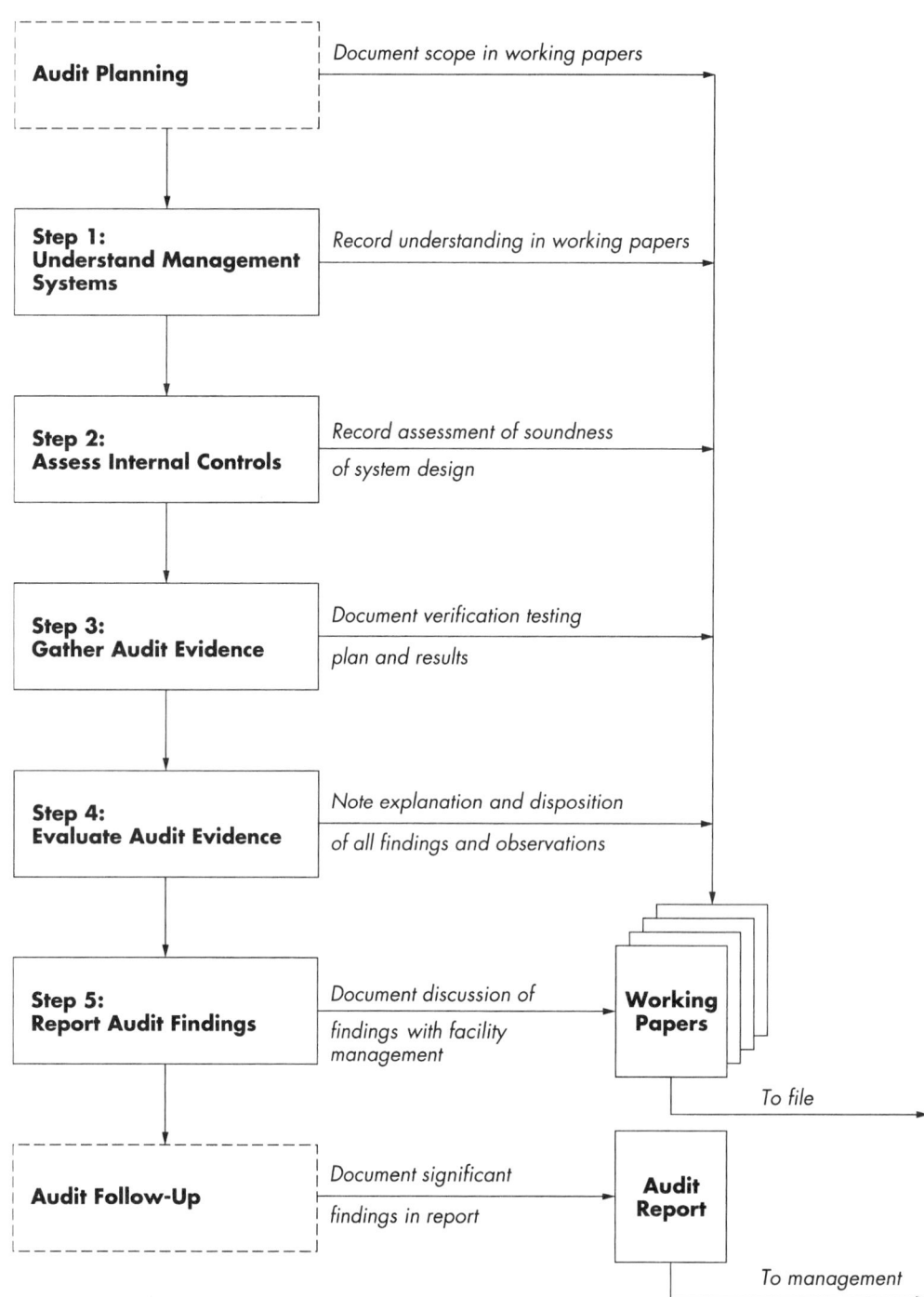

Figure 7–1. Key steps in the audit process.

(e.g., annually or every 2 years). In companies not auditing all facilities on a specific repeat cycle, the facilities that will be audited must be selected and scheduled. A list may be drawn up annually and modified throughout the year.

Companies select facilities for audit by a number of methods, typically on the basis of risk (i.e., perceived hazards, business importance, nature of operations, vulnerabilities, etc.). Initial arrangements relating to a facility audit include scheduling the visit, selecting the audit team, and gathering and reviewing background information. Initial notice of an upcoming audit (audit lead time) varies from company to company and may be anywhere from 1 to 6 months. However, a few companies conduct "surprise" audits to obtain what they believe is a more accurate picture of facility operations.

Background information gathering generally begins well

in advance of the audit and includes regulatory requirements, corporate policies (both written and oral), and facility-specific information (such as organization, processes, and layout). Some companies' audit teams visit the facility before the audit to develop a basic understanding of facility processes and SH&E management systems and to brief the facility staff on the objectives of the upcoming audit.

Field Work

Collection and review of advance information results in an audit plan that outlines the required audit field work steps, how each is to be accomplished, who will do them, and in what sequence.

Step 1: Understand Management Systems. Most onsite activities begin by developing a working understanding of how the facility manages activities that may affect SH&E performance. This usually includes learning about facility processes, internal controls (both management and engineering), facility organization and responsibilities, compliance parameters and other applicable requirements, and any current or past problems. This step allows the team members to understand the actions taken within the organization that help regulate and direct its activities.

In developing this picture of internal management systems, auditors usually draw on information from multiple sources. These may include selected information provided by the facility in advance of the audit, staff discussions, and facility tours. Using questionnaires, discussions, and tours as background, auditors further investigate the more detailed aspects of management systems through in-depth interviews, guided discussions, and additional tours to specific sites. The data-gathering methods most used in Step 1 are inquiry and observation.

Step 2: Assess Internal Controls. After clearly understanding how various aspects of SH&E compliance and performance are intended to be managed, auditors then evaluate the soundness of the facility's management systems to determine whether they are functioning as intended and will achieve the desired level of performance.

For each of the areas or topics assigned, the auditor should ask him- or herself: "If the facility is doing everything the way they say it happens, is that acceptable—will the facility be in compliance with applicable requirements, and is the company adequately protected?"

In assessing the strengths and weaknesses of internal controls, auditors typically look for such indicators as clearly defined responsibilities, an adequate system of authorizations, capable personnel, documentation, and internal verification. It is far easier to identify significant weaknesses in internal controls than to determine adequacy. Each of these indicators usually requires significant judgment on the part of the auditor because there are no widely accepted standards an auditor can use as a guide to what is acceptable internal control. Thus, many auditors look to the audit program objectives, as well as to the company's basic SH&E philosophy, for guidance about what is satisfactory internal control.

This step is especially important in that it determines, to a great extent, how the balance of the audit will be conducted. Where internal controls are judged to be sound, the auditor will spend time confirming the existence of the control systems and testing whether they function effectively on a constant basis. The auditor tends to rely on both the comprehensiveness and objective documentation of these internal controls when looking at the rest of the system. On the other hand, where internal controls are judged to be deficient or lacking, the auditor cannot rely on their presence as an indication that the company is functioning well. In such cases, gathering evidence is limited to scrutinizing the outcomes of individual programs and processes or analyzing current compliance status against federal, state, and company standards.

Step 3: Gather Audit Evidence. Audit evidence forms the basis on which the team determines compliance with laws, regulations, corporate policies, and other standards. Evidence is gathered in a variety of ways, including record reviews, examination of available data, and interviews with facility personnel. Suspected weaknesses in the management system are confirmed in this step. Also, management systems that appear sound are tested to verify that they work as planned and are consistently effective.

Audit procedures (the means by which auditors collect audit evidence) fall into three broad categories:

1. *Inquiry.* The auditor asks questions both formally and informally. Audit questionnaires are common examples of formal inquiry.
2. *Observation.* The auditor collects evidence through what can be seen, heard, or touched. Because physical examination is often one of the most reliable sources of audit evidence, observation is a significant aspect of most SH&E audits.
3. *Verification testing.* The auditor focuses on either the management system or physical equipment and performs systems tests. For example, retracing data would uncover errors in recording original data. Other common types of testing include verifying paper trails and equipment checks.

Programs vary regarding the amount and balance of inquiry, observation, and verification testing required. Some programs depend on inquiry as the primary means of gathering audit evidence. Inquiry is easy, provides rapid feedback, and does not require as many resources. Many of the more sophisticated SH&E audit programs, however, require auditors to conduct a considerable amount of verification testing to determine whether management systems and equipment perform as they are supposed to. For each item to be audited, inquiry typically takes a matter of minutes, observation tens of minutes, and testing a matter of hours. Thus, the more items to be verified, the larger the resource commitment required. In most cases, more items could be verified (and more ways to verify each item are possible) than available audit resources allow. However, SH&E audits usually serve as a check on the SH&E management system rather than as a substitute for it. Therefore, most audit programs do not look at every situation, item, or document.

Step 4: Evaluate Audit Evidence. Once evidence gathering is complete, the audit findings and observations are evaluated. Audit evidence is reviewed in terms of program goals to determine both whether audit objectives are met and the significance of the audit findings.

Although auditors usually make preliminary evaluations of their observations throughout the audit, most audit teams devote a few hours at the end of the audit to jointly discuss, evaluate, analyze, and finalize these tentative audit findings.

Step 5: Report Audit Findings. The goal of an audit is to improve conditions, and that goal is best met by timely feedback to the facility personnel. A meeting at the close of the audit can orally communicate the preliminary findings and clarify, with help from facility personnel, the types of responses that may be needed. This dialogue about corrective actions and deficiencies is more formal than a conversation but more informal than the audit team's written report.

After the onsite audit visit, the company's policy dealing with preparation and handling of audit reports must be followed. In some organizations, the avoidance of generating potential legal liability from audit reports is a serious concern. For these entities, the draft audit report should be handled with the law department or with outside legal counsel, and copies should be carefully controlled so that future claims of legal privilege may be asserted. Even in organizations that do not have a policy protecting the legal status of audit reports, the auditors' choice of wording for an audit report carries serious legal consequences. Selecting what words are to be used when the written audit report describes a dangerous condition or risk scenario will be an important part of the audit team leader's responsibilities.

Most companies prepare a written audit report. The purposes of the report are typically to provide management with information about compliance status; to initiate corrective action; and to document how the audit was conducted, what it covered, and what was found.

Some companies prepare a draft of the audit report on site. Most, however, prepare a draft audit report shortly after the onsite audit is completed. This draft usually undergoes review and comment before a final report is issued. Report reviewers often include the SH&E affairs department, the legal department, facility management, and the audit team.

The content of the audit report varies considerably from company to company. Typically, audit reports contain a background or introduction section that describes the purpose and scope of the audit; outlines the audit approach; and identifies the audit team leader, team members, and other key audit participants. Most audit reports include sections on the facility's overall compliance with regulations, as well as compliance with the company's policies and procedures. Some audit reports identify all applicable facility operations; some include a detailed description of the facility and its history or an impression of the facility management's ability to handle SH&E crises. Still other audit reports contain recommendations for how to address the deficiencies identified.

The content of audit reports is strongly linked to the overall objective of the audit program and the needs of the report recipient. If the goal of the audit program is to provide assurance to management, the audit report often is limited to a factual description of the more significant findings and exceptions. On the other hand, if the goal of the audit program is to assist the facility manager, the audit report often is detailed enough to let the facility manager know precisely what is wrong and may include recommendations on how to improve the situation.

An effective reporting process communicates issues to appropriate persons within the company. Many companies have established a multiple or hierarchical reporting scheme. Under such a reporting process, the type of information and level of detail to be provided in an audit report depend on the problem identified and the individual who has to be notified. Some items may require reporting to corporate management and future follow-up; others may require only the attention of the facility manager.

The distribution that an audit report receives depends on policy decisions concerning the need to know, the power to implement, and the need for future legal defense. Copies should go to managers who have the power to correct undesirable conditions. As a legal strategy, some companies tightly control dissemination and require that attorneys be closely involved with the reporting function. This approach may allow the firm to safeguard its critical assessment of self-identified problems from the risk of civil discovery demands and use as trial evidence by challengers, such as injured persons, downstream neighbors, or government regulators. In any case, strategy for dissemination should be developed with advice from legal counsel. Senior management who need to know of the concerns, and who control budgets for corrective actions, are especially important recipients.

Follow-Up

Most companies have established formal procedures for responding to the audit report. The action-planning process is initiated as audit findings are identified. It typically includes assigning responsibility for corrective action, determining potential solutions and preparing recommendations to correct any deficiencies noted in the audit report, and establishing timetables for completion. Responses to the audit report are generally prepared by the facility manager and sent to line management and the audit program manager for review and, often, for approval. A few companies incorporate the action plan into the final report.

Typically, action plans are monitored by an individual with responsibility for follow-up—generally either operating management, SH&E affairs, or, in a few cases, the auditors. In most instances, follow-up involves a written or oral inquiry about the status of the planned action. In companies where facilities are audited on a repeat basis within a specified time, the auditor or audit program manager is usually involved in action plan follow-up (typically reviewing the status of the action plan implementation during the next audit). Where an audit team is unlikely to return to the facility for some time, operating management or SH&E affairs usually assumes responsibility for the follow-up.

The period of file retention for an audit should be no more than needed—usually 1 month after the scheduled date of the next audit—so that comparisons may be drawn by the subsequent auditors. A longer retention period is not usually necessary, but because audit documents may be sought by challengers in hostile litigation, strict adherence to file retention policies is important.

Basic Audit Tools

The SH&E audit process is most commonly supported by some important tools: the audit protocol and the working papers. Although there is considerable latitude in current practice, most SH&E audit programs use these devices in some form.

Audit Protocols

Names for the various documents that guide the auditor while conducting the audit include audit protocols, audit work programs, review programs, checklists, and audit guides. In this discussion, the term *audit protocol* will be used.

An audit protocol represents a plan of how the auditor is to accomplish the objectives of the audit. It lists the audit procedures that are to be performed to gain information about SH&E practices. An audit protocol also provides the basis for assigning specific tasks to individual members of the audit team, for comparing what was accomplished with what was planned, and for summarizing and recording the work accomplished. A well-designed audit protocol can also be used to help train inexperienced auditors and reduce the amount of supervision required by the audit team leader. Many companies use the audit protocol to help build consistency into the audit, particularly when rotating audit teams are used.

The audit protocol itself is a list of auditing procedures that are to be performed to gain evidence about SH&E practices. Typically, a standard audit protocol is modified by the auditor to suit the circumstances of a particular situation. Modifications are made because of the nonapplicability of certain audit procedures (or regulations/policies), unique SH&E hazards, or risks of potential materiality that may be present. The auditor may also modify the standard audit protocol according to how much he or she relies on the facility's internal control systems.

Each procedure is typically annotated on the audit protocol with the specific working paper page reference. Thus, the completed protocol, combined with the working papers (described shortly), provides a record of the audit steps that were performed in carrying out the audit. The completed audit protocol also provides documentation of the rationale for any modifications. The completed audit protocol should also document the scope (or boundaries) of the SH&E audit by identifying aspects of the SH&E areas under review, which were considered in carrying out the audit. The scope of the audit should also describe the period of time under review.

Working Papers

Working papers (the auditor's field notes) document the work performed, the techniques used, and the conclusions reached by the auditors. Working papers help the auditor achieve the audit objectives and provide reasonable assurance that an adequate audit was performed consistent with audit program goals and objectives. Working papers should include documentation of compliance or noncompliance.

Working papers are usually handwritten and include photocopies of documents selected by the auditor to help substantiate the findings of the audit. These papers are not a report that the auditor prepares from notes after the audit is complete; rather, they are the auditor's field notes to keep track of audit procedures undertaken, results achieved, and items requiring further information.

Working papers can include the following items:

- notes of audit planning
- a completed copy of the audit protocol annotated with working paper references and the auditor's initials
- results of compliance testing and evaluations of internal control
- descriptions of all functional tests (e.g., performance of pollution control equipment) and transactional tests (e.g., documentation of hazardous waste shipments) conducted during the audit
- documentation of all audit procedures and evidence obtained
- notes on any conferences held.

Some working papers also contain a list of the facility's materials and their uses, copies of regulatory permits for the area(s) under review, schedules of documentation and record keeping, and copies of previous audit findings and observations.

The working papers can be divided into two sections: the permanent or continuing file and the current file. The permanent file contains a record of items of continuing interest, whereas the current file contains items pertinent to the most recent audit. For example, the permanent file would contain the following:

- a list of the facility's products and their uses
- copies of relevant regulatory permits and key regulatory correspondence
- copies of any regulatory violations
- internal control questionnaires and flowchart
- schedules of documentation and record keeping
- copies of previous years' statements of findings and observations.

The current file contains the auditor's evidence and final decisions. It includes the following:

- notes of audit planning, including audit leader, team members and assignments, time budgets, rationale for any audit program modifications, and so on
- the completed copy of the audit protocol annotated with working paper references and the auditor's initials
- the completed internal control questionnaire, results of compliance testing, and evaluation of internal control
- schedules for compliance testing
- descriptions of all transactional and functional tests conducted during the audit, as well as an explanation of sampling plans employed
- documentation of all audit procedures and evidence obtained, including favorable as well as unfavorable evidence
- notes on any conferences or meetings held with facility management.

To substantiate the auditor's adherence to the proper protocols in conducting the audit, the working papers should provide sufficient detail of both the testing rationale and resulting evidence for each item in the audit protocol.

Designing Audit Programs

The basic objective of most SH&E audit programs is to verify compliance with corporate, legal, and regulatory policies and rulings throughout the company. This objective can be best served under the following conditions:

- The SH&E audit is conducted independently from those who are responsible for development or implementation of SH&E management policies and programs.
- A mix of SH&E engineering, operating, and auditing expertise is employed in the design and implementation of the audit process.
- Auditors serve as reviewers of the total system rather than as functional, technical experts while conducting a SH&E audit. The company's professional SH&E managers usually verify that the system is managerially and technically correct in design, operated by competent people, supported by management, and capable of operating properly, based on current performance data. However, the depth, detail, extensiveness, and variety of the validation are usually much greater in SH&E auditing than in reviews or assessments conducted by professional SH&E managers responsible for ongoing administration of an SH&E program.
- Definitions of important items are developed, clearly understood, and widely accepted. SH&E auditors should not have to decide unilaterally what is important.
- Working papers document the planning and execution of the audit, describing, for each audit step, (1) what audit review and test activities were carried out, (2) what results were noted, and (3) what conclusions were reached.
- An understanding of the facility's internal management and control system is obtained and described in the working papers. The auditors then test whether the system is actually and consistently in use.
- All observations, findings (both compliance and noncompliance), and recommendations are documented with competent and sufficient evidence in written working papers.
- The evidence cited in the working papers provides a clear flow of logic from the testing rationale to the auditor's discoveries. Evidence is relevant, objective, free from bias, and persuasive.
- Findings are formally communicated to those with the organizational authority to review, evaluate, and, where appropriate, take corrective action.

Those charged with responsibility for a SH&E audit program should keep in mind that in today's corporations, SH&E considerations are likely to have an impact on many of the company's activities. In some matters, a company's SH&E managers have primary responsibility for developing, implementing, and overseeing policies and programs. On the other hand, SH&E performance is also likely to be influenced by a wide variety of other corporate activities. For example, business functions such as marketing, manufacturing, transportation, or finance often have wide influence on corporate performance within salient SH&E areas such as pollution control, occupational health and safety, product safety, and loss prevention.

Corporate activities affecting human health and safety and the environment are regulated by several different federal, regional (state, province), and local agencies and often have been viewed—especially by regulators—as separate (and even unrelated) functions. As a result, there may be considerable variance in the corporation's managerial approach and stage of management system development within each SH&E area. Companies must exercise care to establish effective SH&E auditing approaches rather than attempting to use preconceived auditing systems across various SH&E disciplines. An effective SH&E audit program requires enough flexibility to review a wide variety of existing internal systems and, at the same time, provide meaningful feedback to those charged with developing and operating formal management systems.

Program Organization and Staffing

In deciding where to place the audit program within their organization, some companies emphasize that the audit function should be independent while others emphasize that other groups in the organization need to have easy access to the information gained from audits. Typically, companies place audit programs within a core corporate group, most commonly within the corporate SH&E staff. However, in some companies the group is located in the internal audit department, regulatory affairs department, production or operations department, or legal department.

Companies choosing to place their SH&E audit program in the internal audit department tend to view the program more as a corporate tool than as a SH&E management tool. Production or operations departments are sometimes chosen to reflect or reinforce a company's philosophy that operations are responsible for SH&E management. The legal department may be chosen because of sensitivity to potential legal issues involved in SH&E auditing, such as potential disclosure of sensitive audit information.

Regardless of where the audit program is structured in the organization, the legal department or outside legal counsel must be involved from the outset. If the topic is sensitive and litigation is possible, then the lawyers should be involved with the structure of the audit report, the list of recipients in the circulation of reports, the advice about which actions are or are not "violations" of a regulatory rule, and wording of conclusions about problems at the site. When serious noncompliance is found, the lawyers may defer the writing of an audit report pending corporate evaluation of the liability risks involved.

Companies staff their SH&E audit programs in a variety of ways. Most established SH&E audit teams include individuals with technical expertise, knowledge of SH&E regulations, facility experience, and a strong knowledge of auditing procedures and techniques. Knowledge of SH&E management systems and an understanding of similar companies' hazard control programs are also important staffing criteria. Audit teams usually include SH&E specialists and may also include a facility manager, process or safety engineer, attorney, analytical chemist, internal auditor, industrial hygienist, or outside consultant.

Some companies have a full-time audit program manager; others staff their audit program only on a part-time basis. Of the companies that staff their programs on a part-time basis, some do not conduct enough audits to justify full-time staff; others vary participation in their program by rotating the audit team membership. Some rotate the membership of the audit team to involve a wider range of staff; other companies vary their team membership to get the specific expertise desired for a particular audit.

Using the same auditors for every audit, as opposed to rotating auditors, provides greater continuity from audit to audit and generally greater confidence that the goals and objectives of the program are being met.

Typically, an audit program is funded in a manner consistent with the SH&E and other corporate staff functions. In most companies, costs are absorbed as overhead and included in the budget of the organizational unit responsible for the program. However, in a few companies, audit costs are charged back directly to the audited facilities, especially when a company typically directly charges back the costs of many corporate staff activities. It is not unusual to charge travel and out-of-pocket expenses to an individual's assigned organizational unit when a company uses staff for the audit on a part-time or special-assignment basis.

GOVERNMENT-RELATED REGULATORY AUDITS

The EPA defines an *environmental audit* as "an investigation into the history and current status of a particular piece of property (site)." The purpose of an environmental audit is as follows:

- to identify the presence and extent of environmental contamination or hazardous materials from current or previous site activities
- to determine the level of compliance with current standards or regulations
- to provide a general review of environmental risks associated with the site and its operations.

Auditing

The EPA website, epa.gov, provides the following information on the EPA's current self-policing audit policy.

The EPA Audit Policy, "Incentives for Self-Policing: Discovery, Disclosure, Correction and Prevention of Violations," has been around since 1995. The Audit Policy is designed to provide incentives for regulated entities to comply with applicable federal environmental laws and regulations. These incentives are for regulated entities that voluntarily discover, promptly disclose, and expeditiously correct noncompliance, making formal EPA investigations and enforcement actions unnecessary.

Before agreeing to participate in the EPA Audit Policy Program, it would be a good idea to get the advice of legal counsel to ensure that you feel comfortable with the potential outcome.

In addition, EPA has developed a series of Environmental Auditing Protocols to provide guidance for evaluating compliance with applicable environmental regulations. However, the regulated community's legal obligations are determined by the terms of applicable environmental facility-specific permits and underlying statutes, as well as the applicable state and local laws.

EPA has published the "Interim Approach to Applying the Audit Policy to New Owners" ("Interim Approach"), which describes Audit Policy incentives tailored for new owners who want to make a "clean start" at recently acquired facilities by addressing environmental noncompliance that began before acquisition.

In addition, the EPA is piloting a web-based system to allow companies to electronically self-disclose violations under EPA's Audit Policy eDisclosure.

EPA Audit Policy

The EPA Audit Policy, "Incentives for Self-Policing: Discovery, Disclosure, Correction and Prevention of Violations" ("Policy"), provides several major incentives for regulated entities to voluntarily comply with federal environmental laws and regulations. To take advantage of these incentives, regulated entities must voluntarily discover, promptly disclose to EPA, correct, and prevent recurrence of future environmental violations. Disclosures are often preceded by consultation between EPA and the regulated entity so that they can discuss mutually acceptable disclosure details, compliance, and audit schedules.

If there are compliance issues, these incentives can result in significant penalty reductions. Monetary penalties received under environmental laws generally have two components: an amount assessed based on the severity or "gravity" of the violation, and the amount of economic benefit a violator received from failing to comply with the law. There will be no gravity-based penalties if all nine of the Policy's conditions are met. EPA retains its discretion to collect any economic benefit that may have been realized as a result of noncompliance. In addition, there will be a 75% reduction of gravity-based penalties when the disclosing entity meets all of the Policy's conditions, except detection of the violation through a systematic discovery process. The nine conditions of the Policy are as follows:

1. There is systematic discovery of the violation through an environmental audit or the implementation of a compliance management system.
2. There is voluntary discovery of the violation not detected as a result of a legally required monitoring, sampling, or auditing procedure.
3. There is prompt disclosure in writing to EPA within 21 days of discovery or such shorter time as may be required by law; discovery occurs when any officer, director, employee, or agent of the facility has an objectively reasonable basis for believing that a violation has or may have occurred.
4. There is independent discovery and disclosure before EPA or another regulator would likely have identified the violation through its own investigation or based on information provided by a third party.
5. There is correction and remediation within 60 calendar days, in most cases, from the date of discovery.
6. There is prevention of recurrence of the violation.
7. Repeat violations are ineligible; that is, specific (or closely related) violations that have occurred at the same facility within the past 3 years, or those that have occurred as part of a pattern at multiple facilities owned or operated by the same entity within the past 5 years. If the facility has been newly acquired, the existence of a violation before acquisition does not trigger the repeat violations exclusion.
8. Certain types of violations are ineligible, such as those that result in serious actual harm, those that may have presented imminent and substantial endangerment, and those that violate the specific terms of an administrative or judicial order or consent agreement.
9. Cooperation by the disclosing entity is required.

Corporate Audit Agreements

EPA encourages companies with multiple facilities to take advantage of the agency's Audit Policy to conduct corporate-wide audits and develop corporate-wide compliance systems.

A Corporate Audit Agreement allows an entity (such as a corporation, university, or other organization with many facilities or facility locations within its authority) to plan a corporate-wide or facility-wide audit with an advance understanding between the entity and EPA regarding schedules for conducting the audit and disclosing violations beyond the current 21-day disclosure requirement for single-facility disclosures. In return for the advance agreement of an audit and disclosure schedule, the facility would receive the benefits of EPA's Audit Policy as applicable.

ENTERPRISE-RELATED AUDITS

Internal Audits

Companies with SH&E responsibilities should conduct internal audits on a consistent, company-wide basis and request audits by outside consultants as needed. An audit is an important review of a site's environmental risks and how a company is addressing those risks. A minimum goal for an audit is to determine a company's level of compliance with applicable regulations and standards.

The main purposes of an internal audit are to ensure that a facility or department is safe and that the company is in compliance with environmental standards set by various regulations and statutes. Environmental management programs should institute consistent, comprehensive, internal audits. They can be either "top-down" audits or self-audits. Each has its advantages and disadvantages.

In a *top-down audit*, staff from a centralized environmental management department review procedures, record keeping, and so on in operating departments or facilities throughout the organization. This type of review ensures professional auditing expertise. It also enables corporate staff to ensure company-wide compliance with corporate environmental policy. Its disadvantage is a feeling of a "police action" that can create hostile relationships with operating staff, who often feel that the environmental professionals are not familiar enough with their department or facility to conduct a comprehensive study.

Self-audits enable employees who work in the department or facility to assess problems in the operations with which they are familiar. However, they may be too close to the operations to be able to conduct an objective audit or to spot operational processes that could be improved to go beyond minimal environmental release standards. Operations staff may also lack incentive to identify problem areas, believing that such problems will reflect badly on themselves. Self-audits also do not help corporate staff become more aware of individual operations and areas that could be improved.

A company policy that combines self-audits with top-down audits (periodic department or facility self-assessments with follow-up audits by environmental professionals) can combine the advantages of the two approaches while minimizing their disadvantages. Environmental audits conducted internally, whether top-down or self-audit, should follow some predefined protocol. Records of this protocol should be retained in the company's records for review by environmental regulatory agencies and management.

As part of the procedure to develop the internal audit protocol, the company should develop a procedure for addressing deficiencies or violations detected. There is no point in developing self-examination information if the firm has no program to address remediation of the deficiencies. In fact, an internal audit can present a problem if the regulatory agency detects a deficiency or violation and determines, through a company's internal audit document, that it had prior knowledge of the fact but took no further action.

The primary purpose of any audit program is to translate the information gathered into new environmental programs or change existing programs to help the company resolve problems swiftly.

Consultant Audits

Environmental audits by outside consulting firms are conducted for a variety of reasons. Some companies use these audits as the first step in determining what needs to be done to comply with regulations. The main advantage to a consultant audit is that the company obtains an expert, unbiased opinion of the environmental condition of a facility. Regulatory agencies normally view these audits as more complete, with a higher degree of accuracy than top-down or self-audit protocols. These audits are also used as a periodic review of a company's present overall performance on controlling air emissions, discharges to storm and sanitary sewers, groundwater, storage and handling of hazardous materials, and so on. Such a periodic audit can be integrated with an internal audit program.

An environmental consulting firm can be specifically contracted when there are serious concerns about compliance with environmental regulations and statutes. This can be a complete review of a company's compliance with all pertinent regulations, or it can focus on a specific section of the Clean Water Act (CWA), Superfund Amendments and Reauthorization Act (SARA), and so on. These audits often pay for themselves by reducing future compliance costs and avoiding costly penalties. An environmental consultant also can be engaged to do a site assessment before a real estate transaction.

The Comprehensive Environmental Response, Compensation, and Liability Act (CERCLA) holds current and former property owners liable for environmental contamination of the property, no matter how long the hazard has existed or who caused it. Without an environmental audit before a real estate closing, an innocent buyer can own property that cannot be sold or that must be cleaned up at enormous cost. Many lending institutions have acquired environmentally contaminated property through foreclosure. Because of their potential liability, many of them are requiring an environmental audit before approving a loan.

Federal regulations do not require an environmental audit to be part of every real estate transaction. Environmental audits of commercial, industrial, and some residential property may be required by state law. Whether mandated or not, an environmental audit enables both the buyer and seller to be aware of the condition of the property.

When SARA was passed in 1986, it gave an "innocent purchaser" of contaminated property a narrow way of escaping liability. Liability may be avoided if the buyer can prove that an inquiry into previous ownership and uses of the property (an environmental audit) was made before its purchase. The completed audit form can be submitted as an exhibit in a court case, proving that the buyer did not know that any hazardous substances were released or disposed of at the site before the property was purchased.

Uses of Audits

Environmental audits serve many purposes. Site assessments before a real estate purchase help companies avoid costly cleanup charges or provide a possible defense to potential liabilities when existing environmental issues are discovered in the future. When conducted to audit a company's compliance with regulations, they can prevent or minimize fines. If they include an assessment of environmental hazards to employees, audits promote good labor relations. When their results are reported to the community, they can enhance a company's public image with citizens who may have been suspicious of the facility's activities.

Whenever management plans an audit, the first step is a concise statement of company goals. Once management has determined what information is needed from the audit, a decision can be made to determine whether to conduct an internal audit or hire a consulting firm. Time and expertise are often deciding factors. Complex environmental regulations can require the knowledge of someone other than the environmental manager or department personnel. An extensive environmental audit also can take more time than busy internal staff have to spare.

FUTURE DIRECTIONS

What does the future hold? A number of important trends and further developments in audit program design, content, and coverage will continue to occur over the next few years. These trends are described next.

Emergence of Safety, Health, and Environmental Auditing Standards

As SH&E auditing practitioners have developed a core group of common practices, they have begun to reach a consensus on principles and standards for SH&E auditing. A U.S.-based example of this is ASTM E2107–06, *Standard Practice for Environmental Regulatory Compliance Audits* (2006). The effort to define auditing standards has become more urgent with the advent of international standard-setting initiatives, including the European Union's Eco-Management and Audit Regulation and the International Organization for Standardization's environmental management standards (ISO 14000), described in Chapter 4, Regulatory History.

Although practitioners continue to differ on audit approach and philosophy issues, they generally endorse the following standards:

- *Auditor proficiency.* Audits are conducted by staff who have the qualifications, technical knowledge, training, and proficiency in the auditing discipline to perform their assigned auditing tasks. The organization managing auditing activities and the individual auditor take responsibility for ensuring staff proficiency. Audit team qualifications are suited to the objectives, scope, and complexities of the audit assignment.
- *Due professional care.* Due professional care is the application of diligence and skill in performing audits. Exercising due professional care means achieving accuracy, consistency, and objectivity in the performance of audits; ensuring that the audit examination follows the procedures established for that audit; using good judgment in choosing tests and procedures; reporting audit findings in accordance with the audit's procedures; and developing conclusions and, if necessary, recommendations.
- *Independence.* Auditors are objective and independent of the audit site and/or activity to be audited, free of conflicts of interest in any specific situation, and not subject to internal or external pressures that could influence their findings. When complete independence is not feasible, the audit report should clearly communicate the limiting factors and the client's and auditee's awareness of them.
- *Clear and explicit objectives.* The objectives of an audit are clearly established and fully communicated beforehand to the client and the auditee. The objectives

of specific audits meet the needs of intended recipients of audit results and the provisions of accepted audit standards.
- *Systematic plans and procedures for conducting audits.* Audits are based on the use of systematic plans and procedures that provide uniform, consistent guidance in audit preparation, field work, and reporting. The audit is planned, resources allocated, and procedures selected and supervised to achieve explicit audit objectives.
- *Planned and supervised field work.* Auditors and audit team leaders plan, implement, and supervise field work to foster efficiency, to ensure consistency in all parts of the audit and with the audit plan, and to achieve audit objectives. On site, auditors and team members collect the relevant and accurate information that is needed to meet audit objectives.
- *Thorough review of internal controls.* Audits are conducted to review existing management systems and internal controls and to gather appropriate information for evaluating the reliability of internal controls in achieving environmental performance goals.
- *Audit quality control and assurance.* Audits undergo quality checks to ensure accuracy and to encourage continuous improvement of audit management systems, procedures, and implementation. Quality control measures the extent to which an audit is conducted according to the objectives and scope of the audit and to these standards.
- *Audit documentation.* The auditor prepares documentation of ongoing activities during an audit in "working papers." Each subject reviewed in an audit should be documented sufficiently so that another auditor of similar skill could confirm the conclusions of the auditor without consulting other resources.
- *Clear and appropriate reporting.* For each audit, the auditor prepares a formal report that communicates information in accordance with the audit objectives. The report clearly communicates information and findings to the intended recipients in a timely manner and in sufficient detail and clarity to expedite corrective action.

These emerging standards draw on concepts from other types of auditing and on principles and practices that have emerged in SH&E auditing since the 1970s. In the near future, it is likely that SH&E auditing standards will be much more sharply defined, opening the way for true auditor certification and independent audit verification.

Increased Growth

The numbers of organizations with SH&E audit programs have continued to grow. Although the existence of an SH&E audit program is widespread among larger companies in the manufacturing and process industries, most of the new growth in audit programs has come from other segments of industry and from medium-sized and small companies. These organizations have begun to appreciate that the up-front costs of auditing can significantly outweigh the eventual costs associated with noncompliance and cleanup.

Broader Scope

In the future, organizations with existing or new audit programs will expand the scope of their auditing efforts. The expansion will likely occur both geographically and functionally. Organizations that now conduct audits only of domestic facilities will broaden their scope to include overseas locations. In transporting auditing programs overseas, companies will have to address important considerations such as diverse local regulatory systems, travel and logistical requirements, language and cultural barriers, and different levels of company ownership and control. These considerations, if not fully examined and addressed, will impair the company's ability to audit the systems that have been developed to manage environmental risks.

Organizations whose audit programs focus only on environmental topics will expand them to include health and safety and, in many instances, process safety and product safety issues. Given the many common issues shared by these topics, companies can realize economies in staff and resources if they place different audits under one central management. This expansion in scope, however, will not necessarily mean combining all topics into a single audit. Instead, the scope of the audits will be designed to match audit resources and objectives.

Increased Rigor and Depth of Review

Perhaps the most significant change in the years ahead will be the increased depth and rigor of review. More companies will shift the orientation of their programs from finding regulatory problems to verifying compliance and confirming that management systems are in place and functioning as they were designed to do.
- *Identify problems.* Many companies begin with an assessment orientation, focusing first on identifying problems through inquiry and observation only. The problems identified may relate to regulatory compliance or may be in nonregulated areas.
- *Verify compliance.* Effective environmental audit programs evolve beyond assessment to compliance verification. The focus here is not only on identifying compliance problems, but also on a systematic, rigorous verification of areas, both regulated and nonregulated, that appear to be functioning well and in compliance.

- *Confirm functioning of management control systems.* Because verification of compliance during the audit does not ensure that the facility will stay in compliance, a growing number of companies take a systems perspective. In this approach, the auditing effort extends beyond verification of compliance to include a review of the underlying programs, procedures, and systems that are in place to ensure ongoing compliance.

Increased Effectiveness of Field Resources

There will be a continuing shift in focus toward a more rigorous review and independent verification of the areas believed to be in compliance and operating smoothly. This is particularly true if there is a corresponding examination of the underlying facility-level environmental management systems that will require an increase in the auditor's effectiveness and, perhaps, an increase in the total amount of field resources allocated. It is anticipated that companies will find creative ways to accomplish this goal, despite the fact that budget and staffing constraints will most likely continue.

Increased Emphasis on Basic Skills

More and more companies are beginning to realize that a discrete set of basic skills is required to be an effective auditor. These skills include the following:
- working knowledge of regulatory requirements applicable to the scope of the audit
- training and proficiency in basic auditing skills and techniques
- general familiarity with the type of facility operations being audited
- understanding of the SH&E controls, procedures, and management systems of similar facilities.

Over the next several years, increasing numbers of companies will work toward gaining a better balance among these four skills, with particular emphasis on proficiency in basic auditing techniques.

Expanded Nature of Reporting

As top management continues to request assurance that facilities are operating well and are in compliance, the nature of reporting audit results will shift from simple problem identification reports to performance reports and, ultimately, to assurance reports.
- *Problem reports.* Auditors list a number of problems identified at a facility and, perhaps, provide a list of positive findings as well.
- *Performance reports.* Auditors summarize the overall results of a comprehensive audit not only by listing problems, but also by including a general defensible conclusion regarding the compliance status and overall SH&E performance of a facility.
- *Assurance reports.* Auditors provide top management with statements regarding overall compliance status, which is substantiated by systematic, organized audit documentation.

Self-Reporting and Immunities

Should results of an audit be reported to government agencies as an admission of violations? The organization's lawyers should advise about the federal and state laws that give specific incentives to make reports of detected violations and the corrective actions that follow. In about half of the states, laws encourage audits by granting a legal privilege against the use of an audit report in civil litigation. Some further encourage reporting by granting legal immunity from prosecution. But the several federal programs for audit reporting at EPA, OSHA, and other agencies have important limits and qualifications. A federal prosecution could use the audit report even though the state enforcers could not do so. Details of these legal strategies are beyond the scope of this text. The reader should consult with experienced legal counsel familiar with the SH&E legal defense function before presuming that audits can be shielded from challengers.

SUMMARY

- Safety, health, and environmental (SH&E) auditing is used to verify that a facility's procedures and practices comply with legal requirements and internal policies and conform to good SH&E practices.
- The primary motivation for SH&E audits ranges from the desire to measure compliance to the need to identify potentially hazardous conditions for which standards do not exist. Companies define the scope and focus of an audit in organizational, geographical, locational, functional, media-specific, and compliance contexts.
- The SH&E audit process involves planning, field work, and follow-up stages. Generally, a team approach is used to conduct the auditing activities of gathering background information, conducting the actual audit, and reporting the results. Basic audit tools include audit protocols and working papers.
- The role that auditing programs play in companies determines how much independence the programs are given and where they are placed in the organization. Companies may form audit teams composed of members from a variety of disciplines and departments; they may

also appoint full-time or part-time managers.
- Government agencies or companies themselves can conduct environmental audits of facility operations to (1) identify contamination or hazardous materials from current or previous activities, (2) determine compliance with current standards or regulations, and (3) provide a general review of environmental risks associated with a site and its operations.
- Environmental audits can be internal (conducted by the company) or external, conducted by an outside firm. Internal top-down and self-audits do not always result in objective treatment of a company's environmental problems and risks. Audits performed by outside consultants provide objectivity and may pay for themselves in reduced fines and penalties.
- Environmental audits serve many purposes: avoiding costly cleanup charges, preventing fines, promoting good labor relations, enhancing a company's public relations with the community, and serving as a "report card" on a firm's environmental management functions.
- The number of companies with SH&E audits is expected to increase. In addition, the audits will have increased depth and rigor of review, greater use of field resources, and increased emphasis on basic skills. Careful attention to liability from disclosures to government or to litigation opponents requires close coordination with company lawyers as the audit process is adopted and whenever serious noncompliance is found. SH&E auditing standards that are emerging on both the domestic and international levels will provide guidelines for this increasingly important process.

REFERENCES

American Society for Testing and Materials (ASTM). ASTM E2107–06, *Standard Practice for Environmental Regulatory Compliance Audits*, 2006.
Practice for Environmental Site Assessments. 2 parts: Phase 1 and Transaction Analysis, 1993.
Geltman, E. *A Complete Guide to Environmental Auditing*. Chicago: American Bar Association, 1997.
Greeno, J. L. *Environmental Auditing: Fundamentals and Techniques*. New York: Wiley, 1985.
Hall, R. M., and D. R. Case. *All about Environmental Auditing Edition 2*. Washington DC: Federal, 1992.
McDaniel, T., J. Shih, and E. Ardiente. "Environmental Auditing for Continuous Improvement." Paper presented at the annual meeting of the A&WMA, Pittsburgh, PA, June 1993.
Priznar, F. J. "A Guide to Environmental Auditing, Scrap Processing and Recycling." *Journal of the Institute of Scrap Recycling Industries* 50 (March–April 1993): 101.
———. "Trends in Environmental Auditing." *Environmental Law Reporter*, Environmental Law Institute, 20 ELR 10179, 1990.
U.S. Environmental Protection Agency. *Environmental Auditing Policy*, 60 Fed. Reg. 66706, December 22, 1995.
Incentives for Self-Policing, Discovery, Disclosure, Correction and Prevention of Violations, 65 Fed. Reg. 19618, April 11, 2000.
Young, S. *Environmental Auditing*. Des Plaines, IL: Cahners, 1994.

REVIEW QUESTIONS

1. Briefly define safety, health, and environmental (SH&E) auditing.
2. List six of the nine company objectives in establishing SH&E auditing programs.
   a.
   b.
   c.
   d.
   e.
   f.
3. Name and briefly discuss the six criteria companies can use to define the scope and focus of an audit.
   a.
   b.
   c.
   d.
   e.
   f.
4. What are the seven key steps in the SH&E auditing process?
   a.
   b.
   c.
   d.
   e.
   f.
   g.
5. List the two basic audit tools, and explain why each one is so important.
   a.
   b.

6. Which department of a company is usually involved in the initial development of the audit program and plays a central role in developing the audit reporting process?
   a. internal audit department
   b. production/operations department
   c. legal department
   d. regulatory affairs department
7. In staffing their audit program team, companies should choose which combination of individuals?
   a. facility manager, process or safety engineer, attorney, and an analytic chemist
   b. process safety engineer, internal auditor, industrial hygienist, and an outside consultant
   c. full-time audit program manager, attorney, analytic chemist, and an internal auditor
   d. varies from company to company
8. As companies expand their auditing programs overseas, what types of considerations will they have to address?
   a. diverse local regulatory systems
   b. travel and logistical requirements
   c. language and cultural barriers
   d. different levels of company ownership and control
   e. all of the above
9. List five of the seven trends that will be seen in audit programs over the next 5 years.
   a.
   b.
   c.
   d.
   e.

Workers' Compensation

Pat Lorimer, MPH
Philip E. Hagan, JD, MBA, MPH, ARM, CIH, CET, CHMM, CHCM, CHSP, CEM

Economic Losses

Workers' Compensation in the United States
Early Laws ▸ Compensation Legislation

Objectives of Workers' Compensation
Income Replacement ▸ Accident Prevention and Reduction

Major Characteristics
Covered Employment ▸ Limitations on Coverage
▸ "Exclusive Remedy" for Work-Related Disabilities
▸ Covered Injuries ▸ Occupational Disease
▸ Hearing ▸ Black Lung Disease

Benefits
Income Replacement ▸ Medical Benefits
▸ Rehabilitation

Administration
Objectives ▸ Handling Cases

Rehabilitation
Medical Rehabilitation ▸ Vocational Rehabilitation

Degree of Disability
Temporary Total and Partial Disability ▸ Permanent Partial Disability ▸ Permanent Total Disability
▸ Insurance Incentives ▸ Safety Incentives

Managing a Workers' Compensation Program
Hiring ▸ First Report of an Accident ▸ Physicians and Medical Institutions ▸ Rehabilitation ▸ Follow-Up

Summary

References

Review Questions

Prevention of unintentional injuries and illnesses is the goal of this manual. The ways in which hazards can be identified, categorized, avoided, and mitigated are discussed throughout this text. But despite great efforts, some injuries will occur at workplaces, with the result that workers will incur lost wages, medical bills, and rehabilitation costs. This chapter describes the legal system that states have established to ensure that losses from workplace injuries are compensated and that—through a standardized insurance mechanism—worksites that have greater risks pay a greater proportion of the insurance costs. Ultimately, the workers' compensation system is a "safety net" that alleviates the problems that workplace injuries can cause while providing employers with some relative predictability of the costs that injuries will impose.

This chapter will discuss the following topics:
- the types of economic losses that workers and their families may experience
- a historical review of workers' compensation laws and legislation
- the six primary objectives underlying workers' compensation laws
- major characteristics of workers' compensation and limitations on coverage
- basic benefits covering loss of income, medical payments, and rehabilitation
- how workers' compensation is administered
- issues in rehabilitation of injured workers
- four general classifications of disability
- how to manage a workers' compensation program.

Effective loss control can prevent injuries and accidents and reduce their costs, thus benefiting workers, employers, and the entire economy. Many companies are continuing to show much greater interest in controlling the costs related to such incidents. This interest is sparked by the increasing importance that costs are being given in executive decision making. As a result, safety and health professionals have a great opportunity to influence management to adopt more effective safety and loss control measures.

ECONOMIC LOSSES

Workers and their families may suffer two types of economic losses: (1) loss of earnings and (2) accrual of additional expenses.

If a worker dies because of a work-related injury or sickness, the survivors lose the income the worker would have earned—less the amount spent on personal expenses—over the remainder of the individual's working career and retirement years. This loss can be substantial.

Total and permanent disability causes even greater earnings losses than death because the worker must be taken care of despite being unable to work and contribute income. *Permanent partial disability* accounts for part of the economic losses due to disability, which depend on the proportion of annual earnings lost because the worker cannot function fully. An employee who is totally disabled temporarily loses any income during the time he or she is recuperating. Loss of even a month's earnings can be a serious financial problem for most workers. In addition to these earnings losses, the deceased or disabled worker often is unable to provide valuable household services, which must now be forgone or taken over by someone else at additional cost.

Although not all injured workers are disabled, nearly all will require some form of medical attention. In general, medical expenses usually amount to less than the total earnings lost, but medical expenses for many workers equal or exceed the income lost.

In addition to direct earnings losses, society also loses the taxes that injured employees would have paid and the products or services they would have provided. Some injured employees and their families become public assistance beneficiaries and must be supported by other members of society.

WORKERS' COMPENSATION IN THE UNITED STATES

In the early decades of industrialization in the United States, efforts to implement a system of compensation for industrial injuries lagged far behind the compensation developments in Europe. However, toward the end of the 19th century, as work-related injuries and diseases and their consequences grew more severe and costly, the public and others began to demand radical change. The first tangible evidence of government interest in workers' compensation laws appeared in 1893 when legislators seized upon John Graham Brooks' account of the German system as a guide for their own efforts at reform. Their interest was further stimulated by the passage of the British Compensation Act of 1897.

Early Laws
In 1902, Maryland passed an act providing for a cooperative accident insurance fund. Although this represented the first legislation to embody the compensation principle to any degree, the scope of the act was limited. Benefits, which were quite meager, applied only in cases of fatal accidents. Within 3 years, the courts declared the act unconstitutional. In 1908, a Massachusetts act authorized the establishment of private plans of compensation on approval

of the state board of conciliation and arbitration. This law had no practical significance and proved to be a dead letter from the start.

By 1908, the United States still had no workers' compensation act. Recognizing the injustice, President Theodore Roosevelt, urged passage of an act for federal employees. He pointed out that the burden of an accident fell on the helpless man, his wife, and his children and declared this state of affairs "an outrage." Later in 1908, Congress passed a compensation act covering certain federal employees. Though somewhat inadequate by some standards, it was the first real compensation act passed in the United States.

During the next few years, advocates of compensation continued to press for state laws. A law passed in Montana in 1909, applying to miners and laborers in coal mines, was declared unconstitutional. Nevertheless, many states appointed commissions to investigate the feasibility of compensation acts and to propose specific legislation. A significant number of laws resulted from these commission reports—all of which favored some form of workers' compensation legislation—and were combined with recommendations from various private organizations. However, widespread agreement on the need for compensation legislation did not end all conflict over reform. Special-interest groups clashed over specific bills and over questions of coverage, waiting periods, and state versus commercial insurance.

In 1910, New York adopted a workers' compensation act of general application whose coverage was compulsory for certain especially hazardous jobs and optional for others. None of the early state compensation acts expressly covered occupational diseases, although statutes that provided compensation for "injury" were frequently interpreted to include disability from disease. However, acts that limited benefits to "injury by accident" expressly excluded occupational disease. Every state act except the one passed by Oregon required uncompensated waiting periods of 1 to 2 weeks before benefits were paid; several states provided retroactive payments after a prescribed period.

The 1911 Wisconsin workers' compensation act was the first law to remain effective, and it was quickly followed by laws in Nevada, New Jersey, California, and Washington that same year. In 1916, the U.S. Supreme Court declared workers' compensation laws to be constitutional. Although 24 jurisdictions had enacted such legislation by 1925, workers' compensation was not provided in every state until Mississippi enacted its first law in 1948. Thus, the United States proceeded with a *statewide* workers' compensation system when other countries enacted *nationwide* workers' compensation. As a result, there are 50 different workers' compensation acts in the United States today.

Compensation Legislation

All 50 states, the District of Columbia, Guam, and Puerto Rico have compensation acts. In addition, the Federal Employees' Compensation Act (FECA) covers all employees of the U.S. government, while the Longshore and Harbor Workers' Compensation Act covers maritime workers (other than seamen) and workers in certain other groups. The latter act provides compensation for workers in the "twilight zone" between ship and shore because the U.S. Supreme Court had ruled that they could not be covered under state compensation laws. (Each of the Canadian provinces and territories also has a compensation act or ordinance.)

Although economic changes and public policy have prompted increases in benefits and scope of the laws, the basic concepts have remained relatively unchanged. Employers and the labor force are both dissatisfied with certain aspects of workers' compensation. Labor proponents attack the system for inadequate benefits, coverage limitations, and exclusion of many injuries, illnesses, and disabilities that they consider job related. Employers criticize the system for covering some injuries and diseases that they do not consider job related and for the system's high cost relative to its apparent benefits. Thus, although early advocates of workers' compensation conceived of it as a simple, efficient, equitable remedy to reduce litigation over industrial injuries, both labor and management continue to disagree about what constitutes "efficient" and "equitable."

OBJECTIVES OF WORKERS' COMPENSATION

The U.S. Chamber of Commerce publication *Analysis of Workers' Compensation Laws*, published annually, cites these six basic objectives underlying workers' compensation laws:

1. Provide adequate, equitable, prompt, and reliable income and medical benefits to work-related accident victims or income benefits to their dependents, regardless of fault.
2. Provide a single remedy and reduce court delays, costs, and workloads arising from personal injury litigation.
3. Relieve public and private charities of financial drains resulting from uncompensated industrial accidents.
4. Eliminate payment of fees to lawyers and witnesses as well as time-consuming trials and appeals.
5. Encourage maximum employer interest in safety and rehabilitation through an appropriate experience-rating mechanism.
6. Promote candid study of causes of accidents (rather than concealment of fault), thus reducing preventable accidents and human suffering.

Income Replacement

The first objective of workers' compensation is to replace the wages lost by workers who are disabled because of a job-related injury or illness. According to this objective, the replacement should be adequate, equitable, prompt, and sure.

To be adequate, the program should replace lost earnings (present and projected, including fringe benefits), minus expenses such as taxes and job-related transportation costs that will no longer accrue. The worker, however, should bear a proportion of the loss so that he or she has an incentive to engage in rehabilitation and accident prevention. A two-thirds replacement ratio is used in most state statutes.

To be equitable, the program must treat all workers fairly. According to one concept of fairness, most workers should have the same proportion of their wages replaced. However, workers with a low wage may need to receive a high proportion of their lost wages in order to sustain themselves and their families. High-income workers who can afford to purchase private individual protection may have their weekly benefits limited to some reasonable maximum. However, as long as workers' compensation insurance is regarded primarily as a wage-replacement program, few people should be affected by this maximum.

The first objective also includes medical and vocational rehabilitation and a return to productive employment. To achieve this goal, workers should receive quality medical care—care that will restore them as much as possible to their former physical condition—at no cost. If complete restoration is impossible, workers should receive vocational rehabilitation that will enable them to maximize their earning capacities. Finally, the system should provide incentives for disabled workers to want to return to productive employment as quickly as possible and for prospective employers to help these workers do so.

Another objective of workers' compensation is to distribute the costs of the program among employers and industries according to the degree to which they are responsible for the losses. Such a distribution is considered equitable because each employer and each industry pays its fair share of the costs. In the long run, this distribution shifts resources from hazardous industries to safe industries and from unsafe employers within an industry to safe employers. Eventually, employers with the most unsafe operations will be driven out of the marketplace.

Critics argue that workers' compensation costs account for such a small part of overall operating expenses that they have little, if any, effect on a firm's resource allocation. As a result, unsafe employers would not need to resort to higher prices, and they would remain in the marketplace.

Accident Prevention and Reduction

Occupational accident prevention and reduction is the final commonly accepted objective of workers' compensation. Those who consider this objective important believe that the system can and should provide significant financial and other incentives for employers to introduce safety measures that will decrease the frequency and severity of accidents. More specifically, they believe that the pricing of workers' compensation should reward good safety practices and penalize dangerous operations. By having to share some of the losses, employees are more likely to have some incentive to follow safe work practices. Injured workers should have the opportunity, and should be encouraged, to return to work as soon as they are physically able.

MAJOR CHARACTERISTICS

Compensation laws can be elective or compulsory. Under an elective law, the employer may accept or reject the act. However, if an employer rejects the act, it loses the three common-law defenses—assumption of risk, negligence of fellow employees, and contributory negligence. This means that in practice, all the laws can be considered "compulsory." A compulsory law requires each employer to accept its provisions and provide for the benefits specified.

Most jurisdictions require employers to obtain insurance or to prove financial ability to carry their own risk. Six states, two U.S. territories, and most provinces require employers to contribute to a monopolistic fund operated by the state or provincial agency. In some instances, employers may qualify as self-insurers. Thirteen states permit employers to purchase insurance either from a competitive state fund or from a private insurance company.

Covered Employment

Although most of the state workers' compensation laws apply to both private and public employment, none of the laws cover all forms of employment and occupations. For example, a few states restrict compulsory coverage to so-called hazardous occupations. Many laws exempt employers having fewer than a specified number of employees, usually fewer than three or four in any one location. Most of the laws also exclude workers in farming, domestic service, and occasional employment. Many laws contain other exemptions, such as employment in charitable or religious institutions.

As indicated earlier, federal workers are covered by FECA. Employees of the District of Columbia are covered by the District of Columbia Workers' Compensation Act, which went into effect in 1982. Its provisions closely follow those of FECA.

Two other major groups excluded from coverage by compensation laws are interstate railroad workers and maritime employees. Railroad workers whose duties involve any aspect of interstate commerce are covered by the Federal Employers' Liability Act (FELA). Maritime workers are subject to the Jones Act, which applies provisions of the FELA to seamen.

The Federal Employers' Liability Act is not a workers' compensation law. Instead, it gives an employee the right to charge the employer with negligence and prevents the employer from pleading the common-law defenses that the worker is a fellow servant or assumes part of the risk; moreover, the act substitutes the principle of comparative negligence for the common-law concept of contributory negligence.

It is not known how many state and local employees are covered by workers' compensation or provided with such protection voluntarily by their employers. All states (as well as Puerto Rico, Guam, and the District of Columbia) provide some coverage of public employees, but the extent of the benefits varies widely. Some laws specify no exclusions or exclude only such groups as elected or appointed officials. Others limit coverage to employees of specified political subdivisions or to employees engaged in hazardous occupations. In still others, the extent of coverage is left entirely up to the state or to the city or political subdivision employing government workers. Certain other groups, such as the self-employed, unpaid family members, volunteers, and trainees, generally are not protected by workers' compensation.

Limitations on Coverage

In view of the fact that some of the exemptions or exclusions in many state laws have persisted to this day, it may be helpful to review some of the reasons behind the original limitations. Nearly all state acts were prepared and enacted in the face of constitutional challenges and the outright opposition of certain business or government interests. Thus, each act was the result of political compromises.

Initially, workers' compensation was hailed as an innovation that would introduce greater certainty into the calculation and payment of benefits than existed in the common-law system. Under common law, workers could sue employers and, if successful, might be ensured of adequate payment; however, those who lost would be left with nothing but debts. To reduce this risk, the workers' compensation law specified the benefits that would be paid to all regardless of fault. Although the outcome of workers' compensation cases is far more certain in ordinary suits in which negligence must be shown, the law is not automatically applied.

In part, this remaining uncertainty arises from the wide variety of permanent partial disability cases that the schedules do not cover satisfactorily. Two factors usually prompt compensation litigation. One is the uncertainty about whether an accident arose out of and in the course of employment; the other factor is the extent of disability. As workers' compensation comes to encompass more of the ailments to which the general population may be susceptible, it becomes more difficult to distinguish impairments that are work related from those that are not. In addition, legal skills and medical judgment are required to assess the extent of disability in such difficult cases as occupational diseases, injuries to the soft tissue of the back, heart conditions, and situations in which the only evidence before the commission is a subjective complaint.

"Exclusive Remedy" for Work-Related Disabilities

Before workers' compensation laws were enacted in the states, an employee, in order to recover damages for a work-related injury, had to prove some degree of fault or negligence on the employer's part. Under what is now known as the *quid pro quo* of workers' compensation laws, employers accepted, or were required to accept, responsibility for injuries arising out of and in the course of employment without regard to fault. In exchange, employees gave up the right to sue employers for unlimited damages. These agreements are usually referred to in the state acts as *exclusive remedy* provisions, a term that is quite misleading.

In no state are workers' compensation benefits necessarily the only remedy available to an injured worker. Depending on the wording of the applicable statute, workers may also bring a negligence action against their employer, fellow workers, another contractor on the same job, or some other entity or individual who caused the compensable injury. For example, workers may sue the manufacturer of a piece of equipment that caused an injury. From the employer's viewpoint, the doctrine should be the *exclusive liability rule*. As the employee sees the rule, it remains an "exclusive remedy" for obtaining "workers' compensation" from the employer. However, neither liability nor remedy is perfectly exclusive.

Two concepts that are broadening the exclusive remedy provision are (1) the expansion of the dual-capacity doctrine and (2) the intentional-tort exception.

Under the first concept, an injured employee can sue an employer for an injury—even if it arose out of and in the course of employment—if the injury was caused by the employer's product or a service available to the public. (Examples: A driver of a tire company delivery truck is injured when a defective tire made by the employer causes the truck to have an accident. Or a hospital employee who, after a fall on the job, is further injured as the result of negligent treatment by one of the hospital's medical staff.) In both cases, the injury did not occur as a result of the employer–employee relationship but rather because of a relationship more akin to that of a supplier or service provider and the public.

In relatively few cases, workers have been able to bypass the exclusive remedy approach of workers' compensation laws. These "intentional tort" claims are exceptional, and each state's highest court sets the parameters that satisfy the narrow prerequisites. To generalize, a worker must show that the employer knew of the grave risk, that the employer acted recklessly in sending the worker into danger despite knowing that an injury was substantially likely, and that the worker was in fact harmed. Unlike the conventional workers' compensation remedy, which does not assign blame or allocate responsibility for an incident, these cases usually involve the employer's conscious or willful indifference to a hazard that resulted in a worker's serious injury.

Covered Injuries

Workers' compensation is intended to provide coverage only for certain work-related conditions and not for all of an employee's health problems. Statutory definitions and tests have been adopted that distinguish between conditions that are compensable and those that are not. All jurisdictions, when drafting workers' compensation laws, relied to some extent on the English legal system (or on other statutes based on the English model). Even though the statutory language of these laws is remarkably similar, there are variations in terminology and differences in interpretation; as a result, a condition considered compensable in one state may be held noncompensable in others.

The statutes usually limit compensation benefits to personal injury caused by accidents arising out of and in the course of employment. Although this restriction presents four distinct tests that must be satisfied, in practice these tests are often considered in pairs: the "personal injury" and "by accident" requirements in one set and the "arising out of" and "in the course of" requirements in the other.

For the first pair of tests, if interpreted narrowly, "personal injury" refers solely to bodily harm, such as a broken leg or a cut, whereas the "by accident" test refers to the cause, such as a blow to the body or an episode of excessive or improper lifting. In practice, however, the distinctions are often blurred. The "by accident" concept is a carryover from English law. Early judicial interpretations of English law made it quite clear that the "by accident" requirement is intended to deny compensation to those who injured themselves intentionally. A number of U.S. jurisdictions, however, have applied the test in order to narrow the range of unintentional injuries that must be compensated.

For the second pair of tests, "in the course of" employment generally is an issue of time, place, and circumstance. If an injury occurs on the job or at the employer's place of business during normal working hours, the injury is usually considered to have occurred in the course of employment. Typically, "arises out of" employment hinges on whether there is a causal connection between the injury and the employment. The phrase "arising out of and in the course of employment," which is applied in almost every jurisdiction, is meant to clearly define the relationship between employment and an injury or disease and when an employee is eligible for workers' compensation. The phrase obviously lacks precision. Often, it is quite difficult to determine whether a given set of facts can support an award of compensation.

Different jurisdictions have developed and used four different doctrines to determine whether an injury or disease "arises out of and in the course of employment." These four distinct interpretations are the peculiar-risk, the increased-risk, the actual-risk, and the positional-risk tests.

In the *peculiar-risk* doctrine, which has fallen out of favor, the injured party has to show that the source of the harm was peculiar to the employment. This test looks at how the risk the injured worker faced differed from the risk every other worker faced.

The *increased-risk* test looks at whether the employment resulted in a greater degree of risk than the general public encounters.

The more widely used test is the *actual-risk* doctrine, which requires that the hazard causing the injury be a risk of the particular employment regardless of whether the general public is also exposed to the risk. All the injured employee needs to demonstrate is that the peril may reasonably be considered to arise from the employment.

The *positional-risk* doctrine could also be called the *but for* test. According to this doctrine, if the employment places the worker in a position where he or she is injured ("but for" the employment, the injury would not have occurred), the injury is considered to arise out of the employment. This is the most liberal of the four doctrines.

The "course of employment" aspect of this test refers primarily to the time frame of the injury. Virtually every jurisdiction holds that employees are considered within the course of employment—barring unusual circumstances or unreasonable conduct—from the moment they step onto the employer's premises at the start of the workday to the moment they leave at the workday's end.

Although this test appears to be relatively simple to apply, doing so has often proved difficult. For example, what is meant by the term *premises*? Injuries that occur off premises but that appear to deserve compensation lead plaintiffs to search for exceptions in the laws and encourage courts to modify the basic rules. In addition, many employees do not work at only one premises. Also, even though an injury occurs off premises, as when traveling to and from work, the employee may be compensated if a sufficient employment connection can be established; perhaps the employer paid the worker for the time or expense of travel or provided a company vehicle for transportation. In these circumstances, the travel time to and from the worker's home may be considered "in the course of" employment. In addition, unusual circumstances may exist in which a person was injured while on personal business during the workday and the injury is not covered by workers' compensation. Each instance, therefore, has to be examined on a case-by-case basis.

The "arising out of" segment of the test is intended to establish a causal relationship between the employment and the injury. For example, an employee should not automatically expect to receive compensation after having a heart attack at work. The person must show that the heart attack arose out of the employment. This means that, at minimum (some states have more stringent rules), the employee must show that the stress and strain or exertion of the employment caused the heart attack and that the heart attack was not instead a result of a spontaneous breakdown of the cardiovascular system.

One of the early casualties of the "by accident" requirement was occupational disease coverage. Because the typical judicial finding was that "occupational disease" and "accidental injury" are mutually exclusive concepts, special legislation was required to secure coverage for workers experiencing work-related diseases.

Occupational Disease

Although workers' compensation laws initially had no specific provisions for occupational diseases, all states now recognize responsibility for those diseases. Coverage extends to all diseases arising out of and in the course of employment. Most states do not provide compensation for a disease that is an "ordinary disease of life" or that is not "peculiar to or characteristic of" the employee's occupation. For example, in March 1996, the Virginia State Supreme Court ruled that carpal tunnel syndrome and "trigger finger" are not occupational diseases covered under the state's workers' compensation law. The Virginia Workers' Compensation Act was amended to define "occupational disease" in six parts, but generally as "a disease arising out of and in the course of employment, but not an ordinary disease of life to which the general public is exposed outside of employment" (NSC, *Printing and Publishing Newsletter*, 1996, p.3).

Generally, compensation for occupational diseases is the same as for traumatic injuries, and medical care coverage is unlimited. A few states do not provide permanent partial disability benefits for certain diseases. Occupational diseases usually become evident during employment or soon after exposure; however, certain diseases, such as radiation disabilities, may be latent for a long time. Most states have extended periods in which claims may be filed concerning latent, slowly developing occupational diseases.

Some states impose special restrictions on disabilities resulting from exposure to coal dust, asbestos, silica, cotton dust, or radiation. A number of states have established presumptions for police officers and fire fighters who have heart attacks or respiratory conditions, but no attempt is made to track those states.

Hearing
The difficulty of distinguishing between occupational and nonoccupational hearing loss has led to the enactment of special coverage provisions in many state statutes. These provisions attempt to isolate the occupational component of the hearing loss and to compensate workers accordingly.

Black Lung Disease
One category of occupational ailment, black lung disease, is covered by a federal benefits program under the Federal Black Lung Act, which is part of the Coal Mine Health and Safety Act of 1969, as amended. The tremendous cost of compensation for disabilities arising from this and other occupational diseases has led to pressures to fund such coverage in whole, or in part, through federal programs.

BENEFITS

The three basic types of workers' compensation benefits are (1) income replacement, (2) medical benefits, and (3) rehabilitation.

Income Replacement
Although 70% or more of recent workers' compensation cases have been for temporary total disability, such cases account for only about 25% of cash benefits paid. At the same time, income benefits to workers for permanent partial disabilities have accounted for almost 66% of the total dollar amount.

Basic Features
In general, the cash benefits provided for temporary total disability, permanent total disability, permanent partial disability, and death are payable as a wage-related benefit— that is, the weekly amount is a percentage of the worker's wage. Although the benefit varies by state and by type of disability, it is commonly set at 66% to 100% of current wages. In some states, the statutory percentage, especially for survivor benefits, varies according to the worker's marital status and the number of dependent children.

For many beneficiaries, the benefit rate is limited to less than two-thirds of wages by another statutory provision—the maximum ceiling on the weekly benefits payable. Because of this ceiling, in almost all states disabled workers whose wages are at or above the statewide average receive benefits below the statutory benefit rate. However, for such individuals, benefits may exceed pre-injury take-home pay because the benefits are tax-free.

Other restrictions on benefits set maximum time periods in which to receive compensation or maximum dollar amounts that can be paid out. Such limitations in permanent total disability and death cases may cut off benefits to workers or their survivors even though their need for income continues. Only 14 states limit the duration of total dollar benefits to surviving spouses and children.

To reduce administrative costs and to discourage workers from malingering, all states stipulate in their laws that benefits are payable only after a waiting period following the report of disability. This delay in payment, which ranges from 3 to 7 days, applies only to cash indemnity payments and not to medical and hospital care. In all states, workers who remain disabled beyond the specified minimum waiting period receive payment retroactively for that time. In more than three-fourths of workers' compensation laws, the minimum period before retroactive payment of disability benefits begins is 2 weeks.

Benefits by Type of Disability

Income benefits vary depending on whether an employee's disability is temporary or permanent, partial or total, or fatal. Most compensation cases concern workers who incur temporary disabilities but recover completely. In some cases, dollar ceilings on weekly income benefits mean that many disabled workers are not fully compensated for their earnings lost.

Permanent total disabilities that prevent employees from performing any work in any well-known area or industry in the labor market and are of indefinite duration. Benefits for these disabilities are similar to those paid for temporary total disability benefits. In a few states, the weekly benefits for permanent disability benefits are less than the weekly benefits for temporary disability. A small number of states limit the benefit period, usually from 6 to 10 years, for those with permanent disabilities.

Residual limitations on a worker's earning capacity after recovering from an injury (that is, after recovering from a permanent partial disability) are calculated using a relatively complex formula. Partial disabilities are divided into two categories: "scheduled" injuries, or injuries listed in the law such as loss of specific body members, and "nonscheduled" injuries, or injuries that are of a more general nature such as back and head injuries.

Weekly benefits for scheduled injuries are calculated as a percentage of average weekly wages, and are usually the same as the weekly benefits for permanent total disability. The maximum weekly benefit is usually the same as or lower than the maximum weekly benefit for total disability. Nonscheduled injuries are paid at the same or a similar rate but as a percentage of wage lost. This represents the difference between the wages before the injury and the wages the worker is able to earn after the injury.

Scheduled benefits are paid over fixed periods that vary according to the type and the severity of the injury. For example, most state laws call for payments ranging from 200 to 300 weeks for loss of an arm and 20 to 40 weeks for loss of a big toe. The maximum benefits period for nonscheduled injuries in each state is either the same as or, more commonly, less than the time limit set for permanent total disability. In most states, compensation for permanent partial disability is paid in addition to the benefits a worker received during the healing period or when the worker was temporarily and totally disabled.

Death benefits are intended to furnish income replacement for families who depended on the earnings of an employee who was killed by a work-related incident or disease. As is true for other types of benefits, the amount of survivor benefits and the length of time they are paid vary considerably from state to state. If the surviving spouse is without dependents, the benefits paid, which are a percentage of the deceased worker's wages often are less than the benefits paid for permanent total disability. If the worker had dependent children, the benefits in many states will be augmented to help support those children. In most states, the duration of these benefits is unlimited, although nine states have established limits that range from 7 to 20 years. In several states, payments continue to spouses as long as they do not remarry and continue to children until they are no longer dependent, usually at age 18. In many states, benefits to children continue until they are age 23 or 25 if they are students. In the four states that limit total dollar benefits, death benefits may be terminated earlier.

In addition to paying benefits for surviving spouses and children, some states pay survivor benefits to dependent invalid spouses, parents, or siblings of the deceased worker. Burial expenses are covered in all states.

Medical Benefits

For many years, medical care disbursements provided by workers' compensation have comprised about one-third of the total workers' compensation benefits outlay. Medical care includes first-aid treatment, physician services, surgical and hospital services, nursing services and medications, supplies, and prosthetic devices. Some large employers, in addition to providing first-aid facilities, employ staff physicians for their workers. Most employers insure their medical care responsibilities just as they insure the income benefits of workers' compensation.

Every state law requires employers to provide benefits for medical care to injured workers. In most jurisdictions, such medical treatment is provided without limitations through either explicit statutory language or administrative interpretation. In the few states that limit total medical care by enacting specified maximum dollar amounts or maximum payment periods, management can decide to exceed the initial ceiling if circumstances warrant such an action. However, if specified types of injuries or diseases are denied cash benefits, medical care for these conditions also is denied.

An issue many employers face when providing medical benefits under workers' compensation is choosing the physician who will furnish care. Almost half of the states give the employer the right to designate the physician. In practice, the insurance company of the employer typically selects the physician because it is the insurer who will handle the benefits claims. When the physician is chosen in this way, the medical care furnished may be highly skilled and effective because of the selected physician's specialized experience. On the other hand, workers often feel that their own family physicians place more emphasis on their personal health and well-being. They believe that other considerations, such as company interests or insurance costs, may influence the care of a physician chosen by their employer.

Two other sources of medical benefits for disabled workers are Social Security and private disability programs that are part of fringe benefit packages provided by larger employers in particular. Any disabled worker whose disability is documented as lasting at least 12 months or that results in death may receive (or whose survivors will receive) disability benefits if eligible under Social Security rules. The combination of Social Security benefits and workers' compensation payments cannot exceed 80% of the disabled worker's earnings before disability. These benefits are taken out of Social Security.

Even if they qualify for workers' compensation, disabled workers may not be able to meet Social Security's requirements for proving that they cannot perform any kind of gainful work.

Private disability insurance programs usually coordinate with workers' compensation plans. Frequently, a private disability program requires employees to file workers' compensation and Social Security disability claims before they can qualify for private carrier benefits.

Rehabilitation

Along with industrial safety, medical care, and cash compensation, rehabilitation of workers is recognized, at least theoretically, as one of the primary goals of the workers' compensation system. The most widespread benefits offered through workers' compensation laws to restore a worker to the fullest economic capacity are the special rehabilitation benefits authorized in more than half the states. These benefits usually are paid (sometimes in addition to regular disability compensation) for various training, education, testing, and other services designed to speed an injured person's return to work.

Probably the main source of retraining and rehabilitation is the federal-state vocational program. The federal government, through the Federal Vocational Rehabilitation Act, is the major source of funding for such programs. Facilities operated by this program rehabilitate both individuals with work-related disabilities and individuals who were injured off the job. In all states, these institutions are administered by state vocational rehabilitation agencies and provide medical care, counseling, training, and job placement. Unfortunately, not all workers' compensation cases referred for vocational rehabilitation can be accepted promptly, and many other cases are never brought to the facilities' attention. As a result, workers who may have been able to return to some type of job remain disabled.

Another notable drawback preventing the full use of available rehabilitation facilities are the complex, adversarial proceedings that determine a worker's right to benefits. Because the decision about whether benefits should be awarded for permanent partial or permanent total disability (and how large the benefits should be) is based primarily on a worker's inability to work, the worker may have a strong incentive to put off rehabilitation. Further, in the many compromise settlements, the employer's (or insurer's) main goal is to pay no more than the agreed amount of money and to prevent any future liability for medical, vocational, or other needs arising from the injury. Such settlements also work against a full-fledged effort to restore the worker to full health and productivity.

Advocates say that guidelines for applying medical treatment when managing workers' compensation claims are gaining greater acceptance as more states adopt rules encouraging their use. The guidelines typically describe appropriate treatment to address common work-related injuries and are developed through peer review and consensus among occupational physicians.

ADMINISTRATION

The goals of workers' compensation are to provide quick, simple, and inexpensive decisions on all claims for benefits and to provide the medical care and rehabilitation services needed to restore injured workers to employment. Nearly all states have agencies that carry out these administrative responsibilities.

Objectives

A state agency's responsibilities include closely supervising the processing of cases. The primary objectives of this supervision are to ensure that all parties comply with the law and to guarantee an injured worker's rights under the statute.

A key goal of an agency is to see that the injured worker gets the full benefits due. To do so, the agency must follow an injury case from the first report to the final closing. Some states not only check the accuracy of total payments but also require signed receipts for every compensation payment. To facilitate a complete audit of the individual payments, some states require the employer to file a final receipt that itemizes the purpose of each part of the total benefits outlaid.

Frequently, however, the legislation itself states that a workers' compensation agency must operate on the assumption that each injured worker is responsible for securing his or her rights and that the agency's primary function is to adjudicate contested claims. Even where the law does not favor this policy, a lack of staff may force an agency into having this restricted role.

Many workers are unfamiliar with the provisions of their state's workers' compensation act, and in only a few states does the agency administrator advise the worker (as soon as possible after the injury is reported) of his or her rights to benefits, and medical and rehabilitation services and of the assistance available at the agency's office. Too many states fail to insist that employers report accidents promptly, pay benefits on time, and submit final reports that list the amounts paid and how those amounts were computed. Although prompt reporting is usually required, some states impose no penalty when employers violate even that requirement.

Handling Cases

Workers' compensation claims may be either uncontested or contested. In uncontested cases, the two main payment systems are the direct payment system and the agreement system.

Under the *direct payment system*, the employer or insurer takes the initiative and begins paying compensation to the worker or dependents. The injured worker does not need to enter into an agreement and is not required to sign any papers before compensation starts. The laws prescribe the amount of the benefits. If the worker fails to receive this compensation, the administrative agency can investigate and correct any errors. Jurisdictions whose laws provide the direct payment system include Arkansas, Michigan, Mississippi, New Hampshire, Wisconsin, and the District of Columbia; this feature is also provided in the Longshore and Harbor Workers' Compensation Act.

Under the *agreement system*, which is used in most states, the parties (i.e., the employer/its insurer and the worker) must agree on the settlement before any payment is made. In some cases, the agreement must be approved by the administrative agency before payments start.

In contested cases, most workers' compensation laws provide for a hearing with a mediator or hearing officer. The statutes also provide for either the worker or the employer/insurer to appeal a decision of the mediators or hearing officer to the commission or the appeals board and from there to the courts. Because the administrative agency typically has exclusive jurisdiction over the determination of the facts in a case, appeals to the courts usually are limited to questions of law. However, in some states, in an appeal, the court is permitted to consider issues of both fact and law.

REHABILITATION

Most employees injured in work accidents return to their jobs after minor medical attention and with little if any work time lost. If the effects of the injury are temporary, the incident usually fades from memory. Even those who experience days or weeks of disability and endure substantial medical treatment may find that the injury is not permanent. Although the loss of income and the medical expenses may be distressing, eventually, when workers resume their jobs, they recover economically as well.

Unfortunately, a minority of those injured—as many as 10% of the total—experience injuries that significantly disrupt their lives. Even when these workers receive effective medical care and eventually return to productive jobs, their lives are permanently changed by the event. For some, their injuries are so severe that prolonged medical treatment and convalescence fail to restore them completely to full health and functioning. Residual disabilities prevent them from performing their former jobs, and only retraining and education, combined with special assistance, offer possibilities for future employment.

Some injured employees never return to work. If they do not die from their injuries, they live with such severe disabilities that they barely can manage by themselves. Often, the most that health services can do is to lighten the burden on those who must care for these injured people.

Treatment for workers whose livelihood is threatened by work-related impairments consists of medical rehabilitation and vocational rehabilitation.

Medical Rehabilitation

Each medical rehabilitation program, whether set up by an insurance company or a workers' compensation agency, has its own requirements for treatment, qualifications for eligibility, and definitions of service.

A disabled worker who requires medical rehabilitation receives whatever medical care is necessary to treat the impairment and restore lost function. The worker may report first to a facility nurse or physician for immediate attention. If the injury is serious, the person may then go to a hospital. Workers' compensation laws often obligate employers and insurers to pay the costs of rehabilitative medical care. Costs can be reduced by having health service workers on salary, by contractual arrangement with health personnel, or by paying hospital and doctor bills. The insurer may or may not have much influence on the selection or course of treatment.

For injuries associated with chronic disabilities, the insurer usually attempts to influence the selection of the rehabilitation services, frequently by sending the worker to a particular specialist or facility with an appropriate expertise. Often, the insurer pays for transportation to the specialist or facility as well as for rooms during treatment. Some insurance companies operate their own rehabilitation facilities, under individual or joint ownership, with medical personnel on salary, at least part time. When insurers contract to share rehabilitation programs or facilities, they may pay expenses case by case or through a rental agreement.

When informed of a worker's potential need for rehabilitation, some agencies do little more than notify the worker and the insurer that medical rehabilitation is worth considering. Other agencies conduct formal evaluations of the need for further medical care and recommend action. These agencies seek to convince disabled workers of the wisdom of rehabilitation. When the workers agree, the insurers can be required to finance the care.

Vocational Rehabilitation

Vocational rehabilitation prepares an injured worker for a new occupation or for ways of to continue doing the old one. Usually, vocational rehabilitation is needed when medical rehabilitation fails to restore the abilities the worker had before he or she was injured. The worker's injury may be so severe or the work requirements so arduous that even residual impairment prevents the person from being able to perform the former job effectively. These workers need training to overcome or to compensate for their limitations, and many may even need to enter new occupations. In general, however, the more effective the medical rehabilitation, the less the need for vocational rehabilitation.

The current definition of vocational rehabilitation makes a greater distinction between it and medical rehabilitation than is necessary. Although the difference in the kinds of treatment offered seems clear enough—retraining versus providing medical care—the two categories often overlap to a considerable degree. For example, in the public vocational rehabilitation programs in each state, services include medical diagnosis and evaluation, surgery, psychological support, the fitting of prostheses, and other health-related services along with education, vocational training, on-the-job training, and job placement.

The two programs also blend an employee's medical records. Record keeping by workers' compensation insurers does not separate claimants who receive medical rehabilitation from those who receive vocational rehabilitation, although some insurers distinguish between medical rehabilitation and acute medical care. In contrast, records kept by workers' compensation agencies usually separate vocational rehabilitation from other benefits.

Injured workers who need vocational rehabilitation are served by several means. An employer or insurer may channel the worker to whatever sources the employer/insurer thinks will provide satisfactory rehabilitation. Some workers are referred to the public vocational rehabilitation program, whose services may be financed by taxes, although insurers may reimburse the public agency. Other insurers direct workers to private facilities whose vocational training is conducted by technical schools or on the job. The costs of these services are always paid by insurers.

As with medical rehabilitation, some workers' compensation agencies encourage vocational rehabilitation and often direct a worker into a program if the insurer fails to do so. Several jurisdictions select vocational rehabilitation candidates either in conjunction with screening for medical rehabilitation or separately. Workers who have serious injuries or permanent disabilities or who are receiving extended compensation payments are reviewed by the agency for referral to the state's public vocational rehabilitation agency or to the insurer.

Some workers obtain vocational rehabilitation through their own efforts. If no one refers them, they may go directly to the public vocational rehabilitation office. Since 1920, the federal government and the states have cooperated financially (80% federal and 20% state funding) to support a vocational rehabilitation program that anyone with a vocational disability can use. Rehabilitation counselors, who usually determine a candidate's acceptability, simply verify the disability without regard to its cause and explore ways to overcome the worker's limitations. If the candidate shows relatively good prospects of improvement, the counselor designs a program to help restore the worker's abilities as fully as possible. For those who are unable to return to a paying job, the objective of vocational rehabilitation may be to help them care for themselves and to free other members of the family to earn wages.

Once workers are established in a vocational rehabilitation program, they are assigned whatever resources the counselors think best fit their needs. Generally, the resources are provided not by the vocational rehabilitation agency but by private vendors or other public agencies. For example, a worker may be sent to a private rehabilitation center or school, enter a workshop such as those run by Goodwill Industries of America, or be enrolled in a public institution.

DEGREE OF DISABILITY

Determining the extent of disability is perhaps responsible for more litigation than any other issue in workers' compensation. Making such a determination requires not only the correct application of legal principles but also an evaluation of facts, subjective complaints, and opinions and predictions of what will happen in the future.

As a general proposition (some jurisdictions use different terminology and slightly different classifications), disability can be classified in one of four categories: temporary total disability, temporary partial disability, permanent partial disability, and permanent total disability.

Temporary Total and Partial Disability

Temporary total disability is when an injured worker, although temporarily incapable of gainful employment, has a good chance of improving to the degree that he or she will be able to return to work either with no disability or with only a partial disability. *Temporary partial disability* is similar to temporary total disability in that the worker's physical condition has not stabilized but is expected to improve. The difference lies in the worker's current abilities. When temporarily partially disabled, the worker is capable of some employment, such as duties that are not physically taxing or part-time work, and is expected to improve and regain much of his or her former capabilities.

Judging an injury to be a temporary disability, either total or partial, is the least difficult and requires merely evaluating an employee's present physical condition in light of the work opportunities available. In practice, evaluating a temporary disability is concerned only with the ability of the employee to return to work for the current or previous employer. It is assumed that eventually, the employee will be able to return to work for the current employer. Given the difficulties involved in obtaining employment for workers still under medical care—and realizing that any new employment probably will be temporary—most adjudicators have expressly adopted the position that unless the worker can return to the previous job held, or be supplied with temporary effortless or part-time duties with this employer, he or she will remain temporarily totally disabled. This assumption is made even when the worker is able to perform a different job whose duties are within the individual's temporarily physically limited abilities.

Permanent Partial Disability

An injured worker with a *permanent partial disability* has reached his or her maximum improvement without full recovery. Such a worker has benefited from medical and rehabilitative services as much as possible and still has a partial disability.

Determining the extent of permanent partial disability depends on what the jurisdiction considers a "permanent partial disability." Three theories are used to establish guidelines for the payment of workers' compensation benefits for such a disability: *whole person*, *wage loss*, and *loss of wage-earning capacity* theories. Their underlying philosophies differ somewhat, as do the factors that are considered when applying each theory to a specific situation.

Whole Person Theory
This theory is concerned solely with a worker's functional limitations. Thus, the only considerations are whether the worker has sustained a permanent physical impairment and, if so, to what extent that impairment interferes with the person's usual functions and abilities. Age, occupation, educational background, and other factors are not considered.

Wage Loss Theory
The aim when applying this theory is to determine what wages the worker would have earned had the permanent impairment not occurred. If the worker's earnings dip below the estimated wage figure because of the impairment, he or she is paid compensation that is equal to some percentage of the difference between the wages that would have been earned and those actually earned. Here, the degree of physical impairment is of little or no importance. The only concern is the actual wage loss incurred and whether it is due to the impairment.

Loss of Wage-Earning Capacity Theory
This theory requires a glimpse into the future. After the worker has reached his or her maximum physical improvement, many factors (such as the remaining impairment, occupational history, age, sex, and educational background) are considered in order to estimate, as a percentage, how much of the worker's potential earning capacity has been eliminated by the work-related impairment. The worker is awarded benefits on the basis of this computation. Benefits may be paid at the maximum weekly rate for a limited number of weeks, or they may be based on a percentage of the difference between the wage-earning capacities before and after the disability, which will be paid until a preestablished dollar or time limit is reached.

Combinations

These three theories can also be used in combination. For example, some states use either the wage loss theory or the loss of wage-earning capacity theory and also provide a benefit floor determined by the worker's medical impairment. Thus, an employee who sustains a permanent impairment but no loss of wages or of wage-earning capacity would still receive some permanent disability benefits.

The use of schedules has replaced the tedious and controversial aspects of rating disabilities in a significant proportion of permanent partial disability cases. The typical schedule covers injuries to the eyes, ears, hands, arms, feet, and legs and states that for 100% loss (or loss of use) of that body part, compensation at the claimant's weekly rate will be paid for a specified number of weeks. If the loss or loss of use is less than total, the maximum number of weeks is reduced in proportion to the percentage of the loss or loss of use. Only physical impairment is considered. The effect of the injury on wages or wage-earning capacity is disregarded. If an injury is confined to a body part listed on the schedule, the benefits provided are absolute even when the disability rating on a wage loss or loss of wage-earning capacity basis would result in greater benefits. Even though this is true generally, some states provide additional benefits in the following circumstances: (1) when using one of the other theories would result in higher benefits being paid, (2) when diminished wage-earning capacity would continue after the scheduled amount is paid out, (3) when the scheduled injury resulted in permanent total disability, or (4) when several scheduled injuries are sustained in the same accident.

Use of a schedule may also be prevented by showing that the effect of the injury, such as radiating pain, extends into other parts of the body. A few jurisdictions limit the use of schedules to amputation or 100% loss of use of a body part, as opposed to partial loss of use. Other jurisdictions not only make the schedule absolute for permanent disability awards, but also require that the number of weeks for which benefits were paid during the healing period be deducted from the number of weeks authorized by the schedule before an award is made for permanent partial disability.

Most states primarily use the loss of earning capacity theory. Even when statutory language seems to indicate clearly that only functional impairment should be considered, the courts have managed to hold that loss of earning capacity is the real consideration. The use of schedules has also been justified on an earning-capacity basis as a legislative determination of presumed wage loss that resulted from the impairment listed in the schedule.

Permanent Total Disability

An evaluation of a worker with a permanent total disability is, in most respects, merely an extension of the evaluation of a worker with a permanent partial disability. In fact, the permanent total disability evaluation is a part of the same process as the permanent partial disability evaluation, and the fact finder's only additional task is to determine whether the worker's wage-earning capacity has been eliminated to the extent that he or she will no longer be able to compete in the job market.

Two aspects of the permanent total disability evaluation warrant special attention. First, most states have certain assumptions that make the fact finder's job much easier. For example, a state might assume that the loss of sight in both eyes or the loss of any two limbs constitutes permanent total disability. This relieves the fact finder of the having to evaluate all the factors mentioned for the other forms of disability. In some cases, however, these assumptions may be refuted by providing evidence that the worker has some wage-earning capacity, or the assumptions may be applied only for a limited time.

Second, the concept of permanent total disability must be defined. The definition varies from state to state. However, no state's definition says that to qualify as having a permanent total disability, an injured employee must be completely helpless or unable to earn a single dollar at a job. The person's limitations need only prevent him or her from being able to compete in the open job market and be such that no reliable job market exists for a worker with this disability.

Insurance Incentives

The insurance incentive of workers' compensation is based on a merit-rated pricing policy. *Merit rating* includes both experience rating and retrospective rating systems. All state workers' compensation funds use merit rating of some sort. Most states permit private insurers to rate employers on merit, but do not require them to do so. Under this procedure, a firm is charged a premium based on the dollar amount of claims for which it is liable. Consequently, a merit-rated firm has an incentive to reduce the amount of its claims through loss control measures. The effectiveness of this incentive has been challenged, especially given that only about one-fourth of insured firms, usually large ones, are eligible for merit rating.

The yearly accident rates of firms with only a few employees are not sufficiently reliable indications of their characteristic accidents and cannot be considered when establishing premium rates. On the other hand, merit-rated firms account for 85% of the dollar volume of premiums paid. In addition, self-insured firms, which pay approximately 14% of all benefits, are implicitly merit rated. The incentive effects inherent in experience rating, thus are not available to a large number of small firms and their employers.

Firms not eligible for merit rating are *class rated*. Under this procedure, all employers engaged in similar business operations within a state pay the same rate per $100 of payroll. These employers thus do not have a strong incentive to reduce the rates paid by their industry. In fact, the only accident prevention incentive generated for individual employers within an industry is that, if they are poor risks, they may be unable to obtain workers' compensation coverage. Other considerations of class rating include the relative quality of the safety and claims services provided by insurers, by management service organizations, and by employers themselves; the tax factors; and the opportunity cost of paying an insurance premium instead of losses and expenses as they occur.

Safety Incentives

The greatest contribution a safety professional can make to a firm's success is working with managers throughout the organization to safeguard employees from disabling occupational injuries and diseases. These injuries and diseases arising on the job hamper both the employee and those with whom he or she works in meeting various objectives: workers miss the target dates set for projects, their performance evaluations are disappointing, and they often must forgo opportunities for pay increases and promotions.

The safety professional who seeks to promote safety and health in the workplace should recognize the reasons each worker, supervisor, and executive wants to succeed in his or her occupational efforts. No matter what these reasons are, achieving career goals requires individuals and their colleagues to remain productive and efficient on the job. Thus, the insightful safety professional can demonstrate to all employees that preventing disabling injuries and diseases both on and off the job is in their own—not just their employer's—best interests and that doing one's job safely leads to personal and career success.

MANAGING A WORKERS' COMPENSATION PROGRAM

The total costs of workers' compensation, in terms of both money and human suffering, are each organization's responsibility. Like every other cost, workers' compensation expenditures need to be managed carefully. Cost control should be a top priority of management to ensure that dollars are spent as prescribed by the law. Injured employees are entitled to benefits, but businesses do not have to pay questionable or fraudulent claims.

To be successful within the legal limits requires that companies have a plan of action and treat each individual case fairly and consistently. Each firm should focus on three goals for its workers' compensation program: (1) to prevent accidents, (2) to control costs, and (3) to respond to accidents promptly and efficiently.

When an organization's management style and philosophy encourage cooperation, commitment, teamwork, and communication between labor and management, management lets workers know that accident cases will be handled fairly. Employees in such firms tend to feel fulfilled as both workers and people. Their confidence and self-discipline are reflected in their work habits and productivity. The resulting work environment tends to prevent accidents and promote greater worker safety and productivity.

The organization's philosophy should also be the basis of the workers' compensation program. The following are some factors usually addressed in program management.

Hiring

A comprehensive interview and screening process can generate evidence of applicants' existing physical condition(s) and past medical conditions that may suggest future compensation liabilities. The usefulness of preemployment physical examinations depends on the job, the employer's past experience, the physician's evaluation, and other factors. Although the objective should be to hire the best people—without discriminating and in compliance with the Americans with Disabilities Act of 1990—companies must realize that at times, they will inadvertently hire problematic employees.

During the screening process, the interviewer should be trained to elicit complete answers from applicants and to probe more deeply when the responses are not satisfactory. The interviewer should make sure that all information provided is considered in the hiring decision.

First Report of an Accident

It is vitally important that workers and management report all accidents promptly. Employees must understand the value of letting the employer know when an accident has occurred so that the employer can take immediate, efficient action.

Management should ensure that accident investigation reports are completed promptly and added to any existing files. Employees should be accompanied to the first-aid station when they are injured or ill or when they must be attended to by outside medical services. Doing so demonstrates the company's concern for its employees.

The following are advantages of having a centralized accident-reporting system: it can provide needed information quickly, help select the appropriate medical care for the worker, and offer better monitoring and control of medical treatment.

Physicians and Medical Institutions

The company must choose physicians and medical institutions to provide care to injured employees. In many situations, this care must be from the list of providers approved by the company, but in some cases, the employer allows the employee to see his or her personal physician. When specialized outside or in-house services are needed, the organization might also select the provider. Before selecting medical services, the employer should consult with other firms to find out their experiences with particular physicians or medical institutions and seek second opinions on certain cases. Proper supervisory procedures and open communication about employee injuries and illnesses should eliminate most if not all fraudulent employee claims.

Rehabilitation

Getting a disabled worker back to work as soon as possible is a major goal of medical treatment and workers' compensation coverage. Adequate follow-up of those on leave is often necessary to achieve this objective. The added cost for rehabilitation is usually money well spent. The employer should find out how the prospective physician handles paperwork, bills, reports, patient care and treatment, and patient activities. Some physicians are likely to be more suitable to a company's needs than others.

Follow-Up

The employer should review ongoing bills, claims, and reports to make sure that services are being provided as specified in compensation acts. Management should maintain a file on each case until the case is closed. Accident information learned from each incident must be fed back into the safety and health program.

The employer should treat workers' compensation employees and agents fairly and consistently. If the company buys its own insurance, it should not hesitate to seek bids for each insurance period from several different companies. The organization should frequently keep in touch with injured employees to make sure that follow-up and rehabilitation services are being properly administered.

An effective safety and health program has been shown to prevent accidents, reduce workers' compensation claims, prevent some compensation claims, and reduce an organization's overall costs. In addition, an analysis of high-risk hazards is a valuable tool for accident prevention. Typically, in line with the Pareto principle, 20% of the accidents cause 80% of the costs. Hazard analysis and past accident observations can identify which hazards are the most dangerous and help an organization reduce, control, or eliminate those hazards. At the least, the company will be able to predict with some certainty the costs of accidents resulting from those hazards.

SUMMARY

- All 50 states and all U.S. territories and possessions have workers' compensation laws. Most compensation laws seek to replace wages lost by disabled workers, pay for medical and vocational rehabilitation, reduce litigation, encourage companies to provide safe working conditions, and promote accident prevention and investigation.
- Although compensation laws are compulsory or elective, in practice, all such laws can be considered compulsory. Excluded workers include those in farming, domestic service, self-employment, some interstate commerce, and all federal and many state government employees (who are covered by government programs).
- Under an expansion of the dual-capacity doctrine and the intentional-tort exception, employers can be liable for injuries to their workers caused by the employer's product or service that is available to the public or by the employer's reckless, harmful behavior.
- Workers' compensation covers only work-related injuries or illnesses, as determined by the "personal injury" and "by accident" requirements on one hand and the "arising out of" and "in the course of" requirements on the other.
- Three basic types of workers' compensation benefits are (1) *income replacement*, (2) *medical benefits*, and (3) *rehabilitation*. All employers are required to provide employees with medical benefits that cover immediate and long-term care. Nearly all states have agencies that administer workers' compensation programs.
- Rehabilitation services may be medical or vocational or a combination of the two. Care is usually coordinated among the company, the disabled worker, and various state agencies to select the physician(s), medical facilities, and vocational assistance that will provide the best rehabilitation.
- Worker disability is generally classified into four categories: *temporary total disability*, *temporary partial disability*, *permanent partial disability*, and *permanent total disability*. Various theories are used to help determine the degree of a worker's impairment and the worker's remaining capacity to find employment.
- Many companies are given insurance and safety incentives to reduce their accident and illness rates. Merit-rating systems and occupational health and safety programs are effective ways to prompt organizations to make safety and health concerns a top priority.
- A company's goals for its workers' compensation program should be (1) *to prevent accidents*, (2) *to control costs*, and (3) *to respond to accidents promptly and efficiently*.

REFERENCES

Balge, M. Z., and G. R. Krieger, eds. *Occupational Health and Safety*. 3rd ed. Itasca, IL: National Safety Council, 2000.

English, W. *Strategies for Effective Workers' Compensation Cost Control*. Des Plaines, IL: American Society of Safety Engineers, 1988.

Larson, A., and L. Larson. *Larson's Workers' Compensation Law*, sections 3.01, 3–2, 3–4 (2003).

National Council on Compensation Insurance, 5 Marine View Plaza, Hoboken, NJ 07030.
Rate and rating plan manuals.

National Safety Council, 1121 Spring Lake Drive, Itasca, IL 60143.
Injury Facts (formerly *Accident Facts*). Issued annually.
Printing and Publishing Newsletter, July/August 1996.

U.S. Chamber of Commerce. *Analysis of Workers' Compensation Laws*. Washington DC: U.S. Chamber of Commerce, 1995.

REVIEW QUESTIONS

1. Name two types of economic losses that workers and their families experience when workers are injured at work.
   a.
   b.
2. How can an effective loss control program benefit the entire economy?
3. The Federal Employees' Compensation Act covers all employees of the U.S. government. Which act covers maritime workers?
4. Name the six basic objectives underlying workers' compensation laws.
   a.
   b.
   c.
   d.
   e.
   f.
5. What does a compulsory law require?
6. Under the quid pro quo of workers' compensation law, what were employers required to accept?
7. What two concepts are broadening the exclusive remedy provision?
   a.
   b.
8. What are the three types of workers' compensation benefits?
   a.
   b.
   c.
9. What is the minimum percentage of disability wages that injured workers usually receive?
10. What two types of workers are excluded from workers' compensation?
    a.
    b.
11. Name the four degrees of worker disability.
    a.
    b.
    c.
    d.
12. What is the goal of workers' compensation?
13. What goals does a company hope to achieve through its workers' compensation program?
    a.
    b.
    c.
14. What federal act provides a major source of funding for the retraining and rehabilitating of workers?
15. What are two factors that prompt compensation litigation?
    a.
    b.
16. What is vocational rehabilitation, and when is it necessary?
17. Which three theories are used to help determine the degree of a worker's impairment and the worker's remaining capacity to find employment?
    a.
    b.
    c.
18. When can an employee sue the employer regardless of the exclusive remedy doctrine?
    a.
    b.

Identifying Hazards

John DeLaHunt, MBA
Philip E. Hagan, JD, MBA, MPH, ARM, CIH, CET, CHMM, CHCM, CHSP, CEM

Introduction to Hazard Analysis

System Safety
Profit Motive ▸ Legal Vulnerability ▸ Ethical Motives ▸ Directives ▸ System Safety Terms ▸ How Is System Safety Designed? ▸ Mishap Models

Hazard Analysis
Philosophy ▸ What Is Hazard Analysis? ▸ Formal Methods of Hazard Analysis ▸ Who Should Participate in Hazard Analysis? ▸ What Factors Need to Be Analyzed?

Job Safety Analysis
Benefits of JSA ▸ Selecting the Job ▸ Breaking the Job Down into Steps ▸ Identifying Hazards and Potential Incident Causes ▸ Developing Solutions ▸ Using JSA Effectively

Inspection
Philosophy behind Inspection ▸ Purpose of Inspection ▸ Types of Inspection

Planning for Inspection
Hazard Control Inspection Inventory ▸ What Items Need to Be Inspected? ▸ What Aspects of Each Item Need to Be Examined? ▸ What Conditions Need to Be Inspected? ▸ How Often Must Items Be Inspected? ▸ Who Will Conduct the Inspection?

Conducting Inspections
Preparing to Inspect ▸ Inspection Tools ▸ Relationship of Inspector and Supervisor ▸ Relationship of Inspector and Employee ▸ Recording Hazards ▸ Company Compliance Inspectors ▸ Condemning Equipment ▸ Writing the Inspection Report ▸ Follow-Up for Corrective Action

Regulatory Compliance Inspection
Opening Conference ▸ Inspection ▸ Closing Conference ▸ Postinspection Procedures

Measurement and Testing
Kinds of Measurement and Testing ▸ Measuring and Evaluating Exposure ▸ When to Measure ▸ Who Will Do the Measuring?

Incident Investigation
Why Incidents Are Investigated ▸ When to Investigate Incidents ▸ Who Should Conduct the Investigation? ▸ What to Investigate ▸ Conducting Interviews ▸ Preparing the Incident Investigation Report ▸ Implementing Corrective Action

Summary

References

Review Questions

INTRODUCTION TO HAZARD ANALYSIS

A hazard is an unsafe condition or activity that, if left uncontrolled, can contribute to an unintentional injury or illness. Before hazards can be controlled, they must be identified. This identification of hazards can be accomplished through a systematic hazard analysis program that includes job safety analysis, inspection, measurement and testing, and incident investigation. Including all four functions means that analysis is performed before the operation begins, during the life cycle of the operation, and after indications that the system has broken down. This chapter covers the following topics:
- the system safety process
- the philosophy and methods of hazard analysis
- the benefits and major components of job safety analyses
- planning and conducting inspections as a critical part of hazard identification
- methods of measurement and testing used to identify workplace hazards.

SYSTEM SAFETY

System safety is a deliberate attempt to find patterns of operation that lead to safer, more precise, and predictable results.

The need for system safety is rooted in the economic considerations of production. Unwanted incidents with associated injuries and illnesses incur increased costs and decrease profits. Companies have found that it is more cost effective to correct actions before they lead to injuries and illnesses. The profit motive is the underlying force in system safety, but legal issues, ethical motives, and regulatory directives also play a role.

Profit Motive
Incidents produce direct and indirect losses that must be absorbed by a company. The direct losses of resources affect not only property but personnel as well. The loss of an aircraft due to an incident represents a decrease in the revenue potential of an airline. Fewer aircraft represents fewer passenger-revenue miles. The loss of personnel can leave an airline without qualified workers required for the operations, and training new employees represents a loss in the revenue stream of a company.

The indirect costs of an incident can be devastating to a company. The legal costs related to an aircraft incident, for example, can dramatically decrease the profits earned by a company. But the greatest costs can be to an airline's reputation. One only needs to look at the case of ValuJet, a low-cost airline that had a plane crash into the Everglades due to safety issues in 1996. The company never recovered, and by the end of 1997, it was no longer in business. Absolute safety is not cost effective and perhaps not possible in most cases. For a company to achieve no risk of an incident would require that it cease its operations. People must accept that risk is inherent in any venture. An amount of risk that is tolerable to the consumer and a producer must be achieved.

Legal Vulnerability
Consumers and employers are held accountable in the roles they play in an incident. Criminal and civil law can weigh heavily against those who are responsible. In the 1970s, Ford Motor Company marketed a vehicle with a known design defect: the Ford Pinto. The Pinto had a flawed fuel tank design that would cause it to ignite if the car was hit from behind. To rectify the design for 11 million cars and 1.5 million light trucks with Pinto-like fuel tank design would have cost $11 each, for a total expenditure of $137 million. Ford chose to save the $11 it would have cost to rectify the problem. Because of its decision, the Ford Motor Company was held liable for the deaths caused by Pinto crashes, which led to the most expensive recall in automotive history.

Ethical Motives
Sophisticated societies for the enhancement of professional conduct have existed for centuries. Professional societies and licenses were created to provide uniform methods of scrutinizing the management and behavior of service providers. In the past, these societies lacked the organizational strength to effectively develop systems of safety. Today, organizations such as the American Consulting Engineers Council are guided by codes of ethics:

> Engineers shall hold paramount the safety, health, and welfare of the public in the performance of their professional duties.

Directives
Regulatory directives can specify how companies will proceed in the development of their safety systems. The airline industry, for example, is controlled by Title 14 of the *Code of Federal Regulations* (CFR). These regulations lay out in specific terms how the aviation industry will conduct its business and safety systems. The development of airworthy transport aircraft is specified in Federal Aviation Regulation (FAR) 25.1309 (b):

> The airplane systems and associated components, considered separately and in relation to other systems, must be designed so that:
> 1. The occurrence of any failure condition which would prevent the continued safe flight and landing of the airplane is extremely improbable, and

2. The occurrence of any other failure conditions which would reduce the capability of the airplane or the ability of the crew to cope with adverse operating conditions is improbable.

Organizations can have many directives to follow from many different regulatory schemes. For example, the airline industry must comply not only with 14 CFR, but also with other pieces of legislation such as the Consumer Product Safety Act, the Federal Hazardous Substances Act, and the Flammable Fabrics Act. Federal or organizational directives can influence the emphasis a company places on the development of safety systems.

System Safety Terms

The difficulty in designing safety systems is partly due to the fluid nature of the activities in which workers participate. In order to have a foundation from which to build, the tools and techniques to be used must be identified. Many industries have their own definitions for describing safety, but some are common throughout the safety world:

- *system*—a formation of personnel, procedures, materials, tools, equipment, facilities, and software (the elements of this composite entity are used together in the intended operation or support environment to perform a given task or achieve a specific production, support, or mission requirement) (MIL-STD-882C)
- *safety*—freedom from conditions that cause death, injury, occupational illness, or damage to or loss of equipment or property, or damage to environment (MIL-STD-882C)
- *management*—the process of allocating scarce or limited resources to achieve identified goals
- *system safety*—the application of engineering and management principles, criteria, and techniques to optimize safety within the constraints of operational effectiveness, time, and cost throughout all phases of the system life cycle (MIL-STD-882C)
- *risk*—an expression of the probability/impact of a mishap in terms of hazard severity and hazard probability (MIL-STD-882C)
- *mishap*—an unplanned event or series of events that result in death, injury, occupational illness, or damage of equipment or property or damage to environment (MIL-STD-882C)

How Is System Safety Designed?

The development of system safety should be derived from and supported by the mission statement of a company. Without the full support of management and personnel, a safety program will lack effectiveness. System safety programs must actively identify, verify, and rectify areas of risk, reducing it to a level that is acceptable to both the company and its stakeholders. The risk involved must be identified as early as possible, before it escalates to an unmanageable level. Once the risk is identified, it must be investigated to find its point of origin, solution, and potential for cross-contamination into other areas of operation. The final phase must implement any recommendations in accordance with the findings of the investigation. This final phase is also a starting point because the cycle must continue to maintain an effective and dynamic safety program that will provide optimum solutions for the risks at hand.

Mishap Models

Potential mishaps (incidents or risks) are analyzed using various mishap models. One traditional model is called *Heinrich's dominoes*. Incidents involve a sequence of the following five general factors:

1. environment of the risk
2. fault of a person
3. unsafe act or condition
4. accident
5. injury.

The analogy is that if one of the steps/dominoes is set in motion and they are allowed to continue to fall, then the result will be an accident. Eliminating any of the factors involved should stop the sequence.

Another mishap model assumes that mishaps are the result of multiple causes; it is called the *all/multiple cause model*. This model assumes that a chain or a series of causes must take place before a mishap can occur. Underlying assumptions in this model include the following:

- Single-cause mishaps are extremely rare.
- Identifying single causes provides limited preventive options.
- Mishaps normally have both technical and management causes.
- Technical causes identify deficiencies in the operational system.
- Management causes identify deficiencies in the management system that allowed the operational deficiencies to exist.

A system safety model, however, is in place before a mishap occurs and proactively prevents unintended incidents. One model, the *risk management cycle*, consists of six distinct elements that can be analyzed to provide a means by which system safety can be achieved (Figure 9–1).

1. Identify hazards.
2. Assess risks.
3. Develop and evaluate control measures.
4. Make control decisions and assume residual risk.
5. Implement control measures.
6. Evaluate effectiveness of control measures.

Figure 9–1. A system safety development tree, starting with the overall system and proceeding to specific management of risks.

Because system safety is a deliberate attempt to find patterns of operation that lead to safer, more precise, and more predictable results, it can function only with the cooperation of all who participate within the system. It is a dynamic relationship whereby the interactions of people and their environment must constantly be analyzed to maintain a safe and efficient environment. What is paramount and must be determined is the level of risk that will be accepted by the producer and consumer. Other approaches to identifying and preventing hazards are discussed in the next section.

HAZARD ANALYSIS

Hazard analysis is an analysis performed to identify and evaluate hazards in order to eliminate or control them. Data from hazard analysis can be regarded as a baseline for future monitoring activities. Before the workplace is inspected to ensure that environmental and physical factors fall within safe ranges, hazards inherent in the system must be discovered (i.e., hazard identification). Hazard analysis has proven to be an excellent tool to identify and evaluate hazards in the workplace.

Analyzing a problem or situation to obtain data for decision making is not new. Workers and their supervisors constantly make assessments—even if subconsciously—about their work to guide their actions. Written analyses carry the process one step further by providing the means to document hazard information, providing a historical basis for future decisions.

Philosophy

Written analyses often serve as the basis for more thorough inspections. They can be used to communicate data about hazards and risk potential to those in command positions. They can also be used to educate those in the line and staff organizations who need to know the consequences of hazards within their operations and the purpose and logic behind established control measures. Management can request a formal, written analysis for each critical operation. Such analyses not only gather information for immediate use, but also reap benefits over the long run. For instance, once important hazard data are committed to paper, they become part of the technical information base of the organization. These documents show the employer's concern for locating hazards and establishing corrective measures before an unintentional injury or illness happens.

Traditionally, companies analyzed systems during the operational phase to uncover problems and failures that impaired system effectiveness. Applying hazard analytical techniques returned substantial dividends by reducing both incident and overall operational losses.

Hazard control specialists no longer concentrate solely on operations. They look at the conceptual and design stages of the systems for which they are responsible. They use analytical methods and techniques before the process or product is built to identify and judge the nature and effects of hazards associated with their systems.

This wide assessment has significantly altered the direction of hazard control efforts. When potential problems or failures can be located before the production or onstream process stage of a system's life cycle, specialists can cut costs and avoid damage, injuries, and death. Systems engineering was initially concerned with increasing effectiveness, not profits. Properly applied, however, it can point out profitable solutions to many of management's most perplexing operational problems.

What Is Hazard Analysis?

Hazard analysis is an orderly process used to acquire specific hazard and failure data pertinent to a given system. A popular adage holds that "most things work out right for the wrong reasons." By providing data for informed management decisions, hazard analysis helps things work out right for the right reasons. The method forces those conducting the analysis to ask the right questions and helps answer them. By locating the hazards that are the most probable and/or have the severest consequences, hazard analyses provide information needed to establish effective control measures. Analytic techniques help the investigator decide what facts to gather, determine probable causes and contributing factors, and arrange orderly, clear results.

What are some uses for hazard analysis?
- It can uncover hazards that have been overlooked in the original design, mockup, or setup of a particular process, operation, or task.
- It can locate hazards that developed after a particular process, operation, or task was instituted.
- It can determine the essential factors in and requirements for specific job processes, operations, and tasks.
- It can indicate what qualifications are prerequisites to safe and productive work performance.
- It can indicate the need for modifying processes, practices, operations, and tasks.
- It can identify situational hazards in facilities, equipment, tools, materials, and operational events (e.g., unsafe conditions).
- It can identify ergonomic problems through anthropometrics and workstation design (e.g., worktable heights, chairs, reaching capabilities).
- It can identify work practices responsible for incident situations (e.g., deviations from standard procedures).
- It can identify exposure factors that contribute to injury

and illness (e.g., contact with hazardous substances, materials, or physical agents).
- It can identify physical factors that contribute to incident situations (e.g., noise, vibration, insufficient illumination).
- It can determine appropriate monitoring methods and maintenance standards needed for safety.
- It can determine the possible results of failures/incidents, people or property exposed to loss, and the potential severity of injury or loss.
- It can identify hazards in new equipment or processes before an employee is exposed to them.

Formal Methods of Hazard Analysis

Formal hazard analytical methods can be divided into two broad categories: inductive and deductive.

Inductive Method

The inductive analytical method uses observable data to predict events and outcomes within a particular system. It postulates how the component parts of a system will contribute to the success or failure of the system as a whole. Inductive analysis considers a system's operation from the standpoint of its components, their failure in a specific operating condition, and the effect of that failure on the system.

The inductive method forms the basis for such analyses as failure mode and effect analysis (FMEA) and operations hazard analysis (OHA). In FMEA, the failure or malfunction of each component is considered, including the mode of failure. Management can trace throughout the system the effects of the hazard(s) that led to the failure and evaluate the ultimate impact on task performance. However, because only one failure is considered at a time, some possibilities may be overlooked. Figure 9–2 illustrates the FMEA format used at Aerojet Nuclear Company, Idaho Falls, Idaho. Figure 9–3 illustrates the OHA format used for industrial operations.

Once the inductive analysis is completed and the critical failures requiring further investigation are detected, then the fault tree analysis will facilitate an inspection (see Deductive Method). Job safety analysis (JSA), discussed in a later section in this chapter, also uses the inductive method for determining the safety risks and components of various jobs.

Deductive Method

If inductive analysis reveals what can happen, deductive analysis shows how. It postulates failure of the entire system and then identifies how the components could contribute to the failure.

Deductive methods use a combined-events analysis, often in the form of decision trees. The positive tree shows the requirements for success (see Figure 9–4). Positive trees are less commonly used than fault trees because they can easily become a list of "should," and subsequent moralizing could make it difficult to reach an endpoint.

Fault trees are reverse images of positive trees and show ways troubles can occur. The analyst selects an undesired event, then diagrams in tree form all the possible factors that can contribute to the occurrence of the undesired event. The branches of the tree continue until they reach independent factors. The analyst can then determine probabilities for the independent factors occurring.

The fault tree requires a thorough analysis of a potential event and involves listing all known sources of failure. It is a graphic model of the various parallel and sequential combinations of system component faults that can result in a single, selected system fault. Figure 9–4 illustrates three types of analytical trees.

Analytical trees have three advantages:
1. They accomplish a thorough analysis without wordiness. Using known data, the analyst can identify the single and multiple causes capable of inducing the undesired event.
2. They make the analytical process visible, allowing for the rapid transfer of hazard data from person to person and from group to group, with few possibilities for miscommunication during the transfer.
3. They can be used as investigative tools. By reasoning backward from the incident (the undesired event), the investigator can reconstruct the system and pinpoint the elements responsible for the undesired event.

Cost-Effectiveness Method

The cost-effectiveness method can be used as part of either the inductive or deductive approach. The cost of system changes made to increase safety is compared with the decreased costs of fewer serious failures or with the increased efficiency of the system. Cost-effectiveness frequently is used to decide among several systems, each capable of performing the same task.

Choosing a Method

To decide what hazard analytical approach is best for a given situation, the hazard control specialist will want to answer five questions:
1. What is the quantity and quality of information desired?
2. What information already is available?
3. What is the cost of setting up and conducting analyses?
4. How much time is available before decisions must be made and action taken?
5. How many people are available to assist in the hazard analysis, and what are their qualifications?

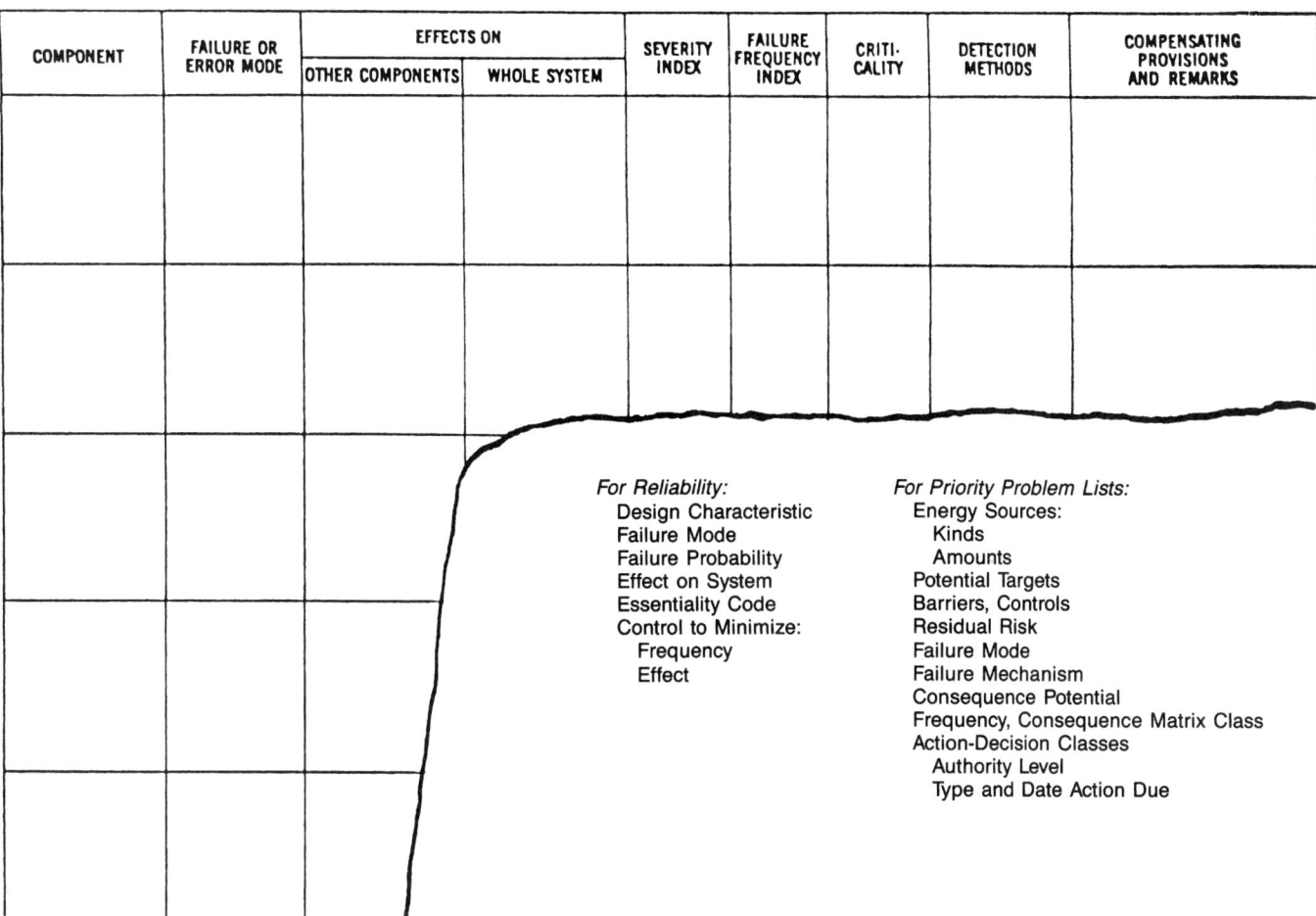

Figure 9–2. A failure mode effect analysis form used by the Aerojet Nuclear Company *(Johnson 1980)*.

Process	Operational Step	Task	Source of Potential Hazard	Triggering Event	Potential Effect on Equip., Material Environment	Personal Injury, Property Damage	RAC	Procedural Requirements	Safety & P.P.E.
Turning steel stock between centers on machine lathe.	Rough turning steel stock.	Select cutting tool and place in tool holder.	Improper tool used for rough cutting operation.	Starting lathe.	Tool jams in stock. Stock comes off centers. Uneven cut. Wasted stock.	Operator is hit in face with flying chips of steel.	2	A right-cut tool or roundhouse tool should be held in a straight tool holder. Lathe located to minimize exposure to other work stations.	Select proper tool for job. Operator to wear face protection while operating lathe.
		Place tool holder in the tool post and adjust cutting tool to proper location.	Tool holder extending too far from tool post.	Starting lathe.	Same as above. Breaking cutting tool.	Operator is hit in face with flying chips of steel from stock and broken cutting tool.	2	Tool post should be at end of T-slot. Face of tool must be on center and turned slightly away from headstock.	Operator to wear face protection while operating lathe.

Figure 9–3. This operations hazard analysis (OHA) form is used for industrial operations. *(Printed with permission from Indiana Labor & Management Council Inc.)*

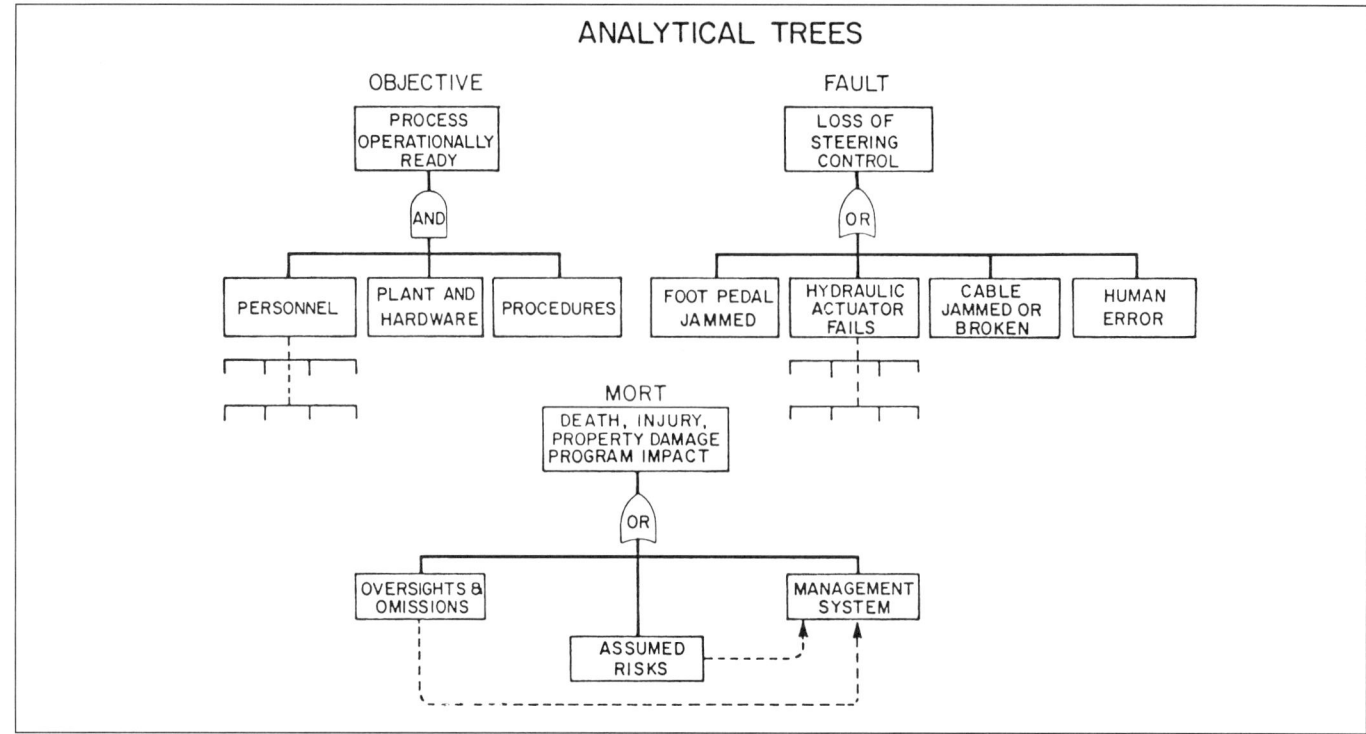

Figure 9–4. Analytical trees are nothing but "structured common sense." Trees are of two major types—objective or positive trees, which emphasize how a job should be done properly, and fault trees, which chart things that can go wrong and produce a specific failure. A fault tree structured for one job can often be generalized to cover a wide variety of jobs. The MORT diagram describes the ideal safety program in an orderly, logical manner. It is based on three branches: (1) a branch dealing with specific oversights and omissions at the worksite, (2) a branch that deals with the management system that establishes policies and makes the entire system go, and (3) an assumed-risk branch visually recognizing that no activity is completely risk-free, and that risk management functions must exist in any well-managed organization *(Printed with permission from* Professional Safety, *February 1977.)*

Conducting a hazard analysis can be expensive. Before a hazard analysis technique is chosen, it is important to determine what information is needed and how important it is.

It is beyond the scope of this manual to go into detail regarding other applications of system safety (see Johnson 1980).

Who Should Participate in Hazard Analysis?

A hazard analysis, to be fully effective and reliable, should represent as many different viewpoints as possible. Each person familiar with a process or operation has acquired insights concerning problems, faults, and situations that can cause unintentional injuries. These insights need to be recorded, along with those of the person initiating the hazard analysis—usually the safety professional. Input from workers and employee representatives can be extremely valuable at this stage.

What Factors Need to Be Analyzed?

All machines, equipment, processes, operations, and tasks in any establishment or facility are good candidates for hazard analysis because they have the potential to cause incidents. Eventually, hazard analyses should be completed for all jobs, but the most potentially threatening should have immediate attention. In determining which processes, operations, and tasks receive priority, those making the decisions should take the following factors into consideration:

- *Frequency of incidents*—Any operation or task with an associated history of repeated incidents is a good candidate for analysis, especially if different employees have the same kind of incident while performing the same operation or task.
- *Potential for injury*—Some processes and operations can have a low incident frequency but a high potential for major injury (e.g., tasks on a grinder conducted without a tool rest or tongue guard).
- *Severity of injury*—A particular process, operation, or task can have a history of serious injuries and be a worthy candidate for analysis, even if the frequency of such injuries is low.
- *New or altered equipment, processes, and operations*—As a general rule, whenever a new process, operation, or task is created or an old one altered (because of machinery, equipment, or other changes), the safety professional or supervisor should conduct a hazard analysis. For max-

JOB SAFETY ANALYSIS	JOB TITLE (and number if applicable): Banding Pallets PAGE 1 OF 2 JSA NO. 105		DATE: 00/00/00	☒ NEW ☐ REVISED
INSTRUCTIONS ON REVERSE SIDE	TITLE OF PERSON WHO DOES JOB: Bander	SUPERVISOR: James Smith	ANALYSIS BY: James Smith	
COMPANY/ORGANIZATION: XYZ Company	PLANT/LOCATION: Chicago	DEPARTMENT: Packaging	REVIEWED BY: Sharon Martin	
REQUIRED AND/OR RECOMMENDED PERSONAL PROTECTIVE EQUIPMENT:	Gloves - Eye Protection - Long Sleeves - Safety Shoes		APPROVED BY: Joe Bottom	

SEQUENCE OF BASIC JOB STEPS	POTENTIAL HAZARDS	RECOMMENDED ACTION OR PROCEDURE
1. Position portable banding cart and place strapping guard on top of boxes.	1. Cart positioned too close to pallet (strike body & legs against cart or pallet, drop strapping gun on foot.)	1. Leave ample space between cart and pallet to feed strapping - have firm grip on strapping gun.
2. Withdraw strapping and bend end back about 3".	2. Sharp edges of strapping (cut hands, fingers & arms). Sharp corners on pallet (strike feet against corners).	2. Wear gloves, eye protection & long sleeves - keep firm grip on strapping - hold end between thumb & forefinger - watch where stepping.
3. Walk around load while holding strapping with one hand.	3. Projecting sharp corners on pallet (strike feet on corners).	3. Assure a clear path between pallet and cart - pull smoothly - avoid jerking strapping.
4. Pull and feed strap under pallet.	4. Splinters on pallet (punctures to hands and fingers.) Sharp strap edges (cuts to hands, fingers, and arms).	4. Wear gloves - eye protection - long sleeves. Point strap in direction of bend - pull strap smoothly to avoid jerks.
5. Walk around load. Stoop down. Bend over, grab strap, pull up to machine, straighten out strap end.	5. Protruding corners of pallet, splinters (punctures to feet and ankles).	5. Assure a clear path - watch where walking - face direction in which walking.
6. Insert, position and tighten strap in gun.	6. Springy and sharp strapping (strike against with hands and fingers).	6. Keep firm grasp on strap and on gun - make sure clip is positioned properly.

Figure 9–5. This sample of a completed JSA shows how hazards and safe procedures are identified to help reduce the occurrence of injuries.

imum benefit, the hazard analysis should be done while the process or operation is in the planning stages. No equipment should be put into regular operation until the safety professional has checked it for hazards, studied its operation, installed any necessary additional safeguards, and developed safety instructions or procedures. Adhering to such a procedure ensures that managers can train employees in hazard-controlled, safe operations and help prevent serious injuries and exposures.

- *Excessive material waste or damage to equipment*—Processes or operations producing excessive material waste or damage to tools and equipment are candidates for hazard analysis. The same problems causing the waste or damage could also, given the right situation, cause injuries.

One of the first steps in hazard and incident analysis is performing job safety analyses. They can be specifically tailored to individual jobs or categories of jobs in the workplace. This next section describes the nature of the job safety analyses and how they can be conducted.

JOB SAFETY ANALYSIS

Job safety analysis (JSA) is a procedure used to review job methods and uncover hazards that (1) may have been overlooked in the layout of the facility or building and in the design of the machinery, equipment, tools, workstations, and processes; (2) may have developed after production started; or (3) resulted from changes in work procedures or personnel.

A JSA can be written as shown in Figure 9–5. In the left column, the basic steps of the job are listed in the order in which they occur. The middle column describes all hazards, both those produced by the environment and those connected to the job procedure. The right column gives the safe procedures that should be followed to guard against the hazards, prevent potential injuries, and perform the job correctly. (See Figure 9–6 for a list of instructions printed on the back of JSA forms published by the National Safety Council.)

For convenience, both the JSA procedure and the written description are commonly referred to as JSA. Health hazards are also considered when making a JSA.

INSTRUCTIONS FOR COMPLETING JOB SAFETY ANALYSIS FORM

Job Safety Analysis (JSA) is an important accident prevention tool that works by finding hazards and eliminating or minimizing them *before* the job is performed, and *before* they have a chance to become accidents. Use your JSA for job clarification and hazard awareness, as a guide in new employee training, for periodic contacts and for retraining of senior employees, as a refresher on jobs which run infrequently, as an accident investigation tool, and for informing employees of specific job hazards and protective measures.

Set priorities for doing JSAs: jobs that have a history of many accidents, jobs that have produced disabling injuries, jobs with high potential for disabling injury or death, and new jobs with no accident history.

Here's how to do each of the three parts of a Job Safety Analysis:

SEQUENCE OF BASIC JOB STEPS

Break the job down into steps. Each of the steps of a job should accomplish some major task. The task will consist of a *set* of movements. Look at the first *set* of movements used to perform a task, and then determine the next logical *set* of movements. For example, the job might be to move a box from a conveyor in the receiving area to a shelf in the storage area. How does that break down into job steps? Picking up the box from the conveyor and putting it on a handtruck is one logical set of movements, so it is one job step. Everything related to that one logical set of movements is part of that job step.

The next logical *set* of movements might be pushing the loaded handtruck to the storeroom. Removing the boxes from the truck and placing them on the shelf is another logical set of movements. And finally, returning the handtruck to the receiving area might be the final step in this type of job.

Be sure to list *all* the steps in a job. Some steps might not be done each time—checking the casters on a handtruck, for example. However, that task is a part of the job as a whole, and should be listed and analyzed.

POTENTIAL HAZARDS

Identify the hazards associated with each step. Examine each step to find and identify hazards—actions, conditions and possibilities that could lead to an accident.

It's not enough to look at the obvious hazards. It's also important to look at the entire environment and discover every conceivable hazard that might exist.

Be sure to list health hazards as well, even though the harmful effect may not be immediate. A good example is the harmful effect of inhaling a solvent or chemical dust over a long period of time.

It's important to list *all* hazards. Hazards contribute to accidents, injuries and occupational illnesses.

In order to do part three of a JSA effectively, you must identify potential and existing *hazards*. That's why it's important to distinguish between a hazard, an accident and an injury. Each of these terms has a specific meaning:
HAZARD—A potential danger. Oil on the floor is a *hazard*.
ACCIDENT—An unintended happening that may result in injury, loss or damage. Slipping on the oil is an *accident*.
INJURY—The *result* of an accident. A sprained wrist from the fall would be an injury.

Some people find it easier to identify possible accidents and illnesses and work back from them to the hazards. If you do that, you can list the accident and illness types in parentheses following the hazard. But be sure you focus on the *hazard* for developing recommended actions and safe work procedures.

RECOMMENDED ACTION OR PROCEDURE

Using the first two columns as a guide, decide what actions are necessary to eliminate or minimize the hazards that could lead to an accident, injury, or occupational illness.

Among the actions that can be taken are: 1) engineering the hazard out; 2) providing personal protective equipment; 3) job instruction training; 4) good housekeeping; and 5) good ergonomics (positioning the person in relation to the machine or other elements in the environment in such a way as to eliminate stresses and strains).

List recommended safe operating procedures on the form, and also list required or recommended personal protective equipment for each step of the job.

Be specific. Say *exactly* what needs to be done to correct the hazard, such as, "lift, using your leg muscles." Avoid general statements like, "be careful."

Give a recommended action or procedure for *every* hazard.

If the hazard is a serious one, it should be corrected immediately. The JSA should then be changed to reflect the new conditions.

Figure 9–6. Use these instructions for preparing a JSA.

Benefits of JSA

The principal benefits of a JSA include the following:
- reducing the frequency of injuries
- reducing the severity of injuries
- providing information to develop effective training programs
- instructing the new person on the job
- preparing for planned safety observations
- giving prejob instruction on irregular jobs
- reviewing job procedures after incidents occur
- studying jobs for possible improvement in job methods.

A JSA can be done in three basic steps. However, before initiating this analysis, management must first carefully select the job to be analyzed.

Selecting the Job

A job is a sequence of separate steps or activities that together accomplish a work goal. Some jobs can be broadly defined by what is accomplished—for example, making paper, building a facility, and mining iron ore. On the other hand, a job can be narrowly defined in terms of a single action, such as turning a switch, tightening a screw, and pushing a button. Such broadly or narrowly defined jobs are unsuitable for JSA.

Jobs suitable for JSA are those assignments that a line supervisor may make. Operating a machine, tapping a furnace, and piling lumber are good subjects for job safety analyses because they are neither too broad nor too narrow.

Jobs should not be selected at random—those with the worst injury experience should be analyzed first if JSA is to yield the quickest results. In fact, some companies make such selections the focal point of their incident prevention program.

Selection of jobs to be analyzed and establishment of the order of analysis should be guided by the following factors:
- *Frequency of incidents*—The greater the number of unintentional injuries associated with the job, the greater its priority claim for a JSA.
- *Rate of disabling injuries*—Every job that has had disabling injuries should be given a JSA, particularly if

the injuries prove that prior preventive action was not successful.
- *Severity potential*—Some jobs may not have a history of injuries but may have the potential for producing severe injury.
- *New jobs*—Changes in equipment or in processes obviously have no history of injuries, and thus their injury potential may not be fully appreciated. A JSA of every new job should be made as soon as the job has been created. Analysis should not be delayed until injuries or near misses occur.

After the job has been selected, the three basic steps in making a JSA are as follows:
1. Break the job down into successive steps or activities and observe how these actions are performed.
2. Identify the hazards and potential injuries. This is the critical step because only an identified problem can be eliminated.
3. Develop safe job procedures to eliminate the hazards and prevent the potential injuries.

Breaking the Job Down into Steps

Before the search for hazards begins, a job should be broken down into a sequence of steps, each describing what is being done. Avoid two common errors: (1) making the breakdown so detailed that an unnecessarily large number of steps results or (2) making the job breakdown so general that basic steps are not recorded.

To do a job breakdown, select the right worker to observe—an experienced, capable, and cooperative person who is willing to share ideas. If the employee has never helped out on a job safety analysis, explain the purpose (to make a job safe by identifying and eliminating or controlling hazards) and show him or her a completed JSA. Reassure the employee that he or she was selected because of experience and capability.

Observe the employee perform the job and write down the basic steps. Consider videotaping the job as it is performed for later study. To determine the basic job steps, ask, "What step starts the job?" Then ask, "What is the next basic step?" and so on.

Completely describe each step. Any possible deviation from the regular procedure should be recorded because this irregular activity may lead to an incident.

To record the breakdown, number the job steps consecutively as illustrated in the first column of the JSA training guide shown in Figure 9–7. Each step tells what is done, not how.

The wording for each step should begin with an action word such as *remove*, *open*, or *weld*. The action is com-

JOB SAFETY ANALYSIS WORKSHEET
JOB: Using a Pressurized Water Fire Extinguisher

WHAT TO DO (Steps in sequence)	HOW TO DO IT (Instructions) (Reverse hands for left-handed operator.)	KEY POINTS (Items to be emphasized. Safety is always a key point.)
1. Remove extinguisher from wall bracket.	1. Left hand on bottom lip, fingers curled around lip, palm up. Right hand on carrying handle palm down, fingers around carrying handle only.	1. Check air pressure to make certain extinguisher is charged. Stand close to extinguisher, pull straight out. *Have firm grip, to prevent dropping on feet.* Lower, and as you do remove left hand from lip.
2. Carry to fire.	2. Carry in right hand, upright position.	2. Extinguisher should hang down alongside leg. (This makes it easy to carry and reduces possibility of strain.)
3. Remove pin.	3. Set extinguisher down in upright position. Place left hand on top of extinguisher, pull out pin with right hand.	3. Hold extinguisher steady with left hand. Do not exert pressure on discharge lever as you remove pin.
4. Squeeze discharge lever.	4. Place right hand over carrying handle with fingers curled around operating lever handle while grasping discharge hose near nozzle with left hand.	4. Have firm grip on handle to steady extinguisher.
5. Apply water stream to fire.	5. Direct water stream at base of fire.	5. Work from side to side or around fire. After extinguishing flames, splay water on smouldering or glowing surfaces.
6. Return extinguisher. Report use.		

Figure 9–7. This JSA worksheet shows how to break down a job and analyze hazards and procedures.

pleted by naming the item to which the action applies—for example, *remove extinguisher* or *carry to fire*.

Check the breakdown with the person observed and agree on what is done and the order of the steps. Thank the employee for helping to enhance workplace safety.

Identifying Hazards and Potential Incident Causes

Before filling in the next two columns of the JSA, Potential Hazards and Recommended Action or Procedure, begin the search for hazards. The purpose is to identify all hazards—both those produced by the environment and those connected with the job procedure. Each step, and thus the entire job, must be made safer and more efficient. To do this, ask these questions about each step:

- Is there a danger of striking against, being struck by, or otherwise making harmful contact with an object?
- Can the employee be caught in, by, or between objects?
- Is there a potential for a slip or trip? Can the employee fall on the same level or to another?
- Can strain be caused by pushing, pulling, lifting, bending, or twisting?
- Is the environment hazardous to safety or health? For example, are there concentrations of toxic gas, vapor, mist, fume, dust, heat, or radiation? (See discussion in the National Safety Council's book *Fundamentals of Industrial Hygiene*, 6th edition.)

Close observation and knowledge of the particular job are required if the JSA is to be effective. The job observation should be repeated as often as necessary until all hazards and potential causes for incident or injury have been identified.

When inspecting a particular machine or operation, ask the question, "Can an injury occur here?" More specific questions include the following:

- Is it possible for a person to come into contact with any moving piece of machine equipment?
- Are rotating equipment, set screws, projecting keys, bolt heads, burrs, or other projections exposed where they can strike at or snag a worker's clothing?
- Is it possible to be drawn into the inrunning nip point between two moving parts, such as a belt and sheave, chain and sprocket, pressure rolls, rack and gear, or gear train?
- Do machines or equipment have reciprocating movement or any motion where workers can be caught on or between a moving part and a fixed object?
- Is it possible for a worker's hands or arms to make contact with moving parts at the point of operation where milling, shaping, punching, shearing, bending, grinding, or other work is being done?
- Is it possible for material (including chips or dust) to be kicked back or ejected from the point of operation, injuring someone nearby?
- Are machine controls safeguarded to prevent unintended or inadvertent operation?
- Are machine controls located to provide immediate access in the event of an emergency?
- Do machines vibrate, move, or walk while in operation?
- Is it possible for parts to become loose during operation, injuring operators or others?
- Are guards positioned or adjusted to correspond with the permissible openings?
- Is it possible for workers to bypass the guard, thereby making it ineffective?
- Do machines, equipment, and appurtenances receive regular maintenance?
- Are machines placed so operators have sufficient room to safely work with no exposure to aisle traffic?
- Is there sufficient room for maintenance and repair?
- Is there sufficient room to accommodate incoming and finished work as well as scrap that may be generated?
- Are the materials-handling methods adequate for the work-in-process and the tooling associated with it?
- If tools, jigs, and other work fixtures are required, are they stored conveniently, where they will not interfere with the work?
- Is the work area well illuminated with specific point-of-operation lighting where necessary?
- Is ventilation adequate, particularly for operations that create dusts, mists, vapors, and gases?
- Is the operator using personal protective equipment?
- Is housekeeping satisfactory, with no debris or tripping hazards or spills on the floor?
- Are there places where employees have access to machines (e.g., the back side)?
- Are energy sources heat controlled for protection?
- Are energy sources controlled for maintenance?

All of these questions will be of most value if they are incorporated into an inspection form that can be filled out at regular intervals. Even though a question may not at first seem to apply to a specific operation, on closer scrutiny it may be found to apply. Using a checklist is a good way to make sure nothing is overlooked.

To complete column 2 of the JSA, the analyzer should list all potential causes for incidents or injuries and all hazards yielded by a survey of the machine or operation. Record the type of potential cause of injuries and the agents involved. For example, to note that the employee may injure a foot by dropping a fire extinguisher, write down *struck by extinguisher*.

Again, check with the observed employee after the hazards and potential causes of injuries have been recorded. The experienced employee will probably offer additional suggestions. You should also check with others experienced with the job. Through observation and discussion, the analyzer can develop a reliable list of hazards and potential injuries.

Developing Solutions

The final step in a JSA is to develop a recommended safe job procedure to prevent the occurrence of incidents. The principal solutions are as follows:
1. Find a new way to do the job.
2. Change the physical conditions that create the hazards.
3. Change the work procedure.
4. Reduce the frequency (particularly helpful in maintenance and materials handling).

To find an entirely new way to do a job, determine the work goal of the job, and then analyze the various ways of reaching this goal to see which way is safest. Consider work-saving tools and equipment.

If a new way cannot be found, then ask this question about each hazard and potential injury cause listed: "What change in physical condition (such as change in tools, materials, equipment, layout, or location) will eliminate the hazard or prevent the potential injury?"

When a change is found, study it carefully to find other benefits (such as greater production or time saving) that will accrue. These benefits are good selling points and should be pointed out when proposing the change to higher management.

To investigate changes in the job procedure, ask the following questions about each hazard and potential injury cause listed: "What should the employee do—or not do—to eliminate this particular hazard or prevent this potential injury?" "How should it be done?" Because of his or her experience, in most cases the supervisor can answer these questions.

Answers must be specific and concrete if new procedures are to be any good. General precautions—"be alert," "use caution," or "be careful"—are useless. Answers should precisely state what to do and how to do it. The recommendation, "Make certain the wrench does not slip or cause loss of balance," is incomplete. It does not tell how to prevent the wrench from slipping. Here, in contrast, is an example of a recommended safe procedure that tells both what and how: "Set wrench properly and securely. Test its grip by exerting a light pressure on it. Brace yourself against something immovable or take a solid stance with feet wide apart before exerting full pressure. This prevents loss of balance if the wrench slips."

Some repair or service jobs have to be repeated often because a condition needs repeated correction. To reduce the need for such repetition, ask, "What can be done to eliminate the cause of the condition that makes excessive repairs or service necessary?" If the cause cannot be elimi-

nated, then ask, "Can anything be done to minimize the effects of the condition?"

Machine parts, for example, may wear out quickly and require frequent replacement. Study of the problem may reveal that excessive vibration is the culprit. After reducing or eliminating the vibration, the machine parts last longer and require less maintenance.

However, reducing the frequency of a job contributes to safety only in that it limits the exposure. Every effort should still be made to eliminate hazards and to prevent injuries through changing physical conditions, revising job procedures, or both.

A job that has been redesigned may affect other jobs and even the entire work process. Therefore, the redesign should be discussed not only with the worker involved, but also with co-workers, the supervisor, the facility engineer, and others who are concerned. In all cases, however, check or test the proposed changes by observing the job after discussing the changes with those who do the job. Their ideas about the hazards and proposed solutions can be of considerable value. They can judge the practicality of proposed changes and perhaps suggest improvements. These discussions are more than just a way to check a JSA. They are safety contacts that promote awareness of job hazards and safe procedures.

Using JSA Effectively

The major benefits of a JSA come after its completion. However, benefits are also to be gained from the development work. While conducting a JSA, supervisors learn more about the job they supervise. When employees are encouraged to participate in job safety analyses, their safety attitudes improve and their knowledge of safety increases. As a JSA is worked out, safer and better job procedures and safer working conditions are developed. But these important benefits are only a portion of the total benefits listed at the beginning of this discussion.

When a JSA is distributed, the supervisor's first responsibility is to explain its contents to employees and, if necessary, to give them further individual training. The entire JSA must be reviewed with the employees concerned so that they will know how the job is to be done—without injuries.

The JSA can furnish material for planned safety contacts. All steps of the JSA should be used for this purpose. The steps that present major hazards should be emphasized and reviewed again and again in subsequent safety contacts.

New employees on the job must be trained in the basic job steps, taught to recognize the hazards associated with each job step, and instructed in the necessary precautions. There is no better guide for this training than a well-prepared JSA. (See Chapter 31, Safety and Health Training, for further discussion of job instruction training.)

Occasionally, the supervisor should observe employees as they perform the jobs for which analyses have been developed. The purpose of these observations is to determine whether employees are doing the jobs in accordance with the safe job procedures. Before making such observations, the supervisor should prepare by reviewing the appropriate JSA to keep in mind the key points to observe.

Many jobs, such as certain repair or service jobs, are done infrequently or on an irregular basis. The employees who do them will benefit from prejob instruction to remind them of the hazards and the necessary precautions. Using the JSA for the particular job, the supervisor should give this instruction at the time the job is assigned.

Whenever an incident occurs on a job covered by a JSA, the JSA should be reviewed to determine whether it needs revision. If the JSA is revised, all employees concerned with the job should be informed of the changes and instructed in any new procedures.

When an incident results from failure to follow JSA procedures, the facts should be discussed with all those who do the job. It should be made clear that the incident would not have occurred had the JSA procedures been followed.

All supervisors are concerned with improving job methods to increase safety and health, reduce costs, and step up production. The JSA is an excellent starting point for questioning the established way of doing a job. In addition, study of the JSA may well suggest ideas for improvement of job methods.

Once the hazards are known, the proper solutions can be developed. Some solutions may be physical changes that eliminate or control the hazard, such as placing a safeguard over exposed moving machine parts. Others may be job procedures that eliminate or minimize the hazard, such as safe piling of materials. If these solutions do not completely or sufficiently control the hazard, personal protective equipment may be necessary to safely perform the job (see Chapter 7, Personal Protective Equipment, in the *Engineering & Technology* volume). A combination of these solutions may also provide a safe work environment.

The first stage of the hazard and incident analysis procedure is inspection. Management must know what problems and potential injury causes may be present in the workplace and what unsafe procedures or practices workers may be performing.

INSPECTION

Inspection should be conducted in an organization to locate and report existing and potential unsafe conditions or activities that, if left uncontrolled, have the capacity to cause injuries in the workplace.

Philosophy behind Inspection

Depending on the conditions surrounding the process, an inspection can be viewed negatively as fault finding, with the emphasis on criticism; or positively as fact finding, with the emphasis on locating hazards and developing plans for eliminating hazards that can adversely affect safety and health.

The second viewpoint is more effective. This viewpoint depends on two factors: (1) performance indicators adequate for measuring a particular situation and (2) comparison of what is with what ought to be (Firenze 1978). Failure to analyze inspection reports for causes of system defects ultimately means the failure of the monitoring function. Corrective action may fix the specific item but fail to fix the system.

Purpose of Inspection

The primary purpose of inspection is to detect potential hazards so they can be corrected before an unintentional injury or illness occurs. Inspection can determine conditions that need to be corrected or improved to bring operations up to acceptable standards, from both safety and operational standpoints. Secondary purposes are to improve operations and thus to increase efficiency, effectiveness, and profitability.

Although management ultimately has the responsibility for inspecting the workplace, authority for carrying out the actual inspecting process extends throughout the organization. Obviously supervisors, foremen, and employees fulfill an inspection function, but so do departments as diverse as engineering, purchasing, quality control, human resources, maintenance, and medical.

Types of Inspection

Inspection can be classified as one of two types, continuous or interval inspection.

Continuous Inspection

This process is conducted by employees, supervisors, and maintenance personnel as part of their job responsibilities. Continuous inspection involves noting an apparently or potentially hazardous condition or unsafe procedure and either correcting it immediately or making a report to initiate corrective action. Continuous inspection of personal protective equipment is especially important.

This type of inspection is sometimes called informal because it does not conform to a set schedule, plan, or checklist. However, critics argue that continuous inspection is erratic and superficial, that it does not get into out-of-the-way places, and that it misses too much. The truth is that both continuous and interval inspections are necessary and complement one another.

As part of their job, supervisors make sure that tools, machines, and equipment are properly maintained and safe to use and that safety precautions are being observed.

Toolroom employees regularly inspect all hand tools to be sure that they are in safe condition. Foremen are often responsible for continuously monitoring the workplace and seeing that equipment is safe and that employees are observing safe practices. When foremen or supervisors inspect machines at the beginning of a shift, a safety inspection must be part of the operation.

Continuous inspection is one ultimate goal of a good safety and health program. It means that each individual is vigilant, alert to any condition having incident potential, and willing to initiate corrective action.

However, in some instances, the supervisor's greatest advantage in continuous inspection—familiarity with the employees, equipment, machines, and environment—can also be a disadvantage. Just as an old newspaper left on a table in time becomes part of the decor, a hazard can become so familiar that no one notices it. The supervisor's blind spot is particularly likely to occur with housekeeping and unsafe practices. Poor housekeeping conditions may be overlooked because the deterioration is gradual and the effect is cumulative. A similar phenomenon may occur with unsafe practices, such as employees not following established production procedures.

In addition, no matter how conscientious the supervisors are, they cannot be completely objective. Inspections of their areas reflect personal and vested interests, knowledge and understanding of the production problems involved in the area, and concern for the employees. A planned periodic inspection of their areas by another supervisor can be used to audit their efforts. Furthermore, the supervisors who inspect another area may return to their own sections with renewed vision. Having looked at the trees day in and day out, they need occasionally to take the long view and see the forest.

Though this section is devoted primarily to discussing planned inspections, continuous inspections should be regarded as a cooperative, not a competitive, activity.

Interval Inspections

Planned inspections at specific intervals are what most people regard as "real" safety and health inspections. They are deliberate, thorough, and systematic procedures that permit examination of specific items or conditions. They follow an established procedure and use checklists for routine items. These inspections can be any one of three types: periodic, intermittent, and general.

Periodic inspection includes inspections scheduled at regular intervals. They can target the entire facility, a specific area, a specific operation, or a specific type of equipment. Management can plan these inspections weekly, monthly, semiannually, annually, or at other suitable intervals. Items such as safety guard mountings, scaffolds, elevator wire ropes (cables), two-hand controls, fire extinguishers, and other items relied on for safety require frequent inspection. The more serious the potential for injury or damage, the more often the item should be inspected.

Periodic inspections can be of several different types:
- inspections by the safety professional, industrial hygienist, and joint safety and health committees
- inspections for preventing incidents and damage or breakdowns (checking mechanical functioning, lubricating, etc.) performed by electricians, mechanics, and maintenance personnel, who are sometimes asked to serve as roving inspectors
- inspections by specially trained certified or licensed inspectors, often from outside the organization (e.g., inspection of boilers, elevators, unfired pressure vessels, cranes, power presses, fire-extinguishing equipment); often are mandated inspections required by regulatory agencies, manufacturers, underwriters, or management
- inspections done by outside investigators or government inspectors to determine compliance with government regulations.

The advantage of periodic inspection is that it covers a specific area and allows detection of unsafe conditions in time to provide effective countermeasures. Measurement data collected at regular intervals indicate degenerative trends. The staff or safety committee periodically inspecting a certain area is familiar with operations and procedures and therefore quick to recognize deviations. A disadvantage of periodic inspection is that deviations from accepted practices are rarely discovered because employees are prepared for the inspectors.

Intermittent inspections are those made at irregular intervals. Sometimes, the need for an inspection is indicated by incident tabulations and analysis. If a particular department or location shows an unusual number of incidents or if certain types of injuries occur with greater frequency, the supervisor or manager should call for an inspection. When construction or remodeling is going on within or around a facility, an unscheduled inspection may be needed to find and correct unsafe conditions before an intentional injury or illness occurs. The same is true when a department installs new equipment, institutes new processes, or modifies old ones.

Another form of intermittent inspection is that made by the industrial hygienist when a health hazard is suspected or present in the environment. This monitoring of the workplace is covered in detail later in this chapter under Measurement and Testing. It usually involves the following tasks:
- sampling the air for the presence of toxic vapors, gases, radiation, and particulates
- sampling physical stresses such as noise, heat, and radiation
- testing materials for toxic properties
- testing ventilation and exhaust systems for proper operation

Intermittent inspections may be initiated because of the following reasons:
- increase in injury or illness rates in an area
- reports from employees in an area
- management directive
- reports of hazards from other departments, companies, manufacturers, or regulatory agencies
- random selection
- incident/severity potential
- reaction to an event (e.g., injury, threat of sabotage, severe weather warning).

A *general inspection* is usually planned and often covers places not inspected periodically. This includes areas no one ever visits and where people rarely get hurt, such as parking lots, sidewalks, fencing, and similar outlying regions.

Many out-of-the-way hazards are located overhead, where they are difficult to spot. Overhead inspections frequently disclose the need for repairs to skylights, windows, cranes, roofs, and other installations affecting the safety of both employees and the physical facility. Overhead devices can require adjustment, cleaning, oiling, and repairing.

Inspections of overhead areas are necessary to make certain all reasonable safeguards are provided and safe practices observed. Inspectors must verify that workers performing overhead jobs have suitable staging and fall-protection equipment. They must also apply this safety directive to themselves during the inspection. They should look for loose tools, bolts, pipelines, shafting, pieces of lumber, windows, electrical fixtures, and other objects that can fall from building structures, cranes, roofs, and similar overhead locations.

Safety conditions change after dark, when illumination consists of artificial light. Therefore, when an organization has more than one shift, it is important to perform inspections at night to make sure illumination is adequate and the lighting system is well maintained. The safety professional should make this inspection, aided by a photometer and camera where necessary. Even in organizations with no regular night shifts, some employees (maintenance personnel, fire fighters, and night security guards) are required to work after dark. The safety professional occasionally needs to check on their night work conditions.

General inspections are usually required before reopening a facility after a long shutdown.

PLANNING FOR INSPECTION

An effective safety and health inspection program requires the following elements:
- sound knowledge of the facility
- knowledge of relevant standards, regulations, and codes
- systematic inspection steps
- a method of reporting, evaluating, and using the data.

An effective program begins with analysis and planning. If inspections are casual, shallow, and slipshod, the results will reflect the method. Before instituting an inspection program, these five questions should be answered:
1. What items need to be inspected?
2. What aspects of each item need to be examined?
3. What conditions need to be inspected?
4. How often must items be inspected?
5. Who will conduct the inspection?

Hazard Control Inspection Inventory

To determine what factors affect the inspection, a hazard control inspection inventory can be conducted. Such an inventory is the foundation on which a program of planned inspection is based. It resembles a planned preventive maintenance system and yields many of the same benefits.

Management should divide the entire facility—yards, buildings, equipment, machinery, vehicles—into areas of responsibility. These areas, once determined, should be listed in an orderly fashion. The analyst may develop a color-coded map or floor plan of the facility. Large areas or departments can be divided into smaller areas and assigned to each first-line supervisor and/or the hazard control department's inspector.

What Items Need to Be Inspected?

Once specific areas of responsibility have been determined, managers should inventory areas of responsibility that could affect safety and health. These would include the following:
- environmental factors (illumination, dust, fumes, gases, mists, vapors, noise, vibration, heat, radiation sources)
- hazardous supplies and materials (explosives, flammables, acids, caustics, toxic or nuclear materials or by-products)
- production and related equipment (mills, shapers, presses, borers, lathes)

- power source equipment (steam and gas engines, electrical motors)
- electrical equipment (switches, fuses, breakers, outlets, cables, extension and fixture cords, grounds, connectors, connections)
- hand tools (wrenches, screwdrivers, hammers, power tools)
- personal protective equipment (hard hats, safety glasses, safety shoes, respirators, hearing protection, gloves)
- personal service and first-aid facilities (drinking fountains, wash basins, soap dispensers, safety showers, eye-wash fountains, first-aid supplies, stretchers)
- fire protection and emergency-response equipment (alarms, water tanks, sprinklers, standpipes, extinguishers, hydrants, hoses, self-contained breathing apparatuses, toxic cleanup, automatic valves, horns, phones, radios)
- walkways and roadways (ramps, docks, sidewalks, aisles, vehicle ways, escape routes)
- elevators, electric stairways, and manlifts (controls, wire ropes, safety devices)
- working surfaces (ladders, scaffolds, catwalks, platforms)
- materials-handling equipment (cranes, dollies, conveyors, hoists, forklifts, chains, ropes, slings)
- transportation equipment (automobiles, railroad cars, trucks, front-end loaders, helicopters, motorized carts and buggies)
- warning and signaling devices (sirens, crossing and blinker lights, klaxons, warning signs, exit signs)
- containers (scrap bins, disposal receptacles, carboys, barrels, drums, gas cylinders, solvent cans)
- storage facilities and areas, both indoor and outdoor (bins, racks, lockers, cabinets, shelves, tanks, closets)
- structural openings (windows, doors, stairways, sumps, shafts, pits, floor openings)
- buildings and structures (floors, roofs, walls, fencing)
- miscellaneous—any items that do not fit in the preceding categories.

There are many sources of information about items to be inspected, especially employees in an organization. For instance, maintenance employees know what problems can cause damage or shutdowns. The workers in the area are qualified to point out causes of injury, illness, damage, delays, or bottlenecks. Medical personnel in the organization can list problems causing job-related illnesses and injuries. Manufacturers' manuals often specify maintenance schedules and procedures and safe work methods.

Gathering information about standards, regulations, and codes is a necessary first step in determining what items need to be inspected. A great deal of work already has been done for the safety inspector.

Building codes, building inspection books, and guides to building and facility maintenance also are useful references. Publications of the NFPA will help safety personnel ensure that fire hazards are being effectively managed. In addition, insurance company surveys often contain checklists to help management evaluate the safety condition of their buildings. Research and reference material is contained in subsequent chapters of this manual. Other publications of the National Safety Council, such as *Injury Facts* (formerly *Accident Facts*), may prove useful as well.

State or provincial and federal governments also publish injury statistics. The *Federal Register* and Title 29 CFR, Parts 1900 through 1950, document OSHA regulations. Figure 9–8 shows some special subjects addressed by the subparts of 29 CFR 1910 (General Industry) and the subparts of 29 CFR 1926 (Construction).

It is important to remember, however, that federal and state or provincial laws, codes, and regulations usually set up minimum requirements only. To comply with company policy and ensure a safe working environment, management must often exceed these regulatory requirements. Some OSHA publications indicate not only what standards are required but also what violations are most frequent. These same sources are helpful in the next step—determining critical factors to be inspected.

What Aspects of Each Item Need to Be Examined?

Particular attention should be paid to the parts of an item most likely to become a serious hazard to health and safety. These parts often develop problems because of stress, wear, impact, vibration, heat, corrosion, chemical reaction, and misuse. Such items as safety devices, guards, controls, work or wearpoint components, electrical and mechanical components, and fire hazards tend to become unsafe first. For a particular machine, critical parts would include the point of operation, moving parts, and accessories (flywheels, gears, shafts, pulleys, key ways, belts, couplings, sprockets, chains, controls, lighting, brakes, exhaust systems). Also to be checked are items related to feeding, oiling, adjusting, maintenance, grounding, attaching, work space, and location.

The most critical parts of an item are not always the most obvious. When the security of a heavy load depends on a cotter pin being in place, then that pin is a critical part.

What Conditions Need to Be Inspected?

The unsafe conditions for each part to be inspected should be described specifically and clearly. A checklist question that reads "Is _____ safe?" is meaningless because it does not define what makes an item unsafe. Inspectors should describe the elements that contribute to an unsafe condition and not simply list unsafe conditions for each item. Usually, conditions described by physical descriptions can be indicated by such

29 CFR 1910
OCCUPATIONAL SAFETY AND HEALTH STANDARDS

Subparts

- C — Access to Employee Exposure and Medical Records
- D — Walking-Working Surfaces
- E — Means of Egress
- F — Powered Platforms, Manlifts, and Vehicle-Mounted Work Platforms
- G — Occupational Health and Environmental Controls
- H — Hazardous Materials
- I — Personal Protective Equipment
- J — General Environmental Controls
- K — Medical and First Aid
- L — Fire Protection
- M — Compressed Gas and Compressed Air Equipment
- N — Materials Handling and Storage
- O — Machinery and Machine Guarding
- P — Hand and Portable Powered Tools and Other Hand-Held Equipment
- Q — Welding, Cutting, and Brazing
- R — Special Industries
- S — Electrical
- T — Commercial Diving Operations
- Z — Toxic and Hazardous Substances

29 CFR 1926
STANDARDS FOR THE CONSTRUCTION INDUSTRY

Subparts

- C — General Safety and Health Provisions
- D — Occupational Health and Environmental Controls
- E — Personal Protective and Lifesaving Equipment
- F — Fire Protection and Prevention
- G — Signs, Signals, and Barricades
- H — Materials Handling, Storage, Use and Disposal
- I — Tools—Hand and Power
- J — Welding and Cutting
- K — Electrical
- L — Scaffolding
- M — Fall Protection
- N — Cranes, Derricks, Hoists, Elevators, and Conveyors
- O — Motor Vehicles, Mechanized Equipment, and Marine Operations
- P — Excavations, Trenching, and Shoring
- Q — Concrete, Concrete Forms, and Shoring
- R — Steel Erection
- S — Tunnels and Shafts, Caissons, Cofferdams, and Compressed Air
- T — Demolition
- U — Blasting and Use of Explosives
- V — Power Transmission and Distribution
- W — Rollover Protection Structures; Overhead Projection
- X — Ladders and Stairways

Figure 9–8. This is a quick reference list of what subpart discusses certain major topics in 29 CFR 1910 and 29 CFR 1926.

words as *jagged*, *exposed*, *broken*, *frayed*, *leaking*, *rusted*, *corroded*, *missing*, *vibrating*, *loose*, or *slipping*. Sometimes exact figures are needed—for example, the maximum pressure in a boiler or the percentage spread of a sling hook.

Checklists serve as reminders of what to look for and as records of what has been covered. They can be used to structure and guide inspections. They also allow on-the-spot recording of all findings and comments before they are forgotten. In case an inspection is interrupted, checklists provide a record of what has and what has not been inspected. Otherwise, inspectors may miss items or conditions they should examine or may be unsure, after inspecting an area, that they have covered everything. Good checklists also help in follow-up work to make sure hazards have been corrected or eliminated.

Many types of monitoring checklists are available, varying in length from thousands of items to only a few. These checklists are useful in determining which standards or regulations apply to individual situations. Once the applicable standards are identified, the organization can tailor a checklist to its needs and uses and enter it into the computer system for action and follow-up.

The Centers for Disease Control (CDC) of the U.S. Department of Health and Human Services (DHHS) have devised a suggested checklist for the safety evaluation of shop and laboratory areas. The worksheet is referenced to the OSHA "General Industry Standards."

Merely running through a checklist, however, does little to locate or correct problems. The checklist must be used as an aid to the inspection process, not as an end in itself. Of course, any hazard observed during inspection must be recorded, even though it is not part of the checklist. Sample inspection checklists from various companies are available at the National Safety Council website (nsc.org). These are used with computer follow-up on inspection results, actions to be taken, and corrections made.

The amount of detail included in the checklist will vary, depending on the inspector's knowledge of the relevant standards and the nature of the inspection. An experienced inspector with thorough knowledge of the standards will need only a few clues as a reminder of items to be inspected. Checklists for infrequent inspection generally will be more detailed than daily or weekly ones.

The format of a checklist should include columns to indicate either compliance or action date. Space also should be provided to cite the specific violation, a way to correct it, and a recommendation that the condition receive more or less frequent attention. Whatever the format of the checklist, space should be provided for the inspector's signature and the inspection date.

Checklists can be prepared by the safety and health committee, by the safety director, or by a subcommittee that includes engineers, supervisors, employees, and maintenance personnel. The safety professional and the department supervisor should monitor checklist development and make sure all applicable standards are covered. In their final form, the checklists should conform to the inspection route.

Choosing the inspection route means inspecting an area completely and thoroughly while avoiding the following:
- time-consuming backtracking and repetitions
- long walks between items
- unnecessary interruptions of the production process
- distraction of employees.

Often, a closed-loop inspection will give good results. Sometimes, it is valuable to follow the production path of the material being processed.

It is important to remember that a checklist is a tool and should be used only to support the safety process. Once unsafe conditions have been identified, it is important to provide workable solutions for addressing the situation in a safe manner.

How Often Must Items Be Inspected?
The frequency of inspection is determined by four factors:
1. What is the loss severity potential of the problem? The inspector should ask, "If the item or critical part fails, what injury, damage, or work interruption will result?" The more severe the loss potential, the more often the item should be inspected. For instance, a frayed wire rope on an overhead crane block has the potential to cause a much greater loss than does a defective wheel on a wheelbarrow. Therefore, the rope needs to be inspected more frequently than the wheel.
2. What is the potential for injury to employees? If the item or critical part fails, how many employees will be endangered and how frequently? The greater the probability for injury to employees, the more often the item should be inspected. For example, a stairway continually used by many people needs to be inspected more frequently than one seldom used.
3. How quickly can the item or part become unsafe? The answer to this question depends on the nature of the part and the conditions to which it is subjected. Equipment and tools used frequently can become damaged, defective, or worn more quickly than those rarely used. An item located in a particular spot can be exposed to greater damage than an identical item in a different location. The faster an item can become unsafe, the more frequently it should be inspected.
4. What is the past history of failures? What were the results of these failures? Maintenance and production records and incident investigation reports can provide valuable information about how frequently items have failed and the results in terms of injuries, damage, delays, and shutdowns. The more often an item has failed in the past and the greater the consequences, the more it needs to be inspected.

The Occupational Safety and Health Administration, the Mine Safety and Health Administration (MSHA), the Federal Aviation Administration (FAA), the Nuclear Regulatory Commission (NRC), the Environmental Protection Agency (EPA), and other federal, state, provincial, and local regulatory agencies require periodic inspections. Consult the regulations and the agency responsible for enforcement for current information regarding inspection criteria and intervals. Frequency of inspections should be described in specific terms—for example, before every use, when serviced, daily, monthly, quarterly, yearly.

Who Will Conduct the Inspection?
Answering the four previous questions—the items to be inspected, the aspects of each item to be inspected, the conditions to be inspected, and the frequency of inspections—will help determine who is qualified to do the inspection. No individual or group should have exclusive responsibility for all inspections. Employees who perform these inspections will benefit from training in hazard recognition. Some items will need to be inspected by more than one person. For example, although an area supervisor may inspect an overhead crane weekly and maintenance personnel inspect it monthly, the operator of the crane will inspect it before each use. When grinding wheels are received, they are inspected by the stockroom attendant, but they must be inspected again by the operator before each use.

As part of the hazard control inspection inventory, management should assign responsibility for each inspection. Figure 9–9 shows how the inventory can designate the proper person by title: area supervisor, operator, foreman, maintenance foreman, and so forth. A suggested guide for planned inspections is as follows:
- *daily*—area supervisor and maintenance personnel, who also can request suggestions from employees at their various workstations
- *weekly*—department heads
- *monthly*—supervisors, department heads, the safety department, and safety and health committees.

The safety department also may be actively involved in monthly, quarterly, semiannual, and annual inspections.

Five qualifications of a good inspector are as follows:

1. knowledge of the organization's incident experience
2. familiarity with incident potentials and with the standards that apply to his or her area
3. ability to make intelligent decisions for corrective action
4. diplomacy in handling personnel and related situations
5. knowledge of the organization's operations—its workflow, systems, and products.

Safety Professionals

Clearly, the safety professional should spearhead the inspection activity. During both individual and group inspections, the professional can educate others in inspection techniques and hazard identification by using on-the-spot examples and firsthand contact. Supervisors, foremen, stewards, and safety and health committees can be shown what to look for when making inspections. The organization's fire protection representative or industrial hygienist usually works with the hazard control specialist in conducting inspections.

The number of safety professionals depends on the size of the company and the nature of its operation. Large companies with well-organized incident prevention programs usually employ a full-time staff. Sometimes, large companies also have designated employees who spend part of their time on inspections.

In organizations where toxic and corrosive substances are present, the industrial hygienist is part of the inspection team (see Measurement and Testing, later in this chapter). When an organization uses chemicals, the chief chemist must cooperate closely with the safety professional and fire protection representative in establishing inspection criteria. If the organization has no industrial hygienist, the safety professional needs to obtain training about the hazardous properties of substances, unstable properties of chemicals, and methods of control. An inspection conducted without this knowledge is incomplete and may miss potentially serious problems.

Company or Facility Management

Safety inspections should be considered part of the duties of company or facility management. By participating in inspections, management demonstrates its commitment to maintain a safe working environment. But the psychological effect of inspection by senior executives goes beyond merely showing an interest in safety. When employees know that management is coming to inspect their area, conditions that seemed "good enough" suddenly appear unsatisfactory and are quickly corrected.

First-Line Supervisor or Foreman

Because supervisors and foremen spend practically all of their time in the shop or facility, they are continually monitoring the workplace. At least once a day, supervisors need to check their areas to see that (1) employees are complying with safety regulations, (2) guards and warning signs are in place, (3) tools and machinery are in safe condition, (4) aisles and passageways are clear and proper clearances maintained, and (5) material in process is properly stacked or stored. Although such a spot check does not take the place of more detailed inspections, it emphasizes the supervisor's commitment to maintaining safety in the area. A supervisor also should conduct regular formal inspections to make certain all hazards have been detected and safeguards are in use. Supervisors can perform such inspections weekly on their own and monthly as part of a safety and health committee.

Mechanical Engineer and Maintenance Superintendent

Either as individuals or as members of a committee, the mechanical engineer and the maintenance superintendent also need to conduct regular formal inspections. They can write necessary work orders on the spot for guards or for correcting faulty equipment.

Employees

As mentioned previously, employee participation in continuous inspection is one goal of an effective hazard control program. Before beginning the workday, the employee should inspect the workplace and any tools, equipment, and machinery that will be used. Any defects the employee is not authorized to correct should be reported immediately to the supervisor. Action resulting from this report must be reported to the employee to encourage further participation.

Maintenance Personnel

Maintenance employees can be of great help in locating and correcting hazards. As they work, they can conduct informal inspections and report hazards to the supervisor, who in turn should encourage the mechanics to offer suggestions.

Joint Safety and Health Committees

Joint safety and health committees (discussed in Chapter 6, Loss Control Programs) conduct inspections as part of their function. They give equal consideration to factors that could cause incidents, fires, and adverse health exposures. By periodically visiting areas, members may notice changed conditions more readily than someone who is there every day. Another advantage provided by the committee is the members' various backgrounds, experience, and knowledge represented.

If the committee is large, the territory should be divided among teams of manageable size. Large groups going

DEPARTMENT Maintenance	UNIT Workshop	SUPERVISOR RESPONSIBLE J. P. Smith	APPROVED BY Ralph T. Welles	DATE 4/16/72	PAGE NO. 1
1. PROBLEMS	2. CRITICAL FACTORS	3. CONDITIONS TO OBSERVE	4. FREQUENCY	5. RESPONSIBILITY	
1. Overhead hoist	Cables, chains, hooks, pulleys	Frayed or deformed cables, worn or broken hooks and chains, damaged pulleys	Daily—before each shift	Operators	
2. Hydraulic pump	High pressure hose	Leaks; broken or loose fittings	Daily	Shift leader	
3. Power generator	High voltage lines	Frayed or broken insulation	Weekly	Foreman	
4. Fire extinguishers	Contents, location, charge	Correct type, fully charged, properly located, corrosion, leaks	Monthly	Area safety inspector	
5. General housekeeping	Passageways, aisles, floors, grounds	Free of obstructions, clearly marked, free of refuse	Daily	Shift leader foreman	

Figure 9–9. A hazard control inspection inventory should list the person responsible for each inspection. *(Reprinted from* Principles and Practices of Occupational Safety and Health, Student Manual, *Booklet Three, U.S. Department of Labor, OSHA 2215.)*

through the facility are unwieldy and distracting.

Other Inspection Teams

If there is no safety and health committee, a planned, formal inspection is still necessary. Management should assign an inspection team that includes the hazard control specialist, production manager, supervisor, employee representative, fire prevention specialist, and industrial hygienist. The important point is that inspections should be directed by a responsible executive who has the authority to ensure that the work is carried out effectively.

Outside inspectors sometimes are needed to perform inspections. For example, insurance company safety engineers and local, state or county, and federal inspectors may lend their expertise to specific inspections.

Contractors' Inspection Services

For some technical systems, notably sprinkler systems, contracting companies furnish inspection services. Companies without either qualified safety professionals or a well-established maintenance program can avail themselves of such services.

For example, a sprinkler contractor may arrange with a customer for periodic inspection and tests of sprinkler equipment. The contractor and the client negotiate how often inspections are to be done. In some cases, the inspection includes other items, such as fire extinguishers, hoses, or fire doors. The contractor furnishes a comprehensive written report. The client can request that the contractor send copies of the report to the insurer.

The basic contract does not include maintenance work or materials required for alterations, repairs, or replacement. However, if the report indicates any maintenance needs, the client can have the contractor perform the work. Contract service does not relieve management of its primary responsibility for inspection and maintenance. Nevertheless, it does provide excellent inspection for small companies, buildings with mixed tenants, and companies with systems too complex for inspection by their own maintenance staffs.

CONDUCTING INSPECTIONS

Companies must not only conduct regular inspections, but also carry them out in a way that emphasizes the company's commitment to safety. Employees should feel part of the team rather than the target of management scrutiny. This section covers how to conduct inspections to maximize their potential for reducing incidents and motivating employees to support a company's safety goals.

Preparing to Inspect

Inspections should be scheduled for times when inspectors will have the best opportunity to see operations and work practices without much interruption. The inspection route should be planned in advance.

Before conducting an inspection, the inspector or inspection team should review all incidents that have occurred in the area. At this brief meeting, team members should discuss where they are going and what they will be looking for. During the inspection, before going into noisy areas, the team will need to discuss what they wish to accomplish in order to avoid arm waving, shouting, and other unsatisfactory methods of communication.

In addition to the regular checklist and incident reports, inspectors should have copies of the previous inspection report for that particular area. Reviewing this report makes it possible to check whether earlier recommendations have been followed and reported hazards corrected.

Those making inspections should wear the protective equipment required in the areas they enter: safety glasses and shoes, hard hats, acid-proof goggles, protective gloves, respirators, and so forth. If inspectors do not have or cannot get special protective equipment, they should not go into the area. They must be careful to "practice what they preach."

Inspectors also should be aware of any special hazards they might encounter. For example, because welding crews and other maintenance crews move from place to place, they may be encountered anywhere in the facility. Inspectors should know what precautions are required where these crews are working.

Inspection Tools

Inspectors should have the proper tools ready before the inspection to make the process more efficient and to gather more precise data. Common tools include the following:
- clipboards
- inspection forms
- pens/pencils
- lockout/tagout supplies
- measuring tape/ruler
- flashlight.

Depending on the inspection area or type, the following equipment may also be useful:
- cameras
- tape recorder
- electrical testing equipment
- sampling devices (air, noise, light, temperature)
- sample containers
- calipers, micrometers, feeler gauges
- ladder
- special personal protective equipment (see Measurement and Testing, later in this chapter)
- stopwatch.

Relationship of Inspector and Supervisor

Before inspecting a particular department or area, the inspector should contact the department head, supervisor, or other person in charge. This person may have important information for the inspection, particularly if conditions are temporarily altered because of construction, maintenance, equipment downtime, employee absence, and so forth.

If no rules prohibit it, the person in charge may want to accompany the inspector. The inspector can agree but should also emphasize that no tour guide is needed. The inspector must preserve independence and the opportunity to make uninfluenced observations.

If the supervisor of the area does not accompany the inspector, the supervisor should be consulted before the inspector leaves the area. The inspector should discuss each recommendation on particular hazards or unsafe conditions with the supervisor. Usually, they can reach an agreement regarding the relative importance of each recommendation. Obviously, an inspector should not focus on numerous trivial items merely to make the report look complete. On the other hand, the inspector does not have the authority to overlook any condition that may cause an injury.

Even minor items that the supervisor can correct quickly should be reported. The inspector can note on the written report that the supervisor promises to correct the particular condition. This keeps the record clear and serves as a reminder to check the condition during the next inspection.

An inspector should not fail to report hazards merely because a supervisor regards such reporting as criticism. If a supervisor becomes defensive or resentful, the inspector can only repeat what the supervisor knows: the purpose of an inspection is fact finding, not fault finding. By retaining objectivity and refusing to let the issue of safety degenerate into a personality conflict, the inspector keeps matters on a proper professional footing and maintains a firm, friendly, and fair attitude.

Sometimes, a supervisor requests the inspector's assistance in recommending new equipment, reassigning space, or transferring certain jobs from one department to another. When these suggestions deal with safety issues, the inspector will want to include them in notes and consider whether to make them part of the report. The inspector must be careful, however, not to promise either a supervisor or an employee more than can be delivered.

Relationship of Inspector and Employee

Unless company policy or departmental rules prohibit conversation with employees, the inspector can ask questions about operations, taking care, however, not to usurp the responsibility of the supervisor. If, for example, a member of a safety and health committee sees an employee who deviates from established safe work practices, it is better to ask the supervisor rather than the employee about the supposed infraction. The committee member may not fully understand the operation and may be incorrect in the assumption.

In another case, the employee may be performing a risky practice that has been sanctioned by those in authority. The employee could become defensive when questioned by the inspector. It is the supervisor's job to require compliance with company safe work procedures; it is the inspector's job to do the inspecting and reporting. If, however, the situation appears to present an immediate danger, the employee and management should be notified immediately.

Recording Hazards

Inspectors should locate and describe each hazard found during inspection. A clear description of the hazard should be written down, with questions and details recorded for later use. It is important to determine which hazards present the most serious threat and are most likely to occur. The hazard-ranking scheme described in Chapter 6, Loss Control Programs, will simplify the job of classifying hazards.

Properly classifying hazards places them in the right perspective. This approach enables the inspector to briefly describe potential consequences and the probability of such consequences occurring. Inspection reports should enable management to quickly understand and evaluate problems, assign priorities, and make decisions.

On the other hand, the inspector should describe unsafe conditions or deviations from accepted practices in detail and identify machines and operations by their correct names. Also, the inspector must accurately name or number locations and identify in detail the specific hazards within them. Instead of noting "poor housekeeping," for example, the report should give the details: "Empty pallets left in aisles, slippery spots on the floor from oil leaks, a ladder lying across empty boxes, scrap piled on the floor around machines." Instead of noting "guard missing," the report should read, "Guard missing on shear blade of No. 3 machine, SW corner of Bldg. D."

Management must adopt some plan to note intermediate or permanent corrective measures. For example, if intermediate safety measures have been taken, the item could be circled. When permanent measures are taken, the item can be crossed out or marked with an X. Such a system identifies items requiring further corrective action (Figure 9–10).

If a committee is performing the inspection, one member assumes the task of keeping notes. Without such notes it is almost impossible to write a satisfactory inspection report.

Company Compliance Inspectors

Chapter 6, Loss Control Programs, differentiated between deviations from accepted practices and workplace-induced human error. The inspection team needs to look for both. Typically, the inspector is not concerned with identifying the person who is responsible for the unsafe behavior (fault finding). The goal is to identify the behavior (fact finding) and see that it is corrected.

Unsafe behaviors will vary from one area to another. Among common items that may be noted are the following:
- using machinery or tools without authority
- operating motorized vehicles at unsafe speeds or in other violation of safe work practice
- removing guards or other safety devices or rendering them ineffective
- using defective tools or equipment or using tools or equipment incorrectly
- using hands or body instead of tools or push sticks
- overloading, crowding, or failing to balance materials or handling materials incorrectly, including improper lifting
- repairing or adjusting equipment that is in motion, under pressure, or electrically charged
- failing to use or maintain (or using improperly) personal protective equipment or safety devices
- creating unsafe, unsanitary, or unhealthy conditions by improper personal hygiene, using compressed air for cleaning clothes, poor housekeeping, or smoking in unauthorized areas
- standing or working under suspended loads, scaffolds, shafts, or open hatches.

Because the inspector's purpose is to locate unsafe practices, not pinpoint blame, the resulting report should not specify any names. When the report states, "An employee in this area was observed," the supervisor has been advised of the need to enforce safe work practices. The inspector should not be seen as a police officer handing out tickets or, worse, as a snoop from "outside." Nor should information derived from inspections be used for disciplinary measures.

Sometimes, it is necessary to closely observe workers on the job to understand their tasks. The inspector should explain to workers why he or she needs to observe the task and should always ask permission to watch. When employees understand that no one is trying to catch them in an error but rather that they have been chosen to demonstrate a task because of their exceptional skills, they probably will agree to being observed.

Condemning Equipment

When a piece of equipment presents an imminent danger, the inspector should notify the supervisor immediately and see that the machine or equipment is shut down, tagged,

Inspection Report

Area Inspected: Building D
Date and Time of Inspection: —11:00 a.m.
Inspector and Title: Ron Baker, Hazard Control Specialist
Date of Report: _____
Names of Those to Whom Report Is Sent: Bob Firenze (Executive Director); Loren Hall (Department Head); file

No. of Items Carried Over from Previous Report: 3
No. of Items Added to This Report: 4
Total No. of Items on This Report: 7

Item (asterisk indicates old item)	Hazard Classification — Consequence	Probability	Hazard Description	Specific Location	Supervisor	Corrective Action Recommended	Corrective Action Taken
*1	II	B	Guard missing on shear blade #2 machine. Work order issued to engineering for new guard 10/16/80. Wooden barrier guard in temporary use 10/23/79. Guard still missing.	S.W. corner, bay #1	Jay Rillo	Contact engineering to replace guard	Engineering says they will have guard by 11/24.
*2	IV	C	Window cracked. Work order issued for replacement 10/30/80.	South wall, bay #3	Joe Whitestone	Have maintenance replace window	Maintenance to replace all broken windows starting next week.
3	II	B	Oil and trash still accumulated under main motor. Was to be cleaned by 10/30/80.	Pump room	Tony Silva	Clean area; have supervisor talk to men	Cleaned out 11/21. Silva told men to keep area clear.
4	III	B	Mirror at pedestrian walk out of line	North end of machine shop	Tom Schroeder	Post temporary warning sign; call maintenance for adjustment	Sign posted 11/21 — Butler has scheduled adjustment for 12/1
5	II	A	Three workers at cleaning tank not wearing eye protection	Electric shop	Hank Beine	Have supervisor give more training and education	Discussed with Beine — he held meeting on 11/25
6	I	A	Cable on jib crane badly frayed	Bay #3	Joe Whitestone	IMMEDIATE ACTION REQUIRED	Tagged crane out of Service Cable to be replaced 11/21
7	II	B	Guard rail damaged on stairway to second floor	Bay #1	Jay Rillo	Issue work order to carpenter shop to make replacement	Work issue order 11/21

Figure 9–10. This inspection report form simplifies procedures and emphasizes carryovers, new items, and responsibilities. The column at right is for noting corrective action taken later. *(Printed with permission from RJF Associates Inc., Bloomington, IN.)*

and locked out to prevent its further use.

When danger tags and locks are used, those persons authorized to condemn equipment must sign them. Only the inspector who places the tag should be permitted to remove it. This step should occur only when the inspector is satisfied that the hazardous condition has been corrected. No equipment or materials should be placed out of service without notifying the person in authority in the department affected.

Writing the Inspection Report

Every inspection must be documented in a clearly written report furnished by the inspector. Without a complete and accurate report, the inspection would be little more than an interesting sightseeing tour. Inspection reports are usually of three types:

1. *Emergency*—Made without delay when a critical or catastrophic hazard is probable. Using the classification system described in Chapter 6, Loss Control Programs, this category would include any items marked IA or IIA.
2. *Periodic*—Covers unsatisfactory nonemergency conditions observed during the planned periodic inspection. This report should be made within 24 hours of the inspection. Periodic reports can be initial, follow-up, final, or a combination of all three.
3. *Summary*—Lists all items of previous periodic reports for a given time.

The written report should include the name of the department or area inspected (giving the boundaries or location if needed), date and time of inspection, the names and titles of those performing the inspection, the date of the report, and the names of those to whom the report was made.

One way to make the report is to begin by copying items carried over from the last report that were not corrected. Each item is numbered consecutively. The item number can be followed by the hazard classification (IB, IIIC, and so on). Carryover items can be marked with an asterisk. The narrative should include the date the hazard was first detected. The inspector should describe each hazard and pinpoint its location. After the hazard is listed, the inspector should recommend corrective action and establish a definite abatement date. There should follow a space for noting corrective action taken later. In addition, a report should show what is right in a work area as well as what is wrong. When the report is from a committee, it should be reviewed by each member of the inspection team for accuracy, clarity, and thoroughness.

Generally, inspection reports are sent to the head of the department or area where the inspection was made. Copies are also given to executive management and the manager to whom the department head reports.

Follow-Up for Corrective Action

After the inspection report has been written and disseminated, the inspection process starts to return benefits. The information acquired and the recommendations made are valueless unless management takes corrective action. Information and recommendations provide the basis for establishing priorities and implementing programs that will reduce unintentional injuries, improve conditions, raise morale, and increase the efficiency and effectiveness of the operation.

Inspectors can list recommendations in the order in which the hazards were discovered or group them according to the individuals responsible for their correction. Recommendations are then sent to the proper member of management for approval. Where possible, management should set a definite time limit for correction for each recommendation and follow up to make sure corrective action has been taken.

Often, the safety professional is authorized to make recommendations directly to the affected foreman, supervisor, or department if such recommendations do not require major capital outlays. Some organizations require that inspection reports be reviewed by the safety and health committee, particularly when recommendations apply to education and training and directly affect employees.

In making recommendations, inspectors should be guided by four rules:

1. *Correct the cause whenever possible.* Do not merely correct the result, leaving the problem intact. In other words, be sure the disease and not just the symptom is cured. If the inspector or supervisor does not have the authority to correct the real cause, the inspector should bring it to the attention of the person who does.

2. *Immediately correct everything possible.* If the inspector has been granted the authority and opportunity to take direct corrective action, he or she should take it. Delays risk injuries.

3. *Report conditions beyond one's authority and suggest solutions.* Inform management of the condition, the potential consequences of hazards found, and solutions for correction. Even when nothing seems to come of a recommendation, it can pay unexpected dividends.

   For example, a company safety and health committee made a detailed proposal about guarding a particularly hazardous location, only to be told that the engineers had planned to move operations to another location. However, instead of feeling that it had wasted its time, the committee pointed out that the organization had serious communication problems, with the right hand not knowing what the left was doing. The committee recommended that effective management techniques be applied to the hazard control program.

4. *Take intermediate action as needed.* When permanent correction takes time, the hazard should not be ignored. Inspectors or supervisors should take any temporary measures they can, such as roping off the area, locking and tagging out equipment or machines, or posting warning signs. These measures may not be ideal, but they are better than doing nothing.

Some of the general categories into which recommendations may fall are setting up a better process, relocating a process, redesigning a tool or fixture, changing the operator's work pattern, providing personal protective equipment, and improving personnel training methods. Recommendations can also call for improvements in the preventive maintenance system and in housekeeping. Cleaning up debris and dirt may be considered the janitor's job, but preventing its accumulation is part of an effective hazard control program.

Management must realize that employees are keenly interested in the attention paid to correcting faulty conditions and hazardous procedures. Recommendations approved and supported by management should become part of the organization's philosophy and program. At regular intervals, supervisors should report progress in complying with the recommendations to the safety department, the company safety and health committee, or the person designated by management to receive such information. Inspectors should periodically check to see what progress toward corrective action is being made. Unsafe conditions left uncorrected indicate a breakdown in management communications and program application.

Sometimes, management must decide among several courses of action. Often, these decisions are based on cost-effectiveness. For example, it may be cost effective as well as practical to substitute a less toxic material that works as well as the highly toxic substance presently in use. On the other hand, replacing a costly but hazardous machine may have to wait until funds can be designated. In this case, the immediate alternative may be to install machine guards. In all cases, action taken or proposed must be communicated to all persons involved.

REGULATORY COMPLIANCE INSPECTION

Each regulatory agency has specific procedures and rules for inspecting an organization's facilities. The safety professional needs to know the specific procedures to be followed when an inspector from a regulatory agency shows up at the front door. The main question to be answered before inspection visits is this: Who will be the person to meet and accompany the inspecting party during the visit? Typical team members include someone from management, a photographer, and someone who has access to pertinent documents. The purpose of an agency inspection is usually to gather evidence that violations are occurring or have occurred, so if photographs are taken by the inspector, a company representative should record the same thing at the same time.

Most regulatory inspectors use similar guidelines when conducting an inspection. Some reasons for initiating compliance inspections are based on the following criteria:
- imminent danger of death or serious physical harm
- reports of noncompliance with regulatory limits for environmental programs
- accidents resulting in a fatality or hospitalization of three or more employees
- employee complaint
- referral by another government agency
- programmed inspection of high-hazard industries
- follow-up inspection to determine whether previously cited violations have been abated.

The following are general guidelines to which the inspecting party can be expected to adhere during the inspection:
- Inspections will normally be made during the regular working hours of the facility.
- The inspecting party will display his or her official credentials and request to meet the employer representative.
- The inspecting party may be required by an employer to obtain a search warrant based on probable cause before the inspection.
- The inspecting party must state the reason for the inspection (e.g., imminent danger, employee complaint).
- Cameras and sampling equipment may be used by the inspecting party.
- Under the "plain view" exception, the inspecting party may issue a citation for any observed violations, even though the subject of the alleged violation is outside the scope of the consent.

Response to the inspection should consider the following actions:
- Immediately on notification of a regulatory inspector, begin a written log of the inspection—time of arrival, comments, opening and closing conference, departure time, and so on. Everything the inspecting party mentions or questions during the inspection should be documented. Forcible interference with the conduct of inspections or other official duties of the inspecting party may be a criminal offense.
- Immediately notify the safety manager/officer of the inspection.
- Request and photocopy the compliance officer's credentials. Ensure that the credentials are valid.
- Request the reason for the visit. If the reason is an employee complaint, a copy of the written complaint should be requested.
- Ensure that the compliance officer is escorted at all times while in the facility.
- Designate an office or area for the inspecting party to examine records.
- Bring any requested records to the officer rather than allowing unrestricted access to files. Provide only those records specifically requested.
- If any sampling is conducted by the inspecting party, try to conduct concurrent sampling.
- If pictures are taken by the inspecting party, additional, duplicate copies should be taken by internal staff.
- Obvious alleged violations pointed out by the inspecting party should be corrected immediately, if practicable.

Opening Conference
The inspecting party will usually conduct an opening conference before any walk-through or physical inspection. Company representatives and/or employees and other interested stakeholders should be in attendance. Separate conferences may be conducted for the company representative and the employee representative, if requested by either party. The inspecting party will then present his or her credentials, furnish a copy of the employee complaint or warrant, if applicable, and state the reason for the inspection. Usually the inspecting party will outline, in general terms, the scope of the inspection, including records review, employee inter-

views, physical inspection, and the closing conference with management to discuss the inspection findings.

Inspection
The inspection usually includes a records review and physical inspection. Sampling, picture taking, and interviews may also be included.

Answer all questions honestly, but do not volunteer any information. If the inspecting party identifies an alleged violation, do not attempt to provide an explanation, state that it is a violation, or defend the occurrence. This could be considered an admission of guilt when that is not the case.

Immediately correct alleged violations that are obvious, if practicable, to demonstrate a good-faith effort to comply with regulations and to ensure the safety of employees.

Closing Conference
Depending on the scope of the visit, a closing conference will be held after either the records review or the physical inspection. The closing conference is conducted to advise the company of any violations observed and abatement requirements. The inspecting party will usually not provide a written summary at the time of the closing conference, so it is important that the company representative take complete and accurate notes.

The company representative should make no statements during the closing conference that could be construed as an admission of the alleged violations at issue or as a limitation of the company's right to contest the citation.

The company representative may brief the inspecting party on any aspects of the company safety, health, and environmental program that specifically relate to any alleged violations. The inspecting party should be encouraged to set reasonable abatement periods for any alleged violations being contemplated.

Postinspection Procedures
The relevant company stakeholders should be updated on the outcome of the inspection, including alleged violations, abatement periods, and potential responses. The notes taken during the inspection should be maintained in a permanent file with any pictures taken during the inspection. Any alleged violations mentioned by the inspecting party that are apparent should be corrected as soon as possible if they were not corrected during the inspection.

MEASUREMENT AND TESTING

Two special sorts of inspection are conducted by the industrial hygienist and the medical staff. Testing for exposures to health hazards requires special equipment not always available to the hazard control specialist. In such cases, management can often obtain assistance from the state labor department's industrial hygiene division or from the provincial department of labor and health. Another source of help can be industrial hygienists employed by consulting firms and by insurance companies.

Conducting physical examinations of employees exposed to occupational health hazards may require medical equipment that the organization's medical staff does not have. The following discussion is a summary of how to anticipate, recognize, evaluate, and control health hazards in the workplace. Those interested in more details should consult *Fundamentals of Industrial Hygiene*, 6th edition (2012). This reference should help safety professionals understand their role in this area of hazard control.

Kinds of Measurement and Testing
Occupational health surveillance monitors chemical, physical, biological, and ergonomic hazards. Four monitoring systems are used: personal, environmental, biological, and medical.

Personal Monitoring
One example of personal monitoring is measuring airborne concentrations of contaminants. The measurement device is placed as close as possible to the site at which the contaminant enters the human body. When the contaminant is noise, the device is placed close to the ear. When a toxic substance could be inhaled, the device is placed in the worker's breathing zone.

Environmental Monitoring
Environmental monitoring measures contaminant concentrations in the workspace or target environment. The measurement device is placed in the general area adjacent to the worker's usual workstation or where it can sample the general environment. Quite often, environmental monitoring is conducted as a worst-case exposure scenario.

Biological Monitoring
Biological monitoring measures changes in composition of body fluid, tissues, or expired air to detect the level of contaminant absorption or the production of metabolic byproducts. For example, blood or urine can be tested to help determine lead exposures. Similarly, the phenol in urine sometimes is measured to document benzene exposure.

Medical Monitoring
When medical personnel examine workers to see their physiological and psychological response to a contami-

nant, the process is termed *medical monitoring*. Medical monitoring can include health and work histories, physical examinations, x-rays, blood and urine tests, pulmonary function tests, and vision and hearing tests. The aim of such monitoring is to find evidence of exposure early enough to identify especially susceptible workers and to detect any damage before it becomes irreversible.

Biological and medical monitoring provide information after the exposure already has occurred. However, such programs also encompass arrangements to treat an identified health problem and to take corrective action to prevent further damage. To understand how industrial hygienists measure for health hazards, it is necessary to define some basic terms, to distinguish between acute and chronic effects, and to see how safe exposure levels are established.

Measuring and Evaluating Exposure

Measuring for toxicity involves several activities. These include determining a substance's capacity for injury or harm, inhalation hazards, influence of solubility, Threshold Limit Values, and permissible exposure limits.

Toxicity

The toxicity of a material is not identical with its potential for being a health hazard. *Toxicity* is the capacity of a material to produce injury or harm. *Hazard* is the possibility that exposure to a material will cause injury or illness when a specific quantity is used under certain conditions and some level of exposure occurs. The key elements to be considered when evaluating a health hazard are the following:
- amount of material to which the employee is exposed
- total time of the exposure
- toxicity of the substance
- availability of personal protective equipment
- individual susceptibility.

Not all toxic materials are hazardous. Most toxic chemicals are safe when packaged in their original shipping containers or contained within a closed system. As long as toxic materials are adequately managed, they can be safely used. For example, many solvents, if not properly handled, will cause irritation to the eyes, mouth, and throat. Some also are intoxicating and can cause blistering of the skin and other forms of dermatitis. Prolonged exposure can cause more serious illness. But if workers use such solvents in a well-ventilated area and are given proper protective equipment to prevent the solvents from contacting skin, the substances can be used safely.

The toxic action of a substance can be divided into acute and chronic effects:
- *Acute effects* involve a short exposure time period and adverse effects occurring within a short time period. They can be the result of sudden and severe exposure, during which the substance is rapidly absorbed. Acute effects can be related to an incident that disrupts ordinary processes and controls. For example, sudden exposure to very high concentrations of methane gas in a confined space can lead to loss of consciousness, coma, or death by asphyxiation.
- *Chronic effects* are usually the result of exposure to a toxic substance over a long period of time. When the chemical is absorbed more rapidly than the body can eliminate it, the chemical begins to accumulate in the body. If the level of contaminant is relatively low, the effects, even if they are serious and irreversible, may go unnoticed for long periods because of a phenomenon known as the latency period. *Latency* refers to an extended time period, usually in years, between exposure and observed health effects.

The action of a toxic substance can result in both acute and chronic toxic effects and exposure.

Inhalation Hazards

Inhalation of harmful materials may irritate the upper respiratory tract and lung tissue or the terminal passages of the lungs and the air sacs, depending on the solubility of the material. Inhalation of biologically inert gases may dilute oxygen levels below the normal blood saturation value and disturb cellular processes. Other gases and vapors may prevent the blood from carrying oxygen to the tissues or interfere with its transfer from the blood to the tissue, producing chemical asphyxia.

Inhaled contaminants that adversely affect the lungs fall into three general categories: aerosols, toxic gases, and gases that produce systemic effects.
- Aerosols (particulates) are substances that, when deposited in the lungs, may produce either rapid local tissue damage, some slower tissue reactions, eventual disease, or only physical plugging.
- Toxic vapors and gases are hazards that produce adverse reactions in the tissue of the lungs themselves.
- Some toxic aerosols or gases do not affect the lung tissue locally but (1) are passed from the lungs into the bloodstream, where they are carried to other body organs, or (2) have adverse effects on the oxygen-carrying capacity of the blood cells themselves.

An example of the first type (aerosols) is particulates that can cause a variety of lung reactions over time, including production of scarring and progressive decrease in lung function.

An example of the second type (toxic gases) is hydrogen fluoride, a gas that directly affects lung tissue. It is a primary irritant of mucous membranes and causes chemical burns. Inhalation of this gas will cause pulmonary edema: after lung tissue is burned, the lungs fill with fluids that directly interfere with the gas-transfer function of the alveolar lining.

An example of the third type is carbon monoxide, a toxic gas that passes into the bloodstream without essentially harming the lung. The carbon monoxide passes through the alveolar walls into the blood, where it preferentially binds to the hemoglobin molecule so it cannot as easily accept oxygen, thus starving the body of oxygen. Cyanide gas prevents cell enzymes from using molecular oxygen; this state disrupts vital cell processes.

Sometimes, several types of lung hazards occur simultaneously. In mining operations, for example, explosives release oxides of nitrogen into the air breathed by miners. These compounds impair the bronchial clearance mechanism so that coal dust (of the particle sizes associated with the explosions) is not efficiently cleansed from the lungs.

Influence of Solubility

A compound that is very soluble—such as ammonia, formaldehyde, sulfuric acid, or hydrochloric acid—may pose less of a hazard. Although it is rapidly absorbed in the upper respiratory tract during the initial phases of exposure, it does not penetrate deeply into the lungs. Consequently, the nose and throat become very irritated, causing workers to leave the exposure area before they suffer serious harm. On the other hand, compounds insoluble in body fluids often cause considerably less throat irritation than do the soluble ones but may penetrate deeply into the lungs. Thus, a serious hazard can be present without workers being immediately aware of it. With less irritation to the lungs, workers have less warning that exposure is building up—the hazardous chemical has poor warning properties. Examples of such compounds (gases) are nitrogen dioxide and ozone. The immediate danger from these compounds in high concentrations is acute lung irritation or, possibly later, chemical pneumonia.

Numerous chemical compounds do not follow the general solubility rule. Such compounds are not very soluble in water and yet irritate the eyes and respiratory tract. They also can cause lung damage, even death in many situations. The supervisor must be sure that all hazardous compounds are identified and workers are properly protected.

Threshold Limit Values (TLVs)

Individual susceptibility to respiratory toxins is difficult to assess. Nevertheless, certain recommended limits have been established. Threshold Limit Values (TLVs) are one commonly used exposure limit. The term *TLV* refers specifically to limits published by the American Conference of Governmental Industrial Hygienists (ACGIH). Because of wide variations in individual susceptibility, however, an occasional exposure of an individual at or even below the TLV may not prevent discomfort, aggravation of a preexisting condition, or occupational illness. The TLVs are reviewed and updated annually. The National Safety Council's *Fundamentals of Industrial Hygiene* (2012) explains this subject in detail. A brief overview follows. There are three categories of TLVs:

1. Time-weighted average (TLV-TWA) is the time-weighted average concentration for a normal 8-hour day or 40-hour week. It is believed that nearly everyone can be exposed day after day to airborne concentrations at these limits without adverse effect.
2. Short-term exposure limit (TLV-STEL) is the concentration to which persons can be exposed for a period of up to 15 minutes continuously without experiencing the following:
   - irritation
   - chronic or irreversible tissue change
   - narcoses of sufficient degree to reduce reaction time, impair self-rescue, increase the likelihood of injury, or materially reduce work efficiency, provided the daily TLV-TWA is not exceeded.

   No more than four 15-minute exposure periods per day are permitted, with at least 60 minutes between exposure periods.

   An STEL is not a separate, independent exposure limit. Rather, it supplements the time-weighted average (TWA) limit in cases where workers experience acute reactions to a substance whose toxic effects are primarily chronic. Short-term exposure limits are recommended only where toxic effects have been reported from high short-term exposures in either humans or animals.
3. Ceiling (TLV-C) is the concentration that should not be exceeded even for an instant.

Nearly one-fourth of the substances in the TLV list are followed by the designation *skin*. This refers to potential exposure through skin absorption. This designation is intended to suggest appropriate measures to prevent absorption of substances through the skin.

Permissible Exposure Limits (PELs)

The first compilation of OSHA health and safety standards appeared in 1970. Because it was derived from then-existing standards, the compilation adopted many of the TLVs established in 1968 by the American Conference of Governmental Industrial Hygienists. Thus Threshold Limit Values—a registered trademark of the ACGIH—became, by federal standards, permissible exposure limits (PELs). These PELs represent the legal maximum level of contaminants in workplace air.

The General Industry OSHA Standards currently list about 600 substances for which exposure limits have been established. These are included in subpart Z, "Toxic and Hazardous Substances," Sections 1910.1000 through 1910.1500. The PELs were updated in 1990, again including many of the newer, revised TLVs. In some cases, OSHA may enforce the "general duty clause" [OSH Act Section 5(a)(1)] when a serious hazard is observed related to a hazardous material exposure that is not covered by a PEL. For current information on exposure limits, refer to current regulatory documentation.

The OSHA action level (AL) is the point at which employers must initiate certain safety provisions: employee exposure measurement, employee training, and medical surveillance. An OSHA action level has not been defined for all workplace exposures. The action level for some OSHA-regulated chemicals such as lead is usually set at about one-half the permissible exposure level (PEL).

Why is an action level set well below the PEL? Setting the action level at one-half the permissible exposure helps protect employees from overexposure with a minimum burden to the employer. When employee exposure measurements indicate that no employee is exposed to airborne concentrations of a substance in excess of the action level, employers, in effect, are exempted from having to initiate certain provisions in the standards.

The action level recognizes that air samples can only estimate the true TLV-TWA. For an extra margin of safety, companies set action levels lower than one-half the PEL. The adequacy of the TLVs to protect workers from illness is controversial, and a lower company action level may be desirable.

When to Measure

The measurements done by the industrial hygienist can be divided into three phases:
1. problem definition phase
2. problem analysis phase
3. solution phase.

Problem Definition Phase

In many instances, inspectors take measurements to determine whether there is a problem in the workplace. In addition, some OSHA regulations require measurement at certain specified intervals or any time there is a change in production, process, or control measures. Measurement often establishes that workers are not experiencing excessive exposure to hazardous materials. Such monitoring of the workplace helps ensure a safe environment.

Monitoring, then, is frequently used to determine that employers are in compliance with OSH Act requirements, state or provincial regulations, commonly accepted standards, TLVs, PELs, and action levels. Newer health standards published by OSHA usually state:

> Each employer who has a place of employment in which [toxic substance name] is released into the workplace air shall determine if there is any possibility that any employee may be exposed to airborne concentrations of [toxic substance name] above the permissible level. The initial determination shall be made each time there is a change in production, process, or control measures that may result in an increase in airborne concentrations of [toxic substance name].

When any hazardous substances are released into the workplace air, the employer must take the first step in the employee exposure monitoring program. For OSHA-regulated substances, there must be an actual exposure measurement to see whether any employee has been exposed to concentrations in excess of the recommended levels. This step should be taken even if there is only a remote chance that employees have been exposed to a substance above recommended levels.

Where does sampling begin? Should the sample be taken at the worker's breathing zone? Out in the general air? At the machine or process that is emitting the toxic substance? Although OSHA requires sampling only in the worker's breathing zone, sampling at all three sites provides a clearer picture of the situation. Should the sample be taken for 2 minutes, 2 hours, or a whole day?

There are two major types of samples: grab samples and long-term samples. The grab sample is taken over so short a period of time that the atmospheric concentration is assumed to be constant throughout the sample. This usually will cover only part of an industrial cycle. A series of grab samples can be taken in an attempt to define the total exposure. However, doing so requires a sound knowledge of statistical sampling techniques.

The long-term sample is taken over a sufficiently extended period of time that the variations in exposure cycles are averaged. Usually, one sample or a series of samples is taken to represent the employee's 8-hour average exposure. OSHA regulations usually require this type of sampling.

An adequate number of tests should be taken to define the TLV-TWA and to relate this level to recommended or regulatory exposure levels. But samples also must be taken to characterize the peak emissions during various portions of the process cycle.

If employee measurements indicate exposure at or above the action level, then OSHA requires that all employees so exposed be identified and their exposure measured. This step clearly determines the population at risk.

When exposure measurements are at or above the action level but not above the PEL or just below the TLV-TWA, the employer needs some statistically reliable means to ensure that exposure levels are not exceeding these values. Management should conduct periodic sampling of the affected area and order medical examinations to determine whether any susceptible individuals are exhibiting effects at these exposures.

If employees are exposed above the PEL or TLV, then a more intensive monitoring program is necessary. Medical staff must examine workers exposed to these excessive levels to measure the effects on their health. Noninhalation exposures—such as skin absorption—also may occur. Therefore, accurate exposure evaluation may require breath, blood, and urine sampling.

The problem definition phase is an orderly progression. At each step of the process, employers can decide whether to proceed to the next higher step.

Problem Analysis Phase

Once the industrial hygienist has defined the problem in the first phase of the measurement process, he or she must determine its causes. Management can identify opportunities for improvements in the workplace, set objectives for solutions, and devise alternative solutions should the initial ones fail to solve the problem.

The following eight methods suggest some ways that exposure hazards can be controlled:

1. substitution of a less harmful material for a hazardous one
2. change or alteration of a process to minimize worker contact
3. isolation or enclosure of a process or work operation to reduce the number of workers exposed
4. wet methods to reduce generation of dust in operations
5. local exhaust at the point of generation
6. personal protective equipment (see Chapter 7 in the *Engineering & Technology* volume)
7. good housekeeping, including cleanliness of the workplace, waste disposal, adequate washing, clean toilet and eating facilities, healthful drinking water, and control of insects and rodents
8. training and education—the OSHA hazard communication standard requires training for employees exposed to hazardous chemicals.

Solution Phase

Once the problem has been analyzed and a number of solutions proposed, the most effective, timely, and practical solution needs to be selected—one that provides optimum benefits with minimal risks. The details of the solution should be carefully worked out. In effect, management needs to develop a blueprint describing what should be done, how and by whom it should be done, and in what sequence the actions are to take place.

Once controls are installed, they must be checked periodically to be sure they are functioning properly. Follow-up monitoring and inspection will determine whether the solution to a given hazardous exposure is controlling it within the specified limits. Thus, managers should regard the monitoring function as a circular, not horizontal, process. If measurements at the solution phase reveal that controls are inadequate, the industrial hygienist must return to the first phase, that of defining the problem.

Who Will Do the Measuring?

Not every organization requires or can afford the services of a full-time industrial hygienist. Independent consultants can be hired to accomplish two major objectives:

1. Identify and evaluate potential health risks and injury hazards to workers in the occupational environment.
2. Design effective controls to protect the safety and health of workers.

Because any person can legally offer services as an industrial hygiene consultant, it is important that the consultant hired is a trained, experienced, and competent professional. A competent industrial hygiene consultant must have detailed knowledge of proper sampling equipment and analytic procedures and will probably hold the designation *Certified Industrial Hygienist (CIH)*.

Good sources of information and assistance regarding consultants are the American Industrial Hygiene Association and the American Society of Safety Engineers, the professional associations related to occupational and health safety, respectively (see the descriptive listing in Appendix 1, Sources of Help). Regional offices of the National Institute for Occupational Safety and Health (NIOSH) usually have lists of consultants in their area. Many insurance companies have loss prevention programs that employ industrial hygienists. The National Safety Council and its chapters with offices in major cities can offer assistance. The NSC offers a full range of consulting services in safety and occupational health management (see Appendix 1, Sources of Help). For a state-by-state listing of governmental consulting service offices, see Appendix 1, Sources of Help under U.S. Government Agencies.

INCIDENT INVESTIGATION

A fourth function of monitoring in the total hazard control system is incident investigation, the subject of Chapter 10,

Incident Investigation, Analysis, and Costs. The following discussion demonstrates how incident investigation fits into the systems approach to hazard control.

Why Incidents Are Investigated

When viewed as an integral part of the total occupational safety and health program, incident investigation is especially important to determine direct causes and uncover contributing causes of injuries, prevent similar injuries from occurring, document facts, provide information on costs, and promote safety. Incident investigation concentrates on gathering all information about the factors leading to the incident.

Determine Direct Causes

Incident investigation determines the direct and contributing causes of incidents. At what points did the hazard control system break down? Were rules and regulations violated? Did defective machinery or factors in the work environment contribute to the incident? Poor machinery layout, for example, or the design of a job process, operation, or task can contribute to an undesirable situation. Chapter 6, Loss Control Programs, outlined the three primary sources of incidents: human, situational, and environmental factors.

Uncover Contributing Incident Causes

Thorough incident investigation is likely to uncover problems that indirectly contributed to the incident. Such information benefits incident reduction efforts. For example, a worker slips on spilled oil and is injured. The oil spill is the direct cause of the injury, but a thorough investigation may reveal other contributing factors: poor housekeeping, failure to follow the maintenance schedule, inadequate supervision, or faulty equipment (such as a lathe leaking oil).

Prevent Similar Incidents

Incident investigation identifies actions and improvements that will prevent similar incidents in the future.

Document Facts

Incident investigation documents the facts involved in an incident for use in any compensation and litigation that may arise. The report produced at the conclusion of an investigation becomes the permanent record of facts about the incident. It may become necessary to reconstruct an incident situation long after the occurrence. To do so, the details of the incident will have to be recorded properly, accurately, and thoroughly.

Provide Information on Costs

Incident investigation provides information on both direct and indirect costs of incidents. Chapter 10, Incident Investigation, Analysis, and Costs, gives details for estimating incident costs.

Promote Safety

Incident investigation yields psychological as well as material benefits. The investigation demonstrates the organization's interest in worker safety and health. It indicates management's sense of accountability for incident prevention and its commitment to a safe work environment. An investigation in which both labor and management participate promotes cooperation between these two groups.

Despite what many people believe, incident investigation is a fact-finding, not a fault-finding, process. When attempting to determine the cause of an incident, the novice investigator may be tempted to conclude that the person involved in the incident was at fault. But if human error is not the real cause, the hazard that produced the incident will go undiscovered and uncontrolled. Furthermore, the person falsely blamed for causing the incident will resent the unjustified accusation, as well as any disciplinary action. The worker will be less cooperative in the future and feel less respect for the organization's safety and health program. Investigators should always stress that the intent of incident investigation is to pinpoint causes of error and defects so similar incidents can be prevented.

Conducting an incident investigation is not simple. It can be difficult to look beyond the incident at hand to uncover causal factors, determine the true loss associated with the occurrence, and develop practical recommendations to prevent recurrence. A major weakness of many incident investigations is the failure to establish and consider all factors—human, situational, and environmental—that contributed to the injury. Reasons for this failure include the following:

- an inexperienced or uninformed investigator
- reluctance of the investigator to accept full responsibility for the job
- narrow interpretation of environmental factors
- erroneous emphasis on a single cause
- judging the effect of the incident to be the cause
- arriving at conclusions before all factors are considered
- poor interviewing techniques
- delay in investigating incidents.

An effective investigator must be ready to acknowledge as contributing causes any and all factors that may have led, in any way, to the incident. What at first may appear to be a simple, uninvolved incident can, in fact, have numerous contributing factors that become more complex as analyses are completed. Immediate, on-the-scene incident investigation provides the most accurate and useful information.

When to Investigate Incidents

The longer the delay before examining the incident scene and interviewing victims and witnesses, the greater the possibility of obtaining erroneous or incomplete information. The incident scene changes, memories fade, and people discuss what happened with each other. Whether consciously or not, witnesses may alter their initial impressions to agree with someone else's observation or interpretation. Further, prompt incident investigation also expresses concern for the safety and well-being of employees.

As a general rule, all incidents, no matter how minor, are candidates for thorough investigation. Many incidents occurring in an organization are considered minor because their consequences are not serious. Such incidents are taken for granted and often do not receive the attention they demand. Management, safety and health committees, supervisors, and employees must be aware that serious injuries arise from the same hazards that produce minor incidents. Usually, sheer luck determines whether a hazardous situation results in a minor incident or a serious injury.

In incident investigation, the investigator must give priority to the health and safety of affected personnel (including any victims). When possible, rescue and first-aid procedures should be used that disturb the incident scene as little as possible. Measures to protect equipment should also preserve evidence. When the area is secure, victims have received medical attention, and appropriate notifications have been made, efforts can be concentrated on investigating the incident.

As with inspections, it is advisable to prepare investigation tools in advance. An investigation kit may include the following:
- camera and film or a memory card (ensure sufficient resolution if using a digital camera)
- tape recorder
- measuring devices
- sample containers
- interview/investigation forms
- flashlight
- barricade markers/tape
- warning tags and padlocks.

Having the necessary equipment ready will facilitate the investigation and certainly help eliminate delays and other difficulties.

Who Should Conduct the Investigation?

Chapter 6, Loss Control Programs, discusses the question of who is to make the investigation: the supervisor or foreman, the safety professional, a special investigative committee, or a company safety and health committee. As a supplement to that discussion, the following section outlines the roles played by physicians and management in incident investigation. It also covers the responsibility of the safety professional in preventing further incidents from occurring during the investigation itself.

Physician

In any case, the assistance of a physician should be garnered whenever there is an issue with a person's health. If a safety professional starts addressing issues that should be dealt with by a physician, then credibility could be lost and never regained.

A physician's assistance is particularly important when human factors have been designated as direct or contributing causes of an incident. The physician can assess the nature and degree of injury and help determine the source and nature of the forces that inflicted the injury. The physician also can (1) determine what special biomedical studies, if any, are needed; (2) establish whether the injured person was physically and mentally fit at the time of the incident and whether the screening, selection, and placement process is adequate; (3) help judge the adequacy of safety and health protection procedures and equipment; and (4) help evaluate the effectiveness of the plans, procedures, equipment, training, and response of rescue, first-aid, and emergency medical care personnel. The physician also can evaluate the effectiveness of measures aimed at early detection of medical conditions, mental changes, or emotional stress.

Management

Management and department heads should help investigate incidents resulting in lost workdays or major property damage. When management actively participates in incident investigation, it can evaluate the hazard control system and determine whether outside assistance is desired or required to upgrade existing structures and procedures. Management also must review incident reports in order to make informed decisions. When incident investigation reveals the need for or desirability of specific corrective actions, management must determine whether the recommended action has been implemented.

Safety during the Investigation

In many cases, the incident scene is a dangerous place. The incident may have damaged electrical equipment, weakened structural supports, and released radioactive or toxic materials.

The safety professional must be particularly alert to the hazards encountered by the investigating team and, when necessary, see that proper protective equipment is provided. Investigators need to be alerted to the hazards they may encounter and emergency procedures they should follow (see details in Chapter 18, Emergency Preparedness).

What to Investigate

The incident investigation must answer many questions. Because of the infinite number of incident-producing situations, contributing factors, and causes, it is impossible to list all of the questions that apply to all investigations. The following questions are generally applicable, however, and will be considered in most incident investigations (Firenze 1978):

- What was the injured person doing at the time of the incident? Performing an assigned task? Performing maintenance? Assisting another worker?
- Was the injured employee working on an unauthorized task? Was the employee qualified to perform the task and familiar with the process, equipment, and machinery?
- What were other workers doing at the time of the incident?
- Was the proper equipment being used for the task at hand (screwdriver instead of can opener to open a paint can, file instead of grinder to remove burr on a bolt after it was cut)?
- Was the injured person following approved procedures?
- Is the process, operation, or task new to the area?
- Was the injured person being supervised? What was the proximity and adequacy of supervision?
- Did the injured employee receive hazard recognition training before the incident?
- What was the location of the incident? What was the physical condition of the area when the incident occurred?
- What immediate or temporary actions could have prevented the incident or minimized its effect?
- What long-term or permanent action could have prevented the incident or minimized its effect?
- Had corrective action been recommended in the past but not adopted?

During the course of the investigation, the preceding questions should be answered to the satisfaction of the investigators. Other questions that come to mind as the investigation continues should be recorded.

Conducting Interviews

Interviewing injury victims and witnesses can be a difficult assignment if not properly handled. The individual being interviewed often is fearful and reluctant to provide the interviewer with accurate facts about the incident. The victim may be hesitant to talk for any number of reasons. A witness may not want to provide information that may implicate friends, fellow workers, or the supervisor. To obtain the necessary facts during an interview, the interviewer must first eliminate or reduce an employee's fear and anxiety by establishing good rapport with the individual. The interviewer must create a feeling of trust and establish open communication before beginning the actual interview. Once good rapport has been developed, the interviewer can follow this five-step method.

1. Discuss the purpose of the investigation and the interview (fact finding, not fault finding).
2. Have the individual relate his or her version of the incident with minimal interruptions. If the individual being interviewed is the one who was injured, ask what was being done, where and how it was being done, and what happened. If practical, have the injured person or eyewitness explain the sequence of events that occurred at the time of the incident. Being at the scene of the incident makes it easier to relate facts that may otherwise be difficult to explain.
3. Ask questions to clarify or fill in any gaps.
4. Repeat the facts of the incident to the injured person or eyewitness. Through this review process, there will be ample opportunity to correct any misunderstanding that may have occurred and clarify, if necessary, any of the details of the incident.
5. Discuss methods of preventing recurrence. Ask the individual for suggestions aimed at eliminating or reducing the impact of the hazards that caused the incident. By asking the individual for ideas and discussing them, the interviewer will show sincerity and emphasize the fact-finding purpose of the investigation, as it was explained at the beginning of the interview.

In some cases, contractual agreements may call for an employee representative to be present during any management interview, if the employee so requests.

Preparing the Incident Investigation Report

Chapter 11, Injury and Illness Record Keeping, Incidence Rates, and Analysis, outlines specific ways to record and classify data: how to identify key facts about each injury and the incident that produced it, how to record facts on a form that facilitates analysis and reveals patterns and trends, how to estimate incident costs, and how to comply with regulatory record-keeping requirements.

An unintentional injury in any organization is of significant interest to employees, who will ask questions that reflect their concerns. Is there any potential danger to those in the immediate vicinity? What caused the incident? How many people were injured? How badly?

Those who investigate incidents should answer these questions truthfully and avoid covering up any facts. On the other hand, they must be certain that they are authorized to release information, and they must be sure of their data.

Because the incident report is the product of the investigation, it should be prepared carefully and adequately to justify the conclusions reached. It must be issued soon after the incident. When a report is delayed too long, employees may feel left out of the process. If a final report must be postponed pending detailed technical analysis or evaluation, then management should issue an interim report providing basic details of the investigation.

Summaries of vital information on major injury, damage, and loss incidents should be distributed to department managers. Such summaries should include information on incident causes and recommended action for preventing similar incidents. Management should maintain incident and statistical report files as dictated by company policy.

Supervisors need to keep employees informed of significant injuries and preventive measures proposed or executed. Posting incident reports is one way to make information available.

Implementing Corrective Action

The preceding section on inspection emphasized that hazard control benefits accrue only after the inspection report is written and disseminated. Until corrective action is initiated, any recommendations—no matter how earnest, thorough, and relevant—remain "paper promises."

The same is true of incident investigation when it is used as a monitoring technique. Viewed from the perspective of hazard control, incident investigation serves as a monitoring function only when it provides the impetus for corrective action.

Whenever management and safety professionals review monthly incident reports, they exercise an essential auditing function. Management (including the chief executive officer) can demonstrate interest in safety by requiring prompt reporting of all serious or potentially serious incidents. They use incident reports to make decisions to prevent similar incidents from occurring, and they look for answers to certain key questions. Are all significant incidents being reported? Are all parts of the organization equally committed to the hazard control effort? Are there trends or patterns in incidents or injuries? What system breakdowns predominate? What supervisors require additional training? Are employees advised of the results of incident investigations and of preventive measures being instituted? What management deficiencies are indicated? Incident investigation as a monitoring function occurs after the hazard control system has already broken down. Although no amount of investigation can reverse the incident, the investigation serves an important monitoring function. Past mistakes can be used to improve future operations. As George Santayana has written, "Those who cannot remember the past are condemned to repeat it."

SUMMARY

- Monitoring, a vital management tool, is a set of observation and data-collection methods used to detect and measure deviations from plans and procedures in current operations. It involves hazard analysis, job safety analysis, inspection, measurement and testing, and incident investigation.
- Inductive or deductive hazard analysis is an orderly process used to acquire specific hazard and failure data pertinent to a given system. Factors such as frequency of incidents; potential for injury; severity of injury; new or altered equipment, processes, and operations; and excessive material waste or damage to equipment determine which tasks or processes are analyzed.
- Job safety analyses are the first step of hazard analysis. JSAs are used to identify and analyze potential hazards and causes of incidents within each job or within specific categories of jobs.
- The primary purpose of inspection is to detect potential hazards so they can be corrected before an incident occurs. Continuous inspections are a routine part of the job. Planned inspections are more formal procedures that may be periodic, intermittent, or general. Inspectors report all hazards—classifying and describing them carefully—and recommend corrective actions.
- Measurement and testing methods are used to monitor chemical, physical, biological, and ergonomic hazards. Four monitoring systems are used: personal, environmental, biological, and medical.
- Standards used to establish health and safety limits include Threshold Limit Values (TLVs), time-weighted averages (TWAs—exposures averaged over a set time, usually 8 hours), short-term exposure limits (STELs—exposure limits are usually higher than TLVs or PELs but are used to evaluate much shorter time periods), permissible exposure limits (PELs), and action levels (ALs). Measurements are divided into the problem definition phase, the problem analysis phase, and the solution phase.
- Incident investigations are conducted to determine direct causes, uncover contributing incident causes, prevent similar incidents, document facts, provide information on costs, and promote safety.
- Incidents must be investigated immediately to ensure accurate details and to preserve evidence. Investigators can be supervisors, safety professionals, special committees, or health and safety committees.
- Investigators must examine all human, situational, and environmental factors in determining incident causes and apply the five-step method for interviewing incident victims and witnesses. An investigation report should be issued as soon as possible and contain recommendations for corrective action.

REFERENCES

Balge, M. Z., and G. R. Krieger, eds. *Occupational Health and Safety*. 3rd ed. Itasca, IL: National Safety Council, 2000.

Ferry, T. *Modern Accident Investigation and Analysis*. 2nd ed. New York: Wiley, 1988.

Firenze, R. J. *The Process of Hazard Control*. Dubuque, IA: Kendall/Hunt, 1978.

Gowen, L. D. "Using Fault Trees and Event Trees as Oracles for Testing Safety-Critical Software Systems." *Professional Safety* 41, no. 4 (1996): 41–44.

Johnson, W. G. *MORT Safety Assurance Systems*. New York: Dekker, 1980. (Also available through the National Safety Council.)

MAS 611 Aviation/Aerospace System Safety. Daytona Beach, FL: Embry-Riddle University, 1998.

National Safety Council, 1121 Spring Lake Drive, Itasca, IL 60143.
  Fundamentals of Industrial Hygiene. 6th ed. 2012.
  Injury Facts® 2014 ed.
  Supervisors' Safety Manual. 10th ed. 2009.

U.S. Chamber of Commerce. *Analysis of WC Laws*. Washington DC: U.S. GPO, 2012.

U.S. Department of Labor.
  Occupational Safety and Health Act of 1970 (Public Law No. 91–596).
  Occupational Safety and Health Act Regulations, Title 29, CFR, Part 1910.
    Section 5(a)(1), "General Duty Clause."
    Subpart Z, "Toxic and Hazardous Substances," Sections 1910.1000 through 1910.1500.
  Occupational Safety and Health Act Regulations, Title 29, CFR, Part 1926.

REVIEW QUESTIONS

1. Define *hazard analysis*.
2. List two formal methods of hazard analysis.
   a.
   b.
3. In determining which hazard analysis approach to use for a given situation, the hazard control specialist will need to answer which five questions?
   a.
   b.
   c.
   d.
   e.
4. In order to decide which processes, operations, and tasks receive priority, what are the five factors that need to be analyzed?
   a.
   b.
   c.
   d.
   e.
5. Define *job safety analysis (JSA)*.
6. After a job has been selected to be analyzed, what are three basic steps in conducting a JSA?
   a.
   b.
   c.
7. Inspecting is the first stage of the hazard and incident analysis procedure. What is the general purpose of an inspection?
8. List and briefly explain two types of inspections.
   a.
   b.
9. The toolroom employee examines all tools before sending them out to be used. What is this type of inspection called?
   a. general
   b. intermittent
   c. continuous
   d. periodic
10. Gathering information about standards, regulations, and codes is the first step in determining what items need to be inspected. What sources of information are available to assist in this process?
11. List four factors that determine the frequency of inspections.
    a.
    b.
    c.
    d.
12. Name and briefly define four kinds of monitoring systems.
    a.
    b.
    c.
    d.
13. The *toxicity* of a material refers to
    a. its potential for being a health hazard.
    b. its capacity to produce injury or harm.
    c. a standard measure of percentage of particles.
    d. all of the above

14. Define *Threshold Limit Values (TLVs)*.
15. Define *permissible exposure limits (PELs)*.
16. List the six main outcomes of an incident investigation.
    a.
    b.
    c.
    d.
    e.
    f.
17. Why is it important to investigate incidents immediately?

Incident Investigation, Analysis, and Costs

James T. O'Reilly, JD

Incident Investigation and Analysis
Types of Investigation and Analysis ▶ Persons Conducting the Investigation ▶ Cases to Be Investigated ▶ Minimum Data Required ▶ Identifying Causal Factors and Selecting Corrective Actions ▶ Classifying Incident Data ▶ Conducting the Analysis ▶ Using the Analysis ▶ Involving Outside Employees and Contractors

Estimating Incident Costs
Definition of Work Incidents for Cost Analysis ▶ Method for Estimating ▶ Example of a Cost Estimate ▶ Adjusting for Inflation ▶ Items of Uninsured Cost ▶ Conducting a Pilot Study ▶ Developing the Final Cost Estimate

Off-the-Job Disabling Injury Cost
Categorizing OTJ Disabling Injury Cost ▶ Estimating OTJ DI Costs ▶ Measuring the Effects of OTJ Safety Programs

Summary

References

Review Questions

This chapter covers the investigation of incidents without injury as well as incidents that result in injuries, illnesses, and/or losses. To increase the understanding that an incident is a potential loss situation and that so-called accidents are not random events but rather preventable events, the term *incident* is used in this chapter wherever possible. Used in its broadest sense, the term *incident* includes events that may lead to injuries, illnesses, and/or property damage. The following definitions are generally used in this chapter:

Accident: An occurrence in a sequence of events that produces unintended injury, death, or property damage. *Accident* refers to the event, not to the result of the event.

Incident: An undesired event that may cause personal harm or other damage. In the United States, OSHA requires that incidents of a certain severity be recorded.

Near-miss incident: For purposes of internal reporting, some employers choose to classify as an "incident" the near-miss incident: an injury requiring first aid; a newly discovered unsafe condition; a fire of any size; or a nontrivial incident of damage to equipment, building, property, or product.

Unintentional injury: The preferred term for "accidental injury" in the public health community. It refers to the result of an accident.

With proper hazard identification and evaluation, management commitment and support, preventive and corrective procedures, monitoring, evaluating, and training, unwanted events can be prevented.

In some contexts, it may be necessary to distinguish between incidents with no injuries and accidents resulting in injuries. To facilitate such record keeping, see the Incident (No-Injury Accident) Report Form (Figure 10–1).

The topics covered in this chapter include the following:
- basic types of incident investigation and analysis
- methods of conducting an investigation and analysis
- types of costs associated with incidents
- how to calculate incident-related costs
- the financial effects of off-the-job incidents.

Successful incident prevention requires at least six fundamental activities:
1. studying of all working areas to detect and eliminate or control the behavioral, physical, and environmental hazards that contribute to incidents (see also Chapter 6, Loss Control Programs, and Chapter 9, Identifying Hazards)
2. studying all operating methods and practices and administrative controls
3. incorporating education, instruction, training, and enforcement of procedures to minimize the human factors that contribute to incidents (see also Chapter 30, Motivation, and Chapter 31, Safety and Health Training)
4. conducting a thorough investigation and causal analysis of every incident resulting in at least an injury that requires medical treatment, with the goal of determining the contributing circumstances. Incidents not resulting in personal injury (so-called *near incidents* or *near misses*) are warnings and should also be investigated thoroughly. This fourth activity is a safeguard against any hazards overlooked in the first three activities, against hazards not immediately obvious, and against hazards resulting from circumstances difficult to foresee.
5. implementing programs to change or control the hazardous conditions, procedures, and practices found in the preceding activities
6. conducting program follow-up and evaluation to ensure that the programs achieve the desired control.

INCIDENT INVESTIGATION AND ANALYSIS

The ultimate purpose of incident investigation and analysis activities is to prevent future incidents. Therefore, the investigation or analysis must produce factual information leading to corrective actions that prevent or reduce the number of incidents. The more complete the information, the more easily management can take effective corrective actions. For example, knowing that 40% of an organization's incidents involve ladders is not as useful as knowing that 80% of the organization's ladder incidents involve broken rungs. A good record-keeping system, as discussed in Chapter 11, Injury and Illness Record Keeping, Incidence Rates, and Analysis, is essential to incident investigation. The system allows the basic facts about an incident to be recorded quickly, efficiently, and uniformly.

All incidents should be investigated, regardless of severity of injury or amount of property damage. The extent of the investigation depends on the outcome or potential outcome of the incident. Unless the outcome could have been disabling injury or death, an incident involving only first aid or minor property damage does not need to be investigated as thoroughly as an incident resulting in death or extensive property damage.

For purposes of incident prevention, investigations must be fact finding, not fault finding; otherwise, they can do more harm than good. This is not to say that responsibility should not be assigned when personal failure has caused injury, nor that such persons should be excused from the consequences of their actions. It does mean that the investigation

Incident (No-Injury Accident) Report Form

(This form must be completed IMMEDIATELY after an accident when there is no injury.)

Exact location of incident: _____ Department: _____

Occurrence date: _____ Time: _____ Date reported: _____

Employee involved: _____ SS#/Employee ID: _____

Job title: _____ Employment date: _____ Time on present job: _____

1. Property damaged: _____

Cost: $ _____ Length of downtime: _____

2. Unsafe condition at time of incident (be specific): _____

   _____

   _____

3. Unsafe practice contributing to the incident (be specific): _____

   _____

   _____

4. Witness(es) to incident: _____

5. Sequence of events (detailed): _____

6. What can be done to prevent a recurrence of this incident? _____

   _____

Supervisor: Department: _____ Date: _____

Immediately forward copies of this report to Department Management.

Figure 10–1. This sample form can be used to report incidents that involve no injuries.

itself should be concerned only with facts. The investigating individual, board, or committee should not be involved with any disciplinary actions resulting from the investigation.

Types of Investigation and Analysis

The investigator may use a variety of incident investigation and analysis techniques, some of them more complicated than others. The choice of method depends on the purpose and the orientation of the investigation. The failure mode and effect approach discussed in Chapter 9, Identifying Hazards, may be useful for investigating situations involving large, complex, and interrelated machinery and procedures, but it may be of limited value for investigating incidents involving hand tools. If management procedures and communications and their relationship to incidents are of great interest, the management oversight and risk tree analysis (MORT; see the References at the end of Chapter 9, Identifying Hazards) could prove to be the best choice.

The incident investigation and analysis procedure outlined in this chapter follows the method recommended in ANSI standards Z16.2, Information Management for Occupational Safety and Health, and Z16.5, Occupational Safety and Health Incident Surveillance. These standards were withdrawn for administrative reasons and are no longer active, but their concepts and methods are still valid. Similar techniques involve investigating within the framework of defects in man, machine, media, and management (the four Ms) or the defects in education, enforcement, and engineering (the three Es). For analysis purposes, these techniques involve placing the data about a group of incidents into various categories. This approach has been referred to as the *statistical method of analysis*. Corrective actions are then designed based on the most frequent patterns of occurrence.

Other techniques discussed in Chapter 9, Identifying Hazards, come under the systems approach to safety. Systems safety stresses a broader viewpoint that takes into account interrelationships between various events that could lead to an incident. Because incidents rarely have just one cause, the systems approach to safety can point out more than one place in a system where effective corrective actions can be introduced. This process allows the safety professional to choose the corrective actions that best meet the criteria for effectiveness, rapid implementation, cost/benefit analysis, and the like. Using systems safety techniques has additional advantages: management can implement them before incidents occur and can apply them to new procedures and operations.

Persons Conducting the Investigation

Depending on the nature of the incident and other factors, the investigation is usually conducted by the supervisor. This person can be assisted by a fellow worker familiar with the process involved, a safety professional or inspector, an employee health professional, the joint safety and health committee, the general safety committee, or a consultant from the insurance company. If the incident involved unusual or noteworthy features, consultation with a state labor department or federal agency, a union representative, or an outside expert may be warranted. If a contractor's personnel were involved in the incident, then a contractor's representative should also be involved in the investigation.

The supervisor should make an immediate report of every injury requiring medical treatment and other incidents that he or she may be directed to investigate. The supervisor is on the scene and probably knows more about the incident than anyone else. Therefore, it is up to this individual, in most cases, to put into effect whatever measures can be adopted to prevent similar incidents.

The Safety Professional

Ideally, the safety professional is an adviser and a guide to the supervisor on incident investigations and should verify the supervisor's findings and the adequacy of his or her investigation because sometimes supervisors attempt to cover up a supervisory error. The safety professional should conduct the investigation on his or her own only in serious cases or when the supervisor is not adequately trained in incident investigation.

Special Investigative or Review Committee

In some companies, a special committee is set up to investigate and report on all serious incidents or to review the quality of incident investigations. To be acceptable to all involved, this committee should be composed of representatives of both management and workers. Workers as well as management would be more likely to accept a report published by such a committee than a report published by a safety professional. (See the discussion of safety committees in Chapter 6, Loss Control Programs.)

The Safety and Health Committee

In many organizations, especially those that are small or moderate sized, a number of safety activities, including incident investigation, are handled by a safety and health committee. Ordinarily, such investigations are conducted in a routine manner, but in important cases, the head of the committee might call an extra meeting to initiate a special investigation.

Cases to Be Investigated

In most organizations, the development, implementation, and evaluation of safety and health programs require information on injury/illness incidents well beyond the record-keeping requirements of OSHA and other government agencies. A comprehensive surveillance system gives the safety professional all the information needed to meet government agency record-keeping requirements and also supplies information on events and exposures that could indicate potential or emerging safety and health hazards before they result in recordable cases.

The key concept is that information should be documented about many different kinds of events and exposures in the workplace, all of which are of interest to the safety professional. From this large database of information, the professional can select cases that must be recorded for various government agencies or workers' compensation authorities (Figure 10–2). This chapter defines the input to the database—that is, what should be documented. The output from this database is determined by the recording requirements of the database users. An important advantage of this database

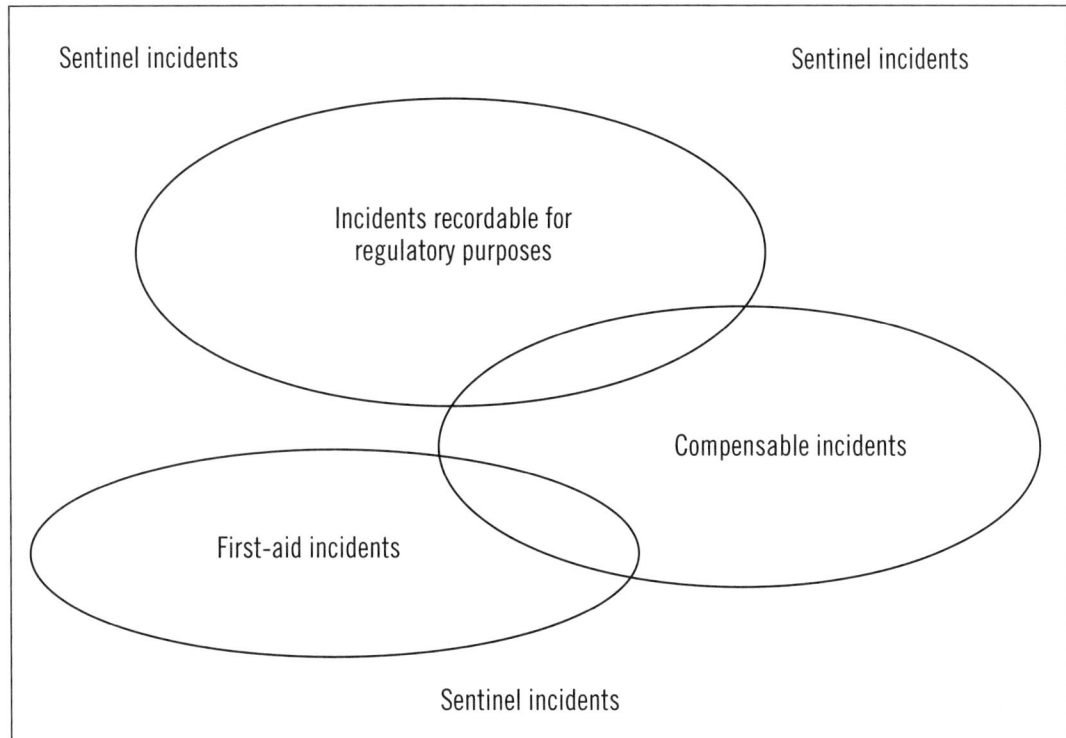

Figure 10–2. Documentable incidents.

is that it puts the employer in control of the data and makes it easier for him or her to respond to the changing data requirements of outside agencies as well as the employer's own needs.

An incident causing death or serious injury obviously should be thoroughly investigated. Also, a sentinel or near-miss incident that might have caused death or serious injury is equally important and should be investigated. Many companies investigate all OSHA-recordable injuries.

Each investigation should be conducted as soon after the incident as possible. A delay of only a few hours may permit important evidence to be destroyed or removed, intentionally or unintentionally. Also, the investigator or committee should present the results of the inquiry as quickly as possible; doing so greatly increases the results' value in the safety education of employees and supervisors.

In addition, any epidemic of minor injuries demands study. A particle of emery in the eye or a scratch from handling sheet metal may seem to be uncomplicated cases: the immediate cause is obvious, and the loss of time is small. However, if these or similar cases occur frequently in the organization or in any one department, they need to be investigated to determine the underlying causes.

The chief value of such an investigation lies in its ability to discover any contributing causes. The safety professional or manager appreciates this type of incident investigation because it can prove more valuable, though less spectacular, to safety efforts than an inquest following a fatal injury.

In any incident investigation, fairness and impartiality are absolutely essential. Otherwise, the value of this tool can be tarnished if employees suspect that its purpose is to place blame or find a scapegoat. No one should be assigned to investigation work unless he or she has a reputation for fairness and is trained and experienced in gathering evidence. The person should clearly understand that incident investigations are conducted solely to obtain information that can help prevent a recurrence of such incidents.

In earlier years, a reduction in incident rates was prompted primarily by humanitarian appeals to management and workers. Now, the most successful methods are aimed at isolating and identifying incident causes in order to permit direct, positive, and corrective actions that will prevent incidents' recurrences.

Like other phases of business management, incident prevention must be based on facts that clearly identify the problem. An approach to incident prevention with this basis not only will result in more effective control over incidents but also will save the organization time, effort, and money. Incident analysis of individual cases identifies the facilities, locations, or departments in which injuries most frequently occur and suggests necessary corrective actions to reduce incidents.

Sometimes, a high incident rate cannot be traced to only one or a few departments but instead represents a high frequency of incidents throughout the facility. Under such circumstances, it is even more important that an analysis of the incidents be made. For example, similar incidents may occur frequently but at widely separated locations, so their high incidence rate is not apparent. Incidents may be more numerous in some machine operations than in others or in certain procedures. Because some unsafe practices that cause incidents may be committed repeatedly but at different times and in different places, their importance as incident causes is not immediately recognized.

Analysis of the circumstances of incidents can produce the following results:

1. Management can identify and locate the principal sources of incidents by determining, from actual experience, the methods, materials, machines, and tools most frequently involved in incidents and the jobs most likely to produce injuries.
2. Investigations may reveal the nature and size of the incident problems in departments and among occupations.
3. Results will indicate the need for engineering revision by identifying the principal hazards associated with various types of equipment and materials.
4. The investigation can uncover inefficiencies in operating processes and procedures such as that poor layout contributes to incidents, or that certain outdated, physically overtaxing methods or procedures should be eliminated.
5. An incident report will reveal unsafe practices that need to be corrected by training employees or changing work methods.
6. By giving supervisors information about the principal hazards and unsafe practices in their departments, the report also will enable supervisors to put their safety work efforts to the best use.
7. Investigation results provide an objective evaluation of a safety program's progress by noting in continuing analyses the effects of corrective actions, educational techniques, and other methods adopted to prevent injuries.

Minimum Data Required

The purpose of an incident investigation is twofold: first, to identify facts about each injury and the incident that produced it and to record those facts; second, to determine a course of action to prevent a recurrence of that incident. In addition, the investigation, instead of focusing solely on the injury and the incident type, should include the entire sequence of events leading to the injury, as far back in time as the investigator feels is relevant. This expanded view of the incident sequence allows an employer to identify and implement a wide variety of corrective actions. Incident investigation records, individually and collectively, serve as guides to the areas, conditions, and circumstances to which incident-prevention efforts can most profitably be directed.

An Incident Investigation Report form is used to help investigators gather, at a minimum, the basic information that should be recorded about each incident (Figure 10–3). Although many report forms are designed mainly for gathering data about injuries, they can also be used to investigate occupational illnesses that arise from a single exposure.

The Incident Investigation Report form provides a minimum data set that can be used to improve the quality of incident investigation and analysis. This minimum data set identifies the *why* of some of the incident characteristics as well as the *who*, *what*, *when*, *where*, and *how*. It acknowledges the existence of multiple causes of incidents by not restricting the investigator or analyst to choosing a single causal act or condition. All questions on the form should be answered as completely and specifically as possible. If there is no answer for a question, or when a question does not apply to the incident, the investigator should indicate this on the form. Supplemental information, such as photographs, drawings, or sketches, should be attached to the report. In a multiple-injury incident, investigators should complete a separate report for each employee who is injured.

The minimum data set that investigators should record includes the following:

- If data will be used to compare one company with another, record data about *employer characteristics*. These characteristics include the type of industry and the size of the company (i.e., the number of full-time-equivalent employees).
- Record *employee characteristics*: the injured employee's age and sex, the department and occupation in which he or she works, and whether he or she is a full-time, part-time, or seasonal employee. Details about the employee's experience are also important. How long has he or she been with the company? How long has he or she been in the current occupation? How long had he or she performed the activity engaged in when the incident occurred? Employee training records may be examined and assessed during this part of the investigation.
- Record the *characteristics of the injury*. Describe exactly the injury or injuries and the part or parts of the body affected by the incident. In the case of an occupational illness, provide the diagnosis and the body part or parts affected. On the form, check the highest degree of severity that applies to the injury.
- Prepare a *narrative description and incident sequence* that provides the exact location of the incident (attach any maps or diagrams to the report); a complete, specific breakdown of the sequence of events leading to the injury or near miss; what objects or substances were involved

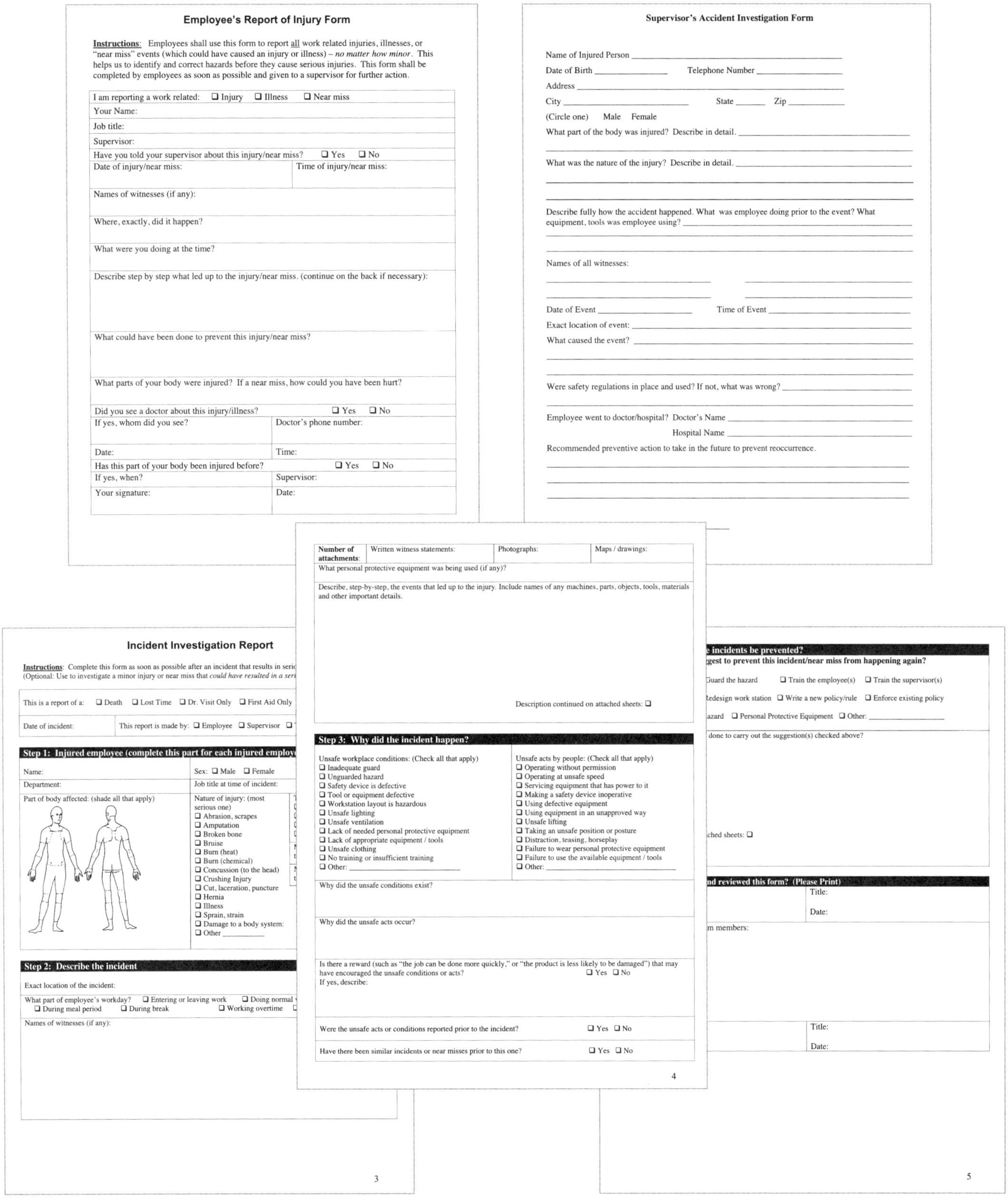

Figure 10–3. These Incident Investigation Report forms can be downloaded from OSHA's website: https://www.osha.gov/dte/grant_materials/fy11/sh-22224-11/3_Accident_Investigation_Form.pdf.

in the incident; the conditions such as temperature, light, noise, and weather pertaining to the incident; how the injury occurred and the specific object or substance that inflicted the injury or was involved in the incident; whether any preventive measures had been in place; and what, if anything, happened after the injury occurred. The investigators should include only the facts obtained during the investigation; they should not record opinions or assign blame.

In most incidents, the incident event and the injury event are different. For instance, an employee might be walking underneath scaffolding when a hammer falls from the platform and strikes the employee on the head. The incident event (hammer falling from the scaffolding) is separate from the injury event (hammer striking the employee's head). Because the events are separate, the form emphasizes this distinction and records other events that led to the incident.

Preceding events can be something that happened that should not have occurred (the hammer falling off the scaffolding) or something that did not happen that should have occurred (no safeguards on the scaffolding platform). To determine whether a preceding event should be included in the Incident Report Form, the investigator should ask whether the event set in motion a sequence of actions that resulted in the incident and the subsequent injury.

- Record the *characteristics of the equipment* associated with the incident. These data should be incorporated into the narrative description on page 4 of the report. Include the type, brand, size, condition, and any distinguishing features of the equipment, and the specific part involved.
- Record the *characteristics of the task* being performed when the incident happened: the general task (such as repairing a conveyor) and the specific activity (such as using a wrench). The description should include the posture and location of the employee (e.g., squatting under the conveyor) and whether the person was working alone or with others.
- Record the *time factors*. The investigator should record the time of day and whether it was the injured employee's first hour of the shift, second hour, or later. Also, what type of shift was it—day, swing, straight, rotating, or other? Other needed information is the phase of the employee's workday: performing work, rest period, mealtime, overtime, or entering or leaving the facility.
- Record *supervision information*. Indicate whether the employee was being supervised directly, indirectly, or not at all at the time of the incident. Also indicate situations in which supervision was not feasible.
- Record the *causal factors*. Record the events and conditions that contributed to the incident. Be as specific and complete as possible.
- Describe the *corrective actions* taken immediately after the incident to prevent a recurrence, including interim or temporary actions.

Investigators may add other information to the report to comply with local or company requirements. These types of data might include the following:

- an estimate or calculation the of costs associated with the incident
- exposure data to be used when calculating incidence rates for injuries associated with certain activities
- management data to be used in performance reviews
- information required for special studies (such as monitoring corrective action)
- information on incident patterns specific to a particular division, company, or industry.

The answers to the questions on this report constitute the minimum information needed to proceed with an analysis. The nature of a company's operations or the interests of the investigator may suggest other questions that should be answered in the investigation.

Two types of analysis can be done. First, the investigator can examine the individual incident to determine the corrective action or actions needed to prevent future occurrences of this specific sequence of events. Second, the investigator can do a statistical analysis and examine a group of similar occurrences for patterns lending themselves to corrective actions. Over time, these statistical analysis can show which corrective actions are more effective than others.

Identifying Causal Factors and Selecting Corrective Actions

In any incident, many factors could have triggered a sequence of events that resulted in an injury. The idea behind selecting corrective actions is to identify all the factors for which a corrective action is possible. Management then selects the corrective actions likely to be most effective, most cost beneficial, most acceptable, and so on, and implements those actions.

Figure 10-4 shows the Guide for Identifying Causal Factors and Corrective Actions. This guide, which should be used with the Incident Investigation Report contains four parts: equipment, environment, people, management, and occupational health. Although these elements usually combine to produce products and profits, at times they result in incidents. The questions in these five sections help the investigator analyze each factor's contribution to the incident.

The structure of the guide makes it easy to identify the causal factors. Questions are answered by placing an X in a circle or in a box. An X in a circle means that the item is a causal factor; an X in a box indicates the item is not a causal factor.

The Comment column provides space to record specific information about the incident being investigated. The Recommended Corrective Actions column has room to write down specific corrective actions for each causal factor. After listing all of the possible corrective actions for the factors identified in the guide and before deciding which ones to implement, each action must be evaluated for its effectiveness, cost, feasibility, reliability, acceptance, effect on productivity, time required to be implemented, and any other component deemed important.

This systematic approach to selecting corrective actions ensures three basic steps: first, that all major actions are considered; second, that the analyst does not stop with familiar and favorite corrective actions; and third, that each corrective action chosen for implementation is carefully thought out.

Classifying Incident Data

Although two incidents rarely happen in exactly the same way, incidents do follow general patterns. Because incidents must be grouped according to patterns for purposes of analysis, finding the patterns and common features of groups of incidents is the statistical approach to incident analysis.

Setting Up Classifications

The Occupational Injury and Illness Classification System (OIICS) provides detailed classification categories and numerical codes for the nature of the injury or illness, the part of the body affected, the source of the injury or illness, the event or exposure, and the secondary source of the injury or illness. This classification system was originally developed in the ANSI Z16 standards and was then adapted and expanded by the Bureau of Labor Statistics for use in its data collection and classification programs. Data coded according to the OIICS are comparable to the benchmarking data published by the Bureau of Labor Statistics and the National Safety Council. The OIICS may be downloaded from the BLS website (bls.gov/iif). For other data elements, classifications must be set up for grouping the various data. For each basic fact, general classifications should be established to group similar data before the actual analysis takes place. For example, among the general classifications for hazardous conditions are the following:

- defects of tools or equipment (which are undesired and unintended characteristics)
- dress or apparel hazards
- environmental hazards
- placement hazards
- inadequate safeguarding
- public hazards.

Within each of these general classifications, more specific classifications are set up. Under "defects of tools or equipment," for example, are the following:

- composed of unsuitable materials
- dull
- improperly constructed or assembled
- improperly designed
- rough
- sharp
- slippery
- worn, cracked, frayed, or broken
- other defects.

It is not always possible to establish classifications before the analysis begins. In this case, management can develop classifications as it reviews reports and notices various incident situations.

For example, if an investigator is analyzing ladder incidents and finds that in a number of cases, broken rungs caused the incidents, a specific category for "broken rungs" should be set up under the classification "defects of tools or equipment."

General and specific classifications should be used for most key elements. For an analysis to be of maximum usefulness, classifications must be set up to encompass the situations pertinent to the particular organization.

Use of a Numerical Code

Regardless of the method eventually used to sort and tabulate the various key elements, the work will be easier if the classifications are assigned numerical codes. A numerical code simply refers to numbers assigned in sequence to a list of similar facts. Each data element should have its own number, but for different elements, the numbering series may be repeated. The OIICS is a good example of how numerical codes may be assigned.

With this numerical code method, the analyst reads each case only once, after which he or she assigns code numbers to the different elements. Subsequent sorting of the various elements, whether by hand or with a computer, can be completed quickly by referring to the code numbers. After the analyst has reviewed the cases and assigned code numbers to the different key elements, he or she can easily and quickly sort or arrange the reports by any of the elements to reveal the principal data involved in all the incidents.

Conducting the Analysis

Experience has proved that the most effective way to reduce incidents is to concentrate on the primary causes of one phase

Guide for Identifying Causal Factors and Corrective Actions

Case Number _____

Answer questions by placing an X in the **"Y"** circle or box for yes or in the **"N"** circle or box for no. A marked **"Y"** circle indicates a possible causal factor.

PART 1 EQUIPMENT

○ □ **1.0 Was a hazardous condition(s) a contributing factor?** If no, proceed to Part 2.
Y N

	Causal Factors	Comment	Possible Corrective Actions	Recommended Corrective Actions
○ □ Y N	1.1 Did any defect(s) in equipment/tool(s)/material contribute to hazardous condition(s)?		Review procedures for inspecting, reporting, maintaining, repairing, replacing, or recalling defective equipment/tool(s)/material used.	
□ ○ Y N	1.2 Was the hazardous condition(s) recognized? If yes, answer A and B. If no, proceed to 1.3.		Perform job safety analysis. Improve employee ability to recognize existing or potential hazardous conditions. Provide test equipment, as required, to detect hazard. Review any change or modification of equipment/tool(s)/material.	
□ ○ Y N	A. Was the hazardous condition(s) reported?		Train employees in reporting procedures. Stress individual acceptance of responsibility.	
□ ○ Y N	B. Was the employee(s) informed of the hazardous condition(s) and the job procedures for dealing with it as an interim measure?		Review job procedures for hazard avoidance. Review supervisory responsibility. Improve supervisor/employee communications. Take action to remove or minimize hazard.	
□ ○ Y N	1.3 Was there an equipment inspection procedure(s) to detect the hazardous condition(s)?		Develop and adopt procedures to detect hazardous conditions. Conduct test.	
□ ○ Y N	1.4 Did the existing equipment inspection procedure(s) detect the hazardous condition(s)?		Review procedures. Change frequency or comprehensiveness. Provide test equipment as required. Improve employee ability to detect defects and hazardous conditions. Change job procedures as required.	
□ ○ Y N	1.5 Was the correct equipment/tool(s)/material used?		Specify correct equipment/tool(s)/material in job procedure.	
□ ○ Y N	1.6 Was the correct equipment/tool(s)/material readily available?		Provide correct equipment/tool(s)/material. Review purchasing specifications and procedures. Anticipate future requirements.	
□ ○ Y N	1.7 Did employee(s) know where to obtain equipment/tool(s)/material required for the job?		Review procedures for storage, access, delivery, or distribution. Review job procedures for obtaining equipment/tool(s)/material.	

Figure 10–4. The Guide for Identifying Causal Factors and Corrective Actions helps the incident investigator consider systematically five contributing factors: equipment, environment, people, management, and occupational health.

10 Incident Investigation and Analysis 247

	Causal Factors	Comment	Possible Corrective Actions	Recommended Corrective Actions
○ □ Y N	1.8 Was substitute equipment/tool(s)/material used in place of the correct one?		Provide correct equipment/tool(s)/material. In job procedures and in job instruction, warn against use of substitutes.	
○ □ Y N	1.9 Did the design of the equipment/tool(s) create operator stress or encourage operator error?		Review human factors engineering principles. Alter equipment/tool(s) to make it more compatible with human capability and limitations. Review purchasing procedures and specifications. Check out new equipment and job procedures involving new equipment before putting into service. Encourage employees to report potential hazardous conditions created by equipment design.	
○ □ Y N	1.10 Did the general design or quality of the equipment/tool(s) contribute to a hazardous condition?		Review criteria in codes, standards, specifications, and regulations. Establish new criteria as needed.	
○	1.11 List other causal factors in "Comment" column.			
PART 2 ENVIRONMENT				
○ □ Y N	2.0 Was the location of equipment/material/employee(s) a contributing factor? If no, proceed to Part 3.			

	Causal Factors	Comment	Possible Corrective Actions	Recommended Corrective Actions
○ □ Y N	2.1 Did the location/position of equipment/material/employee(s) contribute to a hazardous condition?		Perform job safety analysis. Review job procedures. Change the location, position, or layout of the equipment. Change position of employee(s). Provide guardrails, barricades, barriers, warning lights, signs, or signals.	
□ ○ Y N	2.2 Was the hazardous condition recognized? If yes, answer A and B. If no, proceed to 2.3.		Perform job safety analysis. Improve employee ability to recognize existing or potential hazardous conditions. Provide test equipment, as required, to detect hazard. Review any change or modification of equipment/tool(s)/materials.	
□ ○ Y N	A. Was the hazardous condition reported?		Train employees in reporting procedures. Stress individual acceptance of responsibility.	
□ ○ Y N	B. Was employee(s) informed of the job procedures for dealing with the hazardous condition as an interim action?		Review job procedures for hazard avoidance. Review supervisory responsibility. Improve employee/supervisor communications. Take action to remove or minimize hazard.	
□ ○ Y N	2.3 Was employee(s) supposed to be in the vicinity of the equipment/material?		Review job procedures and instructions. Provide guardrails, barricades, barriers, warning lights, signs, or signals.	

Figure 10–4. Continued.

	Causal Factors	Comment	Possible Corrective Actions	Recommended Corrective Actions
☐ ○ Y N	2.4 Was the hazardous condition created by the location/position of equipment/material visible to employee(s)?		Change lighting or layout to increase visibility of equipment. Provide guardrails, barricades, barriers, warning lights, signs or signals, or floor stripes.	
☐ ○ Y N	2.5 Was there sufficient work space?		Review work space requirements and modify as required.	
○ ☐ Y N	2.6 Were environmental conditions (illumination, noise levels, air contaminants, temperature extremes, ventilation, vibration, or radiation) contributing factors?		Monitor, or periodically check, environmental conditions as required. Check results against acceptable levels. Initiate action for those levels found unacceptable.	
○	2.7 List other causal factors in "Comment" column.			

PART 3 PEOPLE

○ ☐ Y N	**3.0 Was the job procedure(s) a contributing factor?** If no, proceed to Part 3.6.			

	Causal Factors	Comment	Possible Corrective Actions	Recommended Corrective Actions
☐ ○ Y N	3.1 Was there a written or known procedure for this job? If yes, answer A, B, and C. If no, proceed to 3.2.		Perform job safety analysis and develop safe job procedures.	
☐ ○ Y N	A. Did job procedures anticipate the factors that contributed to the accident?		Perform job safety analysis and change job procedures.	
☐ ○ Y N	B. Did employee(s) know the job procedure?		Improve job instruction. Train employees in correct job procedures.	
○ ☐ Y N	C. Did employee(s) deviate from the known job procedure?		Determine why. Encourage all employees to report problems with an established procedure to supervisor. Review job procedure and modify if necessary. Counsel or discipline employee. Provide closer supervision.	
☐ ○ Y N	3.2 Was employee(s) mentally and physically capable of performing the job?		Review employee requirements for the job. Improve employee selection. Remove or transfer employees who are temporarily either mentally or physically incapable of performing the job.	
○ ☐ Y N	3.3 Were any tasks in the job procedure too difficult to perform (requiring excessive concentration or physical demands)?		Change job design and procedures.	
○ ☐ Y N	3.4 Is the job structured to encourage or require deviation from job procedures (incentive, piecework, work pace)?		Change job design and procedures.	

Figure 10–4. Continued.

	Causal Factors	Comment	Possible Corrective Actions	Recommended Corrective Actions
○	3.5 List other causal factors in "Comment" column.			
○ □ Y N	**3.6 Was lack of personal protective equipment or emergency equipment a contributing factor in the injury?** If no, proceed to Part 4. Note: The following causal factors relate to the *injury*.			

	Causal Factors	Comment	Possible Corrective Actions	Recommended Corrective Actions
□ ○ Y N	3.7 Was appropriate personal protective equipment specified for the task or job? If yes, answer A, B, and C. If no, proceed to 3.8.		Review methods to specify PPE requirements.	
□ ○ Y N	A. Was appropriate PPE available?		Provide appropriate PPE. Review purchasing and distribution procedures.	
□ ○ Y N	B. Did employee(s) know that wearing specified PPE was required?		Review job procedures. Improve job instruction.	
□ ○ Y N	C. Did employee(s) know how to use and maintain the PPE?		Improve job instruction.	
□ ○ Y N	3.8 Was the PPE used properly when the injury occurred?		Determine why and take appropriate action. Implement procedures to monitor and enforce use of PPE.	
□ ○ Y N	3.9 Was the PPE adequate?		Review PPE requirements. Check standards, specifications, and certification of the PPE.	
□ ○ Y N	3.10 Was emergency equipment (emergency showers, eyewash fountains) specified for this job? If yes, answer A, B, and C. If no, proceed to Part 4.		Provide emergency equipment as required.	
□ ○ Y N	A. Was emergency equipment readily available?		Install emergency equipment at appropriate locations.	
□ ○ Y N	B. Was emergency equipment properly used?		Incorporate use of emergency equipment in job procedures.	
□ ○ Y N	C. Did emergency equipment function properly?		Establish inspection/monitoring system for emergency equipment. Provide for immediate repair of defects.	
○	3.11 List other causal factors in "Comment" column.			

Figure 10–4. Continued.

PART 4 MANAGEMENT

☐ Y ☐ N **4.0 Was a management system defect a contributing factor?** If no, proceed to Part 5.

	Causal Factors	Comment	Possible Corrective Actions	Recommended Corrective Actions
☐ Y ☐ N	4.1 Was there a failure by supervisor to detect, anticipate, or report a hazardous condition?		Improve supervisor capability in hazard recognition and reporting procedures.	
☐ Y ☐ N	4.2 Was there a failure by supervisor to detect or correct deviations from the job procedure?		Review job safety analysis and job procedures. Increase supervisor monitoring. Correct deviations.	
☐ Y ☐ N	4.3 Was there a supervisor/employee review of hazards and job procedures for tasks performed infrequently? (Not applicable to all incidents.)		Establish a procedure that requires a review of hazards and job procedures (preventative actions) for tasks performed infrequently.	
☐ Y ☐ N	4.4 Was supervisor responsibility and accountability adequately defined and understood?		Define and communicate supervisor responsibility and accountability. Test for understanding and acceptance.	
☐ Y ☐ N	4.5 Was supervisor adequately trained to fulfill assigned responsibility in accident prevention?		Train supervisors in accident prevention fundamentals.	
☐ Y ☐ N	4.6 Was there a failure to initiate corrective action for a known hazardous condition that contributed to this incident?		Review management safety policy and level of risk acceptance. Establish priorities based on potential severity and probability of recurrence. Review procedure and responsibility to initiate and carry out corrective actions. Monitor progress.	
☐	4.7 List other causal factors in "Comment" column.			

PART 5 OCCUPATIONAL HEALTH — SUPPLEMENTAL INFORMATION

☐ Y ☐ N **5.0 Was an adverse occupational health environment a potential contributing factor?** If yes, identify adverse environments below and elaborate in "Comment" column. If no, form is complete.

	Causal Factors	Comment
☐ Y ☐ N	5.1 Physical agent. If yes, check applicable causal factor and explain in "Comment" column.	
☐	Noise, vibration	
☐	Temperature extremes	
☐	Ionizing radiation (X, gamma, beta, and alpha rays)	
☐	Nonionizing radiation (microwaves, lasers, ultraviolet rays, infrared radiation, RF)	

Figure 10–4. Continued.

	Causal Factors	Comment
○	Ergonomic (repetitive motion trauma, lighting, glare, incorrect or insufficient tooling)	
○	Other Type of agent: _____	
○ □ Y N	5.2 Chemical agent. If yes, check applicable causal factor, name the chemical agent, and explain in "Comment" column.	
○	Solvents Name: _____	
○	Acids, bases Name: _____	
○	Laboratory reagents Name: _____	
○	Other toxic chemicals Name: _____	
○	Unknown or combination	
○ □ Y N	5.3 Biological agent. If yes, check applicable causal factor, name the biological agent, and explain in "Comment" column.	
○	Microorganism Name: _____	
○	Insect Name: _____	
○	Animal Name: _____	
○ □ Y N	5.4 Medical problem. If yes, explain in "Comment" column.	

Figure 10–4. Concluded.

of the incident problem at a time rather than attempt to prevent all incidents at once. The problem can be approached in different ways, any one of which should prove effective.

The analyst may group reports by the occupations of the injured persons. The analyst would review each group of reports and determine the most prevalent incident types and sources of injury among different occupations. Such information is particularly helpful when planning process or equipment changes, implementing employee training, and developing educational materials and programs.

Injury incidence rates may reveal that injuries occur at significantly higher rates in some departments than in others. If this is the case, the analyst can examine the incident reports in the high-injury-rate departments to find the sources of the incidents and their causes. This method enables management to concentrate efforts on the locations where the most incidents occur.

If injury incidence rates reveal a high rate of occurrence throughout the organization, analysis usually starts with information about the injury, identifies the injury-producing event, and then looks at the circumstances and causal factors. The same procedure can be followed to examine injuries occurring within a high-injury-rate occupation or department.

The analyst can crosstabulate the injury data to show the relationship or interaction between the two categories. Table 10–A illustrates an analysis of the nature of the injury versus the part of the body. This crosstabulation, in addition to suggesting what types of personal protective equipment might be useful, also points out common injury patterns needing further investigation.

A crosstabulation can extend in several directions and produce the need for further, separate crosstabulations. For example, the analyst would use the categories in Table 10–A showing the highest frequency of injuries ("Cut, laceration, puncture, or abrasion to fingers" and "Sprain, strain" of the back) to construct a second (Table 10–B) and a third crosstabulation.

As shown in Table 10–B, by the second tabulation the number of cases in a category usually is small enough that the analyst can read individual incident reports to find common causal factors and determine corrective actions. If the number of cases in a category is still too large, then additional crosstabulations, such as location versus activity or activity versus occupation, can reduce the number enough to allow study of individual incident reports.

The analyst should choose a method of tabulating that is appropriate for the number of reports generated and how the data will be used. For analyzing a small number of reports (up to about 100), hand sorting and hand tallying are effective. This method's principal advantage is that the analyst is using original records and has all the information available for reference.

A computer is best for tabulating large collections of cases. Computers are also useful for tabulating small data sets because they can sort and display cases quickly. Computers' efficiency allows the investigator to concentrate on various incidents and to test alternative hypotheses easily.

Using the Analysis

Merely obtaining the information will not prevent recurrence of incidents. Management must correct the underlying and contributing conditions identified in the information. Thorough analysis of groups of incident investigation reports can point to corrective actions that might not be evident when studying an individual case. In particular, inadequate policies, procedures, or management systems are often apparent after taking in the whole picture.

The statistical evidence revealed by an analysis can guide safety efforts along the most effective path. The analysis will also provide objective support and justification for budget requests, training programs, and other management safety activities.

Involving Outside Employees and Contractors

Owners and employers must also recognize the importance of involving outside employees and contractors in incident investigation and analysis. (For a fuller discussion of this topic, see Chapter 28, Contractor and Customer Safety, in this volume.) This involvement should begin before outside workers arrive on the job site. Employers should insist on the following basic safety requirements, which can be written into a contract or stated in a written policy that is agreed upon by both parties.

TABLE 10–A. Nature of Injury versus Part of Body

Nature of Injury	Part of Body											
	Eyes	Head, Face, Neck	Back	Trunk	Arm	Hand, Wrist	Finger	Leg	Foot, Ankle	Toe	Internal, Other	Total
Amputation	0	0	0	0	0	0	1	0	0	0	0	1
Bruise, contusion	0	2	0	1	0	3	2	3	0	1	0	12
Burn & scald (heat)	0	0	0	0	0	0	0	0	0	0	0	0
Burn (chemical)	2	0	0	0	0	0	0	0	0	0	0	2
Concussion	0	2	0	0	0	0	0	0	0	0	0	2
Crushing	0	0	0	0	0	1	0	0	0	2	0	3
Cut, laceration, puncture, abrasion	0	1	0	0	1	3	18	0	0	0	0	23
Fracture	0	0	0	0	0	2	5	0	1	0	0	8
Hernia	0	0	0	1	0	0	0	0	0	0	0	1
Occupational illness	0	1	0	0	1	2	0	0	0	0	8	12
Sprain, strain	0	0	24	0	0	2	2	3	4	0	0	35
Other	1	0	0	0	0	0	0	0	0	0	5	6
Total	3	6	24	2	2	13	28	6	5	3	13	105

This crosstabulation shows how the nature of the injury and the part of the body interact. In this example, cuts most often affect the fingers, and sprains and strains usually involve the back. Note that bruises and contusions affect several body parts.

TABLE 10–B. Source of Injury versus Type of Accident

Source of Injury	Contact with Electrical Current	Contact with Temperature Extremes	Radiations, Caustics, Toxic and Noxious Substances	Public Transportation Accident	Motor Vehicle Accident	Other	Unknown	Total	Fall from Elevation	Fall on Same Level	Struck against	Struck by	Caught in, under, or between	Rubbed or Abraded	Bodily Reaction	Over-exertion
Machine	0	0	0	0	0	2	1	3	0	0	0	0	3	0	0	0
Conveyor, elev. hoist	0	0	0	0	0	0	0	0	0	0	0	0	0	0	0	0
Vehicle	0	0	0	0	0	0	0	0	0	0	0	0	0	0	0	0
Electrical apparatus	2	0	0	0	0	2	0	4	0	0	0	0	0	0	0	0
Hand tool	0	0	0	0	0	0	0	0	0	0	0	4	0	0	0	0
Chemical	0	0	0	0	0	0	0	0	0	0	0	0	0	0	0	0
Working surface, bench, etc.	0	0	0	0	0	0	0	0	0	0	0	0	0	0	0	0
Floor, walking surface	0	0	0	0	0	0	0	0	0	0	0	0	0	0	0	0
Bricks, rocks, stones	0	0	0	0	0	0	0	0	0	0	0	0	0	0	0	0
Box, barrel, container	0	0	0	0	0	0	0	0	0	0	0	0	0	0	0	0
Door, window, etc.	0	0	0	0	0	0	0	0	0	0	0	0	0	0	0	0
Ladder	0	0	0	0	0	0	0	0	0	0	0	0	0	0	0	0
Lumber, woodworking metals	0	0	0	0	0	0	0	0	0	0	0	0	0	0	0	0
Metal	2	3	0	0	0	4	0	9	0	0	0	9	0	0	0	0
Stairway, steps	0	0	0	0	0	0	0	0	0	0	0	0	0	0	0	0
Other	0	0	0	0	0	0	2	2	0	0	0	0	0	0	0	0
Unknown	0	0	0	0	0	0	0	0	0	0	0	0	0	0	0	0
None	0	0	0	0	0	0	0	0	0	0	0	0	0	0	0	0
Total	4	3	0	0	0	8	3	18	0	0	0	13	3	0	0	0

- Employers should require the use of a system of permits for potentially hazardous activities. This policy ensures that workers have the skills to perform their jobs; understand the hazards of any equipment, chemicals, or processes they use; and know basic safety and emergency measures to follow.
- Employers should require outside employees and contractors to designate a supervisor to be responsible for coordinating safety activities on the job. The supervisor can ensure that workers know and comply with all company and regulatory safety standards that apply to their work. The supervisor should also inform workers about new hazards or changing work conditions that may present additional risks. In addition, this person should report any incidents to the employer and assist in the investigation and analysis of worker injuries or illnesses.
- Employers should provide outside employees and contractors with safety guidelines that personnel must follow. All guidelines for work practices and procedures, use of personal protective equipment, and safe handling and disposal of materials and chemicals should be clearly stated in writing and communicated to all workers.

By including safety as part of the job requirements, employers can establish a spirit of cooperation between their own management and outside employees. This type of cooperation tends to facilitate incident investigation and analysis, and reduce downtime, help to uncover the causes of incidents more quickly, find the best solutions to prevent their recurrence, and minimize the costs associated with incidents.

ESTIMATING INCIDENT COSTS

This discussion concerns the elements of cost most likely to result from a work incident and presents a method whereby an organization can obtain an accurate estimate of the total costs of its work incidents.[1]

Reliable cost information is one basis for making decisions upon which efficiency and profit depend. Even in so obviously desirable an activity as incident prevention, some proposed measures or alternatives must be evaluated on the basis of their potential effects on profits.

Although most executives want to make their company a safe place to work, they also have a responsibility to run their business profitably. Consequently, they may be reluctant to spend money for incident prevention unless they can see the possibility of saving at least as much as they spend.

Estimates of total costs of injuries to the organization may be used in cost/benefit and return on investment (ROI) analyses to justify incident prevention. Without information on the costs of incidents, it is practically impossible to estimate the savings brought about by expenditures for incident prevention.

Reports stressing monetary costs and savings are more meaningful to management than are reports presenting incidence rates. Facts about the costs of incidents also may be used effectively to secure the engaged cooperation of supervisors. Supervisors usually are cost conscious because they are expected to run their departments profitably. Monthly reports showing the costs of incidents or the savings resulting from good incident records can motivate supervisors to ensure that their workers use safe operating procedures.

Definition of Work Incidents for Cost Analysis

Work incidents, for the purpose of cost analysis, are unintended occurrences arising in the work environment. These incidents fall into two general categories: (1) incidents resulting in work injuries or illnesses and (2) incidents causing property damage or interfering with production. The inclusion of no-injury incidents makes "work incidents" roughly synonymous with the type of occurrences a safety department strives to prevent.

Method for Estimating

To be of maximum usefulness, cost figures should represent as accurately as possible the specific experience of a company. It is not useful to have a fixed ratio of uninsured to insured costs that represents many different organizations in many different industries. Estimated costs of incidents in general do not take into account the differences in hazards from one industry to another or the more important differences in safety performance from one company to another.

Because the distinctions between direct and indirect costs are difficult to maintain, the terms have been abandoned in favor of the more precise terms *insured costs* and *uninsured costs*. Using these data, a company can estimate its incident costs with reasonable accuracy.

Insured Costs

Every organization paying workers' compensation insurance premiums recognizes such expense as part of the cost of incidents. In some cases, medical expenses, too, may be covered by insurance. These costs are definite and known and constitute the insured element of the total incident cost.

In addition to these costs, many other costs arise in connection with incidents. Although the expense of damaged equipment is easily identified, other expenses, such as wages paid to the injured employee for rest time on the day of

---

1. This procedure for estimating costs was developed originally by Rollin H. Simonds, PhD, Professor, Michigan State University, under the direction of the Statistics Division, National Safety Council. It has since been modified to be compatible with OSHA record-keeping requirements.

the injury, are not as obvious. These expenses and others constitute the uninsured element of the total incident cost.

Uninsured Costs

Although insured costs can be determined easily from accounting records, uninsured (frequently called "indirect") costs are more difficult to assess. The method described here is one way to calculate uninsured expenses associated with many incidents. The first step is to conduct a pilot study to ascertain approximate averages of uninsured costs in each of the following four classes of incidents:

Class 1—OSHA-recordable cases with lost workdays (days away from work, days of restricted work activity, or both)

Class 2—OSHA-recordable cases without lost workdays (mostly medical treatment cases)

Class 3—Injuries not OSHA recordable (mostly first-aid cases) and less than $100 of property damage

Class 4—Incidents involving no injury or injuries not OSHA recordable (mostly first-aid cases) and more than $100 of property damage

Once average costs have been established for each incident class, they can be used as multipliers to obtain total uninsured costs in subsequent periods (after adjusting for inflation). These costs are then added to known insurance premium costs to determine the total cost of incidents.

Example of a Cost Estimate

An estimate of costs made by one company is given in the following example. First, a pilot study was conducted to obtain the average cost of each class of incident. Included in the study were 20 Class 1 incidents, 30 Class 2 incidents, 50 Class 3 incidents, and 20 Class 4 incidents. Costs were determined and averages developed as shown in Table 10–C.

TABLE 10–C. Average Cost Determined by Pilot Study

Class of Accident	Number of Accidents Reported	Average Uninsured Cost
Class 1	20	$251.10
Class 2	30	80.80
Class 3	50	15.70
Class 4	20	507.10

During the entire year, the company had 34 Class 1 incidents, 148 Class 2 incidents, and 4,000 Class 3 incidents. No record was kept of the Class 4 incidents after the pilot study was completed. Instead, the ratio of the number of Class 4 to Class 1 incidents found in the pilot study was used. This ratio was shown to be about 1 to 1, and because there were 34 Class 1 incidents during the year, it was assumed there were about 34 Class 4 incidents. (A separate record could be kept of the number of Class 4 incidents.)

The average cost for each incident class was applied to these totals to secure the results shown in Table 10–D.

TABLE 10–D. Estimate of Yearly Accident Costs

Class of Accident	Number of Accidents	Average Cost per Accident (from pilot study)	Total Uninsured Cost
Class 1	34	$251.10	$8,537.40
Class 2	148	80.80	11,958.40
Class 3	4,000	15.70	62,800.00
Class 4	34	507.10	17,241.40
Total Uninsured Cost			$100,537.20
Insurance Premiums			54,400.00
Total Accident Cost for the Period			$154,937.20

Because the final total is the sum of many estimates, it should not be assumed that the total figure has absolute accuracy. The estimate should be rounded to three significant digits—in this case, to the nearest thousand dollars. As a result, in this instance, the analyst reported to the facility manager, "During the past year, incidents cost this company about $155,000 in compensation, medical expenses, lost time, and property damage."

The average costs determined in this pilot study represent the actual experience of this particular organization. Until important changes take place in this company's safety program, in the kind of machinery used or workers employed, or in other aspects affecting costs, the same average costs can be used.

Adjusting for Inflation

The effects of inflation can quickly render obsolete the cost figures in a pilot study. To account for these effects, the cost factors should be adjusted for inflation each year. The wage-related cost elements can be multiplied by the change in the general level of wages in the company. Other cost elements can be brought up to date by multiplying them by the general inflation rate as measured by the change in the Consumer Price Index.[2] Because these adjustments are only approximate, the pilot study should be repeated at least every 5 years to establish new benchmarks.

Items of Uninsured Cost

Important to a pilot study is a careful investigation of each incident to determine all the costs arising out of it. The following uninsured costs are clearly the result of work incidents and are subject to reasonably reliable measurement. Less tan-

---

2. Generally, use the CPI-U as reported by the Bureau of Labor Statistics (bls.gov).

gible losses, such as the effects of incidents on public relations, employee morale, or the wage rates necessary to secure and retain employees, are not included in this method of estimating costs but can be important factors in some cases.

Data on costs may be collected by using the Investigator's Cost Data Sheet and Summary Report (Figure 10–5) or the Injury Cost Capturing Tool, a software program developed and published by the National Safety Council.

1. *Cost of wages paid for time lost by workers who were not injured*—These are employees who stopped working to watch or assist after the incident or to talk about it, or who lost time because they needed the equipment that was damaged in the incident or needed the output or the aid of the injured worker.
2. *Cost of wages paid to supervisors for time spent on activities concerning the incident*—The most satisfactory way to estimate this cost is to compute the wages paid to the supervisor during the time spent away from normal activities as a result of the incident.
3. *Cost of time spent by management and clerical workers on investigations or in the processing of compensation application forms*—Time spent by managers or supervisors (other than the supervisor covered in item 2) and by clerical employees for investigating an incident, or for settling claims arising from it, is chargeable to the incident.
4. *Cost of wages paid for time lost by the injured worker, other than workers' compensation payments*—Payments made under workers' compensation laws for time lost after the waiting period are not included in this element of cost.
5. *Wage cost resulting from decreased output of injured worker after returning to work if paid at old rate*—If the injured worker's previous wage payments are continued despite a 40% reduction in the worker's output, the incident should be charged with 40% of the worker's wages during the period of reduced output.
6. *Uninsured medical cost borne by the company*—This cost is usually for the medical services provided at the company dispensary. Estimating an average cost per visit for this medical service is simple. The question may be raised, however, whether this expense can be considered variable. That is, would a reduction in incidents result in lower expenses for operating the dispensary?
7. *Cost of learning period of new worker (temporary or permanent replacement)*—If a replacement worker produces, for the same pay, only half as much in the first 2 weeks as the injured worker would have produced in the same time then half of the new worker's wages for the 2 weeks should be considered part of the cost of the incident. A wage cost for the time spent by supervisors or others in training the new worker also should be attributed to the incident.
8. *Extra cost of overtime work necessitated by the incident*—The charge against an incident for overtime work is the difference between normal wages and overtime wages for the time needed to make up lost production, and the cost of extra supervision, heat, light, cleaning, and other additional services.
9. *Nature and cost of damage to material or equipment*—The validity of property damage as a cost can scarcely be questioned. Occasionally, there is no property damage, but a substantial cost is still incurred for reorganizing material or equipment. The charge should be confined either to the net cost of repairing or reorganizing the damaged or displaced material or equipment or to the current worth of the equipment, less salvage value, if damaged beyond repair. An estimate of property damage should have the approval of the cost accountant, particularly when the current worth of the damaged property differs from the depreciated value established by the accounting department.
10. *Miscellaneous costs*—This category includes less typical costs, the validity of which must be clearly indicated by the investigator on individual incident reports. Among such possible costs are public liability claims, equipment rentals, losses due to canceled contracts or lost orders if the incident caused an overall reduction in total sales, loss of company bonuses, costs of hiring new employees if this expense is significant, cost of above-normal spoilage by new employees, and demurrage. These cost factors and any others not given above need to be well substantiated. Miscellaneous costs were found in less than 2% of the several hundred cases reviewed in connection with this study.

Conducting a Pilot Study

The purpose of the pilot study is to develop for different classes of incidents average uninsured costs that can be applied to future incident totals. Therefore, it is not desirable to include the costs of deaths and permanent total disabilities. Such incidents occur so seldom that the costs should be calculated individually and not estimated based on averages.

The following discussion assumes that the study of costs is made with the injuries grouped in the recommended classes. A different method must be applied to Class 3 injuries, which it will be discussed shortly.

Classes 1, 2, and 4 Incidents

To analyze uninsured costs for incidents in Classes 1, 2, and 4, the supervisor in charge of the department where an incident occurred should secure for each incident the information requested on the Investigator's Cost Data Sheet and Summary Report (Figure 10–5). These data can be

Investigator's Cost Data Sheet and Summary Report

Company name _____

Incident number _____

Date of incident _____

Name of injured or involved worker _____

Department _____ Occupation or job class _____

Incident Class
Class 1 _____
Class 2 _____
Class 3 _____
Class 4 _____

Hourly wage of injured or involved worker $ _____
Hourly wage of supervisor $ _____
Average hourly wage in department $ _____
Average hourly wage of clerical staff $ _____
Average hourly salary of management staff $ _____

Uninsured Costs

1. Cost of wages paid for time lost by workers who were not injured.
   a. Total time lost by employees not involved in the incident by talking, watching, or helping, if paid by employer: _____ hours (or, number of workers × average time lost per worker)
   b. Total time lost by workers because they lacked the equipment damaged in the incident or because they needed the output or aid of the injured worker: _____ (or, number of workers × average time lost per worker)
   c. Cost = (a + b) × average hourly wage in department . . . . (1) $ _____

2. Cost of wages paid to supervisors for time spent on activities concerning the incident.
   a. Supervisor's time lost at time of the incident: _____ hours
   b. Supervisor's time required later: _____ hours
   c. Cost = (a + b) × hourly wage of supervisor . . . . (2) $ _____

3. Cost of time spent by management and clerical workers on investigations or in the processing of compensation application forms.
   a. Time spent by management: _____ hours
   b. Cost of management time = (a) × hourly salary of management = $ _____
   c. Time spent by clerical staff: _____ hours
   d. Cost of clerical time = (c) × hourly wage of clerks = $ _____
   e. Total cost of management and clerical time = (b + d) . . . . (3) $ _____

4. Cost of wages paid for time lost by the injured worker, other than workers' compensation payments.
   a. Time lost on day of injury for which worker was paid: _____ hours
   b. Number of subsequent day's absence for which worker was paid: _____ days
   c. Multiply days (b) times hours per day to obtain hours lost: _____ hours
   d. Number of additional trips for medical attention on employer's time on succeeding days after worker's return to work: _____ trips
   e. Average trip time: _____ hours
   f. Total trip time = (d) × (e) = _____ hours
   g. Additional time lost by worker for which he or she was paid by company: _____ hours
   h. Total cost = (a + c + f + g) × worker's hourly wage . . . . (4) $ _____

Figure 10–5. Investigator's Cost Data Sheet and Summary Report.

5. Wage cost resulting from decreased output of injured worker after returning to work if paid at old rate.
    a. Number of days on moderate work or at reduced output: _____ days
    b. Multiply days (a) times hours per day to obtain hours lost: _____ hours
    c. Worker's average percentage of normal output during this period: _____ %
    d. Total cost = (b) × (c) × worker's hourly wage .... (5) $_____
6. Uninsured medical cost borne by the company.
    (May include first aid, ambulance costs, etc.) .... (6) $_____
7. Cost of learning period of new worker (temporary or permanent replacement).
    (Include hiring costs in item 10.)
    a. Hours spent in training: _____ hours
    b. Time new worker's output was below normal: _____ hours
    c. Average percentage of normal output during this time: _____ %
    d. Total cost = [(a) × worker's hourly wage] + [(a) × trainer's hourly wage] +
        [(b) × (c) × worker's hourly wage]) .... (7) $_____
8. Extra cost of overtime work necessitated by the incident.
    a. Estimated overtime hours of production workers: _____ hours
    b. Wage rate difference for overtime: $_____ per hour
    c. Other overtime costs such as extra supervision, light, heat, etc.: $_____
    d. Total cost = (a × b) + (c) .... (8) $_____
9. Nature and cost of damage to material or equipment.
    a. Describe damage to material or equipment: _____
        _____
        _____
    b. Net cost to repair, replace, or reorganize the above material or equipment .... (9) $_____
10. Miscellaneous costs.
    a. Liability claims $_____
    b. Equipment rentals $_____
    c. Lost contracts or orders $_____
    d. Lost bonuses $_____
    e. Costs of hiring new employee(s) (training costs are in item 7) $_____
    f. Costs of excessive spoilage by new employees $_____
    g. Demurrage $_____
    h. Other costs (explain fully) $_____
    i. Total costs = (a + b + c + d + e + f + g + h) .... (10) $_____
11. **Total Uninsured Costs** (Sum of items 1–10) .... (11) $_____

Insured Costs

12. Workers' compensation indemnity payments .... (12) $_____
13. Workers' compensation medical expense payments .... (13) $_____
14. **Total Insured Costs** (Sum of items 12 and 13) .... (14) $_____
15. **Total Uninsured and Insured Costs** (Sum of items 11 and 14) .... $_____

Figure 10–5. Concluded.

obtained during the supervisor's standard investigation of the incident. As soon as each report is completed, it should be sent to the safety department.

The safety department then assumes the responsibility for securing supplemental information from the accounting department, the industrial relations department, and the other departments where records on lost time and other necessary information are kept.

As an alternative, a member of the safety department could secure all information needed on the data sheet. In this case, the investigator's report form is not used, and the supervisor is required only to report each incident in Class 1, 2, or 4 to the safety department as soon as the incident occurs.

The investigator, before computing averages, should be certain that the pilot study included a sufficient number of cases of Class 1, 2, and 4 incidents to be representative. This number will rarely be fewer than 20 cases. However, more cases should be studied when the costs of the cases in a particular class vary widely. Information should be secured on enough cases of each class so that the average cost per case in each class is fully representative of past experiences and is, by inference, applicable to future experiences.

Once a sufficient number of cases have been accumulated, the investigation of individual cases can be discontinued. For the data thus collected, separate averages should be calculated for the cases in each class. These costs are averages of the uninsured costs only.

Class 3 Incidents

Class 3 incidents are common first-aid cases in which no significant property damage results. They are the most difficult to analyze from the standpoint of cost because the time lost is likely to occur repeatedly and only for short periods. Also, the injuries can occur so frequently that they place an undue burden on the supervisor and safety director when a completed report form and data sheet are required for each case.

The essential information needed includes the average amount of working time lost per trip to the dispensary, the average dispensary cost per treatment, the average number of visits to the dispensary per case, and the average amount of the supervisor's time required per case. For organizations with a dispensary, first-aid cases (Class 3) may be estimated by using the following steps:

1. Estimate the average working time lost per trip to the dispensary for first aid. Departmental time records should be consulted, as they may show the amount of time each worker is absent from the job while receiving first aid. If so, a random sample of 50 to 100 records of workers who have received first aid should be selected from different departments. The average time lost per dispensary visit is calculated by adding the absence time for all visits in the sample and dividing by the total number of visits.
2. If departmental records do not contain this information, an investigator should observe a random sample of 50 or more workers visiting the dispensary. As before, to determine the average time, all the estimated time intervals of absence are added and then divided by the total number of workers observed.
3. Estimate the average cost of providing medical care in each visit by dividing the total cost of operating the dispensary for a year by the total number of treatments given during the year.
4. Calculate the average number of visits to the dispensary per case. This is done by dividing the number of treatments of Class 3 injuries in a representative period, perhaps a month or 6 weeks, by the number of persons with Class 3 injuries reported during the same period of time.
5. Calculate the average amount of supervisor time required per case. Where possible, this can be accomplished by observing the activities of representative supervisors in connection with first-aid cases.

Over time, enough cases will studied to form a representative sample of both the activities of supervisors in different departments and the different types of first-aid cases. The average time spent by a supervisor is then computed by adding all the time intervals recorded and dividing by the number of cases.

In some instances, it may be impossible to make a time study of the supervisor's activities in connection with first-aid cases. The only alternative is to have each supervisor estimate the time spent on a typical first-aid case and then to add the estimates and divide by the number of supervisors to find the average time. Determine the average value of this time by multiplying it by the average hourly wage of a supervisor.

The average total uninsured cost of a Class 3 incident is estimated from the data accumulated earlier as follows: the average amount of time lost for a trip to the dispensary (items 1 and 2) is multiplied by the facility's average wage rate, obtained from the payroll department, to get the average cost per trip for a worker's lost time. To this figure is added the estimated cost of providing medical care in a single visit (item 3). This figure is then multiplied by the average number of dispensary visits per medical treatment case (item 4), and to this result is added the average value of supervisor time required (item 5).

This method of calculating costs is designed to estimate the average uninsured cost per case for incidents causing localized property damage or, at most, a few injuries.

The method of cost estimation for incidents resulting in death, permanent total disability, or unusually extensive property damage is essentially the same as the method of estimating costs of other types of incidents. However, the main difference is that every cost is separately investigated and should be included in the final cost estimate as a separate item. When estimating the cost of a fire, for example, the investigator should bear in mind that the company's fire insurance will probably cover property damage in a major incident, whereas in a less serious incident, fire damage should appear as an uninsured cost.

Developing the Final Cost Estimate

Once the averages of each class of cases have been established, costs for any period in which a sufficiently large number of incidents have occurred can be estimated with considerable accuracy. These costs are calculated by multiplying the average uninsured cost per case for each of the four classes by the number of cases occurring in each class during the period.

If any deaths, permanent total disabilities, or unusually extensive property damage incidents have occurred, the investigator should add the specific uninsured costs of these incidents to the estimated costs of the four classes of incidents.

To these uninsured-cost totals the investigator then adds the costs of workers' compensation and insured medical expenses. For self-insured companies, this figure is the total amount paid out in the settlement of claims plus all the expenses of administering the insurance. For companies not carrying their own insurance, the figure is the amount of their insurance premiums plus deductibles paid.

The method needs to be modified according to the record-keeping systems of different companies. For example, most self-insurers would find it impossible to separate compensated medical expenses from dispensary care expenses. In that case, these items should be combined into one, and the dispensary costs should be omitted from the analysis of noncompensated costs on the data sheets. For an illustration of the development of a final cost estimate, see the example in Tables 10–C and 10–D.

As stated previously, presenting incident costs or estimated cost savings is useful for securing management support. Because incident costs represent possibilities of lost profits, incident dollars may be treated as lost profit dollars. The general method for achieving additional profits is through achieving additional sales. To estimate the sales necessary to recover profits lost because of incident expenses, determine the net profit percentage of gross sales and divide incident costs by the net profit percentage. Doing so provides an estimate of the additional gross sales dollar volume necessary to replace the profits lost because of incident expenses. Table 10–E shows the sales necessary to offset injury costs for different profit margins.

Another method of presenting incident costs to management is to equate lost incident dollars to lost purchasing opportunities for needed supplies or equipment.

OFF-THE-JOB DISABLING INJURY COST

The employer loses the same services whether an employee is injured off the job or on the job and incurs just about the same types of direct and indirect costs. Nevertheless, the costs of off-the-job (OTJ) incidents and illnesses are, at least in part, handled differently. This section explains the difference and presents sample calculations. (See Off-the-Job Safety Programs in Chapter 6, Loss Control Programs; and Off-the-Job Injuries in Chapter 11, Injury and Illness Record Keeping, Incidence Rates, and Analysis.)

When incidents take place off the job, a major portion of the costs is borne by employers. Some of the costs are obvious, such as insurance premiums and the wages paid to absent employees. Some of the costs are not as obvious, such as training new or transferred workers and the extra medical staff time needed by workers returning to work after an incident. For example, a new worker does not produce at the same rate as an experienced worker; thus, the decreased productivity of the new worker indirectly increases the company's overhead.

Other costs, although still very real, are more difficult to assess. As incident rates in a community rise, so do insurance rates and taxes. Not all organizations are aware of the total costs that can result from off-the-job incidents and the impact those costs have on operations and profits. Enough experience has been accumulated, however, to develop a simplified plan for estimating such costs.

Categorizing OTJ Disabling Injury Cost

The cost of off-the-job disabling injuries (OTJ DIs) to an organization falls into the following two categories: insured and uninsured. These are the same categories used for on-the-job incidents that result in disabling injuries, as described in the previous section.

Most uninsured costs are not obvious. Aside from wage costs, most organizations do not keep records of uninsured costs. However, the following costs are associated with all OTJ DI incidents and, therefore, affect profit margins:

- *Insured-worker productivity cost, product loss, and equipment damage*—Costs directly associated with the employee who sustained the OTJ DI injury are included in this expense subcategory.

TABLE 10-E. Sales Necessary to Recover Injury Costs

Injury Costs	Profit Margin				
	1%	2%	3%	4%	5%
$1,000	$100,000	$50,000	$33,333	$25,000	$20,000
$5,000	$500,000	$250,000	$166,667	$125,000	$100,000
$10,000	$1,000,000	$500,000	$333,333	$250,000	$200,000
$50,000	$5,000,000	$2,500,000	$1,666,667	$1,250,000	$1,000,000
$100,000	$10,000,000	$5,000,000	$3,333,333	$2,500,000	$2,000,000
$500,000	$50,000,000	$25,000,000	$16,666,667	$12,500,000	$10,000,000
$1,000,000	$100,000,000	$50,000,000	$33,333,333	$25,000,000	$20,000,000
$5,000,000	$500,000,000	$250,000,000	$166,666,667	$125,000,000	$100,000,000
$10,000,000	$1,000,000,000	$500,000,000	$333,333,333	$250,000,000	$200,000,000
$50,000,000	$5,000,000,000	$2,500,000,000	$1,666,666,667	$1,250,000,000	$1,000,000,000

Injury Costs	Profit Margin				
	6%	7%	8%	9%	10%
$1,000	$16,667	$14,286	$12,500	$11,111	$10,000
$5,000	$83,333	$71,429	$62,500	$55,556	$50,000
$10,000	$166,667	$142,857	$125,000	$111,111	$100,000
$50,000	$833,333	$714,286	$625,000	$555,556	$500,000
$100,000	$1,666,667	$1,428,571	$1,250,000	$1,111,111	$1,000,000
$500,000	$8,333,333	$7,142,857	$6,250,000	$5,555,556	$5,000,000
$1,000,000	$16,666,667	$14,285,714	$12,500,000	$11,111,111	$10,000,000
$5,000,000	$83,333,333	$71,428,571	$62,500,000	$55,555,556	$50,000,000
$10,000,000	$166,666,667	$142,857,143	$125,000,000	$111,111,111	$100,000,000
$50,000,000	$833,333,333	$714,285,714	$625,000,000	$555,555,556	$500,000,000

Injury Costs	Profit Margin				
	11%	12%	13%	14%	15%
$1,000	$9,091	$8,333	$7,692	$7,143	$6,667
$5,000	$45,455	$41,667	$38,462	$35,714	$33,333
$10,000	$90,909	$83,333	$76,923	$71,429	$66,667
$50,000	$454,545	$416,667	$384,615	$357,143	$333,333
$100,000	$909,091	$833,333	$769,231	$714,286	$666,667
$500,000	$4,545,455	$4,166,667	$3,846,154	$3,571,429	$3,333,333
$1,000,000	$9,090,909	$8,333,333	$7,692,308	$7,142,857	$6,666,667
$5,000,000	$45,454,545	$41,666,667	$38,461,538	$35,714,286	$33,333,333
$10,000,000	$90,909,091	$83,333,333	$76,923,077	$71,428,571	$66,666,667
$50,000,000	$454,545,455	$416,666,667	$384,615,385	$357,142,857	$333,333,333

Sales = Injury Costs / Profit Margin

- *Noninjured-worker productivity cost, product loss, equipment damage, and administrative cost*—Costs incurred by personnel other than the employee who sustained the OTJ DI injury are included in this subcategory.
- *Miscellaneous costs*—This subcategory also includes loss of profits from canceled contracts or orders and the costs of demurrage, telephone calls, transportation, and other miscellaneous expenses.

Estimating OTJ DI Costs

Some experts say the ratio of insured costs to uninsured costs is 3 to 2. In order to estimate a company's losses from employee OTJ DIs, management must first determine the insured costs. Next, using the 3:2 factor, it must estimate the uninsured costs to determine the total (insured and uninsured) employee costs. The insured costs for injuries to dependents of employees is then added to the total employee cost to ascertain total losses. The following examples illustrate calculation procedures.

Example 1

Company A is insured by an outside carrier. Twenty percent, or $225,000, of its annual premium charge went to pay for its previous calendar year OTJ DI incidents. Of that total, $75,000 paid for 11 employee injuries, and the remaining $150,000 paid for 22 employee-dependent injuries.

The total cost for employee-dependent injuries is a conservative figure because it does not include the administrative cost incurred by the company's insurance office staff to process claims. If this cost is known, it should be added to the employee-dependent injury expense category. When the $75,000 insured-cost category is added to the uninsured-cost category, the total expenses for employee injuries become $125,000.

Company A Estimated OTJ DI Costs

Insured cost for employee injuries	$75,000
Uninsured cost for employee injuries	$50,000
Insured cost for employee-dependent injuries	$150,000
Total annual estimated OTJ DI cost	$275,000

Example 2

Company B is insured by an outside carrier. Its carrier stated that $850,000 was paid for 138 employee injuries and that $1,800,000 was paid for 279 employee-dependent injuries for the company's previous calendar year OTJ DI incidents. The administrative fee paid to the carrier was 6% of the total cost, resulting in a cost of approximately $900,000 for employee injuries and $1,900,000 for employee-dependent injuries. Adding the cost of employee injuries to the insured-cost category gives a total of $1,500,000.

Company B Estimated OTJ DI Costs

Insured cost for employee injuries	$900,000
Uninsured cost for employee injuries	$600,000
Insured cost for employee-dependent injuries	$1,900,000
Total annual estimated OTJ DI cost	$3,400,000

Example 3

Company C is self-insured. Insurance records indicate that for the previous calendar year OTJ DI incidents, $2,280,000 was paid in medical and health claims for 350 employee injuries, and $4,700,000 was paid for 690 employee-dependent injuries. Insurance staff administrative costs for claim processing should be added to the employee-dependent cost category if the administrative costs are known. Thus, adding the uninsured cost for employee injuries and the insured cost for employee-dependent injuries to the $2,280,000 results in a total cost of approximately $8,500,000.

Company C Estimated OTJ DI Costs

Insured cost for employee injuries	$2,280,000
Uninsured cost for employee injuries	$1,520,000
Insured cost for employee-dependent injuries	$4,700,000
Total annual estimated OTJ DI cost	$8,500,000

Measuring the Effects of OTJ Safety Programs

Calculations of average costs are useful for supporting the initiation or acceleration of off-the-job safety awareness programs. These calculations also can be used to measure the effects of safety programs.

For example, 350 employees of Company C sustained OTJ injuries during the previous year, for a total cost of approximately $3,800,000 and an average cost per incident of approximately $10,860. Based on these OTJ DI losses, top management allocated $50,000 from the budget to initiate a safety awareness program. At the end of the year, the $10,860 average cost figure will be adjusted for inflation, and the new figure will be used to calculate losses. For illustration purposes, it is assumed that the new average cost per incident figure will be $11,500 and that employee injuries will be reduced from 350 to 300. Calculations (300 × $11,500) indicate a loss of $3,450,000, which is a savings of approximately $350,000 over the previous year's total. The estimated net return will be $300,000, which is a 600% return on investment.

To further justify the allocation of operating funds for the safety program, safety personnel can add to the employee savings total the savings realized from fewer employee-dependent injuries. Because this type of analysis indicates the impact of off-the-job disabling injuries on profit margins, it can be used to gain management's commitment to supporting operating budgets for safety awareness programs.

SUMMARY

- Successful incident prevention requires (1) studying all working areas to detect and control or eliminate hazards; (2) studying all operating procedures and administrative controls; (3) incorporating education, training, and discipline to minimize human factors; (4) conducting a thorough incident investigation and analysis; (5) implementing programs to change or control any hazardous conditions, procedures, and practices; and (6) conducting program follow-up and evaluation.
- The primary purpose of incident investigation and analysis is to prevent incidents by uncovering facts about each incident and determining how to prevent a recurrence of those incidents. Several incident investigation and analysis techniques are available to management and can be adapted to the needs of each organization.
- Incident investigation generally should be conducted by the supervisor. The safety professional advises and guides the supervisor. In some companies, a special investigation or review committee reports on all incidents or reviews the investigation.
- All incidents should be investigated, particularly those involving deaths or serious injuries. Near misses and a series of minor injuries also require study. Employers

- should involve outside employees and contractors to support safety on the job and incident investigations.
- Analysis of incident investigations can help management identify sources of incidents, pinpoint incident problems, reveal the need for engineering changes or correcting inefficiencies, disclose unsafe practices, enable supervisors to target safety efforts, and evaluate a safety program's effectiveness.
- Two types of incident analyses can be conducted: (1) to determine corrective actions and (2) to conduct a statistical analysis to uncover incident patterns so that corrective actions can be devised. Statistical analyses can be used to classify various groups of data to identify key factors causing or contributing to incidents. These analyses also can be used to compare current incident rates with prior years' incident rates, with rates of other companies, or with rates for the industry as a whole.
- Most companies have a method for calculating direct (insured) or indirect (uninsured) costs associated with incidents. Pilot studies are used to determine for different classes of incidents the average uninsured costs that can be applied to future incident totals. Workers' compensation costs and insured medical expenses are then added to these totals.
- Investigation, analysis, and cost estimates for off-the-job incidents are handled similarly to investigation, analysis, and cost estimates for on-the-job incidents. These activities can show management the impact of off-the-job incidents on organizational operations and profits and help to gain management support for worker education and safety training programs.

REFERENCES

Blankenship, L. M. *Nonoccupational Disabling Injury Cost Study.* K/DSA–457. Oak Ridge, TN: Martin Marietta Energy Systems, October 1981.

DeReamer, R. *Modern Safety and Health Technology.* New York: Wiley, 1981.

Dougherty, T. "How to Conduct an Accident Investigation." In DiBerardinis, L.J, ed. *Handbook of Occupational Safety and Health,* 2nd ed. New York: Wiley, 1998.

Grimaldi, J. V., and Simonds. *Safety Management—Accident Cost and Control.* 5th ed. Burr Ridge, IL: Irwin, 1988.

Johnson, W. G. *MORT Safety Assurance Systems.* New York: Dekker, 1980.

Kepner, C. H., and B. B. Tregoe. *The New Rational Managers.* Princeton, NJ: Kepner-Tregoe, 1981.

Latino, R. J., and K. C. Latino. *Root Cause Analysis: Improving Performance for Bottom-Line Results.* 3rd ed. Boca Raton, FL: Taylor & Francis, 2006.

National Safety Council, 1121 Spring Lake Drive, Itasca, IL 60143.
Injury Cost Capturing Tool (software), 2008.
The Off-the-Job Safety Program Manual, 2006.

Tompkins, N. C. *Basics of Safety and Health (Revised).* Itasca, IL: National Safety Council, 2004.

Tritsch, S. "Accident Investigation: How to Ask Why." *Safety and Health* 146, no. 6 (2000): 40–43.

Vincoli, J. W. *Basic Guide to Accident Investigation and Loss Control.* New York: Van Nostrand Reinhold, 1994.

REVIEW QUESTIONS

1. Name the six fundamental activities needed for a successful incident-prevention program.
    a.
    b.
    c.
    d.
    e.
    f.
2. What is the primary purpose of an incident investigation?
3. What types of incidents should be investigated?
4. Who is responsible for investigating an incident after it has occurred?
5. How soon after an incident has occurred should an investigation be started?
6. What can a company learn from analyzing incident causes?
7. List the minimum data that should be collected for each incident.
8. What are the three basic steps in a systematic approach to selecting corrective actions?
    a.
    b.
9. After thorough analysis of groups of incident investigations, what corrective actions might be suggested that were not evident when studying an individual case?
10. What are the two general categories of work incidents for the purpose of cost analysis?
    a.
    b.
11. Why is a pilot study of uninsured costs preferred over the use of a fixed ratio such as 4 to 1?

Injury and Illness Record Keeping, Incidence Rates, and Analysis

11

JoAnn H. Dankert, CHMM, CET; Senior Consultant, NSC
Kenneth P. Kolosh, Manager, Statistics, NSC

Introduction

Incident Records
Uses of Records ▸ Incident Reports ▸ Types of Data to Record and Track ▸ Record-Keeping Tools ▸ Periodic and On-Demand Reports

OSHA Record-Keeping Requirements

Injury and Illness Data Analysis
Statistical Measures ▸ Incidence Rates ▸ Severity Measures ▸ Noninjury/Illness Incidence Rates ▸ Control Charts ▸ Implementation and Follow-Up

Off-the-Job Injuries

Summary

References

Review Questions

INTRODUCTION

In this chapter, the terms *incident* and *injury* are restricted to occupational injuries and illnesses or documentable events that don't result in an injury or illness. In other chapters, *incident* is used in its broad meaning: unplanned, undesired events that interrupt or have the *potential* to interrupt the completion of an activity and that may include property damage, injury, or illness.

The Williams-Steiger Occupational Safety and Health Act of 1970 (OSH Act) requires most U.S. employers to maintain specific records of work-related employee injuries and illnesses. Some employers are required to maintain injury and illness records under regulations issued by other federal agencies such as the Mine Safety and Health Administration (MSHA) and the Federal Railroad Administration (FRA). In addition to these records, many employers also are required to make reports to state workers' compensation authorities. Similarly, insurance carriers may require reports for their records. Occupational injury and illness reports and records are now required of nearly every establishment by its leadership group or the government.

Safety personnel are faced with two tasks—maintaining records required by law and by their leadership and maintaining records useful to an effective safety and health program. Unfortunately, the two are not always perfectly aligned. A good occupational safety and health program requires more data than that required by most standard federal and state record-keeping systems.

This chapter covers the following topics:
- reasons for keeping incident records and how to establish an effective system
- how to calculate incident rates for on-the-job injuries and illnesses
- how to analyze injury/illness and sentinel incident data to separate random from caused variation
- how to calculate off-the-job incident rates.

Because the federal record-keeping requirements are subject to change, companies should contact an OSHA regional office for the latest information or consult Title 29 of the *Code of Federal Regulation* (CFR), Part 1904, Recording and Reporting Occupational Injuries and Illnesses (see References at end of chapter). Establishments subject to MSHA or FRA record-keeping requirements should consult the appropriate parts of the regulations 30 CFR 50 and 49 CFR 225, respectively.

Records of incidents and injuries are essential to maintain efficient and successful safety and health programs, just as records of production, costs, sales, and profits and losses are essential to efficient and successful business operations. Records supply the information necessary to transform haphazard, costly, ineffective safety and health practices into a planned safety program that controls both the conditions and the behaviors that contribute to incidents. Good record keeping is the foundation of a scientific approach to occupational safety.

A successful record-keeping program requires a well-defined process that integrates efficiently with other business processes and is understood by all employees. The accompanying flow chart (Figure 11–1) provides an example of one such injury and illness record-keeping process. This example is provided as a benchmark for comparison. No two record-keeping processes will be identical because each unique process will be affected by organizational practices, structure, and the degree of record-keeping tool automation.

INCIDENT RECORDS

Uses of Records

A good record-keeping system can help the safety professional in the following ways:
- provide safety personnel with the means for an objective evaluation of their incident problems and a measurement of the overall progress and effectiveness of their safety and health program
- identify high-incident-rate units, facilities, or departments and problem areas so that extra effort can be made in those areas
- provide data for an analysis of incidents pointing to specific causes or circumstances, which can then be addressed with specific countermeasures
- create interest in safety and health among line leadership (supervisors or team leaders) and area leadership (managers) by furnishing them with information about their departments' incident experience
- provide supervisors and safety teams and committees with hard facts about their safety problems so that their efforts can be concentrated
- measure the effectiveness of individual countermeasures and determine whether specific programs or processes are doing what they were designed to do
- assist leadership in safety performance evaluation.

Incident Reports

To be effective, preventive measures must be based on complete and unbiased knowledge of the causes of incidents. The primary purpose of an incident report is to obtain such information, not to fix blame. The completeness and accuracy of the entire incident record-keeping system depends

11 Incident Records 267

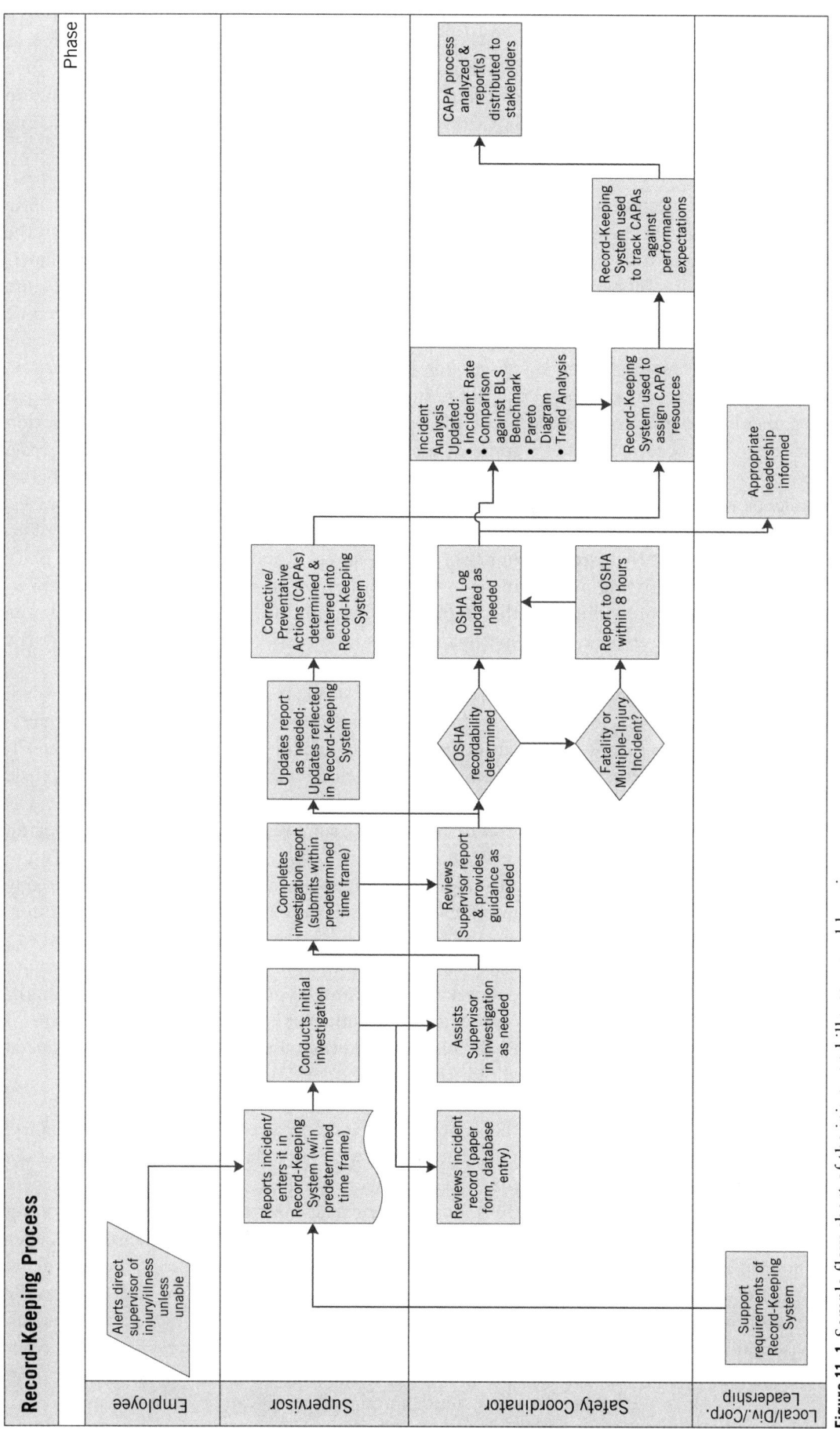

Figure 11–1. Sample flow chart of the injury and illness record-keeping process.

on the quality of information from the individual incident reports. Therefore, leadership must be sure that the necessary tools and their purposes are understood by those who must complete them. Everyone involved in the incident investigation process should be given the necessary training or instruction.

Supervisors or team leaders should set expectations for immediately reporting incidents. Supervisors also have the responsibility to begin the incident investigation as soon as possible after an incident occurs. They need to document their findings in the appropriate reporting tools and forward incident information to the safety department and to other designated persons such as area/department managers and others in local leadership positions. Early reporting and investigation of incidents helps gather information concerning activities and conditions that preceded an occurrence and is important to prevent future incidents. This information is particularly difficult to get unless it is obtained promptly after the incident occurs.

The role of the safety professional is to set the direction and oversee the process of incident reporting and investigation at his or her location. These professionals have the responsibility for collecting incident investigation data, validating it for accuracy, analyzing it for trends, and reporting results back to leadership. In some cases, they may suggest possible interventions or work closely with the safety team/committee to develop next steps for prevention. In many organizations, the safety staff will author or create the data collection tools (report forms, action registers, spreadsheets, databases, etc.). In larger organizations, the division/corporate staff may select and manage an enterprisewide data collection tool (internally created or off-the-shelf product) so that information and statistics can be readily accessed and gathered for routine issuance of management reports.

Management and the organization's leadership also play an important role in incident investigation and data collection. They need to know the investigation process and hold supervisors/team leaders accountable for completing incident investigation reports according to the location's policy/procedure. Management and leadership should review reports generated for accuracy and quality and to ensure that root causes have been uncovered and corrective actions as well as preventive activities have taken place.

Generally, analyses of incidents are made periodically (monthly, quarterly) and often long after the incidents have occurred. Because, after the time of the incident, workers may find it impossible to recall the precise details of what exactly took place, supervisors and others must record details accurately and completely at once or the information may be lost forever.

All facts may not be available when the incident report is being completed. As a result, items such as total time lost and dollar amount can be added later. This should not, however, prevent a detailed investigation from starting immediately after the incident occurs.

The collection of injury data generally begins with first-aid treatment. The first-aid attendant or nurse fills out a report for each new case. Reports are distributed to the safety department or safety committee, the worker's supervisor, and other departments as leadership may wish (Figure 11–2). For additional guidance on OSHA record-keeping requirements, please refer to the *OSHA Recordkeeping Handbook* (see References at end of chapter).

The first-aid attendant or the nurse should know enough about incident analysis, record keeping, and investigation to be able to record the principal facts about each case. The organization's occupational physician or clinic also should be informed of the basic rules for classifying cases. At times, the doctor's or clinic's repeated treatment of an injury or illness—or their opinion of the seriousness of an injury—may be necessary to record the case accurately on the OSHA 300 Log and to help determine whether the specific injury could have been caused by the incident the employee described.

Three different supervisor's report forms are provided. They fulfill all of the information requirements of the present OSHA Form 301, Injury and Illness Incident Report. The forms also include questions about additional basic data that should be known about each incident.

The first supervisor's report form (Figure 11–3) is an open-ended, narrative type of form. The next form (Figure 11–4) is self-coding to allow keying of data items directly into the form without the extra step of recoding this information for data-processing equipment. This method can easily be used to enter information into an electronic data collection system and to produce a variety of summary reports (such as summaries by department and by type of incident). Detailed crosstabulations thus can be produced with little effort.

It is recommended that the supervisor make a detailed report about each incident, even when only a minor injury/illness or no injury/illness is the result. For purposes of the OSHA Log of Work-Related Injuries and Illnesses (Form 300), only reports that meet the minimum severity level need to be separated and recorded. Minor injuries occur with greater frequency than serious injuries, and records of these minor occurrences can help pinpoint problem areas. By working to alleviate these problems, workers and leadership can prevent many serious injuries. Minor injuries should not be regarded lightly, however. Complications may arise from these injuries, and their result can be quite serious.

```
Case No.: _____                                              Date: _____

                              FIRST-AID REPORT
Name: _____   Department: _____
Occupation: _____   Supervisor: _____

                                                          a.m.
Date of Incident: _____   Time: _____   p.m.   Date Treated: _____
Nature of injury/illness: _____
_____

Treatment Given: _____
_____

Sent:   _____ Back to Work    _____ to Doctor/Hospital/Clinic    _____ Home
Estimated Lost Workdays:      _____ Away from Work               _____ Restricted Work Activity
Employee's Description of Incident: _____
_____

Report Prepared by: _____
Signed: _____
```

Figure 11–2. The first-aid report (4 × 6 in. or 10 × 16 cm) is prepared by the first-aid attendant when an injured or ill person comes for treatment. The report serves as a record and permits quick tabulation of such data as department, occupation, and the key facts of the occurrence.

Types of Data to Record and Track

Most safety and health professionals recognize that the OSHA record-keeping system does not provide enough information about what is going on in the workplace to effectively manage a safety and health program. The development, implementation, and evaluation of safety and health programs require information on incidents well beyond government recording requirements. Comprehensive surveillance systems document many different kinds of events and exposures in the workplace, all of which are of interest to the safety and health professional. From this large database of "documentable events," the professional may select those cases that must be recorded on the OSHA 300 Log or reported to workers' compensation authorities and insurance carriers. Broadly speaking, documentable incidents can be classified as either injury/illness or noninjury/illness incidents:

- injuries or illnesses
  - minor (first-aid cases)
  - recordable [see 29 CFR 1904.7(b)(1)]
    - fatalities [1904.7(b)(2)]
    - medical treatment beyond first aid [1904.7(b)(5)]
    - cases resulting in days away from work [1904.7(b)(3)]
    - cases resulting in job transfers or work restrictions [1904.7(b)(4)]
    - loss of consciousness [1904.7(b)(6)]
    - significant injury or illness diagnosed by a physician or other licensed health care professional [1904.7(b)(7)]

- noninjury or illness
  - signs and symptoms not requiring medical treatment or first aid
  - property damage (all severities)
  - substandard conditions/identified hazards
  - at-risk behaviors
  - regulatory deviations (unscheduled inspections, citations, fines)
  - near-miss incident.

Comprehensive surveillance systems document all injuries and illnesses to employees, including first-aid cases and all noninjury/illness incidents. No-injury/illness incidents are events that occur in the work environment that could have, but did not, result in an occupational injury/illness. Noninjury/illness incidents include property damage events (e.g., motor vehicle crash, crane failure, gas explosion, collapse of structure, rupture of pipes or hoses, uncontrolled fires), signs and symptoms not requiring medical treatment or first aid (e.g., complaint of pain, blood lead level greater than the action level in the lead standard [29 CFR 1910.1025]), and "near hits" (sometimes called near misses). A near hit is an occurrence in the work process, including an employee act or behavior, that has an injury, illness, or death potential but where injury, illness, or death did not occur. Such events may include improper operation of a safety device, entrapment of people, or unintentional

INCIDENT INVESTIGATION REPORT

Case Number:

Company:

Address:

Department:

Location (if different from mailing address):

1. Name of injured:

2. Social Security Number:

3. Sex: ☐ M ☐ F

4. Age:

5. Date of incident:

6. Home address:

7. Employee's usual occupation:

8. Occupation at time of incident:

9. Employment category:
☐ Regular, full time ☐ Regular, part time
☐ Nonemployee ☐ Temporary ☐ Seasonal

10. Length of employment:
☐ Less than 1 month ☐ 6 months-5 years
☐ 1-5 months ☐ Over 5 years

11. Time in occupation at time of incident:
☐ Less than 1 month ☐ 6 months-5 years
☐ 1-5 months ☐ Over 5 years

12. Nature of injury and part of body:

13. Case numbers/names of others injured in same incident:

14. Name and address of physician:

15. Name and address of hospital:

16. Time of injury:
A. _____ a.m. / p.m.
B. Time within shift:
C. Type of shift:

17. Severity of injury:
☐ Fatality
☐ Lost workdays—days away from work
☐ Lost workdays—days of restricted activity
☐ Medical treatment needed
☐ First aid needed
☐ Other, specify _____

18. Specific location of incident:

19. On employer's premises?
☐ Yes
☐ No

20. Phase of employee's workday at time of injury:
☐ During rest period ☐ Entering or leaving facility
☐ During meal period ☐ Performing work duties
☐ Working overtime ☐ Other _____

Figure 11–3. The Incident Investigation Report (8 × 11 in. or 22 × 28 cm) captures a record of contributing circumstances to provide a basis for specific remedial action. Users should be trained to properly fill it out.

21. Describe how incident occurred:

22. Incident sequence: Describe, in reverse order of occurrence, events preceding the injury and incident. Starting with the injury and moving backward in time, reconstruct the sequence of events that led to the injury.

A. Injury event:	
B. Incident event:	
C. Preceding event #1:	
D. Preceding event #2, #3, etc.	

23. Task and activity at time of incident:	24. Posture of employee:
General type of task:	
Specific activity:	
Employee was working:	25. Supervision at time of incident:
☐ Alone ☐ With crew or fellow worker ☐ Other, specify	☐ Directly supervised ☐ Indirectly supervised
	☐ Not supervised ☐ Supervision not feasible

26. Causal Factors (Events and conditions that contributed to the incident. Include actions identified using the *Guide for Identifying Causal Factors and Corrective Actions*.):

27. Corrective actions (Those that have been or will be taken to prevent recurrence. Include actions identified using the *Guide for Identifying Causal Factors and Corrective Actions*.):

Prepared by:	Approved:	
Title:	Title:	Date:
Department:	Approved:	
Date:	Title:	Date:

Figure 11–3. Concluded.

SUPPLEMENTARY RECORD OF OCCUPATIONAL INJURIES AND ILLNESSES

SERVICE NO. ▶ 1-9 _____
CASE OR FILE NO. ▶ 10-15 _____

(Meets OSHA requirements when Instruction 1. has been followed.)

THIS REPORT IS
▶ 16. 1 ☐ First report   2 ☐ Revised report

EMPLOYER
1. NAME _____
2. MAIL ADDRESS _____
3. LOCATION, if different from mail address _____

INJURED OR ILL EMPLOYEE
4. NAME _____
   SOCIAL SECURITY NO. _____
▶ EMPLOYEE NO. 17-26 _____
5. HOME ADDRESS _____
▶ 6. AGE 27-28 _____
▶ 7. SEX 29, 1 ☐ Male   2 ☐ Female
▶ 8. OCCUPATION (specify) _____
   30-31, 01 ☐ Manager, official, proprietor
   02 ☐ Professional, technical
   03 ☐ Foreman, supervisor
   04 ☐ Sales worker
   05 ☐ Clerical worker
   06 ☐ Craftsman—construction
   07 ☐ Craftsman—other
   08 ☐ Machinist
   09 ☐ Mechanic
   10 ☐ Operative (production worker)
   11 ☐ Motor vehicle driver
   12 ☐ Laborer
   13 ☐ Service worker
   14 ☐ Agricultural worker
   15 ☐ Other
   16 ☐ Unknown
9. DEPARTMENT _____
   (Enter the name of department or division in which the injured person is regularly employed.)

CLASSIFICATION OF CASE
A. INJURY OR ILLNESS (see code on Log, OSHA No. 300)
▶ 32, 1 ☐ Injury (10)
   2 ☐ Occupational skin disease or disorder (21)
   3 ☐ Dust disease of the lungs (pneumoconioses) (22)
   4 ☐ Respiratory conditions due to toxic agents (23)
   5 ☐ Poisoning (systemic effects of toxic materials) (24)
   6 ☐ Disorder due to physical agents (other than toxic materials) (25)
   7 ☐ Disorder due to repeated trauma (26)
   8 ☐ All other occupational illnesses (29)

B. EXTENT OF INJURY OR ILLNESS
▶ 33, 1 ☐ Fatality
   2 ☐ Lost workday case
   3 ☐ Nonfatal case without lost workdays
▶ C. Number of workdays lost 34-36 _____
D. Permanently transferred or terminated
▶ 37, 1 ☐ Yes   2 ☐ No

INSTRUCTIONS
1. Type or print the narrative where requested.
2. Check the one box that most clearly describes each narrative statement.
3. See also original OSHA No. 301 for more details.

THE ACCIDENT OR EXPOSURE TO OCCUPATIONAL ILLNESS
10. PLACE OF ACCIDENT OR EXPOSURE (mail address) _____
11. WHERE DID ACCIDENT OR EXPOSURE OCCUR?
    a. On employer premises
▶   38, 1 ☐ Yes   2 ☐ No   3 ☐ Unknown
    b. Place (specify) _____
▶   39-40, 01 ☐ Office
    02 ☐ Plant, mill
    03 ☐ Shipping, receiving, warehouse
    04 ☐ Maintenance shop
    05 ☐ General or public area of employer premises (corridor, washroom, lunchroom, parking lot, etc.)
    06 ☐ Retail establishment (store, restaurant, gasoline station, etc.)
    07 ☐ Farm
    08 ☐ Motor vehicle accident
    09 ☐ Other
    10 ☐ Unknown
12. WHAT WAS THE EMPLOYEE DOING WHEN INJURED? (Be specific)
    _____
    _____
    _____
    a. Task performed at time of incident
▶   41-42, 01 ☐ Operating machine
    02 ☐ Operating hand tool (power or nonpower)
    03 ☐ Materials handling
    04 ☐ Maintenance & repair—machinery
    05 ☐ Maintenance & repair—building & equipment
    06 ☐ Motor vehicle driver, operator or, passenger
    07 ☐ Office and sales tasks, except above
    08 ☐ Service tasks, except above
    09 ☐ Other
    10 ☐ Not performing task
    11 ☐ Unknown
    b. Activity at time of incident
▶   43-44, 01 ☐ Climbing
    02 ☐ Driving
    03 ☐ Jumping
    04 ☐ Kneeling
    05 ☐ Lying down
    06 ☐ Lifting
    07 ☐ Reaching, stretching
    08 ☐ Riding
    09 ☐ Running
    10 ☐ Sitting
    11 ☐ Standing
    12 ☐ Walking
    13 ☐ Other
    14 ☐ Unknown

Figure 11–4. A self-coding supplementary record of occupational injuries and illnesses.

13. **HOW DID THE INCIDENT OCCUR?** (Describe fully the events)

_____
_____
_____
_____

a. **AGENCY.** (Object or substance involved)
   INCIDENT AGENCY (1st column). The first object or substance involved in incident sequence.
   INJURY AGENCY (2nd column). The agency inflicting the injury. See also section 15.
   (Example: Worker fell from ladder and struck head on machine. Check "Ladder" under incident and check "Machine" under injury.)

	INCIDENT		INJURY	(Check one box in each column)
▶▶ 45-46,	01 ☐	47-48,	01 ☐	Machine
	02 ☐		02 ☐	Conveyor, elevator, hoist
	03 ☐		03 ☐	Vehicle
	04 ☐		04 ☐	Electrical apparatus
	05 ☐		05 ☐	Hand tool
	06 ☐		06 ☐	Chemical
	07 ☐		07 ☐	Working surface, bench, table, etc.
	08 ☐		08 ☐	Floor, walking surface
	09 ☐		09 ☐	Bricks, rocks, stones
	10 ☐		10 ☐	Box, barrel, container (empty or full)
	11 ☐		11 ☐	Door, window, etc.
	12 ☐		12 ☐	Ladder
	13 ☐		13 ☐	Lumber, woodworking materials
	14 ☐		14 ☐	Metal
	15 ☐		15 ☐	Stairway, steps
	16 ☐		16 ☐	Other
	17 ☐		17 ☐	Unknown
	18 ☐		18 ☐	None

b. **INCIDENT TYPE.** (First event in the incident sequence)

▶ 49-50, 01 ☐ Fall from elevation
       02 ☐ Fall on same level
       03 ☐ Struck against
       04 ☐ Struck by
       05 ☐ Caught in, under, or between
       06 ☐ Rubbed or abraded
       07 ☐ Bodily reaction
       08 ☐ Overexertion
       09 ☐ Contact with electrical current
       10 ☐ Contact with temperature extremes
       11 ☐ Contact with radiations, caustics, toxic and noxious substances
       12 ☐ Public transportation accident
       13 ☐ Motor vehicle accident
       14 ☐ Other
       15 ☐ Unknown

This space may be used for additional information.

OCCUPATIONAL INJURY OR ILLNESS

14. **DESCRIBE THE INJURY OR ILLNESS** in detail and indicate the part of the body affected.

_____
_____
_____

a. **NATURE OF INJURY OR ILLNESS.** (Check most serious one)

▶ 51-52, 01 ☐ Amputation
       02 ☐ Burn and scald (heat)
       03 ☐ Burn (chemical)
       04 ☐ Concussion
       05 ☐ Crushing injury
       06 ☐ Cut, laceration, puncture, abrasion
       07 ☐ Fracture
       08 ☐ Hernia
       09 ☐ Bruise, contusion
       10 ☐ Occupational illness
       11 ☐ Sprain, strain
       12 ☐ Other

b. **PART OF BODY.** (Check most serious one)

▶ 53-54, 01 ☐ Eyes
       02 ☐ Head, face, neck
       03 ☐ Back
       04 ☐ Trunk (except back, internal)
       05 ☐ Arm
       06 ☐ Hand and wrist
       07 ☐ Fingers
       08 ☐ Leg
       09 ☐ Feet and ankles
       10 ☐ Toes
       11 ☐ Internal and other

15. **NAME THE OBJECT OR SUBSTANCE THAT DIRECTLY INJURED THE EMPLOYEE.** Also check one box in injury column under 13a.

_____
_____

16. **DATE OF INJURY OR INITIAL DIAGNOSIS OF OCCUPATIONAL ILLNESS.**

   a. **MONTH**

▶ 55-56, 01 ☐ Jan.   07 ☐ July
       02 ☐ Feb.   08 ☐ Aug.
       03 ☐ March   09 ☐ Sept.
       04 ☐ April   10 ☐ Oct.
       05 ☐ May   11 ☐ Nov.
       06 ☐ June   12 ☐ Dec.

▶ b. **DATE OF MONTH**   57-58 _____

17. **DID EMPLOYEE DIE?**

▶ 59, 1 ☐ Yes   Date of Death_____
      2 ☐ No

OTHER

18. **NAME AND ADDRESS OF PHYSICIAN**_____

19. **IF HOSPITALIZED, NAME AND ADDRESS OF HOSPITAL**_____

DATE OF REPORT_____
PREPARED BY_____
OFFICIAL POSITION_____

National Safety Council
Printed in U.S.A.

Figure 11–4. Concluded.

11 Injury and Illness Record Keeping, Incidence Rates, and Analysis

INCIDENT INVESTIGATION FORM – PAGE 1 OF 3

Privacy: This information will be stored and used in accordance with future incident prevention – Persons named on this form agree to stand in a court of law if requested

1. Location of Incident
Site Address: _____
Supervisor: _____

2. Details of the incident being investigated
Incident Report Number: _____
Name of Injured Person (If Applicable): _____ Date of Incident: __ __ / __ __ / __ __ __ __ (dd/mm/yyyy)
Name of person who reported Incident: _____ Date of Report: __ __ / __ __ / __ __ __ __

3. Details of the Incident Investigation
Name of person completing this form: _____ Date Completed: __ __ / __ __ / __ __ __ __ (dd/mm/yyyy)
Telephone Number: _____ E-mail address: _____

Is this form being completed as part of an "onsite" investigation? Yes / No → Names of Investigation Team → _____

Have any witnesses been interviewed as part of the incident investigation? Yes / No → Names of those interviewed → _____
(Attach witness statements if applicable.)

4. Description of Events
Describe the sequence of events that led to the incident:
_____

Describe the sequence of events following the incident:
_____

Describe the task being performed at the time of the incident:
_____

5. Risk Rating of Incident and Likelihood of Recurrence

Using the Two-Variable Risk Matrix (Right):
- Rate the consequences (Severity) of the incident
- Rate the likelihood of the incident occurring or recurring
- Circle the resultant risk rating on the Risk Matrix

Likelihood Label	Consequences (Severity) Label				
	Negligible	Significant	Moderate	Major	Catastrophic
Almost Certain (A)	Medium	High	High	Very High	Very High
Likely (B)	Medium	Medium	High	High	Very High
Possible (C)	Low	Medium	High	High	High
Unlikely (D)	Low	Low	Medium	Medium	High
Rare (E)	Low	Low	Medium	Medium	High

Figure 11–5. Example of a spreedsheet reporting form.

INCIDENT INVESTIGATION FORM – PAGE 2 of 3

Privacy: This information will be stored and used in accordance with future incident prevention –
Persons named on this form agree to stand in a court of law if requested

5.1. Identify the behavioral causes of the incident

Did any of the following behaviors contribute to the cause of the incident? (Choose below)

Performing task without authority	☐	Distracting, teasing, or abusing a person	☐
Performing task at unsafe speed	☐	Using unsafe or tagged out equipment	☐
Performing task while affected by drugs/alcohol	☐	Using equipment in an unsafe manner	☐
Performing task with improper work technique	☐	Unsafe placement of equipment or objects	☐
Performing task without PPE	☐	Unsafe manual handling technique	☐
Performing task without correct PPE	☐	Unsafe position or posture	☐
Failure to warn of hazard	☐	Unsafe acts of others	☐
Failure to secure hazardous item	☐	Other **(Specify):**	☐
Making safety device inoperable	☐	Not applicable	☐

What are the management systems (procedural) deficiencies that led to the unsafe behaviors? (Choose below)

Deficiency		Deficiency	
Inadequate standard operating procedure/policies	☐	Inadequate workplace inspection	☐
Inadequate supervision	☐	Inadequate equipment provided	☐
Inadequate hazard identification	☐	Inadequate design or construction of workplace	☐
Inadequate assessment of risk	☐	Inadequate task or process design	☐
Inadequate provision of PPE	☐	Unrealistic scheduling	☐
Inadequate operator training	☐	Other **(Specify):**	☐
Inadequate supervisor training	☐	Not applicable	☐

5.2. Identify the physical causes of the incident

Did any of the following conditions contribute to the cause of the incident? (Choose below)

Inadequate or absent guarding	☐	Inadequate fire or explosion risk control	☐
Poor workstation design or layout	☐	Inadequate noise control	☐
Poor condition of equipment or objects	☐	Inadequate ventilation	☐
Equipment or objects with unsafe design	☐	Inadequate temperature control	☐
Unsafe storage of equipment or objects (housekeeping)	☐	Inadequate fall protection	☐
Unsafe walking surfaces	☐	Inadequate signage or warning systems	☐
Unsafe lighting or glare	☐	Inadequately controlled use of chemicals/substances	☐
Unsafe clothing or shoes	☐	Other **(Specify):**	☐
Unsafe task or process	☐	Not applicable	☐

What are the management systems (procedural) deficiencies that led to the unsafe conditions? (Choose below)

Deficiency		Deficiency	
Inadequate standard operating procedure/policies	☐	Inadequate workplace inspection	☐
Inadequate supervision	☐	Inadequate equipment provided	☐
Inadequate hazard identification	☐	Inadequate design or construction of workplace	☐
Inadequate assessment of risk	☐	Inadequate task or process design	☐
Inadequate provision of PPE	☐	Unrealistic scheduling	☐
Inadequate operator training	☐	Other **(Specify):**	☐
Inadequate supervisor training	☐	Not applicable	☐

Figure 11–5. Continued.

Figure 11–5. Concluded.

detonation of explosives. In addition, many injury and illness record-keeping systems also track at-risk behaviors and substandard conditions even if the threat of injury or illness was not imminent.

Noninjury/illness incidents occur much more frequently than injuries/illnesses and because of this disparity, the reporting of noninjury/illness incidents:

- assists in the identification of relationships between non-injury/illness incidents and injury/illness rates
- provides opportunities to prevent or reduce injury/illness incidents through the improved understanding and control of energy sources and transfers in such incidents
- raises staff consciousness regarding the need to reduce or eliminate conditions that may result in injury/illness incidents or property damage
- provides the necessary information to prioritize efforts to eliminate and/or reduce the probability of the occurrence of an injury or illness
- provides information to establish databases on causes and subsequent corrective action(s) to eliminate and/or reduce injuries/illnesses
- assists in helping to identify training needs
- assists in reducing the costs of injuries, illnesses, and non-injury/illness incidents.

Record-Keeping Tools

The system presented in this chapter provides the basic items necessary for good record keeping. It is designed to dovetail with the present record-keeping requirements of the OSH Act and attempts to avoid a duplication of effort on the part of personnel responsible for keeping records and submitting reports. Some of the record-keeping tools presented in this section are also constructed with data analysis methods in mind.

A well-designed record-keeping tool takes into account the person who will use the tool and the way in which the subsequent data will be analyzed and reported. In general, a self-coding check-off or dropdown list can save time for both the person using the data collection tool and the person managing and reporting the data. Regardless of the format, record-keeping tools should accomplish three things:

1. allow for the recording of all causes contributing to documentable events
2. reveal questions the investigator should ask to determine all relevant worksite conditions and human causes
3. provide a means of accumulating documentable incident data.

Tools that accomplish these three goals are more likely to be completed accurately and will present fewer problems for those who manage and analyze the data. Care in the choice and design of tools will pay dividends in better, more reliable data.

Unlike the collection of record-keeping data, which was discussed earlier and is distributed across the organization, the function of data management and analysis has historically been centralized. A centralized data management and analysis process, most often housed in the safety group, helps to improve the reliability of the data and makes it possible to update analysis procedures without retraining as many people. A few key people who work with the system a great deal will have a better understanding of the analysis procedures than will supervisors or managers, who may analyze only a few incident reports each year. Although current technological advances have made it easier for supervisors and others in leadership positions to perform their own analysis as needed, periodic reports developed by the safety group still play a vital role to help ensure information is being collected and interpreted consistently across the organization.

The record-keeping tools in this section are not the only ways to keep records, but are rather provided as examples. The incident problems of individual establishments are unique, and no one set of generic tools can provide every establishment with all the data for solving all of its individual problems. A system that does a good job of collecting the basic facts, however, makes it easier later on to pinpoint the data relating to a specific problem.

In the last several years, many technical advances have emerged that can greatly improve the efficiency of record-keeping activities. The most advanced electronic-based tools provide nearly real-time reporting and support data collection using tablets or smartphones. However, these new technologies have not yet been universally adopted. Because of the cost and complexity of the implementation that is associated with the more advanced systems, many small and mid-sized companies continue to adequately meet their record-keeping needs using either paper- or spreadsheet-based systems. This chapter provides examples of record-keeping tool formats ranging from simple paper-based tools to spreadsheets to electronic systems specifically designed for injury and illness record keeping.

The following sections deal with occupational injuries, illnesses, and other documentable events. Nonemployee incidents are covered in Chapter 28, Contractor and Customer Safety. Specific record-keeping requirements are explained in Chapter 4, Regulatory History. Chapter 10, Incident Investigation, Analysis, and Costs, contains important information on what cases to investigate and the minimum data to be compiled on each case. It also covers some basic data analysis. A statistical approach to data analysis is covered later in this chapter.

Paper-based record-keeping tools are still widely used today either as the primary tool used to collect information during an investigation or as a supplement to other formats. Paper tools have the advantage of working in almost any situation, regardless of circumstances such as power outages or remote field locations, where other technology solutions may not be consistently available. Paper forms also have the advantage of documenting procedures. They can be included as part of an organization's safety policies or be reviewed as part of a vendor vetting process. Figure 11–6 provides a sample of a paper-based form used by the National Safety Council as part of its own safety program. Most forms are used in conjunction with spreadsheets or databases to manage and analyze the data.

Spreadsheet-based systems are one of the most common in-house ways to develop record-keeping formats. Because of the prevalence of spreadsheet software in the workplace, these types of systems can be designed from the ground up within the safety department without external technical support. Spreadsheets also provide the added advantage of being highly flexible and adaptable over time and can be easily updated to reflect organizational structure changes. With the use of macros and formulas, spreadsheets can also be partially automated. The downside of spreadsheets compared with other electronic solutions specifically designed for record keeping is the lack of efficiency, particularly for larger, multi-location organizations. Typically, spreadsheet systems require the manual consolidation of information collected from across the organization. This process, often conducted monthly, can be resource intensive. Because of the manual nature of spreadsheet-based systems, the reporting of results is often slower than in more automated systems.

The adaptability of spreadsheets allows their use in both the collection and analysis of results. Figure 11–5 provides an example of a spreadsheet reporting form, while Figure 11–7 provides an example of a spreadsheet used to analyze data.

Although both paper- and spreadsheet-based systems have supported the record-keeping process for many years, many organizations have moved to electronic solutions specifically designed for record keeping. Electronic injury and illness record-keeping tools can be either custom tools specifically designed to meet an organization's unique needs or an off-the-shelf tool. As the number and sophistication of off-the-shelf tools have increased, the need to develop custom electronic record-keeping tools has decreased. Off-the-shelf tools are currently available with a wide range of features and complexity. On the low end, simple stand-alone systems support OSHA record-keeping requirements as well as report generation. More comprehensive enterprise safety management systems also support the management of audits, inspections, safety training, and other regulatory requirements. Both stand-alone and enterprise tools often support electronic completion of incident reports via portable electronic devices such as tablets or smartphones. These systems often allow the inclusion of photos and videos as attachments to the incident record. Finally, many larger organizations that already have an enterprise resource planning (ERP) system in place utilize that system's built-in environmental health and safety modules.

Compared to spreadsheet- and paper-based tools, electronic tools designed specifically to support injury and illness record keeping greatly improve efficiency in several critical ways. First, this type of system eliminates data-entry duplication, with data entered in one part of an organization automatically integrated into the organization's database. Second, incident data analysis and distribution become a self-service activity, with leadership being able to track incident investigation as well as trends on nearly a

SUPERVISOR INCIDENT INVESTIGATION REPORT

National Safety Council

☐ Injury Incident
☐ Near-Miss Incident
☐ Equipment / Property Damage
☐ Other

Fill out all blocks. Be as specific as possible and include drawings, photos, additional narrative, as needed. **Initial SIIR to be completed within 24 hours. Revised and signed form to be completed within 5 workdays.**

Location	☐ Itasca	☐ Roselle	**Navigator Case #**	Click here to enter text.
	☐ DMC	☐ DC	**Board Review Rating**	
	☐ Alabama	☐ Other:		
	☐ Atlanta	Click here to enter text.		

Person Involved in Incident or Near Miss

Click here to enter text.

Reporting Supervisor/Investigator Name

Reporting Supervisor / Investigator Name	Title	Dept.	Ext.
Click here to enter text.	Click here to enter text.	Click here to enter text.	Click here to enter text.

Date of Incident/Near Miss	Time of Incident	Hours into Shift	Today's Date
Click here to enter a date.	Click here to enter text.	Click here to enter text.	Click here to enter a date.

Employment Category	Length of Employment
☐ Full Time	☐ Less than 1 month
☐ Part Time	☐ 1 – 5 months
☐ Temp/Contractor	☐ 6 months – 5 years
☐ Nonemployee	☐ Over 5 years

Was a Contractor Involved? If yes, name and contact information

Click here to enter text.

INJURED PARTY

	Injured Party's Name & Title	Injured Party's Contact Information
☐ No Injury – Skip this section	Click here to enter text.	Click here to enter text.

Anyone Else Injured?	If yes, names	Contact Information
Yes ☐ No ☐	Click here to enter text.	Click here to enter text.

NATURE OF INJURY/ILLNESS

☐ Strain/Sprain	☐ Burn/Scald	☐ Heat Related	**Treatment**		**Severity of Injury**		
☐ Fracture	☐ Foreign Body	☐ Overexertion	☐ First Aid		☐ Lost Workdays		
☐ Laceration/Cut	☐ Chemical Reaction	☐ Other (Specify)	☐ E.R.		# Days:	Click here to enter text.
☐ Bruising	☐ Allergic Reaction	Click here to enter text.	☐ Dr.'s Office		☐ Restricted Workdays		
☐ Scratch/Abrasion	☐ Concussion	**Body Part(s) Injured**	☐ Hospital Stay		# Days:	Click here to enter text.
☐ Amputation	☐ Slip	Click here to enter text.	☐ NSC First Responder		☐ Fatality		
☐ Dislocation	☐ Trip		**Injury on Employer Premises?**			
☐ Internal	☐ Fall		Yes ☐		No ☐		

ESS AND/OR WITNESS STATEMENT

Witnesses (Name and Contact Information)	Witness Statement Attached?	Yes ☐	No ☐
Click here to enter text.	Click here to enter text.		

PROPERTY DAMAGE

List Property/Material Damaged	Nature of Damage
Click here to enter text.	Click here to enter text.

Object/Substance Inflicting Damage	Approximate Cost
Click here to enter text.	Click here to enter text.

Figure 11–6. Example of a paper-based incident investigation form.

THE INCIDENT

What was employee doing before the event occurred? Click here to enter text.

Describe exactly what happened in detail. (Investigate scene of incident or conditions. Describe who was involved, when and where the incident happened, what happened, and how.)

Click here to enter text.

WHY DID IT HAPPEN? (Root Cause Analysis – What Actually Caused the Illness, Injury, or Incident?)

	Unsafe Acts		Unsafe Conditions		Management System Deficiencies
☐	Safety Rule Violation	☐	Slippery Conditions	☐	Lack of Written Procedures/Rules
☐	Improper Materials Handling	☐	Poor Housekeeping	☐	Safety Rules Not Enforced
☐	Improper Shoes	☐	Poor Workstation Design or Layout	☐	Hazards Not Identified
☐	Horseplay	☐	Level or Surface Change	☐	PPE Unavailable
☐	Drug or Alcohol Use	☐	Fire Hazard	☐	Insufficient Worker Training
☐	Unsafe Act(s) of Others	☐	Hazardous Substances	☐	Insufficient Supervisor Training
☐	Improper Work Technique	☐	Inadequate Ventilation	☐	Improper Maintenance
☐	Improper PPE (not used or used incorrectly)	☐	Improper Material Storage	☐	Inadequate Supervision
☐	Failure to Warn or Secure	☐	Improper Tool or Equipment	☐	Insufficient Job Planning
☐	Operating at Improper Speeds	☐	Insufficient Job Knowledge	☐	Inadequate Hiring Practices
☐	By-Passing Safety Devices	☐	Excessive Noise	☐	Poor Process Design
☐	Servicing/Adjusting Machinery Motion	☐	Inadequate Guarding of Hazards	☐	Inadequate Workplace Inspections
☐	Unnecessary Haste	☐	Defective Tools/Equipment	☐	Inadequate Equipment
☐	Distracted While Performing Task	☐	Insufficient Lighting	☐	Unsafe Design or Construction
☐		☐	Inadequate Fall Protection	☐	Unrealistic Scheduling
☐	OTHER: Click here to enter text.	☐	OTHER: Click here to enter text.	☐	OTHER: Click here to enter text.

Final Root Cause(s) Determined:

Click here to enter text.

List Immediate Actions Taken and Results:

Click here to enter text.

What Should Be Done to Prevent a Recurrence:

☐ Engineering Control	☐ Administrative Control	☐ PPE

Click here to enter text.

Job Safety Analysis Review – JSA Directory Available (N:\Facilities Archive\JSA)

Is there a JSA that applies to the task being performed when the injury or incident occurred? *If yes, review the JSA, answer the following questions, and attach a copy to this report.*		Yes ☐	No ☐
Were hazards sufficiently identified on JSA? If not, explain:		Yes ☐	No ☐
Were identified controls adequate and implemented? If not, please explain:		Yes ☐	No ☐
Were the identified controls not implemented? If not, please explain:		Yes ☐	No ☐

Acknowledgment Sign-Off - Signatures to be obtained AFTER initial Safety Team review
(To ensure management engagement, report MUST be signed by EVERY level up the chain of management of employee)

Manager Chain	Signature	Name/Comments	Date
Employee's Manager	Choose an item.	Click here to enter text.	Click here to enter a date.
Manager's Manager	Choose an item.	Click here to enter text.	Click here to enter a date.
Sr. Manager	Choose an item.	Click here to enter text.	Click here to enter a date.
Executive Management	Choose an item.	Click here to enter text.	Click here to enter a date.

Figure 11–6. Concluded.

Date	Area	Shift	Name	Type	Comments	Reportable	Manager	Corrective action	C. Date	Responsible	Complete
07/12/2011	Office	days	J. Fox	Manual Handling	Person dropped box due to wei	Near Miss	Dominic Maree	Training	1/2/2014	R.Hennelly	Yes

Summary totals:
- Total Slips, Trips, and Falls: 0
- Total Manual Handling: 0
- Total Hit by Falling: 0
- Filtered by type: 1
- Filtered by area: 1
- Reportable: 0
- Non-Reportable: 0
- Near Miss: 1
- Number by manager: 1

Figure 11–7. Example of a spreadsheet summary form.

real-time basis. The use of dashboards and other automated notification tools enables leadership to keep up to date on current safety performance as well as the status of corrective actions without the need to wait for the traditional monthly report.

Figure 11–8 is an excerpt of an incident report form representing a typical stand-alone record-keeping tool. As can be seen, the format of the form is similar to a common spreadsheet. The advantage of this format over a spreadsheet is not the style of the form but rather the ability to better leverage the information collected by automatically compiling data across the organization and then providing real-time review, analysis, and distribution of the information to leadership.

Regardless of the format of the record-keeping tool being used, the same fundamental information needs to be collected for future analysis and reporting. Provided here is a sample of the types of information that can be collected. The actual information collected needs to be tailored to the organization's specific needs as well as regulatory or legal requirements.

1. Employee information
   a. Name
   b. Sex
   c. Age
   d. Marital status
   e. Number of dependents
   f. Occupation
   g. Employment category (regular, full time; regular part time; …)
   h. Length of employment
   i. Time in occupation
2. Description of incident
   a. Date of incident
   b. Time of incident
   c. Location
   d. Narrative of event
   e. Incident sequence (description of the sequence of events preceding the incident)
   f. Task and activity at time of incident
   g. Causal factors
   h. Type of incident
      i. At-risk behavior
      ii. Substandard condition
      iii. Property damage
      iv. Near miss
      v. Injury
      vi. Illness
3. Description of injury/illness
   a. Severity
      i. Signs and symptoms not requiring treatment
      ii. First aid
      iii. Medical treatment
         1. No lost workdays
         2. Lost workdays—days away from work
         3. Lost workdays—days of restricted activity
   b. Nature of injury or illness (sprains or strains, fractures, amputations, …)
   c. Part of body (head, lower back, hand, …)
   d. Source of injury or illness (machinery, floor or ground surfaces, containers, …)
   e. Event or exposure (contact with object, fall to lower level, repetitive motion, …)
4. Corrective actions and preventive actions (CAPA): *corrective actions*—what was done/should be done immediately; *preventive actions*—longer-term actions that need to be considered such as retraining, writing/rewriting procedures, purchase of new tools/equipment, etc.

An additional resource to help guide the type of information to collect and how best to code the information is the Bureau of Labor Statistics' *Occupational Injury and*

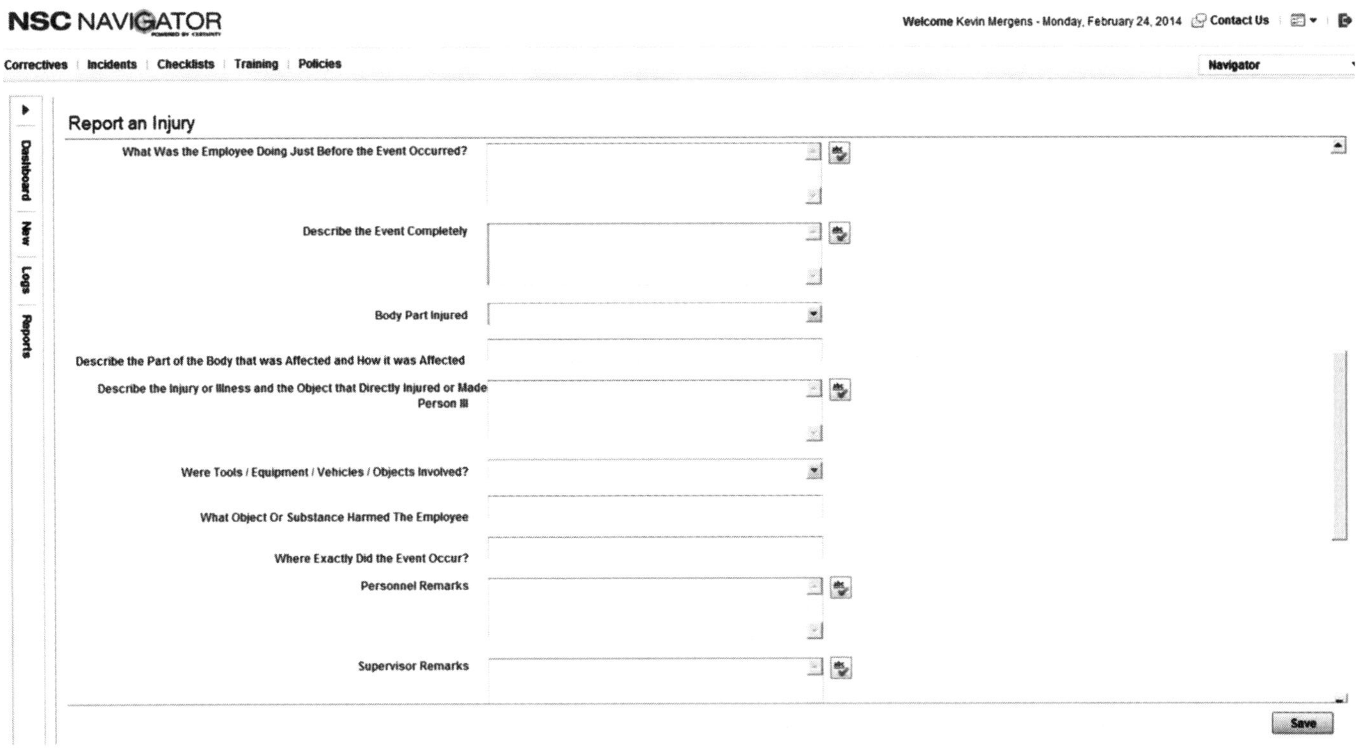

Figure 11–8. Example of an electronic incident reporting form.

Illness Classification Manual, version 2.1, which is available online at bls.gov. An added advantage of using this classification scheme is that data collected will be compatible with the industry benchmarks developed by the BLS each year. This will allow the organization to compare the prevalence of injury or illness types or injury/illness events (i.e., lost-time injuries, restricted work, etc.) against BLS national averages.

Periodic and On-Demand Reports

The incident reporting functions of the tools just discussed are accessed when incidents occur and are used to record the incidents and preserve information about contributing circumstances and factors. The data captured through this process can help to improve an organization's safety performance only if findings generated from the analysis of the data are provided to appropriate employee groups (both line employees and leadership) at regular intervals and in a way that helps to identify hazards as well as track their abatement. Guidelines on how to analyze injury and illness data are provided later in this chapter. The process used to distribute findings will vary depending on the format of the record-keeping tool being used. Paper- and spreadsheet-based tools are generally limited to periodic monthly reports and annual summaries, while the more advanced electronic tools may distribute the majority of information through the use of electronic dashboards, queries performed by individuals, and automated e-mail notifications. Regardless of the record-keeping tool being used and the methods used to distribute the findings, general guidelines are provided for effective distribution. Even when using record-keeping tools that enable leadership to conduct their own analyses, it is also beneficial to develop "official" periodic reports that are consistent throughout the organization and provide a consistent interpretation of the findings and trends.

Monthly Summary of Injuries and Illnesses

At a minimum, monthly incident summaries should be available to leadership (both line and executive) to reveal the current status of incident experience. This monthly summary of injury and illness cases (Figure 11–9) allows for a status check on each incident type, providing monthly totals, cumulative totals, and (possibly) incidence rates.

Because the monthly summary is primarily prepared to reveal the current status of incident experience, the information must be determined and distributed as soon as possible. If an incident report is still incomplete at the time of the report because the employee has not returned to work, or if the classification of an injury or illness is still in doubt, the organization or consulting physician should estimate the outcome. The report must then be included with the completed cases in the monthly summary.

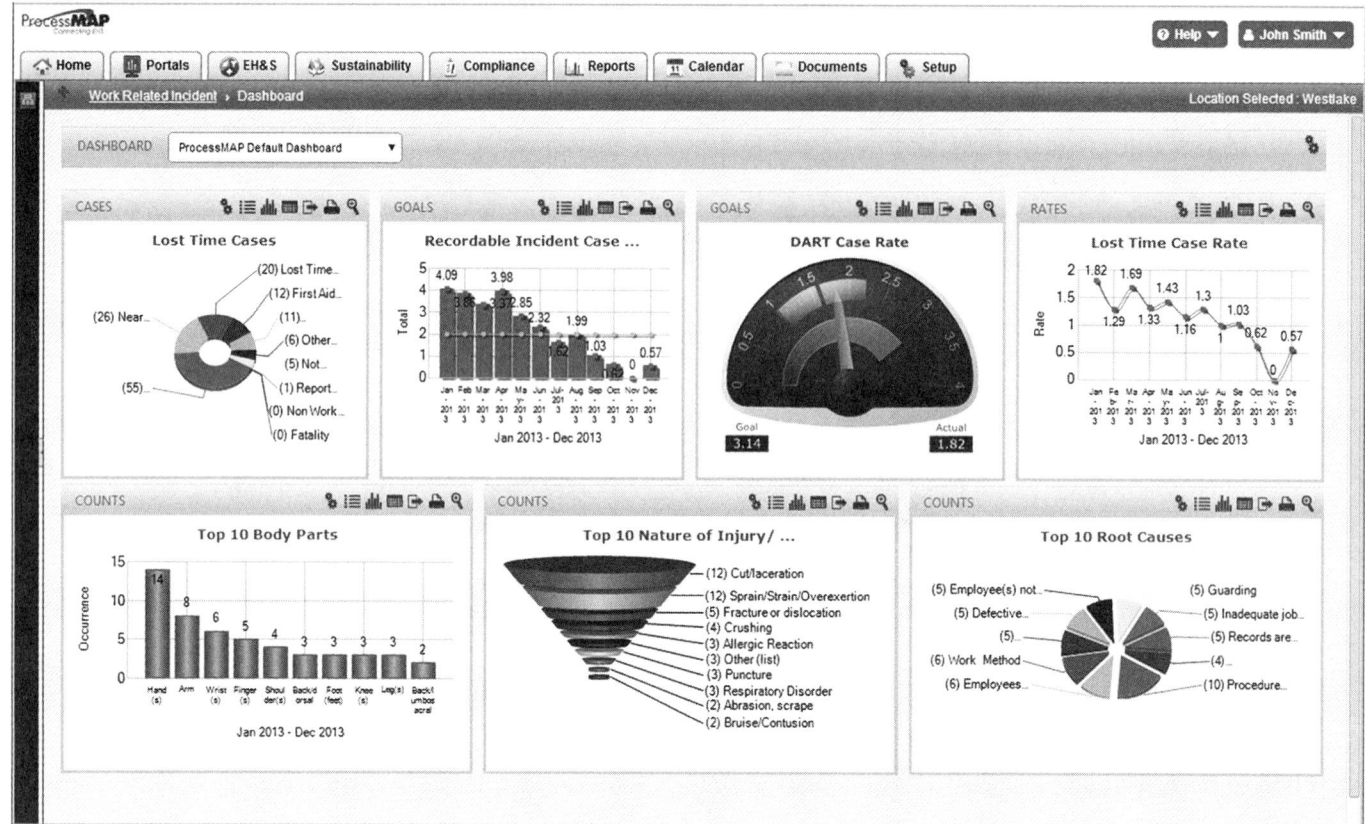

Figure 11–9. Example of an electronic incident report summary.

When definite information becomes available for estimated cases, any change in classification or OSHA lost workdays should be updated in the month of the closing of the case. The adjustment is then included in the cumulative figures for the year through that month. This procedure provides reliable monthly data and an easy method of adjusting cumulative data.

Caution must be used when reviewing monthly injury and illness data, especially in smaller companies. Monthly reports often show wide variations, making it difficult to evaluate safety performance correctly. For example, a small company may have only two or three injuries in a year. Thus, for the months in which these injuries occur, the incident rates will jump to extreme highs; in the other months, the rates will be zero.

Uses of Reports

Prior to distributing any reports or posting data, the safety and health professional should refer to the OSHA resources listed in the References section of this chapter to confirm that confidentiality requirements are addressed.

Reports to Leadership

Leadership must monitor the incident experience of the organization. Therefore, monthly and other periodic summary reports showing the results of the safety and health program should be furnished to the responsible leaders. Such reports do not need to contain details or technical language. They can be supplemented by simple charts or graphs to compare current incident rates with historical trends and the rates of sister facilities or other companies in the industry.

In a large organization, departmental data help leaders visualize incident experience in various facility operations and provide a yardstick for better evaluation of progress made in the elimination of hazards. It can be particularly valuable to compare cost figures, if they can be obtained, for different periods.

Bulletins to Supervisors

Supervisors are primarily interested in their own department and workers. One of the most effective ways to create and maintain the interest of supervisors in incident prevention is to keep them informed about the incident records of their departments. Department injury and illness rates based on sufficient amounts of exposure reflect the effectiveness of the supervisors' safety activities. Because interest increases with knowledge, bulletins containing analyses of the principal causes of incidents in each department can be a very

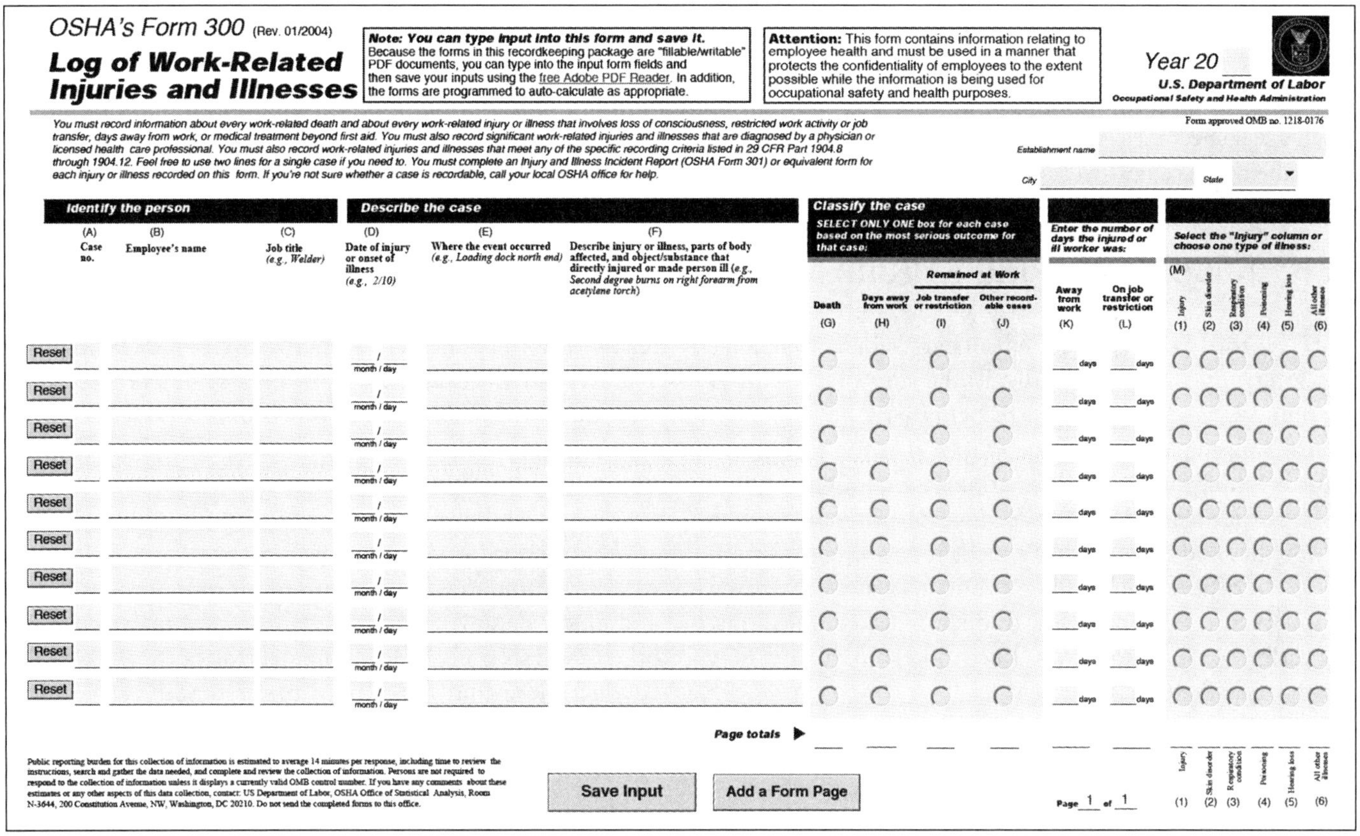

Figure 11–10. OSHA Form 300.

valuable tool. These bulletins not only maintain supervisors' interest at a high level, but also provide the supervisors with information to help them reduce injuries even further.

The agenda for employee safety meetings should emphasize information about unresolved injury and illness problems, frequent hazardous practices, hazardous types of equipment, and similar data disclosed by analysis of incidents that have occurred in the department and facility.

Bulletin Board Publicity

Posting a variety of materials on bulletin boards is one of the best ways to maintain employee interest in safety and health. Incident records furnish many items, such as the following:
- substandard conditions
- at-risk behaviors
- unusual incidents
- causes of incidents
- charts showing trends in incidents
- simple tables comparing departmental records.

Annual Summary

Every establishment subject to the OSH Act is obliged to post its annual summary between February 1 and April 30. The cumulative totals on OSHA Form 300, Log of Work-Related Injuries and Illnesses, serve as the annual report and are recorded and posted on the OSHA Form 300A. Establishments designated as part of the BLS annual survey also must report these figures to the requesting agency. See Figure 11–10 for an example of an OSHA 300 Log.

For leadership purposes, however, the annual report fulfills a more direct function. While monthly summaries of injuries and illnesses are primarily used to show the trend of safety and health performance during the year, the annual summary is used to compare data for the longer periods with data from previous years, with data from similar organizations, and with data from the industry as a whole.

As discussed previously, monthly injury and illness data, especially in smaller companies, often show wide variations, making it difficult to evaluate safety performance correctly. These variations will be smoothed out in annual totals, and the rate for the longer period will become more significant.

Annual summaries should be prepared as soon after the close of the year as information becomes available. In some cases, an injury or illness report may be incomplete because the employee has not yet returned to work. In other cases, the injury or illness occurs in one year but results in lost workdays in the following calendar year. In these cases one should estimate the total number of calendar days the employee is expectd to be away from work. This estimate is used to calculate the total for the annual summary. Then, when a case becomes final, the records can be corrected to show the actual days lost.

OSHA RECORD-KEEPING REQUIREMENTS

OSHA record-keeping requirements are subject to change and interpretation periodically. To be sure that the most current and accurate rules are being used, consult the record-keeping section of the OSHA website (osha.gov/recordkeeping/index.html) or contact a regional or area OSHA office. The *OSHA Recordkeeping Handbook* may be downloaded from OSHA's website or requested from an OSHA office. The handbook contains agency-approved policy, record-keeping rules, frequently asked questions, and letters of interpretation. Record-keeping forms may also be downloaded from the site.

In addition to the record-keeping requirements of 29 CFR 1904, many specific OSHA standards and regulations require maintenance and retention of records of medical surveillance, exposure monitoring, inspections, and other activities and incidents as well as the reporting of certain information to employees and to OSHA. These additional requirements are not covered in this section. (See Chapter 22, Occupational Medical Surveillance, in this volume.

INJURY AND ILLNESS DATA ANALYSIS

The traditional approach to incident investigation has been to expend the greatest amount of time and effort on the most severe and least frequent types of injuries/illnesses. By expanding the database to include "noninjury/illness incidents," an employer takes a big-picture view and can more accurately assess whether the system is stable, improving, or deteriorating.

Noninjury/illness incidents may be seen as precursors to possible injuries and may have the same complex set of root causes as more serious events. Noninjury/illness incidents may include actions or lack of actions that may result in an immediate or future breakdown of a system. For example, if an employee fails to clean up his or her workstation at the end of the shift and leaves a tripping hazard, employees on the next shift will be exposed to a hazard that can cause a potential injury or illness.

Design and development of a noninjury/illness incident reporting system must involve the workers' willingness to report incidents and the supervisor's willingness to gather and use the data. Both sides of the formula have to be adjusted to result in a net gain of mutual positive actions for improvement.

Leaders should use noninjury/illness incident data in conjunction with existing injury/illness incident data to correct system failures. Noninjury/illness incident reporting enhances the ability to monitor the effectiveness of the management system for preventing injuries/illnesses. Such data enable management to translate its own and the workers' insights into more reliable information and communication, to take corrective and preventive actions such as equipment and process redesign, to improve maintenance procedures, to target relevant training, and to solicit cooperation. Noninjury/illness incident reporting, to be effective, has to be dissociated from individual or departmental performance evaluation.

This section provides the mathematical formulas for the most common analyses conducted when exploring injury and illness data. Although the advent of spreadsheets and other analysis tools have eliminated the need to manually conduct these analyses, a basic understanding of how the statistics are calculated is essential to their appropriate use and interpretation.

Statistical Measures
Some common statistical categories of documentable occupational injuries/illnesses are listed here. Other categories may be used for statistical measures as well.
- death
- injuries/illnesses resulting in permanent disability
- injuries/illnesses resulting in days away from work
- injuries/illnesses resulting in days of restricted work activity
- injuries/illnesses involving both days away from work and days of restricted work activity
- medical treatment cases
- organizational levels (e.g., corporate, region, division, location, department, operation, etc.)
- operational function (e.g., maintenance/repair, assembly, paint, etc.)
- day, shift, time
- energy source, energy released, contact with energy.

Incidence Rates
Safety performance is relative. Only when an organization compares its injury and illness experience with that of its entire industry, or with its own previous experience, can it obtain a meaningful evaluation of its safety and health accomplishments. To make such comparisons, a method of measurement is needed that will adjust for the effects of certain variables contributing to differences in injury and illness experience. Injury totals alone cannot be used for two reasons:

1. An organization with many employees may be expected to have more injuries than an organization with few employees.
2. If the records of one organization include all the injuries treated in the first-aid room, while the records of a similar organization include only injuries serious enough to cause lost time, obviously the first organization's total will be larger than the second organization's figure.

A standard procedure for keeping records that provides for these variables is included in the OSHA record-keeping requirements. First, this procedure uses incidence rates that relate injury and illness cases, and the resulting days lost, to the number of employee-hours worked; thus, these rates automatically adjust for differences in the hours of exposure to workplace hazards. Second, this procedure specifies the kinds of injuries and illnesses that should be included in the rates. These standardized rates, which are easy to compute and to understand, have been generally accepted in industry, thus permitting the necessary and desired comparisons.

A chronological arrangement (time series) of these rates for an organization will show whether its level of safety and health performance is improving or worsening. Within an organization, the same sort of time series by departments not only will show the trend of safety performance for each department but will also reveal to leadership other information to make safety and health programs more effective. If it is found, for example, that the trend of incidence rates in an organization is up, a review of the rate trends by department may reveal that this change is accounted for by the rates of just a few departments. With the sources of the highest area/department rates isolated, management can concentrate safety efforts at these points.

A comparison of current incidence rates with those of similar organizations and with those of the industry as a whole serves a critical function. This step provides the safety professional with a more accurate perspective on the organization's safety and health performance than could be obtained by reviewing historical trends.

An "incidence rate" may be calculated for any or all of the categories given earlier as well as other categories of cases including noninjury/illness incidences. The incidence rate, IR, is defined as the number of cases per 100 full-time-equivalent employees per year. It is calculated by multiplying the number of cases (N) occurring in a given employee population by 200,000 and then dividing by the total number of employee-hours (H) worked by all employees in the given population. The number 200,000 is equivalent to 100 employees working 40 hours a week for 50 weeks. The general formula is

(1) $$IR = \frac{(N \times 200{,}000)}{H}$$

If rates are calculated for part of a year, such as quarterly or monthly, the same formula is used and the 200,000 factor remains the same.

It is important that the employee-hours used in the denominator of the incidence rate formula cover the same group of employees (establishment, department, office personnel, temporary employees under direct supervision by the organization, etc.) and the same time period as the injuries/illnesses used in the numerator. For example, to calculate an incidence rate for a manufacturing establishment, the hours worked by the shop-floor employees as well as the office employees should be included in the denominator. In addition, injuries and illnesses from both these groups are also included in the numerator.

The following example gives the formula for the incidence rate for cases involving days away from work:

$$\text{Incident rate of cases involving days away from work} = \frac{(\text{Number of cases involving days away from work} \times 200{,}000)}{H}$$

where H is employee-hours. This is a common incidence rate used for evaluating safety and health programs, for comparing experience between organizations, and for tracking trends over time.

Severity Measures

Measures of severity should be calculated for temporarily disabling injuries/illnesses using the appropriate number of days away from work. The total number of days counted in the particular category should be divided by the number of cases in that category to derive an average severity or average number of days lost per lost-time case.

(2) $$\text{Average days away from work} = \frac{\text{Total number of days-away-from-work}}{\text{Total number of days-away-from-work cases}}$$

An alternative method would be to derive an "incidence rate of days away from work" (either calendar or scheduled) using the following formula:

(3) $$\text{Incidence rate of days away from work} = \frac{(\text{Total number of days-away-from-work} \times 200{,}000)}{H}$$

where H is employee-hours.

For employers with a large work force, a death rate per 100,000 workers should also be calculated. This is not recommended for most employers because fatalities are usually rare events. The formula is:

$$DR = \frac{(D \times 200{,}000{,}000)}{H}$$

where DR is the death rate, D is the number of deaths, and H is employee-hours.

NonInjury/Illness Incidence Rates

Rates for documentable incidents other than occupational injuries/illnesses should be calculated. However, the appropriate exposure measure used in the rate denominator may not be employee-hours and the constant in the numerator may not be 200,000.

For traffic incidents involving highway vehicles, an appropriate rate would be incidents per million vehicle-miles (or kilometers) traveled. The formula is:

(4) $\quad$ Vehicle incident rate $= \dfrac{\text{(Number of incidents} \times 1{,}000{,}000)}{\text{Vehicle-miles (or kilometers)}}$

For other noninjury/illness incidents, the appropriate exposure measure may be hours of operation, or number of units produced (in industry standard units), or some other applicable measure. The general formula for a sentinel incident incidence rate is:

(5) $\quad \dfrac{\text{Noninjury incident}}{\text{incidence rate}} = \dfrac{\text{(Number of noninjury incidents} \times K)}{\text{Number of exposure units}}$

where K is a suitable constant chosen by the user so that the rate is greater than 0 and less than 100.

Examples of industry-specific noninjury/illness incidence rate measures for costs are cents per ton of coal, cents per acre of farm crop, cents per vehicle-mile, or cents per item manufactured.

Small Numbers

When the number of injuries, illnesses, or noninjury/illness incidents in the numerator is small (i.e., less than five per unit of time used in the analysis [week, month, or year]) or when the number of exposure units in the denominator is small, random variations from one period to the next may overshadow any changes in numbers due to real changes in the work environment. In this situation, it is often more meaningful to examine each case individually and to look for similarities by a qualitative analysis rather than through quantitative or statistical analysis. It may be useful to have a trade association or other independent group aggregate data from a number of small employers and do a statistical analysis of the grouped data. One quantitative method that may be useful is a Pareto diagram.

Pareto Diagram

Figure 11–11 shows an example of a Pareto diagram. The bars show the number of cases in each event or exposure category in rank order from largest to smallest. If the number of cases represented by one of the bars is equal to or greater than the arithmetic average (mean) of the bar heights plus three times the square root of the average, then that bar is significantly larger than the others and that kind of event should be given special consideration.

Expressed mathematically, let $b_1, b_2, \ldots b_n$ represent the number of cases in each of n categories (bars). Then the average ($\bar{b}$) is calculated by the formula $\bar{b} = (b_1 + b_2 + \ldots + b_n)/n$ and bar i is said to be significantly higher than the others if $b_i > \bar{b} + 3\sqrt{\bar{b}}$.

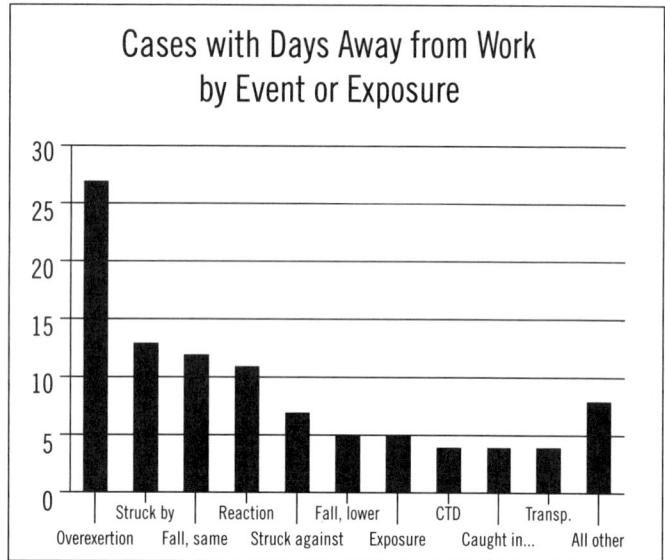

Figure 11–11. Pareto diagram.

There are 100 cases among the 11 bars in Figure 11–11, so the average is 9 cases per bar. The first bar, overexertion, represents 27 cases, which is significantly greater than the other categories: $27 > 9 + 3(\sqrt{9})$. Therefore, overexertion is a category of injuries that deserves special attention.

Tests for Significance of Changes in Rates

A test based on the normal distribution should be used to determine the significance of changes in incidence rates from one time period to another or differences in rates between departments, companies, etc. (Bissell 1994). Let r_1 and r_2 represent two incidence rates. The incidence rate formulas as given earlier in formula (1) are:

$$r_1 = \frac{n_1 \times 200{,}000}{h_1}$$

$$r_2 = \frac{n_2 \times 200{,}000}{h_2}$$

The value of the z-statistic for the difference in rates is given by

$$z = \frac{r_1 - r_2}{\sqrt{\dfrac{r_1}{w_1} + \dfrac{r_2}{w_2}}}$$

(6) where

$$r_1 = \frac{n_1 \times 200{,}000}{h_1}$$

$$r_2 = \frac{n_2 \times 200{,}000}{h_2}$$

Values of z between −3.00 and +3.00 indicate that the difference between the rates is not statistically significant and is likely to have happened by chance. Values of z greater than or equal to +3.00 or less than or equal to −3.00 indicate that there is a very high probability (approximately a 99.7% chance) that the difference between the rates results from a special cause.

An alternative form of formula (6), which may also be used when rates are based on units other than employee-hours (e.g., vehicle-miles), is:

$$z = \frac{\dfrac{x_1}{h_1} - \dfrac{x_2}{h_2}}{\sqrt{\dfrac{x_1}{h_1} + \dfrac{x_2}{h_2}}}$$

where x is the number of events (injuries, crashes, etc.) and h is the amount of exposure (hours, miles, etc.). The statistic z is interpreted as described earlier.

Control Charts

A chart of the kind used for quality control should be developed to evaluate the significance of changes over time in injury/illness as well as noninjury/illness incident experience. There are two kinds of control charts that are appropriate, depending on the situation. If the number of employee-hours worked is reasonably constant from month to month (i.e., the variation is within ±10% of the average), then a "C" chart, based on counts of injuries/illnesses, can be used. If the number of employee-hours fluctuates significantly from month to month, then a "U" chart, based on the incidence rate, should be used.

C Charts

The first step in developing a C chart is to calculate the average number of cases per month. To ensure a stable value, calculate $\bar{c}$ using at least 25 months' experience. After the average number of cases per month has been determined, the upper and lower control limits (UCL and LCL, respectively) are calculated using the formulas:

(7a) $UCL = \bar{c} + 3\sqrt{\bar{c}}$
(7b) $LCL = \bar{c} - 3\sqrt{\bar{c}}$

If the calculated LCL is less than zero, it is reset to equal zero.

U Charts

The first step in developing a U chart is to calculate the average monthly incidence rate, $\bar{u}$, over the base period. Use at least 25 months' experience to get a stable value. To calculate the average monthly incidence rate, do not compute the average of the monthly rates. Instead, determine the total number of cases over the base period (N) and the total number of exposure hours (employee-hours) worked over the base period (H) and enter them into formula (1) as given earlier. The upper and lower control limits are then calculated using the formulas:

(8a) $UCL = \bar{u} + 3\sqrt{\dfrac{(\bar{u} \times 200{,}000)}{\bar{n}}}$

(8b) $UCL = \bar{u} - 3\sqrt{\dfrac{(\bar{u} \times 200{,}000)}{\bar{n}}}$

where $\bar{n}$ is the average monthly employee-hours over the base period (i.e., total employee-hours divided by the number of months in the base period). If the calculated LCL is less than zero, it is reset to equal zero.

In addition to the calculation of the upper and lower control limits that are set at ±3 sigma (σ), one- and two-sigma levels can also be calculated using slightly modified versions of formulas 7a through 8b. An example of a two-sigma calculation for count data is provided here:

(9a) $U2\sigma = \bar{c} + 2\sqrt{\bar{c}}$
(9b) $U2\sigma = \bar{c} - 2\sqrt{\bar{c}}$

Drawing the Chart

When the calculations using formulas (7a) and (7b) or (8a) and (8b) have been completed, draw the chart. The horizontal axis starts with the first month of the base period and extends at least 12 months after the end of the base period. The vertical axis starts at some point below the *LCL* (usually starting at zero) and extends somewhat above the *UCL*. Draw a dotted horizontal line at the $\bar{c}$ or $\bar{u}$ value and solid horizontal lines at the *UCL* and *LCL* values. Finally, plot each of the monthly counts or rates above the appropriate month on the horizontal axis. If any of the data points in the base period fall outside the control limits, find the special cause of variation and correct the problem. Then recalculate the average and control limits and redraw the chart using the new values.

Using the Chart

There are important differences in the way a control chart is used for safety and health purposes as opposed to its use for a manufacturing process. First, if the count or incidence rate falls below the lower control limit, it would not make sense to encourage more incidents so as to return the system to "control." Rather, one looks for changes in the system that may have accounted for the good performance. (It is also possible that a point below the *LCL* could be due to failure to report events, misclassifying events, sabotage of the data, or errors in calculating the statistics or plotting the data.) Corrective and preventive action(s) is required if the system goes out of control by exceeding the upper control limit. Second, while a "steady-state" system (always in control) is desirable from a manufacturing point of view, steady improvement is expected from an occupational safety and health program. The long-term trend in monthly incident frequency (or incidence rates) should be downward, at least initially.

An incidence rate is the best measure for comparing the occupational injury/illness experience among organizations of various sizes. However, an incidence rate alone is not sufficient for determining the statistical significance of month-to-month changes within a company. Manufacturers are already faced with the task of determining the significance of variations over time in such factors as dimensions, weight, or performance of their products. To do this, they frequently employ quality control (QC) charts for process monitoring. A QC chart, when used properly, is a useful tool for distinguishing between random variation in a process (when the process is "in control") and caused variation (when there is lack of control over the process). A process that exhibits only random variation is said to be "in control." By distinguishing between the two types of variation, an organization can concentrate its efforts on reducing caused variations. The QC chart is also a useful tool for monitoring variation in incidence rates. As with any tool, however, its effectiveness depends on the skill of the user. In addition to the material presented here, users of this standard are urged to consult various references on control charts for more information. (See References at the end of this chapter.)

In constructing a control chart, all injury/illness data should be used (often including first-aid cases). This provides a more objective, stable measure for determining the significance of month-to-month fluctuations, especially when there is a seemingly large increase or decrease from the monthly average.

Because the average number of cases per month is calculated from numbers that are both larger and smaller than the average itself, variation from the average is to be expected. The variation can either be random or caused; caused variation is significant in the statistical sense and random variation is not. Caused variations are distinguished from random variations by using a control chart to determine which variations are significant and which are not. As with any statistical method, the use of a control chart can lead to an erroneous conclusion. The important point is that, with a properly constructed procedure, the error rates are controlled.

Figure 11–12 shows an example of a C chart. A C chart was appropriate because employment in the company was fairly constant with no important seasonal variations during the study period. Using 30 months' experience, it was determined that the company had an average of 25 occupational injury/illness cases per month. Substitution of 25 for $\bar{c}$ in formulas (7a) and (7b) yielded the following upper and lower control limits:

$25 \pm 3\sqrt{25} = 25 \pm 15 = 25 + 15 = 40$ for the upper limit

$25 - 15 = 10$ for the lower limit.

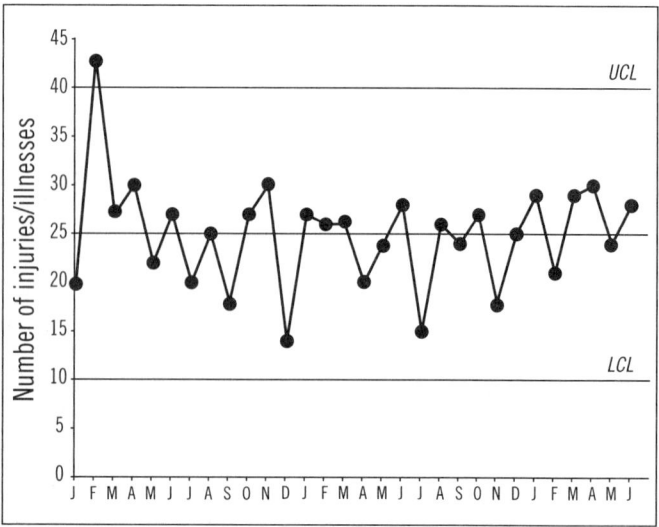

Figure 11–12. C control chart.

The control chart was then constructed, and the company plotted the actual number of cases each month. Note that the actual monthly number of cases, with the exception of the first February, falls within the upper and lower control limits. These variations are random and do not indicate significant changes from the monthly average of 25 cases. The variation for February indicates a lack-of-control condition that may be due to an assignable cause. The cause(s) of the out-of-control condition should be investigated and appropriate corrective measures should be taken. Suppose, for example, an investigation revealed that a guard had been left off a machine after its monthly maintenance and this was the cause of the increase in cases for the month of February. It might be found that the guard was replaced the next month, which corrected this cause and brought the injury/illness experience back in control for March. It is possible, though improbable, that the lack-of-control indication is erroneous, in which case no assignable cause will be discovered by the investigation.

Figure 11–13 shows an example of a U chart. Employment patterns had large seasonal variation, so a C chart was not appropriate. Using 25 months' experience it was determined that the average monthly incidence rate was 8.0 and the average monthly employee-hours was 176,000 (i.e., about 1,000 full-time employees [working 8 hours per day, 22 days per month] with a range of 700 to 1,300). Substituting these values in formulas (8a) and (8b) gives the following upper and lower control limits:

$$8.0 + 3\sqrt{\frac{8.0 \times 200{,}000}{176{,}000}} = 17.04 \quad \text{for the upper control limit}$$

$$8.0 - 3\sqrt{\frac{8.0 \times 200{,}000}{176{,}000}} = 1.05 \quad \text{for the lower control limit}$$

The control chart was then constructed as described earlier, and the company plotted the actual occupational injury/illness incidence rate each month. Note that all of the actual incidence rates fall within the upper and lower control limits, so no corrective action is indicated.

A slightly more rigorous method (and actually easier to implement when using a spreadsheet) is to calculate the UCL and LCL for each time interval (month) using the actual hours (H_i) for that interval. The formulas for the control limits for interval i become:

(10a) $\quad UCL_i = \bar{u} + 3\sqrt{\left[\frac{(\bar{u} \times 200{,}000)}{H_i}\right]}$

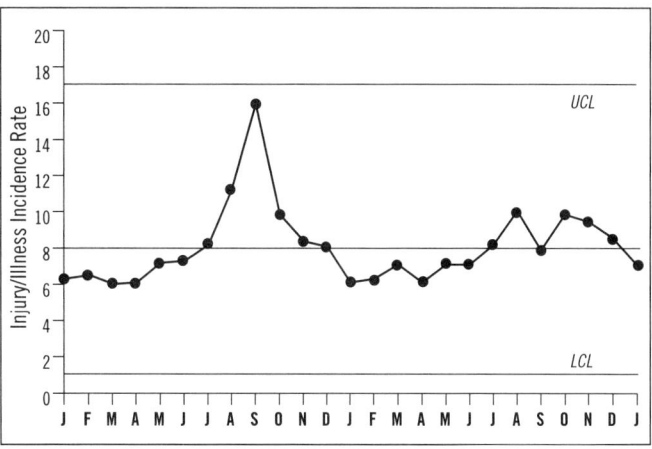

Figure 11–13. U control chart.

(10b) $\quad LCL_i = \bar{u} - 3\sqrt{\left[\frac{(\bar{u} \times 200{,}000)}{H_i}\right]}$

where $\bar{u}$ is as defined earlier.

The control limits will vary inversely with the number of hours worked. More variation is to be expected during months with few hours worked and less variation during months with many worked. (See Duncan 1986 or Wheeler 2004 for more information.)

Additional guidelines for constructing and using control charts:

1. Always use sufficient historical experience when calculating the average to ensure that a stable average will be developed. At least 25 to 30 periods (usually months) is preferable, except as noted in item 3.
2. A control chart can be constructed if the average number of cases per period is less than nine, but the calculated lower control limit will be negative and must be reset to zero. One way to avoid this is to increase the width of the time interval plotted, such as from a month to a quarter, in order to increase the average number of cases per period. If longer time intervals are used, then care must be taken to ensure that data from the earliest time periods are comparable to more recent periods; that is, that no significant changes in external factors have occurred, as discussed in the next guideline.
3. Construct a new control chart if external factors—such as a change in the level of employment, a change in the work environment, or a change in work hazards—bring about a permanent change in injury/illness experience. When calculating a new average in such a situation, do not use experience prior to the change. Verify that the data truly did change as a result of the process change. The existence of a shift in the data can be verified by using either the trend rules given next or a statistical test.

4. A lack of control is indicated (using five rules found in Wheeler 2004) when:
   a. a point falls outside a control limit
   b. two out of three successive values are (i) on the same side of the central line and (ii) more than two σ away from the central line
   c. four out of five successive values are (i) on the same side of the central line and (ii) more than one σ away from the central line
   d. eight successive values fall on the same side of the central line
   e. a pattern (of any length) repeats itself eight times in succession. An explanation should be sought.

Other authors (Bissell, Duncan, etc.) recommend many other rules to detect lack of control.

Implementation and Follow-Up

Having completed the analysis of occupational injuries/illnesses and noninjury/illness incidents and constructed control charts, etc., incorporate the knowledge gained into the organization's decision-making processes and into the leadership policies and practices that will directly affect health and safety.

OFF-THE-JOB INJURIES

In recent years, off-the-job disabling injuries of employees have far exceeded on-the-job disabling injuries. Any unscheduled absence of employees can cause production slowdowns and delays, costly retraining and replacement, or costly overtime by remaining employees. As a result, many safety professionals are very concerned with off-the-job injuries their employees suffer. Moreover, activity to reduce off-the-job incidents should help to promote interest in this area of safety and health. (See also Off-the-Job Safety Programs in Chapter 6, Loss Control Programs, and Off-the-Job Disabling Injury in Chapter 10, Incident Investigation, Analysis, and Costs.)

Definitions and rates used for off-the-job injuries are compatible with those of the OSHA record-keeping requirements. Because the data on off-the-job injuries are not as easy to obtain, however, certain simplifications are introduced. Exposure (for use in rates per 200,000 employee-hours) is standardized at 312 employee-hours per employee per month (equal to 4⅓ weeks less 40 hours per week at work and 56 hours per week for sleeping). Provision is also made for recording home, public, and transportation injuries separately to allow for concentrated effort in problem areas.

SUMMARY

- Organizations must maintain records of work-related employee injuries and illnesses as required by OSHA, state workers' compensation authorities, and insurers. Good record-keeping systems provide data to evaluate incident problems as well as safety and health program effectiveness, identify high-incident-rate areas, create interest in safety, enable the organization to concentrate efforts on more serious incident problems, and measure effectiveness of countermeasures against hazards and at-risk behaviors.
- Record-keeping systems can successfully use a variety of tool formats, including paper-based tools, spreadsheet tools, and electronic tools designed specifically to support the record-keeping process. The main differences among these tool formats are the level of automation that is possible and the degree of integration with other business systems.
- A successful record-keeping process includes all employees. Every employee is responsible for reporting incidences; supervisors are primarily responsible for investigation; safety professionals provide support and expertise; and senior leadership establishes the organizational commitment for the overall safety program.
- The primary purpose of an incident report is to obtain accurate, objective information about the causes of incidents in order to prevent incidents from recurring.
- OSHA incident rates help companies compare their safety performance with the performance of previous years or of the entire industry to evaluate their safety and health programs.
- Quality control charts and other statistical analysis techniques help distinguish between random variations, which are in control, and caused variations, which are out of control. This distinction allows leadership to concentrate safety and health efforts on out-of-control variations.
- Recording and measuring off-the-job injuries is important to help leadership develop comprehensive safety and health programs for employees.

REFERENCES

American Society of Safety Engineers. *Dictionary of Terms Used in the Safety Profession*. 2nd ed. Des Plaines, IL: ASSE, 1981.

Bird, F. E., Jr., and Germain. *Practical Loss Control: Leadership*. 2nd ed. Loganville, GA: Institute Press, November 1992.

Bissell, D. *SPC for TQM*. New York: Chapman & Hall, 1994.

Bureau of Labor Statistics. *Occupational Injury and Illness Classification Manual.* Version 2.01. www.cdc.gov/wisards/oiics/Links.aspx.

Deming, W. E. *Sample Design in Business Research.* New York: Wiley, 1990.

Duncan, A. *Quality Control and Industrial Statistics.* Burr Ridge, IL: Irwin, 1986.

Ferry, T. S. Techniques of Operations Review (TOR). In *Modern Accident Investigation and Analysis: An Executive Guide.* New York: Wiley, 1981.

Juran, J. M. *Juran's Quality Handbook: The Complete Guide to Performance Excellence,* 6th ed. New York: McGraw-Hill Professional, 2010.

McKinnon, R. C. *Safety Management: Near Miss Identification, Recognition, and Investigation.* Cleveland, OH: CRC Press, 2012.

National Safety Council. *Injury Facts.* Issued annually.

Petersen, D. *Techniques of Safety Management: A Systems Approach.* 4th ed. Goshen, NY: Aloray Inc., 2003.

Tufte, E. R. *The Visual Display of Quantitative Information.* 2nd ed. Cheshire, CT: Graphics Press, 2001.

U.S. Department of Labor, Mine Safety and Health Administration. 30 CFR – Part 50, Notification, Investigation, Reports and Records of Accidents, Injuries, Illnesses, Employment, and Coal Production in Mines.

U.S. Department of Labor, Occupational Safety and Health Administration. 29 CFR – Part 1904, Recording and Reporting Occupational Injuries and Illness.

———. *OSHA Forms for Recording Work-Related Injuries and Illnesses.* osha.gov/recordkeeping/new-osha300form1-1-04.pdf.

———*OSHA Recordkeeping Handbook.* www.osha.gov/Publications/recordkeeping/OSHA_3245_REVISED.pdf.

U.S. Department of Transportation, Federal Railroad Administration. 49 CFR Part 225—Railroad Accidents/Incidents: Reports Classification, and Investigations.

Wheeler, D. J. *Understanding Statistical Control,* 3rd ed. Knoxville, TN: SPC Press, 2010.

———*Understanding Statistical Control.* 3rd ed. Knoxville, TN: SPC Press, 2010.

REVIEW QUESTIONS

1. Safety personnel must maintain records for what three reasons?
    a.
    b.
    c.
2. List five of the seven ways a good record-keeping system can help the safety professional.
    a.
    b.
    c.
    d.
    e.
3. In which department does the collection of injury and illness data generally begin?
    a. human resources department
    b. safety department
    c. first-aid department
    d. department in which the accident occurred
4. Which nonsafety professional should make a detailed report of each incident?
    a. human resources director
    b. medical staff supervisor
    c. legal department manager
    d. injured worker's supervisor
5. How often should a summary of injuries and illnesses be prepared?
    a. after every 10 occurrences
    b. weekly
    c. monthly
    d. quarterly
6. In what publication are the OSHA record-keeping requirements found?
7. What is the formula for calculating incidence rates of recordable cases?
8. Why are many safety specialists concerned with the off-the-job injuries their employees incur?
9. How are control charts used in the injury and illness data analysis process?

PART 3

Safety/Health/Environment Program Organization

The chapters in Part 3 build on the topics discussed earlier—the history and future of safety and health as well as controlling loss on many levels. This part directs the safety professional's attention to programs and systems that can prevent accidental injuries and deaths. Taking into account that every workplace has unique needs, the topics in this part offer an overview of a successful SH&E program, the fundamental elements of which are constant.

Core programs that make a positive contribution to the improvement of safety, health, and the environment are described, including developments in indoor air quality, ergonomics, emergency preparedness, violence prevention, and laboratory safety, to name a few.

Occupational Health Programs

12

John F. Montgomery, PhD, CSP, CHMM
Christopher Montgomery
Sherrie C. Wilson

Components of Occupational Health Programs
Objectives ▸ Occupational Health Unit Location
▸ First-Aid Provisions

Occupational Health Professionals
Occupational Health Coordinator ▸ Occupational Health Nurse
▸ Occupational Health Physicians

Physical Examination Program
Preplacement Examinations ▸ Periodic Examinations
▸ Emergency Medical Planning ▸ Employee Health Records
▸ Neck or Wrist Tags for Medic Alert

Special Programs
Health Promotion and Wellness ▸ Medical Surveillance
▸ Health Care for Women ▸ Substance Abuse and Employee Health ▸ Older Workers ▸ Shiftwork

Summary

References

Review Questions

The health and medical services that promote and support employee health in an organization are integral components of every occupational safety and health program. This chapter discusses the important functions and structure of occupational health programs; however, each organization must decide how to provide occupational health services to best meet the needs of its employees. The following topics are covered in this chapter:

- types of employee health services and first-aid provisions companies should provide
- various occupational health professionals and their roles in employee health care
- physical examination programs and emergency medical services provided by companies
- special programs for wellness promotion, female and older workers, substance abusers, and workers coping with unusual work hours.

COMPONENTS OF OCCUPATIONAL HEALTH PROGRAMS

Occupational health programs can range from comprehensive, with a large array of services, to programs with services that meet only mandatory legal requirements. One establishment may have a full-time staff of occupational health nurses, physicians, industrial hygienists, safety specialists, and technicians housed in a model occupational health unit, while another may have only the basic first-aid kit with a trained person to provide first aid.

Ideally, occupational health programs, regardless of size, are composed of elements and services designed to promote and maintain the health of the work force, prevent or control occupational diseases and accidents, and reduce and prevent disability and resulting lost time. A good program should provide the following components:

- maintenance of a healthful work environment through establishing a comprehensive occupational health and safety program
- health examinations, including selective baseline and periodic surveillance for employees performing particular jobs as required by regulatory standards, as well as fit-for-duty, return to work, and job transfer evaluations
- diagnosis and treatment services for occupational injuries and illnesses
- case management services
- immunization programs
- confidential health records kept separate from personnel records
- health promotion, education, and counseling
- open communication between the company's occupational health personnel and an employee's personal physician.

Treatment of ill or injured persons should be provided within the local standards of nursing and medical practice, using appropriate medical guidelines and standards for employee health care.

Objectives

Occupational health programs are concerned with all aspects of the employee's health and the employee's relationship with the workplace environment. The basic objectives of a good occupational health program are as follows:

- to promote health and protect employees against health hazards in their work environment
- to implement audit programs tailored for specific work-related activities
- to facilitate placement and ensure that individuals are assigned work that matches their physical and mental capabilities and that they can perform with an acceptable degree of efficiency without endangering their own health and safety or that of their fellow employees
- to promote adequate health care and rehabilitation of employees injured on the job and to maintain good workers' compensation programs to ensure this
- to monitor the work environment for potential hazards and correct them as needed
- to encourage workers via newsletters, wellness programs, and inexpensive health coverage availability to maintain their personal health.

By applying occupational health principles to the workplace, management can place all employees, including those who are disabled, in jobs according to their abilities to perform the work (see Chapter 23, Workers with Disabilities, in this volume). This approach also promotes continuing health care and rehabilitation of workers who become ill or injured on the job. The achievement of these objectives will benefit both employees and employers by improving health, morale, and productivity.

Industry experience has shown a strong relationship between accident prevention and occupational health. For example, some industrial chemicals, when improperly handled, represent serious hazards to health, property, and the environment. Depending on various environmental and workplace conditions, the vapor from a chemical can ignite, explode, or, if inhaled, cause symptoms ranging from dizziness to death. Dermatitis can be caused by the contact of a chemical with the worker's skin. (For details on the effects of specific chemicals, see the National Safety Council's *Fundamentals of Industrial Hygiene.*)

Safety professionals have demonstrated their ability to reduce the rates of accidental injuries by controlling many phases of the industrial environment through worker education and through improved management techniques (Boylston 1990). As per OSHA industry-specific guidelines for training programs, these two elements must allow for the different physical and emotional characteristics of individual workers. Such characteristics account for variations in workers' job attitudes, productivity, safety and health practices, and personal health management.

Occupational Health Unit Location

Occupational health services can be provided in any clean, private location in the company that is accessible to employees. However, the experience of many organizations indicates that the occupational health unit commands greater respect if management pays careful attention to providing a suitable and efficient location that has a pleasant, comfortable appearance and up-to-date equipment. The location must have adequate space for multiple casualties during emergencies and suitable provisions for disabled employees (Guidotti et al. 1989). Alternatively, occupational health clinics offer more options for treatment and limit liability to the company.

The rooms housing the occupational health services should be easy for workers to find. If management places the department near the greatest number of employees, more workers are likely to report all minor injuries and have them treated. One potential location is near the work area entrance, which is also convenient for loading or unloading supply trucks and for bringing ambulances to the door, if necessary. Also, injured workers who are off duty but under treatment may come and go through a separate entrance. Management should try to put this department in a protected area, outside any potential disaster area.

The occupational health unit should be adequate for the size and needs of the work force and should contain a minimum of two rooms: a waiting room and a treatment/service area. Rooms for special purposes can be added according to the needs and size of the work force and company. The surgical treatment room should be large enough to treat more than one person at a time. Small dressing booths should be provided, which can give workers some degree of privacy.

The health unit will often manage the company's vendor services, using the efficient credibility of hospital-based drug-testing services and other services.

First-Aid Provisions

Good administration of first aid is an important part of every safety program (Workers' Compensation Board 1991). First-aid kits, automated external defibrillators (AEDs), and other supplies approved by the occupational health manager should be stored where they are readily accessible to trained personnel on each shift. Health care personnel should keep a careful record of each injury or illness and the administration of first-aid and send an incident investigation report to the injured worker's supervisor.

In most companies today, the occupational health nurse manages the occupational health unit. This individual should routinely monitor first-aid provisions and train ancillary personnel in first-aid procedures and treatments, including cardiopulmonary resuscitation (CPR) and AED use. Although most companies do not employ a full-time medical doctor, they should maintain a good liaison with a local physician or physicians designated to handle certain injuries. The physician(s) should be invited to the company occasionally to tour the establishment and provide consultation when needed. The company-designated physician(s) should be familiar with the type of work done so that he or she can evaluate injuries, illnesses, and other employee complaints.

In many small organizations and in field operations, it is neither practical nor efficient to maintain qualified professional health care personnel full time. In such cases, the best arrangement is to use trained first-aid responders who follow procedures and treatments (protocols) outlined by qualified health care professionals and who are available on an on-call or referral basis to treat serious injuries. If injured employees have their choice of treating physician, the employer should comply with all requests, if possible.

There are two kinds of first-aid treatments: emergency and prompt attention.

- *Emergency treatment:* Emergency care must be given for immediate, life-threatening conditions. First-aid staff provides care until proper medical treatment can be given (Hau 1994). Proper first-aid measures reduce suffering, lessen the severity of an injury or illness and prevent death.
- *Prompt attention:* This type of first aid is used to treat minor injuries such as cuts, scratches, bruises, and burns. Ordinarily, the injured person would not seek medical attention for these injuries. By requiring that all employees immediately report for treatment when they are injured, regardless of the extent of the injury, the company can help reduce workers' risk of infection, disability, and missed diagnoses.

A first-aid program should include the following:
- properly trained and designated first-aid responders on every shift
- instructions for calling an ambulance, fire, and police
- a protocol to follow for transporting ill or injured employees or calling 911 in an emergency

- protocols for calling a physician and notifying the hospital that a patient is en route
- a first-aid unit or kit, AED, and supplies approved by the health care professional
- a first-aid manual with procedures and protocols
- a list of chemicals, the Safety Data Sheet (SDS), exposure control plan, and other information needed to treat possible reactions to chemicals, including routes of exposure
- an adequate first-aid record system and follow-up.

The first-aid procedures manual should be developed by the occupational health nurse and physician with input from the occupational health staff. All occupational health staff should review and sign off on the manual. Health care staff should render emergency and first-aid treatment in accordance with current science guidelines or protocols and legal parameters of practice. In some situations, physician direction and practice may be required. The consulting physician should specify the type of medication, if any, to be used on injuries such as cuts and burns. The physician should specify the procedures to be followed when medication for temporary relief of nonoccupational ailments, such as for a toothache or headache, is administered. Procedures should also state that in areas where chemicals are stored, handled, or used, emergency flood showers and eyewash fountains (Figure 12–1) should be available and clearly identified. Safety Data Sheets (SDSs) should also be available.

Professional occupational health nurses practice under the state Nurse Practice Act. They must be cognizant of their legal scope of practice as defined by their specific state and the delegatory functions they can handle legally. Anyone responsible for first-aid treatment must understand clearly the limits and scope of practice of this work. Because improper treatment may adversely affect the employee and/or involve the company in serious legal problems, the first-aid attendant should be duly qualified and certified by the Mine Safety and Health Administration (MSHA), the National Safety Council, the American Heart Association, or the American Red Cross. These certificates must be renewed at specific intervals.

First-Aid Training

The National Safety Council first-aid textbooks and the MSHA manual of first-aid instruction are recommended for training employees. Many feel that accidents occur less frequently and are usually less severe among workers trained in first-aid procedures. It is, therefore, advisable that as many industrial workers as possible be given this training (Figure 12–2). The National Safety Council publishes posters and booklets that can be used for training employees. Other valuable sources are included in the References at the end of this chapter.

Figure 12–1. The emergency shower and eyewash should be well identified. Top: Harmful chemicals are quickly washed away and clothing fires doused immediately by the drenching action of the shower. Bottom: First-aid treatment to the eye must be prompt and consists of prolonged irrigation of the exposed eye with low-pressure water. *(Photos reprinted with permission from Haws Drinking Faucet Co.)*

First-Aid Room

It is always advisable to set aside a conveniently located room for the sole purpose of administering first-aid treatment. The person administering first aid should have a proper place to work. A first-aid room should be equipped with the following items:

- examining table
- cot for emergency cases, enclosed by movable curtain
- dustproof cabinet for supplies
- waste receptacle and biohazards-disposable containers
- small table
- chair with arms and one without arms
- magnifying light on a stand
- sink and washing facilities
- dispensers for soap, towels, cleansing tissues, and paper cups
- wheelchair
- stretcher
- blankets
- bulletin board to post all important telephone numbers for emergencies
- bed(s)
- appropriate medications in locked cabinet
- automated external defibrillator (AED) and proper prep kit
- oxygen
- surgical-type mask
- bag-valve-mask.

Health care personnel often use oxygen when treating first-aid cases. Because of the danger of fire or explosion, and smoking, and petroleum products should be prohibited when oxygen is administered. Any type of resuscitating device should be used only by trained persons. In addition, variable-flow oxygen may be restricted to prescription-only uses.

First-Aid Kits

Many types of first-aid kits are available to fill every need, depending on the types of accidents that may occur. Commercial or cabinet first-aid kits, as well as unit kits, must meet regulatory requirements (Workers' Compensation Board 1991). MSHA outlines specific first-aid items required at certain locations in mines and processing facilities.

Kits vary in size from the pocket model to almost a portable first-aid room. The size and contents depend on the intended use and the types of injuries to be treated. For example, personal kits contain articles essential only for the immediate treatment of injuries, whereas departmental kits meet the needs of a group of workers. As a result, the quantity of material each kit contains depends on the size of the work force. Trunk kits, which are the most complete, can be carried easily to an accident site or can be stored near working areas far from well-equipped emergency first-aid rooms. Although a trunk kit can include bulky items, such as a washbasin, blankets, splints, and stretchers, it can still be carried by only two persons.

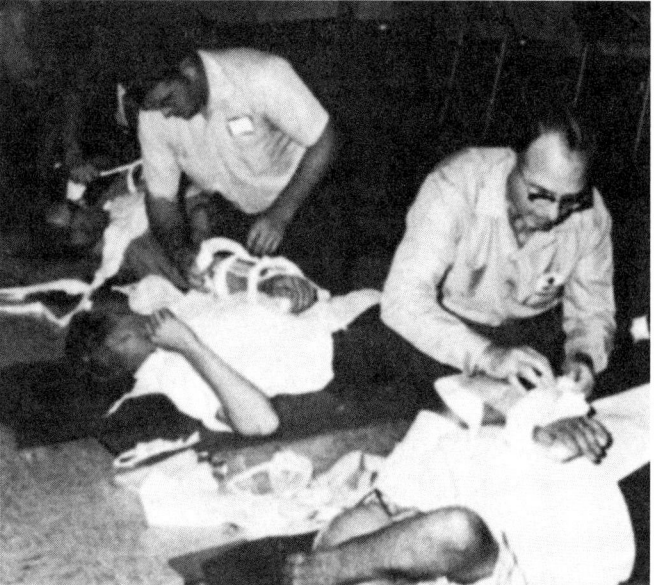

Figure 12–2. Demonstration and practice play an important role in first-aid training.

Keeping first-aid kits strategically located throughout the company seems to work best when such kits are supervised properly (Figure 12–3). A trained individual should be responsible for maintaining each kit, treating serious injuries, and giving first-aid measures for minor cuts and scratches. This employee never provides more than immediate, temporary care. By combining a well-trained first-aid provider and well-maintained first-aid kits, companies can ensure that many minor injuries that may otherwise be ignored receive prompt, proper treatment. All miners, for example, receive first-aid training and annual refresher training.

In some organizations whose activities are widely scattered, employees may need to use first-aid kits and administer treatment themselves. Under these circumstances, the organization can control first-aid service by having the attendant in charge properly instructed in first aid and by seeing that the service as a whole has health care supervision. Many third-party first-aid providers (e.g., Zee Medical) offer free first-aid training to all employees as part of their fee.

In general, qualified health care personnel should supervise the maintenance and use of all first-aid kits and approve all materials provided. A member of the health services group should regularly inspect all first-aid materials and report on their content level and serviceability. Maintaining quantities of materials in the first-aid kit is

 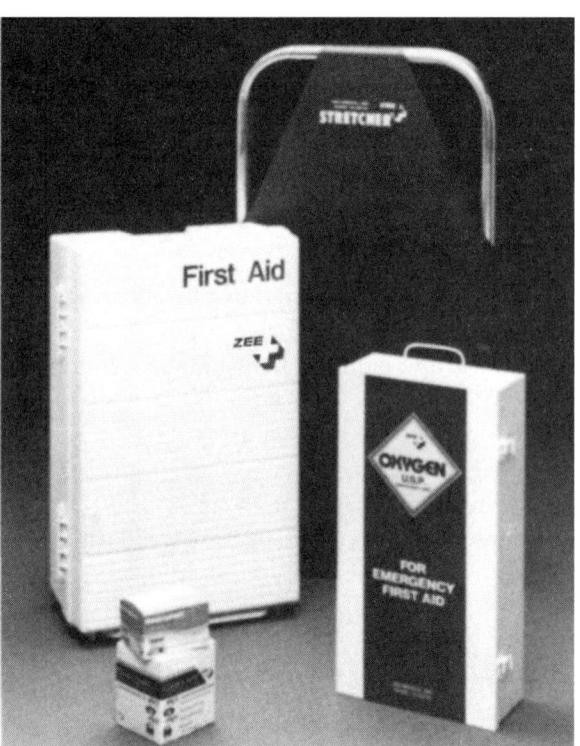

Figure 12–3. First-aid kits should be well distributed throughout a facility. Left: They should be supervised by a trained person so that employees do not try to "doctor" themselves. Right: Stretchers and oxygen kits should also be available in areas distant from the first-aid room. *(Courtesy Zee Medical.)*

easier if each kit contains a list showing the original contents and the minimum quantities needed. Over-the-counter medications made available to employees should be industrial packed. Health care workers should clearly label and date all bottles or other containers in the kit.

Recommended materials for first-aid kits are listed in the textbooks of the National Safety Council, the MSHA, and the American Red Cross. Suggestions are also available from the American Medical Association, the American Petroleum Institute, and manufacturers of first-aid materials.

Stretchers

Company health services must be able to quickly transport a seriously injured person from the scene of an accident to a first-aid room or a hospital (Workers' Compensation Board 1991). Such promptness can help physicians determine the gravity of the injury and perhaps even increase the victim's chances for survival. The stretcher provides the most acceptable method of hand transportation for accident victims. It can be used as a temporary cot at the scene of an accident, during transit, and in the first-aid room or occupational health unit.

Although there are several types of stretchers, the commonly used army stretcher is satisfactory in most cases. However, when workers must hoist or lower the injured person out of an awkward place, it is better to use a lightweight stretcher to which the employee may be strapped and kept immobile.

Stretchers should be conveniently stored near places where employees are exposed to serious hazards. It is customary to keep stretchers, blankets, and splints in clearly marked, prominent cabinets. Stretchers should be clean and ready to use at all times. They should be stored and protected against destructive vapors, gases, dust, or other substances and against mechanical damage. If the stretcher is made of materials that will deteriorate, health care staff should test it periodically for durability and strength.

Any victim of a severe injury—such as a fall, a head injury, a neck injury, a back (spinal) injury, or any injury resulting in unconsciousness—should not be moved until the arrival of emergency medical services (EMS) personnel. Untrained responders may move patients with head, neck, and back injuries only if they are in a life-threatening atmosphere and must be moved to prevent further injury or death.

These patients should be secured and transported on a rigid spine board with proper restraint of head and neck. Regardless of the transportation device, the employee should be secured to the device with straps or cravats. EMS should always be called whenever possible to move a person with a head, neck, or back (spinal) injury.

OCCUPATIONAL HEALTH PROFESSIONALS

Management often needs the services and skills of several professionals to attain the best results in its occupational health and safety programs (Felton 1990; Rogers 1994). These professionals include the following:

- Increasingly, physician specialists are trained in occupational and preventive medicine. They are assisted by specialists in orthopedic surgery, ophthalmology, radiology, surgery, dermatology, psychiatry, and other areas.
- Occupational health nurses use specialized knowledge, not only in basic nursing procedures and health maintenance, but also when dealing with legal, economic, and social issues; labor laws; and occupational health sciences (e.g., toxicology, safety, industrial hygiene). In many instances, nurses are the only full-time health care providers in a company.
- Industrial hygienists apply specialized knowledge in recognizing, evaluating, and controlling health hazards in the work environment.
- Safety specialists use specialized knowledge to design and implement strategies aimed at preventing and controlling workplace exposures that result in unnecessary injuries and deaths.
- Other occupational health professionals may include occupational health coordinators, toxicologists, epidemiologists, and employee assistance counselors.

Working both individually and collectively, these specialists have helped improve the occupational health and safety record of many industries. In some companies, these professionals are so well organized and effective in anticipating and correcting hazards that employees may actually be safer and healthier at work than in their own homes. In fact, one of the current challenges for occupational safety specialists is to educate workers in off-the-job safety at home, in recreational activities, and during travel so that they will be able to return safely to work each day. Employees injured off the job are as much a loss to the operation as if they were injured on the job.

However, safety professionals should be involved in these safety programs without being intrusive. Health care personnel can be influential in extending a safety program beyond the facility to include outside activities of employees and their families.

Some insurance carriers offer a consulting service to help organizations set up an occupational health program suitable for their needs. The consultants usually know which health care providers are available for this kind of service. The basic work, however, has to be done by the organization establishing such a program. Any justification for these services lies in their record of accomplishments and emphasis on an improved quality of life. Prevention is not only better than cure—it is easier and less expensive. In this regard, off-the-job safety programs and activities also can benefit a company of any size.

An effective occupational health service program should be planned by the occupational health staff in cooperation with management. The program must have the full support of top management if it is to be successful. Only with management support is it then possible to establish and maintain an adequate program, professional staff, and facilities for examinations, emergency cases, and record storage.

Over the past 20 years, occupational health services have evolved and expanded considerably. The roles of occupational health professionals who deliver health care have also changed. Today, companies use physicians more as consultants for treating serious illnesses and injuries and to provide limited direct care. Occupational health nursing practice, on the other hand, has greatly expanded. The occupational health nurse is typically the manager of the occupational health unit as well as the sole direct care provider of health services at the work setting.

Occupational Health Coordinator

The medical aspects of health programs depend on the presence of an adequate safety and health program for success. Medical programs would be useless without an effective safety and health program. Managers of the occupational health program must be effective in dealing with safety managers. Sometimes an "occupational health coordinator" position is used to interact with vendors who provide many of the detection and treatment services described in this chapter. The coordinator may work with employee assistance counselors, psychologists, and testing professionals and will aid with the billing and insurance difficulties that are virtually inevitable for today's complex health systems.

Occupational Health Nurse

Occupational health nursing is a specialty practice within the nursing profession. The position requires a registered professional nurse who is licensed to practice in the state where he or she is employed. In addition, the occupational health nurse should know the basic principles of occupational health, health promotion and protection, occupational disease, related work processes and hazards, workers' compensation laws, insurance, health and safety regulations, sanitation, first aid, and record keeping (Guidotti et al. 2002).

The occupational health nurse works with the other human resource and safety professionals to develop a comprehensive health program tailored to the needs of the company.

This health professional has a wide scope of practice directed toward improving, promoting, and protecting worker health. In the work setting, the occupational health nurse has major responsibility for the management and administration of the occupational health unit and for dispensing services. The occupational health nurse may be responsible for direct care services; preplacement, periodic, and other types of examinations; and workplace monitoring surveillance programs. These activities are directed toward identifying potential hazards harmful to workers. Using an interdisciplinary approach, the nurse works closely with the other health care professionals to develop strategies to reduce employee risk.

A major component of the occupational health nurse's practice is the development, implementation, and evaluation of activities that support health promotion and protection strategies through primary, secondary, and tertiary prevention approaches. This process may include conducting employee health needs assessments; administering and interpreting health risk appraisals; designing and developing programs to provide positive lifestyle changes such as exercise, nutrition, and weight control; and initiating programs that support regulatory mandates such as hearing conservation and respiratory protection. The occupational health nurse collaborates with the physician and other health care professionals to provide for rehabilitation and return-to-work programs. This area encompasses the need to pay attention to the physical and psychological needs of the worker and encourages active participation of management in the process.

Today more primary-care services are being provided at the worksite. The occupational health nurse may give workers nonoccupational health-related care for minor health problems and offer chronic disease monitoring and follow-up using established guidelines and protocols. In addition, management services may be available to provide coordinated care for early intervention and reduction in worker disability and further injury. Such services extend employers' liability for potential workers' compensation claims and must be managed carefully.

The occupational health nurse is available to provide counseling to help employees clarify problems and to make informed decisions about their health needs. Referrals for employees in crisis situations are often handled by the occupational health nurse through employee assistance programs. Collaboration with community agencies enables the occupational health nurse to develop a network of resources to more efficiently and effectively develop health care services for employees and the company.

The occupational health nurse may play a major role in implementing many OSHA-mandated programs, such as hazard communication. For example, the nurse should have ready access to Safety Data Sheets (SDSs) for chemicals at the facility and may participate in the overall management of the program and employee training.

The occupational health nurse must maintain a confidential, professional relationship with employees that conforms to legal and ethical codes. The nurse must not divulge information contained in individual employee health records without the employee's written consent. The employee health files should be accessible only to health personnel.

The occupational health nurse is also responsible for maintaining necessary records and reports. These records act as a guide to management and keep both managers and employees informed regarding the success of the health program. Records and reports are necessary to direct and evaluate preventive medical and safety engineering techniques, to chart progress in the reduction of accidents, and to meet the regulatory record-keeping requirements (see Chapter 4, Regulatory History, and Chapter 11, Injury and Illness Record Keeping, Incidence Rates, and Analysis).

Occupational Health Physicians

Physician services in industry depend on such considerations as the hazards in operations, the number of employees, and the type and extent of the occupational health program. The company can arrange for full-time, part-time, or consultant services.

Some large organizations have a full-time corporate medical director with part-time physicians serving their decentralized operations (Felton 1990). Some physicians devote a scheduled number of hours, either daily or weekly, to the medical service needs of a company and are available at other times for emergencies. Others arrange their service on an as-needed basis—for example, when job applicants require examinations, injured employees need medical care, or other medical problems arise.

Management usually makes the on-call physician arrangement with a local physician. This service is most often used by companies (or establishments) with fewer than 500 employees, companies with a low incidence of accidental injuries, or companies with a minimum health service program for employees. The physician usually provides services on referral from the occupational health nurse (if one is on site) when medical care is needed and for illnesses or injuries not requiring hospitalization. Otherwise, the company will make direct referrals to the physician's office. Frequently, the on-call physician has similar arrangements with a number of companies in the area—a particularly convenient system for small-facility or small-company clusters, such as in industrial parks. Also, it is not unusual to find physicians specializing in the care of industrial injuries and diseases and providing occupational medical services in some manufacturing centers or in associated freestanding or hospital-based occupational medicine clinics.

A physician who has a consulting service is not usually called in except to diagnose and treat serious injuries, illnesses, toxicological problems, and special kinds of injuries or disorders (such as eye injuries) that require the services of a specialist. State and local health departments often supply—without charge—medical, nursing, and engineering consultation. They also conduct industrial hygiene and radiological surveys for industries within their areas. Many insurance and private consulting companies offer this survey service to their clients.

Occupational physicians or those providing occupational health services should be familiar with all jobs, materials, and processes that are used in organizations (Levy and Wegman 1995). An occasional inspection visit will help them keep abreast of current technologies, equipment, and processes. During these visits, the physician can develop an inventory of hazards to help diagnose, treat, and prevent occupational health diseases. Physicians can also be involved in other company services that relate to the health of workers, such as food service, welfare service, safety programming, sanitation, and mental health.

Maintaining a true physician–patient relationship (with fairness to both employee and employer) is essential to the success of any occupational health program (Rosenstock and Cullen 1994). For example, employee patients deserve the same courtesy and professional honesty as do private patients. The first meeting of physician and employee usually occurs at the introductory physical examination. The examining physician—within the framework of professional discretion—should tell the worker the results of the examination and, if necessary, refer the worker to a personal physician for further treatment.

The occupational physician should provide emergency medical care for employees who are injured or become ill on the job and require medical intervention. Necessary follow-up treatment for employees with occupational diseases or injuries also should be arranged. The treatment of non-work-related employee injuries or diseases is usually the function of private medical practice. However, more primary-care services are being provided on site by occupational health professionals, including physicians, nurses and nurse practitioners, and occupational health nurses (Rogers, Randolph, and Mastroianni 1996).

Medical and surgical management in every case of industrial injury or disease should be aimed to restore disabled workers to their former earning power and occupation as completely and rapidly as possible. To help achieve these goals, the physician should promptly submit medical reports to those agencies that need to receive them. Furthermore, equitable administration of workers' compensation may depend on a physician's medical testimony regarding the injury and its possible consequences.

Nurse Line Programs

The use of Nurse Line Programs is increasing as insured companies and their insurance providers collaborate to provide immediate medical attention to employees who are injured at any time during the three-shift workday. These programs provide a registered nurse (RN), who follows medical triage guidelines, to direct the employee to the right treatment, at the right time, and in the right setting (ACE/ESIS 2014).

Following an injury and phone contact, the nurse speaks directly with the employee who sustained the injury. As part of the medical triage process, the nurse should ask about the worker's medical history, current medical problems and medications, allergies, and symptoms. The nurse should recommend a course of action for appropriate care to ensure that the employee can return to work as soon as possible. The triage nurse can also provide answers to the injured employee's questions and guidance on care for his or her medical issues. The service is designed to address injuries that are not life-, limb-, or eyesight-threatening. Management should not be involved in the conversation between the nurse and employee; however, the supervisor or manager may speak with the nurse toward the end of the call to be informed of the final diagnosis and required care.

In the event that the nurse prescribes self-treatment, the employee should be equipped with the necessary materials (e.g., cold packs, bandages, etc.). If the injury requires further medical treatment, the employer is responsible for transporting the employee to a medical treatment facility. If the employee has a serious injury that requires immediate care, emergency medical services should be called.

Values of Nurse Lines

- Registered nurses are available 24 hours a day/7 days a week.
- RNs use reliable medical triage guidelines.
- Nurses can provide clear advice and direct the injured employee to the right treatment, including simple first aid.
- Spanish-speaking nurses are available, and language line interpreters are available for all other language needs.

- There is a national relay center for the hearing impaired.
- Self-care or a lower level of care, when appropriate, provides cost savings.
- Network penetration is improved.
- Nurses are available for follow-up.
- Early identification reduces lag time in the first report of injury (i.e., the time between the incident and the report to the insurance company).
- The injury statement is recorded.
- Workplace disruptions are reduced.
- The service helps eliminate unnecessary waiting in the emergency room or doctor's office, reducing the time the employee is off of work.
- Managers can focus on work.
- Managers do not need to make decisions regarding treatment for the injured employee.
- Employee morale improves and managers have peace of mind.
- First report is automatically provided to insurance companies (ACE/ESIS 2014).

PHYSICAL EXAMINATION PROGRAM

Surveillance of workers' health status by qualified personnel is essential if an occupational health program is to provide maximum benefits for both the employee and employer. Therefore, the occupational health service should provide a regular physical examination program for all employees. The health care providers should discuss all significant findings with each worker. With the employee's written consent, a transcript of the data may be supplied to the worker's personal physician or to an insurance company. Courts, workers' compensation commissions, or health authorities may request this information by legal means, but it is preferable to obtain an employee's written consent to disclose his or her medical records. Occupational services personnel must strictly preserve the confidentiality of health examination records. Furthermore, the physician, nurse, or industrial hygienist should inform the employer of potentially harmful work-related exposures or hazards detected through their examination of workers and the environment (Rogers, Randolph, and Mastroianni 1996).

Health care providers determine the scope of a physical examination based on the type of operation involved—the nature of the industry; its inherent hazards; and variations in jobs, physical and mental demands, and health exposures (McCunney 1994). The values of different test procedures must be carefully weighed against their cost in time and dollars. Examinations probably will be different for different jobs; for example, the physical requirements for an ironworker who will be engaged in construction work are different from those of an office worker who will sit at a desk and perform word-processing functions. However, certain basic examination criteria apply to each employee regardless of job assignment.

The various kinds of examinations may be classified as follows—preplacement, periodic, and special examinations such as job transfer, return to work, and termination. Aids to examinations such as x-rays or vision or hearing procedures may be performed to obtain further data about an employee's health status.

Preplacement Examinations

The preplacement examination is done once an employee receives and accepts an offer of employment. This exam is used to determine and record the health condition of the prospective worker in order to match each employee to the right job. The applicant (or the personal physician with the applicant's approval) should be told of any conditions that need attention because follow-up care may be necessary. The primary purpose of the preplacement examination program is to aid in selection and appropriate placement of all workers. The Americans with Disabilities Act requires employers to make reasonable accommodations for people with disabilities. (See Chapter 23, Workers with Disabilities.)

Preplacement examination also provides baseline data for future evaluations. Substance-specific standards (e.g., asbestos, lead, arsenic, benzene) have particular requirements, based on exposure. In addition, the respirator standard requires medical review prior to issuance of negative-pressure respirators.

It is not the health care provider's function to inform applicants whether they are to be employed. This is the prerogative and duty of management. Other factors besides health determine whether a candidate is suitable for employment in the company. Preplacement examinations may lead to future liability if and when an on-the-job injury occurs.

Periodic Examinations

Periodic examinations of all employees may be required or voluntary. These examinations are performed on employees for both preventive health monitoring and health surveillance purposes. Periodic health examinations also are conducted at intervals during employment to determine the worker's continued compatibility with the job assignment and to determine whether adverse health effects have occurred that may be attributable to the work or working conditions (Rogers 1994). Management should institute a program for workers who are exposed to hazardous processes or materials or whose work involves responsibility for the safety of others, such as vehicle operators. In addition, process controls are usually established for substances such as lead, carbon tetrachloride, and many others that can cause occupational diseases. However, caution dictates that the occupational professional periodically examine workers to be certain that the engineering and hygiene controls are effectively safeguarding employees. Periodic examination also permits early detection of highly susceptible individuals and of practices or procedures that workers use to circumvent safety devices and policies.

How often workers are examined varies according to the quality of engineering controls (influenced by how rapidly the hazardous substance acts on the human body), the types of hazards, the findings produced on each examination, and, in some cases, the regulatory standards (Harris and Cralley 1994; McCunney 1994). Thus, worker exposure to substances may require that employees receive examinations or laboratory tests on a weekly, monthly, quarterly, annual, or biannual basis.

In many cases, laboratory tests of blood or urine can serve as the major portion of a periodic examination program, with complete examinations being made less frequently. Deciding which type of special examination (laboratory, x-ray, etc.) is necessary for any exposure and interpreting the results requires expert medical judgment (Jarvis 1992). Appropriate regulatory standards, medical practice, and other sources offer additional guidance for performing tests.

Health History

A carefully taken personal health and occupational history may give as much, if not more, information about the worker's health status as the physical examination. The history also will indicate a worker's need for special tests and perhaps job placement restriction.

The occupational health examiner must know the prospective employee's job title; a description of the job; duties, tasks, and related demands; potential work-related hazards; and any physical and/or emotional capacities required (Felton 1990). The occupational health history is designed to determine previous work experiences and to help the occupational health care provider discover preexisting conditions that may be worsened by conditions on the job (Balge and Krieger 2000; Ginetti and Greig 1981; Rosenstock and Cullen 1994; Stein and Franks 1985).

A synthesis of data obtained from employees' personal and occupational histories and a review of systems can help health professionals perform an individualized risk assessment. The study of individual health data, workplace exposures, and toxicity of the chemicals to which the person is exposed comes together in the risk assessment.

Return to Work

Employees who experience on-the-job health problems often benefit from special examinations. For example, when there is a change in the employee's health status or in work conditions that may place the employee's health at risk, job transfer examinations are conducted to match the employee's capabilities with proposed new jobs.

Some organizations also find it worthwhile to require "return-to-work" examinations of employees who have been absent for more than a specified number of days as a result of an illness or injury. This type of examination is done to determine whether the health status of the employee may require a change in the worker's duties and to ensure proper placement to prevent further illness and injury. The same disease can affect different people in widely different ways.

Return-to-work examinations also serve to evaluate workers after severe injuries or illness. Research has shown that the longer workers are away from their jobs, the less likely they are to return fully to productive employment. Rogers (1994) states that some employees' health problems may require them to have light duty, limited duty, or other work restrictions. This course of action should be determined and recommended by the occupational health professional after a thorough examination of the employee.

Light duty is an adaptation of the employee's original job to reduce the worker's tasks. *Limited duty* is defined as a new job that is appropriate to an injured worker's skills, interests, and capabilities. It is designed for individuals who cannot return to their original work area and is created for either temporary or permanent placement (Rogers 1994).

Another process to rehabilitate workers is known as *work hardening*. This is a progressive, individualized physical conditioning and training program designed to help injured workers return to the workplace as soon as possible. This goal is accomplished by gradually increasing physical and psychological requirements of the job, within the employee's current capabilities, until the employee can perform once again at acceptable work levels.

No matter what assessment for return to work is completed, occupational safety professionals must be sure to match returning workers with appropriate job demands, work conditions, and level of duties performed. The health professional should use job descriptions, job analyses, and discussions with employees in the same positions and the workers' supervisors to identify details and demands of each job.

When an employee returns to work following a serious injury, either occupational or nonoccupational, the safety professional should reevaluate the person's work capacities. Some work restrictions may be necessary. On the other hand, rehabilitation procedures may have improved a worker's ability to perform tasks or increased the range of jobs he or she can do.

Exit Examinations
Upon termination of employment, some organizations give employees a physical examination and document the findings. This procedure is particularly appropriate where operations exposed workers to health hazards such as lead, benzene (benzol), silica, asbestos dust, and excess noise levels, or where required by standards.

Laboratory Tests
Where workers will be exposed to toxic substances, appropriate laboratory tests may be indispensable. The use of general laboratory screening panels (Chem 24) is controversial. These tests are relatively nonspecific and can be affected by many unrelated problems, such as alcoholism and viral infections.

Although not to be used routinely, x-ray tests may be appropriate in alerting a physician to a particular condition. However, x-rays are used much less frequently to detect pulmonary disease. Instead, less potentially hazardous examinations, such as pulmonary function tests, are employed. These tests can detect loss of lung function before the disease reaches the stage where it can be seen on an x-ray.

Vision and Hearing Tests
Special devices have been developed for routine testing of several aspects of vision. Employees' near-sighted or far-sighted vision should be documented. If the company fails to match the visual requirements of a job with the visual abilities of employees, workers may become easily fatigued, inefficient, and frustrated. If the job involves working with colors or color-coded materials, health professionals should test workers' color vision. If the organization does not have vision-testing equipment available on site, management can contract with outside examiners to perform annual eye examinations and to fit prescription safety glasses, if needed.

Workers who will be exposed to high noise levels (generally 85 decibels [dBA] or greater for an 8-hour day) should be examined and tested for hearing acuity before job placement to determine prior hearing loss, if any. Workers with exposures of 85 dBA time-weighted average (TWA) or greater must also be placed in a hearing conservation program (Royster and Royster 1990). The audiometric examination is the accepted method of testing hearing acuity. Employees should be examined periodically thereafter to detect early hearing loss due to noise. Although protective hearing devices can reduce the amount and level of noise reaching the auditory nerve, hearing tests and records are also valuable aids to use when designing programs to prevent premature hearing loss. All hearing booths must meet OSHA requirements.

Emergency Medical Planning
The occupational health physician and nurse and the safety professional should confer with management to plan emergency procedures for handling large numbers of seriously injured employees in the event of a disaster, such as explosion or fire. Management should coordinate these plans with community plans for such events (see Chapter 18, Emergency Preparedness).

Procedures should include the following:
- selection, training, and supervision of allied health and other personnel
- transportation and caring for the injured
- transfer of seriously injured personnel to hospitals
- coordination of these plans with the safety department, security, police, road patrols, fire departments, and other interested community groups.

Employee Health Records
Health records contain an employee's health and medical history, examination and test results, medical opinions and diagnoses, descriptions of treatments and prescriptions, and employee medical complaints. Employee health records should be maintained for the duration of employment plus 30 years. These medical and health records are protected by Health Insurance Portability and Accountability Act (HIPAA) laws and must be maintained separately from employment records.

Regulatory standards require employers to maintain accurate records of work-related deaths, injuries, and illnesses (see Chapter 11, Injury and Illness Record Keeping, Incidence Rates, and Analysis). The OSHA Access to Medical Records standard permits the worker, the worker's representative, and regulatory authorities to have access to employer-maintained health/medical and toxic exposure records. These specific conditions under which access is allowed apply to all employers in general industry, maritime, and construction whose employees are exposed to toxic substances or harmful physical agents. The records that interested parties are allowed to see include exposure records of other employees with past or present job duties or with working conditions related to those of the employee, records containing exposure information concerning the employee's working conditions, and SDSs.

Occupational health records also provide data for use in job placement, in establishing health standards, in health maintenance programs, in treatment and rehabilitation, in workers' compensation cases, in epidemiologic studies, and in helping management with program evaluation and improvement. Such data are collected in the history interview, from the preplacement examination and any subsequent examinations, and from all visits the worker makes to the occupational health unit or first-aid room.

Health records also establish a health status profile of each worker. The key to accurate diagnosis and treatment often lies in the adequacy and completeness of this profile; therefore, record maintenance is a professional responsibility. Further, to compile a complete history, health records may include absences caused by illness or off-the-job injury. Thus, record keeping of nonoccupational and occupational health incidents often reveals chronic or recurrent conditions. In such cases, early treatment (e.g., referral to family doctors) and preventive measures can reduce absenteeism and decrease accident rates.

Maintenance of health records, however, should not be so burdensome that the occupational health nurse or first-aid attendant becomes a file clerk. Recording forms and filing systems must be simple enough that they can be used and interpreted by a physician, nurse, or first-aid attendant. In addition, computerization of records may increase the efficiency of the record-keeping system.

Although the employer needs to know a worker's limitations for job placement purposes, all health/medical records are confidential and protected by HIPAA law. This means that only health care professionals have a legitimate or legal right to see employee health records. If a company does not have a resident medical director or nurse, its medical records are usually filed in the human resources (HR) department and must be kept strictly confidential by HR personnel.

Neck or Wrist Tags for Medic Alert

A universal symbol for emergency medical identification has been developed by the American Medical Association (Figure 12–4). The object of the symbol is to identify its wearer immediately as a person with a health-related condition requiring special attention. If the wearer is unconscious or otherwise unable to communicate, the symbol will indicate that vital medical facts are recorded on a health information card in the bearer's purse or wallet or in another location. The identification tag also gives a telephone number for obtaining more detailed information. These details should be known by all employees before they attempt to help an individual struck down by an accident or sudden illness (see the Emergency Medical Identification and Medic Alert Foundation entries in the References at the end of this chapter).

SPECIAL PROGRAMS

In today's work environment, companies must be aware of the special needs of some employee groups, such as women, employees with special health problems, and older work-

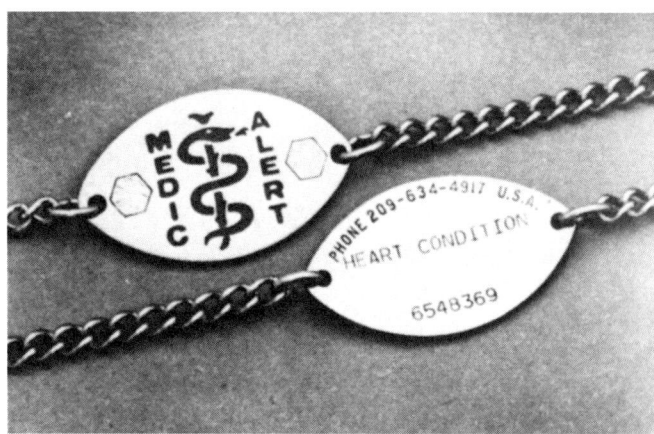

Figure 12–4. Persons with a physical condition for which emergency care may be needed should wear an identification tag. The tag provides a general indication of the problem and a phone number to call for more details if the tag wearer is unable to supply them. *(Reprinted with permission from Medic Alert Foundation, Turlock, CA.)*

ers. Government regulations and professional organizations have developed regulations and guidelines to help companies meet the needs of these groups so that these employees can remain productive members of the work force.

Health Promotion and Wellness

Wellness has been defined as a way of life that promotes a state of health. The health of employees can directly affect the bottom line through decreased workers' compensation and health benefit costs. Maintaining and protecting good health is more effective and less costly than treatment and cure. The government publication *Healthy People 2000: National Health Promotion and Disease Prevention Objectives* (U.S. Department of Health and Human Services 1994) presents a blueprint or plan of action to combat the leading preventable diseases and health-related problems. Priority areas identified include the following:
- health promotion
  - physical activity and fitness
  - nutrition
  - tobacco
  - alcohol and other drugs
  - family planning
  - mental health and mental disorders
  - violent and abusive behavior
  - educational and community-based programs
- health protection
  - unintentional injuries
  - occupational safety and health
  - environmental health
  - food and drug safety
  - oral health

- prevention services
  - maternal and infant health
  - heart disease and stroke
  - cancer
  - diabetes and chronic disabling conditions
  - HIV infection
  - sexually transmitted diseases
  - immunization and infectious diseases
  - clinical preventive services
  - surveillance and data systems.

Achieving these priorities will require a commitment on the part of society and its members to promote and adopt healthful lifestyles and behaviors conducive to optimal health and by business and government to develop the strategies, opportunities, and resources required to support the behavioral and environmental changes needed.

Health Promotion
Pender and Pender (1996) define health promotion as "activities directed toward increasing the level of well-being and actualizing the health potential of individuals, families, communities, and society." Health promotion behaviors are described as continuing activities that are an integral part of one's lifestyle, such as physical exercise or eating nutritious foods, and are directed at achieving the highest level of health.

Health Protection and Prevention Services
Prevention is best described as health-protecting behavior because primary emphasis is placed on guarding or defending an individual or group against specific illnesses or injuries. A major component of occupational health includes activities related to health promotion and protection for both individuals and groups of workers and incorporates primary, secondary, and tertiary prevention strategies (Rogers 1994).

Primary prevention takes place before disease or illness occurs. These types of programs increase awareness and knowledge related to toxic and hazardous exposures and their effects and promote lifestyle interventions such as smoking cessation, stress management, improved exercise and nutrition, use of seat belts, and proper use of prescription drugs.

Secondary prevention begins after a condition or disease is present. Intervention at this stage may lessen the complications and the disability resulting from the disease or illness. Examples of this type of prevention are blood pressure screening; medical surveillance for occupational exposures; and preplacement, periodic, and termination physicals.

Tertiary prevention or *rehabilitation* begins when the disease or condition has stabilized and no further healing is expected. Examples are chronic illness monitoring, onsite therapy, modified duty, or assisting an employee with a work-related injury to reuse an injured hand. The employee may need to learn new skills or new ways to use old skills in order to return to work.

Wellness programs are designed to address and promote health and safety behaviors. According to the September/October 1994 *Occupational Health Nursing Newsletter* (NSC 1995), wellness programs should include the physical aspects of health issues and also should deal with emotional, spiritual, and social dimensions. O'Donnell and Harris (1994) indicate that health promotion and protection programs should be targeted at three levels: awareness, lifestyle and behavioral changes, and supportive environments. Awareness programs are targeted at individuals and groups to increase their understanding of health and work-related risk factors. Uses of pamphlets, health screening activities, or conferences are examples.

Lifestyle change programs are directed at helping individuals change their behavior, such as starting and maintaining exercise programs, eating nutritious foods, and enhancing communication and coping skills. Lifestyle change programs are most successful if they use a multistep process (i.e., introducing one or two programs/activities at a time), include a combination of educational and behavioral modification experiences, and are provided over a number of weeks or months. Supportive environments include families, friends, organizational/work cultures, co-workers, communities, and regulations that help shape these environments (e.g., availability of jogging trails or onsite exercise classes). Fostering a supportive environment or changing an environment to encourage a concept of health will go a long way in improving employees' health behavior outcomes.

Medical Surveillance
The medical surveillance function includes three separate roles: examining overall changes in the workplace population, without regard to individual employee conditions; screening to examine individuals for health concerns such as diminished hearing ability; and examining hired applicants for their qualifications to meet a job's physical parameters. Each of these acts will survey the capabilities of one or more employees.

Health Care for Women
In the last three decades the number of women in the work force has more than doubled (Bureau of Labor Statistics 2008). Many factors such as economics, women's changing roles, increased job opportunities, and gender antidiscrimination laws have influenced both job accessibility and availability in today's job market (Rogers 1995).

Women comprise the largest group of workers in office,

health care, and hospital environments and face various types of hazards, including exposure to chemical and physical agents, unsuitable ergonomic work conditions, and stress. Data suggest that 75% of the hospital labor force is women, and in several occupations, women make up more than 90% of the total work force. This is particularly true of jobs such as nursing (97%), dietary services (91%), and medical records (93%) (Zoloth and Stellman 1987).

In addition, many workers involved in assembly-line production processes, such as in the electronics industry, are women. Besides exposures to chemical cleaning and degreasing substances, these workers are involved in processes that require rapid, repetitive motion of the fingers, hands, wrists, and arms, resulting in cumulative-trauma disorders.

There are more than 18 million clerical workers in the United States, and 75% of them are women. Many jobs that women perform require long hours standing or sitting, sometimes in awkward positions. Sitting at a desk for long periods of time may cause blood pooling and edema in the lower extremities. Poor sitting posture or episodes of standing for long periods can create a postural load and may result in back pain, muscle stress, and general body fatigue. Jobs requiring repetitive tasks such as typing and word processing may result in or aggravate musculoskeletal conditions, including repetitive-motion and carpal tunnel disorders.

The influence of work on reproduction is complicated because of the variety of work-related exposures that can occur. Many chemicals—including pharmaceuticals, anesthetic gases, heavy metals, pesticides, and organic solvents—can disrupt the menstrual cycle and affect the course of pregnancy or the development of the embryo or fetus (Harris and Cralley 1994). Of particular concern is that many women may not know when they first became pregnant and thereby remain at risk to potential exposure.

In the past, many companies used fetal protection policies to prevent women from working in areas where toxic chemicals that might harm a developing fetus were present. A 1991 Supreme Court decision struck down these policies as discriminatory to women (Rogers, Randolph, and Mastroianni 1996) because they may also bar them from higher-paying jobs. Critics of such policies have also argued that they did not deal with reproductive hazards posed to men by chemicals such as lead. The Court decision puts the responsibility on the employer to clean up the work environment and to provide information to the worker on the hazards that exist. Many experts state that as long as there is no negligence in this regard, the employer is not likely to be found liable for any injury to a fetus. Critics of the Supreme Court decision dispute this assertion.

Although reproductive risk is a complex subject, companies must inform women of the hazards they face on the job.

Chemicals such as lead, mercury, some solvents, and therapeutic agents containing synthetic hormones may produce harmful effects on the fetus or lead to spontaneous abortion. Biological agents such as rubella (German measles) or physical hazards such as ionizing radiation may also produce such damage. Pregnancy may limit a woman's ability to do some strenuous physical work. For these reasons, a woman should consult her doctor about any restrictions that should be placed on her work during pregnancy.

In line with the Pregnancy Discrimination Act and the Americans with Disabilities Act, an employer needs to make reasonable accommodations for women during pregnancy. Employers need to consult the latest regulatory interpretations of laws and court decisions in developing programs to deal with issues posed by workers who are pregnant or planning to conceive a child. For example, in the case of employees who desire medical advice regarding lead exposure and the ability to procreate a healthy child, the OSHA lead standard requires that employers offer medical examinations and consultations to both men and women on request.

Substance Abuse and Employee Health

Alcoholism and drug abuse continue to be among the leading illnesses in the work force. Therefore, occupational health professionals and the employer must determine what they can do to help employees who have these illnesses. (See Chapter 17, Employee Assistance Programs.) Managers and supervisors should be alert to employees whose work and performance are deteriorating because of an alcohol or drug problem. These employees should be referred for medical care as quickly as possible.

When alcohol and drugs are combined, they produce a variety of effects that severely impair workers. Concentrations of alcohol and drugs remain in the bloodstream much longer than most users realize, and the effects of this combination may appear unexpectedly. For example, they can make vehicle operators drowsy or impair their reflexes, which can increase the risk of accidents, injuries, and even deaths.

Dependence on alcohol or other drugs is a major cause of family strife, impaired job performance, increased insurance rates, occupational accidents, increased absenteeism, and rising crime rates. These illnesses know no boundaries, nor is there any "generation gap" among abusers—all ages, races, and socioeconomic groups are susceptible to their destructive effects.

To help detect and/or prevent substance abuse and other problems, every time an applicant or employee visits a physician or nurse, some health education and/or counseling should be given. Occupational health units are an ideal way to provide these services. Employers are paying an increasingly large part of health care costs for employees and their dependents. The earlier that health-related problems are detected, the less they are likely to cost. As a result, it makes economic sense for employers to maintain a sound occupational health department to provide proper care for employees and to offer timely referrals for further care, if needed. As a practical and legal matter, drug and alcohol detection measurements are best done by persons trained for and experienced in that role. State law may also require the person performing the testing to have certification or licensing. For that reason, outsourcing this function may be desirable.

Older Workers

It is clear that the American work force is aging. In 2000, the median worker age was 39 years (Campanelli 1990), with nearly half (49%) of the work force between ages 35 and 54. The Bureau of Labor Statistics (2008) states that by 2016, the median worker's age will be 42, and 1 out of every 4 workers will be 55 years of age or older. Nearly all businesses have older workers. As a culture, we have been conditioned to believe that aging means an inevitable decline in cognitive function and an increase in chronic disease, disengagement, nonproductivity, obsolescence, and retirement (Rude and Adams 1993). However, research is showing that these common beliefs are erroneous. As Rude and Adams state, people age differently and at different rates; aging is not a fixed, mechanical experience; and new potential for growth, productivity, and creativity exists in the second half of life. As people age, they become more individual, not more alike. Therefore, it is critical that the occupational health professional take a proactive approach in worker evaluations. Businesses intent on strengthening their future growth will value older workers and recognize their contributions. Industry and the public must shed the belief that older employees are past their prime and begin to realize the incredible resource they represent, if properly managed (Rude and Adams 1993).

For these and other reasons, companies need to offer occupational health programs throughout employment that address both physical and social aspects of life. Although employees now have "work longevity," they also may exhibit specific age-related health problems that must be anticipated and dealt with appropriately. Examples of anticipated health problems include arthritis, hypertension, cardiovascular disease, hearing impairments, diabetes mellitus, and depression. Other lifestyle issues that may need attention are weight control, exercise, smoking, alcoholism, substance abuse, and stress (Hart and Moore 1992).

The occupational health professional must teach employees how to be responsible for making good choices to protect and promote their health and for making sound health care decisions. Information given to employees during employment and special instructions given before retirement should include not only financial aspects but also how to use health care resources wisely; how to find information and resources to continue a healthy lifestyle; and how to locate information about healthy adjustment to retirement, for workers and their spouses, significant others, and family members. Programs targeted at screening for early disease detection and chronic disease monitoring are essential for health promotion in the aging work force. In addition, several points in managing the aging work force can be considered:

- understanding the normal aging process and its effects on health and on job assignments and team building
- addressing work-related stress that may accompany job changes and developing stress management and retraining programs to deal with this issue
- conducting aging programs for management and co-workers to prevent misunderstanding and foster effective employment of older workers' abilities
- capitalizing on older workers' abilities and experiences in job performance.

The result will be an improved quality of life and work life and reduced health and benefit costs.

Shiftwork

Stimulated by the demands in World War II for increased production, the drive for continuous, 24-hour work cycles has been steadily increasing across all industrial service sectors. It is estimated that 25% of the U.S. labor force is engaged in some type of industrial or service activity that requires work around the clock. As the U.S. economy becomes ever more globalized, this trend will accelerate, particularly in the communications and financial services sectors of the economy. Because humans have physiologically and psychologically evolved as "daytime" organisms, it is important to understand the ramifications of the move to a 24-hour society. This type of altered workplace environment must be studied carefully by health and safety professionals to minimize the physiological and psychological disruptions that continuous operations can produce in humans.

Effects on Human Physiology

The human organism adapted over millions of years to an environment that accentuates activity during the daylight hours and sleep (protection) during the hours of darkness. Humans have a 24-hour activity cycle that is constantly reset by the rising and setting sun. This 24-hour cycle is known as "circadian" from the Latin *circa dies* or "about one day." Most physiological functions have cycles of high and low activity during a 24-hour period. Although some processes may have longer cycles, most chemical activities in the body follow this 24-hour period.

Interestingly, the actual human cycle falls within a 23- to 27-hour time frame. In environments that do not depend on artificial or mechanical constraints, most humans have a 27-hour cycle. Apparently, like many other animals, humans are sensitive to the effects of bright lights, which reset their biological clocks back to a 24-hour time cycle. Because shiftworkers are required to sleep during daylight hours and work during the night, it is important that all work activities be conducted with the appropriate level of light. Studies have found that exposing workers to bright lights during the early part of the work shift and again 2 to 3 hours before the shift ends will provide a "boost" in their alertness. The adequate level of light will help overcome the "wall of fatigue" that commonly occurs at about 5:00 a.m.

This constant resetting has important repercussions for the shiftworker because most individuals prefer to add time to the day rather than to subtract time. This phenomenon is frequently compared to east–west jet travel. As an individual flies westward, time is "added," whereas flying eastward "subtracts" time. Because there is a natural resetting in a slight eastward direction by 1 hour per day (to a 24-hour cycle), humans experience more biological disturbances when they fly eastward than when they travel west. North–south movement does not tend to affect the biological clock.

This circadian characteristic is critical in the workplace, where the majority of shiftwork schedules were developed before the current understanding of the biological clock. As a result, most 8-hour shiftwork schedules rotate in an eastward or backward direction such as day shift to night shift to evening shift as opposed to a clockwise rotational pattern. The counterclockwise pattern probably developed because it provides a slightly longer period of continuous time off. This time off is known as the "long change" and is important to most industrial workers. A forward-rotating, 8-hour schedule has a shorter long change at the end of the full cycle rotation. Although the forward rotation pattern appears to make more medical sense, relatively few data demonstrate a true physiological effect. The lack of convincing data may be due to the overwhelming effects of any 8-hour movement in either direction.

The human internal biological clock has a limited ability to adapt to time shifts. Usually, shifts of only 1 to 3 hours per day are tolerated without some noticeable effect. However, the equivalent of one time zone shift that is produced by moving from one 8-hour shift to another (e.g., day or evening to night shift) can overwhelm the body's ability to adapt, regardless of shift change directions. Despite the lack of published data demonstrating physiological improvement in forward (west) rotation versus backward (east) rotation, studies show that many shiftworkers prefer a forward rotational pattern.

The time changes and irregular schedule that shiftworkers experience make them more susceptible to a variety of potential problems that "normal"-scheduled workers do not commonly experience. These potential problems include the following:

- *Reduction in attention span* is common for most shiftworkers and normally begins as early as 20 to 35 minutes into the shift (Krueger 1989). Higher-level thinking skills may also be affected, including decreased memory, recall, logic, reasoning, and decision-making abilities (Balkin and Badia 1998).
- *Chronic fatigue* is due to the failure to obtain sufficient sleep during nonworking hours. Home activities and parental responsibilities often lead to the inability to obtain a full 7 to 9 hours of sleep necessary to work safely. Typically, a shiftworker will sleep only 4 to 6 hours at one time.
- *Sleep debt* is loss of sleep that robs the worker of energy and alertness. Sleep schedules are often disrupted or nonexistent. Length of sleep for night workers may be only 4 to 6 hours compared to the average 7 to 9 hours for day and afternoon workers. Sleep deficit can accumulate after three to four consecutive worknights. Some researchers believe that six consecutive nights may be too exhausting (Scott 1994). When this occurs, the worker can experience brief periods of "microsleep."
- *Microsleep* occurs when the worker slips into light sleep, lasting between 1 and 10 seconds, while continuing normal activities. Microsleep occurs without warning and is a potential safety hazard for these workers.
- *Substance abuse* often begins for the shiftworker with the increase in the use of caffeine and may end with hard drugs or other stimulants to help the worker stay awake and alert. Cases of shiftworkers drinking up to 22 cups of coffee per shift have been reported. The shiftworker may also require alcohol or depressants in an effort to obtain the required sleep. The use of depressants to fall asleep and stimulants to wake up or stay alert often leads to the classic upper/downer scenario.
- *Gastrointestinal and digestive problems* occur because the irregular shifts upset the system in such a way that shiftworkers have poor diets or indulge in "junk foods" that can lead to indigestion, heartburn, and stomachaches. Other problems can be weight gain or loss of appetite.
- *Increased risk of heart attacks* can result from the shiftworker smoking more heavily, exercising less, and participating in fewer leisure activities.
- *Feelings of isolation and depression* can occur because shiftworkers are rarely on the same time schedule as family and friends. They are often not available for family and social events, and children get limited or no contact.

Solutions

Some adverse effects of shiftwork can be reduced by relatively simple administrative solutions. Others require more complex rescheduling plans. Among the easier solutions are the following:

- *Good lighting* for shiftworkers is essential because shiftworkers generally sleep during most of the sunlight hours. Exposing workers to bright light during the early part of the night and 2 to 3 hours before the end of the shift gives them a boost of alertness and has been known to help them overcome the "wall of fatigue" that occurs about 5 a.m. Workers who are exposed to arrays of bright, broad-spectrum lights report that they feel better the night of the treatment, more energetic, less tired, and able to adjust more quickly to shift changes (Eastman 1990).
- *Exercise options* for shift employees could be provided by employers. Preshift or break-time exercises, available exercise equipment, and short exercise sessions during the shift will help workers stay alert during the difficult final hours of their shifts.
- *Napping at work* during breaks is supported by a small, but growing, number of 24-hour companies. A 15- to 20-minute nap can provide an alertness, mood, and energy boost that can last for several hours. However, sleeping too long can bring about sleep inertia or grogginess that can take 5 to 10 minutes to clear.
- *Music in the workplace* can keep the mind alert and improve morale. Careful consideration should be given to the kind of music played. An employer should not have "elevator" music playing for shiftworkers who are doing slow, meticulous duties because it tends to make them sleepy. On the other hand, loud, hard rock tends to agitate some employees and make others nervous and nonproductive.
- *Hot meals* provided by a cafeteria can prevent the potential risks of poor nutrition. To avoid digestive problems, shiftworkers should try to maintain three balanced meals a day, even on the night shift. The first meal should be the largest and the one before bedtime should be the smallest. Shiftworkers should also try to eat low-fat dairy products, raw vegetables, fruits, and other healthy snacks, with their main meal being either broiled or boiled.

Shift Rescheduling Projects

Worker surveys frequently reveal that most employees are dissatisfied with their shiftwork schedules. In response, companies have tried to modify, enhance, or totally redesign rotational schedules. Often, these efforts are initiated by the workers themselves and are frequently based on experience gathered either from nearby facilities or from other sources that report the latest research in circadian rhythms. To their dismay, most workers and management discover that reevaluating and redesigning an operational shift schedule is neither quick nor easy.

Theoretically, the ideal shift schedule would be permanent shifts that did not rotate across the various hours of the 24-hour cycle. In fact, some facilities do have so-called permanent or dedicated shifts. For example, a worker is always assigned to the same shift—day, evening, or night—based on an 8-hour rotational pattern. Although this situation is sometimes successful, it also frequently fails because it is difficult to attract, train, and maintain a dedicated night crew. In addition, unless the night crew continues to maintain their same sleep/wake cycle on their days off, the physiological benefit of a permanent shift is quickly lost as soon as the workers revert back to a daytime schedule. It is critical to realize that the act of rapidly changing sleep–work–leisure patterns produces the physiological and psychological disruption associated with either shiftwork or long-distance east–west jet travel.

Despite the difficulties associated with shift schedule redesign, several guiding principles can be used:

- All schedules are a compromise among three driving forces—medical considerations, work preferences, and operational necessities.
- Schedule redesign is an interactive process among different groups that will never satisfy all parties.
- Worker preference surveys are important and should be conducted in a scientific, confidential manner.
- Detailed industrial hygiene and ergonomics analyses of all chemical and physical hazards must be performed if extended-hour, such as 10- to 12-hour, schedules are considered. This is especially true if the worker is exposed to chemicals that have permissible exposure limits that are time dependent. Any new schedule must be considered temporary and be reevaluated after an appropriate trial period of at least 3 to 6 months.

Shift redesign projects can be a win–win proposition for both workers and management; however, both groups must recognize, define, and discuss certain so-called boundary conditions that may not be easily established. Examples of boundary conditions include management's desire to control costs, the potential for workers to be overexposed to noise or chemical hazards because of extended-hour schedules, a decrease in time off, and changes in contractually mandated work rules.

One important issue that requires in-depth evaluation is the decision to consider extended work schedules. The typical shift schedule is an 8-hour, four-crew configuration that rotates, either forward or backward, through a 28-day calendar and averages approximately 42 hours of scheduled work per week. In this configuration, crews may spend 5 to 7 "days" on a steady shift before rotation to the next scheduled time period. One variation of this pattern that is more common in Europe than in North America is the rapid rotation through the scheduled shifts during a workweek. For example, in a rapidly rotating 8-hour schedule, a crew could be scheduled for two day shifts, two night shifts, and two evening shifts.

The purported advantages of rapid rotation are based on the slow adaptability of the biological clock and the premise that it is easier for workers to keep a daytime internal clock rather than to adapt to constantly changing hours. Unfortunately, there is little or no evidence that this premise is correct. Given the uncertain benefit of one 8-hour schedule versus another, many workers feel that extended-hour schedules offer the best opportunity to improve their shiftwork experience.

Schedules of 10 and 12 hours are extremely popular in certain industrial sectors, such as refineries and chemical facilities. These schedules maximize the amount of time off while still controlling salary costs. However, these schedules have several disadvantages:

- Some jobs are too physically or psychologically demanding to be done for 12 hours.
- Workers may have trouble staying alert and vigilant for extended periods.
- Routine overtime is difficult or impossible to schedule.
- Noise and chemical exposure standards must be modified to reflect the extended exposure period.
- Pay periods including holiday and overtime pay may have to be recalculated if costs are to be controlled.
- Older workers may not be able to handle extended hours.

Despite these disadvantages, none of which is insurmountable, extended work schedules do have strong support in the workplace because of the time-off issue. Instead of working 21 out of 28 days on a routine 8-hour schedule, a typical 12-hour pattern will produce an additional 4 days off. This time off is a powerful incentive because most shiftworkers want to spend more time at home and with their families. Before an operating facility considers using an extended-hour schedule, it should conduct a careful review of health, safety, medical, and ergonomic issues.

Medical Issues

The medical effects of shiftwork have been well described in the occupational medicine and human factors literature. The National Safety Council's *Occupational Health & Safety* (Balge and Krieger 2000) presents some of the issues surrounding the evaluation of medical problems of shiftwork. Some workers, regardless of age, are unable to tolerate shiftwork and drop out of the rotation within 2 to 3 months. Other workers initially tolerate shiftwork well but slowly experience increasing problems after 10 to 20 years on shift. It appears that as workers age, particularly as they enter their 50s, their ability to tolerate shiftwork declines. Both groups show a cluster of problems that has been called shift maladaptation syndrome (SMS).

Shift maladaptation syndrome consists of a variety of medical and psychological complaints such as gastrointestinal problems (ulcers, reflux, chronic indigestion/constipation), sleep disorders (insomnia, fragmented and/or poor-quality sleep), and psychological problems such as extreme mood swings and/or chronic irritability. There is no definitive test for SMS; instead, it is diagnosed by eliminating other, alternative explanations and by removing the worker from the shiftwork for a trial period. Unfortunately, this latter strategy is fraught with potential legal and human resources issues because of workers' compensation and disability considerations. Nevertheless, certain medical problems—such as insulin-dependent diabetes, seizure disorders, and bipolar disease—generally disqualify an individual for shiftwork and should be carefully evaluated by the medical team.

Overall, the safety professional is confronted with a variety of new and rapidly evolving workplace environments. Companies need a multidisciplinary team to respond to these challenges and to develop innovative solutions.

SUMMARY

- Occupational health programs should provide a healthful environment, health examinations, diagnosis and treatment, health records, health education and counseling, and communication between company and occupational professionals. All health services provided by organizations must meet federal, state, and local codes.
- The principal goals of an occupational health program are to protect employee health, to match the right worker to the right job, to treat and rehabilitate injured employees, and to encourage personal health maintenance. Companies should also be involved in off-the-job safety and health programs.
- Management may work with other health professionals such as industrial hygienists and physical therapy specialists in delivering health care to employees. Experience has shown a strong relationship between occupational health and accident prevention.
- First-aid services, generally mandated by regulations, include properly trained personnel; a first-aid unit and supplies; a company manual; a list of reactions to chemicals; instructions for calling physicians, hospitals, and ambulances or rescue squads; methods for transporting ill or injured employees; and an adequate record system of injuries and treatment.
- First-aid treatment can be either emergency (before a phy-

sician can be summoned) or prompt attention (for minor injuries and illnesses). One person in each work area should be responsible for maintaining adequate supplies in good condition and for treating all worker injuries promptly.
- Occupational health nurses and physicians work together to provide occupational health services and health education to workers. Physicians and nurses must maintain a confidential relationship with employees and keep accurate, timely records.
- One of the most important functions of employee health programs is setting up a regular schedule of physical examinations of all workers. The basic purpose is to establish a baseline for worker health at the outset of employment, to detect or monitor any work-related illnesses or injuries, and to match the right worker with the right job.
- Physicians and nurses conduct preplacement, periodic, transfer, promotion, special, and termination examinations. The findings of these examinations are confidential and may be released only if a worker provides written consent. Workers have a right to review these records to determine how they may have been affected by hazards while on the job.
- Employee health records generally are retained for at least 30 years in industries where workers have been exposed to hazardous materials. Such records contain the results of tests along with a complete health and occupational history, medical opinions and diagnoses, treatments, and employee complaints.
- Emergency medical planning is a responsibility of occupational health physicians, nurses, safety professionals, and management. These plans include procedures for training personnel in emergency measures, transporting and treating injured people, transferring the seriously injured to hospitals, and coordinating plans with other units and community departments and agencies.
- Most companies now have some type of substance abuse screening and referral or treatment programs in place. Abuse of alcohol and drugs greatly affects a worker's job performance, places other workers at greater risk, and costs industry several billion dollars a year in lost time, accidents, and lower productivity.
- Shiftwork disrupts the human circadian rhythms, which causes physical and psychological problems for workers. Older workers frequently experience shift maladaptation syndrome or SMS. Rescheduling shiftwork can alleviate some of these hazards.

REFERENCES

ACE Group/ESIS. Triage training material, 2014. http://acegroup.com/esis-en/services-solutions/triage.aspx.

American College of Surgeons, Committee on Trauma. *Emergency Care of the Sick and Injured*. Philadelphia: Saunders, 1982.

Balge, M. Z., and G. R. Krieger, eds. *Occupational Health and Safety,* 3rd ed. Itasca, IL: National Safety Council, 2000.

Balkin, T. J., and P. Badia. "Relationship between Sleep Inertia and Sleepiness: Cumulative Effects of Four Nights of Sleep Disruption/Reduction on Performance Following Abrupt Nocturnal Awakenings." *Biological Psychology* 27, no. 3 (1998): 245–58.

Bonnet, M. H. "Dealing with Shiftwork: Physical Fitness, Temperature, and Napping." *Work and Stress* 4, no. 3 (1990): 261–74.

Boylston, R. *Managing Safety and Health Programs*. New York: Van Nostrand, 1990.

Bureau of Labor Statistics. "Labor Force Statistics from the Current Population Survey." 2008. http://humanresources.about.com/od/worklifebalance/a/business_women.htm.

Campanelli, L. "The Aging Workforce: Implications for Organizations." *Occupational Medicine* 5, no. 4 (1990): 817–25.

Collins, J. G., and O. T. Thornberry, *Health Characteristics of Workers by Occupation and Sex: U.S., 1983–85*. National Center for Health Statistics. Advanced Data from Vital and Health Statistics, Pub. No. 89-1250. Public Health Service, Hattsville, MD, 1989.

Coventry Healthcare. Coventry Worker's Compensation Network, 2013. http://coventryhealthcare.com/products-and-services/workers-comp/index.htm.

Eastman, C. I. "Circadian Rhythms and Bright Light: Recommendations for Shiftwork." *Work and Stress* 4, no. 3 (1990): 245–60.

Emergency Medical Identification. American Medical Association, 535 North Dearborn Street, Chicago, IL 60610.

Felton, J. *Occupational Medical Management*. Boston: Little, Brown, 1990.

Ginetti, J., and A. Greig. "The Occupational Health History." *Nurse Practitioner* (December 1981): 12–13.

Guide for Establishing an Occupational Health and Safety Service. Atlanta: American Association of Occupational Health Nurses, 1994.

Guidotti, T. L., et al. *Occupational Health Services: A Practical Approach*. Chicago: American Medical Association, 1989.

———*Occupational Health Services: A Practical Approach*. Caldwell, NJ: Blackburn Press, 2002.

Harris, R. L., and L. J. Cralley. *Patty's Industrial Hygiene and Toxicology*. Vol. III, Part A *of Theory and*

Rationale of Industrial Hygiene Practice: The Work Environment. 3rd ed. New York: Wiley, 1994.

Hart, B. G., and P. V. Moore. "The Aging Workforce: Challenges for the Occupational Health Nurse." *AAOHN Journal* 40, no. 1 (1992): 36–40.

Hau, M. L. *While Help Is on the Way....* Chicago: Health Products Marketing, 1994.

Jarvis, C. *Physical Examination and Health Assessment*. Philadelphia: Saunders, 1992.

Kemerer, S., and T. Raniere. "Cost Effective Job Placement Physical Examinations." *AAOHN Journal* 38, no. 5 (1990): 236–42.

Krueger, G. P. "Sustained Work, Fatigue, Sleep Loss and Performance: A Review of the Issues." *Work and Stress* 3, no. 2 (1989): 129–41.

Levy, B. S., and D. H. Wegman. *Occupational Health: Recognizing and Preventing Work-Related Disease*. Boston: Little, Brown, 1995.

McCunney, R. J. *A Practical Approach to Occupational and Environmental Medicine*. Boston: Little, Brown, 1994.

Medic Alert Foundation, 2323 Colorado, Turlock, CA 95380.

National Safety Council, 1121 Spring Lake Drive, Itasca, IL 60143.
Fundamentals of Industrial Hygiene. 6th ed. 2012.
Occupational Health Nursing Newsletter, 1995.
OSHA Up-to-Date 24, no. 11 (1995).

O'Donnell, M. P., and H. S. Harris. *Health Promotion in the Workplace*. Albany, NY: Delmar, 1994.

Pender, N. J., and A. R. Pender. *Health Promotion in Nursing Practice*. 3rd ed. Englewood Cliffs, NJ: Prentice-Hall, 1996.

Rogers, B. *Occupational Health Nursing: Concepts and Practice*. Philadelphia: Saunders, 1994.

———"Women in the Workplace." *Women's Health Care*, edited by C. Fogel and N. F. Woods, 363–84. London: Sage, 1995.

Rogers, B., S. A. Randolph, and K. Mastroianni. *Occupational Health Nursing Guidelines: Primary Clinical Conditions*. Boston: OEM Press, 1996.

Rosenstock, L., and M. Cullen. *Clinical Occupational Medicine*. Philadelphia: Saunders, 1986.

Rosenstock, L., and M. R. Cullen. *Textbook of Clinical Occupational and Environmental Medicine*. Philadelphia: Saunders, 1994.

Royster, J. D., and L. H. Royster. *Hearing Conservation Programs: Practical Guidelines for Success*. Chelsea, MI: Lewis, 1990.

Rude, J., and C. Adams. "Doing Business in an Aging Society." *Business News,* August 23–September 5, 1993.

Scott, A. J. "Chronobiological Considerations in Shiftwork Sleep and Performance and Shiftwork Scheduling." *Human Performance* 7, no. 3 (1994): 207–33.

Shiftwork Practices. Steve Mardon, editor@shiftwork.com.

Stein, C., and P. Franks. "Patient and Physician Perspectives of Work-Related Illness in Family Practice." *Journal of Family Practice* 20, no. 6 (1985): 561–65.

Swenson, D. X. *Into the Night: Coping with the Effects of Shiftwork*. Ontario, Canada: Blue Line, 1997.

U.S. Department of Health and Human Services. *Healthy People 2000: National Health Promotion and Disease Prevention Objectives*. DHHS Pub. 91-501212, 94-110. Washington DC: U.S. GPO, 1994.

Wellness Councils of America. *Healthy, Wealthy and Wise: Fundamentals of Workplace Health Promotion*. Omaha, NE: WELCOA, 1993.

Williamson, G. C., and P. V. Moore. "Health Care Cost Containment: A Model for Practice." *AAOHN Journal* 35, no. 11 (1987): 496–500.

Workers' Compensation Board of British Columbia. *Industrial First Aid*, edited by A. Dresser. New York: Van Nostrand Reinhold, 1991.

Zoloth, S., and J. Stellman. "Hazards of Healing. Occupational Health and Safety in Hospitals." In *Women at Work: An Annual Review*, edited by A. H. Stromber, L. Larwood, and B. A. Gutek. Beverly Hills, CA: Sage, 1987.

REVIEW QUESTIONS

1. List five of the eight components of a good occupational health program.
    a.
    b.
    c.
    d.
    e.

2. List the four basic objectives of a good occupational health program.
    a.
    b.
    c.
    d.

3. What is dermatitis caused by?

4. What two methods of controlling the work environment can reduce the rate of accidental injuries?
   a.
   b.
5. The health service office (dispensary) should have a minimum of how many rooms?
   a. two
   b. three
   c. four
   d. It doesn't matter.
6. List and briefly define the two kinds of emergency treatments.
   a.
   b.
7. Why should all employees report for medical treatment immediately when they are injured?
8. List five of the eight elements that a first-aid program should include.
   a.
   b.
   c.
   d.
   e.
9. What federal agency outlines specific first-aid requirements at certain locations in mines and processing facilities?
10. What is the objective of medical and surgical management in cases of industrial injury or disease?
11. What is the primary purpose of preplacement examinations?
12. Explain the difference between light duty and limited duty.
13. What is the primary purpose of exit examinations?
14. Emergency medical planning procedures should include what four items?
    a.
    b.
    c.
    d.
15. How long (maximum) must employee exposure records be kept?
    a. 30 years
    b. 20 years after the employee leaves the company
    c. 30 years after the employee leaves the company
    d. 20 years
16. List five of the eight ways in which the data from occupational health records can be used.
    a.
    b.
    c.
    d.
    e.
17. Define *wellness*.
18. Lifestyle change programs are most effective when they employ a(n) _____ process.
    a. aversive
    b. multistep
    c. baseline
    d. stochastic

Industrial Hygiene Program

David Hibbard, MPH, CIH
Philip Hagan, JD, MBA, MPH, ARM, CIH, CET, CHMM, CHCM, CHSP, CEM

Industrial Hygienist on the Occupational Health and Safety Team
In-House Industrial Hygiene Services
▶ Occupational Health and Safety Team

Practicing Industrial Hygiene
Anticipating and Recognizing Occupational Health Hazards ▶ Evaluating and Controlling Environmental Hazards

Toxicity
Toxicity versus Hazard ▶ Entry into the Body

Industrial Hygiene Consulting Services

Certification and Licensure

Professional Organizations

Summary

References

Review Questions

Industrial hygiene, as defined by the American Industrial Hygiene Association (AIHA), is a "science and art devoted to the anticipation, recognition, evaluation, prevention, and control of those environmental factors or stresses arising in or from the workplace which may cause sickness, impaired health and well being, or significant discomfort among workers or among citizens of the community."

Industrial hygienists are occupational safety and health professionals concerned with the control of environmental stresses or occupational health hazards that arise as a result of or during the course of work. The industrial hygienist recognizes that occupational hazards may endanger life and health, accelerate the aging process, and/or cause significant discomfort. Working with management, and with medical, safety, and engineering personnel, the industrial hygienist can help to eliminate or safeguard against environmental hazards caused by chemical, physical, biological, or ergonomic stresses. This chapter covers the following topics:

- the nature, purpose, and scope of industrial hygiene
- the role of the industrial hygienist in recognizing and controlling environmental hazards
- four general categories of environmental hazards or stresses
- four definitive elements of industrial hygiene practices
- health and psychological problems associated with altered environments
- the inherent medical and physiological issues involved in shiftwork
- toxicity and hazards in the workplace.

Industrial hygiene includes the development of corrective measures to control health hazards by either reducing or eliminating the hazardous exposure. These control measures may include substituting safer materials for harmful or toxic materials, changing work processes to eliminate or minimize work exposure, installing exhaust ventilation systems, practicing good housekeeping (including appropriate waste disposal methods), and providing proper personal protective equipment.

There are five factors necessary to ensure an effective industrial hygiene program: (1) anticipation and recognition of health hazards arising from work operations and processes; (2) evaluation and measurement of the magnitude of the hazards (based on past experience and study); (3) control of the hazards; (4) commitment and support of management because effective controls can be expensive, and the need for these controls must be made clear to company management; and, perhaps most important, (5) the workers' perception of the industrial hygienist as reliable, believable, and honest so that the workers diligently implement recommendations for changes in work practices and believe the industrial hygienist's assurances that conditions are acceptable.

Various reporting structures are used in companies that have an in-house industrial hygiene function. Depending on the company, the industrial hygiene function can be found in the human resources, risk management, medical, environmental, engineering/facilities maintenance, safety, or security department. In today's world of complex regulatory requirements and the need to minimize costs, the industrial hygienist often has collateral responsibilities and spends part of his or her time dealing with environmental and/or safety issues.

For many small and mid-sized companies, employing a full-time industrial hygienist cannot be justified economically. In these cases, an agreement with a qualified local consulting firm or individual consultant might be the most appropriate way for a company to obtain essential services. Ordinarily, the consultant reports to the safety director of the company. Numerous qualified industrial hygienists are consultants, and many companies engage these consultants, as described in the following section. Also, insurance companies and the consulting services offered by OSHA should not be overlooked. Insurance companies (usually the workers' compensation carrier) have a vested interest in improving the workplace environment, in order to reduce the potential for occupational illness.

INDUSTRIAL HYGIENIST ON THE OCCUPATIONAL HEALTH AND SAFETY TEAM

The commitment to the protection of employee health and the industrial hygiene program should be incorporated into the overall safety policy of the company. The industrial hygienist specializes in anticipating, recognizing, evaluating, and controlling health hazards and is principally responsible for controlling chemical, physical, biological, and ergonomic stresses. In large organizations, the control of radiation rests with the radiation specialist or the health physicist. As the need for controlling biological stresses (especially for indoor air quality, such as preventing Legionnaires' disease or mold-related disorders) becomes more widely recognized, many industrial hygienists have taken on this responsibility as well. Depending on their expertise, industrial hygienists may be responsible for ergonomics. Industrial hygienists also maintain lists of all chemical, physical, and biological agents found within a facility; evaluate exposures to those agents; and institute controls to ensure that those exposures are within acceptable limits.

In-House Industrial Hygiene Services

The industrial hygienist must have a good working relationship with the engineering department. Many of the most effective exposure controls involve the industrial hygienist providing input during the design phase of planned construction or renovation projects. Additionally, any hazards identified by the industrial hygienist requiring engineering controls or process modifications to existing facilities will require working collaboratively with the engineering department. Two of the most important departments for the industrial hygienist are the safety and medical departments. The environmental staff, those responsible for controlling facility emissions, should also be part of the industrial hygienist's internal network. Because controls to reduce exposures within the facility can lead to increased emissions, a unified approach to reducing exposures is essential.

Occupational Health and Safety Team

The chief goal of a facility's occupational health and safety program is to prevent occupational injuries and illnesses by anticipating, recognizing, evaluating, and controlling occupational health and safety hazards. The medical, industrial hygiene, and safety programs may have distinct program goals, but all programs must interact with all components of the overall health and safety program. The occupational health and safety team consists of the industrial hygienist, the safety professional, the occupational health nurse, the occupational physician, the employees, senior and line management, and others, depending on the size and characteristics of a particular facility. Frequently, information provided by the medical staff on possible causes of symptoms can be important for developing effective evaluation and sampling strategies. Senior management support can demonstrate to workers that a healthy and safe environment is a company priority. Line management and workers can provide valuable information on which details of production processes should be examined during the industrial hygiene evaluation. All team members must act in harmony to provide information and activities that support the other members to achieve the overall goal of a safe and healthy work environment. Therefore, the separate functions must be administratively linked so that they create a successful and effective program.

Serious commitment to the protection of employee health and to the industrial hygiene program is demonstrated when management is visibly involved in the program and when all personnel comply with health and safety practices. Equally important is the assignment of authority, as well as the responsibility, to carry out the health and safety program. The health and safety function must be given the same level of importance and accountability as the production function.

PRACTICING INDUSTRIAL HYGIENE

The four definitive elements of industrial hygiene are anticipating, recognizing, evaluating, and controlling occupational hazards. Anticipating health hazards before they occur allows a more efficient use of resources by minimizing costly retrofits and other renovations needed to protect the health of workers. An unrecognized health hazard cannot be evaluated or controlled. Upon recognizing a health hazard, the industrial hygienist should be able to identify the measures necessary for proper evaluation. When the evaluation is completed, the industrial hygienist, in consultation with other members of the occupational health and safety team, can implement the controls needed to reduce exposures to tolerable limits.

Anticipating and Recognizing Occupational Health Hazards

The first steps in the process leading to the evaluation and control of exposure to harmful materials and processes are anticipating and recognizing potential occupational hazards. Such recognition can be based on general knowledge of the characteristics of the materials and processes, on clinical findings that disease or discomfort is present in the exposed population, on reports from others in scientific literature, on bulletins from trade associations or governmental agencies, on conversations with peers, or on reports by workers.

Essential to the process of anticipating and recognizing occupational health hazards are identifying and analyzing the potential impacts of the individual parts of a multifaceted process. To achieve this goal, a competent industrial hygienist uses his or her trained power of observation coupled with a fundamental knowledge of communication processes, science, statistics, mathematics, engineering and industrial processes, business, and psychology. This process can include the following:

- reviewing plans for new facilities and renovations
- having conversations with workers about perceived problems
- reviewing historical records
- having discussions with medical personnel
- observing work practices
- reviewing the use of chemical, physical, and biological agents
- checking the effectiveness of control measures
- reviewing production and processes.

The outcome or final product of the recognition phase should be a written industrial hygiene assessment or preliminary survey that identifies the variables observed such as the occupations, the number of employees working the various occupations, the raw materials, the products, the by-products/washes, the potential contaminants/hazards, the representative production levels, worst-case scenarios, and controls available. In addition to the assessment, process-flow diagrams and general layout drawings/sketches of the workplace should be obtained.

Based on the initial assessment, the industrial hygiene exposure monitoring/sampling plan for the workplace should be developed and implemented. The initial or baseline exposure results will, in turn, dictate the frequency of future periodic evaluations and the types of controls necessary.

A health hazard does not exist in isolation from the workplace. A chemical, physical, biological, or ergonomic stress is a concern only if workers can be exposed to it. The industrial hygienist must use all of his or her knowledge, experience, and resources to evaluate the workplace. In addition, discussions with workers and managers are extremely helpful in evaluating their potential for exposure.

Evaluating and Controlling Environmental Hazards

The various environmental factors or stresses that can cause sickness, impaired health, or significant discomfort in workers can generally be classified as chemical, physical, biological, or ergonomic in nature.

Chemical Hazards

Exposure to a variety of chemical substances occurs both on and off the job. Although most chemical substances do not present a hazard under ordinary conditions, all have the potential to be harmful at some concentration and level of exposure. Chemicals present a wide range of potential health hazards (such as irritation, disease, sensitization, and carcinogenicity) and physical hazards (such as flammability, corrosion, and reactivity).

Chemical hazards can be found in the following basic forms:
- *dusts*—particles generated from solid organic or inorganic materials by reducing their size through either mechanical or natural processes
- *liquids*—aqueous substances that flow freely, like water
- *fumes*—aerosolized particles formed when a volatilized solid, such as a metal, condenses in cool air
- *mists*—finely dispersed liquids suspended in the air
- *gases*—formless fluids that can be changed to the liquid or solid state only by the combined effects of increased pressure and decreased temperature
- *vapors*—the gaseous form of substances that are normally in the solid or liquid state at standard temperature and pressure
- *smoke*—carbon or soot particles produced by the incomplete breakdown of carbon-containing materials.

When chemical stresses will be encountered in the workplace, a hazard evaluation should be conducted to identify their potential health impacts. The most effective worksite evaluation looks at all pertinent operations and work activities. The industrial hygienist inspects, researches, or analyzes how the particular chemical hazards at a worksite could affect the workers. When determining the extent of worker exposure, industrial hygienists may review Safety Data Sheets (SDSs), formerly known as Material Safety Data Sheets (MSDSs), and/or conduct environmental monitoring and use applicable analytical methods. Airborne concentration, type and length of exposure, regulatory and consensus exposure limits (both short- and long-term limits) may be some of the issues considered in the evaluation.

If the chemical exposure evaluation indicates hazardous conditions that could affect employee health, the industrial hygienist then recommends appropriate corrective actions. Substituting less hazardous materials is one of the most effective ways of eliminating or reducing exposure to chemicals that are toxic or otherwise hazardous. If substitution is used, one should be careful not to generate a more hazardous situation with the substitute. Engineering controls (ventilation, isolation, enclosure), work practices (procedures, good housekeeping), administrative controls (worker rotation), and other methods (personal protective equipment) can also be used to control potential health hazards.

Physical Hazards

Physical hazards include excessive levels of ionizing and nonionizing radiation, noise, vibration, and temperature extremes.

Ionizing and Nonionizing Radiation. Only industrial hygienists with special training and education should evaluate and recommend procedures to control ionizing and nonionizing radiation. These areas are covered extensively in Chapter 27, Laboratory Safety, in this volume, and in *Fundamentals of Industrial Hygiene* (Plog 2012).

Noise. Noise is one of the most commonly encountered occupational health hazards. If a noise problem is suspected, a noise assessment or survey should be performed to determine the source(s) of the noise, as well as the amount, exposed population, and duration of the exposure. With proper instrumentation, monitoring noise levels is a relatively straightforward procedure for an experienced industrial hygienist.

To prevent adverse outcomes from excessive noise exposure, noise levels should be kept at acceptable levels. After evaluating a workplace, an industrial hygienist can help develop a hearing-protection strategy by recommending engineering and administrative controls and the use of personal protective equipment (PPE) where needed.

The literature documents a variety of control techniques that reduce overall worker exposure to noise. Controls that reduce the amount of sound energy released by the noise source, divert the flow of sound energy away from the receiver, or protect the receiver from the sound energy reaching him or her are frequently used. Types of noise controls include proper maintenance of equipment, revised operating procedures, equipment replacement, acoustical shields and barriers, equipment redesign, enclosures, and administrative controls (NIOSH 1979). When engineering and administrative controls cannot adequately remedy the problem, personal hearing protection can be used (see Chapter 7, Personal Protective Equipment, in the *Engineering & Technology* volume). As always, personal protection should be considered only an interim measure while other means of reducing workplace noise to within acceptable limits are explored and implemented.

Vibration. Vibration exposure occurs in many occupations when a worker comes in contact with vibrating machinery or equipment. Significant exposure to vibration could result from either stationary or portable equipment. The industrial hygiene evaluation should look at the following risk factors: intensity and frequency of the vibration, the duration (time) of exposure to the vibration, and the part(s) of the body receiving the vibration energy. The evaluation is typically described in terms of frequency, amplitude, and acceleration.

- *frequency*—the number of cycles that a vibrating object completes in 1 second. The unit of frequency is hertz (Hz). One hertz equals one cycle per second.
- *amplitude*—the distance from the stationary position to the extreme position of oscillation. The intensity of a vibration depends on its amplitude.
- *acceleration*—a measure of how quickly the speed of a vibrating object changes with time.

Vibration exposure involves contact with a vibrating mechanism that transfers vibration energy to a person's body. The effect of vibration exposure also depends on the frequency of the vibration. Each organ of the body has its own resonant frequency (the tendency to vibrate at one particular frequency depending on the makeup, size, and structure of the organ). If vibration occurs at or near any of these resonant frequencies, the resulting effect is greatly increased.

Adverse effects from vibration exposure can be reduced through engineering controls (isolating the vibrating source), appropriate tool selection (tools with antivibration properties and/or that are vibration-damped), use of appropriate vibration-absorbing materials (such as gloves and shoes), good work practices, administrative controls (limiting contact time), and education programs.

Temperature Extremes. Because both very cold and very hot temperatures can be encountered in numerous occupational settings, these temperatures are frequently evaluated by industrial hygienists. Excessive exposure to heat is referred to as *heat stress*, and excessive exposure to cold is referred to as *cold stress*. An evaluation of temperature extremes includes an examination of environmental conditions, worker behavior, medical records, and body indices.

An industrial hygienist conducts temperature-stress evaluations on operations that involve high and low air temperatures, radiant heat sources, high humidity, or direct physical contact with hot and cold objects. Strenuous physical activities or outdoor procedures conducted in hot-or cold-weather extremes may also require evaluations.

Two types of exposure limits are often used as guidelines for an evaluation: occupational exposure and thermal comfort. Occupational exposure limits are designed to protect industrial workers from temperature-related illness. Thermal comfort limits are used to ensure productivity and quality of work of office workers. An industrial hygienist uses portable heat-stress instruments to measure thermal conditions (temperature, humidity, air velocity). These measurements are then used to calculate the wet bulb globe temperature (WBGT) and effective temperature (ET) indices, two common heat-stress indices.

Measures to control the exposure to temperature extremes include training (in the knowledge of the hazards of thermal stress), fluid replacement when warranted, medical surveillance, engineering controls (ventilation, air cooling or heating, fans, shielding, and insulation), administrative controls (acclimating, scheduling, reducing time of exposure via work/rest regimens), and specialized personal protection. Each method of control should be specific to the type and degree of the hazard.

Biological Hazards

Biological hazards include but are not limited to bacteria, fungi, and viruses. These hazards can result in health effects ranging from skin irritations and allergies to infections and life-threatening diseases. One of the more recognized categories of these hazards is bloodborne pathogen, examples of which are hepatitis B (HBV), hepatitis C (HCV), and the human immunodeficiency virus (HIV). Biological hazards can affect workers in a wide variety of settings including microbiology, public health, clinical, and molecular biology laboratories; hospital and health care facilities; biotechnology facilities; animal facilities and veterinary practices; and agriculture operations. The hazard may be introduced into the work environment via a source (vector) such as plants, birds, animals, insects, or humans. In laboratory settings, exposure can be associated with the direct handling and manipulation of these agents related to the clinical or research operations.

An industrial hygienist completing a biosafety risk assessment examines epidemiology, pathogenicity, susceptible populations, and routes of transmission. Recommended control measures may include education and training in good handling practices, engineering controls (ventilation, biological safety cabinets, isolation), medical intervention and surveillance (immunizations), and use of PPE (see Chapter 7, Personal Protective Equipment, in the *Engineering & Technology* volume). An effective biosafety control measure is using the concept of *universal precautions*. Originally developed for health care settings, the concept has been adapted for a wide range of workplaces. Universal precautions assume that all human blood and certain human body fluids are infectious for bloodborne pathogens. The identification, risk assessment, classification, and control of biological agents are covered extensively in *Fundamentals of Industrial Hygiene* (Plog 2012).

Ergonomic Factors

Ergonomics is the science of matching the job to the worker. A mismatch between the physical requirements of a job and the physical capability of a worker can result in injuries and disorders affecting cartilage, joints, ligaments, muscles, nerves, spinal discs, and tendons. These injuries are generally classified as musculoskeletal disorders (MSDs). Although a qualified health care practitioner should conduct the evaluation and treatment of these disorders, an industrial hygienist can evaluate the workplace factors leading to MSDs and implement corrective actions based on the evaluation.

In analyzing a potential ergonomics problem, an industrial hygienist could perform a job safety analysis (see Chapter 9, Identifying Hazards). Risk factors that are frequently identified during ergonomic evaluations include repetitive motions, awkward work positions, excessive amounts of force used to perform jobs, repeated or improper lifting of heavy objects, cold temperatures, and vibrations. Good engineering and biomechanical principles must be applied to eliminate hazards of this kind.

Ergonomic interventions may include training and educating, adjusting the height of working surfaces, providing the right tool for the job, encouraging short rest breaks, reducing repetitive motions, eliminating forceful or awkward hand exertions, or using equipment for heavy or repetitive lifting. See Chapter 16, Ergonomics Yesterday, Today, and Tomorrow, in this volume, for a more detailed review of ergonomics programs.

TOXICITY

Toxicity is the capacity of a chemical to harm or injure a living organism by other than mechanical means. Because toxicity entails a definite dimension (quantity or amount), a chemical's toxicity often depends on how much of a chemical a worker is exposed to. A toxic effect is any undesirable, reversible or irreversible disturbance of physiological function, including any chemically induced tumor or any mutagenic or teratogenic effect or death resulting from physical stresses or overexposure to a toxic substance via the respiratory tract, skin, eyes, mouth, or any other route into the body. These effects can also arise as side effects of medications and vaccines.

The industrial hygienist is responsible for quantifying levels of exposure to chemical agents and prescribing precautionary measures and restrictions to prevent overexposures. From a toxicological perspective, the industrial hygienist must consider all types of exposures and the possible subsequent outcomes on the living organism. An industrial hygienist should work closely with occupational health practitioners to ensure that recommendations for exposure levels are low enough to prevent adverse impacts on employees.

Toxicity versus Hazard

A distinction must be made between toxicity and hazard. Toxicologists generally consider *toxicity* to be a substance's capability to produce an unwanted effect when that substance reaches a sufficient concentration at a certain site in the body. *Hazard* is the practical likelihood that exposure to a toxic substance will cause harm. Many factors contribute to determining the degree of hazard—route of entry, dosage, physiological state, environmental variables, and other factors. Assessing a hazard thus involves estimating the probability that a toxic substance will cause harm. Toxicity, along with the chemical and physical properties of a substance, determine the level of hazard. Two liquids can possess the same degree of toxicity but present different degrees of hazard.

A chemical stimulus can be considered to have produced a toxic effect when the following are true:
- An observable or measurable physiological deviation has occurred in any organ or organ system. The change can be anatomic in character and may accelerate or inhibit a normal physiological process, or the deviation can be a specific biochemical change.
- The observed change can be passed on from animal to animal even though the dose-effect relationships vary.
- The stimulus has changed normal physiological processes in such a way that a protective mechanism is impaired in its ability to defend against other adverse stimuli.

- The toxic effect does not occur without a stimulus or occurs so infrequently that it indicates that it is a generalized or nonspecific response. When high degrees of susceptibility are noted, equally significant degrees of resistance should be apparent.
- The observation is noted and is reproducible by other investigators.
- The physiological change reduces the efficiency of an organ or function and impairs a physiological reserve in such a way as to interfere with the reserve's ability to resist or adapt to other normal stimuli, either permanently or temporarily.

Entry into the Body

For an adverse effect to occur, the toxic substance must first reach the organ or bodily site where it can cause damage. Common means of entry are inhalation, skin absorption, ingestion, and injection. Depending on the substance and its specific properties, however, entry and absorption can occur by more than one means, such as a solvent that can be inhaled and can also penetrate the skin. When absorption into the bloodstream occurs, a substance may elicit general effects, or, more likely, the critical injury will be localized in specific tissues or organs. When evaluating an exposure, it is therefore important for an industrial hygienist to determine the means of entry and to remember that an exposure could result from several different means.

Inhalation

When industrial exposures to chemicals occur, the most common means of entry is inhalation. Nearly all materials that are airborne can be inhaled.

The respiratory system is composed of two main areas: the upper respiratory tract airways (the nose, throat, trachea, and major bronchial tubes leading to the lobes of the lungs) and the alveoli, where the actual transfer of gases across thin cell walls takes place. Only particles smaller than about 5 μm in diameter are likely to enter the alveolar sac.

The amount of a toxic compound absorbed via the respiratory pathways depends on the compound's concentration in the air, the duration of exposure to the compound, and the pulmonary ventilation volumes, which increase with greater workloads. If the toxic substance is present in the form of an aerosol, deposition and absorption occur in the respiratory tract. For more details, see *Fundamentals of Industrial Hygiene* (Plog 2012).

Gases and vapors of low water solubility but high fat solubility pass through the alveolar lining into the bloodstream and are distributed to the organ sites for which they have an affinity. During inhalation of a uniform level, the absorption of the compound into the blood reaches equilibrium with metabolism and elimination.

Skin Absorption

An important means of entry is absorption through either intact or abraded skin. Contact of a substance with skin results in four possible outcomes: the skin acts as an effective barrier, the substance reacts with the skin and causes local irritation or tissue destruction, the substance produces skin sensitization, or the substance penetrates the blood vessels under the skin and enters the bloodstream.

The cutaneous absorption rate of some organic compounds increases when temperature or perspiration increases. Therefore, absorption can be greater in warm climates or seasons. The absorption of liquid organic compounds may follow surface contamination of the skin or clothes; for other compounds, it may directly follow the vapor phase, in which case the rate of absorption is roughly proportional to the air concentration of the vapors. The process may involve a combination of deposition of the substance on the skin surface followed by absorption through the skin.

Ingestion

Anything swallowed moves into the intestine and can be absorbed into the bloodstream and thereafter prove toxic. The problem of ingesting chemicals is not widespread in industry: most workers do not deliberately swallow the materials they handle.

However, workers can ingest toxic materials as a result of eating in contaminated work areas; contaminated fingers and hands can also lead to accidental oral intake when a worker eats or smokes on the job. Workers also ingest materials when contaminants deposited in the respiratory tract are carried out to the throat by the action of the ciliated lining of the respiratory tract. Workers then swallow these contaminants, which are significantly absorbed in the gastrointestinal tract.

Injection

Material can be injected into some part of the body, such as the bloodstream, the peritoneal cavity, or the pleural cavity. Material can also be injected into the skin, muscle, or any other place a needle can be inserted. The effects produced vary with the location of administration. In industrial settings, injection is an infrequent route of worker chemical exposure.

INDUSTRIAL HYGIENE CONSULTING SERVICES

The ideal circumstance for most companies is to have a full-time industrial hygienist on staff. However, when this is not possible because of economic factors and the lack of fully qualified personnel, industrial hygienists can be employed in a consulting capacity. However, there are drawbacks to this approach. The consultant industrial hygienist is ordinarily called in only when problems arise, which means that the person may not be as familiar with the facility and the personnel as a full-time industrial hygienist would be. This problem can be lessened by scheduling regular industrial hygiene consulting visits to discuss policy issues and inspect the facility during normal conditions. Some of these visits should coincide with visits of the medical consultant. Many companies with industrial hygiene consulting services also have medical consulting services.

Several sources are available to assist in the search for a competent and qualified industrial hygiene consultant. One source is the list of industrial hygiene consultants published by the AIHA. A consultant's name being in the list does not guarantee the consultant's competence, however. For an industrial hygiene consultant to be considered competent, at minimum he or she should be certified (i.e., designated as a CIH) and be familiar with the industry or other occupational setting of interest. As with other professional services, personal recommendations from satisfied users of consulting services are often the best source of information. Recommendations provide differentiation among otherwise apparently equivalently qualified consultants.

In addition to the qualifications of the person being considered as a consultant, the resources of his or her firm can also be important. When quick turnaround of analytical results from air monitoring or other workplace monitoring is required, consulting firm having an in-house laboratory can be significant. In any case, the laboratory considered for use should be accredited by the AIHA. (This accreditation should be required regardless of whether the analysis needed is one for which specific accreditation is offered. The AIHA accreditation process includes evaluating such general areas as quality control and record keeping, as well as evaluating performance on specific analytes.)

In some cases, trade associations have experience with consulting firms and can recommend firms that are familiar with the industry of concern. Formal consulting agreements with the industrial hygiene consultant should stipulate adequate time to discuss with the medical director and the engineering staff problems of mutual interest as well as operations.

CERTIFICATION AND LICENSURE

Certification of industrial hygienists began in 1960, when the AIHA and the American Conference of Governmental Industrial Hygienists (ACGIH) established the American Board of Industrial Hygiene (ABIH) to set up certification requirements. Figure 13–1 shows the industrial hygiene code of ethics. Two classes of certification are currently recognized: diplomates (who are permitted to use the designation Certified Industrial Hygienist—CIH) and Certified Associate Industrial Hygienist (CAIH), which was discontinued in 2006.

Certification is currently offered in comprehensive practice. Beginning in 1993, the specialized competence of diplomates in the two subspecialties of indoor air quality and hazardous materials management and remediation were formally recognized by the ABIH. Before 1992, certification in the specialized fields of acoustical, air pollution, engineering, radiological, and toxicological aspects of industrial hygiene was offered. Certification in the specialized field of chemical practice was discontinued after 2000. Although no new certificates have been offered since then, those certifications remain valid.

Applicants for the Certified Industrial Hygienist (CIH) credential must meet educational and experience requirements as well as demonstrate their knowledge and skills through examination. Qualifications are as follows:

1. Graduation from a regionally accredited college or university, or other college acceptable to the ABIH, with a bachelor's degree in biology, chemistry, physics, engineering, or a program accredited by the Accreditation Board for Engineering and Technology Inc. (ABET) in industrial hygiene or safety. The ABIH may consider and accept any other bachelor's degree from an acceptable college or university provided the degree is based upon appropriate coursework and represents at least 60 semester hours in undergraduate or graduate level courses in science, math, engineering, and science-based technology, with at least 15 of those hours at the upper (junior, senior, or graduate) level.
2. Four years of broad-based professional industrial hygiene experience
3. A total of 180 academic contact hours or 240 continuing education contact hours of industrial hygiene coursework. At a minimum, half of those hours must include coursework in the areas of industrial hygiene fundamentals, measurements, controls, and toxicology. An additional 2 hours of coursework/training are required in the area of ethics.

After certification is achieved via examination, active practice, technical committee work, publications, education, meetings, teaching, retesting, or other approved methods must be documented to ensure that the certification remains current. Maintaining certification requires documenting continued activity in certain defined professional activities and continuing education.

CODE OF ETHICS FOR THE PRACTICE OF INDUSTRIAL HYGIENE

Objective
These canons provide standards of ethical conduct for Industrial Hygienists as they practice their profession and exercise their primary mission: to protect the health and well-being of working people and the public from chemical, microbiological, and physical health hazards present at, or emanating from, the workplace.

Canons of Ethical Conduct

Canon 1
Industrial Hygienists shall practice their profession following recognized scientific principles with the realization that the lives, health, and well-being of people may depend upon their professional judgment and that they are obligated to protect the health and well-being of people.

Interpretive Guidelines
- Industrial Hygienists should base their professional opinions, judgments, interpretations of findings, and recommendations upon recognized scientific principles and practices which preserve and protect the health and well-being of people.
- Industrial Hygienists shall not distort, alter, or hide facts in rendering professional opinions or recommendations.
- Industrial Hygienists shall not knowingly make statements that misrepresent or omit facts.

Canon 2
Industrial Hygienists shall counsel affected parties factually regarding potential health risks and precautions necessary to avoid adverse health effects.

Interpretive Guidelines
- Industrial Hygienists should obtain information regarding potential health risks from reliable sources.
- Industrial Hygienists should review the pertinent, readily available information to factually inform the affected parties.
- Industrial Hygienists should initiate appropriate measures to see that the health risks are effectively communicated to the affected parties.
- Parties may include management, clients, employees, contractor employees, or others, dependent on circumstances at the time.

Canon 3
Industrial Hygienists shall keep confidential personal and business information obtained during the exercise of industrial hygiene activities, except when required by law or overriding health and safety considerations.

Interpretive Guidelines
- Industrial Hygienists should report and communicate information which is necessary to protect the health and safety of workers and the community.
- If their professional judgment is overruled under circumstances where the health and lives of people are endangered, Industrial Hygienists shall notify their employer, client, or other such authority, as may be appropriate.
- Industrial Hygienists should release confidential personal or business information only with the information owner's express authorization, except when there is a duty to disclose information as required by law or regulation.

Canon 4
Industrial Hygienists shall avoid circumstances where a compromise of professional judgment or conflict of interest may arise.

Interpretive Guidelines
- Industrial Hygienists should promptly disclose known or potential conflicts of interest to parties that may be affected.
- Industrial Hygienists shall not solicit or accept financial or other valuable consideration from any party, directly or indirectly, which is intended to influence professional judgment.
- Industrial Hygienists shall not offer any substantial gift, or other valuable consideration, in order to secure work.
- Industrial Hygienists should advise their clients or employer when they initially believe a project to improve industrial hygiene conditions will not be successful.
- Industrial Hygienists should not accept work that negatively impacts the ability to fulfill existing commitments.
- In the event that this Code of Ethics appears to conflict with another professional code to which Industrial Hygienists are bound, they will resolve the conflict in the manner that protects the health of affected parties.

Canon 5
Industrial Hygienists shall perform services only in the areas of their competence.

Interpretive Guidelines
- Industrial Hygienists should undertake to perform services only when qualified by education, training, or experience in the specific technical fields involved, unless sufficient assistance is provided by qualified associates, consultants, or employees.
- Industrial Hygienists shall obtain appropriate certifications, registrations, and/or licenses as required by federal, state, and/or local regulatory agencies prior to providing industrial hygiene services, where such credentials are required.

Canon 6
Industrial Hygienists shall act responsibly to uphold the integrity of the profession.

Interpretive Guidelines
- Industrial Hygienists shall avoid conduct or practice which is likely to discredit the profession or deceive the public.
- Industrial Hygienists shall not permit use of their name or firm name by any person or firm which they have reason to believe is engaging in fraudulent or dishonest industrial hygiene practices.
- Industrial Hygienists shall not use statements in advertising their expertise or services containing a material misrepresentation of fact or omitting a material fact necessary to keep statements from being misleading.
- Industrial Hygienists shall not knowingly permit their employees, employers, or others to misrepresent the individuals' professional background, expertise, or services which are misrepresentations of fact.
- Industrial Hygienists shall not misrepresent their professional education, experience, or credentials.

Figure 13–1. The American Board of Industrial Hygiene Code of Ethics, May 2007. *(Reprinted with permission of the American Board of Industrial Hygiene.)*

Although many industrial hygienists are becoming interested in seeking some form of government licensure, no federal licensure requirements had been established by the end of 2013. However, the CIH designation is often required for industrial hygiene practice on projects or jobs supported by government funds. The CIH designation is increasingly recognized as a minimum requirement to ensure that industrial hygiene work is being performed according to professional standards, especially when litigation is anticipated or feared. As of 2013, 17 states (Alaska, California, Connecticut, Florida, Georgia, Indiana, Minnesota, Nebraska, Nevada, New Jersey, North Carolina, Ohio, Oregon, South Carolina, Tennessee, Texas, and Virginia) had enacted title protection that prohibits the use of the CIH designation without ABIH certification.

PROFESSIONAL ORGANIZATIONS

The AIHA is the predominant U.S. industrial hygiene organization. The AIHA was formed in 1939 by a small group of industrial hygienists and has grown since then to more than 10,000 members. The only restrictions on membership are educational qualifications and the practice of industrial hygiene. A full member must have an appropriate degree (usually in one of the physical or biological sciences) and must have practiced industrial hygiene full time for 3 years. The AIHA provides many services to the profession, including accreditation of laboratories offering analytical services. There are local (regional) sections of the AIHA in all areas of the United States, and many foreign occupational health professionals are members as well.

The American Conference of Governmental Industrial Hygienists previously limited full membership to health and safety professionals practicing in government agencies and educational institutions. Today, the ACGIH offers membership to all practitioners in industrial hygiene, occupational health, environmental health, and safety domestically and abroad. Two publications offered by the ACGIH are of particular value to the profession. One of these is *Industrial Ventilation: A Manual of Recommended Practice* by the Industrial Ventilation Committee of the ACGIH. This manual gives guidance on the design of ventilation systems that control airborne health hazards and is a standard reference source. The other is the annually published *Threshold Limit Values (TLVs) and Biological Exposure Indices (BEIs)*. TLVs and BEIs are widely used measures of allowable exposure limits; in fact, the 1968 TLVs were adopted into law as the permissible exposure limits in the OSH Act.

Many industrial hygienists are also members of organizations that represent major and minor industries. The American Petroleum Institute, the Chemical Manufacturers Association, and the National Agricultural Chemical Association are some industry organizations that have active committees on various aspects of occupational health on which industrial hygienists can serve. Some industrial hygienists also maintain memberships in organizations representing the disciplines of their original training. The American Chemical Society and the American Institute of Chemical Engineers are two societies that claim many industrial hygienists among their members. In addition, industrial hygienists belong to organizations with peripheral interests in industrial hygiene or that represent fields within which industrial hygienists have responsibilities. The Air & Waste Management Association (A&WMA), the American Biological Safety Association (ABSA), the American Management Association (AMA), the American Public Health Association (APHA), the American Society of Safety Engineers (ASSE), the Genetic Toxicology Association (GTA), the Campus Safety, Health and Environmental Management Association (CSHEMA), the National Environmental Health Association (NEHA), and the Society for Epidemiological Research (SER) are a few examples of these organizations.

SUMMARY

- Industrial hygiene is the science and art devoted to anticipating, recognizing, evaluating, and controlling environmental hazards or stresses arising in or from the workplace.
- The industrial hygienist works with management, and with medical, safety, and engineering personnel to help eliminate or guard against environmental pollution and other conditions that can endanger human health.
- The four definitive elements of industrial hygiene are anticipating, recognizing, evaluating, and controlling occupational hazards. These include chemical, physical (excessive noise, radiation, etc.), biological, and ergonomic hazards.
- Toxic effects are undesirable disturbances of physiological function caused by exposure to chemical or physical stresses. Toxicity is the capability of a chemical to harm or injure a living organism by other than mechanical means. Hazard is the practical likelihood that exposure to a toxic substance will cause harm.
- The industrial hygienist is responsible for quantifying levels of exposure to hazardous conditions and prescribing precautionary measures and restrictions to protect workers' health. These professionals can be either full-time or part-time consultants.

REFERENCES

American Conference of Governmental Industrial Hygienists (ACGIH). *Industrial Ventilation: A Manual of Recommended Practice.* Cincinnati, OH: ACGIH. Published biannually, even-numbered years.

———. *Threshold Limit Values (TLVs) and Biological Exposure Indices (BEIs).* Cincinnati, OH: ACGIH. Issued annually.

American Industrial Hygiene Association. "Discover Industrial Hygiene." Accessed July 12, 2014. www.AIHA.org.

Balge, M. Z. and G. R. Krieger, eds. *Occupational Health and Safety.* 3rd ed. Itasca, IL: National Safety Council, 2000.

Burgess, W. A. *Recognition of Health Hazards in Industry.* New York: Wiley, 1981.

Cohrssen, B., and V. E. Rose, eds. *Patty's Industrial Hygiene and Toxicology.* 6th ed. Vols. 1–3. New York: Wiley, 2011.

Cralley, L., and L. Cralley, eds. *In-Plant Practices for Job-Related Health Hazards Control.* Vol. 1, *Production Processes*, and Vol. 2, *Engineering Aspects*. New York: Wiley, 1989.

National Institute for Occupational Safety and Health (NIOSH). *Industrial Noise Control Manual.* Rev. ed. NIOSH Publication No. 79-117, 1979.

Plog, B. A., ed. *Fundamentals of Industrial Hygiene.* 6th ed. Itasca, IL: National Safety Council, 2012.

REVIEW QUESTIONS

1. Define *industrial hygiene*.
2. Whom does the industrial hygienist work with to control environmental stresses or occupational health hazards in the workplace?
   a. management
   b. medical personnel
   c. safety personnel
   d. engineering personnel
   e. all of the above
3. List five control procedures that are used to reduce or eliminate exposure.
   a.
   b.
   c.
   d.
   e.
4. List five factors of an effective industrial hygiene program.
   a.
   b.
   c.
   d.
   e.
5. The industrial hygienist is responsible for monitoring what types of environmental factors or stresses that can cause sickness, impaired health, or significant discomfort?
   a.
   b.
   c.
   d.
6. Which department is the most important to the industrial hygienist?
   a. engineering department
   b. environmental department
   c. safety department
   d. medical department
   e. all of the above
7. Who, at minimum, should be members of the occupational health and safety team?
   a.
   b.
   c.
   d.
   e.
   f.

Environmental Management

14

Anne Germain, PE, BCEE
James Larson, JD, MBA, PE
Philip E. Hagan, JD, MBA, MPH, ARM, CIH, CET, CHMM, CHCM, CHSP, CEM

Environmental Regulations

Proactive Environmental Management
A New Approach ▸ Strategies for Success

Managing for a Healthy Environment
Organizing an Environmental Management Program ▸ Staff Skills and Backgrounds

Waste Minimization

Groundwater Contamination

Global Solutions for Global Problems

ISO 14000:2004
Historical Context ▸ The Next Phase ▸ Certification ▸ Meaning for Organizations ▸ Key EMS Elements ▸ Implementation ▸ Gap Analysis ▸ Environmental Aspects Review ▸ Legal Considerations ▸ Objectives and Targets ▸ Integration and Alignment ▸ Independent Audit Function ▸ Market Drivers ▸ Competition ▸ Disclosures ▸ Influencing Suppliers ▸ Mentoring ▸ Environmental Performance Evaluation ▸ Overview of ISO 14031:2013, the Guideline Document ▸ Environmental Costing and Valuation Methods ▸ Life-Cycle Assessment ▸ Trends—Cause for Optimism

Summary

References

Review Questions

Management of environmental program elements has increasingly become a major focus for safety professionals. Many companies have separate environmental departments; however, as companies streamline and reengineer their professional support services staff, there has been a marked trend toward consolidation of the health, safety, and environmental functions. Therefore, the traditional safety professional has increasingly found that knowledge of environmental affairs is critical. For companies involved in international business, the ISO 14000 standards represent a revolution in corporate environmental management; therefore, the safety professional should become familiar with the ISO 14000 approach to environmental affairs. This chapter covers the following topics:
- a brief synopsis of the major U.S. environmental regulations
- principles of proactive environmental management, waste minimization, groundwater management, and auditing
- a discussion of the International Standards Organization (ISO) 14000 series of standards covering environmental management
- a discussion of the global effects of environmental pollution and issues.

ENVIRONMENTAL REGULATIONS

The U.S. environmental regulatory system has always been characterized by both constant change and complexity. The history and legislative framework of both U.S. and international environmental regulations is presented in the National Safety Council's *Accident Prevention Manual for Business & Industry: Environmental Management* (Krieger 2000). The safety professional should always consult the most current source of legal/regulatory information because environmental laws and regulations are in a state of constant flux, particularly at the state and local levels. In the United States, the major environmental regulations cover the release, treatment, storage, and disposal of potentially hazardous materials into air, water (surface and groundwater), and soil. Some of the most important laws are the following:
- *National Environmental Policy Act (NEPA)*—One of the first laws written, NEPA establishes a broad national framework for protecting the environment using a basic policy to ensure that all branches of government give proper consideration to the environment before undertaking any major federal action that significantly affects the environment. Projects involving airports, buildings, military complexes, highways, parkland purchases, and any other similar federal activities are covered by NEPA. NEPA uses environmental assessments (EAs) and environmental impact statements (EISs) to assess the likelihood of impacts from alternative courses of action.
- *Comprehensive Environmental Response, Compensation, and Liability Act (CERCLA) and Superfund Amendments and Reauthorization Act (SARA)*—CERCLA provides a federal "Superfund" to clean up uncontrolled or abandoned hazardous-waste sites, accidents, spills, and other emergency releases of pollutants and contaminants into the environment. Through CERCLA, the EPA imposes liability on owners or operators of a facility from which there is a release of hazardous substances into the environment. Transfer of a property may not relieve an owner or operator of liability; in addition, the current owner may be held strictly liable. There are complex third-party liability issues and provisions for cleanup cost recovery. The CERCLA or Superfund law has been extremely controversial because of its perceived cost/benefit ratio; that is, the costs of compliance and litigation versus the costs of relatively modest cleanup of sites have been staggering. The last reauthorization of this law was the Superfund Amendments and Reauthorization Act (SARA) of 1986. These amendments cover emergency planning, community right-to-know, and toxic release reporting, among other notification requirements.
- *Resource Conservation and Recovery Act (RCRA)*—This law regulates solid and hazardous management activities. Subtitle C of RCRA regulates the generation, treatment, storage, and disposal of hazardous wastes. It requires an owner/operator of a facility to undertake corrective action to clean up a facility used for the treatment, storage, or disposal of hazardous waste. The owner/operator must also comply with a complex permit program. Subtitle C has become increasingly important because it broadly applies to any operating facility that is a generator of hazardous substances. Subtitle D regulates solid waste facilities.
- *Underground Storage Tank Regulations (under RCRA and state control)*—These laws impose operating, reporting, financial assurance, and potential cleanup obligations on people and companies owning and operating underground petroleum storage tanks (USTs). The rules and regulations covering USTs are quite detailed and prescriptive. In addition, many states have their own regulations and enforcement policies. Underground petroleum storage tanks and petroleum releases have recently been the subject of important studies and recommended standards by the Lawrence Livermore Laboratory (California) and the American Society of Testing and Materials (ASTM). Both the Livermore and the ASTM work evaluated and proposed efficient and cost-effective approaches for the evaluation and risk-based cleanup of petroleum releases from USTs. Because USTs are so common and accidental releases are frequent, the safety professional should carefully review these documents.

- *Clean Air Act (CAA) of 1990 (amended in 2004, PL 108-201, www.epw.senate.gov/envlaws/cleanair.pdf)*—This act is a complex, multifaceted statute that is designed to regulate air emissions from stationary and mobile sources. The act is composed of 11 different titles covering topics such as national ambient air quality standards (Title I), hazardous air pollutants (Title III), acid deposition control (Title IV), operating permits (Title V), and ozone protection (Title VI). Another important CAA topic is the Chemical Safety Information, Site Security, and Fuels Regulatory Relief Act, an amendment to Section 112(r) of the Clean Air Act that addresses reporting and disseminating information on flammable fuels and public access to Off-Site Consequence Analysis (OCA) data. The EPA has drafted new rules that will regulate greenhouse gases including carbon dioxide.
- *Clean Water Act (CWA) (amended November 2002, PL 107-303, www.epw.senate.gov/water.pdf)*—This act prohibits the discharges of pollutants from point sources and storm water into navigable waters of the United States without a permit. The act imposes liability on the person who is responsible for the operation and/or equipment that results in a discharge. The basic thrust of the act is to force compliance with both uniform, technology-based effluent limitations, regardless of the quality of the receiving waters, and with more stringent limitations necessary to meet state-established water quality standards.
- *Safe Drinking Water Act (SDWA) (revised 2002)*—This act imposes federal drinking water standards on virtually all public water systems. The act requires the establishment of drinking water standards for maximum contaminant levels (MCLs) for organic and inorganic chemicals, turbidity, coliform bacteria, and various measures of radioactivity.
- *Toxic Substances Control Act (TSCA) (revised 2002)*—This act governs the manufacture and use of chemical products. The importation and exportation are regulated under the act; in addition, specific regulations control the use, management, storage, and disposal of polychlorinated biphenyl (PCB) materials.
- *Oil Pollution Act (OPA) (revised 2000, PL 106-580)*—This act imposes strict liability on responsible parties for removal costs and damages resulting from discharges of oil into navigable waters of the United States. An owner/operator of an onshore facility that resulted in a discharge would be considered a responsible party.
- *Chemical Facility Anti-Terrorism Standards*—Uncertain times have resulted in new regulations promulgated by the U.S. Department of Homeland Security (DHS). One such rule requires high-risk chemical facilities to complete security vulnerability assessments, develop site security plans, and implement risk-based measures based on DHS-defined risk-based performance standards. The rule establishes risk-based performance standards designed to enhance the security of chemical facilities. Covered chemical facilities are required to prepare security vulnerability assessments and to develop and implement site security plans that include measures that satisfy the identified risk-based performance standards. Risk-based performance standards for the security of chemical facilities use the following criteria:
  - the consequence of a successful attack on a facility (consequence)
  - the likelihood that an attack on a facility will be successful (vulnerability)
  - the intent and capability of an adversary in respect to attacking a facility (threat).

  Some covered chemical facilities can submit alternate security programs in lieu of a security vulnerability assessment, site security plan, or both. The program provides a Chemicals of Interest List (known as Appendix A) to determine whether a facility is required to develop and implement a site security plan.

Other regulations of interest include the following:
Archaeological Resources Protection Act (ARPA)
Asbestos Hazard Emergency Response Act
Coastal Zone Management Act
Disposal of Recyclable Materials
Endangered Species Act
Energy Independence and Security Act (EISA)
Energy Policy Act
Federal Facilities Clean Water Compliance Act of 1999
Federal Facilities Compliance Act of 1992 (FFCA)
Federal Food, Drug, and Cosmetic Act (FFDCA)
Federal Insecticide, Fungicide, and Rodenticide Act (FIFRA)
Food Quality Protection Act (FQPA)
Hazardous Materials Transportation Act
Indoor Radon Abatement Act
Lead-Based Paint Poisoning Prevention Act
Lead Contamination Control Act
Marine Protection, Research, and Sanctuaries Act (MPRSA), also known as the Ocean Dumping Act
Medical Waste Tracking Act
National Defense Authorization Act
National Environmental Education Act
National Historic Preservation Act (NHPA)
National Oil and Hazardous Substances Pollution Contingency Plan (NCP)
Native American Graves Protection and Repatriation Act (NAGPRA)

Noise Control Act
Nuclear Waste Policy Act
Persistent Bioaccumulative Toxic (PBT) Chemical Program
Pollution Prevention Act (PPA)
Pollution Prevention Packaging Act
Safe Drinking Water Act
Shoreline Erosion Control Demonstration Act
Shoreline Erosion Protection Act
Shore Protection Act
Sikes Act
Solid Waste Disposal Act
Surface Mining Control and Reclamation Act
Uranium Mill-Tailings Radiation Control Act
Water Quality Act

One by-product of these environmental regulations is that redevelopment of contaminated land has become prohibitively expensive. The EPA has begun to recognize that the real or imagined fear of being caught in the spider's web of regulation, particularly the CERCLA process, has become a major barrier to redevelopment activities. In response to this problem, the concept of *brownfield* redevelopment has emerged. A brownfield site or property is contaminated land that can be redeveloped as a commercial or industrial site. Many companies have large portfolios of property that may be considered contaminated because of historic activities. The opportunity for focused cleanup and reuse has attracted substantial interest across the United States. The potential for residential usage is not eliminated; however, cleanup standards would be much stricter. The term *brownfield* is wordplay off the concept of *greenfield*, which typically refers to unused open space or farmland. Another term, *greyfield*, has been coined to address commercial shopping sites (i.e., shopping and strip malls) that have fallen into disuse because of competition from newer, better-positioned shopping facilities. For these greyfields, environmental contamination is usually not an issue.

The brownfield movement represents one of the major reforms due to dissatisfaction with the CERCLA (Superfund) process and promises to be a significant activity for environmental managers. This reform process involves use of CERCLA prioritization criteria and associated weighting factors to classify potential contaminant threats. These include the following risks:

- *human population exposed*—population size, proximity to contaminants, likelihood of exposure
- *contaminant stability*—mobility of contaminant, site structure, and effectiveness of any institutional or physical controls
- *contaminant characteristics*—concentration, toxicity, and volume
- *threat to a significant environment*—endangered species or their critical habitats, sensitive environmental areas
- *program management considerations*—innovative technologies, cost delays, high-profile projects, environmental justice, state involvement, brownfield/economic redevelopment.

Each criterion is ranked on a scale of 1 to 5. The highest score for any criterion is 5, representing a current risk–current exposure scenario posing risk to human health and the environment. The lowest score for a factor is 1, representing a future risk–future exposure scenario. These criteria provide a way to prioritize responses to environmental contamination.

The increasing liability and cost of poor or fragmentary environmental management illustrate why most environmental professionals feel that an integrated approach to environmental affairs is necessary. The development of a proactive and comprehensive program for environmental management has moved from a discretionary activity to a virtual requirement if the company is going to prosper and thrive in the 21st century. The next section presents some of the approaches and opportunities that are available for development of proactive environmental management.

PROACTIVE ENVIRONMENTAL MANAGEMENT

With the advent of international, industry-specific, and country-specific standards for managing environmental impacts, the stage has been set for businesses to use these standards for their competitive advantage. Examples of standards include product recall requirements in Germany, recycling regulations in Asia and Europe, and the general use of "green" marketing by a number of consumer product companies. For many multinational corporations, the question remains: How can competitive advantages be realized by proactively identifying and managing safety, health, and environmental (SH&E) issues?

A New Approach

The framework for corporate environmental management has traditionally been "command and control" through regulatory compliance within the borders of a country. As business opportunities have become international in scope, regulations have been enacted by various stakeholder groups that reach beyond a country's borders. This began to occur in the mid-1980s when the chemical industry enacted its Responsible Care program in Canada and subsequently in the United States. Responsible Care was unique in that it advocated voluntary efforts to address the nonregulated environmental aspects of doing business. Since then, similar proactive approaches have been adopted by other industry groups around the world. The benefits extend beyond actual compliance; they add value by improving a company's corporate image, enabling management of suppliers' risk, and cultivating the industrywide credibility needed for expansion beyond national boundaries.

The next major step toward the strategic management of SH&E factors was the issuance of the International Organization for Standardization (ISO) 14000 series of standards. These standards include ISO 14001, which encourages the use of environmental management systems (EMSs). An EMS does not prescribe specific criteria; rather, it provides a process for identifying and managing the environmental aspects of producing, delivering, and using a company's products or services. It calls for setting goals, mission, and policy; defining responsibilities; and developing procedures for valuative and corrective actions.

Although ISO 14000 standards are voluntary and do not require certification or registration, they may become a prerequisite of doing business with many governments and companies. Therefore, their adoption may become a de facto requirement for remaining competitive. Furthermore, the implementation of environmental management systems may also yield product and process improvements that become competitive advantages in their own right. A detailed discussion of ISO 14000 is presented later in this chapter.

Along with Responsible Care and the ISO 14000 series, a number of other standards act as drivers for companies to take action. These include standards such as the European Commission's Eco-Management and Audit Scheme (EMAS) and the British Standard BS-7750. REACH, a European Union regulation on chemicals and their safe use (EC 1907/2006), deals with the **R**egistration, **E**valuation, **A**uthorization, and **R**estriction of **Ch**emical substances and has far-reaching implications for companies that want to do business in the European Union. Although the focus is on environmental aspects, enterprising companies are developing management systems to address a variety of SH&E issues. The challenges associated with developing these programs include the following:

- tracking strategic issues, such as emerging global trends, new product development, and competitors' SH&E initiatives
- elevating SH&E functions to contribute to process and business management decisions
- establishing appropriate responsibility and accountability at all levels of facility and corporate management.

Figure 14–1 illustrates the connections between many of these issues and an overall environmental management.

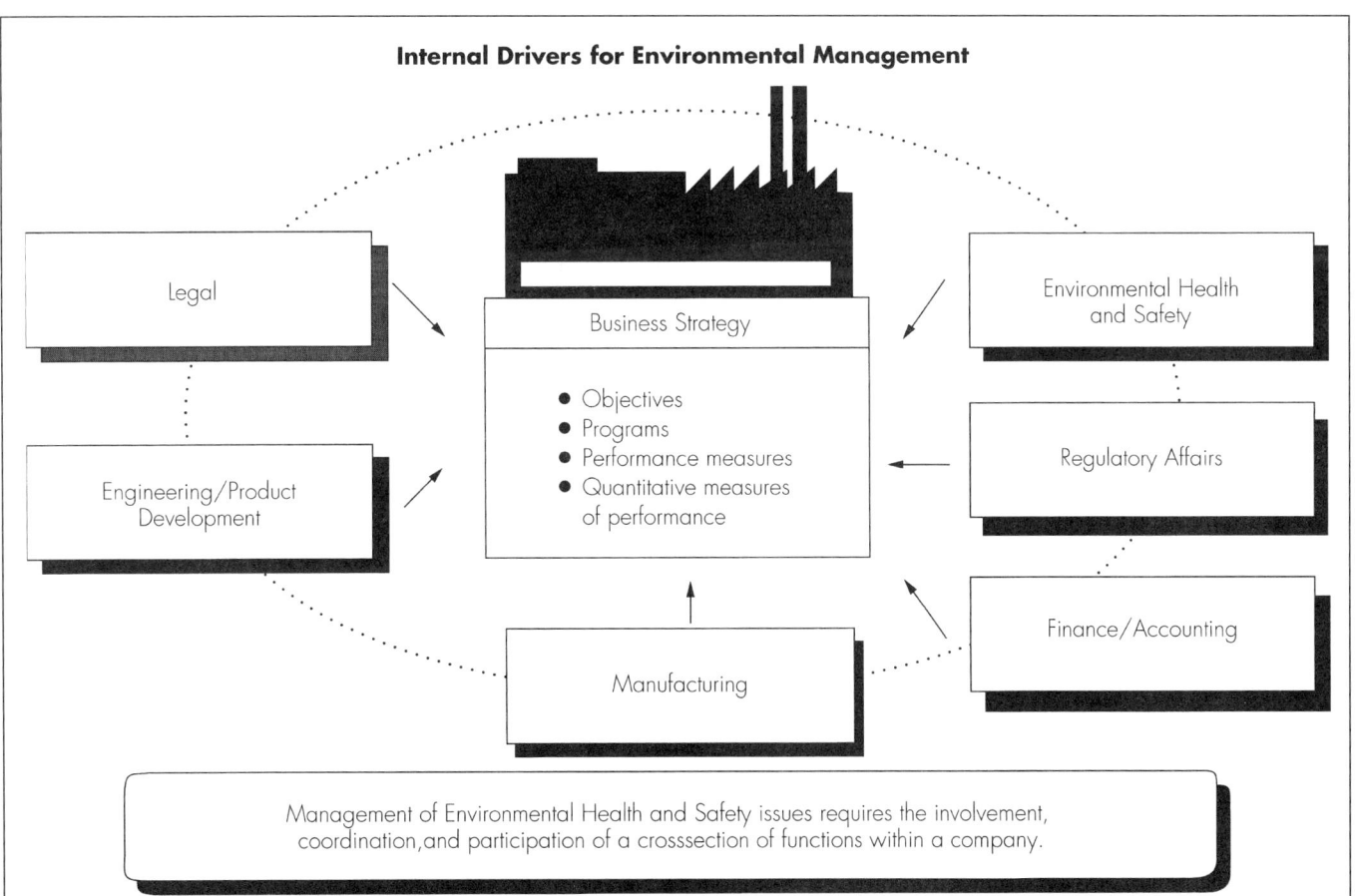

Figure 14–1. Environmental health and safety issues affect overall business operations. Consequently, they must be integrated into the strategic business planning process. *(Courtesy Dames & Moore.)*

Strategies for Success

Competitive advantages can be realized through the alignment of critical SH&E factors with a company's value chain: the process by which products (or services) are conceived, produced, marketed, distributed, used, and recycled or disposed. Examples of implementing an SH&E program in the context of a value chain include the following:

- *research and development*—incorporate design for environment (DFE) or design for disassembly practices for new products or product modifications
- *procurement*—identify suppliers that follow DFE and ISO 14000 practices and encourage other suppliers to adopt them
- *production and manufacturing*—optimize processes to minimize the handling of hazardous materials and reduce consumption of energy and resources
- *marketing and sales*—educate customers and end users as to the effective use and disposal/recycling of products and identify "green" differentiators to gain market share or margin improvements
- *distribution*—design for less packaging to reduce transport costs.

The steps described here are examples of places in the organization where opportunities can support strategic business goals (Figure 14–2).

The value chain approach helps managers understand how SH&E factors affect costs and anticipated revenues across organizational and product lines. It can also help filter out nonstrategic issues so that efforts are directed toward issues that affect goals for growth, diversification, margin improvement, or increased customer satisfaction. The ultimate aim is to move away from command and control and toward the proactive use of SH&E issues in support of broad corporate objectives. Many companies are using the ISO 14000 process to facilitate this evolutionary change; therefore, a detailed analysis of the ISO 14000 process is presented in a later section.

MANAGING FOR A HEALTHY ENVIRONMENT

Industry experts Christopher Hunt and Ellen Auster (1990) describe five distinct stages of environmental management program development. Their survey of corporations revealed departments that range from "beginner" (stage 1) through "proactivist" (stage 5). Small companies or ones that face minimal environmental risks do not require a stage 5 ("proactivist") environmental management program. Companies that use hazardous materials in their manufacturing processes or those with a variety of processes require a more active management approach. A program that is proactive for one company can be inadequate for another. A bakery may operate successfully with a few guidelines and periodic reviews by a consultant. However, large manufacturers or chemical companies often require several hundred staff members who are environmental professionals.

Hunt and Auster describe stage 1 ("beginner") environmental programs as those of either older companies established before the environmental acts were passed or of firms such as banks or real estate developers that do not deal directly with toxic materials, but may encounter related risks from them. Beginners often ignore environmental concerns or deal with them by adding responsibility to an existing job description, such as facility manager or senior engineer. Beginners do not define corporate environmental policy, nor do they consider the potential effects of failing to have such a policy.

Stage 2 ("fire fighter") companies have few people who spend time on environmental concerns or a small group of

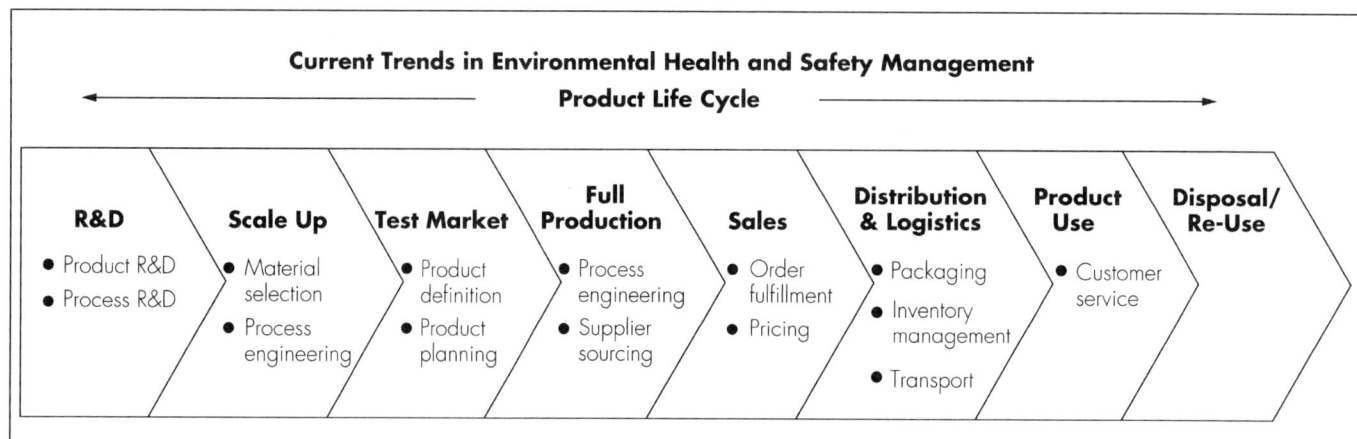

Figure 14–2. With the focus broadening beyond just compliance, environmental health and safety issues and strategy must be managed in the context of the product life cycle. *(Courtesy Dames & Moore.)*

professionals who help individual installations respond to problems. Given this structure, environmental professionals often work at a crisis-intervention level. They do not have the opportunity to consider risks from serious problems that may occur in the future. Small or medium-sized firms that are at this level engage in environmentally hazardous activities and do not consider the benefits of active environmental management.

Hunt and Auster describe stage 3 ("concerned citizen") companies as those whose established environmental departments are either understaffed or not high enough in the corporate structure to make significant changes. They are often staffed by technically competent professionals, such as chemists, biologists, or geologists, whose backgrounds do not prepare them to offer business, legal, and public relations advice for environmental management. These companies, concerned about their environmental responsibilities, have funded programs that operate without serious senior management commitment to integrate them with the operating units.

At stage 4 ("pragmatist") companies, environmental department staffs have sufficient expertise, funding, and authority. Staff members review all facilities and design better ways to limit toxic releases. They look to the future, evaluating potential risks and designing solutions when appropriate. They train key workers in environmental protection and write policy manuals for operating personnel. Formal reporting relationships are established. Although these departments are aggressive, their authority and program funding may be limited. At this stage of development, companies have not made environmental management a top-level corporate concern.

Environmental management is a top priority for companies that have reached stage 5 ("proactivist"). The environmental department is staffed with assertive leaders whose commitment to the environment goes beyond meeting regulations and planning for the firm's future environmental needs. These professionals are active in industry roundtables, sharing information and solutions with competitors to ensure that their performance does not result in polluted resources, adverse publicity, and unnecessary regulations. They participate in local, state, and federal policy-making meetings, aggressively pursuing solutions as they help policymakers determine how best to regulate their industry.

Not every organization needs to budget for a sophisticated, proactive environmental management department. The level of commitment appropriate for each company should be determined by its inherent environmental risks and its size.

Organizing an Environmental Management Program

Establishing an environmental compliance program can be an overwhelming proposition. However, any company whose business is regulated by the EPA, the Occupational Safety and Health Administration (OSHA), the Department of Transportation (DOT), other government agencies, or state or local authorities cannot afford to be without an environmental management program.

Corporate culture—the size of the company, the diversity of its business units, and how it interacts with headquarters—will determine whether the environmental management program will be small and basic or large and sophisticated and centralized or decentralized. Whether the company is large or small, however, senior management must communicate a strong commitment to responsible environmental practices through the ranks, down to the line worker and platform loader. Unfortunately, a sincere environmental management commitment that alters old practices often occurs only after a major disaster or stringent penalty. This does not have to be the case. Top management teams can learn from other companies' misfortunes.

Deciding where to begin may be the biggest obstacle to organizing an environmental management program. The ways to comply with governmental regulations can seem as endless as the thick tomes that transmit those regulations. William Friedman (1988) outlines the following five basic activities essential to any successful environmental compliance program.

Prevent Common Violations

Mislabeling of chemicals and hazardous waste and failing to promptly inform regulators of a release are probably the most commonly cited violations and can most easily result in environmental problems. They are also the easiest violations to prevent. Companies should attach proper labels to all containers of chemicals. In the case of hazardous waste, labels must meet both RCRA regulations and, when prepared for transport, DOT requirements. And, when a release occurs, ensuring that the regulators are informed in accordance with permit and regulatory requirements will establish greater trust and goodwill. OSHA regulations, covered in Chapter 5, Legal and Regulatory Issues for the Safety Manager, also require labeling of chemical containers in the workplace.

Regulations require storage containers to be in good condition. Leaking containers are a common source of environmental fines. Inspect tanks daily and check drums and similar containers weekly. Make sure aisle space is adequate to give inspectors easy access to the facilities. When waste is stored in tanks, regulations require secondary containment to prevent leaks or spills.

Other common violations include the following:
- improper waste disposal
- oil spills
- destruction of wetlands

- dumping into oceans, streams, lakes, or rivers
- improperly handling pesticides or other toxic chemicals
- improperly removing and disposing of asbestos
- falsifying lab data pertaining to environmental regulations
- committing fraud related to environmental crime.

These violations are considered white-collar crimes. They can result in criminal fines, probation, jail time, injunctions, or any combination of these punishments.

Regular audits can identify areas where potential violations exist. See Chapter 7, Safety, Health, and Environmental Auditing, for information on auditing practices.

Record Keeping

Environmental regulations have mandated that organizations document many of their environmental practices during transportation and disposal of hazardous waste. During an inspection by a regulatory agency, records are usually the first target for an inspector. Often, the mere existence of the records and easy accessibility may be all an inspector needs.

It is also important to document all environmental management decisions. For example, whenever a facility has analyzed its waste streams and found them to be nonhazardous, records of the information supporting that determination should be maintained and made readily available for inspection. Relying on a sketchy explanation from a compliance manager (who may or may not be employed or present at the time of the inspection) can lead to an inspection report listing possible violations. Staff must then reconstruct the information and transmit it to the regulatory agency, which may still issue a violation. Accurate, available, and complete records save companies time, money, and legal headaches. It is also important to give the inspector only what is requested. Often, it is a good idea to take copies of records to the inspector at a location away from where the records are kept.

Create a Spill-Reporting Plan

Under numerous statutes and regulations (notably the U.S. federal Superfund law), companies must notify the U.S. National Response Center immediately after release of a reportable quantity of any hazardous substances or pollutants. As regulators decrease the time allowed between an incident and the reporting time, environmental authorities are strictly enforcing existing laws. In some cases, such as air pollution discharges, the permitted time lag for reporting may be only minutes.

An unintentional spill or release of a toxic material can occur during a facility emergency when personnel are literally putting out fires or responding to explosions or other emergencies. For that reason, every facility must have written instructions for employees to follow whenever a release or spill occurs. The reporting plan should detail the facility's procedures, telephone numbers of all agencies that should be notified, and who has the authority to report spills. Employees should be aware of these instructions and where they can be found at various locations. Not only do spill-reporting plans permit companies to respond immediately to emergencies, but their existence is also evidence of a firm's environmental commitment to regulatory agencies that may question the company's ability to respond to an emergency. It is also important to conduct regular training sessions and to practice incident responses so that personnel will feel comfortable dealing with emergency situations.

Set Realistic Limits and Schedules

Environmental managers should suggest realistic compliance schedules and discharge limitations to environmental regulators. Often, in a spirit of cooperation, a manager will offer overly ambitious schedules and limits that ultimately cannot be met. Do not overcommit. Once limits are set in discharge permits and orders, it is difficult to have them changed. Before such limits are written into a company's legal documents, management should consult with government regulators to set realistic dates and limits.

When reporting a release that exceeds set limits, always include the reason and try to solve the problem. Companies have reported pollutant releases above limits, received no response from the regulatory agency, and, over time, assumed that the increased rate is acceptable. The company's own reports then establish its lack of compliance, which can prompt an enforcement action by the agency or serve as evidence in a citizen suit against the company. Whenever it appears that legal requirements cannot be met in the future, the compliance manager should request a change in limits or schedules and detail the reasons why compliance is not possible.

Motivate Employee Action

Spills and releases are almost never caused by the person who writes a company's environmental compliance program. Mistakes that become environmental problems are usually made by the platform loader, line worker, or storage worker. If these employees are not properly educated or trained in environmental management practices, they often focus on doing their jobs quickly at the expense of the environmental program.

Although training for all employees who handle environmentally sensitive substances is essential, it is all too often insufficient. Including their adherence to environmental protection policies in employees' annual reviews and holding them responsible for cited violations brings home the importance of "little details" and "bothersome" reports. It is crucial, as well, to make employees aware that in this age of strict environmental enforcement, it is less costly for a company to adhere to regulations than to defend itself against a lawsuit.

Staff Skills and Backgrounds

A key player in any successful environmental management department is the manager. Whether his or her background is engineering, science, or law is immaterial. The complex and diverse issues that the manager must deal with call for a person with top managerial skills, good internal and external networks, communication skills, rapport with senior management, a diverse background, and the respect of others within the organization. Because of the diversity and sensitivity of day-to-day business, the manager should ideally have some grounding in environmental issues, whether based in law, science, or engineering. Because of the complex nature of managing an environmental program, often a background in business will bring good value to the position.

Most of the environmental management department staff will consist of those with science- and engineering-related backgrounds. Depending on the industry, a company's environmental staff could include chemists, geologists, engineers, biologists, industrial hygienists, safety professionals, and so on. It is important that staff members have appropriate education and credentials. Although science and engineering backgrounds are important in the field, if someone has attained certifications or relevant experiences that support the position, then such information should be part of the hiring decision. Some popular certifications include Associate in Risk Management (ARM), Board Certified Environmental Engineer (BCEE), Certified Environmental Manager (CEM), Certified Environmental Professional (CEP), Certified Environmental, Safety, and Health Trainer (CET), Certified Hazard Control Manager (CHCM), Certified Hazardous Materials Manager (CHMM), Certified Industrial Hygienist (CIH), Certified Safety Professional (CSP), Professional Engineer (PE), Registered Environmental Health Specialist (REHS), Registered Environmental Manager (REM), Registered Hazardous and Chemical Materials Manager (RHCMM), and World Safety Organization-Certified Hazardous Materials Executive (WSO-CHME). This is not an all-inclusive list of credible certifications.

Because many corporate environmental issues are legal in nature, companies also need a close association with an appropriate attorney or law firm. Without such a relationship, the company may overlook key environmental data or legal facts that can plague it far into the future. An attorney can give advice or institute preventive legal measures that will avoid problems that could prove costly both financially and in terms of corporate image and public relations.

Some environmental management departments have an attorney on staff. In others, the corporate general counsel's office includes an attorney who is an environmental specialist. Smaller companies may rely entirely on outside counsel. Whichever approach is taken, it is essential to establish a firm relationship with an attorney skilled in environmental issues who will review environmental data and regulations and oversee company transactions that have environmental elements. Counsel should have a close working relationship with the environmental manager to ensure a timely response to concerns and related questions.

A strong relationship should also be established between the environmental management department and the company's public relations or communications department (or outside public relations firm, if that is the practice for the company). A crisis communication plan should be in place to address potential environmental problems. A collaborative relationship among the environmental manager, public relations function, and legal counsel is the best assurance that the company's environmental efforts meet corporate guidelines and applicable regulations and are appropriately communicated to stakeholders.

WASTE MINIMIZATION

Waste minimization can produce several benefits. It can save money by reducing waste handling, treatment and disposal costs, raw material purchases, and other operating costs. It can reduce potential environmental liabilities. It can protect public health and worker health and safety and protect the environment. Waste minimization supports a goal of sustainability. Waste minimization can be accomplished by source reduction and recycling. Of the two approaches, source reduction is usually preferable to recycling because it has a lower total effect on the environment. RCRA requires generators of hazardous waste to establish a program to reduce the volume and toxicity of waste generated, to the extent that such a program is economically practical.

A waste minimization program must begin with an assessment, which requires a company to evaluate all solid waste streams to determine whether they can be reduced or recycled. Next, the company should calculate the economic impact of reuse or recycling.

Technical evaluation of a waste stream should begin in the production process where the waste is generated. Factors such as changes in raw materials, production process, equipment, and operating conditions could change the characteristics of a solid waste to make it a better candidate for reuse or recycling.

Economic evaluation should consider the capital and operating costs associated with any onsite collection or treatment of the waste. In addition, it should consider disposal fees, transportation costs, raw material costs, and operations and maintenance costs. Economic projections should be as realistic as possible.

In deciding whether to reuse or recycle solid wastes, a company should consider the stability of the minimization technology. If reuse is selected, managers should evaluate the effect on the reuse process of minor changes in waste characterization. Waste is, by nature, inconsistent in composition. Therefore, if a company finds that minor fluctuations in waste composition will adversely affect the finished product or preclude the material from reuse, it may evaluate other options.

In evaluating a solid waste for recycling, bear in mind that recycling is a commodity-driven business, possible only if a market exists for the recycled product. A successful recycling program constantly monitors the market to ensure that the process is cost effective.

An excellent example of a solid waste that is particularly suitable for recycling is the aluminum can. It takes less energy to recycle an aluminum can than it does to produce aluminum by conversion of bauxite ore. In this industry, manufacturers have done an excellent job of establishing facilities capable of recycling aluminum cans. Therefore, there is an excellent market for aluminum cans collected for recycling.

Glass is another material suitable for recycling. Once again, it is less expensive to produce glass through recycling than to manufacture new glass. However, the market for glass recycling is not as well developed. For example, a facility may find that, at times, a recycler is unable to accept its material because the recycled product cannot be sold. This does not preclude the use of recycling as a disposal alternative, however. If a facility selects recycling as a minimization technique, it must be prepared to store waste when the material cannot be recycled or to develop disposal alternatives.

GROUNDWATER CONTAMINATION

One of the most serious and costly forms of pollution is groundwater contamination. When industrial pollution seeps into the groundwater and drinking water supplies, it can, at best, be costly to correct and, at worst, present long-term, chronic health risks for people who live downstream from the facility.

The strata beneath the surface of the earth contain *horizons*, some of which are water bearing. These water-bearing horizons vary in flow, characteristics, and water quality. These water-bearing horizons, known as aquifers, can range broadly in size from local to regional to continental. Some aquifers are high-yield sources of drinking water that serve as the primary drinking water source for a large portion of the population. In evaluating groundwater contamination, a firm should identify an area's geology and hydrology at least to a depth that penetrates the uppermost aquifer.

A company should classify site geology in accordance with the Unified Soil Classification System. This is accomplished by drilling bore holes and collecting soil samples at predetermined depth intervals. In good, cohesive soil conditions, a company can sometimes collect samples from the auger flites. However, when cohesive soil conditions do not exist and there is sidewall sloughing into the bore hole, samples from the auger flites will not allow proper soil classification. To properly classify the soils in noncohesive conditions, it is necessary to use a split spoon or sample tube device to collect samples at predetermined depths. Site hydrology includes the identification of groundwater flow paths, determination of groundwater flow directions (including horizontal and vertical flows), and determination of multihorizontal interconnects.

Horizontal groundwater flow direction is usually found by installing piezometers to determine the potentiometric surface of the groundwater. With these data, a team can identify the horizontal direction of flow by constructing a piezometric surface map.

The vertical flow component is determined by placing vertically nested piezometers in closely spaced, separate bore holes. The piezometers are screened at different depths to measure vertical variations in the hydraulic head. The data are used to construct flow nets, which are graphical representations of the vertical flow components.

After characterizing the site hydrology, an investigative team must determine temporal influences that may alter the piezometric surface. These influences may include location of offsite pumping wells, tidal variations, offsite and onsite land pattern changes, and seasonal variations in groundwater recharge.

Hydraulic interconnects between horizons can affect the horizontal and vertical flow path of groundwater contamination. Hydraulic interconnects are usually detected by installing a pumping well into the lower horizon, pumping at a predetermined rate, and monitoring the potentiometric surface of the upper horizon. Obviously, if there is a drawdown of the potentiometric surface at any of the piezometers during pumping of the lower horizon, hydraulic interconnect may exist.

Groundwater assessments require a great deal of experience in data collection and interpretation to produce meaningful results. If a facility is considering conducting a groundwater assessment, management should seek the assistance of an experienced groundwater assessment firm.

ISO standards promote best practice for sustainable water management and quality and facilitate access to water for the world's population. As of 2012, more than 550 of the 19,000 international standards relate to water. Their solutions feature work in service management and crises, quality and conservation, and infrastructure.

GLOBAL SOLUTIONS FOR GLOBAL PROBLEMS

No part of the globe is immune from chemicals that nations release into the ground and waterways or emit into the air. Water currents, like conveyor belts, deliver toxic chemicals to the Arctic Ocean. Carbon, sulfur, and other pollutants float on air currents from Eurasia to hover over the Arctic. The "imported" pollutants combine with "home-grown" nitrogen oxides from Alaskan oil fields to produce Arctic haze. The Arctic's annual mean level of photochemical smog rivals that of Los Angeles.

In the once-pristine Arctic, high levels of toxic substances have been discovered in seals and polar bears in recent years. These pollutants have made their way to the top of the food chain, where mercury is now found in mammalian milk in many parts of the Arctic. Sulfur dioxide, produced mainly by coal-fueled electrical utilities, and nitrogen oxides, the product of transportation sources and utilities, are chemically transformed in the atmosphere and transported as acid rain over national borders by prevailing winds.

Greenhouse gases (GHGs)—carbon dioxide, chlorofluorocarbons, methane, and nitrous oxide—produced in quantity by industrialized nations over the past 200 years, gather in the atmosphere and absorb the infrared waves emitted from the earth. These gases have contributed to global climate change—the rise of the mean surface temperature of the earth by 0.6°C in the last 100 years. Increases of 2° to 5°C over the next 50 to 100 years are predicted. Those few degrees could turn arable land in the higher latitudes into deserts and could melt polar ice caps, causing a rise in sea levels—potentially flooding large population centers—and other catastrophic consequences.

Greenhouse gas inventories are an accounting of the amount of greenhouse gases emitted to or removed from the atmosphere over some period of time. A greenhouse gas inventory also examines information on the activities that cause emissions and removals, as well as background on the methods used to make the calculations.

In response to the concerns over climate change, organizations have started developing *carbon footprints* in order to develop baselines that can support plans for reducing greenhouse gas impacts on the environment. Footprints can be calculated using a life-cycle assessment (LCA) method or by evaluating emissions attributed to the use of fossil fuels. An LCA is a component of the ISO 14000 environmental management standards and is essentially an analysis of the environmental impacts of a product or service. Carbon footprints can be based on an evaluation of direct emissions or include indirect sources. In the United States, human activities are evaluated to determine potential environmental impacts and measured in units of carbon dioxide. In Europe, the measurements use carbon as a measuring unit. The two methodologies result in essentially equivalent results. There are numerous calculators for determining carbon footprints.

Recent Supreme Court cases have addressed the issue of regulating greenhouse gases. In 2007, the Supreme Court released its ruling in the case of the state of *Massachusetts vs. the Environmental Protection Agency*. Massachusetts and 11 other states, along with several local governments and nongovernmental organizations (petitioners), sued the EPA for not regulating the emissions of four greenhouse gases, including carbon dioxide (CO_2), from the transportation sector. The petitioners claimed that human-influenced global climate change was causing adverse effects, such as sea-level rise, to the state of Massachusetts. In a 5–4 decision, the Court ruled in favor of Massachusetts et al., finding that EPA has the authority to regulate CO_2 and other greenhouse gases. In 2014, the U.S. Supreme Court issued its decision in *Utility Air Regulatory Group v. EPA*, saying that EPA may not treat greenhouse gases as an air pollutant for purposes of determining whether a source is a major source required to obtain a Prevention of Significant Deterioration (PSD) or Title V permit. The Court also said that PSD permits that are otherwise required (based on emissions of other pollutants) may continue to require limitations on greenhouse gas emissions based on the application of best available control technology (BACT) guidelines.

The EPA has responded to climate change issues with regulations covering several different potential sources of greenhouse gases. The EPA's Greenhouse Gas Endangerment Findings are addressed by Section 202(a) of the Clean Air Act, which states that six key, well-mixed greenhouse gases constitute a threat to public health and welfare and that the combined emissions from motor vehicles cause and contribute to the climate change problem.

EPA and the National Highway Traffic Safety Administration (NHTSA) issued Standards to Cut Greenhouse Gas Emissions and Fuel Use for New Motor Vehicles, with coordinated steps to enable the production of a new generation of clean vehicles—from the smallest cars to the largest trucks—through reduced greenhouse gas emissions and improved fuel use.

EPA's Renewable Fuel Standard (RFS) program is also responsible for developing and implementing regulations to ensure that transportation fuel sold in the United States contains a minimum volume of renewable fuel. By 2022, the RFS program will reduce greenhouse gas emissions by 138 million metric tons, about the annual emissions of 27 million passenger vehicles, and is expected to replace about 7% of expected annual diesel consumption and decrease oil imports by $41.5 billion.

In 2010, EPA set greenhouse gas emission thresholds to define when permits under the New Source Review Prevention of Significant Deterioration (PSD) and Title V Operating Permit programs are required for new and existing industrial facilities. This final rule "tailors" the requirements of these Clean Air Act–permitting programs to limit covered facilities to the nation's largest greenhouse gas emitters: power plants, refineries, and cement production facilities.

In 2012, EPA finalized cost-effective regulations to reduce harmful air pollution from the oil and natural gas industries while allowing continued, responsible growth in U.S. oil and natural gas production. The final rules are expected to yield a nearly 95% reduction in volatile organic compound (VOC) emissions from more than 11,000 new hydraulically fractured gas wells each year. The rules will also reduce air toxics and emissions of methane, a potent greenhouse gas.

In 2014, the EPA issued proposals to cut carbon pollution from both existing and new power plants, the largest source of greenhouse gas emissions in the U.S. Also in 2014, EPA proposed updates to its air standards for new municipal solid-waste (MSW) landfills. These updates require certain landfills to capture additional landfill gas, which would reduce emissions of methane, and help further reduce pollution that harms public health.

The international community joined forces to meet the environmental challenge at the 1972 Stockholm Conference and has continued to make giant strides. The Stockholm Declaration articulated several principles of international environmental law. Principle 21, for example, holds that although countries have the right to develop, they have a responsibility not to damage the environment outside their borders. This includes the oceans and Antarctica, as well as other countries. Principle 21 now represents customary international law—law instituted by the agreements and practices of states. Its application to some worldwide environmental problems, however, is difficult. Assessing an individual country's responsibility regarding global warming is not easy.

Serious environmental accidents have spurred nations to cooperate more readily in developing international environmental protection laws. Following several major oil spills in the past 50 years, the International Maritime Organization reached rapid accords on oil spill liability and regulations for oil discharges from ships. When scientific data about ozone layer damage from synthetic chemicals were released, the Montreal Protocol was negotiated and amended in record time. The Basel Convention's agreement was accelerated as soon as the hazardous-waste and incinerator ash shipment to a Nigerian dump site came to light. Immediately after the Chernobyl nuclear accident, the International Atomic Energy Agency swiftly finished new treaties on responsibility for notification and assistance. A key feature codified a country's obligation to notify other nations if the risk of transboundary damage exists. Following the oil spill from Exxon's tanker *Valdez*, Congress passed the Oil Pollution Control Act. The *Deepwater Horizon* disaster in the Gulf of Mexico resulted in new regulatory safety measures that include heightened drilling safety standards to reduce the chances that a loss of well control might occur in the first place as well as a new focus on containment capabilities in the event of an oil spill.

ISO 14000:2004

On July 1, 1995, 39 participating nation members of the International Organization for Standardization's Technical Committee 207 (TC 207), meeting in Oslo, Norway, ushered in a new era of environmental management by approving as Draft International Standards (DISs) five documents of the ISO 14000 series on environmental management. After global voting, these became full-fledged ISO standards in the third quarter of 1996. They dramatically change and broaden the ways in which environmental management is conducted and, potentially, regulated in the global economy. As of 2014, the committee has 21 published standards and normative documents, with an additional nine new or revised documents being prepared. Currently there are only two standards that are published under this TC 207: 14031:2013 and 14033:2102.

The latest version of ISO 14000:2004 is totally nondirective. It tells managers how to identify and manage environmental impacts, not what to manage. It helps organizations (1) examine the complete range of environmental impacts, both adverse and beneficial, associated with their activities, products, and services; (2) logically manage impacts that the organization can control and influence; and (3) balance both environmental and economic goals.

ISO 14000:2004 is a tool for systematically improving environmental management and, consequently, environmental performance. Its focus is on stimulating managers at all levels to consider the environment when making decisions and it seeks to maximize the role of the marketplace in encouraging improvements in environmental performance.

Although ISO 14001:2004 is a voluntary standard, there is nothing to prevent a country, state, or municipality from adopting it as a mandatory standard for performance improvement. ISO 14001:2004 could, in fact, form the foundation for a model regulatory system unique in its reliance on business and market incentives to accomplish positive ends.

Historical Context

The modern era of environmental management in the United States began in 1970 with the first Earth Day, passage of the National Environmental Priorities Act, and the establishment of the U.S. Environmental Protection Agency (EPA). The period since has been marked by development of command and control approaches to regulating end-of-pipe emissions and hazardous waste. Although the system has produced environmental gains, they have been achieved at high economic and social cost from litigation and unpopular regulation. Also, the focus of regulation on toxic releases to air, water, and land has not addressed other, potentially significant global environmental impacts, such as natural resource depletion and nontoxic greenhouse gases, which present arguably greater potential for environmental damage.

ISO 14000 came into existence in the context of Agenda 21, the United Nations Program of Action for Sustainable Development, which emphasizes the development of processes and procedures that balance the needs of both present and future generations of life on the planet.

The Next Phase

ISO 14000:2004 represents an optimistic second phase of environmental management, another beginning step in the necessary transformation of the earth and its finite resources to a point where population, consumption, and the earth's carrying capacity are in balance. The DISs approved in Oslo comprise the fundamentals necessary to establish a new approach to environmental performance and performance verification. The following standards were approved:

- ISO 14001:2004—Environmental Management Systems Standard. This is a straightforward, 19-page document describing the elements necessary to a certifiable environmental management system (EMS).
- ISO 14004:2004—Environmental Management Systems—General Guidelines on Principles, Systems, and Supporting Techniques. This is an elaboration on ISO 14001, providing practical advice for implementing a new EMS or enhancing an existing EMS.
- ISO 14010—Guidelines for Environmental Auditing—General Principles. This is a uniform system for verifying environmental management and performance of all types. (It is superseded by ISO 19011:2002.)
- ISO 14011—Guidelines for Environmental Auditing—Auditing of Environmental Management Systems. This is an audit protocol for objectively evaluating performance of an EMS and is specifically distinguishable from an evaluation of environmental performance. (It is superseded by ISO 19011:2002.)
- ISO 14012—Guidelines for Environmental Auditing—Qualification Criteria for Environmental Auditors. This describes the skills, knowledge, experience, training, and personal attributes deemed necessary to conduct environmental audits. (It is superseded by ISO 19011:2002.)
- ISO 19011—Guidelines for Quality and Environmental Management Systems Auditing. This sets the guidelines for auditing the quality and environmental management systems via a set of principles that help parties to an audit perform efficiently.
- ISO 19011:2011—Guidelines for Auditing Management Systems
- ISO Guide 64:2008—Guide for the Inclusion of Environmental Aspects in Product Standards. This is intended for use by product standards writers to raise their awareness and understanding of products' negative and positive effects on the environment.

Additional guidelines currently approved:
- ISO 14031:2013—Environmental Performance
- ISO 14020 series—Environmental Labels and Declarations
- ISO 14064—GHG Accounting and Verification
- ISO 14065—Accrediting for Greenhouse Gases—requirements for greenhouse gas validation and verification bodies for use in accreditation or other forms of recognition
- ISO 14063:2006—Environmental Communication.

Pending (as of 2009):
14006—Eco-design
14045—Eco-efficiency Assessments
14051—Material Flow Cost Accounting (MFCA)
14067—Carbon Footprint of Products
14069—Calculate Carbon Footprint
14005—Phased Implementation of an EMS
14033—Quantitative Environmental Information
14066—Competency Requirements.

Two guidance documents on environmental labeling, environmental performance evaluation, and life-cycle assessment were developed in the same international process and were issued as DISs: ISO 14001:2004—The Environmental Management System Standard and ISO 14004:2004—The Environmental Management System Guideline.

Certification

One of the most important documents of the ISO 14000 series, ISO 14001:2004, describes what needs to be done either to establish an ISO 14001 EMS in the absence of an environmental management program or to upgrade an existing environmental management program to the ISO 14001 level. Equally important, ISO 14004:2004 parallels ISO 14001 in structure and provides helpful guidance on how an organization may implement an EMS. By 2007, there were more than 129,000 ISO 14001 certifications issued.

An organization may decide that commercial interests or social responsibility dictates certification of the EMS. Certification, under the standard, can be done by the organization (self-certification) or by registrars or certifiers (third-party certification) recognized by the American National Standards Institute (ANSI). Because rational approaches to environmental management are independent of market demands to certify, the decision to certify is made separately from the decision to adopt ISO 14001 as a framework for environmental management.

Meaning for Organizations

A potential end use of ISO 14001 in the United States is to establish a credible third-party environmental audit system, paralleling the principles, standards, and credibility of financial auditing, and to file third-party audit reports with environmental regulators, much as Form 10K is presently created and filed with the Securities and Exchange Commission. This approach would reduce regulatory intrusion on organizations and allow regulators to focus enforcement efforts on organizations not adopting the standard. A challenge for industry lies with doing what it can to help establish the public credibility and acceptability of ISO 14001 certification and the third-party audit as substantive evidence of improved environmental performance.

ISO 14001 presents to U.S. industry and commerce an opportunity to be part of a new, constructive era of environmental/social responsibility that goes beyond regulatory compliance and is creative in its search for new ways of reducing environmental impacts, transparent in its operation, and powerful in its ability to focus attention on our collective environmental future. It also offers U.S. businesses an opportunity to recognize the cooperative, collaborative approaches being taken by the EPA, various state environmental agencies, and many nongovernmental organizations, such as the Environmental Defense Fund, the World Wide Fund for Nature, and the International Network for Environmental Management. The opportunity to form productive alliances to benefit the environment is part of the power of ISO 14000.

Key EMS Elements

The strategic planning process used in developing a successful EMS program generally follows the "plan-do-check-act" (PDCA) cycle used in total quality management efforts. ISO 14001 is based on the same concept. One additional concept highlighted within ISO 14001 is *continual improvement*. The continual improvement concept is aimed at regularly improving the overall environmental management system. ISO 14001 requires that managers evaluate the organization's interactions with and impacts on the environment. The following are some key elements of ISO 14001 that distinguish it from contemporary environmental programs:

- *Requirements*—ISO 14001 requires that top management of the organization commit to continual improvement of the EMS, prevention of pollution, and compliance with applicable environmental regulations in its stated environmental policy. Beyond these three, there are no absolute requirements in the standard.
- *Environmental aspects and impacts*—These are activities, products, or services that the organization can control or influence that can adversely or beneficially interact with (aspects) or change (impacts) the environment. This abridged definition of aspects and impacts suggests how ISO 14001 is markedly different from conventional approaches to environmental management that are rooted in regulatory compliance. ISO 14001 calls for recognition of the following as environmental aspects: emissions that are below permitted levels, emissions that are unregulated (such as CO_2), and consumption of energy and materials. Reference to adverse or beneficial impacts suggests balancing of adverse and beneficial impacts or a net beneficial impact test for new products and services. The concept of impacts that the organization can be expected to influence is a prelude to imposing environmental management standards on the organization's supply and distribution chain, its employees, and the communities in which it operates.
- *Best practices*—Although ISO 14001 encourages use of best practices, it also emphasizes evaluating the cost-effectiveness of best practices and does not require use of best practices that are not also cost effective.
- *Performance audits*—ISO 14001 requires periodic audits and continual improvement of the EMS, leading to

improvements in environmental performance. It does not require environmental performance improvements per se.

Implementation

Implementing an EMS calls for a variety of managerial skills. The objective is to embed an environmental management system so thoroughly into existing, freestanding management systems that environmental management becomes an inherent function of those systems. Knowledge of quality management and risk management principles, in addition to environmental management, is helpful in establishing a number of the principles that are necessary to an EMS. Familiarity with the quality management cycle of plan, do, verify, and assess and the ability to deal confidently with the ambiguities and uncertainties characteristic of risk management are key to setting priorities for dealing with environmental problems. Quality managers and risk managers reading ISO 14001 will recognize practices and tools borrowed from quality and risk management as well as environmental management techniques that can be applied in their domains.

Initial experience with ISO 14001 implementation indicates wide differences in current environmental management practices and approaches to implementation. This variance in implementation is one of the virtues of ISO 14001. It allows and encourages great latitude and interpretation in use of the standard, ultimately resulting in new and creative approaches to improving environmental performance.

Gap Analysis

Organizations that are contemplating installing ISO 14001 as their environmental management system begin by conducting a *gap analysis*, which is a determination of where they are relative to the requirements of ISO 14001, and developing a strategy for migrating from the present position to ISO 14001 conformance.

Gap analysis involves reviewing the major requirements of ISO 14001 to determine the organization's level of conformance. Major requirements include the following:

- *Management commitment*—Does top management show support for environmental management by assigning responsibility and authority for EMS implementation and providing resources, human and monetary, for running the EMS?
- *Environmental policy*—As indicated previously, ISO 14001 calls for an environmental policy that contains organizational commitment to three nonnegotiable elements: (1) regulatory compliance, (2) pollution prevention, and (3) continual improvement of the EMS. Does existing policy reflect these elements?
- *Baseline study (or aspects review)*—ISO 14001 requires that an organization have a procedure for identifying "the environmental aspects of its activities, products, and services to determine which can have significant impacts on the environment." Because most organizations have not been managing the broad ISO 14001 definition of environmental aspects, it is unlikely that a procedure is in place or that a baseline study, known as an environmental aspects review, has been done.
- *Legal requirements*—ISO 14001 requires an organization to have a procedure for assessing legal and other requirements to which it is subject. Generally, satisfaction of this requirement calls for establishment and maintenance of an inventory of environmental laws and regulations affecting the organization's activities, products, and services. Whether that is done by the organization itself or by an outside consultant or law firm is irrelevant so long as access is available. Other requirements include industry standards or environmental principles, such as the American Chemistry Council Responsible Care program or the CERES Roadmap, to which the organization subscribes. If the organization has publicly subscribed to such standards or principles, the standards or principles are regarded by ISO 14001 as having the weight of regulation and become an element of certification.
- *Emergency procedures*—ISO 14001 requires that an organization "establish and maintain procedures to identify potential for and response to accidents and emergency situations, and for preventing and mitigating the environmental impacts that may be associated with them." This requirement overlaps with the typical risk management practice of planning and preparing for emergency events.
- *Training*—ISO 14001 calls for the identification of training needs and ensuring that individuals with significant environmental responsibilities receive appropriate training. Specific training requirements are axiomatic to all management systems and specific to ISO 14001.

Environmental Aspects Review

The environmental aspects review establishes a baseline of significant environmental aspects from which an environmental performance plan, with objectives and targets, can be developed. The aspects review process can involve bringing together a team of inside and outside experts knowledgeable about the organization's activities, facilities, and processes; environmental regulations; relevant geological and hydrogeological conditions; and surrounding plant, animal, and human populations. The team will also review available information on energy and material consumption; transportation; emissions to air, water, and ground; generation of solid wastes; distribution, sale, use, and disposal of products; environmental impacts of services; and environmental impacts of suppliers, employees, communities, and customers that the organization can control or influence.

Informative guidance on the conduct of an environmental aspects review is contained in Annex A to ISO 14001. Annex A indicates that review of aspects should cover four key areas:

1. legislative and regulatory requirements
2. identification of significant environmental aspects
3. examination of existing environmental management practices and procedures
4. assessment of feedback from previous environmental incidents.

The annex further says,

> The process is intended to identify significant environmental aspects associated with activities, products and services. Organizations do not have to evaluate each product, component, or raw material input. They may select categories of services, activities or products to identify those most likely to have a significant impact.

The standard and the annex give organizations latitude to determine what their environmental aspects are. Although significant environmental aspects could be identified through a program of site assessments, site assessments would likely be a time-consuming, expensive way to develop information. Many organizations use a subjective evaluation done by people who are knowledgeable about aspects of the organization's activities, products, and services and briefed on the objectives of Agenda 21, the Rio Declaration on the Environment and Sustainable Development.

Review of the Rio Declaration suggests the following broad categories of environmental aspects:
- water consumption
- water pollution
- energy consumption
- air pollution and ozone depletion
- waste disposal
- transportation
- toxic chemical use
- product mass, use, recovery, and reuse
- biodiversity
- noise, odor, aesthetics, and archeological and architectural heritage
- sudden events and emergency preparedness.

Other approaches to identifying significant environmental aspects are included in ISO 14031, which gives guidance on the design and use of environmental performance evaluations within an organization. They include the following:

- Begin by identifying key activities of the organization (e.g., major manufacturing operations), associating specific environmental aspects with those operations (e.g., emissions into the air of specific constituents), and then identifying the type of impact related to that aspect (e.g., emissions that are potentially harmful to human health).
- Starting from the environmental perspective, identify key attributes in the organization's activities that may contribute to an environmental issue. Examples include raw material inputs, energy resources, discharges, emissions, or wastes. Examples of environmental issues may include global warming phenomena, water quality, or ambient air quality.
- Working from existing data on discharges and emissions, measure and assess these data in terms of quantity and hazards. This approach can be used to prioritize an organization's environmental aspects.
- Analyze environmental aspects (e.g., discharges or emissions) that are regulated and for which data have already been collected by the organization.
- Identify the significant environmental aspects of the organization's products, including manufacture, distribution, use, reuse, and disposal. These products can be traced back through the full range of the organization's activities.

With these evaluations as a starting point, organizations can identify their environmental aspects and determine which ones are significant. In organizations previously committed to regulatory compliance, ISO 14001 aspects reviews indicate that most of the significant environmental aspects are in the areas of energy and materials consumption. Approved toxic emissions have significantly declined because of the impact of waste minimization efforts.

The environmental aspects review is critical to the success of the EMS. Good advice to organizations starting the process is to begin small and slowly expand the review process into all areas of the business.

Legal Considerations

Before the environmental aspects review is conducted, the organization's legal counsel should be consulted on issues of access and confidentiality of the information developed.

Logically, an organization installing an EMS begins by surveying and documenting its environmental exposures and using survey results to establish a baseline for prioritizing objectives and targets. The legal need to protect environmental information from disclosure was of great concern to U.S. companies. Consequently, ISO 14001 contains language that stops short of requiring an aspects review:

> organization[s] shall establish and maintain a procedure to identify ... [their] environmental aspects ... to ensure that the significant environmental aspects are taken into account in establishing, implementing and maintaining [their] environmental management system[s].

In establishing and maintaining a procedure, documentation can be left to the organization. Although this semantic nuance allows some flexibility, a question remains for legal and public relations departments: are attempts at privileging environmental evaluations a useful strategy in view of legal and market factors encouraging environmental transparency?

For U.S. firms, disclosure of environmental aspects should be considered within the context of requirements to report environmental problems and the consequences (i.e., positive or negative) to the organization. Legal considerations prompting voluntary disclosure include the following:

- state and federal prosecutorial guidelines favoring civil or criminal actions against companies that fail to fully disclose their environmental problems
- federal sentencing guidelines that mitigate penalties for companies that do disclose, while increasing penalties for those that do not
- the Securities and Exchange Commission's Staff Accounting Bulletin No. 92, requiring publicly traded companies to disclose environmental liabilities in financial statements
- shareholder suits against directors and officers who withhold environmental information affecting investment decisions
- punitive damages and fines against companies that knowingly violate environmental regulations.

Other competitive market conditions that would further prompt voluntary disclosure include the following:

- environmental management standards such as the European Union's Eco-Management and Audit Scheme (EMAS) requiring a registry of environmental effects (aspects) to be created, validated, and publicly disclosed. This registry would facilitate an inference of similar environmental effects from similar operations in the United States.
- disclosure by one organization of an environmental aspect common to all organizations in a particular industry
- proliferation of "green" reports by organizations seeking to derive competitive advantage by highlighting their environmental stewardship over their operating facilities and products.

Organizations should systematically evaluate their legal exposure and determine a logical strategy for conducting an environmental aspects review. The following evaluative steps can be taken:

- Determine what environmental information is sensitive.
- Determine whether disclosure will result in harmful consequences.
- Determine whether disclosure has already been made or privilege violated (e.g., permit applications or information sharing beyond the scope of the privilege).
- Determine whether information deemed to be sensitive, harmful, and confidential can be protected by attorney/client, attorney work product, or self-evaluative privilege.

With the kind of universal knowledge likely to result from competition and the influence of the EMAS, the standard of care required of managers in carrying out their environmental responsibilities will be increased. With this environmental transparency, the standard of care for managers is greatly increased; the possibilities for successful ignorance, willful or not, on the part of managers becomes remote; and the credibility of nondisclosures becomes suspect.

Objectives and Targets

The ISO 14001 says, "The organization shall establish and maintain documented environmental objectives and targets at each relevant function and level within the organization." It is intended that objectives will conform to the organization's environmental policy, coincide with the findings of the environmental aspects review, and be measurable.

Integration and Alignment

Integration of the EMS refers to the process of embedding the environmental program in other management systems. Alignment refers to specifically coordinating environmental planning on the same time cycle as strategic planning so that strategic business plans reflect environmental issues. In this way, the EMS becomes totally immersed in organizational activities and environmental performance gains come as a function of other management goals. For example, human resources managers should identify all personnel who have environmentally related job functions and include in their performance appraisals a section that reviews environmental performance and encourages creative improvement of environmental performance. Managers should also review the existing training curriculum for opportunities to insert environmental awareness training.

Independent Audit Function

ISO 14001 says,

> The organization shall establish and maintain a program and procedures for periodic environmental management system

audits ... in order to determine whether ... the environmental management system ... conforms to planned arrangements for environmental management ... [and] has been properly implemented and maintained.

In establishing the audit function, the organization can use its own internal auditors or recognized third-party auditors that use either the ISO environmental auditing guidelines or other recognized auditing principles. It is important to recognize that the required audit is of the environmental management system, not the organization's environmental performance.

Market Drivers

ISO 14001 relies on the forces of the marketplace to impose environmental performance improvements. Behind voluntary ideals, however, looms a potential threat—more stringent national environmental management standards could become mandatory if voluntary adoption of ISO 14001 does not result in significant environmental performance improvement.

Market forces are evident in the competition to demonstrate "green" products and performance and to publicly disclose potentially positive aspects of environmental performance. In addition, requirements by government agencies and other large consumers that supplier organizations become ISO 14000 certified will be a major market driver. Similar results have been seen with ISO 9000.

Competition

In many markets, "green" products and processes are thought to confer a competitive advantage. Evidence is found in the proliferation of environmental reports, "green glossies," advertisements extolling the environmental virtues of oil and paper companies, and product claims of favoring sustainable material resources. Some consumer products companies have built their corporate and brand reputations on environmentally responsible products and commitment to the environment (e.g., The Body Shop, Patagonia).

Clearly, many organizations have seen certification to ISO 14001 as an opportunity for competitive advantage. Although certification does not mean that environmental performance is superior, organizational commitment to continual improvement and to the rigors of self-examination required under ISO 14001 can stand for good-faith efforts toward achieving superior environmental performance.

Disclosures

In markets where environmental disclosure can confer an advantage, environmental competitive advantage and performance standards will be established by companies that have a policy of complete disclosure. For example, a paper company publicly acknowledging environmental problems with dioxin would set the disclosure standard for all other companies in the paper industry. The investing and consuming public may assume that other paper companies have similar or worse problems with dioxin. Hence, full disclosure of appropriate and accurate information can suggest management integrity.

SAB 92

In the United States, an effective disclosure tool has been the requirement by the Securities and Exchange Commission (SEC) that publicly traded companies disclose estimates of environmental liability in their annual and quarterly financial statements. Initially, the compliance rate with this requirement, known as Staff Accounting Bulletin No. 92 (SAB 92), was low. The persistence of SEC commissioners and staff members, however, has resulted in improved reporting coming from industry and an indication that companies are looking closely at their environmental liabilities. Although the primary motivation of the SEC was sufficient environmental disclosure to allow investors to make informed investment decisions, the result has been the invocation of the business school maxim, "What gets measured, gets managed." Consequently, this modest staff bulletin from the SEC, interpreting existing accounting principles, is having a great influence on management of environmental problems by companies.

The EMAS

The EMAS requires companies to establish a register of environmental effects verified by a third party and made available to the public. Individual companies that have operations with significant environmental effects in Europe and similar operations in the United States are potentially disclosing the existence of environmental aspects in their U.S. operations. Alert interested parties, including environmental advocacy groups that monitor the environmental performance of industrial organizations, may seek to reconcile the validated existence of environmental problems in Europe with denials or nondisclosures of similar problems in the United States.

Influencing Suppliers

Organizations that have substantial market power are in a position to potentially require environmental performance improvements from their suppliers by buying only from ISO 14001–certified companies. A similar process has occurred with ISO 9000; that is, only ISO 9000–certified suppliers are allowed to bid and compete as suppliers for certain major producers of consumer or industrial products.

Raising the environmental awareness of product specifications writers is the main thrust of the guide for addressing environmental issues in product standards (ISO Guide

64:2008). When specification writers consider the impact of product design requirements on material and energy consumption, they often see opportunities for reducing environmental impacts by using alternatives. This opportunity is clearly present if a government or large organization adds these requirements to a large project bidder's specifications.

For example, as part of a bidding requirement, a purchaser of raw steel may specify that it be produced using the available energy input that has the least impact on atmospheric carbon dioxide levels. Suppliers of steel for this project would have to evaluate and rank the carbon dioxide impact of thermal energy furnished by coal, oil, natural gas, electricity, and the electric renewables: wind, direct solar, and geothermal. Suppliers would be contractually obligated to use the available energy option that their analysis indicates is most carbon dioxide efficient, even though that may add cost to the manufacturing process.

Many large consumers with economic market power will simply begin to require that their suppliers become certified to ISO 14001. Small and medium-sized enterprises, however, are not as aware of the ISO 14000 documents. They will be hard hit when large customers require certification as a condition of doing business. The machine shop whose cash flow depends on sales of a small part to a manufacturer, for example, stands to lose a great deal when that manufacturer will not buy any more parts until the shop is certified to ISO 14001. These businesses will quickly discover the advantage of ISO 14001 certification.

Mentoring

Many small enterprises will be helped by large organizations that agree to mentor them through the ISO 14001 certification process. The EPA has developed a strong record of mentoring for EMS processes. The Merit Partnership of EPA's Region 9 has previously provided incentives for large companies that agree to mentor their supply chain in ISO 14001 certification. Also, EPA had developed a Public Entity EMS Resource Center meant to be a one-stop shop for information regarding EMS. However, the site is no longer available.

Environmental Performance Evaluation

The most critical need of improved environmental performance is the development of consistent methods for measuring impacts on the environment to perform an environmental performance evaluation (EPE). Fundamental questions arise not only about how to measure the present impact of human activities on the environment but also about how we can begin to measure and move toward sustainable development that adequately serves today's needs without depleting resources required by future generations. To make progress on sustainable development, new approaches—environmental performance indicators (EPIs)—must be found.

It is necessary to account more thoroughly for the environmental costs associated with the production of goods and services and, simultaneously, consider the limits of sustainable economic growth and allocation of scarce resources. Although substantial progress is being made on measurement techniques, the concept of integrating sustainable development in economic policy is still in its formative stages.

Overview of ISO 14031:2013, the Guideline Document

ISO 14031:2013, Environmental Performance Evaluation (EPE), is the reference tool developed by TC 207 to help organizations develop reliable, verifiable information to determine whether the organization is meeting the environmental criteria set by management. The 2013 version is the second revision of ISO 14031.

Evaluation of environmental performance at the organizational level raises many collateral issues that impeded the development of the EPE guideline. Many large organizations, particularly manufacturing organizations, have potentially significant environmental impacts and resist the establishment of standards that measure the impact of materials, energy, emissions, effluents, and wastes on the environment. Nevertheless, many organizations evaluated their environmental performance prior to the development of an EPE guideline. Companies recognized that environmental trends have clear implications for how business is conducted in the 21st century. These organizations want to avail themselves of state-of-the-art methods for evaluating their impacts on the environment and also the sustainability of their industry through this century. In the future, estimates of environmental sustainability may be as important to businesses as forecasts of continued market demand are today.

Environmental impacts are always contextual. An organization cannot understand its environmental performance without an understanding of the circumstances surrounding that performance. For example, the environmental impact of water consumption depends on the source of the water. Is it from a basin whose total drawdown is in balance with precipitation into the basin, or is it from an aquifer where drawdown exceeds recharge? Consumption in the former circumstance has virtually neutral environmental impact; consumption in the latter would be of substantial environmental concern.

Environmental impact quantification, in part, involves considering the true costs of externalities. An *externality* is an accounting of indirect costs and can include direct and indirect environmental impacts (such as loss of enjoyment) as well as other economic impacts on employment and human welfare. For example, if permitted air pollutants produce respiratory disease, the costs of medical evaluation and treatment are paid for by society as a whole rather than by the emitter.

Underlying the costs of environmental externalities is an economic and social attitude toward the environment. How are organizations best motivated to make necessary social change? A simplistic answer is that change takes place automatically when environmental impact reduction strategies can be shown to coincide with and enhance economic gain—that is, when it is in the best economic interest of the organization to change. Decision makers react positively to opportunities that increase revenues or reduce costs—hence, the need to express environmental impacts in financial terms and relate their corrective actions to potential for revenue gain or cost reduction. When environmental impacts and corrective actions are expressed financially, the door opens to logical, cost/benefit-based decisions on reduction of environmental impacts and integration of environmental issues with strategic business plans.

Environmental managers and consultants work in terms of EPIs that are different for each contaminant. Concentrations of nitrous oxide, carbon dioxide, or various heavy metals, measured in parts per million/billion of volumes of air, water, or soil, are the EPIs of the environmental scientist. However, for business decision making, the language of environmental science is often insufficient for the following reasons: (1) Scientific measurements of values of carbon dioxide or chlorofluorocarbons are not easily susceptible to full financial valuation. (2) Scientific measurements describe the physical dimensions of contamination; however, business decision making requires an understanding of the uncertainty—that is, what the likely future economic consequences of environmental impacts will be. (3) Scientific language is sometimes foreign to the executive decision maker, who thinks and acts in response to changes in revenues or costs and needs to know the financial cost of environmental impacts.

A variety of analytical tools are used to measure environmental impacts, associated corrective/preventive actions, and the relationship of corrective actions to economic gain. These tools are used so that economically rational decisions can be made. The corrective action of choice becomes the one whose cost, after subtracting the value of measurable economic gain, achieves the largest impact improvement for monies spent. In many instances, the value of economic gain exceeds corrective action costs, producing a win for the environment and for the organization. The approach has several advantages:

- The financial approach is understood by corporate decision makers, who may otherwise wrestle with environmental measurements presented in parts per million or billion.
- Linking corrective action costs to the potential future cost of a problem allows rational choices based on cost-effectiveness among competing demands for limited corrective action funds.
- Measuring impacts (or externalities) rather than outputs captures the true cost of an environmental problem. Impacts consider not only the cost of reducing outputs, but the social costs to environmental receptors as well.

Environmental Costing and Valuation Methods

Conversion of collected data into financial terms and quantifying uncertainty involve a variety of valuation methods, including the following:

- *Direct cost estimates*—An environmental problem is expressed in financial terms at the outset, for example, the cost to clean up a specific regulated contaminated site under Superfund.
- *Fines and penalties*—The amount of fines and penalties is often stated in laws and regulations; what is unknown is the likelihood of penalty assessment and duration of fines. Fines and penalties are not environmental impacts. They are, however, costs to the organization of noncompliance with regulatory requirements and relevant to corrective action strategies.
- *Third-party liabilities*—Estimating third-party liabilities adds a layer of complexity and uncertainty to the valuation process. These liabilities include impacts to people, ecosystems, and property not under the direct control of a company. Both the potential costs and uncertainties are often quite high.
- *Quantifications requiring subjective valuations*—The full cost in damages to the environment is not captured in a wide array of circumstances, particularly where a level of emissions is permitted by law. These costs are transferred to society at large and become part of the price of economic well-being. Under ISO 14001, full costing and attribution of costs to responsible activities, products, or services will increasingly become a component of efforts to improve environmental performance. Although the full amount of contaminating discharges can usually be easily measured, the values of impacts to human health or natural resources are not so easily determined.

  A common method for valuing impacts to natural resources is the contingent valuation method (CVM). This method relies on polling to determine preference or indifference for the value under study. Determining these values facilitates full-cost accounting and subsequent corrective action decisions.

- *Quantifications requiring value establishment*—Value judgments provide an approximation of expected damage that could occur from contaminants that do not have an immediate, direct, and easily determinable environmental impact. For example, carbon dioxide emissions and their long-term impact on global climate are a potential environmental impact that could be quantified using the value judgment approach. This kind of valuation is most useful if conducted and established by an international body, such as ISO, so that there is a uniform approach for documenting aggregate impacts. With an established value scale, carbon dioxide as an emission by-product has a finite value that changes annually depending on the global content of carbon dioxide in the air. The value of emissions and the value of increased forestation (representing ability to absorb greater amounts of carbon dioxide) would then be established. The importance lies not in the absolute accuracy of such projections but in developing some defensible estimate so that organizations have a way of valuing their own emissions and the gains derived from corrective actions.

Life-Cycle Assessment

As a means of relieving government of costly enforcement-based environmental regulations, some countries are establishing product-oriented incentives that are intended to yield environmental benefits. The basis for evaluating such incentives is the province of life-cycle assessment (LCA). ISO 14040:2006, Environmental Management—Life Cycle Assessment—Principles and Framework, defines life-cycle assessment as a systematic set of procedures for compiling and examining the inputs and outputs of materials and energy and the associated environmental impacts directly attributable to the functioning of a product or service system throughout its life cycle.

The impetus behind the efforts to develop a useful LCA standard is the recognition that the European Union may develop and mandate a more stringent standard. The overall process is subject to criticism because LCA is a highly detailed cost evaluation that may be too expensive for use by small firms. In addition, the technical process of performing an LCA is undergoing constant modification. However, several examples of environmental actions that have been analyzed using LCA techniques have produced useful and important results.

France passed a fuel-tax exemption to promote consumer acceptance of biodiesel fuel—that is, diesel fuel produced from agricultural products such as soybeans. Burning biodiesel fuel has been shown to reduce automotive tailpipe emissions. Subsequently, life-cycle assessment established that the process of producing biodiesel fuel resulted in other forms of pollution that exceed that of producing and burning petroleum diesel. In other words, substitution resulted in displacement of pollution from automotive tailpipes to biodiesel production process wastes. Hence, the presumed substitution of environmentally preferable products must be carefully evaluated so that true environmental improvement is demonstrated.

The European Commission's committee on waste management was given the task of developing regulations on packaging waste that would result in both flexibility in consideration of national priorities and environmental improvement throughout member countries. Existing regulations forcing a hierarchy of "reduce, reuse, recycle" resulted in costly investments in Germany and elsewhere that were losing money with no prospect of recovery.

Life-cycle assessment determined that there was a trade-off between recycling and disposal that depended on transportation distances. As distance to the recycling center increased, the environmental benefit of recycling decreased. Beyond certain distances, recycling actually increased pollution over that created by waste disposal nearer the source. As a result of the described study, the European Commission revised its package waste regulations for certain products to favor maximum recyclable content of packaging as opposed to mandatory and, possibly, counterproductive reuse.

Trends—Cause for Optimism

The management changes described in this chapter coalesce into discernible trends at various levels:

- *Capture of relevant environmental costs*—Organizations everywhere, whether driven by social responsibility, market necessity, or regulatory agencies, are moving to include more of the environmental costs of their activities, products, and services in their management accounting systems. This could ultimately lead to a net benefit test for products and services: do the environmental impact costs associated with the design, manufacture, marketing, delivery, use, and disposal of products or services result in social/economic benefits that exceed their environmental costs?

- *Inclusion of beneficial environmental impacts*—Organizations are beginning to see value in creating and measuring beneficial environmental impacts—woodlands, wetlands, creature habitats, and other environmental assets. This could lead to an environmental balance sheet that offsets the adverse consequences of activities, products, and services with the beneficial consequences of other activities, products, or services. Advantages of identifying and quantifying beneficial environmental impacts could include their use as a tool for regulatory negotiations and, possibly, favorable tax treatment. Competitive pressure to produce environmental annual reports may, over time, increase managerial thinking and action toward neutral environmental impact. Environmental balance sheets may be formulated someday to evidence sustainable development practices.

- *Strategic deployment of environmental improvement activities*—Organizations, especially commercial and industrial entities, are increasingly aligning their environmental management planning with their overall strategic planning. For many corporations, this means expressing environmental impacts and corrective/preventive actions in financial terms and considering the potential that each corrective/preventive action has for producing an economic gain. Financial quantification of environmental impacts and ranking of corrective/preventive actions by cost-effectiveness enable rational decisions on sequencing and scoping of environmental activities.

Environmental management has been thrust into a new, dynamic phase as a result of the promulgation of ISO 14001. Based on results to date, the following steps seem appropriate for managers to take now regardless of a need or desire to adopt ISO 14001 or other similar standard or guideline:
- Recognize that environmental management is a high organizational priority.
- Establish a dialogue with internal and external interested parties.
- Determine the regulatory requirements and environmental exposures associated with the organization's activities, products, and services.
- Develop management and employee commitment to protecting the environment, with clear assignment of responsibility and accountability.
- Encourage environmental strategic planning throughout the product or process life cycle.
- Establish a disciplined management process for achieving targeted performance levels.
- Provide appropriate and sufficient resources, including training, to achieve targeted performance levels on an ongoing basis.
- Assess environmental performance against appropriate policies, objectives, and targets, and seek continual improvement where appropriate.
- Establish a management process to review and audit the environmental management system (EMS) and to identify opportunities for improvement of both the system and environmental performance.
- Coordinate EMSs with other systems (e.g., health and safety, quality, finance).

Overall, the opportunities for an integrated, proactive environmental management program are substantial. Although the management activities and time commitment necessary to achieve a fully developed environmental management program are significant, the potential payoffs for both the organization and the environment are equally large.

SUMMARY

- The safety professional must be familiar with the most current environmental regulations covering such areas as the release and recovery of hazardous substances; air, water, and soil pollution; and toxic substances control.
- Environmental standards and regulations in Asia and Europe, such as ISO 14000, are helping companies take proactive steps in managing safety, health, and environmental issues. These standards may be used to help establish international guidelines, develop certification criteria, and encourage voluntary compliance among manufacturers and suppliers.
- ISO 14000 and similar regulations are providing international business with stronger guidelines and regulations for protecting the environment and safeguarding worker and consumer health and safety. These regulations also set standards for environmental protection.
- Industry experts have identified five stages of environmental management program development that can be found in many companies: beginner, fire fighter, concerned citizen, pragmatist, and proactivist. A company's stage often depends on its size, amount of hazardous or toxic material produced, and financial resources.
- For companies that have reached stage 5, environmental management is a top priority. Top management must demonstrate a strong commitment to responsible environmental practices and communicate this priority throughout the company for the program to succeed. Program staff must include skilled professionals with backgrounds in the sciences, regulatory law, and environmental issues.
- Steps in the environmental management program include preventing common violations, maintaining accurate and thorough records, creating a spill-reporting plan, setting realistic limits and schedules for meeting environmental regulations, and training employees in program objectives and procedures. In addition, the company must establish a strong relationship with an attorney skilled in environmental law.
- Efforts to reduce and minimize waste can save companies money, reduce potential environmental liabilities, protect public health and worker health and safety, and protect the environment. Waste can be minimized through source reduction and recycling. Source reduction generally is more cost effective and has less of an impact on the environment than reuse/recycling.
- Groundwater contamination is one of the more serious environmental issues of the past few decades. Seepage of toxic and hazardous materials into aquifers has prompted local and federal agencies to force companies to take stronger measures in assessing and controlling this contamination.

- Complying with environmental regulations and standards can be a competitive advantage to companies that disclose their adherence. The public today is more sensitive to the responsibility of industry to safeguard the environment.
- By buying only from certified companies, organizations with substantial market power are in a position to require suppliers to improve their environmental record.
- The environmental aspects review establishes a baseline of significant environmental aspects to help companies develop an environmental performance plan with objectives and targets. Before conducting a review, a company should consult its legal counsel to protect sensitive company information from disclosure.
- ISO 14031:2013, Environmental Performance Evaluation, is the reference tool to help organizations determine whether they are meeting their environmental protection goals. Companies must weigh the economic and marketing costs and benefits of achieving those goals.
- A variety of valuation methods can help companies estimate the costs of addressing environmental problems and assessing the fines and penalties involved in failing to comply with regulations and standards.
- Life-cycle assessment provides a means of relieving government of costly enforcement-based environmental regulations by establishing product-oriented incentives.
- In the future, companies are likely to take a more proactive role in creating a better environment rather than focusing solely on preventing environmental damage.

REFERENCES

Environmental Management: The ISO 14000 Family of International Standards. ISO Central Secretariat 1, chemin de la Voie-Creuse Case postale 56 CH - 1211 Genève 20 Switzerland.

Friedman, W. J. "Avoiding Environmental Liability in Five Simplified Steps." *Chemical Processing* (April 1988).

Hajost, S. "The Challenge to International Law and Institutions." *EPA Journal* 16, no. 4 (1990).

Hunt, C. B., and E. R. Auster "Proactive Environmental Management: Avoiding the Toxic Trip." *Sloan Management Review* 31 (1990).

Keenan, T. "Why Is Everyone Talking about Environmental Audits?" *Industrial Safety & Hygiene News* 24 (1990).

Krieger, G. R., ed. *Accident Prevention Manual for Business & Industry: Environmental Management*. 2nd ed. Itasca, IL: National Safety Council, 2000.

Leaf, A. "Potential Health Effects of Global Climatic and Environmental Changes." *New England Journal of Medicine* 321 (1989).

Main, J. "Here Comes the Big New Cleanup." *Fortune*, November 21, 1988.

Quarles, J., and W. H. Lewis, Jr. *The New Clean Air Act: A Guide to the Clean Air Programs as Amended in 1990*. Washington DC: Morgan, Lewis, & Bockius, 1990.

Rhodes, D. "Safety, Environmental Crimes and the Tough New Laws." *Professional Safety* 35 (1990).

Title 29 CFR, Labor (OSHA Hazardous Waste Training and Communications Regulations).

Title 40 CFR, Protection of Environment.

U.S. Department of Health and Human Services, Public Health Service, Centers for Disease Control, and National Institute for Occupational Safety and Health. *Occupational Safety and Health Guidance Manual for Hazardous Waste Site Activities*. Washington DC: U.S. Government Printing Office, 1985.

U.S. Environmental Protection Agency, Region 5, 230 S. Dearborn, Chicago, IL 60604.
  EPA Property Searches for Buyers of Real Estate (Fact Sheet), February 1991.
  An Introductory Guide to the Statuary Authorities of the United States Environmental Protection Agency; Our Air, Our Land, Our Water, April 1988.

Young, O. R. "Saving the Arctic: Challenge to Eight Nations." *EPA Journal* 16, no. 4 (1990).

REVIEW QUESTIONS

1. RCRA and CERCLA are used to regulate hazardous materials. What are the major differences?
2. What is a gap analysis when dealing with ISO 14001?
3. What are four aspects that an environmental aspects review should address?
    a.
    b.
    c.
    d.
4. What is the main purpose of the Eco-Management and Audit Scheme?
5. What is the difference between ISO 14010 and ISO 14011?

15 Indoor Air Quality

Philip E. Hagan, JD, MBA, MPH, ARM, CIH, CET, CHMM, CHCM, CHSP, CEM
Lynne Zarate, MSE, CIH, CHMM

Indoor Air Quality Basics
Sick-Building Syndrome ▸ Building-Related Illness ▸ Elements Common to IAQ Problems ▸ Control Strategies

Common Pollutant Sources
Occupant Activities ▸ Outside Pollutant Sources ▸ Building Systems Affecting IAQ

Pollutant Pathways and Driving Forces
Positive Pressure ▸ Natural Forces

Elements of an Effective IAQ Management Plan
IAQ Manager ▸ IAQ Profile ▸ Record Keeping

IAQ Investigations
Developing and Testing Hypotheses ▸ Contaminants

Resolution
Evaluating Solutions ▸ Persistent Problems

Evaluating Potential Consultants

Regulatory Requirements
Pollutant-Related Regulations ▸ Ventilation-Related Regulations

Summary

References

Review Questions

Indoor air quality (IAQ) is a major component of indoor environmental quality (IEQ). IEQ refers to all of the elements found inside a building that could affect occupant well-being and health. IAQ refers to the air inside a building and to its suitability for human consumption. In the last several decades, the potential impact of poor indoor air quality on the work force has become a significant concern. In recent years, comparative risk studies performed by EPA and its Science Advisory Board consistently rank indoor air pollution among the top five environmental risks to public health. As long ago as 1984, a World Health Organization committee report suggested that up to 30% of new and remodeled buildings worldwide may be the subject of excessive complaints related to IAQ.

The condition of a building's indoor environment is based on complex interactions among occupant activities, the surrounding climate and weather, building structure and mechanical systems, and contaminant sources. Since the early 1970s, the construction of energy-efficient (tightly sealed) buildings; decreased ventilation rates to reduce energy costs; and the increased use of synthetic building materials, furnishings, pesticides, and housekeeping supplies have resulted in increased exposures to indoor air pollutants. Poor building design, improperly maintained building mechanical systems, and occupant activities contribute to the pollutant load in a building. On close examination, many buildings reveal some inadequacies of design, construction, operation, and maintenance. Occupant perceptions and susceptibilities can also play a key role in defining the state of the indoor environment.

This chapter will focus on the following aspects of indoor air quality and related topics:
- the nature of IAQ and potential impacts
- identification and discussion of elements that can contribute to IAQ problems: potential pollutant sources, pollutant pathways, and driving forces
- basic elements of an IAQ management plan
- scope of an investigation
- mitigation and resolution of indoor air quality problems
- obtaining and evaluating outside assistance
- regulatory requirements.

Much of the information that follows is based on information contained in the following EPA publications: *Building Air Quality Action Plan*; *An Office Building Occupant's Guide to Indoor Air Quality*; and *Building Air Quality: A Guide for Building Owners and Facility Managers (BAQ)*. Published in 1991, *BAQ* is widely recognized as one of the best available references on developing and managing indoor air quality programs. This publication is available on the Internet at epa.gov/iaq/largebldgs/baqtoc.html. Most of the premises discussed in *BAQ* are based on good building management principles and can usually be implemented with minimal financial impact to an organization. In addition, epa.gov/iaq/ served as a primary source of information and text for this chapter.

EPA has also prepared sector-specific guidance on indoor air quality for schools, homes, and offices. In 2002, EPA released an update to *BAQ*, an interactive software tool called the Indoor Air Quality: Building Education and Assessment Model (I-BEAM). It is a guidance tool designed for use by building professionals and others interested in indoor air quality in commercial buildings. I-BEAM updates and expands EPA's *Building Air Quality* guidance and is designed to be a comprehensive, state-of-the-art tool for managing IAQ in commercial buildings. I-BEAM is available on CD-ROM (EPA 402-C-01-001). To obtain a free CD-ROM copy, contact EPA's National Service Center for Environmental Publications (NSCEP) at PO Box 42419, Cincinnati, OH 45242-0419; online at epa.gov/nscep; or by phone at 800-490-9198. In addition, EPA lists a large number of free publications that can be used to address IAQ issues and that can be either ordered at no cost or downloaded at epa.gov/iaq/schools/pubs.html.

INDOOR AIR QUALITY BASICS

In the past several years, a growing body of scientific evidence indicates that the air within buildings can be more seriously polluted than the outdoor air in most industrialized cities. Other research indicates that people spend approximately 90% of their time indoors. Indoor environments include offices, public buildings, factories, warehouses, buses, automobiles, trucks, trains, subways, airplanes, restaurants, hotels, and commercial buildings.

Many factors that could be encountered in these environments may contribute to complaints or concerns from occupants regarding indoor air quality, including cigarette smoke, dust particles, objectionable odors, airborne compounds generated from a variety of sources, microbial contamination, poor air circulation, thermal discomfort, humidity levels, job pressures, lighting, workstation design, and noise. These factors can cause illness or lead to problems affecting occupant health and comfort. Indoor environmental conditions causing acute problems are usually readily identifiable, and corrective actions can bring resolution in a short time. Chronic problems attributed to poor indoor air quality tend to be a little more difficult to diagnose and are quite often characterized as either "sick-building syndrome" or "building-related illnesses."

Sick-Building Syndrome

Although perhaps a misnomer, because people get sick, not buildings, the term *sick-building syndrome* (*SBS*) is a condition described by general complaints of discomfort including headache; nausea; dizziness; dry or itchy skin; eye, nose, throat, and respiratory irritation; dry coughing; difficulty concentrating; muscle pain; sensitivity to odors; and fatigue. Usually, the specific causes of the symptoms are not known. Complaints may be associated with a particular room or floor or may be widespread throughout a building (Figure 15–1). The symptoms are usually associated with time spent in the building; most affected occupants report relief soon after leaving the building.

Inadequate ventilation, airborne contaminants from indoor and outdoor sources, and biological contaminants have been cited as causes of, or contributing factors to, sick-building syndrome. Sometimes other IEQ concerns regarding thermal discomfort, ergonomics, humidity, noise, or lighting seem to aggravate these conditions.

It is important to ensure that occupants with physical symptoms that could be related to the indoor environment be evaluated by an occupational physician or health care practitioner with experience dealing with similar health care concerns. In some cases, symptoms attributed to SBS are caused by other factors, and it is important that an affected individual receive appropriate treatment.

Building-Related Illness

The term *building-related illness* (*BRI*) is used when symptoms of diagnosable illness are identified and can be attributed directly to building contaminants. Quite often, symptoms such as coughing, chest tightness, fever, chills, and muscle aches can be determined clinically and are the result of clearly identifiable causes. A diagnosis results in a clinically defined illness of known etiology. Allergic reactions, hypersensitivity pneumonitis, and humidifier fever would fall into this category. Affected occupants with BRI may require prolonged recovery times after leaving the complaint area.

Often, the conditions leading to SBS or BRI are temporary, but some buildings have long-term problems requiring identification and mitigation.

Elements Common to IAQ Problems

Although sometimes difficult to ascertain, most IAQ scenarios have several elements in common:
- pollutant source(s)
- a driving force to move the pollutant(s)
- a pathway for the pollutant(s) to travel
- a susceptible population.

However, occupant complaints may not be caused by the indoor air quality of the workplace but could be the result of the following:
- illnesses contracted away from the workplace
- acute sensitivity (allergies)
- stress
- other psychosocial factors.

It is important to enlist the expertise of qualified health care specialists (occupational health nurses and physicians and other specialists) who have special knowledge and training in diagnosis and treatment of workplace-related illnesses when evaluating a potential IAQ-related problem. (See Chapter 12, Occupational Health Programs.)

Control Strategies

Three acceptable strategies for controlling pollutants that contribute to poor IAQ are as follows:
1. managing pollutant sources either by removal, isolation, or controlling use
2. using ventilation to dilute and remove pollutants from the building
3. using filtration to clean the air.

COMMON POLLUTANT SOURCES

A summary of indoor air quality complaints compiled by the National Institute for Occupational Safety and Health (NIOSH) identified the following primary sources of indoor air quality problems:
- inadequate ventilation, 52%
- contamination from inside building, 16%
- contamination from outside building, 10%
- microbial contamination, 4%
- unknown, 13%.

Pollutant sources can include occupant activities such as housekeeping, maintenance, painting, remodeling, and renovation. Building systems, equipment, furnishings, and wall and floor coverings can serve as pollutant sources. Facility personnel should become knowledgeable about IAQ issues and pollutant sources because their perspective can be helpful (Figure 15–2) when IAQ-related issues must be addressed. For example, staff personnel who have not been educated about IAQ issues may observe unsanitary conditions, blocked vents, evidence of water leaks in tenant spaces, or other indicators of potential IAQ problems but fail to recognize their importance and potential adverse impacts. Educating building staff about these issues will allow them to recognize potential problems and address issues before they become a problem.

OCCUPANT DIARY

Occupant Name: _____ Title: _____ Phone: _____
Location: _____ File Number: _____

On the form below, please record each occasion when you experience a symptom of ill health or discomfort that you think may be linked to an environmental condition in this building.

It is important that you record the time and date and your location within the building as accurately as possible, because that will help identify conditions (e.g., equipment operation) that may be associated with your problem. Also, please try to describe the severity of your symptoms (e.g., mild, severe) and their duration (the length of time that they persist). Any other observations that you think may help in identifying the cause of the problem should be noted in the Comments column. Feel free to attach additional pages or use more than one line for each event if you need more room to record your observations.

Time/Date	Location	Symptom	Severity/Duration	Comments

Figure 15–1. Occupants should keep a diary to help identify potential sources and locations of building environmental conditions.

POLLUTANT AND SOURCE INVENTORY

Building Name: _____ Address: _____

Completed by (name/title): _____ Date: _____ File Number: _____

Using the list of potential source categories below, record any indications of contamination or suspected pollutants that may require further investigation or treatment. Sources of contamination may be constant or intermittent or may be linked to single, unrepeated events. For intermittent sources, try to indicate the time of peak activity or contaminant production, including correlations with weather (e.g., wind direction).

Source Category	Checked	Needs Attention	Location	Comments
SOURCES OUTSIDE BUILDING				
Contaminated Ambient Air				
Pollen, dust				
Industrial contaminants				
General vehicular contaminants				
Emissions from Nearby Sources				
Vehicle exhaust (parking areas, loading docks, roads)				
Dumpsters				
Re-entrained exhaust				
Debris near outside air intake				
Soil Gas				
Radon				
Leaking underground tanks				
Sewage smells				
Pesticides				
Moisture or Standing Water				
Rooftop				
Crawl space				
EQUIPMENT				
HVAC System Equipment				
Combustion gases				
Dust, dirt, or microbial growth in ducts				
Microbial growth in drip pans, chillers, humidifiers				
Leaks of treated boiler water				
Non-HVAC System Equipment				
Office equipment				

Figure 15–2. Facility personnel should keep an inventory of potential contaminants or pollutants and actions taken.

POLLUTANT AND SOURCE INVENTORY *(Continued.)*				
Source Category	**Checked**	**Needs Attention**	**Location**	**Comments**
Supplies for equipment				
Laboratory equipment				
HUMAN ACTIVITIES				
Personal Activities				
Smoking				
Cosmetics (odors)				
Housekeeping Activities				
Cleaning materials				
Cleaning procedures (e.g., dust from sweeping, vacuuming)				
Stored supplies				
Stored refuse				
Maintenance Activities				
Use of materials with volatile compounds (e.g., paint, caulk, adhesives)				
Stored supplies with volatile compounds				
Use of pesticides				
BUILDING COMPONENTS/FURNISHINGS				
Locations Associated with Dust or Fibers				
Dust-catching area (e.g., open shelving)				
Deteriorated furnishings				
Asbestos-containing materials				
Unsanitary Conditions/Water Damage				
Microbial growth in or on soiled or water-damaged furnishings				
Chemicals Released from Building Components or Furnishings				
Volatile compounds				

Figure 15–2. Continued.

POLLUTANT AND SOURCE INVENTORY *(Concluded.)*

Source Category	Checked	Needs Attention	Location	Comments
OTHER SOURCES				
Accidental Events				
Spills (e.g., water, chemicals, beverages)				
Water leaks or flooding				
Fire damage				
Special Use/Mixed Use Areas				
Smoking lounges				
Food preparation areas				
Underground or attached parking garages				
Laboratories				
Print shops, art rooms				
Exercise rooms				
Beauty salons				
Redecorating/Repair/Remodeling				
Emissions from new furnishings				
Dust, fibers from demolition				
Odors, volatile compounds				

Figure 15–2. Concluded.

Training and information related to IAQ should also be provided to occupants and contractor personnel whose activities could affect a building's IAQ. Both informal, in-house information sharing (distribution of IAQ information and fact sheets, informal discussions, seminars, or self-training materials) and formal, structured training courses are beneficial.

In-house and contractor personnel whose functions could affect IAQ, such as pest control contractors, housekeeping personnel, and heating, ventilation, and air conditioning (HVAC) maintenance staff, should be identified. It is important to create, keep, and update a list of these personnel and to track their activities so that related activities can be evaluated for potential impacts on IAQ.

Often, occupant activities require the use of chemical substances. Each chemical substance used during occupant activities in the workplace should be evaluated for potential impact to the indoor environment. OSHA's Hazard Communication Standard (29 CFR 1910.1200) requires that information and training be provided to staff who use hazardous chemicals. The training should cover the health effects of the chemicals they use in their duties and how to read, understand, and follow the information contained in label instructions and Safety Data Sheets (SDSs).

OSHA requires manufacturers or distributors of chemicals to provide SDSs for their products. The SDS for every substance used in a facility should be made available to workers and be used to educate the work force on applicable hazards of the chemical substances used in housekeeping and maintenance activities.

The Hazard Communication Standard (HCS) was updated in 2012 and requires chemical manufacturers, distributors, and importers to provide Safety Data Sheets (SDSs) (formerly known as Material Safety Data Sheets or MSDSs) to communicate the hazards of hazardous chemical products. As of June 1, 2015, the HCS requires new SDSs to be in a uniform format and contain information about identification, hazard(s) identification, composition/information on ingredients, first-aid measures, fire-fighting measures, accidental release measures, handling and storage, exposure controls/personal protection, physical and chemical properties, stability and reactivity, and toxicological information. Employers must ensure that SDSs are readily accessible to employees.

Occupant Activities

Building operations and businesses involving occupants (e.g., housekeeping activities, restaurants, renovations, copying, maintenance activities, pest control, renovation, dry cleaners, beauty salons, and printing operations) can serve as a source of indoor pollutants. These operations could be either ongoing or intermittent, making identification and mitigation of potential sources difficult.

Efforts to lower operating costs through reduction of support services, eliminating or reducing training budgets, and deferring maintenance to save money can adversely affect indoor environmental quality. Fewer housekeepers or maintenance personnel left after corporate downsizing are quite often expected to do the same or more work with fewer available resources. These scenarios can result in "cutting corners" to complete assigned tasks. For example, in attempting to do a better job, housekeeping staff will mix cleaning solutions in higher concentrations than recommended by the manufacturer or with other products to increase "cleaning power." Residues left in carpets and upholstery can elicit severe adverse reactions in sensitized individuals. Maintenance workers may postpone preventive maintenance operations in an attempt to save money. Reduced mechanical system efficiency and subsequent breakdowns can be more costly in terms of adverse effects on building occupants and operations than the money saved.

Occupants experiencing thermal discomfort often attempt to regulate their environment by opening a window, adjusting a thermostat, or blocking ventilation supply or exhaust vents. This may bring some immediate relief to the affected occupant. However, quite often, a sequence of events such as these can have a long-term impact of unbalancing the ventilation system as its control system tries to accommodate the unplanned changes occurring throughout a building. The result could be a building environment that is not operating within original design parameters; as a result, the quality of the indoor environment deteriorates. These types of actions can serve as clues of an ongoing problem for an IAQ manager dealing with occupants distressed by their working environment.

Housekeeping

Housekeeping activities can result in occupant exposures to hazardous chemicals, dust, and other pollutants. These activities should be evaluated for potential impacts to the indoor environment:

- cleaning, stripping, and waxing floors
- shampooing carpets
- polishing and cleaning
- furniture vacuuming
- dusting
- removing solid waste.

Suppliers of cleaning products should provide Safety Data Sheets (SDSs). If not, then a request (written or verbal) should be made to the supplier. When feasible, housekeeping staff should use less-hazardous substitutes. Directions and warnings on product labels should be followed. If the label indicates the need for adequate ventilation during preparation, then the product should not be mixed in an unventilated closet.

Training should be conducted for workers on how to properly handle cleaning products. Inappropriate use of cleaning products can adversely affect the workplace environment. For example, the use of ammonia, a popular cleaning product, in an area with inadequate ventilation could result in a potentially hazardous buildup of vapors. Exposure to ammonia vapors can cause eye irritation and headaches. Ammonia cleaners should be used in accordance with the manufacturer's instructions and not mixed with other products. Mixing ammonia with lye, sodium hydroxide compounds, or bleach can create hazardous by-products. A common example of an action resulting in mixing incompatible chemicals would be removal of mold and mildew contamination using bleach, followed by treatment with an ammonia product. If the bleach is not thoroughly rinsed from the surface before using the ammonia product, off-gassing of the resulting by-products could adversely affect sensitive individuals. SDSs should be reviewed to identify incompatible chemicals.

Although housekeeping (vacuuming and dusting) activities are often performed outside normal work hours, dust from cleaning operations can remain suspended for several hours, affecting sensitized individuals in the affected areas. Performing these operations at night or just after workers leave can allow sufficient time for the dust to settle before the beginning of the next workday.

Improperly maintained carpets can serve as a reservoir for microbial contamination, especially after flooding has occurred. Carpets decomposing because of aging and/or excessive wear can produce fibers that, when airborne, could cause eye irritation. Chemical exposure from adhesives or cleaning agents can produce adverse reactions in susceptible populations.

Sanitation also plays an important role in IAQ. Small quantities of food can attract insects and rodents. Birds and bats enter facilities through small penetrations in fascia, soffits, crawl spaces, and unscreened HVAC openings. In addition to carrying diseases and disease vectors, these animals also can elicit a variety of allergic responses and create offensive, lingering odors if they become trapped and die.

Housekeeping staff must be aware of the impact their activities can have on the indoor environment. Simply replacing the lid on a container of cleaning materials can often minimize exposure to hazardous materials.

Maintenance

Preventive maintenance (regular inspection, cleaning, and replacement of worn or broken parts and building systems) is important to ensure that building mechanical systems operate as designed. A written preventive maintenance program is an effective tool for maintaining good IAQ. The plan should include monitoring, inspecting, and cleaning of HVAC system components such as outside air intakes, outside air dampers, air filters, drain pans, heating and cooling coils, the interior of air-handling units, fan motors, belts, air humidification controls, and cooling towers (Figure 15–3). However, a written preventive maintenance plan tends to be ineffective if resources are not allocated to ensure that the plan is implemented.

The frequency of maintenance activities may vary from building to building. It is important to develop a maintenance schedule based on the needs of each building and associated equipment. The schedule should ensure that all equipment is in good, sanitary condition and is operating as close to design set points as possible.

Knowledge and skill levels of workers can also affect operational upkeep of mechanical systems. Reductions in training and education budgets can result in maintenance workers being unable to acquire the necessary knowledge and skills to perform effective repairs. A foreseeable consequence, for example, would be the performance of inadequate repairs on an HVAC system. The outcome would be an HVAC system operating at a reduced capacity, causing building occupants thermal discomfort and possible adverse physical reactions to contaminant buildup.

When unscheduled maintenance events (e.g., equipment failures) require the prolonged deactivation or modification of building HVAC equipment and other associated building systems, maintenance personnel should ensure that necessary personnel are notified so that potential IAQ impacts can be properly evaluated. The individual responsible for managing IAQ issues should review the situation carefully and provide recommendations to maintenance and administrative personnel on how to proceed without compromising the building's IAQ. This information should also be communicated to building occupants and tenants to inform them how their air quality is being protected.

Reductions in preventive maintenance program operating budgets and personnel cutbacks can cause component breakdowns or systems operating below design parameters. These conditions can negatively impact the indoor environment. For more details on preventive maintenance, see Chapter 4, Maintenance of Facilities, in the *Engineering & Technology* volume.

Another important function of maintenance operations is energy management. Some energy efficiency measures can result in IAQ problems, while some IAQ measures have energy impacts. The I-BEAM program, previously mentioned, has a module to help ensure that a building can

HVAC CHECKLIST—SHORT FORM

Building Name: _____ Address: _____
Completed by: _____ Date: _____ File Number: _____

MECHANICAL ROOM
Clean and dry? _____ Stored refuse or chemicals? _____
Describe items in need of attention _____

MAJOR MECHANICAL EQUIPMENT
Preventive maintenance (PM) plan in use? _____

Control System
Type _____
System operation _____
Date of last calibration _____

Boilers
Rated Btu input _____ Condition _____
Combustion air: Is there at least 1 square inch of free area per 2,000 Btu input? _____
Fuel or combustion odors _____

Cooling Tower
Clean? No leaks or overflow? _____ Slime or algae growth? _____
Eliminator performance _____
Biocide treatment working? (list type of biocide) _____
Spill containment plan implemented? _____ Dirt separator working? _____

Chillers
Refrigerant leaks? _____
Evidence of condensation problems? _____
Waste oil and refrigerant properly stored and disposed of? _____

AIR HANDLING UNIT
Unit identification _____ Area served _____

Outdoor Air Intake, Mixing Plenum, and Damper
Outdoor air intake location _____
Nearby contaminant sources? (describe any within 25 feet) _____
Bird screen (mesh < 0.5") in place and unobstructed? _____
Design total cfm _____ Outdoor air (OA) cfm _____ Date last tested and balanced _____
Minimum % OA (damper setting) _____ Minimum cfm OA (total cfm × minimum % OA)/100 = _____
Current OA damper setting (date, time, and HVAC operating mode) _____
Damper control sequence (describe) _____
Condition of dampers and controls (note date) _____

Figure 15-3. This HVAC checklist will assist building personnel in monitoring maintenance of equipment.

HVAC CHECKLIST—SHORT FORM *(Continued.)*

Fans

Control sequence _____
Condition (note date) _____
Indicated temperatures   supply air _____   mixed air _____   return air _____   outdoor air _____
Actual temperatures   supply air _____   mixed air _____   return air _____   outdoor air _____

Coils

Heating fluid discharge temperature _____ DT _____ cooling fluid discharge temperature _____ DT _____
Controls (describe) _____
Condition (note date) _____

Humidifier

Type _____ If biocide is used, note type _____
Condition (no overflow, drains trapped, all nozzles working?) _____
No slime, visible growth, or mineral deposits? _____

DISTRIBUTION SYSTEM

		Supply Air		Return Air		Power Exhaust		
Zone/Room	System Type	Ducted/Unducted	cfm	Ducted/Unducted	cfm	cfm	Control	Serves (e.g., toilet)

Condition of distribution system and terminal equipment (note locations of problems)

Adequate access for maintenance? _____
Ducts and coils clean and unobstructed? _____
Air paths unobstructed? Supply _____ Return _____ Transfer _____ Exhaust _____ Make-up _____
Note locations of blocked air paths, diffusers, or grilles _____
Any unintentional openings into plenums? _____
Controls operating properly? _____
Air volume correct? _____
Drain pans clean? Any visible growth or odors? _____

Filter's Location	Type/Rating	Size	Date Last Changed	Condition (give data)

Figure 15–3. Continued.

HVAC CHECKLIST—SHORT FORM *(Concluded.)*
OCCUPIED SPACE

Thermostat types _____

Zone/Room	Thermostat Location	What Does Thermostat Control? (e.g., radiator, AHU-3)	Set Points		Measured Temperature	Day/Time
			Summer	Winter		

Humidistats/Dehumidistats type _____

Zone/Room	Humidistat/Dehumidistat Location	What Does It Control?	Set Points (% RH)	Measured Temperature	Day/Time

Potential problems (note location) _____
Thermal discomfort or air circulation (drafts, obstructed airflow, stagnant air, overcrowding, poor thermostat location)
_____

Malfunctioning equipment _____
Major sources of odors or contaminants (e.g., poor sanitation, incompatible uses of space) _____

Figure 15–3. Concluded.

provide for energy efficiency without degrading indoor air quality. If they are correctly implemented, the two goals can be mutually compatible.

Remodeling and Renovation

Remodeling and renovation should be planned with IAQ in mind because these activities can create indoor air quality problems by exposing occupants to dust, odors, microorganisms, and volatile hazardous chemicals. Planning issues include the following:

- ensuring that designs and construction activities for all proposed remodeling or renovation activities are reviewed for potential IAQ impacts before initiation
- scheduling work during periods of low occupancy
- isolating work areas by blocking return vents in the work area and/or installing temporary barriers
- configuring ventilation to ensure that air moves into the renovated area in order to prevent migration of pollutants into occupied areas
- using dust suppression techniques during demolition activities
- using specialized cleaning procedures (e.g., HEPA vacuums)
- minimizing emissions from materials processes (e.g., wet-sanding drywall)
- installing products that will have minimal impact on IAQ (e.g., formaldehyde-free cabinetry) or requiring "aging" of products, such as carpet, before installation.

Painting

Painting of interior spaces can produce irritating or harmful vapors. Methods to prevent problems include obtaining and using paints with low odor and chemical emissions, performing work during periods of low occupancy, and arranging ventilation to isolate work areas.

Outside Pollutant Sources

Building HVAC systems are designed to bring in outside air to control temperature, humidity, odors, and air quality. Although HVAC systems filter this outside air to some degree, the weather, season, neighboring structures, and outside pollution sources can contribute to contamination of the incoming air.

Outdoor air also enters and leaves a building by infiltration and natural ventilation. With infiltration, outdoor air flows through breaches in the building envelope to the interior spaces of a building. In natural ventilation, air moves through opened windows and doors.

Pollen, dust, fungal spores, tobacco smoke, industrial emissions, and vehicle emissions are some of the contaminants that can be brought into the building in this manner. Other potential sources include loading docks, Dumpsters, unsanitary debris, or building exhausts located near outdoor air intakes.

Building Systems Affecting IAQ

The quality of indoor air can be affected by building HVAC systems and a variety of other building components, equipment, and furnishings.

HVAC Systems

HVAC systems are used to supply heating, cooling, humidifying, dehumidifying, or otherwise conditioned air for occupant comfort and health. HVAC systems are also used for odor control and the removal of contaminants.

Design parameters for HVAC systems are usually based on building codes, which may vary according to geographic location. However, the American Society of Heating, Refrigerating, and Air-Conditioning Engineers (ASHRAE) ventilation standards are used as the basis for most building ventilation codes. When the HVAC system is operating below design capacity, any or all of these parameters can be adversely affected. NSC's *Fundamentals of Industrial Hygiene* (2012) provides a comprehensive discussion on HVAC systems.

The amount of outdoor air considered adequate for proper ventilation has varied substantially over time. Because updating building codes can take several years, HVAC systems for most buildings probably have a lower amount of air supply than is currently considered adequate. This should be taken into account whenever indoor air quality scenarios are evaluated.

ASHRAE first established recommended ventilation rates for indoor environments in 1973. This standard was amended in 1975 to specify that a minimum of 5 cubic feet per minute (cfm) of outdoor air per person be used in building design, and it has been updated several times. This standard has been incorporated into most building codes. The newest revision, ASHRAE Standard 62.1-2007, recommends minimum ranges of outdoor air in the breathing zone, depending on the use and occupant density of the space, ranging from 6 to 26 cfm per person. A good practice is to ventilate buildings with the maximum volume of outside air that is practical, taking into account HVAC system capacity, use of the space, and current climatic conditions.

Efforts to reduce energy costs or increase the number of people using the building often result in exceeding the capacity of an HVAC system. In addition, HVAC systems that were designed with inadequate supply capacities or that supply air below design parameters can also contribute to IAQ problems.

Reduced airflow rates can occur as a result of the following problems with an HVAC system: uncalibrated controls, inoperable equipment, bad bearings, plugged or dented ducts, slipping fan belts, open access doors, damaged ducts or elbows, closed or opened blast gates to other branches, corroded and stuck blast gates, fans turning in a reverse direction (which occurs when lead wires are reversed and the motor and fan turn backward), worn-out fan blades, clogged air filter systems, or additional branches or filters added to the system since the initial installation. Inadequate maintenance and operating practices (torn or overloaded filters, dirty or damaged insulation, and inoperable dampers/baffles) can contribute to poor IAQ.

HVAC components identified as sources of microbial contamination include wet filters; wet insulation; and wet, clogged, or disconnected undercoil pans, cooling towers, condensers, and evaporative humidifiers. Improper use of biocides and cleaning compounds on HVAC system components can produce airborne contaminants.

Thermal Comfort

Thermal comfort requirements vary for individuals based on clothing, activity level, age, and physiology. A primary goal of HVAC systems is to supply conditioned air to a building to ensure a certain level of occupant thermal comfort. As a result, ranges of temperatures and relative humidities are usually evaluated to determine the appropriateness of thermal comfort for occupants in the indoor environment.

As a source of guidance, ASHRAE Standard 55-2004, "Thermal Environmental Conditions for Human Occupancy," addresses "thermal comfort" in an office environment, which means that an employee wearing a normal amount of clothing feels neither too cold nor too warm. This standard discusses thermal comfort within the context of air temperature, humidity, and air movement and specifies the combinations of indoor space environment and personal factors that will produce thermal environmental conditions acceptable to 80% or more of the occupants:

Recommended Ranges of Temperature and Relative Humidity

Relative Humidity	Winter Temperature	Summer Temperature
30%	68.5°F–75.5°F	74.0°F–80.0°F
40%	68.0°F–75.0°F	73.5°F–80.0°F
50%	68.0°F–74.5°F	73.0°F–79.0°F
60%	67.5°F–74.0°F	73.0°F–78.5°F

Recommendations apply for persons clothed in typical summer and winter clothing, performing light, mainly sedentary activity. ASHRAE addresses ventilation and the removal of air contaminants in a separate standard, ASHRAE Standard 62.1-2007, "Ventilation for Acceptable Indoor Air Quality."

Humidity can play a big part in thermal comfort. A body's cooling mechanism is based on the evaporation rate of perspiration. A high relative humidity reduces the evaporation rate of perspiration. As a result, the effect on humans is similar to raising the environmental temperature. Humidity extremes can also contribute to other IAQ problems. High relative humidities can promote the growth of mold and mildew. Humidity levels that are too low, however, can contribute to irritated mucous membranes, dry eyes, and sinus discomfort.

As a general rule, office temperature and humidity are matters of human comfort. From a regulatory perspective, there are no regulations specifically addressing temperature and humidity in an office setting. However, the *OSHA Technical Manual*, section III, Chapter 2, subsection V, "Recommendations for the Employer," provides engineering and administrative guidance to prevent or alleviate indoor air quality problems. Air treatment is defined under the engineering recommendations as "the removal of air contaminants and/or the control of room temperature and humidity." OSHA recommends temperature control in the range of 68°F to 76°F and humidity control in the range of 20% to 60%.

Hazards for which OSHA does not have a specific standard are governed by section 5(a)(1) of the OSH Act (general-duty clause), which requires that employers provide employment and a place of employment that are free from recognized hazards that are causing or are likely to cause death or serious physical harm. Citations for violations of the general-duty clause are issued when the four components of this provision are present and when no specific OSHA standard has been promulgated to address the recognized hazard. These four components are as follows:

1. The employer failed to keep the workplace free of a "hazard."
2. The hazard was "recognized" either by the cited employer individually or by the employer's industry generally.
3. The recognized hazard was causing or was likely to cause death or serious physical harm.
4. There was a feasible means available that would eliminate or materially reduce the hazard.

Office temperature and humidity conditions are generally a matter of human comfort rather than hazards that could cause death or serious physical harm, and based on current interpretations, as of 2003 OSHA cannot cite the general-duty clause for personal discomfort (osha.gov/pls/oshaweb/owadisp.show_document?p_table=INTERPRETATIONS&p_id=24602).

Non-HVAC Building Components

Contaminants emitted from office equipment, new furnishings, and floor and wall coverings (volatile organic compounds, or VOCs) contribute to IAQ problems. Other indoor contaminant sources include cleaning materials; emissions from trash; pesticides; insects and other pests; and odors and VOCs from paint, caulk, adhesives, and personal care products.

Dry sink traps in plumbing systems allow the release of sewer gas to areas near the affected sinks. This phenomenon occurs in areas where sinks are seldom used or plumbing connections have been disconnected but not sealed during renovation projects. This pollutant pathway can be eliminated by running water in the sink often enough to prevent the water seal in the trap from drying out or by plugging the sink drains. Intermittent sewer gas odors can also result from defective wax seals on toilets.

It is important to control moisture and relative humidity in occupied spaces. Building components contributing to microbiological growth include soiled or water-damaged building materials, furnishings, and floor coverings. Water-damaged materials should be dried promptly (usually within 24 hours) or replaced (ceiling tiles). Equipment with water reservoirs or drain pans (i.e., humidifiers, refrigerators, and ventilation equipment) should be maintained to ensure that standing water drains properly.

POLLUTANT PATHWAYS AND DRIVING FORCES

To adversely impact the indoor environment, pollutants have to be transported from the point of generation to susceptible building populations. This usually requires a pollutant pathway and a driving force. Fundamental to these concepts are airflow patterns in buildings created by the combined action of building mechanical ventilation systems, occupant activity, and natural forces. All building components (walls, ceilings, floors, doors, windows, HVAC equipment, and occupants) interact to affect how air movement distributes pollutants in indoor air quality scenarios. During the investigation of an indoor air quality complaint, the building should be evaluated as a connecting network of many different pollutant pathways.

With the increase in computer networking and telecommunication wiring over the past few years, many pollutant pathways have been created. Other pollutant pathways include stairwells, elevator shafts, utility chases, pipe chases, electrical closets, dumbwaiters, doors, windows, ceiling plenums, and spaces between walls.

Occupants can serve as a driving force for pollutant sources. Colds, influenza, and other communicable diseases can be transmitted by person-to-person contact in the workplace. However, this is not necessarily a function of the workplace indoor air quality but rather of contact with an infected person that could occur anywhere.

Fans vented to the outside that intermittently remove air from a single room, such as bathrooms and kitchens, and portable fans for occupant comfort can act as driving forces because they alter the air movement patterns. Because operation of these fans tends to be intermittent, it is sometimes difficult to identify them as the driving force in an IAQ problem.

Positive Pressure

Positive pressure, a condition that exists when more air is supplied to a space than is exhausted, causes air to move from an area of higher pressure to surrounding areas of lower pressure. The airflow patterns created by these forces also move airborne pollutants from higher-pressure to lower-pressure areas. The most commonly recognized pollutant pathway is the HVAC system and associated components that distribute conditioned air to the building. The system ductwork acts as the pollutant pathway, and the fan component supplies the driving force (source of positive pressure). If ventilation systems and associated components are improperly maintained, they can also serve as breeding grounds for molds and bacteria, which can then be transmitted to the areas served by the system.

Natural Forces

Natural forces can dominate air movement between the interior and exterior of a building. Air movement associated with infiltration and natural ventilation is caused by air temperature differences between an indoor and outdoor environment.

Wind and weather patterns can also profoundly affect the distribution of pollutants in a building. Wind effects tend to be transient, creating localized areas of high pressure (on the wind side) and low pressure (on the opposite side) of buildings. Depending on the size and location of leakage openings in the building exterior, wind can affect the pressure relationships within and between rooms. Entry of outdoor air contaminants may be intermittent or variable, occurring only when the wind blows from the direction of the pollutant source.

Public and commercial buildings are usually designed to be positively pressurized so that unconditioned air does not enter through openings in the building envelope. This is done in part to alleviate indoor air quality problems. Whenever negative-pressure conditions exist, pollutants from outside the building migrate into the building.

Stack effect is the airflow that results from rising warm air creating a positive-pressure area at the top of a building and, subsequently, a negative-pressure area at the bottom of a building. The stack effect can overpower the mechanical system and disrupt ventilation and circulation in a building.

Stack effect exists whenever there is an indoor–outdoor temperature difference, and it becomes stronger as the temperature difference increases (heating season in the North and cooling season in the South). Stack effect usually has a more profound effect for taller buildings.

ELEMENTS OF AN EFFECTIVE IAQ MANAGEMENT PLAN

Effective IAQ management plans typically share several common elements:
- designated primary point(s) of contact
- written guidelines
- identification of potential problems and solutions
- timely communication of relevant issues to all affected parties.

A primary contact or contacts should be identified in an IAQ management plan in order to facilitate the communication process between affected parties. Occupants should be informed about which building management and staff members are responsible for responding to IAQ problems. A plan for maintaining effective lines of communication between management and building occupants should be developed and used on a regular basis (Figure 15–4). Staff should be educated about these procedures. Building occupants and/or tenants should be informed of these procedures and periodically reminded how to locate responsible staff and the methods available for communicating concerns. Operating

IAQ MANAGEMENT CHECKLIST

Building Name: _____ Date: _____

Address: _____

Completed by (name/title): _____

Use this checklist to make sure that you have included all necessary elements in your IAQ profile and IAQ management plan.

Item	Date Begun or Completed (as applicable)	Responsible Person (name, telephone)	Location ("NA" if the item is not applicable to this building)
IAQ PROFILE			
Collect and Review Existing Records			
HVAC design data, operating instructions and manuals			
HVAC maintenance and calibration records, testing and balancing reports			
Inventory of locations where occupancy, equipment, or building use has changed			
Inventory of complaint locations			
Conduct a Walk-Through Inspection of the Building			
List of responsible staff and/or contractors, evidence of training, and job descriptions			
Identification of area where positive or negative pressure should be maintained			
Record of locations that need monitoring or correction			
Collect Detailed Information			
Inventory of HVAC system components needing repair, adjustment, or replacement			
Record of control settings and operating schedules			
Plan showing airflow directions or pressure differentials in significant areas			
Inventory of significant pollutant sources and their locations			
SDSs for supplies and hazardous substances that are stored or used in the building			
Zone/Room record			
IAQ MANAGEMENT PLAN			
Select IAQ manager			
Review IAQ profile			

Figure 15–4. Use this checklist to ensure that you have included all necessary elements in your IAQ profile and IAQ management plan.

IAQ MANAGEMENT CHECKLIST *(Continued.)*			
Item	Date Begun or Completed (as applicable)	Responsible Person (name, telephone)	Location ("NA" if the item is not applicable to this building)
Assign Staff Responsibilities/Train Staff			
FACILITIES OPERATION AND MAINTENANCE			
Confirm that equipment operating schedules are appropriate			
Confirm appropriate pressure relationships between building usage areas			
Compare ventilation quantities to design, codes, and ASHRAE 62.1-2007			
Schedule equipment inspections per preventive maintenance or recommended maintenance schedule			
Modify and use HVAC Checklist(s); update as equipment is added, removed, or replaced			
Schedule maintenance activities to avoid creating IAQ problems			
Review SDSs for supplies; request additional information as needed			
Consider using alarms or other devices to signal need for HVAC maintenance (e.g., clogged filters)			
Housekeeping			
Evaluate cleaning schedules and procedures; modify if necessary			
Review SDSs for products in use; buy different products if necessary			
Confirm proper use and storage of materials			
Review trash disposal procedures; modify if necessary			
Shipping and Receiving			
Review loading dock procedures (Note: If air intake is located nearby, take precautions to prevent intake of exhaust fumes.)			
Check pressure relationships around loading dock			
Pest Control			
Consider adopting IPM methods			
Obtain and review SDSs; review handling and storage			
Review pest control schedules and procedures			
Review ventilation used during pesticide application			
Occupant Relations			
Establish health and safety committee or joint tenant/management IAQ task force			

Figure 15–4. Continued.

IAQ MANAGEMENT CHECKLIST *(Concluded.)*			
Item	Date Begun or Completed (as applicable)	Responsible Person (name, telephone)	Location ("NA" if the item is not applicable to this building)
Review procedures for responding to complaints; modify if necessary			
Review lease provisions; modify if necessary			
Discuss IAQ concerns with architects, engineers, contractors, and other professionals			
Implement IAQ management strategy when work is being conducted in an occupied building; communicate elements to occupants			
Obtain SDSs; use materials and procedures that minimize IAQ problems			
Schedule work to minimize IAQ problems			
Arrange ventilation to isolate work areas			
Use installation procedures that minimize emissions from new furnishings			
Eliminate smoking in the building			
If smoking areas are designated, provide adequate ventilation and maintain under negative pressure			
Work with occupants to develop appropriate nonsmoking policies, including implementation of smoking cessation programs			

Figure 15–4. Concluded.

schedules for activities that could affect IAQ should be provided to building occupants.

The IAQ management plan should contain IAQ policies in written form and a clear complaint–response procedure. Staff, occupants, and contractors should be educated about the potential influence their activities could have on indoor air quality. Lease arrangements and contracts should include provisions for dealing with potential impacts to the indoor environment. A written IAQ management plan should be a dynamic document that is updated as building conditions and occupant activities change. This written plan should outline procedures to alleviate existing IAQ problems and detail preventive actions, investigative procedures, and resolution of IAQ complaints and incidents as they occur.

The plan should address actions that will solve or mitigate specific IAQ problems and prevent them from recurring. General strategies to correct IAQ problems include the following:
- identifying sources and then removing or reducing the source, sealing or covering the source, or modifying the environment
- improving ventilation to provide outside air to occupants and to dilute and/or exhaust pollutants
- improving air filtration to clean air from outside and inside the building
- controlling occupant exposure through administrative approaches such as scheduling contaminant-producing activities during unoccupied periods.

Although an IAQ management plan could cover a number of buildings, certain aspects would have to be specific to a particular site because of differences in structures and building systems. Provisions should be made for operating and maintaining building mechanical systems at design capacity through an effective preventive maintenance program. Activities and projects that could affect air quality should be identified and monitored. Redecorating, renovation, or remodeling; relocation of personnel or functions; and construction projects should be evaluated for potential impact to IAQ.

Effectively implemented IAQ management plans usually prevent problems related to indoor air quality. When they do not, response actions should be in place to enable management to respond and mitigate in an effective manner.

There are value-added benefits to the implementation of the various elements of an IAQ management plan:
- demonstrated ongoing efforts to provide a safe indoor environment and that provide a strong legal and ethical position if problems do arise
- quicker and more cost-effective response if problems occur
- minimized adverse publicity
- greater peace of mind for management and building occupants
- protection of facilities and equipment
- better comfort for occupants.

IAQ Manager

The IAQ management plan should assign the responsibilities of dealing with IAQ issues to a single contact person. This IAQ manager should be an employee of the building owner or manager and should be responsible for coordinating all indoor air quality activities in the building. Having an IAQ manager with overall responsibility as a central point of contact makes it easier to manage building IAQ and to keep occupants informed and involved. This IAQ manager may seek assistance from outside contractors or consultants when faced with problems outside his or her areas of expertise, but should retain primary responsibility. Selection of consultants is discussed later in this chapter.

An effective IAQ manager can come from a variety of backgrounds. Indoor air quality is a field that requires the application of many disciplines to prevent and solve problems because no single field encompasses all the needed principles and skills. The person selected could be the facility manager, the building operating engineer, the health and safety director, the risk manager, or the human resource manager. Whatever his or her job title, the IAQ manager should be given sufficient authority to make decisions and implement improvements.

The IAQ manager will work as a team leader and should possess good communication skills. Where specific technical skills are lacking, another team member may compensate. Nevertheless, the IAQ manager must be familiar with the building's structure and functions and be sufficiently knowledgeable with IAQ issues to communicate effectively with occupants, facility personnel, and the building owner.

IAQ Profile

As part of the IAQ management plan, an IAQ profile for each building should be developed and focus on the following:
- identifying and reviewing records, such as blueprints and operating instructions
- conducting a walk-through inspection to document information on IAQ-related activities and conditions of building systems
- identifying possible pollutant sources throughout the building
- developing a noncomplaint baseline for the building.

Building documents including architectural and mechanical blueprints should be obtained and archived. If these documents are not readily available, the original architects, engineers, and/or equipment suppliers should be contacted for help. These documents are necessary for efficient and effective diagnosing of IAQ problems. If these documents are not available, they should be created either in-house or through an outside contractor.

Record Keeping

An important part of the IAQ management plan is the development and maintenance of a comprehensive, easy-to-use record-keeping system. These records will help the IAQ manager coordinate day-to-day IAQ activities as well as respond efficiently and effectively to IAQ problems. These records will also serve as documentation of program implementation and will provide historical foundation for the IAQ management plan as time evolves and personnel and equipment process change.

To assist in implementation of an IAQ management plan, the forms in Figures 15–1 through 15–4 are provided. Some or all of them may require adaptation to meet specific needs.

IAQ INVESTIGATIONS

A proactive indoor air quality management program should identify many potential impacts to the indoor environment before becoming a problem. And, in many cases, quick response to an IAQ complaint will result in resolution.

When an IAQ complaint warrants investigation past the initial stages, evaluation of a potential hazard can involve various methods including the following:
- direct observation of building activities and work practices
- confidential interviews
- monitoring of basic parameters (CO_2, temperature, relative humidity)
- medical testing or physical examinations
- measurement of contamination levels and determination of the extent of employee exposure
- review of employer's records of injuries and illnesses, medical tests, and job histories.

An IAQ investigation begins with one or more reasons for concern, usually as a result of occupant complaints. For many IAQ complaints, the cause of the complaint is identified and solved with the information collected during the initial review and visual inspection. Some problems could require detailed evaluation and testing by a team of experienced IAQ professionals. Many IAQ problems have more than one cause and may respond to (or require) several corrective actions.

The investigation starts with initial contact followed by a site evaluation. Interviews with occupants, information on potential pollutant sources, driving forces and pollutant pathways, the HVAC system, and any building operations going on in the area (and adjacent areas) should be evaluated. A visual inspection of the complaint area and surrounding areas should be conducted.

Interviews should be conducted in accordance with clearly defined objectives, including determination of the following:
- who has concerns or is exhibiting symptoms
- when the symptoms are occurring
- where in the building the onset of symptoms is associated
- what could be causing the problem.

Management should prepare and follow clear procedures for recording and responding to IAQ complaints, including the following:
- collecting information from the complainant (from interviews and/or daily logs)
- ensuring the confidentiality of information and records obtained from complainants
- determining the response capability of in-house staff
- identifying appropriate outside sources of assistance
- applying remedial action
- providing feedback to the complainant
- following up to ensure that remedial action has been effective.

After the initial gathering of information is complete, the scope of the investigation and potential endpoints should be defined as clearly as possible. The initial walk-through may provide enough information to resolve the problem.

Investigators and building occupants need to be clear about what constitutes a successful resolution to an IAQ problem. The goal of diagnosing an IAQ problem is to discover the cause of the problem so that corrective actions can be implemented. Communicating the findings and explaining the mitigation steps for resolution will be important in convincing building occupants that the problem has been resolved. Regardless of the outcome, building occupants may never be unanimous in their agreement with the findings of an IAQ investigation.

Developing and Testing Hypotheses

As the investigation evaluates identified pollutant sources and pollutant pathway movements, building occupants and staff employees are often a good source of ideas about the potential causes of the problem. A hypothesis should be developed from the gathered information and corrective actions initiated. Remedial action and follow-up evaluation will confirm whether the hypothesis is correct and whether corrective actions are successful.

Under some circumstances, detailed or sophisticated measurements of pollutant concentrations or ventilation quantities may be required. Outside assistance may be needed if repeated efforts fail to resolve the problem or if the information required calls for instruments and procedures that are not available in-house.

Diagnostic outcomes to avoid include evaluations that overemphasize measuring concentrations of pollutants and reports that link all the deficiencies in the building to the problem without considering their actual association with the complaints.

Decisions to perform IAQ measurements should be well justified. A decision to obtain IAQ-related measurements should follow logically from other investigative activities. Nonroutine measurements should not be conducted unless trying to characterize an identified pollutant source or in response to recommendations from a health care practitioner. Before starting to take measurements, investigators and building management need a clear understanding of how the results will be used. Without this understanding, it is impossible to plan appropriate sampling locations and times, instrumentation, and analysis procedures.

Concentrations low enough to comply with industrial occupational standards could still be harmful to building occupants. Also, industrial IAQ problems tend to arise from exposure to high levels of individual chemical compounds, so regulatory standards set limits for individual contaminants or contaminant classes. Exposure standards of this type are rarely encountered in buildings. Instead, IAQ investigators often find a large number of potential sources contributing low levels of many contaminants to the air.

Contaminants

The following pollutants have been associated with indoor air quality episodes. Much of this information is taken from the OSHA IAQ inspection protocol.

Carbon Monoxide

Carbon monoxide (CO) is a colorless, odorless, and tasteless gas produced as a by-product of incomplete oxidation of carbon during combustion processes. Significant sources can be worn or poorly adjusted and maintained combustion devices (e.g., boilers, furnaces) or an improperly sized, blocked, disconnected, or leaking flue. Fossil fuel engine exhausts (auto, truck, bus, or generator exhaust from attached garages, nearby roads, or parking areas) can also be a source.

There are no indoor air standards for CO. The U.S. National Ambient Air Quality Standards for outdoor air are 9 parts per million or ppm (40,000 g/m^3) for 8 hours, and 35 ppm for 1 hour. When CO levels are found above background levels (say, more than a few ppm), an investigation should try to identify potential sources and implement protective measures.

Acute health effects result from the formation of carboxyhemoglobin in the blood, thereby inhibiting oxygen intake. At moderate concentrations, dizziness, headache, nausea, cyanosis, cardiovascular effects, impaired vision, and reduced brain function may result. At higher concentrations, CO exposure can be fatal.

To minimize CO exposure, combustion equipment should be maintained and properly adjusted. Additional ventilation can be used as a temporary control measure when high levels of CO are expected for short periods. Vehicle use next to buildings should be carefully managed, especially in loading dock areas or near potential inlets such as windows and doors, to prevent the intake of the airborne contaminants. Techniques include a combination of signage, no-idling policies, and extending/rerouting exhaust vents. Equipment intended for outdoor use should never be allowed to be used indoors, including generators, grills, propane or kerosene heaters, and powered industrial equipment or trucks.

Formaldehyde

Formaldehyde (CH_2O) is a colorless, water-soluble gas. Materials containing formaldehyde are used widely in buildings, furnishings, carpeting, fabrics, and various consumer products. Formaldehyde-based resins are used in the manufacture of plywood, particleboard, textiles, glues, and adhesives. Urea-formaldehyde (UF) resins are commonly used in interior-grade plywood and pressed-wood furniture, cabinets, and shelving. The walls of some buildings have been insulated with urea-formaldehyde foam insulation (UFFI). Phenol-formaldehyde (PF) resins are normally used in exterior-grade products. Formaldehyde outgasses from all of these products. UF-based products typically emit higher levels of formaldehyde than PF-based materials. Tobacco smoke and other combustion products are secondary formaldehyde sources. Outgassing from materials appears to diminish over time.

Although no federal indoor air quality standard has been set for formaldehyde, OSHA has adopted a permissible exposure limit (PEL) of 0.75 ppm and an action level of 0.5 ppm. OSHA also requires labeling to inform exposed workers about the presence of formaldehyde in products entering workplaces that can cause levels to exceed 0.1 ppm. Some states have established a standard of 0.4 ppm in their codes for residences; others have established much lower recommendations (e.g., the California guideline is 0.05 ppm). Based on current information, it is advisable to mitigate formaldehyde that is present at levels higher than 0.1 ppm.

Formaldehyde has a pungent odor and is detected by many people at levels of about 0.1 ppm. Besides the annoyance, it also causes acute eye burning, hypersensitive or allergic reactions, and irritation of mucous membranes and the respiratory tract. OSHA regulates formaldehyde as a carcinogen.

Increased temperature and humidity accelerate outgassing, so ventilation may not be an effective means for mitigation. Some manufacturers are producing products with lower outgassing rates and surface treatments are being used to seal against outgassing, but long-term effectiveness has not been confirmed.

Volatile Organic Compounds

Volatile organic compounds (VOCs) include trichloroethylene, benzene, toluene, methyl ethyl ketone, alcohols, methacrylates, acrolein, polycyclic aromatic hydrocarbons, and pesticides. Hundreds of other VOCs are found in indoor air; sources include paints, cleaning compounds, mothballs, glues, photocopiers, "spirit" duplicators, signature machines, silicone caulking materials, wood preservatives, waxes, polishes, insecticides, herbicides, solvents, combustion products, asphalt, gasoline or diesel vapors, tobacco smoke, dried-out floor drains, perfumes, cosmetics, dry cleaned clothing, and other personal products. Nausea; dizziness; eye, respiratory tract, and mucous membrane irritation; headache; and fatigue are some of the acute health effects.

Although NIOSH has recommended occupational standards for many compounds, no standards have been set for VOCs in nonindustrial settings.

Several VOCs have been identified individually as causing acute and chronic effects at high concentrations. A few have been directly linked to cancer in humans, and others are suspected of causing cancer.

Where practical, uses of VOC sources should be minimized, and these materials should be stored in properly sealed containers Gasoline-powered lawn mowers and snow removal equipment should never be fueled inside any structure, including sheds.

Nitrogen Oxides

The two most prevalent oxides of nitrogen are nitrogen dioxide (NO_2) and nitric oxide (NO). Both are toxic gases, with

NO$_2$ being a highly reactive corrosive oxidant. NO gradually reacts with the oxygen in the air to form NO$_2$. Primary sources indoors include combustion products from unvented combustion appliances, gas furnaces, and other appliances; vented appliances with defective installations; welding; tobacco smoke; and gas- and diesel-engine exhausts.

No standards have been agreed on for nitrogen oxides in indoor air. ASHRAE and the U.S. National Ambient Air Quality Standards list 100 µg/m^3 (0.053 ppm) as the average long-term (1-year) limit for NO$_2$ in outdoor air.

Oxides of nitrogen have no sensory effect at low concentrations but produce short-term effects on airway activity. High concentrations can lead to acute lung dysfunction. Acute health effects include eye, respiratory, and mucous membrane irritation. Special risks exist for people with chronic bronchitis, emphysema, asthma, and children under age 2. Chronic effects are not well established.

Venting the NO$_2$ source to the outdoors is the most practical measure for existing conditions. Manufacturers are developing devices that generate lower NO$_2$ emissions.

Pesticides

Pesticides are used to control insects (insecticides), termites (termiticides), rodents (rodenticides), fungi (fungicides), and microbes (disinfectants). Widespread use of pesticides associated with food preparation areas, indoor plants, and living spaces can contribute to indoor air pollutants. Pesticides can be highly toxic and short-lived in the environment or less toxic and very persistent. All things being equal, the less-persistent types should be more desirable; those that degrade rapidly, such as the organophosphate insecticides, are extremely toxic and nonselective, encouraging emergence of resistant insects and elimination of their natural enemies. As a result, it is difficult to use pesticides that function without some drawback. Pesticide use entails risks as well as benefits; if used improperly, pesticides can contribute to indoor environmental pollution.

No indoor air concentration standards for pesticides have been set. Pesticide products must be used according to labels and instructions provided by the manufacturer. Pesticide labels also indicate the active ingredients, the potential for environmental damage, first-aid instructions, and storage and disposal directions. If chemical pesticides have been used in an area where an occupant has become sick, then the information should be provided to the physician who is handling the case.

Pesticides are applied in the form of bait, sprays, powders, crystals, and foggers. Because most indoor pesticide exposure occurs via inhalation of spray mists, chemical pesticides should be used only in a well-ventilated area. Instructions should be followed for recommended amounts of a pesticide. Mixing or dilution of pesticides should occur outdoors or in a well-ventilated area. Pesticides should be stored in tightly capped containers. After pesticide use, the area should be ventilated before occupancy.

Most pesticides contain a high percentage of inert ingredients, which are used to make pesticide formulations less concentrated and easier to apply. EPA has identified as toxic more than 100 substances used as inert ingredients in pesticides. When pesticides are used where the indoor environment can be impacted, it is important to identify the inert ingredients.

Nonchemical methods of pest control, also known as integrated pest management (IPM), should be used whenever possible. The primary strategy of IPM is to use a combination of pest prevention techniques, including sanitation, monitoring, exclusion, mechanical control, and structural repairs. It focuses on nonchemical techniques to control pests to minimize the hazard to people, property, and the environment. Routine, "preventive" treatments are not performed; chemical applications are made only when other methods have been exhausted.

Environmental Tobacco Smoke

Environmental tobacco smoke (ETS) from lighted cigars, cigarettes, and pipe tobacco is frequently the source of occupant complaints. Tobacco smoke consists of solid particles, liquid droplets, vapors, and gases resulting from tobacco combustion. Tobacco smoke can irritate the respiratory system and, in people with allergies or asthma, cause eye and nasal irritation, coughing, wheezing, sneezing, headache, and related sinus problems. Tobacco smoke contains several hundred toxic substances, including carbon monoxide, nitrogen dioxide, hydrogen sulfide, formaldehyde, ammonia, benzene, benzo(a)pyrene, tars, and nicotine. Tobacco smoke produces particulates in the respirable range that can adversely affect the indoor environment.

Tobacco smoke is usually controlled by prohibiting smoking in public spaces. Isolation of smokers is potentially effective, but requires careful management of ventilation. ASHRAE Standard 62.1-2007 provides specific construction and operational requirements for buildings with both ETS areas and ETS-free areas. This newest version of the standard also declines to specify what additional ventilation rates should be utilized in smoking areas until a risk-based contaminant level can be determined.

Because the organic material in tobacco doesn't burn completely, cigarette smoke contains more than 4,700 chemical compounds. Although OSHA has no regulation that addresses tobacco smoke as a whole, 29 CFR 1910.1000, Air Contaminants, limits employee exposure to several of the main chemical components found in tobacco smoke. In normal situations, exposures would not exceed these permissible exposure limits (PELs), and, as a matter of prosecutorial discretion, OSHA, as of 2003, will not apply the general-duty clause to ETS (osha.gov/pls/oshaweb/owadisp.show_document?p_table=INTERPRETATIONS&p_id=24602).

Microbial Contamination

Microorganisms and other biological contaminants include viruses, fungi, mold, bacteria, nematodes, amoeba, pollen, dander, and mites. They are also referred to as *microbiologicals* or *microbials*. These particulates range from less than 1 to several microns in size. To put this in perspective, a typical human hair is about 50 to 100 microns wide.

People, animals, and the environment produce biological materials. Dirty cooling coils, humidifiers, condensate drains, insulation, and ductwork can incubate bacteria and molds. Other sources of microbial contamination include air-handling system condensate, water-damaged materials, indoor areas with high humidity, damp organic material and porous wet surfaces, humidifiers, hot-water systems, drapery, bedding, carpet, outdoor excavations, plants, animal excreta, animals and insects, and food and food products.

Acute health effects from exposure to these contaminants range from allergic reactions such as hypersensitivity diseases (hypersensitivity pneumonitis, humidifier fever, allergic rhinitis, etc.) and infections such as Legionellosis and Pontiac fever. Cooling towers can be incubators of *Legionella* bacteria. The organism causing Legionnaires' disease can also be transmitted through airborne exposure. Tuberculosis, measles, staphylococcus infections, and influenza are other diseases known to be transmitted by air.

Molds and mildew are fungi that can cause discoloration and odor problems, deteriorate building materials, and exacerbate health problems such as asthma and allergies in susceptible individuals. Mold spores, nutrients, and moisture are the ingredients necessary for proliferation. In order to control or prevent biological contamination, the two elements essential for biological growth (nutrients and moisture) must be controlled. The likelihood of problems associated with biological growth can be prevented by promptly cleaning up the buildup of dirt, dust, and standing water, and by controlling relative humidity. During an indoor air quality investigation, carpet, furniture, partitions, ceiling tiles, weather stripping/caulk, and exposed cinder block or other foundation materials should be visually examined to determine whether they are in good condition and free from moisture damage.

Molds in the Environment

Molds live in the soil, on plants, and on dead or decaying matter. Outdoors, molds play a key role in the breakdown of leaves, wood, and other plant debris. Molds belong to the kingdom Fungi, and unlike plants, they lack chlorophyll and must survive by digesting plant materials, using plant and other organic materials for food. Without molds, our environment would be overwhelmed with large amounts of dead plant matter.

Molds produce tiny spores to reproduce, just as some plants produce seeds. These mold spores can be found in both indoor and outdoor air and settled on indoor and outdoor surfaces. When mold spores land on a damp spot, they may begin growing and digesting whatever they are growing on in order to survive. Because molds gradually destroy the things they grow on, organizations can prevent damage to building materials and furnishings and save money by eliminating mold growth.

Moisture control is the key to mold control. Molds need both food and water to survive; because molds can digest most things, water is the factor that limits mold growth. Indoors, molds often grow in damp or wet areas. Common sites for indoor mold growth include bathroom tile, basement walls, areas around windows where moisture condenses, and areas near leaky water fountains or sinks. Common sources or causes of water or moisture problems include roof leaks, delayed preventive maintenance, condensation associated with high humidity or cold spots in the building, localized flooding due to plumbing failures or heavy rains, slow leaks in plumbing fixtures, and malfunction or poor design of humidification or dehumidification systems. Uncontrolled humidity can also be a source of moisture leading to mold growth, particularly in hot, humid climates.

Health Effects and Symptoms Associated with Mold Exposure. When moisture problems occur and mold growth results, building occupants may begin to report odors and a variety of health problems, such as headaches, breathing difficulties, skin irritation, allergic reactions, and aggravation of asthma symptoms; all of these symptoms could potentially be associated with mold exposure. This should be verified by examinations by occupational health professionals with experience dealing with such ailments.

All molds have the potential to cause health effects. Molds produce allergens, irritants, and in some cases, toxins that may cause reactions in humans. The types and severity of symptoms depend, in part, on the types of mold present, the extent of an individual's exposure, the age of the individual, and his or her existing sensitivities or allergies.

Potential health effects associated with inhalation exposure to molds and/or the mycotoxins or spores they produce include allergic reactions (e.g., rhinitis and dermatitis or skin rash), asthma, and hypersensitivity pneumonitis. Specific reactions to mold growth can include the following:

Allergic reactions—Inhaling or touching mold or mold spores may cause allergic reactions in sensitive individuals. Allergic reactions to mold are common; these reactions can be immediate or delayed. Allergic responses include hay fever–type symptoms, such as sneezing, runny nose, red eyes, and skin rash (dermatitis). Mold spores and fragments can produce allergic reactions in sensitive individuals regardless of whether the mold is dead or alive. Repeated or single exposure to mold or mold spores may cause previously nonsensitive individuals to become sensitive. Repeated exposure has the potential to increase sensitivity.

Asthma—Molds can trigger asthma attacks in people who are allergic (sensitized) to molds. The irritants produced by molds may also worsen asthma in nonallergic (nonsensitized) people.

Hypersensitivity pneumonitis—Hypersensitivity pneumonitis may develop following either short-term (acute) or long-term (chronic) exposure to molds. The disease resembles bacterial pneumonia and is uncommon.

Irritant effects—Mold exposure can cause irritation of the eyes, skin, nose, throat, and lungs and sometimes can create a burning sensation in these areas.

Opportunistic infections—People with weakened immune systems (i.e., immune-compromised or immune-suppressed individuals) may be more vulnerable to infections by molds (as well as more vulnerable than healthy people to mold toxins). *Aspergillus fumigatus*, for example, has been known to infect the lungs of immune-compromised individuals. These individuals inhale the mold spores, which then start growing in their lungs. *Trichoderma* has also been known to infect immune-compromised children. Healthy individuals are usually not vulnerable to opportunistic infections from airborne mold exposure. However, molds can cause common skin diseases, such as athlete's foot, as well as other infections such as yeast infections.

Spores. Mold spores are microscopic (2–10 micrometers [μm]) and are naturally present in both indoor and outdoor air. Molds reproduce by means of spores. Some molds have spores that are easily disturbed and waft into the air and settle repeatedly with each disturbance. Other molds have sticky spores that cling to surfaces and are dislodged by brushing against them or by other direct contact. Spores may remain able to grow for years after they are produced. In addition, whether or not the spores are alive, the allergens in and on them may remain allergenic for years.

Cleaning and disinfecting with appropriate cleaners and antimicrobial agents provide some protection against mold growth. Mold spores are ubiquitous to the environment, and enough nutrients to support growth can be found in any building environment. If moisture is present, mold will thrive and reproduce. Therefore, moisture control is a primary strategy for reducing mold growth. Mold growth can occur when high relative humidity or the hygroscopic properties (the tendency to absorb and retain moisture) of building surfaces allow sufficient moisture to accumulate. Relative humidity should be maintained below 60% to minimize mold growth.

Good housekeeping practices and maintenance of building systems are important in controlling microbial growth. Adequate ventilation and good air distribution also help. Higher-efficiency air filters can be effective in removing microbial contaminants. Integrated pest management (see the section on pesticides earlier in this chapter) is a viable control measure. Cooling tower treatment protocols exist to reduce levels of *Legionella*. As mentioned earlier, maintaining indoor relative humidity below 60% can be helpful.

Toxic Molds. Under certain circumstances, some molds produce potent toxins called mycotoxins. Although some mycotoxins are well known to affect humans and have been shown to be responsible for human health effects, for many mycotoxins, little information is available and research is ongoing. Some mycotoxins cling to the surface of mold spores; others may be found within spores. More than 200 mycotoxins have been identified from common molds, and many more remain to be identified. Some of the molds that are known to produce mycotoxins are commonly found in moisture-damaged buildings. Exposure pathways for mycotoxins include inhalation, ingestion, and skin contact.

Two species—*Aspergillus versicolor* and *Stachybotrys atra* (chartarum)—have been well studied; some strains of *Stachybotrys atra* can produce one or more potent toxins. Preliminary reports from an investigation of an outbreak of pulmonary hemorrhage in infants suggested an association between pulmonary hemorrhage and exposure to *Stachybotrys chartarum*. Review of the evidence of this association at the Centers for Disease Control and Prevention (CDC) resulted in a published clarification stating that such an association was not established. Research on the possible causes of pulmonary hemorrhage in infants continues. Consult the CDC for more information on pulmonary hemorrhage in infants.

Aflatoxin B_1 is perhaps the most well known and studied mycotoxin. It can be produced by the molds *Aspergillus flavus* and *Aspergillus parasiticus* and is one of the most potent carcinogens known. Ingestion of aflatoxin B_1 can cause liver cancer. There is also some evidence that inhalation of aflatoxin B_1 can cause lung cancer. Aflatoxin B_1 has been found on contaminated grains, peanuts, and other human and animal foodstuffs. However, *Aspergillus flavus* and *Aspergillus parasiticus* are not commonly found on building materials or in indoor environments.

Much of the information on the human health effects of inhalation exposure to mycotoxins comes from studies done in the workplace and some case studies or case reports. Many symptoms and human health effects attributed to inhalation of mycotoxins have been reported, including mucous membrane irritation, skin rash, nausea, immune system suppression, acute or chronic liver damage, acute or chronic central nervous system damage, endocrine effects, and cancer. More studies are needed to get a clear picture of the health effects related to most mycotoxins. However, it is clearly prudent to prevent exposure to molds and mycotoxins.

Some molds can produce several toxins, and some molds produce mycotoxins only under certain environmental conditions. The presence of mold in a building does not necessarily mean that mycotoxins are present or that they are present in large quantities.

Information on ingestion exposure, for both humans and animals, is more abundant—a wide range of health effects has been reported following ingestion of moldy foods, including liver damage, nervous system damage, and immunological effects.

Microbial Volatile Organic Compounds. Some compounds produced by molds are volatile and are released directly into the air. These are known as microbial volatile organic compounds (mVOCs). Because these compounds often have strong and/or unpleasant odors, they can be the source of odors associated with molds. Exposure to mVOCs from molds has been linked to symptoms such as headaches, nasal irritation, dizziness, fatigue, and nausea. Research on mVOCs is still in the early phase.

Glucans or Fungal Cell Wall Components (Also Known as ß-1,3-D-Glucans). Glucans are small pieces of the cell walls of molds that may cause inflammatory lung and airway reactions. These glucans can affect the immune system when inhaled. Exposure to very high levels of glucans or dust mixtures including glucans may cause a flu-like illness known as organic dust toxic syndrome (ODTS). This illness has been primarily noted in agricultural and manufacturing settings.

EPA publication EPA 402-K-01-001, *Mold Remediation in Schools and Commercial Buildings*, provides guidance and more specific details for addressing mold contamination.

Dust

Dust refers to solid particles and aerosols (usually less than 100 µm) generated by grinding, pulverizing, crushing, handling, and wearing of organic and inorganic materials (e.g., rock, metal, wood, microbials). Suspension of particles of any solid material in air can impact the indoor environment. Each specific dust has different characteristics (amount, size, weight, and shape) and requires varying methods of control. Most dust particles are heavier than air and tend to settle out of the environment.

Dust particles are typically classified by size as nuisance or respirable dust. Nuisance dust is typically not a problem in the indoor environment because of its relatively large size. Respirable particles are those that penetrate into and are deposited in the nonciliated portion of the lung. The upper respiratory system filters out most particles larger than 5 µm. Particles greater than 10 µm in aerodynamic diameter are not respirable and are typically filtered by the nose or upper airways.

Air cleaners are used to remove particles from indoor air. Their efficiency depends not only on the airflow rate through the cleaner and the efficiency of its particle capture mechanism, but also on factors such as the following:

- the mass of the particles entering the device
- the characteristics of the particles (e.g., their size)
- the degradation rate of the efficiency of the capture mechanism caused by loading
- whether some of the air entering the unit bypasses the internal capture mechanism
- how well the air leaving the device is mixed with air in the room before reentering the device.

ASHRAE Standard 62.1-2007 recommends that air filters be selected and replaced based on operations and maintenance (O&M) practices specified by the manufacturer. The ANSI/ASHRAE Standard 52.2 User Guide was created by the National Air Filtration Association (NAFA), a group of over 600 air filter distributors, manufacturers, and engineers. The Guide and the application of a particle-based contaminant removal standard prescribed by ANSI/ASHRAE Standard 52.2-2007, "Method of Testing General Ventilation Air-Cleaning Devices for Removal Efficiency by Particle Size," are intended to assist end users and specifiers in selecting appropriate air filtration products and understanding the MERV values in the 52.2 test reporting. Two commonly used methods to evaluate filters are arrestance and atmospheric dust spot tests. Arrestance is used to evaluate low-efficiency filters; the atmospheric dust spot test is used to rate medium-efficiency air cleaners (both filters and electronic air cleaners).

Arrestance is determined by passing a standardized

dust made up of various-size particles through a filter and then determining the weight fraction of the dust removed. Arrestance values are usually high because most of the weight of the standardized dust is attributed to the larger particles. The arrestance values are of little value when determining a filter's capability to remove the smaller particles that are usually implicated in indoor air quality scenarios.

The atmospheric dust spot test is conducted by passing atmospheric dust through a filter. The removal rate is based on the air cleaner's capability to reduce soiling of a clean paper target and is related to the removal of fine particles from the air. As a result, this test does measure smaller particles. However, the variability of atmospheric dusts used for testing may result in different efficiencies for the same filter depending on geographic location and time. Until a new standard is proposed, filters of 25% to 30% as rated by ASHRAE Standard 52.1 are recommended.

For cases in which more efficient particulate cleaning is necessary, the Military Standard 282 (U.S. Department of Defense 1956) test is used to rate high-efficiency air filters (those with efficiencies above 98%). The term *HEPA* (high-efficiency particulate air) is commonly used to describe filters with this level of efficiency. HEPA filters are a subset of high-efficiency filters and are typically rated using the DOP testing method (i.e., the percentage removal of 0.3-μm particles of the chemical dioctylphthalate [DOP]). One standard-setting organization defines a HEPA filter as having a minimum particle collection efficiency of 99.97% using the DOP testing method.

In 2013, OSHA proposed a rule to change the standards for exposure to silica dust. OSHA's proposed rule seeks to lower worker exposure to crystalline silica, which causes silicosis, an incurable lung disease. Leading scientific organizations, including the American Cancer Society, have also confirmed the causal relationship between silica and lung cancer, as well as other lung and kidney diseases. Exposure typically occurs when construction workers create respirable-size particles when chipping, cutting, drilling, or grinding objects that contain crystalline silica. Applying water to a saw blade when cutting materials that contain crystalline silica—such as stone, rock, concrete, brick, and block—substantially reduces the amount of dust created during these operations.

Carbon Dioxide

All combustion processes and human metabolic processes are sources of carbon dioxide (CO_2), including unvented gas and kerosene appliances, improperly vented devices, operations that produce combustion products, and human respiration. Vehicles or other gasoline or diesel engines can be a source of CO_2.

Carbon dioxide is frequently used as a surrogate measure for the effectiveness of an HVAC system in supplying fresh air. Typical values for ambient CO_2 levels in urban settings are about 350 ppm. Concentrations of CO_2 from people are always present in occupied buildings. When indoor levels rise above 1,000 ppm, the effectiveness of the building HVAC system should be evaluated. When CO_2 levels are below 600 ppm, occupant complaints tend to diminish.

Carbon dioxide is a simple asphyxiant, and at concentrations above 1.5% (15,000 ppm), some loss of mental acuity has been noted. These levels would not normally be encountered in an indoor setting. Some acute health effects include difficulty concentrating, drowsiness, and an increased respiration rate.

Miscellaneous Pollutants

Ozone (O_3) is a by-product of smog, electrical arcing, and operation of copy machines and electrostatic air cleaners. Acute health effects include eye, respiratory tract, and mucous membrane irritation and aggravation of chronic respiratory diseases. Although ozone has a short half-life and disperses quickly, it can still be a problem in some scenarios.

Acetic acid can be emitted from x-ray development equipment or silicone caulking compounds and can cause eye, respiratory tract, and mucous membrane irritation.

Miscellaneous inorganic gases include ammonia, hydrogen sulfide, and sulfur dioxide. Sources of inorganic gases include microfilm equipment, window cleaners, acid drain cleaners, combustion products, tobacco smoke, and blueprint equipment. Health effects include eye, respiratory tract, and mucous membrane irritation, and aggravation of chronic respiratory diseases.

Synthetic fibers from fibrous glass and mineral wool can cause dermatitis and irritation to the eyes, skin, and lungs.

Radon, Asbestos, and Lead

Although asbestos and radon are evaluated in some IAQ investigations, acute health effects are not associated with these contaminants and are not usually associated with indoor air quality scenarios. However, building occupants often have concerns about the effects of these contaminants; therefore, these concerns should be addressed.

Radon emanates from the ground beneath buildings, building materials, and groundwater. Radon is a colorless, odorless, and tasteless radioactive gas, the first decay product of radium-226. It decays into solid alpha emitters that can be inhaled directly or attached to dust particles that are inhaled. The unit of measure for radon is picocuries per liter (pCi/L).

Radium is ubiquitous in the earth's crust in widely varying concentrations. Well water can have high concentrations of radon. Masonry building blocks can have radium concentrations. The earth around buildings tends to be the principal source of indoor radon. Radon penetrates cracks and drain openings in foundations into basements and crawl spaces. Water containing radon outgasses into spaces when drawn for use indoors. Some building materials outgas radon, some of which may enter buildings.

EPA recommends taking action to mitigate radon if levels exceed 4 pCi/L. Chronic exposure may lead to increased risk of lung cancer from alpha radiation. In some localities, newer buildings are constructed with passive mitigation systems. If an active radon mitigation system has been installed, electrical/mechanical components (i.e., exhaust fans) need to be inspected regularly to ensure they continue operating as designed.

Asbestos is found in many building materials: insulation, floor tiles, drywall compounds, and reinforced plaster. Renovation and maintenance operations can cause asbestos to be dislodged and become airborne. Evaluation of employee exposure to asbestos is normally covered under the OSHA asbestos standard. An overview of asbestos regulations is provided in Chapter 4, Regulatory History.

Lead is a highly toxic metal. Drinking water, food, contaminated soil and dust, and air are vehicles for lead exposure. Lead-based paint is a common source of lead exposure in the indoor environment. The Consumer Product Safety Commission (CPSC) banned lead in paint after 1978. All other standards are for outdoor air or industrial workplaces.

Lead can cause serious damage to the brain, kidneys, nervous system, and red blood cells. Lead exposure in children can result in delays in physical development, lower IQ levels, shortened attention spans, and increased behavioral problems. Because children are particularly vulnerable to lead, workplaces with day-care facilities should be evaluated to determine that lead will not be a problem. Exposure to lead is estimated by measuring levels in the blood and through neurological testing.

Preventive measures to reduce lead exposure include minimizing operations that generate lead-contaminated dust and ensuring proper cleanup when contamination occurs. Common renovation, repair, and painting activities that disturb lead-based paint (like sanding, cutting, replacing windows, and more) can create hazardous lead dust and chips that can be harmful to adults and children. New EPA regulations require specific work practices during renovation, repair, and painting (RRP) activities in residential locations or where children are regularly present, such as schools and child-care facilities.

RESOLUTION

Solutions to IAQ problems usually include combinations of the following methodologies and approaches:
- pollutant source removal
- increasing ventilation rates and air distribution
- exhausting contaminated air
- air cleaning
- education and communication.

Pollutant source removal or modification is probably the most effective approach to resolving an IAQ problem when sources are known and control is feasible. Examples include routine maintenance of HVAC systems (e.g., periodic cleaning or replacement of filters); institution of smoking restrictions; venting contaminant source emissions to the outdoors; storage and use of paints, adhesives, solvents, and pesticides in well-ventilated areas; use of pollutant sources during periods of nonoccupancy; and allowing time for building materials in new or remodeled areas to off-gas pollutants before occupancy. Several of these options may be exercised at the same time.

Increasing ventilation rates and air distribution often can be a cost-effective means of reducing indoor pollutant levels. HVAC systems should be designed, at a minimum, to meet ventilation standards in local building codes; however, many systems are not operated or maintained to ensure that these design ventilation standards are met. In many buildings, IAQ can be improved by operating the HVAC system to at least its design parameters, and to ASHRAE Standard 62.1-2007, if possible.

When there are strong pollutant sources, local exhaust ventilation may be appropriate to exhaust contaminated air directly from the building. Local exhaust ventilation is particularly effective in removing pollutants that accumulate in specific areas such as restrooms, copy rooms, and printing facilities.

In addition to source control and ventilation, air cleaning can be a useful means of control, but it has certain limitations. Particle-cleaning devices such as furnace filters are inexpensive but do not effectively capture small particles; high-performance air filters capture the smaller, respirable particles but are relatively expensive to install and operate. A shortcoming of mechanical filters is that they do not remove gaseous pollutants. Adsorbent beds can remove some gaseous pollutants. However, this can require frequent replacement of the adsorbent material and can be an expensive undertaking. Air cleaners can be useful but should be used in conjunction with other control methods.

Education and communication can be the two most important elements for both remedial and preventive indoor air quality management programs. When building occupants, management, and building staff understand the causes and consequences of IAQ problems and communicate effectively, problems can be prevented or solved if they do occur.

Usually, the most effective way to improve indoor air quality is to eliminate individual sources of pollution or to reduce their emissions. Some sources can be sealed or enclosed. Others can be adjusted to decrease the amount of emissions. In many cases, source control is usually a more cost-efficient approach to protecting indoor air quality than increasing ventilation because increasing ventilation can increase energy costs.

Evaluating Solutions

The most economical and successful solutions to IAQ problems are those in which the operating principle of the correction strategy makes sense and is suited to the problem. If a specific source of contaminants has been identified, treatment at the source (e.g., by removal, sealing, or local exhaust) is usually a more appropriate correction strategy than dilution of the contaminant by increased general ventilation. If the IAQ problem is caused by the introduction of outdoor air that contains contaminants, increased general ventilation will only make the situation worse (unless the outdoor air is cleaned).

To help ensure that mitigation of IAQ problems is successful, the following criteria should be evaluated:
- permanence
- durability
- operating principles—both initial and operating costs
- control capacity
- ability to institutionalize the solution
- conformity with codes.

Persistent Problems

Even the best-planned investigations and mitigation actions may not resolve the problem. An investigation identifying one or more apparent causes for the problem and implementation of control measures may not have caused a noticeable reduction in the concentration of the contaminant or improvement in ventilation rates or efficiency. In some cases, complaints may persist even though the implementation of control measures was successful.

If attempts to control the problem are unsuccessful, it may be advisable to seek outside assistance. The problem may be fairly complex, occur only intermittently, or cross the borders that divide traditional fields of knowledge. In some cases, it is even possible that poor indoor air quality is not the actual cause of the complaints. Bringing in a new perspective at this point can be effective.

EVALUATING POTENTIAL CONSULTANTS

The IAQ diagnostic process begins when a complaint is registered or an IAQ problem is identified. Many problems can be simple to diagnose, requiring a basic knowledge of IAQ and some common sense. In some cases, outside assistance is needed. Professional help may be necessary or desirable in the following situations:
- Mistakes or delays could have serious consequences (e.g., health hazards, liability exposure, regulatory sanctions, and adverse publicity).
- In-house expertise is not available to effectively evaluate IAQ issues.
- Building management feels that an independent investigation would be better received or more effectively documented than an in-house investigation.
- Investigation and mitigation efforts by in-house staff have not relieved the IAQ problem.
- Preliminary findings indicate the need for measurements that require specialized equipment and training beyond in-house capabilities.
- An examination of affected occupants by a physician or other occupational health practitioner has indicated that indoor air pollutants may be the causative factor in the identified aliments.

There are many commercial consultants that could provide assistance in an IAQ investigation (they will be listed under the following headings: "Engineers," "Environmental Services," "Laboratories—Testing," or "Industrial Hygienists" in phone books or Internet listings for services). Be sure to obtain and check references. Many local, state, or federal government agencies can provide expert assistance or direction in solving IAQ problems, or your local or state health or air pollution agencies may have lists of firms offering IAQ services. Often recommendations from others who have dealt successfully with IAQ situations will serve as a good starting point when trying to identify potential consultants.

It is important to hire consultants with specific expertise in solving difficult IAQ problems. Even certified professionals from disciplines closely related to IAQ issues (such as industrial hygienists, ventilation engineers, and toxicologists) may not have the specific expertise needed to investigate and resolve indoor air problems. Individuals or groups that offer services in this evolving field should be questioned closely about their related experience and methodologies. Firms and individuals working in IAQ may come from a variety of disciplines. Often, a multi-disciplinary team of professionals is needed to investigate and resolve an IAQ problem. Typically, the skills of HVAC engineers and industrial hygienists are useful for this type of investigation, although input from other disciplines such as chemistry, chemical engineering, architecture, microbiology, or medicine may be needed. If problems other than indoor air quality are involved, experts in lighting, acoustic design, ergonomics, interior design, psychology, or other fields may be helpful in resolving occupant complaints about the indoor environment.

It is important to ensure that IAQ consultants conducting investigations do not deal with health-related issues unless they are qualified health care specialists with specialized knowledge of such matters. Often, inappropriate communication regarding clinical diagnoses by unqualified individuals to building occupants can result in problems that can be difficult to address.

A qualified IAQ investigator should have appropriate experience, demonstrate a broad understanding of indoor air quality problems and causative conditions, and use a phased diagnostic approach. Inappropriately designed studies may lead to conclusions that are either false-negative (i.e., falsely conclude that there is no problem associated with the building) or false-positive (i.e., incorrectly attribute the cause to building conditions). Considerable care should be exercised when interviewing potential consultants to avoid those subscribing to the following strategies:

- an evaluation that overemphasizes measuring concentrations of pollutants and comparing them to regulatory standards for industrial workplaces
- a report that lists a series of major and minor building deficiencies and links all the deficiencies to the problem without considering their actual association with the complaints.

Most of the criteria used in selecting a professional to provide indoor air quality services are similar to those used for other professionals:

- experience in solving similar problems
- ability to communicate results effectively and responsibly
- training and experience of the individuals who would be responsible for the work
- quality of interview and proposal
- company reputation
- references
- knowledge of local codes and regional climate conditions
- cost.

REGULATORY REQUIREMENTS

Pollutant-Related Regulations

The federal government has a long history of regulating outdoor air quality and the concentrations of airborne contaminants in industrial settings. At this time, there are few federal regulations for airborne contaminants in nonindustrial settings.

EPA deals with outside air pollution by way of the Clean Air Act (CAA) of 1990 and its Amendments through ambient air quality standards known as the National Ambient Air Quality Standards (NAAQS). These standards address a number of contaminants that could affect the indoor environment and IAQ. In Title 40 CFR, Part 61, EPA lists the current National Emission Standards for Hazardous Air Pollutants (NESHAP). These standards also address a limited number of contaminants that can affect IAQ. These standards are used by HVAC design professionals to categorize outdoor air quality in local regions and to help with their system designs. The EPA and its Office of Air and Radiation, Office of Atmospheric Programs, and Indoor Air Quality Division have no specific regulatory authority over IAQ but provide policy, recommendations, guidelines, and coordination on IAQ issues.

OSHA is the federal agency responsible for enforcing workplace safety and health standards. In the past, OSHA focused primarily on industrial worksites, but it has recently broadened efforts to address other worksite hazards (targeted hazards include ergonomics and indoor air quality). In an industrial environment, specific chemicals released by industrial processes can be present in high concentrations. It has been possible to study the health effects of industrial exposures and establish regulations to limit those exposures. OSHA enforces industrial exposure standards on exposure to chemicals based on a 40-hour week and an 8-hour day to protect workers. The worker permissible exposure limit (PEL) is measured as a time-weighted average (TWA), an action level (usually one-half of the PEL), and a short-term exposure limit (STEL). These standards seldom apply to IAQ issues, but they can provide a point of reference when no other meaningful standard exists. However, it is important to be careful when using these standards as a point of reference in addressing IAQ incidents.

In 1994, OSHA introduced a proposed rule regarding IAQ in nonindustrial environments. In late 2001, OSHA withdrew its indoor air quality proposal and terminated the rulemaking proceedings. Although OSHA does not specifically regulate IAQ, it will investigate building IAQ complaints from workers. The general-duty clause (section 5[a][1] of the OSH Act) is often cited, which requires an employer to "furnish to each of his employees employment and a place of employment which arc free from recognized hazards that are causing or are likely to cause death or serious physical harm to his employees." However, as indicated earlier, OSHA will not use the general-duty clause to cite concerns with personal comfort. OSHA efforts impact IAQ by enforcing existing workplace standards dealing with asbestos, lead, dust, record keeping, hazard communication, and other chemicals that could affect IAQ.

State indoor air quality legislation varies widely. Some states and local jurisdictions have established regulations regarding indoor air quality and specific pollutants, such as radon and lead. Forty-nine states and the District of Columbia have enacted indoor air provisions restricting smoking in public places (offices, restaurants, and government buildings); however, only 28 states and the District of Columbia have comprehensive smoke-free air laws.

Proposed legislation has included the use of commissions, educating residents, and requiring sellers of properties to disclose environmental hazards such as mold, lead, asbestos, and radon. Some states have established government-supported programs for dealing with indoor air quality issues.

Some states have passed specific legislation for addressing IAQ issues and seem to be quite active in pursuing legislation for the future:

- Florida has enacted legislation to regulate the mold inspection industry, including licensing requirements for mold remediators.
- California provides uniform standards for mold, requires education efforts, and requires mold disclosure on commercial and residential property transactions.
- Texas House Bill 2007 requires indoor air quality testing and compliance with mandatory indoor air quality guidelines in newly constructed schools or those undergoing major renovation.
- Illinois, Maine, Massachusetts, New York, and Wisconsin have introduced bills for mold study, assessment, and remediation.
- Several bills relating to indoor air quality in school facilities that will require inspections, evaluation programs, and ongoing maintenance were introduced in Alabama and Indiana.
- In New York, legislation was proposed to (1) amend public health law by establishing guidelines for the operation of a building's HVAC systems in relation to protecting indoor air quality; (2) ensure that people who experience adverse affects from poor indoor air quality have a means for communicating their problems; and (3) research mold, including standards for its prevention, detection, and remediation.
- Many states now require radon testing of certain public facilities, disclosure of radon conditions at the time of sale, radon mitigation, radon standards, and other radon-associated initiatives.
- Many states also require some type of inspection and evaluation of air quality in state buildings and schools.
- A New Hampshire statute requires school principals to conduct an annual IAQ investigation of all school buildings.
- In Maryland, Governor Glendening signed Senate Bill 283, which established a task force on indoor air quality. The task force provided a report that examined the nature, location, and health risks posed to workers by toxic molds, spores, and other substances located in the HVAC systems of office buildings.
- In New Jersey, rules require that employers establish and implement a preventive HVAC maintenance plan, undertake certain prevention and cleanup practices for microbial contamination, protect indoor air quality during renovation, respond to IAQ complaints, and keep and make available records of maintenance activities.
- Statutes require school systems in Nevada, Connecticut, Vermont, New York, Illinois, Hawaii, Iowa, Maryland, and the District of Columbia to use "green" or "environmentally friendly" cleaning products.
- California, West Virginia, and Maryland adopted language regarding carbon monoxide monitoring in schools.

Ventilation-Related Regulations

Ventilation is the other major influence on indoor air quality that is subject to regulation. The federal government does not regulate ventilation in nonindustrial settings. However, many state and local governments do regulate ventilation system capacity through their building codes.

Building codes have been developed to promote good construction practices and prevent health and safety hazards. Professional associations such as ASHRAE and the National Fire Protection Association (NFPA) develop recommendations for appropriate building and equipment design and installation. Relevant IAQ ANSI/ASHRAE standards include the following:

ANSI/ASHRAE Standard 52.2-2007, "Method of Testing General Ventilation Air-Cleaning Devices for Removal Efficiency by Particle Size," establishes a test procedure for evaluating the performance of air-cleaning devices as a function of particle size.

ANSI/ASHRAE Standard 55-2004, "Thermal Environmental Conditions for Human Occupancy," specifies the combinations of indoor space environment and personal factors that will produce thermal environmental conditions acceptable to 80% or more of the occupants.

ANSI/ASHRAE Standard 62.1-2007, "Ventilation for Acceptable Indoor Air Quality," specifies minimum ventilation rates and indoor air quality that will be acceptable to human occupants and are intended to minimize the potential for adverse health effects.

ANSI/ASHRAE Standard 62.2-2007, "Ventilation and Acceptable Indoor Air Quality in Low-Rise

Residential Buildings," defines the roles of and minimum requirements for mechanical and natural ventilation systems and the building envelope intended to provide acceptable indoor air quality in low-rise residential buildings. It is the only nationally recognized IAQ standard developed solely for residences.

These recommendations acquire the force of law when adopted by state or local regulatory bodies. There is generally a time lag between the adoption of new standards by consensus organizations such as ASHRAE and the incorporation of those new standards as code requirements. Local code enforcement officials or consultants can provide information to learn about the code requirements that apply to your building. In general, building code requirements are enforceable only during construction and renovation. Code requirements change over time (as code organizations adapt to new information and technologies), and buildings are usually not required to modify their structure or operation to conform to the new codes. Indeed, many buildings do not operate in conformance with current codes, or with the codes they had to meet at the time of construction.

As discussed earlier in the Dust section, there is a federal standard addressing high-efficiency air filters. Also, standards provided by independent trade associations outside the federal government can be used as guidance in choosing air cleaners for reduction of airborne particles in indoor air.

SUMMARY

- Indoor air quality (IAQ) is a major component of indoor environmental quality (IEQ). IAQ refers to the air inside a building and to its suitability for human consumption. The condition of a building's indoor environment is based on complex interactions between occupant activities, the surrounding climate and weather, building structure and mechanical systems, and contaminant sources.
- Many factors that could be encountered in these environments may contribute to complaints from occupants regarding indoor air quality, including tobacco smoke, dust particles, objectionable odors, airborne compounds generated from a variety of sources, microbial contamination, poor air circulation, thermal discomfort, humidity, job pressures, lighting, workstation design, and noise.
- Chronic problems attributed to poor indoor air quality tend to be a little more difficult to diagnose and are quite often characterized as either sick-building syndrome or building-related illnesses.
- Pollutant sources can include occupant activities such as housekeeping, maintenance, remodeling and renovation, and painting; outside sources; building systems; or equipment, furnishings, and wall and floor coverings.
- To adversely affect the indoor environment, pollutants have to be transported from the point of generation to susceptible building populations. This requires a pollutant pathway and a driving force: airflow patterns in buildings created by the combined action of building mechanical ventilation systems, occupant activity, and natural forces. All building components (walls, ceilings, floors, doors, windows, HVAC equipment, and occupants) interact to affect how air movement distributes pollutants in indoor air quality scenarios. When investigating an indoor air quality complaint, the building should be evaluated as a connecting network of many different pollutant pathways.
- Effective IAQ management plans typically share several common elements, including the designated primary point(s) of contact, timely communication of relevant issues to all affected parties, written guidelines, and identification of potential problems and solutions.
- When an IAQ complaint warrants investigation past the initial stages, evaluation of a potential hazard can involve various methods, including direct observation of building activities and work practices; confidential interviews; monitoring of basic parameters (CO_2, temperature, relative humidity); medical testing or physical examinations; measurement of contamination levels and determination of the extent of employee exposure; and review of the employer's records of injuries and illnesses, medical tests, and job histories.
- Successful solutions to IAQ problems usually include combinations of the following methodologies and approaches: pollutant source removal, increasing ventilation rates and air distribution, exhausting contaminated air, air cleaning, education, and communication.

The following glossary is provided in large part courtesy of the EPA website (epa.gov/iaq/glossary.html).

acid aerosol: Acidic liquid or solid particles that are small enough to become airborne. High concentrations of acid aerosols can be irritating to the lungs and have been associated with some respiratory diseases, such as asthma.

action level: A term used to identify the level of indoor radon at which remedial action is recommended. (EPA's current action level is 4 pCi/L.)

action packet: Refers to the IAQ Tools for Schools Action Kit, which contains numerous products to help school personnel implement an effective yet simple IAQ program in their school.

AHU: See *air-handling unit*.

air cleaning: An IAQ control strategy to remove various airborne particulates and/or gases from the air. The three types of air cleaning most commonly used are particulate filtration, electrostatic precipitation, and gas sorption.

air exchange rate: The rate at which outside air replaces indoor air in a space. It is expressed in one of two ways: the number of changes of outside air per unit of time (air changes per hour [ACH]) or the rate at which a volume of outside air enters per unit of time (cubic feet per minute [cfm]).

air-handling unit (AHU): For purposes of this glossary, refers to equipment that includes a blower or fan; heating and/or cooling coils; and related equipment such as controls, condensate drain pans, and air filters. Does not include duct work, registers, grilles, boilers, or chillers.

air passages: Openings through or within walls, through floors and ceilings, and around chimney flues and plumbing chases that permit air to move out of the conditioned spaces of the building.

allergen: A substance capable of causing an allergic reaction because of an individual's sensitivity to that substance.

allergic rhinitis: Inflammation of the mucous membranes in the nose that is caused by an allergic reaction.

animal dander: Tiny scales of animal skin.

antimicrobial: Describes an agent that kills microbial growth. See *disinfectant*, *sanitizer*, and *sterilizer*.

biological contaminants: Agents derived from, or that are, living organisms (e.g., viruses, bacteria, fungi, and mammal and bird antigens) that can be inhaled and can cause many types of health effects including allergic reactions, respiratory disorders, hypersensitivity diseases, and infectious diseases. Also referred to as *microbiologicals* or *microbials*.

breathing zone: The area of a room in which occupants breathe as they stand, sit, or lie down.

building envelope: Elements of the building, including all external building materials, windows, and walls, that enclose the internal space.

building-related illness (BRI): A diagnosable illness whose symptoms can be identified and whose cause can be directly attributed to airborne building pollutants (e.g., Legionnaires' disease, hypersensitivity pneumonitis). Also, a discrete, identifiable disease or illness that can be traced to a specific pollutant or source within a building. (Contrast with *sick-building syndrome*.)

ceiling plenum: The space below the flooring and above the suspended ceiling that accommodates the mechanical and electrical equipment and that is used as part of the air distribution system. The space is typically kept under negative pressure.

central air-handling unit (central AHU): An air-handling unit that serves more than one area.

cfm: Cubic feet per minute; the amount of air, in cubic feet, that flows through a given space in 1 minute. 1 cfm equals approximately 2 liters per second (L/s).

chemical sensitization: Evidence suggests that some people may develop health problems characterized by effects such as dizziness, eye and throat irritation, chest tightness, and nasal congestion that appear whenever they are exposed to certain chemicals. People may react to even trace amounts of chemicals to which they have become "sensitized."

CO: Carbon monoxide.

CO_2: Carbon dioxide.

combination foundations: Buildings constructed with more than one foundation type, such as basement/crawlspace or basement/slab-on-grade.

commissioning: Start-up of a building that includes testing and adjusting HVAC, electrical, plumbing, and other systems to ensure proper functioning and adherence to design criteria. Commissioning also includes the instruction of building representatives in the use of the building systems. Effective commissioning is an important measure in establishing good indoor air quality in a new building.

conditioned air: Air that has been heated, cooled, humidified, or dehumidified to maintain an interior space within the "comfort zone." (Sometimes referred to as *tempered air*.)

constant air volume system: An air-handling system that provides a constant airflow while varying the temperature to meet heating and cooling needs.

dampers: Controls that vary airflow through an air outlet, inlet, or duct. A damper position may be immovable, manually adjustable, or part of an automated control system.

diffusers and grilles: Components of the ventilation system that distribute and return air to promote air circulation in the occupied space. As defined in this glossary, supply air enters a space through a diffuser or vent, and return air leaves a space through a grille.

disinfectant: One of three groups of antimicrobials registered by EPA for public health uses. EPA considers an antimicrobial to be a disinfectant when it destroys or irreversibly inactivates infectious or other undesirable organisms, but not necessarily their spores. EPA registers three types of disinfectant products based on submitted efficacy data: limited, general or broad spectrum, and hospital disinfectant.

drain tile loop: A continuous length of drain tile or perforated pipe extending around all or part of the internal or external perimeter of a basement or crawlspace footing.

drain trap: A dip in the drain pipe of sinks, toilets, floor drains, and so on, that is designed to stay filled with water, thereby preventing sewer gases from escaping into the room.

environmental agents: Conditions other than indoor air contaminants that cause stress, discomfort, and/or health problems (e.g., humidity extremes, drafts, lack of air circulation, noise, and overcrowding).

environmental tobacco smoke (ETS): A mixture of smoke from the burning end of a cigarette, pipe, or cigar and smoke exhaled by the smoker (also called *secondhand smoke* [*SHS*] or *passive smoking*).

ergonomics: An applied science that investigates the impact of people's physical environment on their health and comfort (e.g., determining the proper chair height for computer operators).

exhaust ventilation: Mechanical removal of air from a portion of a building (e.g., a piece of equipment, a room, or a general area).

flow hood: A device that easily measures airflow quantity, typically up to 2,500 cfm.

fungi: Any of a group of parasitic lower plants that lack chlorophyll, including molds and mildews.

gas sorption: The use of devices that reduce levels of airborne gaseous compounds by passing the air through materials that extract the gases. The performance of solid sorbents depends on the airflow rate, the concentration of the pollutants, the presence of other gases or vapors, and other factors.

general-duty clause: Section 5(a)(1) of the OSH Act, which requires an employer to "furnish to each of his employees employment and a place of employment which are free from recognized hazards that are causing or are likely to cause death or serious physical harm to his employees."

governmental: In the case of building codes, the state or local organizations/agencies responsible for building code enforcement.

green buildings: The building industry is increasingly focused on making its buildings "greener," which includes using healthier, less polluting, and more resource-efficient practices. *Indoor environmental quality* (*IEQ*) refers to the quality of the air and environment inside buildings, based on pollutant concentrations and conditions that can affect the health, comfort, and performance of occupants—including temperature, relative humidity, light, and sound. Good IEQ is an essential component of any building, especially a green building.

HEPA: High-efficiency particulate air (filters).

humidifier fever: A respiratory illness caused by exposure to toxins from microorganisms found in wet or moist areas in humidifiers and air conditioners. Also called *air conditioner fever* or *ventilation fever*.

HVAC: Heating, ventilation, and air conditioning.

hypersensitivity diseases: Diseases characterized by allergic responses to pollutants. The hypersensitivity diseases most clearly associated with indoor air quality are asthma, rhinitis, and hypersensitivity pneumonitis. Hypersensitivity pneumonitis is a rare but serious disease that involves progressive lung damage as long as there is exposure to the causative agent.

hypersensitivity pneumonitis: A group of respiratory diseases that cause inflammation of the lung (specifically granulomatous cells). Most forms of hypersensitivity pneumonitis are caused by the inhalation of organic dusts, including molds.

IAQ: Indoor air quality.

IAQ checklist: A component of the EPA's IAQ Tools for Schools Action Kit containing information and suggested easy-to-do activities for school staff to improve or maintain good indoor air quality. Each activity guide focuses on topic areas and actions that are targeted to particular school staff. The checklists are to be completed by the staff and returned to the IAQ coordinator as a record of activities completed and assistance requested.

IAQ coordinator: An individual who provides leadership and coordination of IAQ activities.

IAQ management plan: A component of the IAQ Tools for Schools Action Kit; specifically, a set of flexible and specific steps for preventing and resolving IAQ problems.

IEQ: Indoor environmental quality, composed of all elements inside a structure that could impact the well-being of an occupant. IAQ is an important component of IEQ and is sometimes referred to as *indoor environmental air quality* (*IEAQ*).

indicator compounds: Chemical compounds, such as carbon dioxide, whose presence at certain concentrations may be used to estimate certain building conditions (e.g., airflow, presence of sources).

indoor air pollutants: Particles and dust, fibers, mists, bioaerosols, and gases or vapors.

IPM: Integrated pest management.

makeup air: See *outdoor air supply*.

"Map of Radon Zones": An EPA publication depicting areas of differing radon potential in both map form and state-specific booklets.

MCS: See *multiple-chemical sensitivity*.

mechanically ventilated crawlspace system: A system designed to increase ventilation within a crawlspace, achieve higher air pressure in the crawlspace relative to air pressure in the soil beneath the crawlspace, or achieve lower air pressure in the crawlspace relative to air pressure in the living spaces, by use of a fan.

mg/m³: Milligrams per cubic meter (a mass-per-volume measurement of air contaminants).

microbiologicals: See *biological contaminants*.

model building codes: The building codes published by the model code organizations and commonly adopted by state or other jurisdictions to control local construction activity.

model code organizations: The following agencies and the model building codes they promulgate:

    Building Officials and Code Administrators International, Inc. (BOCA), used historically on the East Coast and throughout the midwestern United States.

    Comprehensive Consensus Codes (C3), which include the NFPA 5000 building code and companion codes: the National Electrical Code, NFPA 101 Life Safety Code, UPC, UMC, and NFPA 1. NFPA 5000 conforms to ANSI-established policies and procedures for the development of voluntary consensus standards.

    Council of American Building Officials (precursor to the ICC).

    International Building Code (IBC), a model building code developed by the International Code Council (ICC) that has been adopted throughout most of the United States. BOCA, ICBO, and SBCCI consolidated to become the ICC. The IBC is updated on a 3-year cycle.

    International Conference of Building Officials (ICBO), used historically in the western United States.

    Southern Building Code Congress International Inc. (SBCCI), used historically in the southeastern United States.

multiple-chemical sensitivity (MCS): A condition in which a person reports sensitivity or intolerance (as distinct from "allergic" reactions) to a number of chemicals and other irritants at very low concentrations. Medical professionals hold differing views about the existence, causes, diagnosis, and treatment of this condition.

negative pressure: A condition that exists when less air is supplied to a space than is exhausted from the space, resulting in lower air pressure within that space than in surrounding areas. Under this condition, if an opening exists, air will flow from surrounding areas into the negatively pressurized space.

organic compounds: Chemicals that contain carbon. Volatile organic compounds vaporize at room temperature and pressure. They are found in many indoor sources, including many common household products and building materials.

outdoor air supply: Air brought into a building from the outdoors (often through the ventilation system) that has not been previously circulated through the system. Also known as *makeup air*.

PELs: Permissible exposure limits set by OSHA to regulate exposure to chemicals listed in 29 CFR 1910.1000 (Tables Z-1, Z-2, and Z-3), Limits for Air Contaminants, or specifically addressed in OSHA standards in 29 CFR 1910.1001 to 1910.1052.

picocurie (pCi): A unit for measuring radioactivity, often expressed as picocuries per liter (pCi/L) of air.

plenum: An air compartment connected to a duct or ducts.

PM: Preventive maintenance.

pollutant pathways: Avenues for distribution of pollutants in a building. HVAC systems are the primary pathways in most buildings; however, all building components and systems interact to affect how air movement distributes pollutants.

positive pressure: A condition that exists when more air is supplied to a space than is exhausted, so the air pressure within that space is greater than that in surrounding areas. Under this condition, if an opening exists, air will flow from the positively pressurized space into surrounding areas.

ppm: Parts per million (a volumetric measure for chemicals in air).

pressed-wood products: A group of materials used in building and furniture construction that are made from wood veneers, particles, or fibers bonded together with an adhesive under heat and pressure.

pressure, static: In flowing air, the total pressure minus velocity pressure. The portion of the pressure that pushes equally in all directions.

pressure, total: In flowing air, the sum of the static pressure and the velocity pressure.

pressure, velocity: In flowing air, the pressure due to the velocity and density of the air.

preventive maintenance: Regular and systematic inspection, cleaning, and replacement of worn parts, materials, and systems. Preventive maintenance helps prevent part, material, and system failure by ensuring that parts, materials, and systems are in good working order.

psychogenic illness: A group of symptoms that develop in an individual (or a group of individuals in the same indoor environment) who are under some type of physical or emotional stress. This does not mean that individuals have a psychiatric disorder or that they are imagining symptoms.

psychosocial factors: Psychological, organizational, and personal stressors that could produce symptoms similar to those caused by poor indoor air quality.

radiant heat transfer: Heat transfer that occurs when there is a large difference between the temperatures of two surfaces that are exposed to each other, but are not touching.

radon (Rn) and radon decay products: A radioactive gas formed in the decay of uranium. The radon decay products (also called *radon daughters* or *progeny*) can be breathed into the lungs, where they continue to release radiation as they further decay.

reentrainment: A situation that occurs when the air being exhausted from a building is immediately brought back into the system through the air intake and other openings in the building envelope.

reentry: A situation that occurs when the air being exhausted from a building is immediately brought back into the system through the air intake and other openings in the building envelope.

RELs: Recommended exposure limits; recommendations made by the National Institute for Occupational Safety and Health (NIOSH) for exposure to chemicals.

sanitizer: One of three groups of antimicrobials registered by EPA for public health uses. EPA considers an antimicrobial to be a sanitizer when it reduces but does not necessarily eliminate all the microorganisms on a treated surface. To be a registered sanitizer, the test results for a product must show a reduction of at least 99.9% in the number of each test microorganism over the parallel control.

short-circuiting: A situation that occurs when the supply air flows to return or exhaust grilles before entering the breathing zones of building occupants. This is often caused when a supply and return or exhaust grille are located in close proximity. To avoid short-circuiting, the supply air must be delivered at a temperature and velocity that results in uniform mixing throughout the space.

sick-building syndrome (SBS): A set of symptoms that affect some number of building occupants during the time they spend in the building and diminish or go away during periods when they leave the building. SBS cannot be traced to specific pollutants or sources within the building. (Contrast with *building-related illness*.)

soil gas: The gas present in soil that may contain radon.

soil-gas retarder: A continuous membrane or other comparable material used to retard the flow of soil gases into a building.

sources: Sources of indoor air pollutants. Indoor air pollutants can originate within the building or be drawn in from outdoors. Common sources include people, room furnishings such as carpeting, photocopiers, and art supplies.

stack effect: The overall upward movement of air inside a building that results from heated air rising and escaping through openings in the building's superstructure, thus causing an indoor pressure level lower than that in the soil gas beneath or surrounding the building foundation.

static pressure: A condition that exists when an equal amount of air is supplied to and exhausted from a space. At static pressure, equilibrium has been reached.

sterilizer: One of three groups of antimicrobials registered by EPA for public health uses. EPA considers an antimicrobial to be a sterilizer when it destroys or eliminates all forms of bacteria, fungi, viruses, and their spores. Because spores are considered the most difficult form of a microorganism to destroy, EPA considers the term *sporicide* to be synonymous with *sterilizer*.

submembrane depressurization system: A system designed to achieve lower submembrane air pressure relative to crawlspace air pressure by use of a fan-powered vent drawing air from under the soil-gas retarder membrane.

subslab depressurization system (active): A system designed to achieve lower subslab air pressure relative to indoor air pressure by use of a fan-powered vent drawing air from beneath the slab.

subslab depressurization system (passive): A system designed to achieve lower subslab air pressure relative to indoor air pressure by use of a vent pipe routed through the conditioned space of a building and connecting the subslab area with outdoor air, thereby relying solely on the convective flow of air upward in the vent to draw air from beneath the slab.

TLVs: Threshold Limit Values; guidelines recommended by the American Conference of Governmental Industrial Hygienists.

tracer gases: Compounds such as sulfur hexafluoride that are used to identify suspected pollutant pathways and to quantify ventilation rates. Tracer gases may be detected qualitatively by their odor or quantitatively by air monitoring equipment.

TVOCs: Total volatile organic compounds. See *volatile organic compounds (VOCs)*.

unit ventilator: A fan-coil unit package device for applications in which the use of outdoor- and return-air mixing is intended to satisfy tempering requirements and ventilation needs.

variable air volume (VAV) system: An air-handling system that conditions the air to constant temperature and varies the outside airflow to ensure thermal comfort.

ventilation air: The total air, which is a combination of the air brought inside from outdoors and the air that is being recirculated within the building. Sometimes, however, the term refers only to the air brought into the system from the outdoors; this glossary defines this air as *outdoor air ventilation*.

ventilation rate: The rate at which outdoor air enters and leaves a building. It is expressed in one of two ways: the number of changes of outdoor air per unit of time (air changes per hour [ACH]) or the rate at which a volume of outdoor air enters per unit of time (cubic feet per minute [cfm]).

volatile organic compounds (VOCs): Compounds that vaporize (become a gas) at room temperature. Common sources that may emit VOCs into indoor air include housekeeping and maintenance products and building and furnishing materials. In sufficient quantities, VOCs can cause eye, nose, and throat irritations; headaches; dizziness; visual disorders; and memory impairment. Some are known to cause cancer in animals, while others are suspected of causing, or are known to cause, cancer in humans. At present, not much is known about what health effects occur at the levels of VOCs typically found in public and commercial buildings.

zone: The occupied space or group of spaces within a building that has its heating or cooling controlled by a single thermostat.

REFERENCES

American Society of Heating, Refrigerating, and Air-Conditioning Engineers (ASHRAE). ASHRAE Standard 90.1 i-1993. "Addenda to ASHRAE 90.1-1989." New York: ASHRAE, 1993.

———. ASHRAE Standard 52-1976. "Method of Testing Air-Cleaning Devices Used in General Ventilation for Removing Particulate Matter." New York: ASHRAE, 1976.

———. ASHRAE Standard 55-2004. "Thermal Environmental Conditions for Human Occupancy." New York: ASHRAE, 2004.

———. ASHRAE Standard 62-1973. "Standards for Natural and Mechanical Ventilation." New York: ASHRAE, 1973.

———. ASHRAE Standard 62.1-2007. "Ventilation for Acceptable Indoor Air Quality." New York: ASHRAE, 2007.

———. ASHRAE Standard 90-1975. "Energy Conservation in New Building Design." New York: ASHRAE, 1975.

———. ASHRAE/IES Standard 90.1. "Energy-Efficient Design of New Buildings Except Low-Rise Residential Buildings." New York: ASHRAE, 1989.

———. "Guideline 1: Guideline for the Commissioning of HVAC Systems." New York: ASHRAE, 1989.

———. ASHRAE/ANSI Standard 55-2004. "Thermal Environmental Conditions for Human Occupancy." New York: ASHRAE, 2004.

———. ASHRAE/ANSI Standard 62.2-2007. "Ventilation and Acceptable Indoor Air Quality in Low-Rise Residential Buildings." New York: ASHRAE, 2007.

Hansen, S. J. *Managing Indoor Air Quality*. Lilburn, GA: Fairmont Press, 1991.

NAFA User's Guide for ANSI/ASHRAE Standard 52.2-2007, "Method of Testing General Ventilation Air-Cleaning Devices for Removal Efficiency by Particle Size," NAFA, 2007.

Plog, *Fundamentals of Industrial Hygiene*. 6th edition, National Safety Council, 2012.

U.S. Department of Defense. MILSTD282. Military Standard 282. *Filter Units, Protective Clothing, Gas Mask Components and Related Products: Performance Test Methods*. Washington DC: U.S. Government Printing Office, 1956.

U.S. Department of Health and Human Services. Public Health Service. Centers for Disease Control. National Institute for Occupational Safety and Health. *Guidance for Indoor Air Quality Investigations*. Washington, DC: U.S. Government Printing Office, 1987.

U.S. Department of Labor. OSHA Regulations. Title 29 CFR, Part 1910.1000. OSHA Standards for Air Contaminants. Washington DC: U.S. Government Printing Office.

———.Title 29 CFR, Part 1910.1200. *Hazard Communication Standard*. Washington, DC: U.S. Government Printing Office

———. OSHA Instruction CPL 2.103, Field Inspection Reference Manual (FIRM), 1999.

U.S. Environmental Protection Agency. *Building Air Quality Action Plan*. EPA No. 402-K-98-001. DHHS (NIOSH) No. 98123. Washington DC: U.S. Government Printing Office.

———. *Building Air Quality: A Guide for Building and Facility Managers*. EPA No. 0-16-035919-8. Washington DC: U.S. Government Printing Office, 1991.

———. *I-BEAM IAQ. Indoor Air Quality: Building Education and Assessment Model*. Office of Radiation and Indoor Air, Indoor Environments Division (6609J). EPA No. 402-C-01-001. Washington DC: U.S. Government Printing Office, December 2002.

———. *Mold Remediation in Schools and Commercial Buildings*. EPA 402-K-01-001. Washington DC: U.S. Government Printing Office, 2008.

———. *An Office Building Occupant's Guide to Indoor Air Quality*. EPA No. 402-K-97-003. Washington DC: U.S. Government Printing Office, 1997.

———. *Sick-Building Syndrome*. Rev. ed. Indoor Air Quality Fact Sheet No. 4. Washington DC: U.S. Government Printing Office, 1991.

———. TITLE 40, Protection of Environment, Chapter I—Environmental Protection Agency, Subchapter C—Air Programs, Part 61—National Emission Standards for Hazardous Air Pollutants. Washington DC: U.S. Government Printing Office.

———. *Ventilation and Air Quality in Offices*. Rev. ed. Indoor Air Quality Fact Sheet No. 3. Washington DC: U.S. Government Printing Office, 1990.

U.S. Environmental Protection Agency and the U.S. Consumer Product Safety Commission. *The Inside Story: A Guide to Indoor Air Quality*. Washington DC: U.S. Government Printing Office, 1988.

World Health Organization. *Air Quality Guidelines for Europe*. WHO Regional Publications, European Series No. 23. Geneva: WHO, 1987.

REVIEW QUESTIONS

1. Explain the difference between sick-building syndrome and building-related illness.
2. Name three items that can affect a building occupant's health that are not related to IAQ.
   a.
   b.
   c.
3. Name and discuss three control strategies commonly used to control IAQ problems.
   a.
   b.
   c.
4. What are three elements common to every IAQ problem?
   a.
   b.
   c.
5. Occupant comfort is generally related to what two parameters controlled by conditioned air supplied by HVAC systems?
   a.
   b.
6. Describe two natural forces that can have a significant impact on IAQ.
   a.
   b.
7. What are the duties of an IAQ manager?
8. A building IAQ profile should be developed using what sources of information?
9. An IAQ investigation should use interviews to look at what four areas?
   a.
   b.
   c.
   d.
10. What two factors should initiate specialized sampling for contaminants?
    a.
    b.
11. What are two preventive measures for minimizing the possibility of microbial contamination?
    a.
    b.
12. Describe the difference between arrestance and atmospheric dust spot tests.
13. Name and discuss three generally accepted methods for resolving IAQ problems.
    a.
    b.
    c.
14. List five criteria that should be evaluated to determine the success of an implemented IAQ solution.
    a.
    b.
    c.
    d.
    e.

Ergonomics Yesterday, Today, and Tomorrow

Cynthia L. Roth, COHN
Jane Dolezal
Philip E. Hagan, JD, MBA, MPH, ARM, CIH, CHMM, CET, CHCM, CEM

Definition and Scope of Ergonomics

History of Ergonomics

Purpose of Ergonomics

Business of Ergonomics

Ergonomic Risk Factors
Physical Challenges ▸ Environmental Challenges
▸ Mental Challenges

Symptoms of Cumulative-Trauma Disorders

Ergonomics Standards
Occupational Safety and Health Administration

Establishing an Ergonomics Program

Risk Factors
Nonoccupational Risk Factors ▸ Occupational Risk Factors

Ergonomic Risk Mitigation—Worksite Analysis
Passive Surveillance ▸ Active Surveillance

Physiological Demands
Energy Demands ▸ Static Work

Physical Demands
Materials Handling ▸ Repetitive Work

Environmental Demands
Effects of Heat ▸ Effects of Cold

Computer Workstations

Machine Displays and Controls
Visual Displays ▸ Auditory Displays ▸ Controls

Workplace Characteristics
Work Heights ▸ Location of Work Material ▸ Clearance and Accessibility ▸ Design of Accessories

Management of Risk Factors
Engineering and Administrative Controls
▸ Commitment and Involvement ▸ Training
▸ Medical Management

Accommodating Disabilities

Summary

References

Review Questions

When employees and their environments are mismatched, injury levels rise, production and quality are affected, and other problems may result, such as strained labor relations and an increased cost of doing business. All of these detract from worker well-being and organizational efficiency. The goal of ergonomics is to achieve a balance between the demands placed on people during work and their capabilities.

This chapter discusses the definition and scope of ergonomics, ergonomics as an applied science, the ergonomic risk factors that contribute to injury, the signs and symptoms of ergonomic-type injuries, and ways to mitigate and prevent cumulative-trauma injuries. It includes an overview of how to organize and develop an ergonomics program and technical evaluation criteria for various tasks encountered in the workplace. The chapter also covers the following:
- how to establish an ergonomics program that uses ergonomics principles to eliminate or reduce work-related musculoskeletal risk factors
- how to evaluate the workplace to identify potential ergonomic risk factors created by tasks, procedures, or the environment
- how to evaluate and arrange equipment displays and controls
- prevention and control recommendations
- identification of training needs
- medical management issues
- suggestions for accommodations for workers with disabilities.

DEFINITION AND SCOPE OF ERGONOMICS

Ergonomics, also called human-factors engineering, studies the physical and behavioral interaction between people and their environments. This environment could be the workplace, the home, or even the car. Simply put, it fits the job to the employee by addressing human capabilities and limitations.

Many different fields of study have contributed to industry's understanding of work-related risk factors, stresses, and solutions, including the following:
- anatomy and physiology
- anthropometrics
- biomechanics
- engineering
- psychology
- industrial design.

Typically, ergonomists are professionals who have received advanced training in these fields and who, when using applied ergonomic principles, can develop practical solutions for workplace ergonomic risk factors. In approaching a problem, an organization may bring together ergonomists to analyze how people interact with machines, tools, work methods, work organization, and work spaces. Safety professionals, industrial hygienists, and occupational physicians and nurses may add their expertise to find solutions. Management and employees also play important roles in providing budgets and information, assisting in work evaluations, assessing work situations, and giving feedback after solutions have been implemented.

Ergonomics also plays a large role in reducing workers' compensation claims and lost work time by returning employees to jobs that will not aggravate an existing medical condition. In addition, a successful ergonomics program increases productivity and enhances quality, leading to increases in company profits.

In August 2000, the International Ergonomics Association (IEA) adopted the following definition for ergonomics:

> Ergonomics (or human factors) is the scientific discipline concerned with the understanding of interactions among humans and other elements of a system, and the profession that applies theory, principles, data and methods to design in order to optimize human well-being and overall system performance.
>
> Ergonomists contribute to the design and evaluation of tasks, jobs, products, environments and systems in order to make them compatible with the needs, abilities and limitations of people. ...
>
> Derived from the Greek ergon (work) and nomos (laws) to denote the science of work, ergonomics is a systems-oriented discipline which now extends across all aspects of human activity. ...
>
> Domains of specialization within the discipline of ergonomics are broadly the following:
> - Physical ergonomics is concerned with human anatomical, anthropometric, physiological and biomechanical characteristics as they relate to physical activity. ...
> - Cognitive ergonomics is concerned with mental processes, such as perception, memory, reasoning, and motor response, as they affect interactions among humans and other elements of a system. ...
> - Organizational ergonomics is concerned with the optimization of sociotechnical systems, including their organizational structures, policies and processes.

The Human Factors and Ergonomics Society (HFES) has adopted the definition of ergonomics used by the International Ergonomics Association.

Ergonomics deals with the realization and application of worker needs, abilities, limitations, and characteristics to the design of machines, tools, jobs, and workplaces that can result in productive, safe, comfortable, and efficient use. That is, ergonomics is a human-centered discipline that focuses on ways to fit the work to the worker.

The application of ergonomics in any work environment is defined as "the science of fitting the job to the employee." When there is a mismatch between the physical requirements of the job and the physical capabilities of employees, work-related musculoskeletal disorders (MSDs) can result. These disorders—also known as cumulative-trauma disorders (CTDs), repetitive-motion injuries (RMIs), and repetitive-strain injuries (RSIs)—can lead to sprains and similar injuries and can affect all aspects of an employee's life.

HISTORY OF ERGONOMICS

The association between occupations and musculoskeletal injuries was documented centuries ago. Bernardino Ramazzini (1633–1714) wrote about work-related complaints that he saw in his medical practice in the 1713 supplement to his 1700 publication *De Morbis Artificum Diatriba* (*Diseases of Workers*). Wojciech Jastrzebowski (1799–1882) created the word *ergonomics* in 1857 in a philosophical narrative "based upon the truths drawn from the Science of Nature" (Jastrzebowski 1857).

Industrial production in the early 1900s was still dependent on human power, and ergonomic best practices were slowly developing. The results were leading to improvement of production and quality. Scientific management, a method that improved worker efficiency by improving the job process, became extremely popular.

Frederick W. Taylor (1856–1915), a pioneer of scientific management, evaluated jobs to determine the "one best way" they could be performed. At Bethlehem Steel, Taylor dramatically increased worker production and wages in a shoveling task by matching the shovel with the type of material that was being moved (ashes, coal, or ore). This was an "ergonomic" evaluation, although it was not called that at the time.

Frank Gilbreth (1868–1924) and his wife, Lillian (1878–1972), studied jobs and made them more efficient and less fatiguing through time-and-motion analysis and standardizing tools, materials, and the job process. After an application of this approach, the number of motions in bricklaying was reduced from 18 to 4.5, allowing bricklayers to increase their pace of laying bricks from 120 to 350 bricks per hour. This was also the genesis of industrial engineering and the beginning of applied ergonomics.

Human-factors concerns emerged during World War II as a result of the work and experience of a number of specialists involved in the study of then-current human-operated systems. These systems included those operating on the earth's surface, under the sea, and in space. Human-factors studies focused on the following:
- systems performance
- problems encountered in information presentation, detection, and recognition
- related action controls
- workspace arrangement
- work organization
- skills required for the job.

Research in these areas continued with particular emphasis on human operations. This offered the opportunity for early improvements in performance and safety, as significant modifications of equipment were unlikely in wartime. Attention was focused on operations analysis, operator selection, training, and the environment associated with signal detection and recognition, communication, and vehicle control. Concurrently, human-factors work in industry focused on efficiency, task analysis, and time-and-motion studies.

Occupational injuries go back centuries and were well connected to their respective occupations. For example, "telegrapher's cramp" is now totally unheard of, but carpal tunnel syndrome is an all-too-familiar term. As another example, we no longer hear much talk of "seamstress's cramp" in the United States, but it is probably a huge problem in other parts of the world where sewing is still a manual occupation. The following is a description of maladies that illustrate the changing nature of work and the evolution of occupational safety.
- auctioneers' cramp: a professional neurosis affecting mainly the left side of the orbicularis oris muscle (a muscle that encircles the mouth)
- compositor's cramp: an occupation neurosis of the thumb and fingers of compositors resembling writer's cramp
- hammermen's cramp: a spasmodic affection of the muscles of the entire arm
- hephestic cramp: hammermen's cramp (Hephaestus was the Greek god of fire)
- seamstress's cramp: a neurosis of sewing women resembling writer's cramp
- shaving cramp: a neurosis of the hands of barbers resembling writer's cramp
- telegrapher's cramp: a neurosis resembling writer's cramp seen in telegraphers
- watchmaker's cramp: a spasm of the finger muscles peculiar to watchmakers
- writer's cramp: an occupation neurosis due to excessive writing. It is marked by spasmodic contraction of the muscles of the fingers, hand, and forearm together with

neuralgic pain therein. It appears whenever an attempt is made to write.

During World War II, ergonomics also began to emerge and to mature into a distinct discipline through military studies of the aspects of human interaction with complex machines. At first, ergonomics in the United States was concentrated in the military and aerospace programs.

Immediately following World War II, there was rapid growth in the ergonomics discipline. Human-factors requirements were incorporated into government phased-procurement contracts with industry. This led to the use of human-factors specialists by industry and gradually resulted in their involvement in nonmilitary systems and equipment.

After World War II, the focus of concern expanded to include worker safety as well as efficiency and productivity. Research began in a variety of areas:
- muscle force required to perform manual tasks
- compressive low-back disc force while performing lifting tasks
- cardiovascular response when performing heavy labor (oxygen intake/output and heart rate)
- perceived maximum load that can be carried, pushed, or pulled.

Areas of knowledge that involved human behavior and attributes (e.g., the decision-making process, organization design, human perception relative to design) became known as *cognitive ergonomics*. Areas of knowledge that involved physical aspects of the workplace and human abilities (e.g., force required to lift, vibration, reaches) became known as *industrial ergonomics* or simply *ergonomics*.

Today, ergonomics research and application affect all industries, products, and tools. The discipline has continued to broaden its area of concern and activity to include transportation, architecture, environmental "green" designs, consumer products, electronics/computers, energy systems, medical devices, manufacturing, office automation, organizational design and management, aging, farming, health, sports and recreation, oil field operations, mining, forensics, education, and speech synthesis. Well-applied ergonomics leads to enhanced productivity, safety, health and well-being, and job satisfaction.

PURPOSE OF ERGONOMICS

As greater industrialization was occurring globally, applied ergonomics moved into the private sector, addressing health, safety, and productivity in manufacturing settings. From manufacturing, ergonomics has spread through many industries and environments as well as into product and system design. The inclusion of ergonomic factors in the design phase has made products and systems more serviceable, thereby providing a competitive edge in the marketplace.

Having addressed ergonomics issues as they pertain to health and safety in the workplace, many countries are using ergonomics principles and best practices throughout working environments, at home, and during recreation.

The goal of ergonomics is to reduce the physical (and mental) stressors associated with any type of job; to increase the comfort, health, and safety of a work environment; to increase productivity; to reduce human errors associated with a task; and to improve the quality of work life as well as reducing the "costs of doing business." Ergonomics is concerned with interactions among humans and other elements of a system (e.g., tools, equipment, products, tasks, organization, technology, environment) to achieve a seamless interface. The profession applies theory, principles, data, methods, and analysis to design in order to optimize human well-being and overall system performance. It includes the physical, psychosocial, and environmental aspects of the interaction.

Another goal of ergonomics is to optimize the balance between the capabilities (physical and psychosocial) of the employee and the demands required by the work environment to improve comfort, efficiency, productivity, and quality; to prevent musculoskeletal injuries; and to decrease absenteeism and employee turnover. Incorporating effective ergonomics into business processes achieves balance between the capabilities of employees and the demands of the work environment, reducing the potential for bodily harm.

Engineers and others in industry often regard the three elements of machine, raw materials, and end product as composing a complete system. To an ergonomist, however, a system is not complete until the human working within the system is considered. The success of the human/machine interaction often determines the success of the system. To make the most of this process requires identifying demands placed on the employee, also known as *risk factors*.

BUSINESS OF ERGONOMICS

In recent years, employee reports of work-related musculoskeletal disorders (MSDs) have increased. Often, the cost of these work-related injuries is covered by the workers' compensation system. Workers' compensation programs are part of an insurance system through which employers can purchase workers' compensation coverage both to limit employer liability and to ensure that an employee's injuries are addressed. Coverage can be purchased from private insurance companies or state-run agencies, known as state funds. The costs to employers include premiums, payments made under deductibles, and benefits and administrative costs incurred by employers that self-insure or fund their own benefit program. In the mid-1950s, private-sector employers paid an average 0.5% of payroll for workers' compensation. By 1970, this figure was 1%. Employer costs escalated steeply in the 1980s to 2.18% and then declined. In 2001, these costs started to rise again. Now they are still fluctuating.

Estimates by the National Academy of Social Insurance put workers' compensation costs as a percentage of payroll at 1.76% in 2004, up from 1.73% in 2003. However, costs vary widely among states and industries so that the highest-rated (riskiest) groups could pay several hundred times that of the lowest-rated (safest) groups as a percentage of payroll. Also taken into account is the company's own safety record. Workers' compensation claims costs have two components: payments for lost income, which is usually linked to a state's average weekly wage, known as *indemnity costs*, and payments for medical care. Thirty years ago, indemnity costs made up the greater part of total losses. In 1983, for example, indemnity represented 56% of the total. By 2003, indemnity and medical had changed places; indemnity was only 45% of losses as medical costs grew.

As mentioned earlier, employee reports of work-related MSDs have increased. The Occupational Safety and Health Administration (OSHA) reports that these disorders have been increasing both in number and as a percentage of total recorded occupational illnesses. MSDs are found in all kinds of industries (e.g., meatpacking, retail, manufacturing, agriculture, health care, clerical). As employers experience increased costs from these ergonomic-related injuries, workers' compensation premiums also go up.

By understanding the conditions that lead to MSDs and similar ergonomics-related injuries, management and employees can take proper preventive measures to minimize harm to employees and to the employer's profit margin. In addition to workers' compensation costs, MSD-related injuries can affect a company's bottom line in many other areas. The iceberg analogy in Figure 16–1 provides a synopsis for identifying other potential financial impacts to a company.

"Ergonomics" is a general term that has different meanings to different audiences. Most often, this term is applied to work-related MSDs. The U.S. Department of Labor defines an MSD as an injury or disorder of the muscles, nerves, tendons, joints, cartilage, and spinal discs. MSDs do not include disorders caused by slips, trips, falls, motor vehicle accidents, or similar accidents. The Bureau of Labor Statistics (BLS) publishes detailed characteristics

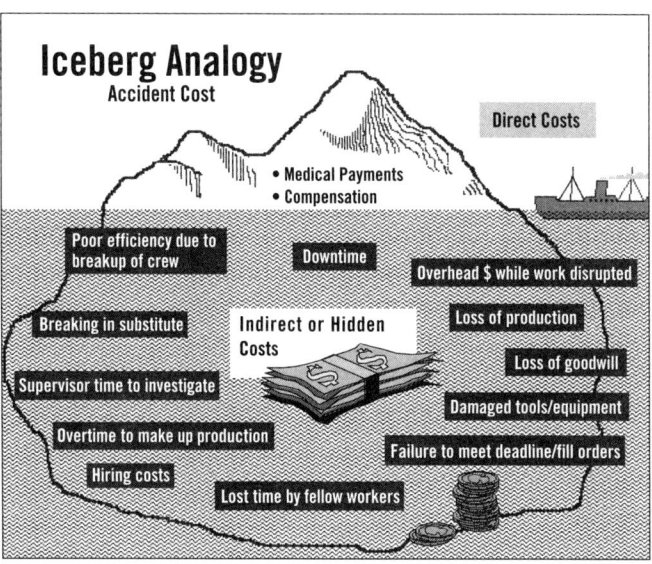

Figure 16–1. Iceberg analogy.

for MSD cases that resulted in at least one lost day from work. However, the BLS changed the case definitions for cases after 2011, so in some instances, it is difficult to make comparisons over time.

Prior to 2011, MSDs included cases where the nature of the injury or illness was sprains, strains, tears; back pain, hurt back; soreness, pain, hurt, except the back; carpal tunnel syndrome; hernia; or musculoskeletal system and connective tissue diseases and disorders when the event or exposure leading to the injury or illness was bodily reaction/bending, climbing, crawling, reaching, twisting, overexertion, or repetition. Cases of Raynaud's phenomenon, tarsal tunnel syndrome, and herniated spinal discs were not included. Although they may be considered MSDs, the survey classified those injuries and illnesses in categories that also included non-MSD cases.

MSDs from 2011 included cases where the nature of the injury or illness was pinched nerve; herniated disc; meniscus tear; sprains, strains, tears; hernia (traumatic and nontraumatic); pain, swelling, and numbness; carpal or tarsal tunnel syndrome; Raynaud's syndrome or phenomenon; or musculoskeletal system and connective tissue diseases and disorders when the event or exposure leading to the injury or illness was unspecified overexertion and bodily reaction, overexertion involving outside sources, repetitive motion involving microtasks, other and multiple exertions or bodily reactions, and body parts rubbed, abraded, or jarred by vibration.

MSD cases (388,060) accounted for 34% of all injury and illness cases in 2012 (http://www.bls.gov/news.release/osh2.nr0.htm). There were 388,060 MSDs in all ownerships (state and local government and private industry), with an incidence rate of 38 cases per 10,000 full-time workers. Both the incidence rate and case count remained statistically unchanged from the previous year; however, the median days away from work increased by 1 day to a median of 12 days. Laborer and freight, stock, and material movers had the highest number of MSD cases and an incidence rate of 164 per 10,000 full-time workers—up from 140 in 2011.

Workers who sustained MSDs required a median of 12 days to recuperate before returning to work, compared with 9 days for all types of cases.

Six occupations together accounted for over 25% of MSD cases: laborers and freight, stock, and material movers; nursing assistants; janitors and cleaners; heavy and tractor-trailer truck drivers; registered nurses; and maintenance and repair workers.

Of these occupations, laborers and freight, stock, and material movers had the highest MSD case count of 26,770. The highest median days away from work in this group was 19 for heavy and tractor-trailer truck drivers.

The most severe MSD cases, for all occupations, occurred to the shoulder, requiring a median of 24 days before returning to work and accounted for 14% of all MSD injuries. MSDs involving the back required a median of 7 days to recuperate and accounted for 41% of the MSD cases. In health care and social assistance, musculoskeletal disorders (MSDs) made up 42% of cases and had a rate of 55 cases per 10,000 full-time workers. This rate was 56% higher than the rate for all private industries and second only to the transportation and warehousing industry.

OSHA (https://www.osha.gov/SLTC/ergonomics/) lists the top 15 occupations with MSDs:
- Nursing assistants
- Laborers
- Janitors and cleaners
- Heavy and tractor-trailer truck drivers
- Registered nurses
- Stock clerks and order fillers
- Light truck or delivery services drivers
- Maintenance and repair workers
- Production workers
- Retail salespersons
- Maids and housekeeping cleaners
- Police and sheriff's patrol officers
- Firefighters
- First-line supervisors of retail sales workers
- Assemblers and fabricators

This chapter will focus on the components of an ergonomics program that can be used to identify and control MSDs.

ERGONOMIC RISK FACTORS

Task demands can be generally classified into three categories:
- physical demands and challenges
- environmental demands and challenges
- mental demands and challenges.

Jobs or tasks are created by engineers and designers who may not have the necessary training and understanding of ergonomics and an employee's capabilities. Physical job demands/requirements (job descriptions) may create the potential for ergonomic injuries because the body has neutral postures for all joints and body parts; these are the postures that are natural or neutral to the body. The job requirements may force the employee out of neutral postures in order to meet the job demands.

Physical Challenges

Challenges based on physical demands are placed on the musculoskeletal system of the body, creating ergonomic risk factors. Some examples of these risk factors are discussed in the following sections.

Neutral Postures

Neutral posture for any body part is a comfortable working posture in which the joints are naturally aligned and relaxed. Working with the body in a neutral position reduces stress and strain on the muscles, tendons, nerves, circulatory system, and skeletal system. Working in a neutral posture reduces the risk of developing an MSD.

The following are important considerations or best practices when attempting to maintain neutral body postures while working in a *seated* posture (e.g., quality control tasks, fine-motor tasks, office environments):
- Hands, wrists, and forearms are straight, in line with and roughly parallel to the floor.
- The head is level or bent slightly forward, forward facing, balanced, and in line with the torso.
- The shoulders are relaxed and the upper arms hang normally at the side of the body, but while seated the arms may be supported by chair armrests for short periods.
- The elbows stay close to the body and are bent between 90 and 120 degrees.
- The feet are fully supported by the floor; a footrest may be used if the desk height is not adjustable.
- The back is fully supported with appropriate lumbar support when sitting vertical or leaning back slightly.
- The thighs and hips are supported by a well-padded seat and generally parallel to the floor.
- The knees are about the same height as the hips, with the feet slightly forward.

Standing neutral posture means that the natural curves of the spine are not stressed or strained, but in a neutral position ready to absorb and distribute loads (e.g., weight) encountered during work and daily activities.
- The feet are apart.
- The lower back maintains the small hollow in the lower back (lumbar curve) by tucking the tailbone (sacrum) in and tilting the pelvis slightly forward. This is done by tightening the muscles of the buttocks and rotating the pelvis into the neutral position.
- The shoulders are pulled back and the chest is raised.
- The head and neck are lightly tilted down, with the chin level and the jaw relaxed.

Manual Material Handling (MMH)
Manual material handling involves physical exertion made without mechanical devices when the tasks require lifting heavy weights (greater than 51 lb, according to the National Institute for Occupational Safety and Health [NIOSH]) repetitively, while in an awkward posture accompanied by other risk factors:
- lifting
- lowering
- pushing
- pulling
- carrying
- holding.

Awkward Postures
Awkward postures are postures that require motion of the joint beyond its natural or neutral limit. These postures increase fatigue and discomfort to the joints and muscles, tendons, ligaments, circulatory system, and nerves. Exposure to awkward postures for prolonged periods can increase the potential for MSDs. The following postures are commonly cited as leading to MSDs:
- awkward arrangement of the body parts, limbs, trunk, and head relative to each other during work
- reaching (all postures including forward, side, overhead)
- staying in one position too long, causing muscles to contract
- prolonged bending (from back to floor, or sideways)
- twisting of limbs
- squatting
- kneeling
- ulnar and radial deviation of the wrist (the ulna is the bone on the little-finger side of the human forearm; the radius is the bone on the thumb side of the human forearm)
  - *Ulnar deviation* is movement of the wrist in the plane of the palm of the hand, away from the thumb and toward the little finger. For most people, ulnar deviation occurs because the shoulder width exceeds the width of the keyboard. In order to position the fingers on the home keys of a standard keyboard, the wrist must be rotated outward.
  - *Radial deviation* is movement of the wrist in the plane of the palm of the hand, toward the thumb and away from the little finger.
- pronation and supination of the forearm
  - *Pronation* is rotation of the forearm and hand so that the palm and inner arm face downward and the radius lies parallel to the ulna. The natural position of our hands on the horizontal plane is not flat, but rather 20 degrees, with the thumb side of the hand higher than the other side. When typing on a conventional keyboard, an individual must force his or her hands to be flat to remain in full contact with the keyboard (pronation).
  - *Supination* is rotation of the forearm and hand so that the palm and inner arm face upward and the radius lies parallel to the ulna.
- flexion and extension of various body parts (e.g., neck, shoulder, hand/wrist, back, ankle)
- abduction and adduction of the shoulders
  - *Abduction* of the shoulders is movement of the shoulders away from the midline of the body.
  - *Adduction* of the shoulders is movement of the shoulders toward the midline of the body.
- twisting the back
- reaching overhead and behind the shoulders.

Force
Force is an action that creates a change in the current state of the object to which it is applied. Examples include the following:
- pounding
- squeezing
- pinching
- excessive gripping
- pressing control buttons
- lifting excessive weights.

Forceful exertion refers to the amount of physical effort required to perform an activity. As muscular effort increases in response to high task load, blood circulation to the muscle decreases, resulting in more rapid muscle fatigue. Symptoms may include spasms or contractions of the affected muscles, resulting in cramping. Insufficient recovery time can result in injury to the soft tissue.

The type of grip used can affect the strength needed to perform a job. *Pinch grip* (using thumb and forefinger) strength is 25% of *power grip* (whole-hand grip) strength. Thus, using a power grip can reduce the strength requirement of a job. The force necessary to do a job depends on the properties of the tool or object being handled (e.g., size and shape, weight, surface frictional characteristics). If gloves are worn, muscles must exert more force to maintain adequate grasp because of reduced tactile feedback. Proper selection of personal protective equipment is important in reducing risk factors. No specific forceful exertion threshold value is associated with the development of MSDs. The association between forceful exertion and potential for MSDs can be modified by other risk factors (e.g., repetition, posture).

Repetition

Repetition is the number of cycles or motions using the same body parts and muscle groups over and over during work. Examples include the following:
- computer keying
- continuous lifting
- excessive walking
- cycle times that are too short in duration.

Repetitiveness is determined by the frequency and/or duration of motion and/or exertion required to perform the job. In a manufacturing setting, the time to complete one unit of assembly or to inspect an item is a *cycle*. Highly repetitive jobs are those with high frequency (a cycle time less than 30 seconds) or high duration of motion or exertion (more than 50% of the cycle time spent performing the same fundamental motions).

Computer task repetitiveness can be estimated by the number of keystrokes per hour performed by the operator. Sustained exertions can be determined as a percentage of time spent in certain postures. Repetitiveness requires more effort by the musculoskeletal system to perform the job and more time for recovery.

Without proper rest breaks, insufficient recovery of soft tissue may occur, which may increase the potential for MSDs. Table 16–A lists some activities of concern and provides some suggestions for improvement. No specific repetition threshold value (cycles per unit of time, movements per unit of time) is associated with the development of MSDs. Repetitiveness is usually a more important factor than force for inducing MSDs (e.g., carpal tunnel syndrome). But forceful activities combined with repetitiveness produce a synergistic effect, increasing the risk for carpal tunnel syndrome by more than five times that of either factor alone. The relationship of repetitiveness and risk of MSDs can be influenced by combination with other risk factors (e.g., force and posture).

Contact Stress

Contact stress occurs when any body part is compressed against a hard or sharp object. Examples include the following:
- leaning on work surfaces (see Figure 16–2)
- resting on chair arms.

Vibration

Vibration is periodic motion of particles away from their position of equilibrium. Examples include the following:
- oscillating equipment

TABLE 16–A. Evaluation for Repetition and Recovery Time

Activity of Concern	Suggestions for Improvement
Identical or similar motions or patterns every few seconds	Enlarge tasks Share the muscle load Use a power tool Job rotation
Short task cycle times (>2/min)	Enlarge the task Job rotation
Low task variety/long duration	Enlarge the task Job rotation
Piece-rate work, machine-paced work	Use accumulating conveyors
Static work	Change positions frequently Use tool balancers/holder Use fixtures/vice

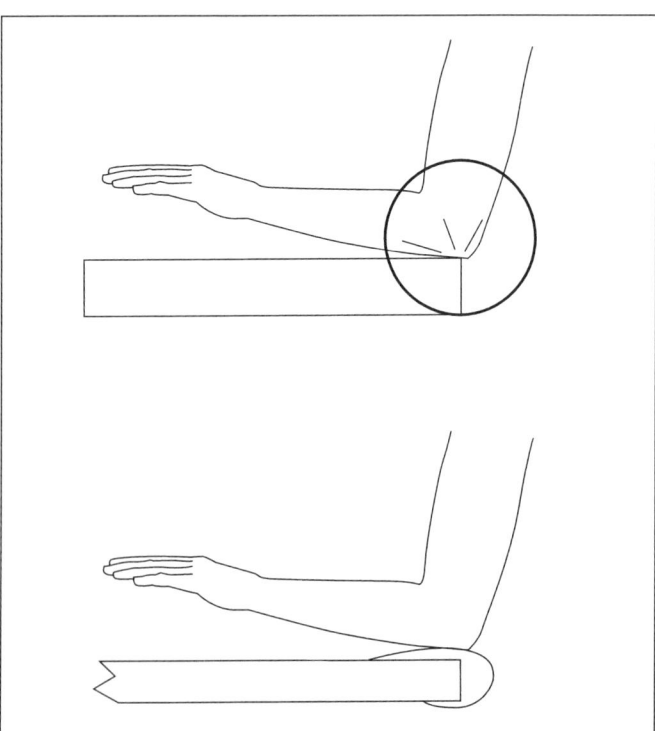

Figure 16–2. Minimizing contact stress.

- pneumatic tools
- riveting/bucking bars
- long-distance driving
    - buses
    - trains
    - cars
    - subways.

Vibration is an external stressor (i.e., has an external source) and, as such, is classified as an environmental factor. At particular frequencies, resonance can be set up in the human body, and this may lead to damage to the musculoskeletal system.

Two forms of vibration cause concern: (1) whole-body vibration from driving trucks, forklifts, or off-road vehicles, and from operating large machinery and (2) segmental vibration or hand-arm vibration from power tools.

Whole-body vibration is considered one of the risk factors for the development of low back pain, early degeneration of the lumbar spinal system, and herniated lumbar discs. Segmental vibration could contribute to the development of a number of disorders, including vibration white-finger syndrome and carpal tunnel syndrome.

NIOSH has published *Criteria for a Recommended Standard for Hand-Arm Vibration* (NIOSH Publication No. 89-106) but has not established exposure limits for such vibrations. However, *Threshold Limit Values (TLVs) and Biological Exposure Indices (BEIs)*, published annually by the American Conference of Governmental Hygienists (2014), provides some guidelines, as do International Organization for Standardization (ISO) standards 5349-1:2001 and 2631-5:2004.

NIOSH recommends engineering and work practice (administrative) controls to minimize exposure to such vibration and to minimize the possibility of developing vibration-related disorders. Suggestions offered by NIOSH include the following:

- selecting tools and machinery with the least amount of vibration
- modifying the vibration at the source
- using vibration isolators for tools, machinery, and seats of transport vehicles
- providing protective clothing and equipment (e.g., anti-vibration gloves)
- using vibration-damping materials on the handles of tools and other contact surfaces
- providing adequate protection from the cold to minimize the possibility of vasoconstriction
- reducing the number of hours/shift of exposure to the vibration
- reducing the number of days/week the vibrating tools are used
- following a regular maintenance schedule for tools to ensure good lubrication and even wear
- decreasing the grip force required to reduce the coupling between the tool and the operator
- restricting the use of piecework and incentive pay.

The ACGIH has recognized Threshold Limit Values (TLVs) for hand-arm vibration (HAV) and whole-body vibration (WBV).

Personal Risk Factors

Personal risk factors are specific to an individual. Examples include the following:
- smoking
- medications
- previous injuries
- aging
- obesity
- rheumatoid arthritis
- hypertensive heart disease
- some cancers
- gender
- genetics.

Because of the aging of the U.S. work force, the physical demands of a job often receive the greatest attention, although environmental and mental demands can be significant factors.

Environmental Challenges

Environmental challenges faced by an employee are a function of the physical designs of the environment in the workplace, such as the following:
- vibration from heating, ventilation, and air-conditioning (HVAC) systems
- workplace layouts
- temperature
- humidity
- noise
- lighting levels
- glare (direct and indirect)
- exposure to dust, gas, fumes, steam, or chemicals
- outdoor work (no effective protection from weather)
- walking on uneven ground (gravel, rocks, mounds)
- working heights (such as on scaffolding or ladders)
- working around moving machinery (forklifts, tractors, mowers)
- protective equipment (e.g., respirator, mask, earplugs, gloves, eyewear, shoes)
- potential exposure to infectious diseases.

There is also the psychosocial environment, with factors such as work organization, pace/cycle times, shift schedule, overtime, perceived work load and job difficulty, and rest schedules.

Mental Challenges

Mental challenges or cognitive responses include the following:
- information needed to perform a particular task
- mental calculations or computations
- short-term memory demands, information processing, decision making
- comprehending and following instructions
- work hours
- maintaining an appropriate work pace (cycle times)
- performing complex or varied tasks
- making decisions
- relating to others (communication skills, written and oral)
- control layouts
- signage.

Physical, environmental, and mental demands are interrelated and expose the employee to ergonomic risk factors. For example, the length of time an employee can work at a high energy level on a hot, humid day is generally less than on a cooler, drier day. Also, the ability to process information and make decisions may be affected by the shift the employee is assigned to work and the level of physical fatigue the body is experiencing. Ergonomic risk factors such as weight, repetition, work organization, and job and workstation design, which all affect the ability of the employee to work safely, productively, and efficiently, have already been discussed.

Special emphasis should be paid to back injuries. These are the most prevalent and expensive of all MSDs in the United States. The following data for 2012 are from the U.S. Department of Labor, Bureau of Labor Statistics (bls.gov):
- total recordable cases: 3,769,100, including state and local government
- cases involving days away from work: 1,154,000
- cases involving sprains, strains, tears: 443,560
  - 279,442 (or 63%) of these cases were due to overexertion and bodily reaction
  - 159,682 (or 63%) of these cases resulted in back injuries
- cases involving injuries to the back: 250,870
- MSD cases (state and local government and private industry): 388,060
- cases involving falls: 234,450.

The spine is made up of stacked bones (*vertebrae*) and cushions of soft tissue (*discs*). When seen from the side they form the three major natural curves of the spine:
- The *cervical curve* is in the neck area.
- The *thoracic curve* is at the chest level.
- The *lumbar curve* is in the lower back (where most workplace low back pain and injuries occur).

The *sacrum* and *coccyx* are two fused bones at the bottom of the spine. Viewed from back to front, the spinal column should be in a straight line or neutral posture. When properly aligned, it keeps the body balanced and comfortable.

Suggested administrative controls to reduce back injury potential include the following:
- Conduct strength testing of existing workers, which one study showed can prevent up to one-third of work-related injuries by discouraging the assignment of workers to jobs that exceed their strength capabilities. Ensure that all Equal Employment Opportunity Commission (EEOC) criteria are met when addressing such issues.
- Train employees to use lifting techniques that place minimum stress on the lower back.
- Provide physical conditioning or stretching programs to reduce the risk of muscle strain.

Suggested engineering controls include the following:
- Reducing the size or weight of the object lifted. These parameters could include maximum allowable weights for a given set of task requirements; the compactness of a package; the presence of handles; and the stability of the package being handled.
- Adjusting the height of a pallet or shelf. Lifting that occurs below knee height or above shoulder height is more strenuous than lifting between these limits. Obstructions that prevent an employee's body contact with the object being lifted also generally increase the risk of injury.
- Installing mechanical aids such as pneumatic lifts, conveyors, or automated materials-handling equipment.

See Table 16–B for a list of occupations most affected by ergonomics-type injuries.

All of the demands mentioned can affect an employee's performance on the job and increase the potential for injury. Demands should be carefully balanced with each person's capabilities. When job demands exceed an employee's physical and mental capabilities, the risk factor potential is increased. If demands exceed workers' capabilities, the first risk factor is usually fatigue. Physical fatigue can result in muscle overuse and can aggravate disorders of the musculoskeletal system. In addition, a tired employee is more likely to have accidents or to produce poor-quality products and is less likely to be efficient.

TABLE 16–B. Occupations Most Affected by Ergonomics-Type Injuries

Occupation	Number of Workers with LWT Injuries	Median Hourly Wage	Median Annual Pay
Nursing aides, orderlies, attendants	41,022	$8.89	$18,491
Registered nurses	10,718	21.56	44,845
Cashiers	8,551	6.95	14,456
Assemblers	6,926	10.32	21,053
Maids and housemen	6,739	7.41	16,190
Misc. machine operators	6,634	7.41	16,190
Laborers (except construction)	5,574	9.04	18,803
Licensed practical nurses	5,232	14.15	29,432
Sales workers, other commodities	5,079	8.02	16,361
Freight, stock, and material handlers	4,530	9.04	18,810

It is just as potentially hazardous and harmful when job demands are too low and the employee becomes bored and easily distracted. Boredom can lead to lapses in concentration and attention, resulting in product defects, quality problems, and accidents.

Other effects of poor ergonomics are high training costs, unnecessary overtime costs, employee complaints, medical restrictions, absenteeism, employee turnover, and increases in workers' compensation claims and disability claims.

SYMPTOMS OF CUMULATIVE-TRAUMA DISORDERS

Ergonomic principles and interventions can be used to identify and control workplace MSDs. MSDs are a group of cumulative conditions that involve injury to the body's soft tissue, muscles, tendons, nerves, blood supply and supporting structures such as the joints, and spinal discs. MSDs are cumulative in nature, which means the root cause of the disorder is difficult to identify.

Acute injuries occur when the root cause is easily identifiable: slips, trips and falls, burns, lacerations, and other events that result in acute injuries.

Musculoskeletal disorders are typically caused or aggravated by risk factors at work. These disorders can cause severe and debilitating injuries with symptoms such as the following:
- pain
- numbness
- tingling
- reduced range of motion
- weakness.

Depending on the severity of the disorder, the employee may experience temporary or permanent disability that prevents him or her from returning to work. These disorders can affect many different parts of the body (e.g., hand, eyes, wrist, arm, elbow, back, shoulder, neck, knee, leg, foot). The following are some of the more common MSDs diagnoses:
- carpal tunnel syndrome (CTS)
- epicondylitis (also known as tennis elbow)
- tension neck syndrome
- tendonitis
- bursitis
- low back pain
- repetitive motion syndrome
- cumulative-trauma disorder
- sprains/strains.

ERGONOMICS STANDARDS

Occupational Safety and Health Administration

The Occupational Safety and Health Administration (OSHA), a division of the Department of Labor, has been working on ergonomic issues for more than two decades. On November 23, 1999, OSHA released a proposed ergonomics standard (Ergonomics Program, Proposed Rule, 64 FR 65768). In proposing this standard, OSHA relied on the experiences of several entities: (1) OSHA programs, (2) private organizations, (3) insurance companies, and (4) research studies conducted by NIOSH and others. These experiences demonstrate that ergonomics programs can be effective in reducing risk, decreasing exposure, and protecting workers against MSDs. OSHA's proposed standard was passed into law in 1999 and terminated almost immediately.

Even with the overturn of the Ergonomics Standard, businesses continue to use the applied science in spite of not having government legislation mandating its use. Ergonomics programs are supported by government agencies, private industries, and companies worldwide. The applied science is accepted as an engineering, health, and safety tool for employees; as a tool for increased productivity and quality; and as a metric for greater company profitability. Table 16–C provides a chronology that documents OSHA's response to addressing ergonomics in the workplace.

TABLE 16–C. OSHA Ergonomics Chronology

Early 1980s	OSHA begins discussing ergonomic interventions with labor, trade associations, and professional organizations. OSHA issues citations to Hanes Knitwear and Samsonite for ergonomic hazards.
August 1983	The OSHA Training Institute offers its first course in ergonomics.
May 1986	OSHA begins a pilot program to reduce back injuries through review of injury records during inspections and recommendations for job redesign using NIOSH's *Work Practices Guide for Manual Lifting*.
October 1986	OSHA publishes a Request for Information on approaches to reduce back injuries resulting from manual lifting (57 FR 34192).
July 1990	OSHA/UAW/Ford corporate-wide settlement agreement commits Ford to reduce ergonomics hazards in 96% of its plants through a model ergonomics program.
August 1990	OSHA publishes *Ergonomics Program Management Guidelines for Meatpacking Plants*.
Fall 1990	OSHA creates the Office of Ergonomics Support and hires more ergonomists.
November 1990	OSHA/UAW/GM sign agreement bringing ergonomics programs to 138 GM plants employing more than 300,000 workers. Throughout the early 1990s OSHA signed 13 more corporate-wide settlement agreements to bring ergonomics programs to nearly half a million more workers.
July 1991	OSHA publishes *Ergonomics: The Study of Work* as part of a nationwide education and outreach program to raise awareness about ways to reduce musculoskeletal disorders.
July 1991	More than 30 labor organizations petition the Secretary of Labor to issue an Emergency Temporary Standard.
January 1992	OSHA begins a special emphasis inspection program on ergonomic hazards in meatpacking industry.
April 1992, 1993	OSHA publishes an Advance Notice of Proposed Rulemaking on ergonomics. OSHA conducts a survey of general industry and construction employers to obtain information on the extent of ergonomic programs and other issues.
March 1995	OSHA begins a series of meetings with stakeholders to discuss approaches to a draft ergonomics standard.
January 1997	OSHA/NIOSH conference on successful ergonomic programs held in Chicago.
April 1997	OSHA introduces the ergonomics web page on the Internet.
February 1998	OSHA begins a series of meetings with stakeholders about the draft ergonomics standard under development.
March 1998	OSHA releases a video entitled *Ergonomic Programs That Work*.
February 1999	OSHA begins small business (Small Business Regulatory Enforcement Fairness Act [SBREFA]) review of its draft ergonomics rule and makes draft regulatory text available to the public.
April 1999	OSHA's Assistant Secretary receives the SBREFA report on the draft ergonomics program proposal and begins to address the concerns raised in that report.
November 1999	OSHA publishes proposed ergonomics program standard.
2001	Ergonomics programs standard is overturned.
2002	OSHA proposes comprehensive ergonomics plan that featured targeted guidelines and enforcement for goals of the plan: decrease ergonomic hazards, reduce injuries and illnesses, ensure flexibility and encourage innovation, and help employers prevent MSDs.
July 2002	Nursing Home National Emphasis program, focused on ergonomic hazards related to resident handling, is published.
2003	OSHA, U.S. Postal Service, postal unions establish strategic partnership to reduce ergonomic injuries.
May 2004	OSHA issues its second set of voluntary ergonomics guidelines since announcing its "comprehensive ergonomics plan" of 2002. The guidelines were for the retail grocery store industry.
2005	OSHA publishes a Standard Interpretation for "Formaldehyde exposure and ergonomic hazards in the embalming/funeral home industry" that addresses workplace ergonomic hazards by providing assistance and information to employers and employees regarding the ergonomic hazards and then applying the "General Duty" portion of the OSH Act in enforcement activities, if necessary.
2007	OSHA publishes CPL 02-00-144—*Ergonomic Hazard Alert Letter Follow-up Policy*—Employers who have received ergonomic hazard alert letters (EHALs) will be asked to provide information on progress in addressing the hazards outlined in the EHAL.

Source: http://www.OSHA-slc.gov

ESTABLISHING AN ERGONOMICS PROGRAM

In the 1990s, the focus of many ergonomics programs was the prevention and management of MSDs. Many industries have and continue to develop ergonomic programs in order to improve worker health, safety, and efficiency. The goal of an ergonomics program is to reduce the human and monetary costs associated with inadequately designed workplaces, work processes, and work environments. To be successful, ergonomics should not be an "add-on" activity but an inherent part of an organization's functions. An effective program combines both a proactive and a reactive approach and involves all personnel. The degree of involvement varies depending on each individual's roles and responsibilities within the organization.

Ergonomics needs to be an integral factor in the design (and execution) of all processes, tools, equipment, jobs, or tasks. However, to make this discipline an inherent part of an organization's functions, it may be necessary first to implement a formal ergonomics process in the organization.

Any ergonomics process must be adopted into the culture of any company in order to be successful. Programs that are "different" will not be accepted, and employees will be skeptical as well as noncompliant. The program has to be supported from the top of the organization down.

The ergonomic work wheel, shown in Figure 16-3, provides a synopsis of the different elements that will be incorporated into the development of a successful ergonomics program. A successful program follows the employee through all phases of employment: entry into a job; preventive maintenance activities, which typically differ from day-to-day job activities; returning to work after an injury has occurred; and changes in job requirements.

First, management commitment and support of ergonomic activities are vital to the success of any program. This commitment can be demonstrated by setting goals for the program and providing the resources required to achieve those goals. Also, incorporating policies and procedures into the management plan is another effective method of supporting the ergonomic process.

Second, case management is another vital component of an effective ergonomics program, especially if the major concerns revolve around MSDs. Timely identification of these disorders and effective treatment will help minimize many of the negative effects of injuries and illnesses. Safety professionals should work closely with health care providers to ensure that they have adequate information on any workplace or job characteristics that may influence the diagnosis and treatment of CTDs. Key components for such a program are provided in Figure 16-4.

Third, training and education of all involved person-

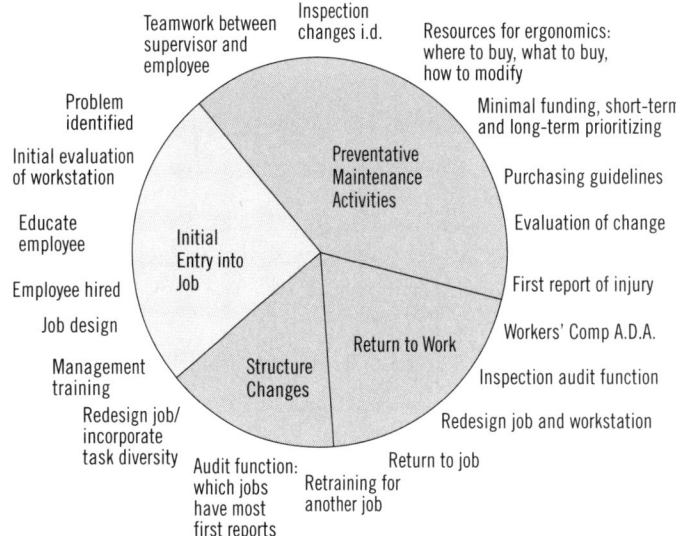

Figure 16-3. Ergonomic work wheel.

nel are critical for the success of such a program. Such training should be appropriate to the role of each employee affected by the program. Prior to employee training, engineers and supervisors must be trained to make and implement changes in the workplace (see Table 16-D).

The fourth, and most important, component is the implementation of workplace improvements. When this process is incorporated into the design stage of new products and processes, it can be an important tool in proactively addressing ergonomic concerns. The basic approach to the workplace improvement process is similar to the traditional quality improvement cycle, as shown in Figure 16-5.

Typical issues addressed during ergonomic evaluations to identify potential problem jobs or tasks include visual task interfaces (including illumination levels), thermal stress (heat/cold), physical workload, duration of work, vibration, and MSDs. The primary factors that are evaluated and assessed can generally be classified into the following categories:
- ergonomic risk factors and workplace characteristics
- physiological demands
- physical demands
- environmental demands
- design of displays, controls, and dials.

Based on these assessments, potential controls and modifications can be implemented to improve the ergonomic impact of these factors on the worker. See Figure 16-6 for an example of issues to evaluate when assessing tool handle and wrist position to enable an operator to work with the wrist in a neutral position.

Figure 16–4. Components of an ergonomics program to manage WMSDs.

TABLE 16–D. Ergonomics Training for Various Categories of Employees

	All Employees	Every Employee in Suspect Problem Jobs	Every Supervisor of Jobs with Suspect Problems	Every Employee Involved in Job Analysis and Control Development	Ergonomics Team or Work Group Members[1]
General ergonomics awareness information[2]	X	X	X	X	X
Formal awareness instruction and job-specific training		X	X	X	
Training in job analysis and controlling risk factors				X	X
Training in problem solving and the team approach					X

[1] If ergonomics teams are formed, added instruction is needed in team-building and consensus-development processes, apart from application of ergonomics techniques.
[2] General ergonomics awareness information for all employees need not require class instruction; it can be disseminated via handouts and all-hands meetings.

Source: NIOSH Publication No. 97-117, pg. 14.

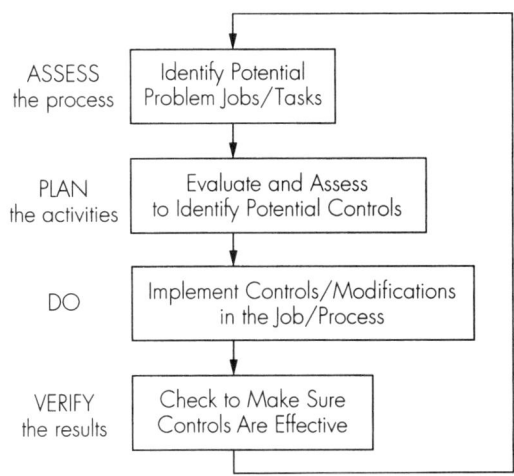

Figure 16–5. Workplace improvement process.

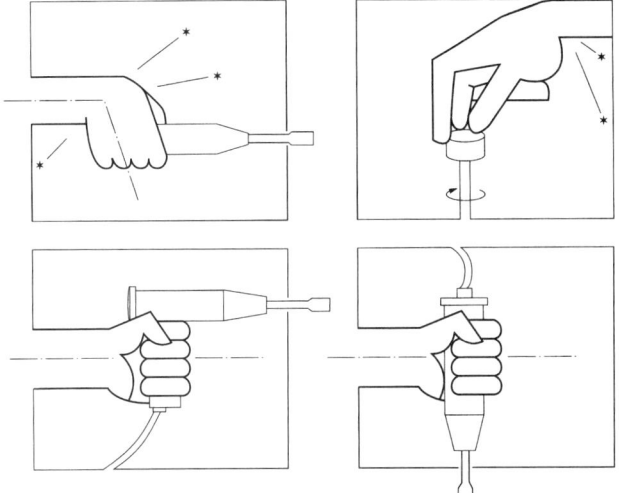

Figure 16–6. Tool handle and wrist position.

Each of these issues is discussed further in the sections that follow.

RISK FACTORS

Nonoccupational Risk Factors

Some frequently reported nonoccupational risk factors that could contribute to MSDs are related to personal risk factors such as the following:

- gender
- pregnancy
- rheumatoid arthritis
- level of physical condition
- certain medications
- prior trauma or injury
- systemic diseases
- smoking
- high blood pressure
- recreational activities
- certain cancers
- weight
- aging
- second jobs.

These personal factors may predispose an individual to MSDs. It is important to obtain a detailed medical and occupational health history to identify predisposing factors. If these factors are present, then the employee should not be placed in a job that may provoke the manifestation of MSDs. However, it is often better to design or redesign the workplace than to rely on the selection of employees. People change through time; for example, they grow older or gain or lose weight, examples of activities that cannot be controlled but may result in personal risk factors. Also, smart companies educate their employees regarding the predisposing nature of personal risk factors. This chapter focuses primarily on preventing MSDs by identifying and controlling occupational risk factors.

Occupational Risk Factors

Occupational risk factors that contribute to the development of MSDs include repetitive or sustained exertions, repetition and recovery time, forceful exertion (see Table 16–E), awkward postures, mechanical stress, vibration, and extremes in temperature.

TABLE 16–E. Evaluating Force

Activity of Concern	Suggestions for Improvement
High gripping, pinching, pressing forces	Provide cushioned grips Optimize handle size for each person/task Provide "return" springs whenever possible Use well-fitting gloves
Concentrated pressure points	Provide cushioned grips Extend handle length Investigate alternative handles Pad sharp edges and corners
Using the body as a brace/vice (static loading)	Use a fixture Provide a tool balancer
Impact forces (i.e., hammering with the palm, or knee)	Use padded, fingerless gloves Consider redesign of task Supplier interaction
High pushing or pulling forces	Minimize the load on the cart Caster maintenance Shelves should have a low friction surface
High rotation, twisting, wringing force	Use appropriate tool/power tool
Reaction torque from a power tool	Use a reaction bar Try an alternative tool

In general, the risk factors that are often reported are repetitive or sustained exertions, forceful exertion, and awkward postures as discussed earlier.

ERGONOMIC RISK MITIGATION—WORKSITE ANALYSIS

To prevent or mitigate the occurrence of MSDs in the workplace, proper methods of identifying jobs possessing ergonomic risk factors are necessary. Passive and active surveillance can be helpful in identifying work areas with potential for MSDs. Then ergonomic job analyses can be performed to assess the presence of risk factors and to prioritize responses (see Table 16–F).

Passive Surveillance

Passive surveillance refers to information that is gathered from existing records to identify potential patterns of disease within a workplace. These records can be used to analyze trends in injury and illness rates in a particular department, job, workstation, or operation. Types of records that can be used for passive surveillance include OSHA Form 300 logs, workers' compensation records, facility medical reports, accident/incident reports, first report of injury, absentee records, and rate of turnover. These and other records should be reviewed periodically, but the frequency of the review will depend on the degree of MSD problems encountered in the workplace.

Active Surveillance

Active surveillance incorporates a system for gathering data that can be used to determine trends of MSDs with greater sensitivity than a passive surveillance system. In contrast to passive surveillance, in which existing information sources are reviewed, active surveillance involves collecting relevant information and data. Active surveillance may help identify precursors that can be indicators of developing MSDs that may have been missed in passive surveillance. It is recommended that data be obtained via periodic worker health surveys. The surveys should collect information on current and past symptoms; affected body parts; and duration, intensity, and frequency of symptoms.

Also, interviews and physical exams can be used to complement the survey. OSHA and Putz-Anderson (1988) present sample questionnaires that can be used for collecting health data. The symptom survey is a good method for identifying areas or jobs that have a potential for MSDs. The major strength of the survey is that it provides data on workers who may be experiencing some form of MSD. Reported symptoms by a group of workers on a specific operation can illustrate the need for further evaluation of the job. If further job analysis is needed to augment the health survey, it can be done via checklist or a systematic evaluation method.

After identifying problematic jobs via passive and/or active surveillance, an effective program requires an ergonomic job analysis of each job identified as having a risk of developing MSDs. An ergonomic job analysis should be performed routinely for jobs that place workers at risk for devel-

TABLE 16–F. Determining Priorities for Job Analyses and Control Actions

Priority and Action	Nature of Available Information			
	Current Cases of WMSDs for Persons in Select Jobs	No Current Cases, but Past Plant Records Indicate WMSDs in Select Jobs or Departments That Have Not Changed	No Current or Past Cases, but Worker Complaints and Symptom Surveys Suggest WMSDs in Select Jobs or Departments	No Cases, Reports, or Complaints of WMSDs, but job Screening and Checklists Suggest Potential High Risk Factor in Select Jobs
Priority for follow-up analysis and control action	Immediate need	Priority is second only to the need	Third in priority, resolving problems	While last in priority, this effort is preventive
Type of follow-up job analyses needed	Perform job analyses to sort and rate job risk factors for observed cases	Perform job analyses to sort and rate risk factors for jobs with highest number or severity of past WMSDs and largest work group at risk	Perform job analyses to sort and rate risk factors for jobs having frequent WMSD complaints and symptoms	Perform job analyses to sort and rate risk factors for jobs with the highest problem potential (based on screening observations)
Focus needed	Control actions should be focused on reducing the highest-rated risk factors in current jobs linked with the greatest number of cases	Control actions should be focused on reducing the highest-rated risk factors in jobs with the highest number or greatest severity of past WMSDs for the largest work group at risk	Control actions should be focused on reducing the highest-rated risk factors in jobs having frequent WMSD complaints and symptoms	Control actions should be focused on reducing the highest-rated risk factors for WMSDs before any are reported

Source: NIOSH Publication No. 97-117, pg. 30.

oping MSDs. The following review of selected evaluation methods is intended as a general descriptor of the process.

Checklists

A checklist is a simple tool that can be used to evaluate the workplace. It can be helpful in identifying potential ergonomic problems in the job. Several types of checklists are available, ranging from general to specific content (see Table 16–G).

General checklists tend to cover a broad spectrum of workplace concerns. They are designed to encompass any situations that may be of concern, including workplace characteristics, physical demands of the job, and environmental conditions of the workplace. General checklists are presented by Eastman Kodak (1986) and Woodson (1981).

A specific checklist has a more refined goal of addressing factors that may contribute to the development of MSDs. It

TABLE 16–G. Tool Checklist

Tool Characteristics	Yes	No	Comments
Does the tool allow the operator to see the work interface?			
Does tool weigh less than 4 lbs (if held with one hand)?			
If weight is excessive, is tool balancer provided/used?			
Can continuous holding of the tool be avoided?			
Can the tool be operated without bending the wrist?			
Can the operator's fingertips close around the tool?			
Does the tool have a spring return?			
Can a pinch grip be avoided when operating tool?			
Can the tool be used with gloves?			
Can the tool be used with either hand?			
Trigger Characteristics	**Yes**	**No**	**Comments**
Can the controls/triggers be used without stretching the fingers?			
Is the force required to activate the tool less than 316 L?			
Can the tool be run without continuous activation of the trigger?			
If it is not a precision tool, are single-finger controls avoided?			
Handle Characteristics	**Yes**	**No**	**Comments**
Are sharp edges on the tool avoided?			
Does the tool handle extend beyond the palm area?			
Is the tool evenly balanced?			
Is there a flange/thumb stop at end of tool to prevent slipping?			
Is the tool comfortable to hold?			
Is the handle slip resistant?			
Is the handle made of/covered with compressible material?			
Is the finger clearance sufficient (>1" for one hand, >2" for two hands)?			
Is the tool handle at a comfortable temperature?			
Power Tools	**Yes**	**No**	**Comments**
Can impact tools/positive clutch tools be avoided?			
Is the force required to keep tool in position minimal?			
Is the reaction torque at acceptable levels?			
Does the tool have an auto shut-off control?			
Is a battery-powered design used to minimize trailing lines?			
Can the operator avoid handling the weight of the air or connection lines?			
Are tools hooked up with swivel-type connections to the lines?			
Are the vibration levels appropriate?			
Are the noise levels within acceptable levels?			
Is air exhaust directed away from the operator?			
General	**Yes**	**No**	**Comments**
Is the tool in good operating condition?			
Are debris and sparks directed away from the operator?			
Is there a guard for moving parts?			

would specifically evaluate the current manufacturing system for risk factors associated with a disorder. Some of these identification techniques are presented by others, such as Lifshitz and Armstrong (1986) and Ridyard and colleagues (1990).

The checklist is, at best, an analytical tool and not intended for synthesizing a new phase of a design stage. It can also help identify potential problems within the system and prioritize the highest to lowest risk. But to quantify the problems, a comprehensive and systematic methodology is required for evaluating the work system. Nevertheless, these and other checklists provide the critical point of departure in initializing the ergonomic analysis and allow a company to organize and prioritize the jobs for analysis.

Table 16–H provides a good example of a checklist that is provided by OSHA for assessing a computer workstation environment and to help with associated purchasing decisions. Table 16–I contains a checklist for workplace characteristics that can be used to evaluate a workstation and the work performed at the station. If any of the answers is no, consider ways to improve the situation.

Systematic Evaluations

Several methods are available to perform a more quantitative evaluation. The evaluation can be done via direct observation or videotaping. An early method by Keyserling (1986) requires videotaping of the job. Later, the videotape is analyzed in real time to determine the percentage of time that various body parts were in neutral and nonneutral postures. The Ovako Working Posture Analysis System (OWAS) method is a direct-observation method that evaluates the work performed by the worker (Karhu, Kansi, and Kuorinka 1977).

Both videotaping and direct observation are still valid methodologies for conducting quantitative evaluations. The ease of using digital cameras and recorders for both video and still pictures has made such evaluations the standard for ergonomic assessments. Use of digital documentation provides an easy venue for additional evaluation by others not participating in the field evaluation and provides an easy method of storage for future comparisons after recommendations are implemented.

PHYSIOLOGICAL DEMANDS

Physiological demands deal with mechanical, physical, and biochemical functions of the human body. The physiological demands placed on a worker performing a given task or job, when affected by the ambient environment (e.g., temperature, humidity), influence the rate at which muscular fatigue develops. The buildup of fatigue can increase the risk of accidents and injuries and can distract operators from the task at hand. This section discusses two different situations that could lead to the buildup of unacceptable levels of fatigue: performing tasks with a high-energy demand or doing static work. However, given the same job and work environment, the rate of fatigue buildup typically depends on the muscular strength of each individual and on his or her level of fitness.

Energy Demands

NIOSH has developed some criteria for the rate of energy expenditure that the average worker can sustain over different periods of time. These levels are shown in Figure 16–7. Similar criteria have been adopted by both the American Conference of Governmental Industrial Hygienists and the International Organization of Standardization (ISO 8996). To determine the average energy expenditure for the day, it may be necessary to estimate the energy demand for each of the tasks performed during the day. The exact level of energy expended for a given task varies from individual to individual. However, an estimate for the energy demand can be calculated by using the NIOSH guidelines shown in Table 16–J. For example, the following data might apply to an operator finishing a metal part with a grinder at a standing workstation.

Basal metabolism	1.0 kcal/min
Standing	0.6 kcal/min
Work with both arms (heavy)	2.5 kcal/min
Total:	4.1 kcal/min

The rate of energy consumption for the whole shift is determined by using a time-weighted average. Thus, short peaks of high-energy demand may be balanced by periods of lower-level activity in order to maintain the average demand for an 8-h day. When estimating the energy demand for a whole day, remember to include rest intervals (breaks and lunch).

For example, suppose this worker's task is to finish parts using a grinder for 6 h of an 8-h day. He spends about an hour of the day performing seated administrative tasks (e.g., tracking parts, updating information on a computer, logging activities) and has an hour total (half-hour lunch, two 15-min breaks) of rest time. The average energy demand for the day would be as follows:

Grinding	4.1 kcal/min × 6 × 60
Administrative tasks	(1.0 + 0.3 + 0.4) × 1 × 60
Breaks	(1.0 + 0.3 + 0.4) × 1 × 60
Average rate for the day	(24.6 + 1.7 + 1.7) × 60/8 × 60 = 3.5 kcal/min

If the job's energy levels are above acceptable limits, the tasks should be modified to decrease the energy demand

TABLE 16–H. OSHA Purchasing Checklist

Monitors

1. Make sure the screen is large enough for adequate visibility. Usually a 15 to 20-inch monitor is sufficient. Smaller units will make it difficult to read characters and larger units may require excessive space. ✓
2. The angle and tilt should be easily adjustable.
3. Flat panel displays take less room on the desk and may be more suitable for locations with limited space.

Keyboards

1. Split keyboard designs will allow you to maintain neutral wrist postures.
2. Keyboards with adjustable feet will accommodate a wider range of keyboard positions and angles. Adjustable feet on the front as well as the back will further aid adjustments. Increased adjustability will facilitate neutral wrist postures.
3. The cord that plugs into the CPU should be long enough to allow the user to place the keyboard and the CPU in a variety of positions. At least 6 feet of cord length is desirable.
4. Consider a keyboard without a 10-key keypad if the task does not require one. If the task does require one occasionally, a keyboard with a separate 10-key keypad may be appropriate. Keyboards without keypads allow the user to place the mouse closer to the keyboard.
5. Consider the shape and size of the keyboard if a keyboard tray is used. The keyboard should fit comfortably on the tray.
6. Consider keyboards without built-in wrist rest, because separate wrist rests are usually better.
7. Keyboards should be detached from the display screen if they are used for a long-duration keying task. Laptop keyboards are generally not suitable for prolonged typing tasks.

Keyboard Trays

1. Keyboard trays should be wide enough and deep enough to accommodate the keyboard and any peripheral devices, such as a mouse.
2. If a keyboard tray is used, the minimum vertical adjustment range (for a sitting position) should be 22 inches to 28 inches from the floor.
3. Keyboard trays should have adjustment mechanisms that lock into position without turning knobs. These are frequently overtightened, which can lead to stripped threads, or they may be difficult for some users to loosen.

Desks and Work Surfaces

1. The desk area should be deep enough to accommodate a monitor placed at least 20 inches away from your eyes.
2. Ideally, your desk should have a work surface large enough to accommodate a monitor and a keyboard. Usually about 30 inches is deep enough to accommodate these items.
3. Desk height should be adjustable between 20 inches and 28 inches for seated tasks. The desk surface should be at about elbow height when the user is seated with feet flat on the floor. Adjustability between seated and standing heights is desirable.
4. You should have sufficient space to place the items you use most often, such as keyboard, mouse, and monitor, directly in front of you.
5. There should be sufficient space underneath for your legs while sitting in a variety of positions. The minimum under-desk clearance depth should be 15 inches for your knees and 24 inches for your feet. Clearance width should be at least 20 inches.
6. Purchasing a fixed-height desk may require the use of a keyboard tray to provide adequate height adjustment to fit a variety of users.
7. Desktops should have a matte finish to minimize glare. Avoid glass tops.
8. Avoid sharp leading edges where your arms come in contact with work surfaces. Rounded or sloping surfaces are preferable.
9. The leading edge of the work surface should be wide enough to accommodate the arms of your chair, usually about 24 inches to 27 inches. Spaces narrower than this will interfere with armrests and restrict your movement. This is especially important in four-corner work units.

Chairs

1. The chair should be easily adjustable.
2. The chair should have a sturdy five-legged base with good chair casters that roll easily over the floor or carpet.
3. The chair should swivel 360 degrees so it is easier to access items around your workstation without twisting.
4. Minimum range for seat height should be about 16 inches.
5. Seat pan length should be 15 inches to 17 inches.
6. Seat pan width should be at least as wide as the user's thighs. A minimum width of about 18 inches is recommended.
7. Chair edges should be padded and contoured for support.
8. Seat pan tilt should have a minimum adjustable range of about 5 degrees forward and backward.
9. Avoid severely contoured seats as these limit seated postures and are uncomfortable for many users.
10. Front edge of the seat pan should be rounded in a waterfall fashion.
11. Material for the seat pan and back should be firm, breathable, and resilient.
12. The seat pan depth should be adjustable. Some chairs have seat pans that slide forward and backward and have a fixed back. On others the seat pan position is fixed and the backrest moves horizontally forward and backward so the effective depth of the seat pan can be adjusted. **Beware** of chairs where the back only tilts forward and backward. These do not provide adequate adjustment for a wide range of users.
13. The backrest should be at least 15 inches high and 12 inches wide and should provide lumbar support that matches the curve of your lower back.

Chairs, continued

14. The backrest should widen at its base and curve in from the sides to conform to your body and minimize interference with your arms.
15. The backrest should allow you to recline at least 15 degrees and should lock into place for firm support.
16. The backrest should extend high enough to support your upper trunk and neck/shoulder area. If the backrest reclines more than about 30 degrees from vertical, a headrest should be provided.
17. Armrests should be removable and the distance between them should be adjustable. They should be at least 16 inches apart.
18. Armrest height should be adjustable between 7 inches and 10.5 inches from the seat pan. Fixed height armrests are not desirable, especially for chairs that have more than one user.
19. Armrests should be large enough (in length and width) to support your forearm without interfering with the work surface.
20. Armrests should be padded and soft.
21. Most chairs are designed for weights under 275 pounds. If the user weighs more than 275 pounds, the chair must be designed to support the extra weight.

Document Holders

1. The document holder needs to be stable but easy to adjust for height, position, distance, and viewing angle.
2. If the monitor screen is your primary focus, purchase a document holder that will sit next to the monitor at the same height and distance.
3. If the task requires frequent access to the document (such as writing on the document), a holder that sits between the keyboard and monitor may be more appropriate.

Wrist Rests

1. Wrist rest should match the front edge of the keyboard in width, height, slope, and contour.
2. Pad should be soft but firm. Gel-type materials are recommended.
3. Wrist rest should be at least 1.5 inches deep (depth away from the keyboard) to minimize contact pressure on the wrists and forearm.

Mouse/Pointing Devices

1. Choose a mouse/pointer based on the requirements of your task and your physical limitations. There really is no difference, other than preference, among a mouse, trackball, or other device.
2. A mouse should match the contour of your hand and have sufficient cord length to allow its placement next to the keyboard.
3. If you choose a trackball, avoid ones that require the thumb to roll the ball—they may cause discomfort and possible injury to the area around your thumb.
4. A smaller mouse may be more appropriate, especially if you have small hands. Caution should be taken if a mouse is used by more than one person.
5. A mouse that has sensitivity adjustments and can be used with either hand is desirable.

Telephones

1. If task requirements mandate extended periods of use or other manual tasks such as typing while using the phone, use a telephone with a "hands-free" headset.
2. The telephone should have a speaker feature for "hands-free" usage.
3. "Hands-free" headsets should have volume adjustments and volume limits.

Desk Lighting

1. Good desk lighting depends on the task you're performing. Use bright lights with a large lighted area when working with printed materials. Limit and focus light for computer tasks.
2. The location and angle of the light sources, as well as their intensity levels, should be fully adjustable.
3. The light should have a hood or filter to direct or diffuse the light.
4. The base should be large enough to allow a range of positions or extensions.

---

associated with them. Another option would be to decrease the proportion of time spent on tasks with a high energy demand.

Static Work

Static work represents another situation in which an unacceptable buildup of fatigue may occur. This type of work is performed by the muscles in any activity that involves holding a position for a certain period of time. Some examples of static work are holding things with the hands, bending or leaning forward for periods of time, standing without moving or sitting in the same position, working with the arms extended, and looking down or sideways or upward for a sustained period. Use this checklist to evaluate the workstation and the work performed at it. If the answer to any of the questions is no, consider ways to improve that element.

TABLE 16–I. Checklist for Workplace Characteristics

Use this checklist to evaluate the workstation and the work performed at it. If the answer to any of the questions is NO, consider ways to improve that element.

When Working, Can the Operator:	Yes	No
Keep the wrists straight?		
Relax the shoulders?		
Keep the back in its naturally curved position?		
Keep the elbows at about 90°?		
Keep their arms close to their sides?		
Keep their neck straight and relaxed?		
Work without twisting their body?		
Easily reach all of the objects they work with?		
Avoid contact with sharp-edged work surfaces?		
Avoid keeping their body or arms or legs in one position for a long time?		
Minimize unnecessary work motions?		
Use a hammer or mallet rather than the hand to pound?		
Shift positions to help circulation?		
Does the Workstation Have:		
Sufficient foot space to minimize any forward lean while standing?		
The flexibility to allow changes in position (e.g., a sit-stand stool)?		
Sufficient leg room to minimize forward lean while sitting?		
A chair with good lower back support, if the operator sits while working?		

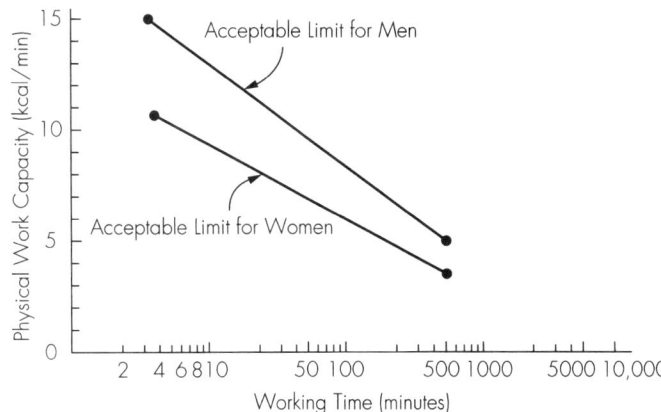

Figure 16–7. Recommended maximum capacities for continuous work.

To minimize the buildup of fatigue, the first priority is to modify the workplace characteristics that contribute to the performance of static work. These include the following:
- installing a fixture or jig to hold the object
- redesigning the workbench or conveyor to minimize forward lean
- providing a sit-stand stool or an adjustable chair
- positioning frequently used objects closer to the employee

TABLE 16–J. Estimating Energy Cost of Work

A. Basal Metabolism	1.0 kCals/min
B. Body Position and Movement	**kCals/min**
Sitting	0.3
Standing	0.6
Walking	2.0-3.0
Walking uphill	add 0.8 per meter rise

C. Type of work:	Average (kCals/min)	Range (kCals/min)
Hand work		0.2 to 1.2
Light	0.4	
Heavy	0.9	
Work one arm		0.7 to 2.5
Light	1.0	
Heavy	1.8	
Work both arms		1.0 to 3.5
Light	1.5	
Heavy	2.5	
Work whole body		2.5 to 9.0
Light	3.5	
Moderate	5.0	
Heavy	7.0	
Very Heavy	9.0	

Source: NIOSH Publication No. 86-113.

- placing displays, dials, and gauges frequently used in front of the employee
- training employees to recognize risk factors before injury.

In situations in which static work cannot be avoided, the work process may need to be evaluated to minimize fatigue buildup. This can usually be achieved through the use of adequate rest and recovery intervals.

PHYSICAL DEMANDS

The physical or muscular demands placed on an operator usually arise from the task the person is performing. These demands can be the result of materials-handling activities (lifting, pushing, pulling, or carrying) or performing repetitive tasks. Demands that exceed the operator's capabilities may result in an injury or an incident or impair the person's ability to perform the task well. The Snook and Ciriello tables of weights and forces published by Liberty Mutual are frequently used during evaluation and design of manual tasks involving lifting, lowering, pushing, pulling, and carrying. In addition, the Liberty Mutual *Manual Materials Handling Tables* list male and female population percentages capable of performing manual materials-handling tasks without overexertion.

Materials Handling

Manual materials handling refers to the human activity necessary to move an object. This often means lifting and carrying but can also describe pushing and pulling. The main issues to consider when performing an ergonomic analysis of a materials-handling task include the following:
- Is there any lifting activity?
- Is material pushed or pulled (even if it is a cart that is pushed or pulled)?
- Is the material carried from location to location?

General guidelines can be used to assess such jobs and identify possible improvements. Each of these issues is discussed in turn.

Lifting

In 1981, NIOSH published guidelines and an equation for analyzing lifting tasks, with the goal of minimizing back injuries. These guidelines were revised and a new equation developed in 1991 (Revised NIOSH Lifting Equation) to incorporate updated information on the biomechanical, physiological, epidemiological, and psychophysical aspects of manual lifting (see Table 16–K and Table 16–L).

In 1994, NIOSH put out the *Applications Manual for the Revised NIOSH Lifting Equation*. This manual provides users of the NIOSH Lifting Equation with methods for accurately applying the equation to a variety of lifting tasks. Use of this manual is limited—for instance, it does not support one-hand lifting, lifting of extremely hot or cold objects, unexpectedly heavy loads, speed lifting, and unstable loads. See the manual for a complete listing of the activities covered by the NIOSH Lifting Equation.

Equation for Lifting

When using the NIOSH equation to analyze tasks, bear in mind that it applies strictly to two-handed lifting tasks and is based on the assumption that other materials-handling activities (e.g., carrying, pushing/pulling) do not consume significant energy. Further, this equation does not apply if the operator is lifting in a constricted space or lifting while sitting or kneeling or using tools (e.g., a shovel). The 1991 Revised NIOSH Lifting Equation assumes that the coefficient of static friction between the worker and the floor surface is at least 0.4—that is, that the worker can obtain a firm footing. The risk of low back injury is presumed to be the same regardless of whether the object is being lifted or lowered, unless the object is lowered by dropping it.

The revised NIOSH equation is used to calculate a recommended weight limit (RWL), which is the weight that a healthy population can handle (for a given set of conditions) without increasing the risk of developing lifting-related low back pain. The RWL is calculated for both the origin and the destination (the starting point of the lift and the end point of the lift). The lower of the two values obtained is defined as the overall RWL for the task. The equation for calculating RWL in English units is as follows:

$$RWL = 51 \times HM \times VM \times DM \times AM \times FM \times CM \text{ lb}$$

where
- HM = Horizontal multiplier = $10/H$ and H is the horizontal distance of the hands from the midpoint between the ankles
- VM = Vertical multiplier = $1 - (0.0075 = |V - 30|)$ and V is the distance of the hands above the floor
- DM = Distance multiplier = $0.82 + 1.8/D$ and D is the vertical travel distance—the absolute difference between the vertical heights at origin and destination of the lift (or lowering task)
- AM = Angle multiplier = $1 - 0.0032 \times A$ and A is the angle the object makes with the frontal (mid-sagittal) plane of the worker
- FM = Frequency multiplier obtained from Table 16–K
- CM = Coupling multiplier obtained from Table 16–L.

To improve the lifting situation, determine which multiplier or factor has the greatest effect on the RWL (i.e., which

TABLE 16-K. Frequency Multiplier

Frequency = Average number of lifts per minute over a 15-minute period
Work Duration = The total amount of time, per shift,* spent lifting

	Work Durations					
	≤ 1 hour		> 1 hour but ≤ 2 hours		>2 hours but ≤ 8 hours	
Frequency	V<30"	V≥30"	V<30"	V≥30"	V<30"	V≥30"
≤0.2	1.00	1.00	0.95	0.95	0.85	0.85
0.5	0.997	0.97	0.92	0.92	0.81	0.81
1	0.94	0.94	0.88	0.88	0.75	0.75
2	0.91	0.91	0.84	0.84	0.65	0.65
3	0.88	0.88	0.79	0.79	0.55	0.55
4	0.84	0.84	0.72	0.72	0.45	0.45
5	0.80	0.80	0.60	0.60	0.35	0.35
6	0.75	0.75	0.50	0.50	0.27	0.27
7	0.70	0.70	0.42	0.42	0.22	0.22
8	0.60	0.60	0.35	0.35	0.18	0.18
9	0.52	0.52	0.30	0.30	0.00	0.15
10	0.45	0.45	0.26	0.26	0.00	0.13
11	0.41	0.41	0.00	0.23	0.00	0.00
12	0.37	0.37	0.00	0.21	0.00	0.00
13	0.00	0.34	0.00	0.00	0.00	0.00
14	0.00	0.31	0.00	0.00	0.00	0.00
15	0.00	0.28	0.00	0.00	0.00	0.00
>15	0.00	0.00	0.00	0.00	0.00	0.00

* Note: The lifting equation only applies for work duration ≤8 hours.

Source: Applications Manual for the Revised NIOSH Lifting Equation (NIOSH Publication 1994-110).

TABLE 16-L. Coupling Multiplier

	Coupling Multiplier	
Coupling Type*	V<30"	V≥30"
Good	1.00	1.00
Fair	0.95	1.00
Poor	0.90	0.90

*Note: If in doubt, choose the more stressful classification

Good coupling—is defined as those situations where the operators can wrap their hands around the handles of the object lifted
Fair coupling—is defined as those situations where the operators cannot comfortably wrap their hands around the handles or hand-hold cutouts or grip the object by flexing their hand about 90°
Poor coupling—is defined as those situations where there are no handles or hand-hold cutouts, or the object is irregularly shaped, or nonrigid (e.g., bags that sag in the middle) or has sharp edges or loose parts or is otherwise hard to handle.

Source: Applications Manual for the Revised NIOSH Lifting Equation (NIOSH Publication 1994-110)

multiplier is the smallest). Focus on these factors to increase the RWL.

Example: An operator is packing cases from a conveyor to a pallet right next to the conveyor. The operator stands in front of the pallet and twists 45 degrees to reach the conveyor. The boxes weigh 25 lb each, arrive at a rate of two boxes per minute, and are placed in a single layer on the pallet. The sole task of the operator for the 8-h shift is to pack these boxes. Because the boxes have no handles or handhold cutouts, the operator holds the boxes by grasping his or her hands around each one. The dimensions of the lift are given in Figure 16–8; the values for each factor are given in Table 16–M.

The RWL is found to be less than 25 lb (the weight of the box) at both the origin and the destination of the lift. In order to rectify this situation, one option would be to lower the weight of the boxes to 18 lb. If this is not possible, then each factor should be examined to determine which one has the greatest impact on increasing the allowable weight of the box. Clearly, the frequency of the lifts, F, has the lowest multiplier (0.65); therefore, slowing down the conveyor would be one solution.

However, in many situations, this may cause an unacceptable decrease in productivity. At the origin, the next lowest multipliers are H and A. At the destination, the lowest multipliers are V and H. Setting up the operator so that he or she does not have to twist to pick up the box from the conveyor and bringing the boxes on the conveyor 1 in. closer to the operator increases the RWL at the origin to 25.5 lb. At the destination of the lift, raising the pallet to the same level as the conveyor (e.g., using a palletizer) increases the RWL to 27.5 lb (Table 16–N). Because the RWL at both the origin and destination are above the actual weight of the box (25 lb), the lifting task would now be acceptable, according to the 1991 Revised NIOSH Lifting Equation.

Pushing and Pulling

As a general rule of thumb, pushing is preferable to pulling because it minimizes the risk of having the cart run over the operator or the operator's feet. Carts, if used, should be designed to be pushed, not pulled.

If material is manually placed on the cart or removed from the cart, it should be handled between elbow and waist height. Objects on the cart should not be stacked so high that the operator cannot see where he or she is going. Finally, cart handles should be approximately at elbow height. Casters should be well maintained and chosen to minimize the force required to move the cart. A well-designed cart also includes a brake system that allows the employee to stop a moving cart in an emergency. The recommended upper limits for the horizontal forces required

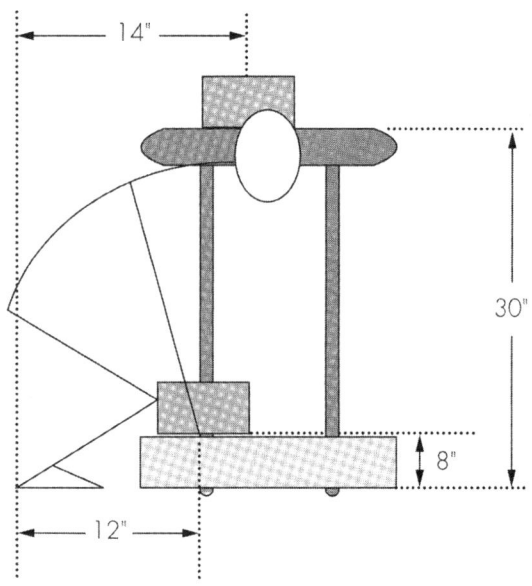

Figure 16–8. Parameters for the lifting task analysis.

TABLE 16–M. Values for the Multipliers for the 1991 NIOSH Lifting Equation

	Origin		Destination	
Factor	Value	Multiplier	Value	Multiplier
H	14"	0.71	12"	0.86
V	30"	1.00	8"	0.84
D	22"	0.9	22"	0.90
A	45°	0.86	0°	1.00
F	2 lifts/min	0.65	2 lifts/min	0.65
C	Fair	1.00	Fair	0.95
RWL	18.2		19.8 lb	

TABLE 16–N. Values for the Multipliers, after the Task Is Modified

	Origin		Destination	
Factor	Value	Multiplier	Value	Multiplier
H	13"	0.77	12"	0.83
V	30"	1.00	30"	1.00
D	8"	1.00	8"	1.00
A	0°	1.00	0°	1.00
F	2 lifts/min	0.65	2 lifts/min	0.65
C	Fair	1.00	Fair	0.95
RWL	25.5 lb		27.5 lb	

to move a handcart and stop a moving cart in an emergency can be found in Eastman Kodak Company (1986) and the Snook tables published by Liberty Mutual.

Carrying

Whenever possible, workers should avoid hand-carrying objects, instead transporting them on either a cart or a conveyor. If workers have to carry objects, they should do so for short distances only. The maximum weight carried is a function of the distance to be covered, the arm and shoulder strength of the individual, the type of carry (one-handed or two-handed), the size of the object, and the design of the handles.

Repetitive Work

The main issue in repetitive work is to minimize the buildup of fatigue in the muscles and to avoid overstressing the muscles. With this in mind, the key factors to be considered when evaluating repetitive work tasks are the following:
- the posture or position of the joints when the task is performed
- the force exerted
- the repetition rate
- the recovery time provided.

Posture

The position of the operator's body when performing different activities affects the force that is exerted (i.e., the stress on the muscles) and, consequently, the rate of buildup of fatigue. For a work environment, the position that places the least stress on the muscles is generally accepted to be the one described later in this chapter under Workplace Characteristics; the suggestions provided in that section can help managers evaluate and improve an operator's posture.

Force

The greater the force exerted, the more stress placed on the muscle. As of yet, there are no generalized or validated numerical limits on the force that can be exerted without overstressing the muscles used. However, some activities to look for and suggestions to address in each situation are given in Table 16–E.

Repetition and Recovery Time

Performing highly repetitive tasks could contribute to a buildup of fatigue, especially if sufficient recovery time is not designed into the task. Although a job with a cycle time of less than 30 seconds is often described as a repetitive task, there is no defined cutoff to classify a particular task as highly repetitive. Further, the exact amount of recovery time required varies from task to task. To identify high-repetition, low-recovery jobs, some key points to look for and suggestions to improve such conditions are given in Table 16–A.

ENVIRONMENTAL DEMANDS

Environmental demands arise from factors external to the operator. These include temperature extremes, relative humidity, and the air movement in the work environment as well as vibration from machines and tools. These factors could impair the ability of a worker to perform his or her job and could contribute to the development of injuries or other incidents. (See Chapter 13, Industrial Hygiene Program.)

Effects of Heat

The human body is designed to work optimally at a temperature of 98.6°F ± 1.8°F. This is the core temperature—that is, the temperature of the internal cavities/organs, not the skin temperature. If the core temperature rises, the operator may experience a heat disorder (e.g., heatstroke, heat exhaustion). The worker's ability to think and reason may also be impaired. The body has its own internal mechanisms that work to keep the body cool. As the heat and humidity levels of the work environment rise above the comfort zone, energy of the operator may be diverted toward keeping the body cool rather than performing the task at hand. In general, the body tries to achieve a balance between the heat gained by the body and the heat lost to the surroundings. This balance is best described by the following equation:

$$S = M \pm C \pm R - E$$

where
S = heat stored
M = heat gained due to metabolic activity (muscular work)
C = heat gained or lost from the surroundings by convection
R = heat gained or lost from the surroundings by radiation
E = heat lost to the surroundings through the evaporation of sweat.

Based on this equation, some methods to decrease the heat stored in the body would be as follows:
- Decrease the metabolic demand or the effort level of the task.
- Decrease the temperature of the surrounding air. If the whole area cannot be cooled, spot cooling may be an option. If the air temperature is less than 97°F (skin temperature), increasing the velocity of airflow could decrease the heat lost through convection. Installing fans or blowers could help increase the air velocity. If, however, the air temperature is more than 97°F, the opera-

tor may be gaining heat from the surrounding air, and increasing the velocity of the airflow may increase the rate of heat gain.
- Reduce the heat gained through radiation. Some options would be to place a reflective shield before the source of radiative heat (furnace, hot surface, etc.). Another option may be to wear reflective clothing.
- Increase the rate of sweat evaporation by decreasing the humidity level of the air (e.g., increase the airflow in an area).

Modifications in work practices could also help address this issue. Some factors to consider include the following:
- If the heat and humidity of the work environment are affected by the weather, try to reschedule heavy or difficult tasks for cooler days or cooler parts of the day. Avoid scheduling such activity during heat waves, unless it is an emergency.
- Provide a cool or air-conditioned area for rest and recovery times (e.g., break times).
- Provide a water fountain or other plentiful source of cool water near the work area.
- Make sure operators pace themselves and take breaks whenever required. One method to determine the schedule of work in hot areas is to use the criteria developed by NIOSH (Publication No. 86-113) (Figure 16–9).

The wet-bulb globe temperature (WBGT) can be determined using a WBGT meter, and the metabolic demand can be estimated by using the method outlined in Table 16–J. For more detailed information, refer to NIOSH Publication No. 86-113, *Occupational Exposure to Hot Environments: Revised Criteria 1986*. If it is not possible to implement these modifications, or if the modifications do not completely resolve the heat-stress issue, the safety professional should investigate the use of protective clothing. A number of devices are available in the market, based on either liquid cooling or air cooling. However, keep in mind that some suits may be restrictive or heavy and could increase the muscular effort or energy expenditure required to complete a task.

A safety professional should work with operators to identify the garment that best suits the work and each worker. When working with the department or division to develop procedures for working safely in a hot environment, it is important to train operators on the key issues around such work.

Effects of Cold

The human body is less capable of coping with heat loss than with heat gain. In cold environments, therefore, workers should be adequately clothed to maintain their body temperature at about 98.6°F. Exposure to cold temperatures (air temperatures less than 61°F) can also reduce manual dexterity. If workers wear gloves to keep their hands warm, the gloves should fit well and should not significantly interfere with grip strength. Suggestions to minimize the effects of working in the cold are discussed in the following sections.

Work Practices
- Work in the sun as much as possible or during the warmest parts of the day if possible.
- Share the workload to avoid overheating and sweating.

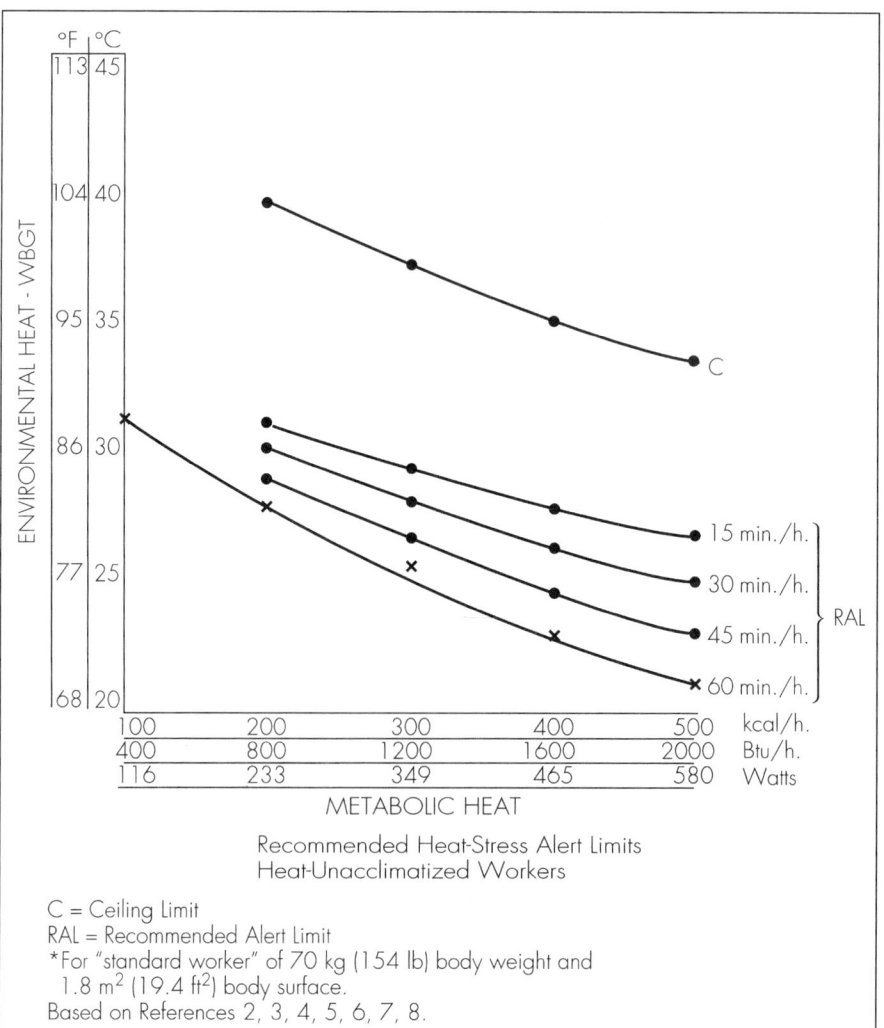

Figure 16–9. Recommended heat stress alert limits.

- When possible, follow a work warm-up schedule.
- Use the buddy system to help detect early signs of frostbite. Avoid working alone. Periodically check under face protectors.
- Pay attention to shivering as a warning sign; workers should get out of the cold if shivering occurs.
- Wear several layers of clothing and remove layers as the body warms up or as external temperatures rise.
- Keep clothing dry. Allow time for changing into dry clothes.
- Brush snow and moisture off clothes whenever possible.
- Evaporate perspiration by opening the neck, waist, arm, sleeves, and so on to provide fresh air circulation.

Protective Clothing

- When considering the insulation requirements of clothes, consider the work to be performed. In general, a person performing heavy work will not require as much insulation to be comfortable, because the body is generating heat.
- Wear insulated underwear and insulated overalls, which are usually sufficient to maintain comfort.
- Ensure adequate moisture control in clothes from both perspiration and precipitation. This is called "wicking" of the fabric. Waterproofed exterior clothing can eliminate absorption from snow, rain, fog, and other sources. Clothing should provide for venting of moisture from perspiration as well.
- Wear a mask or scarf during severe windchill conditions.
- Use extendible hoods to protect the face.
- Wear wind-resistant outer garments to reduce windchill.
- Wear lined, waterproof footwear.
- Overlap pants over boot tops to prevent entry of snow; pants should fit tightly, but not so much that they constrict or restrict movement.
- Use air insole cushions and felt liners with water-resistant boots.
- Ensure that heavy socks and additional dry socks are available when heavy work is to be performed.
- Wear mittens instead of gloves when manual dexterity is not important. Use mitten/glove combinations with liners.
- If using eye protection, use double-layered goggles with foam padding around the edges.
- Ensure that full-face-type respirators have separate respirator channels to prevent fogging and frosting of the facepiece.

COMPUTER WORKSTATIONS

Employees working at computer workstations can minimize the occurrence of MSDs by following a few simple guidelines. The following typing posture is recommended by Cornell University:

- Elbow angle preferably at 95–98 degrees (negatively sloped away from the user). At this angle, the elbow angle is opened to promote circulation to the lower arm and hand. Flexing the elbow (at an angle less than 90 degrees) can compress the median and ulnar nerves at the elbow, restricting blood flow. Working with the forearms sloping up increases the muscle load in the upper arms, shoulders, and neck. Working in this position for extended periods of time can lead to muscle fatigue.
- Chair
  - Has an adjustable seat and backrest that can be tilted back and forward
  - Has a waterfall (rounded) edge to minimize compression on underside of thighs
  - Is upholstered using nonslipping material
  - Has adjustable lumbar support
  - Has increased stability from five legs
  - Has the ability to swivel provided by casters to minimize twisting and turning
- Monitor
  - Sit at a comfortable distance from the monitor where the text can be read with head and torso in an upright posture. Generally, a preferred viewing distance is between 20 and 40 in. (50 and 100 cm) from the eye to the front surface of the computer screen.
  - Place the monitor so the top line of the screen is at or below eye level.
  - As the work force gets older, presbyopia—a gradual loss of flexibility in the natural lens inside the eye—causes a normal, age-related loss of near-focusing ability for workers over age 40. Take this into account when setting up workstations for such workers. Consider reconfiguring workstations for workers who have been around a while and have passed age 40.
- General CTD prevention
  - Keep wrists in neutral position. Place keyboard trays below the work surface with the mouse at the same level.
  - Use wrist rests for resting, not typing. NIOSH studies have found that use of a wrist rest when typing can increase pressure in the carpal tunnel by 140%.
  - Select keyboards based on an individual's own personal characteristics—selecting a single keyboard for a working population is usually not a good decision!
  - Strike keys lightly.
  - Face forward, with neck slightly tilted downward (5–30 degrees). This places the least demand on the static load of neck and shoulder muscles. Relax.
  - Place work materials in front, within easy reach.
  - Position the workstation facing the door, to minimize twisting and turning.

- Support the lower back and rest it by leaning back frequently and by supporting the arms.
- Provide adequate clearance for knees and thighs under desks.
- Position feet flat on the floor, or use a footrest.
- Ensure that work organization allows employee rest breaks.

MACHINE DISPLAYS AND CONTROLS

How workers activate machines and equipment and what information they receive through dials or gauges has a definite effect on work efficiency and worker safety. For example, if a worker must stop a machine in an emergency and pushes the wrong button, the result can be a serious accident.

The interaction between machine and user should be considered when equipment is designed or purchased. This includes reaching and accessing parts of the equipment, setting correct working heights, properly locating controls and displays, and arranging convenient maintenance.

Most organizations use two major types of displays: visual and auditory. A display can be a dial gauge, instrument, or auditory alarm that conveys information to the operator about the status of equipment or processes. Information displayed in a confusing format can cause serious operator errors and possibly disastrous accidents.

Visual Displays

Visual displays are used for one of three purposes:
- *quantitative readings*—to determine the exact quantity involved, such as a scale
- *qualitative readings*—to determine the state or condition at which the machine is functioning, which is usually one of three conditions, such as above, within, or below tolerance
- *dichotomous (check) readings*—to check operations or to identify one of two levels, such as off or on.

The purpose for which the display is to be read will dictate its best design. As a general principle, however, the simplest design is the best. The motion of the display should be compatible with (or in the same direction as) the motion of the machine and its control mechanism. A pointer that moves to the right, up, or clockwise to show an increase should have its corresponding control mechanism designed so that a rightward, upward, or clockwise movement of the control will increase the machine value and the corresponding display output value.

Multiple displays should be grouped according to their function or sequence of use. If dials are arranged in groups on a large control panel and must all be read at the same time, all the indicators should be pointing in the same direction when in the desired range. This method will reduce check-reading time and increase accuracy.

All displays should be labeled so the operator can quickly evaluate the display means, units, and the critical range. If the equipment is located in a dimly lighted area, make sure all displays are adequately illuminated.

Auditory Displays

Auditory displays should follow the principles outlined for visual displays. The question of whether an auditory or visual display should be used depends upon the situation. Table 16–O compares the relative advantages of auditory and visual displays. Other considerations for using auditory displays are found in the References at the end of this chapter.

TABLE 16–O. Visual versus Auditory Presentation of Signals

Use Visual Presentation If
- The person's job allows him or her to remain in one position.
- The message does not call for immediate action.
- The message is complex.
- The message is long.
- The message will be referred to later.
- The auditory system of the person is overburdened.
- The message deals with location in space.
- The receiving location is too noisy.

Use Auditory Presentation If
- The person's job requires him or her to move about continually.
- The message calls for immediate action.
- The message is simple.
- The message is short.
- The message will not be referred to later.
- The visual system of the person is overburdened.
- The message deals with events in time.
- The receiving location is too bright or preservation of dark adaptation is necessary.

The upper part summarizes conditions in the workplace or conditions related to the information to be communicated that make visual presentation the preferred method. The lower part provides a comparable list for auditory presentation of information. Visual presentation is preferred for complex messages in noisy environments where response time is not critical. Auditory presentation is preferred for simple messages in areas where people move around frequently and where response time must be rapid.

To evaluate existing displays (visual or auditory), the following should be considered:
- Is the display intensity higher than the lowest threshold level for sight or hearing? (Each sense has its own threshold level; energy intensities below the level cannot be perceived.)
- Are the cognitive senses overloaded? What other demands are being made on sight or hearing at the same time?

- Is the display compatible with similar displays, controls, and machine movements?
- What environmental factors, if any, could mask the display?

Controls

A *control* is anything—a switch, lever, pedal, button, knob, or keyboard—used by an employee to put information into a system to control or influence its operations. Performance can be enhanced when the controls work as the operator expects them to, when they are designed to fit human dimensions, and when their operating characteristics are within the strength and precision capabilities of the employee.

People expect things to behave in certain ways when they use operating controls; therefore, controls must be intuitive. Most people in the United States, for example, expect to turn on a light switch by flipping the switch up. A clockwise motion generally produces an increase in volume, temperature, or other element. Conversely, people expect the reverse kinds of movements to turn a system off or to decrease a function or flow. Such responses are called *population stereotypes* and are common to nearly everyone in a given population. Any control response that calls for a movement in a direction contrary to the established stereotype is likely to produce errors. The designer is asking for trouble by requiring the operator to change a long-standing habit.

Safety also may be jeopardized if an operator misreads a poorly designed display and operates the wrong control—or the right control in the wrong direction. Despite the fact that many accident reports classify these types of accidents as unsafe procedures or human error, they are, in fact, design errors. Retraining the operator probably will not prevent future incidents because, in an emergency, the operator will tend to fall back on the original stereotypic behavior. The operator should never be required to manipulate controls in unnatural or unexpected ways.

Control Design Principles
Research has established the principles of compatibility and coding in designing effective, safe controls.

Compatibility. As in the design of displays, control movement should be designed to be compatible with the display and machine movement. A lift truck, for example, with lift controls that move right or left to raise or lower the lift is bound to have a number of errors associated with its operation. The correct movement would be up and down.

Coding. Whenever possible, all controls should be coded in some way, such as by a distinguishing shape and texture, location, or color. A good coding system can reduce many errors; consider the following examples:

- Shape and texture to code controls is useful where illumination is low or where a control needs to be identified and operated through touch only.
- Location is another way to code controls. On forklift trucks, for example, all controls pertaining to mast operation can be grouped on the right side of the control panel. Location coding can also be achieved by using a minimum distance between controls.
- Color codes are useful for quick visual identification of various controls and for grouping controls for a particular operation.

All controls require some type of labeling to identify their function.

Arrangement of Controls
The arrangement of controls (and displays) should support interaction between users and the controls or display components. Components of the system should be arranged with these considerations in mind:

- Convenience, accuracy, speed, or other criteria should determine the location of each component.
- Components with related functions should be grouped.
- Controls and displays should be grouped according to their place in a critical set of operations. The important controls should be positioned in the most convenient location for rapid and easy use.
- Controls are frequently used in a sequence or pattern relationship. In these cases, the controls should be arranged to take advantage of such patterns by locating them close to each other.
- Less frequently used controls should be placed in more distant locations.

If there is conflict between control arrangement considerations, compromises will have to be made. Generally, designers should first consider the frequency and sequence of use of a control mechanism. Control or display arrangements in which workers must frequently transfer their entire body, eye, or hand from one place to another should be avoided.

Control Evaluations
There are other questions to consider when designing controls:
- Where are the controls placed?
- Can they be reached easily?
- Are they spaced far enough apart?
- Are they labeled and coded?
- What type of control is used?
- Is it compatible with user needs?

- Do the controls themselves present a hazard?
- Which body limbs are involved?
- Are any of the muscles used to operate the controls being overloaded?
- Are similar control operations alike in design and function?
- How standardized are the controls?

Table 16-P lists stereotypes for switches used in the United States.

TABLE 16-P. United States Stereotypes for Up and Down Switch Settings

UP	DOWN
On	Off
Start	Stop
High	Low
In	Out
Fast	Slow
Raise	Lower
Increase	Decrease
Open	Close
Engage	Disengage
Automatic	Manual
Forward	Reverse
Alternating	Direct
Positive	Negative

Some movement stereotypes for toggle switches used in equipment or production system control panels are given. The expected direction of activation of the switch (when mounted vertically) is given for 13 actions or conditions. These expectations are United States stereotypes and may vary in other countries.

WORKPLACE CHARACTERISTICS

Types of workplace settings include sitting workstations, standing workstations, and sit–stand workstations. Workplaces should be arranged so that the person using them can do so safely and effectively, without exceeding an acceptable level of energy consumption. For instance, a tool and die operator should be able to work in a comfortable position and be able to reach the materials needed easily.

The key to a good workplace setup is to implement a design that accommodates the greatest number of people (90%–95% of the working population). For example, clearances should be designed to accommodate workers with the largest dimensions, whereas reaches should be designed with the smallest workers in mind. The human, or anthropometric, data required for such design decisions can be obtained from a number of sources, such as *Ergonomic Design for People at Work*, Vol. 1 (1986) by Eastman Kodak; *Anthropometric Source Book Vols. I–III* (1978) by NASA; and S. Pheasant's *Bodyspace, Anthropometry, Ergonomics and Design* (1986).

For most people, a comfortable working position is one in which the back is naturally curved, the head is held erect, the shoulders are relaxed, the upper arms are close to the sides of the body, and the wrists are straight. This posture is often referred to as a *neutral posture*. Operators should be able to work with their elbows bent at about 90 degrees, and their feet should be well supported. The edges of the work surface should be smooth and well rounded so that any contact stress is minimized (see Figure 16–2). However, it is not possible or even desirable that workers hold these positions throughout a workday. Remaining in one position for long periods of time can place a static load on the muscle and rapidly build up fatigue. However, postures that operators assume should be close to the neutral positions shown. Any significant deviations should be infrequent and for short durations only. Some factors that influence the postures assumed at work are discussed next.

Work Heights

If the task to be performed requires little force, then the work height should be about elbow height. If the task to be performed involves precision work at close visual range, then the work height should be 3–4 in. (8–10 cm) above elbow level. If the operator has to exert downward force when performing a task at a standing workstation, then the work height should be a few inches (6–8 in., or 15–20 cm) below elbow height.

Elbow height will depend on whether the operator is sitting or standing while working. Further, it will vary with the physical dimensions of the operator. The workstation should preferably be individualized to the person using the area. If more than one person uses the workstation or if the height of the work varies, then an easily adjustable workstation (e.g., pneumatic, crank driven, or electrically powered) is recommended. Either the height of the work surface itself should be adjustable, or the surface the operator is standing on (platform) or sitting on (chair) should be adjustable. If the height of the chair is increased, the operator should still have adequate foot support (e.g., provide a footrest or a foot ring).

Location of Work Material

Work material should be within the reach of the operator. The primary working area (i.e., the workspace together with the material and tools frequently used) should be within the elbow to wrist length of the operator. Secondary objects not often used could be within the shoulder to wrist length (i.e., the operator should be able to reach the items by stretching the arm, but without needing to lean forward). Infrequently used, light objects could be placed beyond this distance but still within reach.

If space is at a premium, management should consider using the air space around a workstation. Some examples are hanging tools with an air or spring balancer or placing parts on a lazy Susan–type shelf arrangement.

Clearance and Accessibility

Clearances always must accommodate the largest population of employees in size. Although clearance and accessibility of work may be issues in a manufacturing situation, they are usually more critical for maintenance workers. Often, equipment and work spaces are arranged to minimize the amount of space used. The engineer may consider the user when designing the machine or work space, but the needs of the maintenance worker are rarely addressed. Maintenance personnel are often seen crouching over a motor or kneeling and reaching to get to the different parts of the equipment.

Access to components that need servicing should be adequate to accommodate both hands and still allow maintenance personnel to see what they are doing. If a tool has to be used, sufficient room should be provided to insert the tool and manipulate it with the required force. An opening of about 8 in. (20 cm) is recommended for one-handed tasks requiring force. If the equipment has to be adjusted, the operator should be able to see the displays or target while making the adjustments. Finally, the maintenance worker should be able to service the equipment without contacting hot surfaces, sharp edges, or electric currents. Note that any protective gear worn by the operator not only will restrict the range of motion of the operator, but may increase the size of the opening required to access the parts.

The position the employee assumes will clearly be influenced by the interaction among the physical characteristics of the workplace, the physical dimensions of the operator, and the range of motion of his or her joints—in other words, by the anthropometrics of the user. The focus should be toward customizing the workstation so that the individual worker can assume a neutral posture while working.

Design of Accessories

The design of the tools used greatly influences the posture of the operator. The joints affected are usually the wrist, hand, and arm. The design of the handle should be such that it enables the operator to work with the wrist in a neutral position.

If workers wear gloves as they perform tasks, the gloves should be comfortable and fit well. An ill-fitting or stiff glove can reduce grip strength and interfere with manipulating the tool.

MANAGEMENT OF RISK FACTORS

Once the probable source of the problem has been determined, the next step is to design measures to control the hazards. Occupational risk factors are prevented primarily by effective job design of workstations, tools, equipment, and jobs. An employer can establish procedures to mitigate these risk factors by using appropriate engineering and administrative controls. Also, commitment and involvement of management and workers, training and educating company personnel, and medical surveillance can be influential in controlling the incidence of MSDs in an organization.

Engineering and Administrative Controls

Engineering controls are the preferred method of control because the primary focus is to design or redesign the job to accommodate the person, rather than making the person adjust to the conditions of the job. This goal can be accomplished by designing or modifying the work methods, workstation, or tools to reduce the job demands of repetitive motion, high force, awkward posture, and any combination therein.

When engineering controls are not feasible or while implementation is occurring, administrative controls are frequently used to limit worker exposures. Administrative controls are actions taken by management to limit the exposure of stressful conditions imposed on the worker. Administrative control is achieved by modifying existing personnel functions. Thus, the control is achieved by having the worker change the risk factors.

Specific recommendations follow for reducing or eliminating occupational risk factors involved in the development of MSDs. The recommendations are not a panacea for all the problems, but an initiation process for providing a better workplace. As management and labor work together in an effort to improve the work environment, more innovative methods of control will evolve. Because highly repetitive motions, high force, and awkward postures contribute to the development of MSDs, it is important to find ways to reduce these risk factors. See the References at the end of this chapter for other sources of recommendations.

Approaches for reducing repetitions include the following:
- Automate highly repetitive tasks.
- Use power tools where applicable.
- Provide short breaks for highly repetitive tasks (5–10 min per hour).

- Design the job to allow the worker to control the pacing.
- Provide job enlargement by increasing the variety of tasks.
- Provide job rotation.

Approaches for reducing forceful exertion include the following:
- Keep tools properly maintained.
- Increase the friction between handles and the hand.
- Reduce the weight to be handled.
- Select gloves to minimize effects on performance.
- Provide handles or handholds on containers and objects that need to be lifted or carried.
- Use jigs and clamps to hold and secure parts.
- Allow for longer standard rest breaks (increase by 5–10 min).
- Provide job rotation.

Approaches for reducing awkward postures include the following:
- Select tools appropriate for the job.
- Move the part or work closer to the worker.
- Move the worker to reduce awkward postures.
- Use mechanical aids to procure parts.
- Allow for longer standard rest breaks (increase by 5–10 min).
- Provide job rotation.

Commitment and Involvement

In an effort to curtail health and safety concerns in the workplace, both management and employees must be committed to and involved in the process of creating safe workplaces. Commitment by management is needed to provide the necessary resources and motivation needed to effectively deal with ergonomics-related risk factors in the workplace. Management should encourage employee involvement in the ergonomics program and decisions that affect worker health and well-being. It is usually a good idea for employees to be involved in the redesigning process for ergonomic fixes. This will ensure that effective methods are developed and implemented to abate the risk factors in the workplace. Both groups must recognize the importance of instilling a concern for safety and health within the organization. Safety should be a part of the company's mission statement and its overall operation. The safety and health of the worker should be seriously considered both as a moral obligation and from the legal position of the organization.

Even if the organization has a small budget and a new program, some actions can still provide a good return on investment by enhancing employees' well-being. Often, it is best to look for "low hanging fruit"—problems that are easily addressed. Inexpensive and simple fixes such as changing the height of a computer screen and enlarging the font size and providing gloves to protect the hands, cushioned insoles for the feet, pads for the knees, rolling stools or chairs, and anti-fatigue matting can provide a new ergonomics program with a positive image and enable you to build support with both management and employees. Develop a list of priorities by using cost/benefit analysis.

Always make sure that any new equipment matches the operation being performed and the workers performing it. Determine whether the product addresses the ergonomic risk factors related to the job, without creating new risks. Document successes and related expenses to justify future program funding.

Training

It is important that all employees, supervisors, engineers, designers, and managers be trained in the principles of ergonomics and its vital role in designing proper workplaces.

If ergonomics teams are formed, added instruction is needed in team-building and consensus-development processes, apart from application of ergonomics techniques. General ergonomics awareness information for all employees need not require class instruction; it can be disseminated via handouts and all-hands meetings. The frequency of training should be determined by the needs of the company's work force, but at minimum, annual retraining is recommended. The proposed OSHA ergonomics standard identifies situations requiring training (e.g., risk factors, initial assignment or reassignment to a job that has risk factors, new hazards or exposure levels, changes in equipment or process) and suggests that training occur periodically as needed or at least every 3 years. The training should address the risk factors in a job or production process, their prevention and control, and their medical treatment. Table 16–D provides guidelines for ergonomics training for various categories of employees.

Medical Management

A proper medical management program is essential in eliminating or reducing the risk of developing MSDs. This program will provide for early identification, evaluation, and treatment of signs and symptoms. The medical management program should address the following issues:
- injury and illness record keeping
- early recognition and reporting
- systematic evaluation and referral
- conservative treatment
- conservative return to work
- systematic monitoring
- adequate staffing and facilities.

ACCOMMODATING DISABILITIES

The ergonomics of a particular task may be altered if the individual employee has a physical disability that limits or impairs his or her ability to perform the task. Title III of the Americans with Disabilities Act (ADA) requires public accommodations to provide goods and services to people with disabilities on an equal basis with the rest of the general public.

Barriers to employment, transportation, public accommodations, public services, and telecommunications have imposed staggering economic and social costs on American society and have undermined well-intentioned efforts to educate, rehabilitate, and employ individuals with disabilities. By breaking down these barriers, the ADA will enable society to benefit from the skills and talents of individuals with disabilities; allow everyone to gain from their increased purchasing power and ability to use it; and lead to fuller, more productive lives for all Americans.

The ADA gives civil rights protections to individuals with disabilities similar to those provided to individuals on the basis of race, color, sex, national origin, age, and religion. It guarantees equal opportunity for individuals with disabilities in public accommodations, employment, transportation, state and local government services, and telecommunications.

Employment discrimination against "qualified individuals with disabilities" is prohibited. This includes applicants for employment and employees. An individual is considered to have a "disability" if he or she has a physical or mental impairment that substantially limits one or more major life activities, has a record of such an impairment, or is regarded as having such an impairment. People who have a known association or relationship with an individual with a disability also are protected from discrimination.

The first part of the definition makes clear that the ADA applies to people who have impairments and that these must substantially limit major life activities such as seeing, hearing, speaking, walking, breathing, performing manual tasks, learning, caring for oneself, and working. An individual with epilepsy, paralysis, HIV infection, AIDS, a substantial hearing or visual impairment, mental retardation, or a specific learning disability is covered, but an individual with a minor, nonchronic condition of short duration, such as a sprain, broken limb, or the flu, generally would not be covered.

The second part of the definition protecting individuals with a record of a disability would cover, for example, a person who has recovered from cancer or mental illness.

The third part of the definition protects individuals who are regarded as having a substantially limiting impairment, even though they may not have such an impairment. For example, this provision would protect a qualified individual with a severe facial disfigurement from being denied employment because an employer feared the "negative reactions" of customers or co-workers. The law requires the employer to make reasonable accommodations to support the conditions under which a task may be done (e.g., lighter loads, hydraulically assisted lifting) if the employer has been given notice that the worker has a disability and that the worker requests an accommodation for that disability.

If expertise is not available on staff, another option is to use a consulting expert who is familiar with current interpretations relevant to the particular task. Opinions change as new problems and new answers are addressed by federal agencies. The Equal Employment Opportunity Commission has a major ADA enforcement role, the Architectural and Transportation Barriers Board advises the federal government about designs for facilities, and the U.S. Department of Justice's Civil Rights Division issues many interpretations of the duty to accommodate people with disabilities.

Safety professionals must examine workplace tasks and safe and reasonable accommodations required for employees with disabilities. Employers must reasonably consider the needs of a worker who uses a wheelchair, particularly in relation to egress in case of fire, or the accommodations required to warn a blind employee about a dangerous blade on a machine. Expert advice from specialists in ADA methods of accommodation should be combined with the advice of safety professionals. It is no longer acceptable to simply refuse to hire a person with a disability, asserting that the disability is an absolute bar to safe performance of workplace tasks. If there are any questions at all regarding an ADA issue, legal counsel should be consulted to ensure compliance.

SUMMARY

- MSDs have increased both in number and in percentage over the years. No occupation is free from this disease. These disorders have a significant effect on the workers' compensation system as well as compromising the health and well-being of workers. Thus, it is important that engineers, managers, designers, and safety professionals, among others, design better workplaces that will improve the safety, health, and productivity of workers.
- Those affected by the workplace (e.g., workers) and those responsible for designing workplaces (e.g., engineers, designers, and managers) should be trained in the principles of ergonomics. Workers who understand and can control the effects of risk factors will be able to function better both at work and at leisure. Participation by the worker should be sought in developing and implementing methods for abating the hazards present at the workplace.

- Further, the employee should be considered by management as an asset rather than an expendable resource.
- Ergonomics deals with the application of worker needs, abilities, limitations, and characteristics to the design of machines, tools, jobs, and workplaces that can result in productive, safe, comfortable, and efficient use. That is, ergonomics is a human-centered discipline that focuses on ways to fit the work to the worker.
- The purpose of ergonomics is to reduce the physical (and mental) stress associated with a given job; to increase the comfort, health, and safety of a work environment; to increase productivity; to reduce human errors associated with a task; and to improve the quality of work life.
- An effective ergonomics program combines both a proactive and reactive approach and involves all affected personnel.
- OSHA has been working on ergonomics issues for about three decades. Because there is no specific federal regulation, OSHA often addresses alleged ergonomics violations with the general-duty clause.
- Some frequently reported, nonoccupational risk factors of MSDs, especially CTS, include wrist size or shape, pregnancy, prior acute trauma or injury, systemic diseases, and age. These personal factors may predispose an individual to MSDs.
- Occupational risk factors that contribute to the development of MSDs include repetitive or sustained exertions, forceful exertion, awkward postures, mechanical stress, vibration, and extreme temperature. In general, the risk factors that are most often reported are repetitive or sustained exertions, forceful exertion, and awkward postures.
- Worksite analysis should focus on repetitiveness of tasks, forceful exertion, and awkward postures. Repetitiveness is determined by the frequency or duration of motion or exertion required to perform the job. Forceful exertion is the amount of physical effort required to perform the activities. Awkward postures require motion of the joint beyond its natural limit.
- To prevent or mitigate the occurrence of MSDs in the workplace, proper methods of identifying problematic jobs are needed. Passive and active surveillance can be helpful in identifying work areas with potential for MSDs.
- The key to designing a good workplace setup is to design it so that it accommodates 95% of the working population.
- The buildup of fatigue can increase the risk of accidents and injuries and can distract operators from the task at hand.
- Excessive physical/muscular demands resulting from materials-handling activities or from performing repetitive tasks can contribute to an injury or incident or impair the worker's ability to perform the task well.
- Excessive temperature extremes, relative humidity, air movement, or vibration from machines and tools can contribute to an injury or incident or impair the worker's ability to do the job.
- Controls to activate or deactivate machinery and information dials or gauges must be designed for worker safety and efficiency.
- All workers, supervisors, engineers, designers, and managers should be trained in the principles of ergonomics and its vital role in designing proper workplaces. Training is an important factor in the control and management of MSDs. The frequency of the training should be determined by the company, but at minimum, annual retraining is recommended.
- A proper medical management program is essential in eliminating or reducing the risk of developing MSDs. This program will provide for early identification, evaluation, and treatment of signs and symptoms.
- The ergonomics of a particular task may be altered if the individual employee has a physical disability that limits or impairs his or her ability to perform the task. The ADA may require the employer to alter the conditions under which the task may be done (e.g., lighter loads, hydraulically assisted lifting) if the employer has been given notice that the worker has a disability and that the worker requests a reasonable accommodation for that disability.

REFERENCES

American Conference of Governmental Industrial Hygienists. *Threshold Limit Values (TLVs) and Biological Exposure Indices (BEIs)*. Cincinnati, OH: ACGIH, 2014.

Armstrong, T. J., R. G. Radwin, D. J. Hansen, and K. W. Kennedy, "Repetitive Trauma Disorders: Job Evaluation and Design." *Human Factors* 28, no. 3 (1986): 325–36.

Bernard, M. L. "Carpal Tunnel Syndrome: Identification and Control." *Occupational Health Nursing* (June 1979): 15–17.

Bureau of Labor Statistics. *Nonfatal Occupational Injuries and Illnesses Requiring Days Away from Work*. Washington DC: U.S. GPO, 2012.

Eastman Kodak Company. *Ergonomic Design for People at Work*. 2 vols. New York: Van Nostrand Reinhold, 1986.

Grandjean, E. *Fitting the Task to the Man*. 4th ed. New York: Taylor & Francis, 1988.

International Organization for Standardization, 1 rue de Varembe, Case Postale 56, CH-1211, Geneva, Switzerland.
Ergonomics—Determination of Metabolic Heat Production. ISO 8996 (1990).

Evaluation of Human Exposure to Whole-Body Vibrations—Part 1: General Requirements. ISO 2631-5 (2004).

Mechanical Vibration—Guidelines for the Measurement and the Assessment of Human Exposure to Hand-Transmitted Vibration. ISO 5349-1 (2001).

Jastrzebowski, W. *An Outline of Ergonomics, or the Science of Work Based upon the Truths Drawn from the Science of Nature.* Nature and Industry 1857. Reprint, Warsaw, Poland: Central Institute for Labour Protection, 1997.

Karhu, O., P. Kansi and I. Kuorinka. "Correcting Working Postures in Industry: A Practical Method for Analysis." *Applied Ergonomics* 8 (1977): 199–201.

Keyserling, W. M. "Postural Analysis of the Trunk and Shoulders in Simulated Real Time." *Ergonomics* 29 (1986): 569–83.

Keyserling, W. M., T. J. Armstrong and L. Punnett. "Ergonomic Job Analysis: A Structured Approach for Identifying Risk Factors Associated with Overexertion Injuries and Disorders." *Applied Occupational and Environmental Hygiene* 6, no. 5 (1991): 353–63.

Kittusamy, N. K., O. G. Okogbaa and A. J. G. Babu. "A Preliminary Audit for Ergonomics Design in Manufacturing Environments." *Industrial Engineering* (July 1992): 47–53.

Konz, S. *Work Design: Industrial Ergonomics.* 4th ed. Scottsdale, AZ: Publishing Horizons, 1995.

Lifshitz, Y., and T. J. Armstrong. "A Design Checklist for Control and Prediction of Cumulative Trauma Disorders in Intensive Manual Jobs." In *Proceedings of the Human Factors Society 30th Annual Meeting,* 837–41. 1986.

Liu, D. *Stay Healthy, Live Longer, Spend Wisely: Making Intelligent Choices in America's Healthcare System.* Stetho Publishing, 2007.

National Aeronautics Space Administration (NASA). *Anthropometric Source Book Vols. I–III* University of Michigan Library. 1978.

National Institute for Occupational Safety and Health, 5600 Fisher Lane, Rockville, MD 20857.

Criteria for Recommended Standard for Hand-Arm Vibration. DHHS (NIOSH) Publication Number 89-106, 1989.

Elements of Ergonomics Programs: A Primer Based on Workplace Evaluations of Musculoskeletal Disorders. Publication No. 97-117, 1997.

Occupational Exposure to Hot Environments: Revised Criteria. Publication No. 86-113, 1986.

Occupational Safety and Health Administration.

Ergonomics Program Management Guidelines for Meatpacking Plants. Publication No. 3123, 1991.

Ergonomics Program, Proposed Rule, 64 FR 65768. November 23, 1999.

Parsons, K. C. *Human Thermal Environments.* Philadelphia: Taylor & Francis, 1993.

Pheasant, S. *Bodyspace, Anthropometry, Ergonomics and Design.* Bristol, PA: Taylor & Francis, 1986.

Putz-Anderson, V. *Cumulative Trauma Disorders—A Manual for Musculoskeletal Diseases of the Upper Limbs.* New York: Taylor & Francis, 1988.

Ridyard, D. T., T. G. Bobick and B. S. Starkman. "Ergonomics Awareness Training for Workplace Design Engineers." *Applied Occupational Environmental Hygiene* 5, no. 11 (1990): 771–81.

Snook, S. H., and V. M. Ciriello. *Manual Materials Handling Tables.* Liberty Mutual, 2012.

Waters, T. R., V. Putz-Anderson, and A. Garg. *Applications Manual for the Revised NIOSH Lifting Equation.* Publication No. 94-110, 1994.

Woodson, W. E. *Human Factors Design Handbook.* New York: McGraw-Hill, 1981.

REVIEW QUESTIONS

1. Ergonomists have received training in which fields of study that have contributed to the industry's understanding of ergonomics problems and solutions?
   a. anatomy and physiology of the human body
   b. anthropometrics
   c. biomechanics
   d. psychology
   e. industrial design and engineering
   f. all of the above
   g. a, c, and e

2. Define *ergonomics*.
3. Define *musculoskeletal disorders*.
4. What disorder is known as "tennis elbow"?
5. What are the five goals of ergonomics?
   a.
   b.
   c.
   d.
   e.

6. Name three general categories of demands placed on the worker, and give examples of each.
   a.
   b.
   c.
7. List the four components of an ergonomics program designed to manage cumulative trauma disorders (CTDs).
   a.
   b.
   c.
   d.
8. What are the primary factors that are assessed during ergonomics evaluations?
9. List six occupational risk factors.
   a.
   b.
   c.
   d.
   e.
   f.
10. Describe passive and active surveillance.
11. List two ways to prevent/control work-related musculoskeletal disorders.
    a.
    b.
12. List three controls for reducing repetitiveness in job activities.
    a.
    b.
    c.
13. List three controls for reducing awkward posture.
    a.
    b.
    c.
14. Describe a "neutral posture."
15. Why can a situation in which the job demands are below a worker's capabilities be just as harmful as one in which the demands are too high?
16. Guidelines and an equation for analyzing tasks to minimize back injuries were revised in 1991 by
    a. the National Institute for Occupational Safety and Health.
    b. the American Conference of Governmental Industrial Hygienists.
    c. the American Industrial Hygiene Association.
    d. the International Organization for Standardization.
    e. the Centers for Disease Control.
17. List four factors that influence the postures assumed at work.
    a.
    b.
    c.
    d.

17 Employee Assistance Programs

James T. O'Reilly, JD

Development of Employee Assistance Programs

How Employee Assistance Programs Work
Staffing and Philosophy of EAPs ▸ Procedures for Obtaining Assistance ▸ EAP and Community Treatment Resources ▸ Benefits of EAPs to Employers

Setting Up an Employee Assistance Program

Internal Employee Assistance Programs
Size and Nature of Internal EAPs ▸ Treatment and Recovery Services ▸ Cost Savings of EAPs to Employers

External Employee Assistance Programs
Choosing an External EAP ▸ Relationship with Employer

Products of Employee Assistance Programs
Treatment Services ▸ Treatment Follow-Up and Monitoring of Recovery ▸ Substance Abuse Professional Services and Other Regulatory Compliance ▸ Critical Incident Stress Debriefing ▸ Work/Family Services ▸ Employee Training ▸ Management Consultation and Organizational Development

Summary

References

Review Questions

All organizations seek to have a highly productive work force with the lowest possible accident and lost-time rates. However, within every workgroup are certain employees whose personal problems are so severe that they pose safety, productivity, and management challenges to the organization. These problems include alcoholism and drug abuse, marital and family problems, and significant psychological and emotional impairment. In addition, the problems of a worker's family often can be an equally severe drain on the worker's emotional stability. To help companies deal with these and other common employee problems, this chapter covers the following topics:

- how employee assistance plans (EAPs) work
- how to establish an EAP tailored to a particular company's needs
- characteristics and benefits of internal EAPs
- external EAPs and their relationships with employers
- the main products that EAPs provide to employees.

In most instances, when employees have serious problems, it is more responsible and cost effective for employers to offer them assistance than to use discipline or threats of termination to try to change their behavior. The goal of employee assistance programs (EAPs) is to enable employers to help troubled employees (or their family members) resolve their personal problems as quickly and cost-effectively as possible and to return them to peak productivity. The cost of ignoring troubled employees can be staggering.

> Estimates are that perhaps 17 percent of our workers ... use alcohol and other drugs on the job. One consequence of this is that every single employee who does use alcohol or drugs on the job costs his or her employer between $4,000 and $5,000 per year above payroll. (EAPA 1991, p. 8)

Much of the cost from troubled employees can be attributed to accidents, theft, and injuries on duty. Workers involved with alcohol or drugs are estimated to be caught up in work-related accidents three to four times as often as other workers. The Employee Assistance Professionals Association (EAPA, eapassn.org) has expressed concern that industrial fatalities and injuries can be linked to alcohol abuse. Worker problems need not be alcohol or drug related to pose safety and productivity risks. A worker who is depressed and exhausted after a night of family strife or distracted because a child has run away from home will certainly have difficulty concentrating on the demands and risks of the job.

EAPs can reduce the costs associated with troubled employees and get workers back on jobs. A 2002 study found significant reduction of work loss and a 43% increase in work productivity (eapassn.org). A survey conducted by *American Management Magazine* in 1985 showed that EAPs were responsible for declines of 33% in use of sickness benefits, 65% in work-related accidents, 30% in workers' compensation claims, and 74% in time spent on supervisor reprimands.

DEVELOPMENT OF EMPLOYEE ASSISTANCE PROGRAMS

The development of today's EAPs began early in the 20th century with occupational health programs designed to treat workers' illnesses. In the 1940s, companies such as DuPont, Eastman Kodak, and Consolidated Edison initiated "alcoholism programs" to address the needs of this particular troubled employee population. In 1962, the Kemper Group expanded its program to include workers with "other health and living problems" and their dependents. This approach gradually grew into what is called today the "broad-brush" approach to EAP, in which the EAP is charged with helping employees with any behavioral problem that may impair their productivity.

Today in the United States, more than 97% of companies with more than 5,000 employees have EAPs, 80% of companies with 1,001 to 5,000 employees have EAPs, and 75% of companies with 251 to 1,000 employees have EAPs. A 2008 National Study of Employers following 10-year trends related to U.S. workplace policies and benefits showed that the EAP industry continues to grow, with 65% of employers providing EAPs in 2008, up from 56% in 1998.

Workplace EAPs providing mental health assistance can reduce productivity losses caused by depression, reduce long-term treatment costs, and improve employee attendance (mentalhealthworks.ca/facts). Although alcohol and drugs are now only two of the many problems EAPs address, these two issues remain central to any program's functioning. Substance abuse is directly correlated to safety problems in the workplace. Further, employee recovery from alcoholism or drug addiction is directly linked to cost savings and productivity improvement for the employer.

HOW EMPLOYEE ASSISTANCE PROGRAMS WORK

> In general, an EAP is a set of professional services specifically designed to improve and/or maintain the productivity and healthy functioning of the workplace and to address a work organization's particular business needs through the application of specialized knowledge and expertise about human behavior and mental health. (EAPA, eapassn.org, 2014)

Assistance is provided through referrals to outside counseling or other treatment services in the community, treatment services provided by the EAP itself, or a wide variety of training and education programs available through the EAP. Services are available to both the employee and members of the employee's immediate family.

Staffing and Philosophy of EAPs

EAPs employ trained professionals whose task is to help employees find the assistance they need. To accomplish this goal in the modern workplace, the EAP will always abide by certain principles and offer certain basic services. The company will need to develop policies to establish the EAP and to encourage employee participation. Joint sponsorship or support of the EAP by both unions and management is a prerequisite for the launch of an EAP.

The principle of confidentiality is vital to any EAP's functioning. Only if employees feel that their employment is not threatened by participation in the EAP and that their problems will remain private will they willingly come forward to seek help. Policies regarding treatment for alcohol and drug abuse must establish what behavior is acceptable in the workplace and what are the consequences for violations.

EAPs are effective only if they attract voluntary participants—and if they generate supervisory or team leader referrals of employees whose performance has noticeably declined. Therefore, all EAPs must include a supervisory training or team leader program about the EAP and its use. Parallel training for union stewards or coordinators will enable unions to assist workers before they come to the attention of the supervisor.

Constructive confrontation is a basic supervisory tool that has been a foundation of EAP work since its inception. In constructive confrontation, the supervisor conveys to the troubled employee the adverse job consequences the employee may experience if his or her deteriorating job performance does not improve. The supervisor then refers the employee to the EAP. In this way, the supervisor becomes a key factor in helping the employee face his or her personal issues.

Finally, the EAP will need a publicity vehicle to inform all employees about the services the EAP has to offer, the policies governing participation, and the details of its operation. A successful EAP will attract large numbers of volunteers, as well as a significant number of directed referrals from supervisors.

Procedures for Obtaining Assistance

Once the employee or dependent (known as a client) uses the EAP's services, either voluntarily or by supervisory referral, the following steps will almost always occur. A professional assesses the client's problem and recommends treatment. The assessment and recommendation are recorded by the professional in confidential EAP files. Treatment or other services will be rendered, either through the EAP itself, by referral to providers under the employer's health plan, or through community services that are either free or must be paid for by the client. If the client is an employee who has taken time off work for treatment, the EAP may assist with reintegration to the work force as soon as practical.

EAP staff will then conduct follow-up sessions with the client to support his or her recovery and to provide additional services if they are needed. When the client's problem (such as a drug problem) may affect safety in the workplace or when the client was a supervisory referral, the EAP may monitor recovery to ensure that safety in the workplace is not compromised.

EAP and Community Treatment Resources

A primary responsibility of EAP staff is to identify and evaluate treatment resources in the community. The staff must locate qualified treatment providers for any and all problems that clients may have and in locations accessible to the employee population. Thus, EAP staff may visit psychiatric hospitals and chemical dependency treatment programs, talk with therapists and counselors, identify credit counseling and legal services, and become familiar with self-help groups and community programs in the area.

The evaluation component of this task is essential. EAP staff must continually assess the quality of treatment being given to the EAP's clients. If clients complain about a particular provider or the quality of treatment begins to decline, the EAP will need to locate new treatment providers that can better meet the needs of the EAP's clients.

Benefits of EAPs to Employers

Using an EAP provides many benefits:
- Morale improves as disruptive employees are helped.
- Managers spend less time working with troubled employees.
- Valued employees with personal problems remain with the company rather than resigning.
- Hiring and training costs are lower.
- Management spends less time and resources on discipline.
- Safety improves and liability declines.

SETTING UP AN EMPLOYEE ASSISTANCE PROGRAM

Regardless of the size of the company, or whether services will be provided by internal staff or by external contractors, the following steps are recommended to ensure the effectiveness of the EAP. First, an EAP advisory committee should be formed. This advisory committee should include representatives from management, human resources, the medical department, supervisory personnel, all labor unions, and the work force (Figure 17-1). The success of the EAP generally depends on the support of all factions within the organization. The committee's functions include evaluating the need for an EAP (and for particular EAP services), planning and implementing the EAP, encouraging its use, setting goals, and evaluating the program and its progress toward those goals. When several small companies use the same EAP, each may place one or more members on a joint advisory committee that oversees all programs.

At the outset, companies must conduct a needs assessment to define the major employee problems within the work force and the kind of EAP services required to address those problems. The needs assessment outlines the following:
- type of organization and industry
- number of worksites
- type of work/jobs
- size of work force and demographics (sex, age, ethnicity, education, special needs)
- major employee problems (from benefits and workers' compensation data)
- risk management issues (from safety, medical, and insurance data)
- management and labor identification of problem issues
- regulatory requirements of government agencies
- resources available to the EAP from the corporation.

This needs assessment may be the joint product of the advisory committee, individuals from management assigned to complete the task, and/or outside consultants brought in to lend objectivity and industry perspective to the study.

Following the needs assessment, a plan should be developed that identifies any barriers to establishing an EAP that need to be overcome, the resources inside and outside the organization that will support the EAP, and the design and structure of the EAP that will best serve the corporation and its workers. For instance, the advisory committee needs to examine the benefits plan to see whether treatments that are likely to be recommended by EAP staff are adequately covered. They should also canvass providers in the community (therapists, hospitals, etc.) to determine what services are available or lacking. Finally, they should draft general policies governing the formation of an EAP and the rights of employees. The most important decision in the plan may be the choice of which EAP delivery system to use.

The major types of EAPs are as follows:
- internal EAP—services delivered by professionals employed by the organization
- external EAP—services delivered by a contracted vendor
- union-based EAP—services delivered by trained union personnel to union members
- consortium EAP—a group of smaller companies band together to jointly contract with an EAP
- blended EAP—any combination of the others.

Management—Management must participate in all stages of assessment and definition to demonstrate support of the program. Open communication between employee and management is essential.

Labor representative—Participation by top-level labor representatives is essential during the entire process. In order to achieve eventual acceptance of the program by employees, labor must be involved from initial information gathering to development planning to program implementation.

Employees—Because they are the users of the EAP services, employees can give information on what is needed and how it can be accessed most easily.

Human resources/benefits coordinator—By monitoring use of current benefits, human resources staff can provide realistic information on the potential use of services and on the costs involved.

Health care—Advice from health care professionals is necessary for assessment and for selection of services.

Safety and health professional—The safety and health professional should be involved in the assessment of health and safety problems that may affect employee performance. If an EAP is adopted, the safety and health professional should perform follow-up evaluations.

Financial officer—Input from the organization's financial officer is necessary to define the feasibility and scope of the program.

Legal counsel—Legal advice is essential to ensure that the EAP structure developed or contracted is fair, accurate, and legally defensible.

Public relations—The public relations department can contribute information on public opinion, from both inside and outside the organizational structure, and on methods of information distribution.

Figure 17-1. Depending on the size and structure of your organization, the EAP task force may include the members discussed here. If these positions do not exist, include persons who do handle pertinent information or seek outside advice.

If the needs assessment and planning have been carefully done, the decision of which EAP type to choose may be quite clear.

Because most EAPs are either internal or external, and because the characteristics of each differ in some significant ways, the two types are described separately in the following sections.

INTERNAL EMPLOYEE ASSISTANCE PROGRAMS

Because the internal EAP is staffed by employees of the corporation, the EAP can interface easily and cooperatively with other departments such as benefits, safety, medical, and all of the operations units. EAP staff are close to the workings of the organization and will have intimate knowledge of the special problems and concerns of employee groups. The placement of the EAP within the organization is important, as managers ranking above the director of the EAP rarely seek help from the program. Yet assisting these individuals can have the greatest positive impact on company productivity. For this reason, many EAP directors/managers report directly to the president of the corporation. Alternatives include placing the EAP under the vice president for human relations, personnel, or risk management.

Size and Nature of Internal EAPs

Internal EAPs range in size from one professional to departments with 60 or more staff members. The Employee Assistance Professionals Association recommends providing one counselor for each 2,500 employees, while the Employee Assistance Society of North America recommends a ratio of one counselor for every 6,000 employees. Professional staff should be appropriately licensed and certified as health professionals in their state of operation. Staff specialties may include psychologists, social workers, and licensed professional counselors. There is now a national credential called Certified Employee Assistance Professional (CEAP), which is given in recognition of the special set of skills required to offer comprehensive EAP services. In addition to general clinical interviewing skills, the ideal employee assistance professional will also be especially knowledgeable regarding alcohol and drug problems, possess marketing skills to help publicize the EAP within the company, and have the political sensitivity to evoke trust in both supervisors and union members.

The perception of confidentiality is especially important for the internal EAP because embarrassment or suspicion on the part of employees about EAP participation can severely hamper its efforts. In addition, EAP offices need to be located at sites convenient enough to attract clients, yet separate enough from other employee areas to protect the privacy of those who seek EAP help. Policies and procedures defining the extent of EAP confidentiality, and guaranteeing that EAP participation will not adversely affect the employee's opportunity for advancement and promotion within the company, must be clearly written and widely distributed to all employee groups. The longer EAP offices can be open during the week, the more accessible the EAP will be. However, this policy may present problems in terms of staffing for evenings and weekends and may sometimes exceed the resources of the company.

The internal EAP has special advantages in its ability to publicize EAP services, train supervisors, and follow employees closely in the workplace after treatment. EAP staff have easy access to company newspapers, company bulletin boards, and other company media vehicles to promote EAP services. The staff can often participate in management meetings to identify new training opportunities and learn quickly about managers' current needs for new EAP "products." The EAP can work in partnership with other departments to provide joint training for supervisors and managers on the best way to use EAP services. Because EAP offices are in or near the workplace, staff can work closely with employees as they complete treatment and return to their jobs.

Treatment and Recovery Services

Research shows that professional follow-up after behavioral treatment improves treatment success. Regular phone calls by EAP staff to troubled employees who are in treatment or have recently completed treatment will help workers integrate back into the workplace as they continue their recovery. If problems arise for these workers, a quick and confidential session with EAP staff can resolve issues or provide an opportunity to revise treatment strategy. Close follow-up and monitoring of recovering alcoholics and drug addicts after treatment can reduce the frequency of relapse and motivate employees to return to sobriety if a relapse does occur. In safety-sensitive industries, the EAP may be responsible for notifying the medical staff to remove an employee from service if relapse occurs.

This type of intervention can be done confidentially (because medical staff, not EAP personnel, contact the supervisor), can strongly motivate the employee to resume recovery, and simultaneously protects the safety of the workplace. In the transportation industry (where the U.S. Department of Transportation encourages, and in some circumstances requires, monitoring of drug addicts), alcohol and drug recovery rates are frequently reported to be much higher than in industries where alcohol and drug monitoring by internal EAPs is less common.

Another strength of internal EAPs is their familiarity with the treatment needs of the employees and their knowledge of local community resources. EAP staff can easily visit a client who has been hospitalized for psychiatric problems or is in a local chemical dependency treatment center. This visit will help the client make a less stressful transition back to the workplace and will allow EAP staff to have firsthand knowledge of any changes in the quality of treatment or staff at the hospital. When many employees show an interest in a common problem or issue in the workplace (such as stress, domestic violence, grief issues, or parenting skills), the internal EAP staff can identify the need and quickly locate community resources to provide training on the issue in question. Because the EAP exists in the same community as its treatment providers, the program staff can establish a rapport with the therapists and hospitals that frequently work with the EAP's clients. This close working relationship may help improve the quality of treatment for clients and may enable the EAP to negotiate reduced fees to benefit both the clients and the company.

Cost Savings of EAPs to Employers

Because the mission of an EAP is to improve productivity and reduce employee costs, the EAP must document its cost savings to the organization. The internal EAP can readily provide this documentation by working closely with workers' compensation, safety, benefits, and medical departments to track the employee costs of clients who come to the EAP. Internal EAPs routinely access benefits data to show that employees treated through the EAP use fewer health benefits for themselves and their family members after completing treatment. Safety and workers' compensation data can be assembled to show that EAP participants are injured less frequently at work and have fewer accidents. Lost-time data should show reductions in lost time after an employee enters the EAP. Similar EAP cost reductions could be shown for employee turnover, theft, arbitration, and legal costs, provided the internal EAP uses its position within the company to assemble this data.

EXTERNAL EMPLOYEE ASSISTANCE PROGRAMS

External EAPs can provide services to a wide variety of companies, from small, privately held corporations to major Fortune 100 organizations. An external provider can focus its efforts on providing a specialty service while spreading the costs over a greater number of client organizations. The external provider can develop sophisticated information systems, hire personnel with greater expertise and experience in EAPs, and develop other systems to better address issues such as quality assurance/quality improvement and provider network development.

Choosing an External EAP

The selection of an external EAP provider should include many considerations, keeping in mind that the external provider will be expected to offer all the services that an internal program would. Companies must take the time to choose an adequate external provider with whom they can build a quality EAP (Figure 17–2).

The key elements to look for in a quality external EAP provider are as follows:

- Does the EAP have the professional expertise on staff (e.g., Certified Employee Assistance Professionals [CEAPs], who have appropriate experience in all areas of the design, implementation, function, and maintenance of EAPs)? Certification as an EAP professional is based on commonly accepted standards of training and experience in the field. As a result, the presence of CEAP personnel on staff also confirms that the provider has the knowledge and ability to handle all basic EAP services and many other types of services.
- Does the EAP have an appropriate and adequate provider network? An appropriate provider network must include credentialed and trained EAP assessors (licensed clinicians) who are available to all employees. All assessors should be subject to a formal credentialing process by the organization that includes verification of personnel's education, licenses, professional liability insurance, and experience. The EAP network should also include adequate means to identify and access all other helping resources in the employer's service areas (i.e., those near all employer worksites).
- Does the EAP staff offer assessment, short-term counseling, and referral services one-on-one with a licensed clinician/assessor, or are these services performed over the telephone?
- How many sessions with a clinician/assessor are available for client assessment and short-term counseling? (Commonly, providers will offer one- to three-session, one- to six-session, or one- to eight-session modules at different prices.)
- Does the EAP follow up with the client and the treatment provider to make sure the client is receiving appropriate and adequate services?
- Does the EAP coordinate with the employer's insurer/managed care organization when the client must be referred for treatment that is paid for using health benefits?
- Does the EAP have an adequate system to maintain confidentiality of all program records?
- Is the EAP's system for quality assurance and quality improvement adequate and appropriate?

Professionals and facilities that provide services must meet at least the following criteria, but you may want to add others:
- Physicians should
    a. hold a degree in medicine (MD, DO)
    b. hold a license to practice in your state
    c. have expertise in dealing with emotional and addiction problems, especially if they will be working on substance abuse cases
    d. be recommended, if possible, by a physician or other reliable source who is familiar with the physician's services
    e. be on staff at a hospital, with an appointment in their specialty
- Psychologists should
    a. hold a state license or certification
    b. hold a doctor of philosophy degree (PhD) in psychology
    c. have two years' supervised clinical practice
- Alternate requirements for psychologists:
    a. a master's degree (MA) in psychology
    b. three years' supervised clinical experience
    c. additional training in the treatment they will provide, such as chemical dependency or marital and family therapy
- Social workers should
    a. hold a master's degree (MA) in social work or a master's of social work (MSW) degree from an accredited school of social work
    b. have two years' supervised clinical experience
    c. be a licensed clinical social worker (LCSW) in states where social workers are licensed or an accredited clinical social worker (ACSW) in other states
- Nurse practitioners should
    a. be licensed by the state as a registered nurse (RN)
    b. comply with state credential requirements for nurse practitioners
    c. hold a master's degree (MS), as recommended by the American Nurses Association
    d. have experience and expertise in dealing with the relevant area
- Professional counselors should
    a. hold a master's degree (MA) in counseling from an approved program at an accredited college or university
    b. be a licensed professional counselor (LPC) in states where counselors are licensed
    c. have two years' supervised clinical experience
    d. have special training in the treatment they will provide, such as marital and family therapy
- Facilities. Visit facilities under consideration and discuss with managers the following criteria:
    a. proof of Joint Commission for the Accreditation of Hospitals accreditation (for inpatient treatment providers)
    b. proof of state licensure
    c. licenses or certifications of personnel
    d. variety of resources
    e. written objectives for program, services, and treatment, and for procedures followed to meet these objectives
    f. individualized program, services, and treatment plans
    g. assurance of confidentiality and compliance with state and federal confidentiality regulations
    h. length of time in business (more than one year)
    i. marketing methods
    j. client references
    k. usage rate of other EAP clients
    l. methods and service consistent with organization's employees
    m. follow-up provided or available (with behavior change programs)
    n. location convenient to employees
    o. service times, length and number of sessions, and frequency convenient to employees
    p. minimum and maximum enrollment for classes
    q. flexibility of service provision logistics
    r. safety precautions consistent with those of the organization
    s. waiver form for participants in the program
    t. evaluation of participants and method of sharing results with the organization
    u. evaluation of program, services, and treatment
    v. willingness to negotiate costs
    w. contract for program, services, and treatment

Figure 17–2. Professionals and facilities that provide services must meet at least these criteria.

- Does the EAP have an adequate information system that is not leased? (Leasing opens another opportunity for breach of confidentiality.)
- Can the EAP produce program activity reports that will meet company information needs?
- Will the EAP provide an organization needs/resource analysis before implementing the EAP, technical assistance in the development of all program policies and procedures, supervisor procedural training, employee orientation, program promotion (posters, wallet cards, brochures, wellness seminars, newsletter articles, etc.), and management consultation regarding the supervisor referrals?
- Does the EAP offer 24-hour access to services, including crisis intervention, via toll-free phone lines?
- Does the EAP offer other desirable services that the purchaser may be interested in, such as critical incident debriefing, fitness-for-duty assessments, and interventions for violence in the workplace?
- Does the EAP cover all personal problems or only selected issues? If some personal issues are excluded, how important are these to a company's work population?
- What additional personal issues and services are offered as optional coverage, and what are the costs?
- Does the EAP cover the immediate family/all household members under the program services?
- How long has the EAP been operating and how stable is its work force? Having to change EAPs unexpectedly can be costly in many ways.
- For what length of time will the EAP guarantee its rates? What is included? Will the EAP offer any performance guarantees?
- What other companies with similar needs have used or are using the EAP? Be sure to request references from these companies.

Relationship with Employer

Although the relationship between the external EAP and the employer is obviously different from that of the internal EAP and the employer, the results are usually similar. The same functions are performed by both the internal and the external programs, but a different approach must be taken by the external provider.

Before the implementation of the EAP, the external consultant must make a special effort to develop relationships with key people in departments such as benefits, safety, and medical, which are essential to the function of the program. This external relationship may seem to be a disadvantage at first glance. However, it may work to the advantage of the external provider and, ultimately, to the EAP itself. Many times, because of internal politics, organizations are willing to give information more freely to outside consultants than they are to internal employees. As a result, in some cases, the external provider wields greater authority as a paid consultant.

In addition, the perception of confidentiality is enhanced by the fact that the external EAP provider is a separate entity from the client organization. Because the people who provide the EAP services are not employees of the client organization, they are less likely to be pressured by management into divulging information about clients to anyone within the employer's organization. On the other hand, this distance between the external EAP and the client organization also creates a greater need for the EAP provider to perform an in-depth organizational analysis to define and clarify the structure, function and resources, and barriers to the successful operation of the EAP.

Both internal and external EAPs have demonstrated their ability to maintain a healthy, productive work force. Each employer must decide which system is the better venue for its needs and situation.

PRODUCTS OF EMPLOYEE ASSISTANCE PROGRAMS

EAPs develop different services, or products, to meet the special needs of the organizations in which they function to improve productivity and safety. All EAPs offer the basic products of assessment of client problems, referrals for treatment, employee training, and record keeping. Among these basic products, different EAPs may offer different levels of service. For instance, one EAP may perform assessments primarily through face-to-face sessions with the client, while another may perform most assessments over the telephone. Some EAPs may offer only yearly supervisor training and some written material for each employee that explains the EAP, while others train hundreds of employees every month to heighten awareness of EAP services.

Clearly, the number of staff and the budget of the EAP will determine the nature of and time spent on these basic products. With the continued development of "broad-brush" EAPs and more sophisticated methods of meeting employers' needs, additional EAP products have become available. The EAP advisory committee will want to consider the following products to find those that best meet the needs of the organization.

Treatment Services

Some EAPs provide clients with three to eight counseling sessions in their offices before deciding whether to refer a client to a provider in the community. Thus, the EAP becomes a service provider itself before it refers the client. This service may include a "crisis-intervention" component in which the

EAP responds quickly to workplace incidents by helping a troubled employee and/or any co-workers who have been affected by the troubled employee's behavior in the workplace. Most EAP counselors have been trained to treat clients by using short-term, problem-resolution-focused therapy. Usually, these sessions are offered at no cost to the client.

The advantages of EAP-provided treatment include the fact that more clients receive help for their problems (with referral, some clients inevitably fail to follow through and do not get help), and fewer providers bill the company's insurance plan(s) for their services. Possible drawbacks to this product include client reluctance to trust an agent of the company to provide their treatment, and the possibility that the EAP counselor is not as skilled in treating the client's problem as an outside specialist might be.

Treatment Follow-Up and Monitoring of Recovery

Although most EAPs offer some form of follow-up after treatment is complete, some EAPs offer more follow-up than a questionnaire or a phone call a few months after services are rendered. Intensive follow-up is most often provided to chemically dependent clients. Research has shown that monitoring after treatment can significantly reduce relapse, especially in the drug-addicted client. For companies where safety is particularly critical, monitoring a drug addict after treatment may be cost effective. Follow-up can consist of regularly scheduled calls or client visits to the EAP, regular reports to EAP staff from a professional aftercare provider, or random drug or alcohol testing coordinated by the EAP.

Some EAPs require all three monitoring tools to ensure recovery and detect relapse by an employee as soon as it occurs. Depending on the employer's policies, relapse by the employee may result in EAP referral for additional treatment, changes in job responsibilities, or termination. In any case, employee monitoring and follow-up will reduce relapse rates and may eventually help the chronically relapsing employee either to change or to leave the organization.

Substance Abuse Professional Services and Other Regulatory Compliance

In 1988, the U.S. Drug-Free Workplace Act was passed, requiring many organizations to provide information and services addressing drug abuse. EAPs are ideally suited to meet many of the act's requirements. In 1991, the Omnibus Transportation Employee Testing Act required drug testing and EAP education for safety-sensitive employees in the transportation industry. This act was revised in 1994 to include alcohol testing and provision of EAP information on alcohol abuse. The Department of Transportation (DOT) now requires substance abuse professional (SAP) assessment, referral, and monitoring recommendations for all employees who test positive on a DOT drug or alcohol test. The DOT lists CEAPs as one of a select group of clinicians who are qualified for SAP work.

As a result, many corporations now expect this SAP service from their EAP. SAPs must adhere to a set of regulations issued by the DOT. They must ensure that all safety-sensitive employees with positive test results receive appropriate treatment and monitoring recommendations before they return to work in DOT-regulated positions. Thus, many EAPs provide a set of substance abuse services required by law, allowing their organization to meet regulatory requirements without having to shop for these services from a variety of vendors.

Critical Incident Stress Debriefing

Traumatic incidents in the workplace, such as employee violence, employee or customer death, fatal or near-fatal accidents, or natural disasters, can cause lasting symptoms or problems for many employees. After an incident, some employees appear confused, anxious, depressed, or angry for an extended period of time. These are normal reactions to trauma and occur in normal employees as well as in troubled personnel. All of these conditions reduce productivity and safety, and may prompt employees who were more severely affected to resign.

In the past 10 years, a technique called critical incident stress debriefing (CISD) has been developed to reduce many of the aftereffects of traumatic incidents and to help identify employees who will need mental health treatment. CISD can be done quickly by trained professionals with groups of 10 to 15 employees, usually in one or two sessions approximately 2 hours long. Many times, the CISD sessions are most effective if previously trained volunteers from the work force assist the professional in the group meeting. These brief interventions have been shown to significantly help employees and reduce the number of PTSD cases that usually develop after an incident.

In many organizations, EAP staff are asked to coordinate the provision of CISD services, or they may actually assume the role of service provider for CISD sessions. The staff may be involved in organizing the training of volunteers to support the CISD effort. After an incident, EAP staff ensure that all employees who need CISD services are encouraged to participate. After each CISD session, the professional conducting the session can involve EAP staff in locating therapists or hospitals for individuals who may require ongoing mental health services.

Work/Family Services

Although some large corporations have separate work/family departments, in many organizations these services are offered by EAP. "Work/family" refers to all issues in the lives of employees that occur outside the job but affect the workplace. Issues include the need to arrange child care, to look after elderly parents, to support a relative who is ill, and to work in a way that supports a healthy lifestyle. An EAP may contract with a day-care network to help working parents with children or may coordinate day-care services offered by the company. The EAP can direct employees to elder-care service agencies in their community or may contract with an elder-care service provider with special expertise in issues of the aging. Again, if the EAP can assist an employee with these tasks, the employee will be more focused on the job, more attentive to safety in the workplace, and more productive.

Employee Training

Training employees to be aware of and to use EAP services is one of the staff's basic priorities. But some EAPs are beginning to widen their training programs beyond the basic services. For example, particular groups of employees may express a need to learn about stress management, parenting skills, couples communication, smoking cessation, or dealing with difficult adolescents. If EAP staff can arrange for experts on these subjects to provide the requested training, or if staff members possess the expertise to offer the training themselves, the EAP can build employee awareness of its services and enhance the level of trust that employee groups feel for the program.

Management Consultation and Organizational Development

Another basic service of EAPs is management consultation, that is, helping managers deal with troubled employees before and after referral to an EAP. In recent years, many EAPs have expanded this basic service to encompass consultation with management on any aspect of the company's operations that seems to be related to an increase in the number, or dysfunctionality, of troubled employees. This service can best be described as organizational development (OD) consultation (which may be the function of a separate department in some larger corporations).

Examples of OD work that may fall to EAP staff could include an analysis of why a certain work group has a sudden increase in carpal tunnel injuries, why a certain work location experiences a much higher rate of employee resignations, or why behavioral care costs in one particular division have risen much faster than they have in other areas of the company. In each case, the EAP's task is not to help a particular employee, but to diagnose a condition that may have caused a group of employees to lose productivity. If an accurate diagnosis is made, then the EAP can work with appropriate operations, safety, or personnel managers to resolve the problem. Companies may look to the EAP for this type of service because of its psychologically sophisticated staff and its access to information about troubled employees that is not available to other departments.

SUMMARY

- The goal of any EAP is to improve employee productivity by reducing lost time, reducing accidents and workplace injuries, and enabling employees to do their best on the tasks assigned to them in the workplace. EAP assistance is provided through referrals to outside counseling or other treatment services in the community, treatment services through the EAP, or a variety of training and education programs available through the EAP.
- To develop an EAP, a company establishes an EAP advisory committee, conducts a needs assessment, develops a plan for establishing the EAP, and drafts general policies governing the formation of an EAP and the rights of employees. The major types of EAPs are internal, external, union-based, consortium, and blended.
- Internal EAPs are staffed by employees of the corporation, which allows it to work easily with other departments of the company. Internal EAPs may also gain employee trust more easily, but workers may have doubts about confidentiality of their records.
- External EAPs often can provide a wider range of services than can internal EAPs. Companies must exercise care in choosing an external provider. In addition, workers may feel that an external EAP will protect the confidentiality of records.
- Products offered by EAPs include counseling sessions, treatment follow-up and monitoring, substance abuse programs, critical incident stress debriefing, work/family services, employee training, and management consultant services.

REFERENCES

American Management Magazine. 1985.

EAPA (Employee Assistance Professionals Association), eapassn.org.
  EAPs: Value and Impact. 1998.

Employee Assistance Society of North America. *EASNA Standards for Accreditation of Employee Assistance Programs.* Berkley, MI: EASNA, 1990, pp. 4, 23.

Keaton, W. H., et al. *Prosser and Keaton on Torts.* 5th ed. St. Paul, MN: West, 1984.

National Institute on Drug Abuse. *National Household Survey on Drug Abuse.* Rockville, MD: U.S. Department of Health and Human Services, 1990, p. 8.

REVIEW QUESTIONS

1. An employee assistance program (EAP) is designed to assist in the identification and resolution of productivity problems associated with employees impaired by personal concerns including, but not limited to:
   a.
   b.
   c.
   d.
   e.
   f.
   g.
   h.
2. How much more often do workers involved with alcohol or drugs become involved in work-related injuries compared with other workers?
   a. 3–4 times
   b. 5–6 times
   c. 7–8 times
   d. 9–10 times
3. How is EAP assistance provided?
4. What principle is vital to the functioning of any EAP in encouraging employees to willingly come forward to seek help, and why?
5. Once the employee (or client) contacts the EAP for services, either voluntarily or by a supervisor's referral, the following steps should occur:
   a.
   b.
   c.
   d.
   e.
   f.
   g.
6. List five benefits that EAPs provide to employers.
   a.
   b.
   c.
   d.
   e.
7. In setting up an EAP, an EAP advisory committee should be formed, composed of members from which group(s)?
   a. high-level management
   b. supervisory management
   c. all labor unions
   d. members of the work force
   e. all of the above
8. What factors should be considered in a needs assessment to define the major types of employee problems and the kind of EAP services needed to address those problems?
9. List the five major types of EAPs.
   a.
   b.
   c.
   d.
   e.
10. What is considered a strength of an internal EAP?
    a. EAP staff can interface easily and cooperatively with other departments.
    b. EAP staff will have more knowledge of the special problems and concerns of employee groups.
    c. EAP staff will become more familiar with the treatment needs of the employees.
    d. EAP staff will have more knowledge of local community resources.
    e. all of the above
11. Stemming from the OSH Act General Duty Clause, employers have legal duties to employees, customers, clients, and third parties in situations involving
    a. family services.
    b. workplace violence.
    c. substance abuse treatment.
    d. a and c
12. List five of the nine EAP products that should be considered by the advisory committee to meet the needs of the organization.
    a.
    b.
    c.
    d.
    e.
13. In monitoring a drug addict after treatment, what are three methods of follow-up?
    a.
    b.
    c.

Emergency Preparedness

Richard Payant, DBA, MA, CFM, CPE, CHS
Philip E. Hagan, JD, MBA, MPH, ARM, CIH, CET, CHMM, CHCM, CHSP, CEM

Management Overview

Developing an Emergency Management Plan

Types of Emergencies
Fire and Explosion ▸ Floods ▸ Hurricanes and Tornadoes ▸ Earthquakes ▸ Civil Strife and Sabotage ▸ Work Accidents and Rumors ▸ Shutdowns ▸ Emergency Management Planning ▸ Hazardous Materials ▸ Radioactive Materials ▸ Weather-Related Emergencies

Plan-of-Action Considerations
Program Considerations ▸ Chain of Command ▸ Training ▸ Hazardous Materials/Spills Emergencies (Hazwoper) ▸ Command Headquarters ▸ Uniform Incident Command System ▸ Emergency Equipment ▸ Alarm Systems ▸ Fire Brigades ▸ Emergency Medical Services ▸ Warden Service and Evacuation ▸ Transportation

Outside Help
Mutual-Aid Plans ▸ Contracting for Disaster Services ▸ Municipal Fire and Police Departments ▸ Industry and Medical Agencies ▸ Governmental and Community Agencies

Summary

References

Review Questions

No industrial, commercial, mercantile, or public-sector organization is immune from disaster. Emergencies can arise at any time and from many causes, but the potential loss is the same—injury and damage to people, the environment, and property. Advance planning for emergencies is the best way to minimize this potential loss. To help management cope with emergencies, this chapter covers the following topics:

- overview of the duties and priorities of management for emergency planning
- discussion of the types of emergency plans available
- the types of emergencies that organizations may encounter
- what elements of emergency planning organizations need to consider
- the need to coordinate outside help during emergencies.

MANAGEMENT OVERVIEW

A written comprehensive emergency management plan is intended to support a planned response to unexpected or disastrous events that can strike any organization, such as a fire or natural disaster. Emergency planning is often assigned to the safety professional. In some instances, this may be acceptable; however, the responsibility ultimately lies with the highest levels of management, who best know a facility's resources, operations, and capabilities. The safety professional should act as the consultant, guiding line management through the process of identifying potential emergency events and developing primary and contingency plans to respond to them.

The safety of employees and the public must be the first concern in planning for an emergency. Planning should take into account the immediate short-term needs people are likely to have (e.g., treatment of injuries) as well as the long-term needs that may result from an emergency event.

Next, management should consider ways to protect the property, operations, and the environment. In new facilities, this may mean locating certain buildings or operations well away from others or locating operational response teams in a more central area to allow for a quicker response. In general, all emergency plans must include salvage, overhaul, and possible decontamination operations.

The final steps in the plan should involve restoring business operations to normal. In emergencies, a facility's operations are likely to be affected or shut down altogether. Management will need to decide how and when to resume operations in the face of such obstacles as temporary wiring, lack of environmental control, or the need for significant rebuilding. Moreover, the probability of environmental damage to the immediate and surrounding community will demand considerable attention from the organization's top management. Operations that could adversely affect the health of employees or community neighbors must meet the emergency-planning requirements of applicable regulatory agencies.

Regardless of the size or type of organization, management is responsible for developing and operating an emergency-planning program designed to address these eventualities. An effective plan requires the same good organization and administration as any business undertaking. No single emergency management plan will address all potential incidents for all organizations. Each organization must decide on a plan that fits its needs and resources.

Emergency plans involve organizing and training small groups or contracting with cohorts of people to perform specialized services, such as evacuation, fire fighting, rescue, spill response, or first aid. These small, well-trained cohorts can serve as a nucleus that can be expanded to meet different kinds of emergencies. Even with outside help available, a self-help plan is the best assurance that losses will be kept to a minimum. Although planning should occur within each unit of the organization, the facility's emergency management staff should develop and implement a large-scale response scheme that encompasses all departments.

This comprehensive emergency management plan should meet all the different types of contingencies. Certain basic elements such as command functions, communications, and emergency staff personnel will be common to all emergency operations. Of course, the comprehensive plan should contain many procedures to deal with natural, technological, and hazardous–materials related hazards.

Organizations should also plan for emergency situations in which they may have to stand alone with no outside help. This is not a remote possibility in the event of a major disaster such as a tornado, hurricane, flood, or brushfire, which may affect a wide area. At times like these, communities will have to either stand alone or join with others in a mutual-aid pact.

Self-help plans should include provisions for recall of off-duty personnel. Many of these callback lists look good on paper, but managers should exercise the lists to see if they really work. Long holiday weekends can present real problems should an emergency occur during these times. Further, management should consider how to contact the required personnel and how long it may take to secure a core group of employees.

Before an organization develops an emergency plan, it is important to identify and evaluate the potential disasters that might occur. Types of emergencies are addressed later, in more detail, in this chapter.

The next step is to assess and prioritize the potential

harm to people, the environment, and property. Time of day and workshift patterns are other factors that should be considered in assessing the potential damage. Planning should take into consideration the impact of a catastrophe that may occur during a weekend or holiday when no one or only a skeleton staff may be on hand.

To estimate potential damage to property, managers should look not only at the general structures but their surroundings as well. For example, a building may be strong enough to resist an earthquake, but a sudden 7-in. (18-cm) rainstorm may cause flooding that could short out all electricity. On the other hand, an exploding boiler located in an adjacent building would probably not harm another building in close proximity.

Next, probable warning time should be considered. For example, although a flood occurs over a period of several days, a telephoned bomb scare may afford only a few minutes' warning. Warning time should give management a chance to alert personnel and to mobilize the plan. It is desirable to have a number of different plans, depending on the nature of the emergency and the actual (or estimated) time available.

Another factor is how much organization operations must be changed to deal with the emergency. For example, in anticipation of a heavy snowstorm, it may be necessary to send employees home early. Some equipment may be left on or idling instead of being shut down completely. Finally, management will need to consider what power supplies and utilities may be needed, particularly those used for fire protection, lighting, ventilation, and communications.

A basic emergency preparedness plan should include a chain of command, warning system, medical treatment plans, communications system, shutdown and evacuation procedures, and auxiliary power systems. Not every element discussed in this chapter will apply to every organization. Also, several functions may be combined and handled by one person, particularly in a smaller company. In general, however, the material in this chapter is directed to the more elaborate and expanded type of organization and planning process.

DEVELOPING AN EMERGENCY MANAGEMENT PLAN

Once management has an idea of the company's risk of various identified hazards, has assessed the response capabilities, and has reviewed existing or previously written plans, it should decide on how to develop the emergency management plan that is best for the facility.

The type of facility and its associated hazards are factors that determine the complexity of the plan. However, other factors also must be considered, such as the availability of qualified personnel to write and maintain the plan and the availability of resources to develop and implement it. Nuclear power facilities, for example, are required to develop extremely sophisticated plans that cost utilities millions of dollars to develop, implement, and maintain. Other facilities, such as small manufacturing plants where few chemicals are located in an area not prone to natural hazards, often require less complicated plans. Developing the correct plan for a particular facility is very important. Action guides, threat assessments, and mutual-aid agreements are often used when developing emergency management plans.

Action guides or checklists are generally short, simple descriptions of basic procedures that must be followed, such as whom to call, necessary emergency information, and basic response functions. These guides and checklists are intended to be more of a reminder and should therefore not be used by an untrained individual.

Threat assessments are used to identify potential problems and to develop potential responses. Threat assessments can be highly detailed and identify all responsible individual actions that must be taken to mitigate the problem at hand. Sometimes, a threat assessment is written for each type of possible hazard at the site. For example, there may be a chemical spill threat assessment plan as well as one for a hurricane.

Mutual-aid agreements with other community stakeholders allow organizations to take advantage of additional resources in times of need.

Developing an effective emergency management plan is important and management should ensure that the plan is flexible enough to address a variety of emergency situations. To ensure effectiveness, the plan's elements should be exercised by affected stakeholders on a regular basis. It is difficult to deal with a crisis situation if participants are trying to implement response measures for the first time.

TYPES OF EMERGENCIES

The first step in the emergency-planning process begins with determining what types of hazards may affect the organization. Targeting specific hazards allows the organization to create a comprehensive planning, organizing, and implementation program. Hazards posing a threat to the organization will vary according to its location, production/engineering processes, and work practices.

Many sources of hazard information can help management determine the likelihood of specific emergency events in a locality. These sources include historical knowledge and records of accidents, fire statistics, National Weather Service logs, U.S. Geological Survey studies, location of rail lines and airports, and local emergency management Hazard Vulnerability Analysis documents. The organization's management should agree on a set of hazards that appear to have some chance of occurring. After this assessment, emergency-response operations planning can begin, based on a realistic study of the potentially disastrous threats facing personnel, property, and environment. A discussion of types of emergencies follows.

Fire and Explosion

Except when fires result from large-scale explosions, warfare, civil strife, or hazardous chemicals, a fire emergency usually allows some time for marshaling fire fighters and organizing an evacuation, if necessary. Many large fires originate as small blazes that, if caught early, could be controlled by trained personnel. Therefore, prompt action by a small, properly trained and equipped group can usually resolve most situations. However, plans should include alerting extensive fire-fighting forces at the first indication of any fire growing beyond the incipient or "small fire" stage. Plans should also include procedures for controlling chemical contamination that may mix with the runoff from water used to extinguish the fire. Also, fire-fighting and evacuation procedures should take into consideration toxic gases, smoke, and fumes that may be produced by burning materials.

The main point is that small fires must be extinguished as soon as they start. The first 5 minutes are considered the most critical. Good housekeeping, prompt action by trained people, proper equipment, and commonsense precautions will prevent a small fire from becoming a disastrous blaze.

Specific information on fire extinguishing and control appears in Chapter 9, Fire Protection, in the *Engineering & Technology* volume.

Both small and large fires have the potential for producing significant environmental problems, including the following:
- production of toxic gas and dust from combustion and decomposition
- thermal plumes that can carry the materials substantial distances
- disposal problems involving large quantities of contaminated water used in extinguishing the fire.

Commonly used construction materials, furniture, carpets, and industrial materials can release significant quantities of toxic combustion products. The combustion of industrial and agricultural chemicals and materials may also be a source of hazardous gases, vapors, fumes, or dusts. The buoyancy effects of the hot gases caused by the fire may enhance the dispersion of these materials. Depending on weather conditions, these plumes may travel substantial distances and remain concentrated. Ash and dusts generated from some forest fires in the western part of the United States have been detected, for example, in the Midwest. To minimize these effects, those responsible for responding to fires must be aware of the potential hazardous airborne materials, be able to track their movement, and be prepared to measure their ground-level concentration. These activities may require access to or on-site collection of meteorological data and air sampling.

An additional environmental hazard in responding to fires is the disposal of the potentially contaminated wastewater produced in extinguishing a fire. For example, 100,000 gal of water can easily be used in fighting a simple residential fire. If 1,000 gal (3,784 L) of a 10% concentrated pesticide is released in the course of the fire, the resultant concentration could be a 100-ppm solution. Introduction of such a liquid waste flow into a sewer system or surface waterway could have significant environmental ramifications. Therefore, diking, chemical neutralization, use of chemical absorbents, or a controlled burn may be considered necessary containment methods.

Floods

When a facility is located in a floodplain, it should have the protection of dikes of earth, concrete, or brick construction. The probable high-water mark can be obtained from the U.S. National Weather Service or the Army Corps of Engineers. The latter group also provides valuable assistance in planning floodwater control.

Floods do not strike suddenly (except flash floods caused by torrential cloudbursts or the bursting of a storage tank, dam, or water main). Ordinarily, there is enough time to take protective measures when a flood seems imminent. Six inches of fast-moving water can knock an adult off his or her feet; a depth of 2 feet will cause most vehicles to float.

Flash floods and floods are the number-one cause of deaths associated with thunderstorms, approximately 140 fatalities each year. Most flash floods are caused by slow-moving thunderstorms, and they frequently occur at night in mid- to late summer. Nationally, 75% of presidential disaster declarations are the results of floods.

Another source of information on flooding in a particular area is the local emergency-planning commission or the U.S. Geological Survey. Many facilities are not located in floodplains. However, the U.S. Geological Survey has classified different types of floods that may hit a region at specified times. One example is a 100-year flood that has a probability of occurring only 1% of the time (once every 100 years). These infrequent events should be included in a company's planning process, even if the chances of their occurring seem remote.

Organizations must make careful preparations for their emergency responses to a flood situation. For example, a list of typical emergency equipment and material could include sandbags, battens for windows and doorways, boats, tarpaulins, fuel-driven generating equipment (such as gasoline-powered arc-welding machines or motor-generator sets), standby pumping equipment, a supply of gasoline in safety containers to fuel this equipment, lubricating oil and grease, rope, life belts, portable battery-operated radio equipment, audio speakers, and so on.

In addition, companies that use tank cars should make some provision for moving these cars to higher ground and anchoring them. They should also move portable containers above the high-water mark, along with buoyant materials and chemicals soluble in water. Storage tanks under the probable high-water mark (including underground tanks) should be specially anchored to prevent floating. Workers can construct auxiliary dikes of sandbags or dirt around key areas.

Other precautions to take in anticipation of flood conditions include the following:
1. eliminating electrocution hazards and providing proper grounding and electrical fault protection
2. bracing or storing important equipment, materials, and chemicals off the ground
3. ensuring availability of pumps and emergency power sources
4. protecting against soil erosion, which may cause serious structural damage
5. provision and protection of potable water supplies.

Hurricanes and Tornadoes

The Atlantic and Gulf coasts are most frequently exposed to winds of destructive hurricane force. However, some inland locations are not immune to this type of disaster.

The National Weather Service and other agencies have developed improved methods of detecting and tracking hurricanes; thus, ample warning can be given for maximum protection of property and evacuation of personnel from threatened areas.

Organizations regularly exposed to this hazard have developed a system of tracking hurricanes on a map. At predetermined locations, a specified alert condition becomes effective, and each supervisor completes a checklist for that alert. As the hurricane progresses toward the facility through the 100-mi (160-km) circle, 50-mi (80-km) circle, and so on, the facility can be shut down in an orderly manner.

Buildings constructed in areas where hurricanes occur should be built to withstand these destructive winds and tides. Basic preventive measures include equipping facilities with storm shutters or battens that can be promptly attached, at least on the side from which the storm is expected to approach. If this is not done, the facilities could have windows shattered, the roof torn away, or other parts of the building destroyed. If the roof is lost or damaged, building contents can be drenched by the heavy rains that accompany the storm and by water from broken sprinkler pipes. Therefore, roofs should be securely anchored and tall structures (e.g., chimneys, water towers, and flagpoles) designed to withstand high winds.

As for tornado emergencies, although the central Mississippi Valley is considered the top tornado area of the United States, almost every state has experienced these destructive storms. These storms can inflict enormous damage quickly, although it is usually restricted to a small area.

In an average year, 1,200 tornadoes cause 70 fatalities and 1,500 injuries nationwide. A tornado is a violent rotating column of air extending from a thunderstorm to the ground. Tornadoes may appear nearly transparent until dusts and debris are picked up or a cloud forms within the funnel. The average tornado moves from southwest to northeast, but tornadoes have been known to move in any direction. The average forward speed is 30 mph (48 kmh) but may vary from nearly stationary to 70 mph (113 kmh).

The National Weather Service cannot give as much advance warning or pinpoint the strike area for tornadoes as accurately as it can with hurricanes. Therefore, an organization must be prepared to protect its personnel on short notice and to take quick action to protect and restore undamaged equipment and materials.

National Weather Service radio bands can be monitored during storm conditions. In one midwestern city, several large companies have set up a cooperative warning network. A lookout is stationed atop the city's tallest building. Through a central network, not only the members but also the city's radio stations and civil defense unit are alerted in the event of an approaching tornado. Private weather consultants are also available.

Organizations need to determine the best location for personnel to seek shelter when threatened by a tornado. A basement or cellar usually affords the best protection. If an underground shelter is not available, identify an interior room or hallway on the lowest level. Closets, small interior hallways, and bathrooms without windows are the best areas. It is always best to conduct periodic tornado safety drills. Decide how and where everyone will gather before and after the storm.

Tornado and hurricane experience indicates that emergency plans should include the following steps:
1. Establish procedures for alerting and getting personnel to a safe place. If the building is not constructed

to withstand these natural forces, emergency shelters should be located close to the work area. All personnel should be instructed in the procedure to follow, with and without advance warning.

2. Assign trained personnel to take care of downed power lines; dangling wires are a serious hazard.
3. Assign trained people to remove wreckage to prevent injury to salvage and repair workers.
4. Schedule regular meals and rest for repair crews.

Earthquakes

Many areas of the United States have the potential to experience damaging earthquakes. In particular, the Pacific Coast contains the San Andreas fault network, and the Midwest contains the New Madrid fault system. Other major fault areas are located in North and South Carolina, Utah, and New England. No reliable earthquake warning system exists, although warning technology is being tested. This factor, coupled with the widespread damage that a quake can cause, makes this geological hazard a difficult phenomenon for planners to address.

"Earthquake-resistant construction" consists of building a structure so that it floats above the bedrock and ballasting it as a ship is ballasted, by making lower stories heavy and upper stories light. Utility lines and water mains should be flexible and laid in trenches that are free of the building, rising in open shafts, and connected to fixtures by flexible joints. Lockers, cabinets, shelves, and the like should be securely installed with seismic bracing and safety restraining strips on shelves containing bottles of chemicals.

The principal dangers of earthquakes come from structural failures of buildings, bridges, and other items; fires; flooding; broken utility lines; hazardous materials releases; water shortages and contamination; health problems; and damage to transportation facilities such as interstate highways, airport runways, navigable waterways, and railways. Water reservoirs or emergency water sources should be available as backup support for fire-fighting operations. Also, a system should be established for shutting down gas mains supplying facilities. The method of turning off the gas supply depends on the size of the main and the pressure on the line that comes into the facility. Organization officials should obtain expert advice when devising these plans.

See Figure 18–1 for recommended steps to take if an earthquake strikes. Be sure the emergency management plan provides for the actions recommended in Figure 18–1 if the facilities are in an earthquake risk area.

Civil Strife and Sabotage

Riot or civil strife is another item on the list of emergencies for which an organization should plan in advance.

Civil Strife

An emergency involving civil strife raises the question of the right to protect property versus individuals' legal right to assemble. An organization should obtain from an elected legal authority in the community (e.g., the local district attorney) a statement explaining the organization's rights in protecting its property and the organization's legal responsibility for the safety of employees and other people, such as customers, supplier salespeople, and visitors, who may be on site during the emergency scenario. An organization's legal department can be helpful in determining such a position, but its opinion does not have the force of law.

Some of the problems involved include disruption of business when an office or facility is invaded by outsiders, protection against a mob intent on destroying property, requests from neighboring businesses for assistance during a riot, and the rights and responsibilities of armed guards. Civil strife emergencies can be just as disastrous as any other type and should receive advance planning by manufacturing, mercantile, or commercial establishments.

Sabotage

Protection against sabotage is also an important consideration. The saboteur may be a highly trained professional or an amateur. He or she may be anyone and, at times, may be one of the least-suspected members of the organization. Because physical sabotage is frequently an inside job or requires the assistance, knowing or unknowing, of someone inside the facility, the principal measures of defense must be denying entry to suspicious persons. Evidence of sabotage should be reported to the U.S. Federal Bureau of Investigation and, if related to the military, to the U.S. Department of Defense.

Work Accidents and Rumors

The "chain reaction" from a so-called routine work accident can result in an emergency situation. For example, a break in a chemical line or accidental emission of toxic vapors may create an emergency.

The potential for work accidents to cause emergencies has increased because of the complexity of processes, the proliferation of industrial and agricultural chemicals, and the often-close proximity of residential areas to industrial activities. For example, a small break in a chemical line could allow hazardous air contaminants to enter a facility's ventilation system, creating a worksite emergency. A larger release of a chemical or toxic vapor may endanger a surrounding neighborhood. The toxic release may cause direct injuries or result in panic and create an emergency situation.

The potential for such events should be investigated and evaluated to protect individuals on site and in the surround-

ACTIONS TO TAKE IF AN EARTHQUAKE STRIKES

The greatest threat during an earthquake is from falling debris. Earthquakes are unpredictable and strike without warning. Therefore, it is important to know the appropriate steps to take when one occurs, and to be thoroughly familiar with these steps to be able to react quickly and safely.

IN THE OFFICE

During the earthquake

Step	Action
1	**Remain inside** the building.
2	**Seek immediate shelter** under a heavy desk/table, or brace yourself inside a doorframe or against an inside wall. • Get at least 15 ft away from windows.
3	**Stay there.** If shaking causes the desk or table to move, be sure to move with it.
4	**Resist the urge to panic.** Organize your thoughts; mentally review the established psychological considerations for earthquake safety. • Don't be surprised if the electricity goes out, fire or elevator alarms begin ringing, or the sprinkler system is activated. • Expect to hear noise from broken glass, creaking walls, and falling objects.

Immediately after the earthquake

Step	Action
1	**Remain in the same position** for several minutes after the earthquake in case of aftershocks.
2	**Do not attempt to evacuate** or leave your immediate area unless absolutely necessary or when instructed to do so by a proper authority.
3	**Check for injuries** and administer first aid. Recognize and assist co-workers who are suffering from shock or emotional distress.
4	**Implement your survival plan.** Establish a temporary shelter if rescue teams are expected to be delayed.
5	**Use stairway** when instructed to exit building.

AT HOME

During the earthquake

Step	Action
1	**Remain inside** your house.
2	**Seek protection** from flying debris or fixtures. Brace yourself inside a doorframe or against inside walls. Seek cover beneath a table, desk, or bed.
3	**Stay in position** until the shaking stops.

After the earthquake

Step	Action
1	**Remain calm.** Organize your thoughts by reviewing your home earthquake survival plan.
2	**Check for injuries** and administer first aid. Be prepared to respond to the psychological aftereffects generated by a major earthquake.
3	**Check water, gas and electric lines.** If you suspect damage, turn off the main valves and leave them off until advised by a utility company representative or other competent source.
4	**Do not use** candles, matches, or other open flames, or turn lights on/off, either during or after the tremor because of possible gas leaks.
5	**Turn on the radio** to receive emergency instructions. Reserve telephone usage for emergency calls only.
6	**Check your house for structural and internal damage.** Wear boots, if possible, to protect against shattered glass. • Chimneys are earthquake prone if not well enforced and should be approached with caution. Look for separation down the sides or for loose bricks. Unnoticed damage could result in a fire. • Check the interior of the house for dislodged items. The continued swaying motion from an earthquake will cause loose doors of medicine and kitchen cabinets/drawers to open and their contents to spill out. • Check closets and bulk storage areas (garage/basements) for items which may have toppled or collapsed during the earthquake, spilling substances which could produce fumes or become potential fire hazards.

Figure 18–1. In earthquake-prone areas, emergency preparedness training should include what to do if an earthquake occurs. *(Courtesy AT&T West.)*

IN PUBLIC	
During the earthquake	
Step	*Action*
	On the street
1	**Enter the closest structure immediately**—do not look up. Enter a store, terminal, office building, etc., just get inside. NOTE: The greatest danger during an earthquake is falling debris.
2	**Remove yourself from windows** that may shatter.
3	**Brace against** an inside doorframe or against inside walls.
	In a stadium, amphitheater or church
1	**Remain in your current location. Do not rush** to exits. The chaotic fleeing of large crowds diminishes the effectiveness of an evacuation procedure and frequently results in unnecessary injuries or deaths.
2	**Seek cover** under a bench or chair. If unavailable, crouch down, and cover your head with your arms to protect against falling debris.
3	**Keep away from overhead electric wires** or anything that might fall.
	In a vehicle
1	**Stop the vehicle** if it is currently in motion. **Avoid** stopping either on or under a bridge or overpass.
2	**Remain inside vehicle** until the shaking stops.
After the earthquake	
Step	*Action*
1	**Remain calm.**
2	**Check for injuries** and administer first aid. Recognize and assist individuals who are suffering from shock or emotional distress.
3	**Await emergency evacuation instructions.**
4	**Watch for hazards** created by the earthquake when traveling to another location such as downed electrical wires, broken or undermined roadways, collapsed freeways, overpasses, or bridge structures.
5	**Stay away** from waterfronts or beach areas. Tsunamis may result as an aftereffect of the earthquake.
6	**Avoid sightseeing.** Emergency vehicles will need ready access to respond to emergency situations.

Figure 18–1. Concluded.

ing community. The methods used in this evaluation may include the application of risk-assessment techniques. These techniques require management to identify hazards, calculate their probability of occurrence, and develop means to respond to these events. Various existing U.S. governmental regulations that require these types of analysis include the following:

- 40 CFR 330 and 335, Extremely Hazardous Substance List and Threshold Planning Quantities; Emergency Planning and Release Notification Requirements (Final Rule. Vol. 52, No. 77, Washington DC, pp. 13378–13410)
- 29 CFR 1910.120, Hazardous Waste Operations and Emergency Response
- 29 CFR 1910.1200, Hazard Communication
- Emergency Planning and Community Right-to-Know Act, Title III, Superfund Amendments and Reauthorization Act, Public Law 99-499.

Panic caused by rumors or lack of knowledge can also create an emergency. Plans for such situations should include establishing auxiliary areas in the building for medical treatment, a method of notifying employees of a current situation, a method of quickly taking a head count of workers, and sources of oxygen supplies available on short notice. A public relations coordinator should deal with the press, the public, and families.

Shutdowns

An emergency situation may occur following an unscheduled action, such as a disaster or strike; hence, a fast shutdown procedure should be covered under an emergency plan. This plan should be based on a priority checklist; that is, all the tasks to be assigned and functions to be performed should be arranged in order of importance ahead of time so that if only short notice is given, at least the most vital precautions are completed. This "crash procedure" is usually an adaptation of the routine procedure used for scheduled shutdowns, such as for vacation or renovation. Naturally, the amount of warning time controls the speed of shutdown.

Whenever a facility or unit must be shut down, safeguards against fire, explosion, or chemical release take on added importance. The extent of these measures will vary with the size and purpose of the facility. It is important to organize a formal program for instructing personnel in emergency shutdown procedures. In particular, workers should do the following:
- remove debris, lint, dirt, and rubbish from the area
- drain and clean dip and mixing tanks and other equipment where flammables have been used
- clean spray booths, ducts, and flammable liquid storage
- close gas, chemical, and fuel line valves; open switches on power circuits that may be out of service
- check serviceable condition of sprinkler systems, fire extinguishers, hydrants, alarms, and other protective apparatus
- anchor cranes.

Before the closing, employees should be alerted by special instructions to keep their workstations clean and fire-safe.

During the shutdown, continuous inspection of any maintenance or special operations, such as remodeling, must be maintained. Gas cutting and welding should be carefully supervised. Employees who remain on duty—the facility protection force, security staff, maintenance workers, supervisors, or executives—should be briefed in effective countermeasures in case a fire breaks out.

If sufficient notice has not been given to conduct a normal shutdown, it may become necessary to allow personnel into the area to perform emergency functions. Emergency shutdown and spill containment procedures should be clearly established for high- and low-staffing situations, and all personnel involved should be appropriately trained.

Organization management should designate someone who can admit the personnel necessary to handle emergencies arising within the area. The emergency coordinator should arrange with local police and fire department officials for assistance if an emergency gets beyond local control. It is especially important that arrangements be completed to speed up admitting fire fighters, security, or cleanup personnel and their equipment to the disaster site.

Some organizations use facility protection service agencies to prevent loss from theft, fire, and accident hazards during shutdowns. Similar plans should be worked out with these people so that police and fire assistance can be obtained quickly when it is needed.

Emergency Management Planning

Emergency management planning (EMP) consists of the plans and preparations of business and industry management to achieve a state of readiness during times of a far-reaching disaster. A far-reaching disaster may require an organization or facility to respond to an emergency disaster scenario alone. In a far-reaching disaster, outside sources of help (e.g., fire and police departments, hospitals and doctors, regular sources of supply for materials and equipment) are not so readily available.

Even if a particular area is not affected, facilities in that area may be asked to furnish transportation to evacuate the injured from damaged areas and to house and feed the evacuees. Facility emergency squads may also be required to assist other stricken buildings or areas. In a major catastrophe, where buildings have been devastated, adequate hospital space may not be available, making it necessary to keep the injured in temporary shelters for some time. In such cases, employees with the proper training may be required to administer sedatives and plasma and to treat injuries.

Hazardous Materials

Because most organizations use a variety of chemical substances, management must be concerned about potential usage, handling, and disposal problems. Although many rules and procedures are in place, the questions organizations should address include the following: What if a safeguard fails? What if the container cracks and substances leak out? In addition to normal hazards, can chemicals react with other substances to create further hazards to people and property? (See the OSHA regulations at 29 CFR 1910.1200, Hazard Communication.)

Chemical hazards are discussed in the National Safety Council's *Fundamentals of Industrial Hygiene*, 6th ed. (2012), and 29 CFR 1910.1200(g).

Radioactive Materials

Fires and other emergencies involving radioactive materials are becoming more common with the widespread, peaceful use of isotopes.

Radioactive Elements and Fire

Giraud (1973) makes the following observations. Radioactivity cannot by itself cause fires, nor can it be destroyed or modified by fire. However, a fire may change the state of a radioactive substance and render it more dangerous by converting it to a gas, aerosol, smoke, or ash and allowing it to contaminate a wide area.

Furthermore, fires can cause structural disruptions in stocks of fissionable materials and in the special equipment for their treatment or use. Such disruptions may, at worst, result in a nuclear chain reaction and initiate a critical nuclear accident.

Radioactive elements are found in various forms, depending on their uses. The human eye can detect no difference between an inactive element and the same element when rendered radioactive. Both appear equally harmless. A fundamental distinction must, however, be made between so-called sealed and unsealed sources.

- In the case of sealed sources, the radioactive substance is not accessible. The container has sufficient mechanical strength to prevent the substance from spreading during normal conditions of use. The capsule is made of stainless steel. The sources are typically of small dimension, approximately ¼ inch (approximately 1 cm).
- In unsealed sources, however, the radioactive substance is accessible. In normal conditions of use, there is no means of preventing it from spreading. Solid substances are kept in aluminum tubes, liquids are kept in flasks, and gases are kept in glass ampules.

The fact that a substance is radioactive does not affect its general physical properties or its behavior when heated to an abnormally high temperature, as, for instance, during a fire. On contact with fire, the substance will undergo the normal transformations, depending on its initial form (i.e., solid, liquid, or gas). Melting, boiling, and sublimation can be expected, with the formation of combustion products corresponding to the chemical properties of the substance: slag, ash, powder, dust, mists, aerosols, fumes, or gases.

These combustion products are generally finer and less dense than the original substance, so they disperse more easily. Although the change in the physical state of the substance will not affect its radioactivity, the radiation hazard will be more difficult to control.

The protective containers currently in use have a widely varying resistance to fire. Therefore, the protection afforded to the contents will depend on the type of container used. In general, sealed sources are strongly fire resistant, and radioactive elements thus contained are well protected (Figure 18–2).

Unsealed sources, however, and solutions or gases in fragile containers easily fall victim to fire. The type of immediate action that must be taken in the event of an accident with radioactive materials can be determined by the fire-fighting staff once the type of container is known. Actions to deal with the hazard will depend on the properties of the radioactive substance. (See the OSHA regulations at 29 CFR 1910.1200(f), Hazard Communication, "Labels and Other Forms of Warning," and 29 CFR 1910.38, Employee Emergency Plans and Fire Prevention Plans.)

When the protective container has been broken as the direct or indirect result of a fire, the radiation hazards for

Figure 18–2. "Radiation Yellow—III" label, which is affixed to each package of highly radioactive material. Different labels are required for different intensities and quantities of radioactive material. For details see 49 CFR 172, Hazardous Materials Table and Hazardous Materials Communications Regulations.

rescue workers at the fire or personnel in the vicinity are likely to be more serious than those from a conventional fire. Accordingly, the person in charge of the rescue work will sometimes be obliged to override the normal fire-fighting procedures to ensure proper confinement of any radioactive elements released. If the elements are already affected by the fire, further hazards may arise.

The release of radioactive elements may result in contamination of surface areas. This may be caused by the spilling or splashing of radioactive substances or by the spreading of solid radioactive substances in paste, powder, or dust form. All possible precautions must be taken to prevent any further spread of the contamination. The means to be used, however, will differ with each case. In the first (spilling or splashing), absorbent materials such as powder, earth, or sand should be used. In the case of spreading, the substances should be slightly dampened with a spray of water—unless it is otherwise specified on the container. (See the OSHA regulations at 29 CFR 1910.1200(g), Safety Data Sheets, now aligned with the UN Globally Harmonized System of Classification and Labelling of Chemicals [*Federal Register*, March 26, 2012.])

Radioactive liquids can be contained by the methods normally used by the fire-fighting brigade. The contaminated area must be clearly marked and roped off to prevent the entry of unauthorized personnel (Figure 18–3).

Contamination of the atmosphere is caused by radioactive elements in the form of dust, aerosols, fumes, and gases. The spreading of such contamination is determined

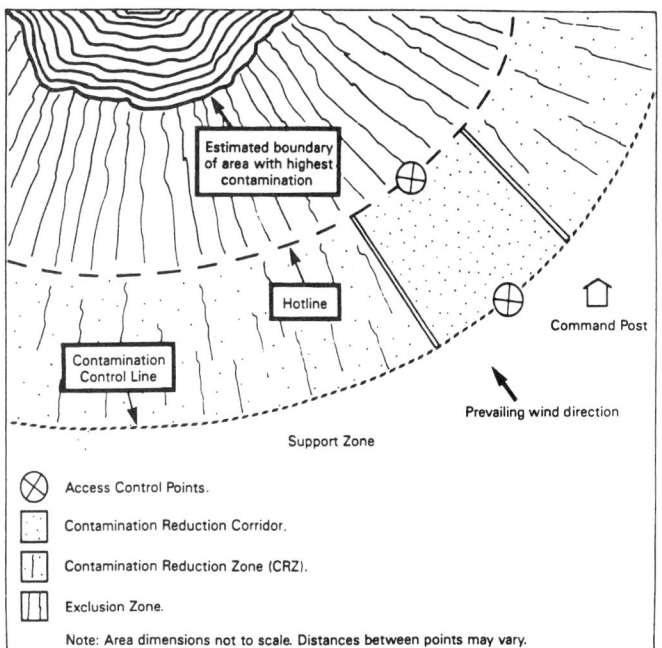

Figure 18–3. This diagram shows site work zones for an emergency situation involving a radioactive hazard. Note that decontamination facilities are located in the Contamination Reduction Zone. (*Courtesy NIOSH/OSHA/USCG/EPA*, Occupational Safety and Health Guidance Manual for Hazardous Waste Site Activities.)

mainly by the prevailing weather conditions, and it is difficult to control. Such atmospheric contamination may lead to other toxic or corrosive hazards associated with the particular chemical. The most serious danger is that of inhaling the substance when it is suspended in the air. Fire fighters, accordingly, should wear a self-contained breathing apparatus.

The danger of internal irradiation is always present whenever there is contamination by a source of penetrating radiation. It may also occur by accidental release of an alpha or beta emitter from its protective container or by the destruction (even partial) of the protective container.

Weather-Related Emergencies

Throughout any year, unusually severe and unexpected weather events can require some changes in normal operations. For example, in North Dakota, the temperature occasionally may drop to −35°F (−30°C), yet most activities and travel are not normally affected because such weather extremes are expected. But if the wind increases in strength or the temperature drops suddenly, people may need help as they travel or participate in other outside activities. (A windchill chart is provided in Figure 18–4.) Employees could be alerted to the danger before leaving work and told when or how they are to be notified about whether the organization will be open in the morning.

Management must plan for a variety of weather-related emergencies appropriate for the locale. In the event of extremely heavy snowfall, for example, what changes may be needed in operations? What should employees be told before leaving for home? An unusually heavy rainstorm may strand hundreds of customers in a store just a few minutes before closing. Are supervisors and clerks prepared to handle the situation? How do they control the crowd? Hail or wind may start breaking glass windows while customers are shopping. What is the immediate action? What if adverse weather causes a power failure or someone suddenly shuts off all power and lights while crowds are shopping? The emergency lighting system may operate as intended, but employees, particularly key supervisors, must understand emergency plans and be prepared to act responsibly.

Other weather emergencies include droughts and extreme heat. Heat kills by taxing the human body beyond its abilities. In a normal year, about 175 Americans succumb to stresses related to summer heat. These harsh conditions require attention to outdoor work schedules, water resources, and health effects.

Estimated Wind Speed (in mph)	Actual Thermometer Reading (°F)										
	50	40	30	20	10	0	-10	-20	-30	-40	-50	-60
	Equivalent Temperature										
Calm	50	40	30	20	10	0	-10	-20	-30	-40	-50	-60
5	48	37	27	16	6	-5	-15	-26	-36	-47	-57	-68
10	40	28	16	4	-9	-24	-33	-46	-58	-70	-83	-95
15	36	22	9	-5	-18	-32	-45	-58	-72	-85	-99	-112
20	32	18	4	-10	-25	-39	-53	-67	-82	-96	-110	-121
25	30	16	0	-15	-29	-44	-59	-74	-88	-104	-118	-133
30	28	13	-2	-18	-33	-48	-63	-79	-94	-109	-125	-140
35	27	11	-4	-20	-35	-51	-67	-82	-98	-113	-129	-145
40	26	10	-6	-21	-37	-53	-69	-85	-100	-116	-132	-148
(Wind speeds greater than 40 mph have little additional effect.)	LITTLE DANGER In < 1 hr with dry skin. Maximum danger of false sense of security.			INCREASING DANGER Danger from freezing of exposed flesh within one minute.				GREAT DANGER Flesh may freeze within 30 seconds.			
	Trench foot and immersion foot may occur at any point on this chart.										

*Developed by U.S. Army Research Institute of Environmental Medicine, Natick, MA
Equivalent chill temperature requiring dry clothing to maintain core body temperature above 36°C (98.8°F) per cold stress TLV®.

Figure 18–4. Windchill factors. The human body senses "cold" as a result of both temperature and wind velocity. The numerical factor that combines the effects of these two is called the *windchill factor*. Because of the extra clothing that people wear in cold weather, their physical size is greater than it is in warm weather. Be sure that equipment and controls are of adequate size and simplicity so that they can be run effectively and safely by persons wearing heavy clothing.

In addition, lightning is a major cause of weather-related deaths in the United States. Lightning results from the build-up and discharge of electrical energy between positively and negatively charged areas. Rising and descending air within a thunderstorm separates these positive and negative charges. A cloud-to-ground lightning strike begins as an invisible channel of electrically charged air moving from the cloud toward the ground. When one channel nears an object on the ground, a powerful surge of electricity from the ground moves up to the clouds and produces the visible lightning strike. Planning should include emergency procedures for halting work in areas vulnerable to lightning strikes or the loss of utilities when this particular danger is imminent.

PLAN-OF-ACTION CONSIDERATIONS

Once the risk assessment of potential emergencies has been completed, the next stage of the planning process should be preparing a plan of action. This plan should be supported by management and include input from both the public and private sectors. It is desirable to include union or labor representatives.

Generally, someone should be appointed emergency-planning director or coordinator, perhaps with help from an advisory committee. Usually, because of their experience and training, the health and safety, medical, fire, and security departments will be involved. Of course, because production and maintenance will be affected, these departments must be consulted. Also, the legal staff needs to be involved in formulating the plan. Finally, management should coordinate the plan with local law enforcement agencies and fire departments.

Some managers may object to the cost and effort involved in giving immediate attention to emergency planning. However, this activity can be justified by weighing the cost of preparedness against the possibility of yearly losses from accidents, fires, floods, and other catastrophes.

For midsize to large organizations, a corporate crisis management team should be developed. This team should be represented by high-ranking officials in finance, human resources, operations, logistics, maintenance, information technology, and safety and others who have the ability to respond to a crisis on a large-scale and/or national basis. The crisis management team should establish a crisis center (e.g., a conference room) on premise and determine an alternate site for scenarios where the premises have been compromised.

Program Considerations

Once management has completed its advance planning and has evaluated the type of emergencies and their potential harm to people and property, the next step is to develop a working emergency management plan within the organization. In some cases, this step requires working with local agencies to protect the organization's operations.

Advance planning is the key. Management must establish a written set of plans for action. The plans should be developed locally within the organization (and corporate structure) and be in cooperation with other neighboring or similar organizations and with governmental agencies. It may not always be possible for every organization or agency to cooperate or participate fully, but through planned action, each one can be aware of the resources and help available. In some instances when outside assistance is limited, an organization may need to depend largely on its own resources to deal with emergencies.

Often, an emergency manual or handbook will be developed for the facility or organization to document emergency response procedures. The following outline covers many of the items that may be included, but other items may be needed as dictated by the expected emergencies and the available resources:

1. delineation of organization policy, purposes, authority, and principal control measures and an emergency organization chart showing positions and functions
2. some description of the expected disasters with a risk statement
3. a map of the facility, office, or store showing equipment, medical and first aid, fire control apparatus, shelters, command center, evacuation routes, and assembly areas
4. a list (which may also be posted) of cooperating agencies and how to reach them
5. a description of the facility warning system
6. a central communications center, including home contacts of employees and corporate functions (if necessary)
7. the shutdown procedure, including security guard
8. how to handle visitors, customers, and contractors
9. locally related and necessary items
10. a list of equipment and resources that would be available and where they can be reached.

Some of these items will be discussed in more detail in the following pages. Management should rehearse the plan under realistic conditions to test the plan's effectiveness. For example, emergency lights may fail when needed, or the telephone service may break down. These are real conditions that may occur in an actual disaster. Therefore, planning should include all possible as well as probable contingencies.

Chain of Command

Once management has decided to establish a disaster plan, it should appoint a director or coordinator and create an advisory committee representing the various departments.

Basic guidelines to follow in establishing the chain of command are (1) keep the chain as small as practical and (2) appoint personnel to crisis management positions based not on their title but rather on their ability to respond to a situation under extreme stress.

Experience has shown that the smaller the chain of command, the more efficient and effective its decisions and actions will be during a crisis situation. No more than three to seven persons should report to each person in the chain of command. Normally, the chain of communication will pattern itself after the chain of command. Communications must be conducted as efficiently as possible. Management should quickly provide accurate information or request the proper support. Long delays could affect the outcome of a given emergency situation.

It is normal to think that the ranking manager should assume the crisis manager's position. However, this may not be the proper action to take. Many people are excellent managers under normal conditions. They are able to give good direction if allowed time to think through the situation and weigh the pros and cons. But given a situation in which extreme stress is introduced and quick, timely decisions are required, the same individual may not be able to perform with the same effectiveness. Therefore, the individuals selected to become a part of the crisis team must be tested, through training exercises, to see if they can perform under emergency conditions. If the ranking manager is found not to be the proper choice for the crisis manager, he or she should be a consultant to whoever is placed in the crisis manager's position. This same philosophy applies to any of the positions established within the chain of command.

The emergency director should be a member of top management, whether for a single building or facility or a national organization. This individual will need to delegate authority and to speak for the organization. The head of the disaster-control organization must be cool and quick thinking and sufficiently healthy to withstand the arduous duties involved in an emergency. The emergency director's regular duties should be such that the greater part of the time will normally be spent at his or her own unit. However, the plan should always name an alternate director. The alternate should be a person who has authority and qualifications similar to the director and should be trained with the director.

The director (and alternate) should be the first to be trained. Management should maintain liaison with local emergency management authorities, if possible, to make sure that the plans are coordinated with those of the community and to keep the organization informed on new developments. (See the OSHA regulations at 29 CFR 1910.156, Fire Brigades; 29 CFR 1910.1200, Hazard Communication; and 29 CFR 1910.120, Hazardous Waste Operations and Emergency Response; and the EPA regulations at 40 CFR 311, Worker Protection.)

The director of emergency management is responsible for the following:
- emergency operations center management
- communications
- fire fighting
- security and law enforcement
- rescue operations
- emergency medical services
- transportation
- damage assessment
- mitigation and investigation
- public information and media briefings
- rumor control
- on-scene safety functions at the emergency site
- warning and evacuation of facility and community personnel
- utilities and engineering functions
- sheltering, feeding, and counseling functions
- morgue establishment and notification of survivors
- notification of SARA Title III, Emergency Planning and Community Right-to-Know Act, authorities in the event of a hazardous materials release.

All of these functions are likely to be essential, although some may be combined. The person (and alternates) responsible for each function should be selected with great care and trained by the director. These managers should be familiar with all parts of the plan and should have experience in the fields in which they are to serve.

Management should train assigned personnel to carry out their duties in accordance with the overall emergency plan. In small operations without regular security staff or fire fighters, operating personnel will be trained to take care of these duties. Of course, the number of members on each of the teams depends on the circumstances of each facility. Each team captain should select personnel from the available volunteers, supervise their training, and procure their equipment.

Because workers' wholehearted cooperation is necessary to the successful operation of an emergency management plan, shop stewards or other employee representatives should take part in the planning. They should understand that measures taken are to protect the lives and jobs of the workers as well as to protect property.

The organization should establish emergency reporting centers so that employees will know where to report should a disaster strike while employees are away from work. Reporting centers give employees a feeling of security and continuity and aid in taking a "roll call." To facilitate these arrangements, each employee should carry an identification card containing specific instructions on where to report, a list of other reporting centers, a basic employment record, and designation of the employee's next of kin in case they must be contacted. The reporting center will keep a duplicate record for each employee assigned to the center. A single facility operation or small organization can consider using the home of a member of management, a supervisor, or an employee.

Training

One of the most important functions of the director and staff, on both the corporate and facility levels, is training. Employee training for each type of disaster is essential in developing a disaster-control plan and keeping it functioning. An emergency plan is vital and real—it has no value if it remains simply an idea (Figures 18–5, 18–6, and 18–7). Training and rehearsals are time consuming, but they keep the program in good working order.

Simulated disaster drills, sometimes called tabletop drills, help key people and employees respond to emergencies with greater confidence and effectiveness. No actual response is required; instead, key personnel operate under the direction of a drill coordinator, who feeds information to them in real time and monitors the response.

Another type of drill is a full-scale dress rehearsal involving all personnel with simulated situations and injuries (Figure 18–8). Management must be careful to prevent injuries to participants and the public while making the full response as realistic as possible. Feedback from either type of drill is essential to improve emergency management plans.

IN CASE OF FIRE OR OTHER EMERGENCY

✓ **KEEP YOUR HEAD**—avoid panic and confusion.

✓ **KNOW THE LOCATION OF EXITS**—be sure you know the safest way out of the building no matter where you are.

✓ **KNOW THE LOCATION OF NEARBY FIRE EXTINGUISHERS**—learn the proper way to use all types of extinguishers.

✓ **KNOW HOW TO REPORT A FIRE OR OTHER EMERGENCY**—send in the alarm without delay; **notify the CHIEF OF EXIT DRILLS.**

✓ **FOLLOW EXIT INSTRUCTIONS**—stay at your work place until signaled or instructed to leave; complete all emergency duties assigned to you and be ready to march out rapidly according to plan.

✓ **WALK TO YOUR ASSIGNED EXIT**—maintain order and quiet; take each drill seriously—it may be "the real thing."

REMEMBER—IT IS PART OF YOUR JOB TO PREVENT FIRES

Figure 18–5. Sample emergency exit notice for general posting.

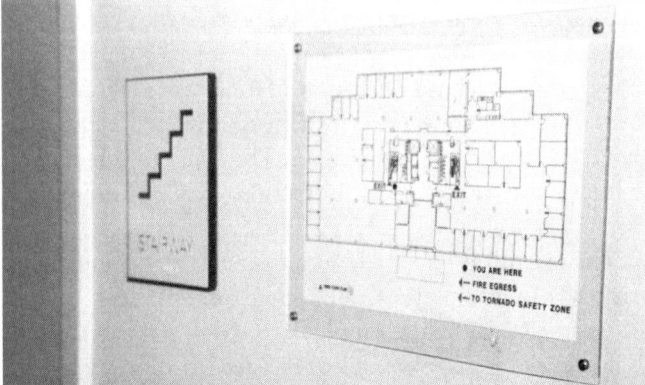

Figure 18–6. Clear identification of evacuation routes from all locations is an important aspect of emergency preparedness training.

EMERGENCY-PLANNING TRAINING REQUIREMENTS

Following are the training programs required to ensure that the XYZ site is prepared in the event of an emergency.

1. Emergency Action Plan (29 CFR 1910.38)
   All employees are trained on the contents of this written plan as a new employee and annually thereafter.
2. Fire Extinguisher Training (29 CFR 1910.157)
   All employees receive annual fire extinguisher training for incipient stage fires. This includes a hands-on demonstration of competency. Incipient stage refers to the severity of a fire in which the progression is in the early stage and has not developed beyond what can be extinguished using portable fire extinguishers.
3. HAZWOPER Training (29 CFR 1910.120)
   Three levels of emergency response training under the OSHA HAZWOPER standard are conducted on an annual basis. All operators and maintenance employees are trained to the First Responder Awareness Level. All supervisors, supervisors on the fire brigade, and select personnel receive Hazardous Materials Technician and Incident Command Level training.
4. Fire Brigade Training (29 CFR 1910.156)
   All members of the voluntary fire brigade attend an annual fire school at the local fire-fighting academy. Other training on site includes the operation of specialized equipment such as the fire truck and remote-controlled fire monitor.
5. Medical Training (29 CFR 1910.146, 1910.151, and 1910.1030)
   All members of the confined-space rescue team receive first-aid, CPR, and bloodborne pathogens training. The refresher cycle is as follows:
   First aid—every three years
   CPR—every two years
   Bloodborne pathogens—annually.
6. Rescue Training Program (29 CFR 1910.146)
   All members of the confined-space rescue team receive annual confined-space entry rescue training and respirator protection training. General confined-space entry training is conducted every three years. In addition, high-angle and technical rescue training is conducted.

Figure 18–7. Sample training outline with associated regulatory references for a manufacturing facility.

DRILLS AND EXERCISES

Drills and exercises are an integral element in the overall training program. The objectives of drills and exercises are to (1) practice emergency procedures and skills; (2) evaluate individual knowledge of emergency response procedures; and (3) determine the overall effectiveness of both emergency response procedures and the Emergency Action Plan.

The following information further describes the role of drills and exercises within the site-training program.

1. DRILL—A planned and defined training session covering a single, specific topic or function. Drills typically reinforce classroom training that has already occurred and are used to demonstrate skill performance.

   In general:
   A. Drills should involve all employees on-site at the time the drill is conducted.
   B. Drills should address the types of emergencies to which the department or site would be expected to respond.
   C. Personal protective equipment required for the event the drill is simulating must be worn/used.
   D. Messages announcing drills should start and end with the statement, "This is a drill."
   E. Facility/system shutdowns should be simulated.
   F. Facility neighbors and appropriate local authorities should be contacted.

   Drills that test the Emergency Action Plan must be conducted per the following guidelines:
   - Department tests must be conducted at least two times per year for each employee.
   - Sitewide emergency tests must be conducted at least two times per year.

   Examples of drills include the donning and doffing of self-contained breathing apparatus, evacuations, shelter-in-place, and telephone notifications.

   The fire brigade should conduct at least quarterly drills.

2. EXERCISE—A mechanism to evaluate both the operational and managerial readiness of emergency response plans and field response capabilities. Exercises may involve internal personnel (e.g., emergency management team) or external personnel (e.g., mutual aid, local public safety agencies, hazmat contractor).

   The site may conduct the following types of exercises as an integral element of its emergency response program:

   A. *Tabletop Exercise*—In this activity, key facility personnel with emergency responsibilities gather to discuss actions to be taken during an emergency based on the site's Emergency Action Plan and the respective operating procedures. The tabletop exercise is designed to elicit constructive discussion by the participants as they examine and resolve problems based on the Emergency Action Plan.

   The purpose of a tabletop exercise is to have participants practice problem solving and resolve questions of coordination and assignment of responsibilities in a controlled environment, under minimum stress. Tabletop exercises typically involve a limited demonstration of operational response and/or internal coordination activities, and can be used in preparation for a functional or a full-scale/field exercise.

   B. *Functional Exercise*—A functional exercise is more extensive than a tabletop exercise in that activities are conducted beyond a conference room atmosphere. It may include both emergency control center operations and concurrent field activities. Generally, this type of exercise will focus on a single function or activity within a function (e.g., notification exercise, implementation and expansion of the facility incident command system).

   C. *Full-Scale/Field Exercise*—A full-scale/field exercise is used to evaluate emergency response operational capabilities in an interactive manner over an extended time frame (typically several hours). The primary purpose of the full-scale exercise is to test a major portion of the Emergency Response Plan functions; it should incorporate a high degree of realism, extensive involvement of resources and personnel, and an increased level of stress. This type of exercise would include mobilization of both personnel and resources to several locations simultaneously, as well as actual movement of personnel, equipment, and resources required, demonstrating a coordinated response and capability. Full-scale/field exercises are always planned with specific written objectives and are normally announced in advance to ensure personnel safety and minimize interruptions of business.

3. CRITIQUES—Critiques of emergency response drills should be held with all participants as soon as possible after a drill, but no later than one working day. The critique report should be issued within 10 working days. Results must be maintained in a file, and action items are to be tracked to completion. The actions learned in a drill should become a part of the site Emergency Action Plan or the unit operating instructions, whichever is more appropriate.

Figure 18–8. Sample of an emergency drill and exercise schedule for a manufacturing site.

Hazardous Materials/Spills Emergencies (HAZWOPER)

The federal OSHA regulation titled Hazardous Waste Operations and Emergency Response (HAZWOPER), Final Rule, became effective in 1990 (29 CFR 1910.120). It addresses the many aspects of health and safety that are legally required at hazardous materials sites; EPA-regulated hazardous waste treatment, storage, and disposal facilities; and other hazardous materials emergency locations. The rule mandates certain requirements for monitoring instrumentation, site safety plans, respiratory and personal protective equipment, medical surveillance, engineering controls, work practices, training requirements, and other operational functions (see Figures 18–9 and 18–10).

Generally, the 29 CFR 1910.120 training standard addresses three categories of employees: hazardous materials site workers (paragraph [e]); treatment, storage, and disposal workers (paragraph [p]); and emergency responders to hazardous substance releases regardless of location (paragraph [q]). (See also the OSHA regulations at 29 CFR 1910.156, Fire Brigades.)

The HAZWOPER team consists of an organized group of employees, designated by the employer, who are expected to handle and control actual or potential leaks or spills of hazardous substances requiring possible close approach to the substance. The team members perform responses to releases or potential releases of hazardous substances for the purpose of control or stabilization of the incident. A HAZWOPER team is not a fire brigade, and a typical fire brigade is not a HAZWOPER team. A HAZWOPER team, however, may be a separate component of a fire brigade or department.

Recommended minimum training requirements for the HAZWOPER team and appropriate courses are discussed in the following sections.

Figure 18–9. A hazmat team transferring material from a tank trailer rolled on its side. *(Courtesy HMHTTC.)*

Figure 18–10. A hazmat team responding to a leaking container in a van trailer. *(Courtesy HMHTTC.)*

Hazardous Materials Site Workers

This course is for general workers who are on site full time and have experience working with hazardous chemicals. They are required to attend 40 hours of classroom, "hands-on" instruction away from the site followed by a minimum of 3 days of actual field experience under a trained supervisor. All site workers must complete 8 hours of refresher training each year, and the training must be documented.

Occasional Hazardous Materials Site Workers

This course accommodates inspectors, engineers, monitoring technicians, and others who are not likely to be exposed over permissible exposure limits. They are required to have 24 hours of offsite instruction followed by 1 day of actual field experience.

Hazardous Materials Site Supervisors

These site managers must have 8 hours of specialized training after completing the 40-hour course. Supervisory training covers site safety plans, personal protective equipment selection, and health monitoring.

Treatment, Storage, and Disposal (TSD) Facility Workers

New employees at these facilities must have at least 24 hours of health and safety training. Current employees must demonstrate the equivalent amount of training from previous experience. All workers must complete 8 hours of TSD refresher training each year.

Hazardous Materials Emergency Responders (Regardless of Location)

These workers are distributed into five categories. The training requirements apply to private facility employees as well as to community emergency services personnel.

First-Responder Awareness. This is a 4- to 8-hour course for responders who will only attempt to identify the involved hazardous material and then notify more qualified personnel.

First-Responder Operations. An 8-hour training requirement exists, but 24 hours of instruction is recommended. This category covers responders who will identify materials, perform basic diking and containment operations, and initiate evacuation. This course requires training in personal protective equipment, decontamination, chemistry, and toxicology. However, employees are taught to respond in a defensive fashion; the course does not qualify them to attempt a patching or plugging operation.

Hazardous Materials Technician. This is a 24-hour course for workers who will perform defensive measures (the actual plugging, patching, or sealing) on a container leaking a hazardous substance. Besides the classroom and "hands-on" training, the hazmat technician must demonstrate certain competencies with regard to the Incident Command System.

Hazardous Materials Specialist. Additional training beyond the 48 hours required for the responder and technician levels must be taken in the specialty area of this skilled responder. No specific hours requirement exists, but this individual should be competent in the chemical, toxicological, and/or radiological behavior of the particular material involved.

Incident Commander (IC). The incident commander must be a graduate of the 24-hour First-Responder Operations course and have additional training (24 hours is recommended) in planning, decontamination, protective clothing, and command systems. The commander does not have to receive instruction in the technician or specialist categories. (See also Uniform Incident Command System later in this chapter.)

Command Headquarters

Although the command headquarters of an organization may not withstand catastrophic disasters, management can plan for emergencies that may occur.

The disaster control organization should be coordinated from a well-equipped and well-protected control room and/or an alternate, offsite command headquarters. The headquarters equipment should include the following:
- telephones
- sound-powered phones
- a public address system
- maps of the facility and surrounding areas
- emergency lighting and electric power
- sanitary facilities
- reference books
- emergency plans
- Safety Data Sheets
- a second exit
- two-way radios for communication both locally and with emergency management authorities, if required.

Good communications are necessary for effective control and flexibility in a disaster situation. Communications include (cellular) telephones, radios, messengers, and the facility's alarm system (discussed separately later in this chapter). The disaster plan should provide for adequate telephones in emergency headquarters to handle both incoming and outgoing calls. In addition, satellite phones and AM radios (better known as "Ham" radios) may be beneficial in some emergencies. Panic and disintegration of the organization will develop quickly if these calls are not handled with dispatch. Operators should keep an accurate log of all incoming and outgoing messages.

Some means of communication independent of normal telephone service must be available during an emergency, such as the one provided by a battery-operated radio. Mobile cellular phones may or may not be operable during or following a disaster. The disaster plan must anticipate the possibility of losing normal telephone communications and electric power. With today's rapidly changing communication venues, it is a good idea to institute the use of a multimedia communication package that will provide cross-media communication networking. At a minimum, this should include texting, phone, e-mail, and other similar communication modalities.

Uniform Incident Command System

The intent of the incident command system (ICS) is to provide a comprehensive management structure that satisfies the requirements set forth in OSHA 29 CFR 1910.120 (Figure 18–11). As stated in the regulations:

> The ICS shall be established by those employers for the incidents that will be under their control and shall be interfaced with the other organizations or agencies who may respond to such an incident.

The command function within the ICS may be conducted in two general ways. *Single command* may be applied when there is no overlap of jurisdictional boundaries or when a single incident commander (IC) is designated by the agency with overall management responsibility for the incident. *Unified command* may be applied when the incident is within one jurisdictional boundary but more than one agency shares management responsibility. Unified command is also used

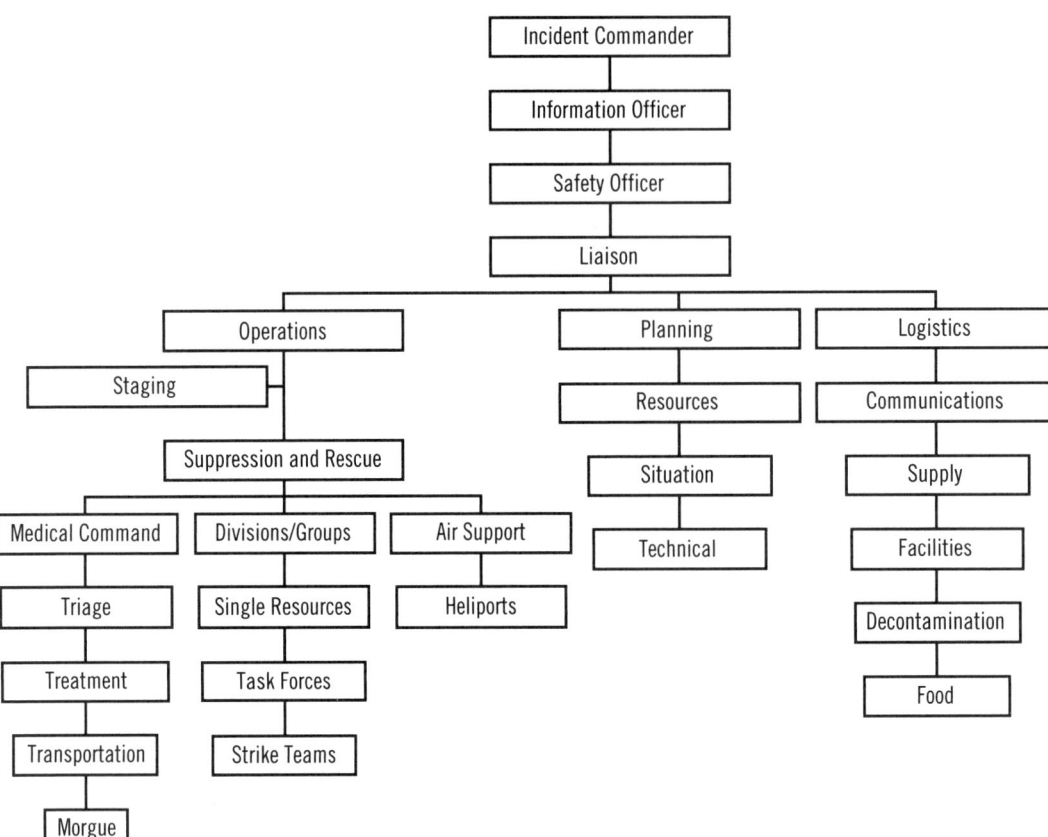

Figure 18–11. This is an example of a basic incident command system for a hazardous materials scenario.

when the incident is multijurisdictional or when more than one individual designated by his or her jurisdictional agency shares overall management responsibility.

Every incident needs some sort of consolidated action plan. Written plans are usually required when resources from multiple agencies are used, when several jurisdictions are involved, or when changes of personnel or equipment are required. The action plan should cover all strategic goals, tactical objectives, and support activities required during the operation. In prolonged incidents it may be necessary to develop action plans covering specific operational periods. Command staff includes the following positions.

Incident Commander

The one function that will always be filled at every incident, regardless of size, is the incident commander. The IC has the responsibility of overall incident management. The incident commander's responsibilities include the following:
1. Assess the incident priorities.
    a. *Safety*—The IC must consider safety issues for all personnel at an incident. No industrial complex or form of property is worth the risk of even one life. Safety comes before all other considerations.
    b. *Incident stabilization*—The IC is responsible for determining the strategy that will minimize the impact that an incident may have on the surrounding area. The size and complexity of the command system developed and implemented by the IC should be directly proportional to the magnitude and complexity of the incident. The ICS structure must match the complexity of the incident, not the size. Situations that may appear hopeless must be managed and ultimately controlled.
2. Determine the incident's strategic goals and tactical objectives. The efforts of the resources available for handling any incident must be properly directed to minimize the damage. The clocks cannot be turned back. Damage that has already occurred cannot be alleviated, but further damage must be minimized. This is accomplished when the IC determines the broad strategic goals for the incident and then transforms these goals into obtainable, practical objectives.

3. Develop or approve and implement the incident action plan. The IC is the primary developer of the incident action plan. On most simple incidents, the action plan will be organized completely by the IC and may not need to be written down. In more complex incidents, the action plan will be a written document developed by a staff headed by the IC. Action plans must be flexible and continually assessed. In the environmental business, conditions rarely remain constant. They are almost always dynamic.
4. Develop an incident command structure appropriate for the incident. The organizational structure is not based on the size or area of involvement; it depends on the complexity of the incident.
5. Assess resource needs and deploy as needed. The IC must continually evaluate and adjust the deployment of resources at all incidents. Initial assessment of the incident and the needed resources is only the first step. As soon as the IC determines the incident's strategic goals and tactical objectives and then evaluates the resource needs to meet these goals and objectives, either the initial action plan will be successful or it will need to be revised. Additional resources may be needed, requiring reorganization. If the IC believes he or she has just enough resources for the required work, it is time to order additional help and/or other resources.

Effective resource management requires that personnel safety be given the highest priority. Although everyone working at an incident must serve as his or her own safety officer, the ultimate responsibility for incident safety rests with the IC. As an incident escalates, the IC will need to assign a person as safety officer, with specific safety responsibilities. As stated in 29 CFR 1910.120 q3 (vii):

> The individual in charge of the ICS shall designate a safety official, who is knowledgeable in the operations being implemented at the incident site, with specific responsibility to identify and evaluate hazards and to provide direction with respect to the safety of operations for the incident at hand.
>
> When activities are judged by the safety official to be an immediately dangerous to life or health (IDLH) condition, the safety official shall have the authority to alter, suspend, or terminate those activities. The safety official shall immediately inform the individual in charge of the ICS of any actions needed to be taken to correct these hazards at an incident site.

6. Coordinate overall site activities. Coordination is essential to effective incident management. Without it, resources will be wasted performing tasks that are not necessary to the overall success of the incident. The IC must constantly monitor the incident activities to ensure that the needed degree of coordination is present and that personnel are not working at cross duplication. The goal of the IC is to obtain the maximum productivity from all on-scene resources. Proper coordination will ensure that personnel and equipment are functioning within the action plan.
7. For incidents of a medium to large scale, a scribe should be assigned to capture incident events as they occur. A logbook should be maintained to ensure that a proper record of the events can be traced.

Note: Under most state emergency response plans, once the fire department arrives on the scene, the fire chief assumes the responsibility of the IC role.

Liaison

A liaison individual is the point of contact for assisting or coordinating agencies. This function is assigned so the IC is not overloaded by questions from the number of assisting agencies that some incidents attract.

One of the most important responsibilities of the liaison individual is to coordinate the management of assisting or coordinating agencies. This is essential to avoid duplication of effort and allows each agency to perform what it does best. Liaison management provides lines of authority, responsibility, and communication.

Operations

Operations is responsible for management of all tactical operations at the incident. Operations is implemented when the IC is faced with a complex incident having major demands in one or more of the remaining major functional areas. For example, the IC may be faced with a rapidly escalating incident with a significant need to evaluate strategy and develop alternative tactical options. Faced with a major functional responsibility in addition to management of tactical operations, the IC may need to staff Operations to maintain an effective span of control.

The most common reason for staffing Operations is to relieve span-of-control problems for the IC. A complex incident, in which the IC needs assistance determining strategic goals and tactical objectives, may also require implementing Operations.

Planning

Planning is responsible for collecting, evaluating, disseminating, and using information about the development of the incident and the status of resources.

When faced with a complex or rapidly escalating incident, the IC may require assistance with the ICS Planning function. A wide range of factors may have an impact on incident operations. Planning must include an assessment of the present and projected situation. Proactive incident management is highly dependent on an accurate assessment of the incident's potential and prediction of likely outcomes. In addition to assessment of the situation status, there is a critical need to maintain information about resources committed to the incident and projected resource requirements.

Logistics

Logistics is responsible for providing facilities, services, and materials for the incident. As incidents grow in size, complexity, and duration, the logistical needs of the operating forces also increase. Even in a relatively simple incident, there are requirements for equipment, drinking water, and emergency medical care. When the organization faces a major incident, the logistical requirements are significant. Long-duration incidents of any type require provisions for feeding personnel, toilet facilities, refueling of vehicles/equipment, lodging, and a myriad of other service and support resources. Acquisition and accurate distribution of materials and equipment is a major functional responsibility for this position. Logistics works very closely with Planning personnel.

Finance

Finance is responsible for tracking all incident costs and evaluating the financial considerations of the incident. During large-scale incidents, many significant purchases and cash transactions are made. Finance personnel must ensure that all disbursements are documented, including but not limited to accurate invoicing for all services and materials used.

Emergency Equipment

An emergency checklist should include equipment and materials to be ordered as well as shutdown actions to be performed. For example, when it is not feasible to keep on hand the necessary emergency equipment and materials, managers should maintain a resource or supply list and names of suppliers. The information should include types of equipment they can supply, after-hours information and phone numbers, and, if possible, prices. These sources must be outside the immediate area.

Items on the shutdown part of the checklist would include closing valves; protecting equipment that cannot be moved; closing and battening doors, windows, and ventilators to keep out looters as well as water; and plugging vents and breather pipes. The checklist should include a list of telephone numbers of supervisors and key employees to be notified.

Other procedures must include shutting off electric and gas utility services at the main line before any water reaches them. Make sure hot equipment is cooled before water reaches it. Operators should coat all machine surfaces liberally with heavy grease, especially around openings to bearings. (This step applies even to machines that may not be under water because dampness can damage equipment.) Shut off all open flames so that any flammable liquid floating on the floodwaters will not be ignited.

If possible, keep a salvage crew at the site to continue preventive operations after the facility has been shut down. This crew can take further steps if the flood exceeds the estimated high-water level.

Alarm Systems

Most industrial operations have a special fire alarm system using existing signaling systems such as a facility whistle. However, to avoid confusing fire alarms with the regularly used signals, some facilities have special codes or other signaling devices for fires. This type of signal also may indicate the location of the fire, or separate signaling devices may be used for each building or working area within the property.

The alarm system activating the emergency plan may or may not go through the communications center, but it should be touched off in the emergency headquarters office. All buildings should contain alarm systems (Figure 18–12).

In hospitals or other locations where both employees and nonemployees can hear an audible page system, a code name can be used to announce a fire and its location, such as "Doctor Red wanted in ———." Employees must be trained to be alert for this subtle signal.

Electric alarms are preferred to mechanical ones, except in a large open shop area with only one alarm-summons station and one alarm-sending device, such as a manually operated gong. Manually operated alarms should supplement electric alarms. Closed-circuit systems of the type specified by NFPA standards are recommended (see the References at the end of this chapter).

Organizations in areas where municipal fire departments are available usually have a municipal alarm box close to the firm's entrance or in one of the buildings. Others may have auxiliary alarm box areas, connected to the municipal fire alarm system, at various points on the premises. Another system often used is a direct connection to the control dispatcher. This system may be touched off by a water alarm in the sprinkler system or be activated manually. If possible, connect the fire alarm system to the local fire-fighting alarm and make sure it has an independent power supply.

In some areas, private central station services are available and provide excellent protection. These central stations receive signals from facility fire alarm boxes, security

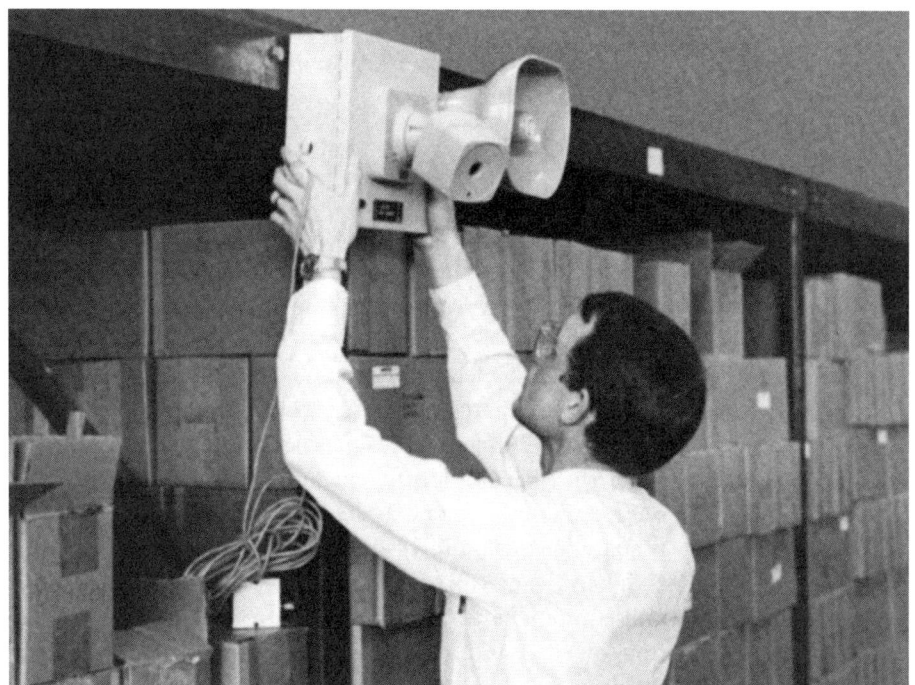

Figure 18–12. Spacing of alarm systems is important for notification of personnel in remote areas of large facilities.

staff, sprinkler head operations, and other hazard control points in the facility. Because these devices monitor facility security, they can relay information to fire or police departments without delay. The signal received at the fire department or assistance agency should pinpoint the site of the fire, or at least the building or area, so the fire can be found quickly (Figure 18–13). Automatic sending stations (thermostatic detectors) may be used but should not interfere with sending a manual alarm.

Regular checks should be made on the alarm system. Checks should include monthly inspections of all stations and a weekly test of the overall system to ensure that it is in proper working condition. These weekly tests should be conducted at a prearranged time and under a variety of wind and weather conditions to determine whether the signal can be heard in all parts of the facility at all times.

Fire Brigades

Fire prevention and fire protection must receive major attention in any emergency program because a fire can start from so many causes. Advance planning is important. (Note: U.S. organizations may elect to have fire brigades, in which case regulatory standards have to be followed.)

The fire chief must be able to command people in addition to having special training in fire prevention and protection. A person who has had experience in city or volunteer fire department work or a military service veteran with experience in fire fighting may be a good choice. In a smaller company or facility, a master mechanic, maintenance department head, or other employee with mechanical experience can be a good part-time fire chief.

Most states in America have a state fire service training program. As part of this service, there are now schools specifically oriented toward industrial fire protection. At the very least, all the fire brigade personnel should have advanced training. They are the ones most likely to use auxiliary fire-fighting equipment. The fire chief may need one or more assistants who have complete knowledge of the facility and equipment, command the respect and obedience of those under them, and are qualified to perform the duties of the chief officer if necessary.

The size of the facility and the fire potential presented by the occupancy will determine the kind of fire-fighting organization (brigade) required. The majority of facilities may need only first-aid fire fighters under the direction of departmental supervisors or managers. In larger facilities,

Figure 18–13. A command console for handling and dispatching 911 emergencies. *(Courtesy Western Will County, Illinois Emergency Communications Center.)*

the fire brigade organization, directed by a full-time fire chief, is composed of full-time and emergency members. The full-time members maintain fire brigade equipment and are responsible for the permanent fire protection of the facility. Emergency members report for fire duty when the alarm is sounded.

Fire brigade apparatus should be selected only after a study has been made of the specific conditions to ensure that the equipment will be adequate for any emergency. Management can obtain advice and assistance from the local fire department, the NFPA, the state fire training organization, or insurance companies.

The large-facility fire brigade is usually organized into squads, each with specific duties. One company's organization chart is shown in Figure 18-14.

The evacuation squad evacuates all employees from the emergency area as quickly and orderly as possible, without injury. They search closed areas, such as washrooms, to determine that everyone has been evacuated.

The environmental monitoring squad identifies and watches environmental conditions. They determine hazardous conditions, make sure sites are not contaminated, and decide when an area can be reentered safely.

Utility control squad members usually are maintenance personnel familiar with facility piping systems and the control of process gases, flammable liquids, and electricity. Sprinkler control squad members must understand the automatic sprinkler system—the direction of rotation of the valve they are to operate, the use of sprinkler stops, and the replacement of sprinkler heads, if this is not a maintenance department function.

Extinguisher Squad

Portable fire extinguishers are frequently operated by designated employees who work in the vicinity. However, as the size of the facility increases, it is advisable that special squads be selected, trained, and equipped for handling fire extinguishers.

Hose squad members are trained to operate fire hydrants and monitor nozzles and hoses. They should drill frequently with wet hose lines so they have the feel of a charged hose. After they become proficient in handling the hose, drills once or twice a year may be sufficient to maintain their preparedness (Figure 18-15).

The salvage squad is trained to protect as much stock and equipment as possible by controlling the directional flow of water and by covering stock with tarpaulins. Their training should include proper methods of throwing tarpaulins and using them to direct the flow of water. Squad members should also be familiar with the location of sawdust or other absorbent material and know how to use it in controlling water on floors.

The brigade-at-large or rescue team consists of specially trained workers. NFPA 600 spells out the qualifications of industrial fire brigade members. The main functions of this unit are to extricate casualties and eliminate hazards that may endanger other workers involved in controlling the emergency. Members respond to all alarms with a utility truck containing such rescue equipment as ropes, chains, block and tackle, ladders, cutting torches, saws, axes, and jacks. The amount of equipment, of course, will depend on the size of the facility and the hazards involved, but every effort should be made to anticipate possible problems.

Under the direction of the brigade officer, personnel in this unit also control utilities, ventilating fans, and blowers. They close fire doors, windows, and other openings in division walls and open windows and doors leading to fire exits. If escape equipment is the swinging section type, rescue personnel should be the first to operate the escapes and to secure the steps to the ground.

Depending on the size and inherent hazards of the facility, it may be nec-

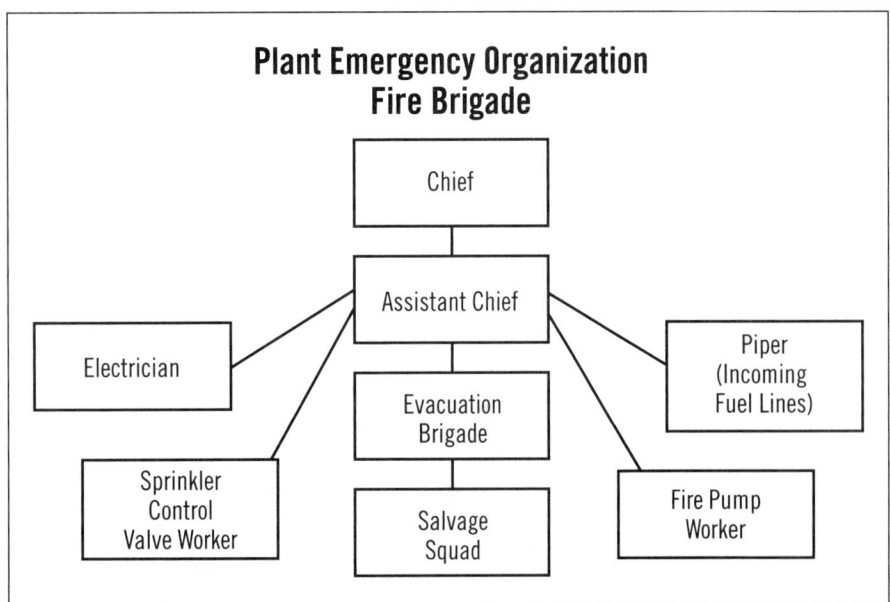

Figure 18-14. Because of the complexity of response needed to cope with an emergency, duties must be divided among facility emergency organization members to avoid confusion. Here is one facility's organizational chart. Current names, phone numbers, and those of alternates should be listed in the boxes. *(Reprinted with permission from Textile Section Newsletter.)*

essary to train squads in handling and erecting ladders, using foam lines, recharging foam generators, using specialized rescue techniques, and following chemical spill response procedures.

Fire Pump Team

There should be at least two competent people for pump duty in the main pump room. Management must assign one trained person to each pump located elsewhere.

Fire-Fighter Training

The brigade should go through complete drills on a routine basis. Drills should be unannounced and thorough in every respect, closely approximating fire conditions (Figure 18–15).

No matter how thoroughly the industrial fire brigade is trained, management should make sure they cooperate closely with adjoining or nearby facilities and the public fire department. It would be a good idea to do mutual training with the local fire department so that each side knows the expectations of the other. This approach builds a better relationship in the short run and helps unite the two groups as a single fire-fighting team for the long term. As stated before, the brigade chief, safety professional, or designated employee should immediately call the municipal fire department whenever a fire is reported at the facility.

In fire emergencies, security staff should open the yard gates and be ready to direct fire apparatus to the blaze. If facilities have railroad tracks, rail crossings that may be needed in an emergency should be kept free of all cars, hand trucks, and the like.

The brigade chief is in command at a fire until the officer in charge of the public fire department takes over. The brigade chief then serves as an advisor on processes and special hazards.

The fire station itself should be centrally located but not exposed to possible fires. It should be built of fire-resistant material or located in a sprinklered part of the facility and protected with portable extinguishers. Larger facilities may require mobile units, such as light hand-drawn trucks outfitted for special hazards.

Emergency Medical Services

First-aid and emergency medical services (EMS) should be headed by the organization's health care component, if available, as discussed in Chapter 12, Occupational Health Programs. In the organization of the medical phase of the emergency plan, those responsible must select and train personnel; decide what measures, equipment, and supplies are needed; and establish first-aid stations and a treatment center.

All employees should be encouraged to enroll in a cardiopulmonary resuscitation (CPR) and first-aid course. People assigned to first-aid and medical units should pass standard and advanced courses in these areas. The National Safety Council's First Aid Institute offers training in CPR and first aid. Local chapters of the American Red Cross also provide training programs. Because more than 140,000 Americans die every year from injuries, and one in three suffers a nonfatal injury, the Council believes that every person at some time will encounter an emergency requiring first aid. The Council trains and certifies instructors who, in turn, teach classes nationwide. The courses meet OSHA requirements for workplace first-aid and CPR training.

In addition, any time an emergency sprinkler system or some type of shower facility is installed in a chemical work area, management should have a light or alarm installed in case of an emergency. In many locations, especially cold climates, people have been known to experience severe cases of hypothermia when they had a chemical accident and drenched themselves to wash the chemicals off. They had no way to summon help and experienced exposure to the cold.

In addition, supervisors must keep copies of Safety Data Sheets (SDSs; formerly Material Safety Data Sheets [MSDSs]) on hand to send with an injured worker when the person is taken to the hospital or medical facility. This

Figure 18–15. Under direction of an instructor (white helmet), an emergency response drill team approaches a 300-gal (1,135-L) oil tank filled with burning fuel oil. Note the shower spray above the lead fire fighters.

information will help emergency and in-house physicians treat the employee. (See also ANSI Z358.1, Emergency Eye Wash and Shower Equipment.)

If a major disaster occurs, there may not be a sufficient number of trained doctors and nurses available to treat victims. In such an eventuality, care beyond the first-aid level will have to be provided by emergency medical technicians (EMTs), paramedics, and first responders who have received additional training. Such a medical team can be developed by recruiting volunteers, preferably with some medical or paramedic experience. In addition to first-aid training, more advanced instruction by regular medical personnel or local hospitals should be provided. Organizations should keep in mind when establishing an EMS program that although most of the medical aid will be given in a central medical station, some may be administered on the job site, possibly under hazardous conditions.

Plans should be made for representatives of the medical team to check all personnel at the disaster scene for trauma. The team can provide written clearance for victims to leave the facility when they are able to do so. Representatives of the investigation team may interview personnel before they leave to ensure that all eyewitness information is recorded.

Welfare and medical services include devising steps to prevent epidemics and to inspect food and sanitation facilities. In the planning stages for these activities, management should designate certain organization trucks as ambulances and make sure they are stocked with the necessary equipment. Two-way radio communication for such ambulance service is essential. Management should also make provisions to have nonperishable food and water rations on hand. There should be close coordination of facility EMS measures with the local civil defense, health, and medical services.

The chemical service responsibility of the EMS unit requires that it provide appropriate respiratory protection and other protective equipment to workers if toxic or corrosive chemical gases are released. The units must also have trained personnel, equipment, and supplies for chemical defense and decontamination. The organization should plan a priority sequence for decontamination—that is, water supply first, power facility second, machinery areas third, warehouse areas fourth, and so on. Medical team personnel will also be responsible for any radiological monitoring required after a nuclear incident. The mere knowledge that such monitoring equipment is available can be a morale builder for workers. After a disaster, no one should be permitted to drink water until it has been examined for contaminants.

Warden Service and Evacuation

The warden service is responsible for maintaining employee control during emergencies, including (1) guiding employees to safe areas; (2) directing employees, including physically disabled workers, away from hazardous areas; and (3) averting panic. In smaller organizations, the wardens may take charge of the shelters. In some cases, the warden service may be responsible for seeing that process and equipment shutdown is carried out smoothly.

This type of service was devised primarily for areas of high population densities, such as commercial structures, factories, and residential areas. For facilities with several hundred or several thousand employees, the use of warden teams can be vital during emergencies. In facilities with only a few personnel, machine operators will have to be trained in shutdown details.

Management, in checking and providing for safe exits, emergency lighting, and evacuation drills, should refer to the NFPA standards, building codes, and appropriate regulatory standards. Smooth, safe functioning of an evacuation plan requires that those in charge have a thorough knowledge of all operations and employees; the number, type, and location of available exits; alternate exits; and location of hazards, as well as knowledge of warning and evacuation facilities. The subject of building exits is covered in Chapter 26, Office Safety in this volume; in Chapter 2, Buildings and Facility Layout, and Chapter 9, Fire Protection, both in the *Engineering & Technology* volume; and in the Life Safety Code, NFPA 101. (See also the OSHA regulations at 29 CFR 1910.38 and 29 CFR 1910.157[a][2].)

Most facilities have a rigid rule that only specially appointed people on the fire brigade should go to the vicinity of the fire. Everyone else should proceed on signal to a safe location or assembly area designated by the evacuation plan.

Transportation

When transportation facilities are disrupted or traffic is temporarily restricted, many employees may find it impossible to get to work. The organization may need to provide transportation with trucks and cars. Advance planning for car pools and pickup stops will greatly facilitate such a procedure.

The transportation responsibility includes arranging for ambulance service, transporting employees to and from work, and moving emergency service crews and supplies as needed. Any planning for adequate transportation service and traffic control requires cooperation with the public police department, emergency management planning authorities, and possibly the military.

An emergency transportation unit might consist of a group of regularly assigned drivers. Company vehicles can be used when it becomes necessary to handle stretcher cases in evacuating any injured. The unit will transport auxiliary fire fighters, first-aid teams, and salvage and rescue workers to the scene of the disaster at the earliest possible instant. The unit can also deliver needed equipment and material from outside suppliers.

When developing emergency transportation plans, the organization must anticipate sources of motor fuel.

OUTSIDE HELP

The chance of survival and recovery from a disaster or major accident is greater when organizations pool knowledge, equipment, and personnel with local community members. Therefore, emergency plans should include a provision for exchanging aid with other organizations in the community.

These plans should be drawn up ahead of time for both groups. Everyone needs to know beforehand what equipment and services will be available and how they can be called in. Plans should include provisions for calling another organization after hours should a problem develop outside normal working times.

Mutual-Aid Plans

A number of industrial communities are organized to assist their members in the event of an emergency or disaster. These organizations include manufacturing facilities, large offices, stores, hotels, utilities, chemical facilities, law enforcement organizations, hospitals, newspapers, and radio and television stations. They operate independently of or as supplements to any individual emergency management teams.

It is usually not possible to have adequate supplies available for a really large disaster such as a major earthquake. The best defense is to have adequate supplies in other areas committed for standby use, with communication channels and a plan for rapid transportation of supplies to the stricken area. Adequate emergency medical supplies and fire-fighting equipment are especially important. Organizations should develop a plan for rapid and accurate communication, as discussed earlier in this chapter.

Mutual-aid plans with neighboring companies and community agencies should include establishing an organizational structure and communication system; standardizing an identification system, procedures, and equipment (such as fire hose couplings); formulating a list of available equipment; stockpiling medical supplies; sharing facilities in an emergency; and cooperating in test exercises and training.

One item often overlooked in most mutual-aid contracts is who will pay any costs, such as repairs to equipment or workers' compensation for injured employees. Most often, workers' compensation is covered by the person who is considered the full-time employer.

Frequently these "cooperatives" establish a task force composed of personnel from each member organization. Their training is supplemented by detailed written instructions. Each facility marks bulldozers, floodlights, and tools for emergency use. Training on a community basis might include instruction by members of the public fire department to facility fire brigade members. In addition, members of a construction or wrecking may provide actual training to show the salvage and rescue teams how to handle heavy weights and to work safely among debris.

Contracting for Disaster Services

Some organizations contract for disaster service. The service may include an annual retainer, plus additional pay for the actual hours worked. For example, a wrecking company can be engaged to supply the workers and equipment necessary to clear debris created by a disaster. Contracting for such a service removes the burden from companies of providing trained personnel and maintaining idle emergency equipment that could easily be damaged in the very disaster for which it was designed.

Sometimes in a major disaster, some of the contractors an organization has on standby may go instead to the highest-bidding firm. In such cases, management must have contingency plans.

Municipal Fire and Police Departments

Fire fighters from the station most likely to respond to an alarm should be fully acquainted with all fire hazards in the facility. The organization may encourage such cooperation by inviting local fire officials to inspect the area before an emergency. The officials can become familiar with the location, construction, and arrangement of all buildings, as well as all special hazards, such as toxic, corrosive, or flammable gases, liquids, and materials. Because public fire department rescue equipment can supplement facility rescue units, they can make sure that equipment is compatible with the municipal equipment. Therefore, the local fire department can formulate an efficient plan of attack before an emergency occurs. Such a procedure is far better than waiting for an incident to happen, and then running the risk that local fire-fighting units will misunderstand the situation and initiate an improper response.

The public police force can help quell large-scale disturbances and assist in evacuating the facility should a major disaster occur. Planning for this outside help should include making arrangements for traffic control, particularly where a parking lot empties immediately onto a public highway.

Industry and Medical Agencies

Details of the following services, as well as a number of other emergency and specialized information sources, are given in Appendix 1, Sources of Help.

In 1970, the American Chemical Council (ACC) (formerly Chemical Manufacturers Association) created the Chemical Transportation Emergency Center (CHEMTREC).

The Toxicology Literature Online network (TOXLINE) has been designed to provide current and prompt information on the toxicity of substances. It is intended for use by health professionals and other scientists involved in antipollution, safety, drug, health, and other disciplines.

Governmental and Community Agencies

During a communitywide disaster, a large number of governmental and private agencies are available to assist industries. These include the Federal Emergency Management Agency (FEMA), the U.S. Army Corps of Engineers, the Salvation Army, the American Red Cross, the U.S. Public Health Service, the National Weather Service, and the National Oceanic and Atmospheric Administration (NOAA). To be effective in coping with a disaster, the efforts of all of these groups must be coordinated and directed toward a common end. Therefore, each organization should have an updated listing of all cooperating agencies; the administrator's name, address, and telephone number; and the task assignment of the agency. If possible, safety professionals and emergency-planning directors should meet periodically with these administrators to discuss mutual problems and disaster control techniques.

The emergency-planning director should become thoroughly familiar with the authority, organization, and emergency procedures that are established by law and that will become effective on declaration of a civil emergency.

It is in the organization's best interests to be active in the county local emergency planning committees (LEPCs). In this way, it can cooperate with other organizations from other industries and groups in the community. The main purpose of an LEPC is handling hazardous materials problems along with natural disasters and other crises.

In wartime, federal, state, and local governments are responsible for relief measures. The American Red Cross has offered to assist the government in providing food, clothing, and temporary shelter on a mass-scale basis during the emergency period immediately following enemy attack. In many communities, local emergency management officials have requested American Red Cross chapters to assume all or part of this responsibility, acting under emergency management authorities.

In natural disasters, the American Red Cross is responsible for helping families and individuals meet disaster-caused needs that cannot be met through their own resources. The relief operations of local American Red Cross chapters are coordinated with the activities of the local, state, and federal governments. The resources of the national organization are available to supplement chapter assistance.

SUMMARY

- Advanced emergency management planning is the best way to minimize potential loss from natural or human-caused disasters and accidents. Emergency planning must provide for the safety of employees and the public, protect property and the environment, and establish methods to restore operations to normal as soon as possible.
- Basic emergency management planning usually includes establishing a chain of command, an alarm system, medical treatment plans, communication system, shutdown and evacuation procedures, and auxiliary power systems.
- Organizations can use a variety of government and private sources to find out what types of hazards are most likely to occur in their area. Once the initial background work has been done, the next step is to develop a working emergency management plan.
- The chain of command established for emergency plans should be kept as short as practical and should be staffed with employees selected for their ability to respond well in high-pressure situations.
- Emergency plans generally call for establishing a command headquarters or center to coordinate communications and disaster response. An incident command system can be established to manage an organization's response to an emergency.
- Most operations have special alarm systems for fire and security. Many organizations establish local fire and hazardous materials brigades known as hazardous materials response teams (HAZWOPER), as defined by OSHA and various statutes. Other services that often organize for emergencies include emergency medical services, warden service, and emergency transportation units.
- In some cases, an organization may need to request outside help to survive and recover from a disaster or major accident. Mutual-aid plans with other firms and organizations in the community can pool the knowledge, resources, equipment, and personnel of many organizations. These cooperatives may establish a task force composed of members from each organization to organize and administer emergency responses.

REFERENCES

American Insurance Services Group, Engineering and Safety Service, 85 John Street, New York, NY 10038.

Fire Hazards and Safeguards for Metalworking Industries, Technical Survey No. 2.
Fire Safeguarding Warehouses, Technical Survey No. 1.

American National Standards Institute, 11 West 42nd Street, New York, NY 10036.
Emergency Eye Wash and Shower Equipment, ANSI Z358.1–2009.

AT&T West. *Earthquake Survival Guide*.

Conference Board, 845 Third Avenue, New York, NY 10022.
Studies in Business Policy, No. 55, "Protecting Personnel in Wartime."

Factory Mutual Research Organization, 1151 Boston Providence Turnpike, Norwood, MA 02062.
Handbook of Industrial Loss Prevention.
Loss Prevention Data.

Federal Emergency Management Agency, 500 C Street SW, Washington DC 20472. (Available through Superintendent of Documents, U.S. Government Printing Office, Washington DC 20402.)
Emergency Planning and Community Right-to-Know Act, Title III, Superfund Amendments and Reauthorization Act, Public Law No. 99-499.

Giraud, R. "Radioactive Elements." *National Safety News* (June 1973.) Adapted from author's article in *Revue Technique du Feu, Entreprise moderne d'edition*, 4 rue Cambon, 75 Paris 1, France.

Kerr, J. W. "Preplanning for a Nuclear Incident." *Fire Command!* (April 1977.)

National Fire Protection Association, 1 Batterymarch Park, Quincy, MA 02269-9101.
Explosion Prevention Systems, NFPA 69.
Facilities Handling Radioactive Materials, NFPA 801.
Fire Protection Handbook. 2008.
Guard Service in Fire Loss Prevention, NFPA 601.
Health Care Facilities, NFPA 99.
Industrial Fire Hazards Handbook, SPP 57A.
Installation of Signalling Systems, NFPA 13.
Life Safety Code Handbook, NFPA 101, 2006.
Professional Competence of Responders to Hazardous Materials Incidents, NFPA 472.
Standard on Industrial Fire Brigades, NFPA 600.

National Petroleum Council, 1625 K Street NW, Washington DC 20006.
Disaster Planning for the Oil and Gas Industries.
Security Principles for the Petroleum and Gas Industries.

National Safety Council, 1121 Spring Lake Drive, Itasca, IL 60143.
First Aid Institute.
Fundamentals of Industrial Hygiene. 6th ed. 2012.
Hazardous Materials (booklet).
(See other appropriate topics treated in both volumes of this Manual, especially those pertaining to organization, training, medical and nursing services, and fire extinguishment and control.)

Noll, G. G., M. S. Hildebrand, and J. G. Yvorra, *Hazardous Materials: Managing the Incident*. Fire Protection Publications. Stillwater, OK: Oklahoma State University

Occupational Safety and Health Administration.
29 CFR 1910.38, Employee Emergency Plans and Fire Preventing Plans.
29 CFR 1910.120, Final Rule for Hazardous Waste Operations and Emergency Response.
29 CFR 1910.156, Fire Brigades.
29 CFR 1910.157, Portable Fire Extinguishers.
29 CFR 1910.1200, Hazard Communication.

Underwriters Laboratories Inc., 333 Pfingsten Road, Northbrook, IL 60062.
Gas Shutoff Valves—Earthquake.
Tests for Fire Resistance of Record Protection Equipment.

UN Globally Harmonized System of Classification and Labelling of Chemicals. *Federal Register* (March 26, 2012).

U.S. Department of Health and Human Services.
Occupational Safety and Health Guidance Manual for Hazardous Waste Site Activities.

U.S. Department of Transportation.
49 CFR 172, Hazardous Materials Table and Hazardous Materials Communications Regulations.

U.S. Environmental Protection Agency.
OS–120, Chemicals in Your Community.
40 CFR 311, 330, 335, Extremely Hazardous Substance List and Threshold Planning Quantities; Emergency Planning and Release Notification Requirements; Final Rule.

REVIEW QUESTIONS

1. What should be the first concern in planning for an emergency?
   a. protecting the property
   b. ensuring the safety of employees and the public
   c. restoring business operations to normal
   d. protecting the environment
2. The first step in the emergency-planning process begins with determining what types of hazards/emergencies may affect the organization. Name 8 of the 11 types of emergencies.
   a.
   b.
   c.
   d.
   e.
   f.
   g.
   h.
3. Many sources of hazard/emergency information can help management determine the likelihood of specific emergency events in a locality. List 5 of the 7 sources.
   a.
   b.
   c.
   d.
   e.
4. The second step in the planning process should be
   a. developing a checklist of emergency equipment and shutdown actions.
   b. designating hazmat team members.
   c. testing all alarm systems.
   d. preparing a specific emergency plan of action.
5. What are 6 of the 12 considerations that should be addressed within the specific plan of action?
   a.
   b.
   c.
   d.
   e.
   f.
6. Why is disaster training one of the most important functions of the director and staff, on both the corporate and facility levels?
7. Which OSHA regulation addresses aspects of health and safety that are now legally required at hazardous materials sites; treatment, storage, and disposal facilities; and other hazardous materials emergency locations?
8. What is the function of the hazmat team?

Workplace Violence

James O'Reilly, JD
Marina Schemmel, JD

Management Overview
Is Workplace Violence a Pervasive Problem? ▶ How Are Government Policies Helping?

What Is Workplace Violence?

Risk Factors for Workplace Violence

Workplace Violence Prevention Programs
Getting Workers and Managers Involved and Committed ▶ Creating Policy Statements ▶ Threat Assessment Team ▶ Workplace Analysis ▶ Hazard Prevention and Control ▶ Special Considerations: Hiring, Discipline, and Termination ▶ General Training Needs ▶ Training on How to Foresee Potential Violence ▶ Assessing Threats of Violence ▶ Incident Reporting Systems ▶ Program Evaluation ▶ Record Keeping

High-Risk Occupations
Late-Night Retail Establishments ▶ Health Care and Social Service Workers ▶ Taxicab Drivers ▶ Law Enforcement

Regulatory Requirements

Domestic Violence

Small Business

Concealed Weapons

Summary

References

Review Questions

MANAGEMENT OVERVIEW

Employers face many challenges in keeping their workplaces and employees safe. Much of this text deals with accidental harms; this chapter addresses the intentional harm that violence can produce. When violence enters the workplace, it can cost the employer money in terms of damages to property, decreased employee security, and potential legal problems for the employer. Workplace violence rarely occurs "at random," or without warning signs (Rugala and Isaacs 2004). Thus, employers who take preventive steps should be more able to mitigate the risk of violence in the workplace. This chapter explains the best practices for reduction or avoidance of violent acts at the workplace.

Is Workplace Violence a Pervasive Problem?

How prevalent is workplace violence in the United States? The U.S. Labor Department's Bureau of Labor Statistics (bls.gov) and several other workplace violence resources compile statistics of workplace assaults and homicides. The dates may be scattered, but altogether they present a cohesive picture on the pervasiveness of workplace violence:

- In 2012, workplace violence was the second highest cause of fatal occupational injuries, causing 767 reported deaths (BLS 2013).
- Workplace homicides accounted for 11% of total occupational deaths in 2012 (BLS 2013).
- The Bureau of Justice Statistics found that between 2002 and 2011, an average of 759,840 private-sector employees were victimized by violence at the workplace each year (528,420 government employees) (Harrell 2013).
- The cost of workplace violence on the economy can be great. The National Institute for Occupational Safety and Health (NIOSH) estimated in 1998 that employers lost a total of 876,000 workdays and $16 million in wages due to nonfatal workplace assaults (NIPWV 2013).
- The number of workplace assaults and homicides committed by customers, clients, and patients has increased in recent years (NIPWV 2013, 1, 5).

OSHA has identified some occupations as "high risk." These occupations have a high number of risk factors for workplace violence. Thus, the statistics for workplace assaults and homicides are higher than for other occupations.

- In 2013, health care patients committed the highest number of workplace assaults against workers (60%) (NIPWV 2013, 9; Restrepo and Shuford 2012, 14).
- In 2009, taxicab drivers had the third highest rate of workplace homicides (down from the highest rate in 2003) (Restrepo and Shuford 2012, 7–8).
- Late-night retail establishments (gas station and convenience store clerks) had workplace assault rates more than double the rate for all retail employees between 1993 and 1999 (OSHA 2009).

The good news is that the numbers of workplace assaults and homicides have been on the decline since the mid-1990s (NIPWV 2013, 3–4; Restrepo and Shuford 2012, 3). In fact, the lowest number of workplace homicides documented was in 2011 (458 total) (NIPWV 2013, 3).

How Are Government Policies Helping?

NIOSH cites increasing efforts to research and raise awareness of workplace violence as well as the passage of government policies at federal, state, and local levels in aiding the decrease (NIOSH 2004, 2–3).

OSHA has published two directives, in 2009 and 2011, and guidelines for high-risk occupations to create workplace violence prevention programs. On average, employers who institute workplace violence prevention programs spend $4.50/employee each year (NIPWV 2013, 12). However, the cost up front could save employers a large amount of money in legal fees, fines from OSHA, and lost productivity and wages. Still, more than 70% of employers do not have workplace violence prevention programs.

Studies have shown that employers can reduce the risk of violence in the workplace by implementing a workplace violence prevention program (OSHA 2011, 4). As OSHA stated in its 2011 directive, "OSHA believes that a well written and implemented Workplace Violence Prevention Program, combined with engineering controls, administrative controls and training[,] can reduce the incidence of workplace violence in both the private sector and Federal workplaces." All employers should be aware that workplace violence can happen in any workplace, no matter what occupation or how large or small the company may be (Rugala and Isaacs 2004, 7).

WHAT IS WORKPLACE VIOLENCE?

Employers must first understand what "workplace violence" encompasses before creating a prevention program. There are many different recognized definitions of workplace violence. A basic but highly cited definition is from NIOSH, which defines *workplace violence* as "violent acts, including physical assault and threats of assault, directed towards persons at work or on duty" (NIOSH 2004, 5). Many organizations have expanded the NIOSH definition to encompass a variety of conduct. Indeed, workplace violence ranges from harassment to homicides. Other examples include:

- stalking
- assaults
- stabbings
- shootings
- rape
- domestic abuse (CCH 2011; NIOSH 2004, 5).

Terrorism is a category of violence that has been increasing at the workplace (Virginia Tech student murders, District of Columbia Navy Yard shooting in 2013). Thus, any conduct that creates anxiety or fear, and "a climate of distrust in the workplace" can be a form of threat or workplace violence (Rugala and Isaacs 2004, 13). Workplace violence is often grouped into four categories based on who the perpetrator or assailant is in relation to the victim/employee. The first category is "criminal intent," or "violence by strangers," and includes people who enter the workplace solely to commit a crime, such as robbery or assault (OSHA 2011, 5; Rugala and Isaacs 2004). Workplaces most vulnerable to this type of violence include occupations classified by OSHA as "high-risk occupations" such as late-night retail, gas station clerks, and taxi drivers or any other occupation with night hours that is located in high-crime neighborhoods and whose employees have access to or carry money (NIOSH 2004, 5; Rugala and Isaacs 2004, 13). Sometimes the criminal will have a gun, which increases the likelihood that the employee may be seriously injured or killed (Rugala and Isaacs 2004, 13). This category has the highest rate of workplace homicides because of robbers, who commit the highest percentage of homicides of male employees and the second highest percent of female employees (BLS 2012).

The second category is "violence by customers, clients, or patients," and includes any person who regularly visits the workplace and who commits violence against an employee while being helped, served, or cared for by the employee (Barling et al. 2009; OSHA 2011, 5). This category has the highest number of workplace assaults, because it includes health care and mental patients, who accounted for 61% of workplace assaults in 2009 (Restrepo and Shuford 2012, 10).

The third category is "violence by co-workers" and includes former, current, or prospective co-workers, as well as violence by supervisors or managers (NIPWV 2013, 2; Rugala and Isaacs 2004, 13). Although violence happens in some cases between managers and employees, assaults between co-workers of the same level are more common (Harrell 2013, 6).

The fourth category is "violence by personal relationships," which includes persons who have or had a relationship with an employee (such as a current or former spouse, friend, or relative) (NIPWV 2013, 2). This category covers domestic violence that enters the workplace. This type of violence, more so than violence by robbers and customers, occurs industrywide, and no industry has a substantially higher risk than others for personal relationship violence (NIOSH 2004, 5). However, there are often warning signs that could cause co-workers and employers to be aware of the risk of potential violence. Thus, a training program as part of a workplace violence prevention program would be especially helpful in reducing the risk of violence from this category (Rugala and Isaacs 2004, 14).

In general, the third and fourth categories have the lowest numbers of workplace assaults and homicides. On the other hand, violence committed by criminals and customers is much more common and thus poses a greater risk to employers. As each type of workplace violence may require different prevention and control strategies, employers should be familiar with the type of violence their employees will encounter.

RISK FACTORS FOR WORKPLACE VIOLENCE

OSHA, NIOSH, and other sources have found that a number of factors can increase the workplace's risk for violence. As mentioned earlier, OSHA's high-risk occupations include late-night retail (gas station and convenience store clerks and bartenders), health care and social service workers (especially home health care), taxicab drivers, and law enforcement employees. These occupations share one or more of the following characteristics:

- are located in areas of high crime
- operate mostly at night or early in the morning
- handle or have access to money
- have customers who are either unstable or volatile (health care patients or people who are under arrest/in jail) or are under the influence of alcohol
- have mobile workplaces (taxicab or police cruiser)
- have employees who work alone or with only one other co-worker (OSHA 2010; OSHA 2011, 3).

Some risk factors are specific to the organization, such as:
- violent incidents in the past
- abusive supervision or inappropriate management behaviors (managers not using fair and consistent disciplinary standards, showing favoritism toward some employees, not treating employees respectfully)
- a high-stress environment (inadequate staffing, employees are overworked/overscheduled, employees feel underappreciated by management)

- improperly trained employees or supervisors
- minimal or ineffective security procedures
- an organizational tolerance for potentially violent conduct (failing to discipline employees who harass or start fights with other employees) (Barling et al. 2009, 680; CCH 2011, 12; Hartsfield 2013, 2; Rugala and Isaacs 2004, 21–22).

These lists are not comprehensive but contain just a few of the risk factors an employer could encounter in evaluating the workplace's potential for violence. However, risk factors will vary for each workplace. Thus, an employer should always include a workplace security analysis as part of a workplace violence prevention program.

OSHA has published guides for establishing workplace violence prevention programs in late-night retail, taxi companies, and health care organizations. Employers in other occupations should consider developing a workplace violence prevention program, especially if the workplace has several risk factors for workplace violence. An effective prevention program will give employers the tools to identify threats before they become violent incidents, and will help the workplace recover faster if violence does occur (CCH 2011, 14).

WORKPLACE VIOLENCE PREVENTION PROGRAMS

Employers are required by federal regulations and state laws to provide a safe working environment for their employees, and a workplace violence prevention program can help aid in that goal (Romano et al. 2011). There are many guides for employers to use in creating a workplace violence prevention program. However, each program must be specifically tailored to the workplace's size, needs (what risk factors face the organization), and available resources (financial and departments within the organization) (CCH 2011, 14). Many employers often have components of a workplace violence prevention program already in place, such as access control, employee standards of conduct, sexual harassment policies, and policies against bullying (Hartsfield 2013, 6). Several sources cite the following components as necessary to create an effective workplace violence prevention program:

- management commitment and employee involvement
- a policy statement
- a threat assessment team
- workplace analysis
- hazard prevention and control
- training
- incident response
- threat assessment
- program evaluations
- record keeping.

Some factors that are important to any workplace violence prevention program are communication, confidentiality, teamwork, and accountability. Communication must occur at all levels: between management and employees and between any workplace violence prevention teams (threat assessment or incident response) and management (NIOSH 2004, 15). Furthermore, a prevention program should focus on being "proactive" rather than "reactive" to incidents of workplace violence. Finally, employers should make sure the plan is implemented and, if possible, practiced so that employees are adequately prepared to deal with potential violence in the workplace (Kelloway et al. 2006, 19).

Getting Workers and Managers Involved and Committed

Just as a workplace cannot function without its employees, a workplace violence prevention program will be ineffective unless everyone is involved and committed to the program's success. Indeed, a workplace violence prevention program begins with support from top management. Top management must commit to provide resources for the program, such as adequate funding for security measures, staffing for threat assessment and/or incident response teams, materials for training, and access to outside resources (local law enforcement, social service and mental health providers), if necessary (CCH 2011, 7; NIOSH 2004, 14). Top management can also develop the workplace's policy statement, stating that violent behaviors will not be tolerated and that all reports of violence will be addressed promptly and confidentially (Philpott and Grimme 2009, 67–68). Management can establish a system of disciplinary procedures for employees who violate the organization's policy. Management can also hold supervisors and employees accountable when necessary.

Senior management can provide support for the program by publicizing the program, recognizing the program's successes, and encouraging employee involvement in implementing and revising the program (CCH 2011, 7; NIOSH 2004, 9). Finally, senior management can "set the tone" for a safe work environment by promoting an atmosphere of respect among co-workers and between employees and management (OSHA 2009, 6).

Managers who supervise employees directly also have a role in workplace violence prevention. Specifically, direct managers should spend enough time with the employees they supervise to be able to detect early warning signs. Managers should also inspect the workplace regularly to ensure that controls for hazard prevention are being used. Managers should make sure all employees have received training on reporting and emergency situation procedures. Managers must know when they should go to top management or seek outside sources for help in defusing or dealing with violence. Managers should make sure each reported threat and incident is investigated and dealt with appropriately. Managers in charge of hiring should use background screening procedures before meeting with prospective employees (Philpott and Grimme 2009, 67–68). Managers must enforce workplace rules fairly and use disciplinary procedures consistently on all employees. During termination, managers should focus on maintaining the employee's respect and dignity.

Employee involvement in the program is vitally important. In developing the program, employees are often the most knowledgeable about potential hazards in the workplace because they see them every day, even when managers are not available. Therefore, employee feedback on hazard assessment and prevention will be more practical. Employees can, and should, participate in any part of developing and implementing a workplace violence prevention program. Specifically, employees should report any and all threats or incidents of violence, even if no one was injured. Employees should always follow the employer's prevention policies and ask management if they have any questions or concerns.

Creating Policy Statements

An essential part of a workplace violence prevention program is a clearly written policy statement (NIOSH 2004, 11). A policy statement should state that violent behavior will not be tolerated in the workplace and outline the procedures that will be implemented to prevent violent behavior and to deal with violence if it arises. A policy statement should include a definition of workplace violence, including specific examples. At a minimum, the definition should include policies on physical violence, harassment, threats, and intimidation (OPM 2013, 10–11). The list should not be exhaustive, however. The policy should state that it covers incidents not only between co-workers, managers, and/or employees, but also violence perpetrated by customers or clients against employees. In addition, if employees often work off site, incidents that occur at different locations should be included. Finally, the policy must state that all potential violent acts, statements, or conduct must be reported, and then the process for reporting such incidents should be described.

The employer must make sure that all employees, even part-time or independent contractors, are aware of the policy statement. The employer must also make sure the policy statement is updated regularly. A final consideration is that no matter how extensive the policy statement, an employer must make sure that the policy is regularly implemented and practiced. Otherwise, the workplace violence prevention program will fail.

Example policy statements are shown in Figures 19–1 and 19–2.

WORKPLACE VIOLENCE POLICY

It is the [insert Department or Agency name]'s policy to promote a safe environment for its employees. The Department is committed to working with its employees to maintain a work environment free from incidents that could be construed as violence, threats of violence, harassment, intimidation, and other disruptive behavior.

Violence, threats, harassment, intimidation, and other disruptive behavior in our workplace will not be tolerated; that is, all reports of incidents will be taken seriously and will be dealt with appropriately.

Such behavior can include oral or written statements, gestures, or expressions that communicate a direct or indirect threat of physical harm. **Individuals who commit such acts may be removed from the premises and may be subject to disciplinary action, criminal penalties, or both.**

Your cooperation is critical to implementing this policy effectively and maintaining a safe working environment. Do not ignore violent, threatening, harassing, intimidating, or other disruptive behavior. If you observe or experience such behavior by anyone on agency premises, whether he or she is an agency employee or not, report it immediately to a supervisor or manager. Supervisors and managers who receive such reports should seek advice from the Employee Relations Office at xxx-xxx-xxxx regarding investigating the incident and initiating appropriate action. [**PLEASE NOTE**: Threats or assaults that require immediate attention by security or police should be reported first to security at xxx-xxx-xxxx or to police at 911.]

I will support all efforts made by supervisors and agency specialists in dealing with violent, threatening, harassing, intimidating, or other disruptive behavior in our workplace and will monitor whether this policy is being implemented effectively. If you have any questions about this policy statement, please contact _____ at xxx-xxx-xxxx.

Figure 19–1. Sample workplace violence policy statement. (*OPM 2013, 10-11*)

> **ABC COMPANY POLICY: WORKPLACE VIOLENCE**
>
> ABC COMPANY is concerned and committed to our employees' safety and health. We refuse to tolerate violence in the workplace and will make every effort to prevent violent incidents from occurring by implementing a Workplace Violence Prevention Program (WPVP). We will provide adequate authority and budgetary resources to responsible parties so that our goals and responsibilities can be met. All managers and supervisors are responsible for implementing and maintaining our WPVP Program. We encourage employee participation in designing and implementing our program. We require prompt and accurate reporting of all violent incidents whether or not physical injury has occurred. We will not discriminate against victims of workplace violence.

Figure 19-2. Sample workplace violence policy statement. (*CCH 2011, 624A*)

Threat Assessment Team

The next step in forming a workplace violence prevention program is to perform a threat assessment for the workplace. Some employers may have the resources to hire a consultant to conduct a threat assessment (Philpott and Grimme 2009, 65–66). If so, the consultant will make recommendations and management will implement the program. Another way to conduct a threat assessment is to create a threat assessment team to evaluate the workplace's risks for violence and provide recommendations to reduce the workplace's risks. The size of the team will depend on the employer's resources. If possible, the team should include a representative from every level or department. On the other hand, the employer may have its management team be responsible for the threat assessment. Regardless of the team composition, the procedures to conduct a threat assessment will be the same.

As part of the workplace analysis, the team should review records of past violent incidents, perform an inspection for security risks, review existing policies, and conduct employee surveys. The team's recommendations should include controls for hazards in the workplace, training programs for employees and supervisors, and plans for responding to violent incidents (ASIS 2011, 20). Management may implement the recommendations or provide the team with resources and allow them to implement the recommendations. In addition to drafting recommendations and implementing them, the threat assessment team should also be responsible for auditing the program.

Workplace Analysis

The first step in a threat assessment is to conduct a workplace analysis. Using records of past incidents, employee surveys, and the results of a security inspection, the threat assessment team will discover what risk factors for violence exist in the workplace.

Records Review

The threat assessment team should gather any documents with information on incidents of violence that occurred in the past 2 or more years. These documents may include police reports, incident reports, medical records, insurance filings, workers' compensation claims, litigation documents, and any forms completed in compliance with OSHA or other regulatory requirements (OSHA 2009, 7). The team should review these documents to look for any patterns such as the type of incident, particular injuries, what response procedures were used, any defects in security equipment, and so forth. The team should use the information from the records review to further investigate possible risks of violence in the workplace or to implement hazard controls.

There may not be many records available, especially if the organization did not previously have an incident reporting system. If so, the threat assessment team may benefit from contacting other businesses in the community, the local police department, and/or other community organizations to see if they have information on workplace violence that could help the team in general in performing a workplace analysis.

Employee Surveys

As mentioned previously, employees are a great source of information for what risks exist in the workplace and how efficient the existing security measures are in preventing violence. The threat assessment team should distribute surveys asking a variety of questions about the workplace's current violence procedures, the employee's experience with violence in the workplace, and if the employee has any recommendations for improvements (OPM 2013, A-2–A-3; Rugala and Isaacs 2004, 22). The surveys may be anonymous if the team is concerned about biased or incomplete feedback.

After the threat assessment team is finished implementing the workplace violence prevention program, employers may decide to continue offering employee surveys, especially after major changes at the workplace or after a violent incident. Employers should also consider providing feedback or following up with employees after reviewing the surveys. Figure 19-3 is an example of a general employee survey.

EMPLOYEE SECURITY SURVEY

This survey will help detect Security Problems in your building or at an alternate worksite. Please fill out this form, get your co-workers to fill it out, and review it to see where the potential for major security problems lie.

NAME: _____

WORK LOCATION: _____ (IN BUILDING OR ALTERNATE WORKSITE)

1. Do either of these two conditions exist in your building or at your alternate work site?	
Work alone during working hours.	
No notification given to anyone when you finish work.	
Are these conditions a problem? If so, when? Please describe. (For example, Mondays, evening, daylight savings time)	
2. Do you have any of the following complaints (that may be associated with causing an unsafe worksite)? (Check all that apply)	
Does your workplace have a written policy to follow for addressing general problems?	
Does your workplace have a written policy on how to handle a violent client?	
Does your workplace have a written policy regarding when and how to request the assistance of a co-worker?	
Does your workplace have a written policy regarding when and how to request the assistance of police?	
Does your workplace have a written policy on what to do about a verbal threat?	
Does your workplace have a written policy on what to do about a threat of violence?	
Does your workplace have a written policy on what to do about harassment?	
Does your workplace have a written policy on working alone?	
Does your workplace have an alarm system(s)?	
Does your workplace have security in and out of the building?	
Does your workplace have security in the parking lot?	
Have you been assaulted by a co-worker? A customer?	
To your knowledge have incidents of violence ever occurred between your co-workers?	
3. Are violence-related incidents worse during shift work, on the road, or in other situations? Please specify:	
4. Where in the building or worksite would a violence-related incident be most likely to occur? lounge—exits—deliveries—private offices—parking lot—bathroom—entrances—other (specify)	
5. Have you ever noticed a situation that could lead to a violent incident?	
6. Have you missed work because of a potential violent act(s) committed during your course of employment?	
7. Do you receive workplace violence-related training or assistance of any kind?	
8. Has anything happened recently at your worksite that could have led to violence?	
9. Can you comment about the situation?	
Where did the incident occur?	
Were the police called?	
Were preventive measures in place at the time of the incident or were they being implemented?	
Were firearms carried or used during the incident?	
How many workers were on duty at the time?	
How often do violent incidents occur?	
10. Has the number of violent clients increased?	

Figure 19–3. Sample employee survey. (*OSHA 2009, 7–8.*)

Workplace Security Analysis

After reviewing records and performing employee surveys, the threat assessment team should perform a thorough workplace security analysis. The team should make sure to follow up on risks identified from the records review and employee surveys. This initial inspection should seek to identify every possible risk that could lead to violence in the workplace. Some questions the threat assessment team may ask during the analysis include:
- Are there any areas in the building that are unsecured, or are there surrounding areas that may put workers at risk?
- Are there security equipment and procedures in use? How effective are they?
- How often are security equipment/procedures tested?
- What are the workplace's emergency plans and procedures? Are all employees aware of them?
- How often are emergency plans practiced (e.g., drills)?
- How often are employees and supervisors trained in identifying and responding to emergency situations?
- How often are assessments conducted, procedures modified, and/or new security equipment purchased? (ASIS 2011, 13; OSHA 2009, 8; Philpott and Grimme 2009, 91)

This list is in no way conclusive. The threat assessment team may benefit from using security checklists, available from multiple resources. Figure 19-4 is an example of a general security analysis checklist. However, the team should make sure to tailor the inspection to known risks in the workplace.

Hazard Prevention and Control

Once the threat assessment team completes their workplace analysis, they should report their recommendations for improvements to management. Management should seek to implement as many of the recommendations as possible, given the available resources. Controls can be implemented to reduce or prevent identified physical hazards. Employers should consider which controls would be most effective but not impose extra burdens on employees. There are two general kinds of controls: engineering and administrative.

Engineering controls will either get rid of the hazard or minimize the employees' contact with a hazard. Such controls include physical changes in the workplace as well as use of security equipment (Hartsfield 2013, 12–13). Employers may prefer engineering controls because they are easier to monitor and do not rely on employee behavior (OSHA 2009, 8). However, some hazards cannot be mitigated by engineering controls. Thus, employers may need to implement administrative controls. *Administrative controls* are changes in procedures that require employee involvement for success. Employers should make sure employees are regularly using the administrative controls. Employers may consider providing constructive feedback to help employees understand the importance of maintaining administrative procedures.

Sometimes, employers will find that a combination of engineering and administrative controls will most benefit the workplace. Again, each workplace will have different controls based on the workplace's risks for violence. The following contains a few general examples of engineering and administrative controls that an employer may consider.

Engineering Controls—General Workplace Design
- Use physical barriers to separate employees from customers and the general public.
- Make sure employees have clear paths to exits.
- Use bright and effective lighting to make high-risk areas more visible.

Engineering Controls—Security Equipment
- Ensure that security devices are tested regularly.
- Implement control access to employee work areas using buzzers, code combinations, or card access.
- Install alarm systems and panic buttons.
- Install closed-circuit television to monitor high-traffic areas inside and outside the building.
- Provide locked drop safes to minimize available cash.
- Provide the ability for cashless transactions (e.g., debit cards) to minimize cash exchanged between customers and workers.
- Install video surveillance to deter and better identify perpetrators.
- Install metal detection systems.
- Provide alarms, trouble lights, or geographic locating devices in taxicabs and other mobile workplaces.
- Provide cellular telephones, beepers, CB radios, or handheld alarms or noise devices for personnel who work in the field.

Administrative Controls—Training
- Train employees to identify hazardous situations and take appropriate responses in emergencies.
- Train employees on awareness, avoidance, and how to prevent crimes (especially robberies and assaults).

Administrative Controls—Procedures
- Base staffing considerations on safety and security assessments (e.g., requiring that employees do not work alone, especially at night and on the weekends).
- Develop internal communication systems to respond to emergencies.

SELF-INSPECTION SECURITY CHECKLIST

Facility: _____

Inspector: _____

Date of Inspection: _____

1. Security Control Plan ____Yes ____No
   If yes, does it contain:
   (A) Policy statement ____Yes ____No
   (B) Review of employee incident exposure ____Yes ____No
   (C) Methods of control ____Yes ____No
   If yes, does it include:
      Engineering ____Yes ____No
      Work practice ____Yes ____No
      Training ____Yes ____No
      Reporting procedures ____Yes ____No
      Record keeping ____Yes ____No
      Counseling ____Yes ____No
   (D) Evaluation of incidents ____Yes ____No
   (E) Floor plan ____Yes ____No
   (F) Protection of assets ____Yes ____No
   (G) Computer security ____Yes ____No
   (H) Plan accessible to all employees ____Yes ____No
   (I) Plan reviewed and updated annually ____Yes ____No
   (J) Plan reviewed and updated when tasks are added or changed ____Yes ____No
2. Policy Statement by Employer ____Yes ____No
3. Work Areas Evaluated by Employer ____Yes ____No
   If yes, how often? _____
4. Engineering Controls ____Yes ____No
   If yes, does it include:
   (A) Mirrors to see around corners and in blind spots ____Yes ____No
   (B) Landscaping to provide unobstructed view of the workplace ____Yes ____No
   (C) "Fishbowl effect" to allow unobstructed view of the interior ____Yes ____No
   (D) Limiting the posting of sale signs on windows ____Yes ____No
   (E) Adequate lighting in and around the workplace ____Yes ____No
   (F) Parking lot well lighted ____Yes ____No
   (G) Door control(s) ____Yes ____No
   (H) Panic button(s) ____Yes ____No
   (I) Door detector(s) ____Yes ____No
   (J) Closed-circuit TV ____Yes ____No
   (K) Stationary metal detector ____Yes ____No
   (L) Sound detection ____Yes ____No
   (M) Intrusion detection system ____Yes ____No
   (N) Intrusion panel ____Yes ____No
   (O) Monitor(s) ____Yes ____No
   (P) Videotape recorder ____Yes ____No
   (Q) Switcher ____Yes ____No
   (R) Hand-held metal detector ____Yes ____No
   (S) Hand-held video camera ____Yes ____No
   (T) Personnel traps ("Sally Traps") ____Yes ____No
   (U) Other _____ ____Yes ____No
5. Structural Modifications
   Plexiglas, glass guard, wire glass, partitions, etc. ____Yes ____No
   If yes, comment: _____
6. Security Guards ____Yes ____No
   (A) If yes, are there an appropriate number for the site? ____Yes ____No
   (B) Are they knowledgeable of the company WPVP policy? ____Yes ____No
   (C) Indicate if they are:
      _____Contract guards (1)
      _____In-house employees (2)
   (D) At entrance(s) ____Yes ____No
   (E) Building patrol ____Yes ____No
   (F) Guards provided with communication? ____Yes ____No
   If yes, indicate what type: _____
   (G) Guards receive training on workplace violence situations? ____Yes ____No
   Comments: _____
7. Work Practice Controls ____Yes ____No
   If yes, indicate:
   (A) Desks clear of objects that may become missiles ____Yes ____No
   (B) Unobstructed office exits ____Yes ____No
   (C) Vacant (bare) cubicles available ____Yes ____No
   (D) Reception area available ____Yes ____No
   (E) Visitor/client sign-in/out ____Yes ____No
   (F) Visitor(s)/client(s) escorted ____Yes ____No
   (G) Barriers to separate clients from work area ____Yes ____No
   (H) One entrance used ____Yes ____No
   (I) Separate interview area(s) ____Yes ____No
   (J) I.D. badges used ____Yes ____No
   (K) Emergency numbers posted by phones ____Yes ____No
   (L) Internal phone system ____Yes ____No
   If yes, indicate:
      Does it use 120-VAC building lines? ____Yes ____No
      Does it use phone lines? ____Yes ____No
   (M) Internal procedures for conflict (problem) situations ____Yes ____No

Figure 19–4. General security analysis checklist.

(N) Procedures for employee dismissal	____Yes ____No	
(O) Limit spouse and family visits to designated areas	____Yes ____No	
(P) Key control procedures	____Yes ____No	
(Q) Access control to the workplace	____Yes ____No	
(R) Objects that may become missiles removed from area	____Yes ____No	
(S) Parking prohibited in fire zones	____Yes ____No	

7a. Off-Premises Work Practice Controls
(For staff who work away from a fixed workplace, such as: social services, real estate, utilities, police/fire/sanitation, taxi/limo, construction, sales/delivery, messengers, and others)

- (A) Trained in hazardous situation avoidance ____Yes ____No
- (B) Briefed about areas where they work ____Yes ____No
- (C) Have reviewed past incidents by type and area ____Yes ____No
- (D) Know directions and routes for day's schedule ____Yes ____No
- (E) Previewed client/case histories ____Yes ____No
- (F) Left an itinerary with contact information ____Yes ____No
- (G) Have periodic check-in procedures ____Yes ____No
- (H) Have after-hours contact procedures ____Yes ____No
- (I) Partnering arrangements if deemed necessary ____Yes ____No
- (J) Know how to control/defuse potentially violent situations ____Yes ____No
- (K) Supplied with personal alarm/cellular phone/radio ____Yes ____No
- (L) Limit visible clues of carrying money/valuables ____Yes ____No
- (M) Carry forms to record incidents by area ____Yes ____No
- (N) Know procedures if involved in incident ____Yes ____No
(see also Training Conducted section)

8. Training Conducted ____Yes ____No
   If yes, is it:
   - (A) Prior to initial assignment ____Yes ____No
   - (B) At least annually thereafter ____Yes ____No
   - (C) Does it include:
     Components of security control plan ____Yes ____No
     Engineering and workplace controls instituted at workplace ____Yes ____No
     Techniques to use in potentially volatile situations ____Yes ____No
     How to anticipate/read behavior ____Yes ____No
     Procedures to follow after an incident ____Yes ____No
     Periodic refresher for onsite procedures ____Yes ____No
     Recognizing abuse/paraphernalia ____Yes ____No
     Opportunity for Q&A with instructor ____Yes ____No
     On hazards unique to job tasks ____Yes ____No

9. Written Training Records Kept ____Yes ____No
10. Are Incidents Reported? ____Yes ____No
    If yes, are they:
    - (A) Reported in written form ____Yes ____No
    - (B) First report of injury form (if employee loses time) ____Yes ____No
11. Incidents Evaluated ____Yes ____No
    - (A) EAP counseling offered ____Yes ____No
    - (B) Other action (reporting requirements, suggestions, reporting to local authorities, etc.) _____
    _____
    _____
    - (C) Are steps taken to prevent recurrence? ____Yes ____No
12. Floor Plans Posted Showing Exits, Entrances, Location of Security Equipment, etc. ____Yes ____No
    If yes, does it:
    - (A) Include an emergency action plan, evacuation plan, and/or a disaster contingency plan? ____Yes ____No
13. Do Employees Feel Safe? ____Yes ____No
    - (A) Have employees been surveyed to find out their concerns? ____Yes ____No
    - (B) Has the employer utilized the crime prevention services and/or lectures provided by the local or state police? ____Yes ____No

Comments _____
_____

General Comments/Recommendations _____
_____
_____

Figure 19–4. Concluded.

- Provide ID cards for all employees and require that they be worn.
- Limit the amount of cash on site and post signs indicating that limited cash is available.
- Develop work practices for opening, closing, money drops/pickups, and other high-risk periods.
- Develop and implement security procedures for working late or off-hours, accounting for field staff, and entering locations where the employee feels threatened or unsafe.
- Include violence prevention in daily procedures, such as replacing lights and checking security cameras regularly.
- Require employees to keep back doors locked at all times and front doors locked after hours (OSHA 2009, 9–10).

- Require employees to change security passwords on software often.
- Develop emergency procedures to use in case of robbery.
- Establish a liaison with local police.
- If resources are available, create an employee assistance program, or contract with an organization to provide similar services (ASIS 2011, 19; OSHA 2009, 9–10).

After implementing controls, management could survey employees to see if the controls are effective and if anything was overlooked in the implementation process. After some time has passed, the threat assessment team should conduct another workplace security analysis to make sure that the controls are effective in preventing hazards and that procedures are being followed.

Special Considerations: Hiring, Discipline, and Termination

Employers are especially vulnerable to violence when bringing in new employees, instituting discipline, and conducting terminations. Thus, a workplace violence prevention program should consider preventive procedures for all three situations.

Hiring

Because individuals who exhibit violent or bullying behavior usually have a history of such actions, a thorough hiring process can serve as one important step in preventing violence in the workplace (Rugala and Isaacs 2004, 22–23). Preemployment screenings will allow employers to identify potentially violent behaviors in prospective employees. Preemployment screening procedures should include:

- confirming the accuracy of everything on the person's application
- conducting a background screening of demographic and background factors
- contacting work references
- holding structured face-to-face interviews
- using a probationary period for new employees (Hartsfield 2013, 9–12; Kelloway et al. 2006, 553; Rugala and Isaacs 2004, 22–23).

Employers should be mindful that all preemployment screening practices must comply with all relevant privacy protection and antidiscrimination laws (Rugala and Isaacs 2004, 20–21). Employers may consider seeking legal guidance in designing hiring procedures.

Disciplinary Procedures

Employers should have firm disciplinary procedures for employees who engage in violence at the workplace. Employers should make sure the procedures are fair and applied consistently for all employees.

Termination Procedures

Employers should also consider setting termination procedures in combination with the workplace violence prevention program. An employee facing termination may, under certain circumstances, act out violently. However, not every termination will result in violence. Termination procedures should be written out and included in employee handbooks. Some termination procedures should include:

- collecting an employee's identification cards, building passes, keys, and parking pass
- blocking a former employee's access to the employer's computer and network systems (usually by disabling accounts or changing passwords)
- if resources are available, offering a severance package
- agreeing in advance with a terminated worker what will be told to prospective employers
- if necessary, implementing additional security measures during termination meetings.

Employers should obtain legal advice regularly to ensure that procedures are consistent with employment laws (ASIS 2011, 31–32; Hartsfield 2013, 9–12; Rugala and Isaacs 2004, 55).

When employers are faced with large-scale terminations, some additional factors may help reduce the risk for violence, such as making sure top management is involved in the process and communicating with employees to reduce rumors (Hartsfield 2013, 9–12).

General Training Needs

A key part of the workplace violence prevention program is to make certain that all employees, both management and nonmanagement, are trained to recognize warning signs of violence and in methods to prevent violence in the workplace (NIOSH 2004, 15). A training program may vary based on the risks in each workplace. However, every program should cover, at minimum, these topics:

- a definition of workplace violence
- general warning signs of workplace violence
- a description of the employer's workplace violence prevention program, including:
    - identified risks/hazards in the workplace
    - what controls were implemented to reduce hazards, and how to use them
    - incident reporting procedures
    - management procedures for incident investigation, response, and follow-up procedures
- the location and operation of safety equipment

- how to defuse potentially violent situations
- what to do in an emergency situation
- how to contact law enforcement during an emergency
- emergency response protocols (bomb threats, hostage situations)
- personal security measures
- employer policies and procedures on how to obtain medical care or counseling after an incident and how to file workers' compensation claims or seek legal assistance after an incident (NIOSH 2004, 17; OPM 2013, 19–20; OSHA 2009, 11).

Employers in high-risk occupations should consult OSHA's recommendations for tips on training programs tailored to the risks faced in the workplace.

Supervisors and managers should receive extra training in addition to all the employee training. Manager training should focus on:
- encouraging employees to report any incidents, including threats of violence
- investigation procedures when a potential problem is identified
- techniques on how to deescalate and resolve conflicts
- when to report incidents to higher management or the police department
- how to address individual employee issues
- how to maintain confidentiality
- how to conduct fair and proper disciplinary procedures
- procedures for screening employees
- procedures for safe employee terminations (NIOSH 2004, 15; OPM 2013, 20–21; Philpott and Grimme 2009, 99–100).

Members of the threat assessment and incident response teams should receive additional comprehensive training on how to complete their tasks. The threat assessment team should receive thorough training on how to recognize hazards in the workplace and conduct workplace security analyses (ASIS 2011, 18). In the same way, training for members of the incident management team should focus on the employer's policies for incident investigation and response, pertinent government regulations and legal considerations, and practices in handling emergencies (OPM 2013, 21).

Training on How to Foresee Potential Violence

In the majority of situations, certain behaviors will precede a violent incident. Employers should cover a number of warning signs or "red flags" in training programs (OPM 2013, 22–25; Rugala and Isaacs 2004, 5). These are the most often cited warning signs of workplace violence:
- direct or veiled threats of harm
- numerous conflicts with supervisors or other employees
- drug and alcohol abuse
- extreme changes in behavior (decreased productivity, increased absences, withdrawal from normal social circles)
- blaming others for problems
- frequent explosive outbursts
- severe personal stressors (divorce, severe financial problems, child custody battle, death in family)
- physical fights with co-workers
- inability to handle criticism
- offensive commentary or jokes relating to violence
- unwanted sexual advances or innuendo
- bullying behaviors
- obsession with weapons
- past incidents of violence (ASIS 2011, 22–24; Hartsfield 2013, 3–4; Kelloway et al. 2006, 556; Romano et al. 2011, 3).

While these factors are in no way conclusive, multiple factors should raise red flags for employers, especially if co-workers begin reporting some of the behaviors. Employers should be careful not to create a "profile" of a violent employee, which can lead to unfair stereotyping (OPM 2013, 17).

New employees should be trained immediately after being hired. In addition, employers should take measures to "refresh" or update current employees on policies regularly. OSHA recommends that high-risk occupations conduct training annually (OSHA 2009, 11).

Assessing Threats of Violence

The workplace violence prevention program should contain procedures for handling threats and incidents of violence as they happen. Although most threats will not become incidents of workplace violence, each report must be taken seriously and investigated for the risk of a violent incident. The same threat assessment team (that performed the workplace analysis) may be responsible for receiving reports of threats and investigating them. However, management may decide to control threat management to maintain confidentiality for employees who report threats or are involved in incidents of violence.

Reports constituting immediate threats will require immediate actions, such as calling law enforcement and notifying management (Rugala and Isaacs 2004, 25–27). A thorough threat assessment should cover the following:
- an analysis of the threat, including when it was made and how (verbally or through threatening behavior)
- the identified target (and the relationship between the person who made the threat and the target)
- whether the threat was accompanied with specific plans
- an analysis into the threatener's background and whether

there are risk factors for violence
- whether the threatener has the means (including weapons) to carry out the threat
- any identifiable reason for the threatener to make the threat (OPM 2013, 35–42; Rugala and Isaacs 2004, 30–32).

This information can be gathered by talking to other employees, interviewing the threatener, and even interviewing the target. However, the investigator should be aware that privacy laws may restrict their analysis.

Teams often will benefit from consulting with law enforcement officials, mental health professionals, emergency response personnel, and other outside specialists or agencies that could become involved in a crisis. To be fully effective, these relationships should be established and maintained *before* an emergency occurs.

Teams should keep good written records of all incidents and interventions, monitor results, and evaluate the actions that were taken.

Incident Reporting System

The first step to addressing violence in the workplace is to establish a system where employees can report threats and/or incidents of violence. A workplace's reporting system will differ depending on the workplace and resources available. The easiest system may be to designate one office or manager to receive the information. An employer could also set up a hotline for employees to call with information. A system that allows for confidentiality may further encourage employees to use it. No matter what kind of system is implemented, employees must be aware of the system for the system to be effective. Management should consider posting information about the system on bulletin boards or on the bottom of paystubs, or sending e-mails about and then hosting additional training sessions to promote awareness of the reporting system (Rugala and Isaacs 2004, 24–25). In addition to providing information about the system, employees should be informed that all threats and incidents, no matter how immediate, should be reported. Every threat that is reported should be analyzed to gauge the risk of the threat and decide whether action should be taken. A reporting system could include procedures to follow up with the employee who made the report to inform him or her of what actions were taken. A follow-up procedure may encourage employees to report because they might then feel that their concerns were promptly addressed (OPM 2013, 8; Rugala and Isaacs 2004, 24–25). In addition, reports should be reviewed occasionally to see if there are patterns of violence or differences in how incidents are handled.

Incident Response

A workplace violence prevention program must include a plan to immediately respond to incidents of violence. Plans should be tailored to the type of workplace and situation involved. The goal of any response plan should be to keep all employees safe. A complete plan will cover prevention efforts (preincident), procedures during an incident, and actions after an incident occurs.

An employer may already have emergency plans in place to deal with natural disasters. However, there should be a separate plan to deal with incidents of violence that could result in injuries to employees or customers in the workplace. An employer can create an "incident response team" consisting of top management, supervisors, security personnel, and human resources and legal counsel (if available), in addition to contacts from law enforcement and medical personnel (ASIS 2011, 25–26). The incident response team can create policies and procedures for preventing violence, dealing with an incident when it occurs, and handling the aftermath if an incident occurs.

In general, if the incident involves a serious emergency, such as an armed person, a bomb threat, or a situation where there are hostages involved, the first step should be to contact law enforcement and emergency medical crews immediately (ASIS 2011, 25–26). The incident response team should create plans for escape in both emergency and nonemergency situations.

Some guidelines for incident response plans include:

Prevention
- Contact law enforcement and involve them in emergency incident response plans.
- Write down all incident response procedures and include them in employee handbooks, post on bulletin boards, or send a copy to employee e-mail accounts.
- Create training programs for employees, management, the threat assessment team, and the incident management team on incident response as well as threat assessment procedures.
- Provide phone numbers for law enforcement and medical personnel and set standards for when law enforcement/medical personnel should be contacted.

During an incident—emergency situations
- Immediately contact workplace security, law enforcement, and medical personnel.
- Secure the building.
- Attempt to alert other employees to the danger (through use of a paging or alarm system).
- If necessary, begin first aid on injured persons (if doing so would not put the employee in danger).

During an incident—nonemergency situations
- Notify management of the incident.
- Attempt to defuse the situation (if doing so will not place the employee in danger).
- Contact security, law enforcement, and medical personnel, if necessary.

After the incident
- Secure the area where the incident took place (to prevent tampering with a potential crime scene).
- Make sure all employees are accounted for (if there was an evacuation).
- Refer employees who were victims or witnesses to the violence to the workplace's EAP or host a debriefing session for all employees.
- Complete any required reports (e.g., OSHA forms).
- Document the incident and response.
- Identify who will be in charge of communication to employees and the media.

The incident response team should evaluate the incident response plans after incidents occur and update them if necessary (ASIS 2011, 35–36; Philpott and Grimme 2009, 71–80).

Incident Investigation
The incident response team should complete an investigation after each violent incident to determine what happened, what factors led to the incident, and whether the responses taken were appropriate for the incident. Investigations should begin as soon as the situation is no longer dangerous so that evidence will not be removed or destroyed. The incident response team should be aware that a criminal investigation could also be going on and take measures not to disrupt those investigators. In addition, each investigation should be conducted professionally and with care so as to not further upset victims or witnesses of the incident (OPM 2013, 27–34). The incident response team can gather information by investigating the location of the incident, talking to victims or witnesses, consulting related reports of the threat/incident, and reviewing any evidence gathered in threat assessments. The team could have victims and witnesses complete incident report forms, such as the sample incident report. The response team should focus on gathering the facts, identifying what led to the incident, and identifying what response plans were used during and after the incident. Based on that information, the team should propose changes to management as needed. The team could also be responsible for reporting the incident to any regulatory agencies as required (e.g., OSHA).

Incident Follow-Up
A workplace prevention program's incident response plan is not complete without procedures for follow-up with employees after an incident. Employees involved in the incident, including victims and witnesses, may have emotional responses to the incident, such as fear of going back to work after the incident and feelings of remorse or helplessness (OSHA 2009, 10). Employers should consider either referring employees involved in incidents to mental health professionals, specifically those specializing in trauma, or providing counseling as part of the incident response program. An employer could offer several types of postincident counseling, such as "trauma-crisis counseling" or "critical incident stress debriefing" for employees. The goal of these programs is to allow employees to talk about the incident and to prevent any negative emotional responses to the incident.

Management should make sure that employees are informed and briefed on the incident and response to both reduce fear and avoid the spread of false information.

If a workplace has the available resources, an employer should consider creating an employee assistance program (EAP) as part of the workplace violence prevention program. An EAP is an effective resource to provide postincident counseling for employees. In addition, the EAP could assist in threat assessment and investigating reports of threats. Professional counselors should be hired for the EAP.

Program Evaluation
A very important part of the workplace violence prevention program is continuous evaluation and updating of procedures. Indeed, after a workplace violence prevention program is created and instituted, the threat assessment team (or management) should occasionally evaluate the program's effectiveness. This evaluation should result in recommendations for changes or new procedures (ASIS 2011, 22). Some things to focus on in an evaluation include:
- whether the number of violent incidents has decreased
- whether the violent incidents have decreased in severity
- whether records are regularly kept of incidents and responses
- if all reports of threats are responded to
- whether hazard controls are effective
- whether OSHA and other regulatory requirements are being met
- how often workplace security analysis take place
- how often training is conducted
- whether law enforcement or consultants are contacted for outside review of the workplace violence prevention program.

Record Keeping

Employers should make sure that detailed, accurate records are maintained at every level of the workplace violence prevention program. A record-keeping program should make sure that all requirements are met for reporting incidents to regulatory and government agencies. In addition, records should be periodically reviewed for potential legal issues that could lead to lawsuits for liability. Records should be kept confidential and be available only to management and members of the threat assessment and/or incident response team. Employers should also keep records for a number of years according to regulatory requirements and for evaluations.

Employers should keep the following records:
- any required OSHA reports (OSHA Form 300)
- investigation reports from an incident of violence
- minutes from meetings of management, the threat assessment team, and/or the incident response team about the workplace violence prevention program
- records from training sessions, including the date, who was trained, and what was covered in the session
- records produced by evaluations and workplace security analyses
- distributed and received employee surveys
- any records of workers' compensation claims
- police reports or follow-up communications
- records from filed insurance claims
- obtained medical records from injuries relating to workplace violence
- records of legal actions (lawsuits, protection orders, etc.) (CCH 2011, 624A; OSHA 2009, 11–12).

HIGH-RISK OCCUPATIONS

As mentioned previously, there are recognized industries that have markedly higher rates of workplace violence. OSHA has created a category for these industries called *high-risk occupations*. OSHA currently recognizes the following jobs as high risk: late-night retail establishments, health care and social service workers, taxicab services, and law enforcement positions. Workplaces in these occupations have a high number of risk factors that increase the chance of workplace violence, such as:
- locations in areas with high crime rates
- operating hours late at night or early in the morning
- employees working by themselves or with only one other person
- customers/clients who may be unstable or dangerous (criminals facing punishment, patients with life-threatening injuries/diseases or mental health conditions)
- employees working with money, especially large amounts
- employees who protect others or valuable property (OSHA 2009, 11).

OSHA has published guidelines specifically for these high-risk occupations to encourage employers in those fields to create workplace violence prevention programs. Although the guidelines are not mandatory for employers in these industries, they should be consulted when dealing with workplace violence. We will briefly discuss what is addressed in these guidelines. All of the OSHA publications can be accessed online at osha.gov/publications.

Late-Night Retail Establishments

Workplaces considered "late-night retail establishments" include convenience stores, liquor stores, and gas stations. These workplaces are often open late at night and/or early in the morning, employees work alone and handle money, and sell alcohol (OSHA 2011, 6). Employers should be aware of conditions that could further increase the risk of violence against employees, such as insufficient lighting and inadequate security measures.

Late-night retail workplaces have had and continue to have high rates of workplace homicides and assaults (OSHA 2009, 4). In 2010, the category of "retail trade" (which includes these late-night workplaces) ranked third for the highest number of homicides at the workplace (26% of total workplace homicides) (NIPWV 2013, 6). In addition, 24% of workplace shootings in 2010 occurred in "retail trade" occupations (NIPWV 2013, 8). Furthermore, between 2005 and 2009, bartenders consistently had the highest rates of incidents of workplace violence (Harrell 2013, 4). OSHA published "Recommendations for Workplace Violence Prevention Programs in Late-Night Retail Establishments" in 2009 for these occupations. The guide contains information about workplace violence, what a workplace violence prevention program should include, sample checklists and incident report forms, and other resources for employers in late-night retail occupations (OSHA 2009, 10; OSHA 2011, B-3–B-4).

Health Care and Social Service Workers

OSHA uses the title "health care and social service workers" for all workplaces that provide physical and mental treatment for patients. This category includes hospitals, specifically emergency departments, and community-care, residential, and long-term-care facilities. In addition, workplaces that offer psychiatric treatment and drug abuse treatment, and even pharmacies, are at higher risk for workplace violence. Employees in these occupations deal directly with patients, many who may be under mental distress or have mental health issues. Furthermore, these workplaces often have direct access to drugs and medications, which may bring persons with criminal intent to the workplace. Therefore, employees in these occupations often have the highest rates of victimization for workplace assaults. In fact, in 2013, about 61% of workplace assaults were committed by patients of health care and social service occupations (NIPWV 2013, 9; Restrepo and Shuford 2012, 14).

This assault risk is not a new issue, however. Indeed, in 2009, assault rates for employees in these occupations were more than six times higher than in the next highest industry (Restrepo and Shuford 2012, 16). In 2004, OSHA published "Guidelines for Preventing Workplace Violence for Health Care and Social Service Workers." In 2010, NIOSH also published a "fast facts" guide, specifically for home health care workers, titled "How to Prevent Violence on the Job." Both guides provide workplace-specific advice on how to create workplace violence prevention programs for these industries.

Taxicab Drivers

Taxicab drivers had the highest rate of homicide in 2003, at 16.4 per 100,000 employees. Taxicab drivers dropped in rank by 2009 to the third highest rate of workplace homicides, after gas station workers and barbers (Restrepo and Shuford 2012, 7–8). Taxi drivers have the same risk factors for violence as late-night retail occupations, such as working alone, at night, dealing with cash, and working with intoxicated or upset customers (OSHA 2010). Furthermore, a taxicab driver has a mobile workplace, which could lead to a lack of hazard controls to prevent violence. In 2010, OSHA published a fact sheet called "Preventing Violence against Taxi and For-Hire Drivers." One of the statistics in the fact sheet states that taxicab drivers "are over 20 times more likely to be murdered on the job than other workers" (OSHA 2010). The fact sheet also lists several control options to reduce the risk of workplace violence.

Law Enforcement

OSHA has not yet published a guide for law enforcement occupations. However, law enforcement officers and security guards (whether government or private employees) often have higher rates of assaults and/or homicides. In fact, between 2005 and 2009, law enforcement employees consistently had an average annual rate of workplace violence incidents higher than the average for other occupations during the same time. In fact, the average number for law enforcement, at 48 incidents per 1,000 employees, was more than two times higher than the occupation with the second highest average (Harrell 2013, 4). The Bureau of Justice Statistics, an agency within the Department of Justice, stated in its 2011 report that "no occupation had workplace violence rates higher than those for law enforcement officers and security guards" (Harrell 2013, 4). Law enforcement and security employees often deal with criminals or violent people, have mobile workplaces (police cruisers), and have access to weapons. Furthermore, employees are often involved in emergency situations where violence has occurred. These risk factors contribute to the high number of assaults and homicides for employees in this occupation. However, law enforcement and security personnel often receive training specific to their job on how to deal with violence. Regardless, the employers should consider additional practices, such as those mentioned in this chapter, to help reduce the risk of violence for their employees, when possible.

REGULATORY REQUIREMENTS

Employers face multiple legal considerations in creating and implementing a workplace violence prevention program. First, employers should be aware of the legal and regulatory requirements that a program should address. There is currently no law requiring employers to have a workplace violence prevention program. However, the OSHA General Duty Clause (29 U.S.C. § 654 [a][1]) is often cited as creating a duty for employers to address potential threats of violence in the workplace and take actions to reduce threats. The General Duty Clause states:

> Each employer shall furnish to each of his employees . . . a place of employment which [is] free from recognized hazards that are causing or likely to cause death or serious physical harm to his employees.

In 2011, OSHA published a directive that provided information about workplace violence and set standards on when OSHA would conduct investigations into violent incidents at workplaces, as well as the procedures investigators will use in conducting the investigations (p. 1). The directive also specifically mentions the high-risk occupations discussed previously (and that have corresponding workplace violence prevention program guidelines). The 2011 directive states that all employers must notify OSHA when a fatal violent incident occurs at the workplace. Employers must also continue reporting incidents as previously required by OSHA directives using Form 300, Log of Work-Related Injuries and Illnesses. Federal agencies are bound by federal regulations that incorporate OSHA standards as well as additional federal regulations (29 CFR 1960). Executive Order 12196 also details additional responsibilities for agency heads, the Secretary of Labor, and the General Services Administration (GSA) (OPM 2013, 3).

An employer may also face a variety of legal liability claims under state statutes if workplace violence occurs. For example, an injured employee could seek payment for injuries and lost wages under state workers' compensation laws. In addition, employees who were threatened or injured could make claims under either Title VII of the Civil Rights Act of 1964 or the Americans with Disabilities Act by claiming they faced harassment in the workplace (Hartsfield 2013, 1–2).

Employees could also use civil litigation to recover damages from a violent incident. Employees have used claims of negligence in hiring and/or not terminating an employee who posed a threat to other employees and failure to provide security measures and/or training for employees and management to reduce the threat of workplace violence (Rugala and Isaacs 2004, 47–48). Some employers must also make sure they are providing protection as required by collective bargaining agreements (ASIS 2011, 6). On the other hand, employers must be cautious of legal restrictions when creating a workplace violence prevention program. For example, when performing background checks for potential employees or when investigating an employee who has made a threat of violence, an employer should consult a lawyer to avoid violating any privacy or anti-discrimination laws (Rugala and Isaacs 2004, 47–48). If an employer decides to terminate an employee as part of the incident response plan, management should be aware of the possibility of a wrongful termination lawsuit from the terminated employee (Rugala and Isaacs 2004, 47–48). Thus, there are many legal considerations in creating and implementing a workplace violence prevention program. Employers would benefit from the help of a lawyer at every stage of the process.

There is a movement to produce laws that would require employers to have programs to prevent workplace violence. The Healthy Workplace Campaign has produced the Healthy Workplace Bill, which seeks to encourage or require employers to institute policies to prevent bullying in the workplace. According to the campaign, there are currently no federal or state antibullying laws. Because workplace bullying is a form of workplace violence, any laws passed in the future prohibiting bullying could have an impact on an employer's workplace violence prevention programs (healthyworkplacebill.org).

DOMESTIC VIOLENCE

Domestic violence is a growing concern for employers. In 2012, "relative or domestic partner" comprised the third highest category of assailant for female victims of workplace homicide, at 21% (3% for male victims) (BLS 2012, 10). About 5% of total workplace homicides are a result of domestic violence at the workplace (Rugala and Isaacs 2004, 41–45). Domestic violence in the workplace consists of any behavior that takes place between an employee and a spouse or other family member that results in the employee being abused, either at home or at the workplace. The employee may be physically beaten, threatened, isolated, emotionally and/or mentally abused, sexually abused, or even controlled by economic means. These behaviors may affect an employee's attendance and performance at work, or the violence may take place at the workplace (in person, through the phone, in the parking lot). The most common forms of domestic violence that employers encounter are stalking and other forms of harassment (Rugala and Isaacs 2004, 41–45).

Employers should be aware of behaviors that could indicate that an employee is a victim of domestic violence, such as:
- excessive tardiness or unexplained absences
- frequent, and often unplanned, use of leave time
- anxiety
- lack of concentration
- a change in job performance
- remaining isolated from co-workers and/or showing reluctance to take part in social events
- receiving disruptive phone calls or e-mails
- unexplained bruises or injuries
- disruptive and recurring visits from a current or former intimate partner
- acting uncharacteristically moody, depressed, or distracted
- being the victim of vandalism or threats (Rugala and Isaacs 2004, 41–45).

Employers may be apprehensive to intervene in potential situations of domestic violence for fear of violating an employee's privacy. However, in the words of the FBI, "domestic violence and stalking that come through the workplace door appropriately become the employer's concern too" (Rugala and Isaacs 2004, 41–45). Thus, employers should consider addressing domestic violence in their workplace violence prevention program. Specifically, employers could make domestic violence prevention a part of the workplace violence prevention program by:
- creating a workplace policy prohibiting any behavior of domestic violence in the workplace. (The policy should

encourage employees to be aware of any signs of domestic violence and also state that any employee who participates in abusive behaviors at work or by using any company resources [such as computer, phones, or vehicles] will be disciplined.)
- requiring, or strongly encouraging, employees to report any suspected behaviors of domestic violence, without the possibility of negative consequences (such as retaliation) and the assurance of full confidentiality
- making a commitment to support employees who are victims of domestic violence, including referrals to workplace and community resources and flexibility in providing time off for reasons relating to the abuse
- stating that threats from domestic violence abusers are treated just as seriously as any other threats of violence in the workplace (ASIS 2011, 39–45; and state law guides)
- providing training on the signs/behaviors related to domestic violence
- posting information on resources for domestic violence in multiple locations around the workplace (ASIS 2011, 39–45; Rugala and Isaacs 2004, 41–45).

Employers can ensure that domestic violence threats are taken seriously by incorporating them into the processes for threat reporting, investigating, and incident response. It is important to make sure all reports of potential domestic violence are confidential to protect the victim. Furthermore, employers should consider contacting legal authorities when dealing with a situation of domestic violence. Legal counsel could ensure that employers are taking the necessary legal steps in the situation.

In 2013, the U.S. Office of Personnel Management published its "Guidance for Agency-Specific Domestic Violence, Sexual Assault, and Stalking Policies" in response to a presidential memorandum in April 2012. The guidance required that federal agencies take measures to prevent discrimination against victims of domestic violence. Many states have also established policies for employers when confronted with domestic violence in the workplace. Employers should be aware of any statutory or regulatory requirements regarding domestic violence before implementing a workplace violence prevention program (OPM 2013, 15–16). A model domestic violence policy is included in Figure 19–5.

SMALL BUSINESS

Small businesses may not have the ability to allocate an adequate number of resources to produce a workplace violence prevention program. Furthermore, small businesses represent the majority of workplaces in the United States. In fact, almost four-fifths of the total 5.6 million employers in the private industry have only one to nine employees (Rugala and Isaacs 2004, 50–52). Small businesses should utilize the numerous sample workplace violence prevention policies available from federal and state regulatory agencies. Here are some additional tips for employers of small businesses:

- Make an effort to keep in touch with employees, and promote an open atmosphere where employees can feel comfortable going to management with issues and know that they will be listened to.
- Pay attention to business cycles, such as periods where business will be heavier than others.
- Focus on accountability: hold each employee responsible for completing his or her job/assignments. This will highlight situations that could create the potential for violence.
- Treat all employees equally.
- Consider programs to provide outlets for aggression, no matter how small (companywide recreational activities/competitions, monthly dinners, etc.).
- Have a consistent, flexible policy for altering work schedules/taking time off.
- Consult community organizations for additional resources (Rugala and Isaacs 2004, 50–52).

CONCEALED WEAPONS

Another issue employers across the United States may confront is the rise in "right to carry," or laws concerning concealed weapons. More than 40 states have laws that allow private citizens to carry concealed weapons, typically guns. However, weapons (specifically shootings) were used in 80% of workplace homicides from 2005 to 2009 (Harrell 2013, 11). In 2010, there were 405 shootings at workplaces in the private sector (NIPWV 2013, 8). Although the number of weapons in the workplace has decreased over time (from 20% in 2002 to 12% in 2011), employers should still be aware of the possibility of weapons being used for workplace violence. There is no question that the presence of a weapon at the workplace can increase the risk of workplace violence. Employers can have policies prohibiting weapons in the workplace. Some state that employees will be terminated if they bring weapons to work. Employers should consult legal advice in forming a weapons policy to make sure it complies with state right-to-carry laws.

MODEL DOMESTIC VIOLENCE POLICY

Purpose

[Employer] is committed to providing a workplace in which employees who are victims of domestic or sexual violence have the support they need at work to address the violence in their lives. This policy is intended to increase awareness of domestic and sexual violence, assist employees who have experienced such violence, and help ensure that the workplace is a safe environment.

Definitions

For the purpose of this policy, a *victim of domestic or sexual violence* is a victim of domestic violence or a victim of sex offenses or stalking, as defined below.

A *victim of domestic violence* is defined as an individual who has been subjected to acts or threats of violence, not including self-defense, committed by a current or former spouse, family member, household member, intimate partner, or person with whom the victim shares a child. Domestic violence may also include other physical, psychological, sexual, economic tactics used to establish and maintain power and control over the victim.

A *victim of sex offenses or stalking* is defined [as] a victim of acts that would constitute violations of article 130 (addressing sex offenses) of the New York State Penal Law or a violation of sections 120.45, 120.50, 120.55, or 120.60 (addressing stalking) of the New York State Penal Law.

Model Policy Language

I. **Protection from Discrimination**

A. *Non-discrimination*: [Employer] will not discriminate against a victim of domestic violence or sexual violence in hiring, staffing, or other terms, conditions, or privileges of employment.

B. *Prohibition on actions against an employee because of an abuser's harassment of the employee at work or disruption of the workplace*: [Employer] will not take any actions against an employee who is a victim of domestic or sexual violence based on the actions of a person who has perpetrated violent acts or threats of violence against the employee, even if such actions disrupt the workplace. [Employer] may take actions, including the filing of a criminal complaint or initiation of a civil action, directly against a perpetrator whose violence affects an employee if the perpetrator's actions disrupt the workplace. Prior to taking any actions against such a perpetrator, [Employer] will consult with the employee to determine whether the action would be likely to put the employee's safety at risk and [Employer] will make reasonable efforts to address these concerns.

II. **Time Off and Other Reasonable Accommodations for Victims of Domestic or Sexual Violence**

A. *Commitment to make reasonable accommodations*: [Employer] will make reasonable accommodations that permit an employee who is a victim of domestic or sexual violence to perform the essential duties of her or his job. Accommodations that impose an undue hardship on [Employer] are not required.

B. *Time off*: [Employer] recognizes that victims of domestic or sexual violence may need time off from work to secure medical or legal assistance or counseling, attend court proceedings, relocate, or make other safety arrangements. [Employer] will try to grant unpaid or paid leave to an employee who is a victim or adjust that employee's work assignments to permit the employee to take steps to address the violence and maintain her employment. When the need for time off is foreseeable, an employee must provide reasonable advance notice of a request for leave, generally at least three days. When the need for time off is not foreseeable, an employee should explain the reason for an absence and request leave if applicable as soon as possible, generally no more than two days after the absence begins.

C. *Other reasonable accommodations*. [Employer] understands that there are other accommodations that could permit a victim to continue to perform her or his job and provide a safe environment for the individual and her or his co-workers. Such accommodations could include changing a phone extension, transferring an employee to a different office, assisting in enforcing a protection order, or taking other steps to ensure that the perpetrator of violence is not allowed to enter the workplace. [Employer] will work with an employee who is a victim to make necessary reasonable accommodations.

D. *Requesting an accommodation:* To request time off or other reasonable accommodations, employees should contact _____ or _____.

[Employer] may ask the employee to provide proof that she or he is a victim of domestic or sexual violence. An employee may satisfy such a request by providing a statement from a victim service provider or other professional assisting the victim to address the violence; a court order; a police report; or other corroborating evidence. Any such documentation will be kept in strictest confidence.

III. **Employee Payroll and Benefit Change Request**

A. If an employee who is a victim informs [Employer] that she or he has separated from a spouse or covered domestic partner, [Employer] will make reasonable efforts to put company benefits in the employee's own name.

B. [Employer] will process an employee who is a victim's requests for changes to his/her electronic payroll transfers.

IV. **Education and Resources Regarding Domestic Violence**

A. *Resources*: [Employer] will make available a list of resources for victims of domestic violence and abusers. The list will be posted in highly visible locations, such as bulletin boards, break rooms, rest

Figure 19–5. From the New York State Office on Prevention of Domestic Violence's Model Policy.

rooms, health or first aid offices, company phone directories, and online information data bases.

B. *Referrals*: [Employer]'s human resources department [or other relevant department] will provide information regarding local or national domestic violence or sexual violence service providers that may be able to provide support and advice to victims seeking to address an abusive or violent relationship. [Except as relevant through an Employee Assistance Program], [Employer] will not counsel or advise employees regarding safety planning or other aspects of addressing violent relationships except as is directly relevant to protecting the safety of the workplace.

V. **Performance Issues Related to Being a Victim of Domestic Violence**

A. [Employer] is aware that victims may have performance problems such as chronic absenteeism or trouble concentrating as a result of domestic or sexual violence.

B. [Employer] will make reasonable efforts to consider all aspects of the employee's situation and, to the extent possible, utilize all reasonable options to attempt to resolve the performance or conduct problem.

C. If reasonable attempts to resolve the performance problems are unsuccessful, [Employer] may decide to terminate the employee or the employee may decide to resign. In that event, [Employer] will inform the employee that she or he may be eligible for unemployment insurance.

VI. **Confidentiality**

If an employee reports that she or he is a victim of domestic or sexual violence, requests a reasonable accommodation because of such violence, or seeks a benefits or payroll change related to such violence, [Employer] will, to the extent allowed by law, take reasonable steps to respect the confidentiality and autonomy of the reporting employee, informing other employees or other persons on a need to know basis only, and only to the extent reasonably necessary to protect the safety of the employee or others and to comply with the law. Wherever practicable, advance notice will be given to the reporting employee if [Employer] needs to inform others about the domestic or sexual violence situation or if [Employer] receives a subpoena ordering the disclosure of such information.

VII. **Employees Who Commit Acts or Threats of Domestic Violence, Sex Offenses, or Stalking**

A. Any employee who threatens, harasses or abuses a family, household member, or intimate partner, or who threatens or conducts activity that constitutes a sex offense or stalking, at, or from, the workplace may be subject to disciplinary action up to and including dismissal. This includes employees who use workplace resources such as phones, fax machines, e-mail, mail, or other means to threaten, harass, or abuse a family or household member.

B. Some job positions may give an employee access to certain types of information or resources. If he or she uses this access to enable an abuser to harm the victim, that employee may be subject to corrective or disciplinary action.

C. Some employees may be licensed to possess firearms as a condition of employment. If such employee is arrested, convicted, or the subject of an order of protection in a domestic violence related offense, the employee's authority to possess a firearm may be unlawful under federal law, 18 U.S.C. § 922(g)(8)& (9), or suspended/revoked under New York State Penal Law § 400.00(11), Criminal Procedure Law § 530.14 or Family Court Act § 842-a. [Employer] shall be notified by the employee in the event any of these circumstances occur.

VIII. **Complaints Related to Violation of This Policy**

A. [Employer] encourages individuals who believe that a violation of this policy may have occurred to report the circumstances to _____ or _____. We encourage prompt reporting of complaints so that rapid and appropriate action may be taken.

B. [Employer] will not retaliate in any way against an individual who makes a report alleging a violation of this policy, nor will [Employer] permit any supervisor or other employee to do so. Retaliation is a serious violation of this policy and anyone who feels subject to any acts of retaliation should immediately report such conduct to _____ or _____.

C. Any allegation of a violation of this policy will be promptly investigated. Reports will be kept confidential to the extent practical and appropriate under the circumstances. [Employer], however, has an obligation to act on all information it receives if it believes an individual may be engaging in wrongful conduct or violation of the law. Our immediate goal is to take prompt remedial action to stop inappropriate or unlawful conduct. Our second goal is to assure violations do not recur. Even where a violation is not found, it may be appropriate to counsel individuals regarding their behavior.

IX. **Disciplinary Action for Violating This Policy**

A. If [Employer] finds that this policy has been violated, the violator will be subject to appropriate disciplinary action, which may include verbal or written reprimand; referral to appropriate counseling; temporary suspension; and/or discharge.

Legal Momentum, Model Domestic, and Sexual Violence Policy for Private Businesses, available at legalmomentum.org/resources/model-employment-policy

Figure 19–5. Concluded.

SUMMARY

Employers should consider implementing a workplace violence prevention program in their workplaces. A program, no matter the size, can help reduce the risk of violence from inside and even outside from customers and clients. In addition, a prevention program will help employees deal with threats and incidents of violence if they happen. Each program must be tailored to the specific workplace. Workplace violence affects not only the employees involved but the entire workplace and the community surrounding it (Rugala and Isaacs 2004, 15). As the FBI stated in its report, violence "damages trust, community, and the sense of security every worker has a right to feel while on the job. In that sense, everyone loses when a violent attack takes place, and everyone has a stake in efforts to stop violence from happening" (Rugala and Isaacs 2004, 15).

REFERENCES

29 CFR 1960. Washington DC: U.S. Government Printing Office.

ASIS International. "Workplace Violence Prevention and Intervention," ASIS/SHRM WVPI.1-2011. Washington DC: ASIS, 2011.

Barling, J., et al. "Prediction of Workplace Aggression and Violence." *Annual Review of Psychology* 60 (2009): 671–92.

Bureau of Labor Statistics (BLS), "Census of Fatal Occupational Injuries," Table A-7. Washington DC: U.S. Department of Labor, 2013.

———. "National Census of Fatal Occupational Injuries in 2012" (Preliminary Results), USDL-13-1699. Washington DC: U.S. Department of Labor, 2012.

CCH Inc. "Violence Prevention Policy." Chap. 4267V in *HR Practices Guide Explanations*. Chapter 4267V, "Violence Prevention Policy," CH-HRMHRPG, 2011. Westlaw database (2011WL 1831381).

Harrell, E. "Workplace Violence against Government Employees, 1994–2011." Washington DC: Bureau of Justice Statistics, 2013.

Hartsfield, W. E. *Violence in the Workplace, Investigation of Employee Conduct*. Section 7.15. 2013.

Healthy Workplace Bill (accessed November 25, 2013). healthyworkplacebill.org

Kelloway, E. K., et al. *Handbook of Workplace Violence*. Thousand Oaks, CA: Sage Publications, 2006.

National Institute for Occupational Safety and Health (NIOSH). "Workplace Violence Prevention Strategies and Research Needs." Washington DC: U.S. Department of Health and Human Services, 2004.

National Institute for the Prevention of Workplace Violence (NIPWV). "The Workplace Violence Fact Sheet." 2013.

Occupational Safety and Health Administration (OSHA). "Enforcement Procedures for Investigating or Inspecting Workplace Violence Incidents," Directive CPL-02-01-052. Washington DC: U.S. Department of Labor, 2011.

———. "Preventing Violence against Taxi and For-Hire Drivers" (OSHA Fact Sheet), CPL 02-01-052. Washington DC: U.S. Department of Labor, 2010.

———. "Recommendations for Workplace Violence Prevention Programs in Late-Night Retail Establishments." Washington DC: U.S. Department of Labor, 2009.

Office of Personnel Management (OPM), Interagency Security Committee. *Violence in the Federal Workplace: A Guide for Prevention and Response*. Washington DC: OPM, 2013.

OSHA General Duty Clause, 29 U.S.C. § 654 (a)(1). Washington DC: U.S. Government Printing Office.

Philpott, D., and D. Grimme. *The Workplace Violence Prevention Handbook*. Washington DC: Government Institutes, 2009.

Restrepo, T., and H. Shuford. "Violence in the Workplace." NCCI Holdings Inc., 2012.

Romano, S. J., et al. "Workplace Violence Prevention: Readiness and Response", 2011.

Rugala, E. A., and A. R. Isaacs eds. *Workplace Violence: Issues in Response*. Washington DC: Critical Incident Response Group, National Center for the Analysis of Violent Crime, FBI Academy, 2004. fbi.gov/stats-services/publications/workplace-violence.

REVIEW QUESTIONS

1. What sources of violence can be risks for the workplace?
2. List three steps a small business should take to plan for possible violence.
   a.
   b.
   c.
3. How is domestic violence defined?
4. What are four signs that domestic violence may be occurring?
   a.
   b.
   c.
   d.
5. Name four elements that should be included in a workplace violence prevention strategy.
   a.
   b.
   c.
   d.
6. What elements should be covered in workplace violence training?

Product Safety Management 20

Richard Hackman, CIH, QEP

Understanding Product Safety Management

Key Elements of Exemplary Product Safety and Compliance Programs
Top Leadership Commitment ▶ Safety and Regulatory Assessment ▶ In-Market Surveillance ▶ Responsiveness and Transparency with Consumers ▶ Working Proactively with Retail Customers to Promote Product Safety ▶ Building Open and Transparent Relationships with Government Agencies ▶ Partnering with External Thought Leaders/Experts to Review Programs ▶ Building Relationships with Nongovernmental Organizations and Critics of the Company ▶ Program Evaluations and Continuous Improvement ▶ Doing the Right Thing

Summary

References

Review Questions

UNDERSTANDING PRODUCT SAFETY MANAGEMENT

Product safety and regulatory compliance are the foundation for any business enterprise. While this is especially true of physical products sold to consumers, it is also very applicable to the services sector and software industry. If done correctly, product safety and compliance provide the license for a company to operate and can become a source of competitive advantage. If done incorrectly, poor product safety/compliance can result in consumer injuries and illnesses, product recalls, damaged corporate reputation, lawsuits, and overall business disruption. For small and medium-sized companies, a single product safety incident can be so damaging that the corporation never recovers.

In the service industry, excellent and responsible service builds trust with customers, protects the company from liability, and enables future growth. Poor or harmful service has the exact opposite effect, resulting in lost business and, often, litigation. For the software industry, consumers and their advocates (government and nongovernmental organizations [NGOs]) routinely evaluate consumer software and video games through a product safety lens. Video games with harmful content may become subject to market restriction or consumer campaigns.

The goal of a robust product safety/compliance program is to provide products and services that are enjoyed by consumers and meet applicable laws and regulations. Doing so will build loyalty, create respect and admiration for the company, and enhance future growth.

KEY ELEMENTS OF EXEMPLARY PRODUCT SAFETY AND COMPLIANCE PROGRAMS

Any exemplary product safety and compliance program may contain multiple elements. The most critical of these include:
- top leadership commitment and corporate expectations
- safety/regulatory assessment
- in-market surveillance
- responsiveness and transparency with consumers
- working proactively with retail customers
- building open and transparent relationships with government agencies
- partnering with external thought leaders
- building relationships with nongovernment organizations and critics
- program evaluations and continuous improvement
- doing the right thing.

Top Leadership Commitment

Top leadership support and commitment is absolutely essential to the success of a company's product safety and compliance program. Without it, the other elements will never achieve their desired results. Top leadership provides the following critical functions:
1. funding and resources to implement a successful product safety and compliance program
2. a foundation of internal support for product safety and compliance activities
3. an environment of trust and respect that enables candid discussion of product/service issues. This also provides support for tough decisions that often need to be made (e.g., product recall, customer notification, consumer outreach, etc.).

The CEO and his or her top leadership team need to show visible support and resources for any product safety and compliance program to be successful. This support needs to be grounded in a company's corporate values and expectations. These serve as the bedrock for the product safety and regulatory compliance program and transcend changes in individual leaders.

Safety and Regulatory Assessment

Every existing and new product must go through a rigorous safety and regulatory assessment. This review needs to include:
- evaluation of all ingredients, all constituents, all "formed" materials as a result of chemical reactions, possible contaminants, and the finished product as a whole
- consideration of intended uses, reasonably foreseeable unintended uses, and possible misuses
- assurance that all ingredients/finished products meet all regulatory requirements and are on approved country inventory lists
- evaluation of new safety compliance risks posed in the manufacturing process of new products
- consideration for the "perceived" risks posed by a new product or technology. This will have to take into account possible consumer reactions and concerns. Unaddressed consumer complaints and allegations can spread quickly in electronic media and destroy a brand in a short time.

In-Market Surveillance

Regardless of how thorough a company's pre-market safety and compliance assessment process is, there will always be unanticipated exposures and consumer incidents. The only way to monitor the external environment for these types of issues is to implement a robust in-market surveillance program. An in-market surveillance program will be the primary way to detect product safety concerns and issues that may be brewing within the organization. Such a program needs to carefully evaluate consumer comments, inquiries, allegations of injury/illness associated with the product, any defects, and reports of possible misuse. The in-market surveillance process also needs to consider external quality issues reported by consumers and suggestions for improvements, as well as internal quality assurance evaluations.

As a part of this process, trends in consumer comments/complaints need to be considered. This may stimulate the need to revisit the initial safety/regulatory assessment in light of new conditions and information. This in-market surveillance program will contribute to overall knowledge of consumer uses and practices. Ideally, this will allow an objective analysis of known uses, unexpected uses, and reasonably foreseeable misuses. Any significant product defect or confirmed serious injury/illness reported by consumers will need to be evaluated and considered for reporting to the appropriate government agency.

Responsiveness and Transparency with Consumers

It is absolutely essential that companies and organizations be responsive and transparent with consumers. Consumers, and the organizations that represent them, can be great advocates for an organization's product or services. Conversely, with the speed of Internet communications and social media, an adverse report on a product or service can become a global issue within minutes. Building these relationships will be critical to:
- ensuring the smooth launch of new products
- containing false allegations or rumors
- understanding consumer perspectives and building them into product planning
- having consumers/consumer associations be ambassadors for the organization's product/services.

Many companies now use prominent consumer bloggers to evaluate new products/services and ask them to provide critique prior to commercialization.

It is important to note that in addition to responding to consumer needs and reports of product safety issues, attention must also be paid to consumer perceptions of product safety. A perceived product safety issue will have real consequences and must be addressed promptly.

Working Proactively with Retail Customers to Promote Product Safety

The organization's retail customers are the face to the consumer, and the conduit to bring the organization's products to the market. Developing positive, proactive relationships with key retail customers will carry the following benefits:
- The organization will hear about consumer concerns/complaints much more quickly and be able to respond more effectively.
- Positive relationships with retailers will allow joint promotions or other activities that build the business for both and provide the opportunity to advance beneficial causes, including consumer safety programs.
- In the event of a recall, it is critical that effective working relationships are in place with key retailers. This will facilitate the identification and removal of the recalled product and will be essential in the development of return procedures.

Building Open and Transparent Relationships with Government Agencies

Building open and trusting relationships with key government agencies is a critical element of an exemplary product safety and compliance program, and this should be done well before a crisis occurs. This will provide the basis for sharing information that is helpful to each organization, such as new product safety methodology, advances in testing protocols, and other relevant new learnings. Such a relationship will also allow a company to share preliminary information on an emerging issue with the appropriate government agency and to provide the agency with the assurance that it will be kept apprised of any new developments. In the event of a product defect that presents a significant risk to consumers, having a positive relationship with the governing regulatory agency will result in more effective and collaborative discussions and a higher likelihood of achieving solutions that meet both organizations' needs.

It is important to note that, increasingly, government agencies globally are collaborating on common product safety issues. In many cases, a product safety hazard occurring in one part of the world may be reportable to government agencies in other countries if the same or a similar product is sold in those locations.

The decision to report a potential product defect or safety risk to a government agency can be a very tough decision. There is the inherent fear that the agency will demand an immediate product recall upon hearing of the issue. Having a well-established relationship with applicable government agencies provides the foundation for honest debate about the proper course of action.

Partnering with External Thought Leaders/Experts to Review Programs

The importance of using external thought leaders cannot be overstated. External experts provide valuable support in many areas:
- They often have the best knowledge of the science and technology critical to the organization's product.

- They can provide critique of new product concepts in the early stages of development—typically when it is least costly to make changes.
- They can provide wise counsel on emerging issues.
- They can serve as external spokespersons supporting the company during a crisis.
- They can provide the external perspective that often gets lost during internal debates.
- They can help develop new standards or expectations for the safety of the company's products.
- They will stretch conventional thinking in ways others within the organization cannot imagine.

Building Relationships with Nongovernmental Organizations and Critics of the Company

Many companies are hesitant to build relationships with nongovernmental organizations (NGOs) or other groups that may be critical of them. However, doing so is absolutely necessary in today's world. These types of groups often can provide honest and unfiltered feedback to a company that is beneficial in so many ways. Such candid feedback is often not found with paid consultants or internal members of the company. Building relationships with NGOs can allow for positive compromise on many areas of mutual interest—for example, new legislative/regulatory initiatives, public safety campaigns, joint awareness programs, and the like. These relationships can also provide early indicators of emerging issues and may result in future areas of collaboration. Failure to build these types of relationships can result in NGOs not trusting the company and other potentially negative consequences. In many cases, NGOs are well-funded organizations and sophisticated users of the media, as well as grass-roots advocates. It is far better to build a relationship and explore common areas of collaboration than to have external groups continually suspicious of company activities and products.

Program Evaluations and Continuous Improvement

Like any effective work process, it is essential to conduct regular evaluations of the organization's product safety and compliance programs to ensure that they are delivering the desired results. Efforts must then be focused on continuously improving areas of opportunity. In addition to periodic evaluations/audits, the following issues likely indicate these programs are faltering:

- frequent consumer complaints
- product recalls
- increasing consumer incidents
- poor quality assurance results
- heightened media attention (social and traditional)
- regulatory agency violations and penalties.

The impact of these types of issues on the company's reputation and market share can be immeasurable. Best-in-class companies often benchmark with peer or other high-quality companies to stretch their thinking and observe new approaches that may improve their programs.

Doing the Right Thing

There is no standard definition for "doing the right thing." Different individuals may have completely different viewpoints while observing the same set of circumstances. It is important, however, that each company establish a set of principles and values that govern overall behavior—especially expectations for product safety and compliance. These principles and values become the moral compass for the organization. Doing the right thing can be difficult and painful in the short term, but it will establish the foundation for long-term success. Over time, consumers, government agencies, NGOs, shareholders, suppliers, and other critical stakeholders will form an opinion of a company. Those opinions then become that company's corporate reputation. Good reputations can only be forged over time with demonstrated exemplary performance. Good reputations can be ruined by a single set of bad actions. If an organization is perceived as trying to do the right thing for consumers and constituents, its reputation will survive and flourish—withstanding the inevitable set of issues and crises. Product safety and compliance are key indicators of corporate behavior. These areas should be used to enhance the company's reputation for the betterment of its consumers and customers!

SUMMARY

Product safety requires planning, awareness, and considerable sustained effort by a company's management, operations, and development staff. Focusing in on safety as a key principle for all employees in the product chain will produce a positive image, consumer loyalty, and commercial success.

REFERENCES

Consumer Product Safety Commission, www.cpsc.gov
National Highway Traffic Safety Administration, www.nhtsa.gov
U.S. Federal Aviation Administration, www.faa.gov
U.S. Food and Drug Administration, www.fda.gov

REVIEW QUESTIONS

1. What are the liability and publicity consequences of failure in designing safe products?
2. How will global connectivity of safety regulators affect local product makers?
3. How should a product safety assessment for a brand-new product be structured?

Industrial Sanitation and Personnel Facilities

Stephen R. Blackwell, MS, REHS
Philip E. Hagan, JD, MBA, MPH, ARM, CIH, CET, CHMM, CHCM, CHSP, CEM

Drinking Water
In-Facility Contamination ▸ Plumbing ▸ Private Water Supplies ▸ Water Quality ▸ Wells ▸ Disinfecting the Water System ▸ Water Purification ▸ Water Storage

Waste Disposal
Building Drains and Sewers ▸ Wastewater Disposal ▸ Solid-Waste Disposal ▸ Refuse Collection ▸ Insect, Rodent, and Nuisance Bird Control

Personal Service Facilities
Drinking Fountains ▸ Washrooms and Locker Rooms ▸ Showers ▸ Toilets ▸ Janitorial Service

Food Service
Nutrition ▸ Types of Service ▸ Equipment Installation and Maintenance ▸ Eating Areas ▸ Kitchens ▸ Controlling Food Contamination

Summary

References

Review Questions

This chapter covers both the essential areas involved in employee health and the basic facilities used for employee sanitation. The topics include:
- issues of drinking water quality at the worksite
- appropriate disposal of company sewage, waste, and garbage
- how to keep employee facilities clean and sanitary
- issues of nutrition and sanitation in employee food service.

Companies must give adequate attention to these areas if employees are to work efficiently, with the assurance that their health and welfare are well protected.

Management should provide employees with a sanitary work environment. To make sanitation safe, efficient, orderly, and economical, management must supervise it properly and integrate it effectively with production and maintenance. The general rules for sanitation include:
- personal cleanliness—hand-washing, grooming, etc.
- good housekeeping—as clean as the nature of the work allows
- an approved water and wastewater distribution and storage system
- a regular, complete self-inspection system.

Where wet processes are used, workers and managers must have proper drainage in the work area.

The department director or supervisor responsible for maintaining the work environment must have appropriate authority within the organization. This position should give the individual the authority to monitor the entire company or facility environment to ensure that it meets appropriate levels of cleanliness and order. Some firms appoint a director of maintenance or facility services (which may include safety and health).

DRINKING WATER

Most facilities receive water for drinking, washing, and food preparation from an approved municipal supply. When delivered to the facility meter, this water must meet all applicable regulatory standards and meet the criteria for classification as potable water (i.e., that it is safe and satisfactory for drinking and cooking).

In-Facility Contamination

Simply because water is potable when delivered to the facility meter does not necessarily mean that it remains so. Many opportunities exist within a facility for water to become contaminated.

Contamination of water supplies can occur due to microbiological, chemical, or physical adulteration. Waterborne disease outbreaks occur from exposure to microbial agents such as salmonella, campylobacter, giardia, legionella, hepatitis, listeria, and cryptosporidium (see Table 21–A for several examples). Mercury, benzene, lead, copper, and asbestos are some common chemical contaminants. Physical contamination can result in discoloration, bad taste, and high turbidity (cloudiness). Although cloudy water may have no direct adverse health effects, it could interfere with the water treatment process and provide a medium for microbial growth.

One of the most common ways a water supply becomes contaminated is by mixing potable with nonpotable water. This can occur through a physical connection between a potable water system and an unapproved water supply or other source of contamination. A cross-connection provides a pathway and pressure differences causing backflow and provides the driving force that delivers the contaminant to the potable water system. Contamination risk from cross-

TABLE 21–A. Infectious Waterborne Diseases Caused by Drinking Contaminated Water

Diseases	Incubation Period	Symptoms	Frequency	Mode
Gastroenteritis	Variable	Lethargy, nausea, diarrhea, cramps, and other stomach ailments	Implicated in over ½ of waterborne diseases	Sewage or chemicals in water
Bacterial Shigellosis	1-7 days	Fever, vomiting, stomach cramps, diarrhea	Serious in some cases. Common	Sewage in water
Salmonellosis	6-72 hours	Abdominal pains, fever, vomiting, and nausea	Less common	Sewage in water
Typhoid fever	1-3 days	Abdominal pains, fever, chills, diarrhea or constipation, and tearing of the intestines	Rare occurrences	Sewage in surface water
Giardiasis	1-7 days	Chronic diarrhea, weight loss, intestinal and stomach gas, bloating, and anorexia	Outdoor enthusiasts commonly affected	Surface water and food
Hepatitis A	14-45 days	Jaundice, nausea, anorexia, fever, and general physical discomfort	Rare in the U.S.	Drinking and swimming

Infectious Waterborne Diseases Caused by Drinking Contaminated Water

connection scenarios is greatest when utilities are improperly installed, poorly maintained, or aging.

Other causes of contamination include improper maintenance of drinking and cooking facilities and improper installation of plumbing facilities, allowing back-siphonage of used water. Back-siphonage is a form of backflow to parts of the system that have fallen below atmospheric pressure. For example, a company pumps well water into an application tank to dilute stock chemicals, such as pesticides. If the pumping is interrupted, perhaps by a power outage, a vacuum is created by the pumping interruption. This situation could back-siphon the solution—now a toxic chemical—into surrounding wells. Such back-siphonage incidents can affect users of both company-maintained and publicly maintained water supplies. This type of contamination rarely occurs because required air gaps are usually in place to prevent this from happening. An air gap is a physical separation between a potable water supply and a nonpotable water system that prevents the contamination of drinking water by back-siphonage or backflow because the nonpotable water cannot reach the potable water. An air gap is often required by code. In instances where the use of an air gap is not feasible, various mechanical backflow prevention devices are available for use, depending on the application. The impact of a back-siphonage incident could be much worse if it happened at an establishment such as a food-processing facility. Just a few disease organisms can infect an entire piping network if they are back-siphoned to the community well or pump station.

Because of the serious consequences of contamination, facility management must ensure that the integrity of the facility's drinking water system is maintained. If the facility has piping systems containing water that is used for nondrinking purposes (e.g., sprinklers, fire hydrants, or manufacturing processes), these systems should be clearly identified, particularly at all outlets. Facility services should see that no direct connection exists between drinking water and other water systems except through a properly installed, approved, and regularly inspected backflow prevention device. In addition, facility services should have eliminated any long, dead-end runs of pipe that cannot be flushed or drained and that might serve as a reservoir for contaminated water. Finally, facility services should ensure that the location of piping approved for potable water use is easily identifiable.

Where the possibility of misusing or cross-connecting pipelines exists, all nonpotable water lines should be clearly marked as unsafe for drinking; for use in food areas or personal service rooms; or for washing of utensils, clothes, or people. Nonpotable water may be used for cleaning work premises (other than food preparation and personal service rooms), provided it does not contain harmful concentrations of chemicals or microbial contamination.

Plumbing

Construction workers should install fixtures and faucets to prevent back-siphonage of contaminated water if the pressure drops in a supply line. Faucets and similar outlets should be at least 1 in. (2.5 cm) above the flood-level rim of the receptacle. To prevent backflow into the drinking water supply, it may be necessary to place surge tanks and air gaps in the fill lines to process equipment.

Another common source of water supply contamination is open joints in underground supply lines, which may allow seepage from groundwater or water from leaky sewers to enter the pipes. This condition can arise where pipes are subject to vibration or corrosion and the joints between pipes open mechanically or the pipe sections crack. Codes usually prohibit installation of sewer and drinking water lines in the same trench, unless the sewer line is placed at a much greater depth and angled to prevent backflow into the water pipes.

If the supply for sprinklers and fire hydrants is the same as that for drinking water, hydrant drains or "weeps" connected directly to sewer lines may be a source of contamination. An open standpipe or reservoir may also allow contamination.

Frequently, contamination of the water supply results when a system is opened for repair or alteration. The new pipe must be disinfected and properly flushed with clean water before being put back into service. In addition, to reduce another source of contamination, the U.S. Safe Drinking Water Act restricts the use of lead pipe, solder, or flux in the installation or repair of any public water system or in residential plumbing connected to a public water system.

Private Water Supplies

Industrial establishments in outlying districts commonly supply and treat their own water from private sources. Such water treatment installations should be built and operated under the supervision of a thoroughly trained and experienced water treatment engineer. Information in this and subsequent sections provides general background but is not a substitute for appropriate professional staff and/or consultants.

All underground and surface waters considered for drinking purposes should be viewed as contaminated until proved otherwise. Water supplied from private sources for the personal use of facility personnel must meet the appropriate health and environmental regulations. As a rule, groundwater collected from deep-drilled wells will be free of biological contamination but may be affected by various minerals at levels that degrade taste, odor, or other aesthetic characteristics. In contrast, shallow wells are more likely to have biological and synthetic chemical contamination.

Water drawn from surface sources, such as lakes and

streams, should always be treated with disinfectant and filtered to ensure potability before use. Many types of both biological and chemical contamination can come from surface water sources; therefore, companies should obtain professional advice in the design and operation of a surface water treatment system.

Because a company may have several different choices for its water supply, the supply choice will depend on daily water requirements, the amount of treatment required so that water from each source meets purity standards, and the potential each source has for additional contamination.

The daily per-person water requirements of an industrial facility can be estimated as follows: 15 to 20 gal (55 to 75 L) for drinking, lavatory, and toilet usage; 20 to 25 gal (75 to 95 L) per shower; and 5 to 10 gal (20 to 40 L) per meal if food is prepared on the premises.

Water Quality

An organization must evaluate its water supply source on the basis of the contaminants it may contain. Tables 21–B through 21–H list the limiting concentrations of two classes of contaminants. U.S. regulations have established two sets of standards relating to the quality of drinking water: primary and secondary. Primary standards are mandatory and cover all contaminants considered health hazards. These contaminants fall into the general categories of chemicals, organic chemicals, physical parameters, microbial agents, and radioactivity. Secondary standards cover the aesthetic qualities of water, such as taste, odor, and color. Although compliance with secondary standards is generally not mandatory, the EPA has given states the choice to adopt them as enforceable standards, and a prudent course of action would be to check the status in the appropriate jurisdiction. In any case, it is strongly recommended that all potable water meet these criteria as well.

Note: Because U.S. regulations are constantly being revised, managers of all potable water supplies should request updates of regulations and guidance materials from the appropriate agencies.

The equipment needed to treat water and make it potable depends on the degree of contamination and the potential for heavier contamination of the water source in the future. These factors can be evaluated only on the basis of a thorough survey of the water source. Such a survey will determine not only the type of treatment necessary but also the nature and frequency of periodic laboratory tests of the source water and the treated water.

Wells

The safest source of water is usually a drilled well with intake below the water table. Typically, such wells show a reliable yield and are free from bacterial and chemical

TABLE 21–B. MCLGs and MCLs for Inorganic Contaminants

	MCLGs	MCLs
(1) Asbestos	7 million fibers/liter (longer than 10 μm)	7 million fibers/liter (longer than 10 μm)
(2) Cadmium	0.005 mg/L	0.005 mg/L
(3) Chromium	0.1 mg/L	0.1 mg/L
(4) Lead	Zero action level = 0.015	—
(5) Mercury	0.002 mg/L	0.002 mg/L
(6) Nitrate	10 mg/L (as N)	10 mg/L (as N)
(7) Nitrite	1 mg/L (as N)	1 mg/L (as N)
(8) Total Nitrate and Nitrite	10 mg/L (as N)	10 mg/L (as N)
(9) Selenium	0.05 mg/L	0.05 mg/L

MCL: Maximum Contaminant Level (expressed as mg/L)
MCLG: Maximum Contaminant Level Goal

TABLE 21–C. MCLGs and MCLs for Volatile Organic Contaminants

	MCLGs (mg/L)	MCLs (mg/L)
(1) o-Dichlorobenzene	0.6	0.6
(2) cis- 1,2 Dichloroethylene	0.07	0.07
(3) trans 1,2 Dichloroethylene	0.1	0.1
(4) 1,2- Dichloroethylene	0	0.005
(5) Ethylbenzene	0.7	0.7
(6) Monochlorobenzene	0.1	0.1
(7) Styrene	0.1	0.1
(8) Tetrachloroethylene	0	0.005
(9) Toluene	1	1
(10) Xylenes (total)	10	10

TABLE 21–D. MCLGs and MCLs for Pesticides/PCBs

	MCLGs	MCLs (mg/L)
(1) Alachlor	Zero	0.002
(2) Atrazine	0.003 mg/L	0.003
(3) Carbofuran	0.04 mg/L	0.04
(4) Chloradane	Zero	0.002
(5) 1,2-Dibromo-3-chloropropane (DBCP)	Zero	0.0002
(6) 2,40D	0.07 mg/L	0.071/L
(7) Ethylene dibromide (EDB)	Zero	0.00005
(8) Heptachlor	Zero	0.0004
(9) Heptachlor epoxide	Zero	0.0002
(10) Lindane	0.0002 mg/L	0.0002
(11) Methoxychlor	0.04 mg/L	0.04
(12) Polychlorinated biphenyls (PCBs) (as decachlorobiphenyl)	Zero	0.0005
(13) Toxaphene	Zero	0.003
(14) 2,4,5-TP (Silvex)	0.05 mg/L	0.05

TABLE 21–E. MCLGs and Treatment Technique Requirements for Other Organic Contaminants

	MCLGs	MCLs
(1) Acrylamide	Zero	Treatment technique
(2) Epichlorohydrin	Zero	Treatment technique

TABLE 21–F. Secondary Maximum Contaminant Levels (SMCLs)

(1) Aluminum	0.05 to 0.2 mg/L
(2) Silver	0.1 mg/L

TABLE 21–G. Best Available Technologies to Remove Inorganic Contaminants

Inorganic Contaminant	Activated Alumina	Coagulation/ Filtration[2]	Corrosion Control	Direct Filtration	Diatomite Filtration	Granular Activated Carbon	Ion Exchange	Lime Softening[2]	Reverse Osmosis	Electro-dialysis
Asbestos		X	X	X	X					
Barium							X	X	X	X
Cadmium		X					X	X	X	
Chromium III		X					X	X	X	
Chromium VI		X					X		X	
Mercury		X[1]				X		X[1]	X[1]	
Nitrate							X		X	X
Nitrite							X		X	
Selenium IV (Selenite)	X	X						X	X	X
Selenium VI (Selenate)	X						X	X	X	

[1] BAT only if influent mercury concentrations exceed 10 µg/L. Coagulation/filtration for mercury removal includes PAC addition or post-filtration GAC column where high organic mercury is present in source water.
[2] Not 1415 BAT for small systems for variances unless treatment is currently in place.

TABLE 21–H. Secondary Maximum Contaminant Levels

Contaminant	Level
Aluminum	0.05-0.02 mg/L
Chloride	250 mg/L
Color	15 color units
Copper	1.0 mg/L
Corrosivity	Noncorrosive
Fluoride	2.0 mg/L
Foaming agents	0.5 mg/L
Iron	0.3 mg/L
Lead	0.015*
Manganese	0.05 mg/L
Odor	3 threshold odor number
pH	6.5-8.5
Silver	0.1 mg/L
Sulfate	250 mg/L
Total dissolved solids (TDS)	500 mg/L
Zinc	5 mg/L

* Action Level

contamination. If a company uses both well and city water, facility services must be sure that no cross-connection occurs between the two systems. Local codes provide specific requirements for each jurisdiction.

The wellhead should be located as far as possible from sewage lines, septic tanks, and sewage drainage fields or process waste-disposal systems. A 200-ft (61-m) separation is usually considered a minimum distance. Toxic chemicals that have leaked from pipelines, tanks, or lagoons or that have been spilled or disposed of on the land surface can contaminate a well at much greater distances. Such contaminants may present a serious health threat. Pesticides and herbicides spilled on or applied to land in the area of a well can also contaminate water at a considerable distance, depending on local hydrogeology.

To prevent contamination of underground water caused by seepage of surface waters, the space between the well casing and the surrounding area can be sealed with a cement grout to a minimum depth of 10 ft (3 m) below the finished ground level or floor. As a further precaution, the casing should be grout-sealed to the lowest impervious stratum it passes through.

Well installations must meet the current state well code or standards for such design considerations as siting and construction materials. Most state codes require that the top of the well casing project at least 12 in. (30.48 cm) above the ground surface. The wellhead should not be covered over by paving or other material that would make access difficult.

Submersible, turbine, or jet pumps may be considered in any particular well. Submersibles are located in the well and do not require a pump house. Local water or sanitation agencies can add a safety factor by requiring two wells and two pumps for a water supply. In this way, if either unit fails, an automatic alternator can switch over to the other pump.

Disinfecting the Water System

The pipes, reservoirs, standpipes, pump, and well casing of a new system should be thoroughly disinfected before being placed into service. Old systems carrying treated water for the first time following extended disuse should also be disinfected on the discharge side of the treatment facility. Likewise, a system that has been opened for repairs should be disinfected before being placed back into service.

Workers can disinfect a drinking water system more easily by filling it with water containing not less than 100 mg/L of available chlorine. The solution should remain for 24 hours in either a new system or one that has not previously carried treated water. For systems that have carried treated water and are being put back into service following minor repairs, 12 hours will be sufficient. However, if officials suspect that giardia lamblia or cryptosporidium are present in the water, other measures should be taken, such as filtering. Using chlorine-containing products, ozone, and ultraviolet radiation can disinfect the water. Each type of disinfection has unique advantages and disadvantages. Overall effectiveness is based on such parameters as dose of disinfection product, temperature, contact time, and quality of the supply water.

A company can determine the success of a disinfecting job by measuring the residual chlorine in the solution at the end of the required time. Test kits for this purpose are commercially available and easy to use. Tests will show residual chlorine if the biological chlorine demand of the system has been met. Workers may connect the system to the drinking water supply, flush it out, and put it into service. If no residual chlorine is present, workers should drain the system and recharge it with new disinfectant solution and then repeat the procedure.

If the system contains a standpipe or reservoir, workers should add the disinfectant solution through it. Otherwise, workers can add the solution in a temporary reservoir on the supply side of the system pump and inject the solution through it. A solution containing 500 mg/L available chlorine, applied with a fog nozzle, will disinfect standpipes and covered reservoirs.

Underground water supplies can become contaminated while being developed. If so, they will have to be disinfected in a sequential fashion. After determining the 24-hour yield of the well, workers should run the test pump to clear the well of turbidity (cloudiness). They should then add a chlorine solution to the well that, with the 24-hour yield, will make a chlorine solution of 500 mg/L. The permanent pumping equipment should be connected to the wellhead and operated until the discharge has a distinct smell of chlorine.

Several methods are available to distribute the disinfectant solution uniformly throughout the water system. Workers may seal the well casing and inject the solution under pressure, or they may add the solution from a hose or a small pipe several levels beneath the surface of water in the well. The chlorine solution should remain in the well for 24 hours.

Water Purification

Of the several methods of water purification available, filtration and chemical disinfection are the most practical for industrial private water supplies.

Filtration

This method of water purification primarily clears turbid waters; however, it can also remove some bacterial contamination. Filtering facilities are usually two types: slow filters and rapid filters.

Slow sand filters will clarify turbid waters when operated at a rate of 25 to 50 gal/day/ft^2 (100 to 200 L/day/m^2) of filter area. Such filters should be made with 0.015 to 0.019 in. (0.25 to 0.35 mm) of sand and should be at least 20 in. (50 cm) deep, but preferably 36 to 40 in. (90 to 100 cm) deep. While the filters are operating, the film that accumulates on the surface of the sand must be kept below water level because this film increases the effectiveness of the filter. Rapid sand filters, made using a uniform 0.015 to 0.019 in. (0.4 to 0.5 mm) sand with a depth of 30 in. (75 cm), will handle about 3,000 gal/day/ft^2 (12,000 L/day/m^2) of turbid water based on available filter surface.

Both filters should be made under competent engineering supervision and operated under continuous inspection. Depending on the water source, filters can require a pre-sedimentation basin for preliminary water treatment. Facility services should purchase filters in pairs so that workers can remove one for cleaning and maintenance without disrupting the supply of filtered water.

Disinfection

Chlorine is generally the best available disinfecting agent for drinking water. It can be added to the water directly as a gas or as a soluble salt (calcium hypochlorite or chlorinated lime—refer to the National Safety Council's *Fundamentals of Industrial Hygiene*, 6th ed. [2012], for hazards of these chemicals).

Small-capacity chlorinators, which inject gaseous chlorine into a water system, are available and easy to operate. However, it is better to install injection pumps that supply high-concentration chlorine solutions to the system at a proper rate because they are safer and easier to operate.

Companies must maintain standby equipment at all chlorinating stations, along with an adequate supply of spare parts. Supervisors should keep on hand gas masks that are effective against chlorine and a small bottle of ammonia to test for leaks outside areas where chlorine is stored or used. Make sure masks are inspected regularly, and train authorized employees in emergency procedures.

The chlorinator should be adjusted to leave a chlorine residue of about 0.2 mg/L in the water after 20 minutes of contact between chlorine and the untreated water. Test kits are available that can quickly measure residual chlorine.

It is easy to disinfect small quantities of water for emergency use in one of several ways. Commercial preparations provide good protection but must be used according to the manufacturer's instructions. Workers can use point-of-use treatment such as boiling, chlorine, iodine, and filtering. Safe drinking water can be produced by vigorous boiling for 1 minute. As long as contamination from giardia lamblia or cryptosporidium is not a problem, chemical disinfection (chlorine or iodine) can be used to make water safe to drink. Safe drinking water can also be produced by boiling water for 5 minutes or by adding four drops of household bleach (hypochlorite solution, 4% available chlorine) or two or three drops of common tincture of iodine to a quart of water and allowing it to stand for 30 minutes. If the water is cloudy, the chemical disinfection dose should be doubled. Workers can lessen the water's flat, medicinal taste after disinfection by pouring it from one container to another to aerate it, by adding a pinch of salt, or by letting it sit for awhile.

Water Storage

Reservoirs or standpipes for treated water should be enclosed completely and located to prevent accidental contamination. A reservoir large enough to hold a 48-hour reserve supply of treated water will provide adequate supplies in the event of unusually heavy water use or supply failure. Fit all vents with screened downspouts well above floor level. Entrance manholes should be enclosed by watertight frames at least 6 in. (15 cm) higher than the surrounding surface and fitted with watertight covers extending at least 2 in. (5 cm) down the outside of the frames. Make sure the cover is closed and locked when the entrance is not in use.

A reservoir can permit full use of a small well and pump, and it still provides a buildup for peak demand.

WASTE DISPOSAL

Industrial facilities must ensure that a careful evaluation of all waste streams (wastewater, solid, and hazardous waste) is conducted to ensure that disposal is in accordance with regulatory requirements. Although many state and local agencies regulate the disposal of solid and hazardous wastes, the EPA has combined regulations governing solid and hazardous waste programs. The disposal of process waste requires special handling and packaging provisions if the material meets the definition of regulated hazardous waste, including documentation from "cradle to grave" (see Chapter 8, Hazardous Wastes, in NSC's *Accident Prevention Manual for Business & Industry: Environmental Management*, 2nd ed. [2000]).

Some industrial facilities must provide their own wastewater disposal systems, and all must adhere to applicable regulations governing disposal. The company must also provide for proper storage and collection or disposal of solid waste.

Building Drains and Sewers

A firm's in-facility sanitary sewage collection system should conform to regulatory standards. Every fixture should be properly trapped and vented by drain(s) and stack(s) to prevent discharging sewer gases into the building and to ensure proper drainage. Facility services must see to it that traps and especially grease interceptors (such as those placed in waste pipes serving the facility cafeteria and kitchen) and interceptors designed to collect other particulate foreign materials are large enough, located for easy access, and cleaned periodically.

The building drain and sewer may be constructed of extra heavy cast-iron bell-and-spigot pipe with drainage fittings. This material is less susceptible to clogging and much easier to clean than pipes made from other materials. It also provides "insurance" when installed under floors that would be expensive to tear up. The sewer should be tight under a 10-ft (3-m) head of water. Plastic and copper plumbing materials are generally approved for drainage.

A company should provide a cleanout where the building drain passes through the building wall and at other selected locations. Local codes for trap size, permitted locations, and strainer requirements should be checked. Codes may call for installation of backwater valves to prevent backup in the sewer line. These valves should be placed where they will be accessible for inspection and cleaning.

Wastewater Disposal

The most common form of private sewage treatment involves use of a septic tank. The septic tank is comparable to the primary treatment phase of a municipal sewage treatment facility. Septic tanks are buried, watertight receptacles designed and constructed to receive wastewater; separate solids from liquids; provide limited digestion of organic matter; store solids; and allow the effluent, clarified liquid to be discharged for further treatment. To retain the solids, the septic tank uses a baffle on the discharge line. The retention of solids is essential to prevent malfunctioning of the secondary treatment component.

The effluent from the septic tank should pass through a watertight sewer that meets regulatory codes, over as short a distance as possible to the secondary treatment phase of the private sewage disposal system. The type of secondary treatment component installed may depend on the soil conditions of the proposed site. Determining the soil conditions may involve performing a soil permeability evaluation or a percolation test by a qualified individual or using some other diagnostic method approved by the state or local agencies. The secondary treatment component may consist of a subsurface seepage field, a waste oxidation lagoon, an aerated treatment system, or another type of sewage treatment system meeting state or local codes.

Sewage disposal systems should be located and constructed to prevent contaminating the groundwater or polluting the surface water (Figure 21–1). All sewage systems must be maintained to avoid causing a nuisance or health hazard to the community. For more information concerning acceptable secondary treatment and proper location, contact the local or state health department or other regulatory agencies with jurisdiction.

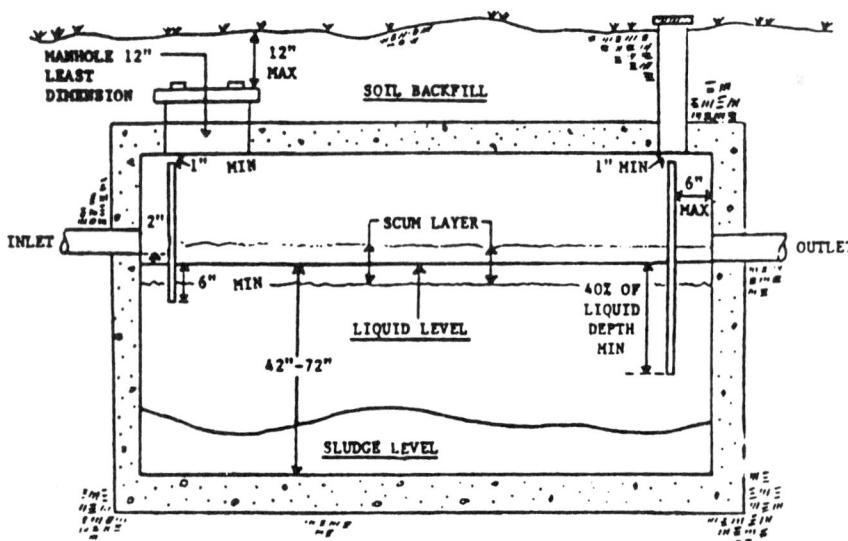

Figure 21–1. Septic tank with slip-in baffles.

Solid-Waste Disposal

Companies and facilities providing food services for their employees must arrange for proper disposal of food wastes and other refuse. Several methods of disposal are available; follow local ordinances in each case.

Many facilities collect garbage and store it for pickup and disposal by municipal or private collection services. Companies should fence in any large outside garbage receptacles and compactors and lock the gates to keep people out and to prevent animals from strewing refuse about.

Metal or plastic garbage and refuse containers with tight-fitting covers should be used to keep out insects and rodents. Containers should be easy to clean and handle and should be washed periodically with a detergent-deodorant solution. Many waste-hauling vendors require the garbage in containers to be kept in plastic bags, paper bags, or liners because bags or liners allow the refuse to be removed easily and help keep containers clean. Place enough garbage containers in the facility so employees will not throw food waste in wastebaskets or other unsuitable receptacles. Proper separation of different waste products may be required. Management should schedule frequent collections to prevent garbage from accumulating.

With the reduction of available landfill space, some states and communities are requiring waste reduction and recycling programs. The goal of these efforts is to remove recyclable materials from the waste stream and to reduce the overall amount of garbage generated. This goal can be accomplished through pollution prevention efforts such as source reduction, reuse, substitution, and composting. Management should work with the company's waste-hauling vendor to develop feasible programs that reduce waste and recycle materials and chemicals that can be recycled.

In areas where recycling efforts are focused, unique hazards may be present from sorting and collecting recyclable materials. Glass, metals, paper, batteries, chemicals, and even cardboard collection can pose a safety and health risk to the employees handling those materials. Use of mechanical devices is encouraged to avoid injuries from improper lifting or awkward postures. Baling or compacting machines are often used to compress and tie together cardboard or polystyrene plastic. Safeguards should be maintained to avoid eye or compression injuries. Cut hazards arise from glass and metal collection. Proper handling and storage and proper disposal of waste batteries and chemicals are important to prevent exposure, injury, and environmental pollution.

Many localities allow firms to discharge their ground food waste into the municipal sewage system. Local ordinances governing this practice will provide details. Installation of food-waste disposers in the facility kitchen can provide a convenient, efficient way to reduce the amount of garbage that must be stored and collected. Supervisors should ensure that disposers are located, grounded, and installed according to approved plumbing practices and local code requirements.

Kitchen employees should be instructed to use nonmetallic tampers, keep silverware out of the disposer, and clean the metal trap daily. Before workers clean or clear the disposer, they should make sure that it is completely stopped and disconnected from its power source.

Refuse Collection

Hazards in refuse collection vary with the type of equipment used and the conditions surrounding the operation. A frequent cause of accidents involves packing blades that cause injury to fingers, hands, arms, and feet. Other hazards arise from "booby traps" unwittingly laid by the companies that refuse collectors serve—loose, broken glass in a refuse container; lightweight trash cans filled with heavy objects (like chunks of concrete); heavy objects concealed by paper or other trash; or hose or other obstacles strewn along the pathway to a rubbish can. Containers that are rusted through or have unserviceable handles increase the risk of injury to refuse workers.

The National Safety Council's *Work Injury and Illness Rates* indicates that cuts, lacerations, and punctures account for about 14% of the lost-time injuries in the refuse collection industry; this statistic compares favorably with the industry in general, where 17% of workers suffer such injuries. Wearing heavy work gloves minimizes these types of injuries.

Unique hazards may arise from sorting and collecting recyclable materials. Collecting waste glass, metals, batteries, chemicals, and even cardboard can pose a safety and health risk to the employees handling such materials. Use of mechanical devices is encouraged to avoid injuries from improper lifting or awkward postures. Baling or compacting machines are often used to compress and tie together cardboard or polystyrene plastic. Refuse collectors should maintain safeguards to avoid eye or compression injuries. Cut hazards arise from glass and metal collection. Proper handling, storage, and disposal of waste batteries and chemicals are important to prevent exposure, injury, and environmental pollution. If employees are exposed to blood products, then a bloodborne pathogens program must be in place.

Insect, Rodent, and Nuisance Bird Control

In industrial and commercial sites where insect, rodent, and nuisance bird infestations are a problem, it is always best to employ a certified professional pest-control operator. To obtain a list of licensed pest-control operators, companies should contact their local health agencies.

The hazards of poisonous chemicals and applications should not be risked by personnel with little or no previous experience in pest control. In some places, toxic insect-, rodent-, and bird-control chemicals may be handled only by persons with appropriate training and licensing. Because of the toxicity and resulting hazards, firms should consider or use alternative methods of pest control. These methods of pest control include removing harborages and feeding sites, pest-proofing structures, and so on. Pests contribute to the bacterial problem of waste disposal and are often the limiting factor for obtaining good sanitary practices.

Integrated pest management (IPM), a very popular method used to control pests and minimize the use of pesticides, is based on a collaborative approach among housekeeping, maintenance, and pest-control services. Operational and administrative intervention strategies are used to reduce the amount of pesticides needed to control pests. The implementation of an IPM program tends to be site specific, and cost and applicability should be examined before implementation. In some settings, IPM may not be feasible.

Typical elements to be included in the implementation of a successful IPM program include sanitation practices, facility design and maintenance, record keeping, use of nonpesticide methods for pest control, minimizing the use of preventive applications of pesticides, and program monitoring and evaluation on a continuous basis. Sources of information on the implementation of IPM include state and local agricultural departments, pest management consultants, or pest management firms.

Management should notify all departments in the company of a pest-control operator's visit well in advance. When ordering the fumigation of a large facility, such as a group of buildings, railroad cars, or grain elevators, management should advise the local fire department in advance. Outside signs can be used to warn neighboring complexes or residents to keep children and pets away from treated areas.

PERSONAL SERVICE FACILITIES

The company should conveniently locate all personal service facilities contributing to employee comfort, such as drinking fountains, washrooms, locker rooms, showers, and toilets. These facilities make up an essential part of the occupational health program in most industries.

Drinking Fountains

Sanitary drinking fountains, one to about every 50 persons, should be installed at convenient places throughout an industrial facility, in accordance with local code requirements. The fountain should have an angle jet and a lip guard (Figure 21–2) and may have a waste container for employees to discard papers or cups. It is important to be sure that the stream projector cannot be flooded or submerged should the water stream be stopped. Also, have the installer direct and project the stream so that users cannot contaminate it. In dusty areas, fountains should be covered.

The water temperature should be 50° to 55°F (10° to 13°C) for heavy manual labor or 45°F (7°C) for less-active office work. An ice dispenser should be kept separate from water, or ice can be provided from an ice maker.

Where an approved municipal source of water is available on construction worksites, the company can extend a water line to upper floors as the building is being erected. On each floor, a standard drinking fountain can be installed.

On worksites involved in highway, pipeline, and power-line construction, timber clearing, and the like, the drinking water source is usually so remote that it is impractical to pipe water to the job. Some companies have successfully solved this problem by using portable drinking fountains. These fountains have an insulated tank equipped with an angle jet drinking nozzle. The tank has an air pump and pressure-release valve so that it can be pumped up to the necessary operating pressure. Keep containers scrupulously clean, and sanitize them daily with steam, boiling water, or chlorine solution.

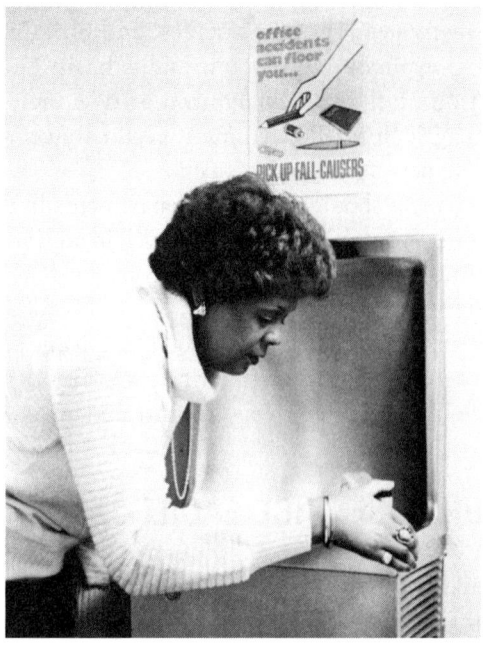

Figure 21–2. Fountains should be conveniently located so that employees can maintain their daily water intake. A safety poster can be placed near the fountain.

Under no circumstances should workers use a common drinking cup or ladle. If drinking cups are required, they should be single-service paper cups kept in a sanitary container at the drinking faucet, with a receptacle provided for disposal. ANSI standard ANSI A117.1–2009, Accessible and Usable Buildings and Facilities, mentions modifications necessary for use by the physically challenged.

Codes generally do not permit installation of drinking fountains in any toilet room. Nor do they permit installation of bubblers as an integral part of or as an accessory to another fixture, such as a lavatory or sink.

Carafes (vacuum bottles) that are frequently used in private offices are a potential source of bacterial contamination. Employees should rinse and refill them daily and clean them periodically, using a sanitizer such as a cationic quaternary ammonium germicide.

Washrooms and Locker Rooms

ANSI Z4.1–1986 (R2005), Minimum Requirements for Sanitation in Places of Employment, serves as a guide to the types and sizes of washrooms, locker rooms, and accessories. A large, single washroom and locker room for each gender may be sufficient for a compact facility or an establishment employing fewer than 500 people. Washrooms in a large one-story facility are generally scattered throughout the building. If the facility consists of a series of separate buildings with only a few people working in each, the company can place all the facilities in a centralized building. This arrangement has proved successful in such establishments as chemical facilities, oil refineries, and railroad yards.

If the facility is relatively small, dressing rooms, lockers, and washrooms should be near the entrance. In a larger facility, it is better to place these facilities in a single, centrally located building or in several buildings near the work areas. In some industries, washing facilities are also used to prevent product contamination (Figure 21–3).

Washing Facilities

The company must maintain all washing facilities in a sanitary condition. Each lavatory should have hot and cold water, or at least tepid running water, and liquid hand soap or a similar cleansing agent. Waterless skin cleansers/sanitizers are not substitutes for soap and water, but they are convenient for special use or where water is scarce.

For industrial occupancies of up to 100 employees, one lavatory for every 10 employees is recommended; for more than 100 employees, one lavatory for every 15 additional employees is considered adequate. In industries where workers need additional washing time, one lavatory for every 5 employees is recommended.

Circular washbasins (Figure 21–4) of stainless steel,

stoneware, enameled iron, or other materials impervious to moisture permit a number of employees to wash at the same time by using central water sprays that are either continuous or controlled by a treadle. These basins are easy to keep clean and sterile. Their construction prevents water from splashing and spilling.

To eliminate standing water, which can transmit disease from one employee to another, make sure lavatories have no stoppers. A mixing faucet or a spray will permit employees to wash in a flowing stream with controlled temperature. Some lavatories also feature knee-actuated controls.

Wherever practicable, a thermostatic control should be installed in the hot-water supply system to keep water temperature below 140°F (60°C). Injecting live steam into tanks or lines of a cold-water system (to make warm water) is dangerous because pressure failure in such a system could release steam through the taps.

Management should establish a regular maintenance program for equipment and train employees to report defective machinery, tools, and other equipment. For example, broken faucets and valve handles may cause serious cuts or lacerations. Handles should be made of metal, not a breakable material, such as porcelain. If leaky faucets are repaired at once, employees will not develop the bad habit of shutting valves too tightly. Also, require employees to wash their hands after using the restroom.

Figure 21–3. These employees must "prep" before they can enter critical, super-clean areas. Washing hands and rubber gloves, then drying them under air dryers, prevents lint and dust contamination. *(Printed with permission from Western Electric Company, Allentown, PA.)*

Soaps

The proper type of soap is important, not only for ordinary hygiene, but as a protection against dermatitis caused by cleaning agents. The soap used should have no free alkali and should have a pH less than 10.5. It should be free of mineral abrasives. Management should provide individually dispensed paste, liquid, or powder (not bar soap) for common use. Liquid or powdered soaps are preferable because they are easy to dispense and serve as an aid to housekeeping.

Management should strongly discourage workers from removing paint, dye, and other stains with solvents or other chemicals and especially from removing grease from their hands with naphthas. Those solvents may cause a severe skin irritation.

Dryers

One or more of the following means of drying hands and/or face must be convenient to the lavatories: individual paper or cloth hand towels or sections, clean individual sections of continuous cloth toweling, or warm air blowers. The firm should not allow common-use towels but should supply paper towels instead. Paper towels should be soft enough not to cause irritation and should be kept in a covered container with a disposal receptacle nearby.

Hot-air hand dryers should be tightly secured either to the floor or to the wall to prevent the fixtures or the electric element from coming loose. The equipment must be grounded and permanently installed without extension cords or plugs. Blowers must provide air at greater than 90°F (32°C) and no more than 140°F (60°C).

Barrier Creams

Protective or barrier creams, if properly used and reapplied frequently, provide limited protection against hand and arm irritants. There are several common types of creams, but no one cream is effective against all irritants. One principal benefit of using barrier creams is that workers must wash their hands repeatedly to remove the cream.

Figure 21–4. Lavatories should be supplied with running water at a controlled temperature. A sufficient number of wash-up facilities should be available. In a multiple-use lavatory, 24 in. (60 cm) of wash sink or 20 in. (50 cm) of circular basin, with water outlets for each space, is considered to be equivalent to one lavatory.

Lockers

Lockers should be perforated for ventilation and be large enough to permit clothing to be hung to dry. If the clothing is heavy or wet, management should provide forced circulation of hot air through the base of the lockers and out through the top or provide hangers on elevating chains so the work clothing can be dried between shifts.

Lockers should have sloped tops to prevent workers from storing material on them (see Figure 21–4). The multiple legs of lockers are serious impediments to floor cleaning; place lockers on metal frames with a minimum of floor supports. Also, anchor lockers together to prevent them from being overturned.

Persons working with toxic materials that can contaminate clothing should have separate lockers for work clothing and street clothing. These lockers preferably should be in rooms on opposite sides of the shower room so employees will have to pass through the shower room when changing from work clothing to street clothing, and vice versa.

Benches

Benches in front of the lockers should be permanently fastened to the locker base, preferably on a hinged support so they can be turned up against the faces of the lockers while the aisles are swept. Maintenance workers should check the benches regularly and keep them in good repair—free from splinters, breaks, and other imperfections.

Floors

Washrooms and locker rooms should be kept well ventilated, warm, and at 30% to 60% relative humidity. Make sure heating equipment is installed to protect against burns and in compliance with all relevant codes.

The floors of washrooms and locker rooms should be made of nonabsorbent material such as glazed brick, tile, or concrete. The floor material should continue up into the walls as a cove for at least 6 in. (15 cm) before there is a joint. The walls must connect to the floor cove with a tight joint and should be watertight to a height of at least 5 ft (1.5 m).

Select flooring material for durability and sanitation and to minimize the hazard of slipping and falling. Terrazzo, tile, marble, and polished concrete floor surfaces are particularly hazardous when wet. For safety, maintenance should establish a strict cleaning and mopping schedule so that the flooring is dry by the time workers arrive.

Concrete floors can be made much less hazardous by covering the surface with a finishing layer of abrasive grain concrete. Abrasive strips can reduce slipping hazards on old concrete floors that have been worn smooth. Ceramic tiles are available with a skid-resistant, nonabsorbent, and watertight surface, or mats can be used.

Floors need to be inspected to be sure they are watertight. Leaky floors cause damage to joists and other structural members of the building, and if organic materials collect in them, they can attract vermin. Plastic material can cover worn wood or concrete floors to obtain a watertight surface.

Showers

Companies should install showers in work areas where employees become dirty, become wet with perspiration, or are exposed to dust or vapors on the job. Showers can also be provided to encourage employees to bicycle or walk to work. The showers should be as close to the job as possible, preferably in a separate room adjacent to the dressing rooms and locker rooms. Workers exposed to high temperatures who come off the job wet with perspiration should not have to walk through cold air when going to the shower and dressing rooms.

A company should provide one shower for every 10 employees of each sex who are required to shower during the same shift. Each shower should be supplied with hot and cold water through a mixing fixture that the user can regulate. Use an automatic regulator to maintain the maximum hot water temperature at 140°F (60°C). Deluge showers, eyewash fountains, and similar installations for emergency use are discussed in Chapter 12, Occupational Health Programs, in this volume.

Place body soap or other appropriate cleaning agents conveniently near the showers. Hot and cold water should feed a common discharge line. The company should provide employees who use showers with individual, clean towels.

If a particular standard requires employees to wear protective clothing because of possible contamination with toxic material, make sure change rooms have storage facilities for street clothes. As discussed earlier under Lockers, protective clothing should be stored separately.

When the company provides clothes for workers, it should ensure that all wet clothes or those washed between shifts are dry before reuse.

The floor of the shower room and of the individual compartments should be made of nonskid material to provide good footing when wet. The surface can be either abrasive grain concrete or concrete with a wood-float finish.

Existing floors that were made smooth or have become smooth through long wear can be given a nonskid surface. Maintenance staff should thoroughly scrub concrete floors, using an abrasive pad and a synthetic detergent. They can apply strips of abrasive material to other types of floors to provide a slip-resistant surface. The floor throughout the shower room area should slope toward drains, preferably at the back of the shower stalls. It is not necessary to have curbs around the individual shower stalls if the floor is properly sloped. The curbs can be a tripping hazard; if used, they should be dyed or painted a contrasting color.

Do not use wood mats on shower room floors. They present a tripping hazard and may expose a worker to splinters and loose-joining members.

As an item of general sanitation, shower rooms and stalls should be well ventilated and adequately lighted to prevent the formation of mold. The floor of the shower should be mopped daily with detergent, hot water, and disinfectant to combat athlete's foot (fungus and ringworm infection).

Toilets

Wall-hung, elongated-bowl flush toilets with open-front seats should be provided according to the number of employees (Table 21–I). If people other than employees are allowed to use toilet facilities, the number of toilets should be increased accordingly. There must be adequately supplied toilet paper holders in every water closet. For every three toilet facilities, the company should provide at least one lavatory in the toilet room or adjacent to it.

TABLE 21–I. Minimum Toilet Facilities

Number of Employees	Minimum Number of Water Closets*
1-15	1
16-35	2
36-55	3
56-80	4
81-110	5
111-150	6
More than 150	One additional fixture for each additional 40 employees.

* Where toilet facilities will not be used by women, urinals may be provided instead of water closets, except that the number of water closets in such case shall not be reduced to less than 2/3 of the minimum specified.

Printed from OSHA Regulations, 1910.14 (c) (1)

Wall-hung units are easier to keep sanitary and to clean. Codes prohibit any type that is not thoroughly washed at each discharge or that might permit siphonage of bowl contents back into the tank. Water supplied to tanks must have vacuum breakers or a positive air gap between the surface of the water in the tank and the water supply inlet.

Toilets should be no more than 200 ft (60 m) from any workstation. In multistory buildings, toilets should be only one floor above or below the work area. With toilets and lavatories placed at various points throughout the facility, the main locker room and shower room can be closed for cleaning during the work period—an advantage for the janitorial crew.

Some states require that women work no more than a certain distance from a women's restroom. Management should check for this requirement during design stages. In addition, some states require a company to install cots so women can lie down.

Ventilation is required for toilet rooms. If natural ventilation is used, windows or skylights must have a ventilation area of 6 ft^2 (0.5 m^2) for a room with one toilet, with an additional 1 ft^2 (0.1 m^2) of window ventilation space for each additional toilet. If this amount of window space is not available, the company should supply forced ventilation at a rate of three to four air changes per hour in the room.

Because windows and skylights generally do not afford sufficient light, companies should install light fixtures in all toilet rooms and washrooms. Place switches for lights, electric dryers, or other equipment where they cannot be operated by anyone who is also in contact with piping or other grounded conductors. Ground-fault circuit interrupters (GFCIs) should be used whenever electrical receptacles are located within 6 ft (2 m) of a sink.

Individual wall-hung urinals should be provided in the men's room. These may be substituted to the extent of one-third or less of the number of stools specified. Approved urinals must be designed so that all surfaces subject to soiling can be easily cleaned. Integral screens over the discharge openings are the major cause of chronic toilet room odors. The decomposing soil under the screen cannot be removed by any practicable method. Blowdown washout urinals are the only acceptable type. Floor-type urinals, in which the drain pipe becomes chronically offensive, and wall-hung urinals with integral screens should be replaced by the approved sanitary type (blowdown washout), thus making room deodorants unnecessary.

Management should prohibit employees from lunching in toilet rooms or in process areas where toxic or noxious materials are present. The habit of some workers to heat foods in molten lead reservoirs or other process heating equipment can be dangerous to their health and should be prohibited. Instead, companies must provide proper lunchrooms or other eating facilities outside the toilet rooms or process areas.

Covered receptacles in facility lunchrooms are the proper places to dispose of waste food and papers. Employees should never discard such refuse in the toilet rooms. If they carry cups of coffee or other drinks from the lunchroom, make sure the drinks are in covered containers or on trays to prevent spills that could create unsanitary or slippery conditions.

Portable toilets are often necessary on construction jobs. The supplier can provide waste removal and maintenance.

Janitorial Service

As a part of an overall managed facility sanitation function, the company should set up a minimum daily janitorial service for all personal service facilities. When properly designed, washrooms, shower rooms, and toilets can be thoroughly cleaned with little personal involvement in the process. Janitorial workers should mop and clean floors and fixtures with detergent and hot water at least once daily and use a sanitizing cleaner as often as necessary. The occasional use of an acid-type cleaner may be required on toilet bowls and urinals. Workers should wear proper personal protective equipment when using chemical cleaners and thoroughly flush the fixtures following use.

When floors are being mopped, block off the area with signs reading CAUTION—WET FLOOR to prevent possible slipping accidents. This subject is also covered in Chapter 4, Maintenance of Facilities, in the *Engineering & Technology* volume. Employee exposure to bloodborne products should be considered. If such exposure is likely, then the company should institute a bloodborne pathogens program in accordance with OSHA's 29 CFR 1910.130, Bloodborne Pathogens.

FOOD SERVICE

Nutrition

Nutrition, another factor in industrial safety and health, concerns the medical and safety departments of any company or facility. Some facilities may establish food-service facilities. With care and thought, the company can deliver adequate and balanced in-facility meals. Food should be properly prepared and attractively served, with strict adherence to sanitary practices.

The nurse, working with the physician, can provide information to employees on better nutrition. The American Medical Association and the Academy of Nutrition and Dietetics (formerly the American Dietetic Association; see References) offer many excellent articles and other materials providing up-to-date nutritional information. A good breakfast, high in protein for timed energy release during the day, lessens fatigue and consequently lessens the chance of accidents.

Some organizations hire full-time dieticians to review their food-service menus and even talk to employees about nutrition, job performance, and health.

Types of Service

There are six main types of industrial food service:
1. cafeterias preparing and serving hot meals
2. canteens or lunchrooms serving sandwiches, other packaged foods, hot and cold beverages, and a few hot foods
3. mobile canteens that move through the work areas, dispensing hot and cold foods and beverages from insulated containers
4. box-lunch service
5. vending machines
6. food trucks.

Even using a mobile canteen to provide a midshift snack adds considerably to the nutrition of the average worker. If lunches are also served, the mobile canteen should carry both hot and cold foods and beverages.

The central cafeteria, with a kitchen wherein full meals can be prepared and served, is often the most satisfactory form of food service. In large facilities, it may be economical to supply several cafeterias from a central kitchen.

Self-service vending machines offer a wide variety of packaged, ready-to-eat foods. Some machines have ovens that let the user quick-cook meals.

Four important safety and health precautions should be followed in the use of microwave ovens:
1. All repairs should be made by manufacturer-authorized repair personnel.
2. Persons with cardiac pacemaker units should be warned against coming too close to microwave ovens. Most pacemakers, however, are protected against microwave radiation.
3. Interlock and leakage testing should be performed on a regular basis.
4. Vending machines or microwave ovens should not be plugged into extension cords.

Details are given in the National Safety Council's *Fundamentals of Industrial Hygiene*, 6th ed. (2012).

Equipment Installation and Maintenance

Proper installation and maintenance of food-heating and refrigeration equipment is essential to employee health. For example, maintenance staff must be sure that vending machines are kept sanitary and that can openers are routinely cleaned, sanitized, and kept in good working order. The company must supply sufficient utensils and adequate waste-disposal facilities in kitchens and eating areas as a necessary part of a self-service operation.

Eating Areas

The cafeteria or lunchroom should be clean and attractive to present a pleasant environment to the employee, thus encouraging its use. Employees should eat only within designated areas to help maintain proper sanitation in the workplace. This requirement will prevent insect and rodent infestations.

Minimum floor spaces for the number of people using the eating area at one time are given in Table 21–J. Where space is limited, managers can stagger lunch periods so that employees do not have to eat on the job. Employees should wash their hands before and after eating or smoking.

TABLE 21–J. Minimum Floor Space in Eating Areas

Number of People	Sq Ft per Person
25 or less	13
26 through 74	12
75 through 149	11
150 and over	10

Kitchens

When a company sets up a cafeteria kitchen, management should pay the same attention to proper equipment and working conditions as in any other part of the facility. Although not required by some regulatory authorities, it is a good idea to use food-service equipment that complies with standards of or receives approval from nationally recognized testing agencies [National Sanitation Foundation International (NSFI) and Underwriters Laboratories (UL)]. Make sure the layout conforms to all relevant public health food-service codes. To ensure that health codes are met, always consult with the local public health jurisdiction early in the planning stages.

Floors made of impervious, water-resistant, easily cleaned, skid-resistant material will minimize slipping and falling hazards if water or grease is spilled. Walls and ceiling should be constructed of a durable, smooth, easily cleanable, light-colored surface. Because areas near the dishwasher and/or sink have high moisture levels, they may need special waterproof materials. The area constructed near the cooking equipment should be approved by the local fire marshal having jurisdiction.

Artificial lighting in the food-service areas should provide at least 20 footcandles of light on all food preparation surfaces and at equipment or utensil-washing areas. All artificial lighting should have shielding to prevent broken glass from falling into the food.

Ranges and other heat-producing equipment require hoods and ventilation to carry away heat, grease, odors, and vapors. Because the ventilation system gets greasy, the ductwork should be easily accessible for cleaning. The ventilation system must meet local fire and building codes.

Sprinkler systems and portable extinguishers should be installed for fire protection (see Chapter 9, Fire Protection, in the *Engineering & Technology* volume). Place a fire blanket and extinguisher near any area where workers' clothing could be ignited by an open flame.

The company should install easily changeable and cleanable racks for such hand tools as knives, cleavers, and saws and place storage racks or cabinets for utensils in convenient locations. All wall- and ceiling-mounted equipment must be easy to clean and maintain in good repair.

Controlling Food Contamination

Many communities and counties have detailed food sanitation regulations governing the installation and operation of an industrial food-service facility. The company must closely adhere to regulations of the health authority that has local jurisdiction. In areas in which no local health authority exists, the organization must follow the other regulatory standards or the recommendations of the U.S. Food and Drug Administration (FDA) Model Food Code (see References).

Because of the potential for contamination during the food preparation and serving process, the FDA recommends the use of a proactive methodology. This methodology, known as the Hazard Analysis Critical Control Points (HACCP) system, is used to identify and eliminate food safety problems caused by biological, chemical, and physical contamination. Effective use of the HACCP system enables one to establish a list of critical control points that can be controlled to reduce or eliminate foodborne illness hazards.

Potentially hazardous foods are those that contain milk or milk products, eggs, meat, poultry, fish, shellfish, edible crustaceans, or other ingredients, including synthetic ingredients, in a form capable of supporting rapid and progressive growth of infectious or toxigenic microorganisms. Potentially hazardous foods should be kept at or below 40°F (4°C) unless being prepared or served. Reheating these foods after they have remained for some time at room temperature will not protect employees from foodborne illnesses.

These illnesses can be caused by a variety of bacteria, viruses, parasites, and chemicals. An example of one of these is the bacteria Staphylococcus, which produces a toxin in foods that is not destroyed by normal cooking temperatures and can therefore cause illness. For example, if the bacteria have been growing in food during storage at about 40°F (4°C), they will be killed by heating, but their toxins will still be capable of causing a foodborne illness. The contaminating agent of a foodborne illness usually cannot be detected through the taste, odor, texture, or appearance of the food. Therefore, proper sanitation is essential to keep food safe for consumption.

To prevent a foodborne illness, workers should observe the following guidelines:

- Wash hands frequently during food preparation and after any interruption in food preparation, after smoking, or after using the restroom. Hand-washing must involve using soap and warm water for at least 20 seconds to be effective. Do not use common towels for hand drying because these towels can recontaminate the hands.
- Keep hot foods hot (140°F [60°C] or above) and cold foods cold (40°F [4°C] or below).
- Exclude from food preparation and handling personnel with open wounds, infections, or communicable diseases transmittable by food.
- Promptly place leftover food under refrigeration.
- Reheat leftovers rapidly to 165°F (73.8°C) before serving.
- Refrigerate all potentially hazardous foods.
- Eliminate flies, roaches, rodents, and other pests from the food-preparation and -storage areas. Consult a certified professional pest-control operator for assistance.
- Never use galvanized or cadmium-plated containers for storage of moist or acid foods.
- Consult the local health authority for answers to questions or concerns on sanitation.

As a further precaution in preventing a foodborne illness, no first-aid material or personal medication should be permitted in the food-preparation area. All workers requiring immediate first aid should be seen by the company or facility nurse or the physician; injured workers should continue on the job only at the doctor's discretion.

To prevent cross-contamination—that is, the contamination of clean items by soiled sources, such as unwashed hands or dirty shelves—it is important to handle and store containers, utensils, glassware, dishes, and silverware properly.

It is generally easier to clean utensils, dishes, and silverware by machine washing than by manual washing as long as the machine is kept clean and properly maintained. In either case, one of the main requirements in ensuring adequate sanitation is proper training of the food staff.

Utensils, dishes, and the like should be carefully scraped and preferably rinsed before washing. The wash cycle includes cleaning with warm water containing soap or detergent and a clear water rinse to remove any soap/detergent residue. The final stage properly sanitizes the utensils using a chemical method (hypochlorite or quaternary ammonium compounds), or 180°F (76.6°C) water, measured at the manifold. All utensils should be allowed to air dry—towel drying is not an acceptable practice in any case.

For effective machine washing, workers must stack the utensils in the trays loosely enough to allow the cleansing agent to reach every part. The machine should maintain the proper concentration of soap/detergent and the appropriate water temperature. The wash water must be changed periodically to prevent excessive buildup of food debris and grease, which impairs cleaning the utensils. To ensure successful operation, workers should clean the dishwashing machine, including the spray nozzles, daily to allow proper flow and distribution of the water.

Requirements for manually washing utensils are the same, except that workers must carefully scrub each utensil individually rather than simply stacking them in a tray. For manual washing of utensils, the dishwashing area must have a sink with no fewer than three compartments (Figure 21–5). The compartments should be large enough to accommodate the largest utensils and have hot and cold potable water under pressure. The sink should be supplied with attached drain boards or movable dish tables large enough to accommodate the freshly washed utensils and dishes. All utensil washing requires the same steps: prewash/scrape, wash, clear water rinse, sanitize, and air dry.

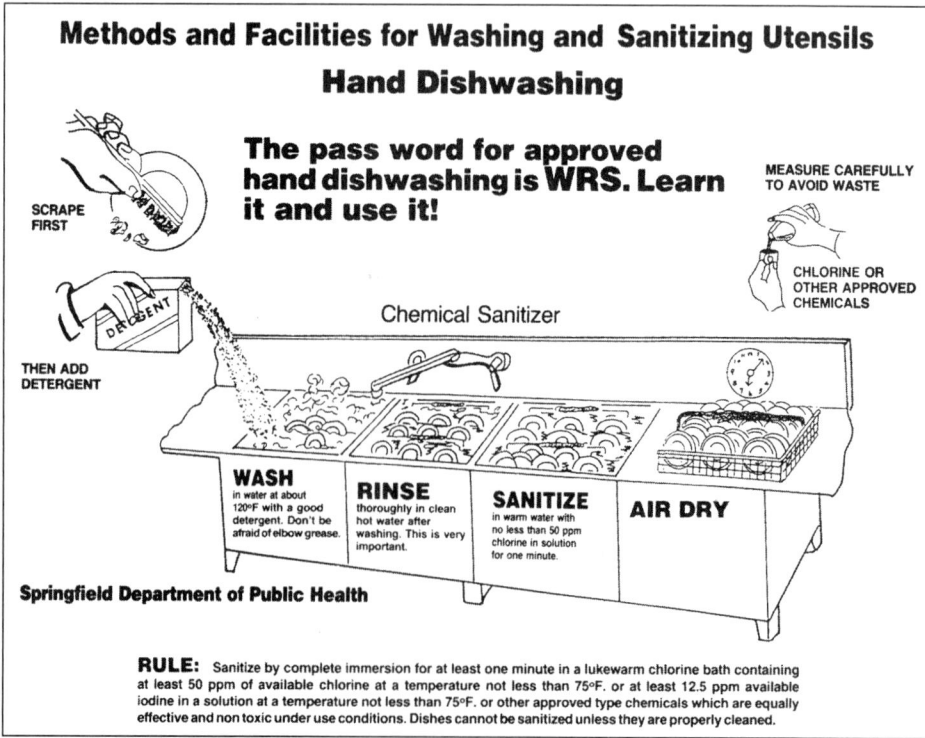

Figure 21–5. A sink with a minimum of three compartments must be provided for hand dishwashing.

Although several methods of sanitizing utensils are acceptable, the most common methods involve hot water or one of numerous chemicals. If using the chemical method, make sure the chemical chosen is approved for use on food-contact surfaces such as silverware, dishes, and so on.

To use the hot-water method of sanitizing, workers must maintain the temperature at 170°F (76.6°C) or above and keep utensils in the water for at least 30 seconds. The main problem is ensuring sufficient amounts of hot water. Long sessions of dishwashing require large amounts of hot water. Thus, the company will need a large hot-water heater with quick recovery, or it should install a booster heater on the water line or in the sanitizing tank.

Less hot water will be necessary if a chemical method is used. The most common chemical method involves immersing the utensils in a solution of hypochlorite containing at least 50 ppm available chlorine in water at least 75°F (23.8°C) for a period of 1 min. Workers can use cationic quaternary ammonium and iodine germicides if this method achieves the same sanitary specifications as a hypochlorite solution. Any time chemicals are used, employees should have a test kit or other device that accurately measures the concentration of the chemical in solution as parts per million.

After utensils have been properly cleaned, they must be properly stored and handled to prevent recontamination by workers or by ordinary dust and dirt. Use of single-service containers and utensils can eliminate washing and handling. However, these items must be carefully stored to prevent accidental contamination.

SUMMARY

- Four industrial health areas must be kept sanitary and well equipped for employee health and convenience: potable water supplies; sewage and garbage disposal; personal service facilities; and food service.
- Management should consider establishing a separate department with its own director to ensure a sanitary work environment. General rules for sanitation include an approved piping and storage system, good housekeeping, personal cleanliness, and a good inspection system.
- Drinking water must meet all regulatory standards of purity to avoid exposing workers to waterborne contaminants. Private water sources must be surveyed routinely to protect workers' health.
- All plumbing and fixtures should be disinfected before being put into service either for the first time or after repairs.
- Filtration and chemical disinfection are the most practical methods for purifying industrial private water supplies.
- Organizations must adhere to all relevant ordinances in disposing of and recycling sewage, waste, and garbage.
- Companies providing food services for their employees must arrange for proper disposal of food refuse. All garbage-collection areas and receptacles must meet state and local codes.
- In industrial and commercial sites where insect, rodent, and nuisance bird infestations are a problem, companies should employ certified professional pest-control operators. Companies must ensure that refuse receptacles and liners comply with local codes.
- All personal service facilities contributing to employee health, such as drinking fountains, washrooms, locker rooms, showers, and toilets, should be kept clean and safe for workers' use.
- Companies must maintain washrooms, showers, locker rooms, toilets, hand-washing facilities, and hand-drying mechanisms in a sanitary condition.
- Company safety and health staff can educate workers on proper nutrition and its relation to job performance and health. There are six main types of industrial food service: cafeterias, canteens or lunchrooms, mobile canteens, box-lunch services, vending machines, and food trucks.
- Food-service staff must receive careful training in sanitation and food-handling methods to prevent food contamination, cross-contamination, and the transmission of foodborne illnesses.

REFERENCES

Academy of Nutrition and Dietetics, 120 South Riverside Plaza, Suite 2000, Chicago, IL 60606-6995.

American Medical Association, AMA Plaza, 330 N. Wabash Ave., Chicago, IL 60611-5885.

American National Standards Institute, 1899 L Street NW, Washington DC 20036.

    Accessible and Usable Buildings and Facilities, ANSI A117.1–2009.

    Gas-Burning Appliances, Z21 Series.

    Minimum Requirements for Sanitation in Places of Employment, ANSI Z4.1–1986 (R2005).

    Pipe Flanges and Fittings, ASME/ANSI B16 Series.

American Public Health Association, 800 I Street NW, Washington DC 20001-3710.

    Standard Methods for the Examination of Water and Wastewater. 22nd ed.

American Water Works Association, 6666 West Quincy Avenue, Denver, CO 80235.

Bennett, G. W., and J. M. Owens, eds. *Advances in Urban Pest Management.* New York: Van Nostrand Reinhold, 1986.

Food Service Sanitation Code, 1989, 77 Ill. Adm. Code 750. Private Sewage Disposal Licensing Act and Code, 1986, pp. 50, 51.

Freedman B., *Sanitarian's Handbook, Theory and Administrative Practice for Environmental Health.* 4th ed. New Orleans: Peerless Publishing, 1977, p. 525.

Illinois Department of Public Health, Office of Health Protection, Division of Food, Drugs, and Dairies, 525 West Jefferson Street, Springfield, IL 62761.

National Fire Protection Association, 1 Batterymarch Park, Quincy, MA 02169-7471.
   National Electrical Code, NFPA 70, 2014.

National Restaurant Association Educational Foundation, 2055 L St. NW Washington DC 20036.
   A Safety Self-Inspection Program for Foodservice Operators.

National Safety Council, 1121 Spring Lake Drive, Itasca, IL 60143.
   Accident Prevention Manual for Business & Industry: Environmental Management. Edited by Gary R. Krieger. 2nd ed. 2000.
   Fundamentals of Industrial Hygiene. 6th ed. 2012.
   Work Injury and Illness Rates. Annual.

NSF International, P.O. Box 130140, 789 N. Dixboro Road, Ann Arbor, MI 48113-0140.

Robinson, W. H. *Urban Entomology: Insect and Mite Pests in the Human Environment.* New York: Chapman and Hall, 1996.

U.S. Environmental Protection Agency, Washington DC 20460.
   Drinking Water Academy. Online training modules.

U.S. Environmental Protection Agency, Office of Water Program Operations, Office of Research and Development, Municipal Environmental Research Laboratory, Cincinnati, OH 45268.
   Design Manual Onsite Wastewater Treatment Disposal Systems, October 1980, 98.

U.S. Food and Drug Administration, U.S. Department of Health and Human Services, 200 Independence Avenue SW, Washington DC 20201.
   FDA Model Food Code. 2013.

REVIEW QUESTIONS

1. The general rule for sanitation includes:
   a. an approved piping and storage system.
   b. good housekeeping as clean as the nature of the work allows.
   c. personal cleanliness.
   d. a routine, complete self-inspection system.
   e. all of the above
   f. only b and c
2. How many infectious waterborne diseases are commonly caused by drinking contaminated water?
   a. 2
   b. at least 5
   c. 9
   d. at least 12
3. Name the three adulterations that can contaminate a water supply.
   a.
   b.
   c.
4. Faucets and similar outlets should be at least how far above the flood-level rim of the receptacle?
   a. 1 in.
   b. 2 in.
   c. 3 in.
5. How can a worker easily disinfect a drinking water system?
6. Name the two methods of water purification that are the most practical for industrial private water supplies.
   a.
   b.
7. List six recyclable materials that can be hazardous to employees when sorting and collecting them.
   a.
   b.
   c.
   d.
   e.
   f.
8. What are the five personal service facilities that contribute to employee comfort and are essential to the occupational health program in most industries?
   a.
   b.
   c.
   d.
   e.
9. Codes require how many drinking fountains to be installed for every 50 people throughout an industrial facility?
   a. 1
   b. 2
   c. 3

10. List four factors to consider when selecting a floor for washrooms and locker rooms.
    a.
    b.
    c.
    d.
11. Janitorial service workers should mop floors and clean fixtures in all personal service facilities with detergent and hot water at least:
    a. once weekly
    b. twice weekly
    c. once daily
    d. twice daily
12. Two important safety and health precautions that should be followed in the use of microwave ovens are:
    a.
    b.
13. List the six main types of industrial food services.
    a.
    b.
    c.
    d.
    e.
    f.
14. What is the best way for workers to combat fatigue and maintain normal blood sugar?
    a. Drink at least two cups of coffee.
    b. Eat foods high in sugar.
    c. Eat a breakfast or snacks high in protein.
15. Food-service equipment should either receive approval from or meet established standards of which organizations?

Occupational Medical Surveillance

Antonello Pileggi, MD, PhD
Philip E. Hagan, JD, MBA, MPH, ARM, CIH, CET, CHMM, CHCM, CHSP, CEM

Occupational Surveillance
Combined Effects of Exposure ▸ Surveillance of Working Populations ▸ Types of Surveillance Systems

Occupational Medical Screening
Prevention ▸ Biomarkers ▸ Ethical and Legal Issues ▸ Medical Screening Programs

Summary

References

Review Questions

OCCUPATIONAL SURVEILLANCE

Epidemiological surveillance is the systematic collection and analysis of health data for planning, implementing, and evaluating public health programs, with the timely dissemination of these data to those who need to know (Langmuir 1976). The ultimate goal of surveillance is prevention of illness or injury due to ambient risk. Applied to the workplace, occupational surveillance focuses on monitoring the health of working populations and the exposure to hazards in the workplace.

The four essential components of an occupational surveillance system include:
- gathering information on adverse health events and exposure circumstances
- distilling and analyzing the data
- disseminating data to interested parties
- intervening on the basis of the evidence provided by the data to alter the factors that produced the hazards and adverse health outcomes.

How can occupational surveillance be used? It allows public health officials, employers, researchers, enforcers, and other stakeholders to (1) become familiar with the magnitude and distribution of occupational illnesses and injuries, (2) monitor trends over time, (3) identify emerging injury and exposure problems, (4) flag specific cases or situations for follow-up investigations, (5) set intervention priorities, and (6) evaluate intervention activities.

Occupational surveillance can be applied to working populations, overall, as well as to individual workers employed by individual companies. In the realm of public health, federal, state, and local officials may use the results to monitor and set policy to mitigate workplace risks within their jurisdictions. In individual workplaces, employers may use surveillance to establish and monitor prevention programs that are specific to the processes and conditions in their own work environments.

Combined Effects of Exposure

OSHA has medical screening requirements for certain chemical exposures (Table 22–A) and for certain working populations (e.g., hazardous-waste workers, noise-exposed workers, those who wear respiratory protection, and fire brigade members). However, little is known about how the human body responds when exposed to more than one toxin or hazardous substance at the same time. The usual assumption made by health care professionals is to consider the effects of each chemical or substance independently. In fact, exposure to a single chemical rarely, if ever, occurs. As a result, researchers and occupational health care professionals are beginning to consider the synergistic effect of substances—that is, with each chemical or substance to which a worker is exposed, the total effect becomes more than the sum of the parts. Many chemical exposures do not occur from a single substance but as part of a hazardous mixture. Hence, as chemicals or substances combine in the body, the potential for an individual developing an injury or disease can increase significantly.

Perhaps the best-known example of an exposure with a synergistic impact is that of smoking combined with asbestos. When workers are exposed to these two substances simultaneously, they greatly increase their chances of developing lung cancer. Other studies have found that carbon disulfide and environmental noise act synergistically to cause significant hearing loss in workers (Ryback 1992).

OSHA's permissible exposure levels (PELs) and the Threshold Limit Values (TLVs) of the American Conference of Governmental Industrial Hygienists (ACGIH) were developed under a similar assumption that workers are exposed to chemicals one at a time. To determine if overexposure has occurred, health professionals can add concentrations of these substances as a fraction of their respective TLVs. If the total equals or exceeds 1, then an overexposure has been detected. This is known as the mixture rule and has been published by ACGIH. The underlying assumption is that the "combined" chemicals act on the same end-organ. For example, three similar solvents could be evaluated using this principle if they act in mechanistically (toxicologically) similar ways and have the same target end-organ (e.g., central nervous system, liver).

The fact that workplace exposure involves chemical, physical, biological, and psychological hazards all interacting simultaneously complicates the health professional's efforts to determine the causes of worker complaints. As researchers develop more sophisticated models exploring the connection between synergistic effects and human health, it may be possible to devise more effective preventive measures and treatments.

Surveillance of Working Populations

There are several data collection systems that may be used for occupational surveillance. These systems may provide information on:
- what illnesses and injuries are occurring (occupational sentinels)
- why they are occurring (workplace conditions)
- how they are occurring (mechanism of injury)
- where they are occurring (industry type, economic sector, establishment size)
- when they are occurring (day of the week, time of day, seasonal variations; changes over time).

TABLE 22–A. Medical Screening Mandated by the Occupational Safety and Health Administration, 1998

Class/Agent	Regulation	History (**standard questionnaire)	Physical Exam	Chest X Ray	Pulmonary Function Test	Laboratory Tests
Respirator program	29 CFR 1910.134	X				
Chemical agents—metals						
Arsenic (inorganic)	29 CFR 1910.1018	X	X	X		Sputum cytology
Beryllium	No OSHA requirement; under review					
Cadmium	29 CFR 1910.1027	X**	X	X	X	BUN, creatinine, CBC, blood cadmium, urinalysis, urine cadmium, beta 2 microglobulin
Lead	29 CFR 1926.1025 29 CFR 1926.62 (construction)	X	X			CBC, BUN, creatinine, blood lead, ZPP, urinalysis
Chemical agents—organics						
Acrylonitrile	29 CFR 1910.1045	X	X	X		Fecal occult blood
Benzene	29 CFR 1910.1028					CBC, WBC differential, blood smear, urinary benzene, phenol
1,3 butadiene	29 CFR 1910.1051	X**	X			CBC
Dibromochloropropane	29 CFR 1910.1044	X	X			CBC, estrogen (women) LH, FSH (women); sperm count (men)
Ethylene oxide	29 CFR 1910.1047	X	X			CBC, pregnancy/fertility if indicated
Formaldehyde	29 CFR 1910.1048	X**	X	X	X if using respirator	
Methylenedianiline	29 CFR 1910.1050	X	X			liver function tests, urinalysis
Methylene chloride	29 CFR 1910.1052	X	X			
Vinyl chloride	29 CFR 1910.1017	X	X			SGOT, SGPT, GGTP, bilirubin, alkaline phosphatase
Dusts						
Asbestos	29 CFR 1910.1001 29 CFR 1915.1001 29 CFR 1926.1101 (OSHA interpretation letter, 12/31/86)	X	X	X	X	
Cotton dust	29 CFR 1910.1043	X**	X		X	
Physical agents						
Noise	29 CFR 1910.95 DOD Instruction 6055.12					Baseline and annual audiogram
Radiation*	OSHA 29 CFR 1910.1096 DOE 10 CFR Part 835 NRC 10 CFR 20 MSHA 30 CFR 57.5038					
Infectious agents						
Bloodborne pathogens	29 CFR 1910.1030	X	X			HIV, HBV, HCV (postexposure)
Tuberculosis	DOD 60555.5-M (CDC guidelines)					
Occupational groups						
Hazardous-waste workers	29 CFR 1910.120 (5 CFR 339)	X	X			CBC, BUN, creatinine
Coke oven workers	29 CFR 1910.1029 (coal tar pitch) 29 CFR 1910.1002 (coke oven emissions)	X**	X	X	X	Urinalysis, urine cytology, sputum cytology

*Radiation workers regulated by four federal agencies
**OSHA requires use of their standard questionnaire for taking a history (OSHA 1994)

What: Occupational Sentinel Health Events

An occupational *sentinel health event* is "a disease, disability, or untimely death which is occupationally related and whose occurrence may:

1. provide the impetus for epidemiologic or industrial hygiene studies
2. serve as a warning signal that materials substitution, engineering control, personal protection, or medical care may be required" (Rutstein et al. 1983).

The occurrence of mesothelioma is evidence for prior asbestos exposure; hearing loss provides evidence for noise exposure; the development of lung cancer is caused by exposure to carcinogens. Mesothelioma, noise-induced hearing loss, and lung cancer are three occupational disease sentinels that signal a failure of preventive practices. Some of the original 50, and later, 64, listed occupational sentinels are specifically caused by workplace exposures; some of these may have nonoccupational causes, as well (Mullan and Murthy 1991).

The usefulness of the sentinel health event concept is that managers, health care workers, and public health personnel are able to review health data for individuals or populations to look for patterns that may be related to occupational exposure. In addition, a disease or injury in an individual may be an indication that his or her co-workers are at risk. Steps may then be taken—not only to reverse disease in the individual, but to prevent disease in co-workers. The occurrence of an occupational illness or injury should be regarded as evidence of a failure of prevention in the workplace.

Why/How: Occupational Sentinel Exposures

A more effective approach to prevention of occupational illness and injury is hazard surveillance and remediation. Much is known about chemical, physical, biological, biomechanical, and psychosocial hazards associated with work. Periodic characterization of these is essential to protecting workers from harm. Exposure hazards may be evaluated on an industrywide scale or in an individual company. In an ongoing, population-based hazard surveillance program, samples of workplace types can be investigated.

In addition to exposure hazards that lead to illness, injury mechanisms may be coded in certain health surveillance systems so that mechanisms and circumstances of injury may be determined. For example, clinical care of an injured worker may be coded for the mechanism of injury, in addition to the diagnosis code. External injury codes (E-codes) are used for this purpose and may provide information about how the injury came about (e.g., motor vehicle crash, fall, drowning, fire). In addition, some surveillance systems have narrative fields, where the health care provider may write a description of the injury mechanism or workplace conditions. These do not tend to be standardized, but they may also give important information that can lead to prevention.

Where: Economic Sector, Industry Type, Job Titles

Some surveillance systems code for economic sector. Until the 21st century, Standard Industrial Classification (SIC) codes were used, and they still are used in most workers' compensation reporting systems. However, more recently, the North American Industrial Classification System (NAICS) is being used because it breaks down industry and job tasks more finely and gives a clearer picture of the work tasks leading to injury.

When: Day, Date, Time of Injury

Systems that record information on when an injury occurred may give clues to reasons for certain types of injuries and may provide a focus for preventive activities. For example, outdoor construction and agricultural work take place during warmer seasons, and injuries in those sectors are more likely to occur in the summer months. Focusing preventive and enforcement efforts during these months may be a better use of resources; deviations from expected injury patterns may call attention to unexpected workplace circumstances. The time of day injuries occur can give clues about the impact of shiftwork on workplace health and safety.

Types of Surveillance Systems

Occupational surveillance systems may be active: an agency specifically selects a sample of workplaces, surveys those workplaces, analyzes the data, and extrapolates the results to U.S. industry as a whole. The U.S. Bureau of Labor Statistics (BLS) collects data on workplace deaths through its Census of Fatal Occupational Injuries. In the BLS Survey of Occupational Illnesses and Injuries, a representative sample of 6,000 workplaces across the United States is selected each year to provide illness and injury reporting for OSHA 300 logs. Alternatively, questionnaires may be sent to the workers themselves in order to ascertain illness and injury trends. The National Health Interview Survey collects general health data as well as occupational data. There have been several national surveys of farmers and farm workers.

Surveillance systems also may be passive. For example, data that are collected for purposes other than occupational surveillance may be used to inform our understanding of occupational illness and injury. Clinic and hospital records, emergency room visits, and workers' compensation reports are all examples whereby specific occupational injuries or illnesses (occupational sentinels) may be examined.

Data from these sources can be analyzed to determine the numbers, rates, trends, circumstances, and outcomes of workplace illnesses and injuries. There are a number of sources that provide data on injury and illness numbers, rates, and trends. Each data source has strengths and limitations, as shown in Table 22–B. Each has been demonstrated to have limited capture due to a variety of disincentives to report (Azaroff, Levenstein, and Wegman 2002).

OCCUPATIONAL MEDICAL SCREENING

Screening is the performance of medical testing for the purpose of detecting organ dysfunction or disease in an individual before he or she would seek medical care and at a time when intervention could be beneficial (Murthy and Halperin 1995). Screening of individual workers or worker populations is one source of health data that may be used to provide surveillance information on populations and identify sentinel cases or monitor trends. However, the main goal of screening is to benefit the individual worker. This secondary preventive approach allows detection of illness at a point in time where removal from exposure or medical treatment can either reverse or minimize long-term health effects. Table 22–A, provided earlier, shows requirements by the U.S. Occupational Safety and Health Administration to provide screening for employees who are exposed to the specific hazards listed.

Medical screening programs take a systematic approach to data collection and require an understanding of many disciplines in order to be accurate, informative, and preventive. Examples include the principles of toxicology and

TABLE 22–B. Sources of Data for Occupational Illness and Injury Surveillance

Data Type (A = Active; P = Passive)	Examples	Strengths	Limitations
Preexisting health care and vital records (P)	Death certificates Hospital discharge records Clinic records Trauma registries Insurance claims	Available at low cost Coding schemes standardized	Occupation rarely listed Physicians don't recognize condition as occupational Misclassification
Health care provider (A)	Sentinel event notification system for occupational risks	In-depth reporting for six sentinel health events	Conducted in only seven states Underreporting
Employer reports (P)	Bureau of Labor Statistics: conducts Survey of Occupational Illnesses and Injuries—yearly survey of OSHA 300 logs	Provides national estimate	Usually completed by nonmedical personnel Shown to underreport
	Workers' compensation data	All records are occupational Narrative fields describe circumstances of injury/illness	Disincentives to report Differences in laws between states Statute of limitations
Employee reports (P)	Workers' compensation claims	Adjudicated cases Elaborate data on workplace, health outcomes, cost	Disincentives to report Only adjudicated cases
Biological monitoring data (A)	Adult blood lead registry	Laboratory reports Reporting mandatory to maintain lab certification Easy to obtain data	Workers may choose not to be tested Few lab tests require mandatory reporting
National health surveys (A)	U.S. Health Interview Survey—sample of 50,000 households and 130,000 individuals; contains occupational health portion	Well funded Sampling strategy allows for generalization of results	Self-reported Occupation is not main emphasis
	National Health and Nutrition Examination Survey—questionnaire and blood tests on 30,000–40,000 people	Well funded Sampling strategy allows for generalization of results Body fluid testing is objective	Questionnaire is self-reported Limited occupational section

clinical medicine in the selection and application of clinical tests, industrial hygiene in applying exposure measurements to the situation, epidemiology and clinical decision making in interpreting results, and policy and ethics in making recommendations.

Medical screening entails conducting periodic clinical examinations on workers who are at risk for work-related diseases and conditions. In workplace screening, the goal is to prevent work-related injuries and illnesses in individuals. A program that focuses on prevention in individuals should reduce injury and illness for the working population as a whole.

Recent events on the global stage have added new parameters to medical screening efforts, with the onset of outbreaks of Ebola, H1N1, and swine flu. Due to the flow of travelers across international boundaries, it is important to stay current with world affairs with regard to infectious disease outbreaks. In the United States, the Centers for Disease Control and Prevention (CDC) takes the lead in providing parameters for screening of infectious diseases. Whenever there is concern regarding a communicable disease exposure in the workplace, safety personnel should obtain information from their local health department and from the CDC at CDC.gov.

Prevention

If there are hazardous conditions in the workplace, they can be ameliorated before they adversely affect the health of employees. Removing the hazard before exposure occurs is an example of *primary prevention*. Optimally, the workplace should be monitored for hazardous conditions on an ongoing basis, and action should be taken when conditions are hazardous for employees. According to the industrial hygiene hierarchy of controls, it is most effective to implement engineering controls or to substitute hazardous agents, or unsafe working conditions, with nonhazardous ones in order to have the greatest effect on health and safety. It is optimal if the workplace can be modified so that hazards are contained; however, administrative controls and personal protective equipment may be implemented in situations in which engineering controls cannot be used. Rotating workers so that they do not contact a hazard frequently or for prolonged periods is an example of an administrative control, as is training. Use of personal protective equipment—gloves, goggles, masks, aprons, hats, and earplugs—is the last recourse for prevention because such equipment may be uncomfortable to use, may create its own hazards, and can be bypassed by workers.

Medical surveillance can occasionally be used for primary prevention if it entails detecting hazardous agents in the tissues of workers before irreversible clinical effects occur. For example, lead may be tested in the blood and the employee may be removed from exposure if his or her level is elevated, but before he or she becomes ill. Medical removal allows the individual time to excrete the lead before he or she becomes sick from it.

Clinical testing may be done for *secondary prevention*, as well. For example, in workers who are exposed to toxins such as solvents, it is possible to detect abnormal elevations in liver enzymes in the bloodstream, signaling an adverse effect of solvents on the liver. To protect the worker from irreversible liver disease, he or she may be removed from exposure early on, when the liver enzyme levels, and the liver itself, will revert to normal. To effect secondary prevention, the abnormal health effect must be reversible at the time it is detected.

Finally, *tertiary prevention* is undertaken when an irreversible injury or illness has occurred; the goal is to limit the damage. For example, if an individual develops asthma as a result of an occupational exposure, he or she can be removed from the aggravating exposure and treated in order to prevent exacerbations and limit the progression of the disease. In the realm of injury, a plan for first aid, emergency transport, and aggressive hospital care can limit adverse health outcomes. Figure 22–1 shows the timing of primary, secondary, and tertiary preventive actions.

Biomarkers

Biological markers are substances, structures, or processes that can be measured in human tissues that may predict disease. Biomarkers of *exposure* measure internal dose. Blood lead level is one example of the more than 100 such available laboratory tests. Biomarkers also may be used to determine *susceptibility* to disease. This type of biomarker may test DNA, immunologic response, or specific gene sequences that are more commonly associated with certain

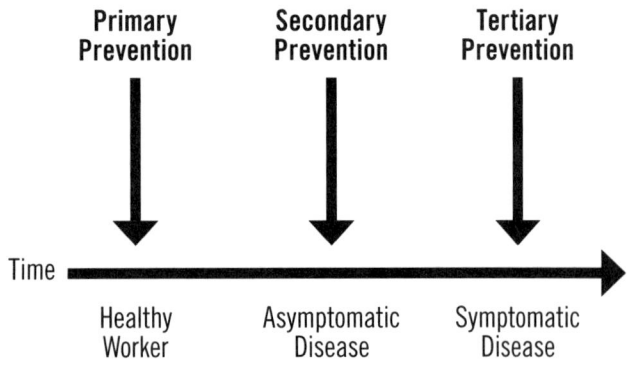

Figure 22–1. Model for prevention of occupational diseases.

diseases. For example, it is possible to detect susceptibility to beryllium-induced lung disease by determining human leukocyte antigens (genetically determined proteins that are inherited). There are very few such tests available at this time. Development of disease tends to be multifactorial—having to do with the genetic makeup of the individual as well as the agents and conditions of exposure; interpretation of these tests must be done with an understanding of gene-environment interaction.

The *limitations* of biological monitoring include:

- difficulty correlating a health risk with exposure once the exposure information is known
- short biological half-lives of some substances, which prevent accurate exposure assessment except within a limited time of exposure
- ineffective monitoring for surface active agents (e.g., hazards causing skin or upper airway irritation)
- interference of tobacco, alcohol, and other agents with some test results
- measurement that may reflect multiple exposure sources (air, food, water, soil, and skin contact), preventing accurate determination of occupational exposure.

Given our primitive understanding of the phenomenon of genetic propensity, excluding individuals from workplaces because of potential susceptibility to disease is unethical and, in most cases, illegal, at this time.

Ethical and Legal Issues

The use of biological markers raises important ethical and legal concerns. Markers of susceptibility involving genetic screening in the worksite, for example, have provoked a serious debate over who has a right to the information and for what purpose. Concerns include the potential for discrimination against workers on the basis of racial or cultural characteristics and acquired or inherited genetic susceptibility.

The central legal issues address the rights of those monitored and the use of monitoring as a primary control strategy. Thus, before companies approve the use of biomarkers in health surveillance, they must carefully consider how they will handle and communicate personal health data beyond the physician–patient relationship. In addition, occupational medical surveillance raises difficult issues related to labor–management relations, labor law, and discrimination law (Ashford 1994; Van Damme et al. 1995).

These ethical and legal issues are forcing companies to exercise great care when they develop their medical surveillance and monitoring programs. They must balance the benefits to be gained from these programs with the risks involved in handling confidential information.

Medical Screening Programs

Matte et al. (1990) proposed 13 steps to designing and conducting a screening program:

1. *Assess the hazards:* evaluate routes of absorption and exposure levels, either qualitatively or quantitatively.
2. *Identify target organ toxicity:* review toxicology literature; this may be available in a distilled or simplified form in secondary sources, like regulations from federal agencies.
3. *Select tests:* these should be able to detect disease at an early stage when it is possible to limit or prevent disease/injury.
4. *Develop action criteria:* before undertaking screening, there must be a plan for how to interpret data and act on results.
5. *Standardize the testing process:* a quality control protocol and standardization of testing are critical to ensuring reliability and comparability.
6. *Perform the test:* informed consent should be obtained from workers and confidentiality of results maintained.
7. *Interpret results:* predetermined action criteria guide decision making when intervening on behalf of individuals; data from individuals [are] aggregated to look for patterns.
8. *Confirm results:* abnormal tests should be confirmed by performing a second test before rendering a decision about removal from work.
9. *Determine work fitness:* appropriate tests should be used before recommending removal from the job; retention of salary and benefits may be of concern.
10. *Notify workers:* notification of results of screening tests should be delivered to the worker promptly and in strict confidence, with an interpretation by a health professional.
11. *Diagnostic evaluation:* abnormal results should result in a sound, medical evaluation.
12. *Evaluate and control exposure:* if an occupational health sentinel is detected, the work environment should be evaluated and hazardous exposures controlled.
13. *Record keeping:* medical records should contain all testing results, interpretation, and copies of employee/employer notifications.

Aggregate test results from workers in a screening program that also monitors trends over time may provide information about changes in exposure of the working population as a whole. This information may prove to be more informative than results for individuals. For example, while small changes in an individual's result may fall into the normal range, changes in averages and rates over time may provide evidence of an increasing exposure hazard.

SUMMARY

- Public concern about human exposure to environmental and occupational hazards has grown steadily over the past three decades. As a result, epidemiological surveillance has become an important tool in detecting, monitoring, and preventing injury and illness in the workplace.
- The four essential components of an occupational surveillance system include gathering information on adverse health events and exposure circumstances, distilling and analyzing the data, disseminating data to interested parties, and intervening on the basis of the evidence provided by the data to alter the factors that produced the hazards and adverse health outcomes.
- The goals of medical surveillance include the detection of overexposure to a hazardous agent and the early detection of disease in order to improve clinical outcome. Environmental and medical monitoring is an important element because by the time a disease can be detected in a person, it is often very difficult to alter the course of the disease.
- Biological monitoring, which consists of testing for biomarkers of exposure, effect, and susceptibility, can assist the clinician in assessing the disease-exposure relationship. Measuring or evaluating the combined effects resulting from exposure to multiple hazards is difficult with the current level of knowledge; this is an active area of research.
- Steps for designing and conducting a successful screening program could include the following elements: assessing hazards, identifying target organ toxicity, selecting appropriate tests, developing action criteria, standardizing the testing process, performing the test, interpreting results, confirming results, determining work fitness, conducting diagnostic evaluations when necessary, evaluating and controlling exposures, and keeping records of all monitoring and treatment efforts.
- There are ethical and legal issues associated with confidentiality of medical data derived from the administration of occupational medical surveillance programs. All issues dealing with the dissemination of medical data should be addressed after consultation with legal counsel.

REFERENCES

Ashford, N. A. "Monitoring the Worker and the Community for Chemical Exposure and Disease: Legal and Ethical Considerations in the U.S." *Clinical Chemistry* 40, no. 7, pt. 2 (1994): 1426–37.

Azaroff, L., C. Levenstein, and D. H. Wegman. "Occupational Injury and Illness Surveillance: Conceptual Filters Explain Underreporting." *American Journal of Public Health* 92, no. 9 (2002): 1421–29.

Baker, E. L., J. M. Melius, and J. D. Millar. "Surveillance of Occupational Illness and Injury in the United States: Current Perspectives and Future Directions." *Journal of Public Health Policy* 9 (1988): 198–221.

Fine, L. J. "Surveillance and Occupational Health." *International Journal of Occupational and Environmental Health* 5 (1999): 26–29.

Halperin, W. E. "The Role of Surveillance in the Hierarchy of Prevention." *American Journal of Industrial Medicine* 29, no. 4 (1996): 321–23.

Langmuir, A. D. "William Farr: Founder of Modern Concepts of Surveillance." *International Journal of Epidemiology* 5 (1976): 13–18.

Matte, T. D., L. J. Fine, T. J. Meinhardt, and E. L. Baker. "Guidelines for Medical Screening in the Workplace." *Journal of Occupational and Environmental Medicine* (1990): 439–56.

Mintz, B. W. "Medical Surveillance of Employees Under the Occupational Safety and Health Administration." *Journal of Occupational and Environmental Medicine* 28, no. 10 (1986): 913–20.

Mullan, R. J., and L. I. Murthy. "Occupational Sentinel Health Events: An Updated List for Physician Recognition and Public Health Surveillance." *American Journal of Industrial Medicine* 19 (1991): 775–99.

Murthy, L. I., and W. E. Halperin. "Medical Screening and Biologic Monitoring. A Guide to the Literature for Physicians." *Journal of Occupational and Environmental Medicine* 37 (1995): 170–84.

Occupational Safety and Health Administration. Medical Questionnaires; Mandatory, 1910.1001 App D (accessed May 26, 2008). http://www.osha.gov/pls/oshaweb/owadisp.show_document?p_table=STANDARDS&p_id=9999.

———. *Occupational Medical Surveillance Manual.* May 1998. Accessed May 27, 2008. http://www.dtic.mil/whs/directives/corres/html/605505.htm.

Rutstein, D. D., R. J. Jullan, T. M. Frazier, W. E. Halperin, J. M. Meliu, and J. P. Sestito. "Sentinel Health Events (Occupational): A Basis for Physician Recognition and Public Health Surveillance." *American Journal of Public Health* 73 (1983): 1054–62.

Ryback, L. P. "Hearing: The Effects of Chemicals." *Otolaryngology–Head and Neck Surgery* 106 (1992): 677–85.

Van Damme, K., et al. "Individual Susceptibility and Prevention of Occupational Diseases: Scientific and Ethical Issues." *Journal of Occupational and Environmental Medicine* 37, no. 1 (1995): 91–99.

REVIEW QUESTIONS

1. What is the ultimate goal of surveillance in the workplace?
2. What are the four components of an occupational surveillance system?
   a.
   b.
   c.
   d.
3. What are the specific uses of a surveillance system?
4. What is the difference between surveillance and screening?
5. What are some components to consider when establishing a medical screening program?
6. What is a "sentinel health event," and what is its significance?
7. What are biomarkers? What are their limitations? What are associated ethical and legal concerns?
8. Explain the synergistic effect of substances and give an example.

23 Workers with Disabilities

James T. O'Reilly, JD
Janice Comer Bradley, CSP

Overview and Background
How Disabilities Relate to Safety Issues ▸ What Employees Are Covered? ▸ What Does the Law Require? ▸ Accommodations ▸ Safety Issues ▸ Nonemployment Protections ▸ Enforcement

Americans with Disabilities Act

History and the Law
Rehabilitation Act of 1973 ▸ U.S. State and Local Laws ▸ Insurance Considerations ▸ Discrimination

Who Are Disabled Job-Seekers?
Disabled Individual ▸ Disabled Veteran ▸ Qualified Individual with a Disability

Reasonable Accommodation
Three Examples ▸ Accommodation Is Not New

Role of the Safety and Health Professional
General Responsibilities ▸ Specific Responsibilities

Job Placement

Analysis of the Job
Physical Classification ▸ Job Appraisal/Job Description

Access to Facilities
General Access ▸ Access to Buildings ▸ Interior Access ▸ Office Accommodations

Emergency Procedures

Summary

Sources of Help

References

Review Questions

OVERVIEW AND BACKGROUND

This chapter is intended to assist companies and organizations in the safe and productive inclusion of disabled individuals into the work force. This chapter will cover policy and liability issues and will address the health and safety professional's role in making the workplace more accessible to disabled persons.

Many surveys have demonstrated that it is good business to hire persons with permanent disabilities because they are productive contributors to the workplace and have lower turnover than other workers. The increasing enforcement of laws preventing discrimination based on disabilities has given another important reason for hiring persons with disabilities.

How Disabilities Relate to Safety Issues

The health and safety professional is a team player, and the teams involved with compliance on disability issues will very much need the expertise of the professional. Human resources and equal employment opportunity (EEO) managers of the organization will be interested in the degree to which safe operations of the facility can be maintained as disabled workers join the production team. The professional will be involved with such issues as safe mobility in the vicinity of fast-moving machinery, visibility of warnings and instructions for impaired persons, and adapted warning signals for hearing-impaired employees. Integrating the worker into the workplace may require creative ways of addressing the "fit" of assistance to needs and aiding the productivity of the worker by reducing barriers to his or her performance of the job. As team players, the professional, the line supervisors, and the Human Resources/Personnel group members will take into account the most practical ways to do the best possible arrangements. Many managers are already aware of the legal liability the organization faces if it fails to adequately accommodate a legitimate need of the disabled person, so there is a strong incentive to get it done correctly and promptly.

The 1990 Americans with Disabilities Act requires employers to hire workers with disabilities if, absent the disability, the worker would be the best qualified applicant for that job. Reasonable accommodations to deal with a disability are required, so long as the person is otherwise capable of performing the job tasks.

What Employees Are Covered?

Disabled persons have a mental or physical impairment, or a history or record of having an impairment, or are perceived as having an impairment; the impairment also substantially limits one or more of the major activities of the person's life. Having an impairment is the first part of the two-part test; it must also interfere with the ability to lead a normal life. Examples of impairments would include loss of leg functions, blindness, mental retardation, and emotional illnesses. Examples of substantial limits on life activities include paralysis, amputation, dependence on insulin, and so forth. A person is substantially limited if, compared with the average person having comparable training or skills, this individual is significantly restricted in the ability to perform a class of job or a broad range of tasks. Permanent, severe, and long-term effects are part of the disability, so a few weeks' trouble with a broken arm is not legally protected as a disability. Broken or sprained body parts, flu, appendicitis, obesity, and so forth, are not impairments that trigger disability law protections.

Employers only have to provide accommodation of a disability after the employer knows that the worker is disabled. The employer has the right to ask if the worker needs a reasonable accommodation of his or her disability. The worker has the responsibility to inform the employer that an accommodation is needed.

What Does the Law Require?

The law does not compel employers to hire or retain persons who are not qualified. A qualified person with a disability is a person who can perform essential functions of the job with or without reasonable accommodation. Essential functions are the fundamental duties of the job. This must be based on qualifications such as a job description, union contract, work experience of others holding similar jobs, and so forth, if the employer's assertions about the needed qualifications are realistic for that job. The capability of the person as of the time of hiring is measured, not speculation about future ability to perform the work.

The definition also requires that the disability cause substantial limitations on the ability to work. Limitations include factors that the federal Equal Employment Opportunity Commission (EEOC) has listed, including the numbers and types of jobs within the same geographical area that call for similar skills or knowledge, from which the impairment has disqualified the worker. For example, a utility pole worker who develops a fear of heights does not have a substantial limitation on his or her employability. If there is a range of other jobs available, the person might not be considered disabled.

Some persons with a history or record of disabling conditions are also protected. A former lung cancer patient cannot be discriminated against because of fear that others would "catch" cancer, for example. Others may be regarded or perceived as having a disability, and they receive protection against employment discrimination. In a U.S. Supreme

Court decision, a teacher was given protection and reinstated after the employer fired her because of fear that her tuberculosis would infect school children.

Accommodations

Once a person is both disabled and otherwise qualified, the employer may need to make a "reasonable accommodation." Such accommodations include making the workplace more accessible, restructuring work tasks, acquiring assistive equipment, and so forth. The employer is not required to provide accommodations that would be for off-work as well as work-related aspects; for example, there may be need for a height-adjustable desk but the employer need not provide a wheelchair or prosthetic limb for the worker. The law does not require an accommodation that would impose "undue hardship" on the company. This claim of hardship requires that the employer show a court or regulatory agency that the costs, financial resources, size of facility, type of operation, size of the employer organization, and so forth, make the impact of this accommodation "unduly" hard to bear.

When a request for reasonable accommodation is received from a current worker or applicant, the job's essential functions should be analyzed. The employer has an obligation to consult with the employee, to find the precise limitations that need to be accommodated, and then to identify effective means of providing the accommodation. The employer has discretion to select among several effective accommodation methods.

Safety Issues

In some cases, an employee will be a risk to him- or herself and others. Removal from a task that includes conduct inconsistent with the disabling condition (e.g., back injury survivors hauling steel drums by hand) can sometimes be justified. The employer who wants to claim that the disabled person cannot have a particular assignment because of its risks must consider duration of the risk, severity of potential harm, likelihood harm will occur, and imminence of potential harm. But the supporting information must be gathered carefully for presentation in a future penalty proceeding or civil trial.

Employers who claim a safety reason for not hiring or assigning a disabled person to a task must justify their action if challenged before the EEOC or in a lawsuit, so the health and safety professional should participate actively in making good decisions that will affect the organization and its people. The employer may need to show that the particular job would put an individual in a hazardous situation and that job requirements cannot be met by an accommodation for that person's limitations. The health and safety professional's data, advice, and expertise will be valuable to the decision on whether such a denial of a job would be able to be successfully defended. As a critical resource person, the health and safety professional should expect to have a role in the decisional process (e.g., reporting on the safety of machinery used by a disabled worker or assessing the fire evacuation potential for a disabled worker).

Nonemployment Protections

Beyond employment situations, the disabilities laws protect persons in hotels, restaurants, and other public places against discrimination based on disability. A narrow exception applies if the person poses a direct threat to safety of other persons that cannot be eliminated by modifying practices or providing assistance to the disabled person.

Enforcement

Health and safety professionals will most likely be involved in decisions about workplace accommodations that serve the needs of the disabled worker and that maintain productivity. It would be unfortunate if the organization were to refuse to make reasonable accommodations where requested by a worker who has a recognized disability. Such a refusal would perhaps result in a penalty from the federal EEOC or the counterpart state discrimination agency.

It would also be a source of regulatory problems if the firm aggressively inquired of slower workers whether they have a disability, if this questioning occurred in a context that was not "job-related and consistent with business necessity." Gossip and speculation are not justification for a supervisor to inquire about the physical or psychological disability of a worker. The professional may, however, aid the supervisor in determining what testing or evaluation would properly evaluate the worker's abilities to deal with the hazards of the particular job assignment. It would be illegal and unwise for an employer who dislikes homosexuality to insist on testing for HIV/AIDS for a small, selected group of workers. Testing must be job related and consistent with a business need.

AMERICANS WITH DISABILITIES ACT

The 1990 U.S. Americans with Disabilities Act (ADA) encompasses the following areas: Title I, Employment Provisions; Title II, State and Local Government Provisions; Title III, Public Accommodations and Services Operated by Private Entities; Title IV, Telecommunications; and Title V, Miscellaneous Provisions.

Employment (Title I) makes it illegal to discriminate against an individual with a disability in hiring or promotion if the person is otherwise qualified for the job and takes into account all aspects of the employment process—application procedures, the type of questions asked during an interview, the identification and delineation of essential functions of a job, accommodations, and preemployment exams.

Transportation (Title II) provides equal access for individuals with disabilities who use public transportation and other public services that have a major impact on the "employability" of people with disabilities.

Public Accommodations (Title III) makes it illegal to discriminate against individuals with disabilities in the full and equal enjoyment of goods, services, facilities, privileges, advantages, and accommodation of any place of public accommodation.

Telecommunications (Title IV) requires common carriers of telecommunications services to provide telecommunication relay services to hearing-impaired and speech-impaired individuals.

Miscellaneous Provisions (Title V) prohibits discrimination by state and local governments against qualified individuals with disabilities and mandates that all government facilities, services, and communications be accessible consistent with Section 504 of the Rehabilitation Act of 1973. It also gives individuals the right to file complaints and bring private lawsuits.

Title I of the act requires that employers make reasonable accommodations for the disabled. An employer is required to provide sufficient accommodation to allow qualified individuals with a disability to attain the same level of job performance as co-workers having similar skills and abilities. An employer is not required to employ an individual where to do so would pose a "direct threat" to the health or safety of others. The determination of direct threat must be based on the actual condition of the individual and not upon generalizations or stereotypes about the disability.

A significant aspect of Title III is that public accommodations and services operated by private entities are not required to permit individuals to participate in their services when the individual poses a direct threat to the health and safety of others that cannot be eliminated by modification of policies or practices or provision procedures or by providing auxiliary aids or services.

The law requires that companies establish affirmative action guidelines for the hiring, upgrading, promotion, award of tenure, demotion, transfer, layoff, termination, right-of-return from layoff, and rehiring of qualified disabled individuals. The law also requires employers to provide "reasonable accommodation" to modify the work environment or the job for these workers when necessary.

If an employer denies a disabled individual a specific job, the burden of proof is upon management to show that the person is unqualified because of one or more of the following reasons:
- The job would put the individual in a hazardous situation.
- Other employees would be placed in a hazardous situation if the person were on the job.
- The job requirements cannot be met by an individual with certain physical or mental limitations.
- And (for all of the preceding) accommodation of the job cannot reasonably be accomplished.

Developing affirmative action programs, including those for hiring workers with disabilities, is usually the responsibility of personnel other than the safety and health professional. The health and safety professional should serve as a resource person and play a critical role in evaluating the job and work environment in order to establish whether the employee can perform the essential functions of the job. Safety evaluations for the worker must include adequate access to and exit from the workplace as well as safety on the job. Health and safety professionals must be sensitive in their use of language to describe the disabled and the tasks that the disabled may be asked to perform (Figure 23–1).

Under the ADA, an employer's written job descriptions are considered evidence of the essential functions of a job if the job description existed before the job was advertised or the applicant/employee was interviewed for the job, considered for promotion, or other job-related action was taken. It is important that all essential job-related functions be defined and contained in the job description. When developing these descriptions, keep in mind intellectual as well as physical demands.

Companies should not be too quick to decide that health and safety problems are insurmountable. There are many organizations available that can help management find solutions. (See Sources of Help at the end of this chapter.)

HISTORY AND THE LAW

In the 1940s, special attention was given to employing workers with disabilities by a number of large companies that realized hiring these people was smart and sound business practice. Although some companies employed such workers before the 1940s, three events occurred in that decade to stimulate these programs and encourage other companies to institute hire-the-disabled programs.

Many individuals with disabilities were hired to help fill job vacancies left by employees who joined the military. Also in the early 1940s, one company undertook an affirmative action program to help each returning disabled veteran to become an employable person. Other companies established similar programs.

The third event was a study published by the U.S. Department of Labor that debunked some myths about workers with disabilities being less productive, suffering more injuries, and losing more time from work than other workers. On the contrary, the Department of Labor study showed disabled workers were as productive as other workers, had lower frequency and severity of injury rates, and were absent from work only 1 day more per year than other workers. In researching more than 11,000 workers with disabilities for almost 2 years, the study's authors did not find a single serious injury caused by a disabled worker, to himself/herself or to a fellow worker, that was a direct result of the disability.

One company, in a study in 1958 and another in 1981, found that of its more than 2,700 disabled workers, 96% rated average or better on safety performance; 92% rated average or better on turnover; and 85% rated average or better on attendance. After a decade or more of direct experience in hiring the disabled worker, the personnel files of many companies contain indisputable proof of the value of these employees—and of affirmative action programs—to their companies.

Rehabilitation Act of 1973

The U.S. Vocational Rehabilitation Act of 1973 (Public Law 93-112), commonly referred to as the Rehabilitation Act, was the first major civil rights law protecting the rights of persons with disabilities. This law applies to federal contractors (Section 503) and recipients of federal assistance programs (Section 504). Therefore, all employers who do work for the federal government, or receive funds from the government, are subject to this law.

Section 503 of the act requires employers to take "affirmative action" to recruit, hire, and advance qualified individuals with disabilities. The law applies only to employers who have federal government contracts or subcontracts of $2,500 or more. Those holding contracts or subcontracts of $50,000 or more, with at least 50 employees, must prepare and maintain (review and update annually) an affirmative action program at each establishment. The program—which sets forth the employer's policies, practices, and procedures regarding disabled workers—must be available for inspection by job applicants and employees. Section 504 of the act forbids acts of employment discrimination against qualified disabled persons by employers who receive federal funds. This section is enforced by each department or agency that provides federal funds.

In the Rehabilitation Act Amendments of 1974, the defi-

Negative Language	Positive Language
handicapped/handicap	disabled/person with disability
cripple/crippled by	person who has (whatever)
victim	person who uses (assistive device)
spastic	
paralytic	
afflicted/afflicted by	caused by …
deformed/deformed by	as a result of …
suffering from	
confined/restricted to a wheelchair	person who uses a wheelchair
wheelchair bound	wheelchair rider, user
deaf and dumb	pre-lingually deaf (at birth)
deaf mute	post-lingually deaf (after birth)
poor, unfortunate and similar words	
Normal—When used in the statistical sense or to express "average," the term is fine. However, this word should never be used to refer to people without apparent physical, emotional, or mental disabilities. Because most people are disabled at some time in their lives, the average person can be a disabled person. The disabilities may be inability to control one's temper, effects of past broken bones, strong prejudices, substance abuse, and other problems.	**Deaf/hearing impaired**—All deaf people are hearing impaired, but not all hearing-impaired people are deaf. This distinction is important as the needs of a deaf employee are very different from those of a hearing-impaired employee. Some deaf people use sign language, while others lipread. Some hearing-impaired people need relative quiet to understand verbal communication; others need a person with a deep voice to relay messages from a person with a higher-pitched voice. Ask what a particular hearing-impaired person needs.
Wheelchair—People who are able to walk and run usually see a wheelchair as a confining device. In reality, it gives mobility to people who would otherwise be confined to bed. Wheelchairs come in many types, including some that have variable height controls, extra-narrow axle widths, or other features. It is often possible to build a wheelchair or other mobility device to suit a particular worker and his or her job duties. Employers should explore several options if a workplace cannot be safely or easily adapted to a standard wheelchair.	**Blind/visually impaired**—Likewise, not all visually impaired people are blind or even legally blind. Some have tunnel vision, others have peripheral but no central vision. Some need strong light while others require dimmer light. Simply because a visually impaired person wears glasses does not mean their vision is 20/20. Ask what a particular visually impaired employee needs to be able to see comfortably.

Figure 23–1. Language issues regarding disabilities. *(Copyright by the Right to be Proud Campaign, St. Louis, MO.)*

nition of *handicapped (disabled) individual* was broadened for purposes of Section 504. With this amended definition, it became clear that Section 504 was intended to forbid discrimination against all disabled individuals, regardless of their need for or ability to benefit from vocational rehabilitation services. Thus, Section 504 reflects a national commitment to end discrimination on the basis of disability and establishes a mandate to bring persons with disabilities into the mainstream of American life.

Other U.S. federal departments and agencies also have issued regulations similar to those of the Department of Education. For example, the U.S. Department of Labor has issued regulations (29 CFR 32), effective November 6, 1980, that implement Section 504 of the act for the department. All of these regulations require federal contractors and recipients of federal funds to make reasonable accommodations to the workplace when necessary for employing people with disabilities.

All records pertaining to compliance with the act, including employment records and any complaints and actions taken as a result, must be retained by the employer for at least 1 year. If the company fails to maintain complete and accurate records or fails to update the affirmative action program each year, the government can impose "appropriate sanctions" against the employer. When complaints are brought against the employer or there is some question about affirmative action programs, the employer must allow investigators to have access during normal business hours to its place of business; its books, records, rules, and regulations; and its accounts pertinent to compliance with the act.

In 2010, final regulations revising the Department of Justice's ADA regulations, including its ADA Standards for Accessible Design, were implemented. The official text was published in the *Federal Register* on September 15, 2010, with corrections to this text published in the *Federal Register* on March 11, 2011. The revised regulations amend the DOJ's 1991 Title II regulation (state and local governments), 28 CFR 35, and the 1991 Title III regulation (public accommodations), 28 CFR 36. Appendix A to each regulation includes a section-by-section analysis of the rule and responses to public comments on the proposed rule.

These final rules went into effect on March 15, 2011, and were published in the 2011 edition of the *Code of Federal Regulations (CFR)*. The DOJ has assembled an official online version of the 2010 standards to bring together the information in one easy-to-access location. It provides the scoping and technical requirements for new construction and alterations resulting from the adoption of revised 2010 standards in the final rules for Title II (28 CFR 35) and Title III (28 CFR 36). These standards can be found online at www.ada.gov/regs2010/2010ADAStandards/2010ADAstandards.htm.

If a business facility was built or altered in the past 20 years in compliance with the 1991 standards, or barriers to specific elements were removed to ensure compliance with those standards, further modifications to those elements are not necessary—even if the new standards have different requirements for them—to comply with the 2010 standards. This provision is applied on an element-by-element basis and is referred to as the "safe harbor." The following examples illustrate how the safe harbor applies:

- The 2010 standards lower the mounting height for light switches and thermostats from 54 in. to 48 in. If the light switches are already installed at 54 in. in compliance with the 1991 standards, companies are not required to lower them to 48 in.
- The 1991 standards require one van accessible space for every eight accessible spaces. The 2010 standards require one van accessible space for every six accessible spaces. If the facility has complied with the 1991 standards, it is not required to add additional van accessible spaces to meet the 2010 standards.
- The 2010 standards contain new requirements for the input, numeric, and function keys (e.g., "enter," "clear," and "correct") on automatic teller machine (ATM) keypads. If an existing ATM complies with the 1991 standards, no further modifications are required to the keypad.

U.S. State and Local Laws

All 50 states and many local governments have now adopted building codes or legislation requiring barrier-free design or removal of barriers preventing access to the building by disabled persons. Many of these codes mandate that any public building or facility must be barrier free if the public is invited to use it for any normal purpose such as shopping, employment, recreation, lodging, or services. If accessible facilities need to be identified, the organization should use the international symbol of accessibility (Figure 23–2).

Insurance Considerations

Some companies erroneously believe that employing disabled workers will raise their workers' compensation premiums. This is not true. Rates are based on experience by the class of industry and modified in most cases by individual plant experience. There is no indication that losses are increased when persons with disabilities are properly placed because this should not put the worker at risk for injury.

Discrimination

Under the ADA, discrimination includes such actions as limiting, segregating, or classifying a job applicant or employee in a way that adversely affects the person's opportunities or status. Discrimination also includes not making reasonable accommodation for the known physical or mental limitation of an otherwise qualified person with a disability. It also includes the denial of employment because a qualified person with a disability needs reasonable accommodation.

WHO ARE DISABLED JOB-SEEKERS?

The law defines three types of disabled persons seeking employment: the disabled individual, the disabled veteran, and the qualified disabled individual.

Disabled Individual

ADA defines a *disabled individual* as a person who has one of the following:
1. a physical or mental impairment that substantially limits one or more of the person's major "life activities," such as:
   - ambulation
   - communication
   - education
   - employment
   - housing
   - self-care
   - socialization
   - transportation
   - vocational training
2. a record of such impairment
3. a perception of having such an impairment.

The term *substantially limits,* as used herein, has to do with the degree to which the disability affects the person's employability. A qualified disabled person who, because of the disability, finds it difficult to obtain an appropriate job or advance in a job would be considered substantially limited.

The term *physical or mental impairment* would include, but not be limited to, these conditions:
- diseases and infections
- orthopedic impairments
- visual, speech, and hearing impairments
- cerebral palsy
- epilepsy
- muscular dystrophy
- multiple sclerosis
- HIV
- cancer
- heart disease
- diabetes
- mental retardation
- emotional illness
- drug addiction
- alcoholism.

It should be noted that homosexuality and bisexuality are not defined as impairments. Compulsive gambling, illegal drug use, kleptomania, and other such activities are also not considered to be impairments.

Figure 23–2. International symbol designates access for the disabled. The symbol is in white on a blue background. *(Printed with permission from the American National Standards Institute.)*

Disabled Veteran

A *disabled veteran* is a "special handicapped individual" who:
1. is entitled to disability compensation under laws administered by the Veterans Administration for disability rated at 30% or more
2. was discharged or released from active duty due to a disability incurred or aggravated in the line of duty.

A veteran with non-service-connected disabilities is not considered a special disabled veteran but may still qualify as a disabled individual under Sections 503 and 504 of the Rehabilitation Act of 1973.

The Vietnam War had the highest proportion of disabled service personnel of any war in U.S. history. A disabled veteran of the Vietnam War is a person who was discharged or released from active duty for a service-connected disability if any part of such duty was performed between August 5, 1964, and May 7, 1975.

Qualified Individual with a Disability

Not every disabled person is covered by the ADA. The crucial word is *qualified*. A person must be capable of performing the essential functions of a job—with reasonable accommodation to the disability.

Nor is every disabled veteran and every Vietnam Era disabled veteran covered by the Rehabilitation Act of 1973 or the Vietnam Era Veterans' Readjustment Assistance Act of 1974. As with the ADA, the veteran also must be qualified, that is, capable of performing a particular job, with reasonable accommodation to the disability. Organizations should be aware that the Americans with Disabilities Act requires that a certain number of deadlines be met.

REASONABLE ACCOMMODATION

According to the ADA, an employer should make "reasonable accommodation" to the known physical or mental limitations of an otherwise qualified disabled applicant or employee, unless the employer can demonstrate that the accommodation would impose an undue hardship. Accommodations can include modifications of equipment or facilities and alterations in processes or job descriptions. The employer may not deny any employment opportunity to a qualified disabled employee or applicant if the only basis for the denial is the need to make a reasonable accommodation.

Undue hardship means an action requiring significant expenses or difficulties and is determined by considering the following factors:
- overall size of the employer's operation with respect to number of employees, number and type of facilities, and size of budget
- type of operation, including the composition and structure of the work force
- nature and cost of the accommodation needed.

Reasonable accommodation may include but is not limited to:
- making facilities used by employees readily accessible to and usable by disabled persons
- job restructuring, part-time or modified work schedules, acquisition or modification of equipment or devices, provision of readers or interpreters, and other similar actions.

The concept of reasonable accommodation has compelled employers to have very detailed descriptions of job tasks, performance expectations, and physical requirements necessary to perform the job successfully. (See also Chapter 16, Ergonomics Programs, for a discussion of accommodations in relation to ergonomics factors.)

Three Examples

Reasonable accommodation is demonstrated in these three examples:
- A construction equipment salesperson, whose job description required him to climb onto the equipment and demonstrate its operation during sales presentations, was given a desk job after he suffered the amputation of his arm during an off-the-job motor vehicle collision. Although his prosthetic device enabled him to operate the equipment controls, the employer had considerable concern about the man's ability to climb on and off the equipment using the prosthetic device. This resulted in the job change. Upon enactment of the Rehabilitation Act of 1973, and its amendments, his employer reinstated the man as a salesperson, making a reasonable accommodation for his disability. The accommodation consisted of providing him with a portable climbing device so that he could get on and off the equipment safely.
- A forklift (powered industrial truck) mechanic became blind in one eye due to a nonindustrial health problem. The man's job description required him to test drive each forklift truck after completing maintenance or repair work on it. Because the employer's standard safety policy required that all drivers of powered industrial trucks must have binocular vision (use of both eyes), the employer at first was going to switch the man to another job, which would have lowered his earnings. Upon reviewing the requirements of the U.S. Rehabilitation Act of 1973, however, the employer provided the mechanic with a reasonable accommodation. The company altered the mechanic's specific job description to eliminate the requirement to test drive the vehicles. Instead, the job descriptions of the other mechanics were broadened to include the test driving of any vehicle repaired by the disabled mechanic.
- A disabled individual working for an electrical appliance firm was provided with a reasonable accommodation to assemble small parts. It consisted of minor adjustments to the workbench to accommodate a wheelchair.

What might be considered an unreasonable accommodation is a company completely redesigning or altering the circuitry or operating levers of a machine to accommodate a physically disabled individual.

Accommodation Is Not New

Accommodation in employment is not a novel concept. The first applications of machine guards and ventilating fans were job accommodations. Also, the first hod carrier who lacquered and reinforced his bowler as a hard hat made a job accommodation. Job placement of employees based on medical examinations, when newly hired or returning to work after an illness or injury, is again an application of accommodation. This experience is common to every employer.

Until individuals receive adequate training in the field of rehabilitation medicine, they cannot be qualified to evaluate reasonable accommodations of the workplace, its procedures, and access for the physical or mental limitations of a disabled worker. The health and safety professional without such expertise should be one member of the team consisting of the worker's physician, the occupational physician, and a rehabilitation specialist; in some situations, other individuals with similar disabilities may be included as part of the team. A team approach will greatly enhance the organization's commitment to an affirmative action program and nondiscrimination. It will also contribute to an affirmative action program and nondiscrimination against disabled employees and applicants in the organization.

Job safety analysis and safe work procedures are a means of eliminating or reducing work hazards to minimize worker risk. They directly transfer to the process of accommodation. Training in safe work procedures will be important in accommodating the person with a disability to the job.

ROLE OF THE SAFETY AND HEALTH PROFESSIONAL

Affirmative action programs required by the U.S. government usually come under the responsibility of EEO managers or coordinators (or labor affairs personnel). As a result, the placement of qualified disabled individuals is normally under their jurisdiction.

The health and safety professional, nonetheless, should be a resource person to those responsible for job placement of qualified disabled individuals. This professional should be consulted before workers are placed and asked to evaluate any proposed reasonable accommodation. The following sections list some of the responsibilities of the health and safety professional in relation to disabled employees.

General Responsibilities

The general responsibilities of the safety and health professional include:

- Maintain close liaison with the EEO manager-coordinator and with medical and personnel departments when they are placing disabled employees. Rehabilitation specialists and people with similar disabilities may also be necessary members of the placement team.
- Perform a job safety analysis of existing work based on the job responsibilities and the abilities and limitations of the disabled employee or applicant when employing, promoting, transferring, and selecting workers with disabilities.
- Make recommendations for safety modifications of machine tools, established processes and procedures, and existing facilities and workplace environment when the company must make reasonable accommodations for disabled employees.
- As required, cooperate with the plant or building engineer or mechanical engineer and the planning, production, and maintenance departments when disabled employee accommodations are being evaluated.

Specific Responsibilities

In addition to reviewing the company's affirmative action program, the health and safety professional's specific responsibilities include establishing specific communication channels, pertaining to disabled employees, with the:

1. EEO manager-coordinator. Make sure this person knows the safety and health professional is part of the team and available when a job needs a safety evaluation.
2. Medical department. Let them know the professional will be requesting their help when evaluating a job.
3. Personnel department. Let them know the health and safety professional is ready when necessary to help them evaluate a job's safety and remind them of safety considerations such as:
   a. Don't place a worker with a coronary condition in a job that would aggravate that condition, which is a medical evaluation process.
   b. Don't place a person with a back problem in a job requiring heavy manual lifting, unless other considerations are given.
   c. Make certain to place a disabled employee in a job that would be safe for that person and that will not cause a hazard to others. Note: Individual judgments are based on a physician's evaluation with input from the health and safety professional.
   d. A person who has been treated for carpal tunnel syndrome should not be placed in a job that involves repetitive motion of the arm/wrist.
4. Plant and mechanical engineers. Reasonable accommodation does not necessarily mean reinstalling machines; rather, it could mean minor relocation of a machine's controls so a disabled employee can operate them properly and safely. Therefore, advise the engineers that the health and safety professional will evaluate all safety aspects of such an accommodation. Also advise them that the professional will be available for safety evaluations when they design reasonable accommodations into future facilities such as:
   a. ramps for wheelchairs
   b. wider door passages for wheelchairs
   c. grab bars in accessible washroom facilities
   d. Braille numbers (within a disabled person's reach) on elevators (Figure 23–3)
   e. easy access to company facilities such as lunchrooms
   f. elimination of curbed crosswalks.
5. Production and maintenance departments.
   a. Because reasonable accommodations also refer to job restructuring and modifications, the health and safety professional should tell the production and maintenance departments that he or she will help by evaluating the safety aspects of such changes. Experienced rehabilitation experts should be consulted before the company declares it is unable to make the job suitable for a particular applicant.

Often a fresh look at the situation and job from an outsider's perspective will reveal new strategies that have been missed.

b. Upon the deletion of any job specification that would arbitrarily and without justification screen out disabled individuals, the health and safety professional will be available if a safety evaluation is needed.

A critical and often overlooked component of ensuring success in dealing with ADA issues is comprehensive and ongoing staff training for dealing with customers. If frontline staff are not aware of ADA policies or do not know how to implement them, problems can arise. Businesses of all sizes should educate staff about the ADA's requirements. Staff members need to understand the requirements on modifying policies and practices, communicating with and assisting customers, and accepting calls placed through the relay system.

The health and safety professional should conduct a safety evaluation of a disabled employee (in relation to the specific job or prospective job) and perform an entire job hazard analysis if needed. Rehabilitation specialists and others may be consulted as appropriate, particularly if it appears at first that the disabled employee is simply unable to perform the job. If the consensus is that the job and employee are incompatible, the company's decision, made along with the rehabilitation and other specialists, is easier to defend.

In addition, the health and safety professional should:
- Conduct a safety evaluation whenever a reasonable accommodation is being planned for a disabled employee.
- Coordinate with both the EEO manager-coordinator and the employment department to make certain any disabled employee being considered for a new position is qualified to safely and capably perform the job. This usually requires the health and safety professional to make a safety evaluation and observe the employee during training.
- Evaluate any reported harassment of a disabled employee that affects safety. For example, name-calling would not normally jeopardize the employee's safety, but pranks by other employees could result in incidents and injuries. Verbal harassment of any worker for any reason should not be tolerated. It is particularly inexcusable for a health and safety professional to stand by, claiming that such harassment does not generate a safety hazard. Angry, defensive, or depressed workers do not make sound safety judgments on the job. Further, when verbal harassment is tolerated by management, it often escalates into physical harassment.
- Conduct safety evaluations of areas accessed by customers and visitors to the site using an ADA compliance perspective.

Failure to discipline workers who are harassing others can lead to serious consequences. These can range from workplace deaths and injuries, to a decline in production, to a loss of valued employees who may quit in disgust and resort to lawsuits that the employer will find difficult to defend. The health and safety professional should work with the EEO manager-coordinator and other pertinent personnel to appropriately and effectively handle such situations as soon as such incidents come to their attention.

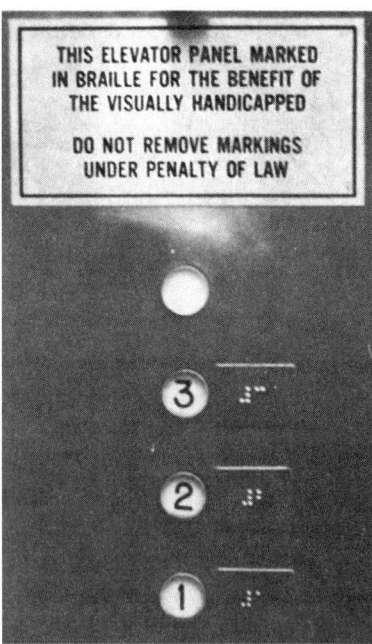

Figure 23–3. Reasonable accommodation for the disabled includes marking elevator buttons in Braille. *(Courtesy Governors State, University Park, Ill.)*

All questions about interpretation of government requirements should be referred to the personnel responsible for implementing the affirmative action program (usually the EEO manager-coordinator) or to legal counsel (either in-house or external, as appropriate).

The evaluation form (Figure 23–4) can help the health and safety professional perform safety evaluations for disabled employees, especially when reasonable accommodation is involved. A written report and supporting material (such as memos, blueprints, and photos) can be attached to the form to provide detailed information explaining why certain decisions were made.

Such evaluations should be kept for at least 1 year after the employee leaves the company. Records can be destroyed only after approval from the EEO manager-coordinator or other personnel responsible for the government-required affirmative action program.

JOB PLACEMENT

Time and again, companies have found that disabled workers make for a first-class work force. When properly placed and trained so they can compete on an equal basis with others, workers with disabilities usually equal or prove to

DISABLED EMPLOYEE SAFETY EVALUATION

☐ Applicant: _____

☐ Employee: _____ _____
        (Last Name)   (First Name)   (M.I.)   (Clock No.)

Disability: _____

Evaluation of   ☐ Current job   ☐ Prospective job

Job Title: _____

Job Description (primary duties): _____

Hazards to This Employee:
(State "none" if none) _____

Hazards to Other Employees:
(State "none" if none) _____

Proposed "reasonable accommodation" (if any): _____

CONCLUSION (Based on all known factors at this time):

☐ Job is safe for this employee:
    ☐ as is   ☐ with proposed reasonable accommodation

☐ Job is unsafe for this employee:
    ☐ as is   ☐ with proposed reasonable accommodation

☐ No hazard to other employees:
    ☐ as is   ☐ only with proposed reasonable accommodation

☐ Hazard to other employees:
    ☐ as is   ☐ with proposed reasonable accommodation

Location:	Safety Supervisor (Print Name)
Date:	Safety Supervisor (Signature)

Note: Complete two copies of this form and give one copy to local EEO manager-coordinator. Second copy is for the safety file.

Figure 23-4. This evaluation is an excellent administrative tool, especially when a written report and other supporting documentation are attached.

---

be slightly better than other employees in production and safety. Their overall attendance and job turnover records are usually superior to those without disabilities.

To place a disabled worker properly, the following requirements should be observed, where applicable, after receiving a physician's evaluation of the individual:

- The worker should meet the physical demands of the job. When necessary, the worker should receive the support of reasonable accommodation.
- The worker should not be a hazard to him- or herself. For example, a person subject to dizzy spells should not work on a ladder or scaffold or around moving machinery, where injury or death could occur.
- The worker should not be a hazard to others. For example, a person with severe vision impairment should not drive a bus or operate an overhead crane because of the potential for personal injury or injury to others.
- The task should not aggravate the known degree of dis-

ability. A person with skin disease should not be exposed to substances that may aggravate this condition. Another example might be a worker with impaired lung function. These workers should not be exposed to substances such as smoke that will further impair lung function.
- To obtain valuable input, a conference with the individual should be held before job placement is made.

Proper placement matches the right worker to the right job on the basis of the person's ability to meet the job qualifications (Figure 23–5). This has become known as fitness for duty and is an important placement criterion for employers to hire workers that are appropriately suited for the position. As a result, the impairment virtually disappears as a factor of job performance. Moreover, employers should be aware that most disabled persons have more ability than disability because few jobs actually require all of a worker's abilities. The job–employee match forms shown in Figure 23–5 are not only used to place disabled persons; some companies use them to place all new and transferred workers.

On the other hand, employers should also remember that each impairment can impose limitations on the type and number of activities in which the disabled person can engage. The impairment will also limit the working conditions and hazards to which this person can be exposed.

Many workers with disabilities are particularly vulnerable to tobacco smoke. Quadriplegics are endangered because they cannot cough to clear their lungs; people with heart or respiratory disease should not be exposed to environmental tobacco smoke (ETS), which contains many irritants. Workers with circulatory problems are further compromised by being exposed to nicotine, a compound that constricts the blood vessels. Other workers' disabilities may also make them hypersensitive to tobacco smoke. Clearly, prohibiting smoking in the workplace is not only reasonable but can benefit all employees, not just the disabled worker.

The health and safety professional should be aware that there are regulatory standards that, although promulgated for the protection of the average employee, are not sufficient to protect employees with disabilities.

PHYSICAL CAPACITIES FORM

Leg Amputation 5" below knee
Artificial leg—good fitting
Name _____ Sex M Age 31 Height 5'9½" Weight 155

PHYSICAL ACTIVITIES		WORKING CONDITIONS	
x 1 Walking	16 Throwing	51 Inside	x 66 Mechanical Hazards
0 2 Jumping	x 17 Pushing	52 Outside	x 67 Moving Objects
0 3 Running	x 18 Pulling	53 Hot	0 68 Cramped Quarters
x 4 Balancing	19 Handling	54 Cold	69 High Places
x 5 Climbing	20 Fingering	55 Sudden Temp. Changes	70 Exposure to Burns
0 6 Crawling	21 Feeling	x 56 Humid	71 Electrical Hazards
x 7 Standing	22 Talking	57 Dry	72 Explosives
8 Turning	23 Hearing	x 58 Wet	73 Radiant Energy
9 Stooping	24 Seeing	59 Dusty	74 Toxic Conditions
0 10 Crouching	25 Color Vision	60 Dirty	75 Working With Others
x 11 Kneeling	26 Depth Perception	61 Odors	76 Working Around Others
12 Sitting	27 Working Speed	62 Noisy	77 Working Alone
13 Reaching	28	63 Adequate Lighting	78
x 14 Lifting	29	64 Adequate Ventilation	79
x 15 Carrying	30	65 Vibration	80

Blank Space=Full Capacity   x=Partial Capacity   0=No Capacity
May work __ hours per day __ days per week. (IF TB, cardiac or disability requiring limited working hours.)
May lift or carry up to __**__ pounds.
DETAILS OF LIMITATIONS FOR SPECIFIC PHYSICAL ACTIVITIES:
Should not be required to walk, balance, climb, stand, kneel for prolonged periods of time.
**Should not lift heavy weights continuously.
Should not carry long distances.

PHYSICAL DEMANDS FORM

Job Title Data Processing Entry   Occupational Code 4-44.110
Dictionary Title _____ Data Processing Entry
Firm Name & Address _____
Industry _____   Industrial Code _____
Branch _____   Department _____
Company Officer _____   Analyst Wetzel   Date ____

PHYSICAL ACTIVITIES		WORKING CONDITIONS	
x 1 Walking	16 Throwing	x 51 Inside	66 Mechanical Hazards
2 Jumping	17 Pushing	52 Outside	67 Moving Objects
3 Running	18 Pulling	53 Hot	68 Cramped Quarters
4 Balancing	x 19 Handling	54 Cold	69 High Places
5 Climbing	x 20 Fingering	55 Sudden Temp. Changes	70 Exposure to Burns
6 Crawling	21 Feeling	56 Humid	71 Electrical Hazards
7 Standing	22 Talking	57 Dry	72 Explosives
8 Turning	23 Hearing	58 Wet	73 Radiant Energy
9 Stooping	x 24 Seeing	59 Dusty	74 Toxic Conditions
10 Crouching	25 Color Vision	60 Dirty	75 Working With Others
11 Kneeling	26 Depth Perception	61 Odors	x 76 Working Around Others
12 Sitting	27 Working Speed	62 Noisy	77 Working Alone
x 13 Reaching	28	63 Adequate Lighting	78
x 14 Lifting	29	64 Adequate Ventilation	79
x 15 Carrying	30	65 Vibration	80

DETAILS OF PHYSICAL ACTIVITIES:
Sits at computer most of the day, reads copy, and fingers keyboard to enter data. Periodically walks short distances, reaches for, lifts, and carries small stacks of billing materials. Reaches for, handles, pushes, and pulls when organizing and filing paperwork at desk level.

Figure 23–5. Example of an employment service form used when matching workers to jobs.

For instance, standards referring to storage of flammable and combustible liquids, 29 CFR 1910.106 (d)(6)(iii), include a requirement for a curb to capture spilled liquids. Disabled employees are just as vulnerable as anyone else to the hazard of chemicals spilled or escaping from a storage room. The curb can be equipped easily with a portable ramp, which not only permits a wheelchair to get in and out, but allows carts to enter and exit the room. This reduces the hazard of workers transferring chemicals one-by-one from the cart and possibly tripping on the curb while doing so.

The means of egress standards, 29 CFR 1910.37, are based on the ability of an individual to move 100 ft (30 m) in 30 seconds (s). Perhaps some employees with disabilities cannot move that fast.

Respirators are required by the standards (e.g., 29 CFR 1910.134) for certain jobs. Some individuals may have a physical impairment that can be affected by restricted breathing. If there is some indication of this problem, such employees must not wear a respirator until it is determined by a physician that it can be worn safely. This may preclude the individual from performing a specific job where a respirator is necessary.

The permissible exposure levels (PELs) listed in the OSH Act air contaminants standards, 29 CFR 1910.1000, are based on the susceptibility of persons with normal breathing capacities to such contaminants. Some disabled individuals do not have normal breathing capacities and, therefore, are susceptible to lower levels of contaminants.

The health and safety professional must consider whether existing safety standards are sufficient to protect a particular disabled worker or candidate. More stringent standards, such as less exposure to an air contaminant than the average employee can tolerate without harm, may be necessary to protect the disabled worker or applicant. Such decisions must be made with the assistance of medical and rehabilitation consultants.

ANALYSIS OF THE JOB

Each job must be evaluated to make sure the individual being considered can do it safely. The following areas should be taken into account when making the analysis.

Physical Classification

The labor market simply does not supply only "physically perfect" workers. In fact, the percentage of workers in perfect health is relatively low—the working population now includes many persons with disabilities. Advances in medical science prolong the lives of many who would have died of war injuries or such illnesses as smallpox, tuberculosis, diabetes, and heart disease. Injuries in industry, in traffic, and in the home continue to increase the number of persons with physical disabilities.

Because the company's physician conducts physical examinations of prospective employees and makes regular plant inspections, he or she often has a better understanding of various physical and mental requirements for company jobs than do noncompany physicians. The company physician's responsibility should be to provide management with clear evaluations of each applicant's fitness for a particular job. The physician's determinations must be made only on the basis of job–worker compatibility. Although the company physician is better able to understand job requirements, he or she may need to consult with a rehabilitation specialist.

On the other hand, safety, medical, and even many rehabilitation professionals often cannot review a work situation and ask the right questions to develop a workable compromise or accommodation for a particular disabled worker or applicant. For this reason, it is wise to consult with those who are experts in a particular disability involved.

Many support groups for disabled people exist, and the members can share experiences and insights that only those living with a particular disability can provide. Such groups can help the safety, medical, and rehabilitation specialist find creative ways to accommodate a particular job to a disabled person or suggest accommodation devices that can be made or purchased.

To locate such support groups, companies should contact the national headquarters of those groups, which can supply the names and telephone numbers of local chapters. The service directories have Rehabilitation Services and Vocational Rehabilitation listings of companies and agencies that can provide contacts for local groups. Local hospitals and physicians are also good sources, along with governmental agencies. It may be a good idea for the health and safety professional to call the groups to find out if they can provide assistance.

Many systems of employee disability classification are now in use. Generally, however, these systems use broad statements, such as "physically fit for any work," "defect that limits applicant to certain jobs" (the defect may or may not be correctable, but may require medical supervision), and "defect that requires medical attention and is presently handicapping."

In yet another method, which approaches an ideal functional evaluation of the individual, the physician documents an employee's capacities on a form that uses identical terminology to evaluate both the physical (functional) factors and the working conditions of jobs (Figure 23–6). This effective method of presenting information from a physical examination clearly indicates the specific work capability and limitations of the individual.

JOB ANALYSIS FOR PHYSICAL FITNESS REQUIREMENTS

TITLE OF POSITION | GRADE

NAME AND LOCATION OF ESTABLISHMENT | AGENCY

Does establishment have medical supervision ☐ Yes ☐ No
Is there an industrial safety branch ☐ Yes ☐ No

Refer to the manual for job analyses before using this form. Check all functional and working condition factors as well as acceptable disabilities whenever appropriate.

I. FUNCTIONAL FACTORS
L - Little   M - Moderate   G - Great   O - None

Hands - Fingers	L	M	G	O	Arms	L	M	G	O	Legs - Feet	L	M	G	O	Body - Trunk	L	M	G	O
1. Reaching					8. Reaching					14. Walking or running					22. Sitting				
2. Pushing or pulling					9. Lifting					15. Standing					23. Bending				
3. Handling					10. Pushing or pulling					16. Sitting					24. Reaching				
4. Fingering					11. Carrying					17. Carrying					25. Lifting				
5. Climbing					12. Climbing					18. Climbing					26. Carrying				
6. Throwing					13. Throwing					19. Jumping					27. Jumping				
7. Touching					**Eyes**					20. Turning					28. Turning				
Voice					30. Near vision					21. Lifting									
29. Talking					31. Far vision					**Ears**									
					32. Color vision					33. Hearing									

II. WORKING CONDITION FACTORS

34. Inside					41. High humidity					48. Odors					55. Toxic conditions				
35. Outside					42. Low humidity					49. Body injuries					56. Infections				
36. High elevations					43. Wetness					50. Burns					57. Dust				
37. Cramped body positions					44. Air pressure					51. Electrical hazards					58. Silica dust				
38. High temperature					45. Noise					52. Explosives					59. Moving objects				
39. Low temperature					46. Vibration					53. Slippery surfaces					60. Working with others				
40. Sudden temperature changes					47. Oily					54. Radiant energy									

III. ACCEPTABLE DISABILITIES   *Check appropriate square if acceptable*
A - Amputation   D - Disability   Y - Yes   N - No

Hands - Fingers	A	D	Arms	A	D	Legs - Feet	A	D	Body - Trunk	D
1 or 2 on primary hand			1 Arm			1 Leg (high)			1 Hip	
1 or 2 on secondary hand			2 Arms			2 Legs (high)			2 Hips	
More than 2 on primary hand			None ☐			1 Leg (low)			1 Shoulder	
More than 2 on secondary hand						2 Legs (low)			2 Shoulders	
1 Hand						1 Foot			Back	
2 Hands						2 Feet			None ☐	
None ☐						None ☐				

Eyes	Y	N	Ears	Y	N	Cardio - Vascular	Y	N	Tuberculosis	Y	N
Blind			Deaf			Moderate tension			Minimal (healed, stable or arrested)		
Industrially blind			Hard of hearing, 1 ear			High tension			Moderate (healed, stable or arrested)		
Blind one eye			Hard of hearing, 2 ears			Organic heart disease compensated			Far advanced (healed, stable or arrested)		
Color blind			Hearing aid acceptable						Collapse therapy		
Color blind for shades											

Figure 23-6. This form is used when analyzing jobs for fitness requirements.

Thus, the medical report is more meaningful to the placement manager because the examining physician is responsible for determining the occupational significance of physical disorders. This method, in turn, makes proper analysis of each job essential.

Job Appraisal/Job Description

Employers must know the physical requirements of jobs and the unintentional injury and health hazards involved in each one. Each job-appraisal factor can make the position unsuitable or potentially undesirable for employees with one or more types of disability. The factors to be considered in job appraisal are physical requirements, working conditions, health hazards, and injury hazards.

Physical requirements include agility, strength, exertion, vision, hearing, talking, sitting, standing, walking, running, climbing, crawling, kneeling, squatting, stooping, twisting, lifting, and handling. They should be evaluated according to quality of ability and duration of activity. For example, a job involving a considerable amount of stair climbing is unsuitable for workers with heart disease, respiratory diseases, obesity, or lower limb orthopedic disorders. Some of these people can tolerate only a small amount of stair climbing.

Both indoor and outdoor working conditions can include excessive heat or cold; excessive wetness or dryness; sudden temperature changes; and ventilation, lighting, and noise problems. Also consider whether the work is performed alone, near others, with others, or as shift work or piecework. Some of these conditions could be harmful for individuals with certain disabilities. For example, work in excessive heat is generally unsuitable for persons who have had malaria; for those with high blood pressure, heart disease, or skin disease; and for older or obese workers.

Health hazards include air pressure extremes; radiant energy (ultraviolet, infrared, radium emanations, and x-rays); silica, ETS, asbestos, dusts, and skin irritants; respiratory irritants; systemic poisons; and asphyxiants. These hazards have serious effects and can aggravate a preexisting disorder. For example, a job might involve exposure to respiratory irritants of insignificant quantities to most people, yet this condition might aggravate the disability of a person who has chronic bronchitis.

The job description needs to spell out how much lifting, how much standing, how much vision is required to do the job, and so forth. Specific details will be of benefit to the employer and the applicant. (These issues were discussed in the previous section, Job Placement.)

Hazards include danger of falls from elevations; working while on moving surfaces; slipping and tripping hazards; exposure to vehicles or moving objects; objects falling from overhead; exposure to sources of foot injuries, eye injuries, cuts, abrasions, bruises, and burns; mechanical and electrical hazards; and fire and explosion hazards. These hazards could present greater dangers to workers with particular disabilities. For example, a job that may involve foot injury hazards is unsuitable for the diabetic because of his or her impaired circulation, which reduces sensation in the extremities, slows the healing of wounds and fractures, and increases susceptibility to gangrene of the foot.

Preemployment Medical Evaluation

Traditional preemployment medical examinations are prohibited under the ADA. Examinations can only be administered after an offer of employment is made and when all employees take the examination. The examinations must be restructured to comply with the ADA.

Post-Offer Preemployment Medical Examinations

These medical examinations may be allowed after an offer has been made. They also are made to determine the nature of any accommodations that may be needed.

Drug testing is not considered a medical examination and may be required before offering employment. Other regulatory tests may also be required.

ACCESS TO FACILITIES

One of a company's primary safety considerations is providing a safe and accessible parking lot for disabled employees. If people cannot find a safe place to park in the lot, they will not make it to the building. Parking spaces need to be wider than normal; details on requirements can be obtained from architects or websites dealing with disabilities. Both a painted symbol on the space itself and a blue-and-white sign above it should mark the parking space. The company must make sure that nondisabled employees, visitors, and others do not use these spaces. This may require rigorous enforcement.

People with mobility impairments are not the only ones who need such spaces. Employees with severe heart and lung disease, arthritis, or other chronic conditions that restrict movement will also need specially marked spaces. Hearing-impaired people, although able to walk, are at great risk walking through a parking lot and should be given special parking considerations. Finally, assistants of seriously disabled employees will need additional space in which to maneuver wheelchairs, walkers, and other equipment.

In some instances, employees may have new wheelchairs that can climb curbs and steps without assistance. However, this does not mean that curbs should not be eliminated or that some wheelchair users will not require assistance.

Safety considerations for hiring a disabled person begin with the first step in the employment process—and on the first day of work for that employee. For example, is there reasonable, safe access to the reception area, applicant-processing area, or workplace for the new employee? Curbs and stairs present barriers to those who use canes, crutches, walkers, and wheelchairs and also increase their chances of falling. A wheelchair user cannot safely negotiate even one step without assistance, which can cause a slight risk to both the wheelchair user and the person assisting him or her. Wheelchair ramps are discussed under General Access later in this section.

Can disabled job applicants safely proceed to where they must go to complete an application? If already employed, can such individuals safely proceed to their workstations? Access and safety are interdependent factors that need to be reconciled when employing these people. A lack of access to the company premises has been the principal factor preventing disabled persons from seeking jobs with some organizations.

General Access

In designing access for disabled workers, do not overlook cafeteria, washroom, and restroom facilities; width of doors; and height of plumbing fixtures, electrical controls, phones, and drinking fountains. Facilities designed to be truly accessible to disabled employees enhance their feelings of dignity and independence—which, in turn, can raise the morale of the entire work force.

With minimum expense, improvements and special considerations made for workers with disabilities also will benefit other employees. For instance:
- Wheelchair ramps are safer than steps—for everyone. However, the slope should be correct, not too steep and without sharp turns, to make negotiating the ramp safe and easy for wheelchair users. Ramp surfaces should be slip-resistant and kept free of obstacles. All ramps should be cleaned of mud, snow, and ice, and railings must be in place where needed.
- Clean and unobstructed aisles are necessary for safe wheelchair, cane, and crutch use. They also make the workplace safer for all employees and visitors. Good housekeeping improves traffic patterns and eliminates hazards.

Access to Buildings

Parking spaces 8 ft wide, next to a 5-ft-wide access aisle, should be reserved for automobiles driven by disabled personnel and visitors (Figure 23–7a). Two accessible parking spaces, however, can share a common access aisle (Figure 23–7b). Workers need room to remove and replace their wheelchairs in an auto and also to ride the chairs between aisles.

Parking spaces for disabled persons should be marked with an upright marker; symbols painted on the ground are often difficult to see or can be covered by snow or debris. At least one accessible route and entrance to the building must be provided. The entrance width should be 32 in. (80 cm); if a ramp leads to the entrance, it should be at least 36 in. (90 cm) wide. The ramp should be a maximum of 30 ft (9 m) long with an open, level area of at least 5 ft (1.5 m) at the bottom. Employees (and visitors) using wheelchairs, crutches, or canes can then move in and out of the building completely on their own. Protective side rails must be at least 36 in. (90 cm) apart if not adjacent to a wall to prevent persons in wheelchairs from running or falling off the side.

Revolving doors are inaccessible for anyone in a wheelchair and for most people using crutches, wearing a leg cast, or even carrying bulky packages. These people need entry doors that preferably open automatically. When a doorknob is necessary, it should be 36 in. (90 cm) from the floor. It is better, however, to have a vertical grab handle on the door.

The space requirements of the average wheelchair are as follows:
- Most wheelchairs are 36 in. high, 26 in. wide, and 42 in. long (Figure 23–8a). They require at least 60 in. to make a 180- or 360-degree turn. However, 78 in. is preferred to make a smoother U-turn.
- All access aisles should be wide enough to allow a wheelchair user to make a smooth turn. *Accessibility Standards Illustrated,* published by the State of Illinois, contains illustrated information on maneuvering space requirements for wheelchairs (Capital Development Board 1985).
- The average arm reach of people who use wheelchairs is usually 48 in. on the diagonal and 64 in. to the side. The average reach directly upward is 60 in. (Figure 23–8b). The usual maximum downward reach from the chair is 10 in. (Figure 23–8b). Shorter distances may be needed, however, to accomplish certain tasks (Figure 23–8c).

Interior Access

Both entry and interior doors should be a minimum of 32 in. (80 cm) wide. Interior doors should open by a single effort and have thresholds as level with the floor as possible. To make a 180- or 360-degree turn, persons in wheelchairs need an open area of 60 × 60 in. (1.5 m^2) in a typical building corridor.

Restrooms should have at least one stall wide enough for wheelchair entry. The stall should be equipped with grab bars and other fixtures no higher than 36 in. (90 cm) above floor level. The grab bars should be at 33 in. (83 cm). Stall doors should be at least 32 in. (80 cm) wide. Controls, switches, fire alarms, and other devices that might be used by a disabled individual must be within convenient reach of a wheelchair user. Restrooms should be located on each of the floors where disabled persons work.

Office Accommodations

Desktops should be no less than 28 in. (70 cm) above the floor to accommodate wheelchairs. Metal desks usually have adjustable feet that workers can raise to the maximum. If more room is needed, the desk can be raised with additional blocks.

Chairs can be regular height, but they should be sturdy and have arms to help disabled people lift themselves up. Some individuals need a chair that will not move easily so they can stand without the chair sliding out from under them. For other disabled workers, however, casters placed on the bottom of chairs may be desirable to help them slide the chair in and out. In addition, they allow a worker to pull him- or herself from one piece of furniture to another,

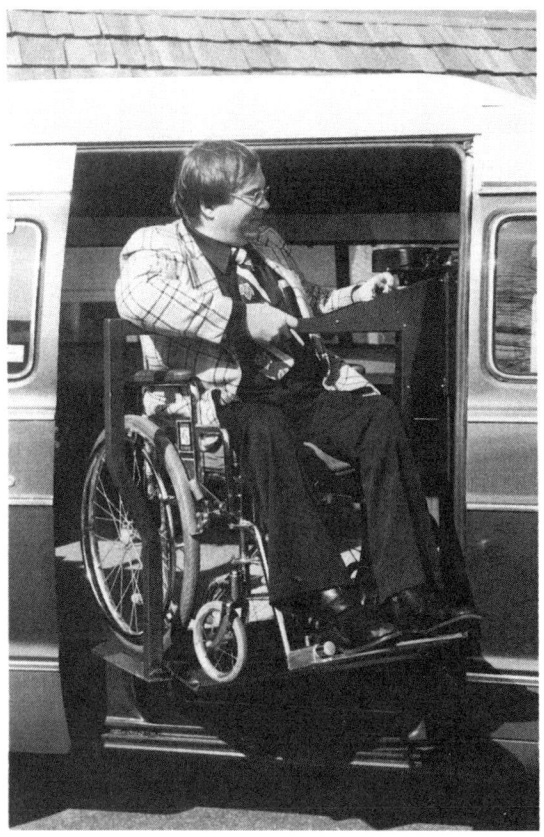 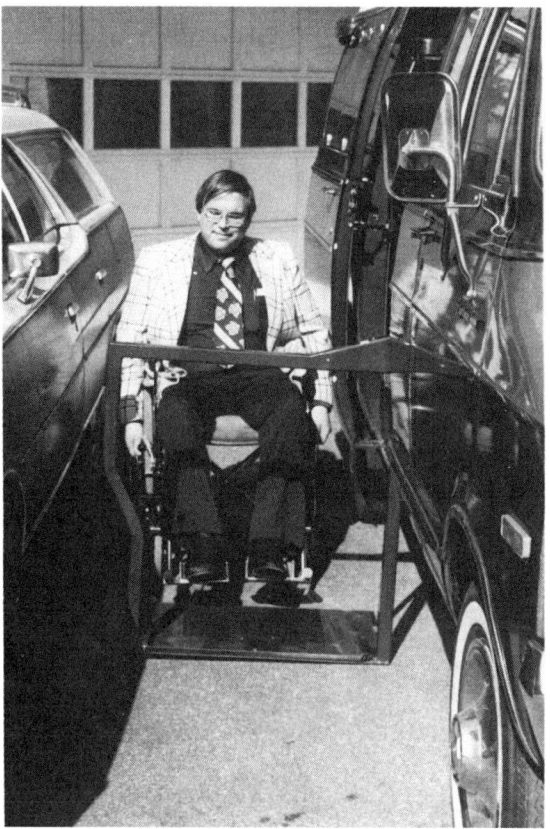

Figure 23–7a. Access aisle adjacent to a normal-width parking space is required for wheelchair clearance. Shown here is a lift that swings alongside a van. For return, a locking, outside control box opens sliding doors and allows users to control the lift. *(Courtesy ABC Enterprises Inc.)*

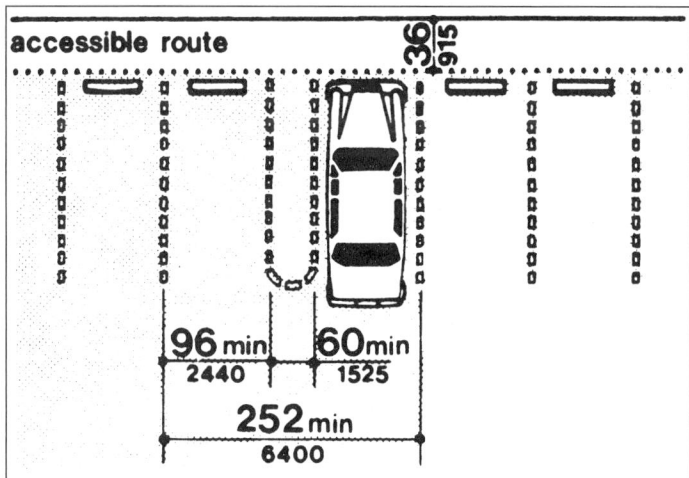

Figure 23–7b. Parking spaces for disabled persons can share a common access aisle. Aisle should be part of the accessible route to the building entrance. Overhangs from parked vehicles must not reduce the clear width of the accessible route, which must be the shortest possible distance to the entrance. *(Printed with permission from American National Standards Institute.)*

thereby avoiding the need to continually move in and out of the chair.

Business machines should be placed, if possible, where they will not become a barrier or obstruct traffic. This precaution is particularly important for wheelchair users. The following accommodations may sometimes apply, depending on the individual worker and the job requirements.

- File cabinets should be placed so that workers can reach drawers from both the front and side. This eliminates awkward reaches by those who must use crutches, a cane, or a wheelchair.
- If books, reports, or other bulky objects must be carried from place to place, make sure a wheeled cart is kept handy so workers can load materials on the cart.

23 Workers with Disabilities

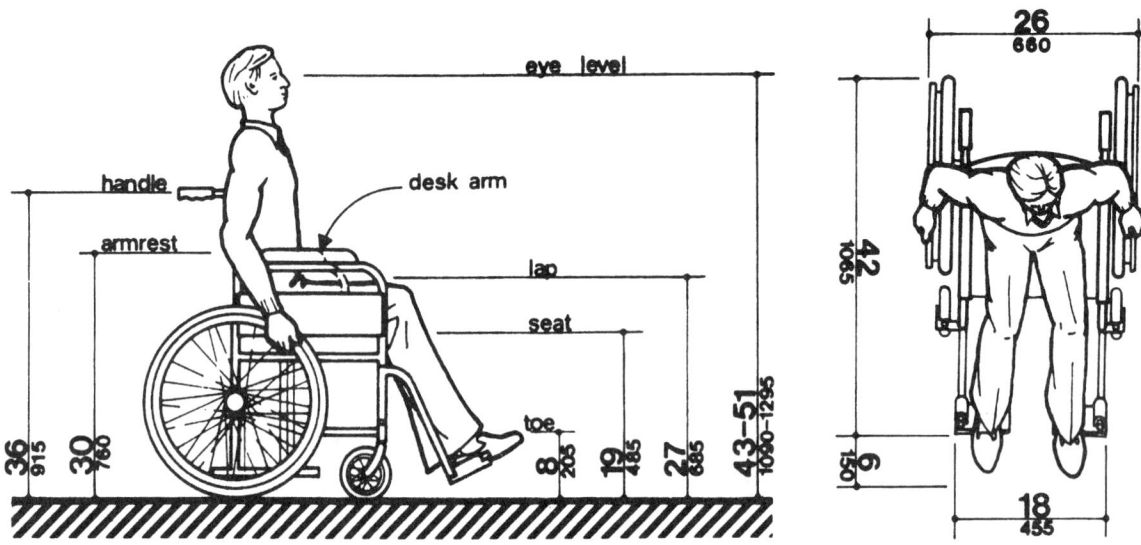

Figure 23–8a. Dimensions of adult-sized wheelchairs. Footrests can extend farther for very large people. Dimensions in this figure and in Figures 23–8b and 23–8c are in both inches and millimeters. *(Printed with permission from American National Standards Institute.)*

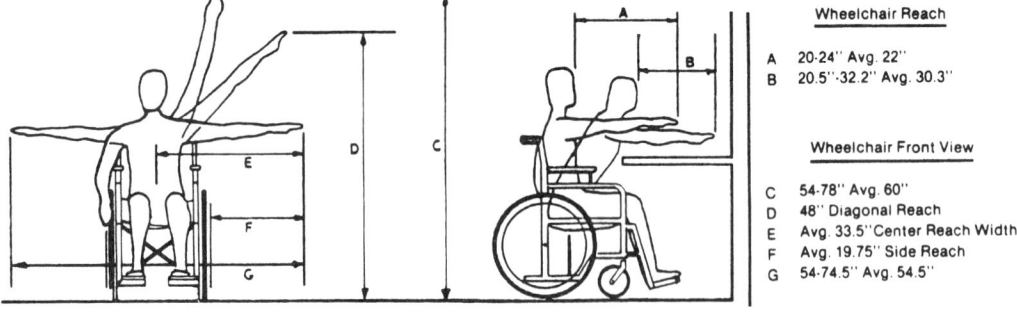

Figure 23–8b. Maximum reach from wheelchair—left: to sides; right: to front. (To convert to millimeters: 1 in. = 25.4 mm.) *(Printed with permission from State Board of Barrier-Free Design, Columbia, SC.)*

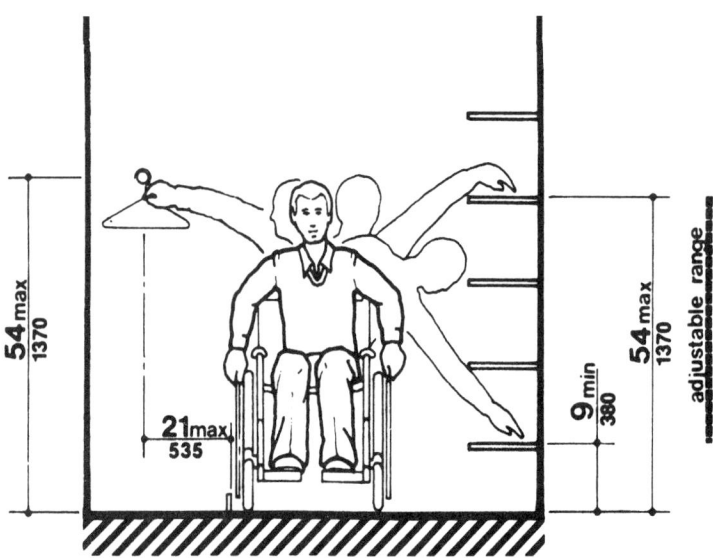

Figure 23–8c. Shorter distances are required when tasks are done. Shown here are the suggested dimensions for storage shelves and clothes racks. *(Printed with permission from American National Standards Institute.)*

- Cords for Venetian blinds, window shades, and draperies should be long enough that disabled workers can reach them easily. Floors must be free of extension cords, raised box receptacles, or any other raised items.

If a disabled employee has an assistive dog, then space, exercise facilities, and water for the dog will have to be arranged. The employee will know exactly what is needed. Other employees must understand that the dog is a working animal; they should not pet or feed it without express consent from the owner. Otherwise, the dog can become confused about its role and stop being dependable as the eyes or ears of the disabled employee.

EMERGENCY PROCEDURES

One concern within the facility is adequate means of escape for all workers in an emergency. This safety requirement frequently restricts or denies disabled persons the freedom to use premises as they would wish. However, many establishments employing workers with mobility disabilities have successfully devised safe evacuation plans.

Safety management of the disabled in many firms is well established and provides a freedom of movement that is entirely compatible with principles of general safety. These measures include supervised use of the stairs for means of escape (discussed later), designated staff to assist in an emergency, strategic ramping of entrances and exits, and alarm systems suitable for vision- and hearing-impaired workers.

Employees with impaired hearing and/or vision may need additional devices to perceive an alarm. Some people with impaired hearing can hear an alarm in a lower pitch than normal; others do best with a buzzer or a device that vibrates against their skin. These employees need to be interviewed separately and a workable notification system developed. Often, all that is necessary is a "buddy" system in which a worker who can see or hear the alarm is assigned to notify the disabled employee whenever there is an emergency or drill.

One means of safely evacuating wheelchair users and permanently or temporarily disabled persons is through use of an evacuation chair (Figure 23–9). This chair is designed to ride on the ends of stair treads so one person can easily guide the disabled worker down fire stairs without putting either person in additional jeopardy during an emergency evacuation. The evacuation chair is lightweight, folds flat, and can be safely and easily stored on a wall bracket.

Without these measures, many disabled persons would be denied access to their places of employment. It is suggested that employers discuss their safety problems with fire prevention specialists and check regulatory requirements.

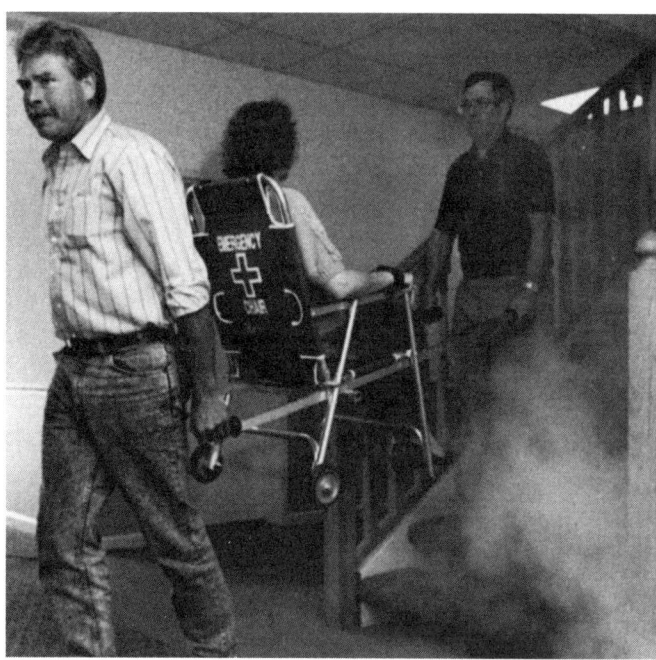

Figure 23–9. Using an evacuation chair, these workers safely transport a disabled worker down fire stairs without putting themselves or the disabled worker under additional risk during an emergency. *(Courtesy Safety Chair Co., Benicia, CA.)*

There are many possible ways to enable disabled people to escape safely in an emergency.

SUMMARY

- The Americans with Disabilities Act of 1990 mandates that employers provide "reasonable accommodation" for workers with disabilities when necessary. Supervisors and other management personnel must be trained to accommodate this group of workers.
- Employers who deny disabled workers a job must prove that these workers would endanger themselves or others, cannot meet job requirements, or reasonable accommodation of the workplace or job cannot be made.
- Various federal, state, and local laws also mandate affirmative action programs for the hiring and advancement of disabled persons. The law defines three types of disabled job seekers: disabled individual, disabled veteran, and qualified disabled individual.
- "Reasonable accommodation" to the known physical or mental limitations of an otherwise qualified disabled applicant or employee can include modifying equipment or changing job descriptions.
- The health and safety professional can work with the EEO coordinator to evaluate reasonable accommodation and

ensure compliance with government affirmative action regulations.
- Management must evaluate each disabled job applicant to place the right person in the right job so the impairment is not a factor of job performance. More stringent standards may need to be developed to ensure safe working conditions for these employees.
- Employers must evaluate jobs in terms of physical requirements, working conditions, health hazards, and injury hazards to eliminate or control job risks that might endanger disabled workers.
- Companies can also contact self-help and support groups, governmental agencies, and private agencies to learn more about how to accommodate a job for a disabled employee and how to work with disabled persons.

SOURCES OF HELP

The overall goal of all employers should be to hire qualified disabled individuals and place them in available and safe occupations. One of the goals of the safety and health professional is to assist the employer in this worthwhile endeavor. The following list includes government and private agencies that can help achieve this goal.

Alabama Institute for the Deaf and Blind, Talladega, AL 35160.

American Council for the Blind, 1115 Fifteenth Street NW, Suite 720, Washington DC 20005; 202-467-5081 or 800-424-8666.

Architectural and Transportation Barriers Compliance Board, 1111 18th Street NW, Suite 501, Washington DC 20036; 800-USA-ABLE (voice) or 800-USA-ABLE (TDD).

Arthritis Foundation, 1314 Spring Street NW, Atlanta, GA 30309; 404-872-7100.

Association for Children and Adults with Learning Disabilities, 4156 Library Road, Pittsburgh, PA 15234; 412-341-1515.

Asthma and Allergy Foundation of America, 1717 Massachusetts Avenue NW, Suite 305, Washington DC 20036; 202-265-0265.

Cystic Fibrosis Foundation, 6931 Arlington Road, Suite 200, Bethesda, MD 20814; 800-344-4823; 301-951-4422.

Department of Justice, Office on the Americans with Disabilities Act, Civil Rights Division, P.O. Box 66228, Washington DC 20035-6118; 202-514-0301 (voice); 202-514-0381 (TDD); 202-514-6193 (electronic bulletin board).

Department of Transportation, 400 Seventh Street SW, Suite 501, Washington DC 20590; 202-366-9305 (voice); 202-755-7687 (TDD).

Easter Seal Society, 70 East Lake Street, Chicago, IL 60601; 312-726-6200.

Epilepsy Foundation of America, 4351 Garden City Drive, Landover, MD 20785; 800-332-1000; 301-459-3700 (voice); 301-577-2684 (fax).

Equal Employment Opportunity Commission, 1801 L Street NW, Washington DC 20507; 202-663-4900 (voice); 800-800-3302 (TDD); 202-663-4494 (TDD).

Federal Communications Commission, 1919 M Street NW, Washington DC 20554; 202-632-7260 (voice); 202-632-6999 (TDD).

George Washington University, Job Development Laboratory, Rehabilitation Research and Training Center, 2300 I Street NW, Washington DC 20052.

Internal Revenue Service, Department of the Treasury, 1111 Constitution Avenue NW, Washington DC 20044; 202-566-2000.

Job Accommodation Network, West Virginia University, 809 Allen Hall, P.O. Box 6123, Morgantown, WV 26506-6123; 800-526-7234 (U.S. other than WV—voice and TDD); 800-526-4698 (WV—voice and TDD); 800-526-2262 (Canada—voice and TDD).

Lupus Foundation of America, 1717 Massachusetts Avenue NW, Suite 203, Washington DC 20036; 800-558-0121; 202-328-4550.

Mainstream Inc., 1200 15th Street NW, Washington DC 20005.

Multiple Sclerosis Society, 205 East 42nd Street, New York, NY 10017; 800-624-8236; 212-986-3240.

Muscular Dystrophy Association, 801 Seventh Avenue, New York, NY 10019; 212-586-0808.

Myasthenia Gravis Foundation Inc., 53 West Jackson Boulevard, Suite 1352, Chicago, IL 60604; 800-541-5454; 312-427-6252.

National Association of the Deaf, 814 Thayer Avenue, Silver Spring, MD 20910-4500; 301-587-1785 (voice); 301-587-1789 (TDD).

National Association of Protection and Advocacy Systems, Client Assistance Program, 220 I Street NW, Suite 150, Washington DC 20002; 202-546-8202.

National Association for Sickle Cell Disease, 4221 Wilshire Boulevard, Suite 360, Los Angeles, CA 90010; 213-936-7205.

National Council on Disability, 800 Independence Avenue SW, Suite 814, Washington DC 20591; 202-267-3846 (voice); 202-267-3232 (TDD); 202-453-4250 (fax).

National Council of Persons with Disabilities, P.O. Box

29113, Washington DC 20017; 202-529-2933.

National Down Syndrome Society, 666 Broadway, New York, NY 10012; 800-221-4602; 212-460-9330.

National Head Injury Foundation, 333 Turnpike Road, Southborough, MA 01772; 800-444-NHIF; 508-485-9950.

The National Institute for Rehabilitation Engineering, 97 Decker Road, Butler, NJ 07405.

National Kidney Foundation, 30 East 33rd Street, New York, NY 10016; 800-622-9010; 212-889-2210.

National Organization on Disability, 910 16th Street NW, Suite 600, Washington DC 20006; 202-293-5960 (voice); 202-293-5968 (TDD); 202-293-7999 (fax).

National Spinal Cord Injury Association, 600 West Cummings Park, Suite 2000, Woburn, MA 01801; 800-962-9629; 617-935-2722.

Paralyzed Veterans of America, 801 18th Street NW, Washington DC 20006; 202-872-1300.

President's Committee on Employment of People with Disabilities, 1331 F Street NW, Washington DC 20004-1107; 202-376-6200 (voice); 202-376-6205 (TDD); 202-376-6219 (fax).

The Rehabilitation Institute of Chicago, 345 East Superior Street, Chicago, IL 60611.

Retinitis Pigmentosa Foundation Fighting Blindness, 1401 Mt. Royal Avenue, Baltimore, MD 21217-4245; 800-638-2300.

Self Help for Hard of Hearing People, 7800 Wisconsin Avenue, Bethesda, MD 20814; 301-657-2248.

Spina Bifida Association of America, 1700 Rockville Pike, Suite 540, Rockville, MD 20852; 800-621-3141; 301-770-SBAA.

Support Dogs for the Handicapped, P.O. Box 966, St. Louis, MO 63044; 314-487-2004.

REFERENCES

ADA Compliance Guide, Thompson Publishing Group, 1725 K Street NW, Suite 200, Washington DC 20006.

American National Standards Institute, 11 West 42nd Street, New York, NY 10036.

BNA's Americans with Disabilities Act Manual, The Bureau of National Affairs Inc., 1231 25th Street NW, Washington DC 20037.

Capital Development Board. *Accessibility Standards Illustrated.* 1978. Reprinted edition with all revisions to March 1, 1985, and Environmental Barriers Act, Public Act 84–948. Springfield: State of Illinois, 1985.

Epstein Becker & Green, PC [Law firm]. *What You Absolutely Must Know about the Americans with Disabilities Act.*

The Gallaudet Survival Guide to Signing. Washington DC: Gallaudet University Press.

Hill, N., et al. "Hiring the Handicapped: Overcoming Physical & Psychological Barriers in the Job Market." *Journal of American Insurance* (Third quarter, 1986.)

———. "The Merits of Hiring Disabled Persons." *Business & Health* (February 1987).

Mental and Physical Disability Law Reporter. American Bar Association, Commission on Mental and Physical Disability Law, 1800 M Street NW, Washington DC 20036-5886.

President's Committee on Employment of the Handicapped. "Affirmative Action to Employ Handicapped People—A Pocket Guide."

"Supervising Handicapped Employees."

The Registry of Interpreters for the Deaf Inc., RID Publications, 8719 Colesville Road, Suite 310, Silver Spring, MD 20910-3919.

U.S. Department of Health and Human Services. "Nondiscrimination on the Basis of Handicap in Programs and Activities Receiving or Benefiting from Federal Financial Assistance" (45 CFR 84).

U.S. Department of Labor.
  Occupational Safety and Health Act of 1970 (Public Law No. 91–596).
  Occupational Safety and Health Act Regulations, Title 29, CFR, Chapter XVII, Part 1910.
  Rehabilitation Act of 1973 (Public Law No. 93–112).
  Rehabilitation Act Amendments of 1974 (Public Law No. 93–516).
  Vietnam Era Veterans' Readjustment Assistance Act of 1974 (Public Law No. 93–508), Section 38 USC 2012.

U.S. Department of Labor, Employment Standards Administration, Office of Federal Contract Compliance Programs. "Affirmative Action Obligations of Contractors and Subcontractors for Handicapped Workers" (41 CFR 60–741).

West, J., ed. *The Americans with Disabilities Act—From Policy to Practice.* New York: Milbank Memorial Fund.

Woodward, R. E. *Comprehensive Barrier-Free Design Standard Manual.* Columbia, SC: State Board for Barrier-Free Design, 1979.

———. "Industry Unlocks Its Doors to the Handicapped." *Plant Facilities* 2, no. 2 (February 1979).

REVIEW QUESTIONS

1. The 1990 U.S. Americans with Disabilities Act (ADA) encompasses what five general areas?
   a.
   b.
   c.
   d.
   e.
2. If an employer denies a disabled individual a specific job, management must prove that the person is unqualified because of what four reasons?
   a.
   b.
   c.
   d.
3. What three occurrences in the 1940s stimulated hire-the-disabled programs?
   a.
   b.
   c.
4. List the three types of disabled persons seeking employment as defined by law.
   a.
   b.
   c.
5. Examples of "reasonable accommodation" include:
   a. making employee facilities readily accessible to and usable by disabled persons.
   b. restructuring a job, modifying work schedules, and modifying equipment or devices.
   c. providing readers or interpreters.
   d. completely redesigning or altering the circuitry or operating levers of a machine.
   e. all of the above
   f. only a, b, and c
6. Explain four general responsibilities of the safety and health professional in relation to disabled employees.
   a.
   b.
   c.
   d.
7. The safety evaluation form for disabled employees should be kept for at least _____ after the employee leaves the company.
8. What are the four factors employers must consider when appraising a job for employees with one or more disabilities?
   a.
   b.
   c.
   d.
9. What are six important factors to consider when designing access to facilities for disabled workers?
   a.
   b.
   c.
   d.
   e.
   f.
10. To accommodate disabled persons, the entrance width to a building should be:
    a. 28 in. (70 cm)
    b. 32 in. (80 cm)
    c. 36 in (90 cm)
    d. none of the above
11. Desktops should be no less than how many inches above the floor to accommodate wheelchairs?
    a. 28 in. (70 cm)
    b. 32 in. (80 cm)
    c. 36 in. (90 cm)
12. One means of safely evacuating wheelchair users and permanently or temporarily disabled persons is through the use of a(n) _____ .

Retail/Service Facilities Logistics

24

Wendy Key, BS, MS, CS
Nic Valter, JD, MBA
Richard Payant, DBA, MA, CFM, CPE, CHS

Importance of an Occupational Health and Safety Management Program
Management Leadership and Employee Participation ▸ Planning, Risk Assessment, and Prioritization ▸ Implementation and Operation ▸ Evaluation and Corrective Action ▸ Management Review ▸ Corporate Safety Culture

OSHA Regulations
General Duty Clause ▸ Posting Requirements ▸ Reporting and Record Keeping ▸ Means of Egress ▸ Hazard Communication ▸ Medical Services and First Aid ▸ Walking and Working Surfaces ▸ Lockout/Tagout ▸ Electrical Installations and Equipment ▸ Machine Guarding ▸ Powered Equipment

General Liability Issues
Hiring and Placement ▸ Risk Control ▸ Audit and Inspection Program ▸ Materials Handling ▸ Fleet Safety Programs ▸ Ergonomics ▸ Emergency Preparedness ▸ Violence and Crime in the Workplace

Summary

References

Review Questions

This chapter provides an overview of the safety management system that should be established for employees in retail, service, and warehouse facilities and the special safety-related issues confronting employers in these facilities. For more detailed discussions of each topic, see the specific chapters referenced within each section. Chapter 28, Contractor and Customer Safety, discusses issues relating to the safety of outside contractors, vendors, and customers. Topics covered in this chapter include:
- establishing an occupational health and safety management system consistent with ANSI Z10–2012 or other safety management systems
- OSHA regulations relevant to service and retail industries
- ways to limit liability.

An effective safety management system that is incorporated into the operations of an organization not only reduces health risks, injuries, and accidents, but substantially increases the quality of services offered and ensures better management of products and services. Effective safety management will lower the overall costs of running a business.

IMPORTANCE OF AN OCCUPATIONAL HEALTH AND SAFETY MANAGEMENT PROGRAM

Service and retail facilities invite the public into their facilities and can have a high exposure of risk of injury to both customers and employees. The direct cost of unintentional injuries is only attributable to a small part of the total expenditures (insurable cost), while indirect costs can range between 5 to 20 times the direct costs, according to the National Safety Council. No organization can efficiently sustain a business with a high volume of injury losses. Since all organizations must pass the cost of such losses to their customers, and these losses cause services and products to become more expensive, their customers are negatively impacted.

One key component of a company optimizing income while minimizing expenditures is for the company to establish an effective and comprehensive occupational health and safety management system. One example of a widely used management system is the ISO series Continual Improvement Model, "Plan, Do, Check, Act." The ISO Continual Improvement Model includes the following elements:
- management leadership and employee participation
- planning, risk assessment, and prioritization
- implementation and operation
- evaluation and corrective action
- management review.

Management Leadership and Employee Participation

Senior management should provide leadership for safety activities and assume overall responsibility for the safety management process. Managers at every level of the company need to instill the importance of safety management and create a culture that embraces safety.

Management Leadership—Policy Statement

Each company should have a clear statement of policy regarding its position on safety, health, and environmental matters. This policy should be posted or accessible for all employees to review. It should identify those responsible for safety matters and the steps employees should take if they have questions and/or concerns regarding any environmental or safety matter.

Employee Participation—Safety Committee

The company should establish a safety committee to help define safe working conditions. The committee's focus should be to listen to employee concerns, monitor operating conditions, and serve as a communications liaison between management and employees. This committee should also have the authority to enforce the safety policy with the cooperation of management. Management must be actively involved with the committee and must meet with it regularly.

Planning, Risk Assessment, and Prioritization

A certain amount of risk is inherent in any facility that accepts the public during its normal course of operation. Inviting the public into a facility during rapidly changing weather conditions, for example, increases the probability of an accident or injury. Although an organization does not have control over certain variables, such as the physical condition of customers or the severe weather, control measures can be implemented to reduce this risk. It is difficult to effectively manage risk without first assessing and measuring it. After risk is assessed and measured, prioritizing risks can help a facility utilize its limited resources in the most efficient way. Resources must be allocated so that there is a good balance between the cost of loss prevention efforts and costs of potential losses.

Implementation and Operation

Employee Work Procedures

Each company should develop employee work procedures

that not only define the level of safety expected from employees but also provide specific instructions. An example would be a statement on safe practices for forklift operators, which might read as follows:

> Only trained and certified forklift drivers will drive any forklift on company property.

In this example, the policy should specify (1) safe work practices regarding the use of forklifts; (2) do's and don'ts regarding use of forklifts at the company; (3) maintenance and emergency procedures for forklift safety; and (4) disciplinary actions, with management enforcement, for employees in violation of the safety rules.

Training
Each company should conduct a general safety orientation and initial training for all employees, along with task-specific safety training. Workers in high-risk jobs (e.g., chemical handling, heavy lifting, or operation of heavy equipment) need specialized safety training relevant to the specific high-risk jobs. Training manuals or computer-based training can be used to ensure consistency across the company and within particular jobs. Effective training requirements and training manuals should be created with the input of employees who have a high level of knowledge and work experience within their areas of expertise. The safety manuals should be well written, concise, and easy to understand, so that safety policies and safe work practices may be easily taught to other employees. The company must document all training that is conducted. (For more details, see Chapter 31, Safety and Health Training.)

Evaluation and Corrective Action

Safety Audits
A critical part of any safety management system is safety auditing. A formal procedure should be established to systematically and comprehensively audit operating tasks within a company to identify areas needing improvement. The auditors should have expertise in the type of operation being audited (e.g., warehousing, trucking, retailing). Often, establishing a team composed of an independent outside auditor, a management auditor, and a facility manager works the best. (See Chapter 7, Safety, Health, and Environmental Auditing.)

Investigation and Action
A vital part of an effective safety management system is that it continues to improve over time and learn from mistakes and modify practices to reduce future exposures. Thus, companies should thoroughly investigate incidents, identify root causes, recommend changes, and take corrective actions. Employees should be encouraged to submit ideas to management if they have a new idea to improve the safety management system. Implemented ideas could be rewarded by an incentive program. Active employee involvement in these systems and investigations can reduce the chance of recurrence and strengthen corporate commitment to safety. (For more details on this topic, see Chapter 10, Incident Investigation, Analysis, and Costs.)

Management Review
The organization should establish and implement a process for top management to review the occupational health and safety management system at least annually and recommend improvements to ensure its continued suitability, adequacy, and effectiveness.

Corporate Safety Culture
Effective implementation of the steps previously described will greatly enhance employee awareness of safety in the workplace. This heightened awareness is the building block that creates the safety culture of a company. A company that possesses a high-level "safety culture" will reduce work-related losses. The service industry, like other industries, also must be concerned about government regulations that pertain to health and safety both in the workplace and in the community where the company operates. These regulations include Occupational Safety and Health Administration (OSHA) standards, state workers' compensation laws, Environmental Protection Agency (EPA) regulations, and Department of Transportation (DOT) regulations.

OSHA REGULATIONS

Several OSHA regulations address specific safety issues within the service industry. The major relevant OSHA regulations include the following:

29 CFR 1910—Occupational Safety and Health Standards
- Subpart C—General Safety and Health Standards
- Subpart D—Walking-Working Surfaces
- Subpart E—Means of Egress
- Subpart F—Powered Platforms, Manlifts, and Vehicle-Mounted Work Platforms
- Subpart G—Occupational Health and Environmental Control
- Subpart H—Hazardous Materials
- Subpart I—Personal Protective Equipment
- Subpart J—General Environmental Controls

- Subpart K—Medical and First Aid
- Subpart N—Materials Handling and Storage
- Subpart O—Machinery and Machine Guarding
- Subpart P—Hand and Portable Powered Tools and Other Hand-Held Equipment
- Subpart R—Special Industries
- Subpart S—Electrical.

Companies should comply with these regulations and make sure that their policies, written procedures, and safety notices conform to these requirements.

General Duty Clause

When a particular hazard is not addressed by a specific OSHA standard, the General Duty Clause may be applicable. The General Duty Clause, section 5(a)(1) of the Occupational Safety and Health Act of 1970, applies to all employers and requires each employer to provide employees with a workplace that is free of recognized hazards that may cause death or serious physical harm. The general duty provisions are used only when there is no specific standard available that applies to the particular hazard involved. In general, OSHA case law has established that the following elements are necessary to prove a violation of the General Duty Clause:

- The employer failed to keep the workplace free of a hazard to which employees of that employer were exposed (exposure threatens physical harm to employees).
- The hazard was recognized (by the employer's industry; where there is evidence of actual employer knowledge; or in flagrant cases, when any reasonable person would have recognized the hazard).
- The hazard was causing or was likely to cause death or serious physical harm.
- There was a feasible and useful method to correct the hazard. (The information should indicate that the recognized hazard, rather than a particular incident, is preventable.)

The General Duty Clause would not be used for nonserious citations but is limited to alleged serious violations. OSHA may use this clause to cite a company in the service industry for the following:

- Cumulative-trauma disorders such as carpal tunnel and tendinitis would generally fall under the General Duty Clause because for most industries, there are no comprehensive, formally approved OSHA standards on ergonomics.
- A violation of the General Duty Clause relating to a serious training deficiency may be present if the training is not addressed under another general or specific training requirement.

In summary, employers have a duty to take necessary action to protect workers from known hazards. This duty extends beyond the hazards addressed by existing written standards and includes hazards that are associated with recognized employee exposure and may or may not be included in current OSHA safety and health regulations. Failure to adhere to this duty may result in OSHA citations and/or fines under the General Duty Clause. OSHA has ergonomic guidelines for retail grocery stores on its website (osha.gov/SLTC/ergonomics/guidelines.html#retail), as well as several useful tools such as the Grocery Warehousing eTool (osha.gov/SLTC/etools/grocerywarehousing/index.html).

Posting Requirements

Each employer must post a notice informing employees of their rights under the OSH Act and the safety requirements and protection mandated by OSHA. The notice must be posted in each store or work area in a conspicuous place where other employee notices are posted. The notice can be obtained by contacting the nearest Department of Labor office.

Reporting and Record Keeping

Employers are required to prepare and maintain two forms for OSHA record keeping. OSHA Form 300 serves as the log and summary of occupational injuries and illnesses. Recordable injuries are placed in the log, and the summary of occupational injuries and illnesses is used at the end of each year. OSHA Form 301, Injury and Illness Incident Report, provides detailed information on each of the cases recorded on OSHA Form 300. (See also Chapter 4, Regulatory History, and Chapter 11, Injury and Illness Record Keeping, Incidence Rates, and Analysis.)

Means of Egress

The general requirements for means of egress from a building include regulations that provide for the following:

- exits sufficient to permit the prompt escape of occupants in case of fire or other emergency
- a building constructed, arranged, equipped, maintained, and operated to avoid undue danger to the lives and safety of its occupants
- exits of kinds, numbers, locations, and capacities appropriate to the individual building or structure and occupancy
- exits arranged and maintained to provide free and unobstructed egress from all parts of the building or structure at all times when it is occupied
- clearly visible exits that are indicated in such a conspicuous manner that every occupant will readily know the direction of escape from any point
- adequate and reliable illumination for all exits
- fire alarm facilities in every building or structure of such

size, arrangement, or occupancy that a fire may not itself provide adequate warning to occupants
- at least two means of egress remote from each other in every building or structure.

NFPA 101, Life Safety Code, is a good reference for proper exit capacity and arrangements. Compliance with these conditions does not eliminate or reduce the need for other provisions for the safety of people using a structure under normal occupancy conditions.

Maintenance and construction of exits are addressed by three requirements:
1. Doors, ramps, passages, signs, and all other components of the exit should be of substantial and reliable construction.
2. All exits should be continuously maintained free of all obstructions or impediments.
3. Any device or alarm installed to restrict the improper use of an exit should be so designed and installed that it cannot, even in cases of failure, impede or prevent emergency use of such exit. This prohibits the use of a lock and hasp on exit doors and sets requirements for push-bar alarm systems. Check with the local fire official for specific requirements within the jurisdiction.

Companies that handle large quantities of materials and have limited storage space for products and merchandise need to establish specific procedures on maintaining clear exits. Management and supervisors should inspect exits on a regular basis to ensure that they remain accessible at all times. Management needs to pay particular attention to proper maintenance of the components and hardware of exits, and to enforce strict rules prohibiting blocking exits with stored products and merchandise, equipment, mop buckets, hand trucks, or other obstacles.

Hazard Communication

The hazard communication standard (HAZCOM) is based on the concept that employees have a need and a right to know the identities and hazards of the chemicals to which they are exposed in the workplace (29 CFR 1910.1200). Employees also have the right to know what protective measures are available to prevent exposure to these chemicals. With proper information, employees have the responsibility to handle chemicals in a safe manner and use required personal protective equipment.

The main requirements of the hazard communication standard are as follows:
- a written hazard communication program
- labels and other forms of warning on containers of chemicals
- Safety Data Sheets (SDSs; formerly Material Safety Data Sheets [MSDSs]) available for review by employees (available electronically or in an SDS book within the facility)
- information and training provided for potentially exposed employees
- documented hazard communication training for all employees
- provisions for communicating with contractors on chemicals their employees may be exposed to in the facility and providing chemical information on chemicals brought into the facility by contractors for work conducted, such as repairs.

Medical Services and First Aid

The OSHA standard at 29 CFR 1910.151–153 (Subpart K) also requires that employers provide access to and information about medical services and first aid. The standard states the following:

a. The employer shall ensure ready availability of medical personnel for advice and consultation on matters of facility health.
b. In the absence of an infirmary, clinic, or hospital close to the workplace, which is used for the treatment of all injured employees, a person shall be adequately trained to render first aid. First-aid supplies approved by the consulting physician shall be readily available.
c. When the eyes or body of any person may be exposed to injurious corrosive materials, suitable facilities for quick drenching or flushing of the eyes and body shall be provided within the work area for immediate emergency use. Emergency showers and eyewash stations must be kept clear and accessible at all times and should be tested regularly.

Emergency service telephone numbers should be available to workers so they can obtain help quickly. Posting emergency phone numbers at company telephones is sufficient and emphasized in areas without 911 capability. All employees should be trained in emergency-response procedures to minimize the damage or severity of any injury or illness.

Walking and Working Surfaces

The OSHA standard at 29 CFR 1910.21–32 (Subpart D) also requires development of a comprehensive program to identify and address the physical hazards of a facility. The following elements are components of the program:
- *Housekeeping*—All places of employment, passageways, storerooms, and service rooms should be kept clean, orderly, and in a sanitary condition.
- *Floor loading protection*—Elevated floors in buildings

used for mercantile, business, industrial, or storage purposes should be posted to show maximum safe floor loads.
- *Stairway railings and guards*—Every flight of stairs with four or more risers should have a standard railing on all open sides. Handrails should be provided on at least one side of closed stairways, preferably on the descending right side.
- *Fixed stairways*—Fixed stairways should have a minimum width of 22 in. (56 cm). Fixed stairways should be provided for access from one elevation to another where operations necessitate regular travel between levels.
- *Dockboards (bridgeplates)*—Dockboards should be strong enough to carry the load imposed on them.
- *Portable ladders (stepladders and straight ladders)*—Stepladders should be equipped with a metal spreader or locking device of sufficient size and strength to securely hold the front and back sections in the open position.
- *Fixed ladders*—All rungs should have a minimum diameter of ¾ in. (1.9 cm) if metal, or 1⅛ in. (3.5 cm) if wood. They should be a minimum of 16 in. (41 cm) wide and should be spaced uniformly no more than 12 in. (30 cm) apart.
- *Aisles and passageways*—Where material-handling equipment is used, sufficient safe clearances should be allowed for aisles, at loading docks, through doorways, and wherever turns or passage must be made.
- *Floors, general condition*—All floor surfaces should be kept clean, dry, and free from protruding nails, splinters, loose boards, holes, or projections. Where wet processes are used, drainage must be maintained.
- *Open-sided floors*—Every open-sided floor, platform, or runway 4 ft (1.2 m) or more above the adjacent floor or ground level should be guarded by a standard railing with a toeboard on all open sides, except where there is an entrance to a ramp, stairway, or fixed ladder.
- *Railings*—A standard walking/working surface should consist of top rail, intermediate rail, posts, and toeboard. The footwalk should have a railing with the top edge height of the top rails, or equivalent guardrail system members, 42 in. (1.1 m) ± 3 in. (8 cm) above walking/working levels. Midrails should be installed at a height midway between the top edge of the guardrail system and the walking/working surface, and on the exposed side, toeboards should be 4 in. (10 cm) high. (See 29 CFR 1910.23 and 29 CFR 1926.502[b][1] and [b][2][i].)
- *Toeboards*—Railings protecting floor openings, platforms, scaffolds, and the like should be equipped with 4-in.- (10-cm-) high toeboards wherever persons can pass beneath the open side, when there is moving machinery, or when there is equipment from which falling material could cause a hazard.

Lockout/Tagout

The OSHA standard at 29 CFR 1910.147 helps safeguard employees from hazardous energy while they are servicing or performing maintenance on machines and equipment. According to the standard, employers must do the following:

- Develop a control program to prevent unintentional start-up of machinery or equipment being repaired or serviced.
- Use locks when equipment can be locked out.
- Ensure that new equipment and overhauled equipment can accommodate locks.
- When lockout procedures cannot be employed, tagout procedures should be in place. Tagout procedures require that all switches, valves, levers, and other similar devices be tagged to instruct all parties not to open or operate such controls until the tag is removed by the individual who placed it.
- Identify and implement specific procedures (generally in writing) for the control of hazardous energy, including preparation for shutdown, equipment isolation, lockout/tagout application, release of stored energy, and verification of isolation.
- Institute procedures for release of lockout/tagout, including machine inspection, notification, and safe positioning of employees and removal of the lockout/tagout device.
- Obtain standardized locks and tags that identify the employee using them. The locks and tags should be of sufficient quality and durability to ensure their effectiveness.
- Conduct inspections of energy control procedures at least annually.
- Train employees in the specific energy control procedures, and provide refresher training as part of the annual inspection of control procedures. Document all training.
- Adopt procedures to ensure safety when equipment must be tested during servicing, when outside contractors are working at the site, when a multiple lockout is needed for a crew servicing equipment, and when shifts or personnel change.

The following machines and equipment are covered by the OSHA lockout/tagout standard: dough mixers, meat saws, deli slicers, meat wrappers, meat grinders, balers, and compactors. Also included are carpentry and maintenance shop equipment such as radial saws, table saws, lathes, drilling machines, bench grinders and wire buffers, and similar equipment that could cause injury to a person during maintenance or servicing of such equipment if it could be energized by another employee or by stored energy. All persons using such equipment should be trained in their safe use and such training documented. Employees under 18 years old should not operate hazardous equipment. (See also Chapter 6, Safeguarding, in the *Engineering & Technology* volume.)

Electrical Installations and Equipment

According to OSHA citations, the electrical standards most frequently violated in the service industry are the following:

1. *General electrical standard* (29 CFR 1910.301–399, Subpart S)—This standard states that companies must maintain sufficient access and working space around electrical panels, service entrances, circuit breakers, and the like to permit ready and safe operation and maintenance. A good rule of thumb is to keep all storage and materials at least 3 ft (0.9 m) away from major electrical equipment.

   The standard also requires that electrical equipment operating at 50 volts or more (most appliances and equipment) be guarded against unintentional contact. This means that the covers, boxes, and cabinets provided with most equipment must be properly installed and in place at all times. Proper warning signs are required to be in place where exposed live parts are available, such as during maintenance and repair.

2. *Flexible cords and cables*—This standard is cited for improper use of extension cords. Long runs of extension cords running through walls and ceilings can be both fire and tripping hazards. Extension cords cannot be permanently attached or used for permanent wiring (no longer than 90 days).

3. *Cabinets, boxes, and fittings*—This standard offers guidelines for physically protecting electrical installations.

4. *Examination, installation, and use of equipment*—This section requires that electrical equipment be examined and found free of recognized hazards before use.

5. *Identification of disconnecting means and circuits*—This section requires that each means of disconnecting equipment and appliances (circuit breakers, etc.) be legibly marked to indicate its purpose.

Machine Guarding

A company must be sure to guard any machine part function or process that may cause injury. Where the operation of a machine or unintentional contact with it can injure the operator or others in the vicinity, the hazard must be either controlled or eliminated. Management should conduct routine audits to ensure that such guards are in place and being properly used by trained employees. Power tools or hand tools should never be loaned to contractors, clients, untrained employees, or other unauthorized persons.

Powered Equipment

The OSHA standard at 29 CFR 1910.241–247 (Subpart P) requires that all companies using powered industrial trucks and equipment establish a safety program that includes the following:

- General rules for safe operation have been developed and implemented.
- Only trained and authorized operators are permitted to operate powered industrial trucks and equipment.
- Employers must provide training for operators in safe work practices and equipment operation.

Although the practice is not regulated by OSHA, spotters should be used when power equipment is operated in areas when guests or customers are present.

GENERAL LIABILITY ISSUES

Each state has its own workers' compensation and general liability laws. These laws have considerable variability and must be consulted for specific guidelines because a business is responsible for compliance with applicable state rules and regulations. Companies in the service industry must follow the laws of the state(s) in which they operate. Not surprisingly, management commitment to compliance has increased in the past few years because the industry realizes that to maintain competitive pricing in addition to ensuring good customer service, controlling safety, health, and environmental costs is critical.

Hiring and Placement

The logistics and distribution sector is the industrial part of the service and retail industry that requires employees to be physically fit. Consistent with the Americans with Disabilities Act (ADA), once a company makes a tentative job offer to an employee, a drug screening and physical examination may be conducted. The physical examination also includes a review of the individual's past medical history.

Many warehousing environments now conduct a physical capacity screening process, which measures a person's strength and endurance. The screening helps an employer determine whether an applicant is physically capable of performing a job that requires a significant amount of manual labor. The results of the screening are matched against the specific demands of the job. When the applicant's capabilities meet or exceed the job demands, the person is considered physically qualified to do the job. If the screening indicates the applicant's capabilities are insufficient to perform the job, the employer must determine whether a reasonable accommodation can be made so that the applicant can perform the essential functions of the job. (See Chapter 16, Ergonomics Yesterday, Today, and Tomorrow, and Chapter 23, Workers with Disabilities, both in this volume.) If the employer's efforts meet the specifications set by the ADA and the individual still cannot perform the work, the employer is justified in turning down the applicant.

Service, retail, and logistics organizations should also conduct new-hire orientations to ensure that the new employee is made aware of company safety and health regulations and procedures. Most orientation programs cover the following topics:
- employee responsibilities and discipline
- safety rules
- incident and injury reporting
- general fire protection
- emergency procedures
- security alarms and inventory controls.

(See also New Employee Training and Orientation and Job Instruction Training in Chapter 31, Safety and Health Training.)

Risk Control

Many jobs in the distribution industry involve manually handling materials. If not properly managed, this type of work can result in a wide variety of injuries. To help reduce the frequency and severity rates of occupational injuries and illnesses, it is common for a company to use a risk control action plan. This plan typically outlines the specific efforts necessary to obtain the risk control objectives of the company. After a thorough analysis of individual risks in company operations, the company uses the plan to establish measurable, specific risk control efforts and numerical goals that will be used in performance reviews for management personnel.

The service industry has emphasized employee safety training and management operational accountability as effective ways to control risk exposures. Managers are given training in risk control management that includes the following:
- cost and efficiency consequences of incidents and injuries
- modified duty (light duty)
- injury trend analysis
- how to establish safe work practices
- how to approach and care for injured employees and customers' compensation laws
- OSHA regulations
- ergonomics
- fire protection and prevention.

One area of special emphasis has been active training of supervisory personnel, safety committees, and general employees in proper incident investigation techniques.

Because of the number of customers who shop at retail establishments, these stores are susceptible to customer incidents. For this reason, retail companies should analyze in detail the potential causes of customer incidents (falls, slipping, falling merchandise, cuts, etc.) and then take measures to reduce the risks. This process is particularly critical because third-party lawsuits can be generated as a result of these types of incidents.

On the other hand, the general liability risk for the distribution side of the industry is posed by incidents involving outside vendors or drivers who routinely visit distribution centers to monitor or deliver products. To reduce the risk of this type of loss, many companies restrict outside vendors and drivers to certain areas, such as the side or back of a building or a special delivery entrance. In addition, companies can require written review and compliance with in-house policies and procedures. Outside driver ignition key controls can help reduce unintentional pullouts at dock locations.

Because property losses can represent a significant cost, many companies train their employees during orientation about the importance of eliminating property damage. The warehousing industry has a risk of different types of property damage because of the volume of heavy equipment used to move material and products in the warehouse. Requiring suppliers to place the product within the perimeter of the pallet and secure the product with shrink-wrap can assist in reducing product damage during shipment.

For examples of incident report forms that can be used to report and analyze the causes of employee and customer losses, see Figures 24–1 and 24–2. (See also Chapter 10, Incident Investigation, Analysis, and Costs.)

Audit and Inspection Program

As discussed in Chapter 7, Safety, Health, and Environmental Auditing, auditing and inspections are an important part of a company's health and safety program. The service industry has developed audit and inspection programs to ensure that risk control initiatives are implemented and followed. Many companies are beginning to use both external and internal auditors to conduct a comprehensive evaluation of loss control initiatives. Typically, these are annual audits that examine the entire company's risk control issues. For example, an external audit could include the following topics:
- management direction and compliance with corporate policies
- compliance with federal and state regulations
- management accountability and compliance with corporate safety policies
- hiring and placement
- physical conditions

24 General Liability Issues 555

LIBERTY PERSONNEL: PLEASE FORWARD TO DIVISION STATISTICAL

Employee Injury Data Entry Report - Distribution Center

DIVISION _____ LOCATION CODE _ _ _ _ _ _

DIVISION ADDRESS _____ STATE _____ ZIP _____

THIS IS AN EMPLOYEE INJURY REPORT BASED UPON THE SUPERVISOR'S INTERPRETATION OF INITIAL ALLEGATIONS OF FACTS AS RELAYED BY OTHERS.

WHO? INJURED EMPLOYEE'S NAME _____ MALE _____ FEMALE _____

AGE _____ LENGTH OF EMPLOYMENT: YEARS _____ MONTHS _____ FULL TIME _____ PART TIME _____

WHEN? DATE OF INJURY ___/___/___ (MONTH/DAY/YEAR) TIME OF ACCIDENT ___:___ ☐ AM ☐ PM HOURS WORKED BEFORE INJURY _____ SHIFT _____

WHAT REPORTEDLY HAPPENED? BRIEF FACTUAL DESCRIPTION _____

WHERE DID THE ACCIDENT OCCUR (CHECK ONE)

OUTSIDE
- ☐ 70 Parking Lot
- ☐ 75 Personnel Entrance/Sidewalk
- ☐ 97 Trailer
- ☐ 69 Yard
- ☐ 10 Transportation-While Driving
- ☐ 95 Not Otherwise Classified

INSIDE
- ☐ 68 Battery Recharge
- ☐ 96 Bindery Area
- ☐ 11 Building Maintenance
- ☐ 87 Cake Production
- ☐ 94 Camera Area
- ☐ 22 Dairy
- ☐ 88 Dough Production
- ☐ 01 Dry Grocery
- ☐ 05 Freezer
- ☐ 09 Garage
- ☐ 02 General Merchandise

OCCUPATION _____ **CODE** _____
(See back of this form for Occupation Code Numbers)

- ☐ 78 In-Plant Vehicle Shop
- ☐ 71 Lunchroom
- ☐ 82 Mailroom
- ☐ 21 Meat
- ☐ 12 Office
- ☐ 67 Packaging
- ☐ 83 Print Shop
- ☐ 04 Produce
- ☐ 06 Railcar Receiving
- ☐ 13 Recoup/Salvage
- ☐ 14 Repack Department
- ☐ 90 Sanitation
- ☐ 74 Transportation at Store or Backhaul
- ☐ 07 Truck Receiving Dock
- ☐ 08 Truck Shipping Dock
- ☐ 95 Not Otherwise Classified

TYPE OF ACCIDENT (CHECK ONE)

FALL OR SLIP
- ☐ BS Fall - From Racking
- ☐ A3 Fall - From Elevation (Other Than Rack)
- ☐ AA Fall - Same Level
- ☐ A2 Slip or Trip (Not Fall)
- ☐ A7 Falls Not Otherwise Classified

MATERIAL HANDLING (Complete Backside Of Pink Copy If Coded For Order Selector as Occupation)
- ☐ HD Carrying, Holding
- ☐ HB Lifting/Lowering
- ☐ HC Pulling, Pushing
- ☐ WY Repetitive Motion
- ☐ H3 Twisting
- ☐ H4 Reaching
- ☐ JC Not Otherwise Classified

STRUCK BY OR AGAINST
- ☐ RP Powered Object
- ☐ RS Stationary Object
- ☐ RR Falling, Flying Object
- ☐ RM Not Otherwise Classified

MISCELLANEOUS
- ☐ NL Caught In Between or Under
- ☐ Y1 Contact with Sharp Object
- ☐ YD Electrical Contact/Shock
- ☐ XN Explosion or Flash Fire/Back
- ☐ XS Friction
- ☐ X1 Hot Liquid, Steam
- ☐ XZ Hot Surfaces, Objects
- ☐ X9 Misconduct
- ☐ GD Motor Vehicle Accident
- ☐ YA Not Otherwise Classified

TYPE OF INJURY/ILLNESS (CHECK ONE)

- ☐ 300 Abrasion
- ☐ 100 Amputation
- ☐ 110 Asphyxiation/Choking/Drowning
- ☐ 160 Bruise/Contusion
- ☐ 120 Burns (Other Than Chemical)
- ☐ 130 Chemical Burns
- ☐ 140 Concussion
- ☐ 165 Crushing Injury
- ☐ 170 Cut/Lacerations
- ☐ 190 Dislocation
- ☐ 200 Electric Shock
- ☐ 235 Foreign Object/Body (Eyes)
- ☐ 210 Fracture
- ☐ 220 Frostbite or Freezing
- ☐ 230 Hearing Loss
- ☐ 420 Heart Attack
- ☐ 240 Heat Exhaustion
- ☐ 250 Hernia
- ☐ 350 Infection
- ☐ 150 Infectious Disease
- ☐ 260 Inflammation (Irritation of Muscles, Joints)
- ☐ 145 Loss of Consciousness
- ☐ 400 Multiple Injuries
- ☐ 175 Puncture
- ☐ 180 Rash/Dermatitis, Skin Inflammation
- ☐ 311 Spasms/Muscle Cramps/Sprains/Strains
- ☐ 994 Not Otherwise Classified - Illness
- ☐ 995 Not Otherwise Classified - Injury

BODY PART (CHECK ONE)

- ☐ 100 Head (Multiple)
- ☐ 150 Scalp
- ☐ 140 Face
- ☐ 120 Ear
- ☐ 146 Nose
- ☐ 130 Eye
- ☐ 144 Mouth/Tongue/Teeth
- ☐ 201 Neck/Throat
- ☐ 300 Upper Extremities (Multiple)
- ☐ 310 Arm
- ☐ 313 Elbow
- ☐ 320 Wrist
- ☐ 330 Hand (Multiple)
- ☐ 350 Thumb
- ☐ 340 Fingers
- ☐ 400 Trunk (Multiple)
- ☐ 410 Abdomen/Stomach
- ☐ 412 Pelvis
- ☐ 420 Back
- ☐ 430 Chest
- ☐ 440 Hips
- ☐ 441 Buttocks
- ☐ 411 Groin
- ☐ 450 Shoulder
- ☐ 500 Lower Extremities (Multiple)
- ☐ 510 Leg (Multiple)
- ☐ 560 Thigh
- ☐ 513 Knee
- ☐ 550 Calf
- ☐ 520 Ankle
- ☐ 530 Foot
- ☐ 540 Toes
- ☐ 700 Multiple Body Parts - Exterior
- ☐ 860 Internal Body (Multiple)

POSSIBLE CONTRIBUTING FACTORS (AGENCY)
(CHECK ONLY ONE)

- ☐ 0200 Animal/Insect
- ☐ 0620 Bag
- ☐ 5602 Bike
- ☐ 0600 Box(es)
- ☐ 2246 Box Cutter/Utility Knife
- ☐ 0700 Building Structure/Walls, Pillars
- ☐ 5920 Building Heating/Refrigeration Equipment
- ☐ 5600 Car/Auto
- ☐ 5610 Cart
- ☐ 0900 Chemical
- ☐ 5620 Compactor/Bailer
- ☐ 1300 Conveyor
- ☐ 0761 Dock Leveler, Dock

- ☐ 0762 Dock Plate
- ☐ 0780 Dolly Crank
- ☐ 0771 Door - Dock
- ☐ 0770 Door - Personnel
- ☐ 0773 Door - Trailer
- ☐ 2620 Elevator
- ☐ 4110 Fastener, Nail, Screw
- ☐ 1710 Fire/Flame/Smoke
- ☐ 1951 Floor (Uneven/Holes)
- ☐ 1952 Floor (Slippery Surface)
- ☐ 1960 Floor Sweeper
- ☐ 8000 Foreign Object
- ☐ 5640 Fork Lift - Stand Up
- ☐ 5641 Fork Lift - Sit Down
- ☐ 2000 Glass

- ☐ 2200 Hand Tools - Non-Powered
- ☐ 2300 Hand Tools - Powered
- ☐ 2600 Hoisting Equipment
- ☐ 6250 Ice/Snow
- ☐ 2245 Knife
- ☐ 2800 Ladder
- ☐ 2900 Liquids
- ☐ 3000 Machinery
- ☐ 2240 Meat Slicer
- ☐ 5930 Noise
- ☐ 7700 Other Person
- ☐ 7300 Pallet/Slips
- ☐ 7320 Pallet Jack - Non-Powered
- ☐ 7310 Pallet Jack - Powered
- ☐ 4510 Paper/Cardboard

- ☐ 5940 Product Spill
- ☐ 5650 Rack
- ☐ 9700 Rope
- ☐ 5826 Scaffold
- ☐ 5603 Scooter
- ☐ 5840 Stairs/Steps
- ☐ 0601 Tote/Tub/Pan/Tray
- ☐ 5625 Trailer
- ☐ 6450 Truck/Tractor/Shuttle
- ☐ 2550 Welding Equipment
- ☐ 4030 Wire/Banding
- ☐ 5700 Wood
- ☐ 6700 Not Otherwise Classified

Date Report Completed _____ Completed By _____ Title _____

PS 1133 (Rev. 8/95)   WHITE COPY TO INSURANCE CARRIER   YELLOW COPY (If Completed) TO HOME OFFICE RISK MANAGEMENT   PINK COPY - KEEP FOR YOUR RECORDS

Figure 24–1. This sample form can be used to report and analyze employee injuries.

CUSTOMER STORE INCIDENT REPORT

Store #: _____ Manager completing this report: _____

Date: _____ Time: _____

Name of injured: _____ Age: _____

If a minor, parent's name: _____

Address: _____

Phone number (home): _____ (Work): _____

Personal injury: _____ and/or property damage: _____

Describe fully how incident occurred: _____
_____
_____
_____
_____

Nature of injury/damage (describe fully): _____
_____
_____
_____
_____

Witnesses (name, address and phone number): _____
_____
_____

Figure 24–2. A report such as this sample form should be completed whenever a customer is involved in an incident that involves personal injury or property damage.

- risk assessment
- incident follow-up
- ergonomic control
- fleet operations
- business interruption.

Internal audits can be just as comprehensive but are more commonly conducted by supervisors or employees on a daily or weekly basis. Observation audits are more focused on processes that can prevent injuries and include operational safety issues such as the following:

- materials handling
- body positioning
- environment
- equipment
- racking
- employee exposure potentials
- fire and life safety
- motorized equipment
- chemical management
- customer exposure potential/controls.

See Figure 24–3 for a sample internal audit form.

Materials Handling

Materials handling carries the most common risk for injuries and property damage in both the retail and distribution industries. This is due to the volume of product that has

Safety/Sanitation Checklist

Completed By:	Date:
Location:	Time:
Employees Observed:	Shift:

Directions: For each item, check entire department/location. If satisfactory, make a checkmark in the "Safe" column. If not, make a checkmark in the "Unsafe" column and explain why in the "Comments" column. Leave blank if item is not applicable.

SAFETY

	SAFE	UNSAFE	COMMENTS
Manual Materials Handling			
Loads handled as close to the body as possible.			
Pivots feet and avoids twisting of trunk.			
Unnecessary awkward postures avoided.			
"Z-pick" selecting pattern followed.			
Avoids walking on inappropriate surfaces (i.e., pallets).			
Places feet properly when climbing up or down racks.			
Watches for obstacles when climbing up or down steps.			
Uses hook to select products when necessary.			
Body Positioning			
Avoids putting hands or feet under pallets.			
Avoids walking under forklift's raised head.			
Works in a deliberate motion when in confined space.			
Environment			
All exit doors accessible for emergency exit.			
No plastic in slots.			
All aisles/floors free from physical debris or spills.			
When debris/spill is detected, it is removed/guarded.			
Items slotted according to ergonomic specifications.			
First-aid kit well stocked.			
Product stacked to prevent sliding/falling/collapse.			
Smoking rules observed.			
Sprinkler heads have 24" clearance above stock.			
Equipment			
Material-handling equipment in safe operating condition.			
Faulty equipment not used and referred for repair.			
Eyewash station in good order.			
Racking			
Rubber steps installed on racks as needed.			
Rubber pads installed on sharp edges of racks.			
All racks are structurally sound.			
Additional Comments:			

Figure 24–3. This sample checklist can be used for internal audits.

Safety/Sanitation Checklist			
SANITATION			
Warehouse Inspection	SATISFACTORY	UNSATISFACTORY	COMMENTS
No open product indicating pilferage.			
No damaged product in slots.			
No cigarette butts found.			
No product stored directly on the floor.			
No cleaning or processing chemicals near food.			
No dead, crawling, or flying insects encountered.			
If rodent evidence is observed, are rodent traps set?			
Open bait.			
Floor is clean.			
No objectionable odor.			
Stock is properly rotated in all departments.			
No damaged product in overstock and/or picking slots.			
White lines and curbs are clean.			
Areas under cart rack have been cleaned.			
All empty pallets have been removed from the area.			
Docks (Inside)			
Floor and plates clean.			
No cracks in wall openings.			
Doors in good repair, closed, and locked.			
Lights operating.			
Blacklight available and working in receiving office.			
Receiving crews monitor receipts for insect/rodent signs.			
Plastic wrap removed at time of receiving.			
Additional Comments:			

Figure 24-3. Concluded.

to be moved by manual handling or by motorized equipment such as forklifts and power pallet jacks. To reduce the risk, the industry has improved freight flow to minimize the number of times product is touched. Training employees on proper handling technique is critical. The training starts with new employee selection and orientation and continues with job-specific training regarding proper material handling and proper use of motorized equipment. To reduce the risk of injuries that occur while lifting materials or product, the industry offers many types of training programs in proper materials-handling techniques. (See Chapter 12, Materials Handling and Storage, and Chapter 15, Powered Industrial Trucks, both in the *Engineering & Technology* volume. See also Swartz 1999a, 1999b.)

In order to reduce injuries involving forklift and pallet jack operations and to comply with OSHA requirements, the distribution industry has implemented a certification program that requires forklift and pallet jack operators to have both initial and recurrent training (Figure 24-4). (OSHA now requires refresher training based on five criteria and that takes place no less than once every 3 years. See 29 CFR 1910.178[1][4][ii].)

Forklift Training Outline

I. Safety
   A. Daily safety check
   B. Personnel protection
   C. Starting and stopping
   D. Travel
   E. Loading
   F. Accidents
   G. Parking
II. Forklift Responsibilities
   A. Letdowns
   B. Incoming freight
   C. Prevents
   D. As you work
   E. Pallet handling and sorting
   F. Shift-end
III. Operational Procedures
   A. Common practices
   B. Grocery specific practices
   C. Perishable specific practices
   D. Freezer specific practices
IV. Telxon Procedures
   A. General
   B. Putaways
   C. Transfers
   D. Date rotation
V. Reserve practices
   A. Receiving
   B. Putaways
   C. Single selection bays
   D. Drive-in bays
VI. Housekeeping
   A. Common practices
   B. Grocery specific practices
   C. Perishable specific practices
   D. Freezer specific practices
VII. Productivity
   A. Service
   B. Time management
   C. Commitment to continuous improvement

Figure 24-4. Topics that should be covered in a forklift training program.

Fleet Safety Programs

Operating motor vehicles on public highways represents the most serious off-premise risk exposure in the retail, services, and logistics industry. Although the DOT regulates truck fleets, companies should look at ways to implement safety controls to manage light vehicle fleets, including company cars. The high costs, both direct and indirect, of vehicle collisions can affect companies in several ways. The company usually pays out more than it recovers from any insurance policy, must pay higher insurance premiums, and experiences a negative impact on profitability. The logistics industry in particular can have a high risk of vehicle collisions because of the number of miles driven to deliver product.

A defensive driving course can reduce light vehicle fleet driver accidents. To comply with DOT regulations for large trucks (more than 26,000 lb [11.8 metric tons] Gross Vehicle Weight) and to address this loss potential, companies should develop a fleet safety program that includes the following:

1. Driver selection (application review)
   - past employers
   - driver's license
   - driving experience
   - prior collisions
   - prior traffic violations
2. Motor vehicle reports (MVRs)
   - MVRs obtained on applicants before hiring
   - MVRs obtained annually on existing drivers
   - objective standards set and used for the analysis of MVRs, both preplacement and annually
   - new drivers' preplacement physical examinations that incorporate DOT-required parameters
3. Driving test
   - determination of applicant's ability to drive (DOT 391.31)
   - road tests with a preplanned route of 10 mi (16 km) or more, requiring a minimum of 30 min
4. Documented orientation and training
   - defensive driving courses, including online courses for remotely located drivers, with the goal of getting home safely to their families
   - steps to take in the event of a collision (company-specific forms and checklists should be available in all company vehicles)
   - state and federal regulations involved in fleet operations
   - expected safe-driving practices and following distances
   - proper care and operation of the vehicle
   - use of seat belts for all occupants of the vehicle
   - restrictions on the use of cell phones and other electronic devices
   - inspection responsibilities (before and after each trip)
   - safe cab entry and exit procedures
   - procedures for ceasing operations
   - emergency response for tank rupture or fuel leak
   - written fleet safety manual reviewed with and provided to new drivers
5. Follow-up training; procedure for taking a defensive driving course after a preventable accident
   - annual safety-related training for all drivers
   - corrective action, including training, given after a chargeable collision
6. Incident reporting

- established procedures for drivers to follow at the scene of an incident in which they are involved
- investigation of serious collisions and a detailed analysis completed by the safety manager and the transportation manager at the scene of the collision
- review of incidents by the accident review committee to determine the underlying and immediate causes of the incident
7. Vehicle inspections
   - daily documentation of pre- and post-trip inspection reports.

This comprehensive fleet safety program can help companies prevent serious injuries to workers, pedestrians, and other drivers and reduce the costs associated with all accidents. (See also Chapter 25, Transportation Safety Programs.)

Ergonomics

Ergonomics is the study of efficiency of people in their work environment. There are various methods to create a "good fit" between the employee and the job. The goal of a good ergonomics program is to prevent and minimize accidents and illnesses due to repetitive physical and psychological stresses on the body. (See also Chapter 16, Ergonomics Yesterday, Today, and Tomorrow.)

The service industry, especially the distribution and warehousing section, can benefit greatly by using ergonomics to design a job. Many companies are beginning to develop a variety of ergonomics programs to safeguard worker health and to contribute to a more efficient operation. A good ergonomics program must have strong support from senior management. Additionally, an in-house ergonomics team should be established and include supervisory personnel and personnel from engineering, human resources, operations, purchasing, health and safety, and appropriate employees. Ergonomic programs need to be site-specific and include at least four basic activities:

1. *Identify the problems*—Management must know which jobs require the most study and which type of injury patterns occur most frequently on those jobs. Past incident and injury records can be used.
2. *Study the physical demands of high-priority jobs*—The company can develop checklists to help identify specific activities that give rise to particular injury patterns.
3. *Formulate a written action plan*—The plan should include a description of the action to be taken and a realistic completion date.
4. *Maintain the effort*—This goal can be accomplished through ongoing, periodic reviews of job demands and through obtaining management and employee comments regarding emerging problems, changes in processes, or other ergonomic issues that need to be addressed.

OSHA has developed guidelines for retail grocery and warehouse operations that are available on its website (osha.gov/SLTC/ergonomics/guidelines.html).

Emergency Preparedness

Retail and service operations need a plan to handle emergencies to protect their guests and employees. The recovery plan allows a company to get back into operation as soon as possible, which also helps the community in its recovery process.

The two essential steps in preparing for and successfully managing responses to emergencies are creating a plan and exercising the plan. Many retail, service, and warehouse facilities have substantial inventories. Because of the potential impact of wholesale losses due to sudden, unexpected problems, companies must be adequately prepared and practice for a variety of emergencies. Fire, tornado, hurricane, earthquake, and other emergency drills are an important way to learn how to manage crisis situations, reduce recovery time after such disasters, and better protect employees and property. During or after practice drills, employees often raise critical questions that could help save lives and preserve property in the event of a real emergency. The company should document all such exercises and their results.

An emergency management plan should be established and used to prepare for and react to possible disasters. The plan allows authorized personnel to find the right information quickly and to respond to the disaster. The plan can also serve as a training document to acquaint new personnel with the company's procedures for coping with such emergencies.

The geographic location of a facility often determines the types of emergencies that will be included in the plan. It is important that a cross-section of employees conduct a detailed analysis of the company's needs so that an appropriate plan is developed for all potential emergencies. Contingency plans should be developed for the following potential emergencies:

- security breaches in facilities and inventory
- fires in the workplace or on the grounds
- chemical releases or spills
- riots/strikes
- bomb threats
- power failures
- product recalls/tampering
- violence in the workplace
- natural disasters, such as tornadoes, earthquakes, hurricanes, floods, and fires.

(See also Chapter 18, Emergency Preparedness, for more information on developing a plan and conducting drills.)

In addition to the emergency preparedness plan, a business continuity plan should address recovery procedures that facilitate return to normal business operations as soon as possible.

Although no recovery plan can cover all aspects of a situation, it should provide the following minimum recovery procedures:
- definition of responsibilities of those involved in the recovery
- computer backups to allow for the continuance of operations
- a guide for successful recovery of operations after an emergency
- procedures to ensure continued review and update of the emergency preparedness and business continuity plans.

Violence and Crime in the Workplace

The retail trade and service industries account for a large percentage of workplace homicides and nonfatal workplace assaults. Under NIOSH *Violence in the Workplace: Risk Factors and Prevention Strategies* (1996), employers can be cited for violating the General Duty Clause of the OSH Act if there is a recognized hazard of workplace violence in their establishments and they do nothing to prevent or abate it.

Service and retail employers should review their risk exposure and consider the need for procedures to prevent or appropriately respond to violence in the workplace. There are three major types of exposure to workplace violence: violence in the course of a crime, violence by a current or former customer, and violence that is employment related. Workers in the retail and service environment are at risk for all three types of workplace violence. In the distribution industry, the greatest risk is from employment-related violence. This type of violence may be caused by a current or former employee, supervisor, or manager who is a spouse, friend, or acquaintance of an employee. The primary target of employment-related violence is a co-worker, supervisor, or manager. In committing the assault, the individual generally is seeking revenge to settle a personal score or for what he or she perceives as unfair treatment. This could include an unsatisfactory review, some type of disciplinary action, loss of pay or benefits, demotion, or termination.

Studies have shown that many cases of employment-related violence can also involve domestic or romantic disputes. For these reasons, most retail and distribution industries are beginning to formulate plans and provide specific training to address this issue. Such training should include how to handle personnel issues before they escalate and what to do when an incident occurs.

Risk Factors
The National Institute for Occupational Safety and Health (NIOSH) report notes factors that may increase a worker's risk for a workplace assault:
- contact with the public
- exchange of money
- delivery of passengers, goods, or services
- having a mobile workplace, such as a taxicab or police cruiser
- working with unstable or volatile persons in health care, social services, or criminal justice settings
- working alone or in small numbers
- working late at night or during early morning hours
- working in high-crime areas
- guarding valuable property or possessions
- working in community-based settings (1996, 14).

Obviously, employees who have the greatest exposure to violence committed in the course of a crime are those who have face-to-face contact and exchange money with the public. They often work alone or in small numbers and work late at night. Companies—particularly those whose employees have these risk factors—should develop formal plans and training programs for employees to reduce employee risk from violent assault. These plans should include using methods of crime deterrence whenever possible.

Prevention Strategies
According to *Armed Robbers and Their Crimes* (Erickson 1996), a report based on interviews with convicted armed robbers, robbers are generally not deterred by surveillance cameras, the threat of arrest, police patrols, or bullet-resistant barriers. Therefore, prevention plans that rely on only these strategies may not prevent robberies. The NIOSH report suggests a variety of environmental designs, administrative controls, and behavioral strategies for a comprehensive approach to reducing workplace violence:
- environmental designs
    - using time-controlled drop safes
    - carrying small amounts of cash
    - posting signs and printing notices that limited cash is available
    - implementing cashless transactions using automatic teller account cards or debit cards, where possible
    - physical separation of workers from customers, clients, and the general public
    - making high-risk areas visible to more people (e.g., by clearing window areas)
    - increasing external lighting
    - controlling the number of entrances and exits, escape routes, access to work areas, and hiding places

- using numerous security devices (closed-circuit cameras, alarms, two-way mirrors, card-key access systems, panic-bar doors locked from the outside only, trouble lights, geographic locating devices for mobile workplaces)
- body armor
• administrative controls
  - staffing plans and work practices that prohibit unsupervised movement
  - increasing the number of staff on duty
  - using security guards or receptionists to screen persons entering
  - photo identification badges
  - controlling access to work areas
  - establishing policies and procedures for reporting and assessing threats to employees
  - training employees on recognition of potential for violence, methods for defusing violent situations, use of security devices and protective equipment, and procedures for obtaining medical care and post-trauma care
• behavioral strategies
  - training employees in nonviolent response and conflict resolution
  - establishing hazard recognition and prevention strategies.

(See also NIOSH, *Violence in the Workplace: Risk Factors and Prevention Strategies*, and National Safety Council, *Accident Prevention Manual for Business & Industry: Security Management*.)

Theft Prevention

Companies that have direct contact with customers who visit the place of business to purchase a product may be at risk for both product losses and potential injuries due to shoplifting. To reduce this risk, these establishments should implement a theft-prevention program that includes a comprehensive policy that employees are trained to consistently follow. An organization that does not have such a policy or does not consistently follow that policy can significantly increase its liability. The policy should include the following:
- techniques used by shoplifters
- how to detect a potential shoplifter
- methods of surveillance
- how to apprehend a shoplifter.

Employers should emphasize in the policy statement what specific activities are to be followed in case of a robbery. It is important that all employees understand that their main goal should be to avoid violence and to prevent anyone from being injured. The important message that must be understood by all employees is, "Don't do anything that would jeopardize personal safety. Don't be a hero."

SUMMARY

- Good safety management in the service and retail industry is based on a strong corporate commitment to safety and an investment in time to develop specific guidelines and policies to address trends identified during the risk assessment process. This investment pays off through better customer service, a better work environment for employees, reduced injury rates, reduced workers' compensation problems, reduced regulatory fines, higher-quality service, and increased profit margins.
- A company can use ANSI Z10–2005, Occupational Health and Safety Management System, to reduce its risk of losses and to enhance productivity. The system includes developing a clear policy statement and specific documents, employee training and disaster preparedness, management and employee accountability for action and safety records, internal and external safety audits, and the establishment of safety committees or other oversight groups.
- OSHA regulations that particularly apply to the service industry include the General Duty Clause, posting requirements, reporting and record keeping, means of egress, hazard communication, medical services and first aid, walking and working surfaces, lockout/tagout procedures, electrical installations and equipment, machine guarding, and powered equipment.
- Because the service industry can be physically demanding, a company must make sure that employees are physically fit to do their work. The company should also provide employees with proper training in safe work practices and procedures.
- A company in the retail and distribution industry can manage its risk of loss by training employees in proper incident reporting and analysis procedures, introducing motor fleet safety programs, conducting safety audits, training workers in materials handling, introducing ergonomics to the workplace, planning for emergencies and disasters, and ensuring greater security against crime and violence on the job.

REFERENCES

Erickson, R. *Armed Robbers and Their Crimes*. Seattle, WA: Athena Research, 1996.

National Safety Council. *Accident Prevention Manual for*

Business & Industry: Security Management. Itasca, IL: National Safety Council, 1997.

NIOSH. *Violence in the Workplace: Risk Factors and Prevention Strategies.* Current Intelligence Bulletin 57. DHHS NIOSH Publication No. 96-100. Washington DC: U.S. Department of Health and Human Services, National Institute for Occupational Safety and Health, June 1996.

Swartz, G. *Forklift Safety: A Practical Guide to Preventing Powered Industrial Truck Incidents and Injuries.* 2nd ed. Rockville, MD: Government Institutes, 1999a.

———. *Warehouse Safety: A Practical Guide to Preventing Warehouse Incidents and Injuries.* Rockville, MD: Government Institutes, 1999b.

REVIEW QUESTIONS

1. The indirect costs of incidents are ____ times the direct costs.
    a. 2
    b. 5 to 20
    c. 2 to 4
    d. 10 to 50
2. List five of the eight components of a sound risk management program.
    a.
    b.
    c.
    d.
    e.
3. An employee manual should not only define the required level of safety, but also _____.
4. Briefly define the General Duty Clause.
5. What is the difference between OSHA Form 300 and OSHA Form 301?
6. List five of the eight OSHA requirements for all means of egress.
    a.
    b.
    c.
    d.
    e.
7. Briefly define HAZCOM.
8. List the four main requirements of the hazard communication standard.
    a.
    b.
    c.
    d.
9. Briefly describe the tagout procedure.
10. Which of the following machines/equipment are not covered by tagout standards?
    a. balers
    b. meat wrappers
    c. ventilation hoods
    d. wire buffers
11. Company posting of emergency phone numbers is a sufficient action under OSHA regulations.
    a. true
    b. false
12. All storage and materials should be kept at least ____ ft from all major electrical equipment.
    a. 3
    b. 5
    c. 10
    d. 15
13. During physical capacity screening, an applicant is considered qualified when his or her capabilities _____ the job demands.
14. List four of the six topics that an orientation program should cover.
    a.
    b.
    c.
    d.
15. _____ is the area that has the greatest risk for injuries.
16. Briefly define ergonomics.
17. Ergonomics programs should contain which four basic activities?
    a.
    b.
    c.
    d.
18. The leading cause of death in the workplace is
    a. violence.
    b. chemical exposure.
    c. asbestos exposure.
    d. accidents.
19. What are the two major types of exposure to workplace violence?
    a.
    b.

Transportation Safety Programs

John F. Montgomery, PhD, CSP, CHCM, CHMM
Thomas Bush
Von M. Griggs-Laws

Cost of Vehicle Collisions

Vehicle Safety Program
Responsibility ▸ Driver Safety Program ▸ Collision-Reporting Procedures ▸ Interviewing Drivers ▸ Driver Record Cards ▸ Fleet Collision Frequency ▸ Selection of Drivers ▸ Selection Standards ▸ Information-Gathering Techniques ▸ Legal and Social Restrictions ▸ Driver Training ▸ Safe-Driving Programs ▸ Safety Devices ▸ Preventive Maintenance

Repair Shop Safe Practices
Servicing and Maintaining Equipment ▸ Tire Operations ▸ Fire Protection ▸ Lubrication and Service Operations ▸ Wash Rack Operations ▸ Battery Charging ▸ Gasoline Safety ▸ Other Safe Practices ▸ Training Repair Shop and Wash Rack Personnel

U.S. DOT–Required Drug and Alcohol Testing
Drug Testing and DOT Requirements ▸ Alcohol Testing and DOT Requirements ▸ Setting Up a Testing Program

Summary

References

Review Questions

Transportation safety programs cover a range of vehicles from company cars to corporate jets, each type with its own safety issues and problems. The fleet equipment discussed in this chapter includes trucks, passenger cars, buses, and motorcycles. Powered industrial trucks and hand trucks are covered in Chapter 12, Materials Handling and Storage, and in Chapter 15, Powered Industrial Trucks, both in the *Engineering & Technology* volume. Haulage and off-road equipment are also discussed in Chapter 16 of the *Engineering & Technology* volume. This chapter covers the following topics:
- cost of vehicle collisions
- basic elements of a vehicle safety program
- safe practices and procedures in the repair and maintenance shop
- U.S. Department of Transportation (DOT) drug and alcohol testing requirements.

Many readers of this manual will have overall responsibility for the safety program of their entire company. However, most will not consider themselves the fleet safety manager. In fact, many do not even consider the fact that their company has a fleet of vehicles. They often overlook such vehicles as those used by maintenance workers, security personnel, company management, and so on. Yet, for many industries, these types of vehicles represent their firm's greatest risk of work-related injuries and deaths and civil liability losses. This chapter cannot address the entire range of vehicles that a company or industry uses—only the highlights of a total fleet safety program. For more details, see the National Safety Council's *Motor Fleet Safety Manual*, 4th edition.

Safe vehicle operation is not the result of chance, but of training, skill, planning, and action. Unfortunately, many companies fail to pay enough attention to the safe operation of motor vehicles. The reason for this lapse may be the difficulties of organizing an adequate safety program and of providing good driver and fleet supervision.

Most motor vehicle collisions are caused by driver error or poor operating practices—both factors that can be controlled. Only small percentages are due to mechanical failure of vehicles or improper maintenance of equipment. As a result, an organization's vehicle collision-prevention efforts should focus primarily on two principal collision factors: driver error and vehicle failure.

Companies can control driver error by implementing a program of driver selection, training, and supervision; vehicle failure can be reduced by a systematic preventive maintenance program. Experience has shown that an unsupervised fleet usually has higher collision costs than a supervised one.

COST OF VEHICLE COLLISIONS

The total cost of a vehicle collision almost always exceeds the amount recovered from the insurance company. Collision control in a large motor vehicle fleet is critical because increased insurance premiums and indirect costs reduce profits. (See Table 25–A.) Insurance premiums can fall or rise with the collision frequency and costs.

As discussed in Chapter 10, Incident Investigation, Analysis, and Costs, medical and property damage costs are only part of the costs resulting from a collision. There are also indirect costs, which, as with most work-related unintentional injuries, may be several times the direct costs. (Table 25–A). Typical indirect costs include the following:
- salary paid and loss of service of employees injured in a collision
- added workers' compensation costs resulting from a disabling injury
- loss of vehicle's commercial value while it is out of service, the cost of the replacement vehicle, or rental costs
- cost of supervisory time spent in investigating, reporting, and cleaning up after the collision
- poor customer and public relations resulting from involvement of a company vehicle in a collision
- cost of replacing or retraining an injured employee
- time lost by co-workers while discussing the nature of the collision and the extent of the victims' injuries.
- lost clients/customers and sales
- missed meetings
- salaries paid to injured employee
- medical costs paid by company
- any government agency costs
- possible overtime hours worked by others while covering duties of the injured employee.

Besides appealing to the company's profit motive, the safety professional can make the most telling argument for controlling collisions by pointing out the company's moral obligation to act reasonably and responsibly toward its employees and the public. The employer, through the authority to hire, supervise, discipline, and discharge employees, exercises considerable control over their driving performance. Management can require employees to enroll in driver safety education programs and provide proper supervision on the job.

VEHICLE SAFETY PROGRAM

A vehicle safety program should provide for the following:
- a written safety policy developed, supported, and enforced by management

TABLE 25–A. Revenue Necessary to Pay for Accident Losses

This table shows the dollars of revenue required to pay for different amounts of costs for accidents:

It is necessary for a motor carrier to generate an additional $1,250,000 revenue to pay the cost of a $25,000 accident, assuming an average profit of 2%. The amount of revenue required to pay for losses will vary with the profit margin.

YEARLY ACCIDENT COSTS	PROFIT MARGIN				
	1%	2%	3%	4%	5%
	REVENUE REQUIRED TO COVER LOSSES				
$1,000	$100,000	$50,000	$33,000	$25,000	$20,000
$5,000	$500,000	$250,000	$167,000	$125,000	$100,000
$10,000	$1,000,000	$500,000	$333,000	$250,000	$200,000
$25,000	$2,500,000	$1,250,000	$833,000	$625,000	$500,000
$50,000	$5,000,000	$2,500,000	$1,667,000	$1,250,000	$1,000,000
$100,000	$10,000,000	$5,000,000	$3,333,000	$2,500,000	$2,000,000
$150,000	$15,000,000	$7,500,000	$5,000,000	$3,750,000	$3,000,000
$200,000	$20,000,000	$10,000,000	$6,666,000	$5,000,000	$4,000,000

Source: Federal Motor Carrier Safety Administration, "Revenue Necessary to Pay for Accident Losses," Accident Cost Table (Washington DC: U.S. Department of Transportation, 2010), January 1, 2011.

- a person designated to create and administer the safety program and to advise management
- a driver safety program, including driver selection procedures, driver training (to include classroom instruction and demonstration), and safety-motivating activities (Figure 25–1); proper supervision and implementation are mandatory for success
- an efficient system for collision investigation, reporting, and analysis; determination and application of appropriate corrective action; and follow-up procedures to help prevent future collisions
- a vehicle preventive maintenance program with documentation of inspections and completed maintenance.

Responsibility

The first requirement of an effective driver safety program is that all employees from the president down to the individual vehicle operator accept responsibility for safe operation of company vehicles. Management must define standards of acceptable driver and vehicle performance and establish criteria and procedures to evaluate and correct job performance to meet these standards. Companies must make sure all employees understand that collision prevention is a requirement of continued employment.

In larger companies, a fleet safety professional is usually responsible for supervising the program for the safe operation of all automotive equipment. However, in a small industrial concern, this function may be assigned part-time to a supervisor or manager. Some of the duties of the person include the following:

- Advise management on collision prevention and safety matters.
- Develop and promote safety activities and work-injury prevention measures throughout the fleet.
- Study and recommend fleet safety programs regarding equipment and facilities, personnel selection and training, and other phases of fleet operation.
- Evaluate driver performance and skills requirements.
- Conduct or arrange for effective safety training, and procure or prepare and disseminate safety education material.
- Review collisions to determine their causes, and recommend corrective action to management.
- Compile and distribute statistics on collision-cause analysis and experience; identify problem persons, operations, and locations.
- Maintain individual driver safety records, and administer the safe-driver award incentive program.

Driver Safety Program

A driver safety program should include the following five basic collision-prevention procedures:

- Initiate a driver training certification and recertification program.
- Develop standards to determine ways collisions can be prevented.
- Require immediate reporting of every collision and near miss.
- Recommend performance goals to management; compute and publish the fleet collision record to all affected drivers.
- Establish competency, physical capacity, and skills levels; set objectives; and maintain a collision record for each driver.

XYZ Corporation—Safe Driving Program

1. **PURPOSE**
   To provide motor vehicles that meet company and government standards and to define the minimum requirements for their use by XYZ employees and its subsidiaries.

2. **SCOPE**
   This procedure applies to XYZ Corporation and its subsidiaries.

3. **DEFINITIONS**
   3.1 **Commercial Driver's License ("CDL")**
       License required in different states to drive a commercial (e.g., non-passenger) vehicle that may impose specific requirements and limitation regarding:
       - Driver proficiency
       - Type of vehicle
       - Number of passengers
       - Permissible cargo, and/or
       - Weight of vehicle
   3.2 **Passenger Vehicle**
       A vehicle designed for transporting personnel; e.g., cars, sport utility vehicles, mini-vans, pick-up trucks.
   3.3 **Truck**
       A vehicle designed for transporting materials (e.g., flatbeds and dump trucks), not including tractor-trailers or semi-trucks.

4. **PROCEDURE**
   4.1 **General Requirements**
       4.1.1 **Minimum Equipment**
             All XYZ contracts for purchase or lease of motor vehicles should require, at a minimum, basic safety equipment (e.g., seat belts, passenger and driver side air bags, spare tire and jack).
       4.1.2 **Additional Equipment**
             Additional specifications are listed below for purchased/leased vehicles that are acquired after the date of this procedure:
             - First-aid kits
             - Anti-lock brakes
             - Three-point restraint type seat belts
             - Rear window wipers and rear defrosters on all vans and station wagons (as technically feasible for the vehicle)
             - Outer side view mirrors on both sides
             - A barrier or safety net in cargo vans to stop materials from moving forward in case of accidents or sudden stops, and
             - Whiplash head restraints on all seats unless technically infeasible.
       4.1.3 **Other Requirements**
             - Individual states or countries may require additional items of safety equipment.
             - Leased vehicles must have the safety equipment specified by the manager of the contract.
       4.1.4 **Prohibited Equipment and Cargo**
             Nothing is to be installed or used in XYZ vehicles:
             - That violates any applicable laws,
             - Is designed to thwart law enforcement activities (e.g., radar detectors), or
             - Creates a hazard.
             Nothing is to be transported in XYZ passenger vehicles or personal vehicles used in XYZ business that violates laws or creates a hazard (e.g., improperly packaged chemicals).
       4.1.5 **Seat Belts**
             Seat belts **must be worn at all times** by all occupants in XYZ owned, leased or rented vehicles.

Figure 25–1. Here is a sample of a company's safe driving program.

4.1.6 Cell Phones
The use of cellular phones is **prohibited** while operating an XYZ owned, leased or rented vehicle. This includes text messaging.

4.2 Communication
Each XYZ terminal must have a means to inform common carrier drivers and leased drivers of terminal requirements for traffic safety and emergencies on site.

4.3 Driver Education and Training
4.3.1 Program Development
Each terminal, operation or department should, based on an assessment of needs, develop and implement a driver education and training program.
- Candidates for training may include employees who will be assigned a company vehicle or who will drive extensively in their work.
- Initial and refresher training should be included if deemed necessary based on observation of accident/incident reporting and observed competencies.

4.3.2 Driving Record
- Employees who will be assigned a company vehicle or who will drive extensively in their work must:
  - Have their driving record checked as a part of their pre-employment screening, and
  - Maintain a valid driver's license.
- Driving record checks may be done periodically at the discretion of XYZ.

4.3.3 Commercial Driver's License (CDL)
- A CDL license must be obtained in accordance with individual state or country laws.
- A thorough knowledge of state or country laws applicable to the operation of commercial vehicles is required for XYZ employees who drive with a CDL.

4.4 Accident Investigation
All vehicular accidents must be investigated and documented using the appropriate insurance company, leasing company and XYZ Incident Investigation forms and will be reported to the appropriate functions within XYZ (e.g., EHS Group, business function [sales, etc.] Insurance Department).

4.5 Vehicle Breakdowns
4.5.1 Assistance
Assisting non-XYZ personnel who have vehicle breakdowns along the road should be limited to calling for aid.

4.5.2 Good Samaritan
Assisting a person who is in distress due to an illness or accident is left to the reasonable good faith judgment of the employee. The "Good Samaritan" may incur personal liability if his or her actions are determined to have been grossly negligent or willful.

4.6 Authorized Use
Authorized drivers of company assigned vehicles are limited to properly designated employees and their legal spouses and contractors.

4.7 Passengers
4.7.1 Transportation of Passengers
Transportation of passengers in company vehicles is discouraged unless such passengers are on XYZ business.

4.7.2 Hitchhikers
Hitchhikers should not be permitted to ride in company-owned, -leased or -rented vehicles.

4.8 Personal Cars on Extended Business
Use of personal cars on extended business is discouraged. The immediate supervisor must be notified in the event that employees wish to drive their own cars on XYZ business.

4.9 Trailers
Personally owned or rented trailers, boats, etc. should not be attached to or drawn by company vehicles unless properly authorized by immediate supervision.

4.10 Motor Vehicle Maintenance
All XYZ owned or leased vehicles should be maintained and serviced according to the manufacturer's recommendations.

Figure 25–1. Concluded.

The planning and administration of a safety program for motor fleets are described in greater detail in the National Safety Council's *Motor Fleet Safety Manual*, 4th edition (see the References at the end of this chapter).

A motor vehicle collision can be defined as any incident in which the vehicle comes in contact with another vehicle, person, object, or animal, in a way that results in death, any degree of personal injury, or any extent of property damage, regardless of where the incident took place or who was responsible.

This definition includes even minor collisions such as fender scratches. All vehicle collisions, major and minor, are important to the person responsible for vehicle safety, whose major concern is eliminating poor driving habits or careless attitudes. Even minor collisions, not just spectacular ones, provide clues that can help prevent recurrences.

Collision-Reporting Procedures

Drivers should be required to complete a standard collision report form for every incident and near miss in which they are involved. Figure 25–2 is a good example of a collision report form. (See also the National Safety Council's *Motor Fleet Safety Manual*, 4th edition, for illustrations of other helpful forms.) Make sure drivers know how to fill out collision report forms accurately and thoroughly. Drivers should complete the report at the scene of the collision, if possible, and then send the form to their supervisor no later than their next assigned shift.

Companies can supply drivers with an incident report packet to be carried in the glove box. This packet (Figure 25–3) should provide for a memorandum report of the collision (such as Figure 25–2), an incident report checklist, pencils, plain paper, courtesy cards, telephone numbers of insurance representatives, and a disposable camera. Drivers will then have what they need to record all pertinent information at the scene of a collision. Smartphones may be permitted for photos and voice recordings of witnesses and the photos of the accident scene. Check with your insurance carrier for local and state laws.

Specially printed courtesy cards can help the driver quickly get names and addresses of key witnesses at the collision site (Figure 25–4). Drivers should be told that it is important to identify as many witnesses to the collision as possible.

Failure to report a collision, no matter how slight, or falsification of data on an incident report should be cause for disciplinary action against the driver. Management can ensure timely collision reporting by requiring a documented vehicle inspection before and after each trip.

Some companies have developed supplemental reports to ensure that all pertinent information and documents have been received and a thorough investigation can be accomplished (Figure 25–5).

In the case of serious collisions, especially those resulting in a fatality or severe personal injury, a representative of the company—the supervisor, safety director, or claim agent—should investigate the collision at the scene as quickly as possible. The purpose of such an investigation is to verify the driver's information on the collision report form and to obtain other data that may prove valuable for collision-prevention work or for defense against liability claims.

Interviewing Drivers

As soon as possible after a collision, the investigator should interview the driver involved to determine whether the collision could have been prevented. If this is determined to be the case, based on the interview and written reports, the collision should be classified as preventable. The fleet professional must then explain to the driver what actions contributed to the collision and make sure the driver understands what to do to prevent similar collisions in the future. For example, if the driver's lack of skill contributed to the collision, the safety investigator should recommend remedial driver training.

In some cases, the driver may disagree with the decision of the supervisor. For such instances, many firms have established an internal collision review committee to referee contested decisions. Should the driver wish to appeal the employer's determination that a collision was preventable, the National Safety Council maintains an Accident Review Committee, which will render an unbiased opinion about the case.

If, on the other hand, the investigator believes the driver did everything possible to prevent the collision, it should be classified as non-preventable. Responsibility for preventing collisions involves more than carefully observing traffic rules and regulations. Drivers must practice "defensive driving" to prevent collisions; that is, they must assume that other drivers will do the unexpected and will fail to observe normal traffic laws and safe-driving practices.

Smartphones for recording may be permitted. Check with your insurance carrier for local and state laws.

Driver Record Cards

Management should maintain a record card or computerized record for each employee who drives a company vehicle. This record not only furnishes a history of all collisions a driver has been involved in, but also provides information needed for safe-driver awards or other forms of safety recognition (Figure 25–6). The date of each collision and the collision category (preventable or non-preventable) should be entered on this record. The investigator should

MOTOR TRANSPORTATION
DRIVER'S COLLISION REPORT
READ CAREFULLY – FILL OUT COMPLETELY

(For office use)
File No. _____
☐ Preventable
☐ Not Preventable
☐ Reportable
☐ Not reportable

COMPANY _____
DIVISION _____ ADDRESS _____

COLLISION INFORMATION

DATE OF COLLISION _____, 20 ____
DAY OF WEEK _____
TIME _____ A.M. P.M.

MOVING
☐ Another com'l vehicle
☐ Passenger car
☐ Pedestrian
☐ _____

FIXED
☐ Building or fixture
☐ Parked vehicle
☐ _____

TYPE
☐ Head On   ☐ Rear End   ☐ Front End
☐ Sideswipe   ☐ Other (Describe) _____
☐ Right Angle

☐ Non Collision (Describe) _____

LOCATION

PLACE WHERE COLLISION OCCURRED:
ADDRESS OR STREET ON WHICH COLLISION OCCURRED: _____
☐ AT INTERSECTION WITH: _____
CITY/TOWN, STATE _____
☐ NOT AT INTERSECTION: _____ FEET N S E W ☐☐☐☐ OF _____
Nearest Intersecting Street/Road; House Number, or Landmark: Bridge, Milemarker, etc.

POLICE PRESENT? ☐ YES ☐ NO
Name of Force _____
☐ Local ☐ County ☐ State
Officer's Name _____
Badge No. _____
Report No. _____
Tickets/Arrests   Driver 1 ☐ 2 ☐
Other _____

DRIVER / PASSENGER / PEDESTRIAN INFORMATION

DRIVER VEHICLE ONE
Driver's Name: _____ Company I.D. Number: _____
Driver's Address: _____ City/State: _____
Phone No.: _____ Driver's License Number: _____
Vehicle Number(s): _____ Run: _____ Route: _____
Date Of Birth: _____ Employment Date As Driver: _____ Hours Since Last 8 Hours Off: _____
Parts Of Vehicle Damaged: _____
Was Street Lighted? ☐ Yes ☐ No
Was Vehicle Lighted? ☐ Yes ☐ No

DRIVER VEHICLE TWO
Driver's Name: _____ Phone No.: _____
Driver's Address: _____ City/State: _____
Driver's Occupation: _____ Driver's License Number: _____
Age: _____ Insurance Co.: _____
Make Of Vehicle: _____ Year: _____ Type: _____ Vehicle License No.: _____
Registered Owner: _____
Owner's Address: _____
Parts Of Vehicle Damaged: _____
Others In Vehicle: _____
Was Vehicle Lighted? ☐ Yes ☐ No

PASSENGER / PEDESTRIAN

PASSENGER ☐
PEDESTRIAN ☐
NO. 1
Name _____ Address, City, Zip _____ Home Phone _____
Passenger In Vehicle # _____ Date Of Birth _____ Sex _____ What Was Pedestrian/Passenger Doing: _____

PASSENGER ☐
PEDESTRIAN ☐
NO. 2
Name _____ Address, City, Zip _____ Home Phone _____
Passenger In Vehicle # _____ Date Of Birth _____ Sex _____ What Was Pedestrian/Passenger Doing: _____

INJURIES	INJURED: YES / NO	DESCRIBE INJURIES	INJURED TAKEN TO/BY
DRIVER VEHICLE 1			
DRIVER VEHICLE 2			
PASSENGER VEH ____			
PASSENGER VEH ____			
PEDESTRIAN 1			
PEDESTRIAN 2			
OTHER			

Figure 25–2. Driver's Collision Report.

VEHICLES/PEDESTRIANS/PASSENGERS

(CHECK ONE OR MORE FOR EACH DRIVER)—YOU ARE No. 1	PREVIOUS TO ACCIDENT		WHEN FIRST IN DANGER		AT IMPACT	
	No. 1	No. 2	No. 1	No. 2	No. 1	No. 2
Going straight ahead	☐	☐	☐	☐	☐	☐
Slowing	☐	☐	☐	☐	☐	☐
Stopped in traffic	☐	☐	☐	☐	☐	☐
Park or stopped in zone	☐	☐	☐	☐	☐	☐
Backing	☐	☐	☐	☐	☐	☐
Starting	☐	☐	☐	☐	☐	☐
Passing	☐	☐	☐	☐	☐	☐
Being passed	☐	☐	☐	☐	☐	☐
Changing lanes	☐	☐	☐	☐	☐	☐
Turning left	☐	☐	☐	☐	☐	☐
Turning right	☐	☐	☐	☐	☐	☐
Entering zone/Pulling to curb	☐	☐	☐	☐	☐	☐
Leaving zone/Pulling from curb	☐	☐	☐	☐	☐	☐
Other (explain)	☐	☐	☐	☐	☐	☐

YOUR SPEED _____ MPH _____ MPH _____ MPH
SPEED OF OTHER VEHICLE _____ MPH _____ MPH _____ MPH

DISTANCE YOUR VEHICLE FROM OTHER VEHICLE _____ FEET _____ FEET

DID YOU SOUND HORN? ☐ YES ☐ NO HOW FAR AWAY? _____ FEET
DID YOU APPLY BRAKES? ☐ YES ☐ NO HOW FAR AWAY? _____ FEET

AFTER IMPACT—VEHICLE MOVED _____ FEET
AFTER IMPACT—OTHER VEHICLE MOVED _____ FEET

GIVEN CONDITIONS, WHAT WAS SAFE SPEED FOR:
VEH. 1 _____ MPH   VEH. 2 _____ MPH

VEHICLE
1	2	
☐	☐	Did not have right-of-way
☐	☐	Following too closely
☐	☐	Failure to signal intentions
☐	☐	Speed too fast for conditions
☐	☐	Disregarded traffic signs or signals
☐	☐	Improper passing
☐	☐	Improper turning
☐	☐	Improper backing
☐	☐	Improper traffic lane
☐	☐	Improper parking
☐	☐	No improper driving
☐	☐	Defective brakes
☐	☐	Defective steering
☐	☐	Defective lights
☐	☐	Defective tires
☐	☐	No defects
☐		_____
☐		(Specify other)

PEDESTRIAN
☐ Walking with traffic
☐ Walking against traffic
☐ Coming from behind parked vehicle
☐ Crossing at intersection
☐ Crossing not at intersection
☐ Alighting from a vehicle
☐ Working in roadway
☐ Playing in roadway
☐ _____
(Specify other)

PASSENGER
☐ Boarding vehicle
☐ Alighting from vehicle
☐ Caught in doors
☐ Seated
☐ In motion in vehicle
☐ Other (describe)
_____
_____

ENVIRONMENTAL CONDITIONS (CHECK ALL THAT APPLY)

WEATHER (check one)
☐ CLEAR
☐ CLOUDY
☐ RAINING
☐ SNOWING
☐ FOGGY
☐ OTHER _____

SURFACE (check one)
☐ DRY
☐ WET
☐ ICY
☐ SNOWY
☐ OTHER _____

TRAFFIC CONTROL (check one)
☐ STOP SIGN
☐ YIELD SIGN
☐ TRAFFIC SIGNAL
☐ FLAGMAN
☐ NO CONTROL
☐ OTHER

LIGHT (check one)
☐ DAWN
☐ DAY
☐ DUSK
☐ DARK-NO LIGHTS
☐ ARTIFICIAL LIGHT
☐ OTHER

ROADWAY No. of Lanes
☐ DIVIDED _____
☐ UNDIVIDED _____
☐ ASPHALT _____
☐ CONCRETE _____
☐ GRAVEL _____
☐ OTHER _____

ALIGNMENT (check one)
☐ STRAIGHT
☐ CURVE
☐ BRIDGE
☐ INTERSECTION
☐ RAMP
☐ RAILROAD

☐ OVERPASS
☐ UNDERPASS
☐ LEVEL
☐ UPHILL
☐ DOWNHILL

WITNESSES

NAME	ADDRESS	PHONE

COLLISION DESCRIPTION

INDICATE ON THIS DIAGRAM WHAT HAPPENED
Use one of these outlines to sketch the scene of your accident, writing in street or highway names or numbers.

1. Number each vehicle and show direction of travel by arrow:
2. Use solid line to show path before accident ____; dotted line after accident ----
3. Show pedestrian by: ──────○
4. Show railroad by: ┤┤┤┤┤┤┤┤┤┤
5. Show distance and direction to landmarks; identify landmarks by name or number.
6. Indicate north by arrow, as: ↑

INDICATE NORTH BY ARROW

DRIVER'S ACCOUNT OF ACCIDENT _____
_____
_____
_____
_____
_____

This report is accurate to the best of my knowledge   DRIVER(S) _____   DATE _____

Figure 25–2. Concluded.

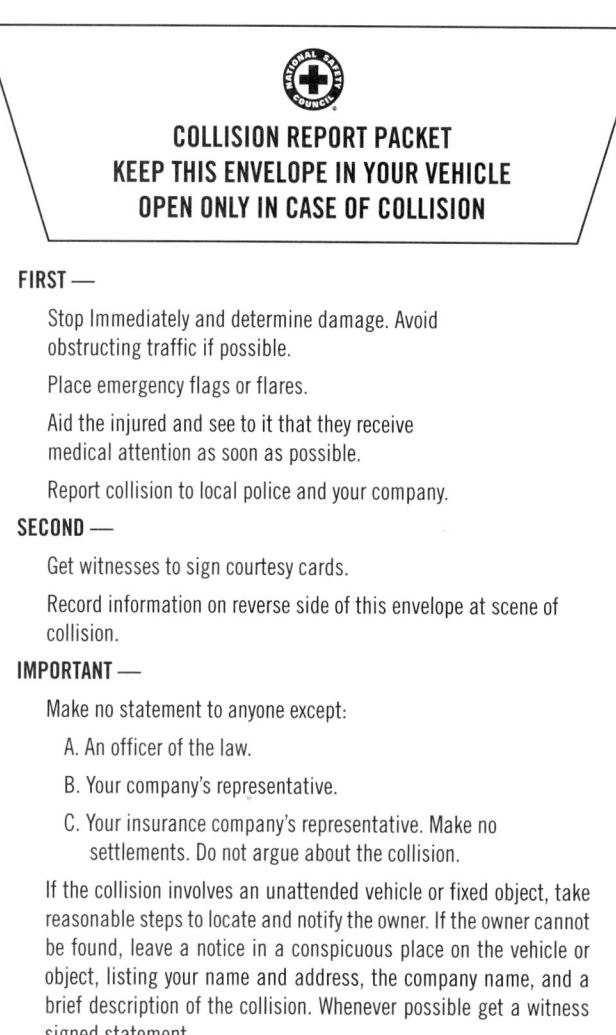

Figure 25–3. This Collision Report Packet contains almost everything a person needs to complete an accurate report and obtain the identification of witnesses. Left: back side of the envelope. Right: front side of the envelope.

WILL YOU KINDLY FAVOR THE DRIVER BY FILLING IN THIS CARD?

ACCIDENT AT _____
(show street no. or intersection)
DATE _____ TIME _____ ☐ A.M. ☐ P.M.

DID YOU SEE THE COLLISION HAPPEN?   ☐ YES ☐ NO
DID YOU SEE ANYONE HURT?   ☐ YES ☐ NO
WERE YOU RIDING IN A VEHICLE INVOLVED?   ☐ YES ☐ NO

IN YOUR OPINION WHO WAS RESPONSIBLE?
☐ OUR DRIVER ☐ OTHER DRIVER ☐ PASSENGER ☐ PEDESTRIAN

NAME _____
ADDRESS _____
CITY & STATE _____ TELEPHONE _____

THANK YOU

Figure 25–4. When this courtesy card is distributed, filled in, and collected, it will help determine who saw the collision, in case witnesses are needed later.

Supplemental Incident Report
Must be completed and attached to each Incident Report

Direct Cause of the Incident (unsafe behavior or condition): _____

_____

Indirect Cause of Incident: _____

Preventable Measures: (Recommendations) _____

Follow-up action requested: _____

Additional Information:
Length of Service with XXX: _____
Years of Professional Experience: _____
Number of Previous Preventable Losses: _____
Date of Last Preventable Accident: _____
Name of Co-Driver: _____

Incident Classification:
Preventable: _____   Non-Preventable: _____

Disciplinary Action to be taken:
Written warning (attach a copy)
Suspended from _____ to _____
Terminated   Yes: _____ No: _____
Lease Terminated   Yes: _____ No: _____
Safety Incentive Loss: _____
Other: _____

_____   _____
Terminal Manager's Signature   Date

Attachments:
Initial each and attach required reports

Telephone Report of Incident or Incident Report	Yes: _____	No: _____
Police Report(s)	Yes: _____	No: _____
Photographs (digital, film, etc.)	Yes: _____	No: _____
Driver's Logs (8 days, including day of Incident)	Yes: _____	No: _____
US DOT Incident Report	Yes: _____	No: _____
Spill Report	Yes: _____	No: _____
5800.1 Spill Report (if applicable)	Yes: _____	No: _____
Driver's Statement	Yes: _____	No: _____
Travel Order	Yes: _____	No: _____
Customer Bill of Lading	Yes: _____	No: _____
Worker Compensation Report (if required)	Yes: _____	No: _____

EHS Department Review: _____

_____

Figure 25–5. This is a sample of a Supplemental Incident Report that details all pertinent information necessary to conduct an incident investigation.

MOTOR TRANSPORTATION
AWARD AND INCIDENT RECORD

JAN	FEB	MAR	APR	MAY	JUN	JUL	AUG	SEP	OCT	NOV	DEC
			X								

Company Highway Express

Address 123 (Number) Home St., (Street) Any Town

Name Smith, (Last) Bill (First) E. (Middle)

Location Midwest Division

Badge Number 123 Date Employed Feb. (Month) 1 (Day) 20-- (Year) Age 29 (At Emp.)

SAFE DRIVER AWARD RECORD

Earned 2 Year N. S. C. Award during period from 3-2-20-- to 3-2-20-- with Company Overnight Freight, Inc.
Certified by F. J. O'Connell, Supt.

DRIVING TESTS

Date	Score	Remarks
2-4-20--	90	Slow reaction time

Award earned	Certificate Number	Date Award earned	Preventable Accidents: File No	Preventable Accidents: Date	Non-Driving Time: From	Non-Driving Time: To
3	468950	3/2/20--	1016	4/10.20--	12/1/20--	1/1/20--
4	501850	3/2/20--				
5	600100	4/2/20--				

REMARKS: 12-1-20-- took on month leave of absence.

Form Vehicle 6 PRINTED IN U.S.A. STOCK No. 229.36

Figure 25–6. An Award and Incident Record should be kept as part of each driver's personnel record. This form is useful in administering a company award plan and in counseling collision repeaters.

review the records regularly to determine trends and recurring incidents.

When a driver becomes a collision repeater, management should make every effort to rehabilitate the person through counseling, retraining, closer supervision, or reassignment. When all efforts fail to reduce preventable collisions, the driver should be discharged or assigned to non-driving duties in the best interests of both the firm and the employee.

Fleet Collision Frequency

A useful collision control tool is monthly or quarterly computing of the fleet's collision frequency rate per million vehicle miles driven. Vehicle miles should be computed from odometer readings of all vehicles and not left to rough guesses based on route mileages unless the operations of the fleet are stable from day to day. The standard formula for figuring a fleet collision rate is as follows:

$$\text{Fleet accident frequency rate} = \frac{\text{Number of accidents} \times 1{,}000{,}000 \text{ actual miles driven}}{\text{Millions of miles driven}}$$

By keeping a monthly record of frequency rates, the investigator can do the following:

- Analyze changes in group safety performance.
- Compare records from several years to find seasonal or other trends.
- Compare the fleet's performance with that of similar fleets.
- Establish year over year safety performance goals.

The fleet safety program can be planned based on an analysis of the collision frequency rate, collision causes, and trends. Collision frequency rates also are of interest to management because they can prove whether the safety program is effective.

The National Fleet Safety Contest, conducted annually by the National Safety Council, issues a bulletin giving the collision frequency rates of all participating fleets, thus providing a means of comparison among similar companies.

Selection of Drivers

For some jobs that require driving, such as sales representative or technician, other qualifications need to be considered besides a safe driving record. However, if the interviewer does not regard driving as an important part of the job, the prospective employee's competence as a driver may not be investigated completely. As a result, when a person is to be hired for a job in which driving is a regular or even occasional function, the company should make every effort to select an individual who can be expected to drive safely. In making that selection, the following factors should be considered.

The company should set up a pre-hire driver profile that identifies the type of driving violations or point system that would assist it in determining if the driver is a safe driver with no serious or repeated traffic violations. A potential driver with numerous small violations may place the company at a greater risk than a driver with a ticket for driving in excess of 20 miles per hour over the speed limit. In all cases, the prospective employee's motor vehicle record (MVR) must be reviewed to ensure that there are no serious traffic violations of the pre-hire driver profile. The driving record should be reviewed by the safety manager. Following initial hire, an annual check of the MVR should be conducted.

Experience
Individuals who have a record of frequent vehicle collisions should not be assigned to drive company vehicles. An individual's safety record should be investigated (1) in a personal interview, (2) by consulting with former employers, and (3) by checking the state motor vehicle department or license bureau for collision reports and the National Driver Register service for either out-of-state or two or more in-state license violations. A number of private services can provide motor vehicle record information electronically via computer within 24 hours.

Personal Traits
Researchers have reported a close relationship between one's ability to drive safely and such personal traits as dependability, good judgment, courtesy, pleasant personality, and the ability to get along with other people. Conversely, those who tend to be antisocial, argumentative, and impulsive often prove to be problem drivers. Likewise, dissatisfied, timid, cocky, troublesome, or otherwise temperamental or unstable individuals often are not consistently good drivers (Figure 25–7).

Selection Standards
In determining standards of driver selection, management should conduct a careful analysis of the particular job and of pertinent driver qualifications. The next step is to decide what qualifications the applicant must have to perform the job effectively. Management must be able to justify each qualification imposed. It is helpful to study the employees who are performing the job in an average or better-than-average manner. Their skills and abilities should serve as the basis for what is expected of new employees. In addition, interstate/intrastate and hazardous materials carriers in the United States must comply with the driver qualification regulations of U.S. Department of Transportation (DOT), which include drug and alcohol testing.

Canadian provinces are encouraged to adopt the National Safety Code. Canadian carriers are required to comply with the provincial regulations adopted from the National Safety Code.

Management should always rank a candidate's safe-driving skills as critically important on the job. Otherwise, collision losses may completely offset any advantages gained through a driver's other special abilities (e.g., sales abilities).

Information-Gathering Techniques
After essential job tasks and employee qualifications have been determined, management's next step is to develop methods for gathering and evaluating data about each applicant. Standard employment procedures include the following:
- application forms
- personal references
- interviews
- psychological tests
- driving tests
- motor vehicle records
- physical examination.

Illustrations of the forms and more descriptive detail can be found in the National Safety Council's *Motor Fleet Safety Manual*, 4th edition.

The application form is used to record details of an individual's past employment and other personal data to help the interviewer determine the best candidate for the job. Because of the Right to Privacy Act, no personal information can be requested on the written application or during the interview that does not have a direct bearing on a position's occupational requirements.

Questions on the form should cover the basic qualifications for the job and should be arranged in logical sequence. Specifically, questions about driving experience should include mileage, number of years applicants have been drivers (long gaps in employment history should be questioned), and types of vehicles they operated; seasons of the year and geographical areas in which candidates operated vehicles; preventable and non-preventable collisions they experienced; number of convictions received for traffic violations; number and type of driver's licenses

Structured Telephone/Walk-In Interview

Hi and thanks for calling/stopping in! My name is _____. I'm sure you would like to hear about our organization and the job we have to offer. Let me ask you some questions and then we can discuss the job in detail.

Name: _____

Address: _____

City: _____ State: _____ Zip Code: _____

Telephone Number: _____ Cell Number: _____

How did you hear about us?

Newspaper: _____ Internet: _____

Trucks: _____ Magazine: _____

Driver: _____ Walk-In: _____

Other: _____

Are you over 21 years of age?   [ ] Yes [ ] No
Are you an experienced Commercial Driver?   [ ] Yes [ ] No
Do you have experience in driving Tankers? [ ] Yes [ ] No
If yes, please describe your driving experience: _____

Employer & Address	Employment Dates	Miles Driven	Contact	States

Driver's License Number: _____ State Licensed In: _____

Are you seeking Over the Road or Regional/Local employment? _____

What home time do you expect? _____ Are you available for work 7 days a week? [ ] Yes [ ] No

Have you ever been denied a License, Permit or Privilege to operate a Motor Vehicle? [ ] Yes [ ] No
If yes please explain: _____

Have you ever had your Driver's License, Permit or Privileges suspended or revoked? [ ] Yes [ ] No
If yes please explain: _____

Have you ever been disqualified to drive a Commercial Vehicle in accordance with Section 391.15 of the Federal Motor Carrier Safety regulations? [ ] Yes [ ] No

As part of our Security Program, **XXX** conducts Criminal Backgrounds checks on all applicants. Have you ever been convicted of a misdemeanor or felony that could prevent you from driving in Canada?
[ ] Yes [ ] No If yes please explain: _____
_____

Figure 25–7. A sample Structured Telephone/Walk-In Interview form for potential driver candidates.

How many accidents have you been involved in during the past five (5) years (regardless of fault or preventability)? _____

How many violations have you incurred? _____

Prior to hiring an applicant, we review their State Motor Vehicle Report. What will we find listed on yours? _____

Prior to hiring an applicant, we require a Physical and a Drug Test in accordance with DOT regulations. Are YOU willing to take and can you pass the tests? [ ] Yes [ ] No

How would you handle being late for a pickup or delivery? _____

Describe a time when you were asked to take another load at the last minute. How did you respond to this request? _____

How would you handle conflicts?
- With your Dispatcher: _____
- At a customer's location: _____
- On the roadway: _____

Do you consider yourself to be a safe driver? [ ] Yes [ ] No

Why? _____

What do you want from this job that you were not getting from your current or past employers? _____

Okay. I have asked enough questions for now. Let's give you a turn. What can I tell you about our organization?

Do you feel the position I described is something that is of interest to you? If applicant says yes invite him/her to come in and complete an application. Provide them with two options to come in. Example: we can get together today after 1 PM or would tomorrow at 10 AM be more convenient? I am really looking forward to having you on the **XXX** Team.

Application Mailed: _____ Interview Set Up For: _____

Interviewer: _____ Date: _____

Recommendation: _____

Figure 25–7. Concluded.

held; and safe-driver awards they received. The Office of Motor Carriers requires such specific information on the application form; other regulatory agencies may have other requirements.

Personal References
The applicant should furnish the names and addresses of previous employers in the space provided on the application form; the interviewer should check these references. The interviewer can also look up an applicant's record with the appropriate state motor vehicle department. Organizations should consult their legal adviser to guard against any liability resulting from personnel investigation activities or any other aspects of the hiring process.

A properly conducted interview is intended to reveal additional facts about the applicant's employment experience, knowledge of traffic regulations, attitude, personality, appearance, family life, and general background in order to help the interviewer make a good decision. The interviewer should conduct the session in private and make sure the applicant is seated and put completely at ease. The interviewer should keep in mind at all times the inventory of basic job qualifications; a written checklist can serve as a guide. After the interview, the applicant can be evaluated for each of the qualifications listed (Figure 25-7).

Driving Tests
Each applicant should undergo an actual driving test or a written examination on traffic regulations as part of the selection procedure. Some firms use a written examination to evaluate driving ability. After an individual is hired, the examination can again be used to monitor his or her developing skills when undergoing an extensive driver training course.

There are two types of driving tests: driving range (off the road) and in traffic (on the road). The requirements of a good driving test in traffic can be listed as follows:
- The test should be designed to sample a number of typical driving situations.
- The test should include typical maneuvers to test a driver's ability.
- The test should use a standard scoring procedure and predetermined test route so that all drivers are evaluated equally.

The examiner should have an objective checklist concerning the driver's performance (to reduce subjective judgment) and determine whether driving errors may be corrected by proper training.

Driver Performance Measurement Test
Michigan State University (MSU) has developed the Driver Performance Measurement Test. This test can help assess safe-driving ability during selection by pointing out the precise unsafe habits likely to result in collisions. Information can be obtained from Michigan State University's Lifelong Education Highway Traffic Safety Program at 517-355-3270.

The MSU system requires the preparation of a thoroughly calibrated test course on public roads. As many as 50 traffic situations may be charted and correct/incorrect driving procedures determined for each. After MSU validates the course, nearby companies can use it to test their drivers once individual company observers have been trained and certified into the MSU system.

Although the driving range test course obviously requires considerable organization and test time, it does give a standardized measure of driving performance.

Some companies use a truck "rodeo" type of driving course in which maneuvering skill can be measured. Skills may be important to measure in a city driver who, for instance, may be backing into tight alleys and loading docks.

In some trucking operations, companies may test job applicants' skills in nondriving tasks, including required paperwork and reports. Experienced tank truck driver applicants may be asked to hook up hoses or load trucks. A driver claiming experience with hauling doubles (trailers) may be required to hook up a set of trailers. The applicant's driving test performance will indicate whether driver training is needed and whether weaknesses can be corrected during the training period (Figure 25-8).

Tests for interstate/intrastate and hazardous materials drivers must meet DOT federal requirements. The DOT has developed regulations covering driver licensing, drug testing, training, physical examinations, and other qualifications. Depending on the type of motor vehicle operated, organizations must comply with all or some of these requirements. As mentioned earlier, Canadian operations need to check the provincial regulations adopted from the National Safety Code.

Medical Examination
Before being hired, all applicants should be examined by a physician who is familiar with the various governmental requirements. For U.S. firms engaged in interstate/intrastate commerce or hazardous materials transport, medical examination of all new drivers is mandatory, and drivers must be reexamined at least every 2 years.

Drivers should also meet certain psychophysical, vision, depth perception, and hearing qualifications. Drivers should be advised of substandard results and the best way to compensate for them.

RECORD OF ROAD TEST

Instructions to Evaluator: Use a checkmark on items that the driver performs satisfactorily, use "X" where performance is unsatisfactory. Any items not evaluated, leave blank.

Driver's Name _____ Home Address _____

Social Security No. _____ License No. _____ State Class _____

Equipment Driven: _____ Tractor Trailer(s) _____
　　　　　　　　　　(Make & Model)　　　　　　　　　　　　　　　　　(Body Type & Length of Each)

Length of Test _____ Mi.From/In _____ To _____

Start Time _____ Finish Time _____ Weather Conditions _____

PART 1—PRE-TRIP INSPECTION AND EMERGENCY EQUIPMENT
Checks general conditions approaching unit　　　_____
Checks fuel, oil, water and for excessive oil
on engine　　　_____
Checks around unit-Tires, lights, trailer hookup,
brake and light line, doors and inspects for body damage　　　_____
Tests steering, brake action, tractor protection
valve, and parking brake　　　_____
Checks horn, windshield wipers, mirrors,
emergency equipment; reflectors, flares, fuses,
tire chains (if necessary), fire equipment　　　_____
Checks instruments for normal readings　　　_____
Checks dashboard warning lights for proper functioning　　　_____
Cleans windshield, windows, mirrors, light and reflectors　　　_____
Reviews and signs previous report　　　_____

PART 2—COUPLING AND UNCOUPLING
Connects glad hands to trailer to apply trailer
brakes before coupling　　　_____
Connects glad hands and light line properly　　　_____
Couples without difficulty　　　_____
Raises landing gear fully after coupling　　　_____
Visually checks king pin assembly to be certain
of proper coupling　　　_____
Checks coupling by applying hand valve or tractor-protection
valve (trailer air supply valve) and gently applying pressure
by trying to pull away from trailer　　　_____
Ensures that surface will support trailer before
uncoupling　　　_____

PART 3—PLACING VEHICLE IN MOTION AND USE OF CONTROLS
A. MOTOR
Places transmission in neutral before starting engine　　　_____
Starts engine without difficulty　　　_____
Checks instruments at regular intervals　　　_____
Maintains proper engine rpm while driving　　　_____

B. BRAKES
Knows proper use of and checks tractor-
protection valve (trailer air supply valve)　　　_____
Tests service brakes　　　_____
Builds full air pressure before moving　　　_____

C. CLUTCH AND TRANSMISSION
Starts unit moving smoothly　　　_____
Uses clutch properly　　　_____

D. LIGHTS (If tested at night)
Adjusts speed for range of headlights　　　_____
Dims lights when approaching another
vehicle or following traffic　　　_____

PART 4—BACKING AND PARKING
A. BACKING
Gets out and checks area before backing　　　_____
Understands and utilizes mirrors properly　　　_____
Signals when backing (if appropriate)　　　_____
Avoids backing from blind side　　　_____

B. PARKING (CITY)
Parks without hitting any other vehicles or
stationary objects　　　_____
Parks correct distance from curb　　　_____
Secures unit properly—sets parking brake,
transmission in correct gear, shuts off
engine, blocks wheels (when necessary)　　　_____
Carefully enters traffic from parked position　　　_____

C. PARKING (ROAD)
Parks off pavements　　　_____
Secures unit properly　　　_____
Uses emergency warning signal or devices when
necessary　　　_____

Figure 25–8. A sample driver road test evaluation form. The road test is usually conducted by a company representative or certified driver trainer.

PART 5—SLOWING AND STOPPING
Uses clutch and gears properly _____
Gears down properly before descending hill _____
Starts without rolling back _____
Tests brakes before descending grades _____
Uses brakes properly on grades _____
Makes proper use of mirrors _____
Plans stop far enough in advance to avoid hard braking _____
Stops clear of crosswalks _____

PART 6—OPERATING IN TRAFFIC, PASSING, AND TURNING
A. TURNING
Signals intention to turn well in advance _____
Gets into proper lane well in advance of turn _____
Checks traffic conditions and turns only when intersection is clear _____
Restricts traffic from passing on right when preparing to complete right hand turn _____
Completes turn promptly and safely and does not impede other traffic _____

B. TRAFFIC SIGNS AND SIGNALS
Plans stop in advance and adjusts speed correctly _____
Obeys all traffic signals _____
Comes to a complete stop at all stop signs _____

C. INTERSECTIONS
Yields right of way _____
Checks for cross traffic regardless of traffic controls _____
Enters all intersections prepared to stop if necessary _____

D. GRADE CROSSINGS
Stops at a minimum of 15 feet but not more than 50 feet before crossing if stop is necessary _____
Selects proper gear and does not shift gears while crossing _____
Knows and understands federal and state rules governing grade crossings _____

E. PASSING
Allows sufficient space ahead of passing _____
Passes only in safe locations _____
Signals changing lanes before and after passing _____
Warns driver ahead of his or her intention to pass _____
Passes with sufficient speed differential to minimize obstructing traffic _____
Returns to right lane promptly, but only when safe to do so _____

F. SPEED
Observes speed limits _____
Drives at speed consistent with ability _____
Adjusts speed properly to road, weather and traffic conditions _____
Slows down in advance of curves, danger zones, and intersections _____
Maintains constant speed where possible _____

G. COURTESY AND SAFETY
Yields right of way _____
Consistently strives to drive in safe manner _____
Allows faster traffic to pass _____
Uses horn only when necessary _____

PART 7—MISCELLANEOUS
A. GENERAL DRIVING ABILITY AND HABITS
Consistently alert and attentive _____
Consistently aware of changing traffic conditions _____
Anticipates problems _____
Performs routine functions without taking eyes from road _____
Checks instruments regularly while driving _____
Remains calm under pressure _____

B. USE OF SPECIAL EQUIPMENT (SPECIFY)
_____
_____
_____

REMARKS:
_____
_____

GENERAL PEFORMANCE: Satisfactory ☐   Needs Training ☐   Explain _____

QUALIFIED FOR: Straight truck ☐ ;   Tractor-Semitrailor ☐ ;   Twin Trailers ☐ ;   Other Combinations ☐ ;

Special Equipment _____
(Specify)

_____ Date: _____
Signature of Examiner

Figure 25-8. Concluded.

Acceptance Interview

Management should contact all applicants who pass the various requirements and tell them whether they have been hired or placed on a waiting list. The interviewer should reinforce applicants' enthusiasm for the job. Applicants should feel that although company employment standards are rigorous, they have attributes the company sincerely wants. When applicants are accepted, the interviewer and others should welcome them into the company as valued new employees. Their new boss or supervisor should give further orientation about the company, what training they will receive, when and where to report for work, and an explanation of wages and working conditions.

Legal and Social Restrictions

The employer must abide by certain state and federal hiring requirements. Laws forbid discrimination against any job applicant on the basis of characteristics such as sex, creed, race, ethnic background, and disability. These are not unreasonable restrictions and certainly should not force the employer to hire the unqualified. No applicant should be hired for a position for which he or she is not qualified. Such an individual runs the risk of being involved in collisions, incurring possible injuries, and creating an unfavorable work record.

Driver Training

Companies should strive to provide individual training by a skilled instructor to all employees assigned to drive company motor vehicles. Various types of training can be offered:
- basic—for new employees
- remedial—for drivers who get into trouble
- refresher—for periodic updating of all drivers
- special—for operators of specialized equipment.

The company must plan a course to fit each job, assemble appropriate training materials, and provide training facilities. The driver training course should cover the following points (see Figure 25–9).

State and Federal Driving Rules

Most state and provincial motor vehicle departments publish the rules and regulations for those seeking driver's licenses. The training course should cover the salient points found in such booklets. Employees subject to U.S. licensing requirements and U.S. DOT examinations should attend appropriate training courses. The same is true of employees of Canadian operations as well as those of other countries.

Company Driving Rules

Each company has rules governing the use of company vehicles: for instance, how the vehicles may be obtained, who may drive them, where they may be operated, where they may be parked and under what conditions, what speed should be observed, and so on. Instructors should cover these rules thoroughly in the training course and give drivers written guidelines to take with them on the job.

What to Do in Case of a Collision

This topic should cover how to prepare company collision report forms, how individual driver records affect the employee, what to do in case of a vehicle collision away from the facility, mandatory reporting requirements, and so on.

Defensive Driving

This concept embraces all the commonsense rules of safe and courteous driving. Defensive driving instruction seeks to impart to prospective drivers a strong sense of responsibility not only for their safety, but also for the safety of others who are less skilled and who have had less training and practice (Figure 25–10). The company's safe-driver award or incentive plan should be explained, along with policies for disciplinary action and criteria for determining preventable collisions.

Safe-Driving Programs

Most fleet professionals regard it as part of their job to provide effective safety motivation so drivers will use their driving skills all of the time. To provide this motivation directly or indirectly, the safety program should include the following:
- a required detailed report of every collision
- driver interviews after each collision to determine whether it could have been prevented
- a record of each driver's safety performance
- continuous safety instruction and reminders through the use of all media: company newsletters, bulletins, booklets, posters, bulletin board displays, meetings, and direct conversation
- safe-driving performance recognition.

A cautionary word about safe-driver awards and prizes: although these awards and cash or merchandise prizes for safe driving are strong motivators, company management must understand that a recognition program supports a safety program and is not a substitute for it.

Safety Devices

Some motor vehicle collisions are due to lack of safety devices and/or inadequate vehicle maintenance. Therefore, one of

Student Trip Report

Trainee_____ Tractor#_____
Trailer#_____
Entry Level: [ ] Yes [ ] No Date of Training_____ Day of training_____
Shipper_____ _____ Consignee_____
Origin_____ _____Destination_____

Grade this report by placing P – Pass or F – Fail in the blanks that apply.

A. PERSONAL		D. TERMINAL & PRE-TRIP PROCEDURES	
1. Appearance	_____	1. Travel Order	_____
2. General Attitude	_____	2. Verify Trailer #	_____
3. Overall Conduct	_____	3. Accident Packet	_____
		4. Pre-trip Inspection Tractor/Trailer	_____

B. ATTITUDE **E. DRIVING ABILITY**
1. Towards Safety _____ 1. Tanks _____
2. Towards Customers _____ 2. Dry Bulk _____
3. Towards Co-Workers _____

C. ABILITY TO LEARN/UNDERSTAND **F. POST TRIP PROCEDURES**
1. Directions _____ 1. Completing Paperwork
2. Loading Procedures _____   Travel Order _____
3. Plant Procedures _____   Shipping Papers _____
   Lab Samples _____   Logs _____
   Shipping Papers _____   Vehicle Condition Report _____
4. Followed Consignee rules _____ 2. Call-in procedures
   Authority Signed on TO _____   Qualcom _____
   Open/Close Dome Lids _____   GeoLogic _____
   Hook/Unhook hoses _____   Two-way Phone _____
   Open/Close Trailer Valves _____ **G. SAFETY**
   Clears lines and hose _____ 1. PPE used _____
   No spills occurred _____ 2. 3-point contact used (tractor/ladders) _____
  3. Proper lifting – hoses _____
  4. Understands Emergency Procedures _____

If any of the following items were used on this trip, place a check in the appropriate blank:

Truck pump _____   Plant Pump _____   Compressor _____   Plant Air _____   Gravity _____
Hoses _____   Drum Nozzle _____   Fittings _____   Bug Screens _____   Pneumatic Blower _____
Valves: Hydraulic _____   QRB _____   Dairy _____   Butterfly _____   Emergency _____

Trans-loading from rail cars: Liquid_____ Dry _____
PD rail car_____ Air Slide car _____ Sifter _____ Spill containment_____ Grounding_____
Product _____ UN #_____ Hazard _____

Please answer all questions as if you were hiring this person to drive your truck!
Comments:_____
_____

Student signature _____ Date _____
_____ has met all the conditions of the Student Driver Report period and is ready
          Student
to drive for **XXXX** effective _____Date

_____   _____
       Driver Trainer/Date   Terminal Manager/Date

Figure 25–9. A sample Student Trip Report for tank/dry bulk trailers.

the fundamental requirements for safe operation is that all vehicles be equipped with the necessary safety devices. Tires, brakes, steering mechanisms, headlights, taillights, horns, and safety restraints must be maintained in first-class condition and used when required (Figures 25–11a and 25–11b). Additional safety devices include the following:
- directional signals
- windshield wipers
- windshield defroster
- fire extinguisher
- power steering/brakes
- antilock brakes/traction control
- low air-pressure warning system
- rock guards over the drive tires/mud guards on other tires
- adequate outside mirrors
- backup lights/required warning lights
- audible backup signal for trucks/power equipment
- slip-resistant surfacing on fenders, floors, and steps
- safety belts/restraint/protection devices/air bags/tires/chains (when required)
- automatic sander (buses)
- antijackknife device (where required)
- reflective markings.

Figure 25–10. Training sessions in the National Safety Council's Defensive Driving Course use audiovisual materials to aid retention of good driving tips.

In addition, the following devices are recommended for dump trucks:
- light or indicator to show when the body is raised
- CAUTION sign on the rear of packer-loader trucks
- cab protector or canopy
- built-in body prop.

Certain hazardous materials and cargo require placards and special precautions when they are loaded and unloaded. U.S. DOT, Environmental Protection Agency (EPA), and/or other regulatory requirements must be followed where applicable.

Unintentional injuries incurred during the loading and unloading of materials, such as lumber, pipe, equipment, and supplies, are especially numerous but can be avoided if these precautions are followed:
- The bulk and weight capacity of the truck should be observed.
- Loads that may shift should be blocked or lashed. Tie downs (ropes, chain, boomers) should be tightened on the right side or top of the load.
- If material extends beyond the end of the tailgate, a red flag (or, at night, a red lamp) should be fastened to the end of the material. No material should extend over the sides.
- Before loading or unloading a truck, the brakes must be securely set and the wheels blocked to protect the workers both on the truck and on the ground (Figure 25–12).
- To reduce the danger to the driver from falling material while the truck is being loaded or unloaded, the truck should be spotted so the load does not swing over the cab or seat.
- If a truck cannot be so located and does not have a protective canopy over the cab, the driver should dismount and stand clear of the truck.
- A truck should not be moved until all workers are either off the truck or properly seated on seats provided and are protected from injury if the load should shift during transit.
- To avoid falling when unloading a flatbed truck, employees should keep away from the sides of the truck, especially when shoes, floors, and loads are wet or muddy.
- Workers must be alert for pinch points when loads are being pulled, hauled, or lifted.
- All safe practices for materials handling, such as using mechanical handling equipment and getting sufficient help, should be observed (Figure 25–13; see also Chapter 12, Materials Handling and Storage, and Chapter 16, Haulage and Off-Road Equipment, both in the *Engineering & Technology* volume).
- Specialized training depending on the classification of the cargo (e.g., flammable, corrosive, or radioactive) should be provided, as appropriate.
- When loading and unloading detached trailers with a lift truck, be sure the wheels are properly blocked/chocked.

It is important to place the chock properly when trailers are being loaded and unloaded at the dock. The preferred location of chocks is at the rear set of wheels. If a trailer is not properly blocked, the vehicle may move because of an incline or be set in motion by the loading or unloading operation (Figure 25–12).

- Under a heavy forklift load, landing gears have collapsed from the weight because of dolly metal rust or fatigue, defective struts, or some other cause. If inspection indicates this possibility, the trailer nose can be supported by screw or hydraulic jacks—one on each side of the nose—to provide additional support for the landing gear assembly (Figure 25–13).

Preventive Maintenance

A well-managed motorized equipment program, whether highway or off-road, includes a properly designed and implemented preventive maintenance (PM) program. Such a program, based on either the mileage or the operating hours of the equipment (as recommended by the manufacturer or company procedures), determines when the employer should change the oil, rotate or replace tires, and undertake minor and major routine maintenance. The objectives of such a program are as follows:

- to prevent collisions and delays
- to minimize the number of vehicles down for repair
- to stabilize the workload of the maintenance department
- to save money by preventing excessive wear and breakdown of equipment and unscheduled downtime.

DRIVER'S VEHICLE INSPECTION REPORT

AS REQUIRED BY THE D.O.T. FEDERAL MOTOR CARRIER SAFETY REGULATIONS

CARRIER: ____________________

DATE: __________ TIME: __________ A.M. __________ P.M. __________

Check any defective item and give details under "Remarks"

TRACTOR NO. ____________________

☐ Air compressor	☐ Front Axle	☐ Springs
☐ Battery	☐ Fuel Tanks	☐ Steering
☐ Body	☐ Heater	☐ Tachograph
☐ Brake Accessories	☐ Horn	☐ Tires
☐ Brakes	☐ Lights	☐ Transmission
☐ Clutch	☐ Loses Water	☐ Wheels
☐ Defroster	☐ MIrrors	☐ Windows
☐ Door Handles	☐ Oil Pressure	☐ Windshield Wiper
☐ Drive Lane	☐ Radiator	
☐ Engine	☐ Rear End	☐ Other
☐ Fifth Wheel	☐ Safety Equipment	

TRAILER(S) NO.(S) ____________________

☐ Brake Connections	☐ Roof
☐ Brakes	☐ Springs
☐ Coupling Chains	☐ Tarpaulin
☐ Coupling (King) Pin	☐ Tires
☐ Doors	☐ Wheels
☐ Hitch	
☐ Landing Gear	
☐ Lights	☐ Other

☐ CONDITION OF THE ABOVE VEHICLES IS SATISFACTORY

REMARKS: ____________________

Driver's Signature ____________________

☐ Above defects corrected ☐ Above defects need not be corrected for safe operation of vehicle

Mechanic's Signature ____________________

Driver reviewing repairs ____________________ (Signature)

Figure 25–11a. Although maintenance personnel are responsible for giving the driver a vehicle that is in top mechanical condition, the driver must inspect the vehicle at the start of each day. Using a checklist ensures that no point is forgotten. *(Courtesy JJ Keller & Associates, Inc., Neenah, WI. Used with permission.)*

The PM program should cover all mechanical factors relating to safe operation of motorized equipment, such as brakes, headlights, rear and stop lights, turn signals, tires, wheel rims, coupling devices, emergency equipment, windshield wipers, muffler and exhaust systems, steering mechanisms, glass fixtures, horns, and rearview mirrors. The manufacturer of the equipment and lubricant supplier can help with maintenance specifications.

Drivers and operators can play an important role in a PM program if they are properly instructed and motivated. Because they are most familiar with the vehicle and how it normally operates, they are usually the first to notice when minor or major mechanical defects develop. If possible, each driver or operator should be assigned a specific vehicle and be given responsibility for reporting defects.

9-Step Safety Check

(FOR USE AS A PRE-TRIP VEHICLE INSPECTION)

Inspect the following items that apply to the vehicle before leaving the terminal. Report any unsatisfactory condition to your supervisor.
ITEMS IN ITALICS REFER TO COMBINATION VEHICLES ONLY.

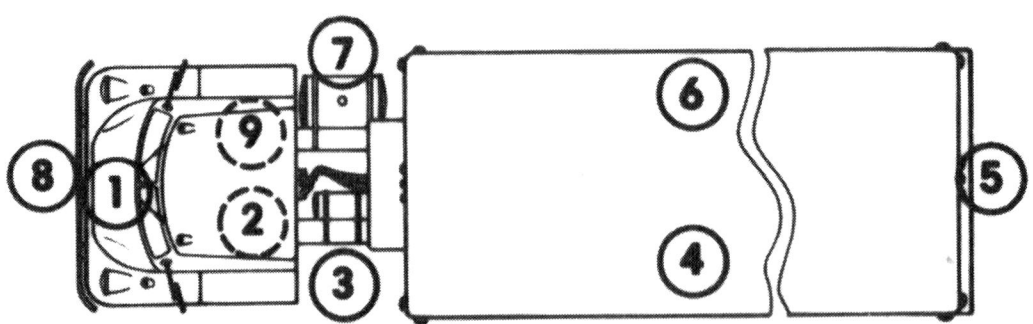

Wind tachograph, if necessary, and insert tach chart before starting nine point check.

STEP 1 – UNDER HOOD
A. Check engine oil level before starting engine.
B. Check radiator coolant level.
C. Check drive belts.
D. Check fluid level in alcohol evaporator or injector (where applicable).
E. Check for oil and water leaks.
F. Secure hood.

STEP 2 – IN CAB
A. Ensure parking brake is applied.
B. Adjust driver's seat.
C. Check vehicle for registration and accident forms also driver's log book (where applicable).
D. Inspect emergency equipment.
E. Clean and inspect windshield, mirrors, and windows.
F. Start engine. Note operation of low air pressure indicator.
G. Check oil pressure, fuel gauges, alternator charging rate.
H. Check operation of horns, windshield wipers, defroster, and fans.
I. Turn on all external vehicle lighting devices.

STEP 3 AND 7 – AT REAR OF CAB
A. Check that correct hazardous material placard is intact and clean (where applicable).
B. *Inspect glad hands to ensure they are properly locked.*
C. *Ensure electrical lines between tractor and trailer(s) are properly connected.*
D. Check air dryer and drain water from air tank(s).

E. Clean and inspect operation of lighting devices and reflectors.
F. *Inspect tractor rear tires (4/32" tread or more), proper inflation, valve caps, rim lugs, wheel slippage, leaking grease seals.*
G. Check for wheel chocks (where applicable).
H. Inspect emergency equipment (where applicable).
I. Check fuel supply and inspect fuel tank(s) for leaks.
J. *Check D.O.T. Visual inspection date (okay if within last two years).*
K. *Check that fifth wheel locking device is properly closed.*
L. Check exhaust system for leaks.
M. Check tractor license plate.

STEP 4 AND 6 – TRAILER SIDES
A. Check that correct hazardous material placard is intact and clean (where applicable).
B. Check fire extinguisher seals.
C. *Check that trailer support legs are properly stowed.*
D. Clean and inspect operation of lights and reflectors.
E. *Drain water from trailer air tanks.*
F. Inspect tires (4/32" tread or more), proper inflation, valve caps, rim lugs, wheel slippage, leaking grease seals or proper oil level in wheel bearings.
G. Check that horns are properly stowed.
H. Check for appropriate delivery fittings (where applicable).
I. Ensure all delivery valves and can box doors are closed.
J. Check vapor recovery lines and ensure that interlock valves are working.

STEP 5 – AT REAR OF VEHICLE
A. Check that correct hazardous material placard is intact and clean (where applicable).
B. Clean and inspect operation of lights and reflectors.
C. Secure cargo – check cargo fastening devices, doors, valves (where applicable).
D. *Check tow-bar and safety chain or cable (where applicable).*
E. Check vehicle license plates, and highway use plates.
F. Check tailgate operation.

STEP 8 – FRONT OF VEHICLE
A. Check that correct hazardous material placard is intact and clean (where applicable).
B. Clean and inspect operation of lighting devices and reflectors.
C. Check vehicle license plates, and highway use plates and decals.
D. Inspect vehicle front tires (4/32" tread or more), proper inflation, rim lugs, wheel slippage, valve caps, leaking grease seals or proper oil level in wheel bearings.

STEP 9 – IN CAB
A. Test service brake system for air leaks.
B. *Check operation of tractor protection valve.*
C. Apply service brakes to check stop – lamp operation.
D. Fasten seat belt.
E. *Test fifth-wheel connection under power with trailer brakes applied.*
F. Make a test stop with service brakes.
G. Check steering mechanism for ease of operation.

This form complies with the intent of Department of Transportation (D.O.T.) Regulation 392.7 and 392.8.

Figure 25-11b. Using this truck inspection chart will ensure that no items are overlooked.

This encourages drivers and operators to take better care of their vehicles.

Drivers regulated by U.S. DOT (40 CFR 392.7) must perform a pre-trip and post-trip vehicle inspection. However, all drivers should conduct, at a minimum, a walkaround inspection of their vehicles after every stop (food, fuel, sleeper berth, etc.). Any major or safety-related defects must be reported and corrected before that vehicle

Figure 25–12. This 8-in. (20-cm) -high heavy-duty cast aluminum chock provides secure chocking for trucks or trailers. *(Courtesy Worden Safety Products Inc.)*

Figure 25–13. Properly placed trailer stabilizing jacks will help prevent trailer tipping or falling. *(Courtesy The Aldon Company.)*

is used. Minor items not affecting the vehicle's operation can be corrected at the next scheduled PM.

The company should provide vehicle condition report forms to all drivers and operators for the inspection, and a copy of the form should be kept in the vehicle cab. The form should cover the following items:

- Brakes should apply evenly to all wheels to prevent a vehicle from swerving when coming to a stop. This even pressure also gives maximum braking effectiveness.
- Headlights should function and be properly aimed to avoid blinding other motorists and to provide maximum lighting of the road ahead and to each side. The dimming switch and upper and lower beams should work properly at all times.
- Connecting cables on a combination vehicle should be fastened securely enough to be unaffected by vehicle vibration. All other cables, such as brake and electric ("glad hands"), should be free of defects.
- Make sure all stop lights, turn lights, rear lights, warning lights, and side-marker lights function properly.
- Tires should be inflated to the recommended pressure and regularly checked for adequate tread and any cuts, breaks, or other defects. Dual tires should be well matched.
- Windshield wipers must wipe clean and not streak or skip any of the glass, which can obscure or blur a driver's view of the road.
- Steering wheels should be free from excessive play. Front wheels should be properly aligned.
- All window glass should be free from cracks, discoloration, dirt, or unauthorized stickers that may obscure vision.
- Horns should respond to a light touch and be easy to locate.
- Rearview mirrors should give the driver a clear view. Portions of outside rearview mirrors can be conventional and convex to provide maximum sight advantage.
- Any stalling or lugging problems of the engine should be investigated and corrected immediately.
- All instruments must work properly.
- Exhaust systems should be checked to protect the driver against carbon monoxide gas leaks. The exhaust manifold, pipe connections, and muffler should be inspected periodically.
- Load securement devices such as wing nuts on dome lids; straps on flatbeds; bracing in van trailers; and valves on tank/dry bulk trailers must be secured and verified prior to any movement of the shipment.
- The cab should be kept free of unnecessary clutter and debris.
- Coupling devices (such as on a fifth wheel) should be securely mounted to the frame, with the locking jaws around the shank and not the head of the kingpin. The kingpin should not be worn, bent, or damaged.

Emergency equipment in every vehicle must include a fire extinguisher, essential tools for road repairs, spare bulbs, flares, reflectors, flags, and such other equipment deemed necessary in case of fire, collision, or road breakdown. These items should be checked regularly to ensure that they are in good working order and readily available. Interstate and hazardous materials vehicles must be equipped with emergency items required by the appropriate regulatory agency.

Many governmental agencies require periodic safety tests and inspections for all vehicles. The maintenance facility must know the applicable inspection standards, and the company should ensure that all vehicles meet these requirements.

REPAIR SHOP SAFE PRACTICES

Vehicle maintenance facility supervisors must constantly evaluate the work habits of automotive repair employees, determine their safety attitudes, and cooperate with management to create a safe work environment. An effective safety program can help ensure that all employees engage in safe work practices and that all federal and state or provincial regulations are followed.

Servicing and Maintaining Equipment

Workers may incur a wide range of injuries while servicing and maintaining trucks. The equipment often requires mechanical aids for handling heavy parts.

Serious injuries are likely to occur when equipment undergoing repair shifts or moves unexpectedly. Workers should set equipment brakes and block the wheels. If work must be done under a raised vehicle, the vehicle must be propped up in such a way that it cannot come down should someone or something inadvertently strike the hoist or jack control levers or pedals and release the load.

Workers often use a jack to raise equipment and then leave it in place as an unstable support. Because serious injuries may occur when a vehicle falls off a jack, workers must set the jack on a firm foundation and make sure it is exactly perpendicular to the load. When the vehicle has been raised to the desired height, it should be supported by stanchions, blocking, or other secure support.

Employees should never work near the engine fan or other exposed moving parts until the engine has been stopped. If they must run the engine to inspect or check on moving parts, they should keep a safe distance away and not attempt to make an adjustment. Workers should wear close-fitting clothing, safety shoes, and goggles, but remove jewelry, especially rings, while repairing equipment.

Lockout procedures must be used when a mechanic is working under the hood or anywhere on the tractor or hooked trailer to prevent someone from inadvertently starting and/or moving the vehicle.

For internal tank trailer integrity testing or repairs inside a trailer (especially hot work activities), a permitted confined-space entry program (29 CFR 1910.146) must always be enforced.

Mechanics working on the top of trailers must be part of a fall-protection program. The program can include a walking/working (catwalks, ramps) rack system but must include a full fall-arrest system (engineered trolley system, harnesses, lifelines, etc.) if a worker is required to work on top of a trailer.

Employees risk burn injuries when servicing vehicle engines that have been running. To open a radiator, the employee should use a heavy work glove, bleed off any steam, and then remove the cap. Gasoline or alcohol spilled on hot engines can cause a serious fire. Suitable funnels and safety containers should be used when transferring liquids.

Tire Operations

Workers face a particularly serious hazard when inflating vehicle (i.e., truck) tires should the locking ring blow off at high pressure (Figure 25–14). Using a tire safety rack will greatly reduce the potential for injury. Workers need to check all tires to make sure the valve has been removed and the tire fully deflated before disassembling it. When removing a tire from a dual wheel, fully deflate both tires before removing the one to be repaired. Workers must inflate tires in steel cages that will restrain flying objects should a blowout occur. A locking ring must be seated properly and must not be yanked free by being twisted. Always replace a defective locking ring or rim with a new one and keep ring and rim seats clean. Carefully separate parts, rims, and rings for various types of wheels to eliminate a possible mismatch.

Workers should use inflators that they can preset and that have locking attachments so hands or arms are not in the danger area, even if the tire is in a cage. Blowouts can occur because of overinflation, improper placement of the

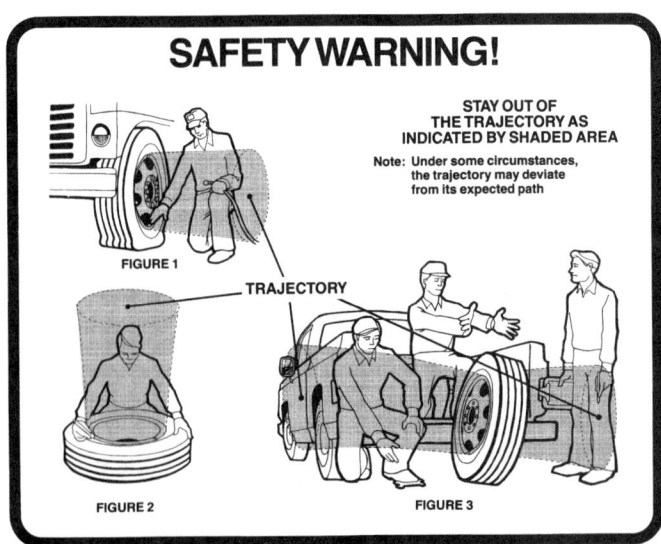

Figure 25–14. Stay out of the trajectory area while inflating tires. (*Courtesy Rubber Manufacturers Association.*)

tire on the rim or wheel (which may pinch or chafe the tire or tube), or improper mounting of lock rings or rims. Only employees trained in tire repair and thoroughly familiar with the hazards and safe methods involved in handling tire equipment should inspect, install, repair, and replace tires and rims.

Other potentials for injury include strains or hernias resulting from lifting heavy tire assemblies. Companies should provide mechanical lifting and moving devices so workers are not required to lift heavy tires.

Casing compounds used for filling tire cuts and rubber cement and flammable solvents used for patching inner tubes should be kept in safety cans. Routinely inspect electric heating elements used for vulcanizing or branding tires and replace any defective wiring.

Where power-driven rasps or scrapers are used for casings or inner tubes, operators must wear eye protection and an approved particulate filtering respirator. Attaching a local exhaust or vacuum system to these machines will keep the fine rubber dust out of the workroom air.

Fire Protection

Because of fire hazards, heavy-duty trucks should be equipped with type ABC fire extinguishers listed by Underwriters Laboratories for use on combustible materials (paper, cardboard), burning oil, gasoline, grease, and electrical equipment. Place the extinguisher in a convenient location in the cab or mount it on the vehicle, and train drivers to operate it. Some regulatory agencies require a monthly inspection of fire extinguishers.

The maintenance facility should also have an adequate number of type ABC fire extinguishers. Trained personnel must show all employees how to use the equipment.

Workers who are performing cutting, burning, or welding tasks near fuel oil tanks should have an extinguisher on hand. A wet tarpaulin may be used to cover fuel, oil tanks, or combustible materials to protect them against sparks and excessive heat. Do not perform such work on a fuel tank or other fuel container until it has been drained and thoroughly purged of vapors. Properly calibrated and adjusted combustible gas indicators (CGIs) will determine vapor levels. Use of the "hot work" permit system is recommended.

Fueling requires certain precautions to avoid fires. Workers must shut off the engine and extinguish all smoking materials. Use safety containers and a grounded fuel hose for fueling. When the tank is being filled, the metal spout of the hose should firmly contact the tank to ensure bonding and to dissipate static charges strong enough to ignite fuel vapors and cause an explosion or fire.

Fires occur in shops each year because gasoline and similar flammable solvents are used for cleaning parts. Therefore, workers should use only safe, nontoxic, and nonflammable cleaning liquids. Parts-cleaning equipment should be provided with a self-closing lid in case a fire occurs. The risk of fire in a shop can be reduced greatly by good housekeeping, especially proper disposal of oil-soaked combustible waste and similar materials in covered metal containers.

Lubrication and Service Operations

In lubrication operations, make sure floors are kept free of grease and oil to prevent slips and falls. Spills should be wiped up immediately or covered with an oil-absorbent compound.

Advise workers to keep their hands away from sharp or rough edges and to obtain immediate first aid for all cuts and scratches. Also, train workers not to put their hands or face in front of the grease gun nozzle when the handle is pulled. In some cases, quantities of grease have been forced under workers' skin by high-pressure grease guns. Tops of grease cylinders should be securely fastened into place; otherwise, covers may blow off and seriously injure anyone who is nearby. Inspect all equipment weekly and make repairs when needed.

Workers should be warned of the danger of inhaling sprayed or atomized oils. They should stand clear of the lubricant spray, which settles quickly, and must not direct the spray at other employees. Many state and federal regulations require that used oil drained from machinery be collected for recycling or properly disposed of by an approved waste-handling company. Drained antifreeze and other drained lubricants such as car oils can be considered hazardous waste and should be treated accordingly. Most companies set up a contract with a certified oil and/or glycol recycler.

Wash Rack Operations

When washing vehicles, a person can slip and fall on wet floors, suffer cuts or abrasions from the sharp or rough edges of the vehicle, or experience burns from careless use of hot water or steam. The concrete floor of the wash rack should be rough troweled to produce a slip-resistant surface. While washing vehicles, employees should wear safety-toe rubber boots, preferably with slip-resistant soles and heels, and a rubber coat or apron and gloves.

Workers should never point high-velocity streams of hot or cold water at another person because serious injuries can result. They should direct the hose, particularly when washing under the vehicle, in such a way as to avoid being struck by a backlashing stream of water and dirt.

Supervisors should provide workers with product safety information on the washing solutions used. Where a hot water or steam hose is used, employees must wear personal protective equipment to avoid skin contact and to prevent burns. Heavy-duty gloves and face shields should be used when necessary. A portable fan can blow steam away to enable the operator to see the work. Supervisors should schedule a periodic cleanup of the entire wash rack and associated equipment.

For internal tank trailer/intermodal cleaning operations, a permitted confined-space entry program (29 CFR 1910.146) must always be enforced. Tank washers working on the top of trailers must also be part of a fall-protection program as mentioned earlier in this chapter under Servicing and Maintaining Equipment.

Battery Charging

Although most battery-operated vehicles are recharged with on-board chargers and plug-in cords, some batteries still require recharging outside the vehicle. Other circumstances occasionally call for battery removal and charging, such as battery servicing, replacement, or emergency jump starts. The principal hazards of battery-charging operations are acid burns, back strains from lifting, electric shocks, slips, falls, and explosions from buildup of hydrogen gas. Management should train workers to follow proper safety procedures whenever they are servicing or charging a battery. They should not charge a frozen battery—most of the charge will go to heat the battery and not to energize it.

Employees should wear safety apparel suitable for battery work. This includes splashproof eye and face protection and acidproof gloves, aprons, and boots with slip-resistant soles. Workers should wear rubber boots and aprons when filling batteries. Goggles and face shields should be used when working around batteries to prevent acid burns to the eyes and face.

Flooring should be constructed of wood slats and kept in good condition to prevent slips and falls and to protect against electric shocks from batteries being charged. Install fire doors between battery-charging rooms and adjacent areas where flammable liquids are handled and stored.

Always follow the manufacturer's recommendations about charging rates for various-sized batteries to prevent rapid generation of hydrogen. Potentially explosive quantities of oxygen and hydrogen accumulate in battery cells. This is particularly true if the battery is defective or if a heavy charge has been or is being applied. The lower the water level in the battery, the larger the cavity in which gas can accumulate. Care should be taken to prevent electric arcing while batteries are being charged, tested, or handled. Make sure tools and loose metal (and even lifting-hoist chains) cannot fall on batteries and cause a short circuit, which, in turn, can inflict serious burns on workers or cause an explosion.

When manual lifting is necessary, tell employees to request sufficient help to prevent strains, sprains, or hernias. Use hand carts for transporting batteries.

Workers must handle acid carboys with special care to prevent breakage and possible injury caused by splashing acid. Acid carboys should never be moved without their protecting boxes nor stored in excessively warm locations or in the direct rays of the sun. Instead, use carboy tilters. A safety shower and an eyewash fountain are required by certain regulatory agencies wherever acids or caustics are used (see Chapter 12, Occupational Health Programs, in this volume).

First Aid for Chemical Burns

Although many batteries are sealed, some batteries require checking fluid levels and contain an acid electrolyte; some, such as the nickel-iron battery, contain an alkali solution. Whether it is acidic or alkaline, if electrolyte gets on a person's skin or eyes, it must be immediately washed off with large quantities of running water for 15 minutes or more. Supervisors and workers should understand that because neutralizing agents are mishandled so often, these agents can do more harm than good. Use them only if first-aid directions are printed clearly on labels or on company or facility directions. Get medical aid for the injured employee at once.

First Aid for Burns of the Eye

Flush the eye thoroughly with large amounts of clean, cool water for 15 minutes or more. Place a sterile dressing over the eye to immobilize the lid and get medical aid at once. Check with the company physician for emergency help. Smartphones for recording may be permitted. Check with your insurance carrier for local and state laws.

Gasoline Safety

Handling and storing of gasoline should comply with the provisions of the National Fire Protection Association's Flammable and Combustible Liquids Code, NFPA 30 and NFPA 30A, which is discussed in Chapter 9, Fire Protection, in the *Engineering & Technology* volume. Never use gasoline for any cleaning tasks. Solvents with higher flash points are equally effective and much safer. Even when higher-flash-point solvents are used, if carburetors or gas-line parts are cleaned, the solution should be changed regularly. The admixture of even small quantities of gasoline will lower the solvent's flash point and increase the danger of fire and explosion.

Workers can remove grease, oil, and dirt from metal parts with nonflammable solutions or with high-flash-point solvents in special degreasing tanks with adequate ventilating facilities and fire protection. Advise workers not to use gasoline to remove oil and grease from garage floors.

Nonflammable commercial cleaning compounds are available for such tasks.

Workers should never use gasoline to remove oil and grease from their hands. Strong commercial soaps will effectively remove greasy dirt from the skin without risk of injury. In addition, protective creams and ointments, if applied before starting work, will protect an employee's skin from dirt and grease. Likewise, workers should never use gasoline to clean work clothes.

Gasoline spills should be cleaned up immediately; many absorbent substances can be used to soak up spills. Dirt, sand, or other material should be used to prevent gasoline from entering sewers, drains, and so on. If quantities of gasoline get into the sewage system, notify the fire department and state EPA at once so the hazard can be handled properly. Because gasoline vapor is heavier than air, it collects in low spots, such as basements, elevator pits, and sumps. Ventilate these places until gasoline vapors are dissipated to a safe level. Do not dispose of spilled gasoline by flushing it down a sewer drain, sink, toilet, or the like. Instead, collect spilled gasoline into an approved container and dispose of it through a waste-handling company or other approved means.

Other Safe Practices

Management should also make sure that workers understand and use safe practices for other types of equipment that pose safety and health hazards. Workers should receive training in these practices during orientation as well as on the job whenever equipment is changed or replaced, jobs are rotated, employees return after leaves of absence, and refresher training is needed. More details on mechanical and chemical safety can be found in the *Engineering & Technology* volume and in *Fundamentals of Industrial Hygiene*, 6th edition (NSC 2012), respectively.

Using Jacks and Chain Hoists

Vehicles jacked up or hung on chain hoists should always be blocked with stanchions or pyramid jacks (make sure these items are rated for load capacity and carefully inspected before use). The best jack for maintenance garage work is the hydraulic-over-air type—if one system fails, the other will operate. Do not use ordinary pedestal jacks, especially the type supplied for passenger cars, because the vehicle can be tipped or jarred off the jacks and cause injury.

When someone is working under a blocked-up vehicle, other employees should not do any work on the car that may knock it off the blocks. Employees who work under vehicles should be safeguarded from danger when their legs protrude into passageways. Erect barricades around them or make sure the worker's entire body is under the vehicle. Advise employees to wear suitable eye protection such as safety glasses, goggles, or plastic eye shields to avoid injury from dirt and metal chips falling into their eyes. When necessary, fog-resistant eye protection should be used.

Removing Exhaust Gases

Local exhaust and ventilating equipment can prevent accumulation of vehicle exhaust gases within a repair shop.

Repairing Radiators

Where radiators are boiled out or tested for leaks, the operator should have both chemical goggles and a face shield of clear plastic. The entire face should be protected.

Cleaning Spark Plugs

All mechanics using blast-type spark plug cleaners should wear goggles or face shields.

Replacing Brakes

Workers should not use air pressure when cleaning around brake drums and backing plates while replacing brakes. The pressure can send asbestos, metal particles, or dust from the brake area into the air and cause respiratory problems. Instead, workers can clean brakes with high-efficiency particulate air (HEPA) vacuums, chemical wash solutions, or steam cleaners. Train workers to wear approved respiratory protective equipment until brake components are free of dust. Often, the work area may not be free of dust, and respiratory equipment may still be needed.

Controlling Traffic

Supervisors should regulate movement of vehicles inside maintenance facilities. Traffic lanes and parking spaces should be painted on the floors, with arrows and signs indicating the direction of traffic flow. Vehicles with air brakes should not be moved until sufficient air pressure has been built up. Move vehicles using low gear and low speeds inside shop areas, especially up and down ramps.

Drivers must stop their vehicles at entrances or exits and make sure the way is clear or sound their horn before passing through. Signs requiring this procedure should be posted in conspicuous places. Install mirrors at blind corners.

Other potential sources of injury include jumping across open inspection pits, falling off ladders, materials handling, and using hand tools improperly. Prevention of all work injuries requires proper selection and training of employees, careful supervision of their work habits, review of all injury causes, and compliance with safety programs.

Training Repair Shop and Wash Rack Personnel

Supervisors and managers should train apprentices and new employees to do each job in the most efficient manner. Make sure job instruction includes the safety rules and regulations pertaining to each job and the reasons for such rules. Job safety analysis (as outlined in Chapter 9, Identifying Hazards, in this volume) should be developed for all repair shop and wash rack job tasks and used as a tool during the orientation process. New mechanics and wash rack operators must be thoroughly indoctrinated concerning the company's policy toward safety and the safety program.

The new employee, having been indoctrinated and trained to work safely, must be motivated to observe accepted safe practices while performing job tasks. Many motivational methods are available to accomplish this goal, including safety supervision; safety contests; and meetings, posters, safety bulletins, and pamphlets (see Chapter 30, Motivation, and Chapter 31, Safety and Health Training, both in this volume).

U.S. DOT–REQUIRED DRUG AND ALCOHOL TESTING

The Omnibus Transportation Employee Testing Act of 1991 requires alcohol and drug testing of safety-sensitive employees in the motor carrier, railroad, aviation, and mass transit industries. The DOT published rules mandating antidrug programs and alcohol misuse prevention programs in February 1994. These rules also expand and supplement the U.S. Drug-Free Workplace Act of 1988, which mandated drug testing of aviation, interstate motor carrier, railroad, pipeline, and commercial marine employees. The February 1994 rules generally required implementation beginning on January 1, 1995, for employers with 50 or more employees in safety-sensitive jobs, and January 1, 1996, for employers with fewer than 50 employees in such jobs. The Omnibus Transportation Employee Testing Act of 1991 is updated regularly, and the most recent levels of drug concentrations should be used in the development and implementation of a company drug program.

The DOT rules include a drug and alcohol testing rule (49 CFR 40) that establishes procedures for urine drug testing and breath alcohol testing. The urine drug-testing procedures rule governs drug-testing programs mandated by the Federal Railroad Administration (FRA), Federal Transit Administration (FTA), Research and Special Programs Administration (RSPA), and the U.S. Coast Guard. Amendments to Part 40 added breath alcohol testing procedures and additional urine specimen collection procedures that provide for split urine specimens. The DOT rules cover safety-sensitive employees in commercial transportation as defined by each DOT agency.

Drug Testing and DOT Requirements

Transportation employers, even self-employed individuals who must conduct urine drug-testing programs under regulations issued by the various agencies of the Department of Transportation, must ensure that all testing is conducted in accordance with 49 CFR 40. Under Federal Motor Carrier Safety Administration (FMCSA) rules, employers are required to test employees for five drugs: marijuana, cocaine, opiates, amphetamines, and phencyclidine (PCP). The rule requires five types of drug tests: pre-employment, post-incident, reasonable suspicion,[1] return-to-duty, and random or unannounced testing. Random testing must be given to a certain portion of the employer's safety-sensitive pool.[2] The random testing works like a lottery, and the percentages of employees to be tested are prescribed by the DOT.

A urine drug-testing custody and control form must be completed at the time of a DOT urine collection and accompany all specimens to the laboratory. This form must conform to the requirements of 49 CFR 40. Employees should be provided a wrapped or prepackaged specimen bottle or collection container in which to urinate. The specimen bottle must be sealed with a tamperproof seal, signed, and dated by the collector.

Collection site personnel must be trained to carry out the required collection procedures, or, if they are licensed medical professionals or technicians, they must have instructions for conducting the required collection procedures. Generally, supervisors of employees being tested should not collect the specimen. Collection site personnel must be the same gender as the donor when a collection is conducted under direct observation. When a collection is conducted in a public restroom or other facility that does not afford the donor complete privacy, a medical professional or technician of either gender may collect the specimen; however, if a nonmedical collector is used, the collector must be the same gender as the donor.

Written instructions and/or procedures concerning specimen collection must be provided to collection site personnel, the employer representative, and employees. Any specimen, if used, must be transported along with associated paperwork from the collection site to a Department of Health and Human Services (DHHS)–certified laboratory

---

[1] Reasonable suspicion exists when a trained supervisor or company official observes behavior or appearances that are characteristic of alcohol and/or substance abuse. All employees designated to supervise drivers must receive, at a minimum, 1 hour of training on recognizing alcohol abuse and 1 hour of training on recognizing controlled substance abuse.

[2] A "safety-sensitive function," as defined by the Federal Motor Carrier Safety Administration (FMCSA), includes all of the time a driver is working and/or is required to be in readiness to work until the time that he or she is relieved from work and all work-related responsibility. This means a driver (or other terminal employees, such as mechanics, loaders, etc.) has to be able to perform the job functions associated with FMCSA-related tasks, including driving a commercial vehicle, performing vehicle inspections, loading or unloading, handling paperwork, etc., while in attendance with the commercial motor vehicle.

in a shipping container that can be sealed with tamperproof tape and initialed to prevent undetected tampering. The collection site should arrange to ship the specimen to the laboratory in shipping containers that minimize the likelihood of transport damage and are sealed with tamperproof tape. The collection site person should sign and enter the date of shipment on the tape.

Employers subject to 49 CFR 40 must use laboratories certified under the DHHS Mandatory Guidelines for Federal Workplace Drug Testing Programs (53 CFR 11970 and subsequent amendments). All DOT tests must be performed in a certified laboratory to meet DOT requirements. The laboratories are required to have internal quality control procedures to monitor each step of the drug-testing process. At least 10% of all specimens tested in the laboratories must be quality control specimens introduced into the testing process by the lab's quality assurance section. Additionally, employers are required to submit three blind quality control specimens to the laboratory for every 100-employee specimens sent for testing. These quality control specimens are called *blind performance tests* because they are not known to the laboratory. Blind quality control specimens can be either blanks (negatives) or spikes (positives).

If a false-positive performance test error is determined to be a technical or analytical procedural error, the employer will, in addition to notifying the DOT agency concerned, instruct the laboratory to submit to this DOT agency all quality control data from the batch of specimens that included the false-positive specimen. Depending on the information provided, DHHS will take appropriate action on the laboratory's certification.

A medical review officer (MRO) should review all laboratory-confirmed positive results. The MRO should be a licensed physician, should have knowledge of substance abuse disorders, and may be on the employer's staff or be a private physician under contract to the employer. The MRO should not be an employee of the laboratory conducting the drug testing unless there is a clear separation of functions to preclude any conflict of interest. The MRO will review positive results reported by the laboratory, will examine alternative medical explanations for any positive test results, and may include a personal interview or physical examination of the employee.

A positive test is not positive until the MRO confirms the test results. The MRO must not discuss the results with the employer or report any findings until his or her verification is completed. Disclosure of any medical information provided by the employee to the MRO as part of the test-verification process is prohibited except under the following circumstances:
- A DOT regulation requires such disclosure.
- The MRO believes that the information could result in the employee being found medically unqualified under a DOT agency rule.
- The information indicates that continued performance by the employee could pose a significant safety risk.

Alcohol Testing and DOT Requirements

Because alcohol is a legal substance, the DOT rules define specific prohibited alcohol-related conduct. Performance of safety-sensitive functions is prohibited under the following conditions:
- when an employee has a breath alcohol concentration of 0.02% or greater as indicated by an alcohol breath test
- when the employee is using alcohol
- within 4 hours, and 8 hours for aviation crew members, after an employee has used alcohol.

In addition, employees are prohibited from refusing to take an alcohol test and from using alcohol within 8 hours after a collision or until tested.

Testing procedures that ensure accuracy, reliability, and confidentiality of test results are outlined in 49 CFR 40. These procedures include training and proficiency requirements of breath alcohol technicians, quality assurance plans for the breath-testing devices, requirements for a suitable test location, and protection of employee test records.

The DOT requires alcohol testing to be performed for pre-employment, post-collision, reasonable suspicion, return-to-duty, follow-up, and random tests. Random alcohol testing must be conducted just before, during, or just after an employee's performance of a safety-sensitive duty. The employee is randomly selected for testing. As with drug testing, the testing dates and times are unannounced and are given with random frequency throughout the year, similar to a lottery. The random rate is set for each industry each year.

DOT rules require breath testing using evidential breath-testing devices (EBTs) approved by the National Highway Traffic Safety Administration (NHTSA). Two breath tests are required to determine whether a person has a prohibited breath alcohol concentration. A screening test is conducted first. Any result less than 0.02% alcohol concentration is considered a negative test. If the alcohol concentration is 0.02% or greater, a second or confirmation test must be conducted. The employee and the individual conducting the breath test (called a breath alcohol technician) complete the DOT-prescribed form to ensure that the results are properly recorded. The confirmation test, if required, must be conducted using an EBT that prints out the results, the date and time, a sequential test number, and the name and serial number of the EBT to ensure the reliability of the results. The confirmation test result determines any actions taken by the DOT and/or the company.

Employees who violate the alcohol misuse rules will be referred to a substance abuse professional for evaluation. Any treatment or rehabilitation should be provided in accordance with the employer's policy or labor/management agreements. The employer is not required under these rules to provide rehabilitation, pay for treatment, or reinstate the employee in his or her safety-sensitive position. Any employer that returns an employee to a safety-sensitive duty must ensure that the employee has been evaluated by a substance abuse professional, has complied with any recommended treatment plan, has taken a return-to-duty alcohol test with a result less than 0.02%, and is subject to unannounced follow-up alcohol tests.

All prescribed and over-the-counter medications that will alter the driver's motor skills, cognitive actions, or response time must be reported to the immediate supervisor for consideration of the driver's operating limitations.

Consumption of energy drinks and use of power aids must also be considered by management.

Setting Up a Testing Program

Many service companies across the United States provide both drug and alcohol testing services. Employers should interview several companies, comparing the pricing and services offered to determine the best provider for their company's needs. In addition to testing, DOT regulations require companies to keep records, offer employee training, and provide employee information on drug and alcohol testing and results (Table 25–B).

SUMMARY

- Most motor vehicle collisions are caused by using improper driving procedures; only a small percentage are the result of mechanical failure. Companies can control driver error by introducing a proactive program of driver selection, training, and supervision; vehicle failure can be reduced by implementing a preventive maintenance program.
- The total cost of vehicle incidents almost always exceeds the amount recovered from the insurance company and includes direct and indirect expenses of collisions. The costs of vehicle collision-prevention programs are more than justified when compared with potential losses related to collisions.
- A vehicle safety program should include a written safety policy, a designated safety program manager, efficient accident investigation and reporting systems, and a preventive vehicle maintenance program.
- The safety professional or fleet manager is responsible for supervising the program and reporting on safety issues to top management.
- A driver safety program should include a training program, collision-prevention measures, reporting procedures, driver performance goals and incident reports, and a method for establishing competency and skills levels and collision/safety records for each driver.
- A motor vehicle collision can be defined as any incident in which the vehicle comes in contact with another vehicle,

TABLE 25–B. Cutoff Concentrations for Drug Tests

U.S. Department of Transportation   49 CFR Part 40 Section 40.87

Initial Test Analyte	Initial Test Cutoff Concentration	Confirmatory Test Analyte	Confirmatory Test Cutoff Concentration
Marijuana metabolites	50 ng/mL	THCA[1]	15 ng/mL
Cocaine metabolites	150 ng/mL	Benzoylecgonine	100 ng/mL
Opiate metabolites			
Codeine/morphine[2]	2000 ng/mL	Codeine	2000 ng/mL
		Morphine	2000 ng/mL
6-acetylmorphine	10 ng/mL	6-acetylmorphine	10 ng/mL
Phencyclidine	25 ng/mL	Phencyclidine	25 ng/mL
Amphetamines[3]			
AMP/MAMP[4]	500 ng/mL	Amphetamine	250 ng/mL
		Methamphetamine[5]	250 ng/mL
MDMA[6]			
	500 ng/mL	MDMA	250 ng/mL
		MDA[7]	250 ng/mL
		MDEA[8]	250 ng/mL

[1] Delta-9-tetrahydrocannabinol-9-carboxylic acid (THCA).
[2] Morphine is the target analyte for codeine/morphine testing.
[3] Either a single initial test kit or multiple initial test kits may be used, provided the single test kit detects each target analyte independently at the specified cutoff.
[4] Methamphetamine is the target analyte for amphetamine/methamphetamine testing.
[5] To be reported positive for methamphetamine, a specimen must also contain amphetamine at a concentration equal to or greater than 100 ng/mL.
[6] Methylenedioxymethamphetamine (MDMA).
[7] Methylenedioxyamphetamine (MDA).
[8] Methylenedioxyethylamphetamine (MDEA).

person, object, or animal with resulting injury or property damage. All collisions must be reported, and reports provide information that may help to prevent similar accidents in the future.
- Collision-reporting procedures should enable all employees to follow the same steps and use the same forms to document details and to record names and addresses of witnesses. An investigator should interview drivers immediately after a collision to help the company calculate collision frequency rates and identify unsafe drivers.
- The selection of drivers is the most important function of a good transportation safety program. Companies must follow careful procedures when hiring employees who will be driving vehicles as part of their jobs. Applicants can be given driving tests to evaluate their driving skills and provide either basic, remedial, refresher, or special training, as required.
- Vehicles should have proper safety devices in good working order to help prevent collisions. Workers must take precautions to avoid being injured when loading or unloading trucks or when handling hazardous or toxic materials.
- A preventive maintenance program can be based on either mileage or operating hours of the equipment. Drivers should report any malfunctions or problems they encounter when driving their vehicles.
- Management and workers must adhere to proper procedures for working on vehicles to maintain safe practices in a maintenance facility. Repair shop workers should receive special training in safe work practices, and new employees should work under close supervision.
- The DOT mandates drug and alcohol testing of all employees who work in safety-sensitive jobs, such as operating motor vehicles. Employers must establish testing programs in compliance with DOT rules and regulations.

REFERENCES

American Automobile Association, 1000 AAA Drive, Heathrow, FL 32746.
    "Driver Training Equipment" (catalog).
American National Standards Institute, 11 West 42nd Street, New York, NY 10036.
    Powered Industrial Trucks, ANSI B56.3/NFPA 505/UL 583
American Society of Safety Engineers, 1800 East Oakton Street, Des Plaines, IL 60016.
    Dictionary of Terms Used in the Safety Profession.
    Photographic Techniques for Accident Investigation.
    Profitable Risk Control.
    Safety Law—A Legal Reference for the Safety Professional.
American Trucking Associations, Inc., 950 North Glebe Road, Suite 210, Arlington, VA 22202–4181.
Associated General Contractors of America, 1957 E Street NW, Washington DC 20006.
    Manual of Accident Prevention in Construction.
Association of American Railroads, 50 F Street NW, Washington DC 20001-1564.
    Rules Governing the Loading, Blocking and Bracing of Freight in Closed Trailers and Containers for TOFC/COFC Service, 2002 Circular 43-D.
Association of Casualty and Surety Companies, 110 William Street, New York, NY 10038.
    Guide Book, Commercial Vehicle Drivers.
    Truck and Bus Drivers Rule Book.
Baker, J. S. *Traffic Accident Investigation Manual.* 9th ed. Chicago: Traffic Institute, 1986.
Current, W. F. *Does Drug Testing Work?* Vienna, VA: Institute for a Drug-Free Workplace, 1992.
DHHS Mandatory Guidelines for Federal Workplace Drug Testing Programs (53 CFR 11970).
Federal Motor Carrier Safety Administration, www.fmcsa.dot.gov/facts-research/facts-figures/analysis-statistics/revenue.htm.
Kirk-Othmer Encyclopedia of Chemical Technology. 5th ed. New York: Wiley Interscience, 2005.
National Committee for Motor Fleet Supervisor Training, A 364, Engineering Bldg., Michigan State University, East Lansing, MI 48824.
    Motor Fleet Safety Supervision, Principles and Practices.
National Fire Protection Association, 1 Batterymarch Park, Quincy, MA 02269.
    Fire Prevention Code, NFPA 1, 2006.
    Fire Protection Handbook. 20th ed. 2008.
    Flammable and Combustible Liquids Code, NFPA 30, 2008.
    Life Safety Code Handbook. 2006.
    National Electrical Code Handbook. 2008.
National Private Truck Council of America. *Driver Training Manual.* Washington DC: National Private Truck Council of America, 1981.
National Safety Council, 1121 Spring Lake Drive, Itasca, IL 60143.
    Defensive Driving Program materials.
    First Aid Program materials.
    Fleet Accident Rates—Manual. Published annually.
    Fundamentals of Industrial Hygiene. 6th ed. 2012.
    Injury Facts (formerly *Accident Facts*). Issued annually.
    Motor Fleet Safety Manual. 4th ed.

National Fleet Safety Contest.
Safe Driver Award Program.
Standards for School Buses and Operations Manual.
"Street and Highway Maintenance."
Supervisors' Safety Manual. Current ed.
"Vehicular Equipment Maintenance."

U.S. Department of Defense, Department of the Army, The Pentagon, Washington DC 20310.
Driver Selection and Training, TM 21–300.
Drivers' Manual, TM 21–305.
General Safety Requirements, EM 385–1–1, U.S. Army Corps of Engineers.
"Methods of Teaching."
Motor Transportation, Operation, FM 25–10.

U.S. Department of the Interior, Bureau of Mines, 2401 E Street NW, Washington DC 20241.
Minerals Yearbook.

U.S. Department of Labor, 200 Constitution Avenue NW, Washington DC 20210.
29 CFR 1910, The Occupational Safety and Health Standards.
29 CFR 1926, The Construction Safety and Health Standards.
OSHA Compliance Guide. Current ed.
OSHA Compliance Operations Manual.
OSHA Job Hazard Analysis.
OSHA Recordkeeping Requirements.

U.S. Department of Transportation, 400 Seventh Street SW, Washington DC 20590.
40 CFR Part 40, Section 40.87.
49 CFR 40, Procedures for Transportation Workplace Drug and Alcohol Testing Programs (53 CFR 11970).
49 CFR 390–397, Motor Carrier Safety Regulations. (U.S. DOT 40 CFR 392.7).
Federal Motor Carrier Safety Administration Regulations.
Hazardous Materials Emergency Response Guidebook. Current ed.
Manual on Uniform Traffic Control Devices for Streets and Highways, ANSI D6.1.
Model Curriculum for Training Tractor-Trailer Drivers.

U.S. Government Printing Office, North Capitol and H Streets NW, Washington DC 20401.
Title 29 CFR, Labor.
Title 40 CFR, Protection of the Environment.
Title 49 CFR, Transportation.

REVIEW QUESTIONS

1. An organization's vehicle collision-prevention efforts should focus primarily on what two factors?
   a.
   b.
2. Which of the following statements is true?
   a. The total cost of a vehicle collision is usually more than the amount recovered from the insurance company.
   b. The total cost of a vehicle collision is usually less than the amount recovered from the insurance company.
   c. The total cost of a vehicle collision is usually equal to the amount recovered from the insurance company.
3. List the five basic elements that a vehicle safety program should provide.
   a.
   b.
   c.
   d.
   e.
4. What two factors should be considered when first screening an individual who is applying for a job that requires driving?
   a.
   b.
5. List the four objectives of a preventive maintenance program.
   a.
   b.
   c.
   d.
6. Repair shop workers should receive training in safe practices whenever:
   a. equipment is changed or replaced.
   b. jobs are rotated.
   c. employees return after leaves of absence.
   d. refresher training is needed.
   e. all of the above
7. Which regulation has established procedures for urine drug testing and breath alcohol testing of all employees who work in safety-sensitive jobs?
8. Why should companies implement their own internal safety education programs?

Office Safety

Joy Prescott, MS
Wendy Key, BS, MS, CS
John F. Montgomery, PhD, CSP, CHCM, CHMM

Office Injuries
Who Gets Injured? ▶ Types of Disabling Injuries

Office Hazard Prevention
Safe Office Equipment ▶ Printing Services ▶ Basic Office Safety Procedures ▶ Fire Protection

Ergonomics in the Office
Visual Demands ▶ Posture and Reach ▶ Muscle Exertion and Repetitive Motion ▶ General Considerations

Safety Organization in the Office
Safety and Health Training ▶ Safety and Health Committee ▶ Incident and Illness Records

Summary

References

Review Questions

Many large organizations today, such as insurance, governmental, and financial companies, consist almost entirely of office workers. Injuries can be just as painful, severe, and expensive when they happen to office workers as when they happen to production workers. Through an effective office safety and health program, incidents and injuries can be managed. This chapter covers the following topics:
- statistics on types and rates of office incidents
- how to control office hazards through an effective safety and health program that covers the work environment, equipment, and procedures
- how to set up computer workstations that meet the ergonomics needs of the operators
- programs that can be established to promote and maintain interest in office injury prevention.

The risk of an occupation-related injury for office workers is lower than the risk to employees involved in manufacturing or transportation. However, office risks often go unrecognized and unmanaged, and some could eventually lead to serious injuries and property loss.

A company safety program cannot be fully effective if there is only partial participation by employees and management. A safety program that is not vigorously pursued in company offices probably will not be vigorously pursued in the factory or shop. If office workers are exempt from safety and health policies, then production workers may feel that following rules to avoid hazards is an unnecessary burden and, perhaps, an unfair exercise of authority by management. Exempt office workers may not understand the importance of safety and health policies and procedures and may scoff at production-oriented safety and health activities.

The safety professional who expects to sell safety to management must get management involved in a total safety program. The emphasis must be on preventing injuries and illnesses in offices as well as in production areas.

OFFICE INJURIES

One reason office safety and health programs are not more widespread is that many people believe office injuries are minor. This is a serious mistake. One aerospace firm, for example, paid out $102,000 over 8 years at one facility just for injuries incurred by people falling out of chairs. Approximately 25,000 people worked at the facility, and more than half were office workers. Of the 14 chair incidents, the two worst cost the company a total of $97,000. Not only are medical and wage replacement benefits expensive for disabling incidents, but there are also hidden costs such as the loss of productivity. (See Chapter 10, Incident Investigation, Analysis, and Costs.)

Studies made by the State of California Department of Industrial Relations (see the References at the end of this chapter) and the Equitable Life Assurance Society of the United States show that this is not an isolated example (Figure 26–1). The California State Department of Industrial Relations analyzed reports filed by more than 3,000 California employers on disabling injuries to employees. Those employers together employed more than 1 million office workers. "An office worker" was defined in the California study as "a person primarily engaged in performing clerical, administrative, or professional tasks indoors in an office at the employer's place of business." The definition did not include salespersons, claims adjusters, social workers, medical and teaching personnel (other than clerical or administrative), and certain stock, order, and inventory clerks.

When the results of the study were extrapolated nationwide, investigators found that on-the-job office incidents amounted each year to about 40,000 disabling injuries at a direct cost (indemnity benefits and medical expenses) of about $100 million. That figure does not include any indirect cost for employers, workers, or the nation.

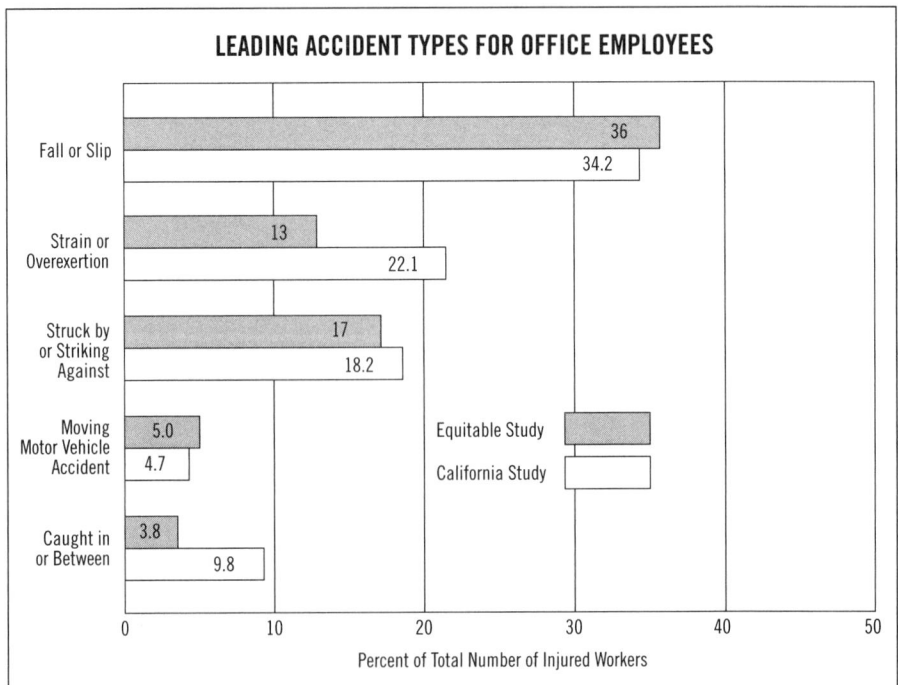

Figure 26–1. A comparison of two studies of incidents occurring in the office worker population.

Of unintentional fatalities among office workers, approximately half are due to work-related automobile crashes. That amount can be significantly reduced by using defensive driving courses (DDCs) such as those developed by the National Safety Council.

The California study did not include employees of the federal government, maritime workers, and railroad workers in interstate commerce. An Equitable Life Assurance Society study, on the other hand, did include salespeople, claims adjusters, medical personnel, and supply and warehouse personnel. The insurance company study included approximately 8,000 employees of one company working in one building, about 5% of whom were maintenance personnel. (See Chapter 11, Injury and Illness Record Keeping, Incidence Rates, and Analysis.) The average number of days charged per disabling injury was 5.9. See Tables 26–A and 26–B for details of the studies.

Who Gets Injured?
The California and Equitable Life studies described to whom most injuries occurred, and how they occurred.

New Surroundings
For employees beginning to work in a new environment, studies showed the importance of teaching office employees how to look for hazards and correct them. Studies revealed a substantial increase in the number of injuries in the first year after a company moved into a new office building. The change upsets established routines and presents unknown hazards. Even going to and from work becomes more hazardous as employees explore new driving and transportation routes.

New or Young Employees
The California study found that new and younger employees do not have a higher incident rate than lon-

TABLE 26–A. Disabling Injuries from Falls in Offices

Equitable Life—800 employees (an 8-year study)	Disabling Accidents	Days Lost
In hallways and work areas, caused by running, slipping, tripping over wires, desk drawers, file cabinet drawers, etc.	53	553
From chairs	21	120
Stairs	16	117
Escalators or elevators	8	55
Total	98	845
California survey—100,000 employees (a 1-year study)		**Disabling Accident Totals**
Falls (all categories)		4,360
Falls or slips on stairs or steps		752
Falls from other elevations		370
Falls on the same level		3,238

TABLE 26–B. Percentage of Disabling Work Injuries* to Office Workers by Occupation and Length of Service

Occupation	Total Disabling Work Injuries	1 mo.	2 mo.	4-6 mo.	3-5 yr.	6-10 yr.	11-20 yr.
Clerical and kindred	12,858	2.1	4.3	4.0	17.1	15.7	9.2
Professional, technical, and kindred	1,418	1.5	2.5	7.5	17.7	15.7	12.1
Managers and administrators	2,000	1.4	2.4	3.4	18.1	16.0	18.9
Totals	16,276	2.0	3.9	8.2	17.2	15.7	11.0

*These figures are true only for the California study.

ger-employed and older workers (Table 26–B). Office employees who had been on the job for 1 and 2 months had an injury percentage rate of 2.0% and 3.9%, respectively. Those who had been employed for 3 to 5 years had the highest incident percentage rate (17.2%). The study showed that only 3% of the injuries occurred in the 18-to-19 age bracket.

Gender of Employee
In the California study, 70% of the disabling injuries occurred to women, who comprised 68% of the office labor force. For office occupations, the estimated disabling work injury rate for women was also very close to the injury rate for men. The rate was approximately 7.8 disabling injuries per 1,000 women employed in office work compared with 7.1 for the same number of men.

The injury statistics compiled from the Equitable Life study were similar to the California study. The rate of injury incidents per thousand male employees found in the Life study was about the same as it was for female employees. However, the rate of total days lost from disabling injuries was two and one-half to three times as high for men as for women.

Types of Disabling Injuries
Falls are the most common office incident and account for the most disabling injuries, according to both surveys (Figure 26–1). The disabling injury rate for falls among office workers is two to two and a half times that for non-office employees. Falls were the most severe office incident and were responsible for 55% of the total days lost because of injuries.

Most chair falls occurred when a person was sitting down, getting up, or moving about on a chair. A few were caused by people leaning back and tilting their chairs in the office or cafeteria, or putting their feet up on the desk. Although stairs would seem to be more hazardous than chairs, people recognize the stair hazard and are more cautious. Furthermore, people are not as often exposed to the stairs as they are to chairs.

Another major incident category is falls occurring on the same level. That includes slipping on wet or slippery floors and tripping over equipment, cords, or litter left on the floor. Good housekeeping procedures can reduce incidents in this category.

A final category was falls from elevations, caused by standing on chairs or other office furniture, and by falls from ladders, loading docks, or other miscellaneous elevations. Those falls (not including stairs) accounted for approximately 2% of the disabling injuries in the California study. Falls on stairs accounted for almost 5%.

Overexertion
Almost three-fourths of the strain or exertion mishaps occurred while employees were trying to move objects—carrying or otherwise moving office machines, supplies, file drawers and trays, office furniture, heavy books, or other loads. Often, the employees were moving office equipment or furniture without authorization from their supervisors. A significant number resulted when the employee made a sudden or awkward movement that did not involve lifting or carrying any outside object. Reaching, stretching, twisting, bending down, straightening up, and cumulative trauma were often associated with these injuries.

Objects Striking or Struck by Workers
Objects striking office workers accounted for about 11% of the injuries in both studies. Most of these injuries were sustained when the employee was struck by a falling object—file cabinets that became overbalanced when two or more drawers were open at the same time, file drawers that fell out when pulled too far, office machines and other objects that employees dropped on their feet when attempting a move, or equipment that fell from a folding pedestal or rolling stand. In addition, a number of employees were struck by doors being opened from the other side. Office supplies or other material and equipment sliding from shelves or cabinet tops caused a few injuries in this classification.

Striking against objects caused approximately 7% of the office injuries in both studies. Two out of three of these injuries were the result of bumping into doors, desks, file cabinets, open drawers, and even other people (Figure 26–2 and Figure 26–3) while walking. Hitting open desk drawers or the desk itself while seated at a desk, or striking open file drawers while bending down or straightening up caused most of the rest of these injuries. Other incidents included workers bumping against sharp objects such as office machines, spindle files, staples, and pins. This category also comprised infected cuts incurred when employees handled paper, file drawers, and supplies.

Caught in or Between
The final major classification was incidents in which the worker was caught in or between machinery or equipment. Mostly, this was getting caught in a drawer, door, or window. However, a number of employees got caught in duplicating machines, copying machines, addressing machines, and fans. Several injured their fingers under the knife edge of a cutter.

Miscellaneous office incidents included foreign substances in the eye, spilled hot coffee or other hot liquids, burns from fire, insect bites, electric shocks, and paper cuts.

Figure 26-2. Desk and file drawers should always be closed when not in use. Labels on drawers help relieve stress and strains by reducing time spent bending, reaching, and searching for items. *(Courtesy Zee Medical.)*

Figure 26-3. Office workstations should be arranged so that required equipment is easily accessible. Provide wheeled carts for transport of light loads, such as computer printouts or books. *(Courtesy Zee Medical.)*

OFFICE HAZARD PREVENTION

Strategies for a Safe Office Environment

Office incidents can be controlled by eliminating hazards or, when they cannot be eliminated, by reducing exposure to them. Management can eliminate or reduce hazards most easily when the office is in the planning stage, before equipment and furniture are purchased, or when new office procedures are set up.

Layout and Ambience

Offices should be laid out for efficiency, convenience, and safety. The principles of work flow apply to offices as well as to factories.

Stairways and exits (including access and discharge) should comply with NFPA 101, Life Safety Code; floor and wall openings should comply with ANSI Construction Safety Requirements for Temporary Floor and Wall Openings, Flat Roofs, Stairs, Railings, and Toeboards, A10.18. Handrails not less than 30 in. (0.8 m) or more than 34 in. (0.9 m) above the upper surface of the tread are specified for one side of stairs up to 44 in. (1.1 m) wide, and both sides for stairs wider than 44 in. Stairs wider than 88 in. (2.2 m) require an intermediate (center) rail. Staff should be trained to always keep a hand on the handrail while ascending or descending the stairway. See Stairways in Chapter 2, Buildings and Facility Layout, in the *Engineering & Technology* volume.

Exits, particularly stairways, should be checked frequently to be sure that they are unobstructed and well illuminated. Exit hallways or paths should have emergency lighting. Exit doors, if locked, should not require the use of a key for operation from inside the building.

Doors are another frequent source of incidents in offices. Glass doors should have some conspicuous design, either painted or decal, about 4 ft (1.2 m) above the floor and centered on the door so that people will not walk into them (see Chapter 28, Contractor and Customer Safety, for details). Safety glass complying with ANSI standard Z97.1 should be installed rather than plate glass. Sometimes local codes specify the type that must be installed.

Frosted safety glass windows in doors provide visual clues to prevent incidents while preserving privacy. Solid doors present a hazard because they can be approached from both sides at the same time and one person can be struck when the door opens. Employees should be warned of this hazard and instructed (1) to approach a solid door in the proper manner, that is, away from the path of the opening door; (2) to reach for the doorknob so that if the door is suddenly opened from the other side, the hand receives the force of the impact rather than the face; and (3) to open the door slowly if it opens outward.

Another hazard is a door that opens directly onto a passageway. If the door opens directly into the path of oncoming traffic, somebody might bump into the edge of the door. If doors that open onto hallways cannot be recessed, they should be protected with short-angled deflector rails or U-shaped guardrails that protrude about 18 in. (46 cm) into the passageway. As an alternative, the area they swing over can be marked as suggested in the following paragraph. Another procedure is to place storage lockers or benches along the wall near the door to provide the safety of a recessed door.

Some offices with tile floors have white or yellow stripes or tape on the floors to mark traffic flows or to guide people away from a rapidly opening door. The floor in front of a swinging door also can be marked or painted as a warning, or a warning sign can be posted. As a final precaution, it is good practice to have the door hinges on the upstream side of the traffic; that is, on the right-hand side as one faces the door from the hallway. Doors are covered in the Life Safety Code, NFPA 101.

When designing the layout of an office and allocating office space, the designer should consider the type of work that is being performed as well. For example, those individuals who are working in marketing, trading, and scheduling for gas pipelines communicate constantly with multiple people at the same time on trading days. Open offices with smaller footprints often work well for those who require constant communication. This openness promotes easy communication and reduces the number of electronic devices required to communicate. However, accountants must often concentrate on detailed information for long periods of time. The noise in open offices interferes with concentration and often creates an environment that is less productive and less accurate. As well, inappropriately designed office space creates stress and can cause injury reporting to escalate.

Lighting

Adequate light, ventilation, and other employee services have an important influence on employee morale. Business growth often requires installing more desks and other equipment than originally planned. Overcrowding is undesirable for both appearance and the physiological effect on employees, especially if it overtaxes ventilation facilities. Smaller offices can be made to appear larger and less crowded if walls, woodwork, and furniture are the same color.

Illumination levels recommended by the Illuminating Engineering Society for an office are listed in Table 26–C. Also see Chapter 2, Buildings and Facility Layout, in the *Engineering & Technology* volume.

Some incidents can be attributed to poor lighting. However, many other factors associated with poor illumination are contributing causes of office incidents. Some of these are direct glare, reflected glare from the work, and harsh shadows, all of which hamper sight.

Excessive visual fatigue can be an incident-causing element. Incidents also can be prompted by delayed eye adaptation when moving from bright surroundings into dark areas and vice versa. Some incidents attributed to an individual's carelessness can actually be traced to difficulty in seeing.

Office design can facilitate good lighting. If offices depend primarily on daylight, for example, employees engaged in visual tasks should be located near windows. North light is preferred by drafters and artists working on paper or canvas. However, employees generally should not face windows, unshielded lamps, or other sources of glare. Windows should be perpendicular to the line of sight with the monitor. Indirect lighting can produce high levels of illumination without glare. Furthermore, walls and other surfaces should avoid annoying reflections. Ceilings, walls, and floors act as secondary large-area light sources and, if finished with the recommended reflecting paints or wall coverings, will increase light and reduce shadows. Finally, modern office lighting must be designed to accommodate computer users.

For further information on lighting, consult the *IESNA Lighting Handbook*. One of the best guides for reducing glare and reflections in the computer environment is "Solving the Problem of VDT Reflections" by Mark Rea in the October 1991 issue of *Progressive Architecture*.

Ventilation

For large interior spaces, forced ventilation is needed if the space is to be used as an office area. All mechanical ventilation and comfort conditioning systems require careful planning and installation by qualified specialists. Private offices installed around the outer walls of a large office space should not cut off light and ventilation to other employees. If fans are used in an office, they should be guarded, secured, and installed where they cannot fall.

The designs of many office buildings seal in office air—odors, smoke, office solvents and chemicals, molds, fungus, and other contaminants. In these buildings, it is particularly important to design and install adequate ventilation systems. Maintenance should regularly inspect the systems, change filters, and update them if problems appear.

Electrical

Management must protect workers against electrical equipment hazards in an office. In some cases, the hazard can be avoided completely, such as not using electric-key switches. In other cases, hazards can be reduced by using UL-listed equipment, using ground-fault circuit interrupters (GFCIs), providing sufficient well-located receptacles, and arranging cords and outlets to avoid tripping hazards. The National Electrical Code, ANSI/NFPA 70, requires GFCIs in restroom areas.

TABLE 26–C. Levels of Illumination for Offices

	Recommended Illumination* (Footcandles)
Cartography, designing, detailed drafting	200
Accounting, auditing, tabulating, bookkeeping, business machine operation, reading poor reproductions, rough layout drafting	150
Regular office work, reading good reproductions, reading or transcribing handwriting in hard pencil or on poor paper, active filing, index references, mail sorting	100
Reading or transcribing handwriting in ink or medium pencil on good quality paper, intermittent filing	70
Reading high-contrast or well-printed material, tasks and areas not involving critical or prolonged seeing such as conferring, interviewing, inactive files, and washrooms	30
Corridors, elevators, escalators, stairways	20 (or not less than ⅕ level in adjacent areas)

*Minimum on task at any time.

Employees should not use poorly maintained, unsafe, or poor-quality, non-UL-listed electrical equipment such as coffeemakers, radios, personal heaters, and lamps. Such appliances can create fire and shock hazards. Electrical holiday decorations must be UL-listed and approved for use.

The company should make sure that a sufficient number of outlets are installed to eliminate the need for extension cords. Those that are necessary should be clipped to the backs of desks or taped down. If cords cannot be dropped from overhead and must cross the floor, cover them with rubber channels designed for this purpose. Cords should not rest on steam pipes or other hot or sharp metallic surfaces.

Outlets should accommodate three-wire grounded plugs to help prevent electric shock to operators. Floor outlets should be located so that they are not tripping hazards and cannot be kicked or used as a footrest. A floor outlet protruding above floor level is frequently shielded by a desk or some other piece of furniture. However, when the desk is moved, the outlet can become an immediate tripping hazard unless it is appropriately covered. Such floor design is not common any longer; underfloor or cellular floor raceways are usually used in new construction.

In all areas, wall receptacles should be designed and installed so that no current-carrying parts will be exposed, and outlet plates should be kept tight to eliminate the possibility of shock or collision injury.

Cords for electrically operated office machines, fans, lamps, and other equipment should be properly installed and frequently inspected for any defects that could cause shocks or burns. Extension cords cannot be used in lieu of permanent wiring. Switches should be provided, either in the equipment or in the cords, so workers do not have to remove the plugs to shut off the power. (See Chapter 8, Electrical Safety, in the *Engineering & Technology* volume.)

Office electrical service should be designed to accommodate changes in equipment and technology. Electrical, coaxial, computer, and telephone cables are easily serviced and moved when run through modular channels dropped from the ceiling.

Installation or repair of any electrical equipment should be done by qualified workers using only approved materials. Because defective wiring may constitute both shock and fire hazards, management and workers must follow all recommendations of the National Electrical Code, ANSI/NFPA 70. Daisy-chaining of multiple outlet strips is prohibited.

Equipment

Workers should never place an office machine on the edge of a table or desk. Maintenance staff or manufacturer representatives should secure machines that tend to creep during operation. They can either affix the machines directly to the desk or table or place them on a nonslip pad.

Workers should place heavy equipment and files against walls or columns; files also can be placed against railings. File cabinets should be bolted together or fastened to the floor or wall so workers cannot tip them over.

Floors

Improper floor surfaces are one of the major causes of office incidents. They should be as durable and maintenance free as possible. Management should select floor finishes for slip-resistant qualities. Well-maintained carpet provides good protection against slips and falls. Maintenance staff must repair defective tiles, boards, or carpet immediately. They should also replace or repair worn or warped mats under office chairs and rubber or plastic floor mats with curled edges or tears. These conditions create tripping hazards.

Highly polished and extremely hard but unwaxed surfaces such as marble, terrazzo, and steel plates represent slipping hazards. Slip-resistant floor wax can increase the coefficient of friction and reduce the slipping hazard. Maintenance staff, however, must not apply wax so thickly that a smeary coating results. They should not use an oil mop on a waxed floor because it creates a soft, smeary coating that could become a slipping hazard (see Chapter 28, Contractor and Customer Safety).

Special slip-resistant protection should be used on stairways and at lobby or elevator entrances. Be sure these especially hazardous areas are always maintained in the best possible condition. Floor mats and runners often provide a better, more slip-resistant walking surface. Cracks, gaps, loose pieces, and raised surfaces more than ¼ in. (6.4 mm) high require special attention and repair. A well-planned, routine maintenance program is needed to keep entrance, cafeteria, and vending area floors dry. Consider appropriate signage on rainy days, such as CAUTION WET FLOOR. Refer to Chapter 28, Contractor and Customer Safety.

Each person should be encouraged to maintain proper footwear. Shoes should not be worn if they are slick on the bottom or damaged in a way that could catch on a surface and cause the individual to fall.

Outdoor Areas

Parking lots and sidewalks represent major problem areas. Many slip-and-fall incidents occur in the company parking lot. To reduce the hazards, maintenance workers must keep the lot clean, remove debris, fill potholes, and correct uneven surfaces. In colder climates, effective snow and ice removal controls should be used during the winter months. Chapter 28, Contractor and Customer Safety, covers the subject in greater detail.

Aisles and Stairs

The suggested minimum width for aisles is 4 ft (1.2 m). All aisles and passageways must meet the width requirements of NFPA 101, Life Safety Code. Keep passages through the work area unobstructed, and place wastebaskets where people will not trip over them. Install telephone cables and electrical outlets so their wires do not create a tripping hazard in passageways. These and other obstructions, such as low tables and office equipment, should be placed against walls or partitions, under desks, or in corners. Avoid building step-offs from one level to another in an office; if one exists, it should be well marked and guarded with a railing.

File drawers should not open into aisles, particularly narrow ones, unless extra space is provided.

Storage

Materials stored in offices sometimes cause problems. In general, materials should be stored only in areas specifically set aside for the purpose. Where possible, the storage area should be located so general traffic patterns do not have to be crossed to reach the stored items. Workers should not store or leave anything on the floor in a passageway where it could become a tripping hazard.

Train workers to stack materials in stable piles that will not fall over. They should put the heaviest and largest pieces on the bottom of the pile. When materials are stored on shelves, the heavy objects should be stored so that they can be handled within the power zone (i.e., closest to the body between mid-thigh and mid-chest), where the body has the most strength and comfort.

Workers should not stack objects on windowsills if there is a danger that the objects may break the window or fall through it. Refer to NIOSH lifting guidelines for additional information.

Supervisors and managers should plan storage areas to make items easily accessible. Appropriate stepladders should be provided where necessary. Office falls can occur when workers stand on chairs, counters, or shelves to reach inconveniently stored items. Rolling ladders are discussed in this chapter under Safe Office Equipment.

Companies should prohibit smoking in mailing and shipping areas, print shops, and receiving rooms. They should also ban smoking in other areas where large quantities of loose paper and other combustible material may be stored and in areas where flammable fluids are used, such as duplicating rooms or artists' supply areas.

Workers should store flammable and combustible fluids and similar materials in safety cans, preferably in locked and identified cabinets. Only minor quantities should be left in the office, and bulk storage should be in properly constructed fireproof vaults. (See Chapter 10, Flammable and Combustible Liquids, in the *Engineering & Technology* volume.)

Safe Office Equipment

Good-quality office furniture not only contributes to the safety of the office but also enhances its appearance. This, in turn, improves the attitudes of both employees and visitors.

Chairs, especially, should be comfortable and sturdily built with a wide enough base to prevent easy tipping. Five-legged chairs are more stable and discourage employees from tilting back on their chairs. (See Chapter 16, Ergonomics Yesterday, Today, and Tomorrow.) The casters on swivel chairs should be on at least a 20-in.- (0.5-m-) diameter base, but a 22-in. (0.6-m) base is preferred. However, the maximum adjustable height of the chair ultimately determines the base size that is adequate to avoid tipping. The casters should be securely fixed to the base of the chair and well constructed. Loose or broken casters are a frequent cause of chair falls. About 20% of the chair falls in the California study were due to chair defects. Instruct all individuals to report defective chairs to the appropriate supervisor or facilities group and to remove them from service.

Companies should purchase chairs with easy-to-adjust seat heights and back supports. Show employees how to properly adjust their chairs in relation to their workstation height. The correct fit will make the employee more comfortable and help reduce acute and chronic back strain—enabling office workers to work more safely and productively. Details are given in the standard for general-purpose office chairs in ANSI/BIFMA X5.1–2012.

Even if good-quality desks and file cabinets are purchased, it is possible that one will occasionally have a sharp burr or corner on it. Supervisors or maintenance staff should inspect office furniture when it is received and remove such burrs or corners immediately. Drawers on desks and file cabinets should have safety stops to prevent workers from pulling them out of the drawer slot.

Other safety tips can help prevent incidents and injuries. Purchase office machines, such as rotary files, copying machines, paper cutters, and paper shredders, with well-designed guards. Glass tops on desks and tables can crack and cause safety hazards. Durable, synthetic surfaces are safer. Make sure workers have enough noncombustible wastebaskets. Where smoking is permitted, provide safety-type ashtrays that are large and stable enough to safely contain smoking materials.

Office fans should have substantial bases and convenient attachments for moving and carrying. They must be well guarded, front and back, with mesh to prevent workers' fingers from getting inside the guard. Many cut fingers result when people try to move fans by grasping the guard or try to catch falling fans. Train workers not to handle fans until they shut off the power and the blades stop turning.

Rolling ladders and stands used for reaching high storage should have brakes that operate automatically when weight is applied to them. Steps with four or more risers should have a handrail. All stepladders should have non-skid feet.

Computers

If computers are installed in a building with overhead sprinklers, keep sprinkler protection in service, but get advice on necessary protection against both fire and water damage to computer hardware. Actually, water damage is not to be feared as much as previously thought. Most new computers are less susceptible because of their solid-state circuitry. In addition, tests have shown that water does not harm magnetic tape. Most of the damage suffered by computers in a fire results from the heat. One of the best ways to prevent a damaging fire is to keep combustible materials such as paper, tapes, and cards at an absolute minimum in the room with the computer. When safeguarding such an investment, call in a fire protection adviser as well as a computer installation expert (see Chapter 23, The Computer as a Safety Information Tool, in the *Engineering & Technology* volume).

Chemical Products

Organizations often underestimate the number and types of hazards represented by office chemicals. The safety professional should assess all chemical products used in copying and duplicating machines and in print shops, as well as all adhesives and cleaning materials. Workers must be informed of any dangers and instructed in the safe use of hazardous chemical products. Workers should also know the location of Safety Data Sheets (SDSs) and be familiar with them. All containers and bottles should be clearly labeled.

If possible, substitute nontoxic and nonflammable solvents for those used in printing and duplicating or other operations. (Details are given in the National Safety Council's *Fundamentals of Industrial Hygiene*, 6th edition.) If chlorinated bleaches are purchased for cleaning purposes, make sure that they will not be mixed with strongly acidic or easily oxidized materials. Purchase a good grade of slip-resistant floor wax.

Purchasing Equipment

The company safety professional should work with the purchasing agent in buying office furniture and equipment. Both should be aware that although advertisers sometimes stress the safety features of office equipment, the machines may be delivered without these important safeguards. Mechanical hazards of heavy office equipment can be determined by careful, expert inspection before purchase. These hazards can almost always be eliminated or minimized, although sometimes at substantial expense.

The safety professional should also inform the purchasing department of precautions to be taken in connection with chemicals, dyes, inks, and other supply items. Particular attention should be paid to toxic, irritant, or flammable properties. When hazards are unavoidable, manufacturers, suppliers, or the safety, health, and environmental department should supply labels and specific instructions for careful handling or issue instructions when workers receive the material.

The purchasing department should gather all pertinent information from the manufacturer on equipment design and electrical and space requirements and should try to determine the composition of proprietary compounds. They can forward this information to the safety professional (or safety and health department) for an opinion concerning inherent safety hazards before purchasing new equipment or supplies.

Office Machines

Machines that have external moving parts that could be hazardous should have enforced safety procedures and constant training and retraining of operators as necessary. Tell employees that if any office machine gives a shock, appears defective, sparks, or smokes, they should turn it off, pull the plug, and inform the supervisor.

Some office machines may be noisy, especially the telex, computer printers, and printing equipment. This noise is usually more of an annoyance than a health hazard but may need to be evaluated by a professional for possible adverse effects. Various covers are available to dampen machine noise.

Printing Services

Larger Offset Presses

Only qualified operators should operate presses. Check the operation of offset presses. Is the operator putting his or her fingers on the blanket while the press is in motion? One offset press department had seven finger injuries in the first 2 weeks of operation, all caused by press operators who put their fingers in the running press to remove dirt or other particles from the plate. Presses should conform to all guarding regulations imposed by local, state or provincial, and federal agencies.

Make sure the area around the presses is free from clutter and well lighted. The flooring should be resilient, or rubber mats should be provided to minimize operator fatigue and to prevent slipping. Loose clothing and long hair are hazardous around these machines. Use a safe, nontoxic substance to clean the presses; office supervisors and press operators should understand the fire and possible health hazards involved and follow all instructions for safe use and storage. Cleaning materials should be disposed of in a safe, acceptable manner.

Gathering and Stitching Machines

Supervisors should make sure guards are installed on open sprockets and collector chain drives of gathering and wire-stitching or stapling machines to protect employees from hand and body injury. The operating arm on the end of the gathering machine should be guarded.

Install hinged drop guards to cover any exposed operating mechanism that creates nipping hazards under the machine and along the working area where operators fill the pockets. Nonskid material should cover the floors and work platform at this area. Supervisors should train operators to open signatures in the middle and place them on the saddle or rod between the hooks on the moving chain. If the hook is not put on the rod or chain correctly, workers must shut off the machine before attempting to straighten the hook out. Operators should also shut down the machine when threading stitcher heads, making any adjustments, or removing jams.

Folding Machines

Here are points to be stressed for safe operation of folding machines:

1. Before jammed paper is pulled from the machine, shut the motor off to avoid getting hands in the feed rollers.
2. Finger clearance at the folding knife should be checked before pulling out paper, putting tape on rollers, or adjusting plates and roller pressure.
3. Workers should walk down the steps of folder feeder platforms facing forward, never backward.
4. On large-sized folders, all steps and platforms should be protected by railings.

Defective staples protruding from reports or booklets should be removed to avoid cuts from them while books are being jogged, trimmed, or wrapped. Workers should be trained to cup their hands over the work when removing defective wire staples. Employees engaged in this operation should wear eye or face protection, and passersby should be protected against flying staples by screens or by isolation of this work.

Basic Office Safety Procedures

Because the major category of office incidents is slips and falls, employees should never run in offices. Also, a number of office incidents can be prevented if everyone walking in passageways would keep to the right. Convex mirrors should be placed at corners and other blind intersections. Collisions at a door, as discussed earlier in this chapter, can be prevented if people stand away from the path of its swing when they go to open it. Texting on a mobile device while walking down the hallway of the office should be strongly discouraged.

People carrying material must be able to see over and around it when walking. They should not carry stacks of materials on stairs, but use the elevator instead. If one is not available, the person should make more trips, if necessary.

People should not have both arms loaded when using stairs; one hand should be free to use the handrail.

When using stairs outside at night or in a dimly lit area, workers should go single file, keep to the right, and always hold the handrail. People should not crowd or push on stairways. Falls on stairs often occur when the person is talking, laughing, and turning to friends while going downstairs. Employees should not congregate on stairs or landings or stand near doors at the head or foot of stairways.

Good housekeeping is essential. Employees should not be permitted to litter in their work areas and should wipe up all spilled liquids immediately. Pieces of paper, paper clips, rubber bands, pencils, and other loose objects must be kept off the floor.

Broken glass should be swept up at once. Do not allow employees to discard loose broken glass in a waste container. It should be wrapped in heavy paper and marked BROKEN GLASS FOR DISPOSAL. Glass that has shattered into fine pieces can be picked up with damp paper towels. Any sharp objects (e.g., razor blades) should be disposed of properly.

All tripping hazards, such as defective floors, rugs, or floor mats, should be reported to the maintenance department and immediately repaired or replaced. Avoid using multiple small floor mats, which can be a tripping hazard. Use one long mat for maximum foot contact length. Long runners (8–12 ft [2.4–3.7 m]) are preferred over shorter mats, in most cases. Many falls could be prevented if employees wore supportive footwear with nonslip soles; high heels should be discouraged.

Chair Falls

Some habits can lead to chair falls. Supervisors should instruct employees not to scoot across the floor while sitting on a chair or lean sideways from the chair to pick up objects on the floor. Discourage workers from leaning back in the chair and placing their feet on the desk. It is possible to fall over backward in this position.

People should properly seat themselves in their chairs. They should form the habit of placing a hand behind them to make sure the chair is in place. Sitting down on the edge of the seat rather than in the center, backing up too far without looking, or kicking the chair out from under can result in a sudden fall to the floor. Chairs with arms allow the user to easily grasp the chair with the chair-arms before sitting in the seat. Standing on a chair with casters to reach an overhead object is particularly dangerous and must be forbidden.

Filing cabinets, as discussed earlier in this chapter, are a major cause of injuries. These include bumped heads from getting up too quickly under open drawers, mashed fingers from improperly closing drawers, and hand injuries and strains from moving the cabinets.

Some precautions are necessary against these incidents:
- People should never close file drawers with their feet or any other part of their body. They should use the drawer handle to close the cabinet, making sure their fingers are not curled over the edge when the drawer closes. File drawers should be closed immediately after use. File cabinets should be designed to allow only one drawer to be opened at a time.
- Employees should open only one file drawer in the cabinet at a time to prevent the cabinet from toppling over. As previously stated, where possible, have the file cabinets bolted together or otherwise secured to a stationary object to safeguard against this chance of human failure.
- Do not open a file drawer if someone else is close by or underneath and could be injured by the drawer. Do not leave open drawers unattended—not even for a minute. Whoever opens a file drawer should warn others working in the area so they do not turn around or straighten up quickly and bump into or trip over an open drawer.
- No one should ever climb on open file drawers.
- Small stools used in filing areas are tripping hazards when left in passageways. They should be stored where they cannot cause falls.
- Filing personnel should wear rubber finger guards to prevent finger cuts from metal fasteners or paper edges.

Office personnel should never move desks or files; they should be moved by maintenance workers, preferably using special dollies or trucks. In general, furniture should not be rearranged without authorization from office management. When desks or cabinets are moved, workers should consider whether they will obstruct floor space or aisles before making the move. If a telephone terminal box on the floor or electrical outlet box is exposed after moving furniture, the box should be marked with a tripping hazard sign until it is removed. Maintenance staff must remove the outlet and, if it is needed, relocate it; this step is far cheaper than the medical costs of a fall.

Do not run electric cords under rugs; they sometimes come out because of traffic movement and form tripping hazards. They also are fire hazards. New outlets should be installed to eliminate the need for extension cords.

Materials Storage

A number of precautions should be taken when storing materials. Neat storage makes it easier to find and recover materials without dropping or knocking over other items. Supervisors must keep employees from stacking boxes, papers, and other heavy objects on file cabinets, desks, and window ledges or placing these materials carelessly on shelves where they could tumble down. If heavy objects fall toward a window, the glass may break and cause an injury.

Instruct workers not to place card index files, dictionaries, or other heavy objects on top of file cabinets and other high furniture. Caution workers not to throw loose razor blades, thumbtacks, and other sharp objects into their drawers but to store them in small boxes. Tools with blades and points should have the cutting or sharp end stuck in foamed-polystyrene blocks.

Lifting

Occasionally, it is necessary for office personnel to lift heavy objects, such as files, books, boxes, and computer tapes. For these times, make sure office workers are trained in proper lifting techniques; use carts, hand trucks, and the like as needed; or ask for help. When lifting, consider the following guidelines: test the load for stability and weight, plan the route, ask for help or use assistive devices as needed (e.g., carts, dollies), get a firm grip with both hands whenever possible, use smooth movement (avoid jerking motions), keep the load close to the body, maintain the natural curves in the back while pushing up with the legs, and avoid twisting. Consider labeling items with corresponding weights. Refer to NIOSH lifting guidelines for additional information.

Other Hazards

Some additional precautions follow: (1) never use a spindle (spike) file in the office; (2) never store pencils in a glass on the desk with points outward; (3) never leave a knife or scissors on a desk with the point toward the user or hand sharp-pointed objects to anyone point first; (4) equip paper cutters with guards that afford maximum protection (bar guards or single-rod barriers found in some cutters are not considered full protection); and (5) do not leave glass objects on the edges of desks or tables where they can easily be pushed off. Make sure that office machinery is operated only by authorized persons.

Some offices have an employee lounge or eating area with a hot plate for brewing coffee, a microwave oven for warming lunches, or a sink to rinse/wash utensils. In these areas, spilled liquids can be a burning and slipping hazard. (See Chapter 21, Industrial Sanitation and Personnel Facilities, in this volume.)

If employees travel on company business, a safe driving program should be part of the company's safety program. Discourage cell phone usage and texting while driving. Consider developing safety policies and procedures for international travel.

Supervisors should encourage employees to report all broken chairs, missing casters, stuck drawers, cracked glass, and other hazards for correction. Management should establish a policy for immediate correction of these defects and set up a formal program requiring quarterly office safety inspections.

Fire Protection

Fire Hazards

To prevent spontaneous-combustion fires, store all solvent-soaked or oily rags used for cleaning duplicating equipment in a metal safety container. Management should prohibit smoking within 10 ft (3 m) of where flammable solvents are used in duplicating or any other office operation. Workers should be trained in handling solvents to prevent eye injuries from splashes and wear proper protective equipment.

In recognition of the health hazards caused by smoking, many organizations do not allow smoking anywhere in the facility. Other safety rules regarding smoking include (1) never allow smoking on elevators and (2) do not throw matches or cigarettes into wastebaskets; the contents usually are highly combustible. Supervisors or department heads should establish procedures so cleaning and maintenance personnel do not collect possible smoldering combustible material from ashtrays and throw it into combustible containers, such as cardboard boxes or cloth bags.

Some waste containers made from plastic or other flame-resistant material may actually be combustible if subjected to fire or intense heat. Such fires can generate dangerous toxic gases and dense smoke, which can easily endanger a whole office. To control for this hazard, use only metal or fire-safe tested materials designed to contain fire.

Fire Extinguishers

Portable fire extinguishers in a fully charged, operable condition must be kept in their designated places at all times when not in use (Figure 26–4). See Chapter 9, Fire Protection, in the *Engineering & Technology* volume, for the correct type of extinguishers for specific office hazard areas.

Employees in general should know what to do in case of fire. Supervisors must train workers to operate extinguishers and fire hoses, if provided, and show them how to react in case of fire or other emergency. (Panic and confusion can be as dangerous as flame and smoke.)

When a fire is discovered, the employee should do three things: (1) turn on the alarm (no matter how small the fire is), (2) alert fellow workers, and (3) use the proper fire-fighting equipment, but only if the employee has been trained to do so and always has a safe path of escape while fighting the fire.

Figure 26-4. Note that this fire extinguisher is placed in an easily accessible location, has nothing stored in front of it, has a sign indicating its location and other instructions regarding replacement, and carries an inspection tag. *(Courtesy Signode.)*

Office employees should receive annual fire and other emergency training. The training should include the use of portable fire extinguishers, procedures for reporting emergencies, and location of escape routes and shelters.

Emergency Plan
Every office should have a written emergency plan that includes first-aid and CPR training as well as training in use of automated external defibrillators (AEDs). Employees should know the location of the AED and first-aid kit. Supervisors should be appointed in every area to safely guide people out of the building. Every department should be assigned a specific route and an alternate route in case that exit is blocked (see Chapter 18, Emergency Preparedness). A fire drill may save lives in an emergency situation in the future. When the alarm sounds, supervisors should direct the show, but every employee must play a part. The group should move calmly along, without hurrying or pushing, and wait on a different floor or outside the building for the signal to return. In a real emergency, the officials in charge would authorize return to the building. (See Chapter 18, Emergency Preparedness.)

ERGONOMICS IN THE OFFICE

Ergonomics is the science of optimizing a system by designing for the capabilities and limitations of the human interacting with it. When the system under consideration is the modern office, the primary components of the "system" are usually a computer (i.e., monitor, keyboard, mouse or other input devices), the table or desk on which it rests, a telephone, and a chair (because most offices are seated workstations). When designing a seated or sit-stand workstation, the factors to be considered are (1) visual demands, (2) posture and reaches required, and (3) the muscular strength and repetition exerted to perform the task. The following section discusses ways of optimizing these factors. Because each individual working in an office environment is different, layout suggestions are discussed in terms of anthropometric dimensions, not absolute measurements. (See also Workplace Characteristics and Accessories in Chapter 16, Ergonomics Yesterday, Today, and Tomorrow.)

Visual Demands

In order to optimize the visual interface, the top of the computer screen should be placed at or slightly below seated eye height and at a comfortable distance for reading. This may necessitate raising or lowering the monitor. If the monitor is placed on top of the CPU, it can often be lowered by taking it off the CPU and placing the monitor on the tabletop. Then, if necessary, it can be raised a couple of inches by using risers or monitor holders, although most flat-screen monitors are independently adjustable and no longer need risers. However, when multiple monitors are being used, monitor-arms are helpful in optimum placement of the monitors by allowing them to be easily repositioned. To alleviate some of the frequent mouse usage, some individuals are using multiple monitors, oversized monitors, or a combination of oversized multiple monitors. This creates a unique situation where the individual has to place the "window" he or she is working on in the best visual field on the screen rather than be concerned with the actual placement of the monitor (these are situations where the screens cover so much area that, even though the screen is directly in front of the user, if the work is on the far left of the screen, the user may be forced into an awkward, twisted posture). A good rule of thumb is that the information on the monitor that the individual is viewing most of the time should be placed at eye level or just below eye level and directly in front of the individual.

The goal is to have the operator's head positioned comfortably on his or her neck without needing to hold the head back, or to look down or to the side to see the screen. This goal has implications for bifocal wearers whose eyeglass prescription may require that they look at the screen through the bottom half of their lenses. They may be forced to tilt the head back to read a screen that is placed at eye level.

There are a number of options to accommodate such situations. For example, operators can use "computer glasses," a full-frame prescription for the distance they prefer to sit from the screen. These glasses also give a larger field of vision through which to see, so less head movement is required to view all parts of the screen. An option for those who need or strongly prefer to use bifocals while at the computer is to lower the monitor so that they can see it comfortably through the bottom half of their bifocals without having to tip their head back.

The monitor should be at a distance that suits the visual acuity of the operator. The operator should be able to read the information on the screen without leaning forward. The monitor should be far enough from the operator's head so that he or she does not have to move it to read the whole screen. The operator should be able to scan the screen simply by moving the eyes. The optimal distance is usually 18 in. to 36 in. (5.5 to 11 m) from the operator. Finally, the monitor should be placed directly in front of the operator to minimize twisting of the neck or trunk.

Lighting needs for computer users can vary greatly depending on the visual abilities of the individual and job demands. In the office, it is best to allow the user to control the light source. Window blinds should be available to control outside light. Ideally, each individual should have the ability to control the lighting in his or her individual workspace. Rheostats allow for maximum control of the lighting level, but when not available, the individual should be able to turn his or her lights off when operating a computer for long periods of time. Task lighting should be provided to view paper documents. Task lights should be located so that they do not create glare on the computer screen or in the individual's eyes. This is best achieved by providing task lighting on an adjustable arm.

Glare can sometimes be a concern—from a window in the office, inappropriately placed task lighting, or overhead fluorescents. Glare from the windows can be eliminated by placing the monitor at right angles to the window or by covering the window. Glare from overhead lights can be minimized by reorienting the monitor perpendicular to the light source and by slightly tilting the screen downward. The contrast and brightness should be adjusted to optimize for clear visibility and minimal reflections from the surroundings. If it is a color monitor, a color scheme that emphasizes contrast while minimizing reflections is often based on a pale background with dark lettering. Keep monitor screens clean and free of fingerprints and dust, which can create glare. A final option would be to reconfigure the lighting in the work area.

Hard-copy material ideally should be located at the same distance from the operator's eyes as the screen and at the same height. This position would minimize muscle fatigue caused by eye movement from copy to screen and minimize the head/neck motion that is often necessary when referring to copy that may be placed flat on the desk. Another option would be to place the copy between the keyboard and the monitor.

Studies have shown that the rate of blinking when reading a screen is decreased, thus increasing the chances of drying out the eyes. Supervisors or trainers should remind computer operators to blink frequently and to keep their eyes hydrated, especially if they wear contact lenses. Computer workers who need glasses should be reminded to get their eyes checked regularly to make sure their prescriptions remain current. Inadequately corrected eyesight may promote poor posture and eyestrain. It is also a good idea to periodically relax the eyes by closing them for a few seconds, doing eye exercises, or refocusing them on a distant object. To help prevent eyestrain, use the 20/20/20 rule: focus away from the screen every 20 min at an object 20 ft away for about 20 s. Move eyes in all directions—up, down, side to side, clockwise, and counterclockwise.

Posture and Reach

The objects most frequently used by computer operators are the keyboard and the mouse. In order to minimize "reaches," the keyboard and mouse should be placed so that operators can use both of them without stretching out the arms. In other words, the keyboard and mouse should be placed close (toward the edge of the desk) to the body to avoid forward reach. The mouse should be placed adjacent to the keyboard to avoid sideways reach. Workers should be able to work with the shoulders relaxed and the upper arms close to their body. Other objects that may be often reached for should be placed close to the operator (e.g., the telephone, writing materials). An often-overlooked culprit in shoulder problems is the coffee cup. If computer workers are frequently stretching out their arms to pick up a coffee cup and put it back down, they may build up fatigue in the shoulder muscles.

Back and leg concerns are related primarily to the chair height and seat pan depth, as well as the space under the work surface. There should be sufficient room for the operator's legs under the work surface so that the operator isn't required to lean forward to reach the keyboard or read the monitor. Operators should be able to sit all the way back in their chairs without the backs of their knees pressing up against the edge of the chair. There should be a distance of the width of two to four fingers between the back of the knee and the front of the seat pan to allow for adequate blood flow to the lower legs. The operator's weight should be evenly supported by the seat pan along the length of the thighs. The feet should be well supported while sitting. If the operator's feet cannot be placed flat on the floor when sitting back in the chair, a footrest may be required.

Another factor to be considered is the stress placed on the back while sitting. Some studies have shown that sitting increases intradisc pressure and so, possibly, the risk of back injury. The natural curves of the back should be maintained while sitting so that the pressure on the discs remains even from front to back. This means that the chair should have a convex curve that matches the "small" of the back. If the chair does not have such a curve, a back support pillow (similar to those used for long-distance driving) can help.

An alternative to sitting at the workstation is standing. An electric adjustable desk may provide the computer user the ability to "rest" from sitting by standing for a few minutes at a time throughout the day. By changing to a standing position, the individual can increase blood flow and decrease sitting fatigue, which in turn can increase productivity. When standing, the height of the keyboard surface should be about the same height as the user's elbow rest height. The top of the monitor should be, as when sitting, at or just below eye level. A standing footrest may be used to allow the individual to stand longer without fatigue.

Muscle Exertion and Repetitive Motion

The main muscles that are used when typing on the keyboard are those of the forearm, hand, and fingers. Because the strength of the fingers and hand is greatest when the wrist is straight (not bent side to side or up and down), the keyboard and the mouse should be placed so that operators can use them with their wrists straight. In addition, the tendons that slide back and forth in the hand experience less friction, and thus less irritation, when the wrists are straight during keying. Because keyboards are usually placed on a flat work surface, the operator would have to bend the elbows in order to flex and extend the wrist. The muscles used to bend the elbow (biceps) are strongest at about 90 degrees of flexion. Therefore, both elbows should rest at the sides of the torso and the keyboard should be placed at approximate elbow height. To achieve this support, the surface for the keyboard and mouse may need to be raised or lowered. Another option would be to raise or lower the chair. Using a wrist rest in conjunction with the keyboard may also help minimize wrist extension. If the operator types by moving the whole arm across the keyboard, there may not be significant radial or ulnar deviations (i.e., lateral/sideways bending of the wrist). However,
if such deviations are placing an unacceptable stress on the operator's wrists, a curved or split keyboard may help reduce such stress.

In general, using a keyboard is not a high-force task; however, some studies have shown that the level of force exerted while typing may be associated with the development of wrist-related musculoskeletal disorders. A light touch on the keyboard is therefore recommended. High forces and power grips are often required to perform stapling and manual stamping. Consider electric or power staplers and stamps.

General Considerations

Avoid resting the elbows, forearms, or wrists on sharp edges because this could place pressure on the ulnar nerve and possibly on the muscles and ligaments in this region. Such "contact stress" can be reduced by rounding and/or cushioning edges on which the elbow, forearm, or wrists are supported. Consider an articulating keyboard tray to adjust arms to a neutral position.

Balancing the telephone between the shoulder and the ear could require an office worker to bend the neck to keep the phone in place. If office workers spend a significant portion of their work hours communicating via telephone, they may assume this position frequently. In order to enable the operator to work with the neck and shoulders relaxed, a shoulder rest may be necessary for infrequent phone users. If phone use is frequent and/or the operator's hands need to be free for using the keyboard and mouse while on the telephone, a headset or speakerphone may provide a better option.

Because of the multifunctional capacities of computers, office workers may spend many hours without leaving their workstation. "Micro-breaks" (a break of 30 to 45 s every 30 to 45 min) have been found to minimize the buildup of fatigue in the muscles.

To summarize, the computer workstation should be laid out so that operators can work with their:
- neck at a comfortable angle for viewing the monitor
- back firmly against the chair backrest
- elbows close to their sides and bent at about 90 degrees
- wrists straight and not resting against any sharp edges
- legs evenly supported by the seat pan
- feet flat on the floor or on a footrest.

See Table 26–D for a computer workstation checklist. Any modifications to a computer workstation should be discussed with the workstation operator and a qualified person with training in ergonomics. Items in the checklist may be interrelated. For example, if a monitor is lowered but not tipped back, greater discomfort could result.

TABLE 26-D. Computer Workstation Checklist*

Monitors	Yes	No
1. Is the top of the monitor at least 15° below horizontal eye level?		
2. Is the top of the monitor farther from the eyes than the bottom of the monitor?		
3. Is the screen free of glare and reflections?		
4. For non-color-sensitive work, can the screen be set with dark letters on a white background?		
5. Is the front of the screen at least 25 in. (63.5 cm) from the user's eyes?		
6. Can the font size of the software be increased if it is not large enough to allow for at least a 25-inch viewing distance?		
7. Is the screen free from perceptible flicker?		
8. Are the screen contrast and brightness set for maximum clarity?		
9. Do the characters on the screen appear sharp and well defined?		
10. Is the field of view of the operator free from bright light sources that are causing discomfort?		
11. Is a copy holder available so hard copy (if frequently referenced) does not have to be laid flat on the work surface?		
Seating		
1. Does the seat height allow the operator to rest his/her feet flat on the floor or a footrest without requiring a thigh-torso angle of less than 90°?		
2. Can the operator sit in full contact with the backrest without the back of the legs contacting the front of the seat pan?		
3. Is the seat wide enough to support the user's thighs?		
4. Does the chair have a lumbar support?		
5. Are adjustable armrests available, if requested?		
6. Are the casters appropriate for the flooring surface?		
Keyboard and Mouse Support Surfaces		
1. Does the height of the keyboard or mouse support surface allow the operator to work without bending the wrist backward?		
2. Does the keyboard or mouse support surface allow for elbow angles between 75° and 135°?		
3. Are the keyboard or mouse support surface and other surfaces free of sharp edges that are likely to contact the operator's wrists or forearms?		
4. Can the elbow be close to the body while keying or mousing?		
5. Are the shoulders relaxed while keying or using a mouse or other input device?		
6. Are the wrists held off the wrist surface of the palm rest while keying?		
Administrative		
1. Is the operator able to intersperse noncomputer work (e.g., filing, copying) with computer work?		
2. Can the operator take "micro-breaks" to stand up, stretch, or focus the eyes at a far or intermediate distance?		
3. Is consideration given to reducing stress in the workplace?		
4. If a phone is used at the same time as computer work is done, is a headset provided?		

*This checklist should be used in consultation with a qualified person who knows the ways in which items are interrelated.

The focus of any modification of a work area should be to customize the workstation to fit the individual(s) using it. People come in all different sizes, the furniture they are given varies, and the components of their computer system may be different as well. What works for one individual with a given workstation may not work for another. Suggestions for possible modifications to make a computer workstation more comfortable include the following:

- Monitor risers or even old phone books can position the monitor at the correct height for comfortable viewing.
- "Computer glasses" can help eliminate neck bending for bifocal or trifocal wearers and allow a larger field of corrected vision.
- Reorienting the monitor in the work area can help eliminate glare from windows or overhead fluorescent lights.
- Glare screens can be fitted on monitors if glare is still a

problem. Parabolic louvers or retrofits for overhead fluorescents can eliminate glare. Usually, however, the whole room would need to be retrofitted.
- A copyholder can position hard copy at the same distance and plane from the eyes as the monitor.
- Adjusting the height of the chair or table can help achieve the correct height in relation to the keyboard.
- A computer table or under-table adapter, such as an articulating keyboard arm, can also help put the keyboard at the correct height for keying.
- A wrist or forearm rest can help minimize wrist extension and eliminate sharp edges against soft tissue.
- Maintaining the natural curves of the back while sitting is very important. If the design of the chair does not support this, lumbar pillows may help.
- A footrest can provide support for the legs if they are not fully supported on the floor.
- To the extent possible, noncomputer work should be interspersed with computer work.
- "Micro-breaks" to stand up, stretch, and refocus the eyes on something in the distance can help minimize the buildup of fatigue.

SAFETY ORGANIZATION IN THE OFFICE

The supervisor is, of course, the key person in the office safety program. However, even the hardest-working supervisor will have difficulty maintaining full-time interest in safety all alone. On the other hand, the office safety committee can help maintain interest in the incident-prevention program, but it cannot substitute for good management.

Safety and Health Training

Safety training has a tendency to be overlooked in an office environment. All office workers should be provided with safety training that focuses on incident prevention, fire prevention, fire emergency response, and medical emergency response. Depending on the specific nature of the office environment, hazard communications training may be required. If an office environment contains an art department, print shop, or duplicating center, hazard communications training should be given to all employees.

To develop proper safety behavior, managers must provide safety instructions for all new office employees. The personnel or industrial relations department can supply an incident-prevention brochure or a set of printed rules. They also should arrange for all explanations of procedures as quickly as possible during the employee's early workdays. (See also Chapter 30, Motivation, and Chapter 31, Safety and Health Training.)

Unfamiliar surroundings, new equipment, and altered work tasks increase the likelihood of incidents, even among veteran employees. Therefore, these people also should be trained when beginning a new job and given specific instructions for each piece of equipment. No one should ever be permitted to use a machine unless fully instructed in its operation and shown the location of fire equipment, how to use it, and how to summon medical aid.

Managers should have safety instruction in safe office operation because they are likely to be as unaware of hazards as the employees. However, prevention of incidents requires the dedicated vigilance of the supervisor throughout every working day. If he or she fails to carry out this function, the number of incidents due to unsafe practices by employees will continue undiminished.

Routine safety and health training should be part of company policy, and employees should receive continuing information on work-related hazards and safe practices. Training meetings are recommended on such topics as slip and fall prevention, proper lifting, fire safety, emergency procedures, office chemical safety, and off-the-job safety. Safe attitudes and behavior are not merely put on when an employee enters the office and taken off when the individual walks out the front door (see Chapter 31, Safety and Health Training, for more detailed discussion).

All office employees who must enter production areas where safety hats, eye protection, and hearing protection are necessary must be provided with these items and should be required to wear them. Every employee who visits the facility should have a card outlining the general safety rules that apply to the facility and should be familiar with them. The same requirement should be enforced for all visitors. Safety rules should apply to everyone if the program is to be successful.

Another critical aspect of training focuses on ergonomic and stress-related areas. Supervisors need to educate workers to recognize common physical complaints that could be caused by the ergonomic design of the workstation or how they use equipment at the workstation. Common physical complaints include eyestrain; back pain; neck pain; shoulder and arm symptoms; wrist, hand, and finger symptoms; leg and foot symptoms; headache; and fatigue. Training materials should emphasize how employees can change their work habits or adapt equipment and furniture to reduce these complaints. This training MUST include how to operate and adjust the office chair and desk.

Office workers tend to work under a high level of stress. One definition of stress is an employee's physical and emotional reaction to change. A definite focus of training should be stress reduction and occupational wellness. Workers can learn how to recognize stressors in the work environment, detect physical symptoms of stress overload, and use various methods to alleviate stress. Stress reduction programs can include progressive relaxation, positive imagery, values clarification, proper nutrition, and exercise.

Many other programs can be developed that focus on occupational wellness. Chapter 31 contains additional information on safety and health training.

Finally, a company safety program cannot succeed unless it has the wholehearted backing of its top management. Supervisors must know that their incident-prevention performance is watched and that good performance is appreciated. The NSC's *Supervisors' Safety Manual* contains information on this topic that can guide both top management and the safety professional.

Safety and Health Committee

In planning an office safety program, management should ensure that worker representation on the safety and health committee reflects the composition of the work force in company departments or divisions. The organization of the office safety committee can be the same as that of the company and joint safety committees discussed in Chapter 6, Loss Control Programs.

The office should be on the inspection itinerary of the company's safety professional. The office supervisor should accompany the safety inspector on every inspection, along with an office safety committee member. (See Chapter 9, Identifying Hazards, for inspection procedures.)

The committee's responsibilities include helping all supervisors maintain safe working conditions in office areas. The committee reports directly to the safety director or whoever is in charge of the program. Along with the department head, it can make periodic inspections of the office to look for hazards. The committee makes recommendations, many of them based on suggestions from supervisors and other employees. It also can help prepare and revise company safety rules.

Often, the committee is in charge of officewide communication, training, and incentive programs. These are designed to maintain peak interest in safety and use such means as posters, bulletins, and contests.

Incident and Illness Records

If a safety program is to succeed, the company needs to keep accurate incident records. Not only do incident investigations and analysis of records spotlight problems that must be corrected, but the records show whether the company is making progress in incident prevention.

Office employees, like facility workers, should report every incident, no matter how minor the injury. The reports should be detailed and made as soon as possible following the injury or near miss. Unsafe conditions or procedures indicated in the reports should be corrected as soon as possible, because near-miss incidents are warnings of worse incidents to come.

Records are the concrete foundation of the safety structure. They tell the "who, what, when, why, and how" of incidents in the office—and help supervisors and employees prevent repeat performances. Accurate records also provide guidelines on which company insurance rates are based.

The average office will not have enough major injuries to warrant extensive investigation and analysis. However, records must be kept to pinpoint problems and prevent future incidents. If, for instance, a large number of falls are injuring workers, supervisors can double-check possible hazards and devote special attention to the problem in meetings and other communications. Standard report forms are available. (See Chapter 10, Incident Investigation, Analysis, and Costs, and Chapter 11, Injury and Illness Record Keeping, Incidence Rates, and Analysis.)

SUMMARY

- Because the risks of serious injury are as great for office workers as they are for production workers, a company's safety policy must include office workers in its program. Office workers and managers must be informed of the hazards and safe work procedures that apply to their jobs.
- Researchers have found that new surroundings and length of service increase the chances of worker incidents and injuries. The most common categories of major injuries are falls, strains, and other injuries related to overexertion; workers either being struck by objects or striking them; and workers being caught in or between machinery or equipment.
- Office injuries can be controlled by eliminating hazards or reducing exposure to them. Offices should be laid out for efficiency, convenience, and safety.
- Adequate lighting, proper office design, and proper ventilation can reduce or eliminate eyestrain and visual fatigue and ensure that contaminants do not accumulate in office areas. Management must protect office workers against hazards from electrical and movable equipment and from hazardous flooring.
- Supervisors and managers should plan storage areas for safety and to make items easily accessible. Workers should be trained in how to stack materials in stable rows or piles and to keep passageways free.
- Workers must also be trained in the safe use and storage of chemicals used in copying, printing, and duplicating

machines and for cleaning purposes. Office machines must be properly guarded and designed with built-in safeguards.
- The safety professional should work with management to ensure that office furniture and workstations are designed according to good ergonomic principles. Following these principles can reduce or eliminate worker fatigue and injury.
- The supervisor and safety professional should ensure that workers receive safety training and know safety procedures for their jobs. Employees should understand how to prevent fires in office work areas and what to do during a fire drill or outbreak of fire. Every company should draw up written emergency plans in the event of a natural or human-caused disaster.
- The company should establish a formal safety program for office workers and establish training sessions for all employees, including training in electrical and fire safety.
- The company needs to keep accurate records on all incidents and injuries and to help pinpoint major problems so that management can design effective solutions.

REFERENCES

American National Standards Institute, 11 West 42nd Street, New York, NY 10036.
    American National Standard for Office Furnishings—General Purpose Office Chairs—Tests, ANSI/BIFMA X5.1–2012.
    Construction Safety Requirements for Temporary Floor and Wall Openings, Flat Roofs, Stairs, Railings, and Toeboards, ANSI A10.18–1983.
    Practice for Office Lighting, ANSI/IES RP1–1992.
    Safety Performance Specifications and Methods of Test for Glazing Materials Used in Buildings, ANSI Z97.1–1984 (R1994).
The Illuminating Engineering Society of North America. *IESNA Lighting Handbook.* 10th ed. 2014.
National Fire Protection Association, 1 Batterymarch Park, Quincy, MA 02269-9101.
    Life Safety Code, NFPA 101, 1994.
    National Electrical Code, NFPA 70, 1993.
National Safety Council, 1121 Spring Lake Drive, Itasca, IL 60143.
    Fundamentals of Industrial Hygiene. 6th ed. 2012.
    Motor Fleet Safety Manual, 4th ed. 1995.
    Starting an Office Safety Program. 1990.
    Supervisors' Safety Manual. 9th ed. 1997.
Rea, M. "Solving the Problem of VDT Reflections." *Progressive Architecture* (October 1991).
Scott, D. *Sitting on the Job: How to Survive the Stresses of Sitting Down to Work—A Practical Handbook.* Boston: Houghton-Mifflin, 1989.
State of California, Department of Industrial Relations, Division of Labor Statistics and Research, 525 Golden Gate Avenue, San Francisco, CA 94102.
    Disabling Work Injuries to Office Employees. 1963 and 1978 eds.
    "Work Injuries and Illness in California." Issued annually.
U.S. Department of Commerce. *Statistical Abstracts of the United States.* Latest ed.

REVIEW QUESTIONS

1. What is the most common and most severe type of office incident?
   a. objects falling on worker
   b. overexertion by worker
   c. worker falling
   d. worker caught in or between machinery
2. List five of the nine different elements of an office layout that can be hazardous to workers.
   a.
   b.
   c.
   d.
   e.
3. In addition to contributing to the safety of an office, which of the following improves the attitudes of both employees and visitors?
   a. eliminating overcrowding of desks and equipment
   b. adequate lighting and ventilation
   c. good-quality office furniture
   d. all of the above
   e. a and c
4. Improper floor surfaces are one of the major causes of office incidents. List five ways to make floors safer.
   a.
   b.
   c.
   d.
   e.

5. Define *ergonomics*.
6. What three basic factors should be considered when designing a seated workstation?
   a.
   b.
   c.
7. List four safety guidelines that help prevent visual and/or muscular fatigue when looking at the computer screen.
   a.
   b.
   c.
   d.
8. Which of the following tells the "who, what, when, why, and how" of incidents in the office?
   a. the safety professional
   b. incident records
   c. office management
   d. all office workers

27
Laboratory Safety

Jairo Betancourt
Philip Hagan, JD, MBA, MPH, ARM, CIH, CET, CHMM, CHCM, CHSP, CEM

Introduction

Laboratory Safety Management
Scale of Laboratory Hazards ▶ The Challenges of Laboratory Safety Management ▶ Laboratory Safety Management Tools ▶ A Sample Safety Management System for Laboratories ▶ Safety Management and Safety Culture

Chemical Safety
Chemical Safety Information ▶ Chemical Safety Practices ▶ Laboratory Design ▶ Laboratory Chemical Safety Regulations

Biological Safety
Biosafety Guidelines ▶ Containment Practices ▶ Primary Barriers ▶ Personal Protective Equipment ▶ Secondary Barriers ▶ Bloodborne Pathogens ▶ The Bloodborne Pathogens Standard—Infection Control Plan (29 CFR 1910.1030)

Radiation Safety
Principles of Radiation ▶ Radiation Hazards ▶ Radiation Protection Methods ▶ Radiation Standards

Nonionizing Radiation Safety
Radiation Safety Program ▶ Radiation Safety Officer ▶ Exposure Control ▶ Contamination Control ▶ Radiation Detection Instrumentation ▶ Other Program Elements

Laser Safety
Principles of Lasers ▶ Hazards of Laser Radiation ▶ Hazard Controls ▶ Recommendations for Laser Safety Programs

Physical Hazards and Other Considerations
Cryogenic Liquids ▶ Fiber-Optic Cables

Altered Environments
Clean Rooms ▶ Animal Facilities ▶ Dark Rooms ▶ Warm and Cold Rooms ▶ Potential Psychological Problems

Laboratory Ergonomics
Biosafety Cabinets, Fume Hoods, and Laboratory Workbenches ▶ Computer Workstations ▶ Microscopy ▶ Repetitive Pipetting ▶ Microtome and Cryostat Work ▶ Glove Boxes ▶ Centrifuges

Emergency Planning

Laboratory Waste and Closing Procedures

Summary

References

Review Questions

INTRODUCTION

Laboratories are small (one- or two-room) general-use workplaces designed to provide as much flexibility as possible to manage potentially hazardous operations. These operations can vary frequently (often from week to week) and are generally managed on an individual basis by many different individuals. Laboratories can be found in clinical, industrial, and academic settings. They also have varied purposes, such as teaching, research, and routine quality control testing of industrial products.

The hazards associated with laboratory settings vary widely, but they share some common characteristics. Hazardous materials are present in small quantities. For example, according to the OSHA laboratory standard (29 CFR 1910.1450), "laboratory scale" refers to hazardous chemicals in quantities that are easily and safety manipulated by one person. Engineering controls are usually limited to general ventilation of the room, some generic local ventilation units (e.g., fume hoods and biological safety cabinets), and chemical storage cabinets. This means that laboratory safety depends to a large extent on the professional expertise of the workers and the administrative controls this expertise supports.

Regulatory requirements and consensus guidelines, when properly implemented, can help ensure that these operations are conducted in a manner safe to the worker and environment. Unfortunately, many regulations that apply to the materials used in laboratories are written with industrial-scale use of chemicals in mind. In many cases, this leaves significant ambiguity as to how these regulations apply in laboratory settings. For this reason, each laboratory process should be assessed for potential risks, and appropriate safeguards should be put into place. These safeguards will often go beyond regulatory requirements.

Some of the more common hazards encountered in laboratories that should be evaluated include chemical hazards, biohazards, ionizing and nonionizing radiation, and physical hazards. To provide an overview of identification and control of these hazards, this chapter discusses the following topics:
- laboratory safety management
- chemical safety practices and applicable regulations
- biological safety, biosafety levels, and appropriate containment measures
- the nature and hazards of ionizing, nonionizing, and laser radiation and how these hazards are controlled
- special considerations for working in altered environments and other unusual laboratory hazards.

LABORATORY SAFETY MANAGEMENT

Laboratories are general-purpose workplaces designed to permit the safe use of a wide variety of hazardous materials and processes using generic safety equipment such as fume hoods and flammable storage cabinets. In general, safe laboratory practice uses few engineering controls, and relies heavily on the technical training of laboratory staff to develop and follow administrative controls specific to the work at hand. The OSHA Lab Standard (29 CFR 1910.1450) addresses the importance of these administrative controls when it requires the development and implementation of a chemical hygiene plan (CHP) and the appointment of a chemical hygiene officer (CHO) for laboratories. To understand why administrative controls are the preferred approach to laboratory hazards, it is helpful first to discuss the nature and magnitude of these hazards.

Scale of Laboratory Hazards

Chemical Hazards

Laboratories use a wide variety of chemicals in varying amounts and ways. The hazards associated with these chemicals vary similarly. For example, in biomedical laboratories, flammable liquids in gallon quantities and other auxiliary chemicals with a range of physical and health hazards in similar and smaller quantities are likely to be found. However, which specific chemicals will be used on a day-to-day basis is quite unpredictable. On the other hand, chemistry and engineering research laboratories may routinely work with highly reactive materials in quantities that present explosion or other reactivity hazards.

Generally, laboratory chemicals include flammable chemicals, corrosive chemicals (acids and bases), reactive chemicals (which can be water reactive or explosive under certain conditions), and acutely toxic chemicals. Many of these chemicals are in weak concentrations that present a much lower hazard than those described on Safety Data Sheets (SDSs) for those chemicals. In many laboratories, unnamed chemicals are produced. Although generally the flammability or corrosivity of these new chemicals may be predictable, their toxicity is not.

Biological Hazards

Biological laboratories work with specific biological agents in the course of their research. These laboratories are proliferating rapidly as biotechnology applications expand. The risk of exposure to these biological agents is rated as biosafety levels on a scale of 1 to 4 by the scientific community, with 1 representing biological agents not known to cause illness in healthy humans and 4 representing biological agents that present high individual risk of an aerosol-transmitted laboratory infection for which there is no cure. For example, human blood is a Biosafety Level 2 agent, which means it can cause illness in healthy people if the person comes into mucous membrane or injection contact with the material and a disease is present. Biosafety levels refer not only to the biological agent, but also to the laboratory facility and to the practices to be implemented.

Radiation Hazards

Many laboratories use radioactive isotopes and/or electrically powered sources of radiation of various kinds in their work. The strength of unsealed chemicals marked with radioactive isotopes is barely above natural background levels in the environment. The primary health concern associated with such unsealed sources is the ingestion of these materials, in which case they could cause health concerns for exposed individuals. Radiation from x-ray devices and other electrically powered radiation sources is a concern when these units are turned on. These devices are marked with radiation hazard signs similar to those for radioactive chemicals.

Physical Hazards

Laboratories also include many physical hazards, such as strong magnets, lasers of all classes, gas cylinders, and high-voltage electricity. Many of these are permanently located in specific rooms designed to control these hazards. Other hazards are mobile, depending on the needs of particular processes, so some rooms may not be designed for them.

The Challenges of Laboratory Safety Management

Laboratory safety management involves many different aspects. The safety manager not only must master a significant amount of technical, scientific, and safety information, but also must understand government regulations and how they are enforced. He or she must be comfortable with management information (such as budgets and deadlines) that helps him or her to understand how decisions are made within the organization. The safety manager must also use communication and leadership skills to develop a set of safety recommendations and present this information in an effective way. This section provides a brief review of these different issues.

Scientific Information

Understanding the safety implications of work under way in a laboratory requires at least a general understanding of the work itself. For example, identifying the hazards associated with the chemicals in use requires looking at specific physical properties such as flash point, pH, and known incompatibilities with other chemicals. It is also important to understand how the use of this information is affected by the concentrations and amounts of the chemicals and how the hazards associated with these properties will change as the process proceeds. The laboratory safety program must also address situations in which this information is unavailable. One way is by including provisions in the safety program to ensure that scientists will discuss the properties of the chemicals they are using to elicit whatever information is available to make reasonable predictions of these factors.

Regulatory Information

Although the OSHA laboratory (lab) standard is a reasonable and flexible approach to the laboratory work setting, the same cannot be said of all government regulations that apply to laboratories. For example, the lab standard acknowledges that the chemical hygiene plan must address issues such as proper disposal of the chemicals involved in laboratory processes. This issue will be determined by EPA's Resource Conservation and Recovery Act (RCRA) regulations (40 CFR 239–299) and State variants of these regulations. A chemical waste disposal plan must address the Department of Transportation's requirements for shipment of hazardous materials (49 CFR 100–185), and the transport of threshold quantities of certain waste chemicals is covered by Department of Homeland Security requirements for security plans (6 CFR 27).

Combining all of these requirements, with their varying jargon and expectations, requires understanding not only the text of the appropriate regulations, but also the enforcement policies of the relevant agencies. These factors must be connected with each other to develop a coherent plan that answers the question, "What do I do with this stuff now that I'm finished with it?" Similar chains of regulatory responsibility can be developed for many other laboratory safety questions.

Management Information

Laboratory safety is significantly affected by the nature of the work to be carried out, the facilities available in which to conduct the work, and the people available to do the work. These choices are made by the management of the organization that oversees the laboratory. The laboratory safety program must consider any such factors that may compete with its recommendations for prudent laboratory practice. Such factors can include budget limitations, deadline requirements, available facilities and equipment, and the level of training of laboratory workers.

Communication and Leadership Skills

The appropriate guidelines for conducting safe work in the laboratory established by the organization must be communicated to the affected populations. This often requires translating the regulatory requirements for the scientist while describing the requirements of the laboratory process in the terms of the regulatory standard. In effect, both the regulatory parties and the laboratory parties must have confidence in the abilities of the laboratory safety program manager's competence for them to accept his or her guidance related to regulatory issues.

To successfully transmit their recommendations, laboratory safety managers must have good communication skills, both oral and written, and the ability to provide leadership around laboratory safety issues for groups with multiple competing priorities.

The Unexpected

Laboratory safety would not be an issue if all work went according to plan. No one plans to create situations that risk their own or others' health or put their work in danger of being lost. However, laboratory safety issues inevitably arise when an adopted procedure is inadequate to meet events that transpire. These events can become emergencies that involve assistance from outside agencies such as fire departments and emergency medical services. The laboratory safety program must address these issues—not only in terms of emergency planning and response, but also by using these events as opportunities for developing information from "lessons learned" that can be incorporated into future work practices. Again, this role requires not only technical knowledge, but also effective use of communication supported by leadership skills, both during and after the emergency.

Laboratory Safety Management Tools

The challenges just described may seem daunting. However, important tools are available for use in laboratory safety programs. The safety manager has the opportunity to connect with workers' self-interest when developing and implementing safety guidelines. Not only are laboratory workers' health and safety involved, but following safety protocols significantly enhances the protection of their work and its validity. Laboratory safety programs can take advantage of two other key assets: peer sharing and support and development of a management system approach to the work.

Peer Sharing and Support

Safety professionals often lack the resources they need to meet the challenges and opportunities presented by their role and the variety of skills required to succeed in their goals. Fortunately, they are not alone; many other safety professionals face similar challenges every day. These professionals form an important peer-to-peer support network for both technical and management advice. In particular, the Internet has developed into a professional communication medium that provides many low-cost opportunities to learn and elicit advice from peers in other organizations. Sharing information resources such as training materials, forms, and experiences with similar situations via websites and e-mail lists is a common and often critical strategy to support a laboratory safety program as new technologies and related hazards arise.

A Management System Approach

Because there is no "bright line" between safe and unsafe conditions in the laboratory, developing laboratory safety guidelines quickly leads to larger questions such as, "How does an organization as a whole effectively manage laboratory safety?" and "What tools are available to make this process easier and more effective?" The answers often revolve around the amount of resources an organization is willing (or able) to put into safety response efforts. For this reason, these questions fall into the management category, not only for the laboratory safety manager, but also for the upper management of the organization.

A key strategy for approaching these questions is to set goals for continuous improvement of the safety program by establishing indicators of the program's success that can be tracked over time. Fortunately, significant research into how to implement this approach has taken place over the last few decades. For an example of such work, see OSHA's Voluntary Protection Programs.

This research has led to a focus on the "Plan, Do, Check, Act" management system approach to health and safety issues based on Dr. Edward Deming's theories of quality control. More specifically, occupational safety management systems (SMSs) have come to the fore as an important organizational tool in identifying ways to continuously improve the safety program. Figure 27–1 provides a conceptual outline of how this approach works. In the case of the laboratory safety program, the laboratory safety plan and its associated procedures and documentation fill the "standardization" role of the "chock" that maintains safety program improvements over time.

A Sample Safety Management System for Laboratories

A safety management system does not provide a cookie-cutter approach to laboratory safety. The specifics of the safety program developed will depend on the specific resources and needs of the organization involved. The SMS model is

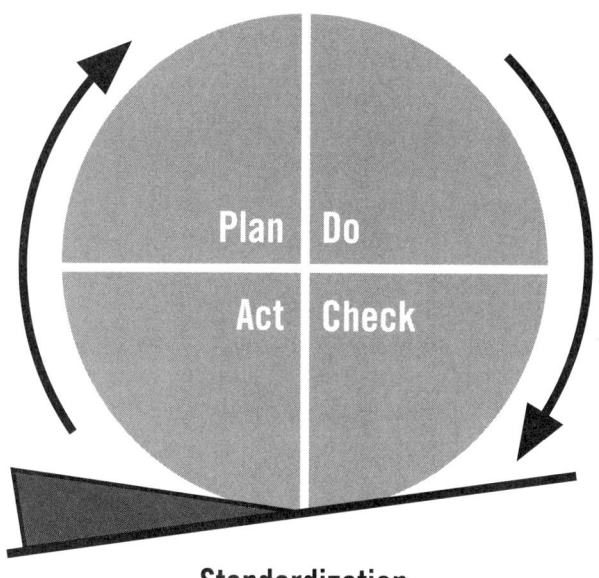

Figure 27-1. Total quality management principles can be used to formulate a chemical hygiene management system.

best used to assess whether the safety program is complete and balanced from both a management and a compliance point of view. This means expressing the program activities in terms of performance indicators and goals for those indicators. These indicators may be qualitative or quantitative, but quantitative indicators are usually easier to work with and provide a better framework for developing a safety program, particularly in a scientific setting.

As an example, one can outline a generic chemical hygiene management system based on the "Plan, Do, Check, Act" framework with this goal in mind. This system could be one of several that work together to manage overall laboratory safety for a particular organization. Such a plan could consist of the following components.

Hazard Identification (Plan)
The rapidly changing nature of work in laboratory settings often makes the hazard identification step a significant challenge. Regular reviews of the work being conducted to ensure that hazard evaluations are still appropriate would be an important sign of a healthy chemical hygiene program. A numerical indicator of the health of this step could be the frequency at which these evaluations are reviewed for completeness. OSHA's requirement of an annual review of the CHP may not be adequate in many research laboratories.

Training and Consultations (Do)
After hazards are identified, specific precautions to control those hazards must be established and disseminated to laboratory workers. In many organizations, counting the number of training sessions and safety consultations that the laboratory safety program conducts is one convenient way to monitor this aspect of the program. Another way is to track contact hours.

Laboratory Audits (Check)
Training and consultations are not likely to be effective without follow-up to determine whether recommendations are implemented. The most common approach to determining the strength of this part of the program is development of a safety checklist specific to each laboratory that tracks conditions and habits that affect safety conditions. This checklist gives each laboratory a numerical grade that can be tracked over time to measure improvement in safety practices. The checklist must be used regularly and also routinely modified to reflect changes noted during the hazard evaluations conducted in the "Plan" stage of this system.

Incident Review and Waste Quantities (Act)
Two potential indicators for the "Act" stage of the system present themselves: review of safety incidents in the laboratory (not only accidents, but near misses as well) and tracking of the laboratory wastes generated in the course of the laboratory work. Clearly, over time the first of these indicators should drop in number as the review of safety incidents results in implemented recommendations. For this reason, the waste analysis information should be thought of as a complement to the incident review indicator. For example, if the degree of hazard associated with the waste decreases with time (e.g., less waste is generated, less-hazardous materials are substituted), this can indicate the success of the safety program, not only for improving the safety conditions in the laboratory but also for pollution prevention.

Safety Management and Safety Culture
Laboratory safety management must be done in the context of the culture and priorities of the organization that hosts the laboratory. The SMS approach requires a significant allocation of resources and continued attention over time. As described in *Prudent Practices in the Laboratory* (National Research Council 1995), the ultimate goal of the management system is to foster the development of a "safety culture" throughout the laboratory organization. The key purpose of this culture is to support laboratory workers' ongoing efforts to understand and manage the hazards of their work. These hazards can be expected to change continuously, and the development of an effective safety culture is critical to maintaining a safe workplace as these changes occur.

CHEMICAL SAFETY

Nearly all laboratories use chemicals in one form or another. And nearly all chemicals are considered hazardous by one regulatory agency or another. Chemicals often present the most significant hazards associated with laboratory work. Therefore, chemical safety is a leading issue in laboratory safety management.

In this context, *chemicals* are materials whose use does not depend on their shape. Thus, chemicals include gases, liquids, powders, and dusts. Clearly, a wide variety of materials fall into this category, so developing blanket chemical safety rules to cover all chemical processes is a significant challenge. This section describes the overall process of chemical safety management in the laboratory setting. Specific details of chemical safety hazards can be found in the References at the end of this chapter.

Chemical Safety Information

A primary challenge in dealing with chemical safety questions is gathering relevant information. Once this information is gathered, it must be interpreted in light of the specific situation in which it will be used. Two sources of information for chemical safety are labels and Safety Data Sheets. OSHA's hazard communication standard (29 CFR 1910.1200) requires manufacturers, distributors, and importers of chemicals to provide extensive safety information that can be used to train end users on how to use the chemical safely.

However, OSHA's hazard communication requirements for laboratory workplaces are simpler: they require only that labels on incoming containers of hazardous chemicals should not be removed or defaced and that employers should maintain any Safety Data Sheets that are received with incoming shipments of hazardous chemicals and ensure that they are readily available and accessible during each work shift to laboratory employees when they are in their work areas.

Safety Data Sheets

The Hazard Communication Standard (HCS) requires chemical manufacturers, distributors, or importers to provide Safety Data Sheets (SDSs) (formerly known as Material Safety Data Sheets or MSDSs) to communicate the hazards of hazardous chemical products. As of June 1, 2015, the HCS requires new SDSs to be in a uniform format. This format includes 16 sections that cover topics such as hazard identification, storage and handling, required personal protective equipment, chemical and physical characteristics, disposal considerations, and the like. These changes respond to the adoption of the Global Harmonization System (GHS), which has been adopted by OSHA and EPA:

- Global Harmonization System (GHS) for Classification and Labeling of Chemicals (Environmental Protection Agency [EPA])
- osha.gov/pls/oshaweb/owadisp.show_document?p_table=INTERPRETATIONS&p_id=28880
- osha.gov/dsg/hazcom/HCSFinalRegTxt.html.

Training requirements from the new standard are already in place.

Laboratory Chemical Safety Summaries

The National Research Council committee that wrote *Prudent Practices in the Laboratory* (1995) recognized the problem of using SDSs in laboratories. In response, it developed a format for safety information that is more appropriate for the laboratory setting. This format described Laboratory Chemical Safety Summaries (LCSSs), which can be used to highlight the most pertinent information for the safe laboratory use of chemicals. The committee also provided 88 sample LCSSs for the most commonly used chemicals. In contrast to SDSs, which may be longer than 10 pages, LCSSs are no longer than 2 pages. The LCSS identifies the major hazards associated with the chemical in the laboratory setting in straightforward language and is intended to be used in the context of an effective CHP by workers well trained in chemical terminology.

Other Sources

In addition to SDSs and LCSSs, which are specific to certain chemicals, safety information about chemical processes commonly used in laboratories should be reviewed. Sometimes, these processes create hazards that cannot be predicted by simple review of the hazards of the various chemicals involved. Fortunately, in the 21st century, Internet search engines provide a convenient way to find such information. All such safety information is written for circumstances that may not match those in a specific laboratory. Therefore, any information collected from other sources must be reviewed with this limitation in mind.

Unfortunately, adequate health and safety information is not available for many laboratory chemicals. This may be because the chemicals have not been tested yet or because the chemical has not been produced anywhere else. In these cases, the laboratory workers using the chemical must work closely with safety professionals to develop appropriate operating procedures for its use. The goal of these procedures should be similar to the core principle of radiation safety, which is to keep exposures to new chemicals as low as reasonably achievable (ALARA).

Training and Documentation

The OSHA lab standard recognizes that laboratory workers generally have more education in chemistry than typical workers. This is one reason a separate regulation was written for the laboratory workplace. However, OSHA recognizes that this education does not necessarily directly translate into a familiarity with good chemical practice. Therefore, OSHA requires that laboratory workers be provided with specific safety training for the chemical procedures they are using. Because of the changing use of chemicals inherent in the laboratory environment, this training should be conducted on an ongoing basis.

Training should provide both general information about chemical hazards, such as might be found on an LCSS or SDS, and information about how these hazards are affected by the specific chemical process conducted in the laboratory. An explanation of what safety equipment, such as fume hoods, and personal protective equipment such as gloves and eye protection should be used as well as the limitations of these safety precautions should be provided to laboratory workers before they begin work. This training should be documented so that it is clear who is authorized to conduct what work within the laboratory.

Chemical Safety Practices

Although chemical hazards vary widely, certain generic practices that are common to safe use of hazardous chemicals in laboratories can often be identified.

Housekeeping

Housekeeping is the daily practice of maintaining an organized and neat workplace. Housekeeping protects not only the workers in the laboratory, but also the wide variety of people who provide support services to the laboratory, including facility maintenance people, chemical waste technicians, cleaning staff, and emergency response personnel such as fire fighters and emergency medical technicians. Any of these people can be put at risk by bad housekeeping when they are required to enter the laboratory in the course of their duties.

In this context, "good housekeeping" means that chemical work surfaces are routinely cleaned. Sharp instruments are stored in puncture-resistant containers. All chemical containers must be stored in the correct storage cabinets and/or shelves. Aisles are kept clear of equipment and chemicals to allow easy access to all parts of the laboratory.

In larger laboratory organizations, housekeeping becomes a significant challenge in shared areas. When specialized laboratory spaces and equipment are shared among several laboratory groups, a conscious commitment to maintaining a group approach to housekeeping is necessary to maintain a safe workplace.

Labeling

Clear, accurate labeling of chemicals is a critical safety step for both laboratory workers and others who work in or pass through their work area. All containers must have the appropriate labels at all times. Chemical labels should include the name of the chemical written in English (rather than scientific notation or abbreviations) with a description of the hazards associated with that chemical. In general, common words such as *flammable*, *corrosive*, *acutely toxic*, or *reactive* are appropriate for these hazard warnings. Ideally, the label should also include the date the chemical was received and the date it was opened.

When transferring chemicals from an original container into a laboratory container, all relevant information about the chemical hazards must be reflected in the new label.

In addition to enabling laboratory workers to identify the contents of a chemical container, this information is also critical to ensure correct disposal of the chemicals and to enable emergency responders to assess the risks associated with responding to an emergency in the laboratory. Emergency responders may appropriately refuse to respond to rooms in which unlabeled, potentially hazardous materials are present. Appropriate warning labels should be attached to the entry door to a laboratory. This gives emergency responders ample warning of possible laboratory hazards that could be encountered during a response.

Storage of Chemicals

Chemicals can present significant hazards if good storage practices are not followed. Incompatible chemicals must not be stored in a way that they will interact with each other. In addition, hazardous liquids should be stored in secondary containment in case their containers are broken or leak.

Many segregation schemes for chemical storage have been developed over the years. Each should be evaluated on a case-by-case basis for use in a particular laboratory. The best arrangement will depend on the specific chemicals included in a laboratory's inventory. The following guidelines present the most important considerations involved:

- *Flammables*—Materials with a flash point of less than 100°F must be stored in a flammable liquid storage cabinet; organic materials must be kept separate from oxidizing materials; if they mix, a fire is likely to result (examples: acetone, ethanol, xylene).
- *Nonflammable solvents*—Organic materials with flash points of more than 100°F can be stored with flammable liquids, still separate from oxidizing materials (examples: carbon tetrachloride, ethylene glycol, mineral oil).
- *Inorganic acids*—Keep separate from other acids; store in a cabinet of acid-resistant material; keep separate from caustics, cyanides, sulfides (example: hydrochloric acid).

- *Oxidizing acids*—Keep separate from other acids and flammables (examples: nitric acid, sulfuric acid, perchloric acid).
- *Organic acids*—Keep separate from oxidizing acids, caustics, cyanides, sulfides (examples: acetic acid, formic acid).
- *Bases*—Store in a dry area, separate from acids (examples: ammonium hydroxide, sodium hydroxide, amines).
- *Poisons*—Keep separate from all other chemicals, and secure from unauthorized access.
- *Water-reactive chemicals*—Store in a cool, dry location, separate from aqueous solutions; protect from fire sprinkler water (examples: sodium, calcium hydride, lithium).
- *Oxidizers*—Store in a cabinet of noncombustible material, separate from flammable and combustible materials (examples: potassium permanganate, sodium nitrate, sodium hypochlorite).
- *Nonvolatile, nonreactive solids*—May be stored in general-use cabinets or open shelves (examples: agar, sodium chloride, sodium bicarbonate).

A more detailed consideration of chemical storage schemes can be found in *Prudent Practices in the Laboratory* (*Prudent Practices in the Laboratory: Handling and Management of Chemicals* [March 25, 2011]). It is also important to review SDSs concerning specific reactivity information for chemicals being stored together.

In addition to appropriate segregation of stored chemicals, a well-maintained chemical inventory can be an important safety tool. Many chemicals degrade over time and can create serious hazards if they undergo spontaneous reactions such as polymerization or dehydration. A good example would be opening a container of peroxide. The container should be dated on opening and contain an expiration date that would preclude the formation of explosive peroxide crystals. All chemical containers should have shelf lives marked or coded on their label. Routine culling of aging chemicals is an important safety practice. An even more important practice would be obtaining only the chemicals that will be used within an appropriate time frame.

Chemical Waste Management

When chemical wastes are considered regulated hazardous wastes by federal or local regulations, their disposal must be considered carefully and managed appropriately. EPA regulations form the basis for the hazardous waste disposal system in the United States and are tracked "cradle to grave." However, many laboratories, both academic and industrial, have been fined by EPA and State agencies for improper management of hazardous wastes within the laboratory. To ensure a good understanding of how these regulations work, seek professional guidance in addressing this issue. In most cases, a laboratory that uses chemicals will produce chemical waste. Chemical waste must be treated as the original chemicals, considering the chemical hazard class as in flammables, corrosives, and so on. All laboratories generating regulated chemical waste must select a satellite accumulation area (SAA) to store and manage the waste following basic principles:

Label all chemical waste containers properly (name, hazard, and date).

Do not leave the funnel on the waste container; always cap it.

Dispose properly of empty chemical containers.

Buy only the chemicals needed for an activity—extra, unused chemicals could become hazardous waste.

Remember, nothing goes down the drain!

Laboratory Design

The design of the laboratory has a direct impact on the safety practices necessary to work effectively with laboratory chemicals. As mentioned earlier, many safety design considerations for laboratories are quite generic; some of the key elements are as follows:

Ventilation

Heating, ventilation, and air conditioning (HVAC) systems are designed to maintain temperature control in the laboratory. They are also used as an engineering control measure to protect laboratory workers by providing dilution air to the space and removing airborne contaminants. Laboratories should be designed so that the HVAC system maintains a negative air pressure inside the lab relative to the surrounding environment. This pressure relationship ensures that air moves from nonlaboratory spaces into the laboratory, thus preventing laboratory processes from contaminating adjacent spaces. Air moved by laboratory HVAC systems should be entirely exhausted from the building to reduce the possibility of a buildup of contaminated air.

Fume Hoods

In general, most work with chemicals in the laboratory has to be done in a chemical fume hood. Ventilation of specific processes that require more control than the general HVAC system can provide is managed by laboratory fume hoods. For a hood to be effective at containing chemical vapors, the hood must be used carefully. Variables that impact the effectiveness of a hood include the height of the sash opening, the amount of equipment in the hood, chemical storage within the hood, velocity of the air moving through the hood, and hood location. These factors should be evaluated routinely to ensure that the hood is performing as expected. Chemical fume hoods are designed to work with toxic chemicals, not to store chemicals.

Traditionally, the optimum performance of a fume hood is considered to be at a face velocity of between 80 and 120 feet per minute (fpm). In most situations, the target average face velocity is 100 fpm. However, in recent years, this target face velocity has been more carefully examined because of the expenses associated with heating and cooling the air exhausted by the fume hood. Because laboratory HVAC systems are designed to exhaust and not recirculate air because of potential contamination issues, a large cost is usually associated with conditioning the air. This cost can be reduced when a low-flow fume hood system is used. Fume hoods can be operated over a range of lower face velocities and still achieve effective containment when properly designed and installed. Containment of contaminants in a low-flow hood is demonstrated by the ASHRAE 110 tracer gas test. Experienced ventilation engineers should be involved in the placement and installation of low-flow fume hoods to ensure that they meet the requirements of this test. In addition, strict work practices may have to be implemented to take advantage of these types of fume hood systems.

Emergency Eyewashes and Showers

OSHA regulations [29 CFR 1910.151(c)] require employers to provide suitable facilities for quick drenching or flushing of the eyes and body when employees are potentially exposed to corrosive materials. Minimum performance and use requirements for emergency eyewashes and showers are provided by ANSI Z358.1–2007. In summary, emergency eyewashes and showers should be in accessible locations that require no more than 10 seconds to reach. A shower or eyewash station should be capable of providing a flow of tepid water for at least 15 min. Eyewashes may be located at sinks or at other readily accessible areas. Eyewashes must be capable of hands-free use after starting the water flow, so that the hands can be used to keep eyes open while flushing.

Stagnant water in both safety showers and eyewashes can be breeding grounds for environmental microbes such as *Legionella* and *Acantha amoeba*. Both of these can present serious health concerns to exposed people; therefore, showers and eyewashes must be routinely flushed. The general recommendation is to flush them weekly. This flushing also ensures that the equipment is ready for use if needed.

Equipment Hazards

Regular inspections and preventive maintenance should be performed on equipment. Whenever preventive maintenance or repairs are conducted on equipment, lockout/tagout procedures should be practiced in accordance with 29 CFR 1910.147 to prevent inadvertent start-up.

Heating Devices

Uncontrolled heat sources (e.g., Bunsen burners) should not be used near flammable substances and must not be left unattended. Hot plates, heating mantles, and other heaters should have enclosed elements, and controls should have a thermal shutoff safety device. Heating devices should have cutoff points to preclude overheating.

Electrical Safety Considerations

All electrical equipment used in the laboratory should be grounded. Ground-fault circuit interrupters must be used whenever equipment is exposed to a wet environment (e.g., near a sink or in a cold room). The use of extension cords should be avoided when possible. When absolutely necessary, extension cords should be used only temporarily (less than a workshift). Electrical cords should be placed in such a way to minimize the risk of tripping. Frayed or damaged electrical cords should not be used. Equipment with electrical plugs and cords should be kept in good repair.

Compressed-Gas Cylinders

Compressed-gas cylinders present hazards in a variety of ways:

- They are heavy, clumsy containers that can be awkward to handle without the correct equipment.
- The gas inside them is stored at high pressure; if the cylinder is ruptured, it can become a high-speed projectile.
- Some gases stored in them can create toxic atmospheres, either by the nature of the gas itself or by displacing oxygen from the room it is in.

For these reasons, gas cylinders must be carefully managed. Cylinders transported by truck must be fastened securely in an upright position so that they will not fall or strike each other. Always secure gas cylinders upright to a wall, cylinder truck, or cylinder rack. A clamp and belt or chains are used for this purpose and should be attached around the body of the cylinder, not around the valves or caps.

Caps used for valve protection should be kept on cylinders at all times, except when the cylinder is in use. Cylinders should not be transported without safety caps in place and screwed all the way down on the cylinder's neck ring. Cylinder valves should be opened slowly. Always use a cylinder wrench or another tightly fitting wrench to tighten the regulator nut and hose connections unless the cylinder is a handwheel type.

A handwheel-type cylinder valve should be operated

only by hand. Before attaching a cylinder to a connection, be sure that the threads on the cylinder and the connection are compatible. If the connections do not fit together readily, then the wrong regulator is being used. Do not permit oil or grease to come in contact with cylinders or their valves. The threads and mating surfaces of the regulator and hose connections should be cleaned before the regulator is attached. Ensure that the regulator is attached securely before opening the valve wide.

If using cylinders containing toxic gases, use the smallest size possible. Appropriate monitoring devices, soapy water, or solutions recommended by the gas supplier should be used to check for leaks. Emergency procedures should be in place for each type of cylinder so if a leak does occur, there will be minimal impact to laboratory personnel and the surrounding environment.

Portable Fire Extinguishers
Fire extinguishers should be available, unobstructed, and inspected on a schedule that complies with NFPA 10, Standard for Portable Fire Extinguishers. This should include a monthly "quick check," verifying that the fire extinguisher is available and will operate (fully charged) and that no obvious physical damage or condition to prevent operation is noted. The inspection needs to be documented with the date and initials of the person performing the inspection. This documentation can be either through the use of a tag attached to the fire extinguisher or in an electronic system (e.g., bar coding) that provides a permanent record.

Laboratory Chemical Safety Regulations
As mentioned earlier, a variety of federal regulations address chemical hazards in the laboratory. The most important of these is the OSHA laboratory standard, which covers workplace laboratories where relatively small quantities of hazardous chemicals are used on a nonproduction basis.

The OSHA Laboratory Standard (29 CFR 1910.1450)
All facilities engaged in the "laboratory use of hazardous chemicals" must comply with the provisions of the standard. "Laboratory use of hazardous chemicals" means use of chemicals in which all of the following conditions are met:
- Chemical manipulations are carried out on a "laboratory scale," which means work in which the containers used are easily and safely manipulated by one person.
- Multiple chemical procedures or chemicals are used.
- The procedures involved are not part of a production process, and do not in any way simulate a production process.
- "Protective laboratory practices and equipment" are available and in common use to minimize the potential for employee exposure to hazardous chemicals.

Chemical Hygiene Plan
To comply with the laboratory standard, the employer must implement a chemical hygiene plan (CHP). The CHP documents the work practices, procedures, and policies to ensure that employees are protected from hazardous chemicals in their work area. Thus, this plan provides a context for the standard operating procedures associated with specific chemical processes. The CHP must be available to employees, employee representatives, and OSHA inspectors.

The employer must provide employees with information and training to ensure that they are aware of the hazards of the chemicals present in their work area and how to protect themselves from those hazards. The training must cover the following:
- the location and contents of the employer's chemical hygiene plan
- the physical and health hazards of chemicals in the work area
- the location and availability of reference materials on the hazards
- methods and observations that may be used to detect the presence or release of a hazardous chemical
- signs and symptoms associated with exposures to hazardous chemicals used in the laboratory
- OSHA's permissible exposure limits (PELs) for chemicals that have such limits established
- the measures employees can take to protect themselves from these hazards.

Training should be conducted at the time of an employee's initial assignment to a work area where hazardous chemicals are present and before assignments involving new exposure situations.

Employee Monitoring
The employer must determine which hazardous chemicals are present in the laboratory. This includes both chemicals delivered to the laboratory and chemicals produced in the laboratory. Based on this inventory, the employer must evaluate employees' potential exposure periodically and monitor for exposure to any regulated substance if there is reason to believe that exposure levels for that substance routinely exceed the action level (or in the absence of an action level, the PEL). The employer must notify the employee of the results within 15 working days after receipt of the monitoring results.

All employees who work with hazardous chemicals must

be given the opportunity to receive medical attention when they develop symptoms associated with the chemicals they work with or after a chemical release in the laboratory. Medical examinations and consultations must be provided without cost to the employee, without loss of pay, and at a reasonable time and place.

The employer must provide the physician with the identity of the hazardous chemicals, a description of the conditions under which the exposure occurred, and a description of the signs and symptoms of exposure that the employee is experiencing.

A system must be in place for informing employees of the hazards associated with chemicals they use in the laboratory. At a minimum, labels on incoming containers of hazardous chemicals must not be removed or defaced. SDSs received with shipments of incoming hazardous chemicals must be retained and made available to lab employees.

BIOLOGICAL SAFETY

Biosafety can be described in lay terms as a set of principles designed to work with infectious and potentially infectious biological agents in a safe manner and avoiding exposure to laboratory employees. The principal mechanisms to handle safely a biological agent are based on a risk assessment process. This process will necessarily result in a risk management plan, which translates into the following basic elements of containment:
- laboratory design
- safety equipment
- good laboratory practices
- personal protective equipment.

Laboratory-acquired infections (LAIs) have been documented since the late 1800s. However, with the emergence of new and reemergent diseases and the growing interest in applying biotechnologies in a variety of settings, an increasing percentage of the laboratory community is working with materials that present a risk of infection.

Biosafety Guidelines
The CDC and NIH have developed and published guidelines for different levels of protective measures appropriate when working with biohazards; these guidelines are published in *Biosafety in Microbiological and Biomedical Laboratories*, 5th ed., commonly known as the BMBL. These guidelines outline four biosafety levels (BSLs) that consist of combinations of laboratory practices and techniques, safety equipment, and laboratory facilities that combine to provide appropriate containment for the agents being used. Each level is appropriate for the use of specific procedures with specific agents, based on the risk presented by the operations performed, the documented or suspected routes of transmission of the infectious agents, and the laboratory activity. These levels of protection are designated Biosafety Levels 1, 2, 3, and 4. Biosafety Level 4 practices and engineering controls offer the highest level of protection to a lab worker and the surrounding environment.

Biosafety Level 1
Biosafety Level 1 (BSL-1) represents a basic level of containment that relies on standard microbiological practices with no special primary or secondary barriers recommended, other than a sink for hand-washing. Biohazards that are managed at this level are not known to cause disease in healthy humans.

Biosafety Level 2
Biosafety Level 2 (BSL-2) practices, equipment, and facility design and construction are applicable to clinical, diagnostic, teaching, and other laboratories in which work is done with a broad spectrum of moderate risk agents that are present in the environment and associated with human disease of varying severity. The primary hazards to personnel working with BSL-2 agents are accidental percutaneous or mucous membrane exposures or ingestion of infectious materials. With good microbiological techniques, these agents can be used safely in activities conducted on the open bench, provided that the potential for producing splashes or aerosols is low. Hepatitis B virus, HIV, and *Toxoplasma* spp. are representative of microorganisms assigned to this containment level.

BSL-2 is also appropriate when work is done with any human blood, body fluids, tissues, or primary human cell lines in which the presence of an infectious agent may be unknown. Laboratory personnel working with human-derived materials should also follow the OSHA blood-borne pathogen standard for specific precautions required by this regulation.

Caution should be taken with needles or sharp instruments contaminated with these agents. Other elements of personal primary protective equipment such as splash shields, face protection, gowns, and gloves should also be used as appropriate. Even though organisms manipulated at BSL-2 are not known to be transmissible by the aerosol route, work that presents aerosol or high splash potential must be conducted in primary containment equipment or in devices such as a biological safety cabinet or safety centrifuge cups. A risk assessment of the procedure being conducted will determine which of these forms of containment are necessary.

In addition to these steps for worker protection, secondary barriers such as hand-washing sinks and waste decontamination facilities must be available for BSL-2 work in order to reduce potential environmental contamination.

Biosafety Level 3

Biosafety Level 3 (BSL-3) practices, safety equipment, and facility design and construction are applicable to clinical, diagnostic, teaching, research, or production facilities in which work is done with "indigenous or exotic agents that may cause serious or potentially lethal disease through the inhalation route of exposure." *Mycobacterium tuberculosis*, St. Louis encephalitis virus, and *Coxiella burnetii* are representative of the microorganisms assigned to this level. The primary hazards to personnel working with these agents are exposure to infectious aerosols.

BSL-3 requirements place more emphasis than BSL-2 on secondary barriers to protect personnel in contiguous areas, the community, and the environment from release of potentially infectious aerosols. All manipulations of infectious agents at this level must be done in the biosafety cabinet or other enclosed equipment, such as a gas-tight aerosol generation chamber. Secondary barriers for this level include controlled access to the laboratory and ventilation design that prevents the release of infectious aerosols from the laboratory.

Biosafety Level 4

Biosafety Level 4 (BSL-4) practices, safety equipment, and facility design and construction are applicable for work with dangerous and exotic agents that pose a high individual risk of life-threatening disease, which may be transmitted via the aerosol route and for which there is no available vaccine or therapy. Agents with a close or identical antigenic relationship to Biosafety Level 4 agents also should be handled at this level. Viruses such as Marburg or Congo-Crimean hemorrhagic fever are manipulated at BSL-4.

The primary hazards to personnel working with BSL-4 agents are respiratory exposure to infectious aerosols, mucous membrane or broken skin exposure to infectious droplets, and autoinoculation. All manipulations of potentially infectious diagnostic materials, isolates, and naturally or experimentally infected animals pose a high risk of exposure and infection to laboratory personnel, the community, and the environment.

Complete isolation from aerosolized infectious materials is accomplished by working in a Class III BSC or in a full-body, air-supplied, positive-pressure personnel suit. The BSL-4 facility itself is generally a separate building or completely isolated zone with complex, specialized ventilation requirements and waste management systems to prevent release of viable agents to the environment.

At all of the biosafety levels, the laboratory director has primary responsibility for the safe operation of the laboratory. His or her knowledge and judgment are critical in assessing risks and appropriately applying these recommendations. Many institutions supplement this judgment by appointing a biosafety officer, who assists the supervisor in assessing risk levels and implementing the appropriate procedures. Appointment of such an officer is required to oversee BSL-3 and BSL-4 work.

Four biosafety levels are also described for activities involving infectious disease work with experimental animals. These four combinations of practices, safety equipment, and facilities are similar to, but distinct from, those just discussed, and are designated Animal Biosafety Levels 1, 2, 3, and 4 in order of increasing levels of protection to personnel and the environment. These guidelines describe special handling techniques, personal protective equipment, and facility design that should prevent the occurrence of LAI or the release of biological agents to the environment.

Biological hazards, or biohazards, are pathogenic (disease-causing) microorganisms that pose a risk to the health of humans, animals, or other biological organisms. In the past, this definition has focused on infectious pathogenic microorganisms responsible for common communicable diseases. However, the realm of biohazards has now expanded to include other agents with the potential for causing disease, including cancer-causing viruses, recombinant DNA molecules, animals and plants and their by-products, microorganisms (including fungi, bacteria, yeasts, and algae), and some proteins causing disease such as Prions.

Containment Practices

The general term *containment* describes methods for controlling the location of infectious materials in the laboratory environment. The goal of containment is to reduce or eliminate exposure of laboratory workers, other persons, and the outside environment to potentially hazardous biological agents. The three key elements of containment are facility design, safety equipment, and laboratory practices and techniques.

Primary containment (primary barriers) focuses on the protection of workers and the immediate laboratory environment from exposure to infectious agents. Good microbiological technique and the use of appropriate safety equipment are associated with primary containment. Agent-specific vaccines can be used to provide an increased level of personal protection in some cases.

Secondary containment (secondary barriers) protects the environment outside the laboratory from contamination with infectious materials. It is achieved by a combination of facility design and operational practices.

To determine the appropriate combination of these elements necessary to achieve containment, a person experienced in biosafety should conduct a risk assessment of planned work before work begins with a new biohazard.

Standard Microbiological Technique

The most important element of containment is adherence to standard microbiological practices and techniques. These are outlined in the BMBL. This publication defines four different biosafety levels (1 to 4). Biosafety levels refer not only to the biological agent but also to the laboratory facility and to the practices to be implemented.

The essential elements of the four biosafety levels (BSLs) for activities involving infectious microorganisms and laboratory animals are summarized in Tables 27–A and 27–B. The levels are designated in ascending order, by degree of protection provided to personnel, the environment, and the community. In summary, these practices require that persons working with infectious agents or potentially infected materials must be aware of the hazards and must be trained and proficient in the practices and techniques required in handling such material safely. The person in charge of the laboratory is responsible for arranging the appropriate training of personnel.

In addition, each laboratory should develop a biosafety manual that documents potential hazards that may be encountered and control methods to be used. A scientist trained and knowledgeable in appropriate laboratory techniques, safety procedures, and hazards associated with handling infectious agents must be responsible for the development of this manual and oversight of affected work. This individual should consult with biosafety or other health and safety professionals with regard to risk assessment for new procedures. Institutions receiving NIH funds are required to have this risk assessment process overseen by an Institutional Biosafety Committee (IBC).

TABLE 27–A. Selection of a Safety Cabinet through Risk Assessment
(Adapted from the Office of Health and Safety, Centers for Disease Control and Prevention)

Biological Risk Assessed	Protection Provided			BSC Class
	Personnel	Product	Environmental	
BSL 1,2,3	YES	NO	YES	I
BSL 1,2,3	YES	YES	YES	II (A, B1, B2, B3)
BSL 4	YES	YES	YES	III (B1, B2)

TABLE 27–B. Comparison of Biosafety Cabinet Characteristics
(Adapted from the Office of Health and Safety, Centers for Disease Control and Prevention)

BCS Class	Characteristics		Applications	
	Face Velocity	Airflow Pattern	Nonvolatile Toxic Chemicals and Radionuclides	Volatile Toxic Chemicals and Radionuclides
I	75	In at front; exhausted through HEPA to the outside or into the room through HEPA (see Figure 27-2b)	YES	YES[1]
II, A	75	70% recirculated to the cabinet work area through HEPA; 30% balance can be exhausted through HEPA back into the room or to the outside through a thimble unit	YES	NO
II, B1	100	Exhaust cabinet air must pass through a dedicated duct to the outside through a HEPA filter	YES	YES (minute amounts[2])
II, B2	100	No recirculation; total exhaust to the outside through hard-duct and HEPA filter	YES	YES (small amounts)
II, B3	100	Same as II, A, but plenums are under negative pressure to room; exhaust air is thimble-ducted to the outside through a HEPA filter	YES	YES (minute amounts[2])
III	N/A	Supply air inlets and hard-duct exhausted to outside through two HEPA filters in series	YES	YES (small amounts)

[1] Installation may require a special duct to the outside, an in-line charcoal filter, and a spark-proof (explosion-proof) motor and other electrical components in the cabinet. Discharge of a Class I cabinet into a room should not occur if volatile chemicals are used.

[2] In no circumstances should the chemical concentration approach the lower explosion limits of the compound.

Primary Barriers

Safety Equipment
Safety equipment includes biological safety cabinets (BSCs), enclosed containers, and other engineering controls designed to remove or minimize exposures to hazardous biological materials. The biological safety cabinet is the principal device used to contain infectious splashes or aerosols generated during microbiological procedures. High-efficiency particulate air (HEPA) filters are incorporated by design into BSCs to filter and control the movement of aerosolized particles outside of the cabinet. BSCs do not provide any containment of chemical vapors; in fact, because the air in the cabinet is recirculated, chemical vapors can be released outside the cabinet into the laboratory air.

Three types of biological safety cabinets are used in microbiological laboratories (Figures 27–2a, 27–2b, and 27–2c). Class I biological safety cabinets are primary barriers that protect laboratory personnel and the environment when used with good microbiological techniques. The Class II biological safety cabinet also protects biological materials from external contamination. The gas-tight Class III biological safety cabinet provides the highest attainable level of protection to personnel and the environment. Tables 27–A and 27–B provide information to assist in selection of a safety cabinet.

Personal Protective Equipment
Personal protective equipment (PPE) may include clothing for personal protection, such as gloves, coats, and face shields. Personal protective equipment is often used in combination with biological safety cabinets and other devices that contain the agents, animals, or materials being handled to protect both the worker and the experimental material. In some situations in which it is impractical to work in biological safety cabinets, personal protective equipment may form the barrier between personnel and the infectious materials.

Autoclaves
Autoclaves are important pieces of biosafety equipment because they are used to steam-sterilize equipment and materials that are potentially contaminated with either environmental or experimental biohazards. However, they can present significant hazards to the worker and/or the environment if they are not used correctly.

Autoclaves should be operated in accordance with the manufacturer's instructions. PPE such as eye protection, heat-resistant gloves, and aprons should be worn when loading and unloading a hot autoclave. To minimize exposure to hot fluids and noxious vapors, the operator should not open autoclave doors too soon after a run is finished (20 min is commonly required for the autoclave to cool down).

An autoclave should have an easily accessible manual control that can run a complete cycle in the event of a power failure. Each autoclave should have alarms that indicate when the chamber is flooded and that the steam and pressure parameters were within acceptable ranges during the cycle. Common safety interlocks include control lockout switches that prevent cycles from starting if the door is not closed and locked and mechanical steam pressure locks to prevent operators from opening a door while the chamber is under positive pressure. Before installation of an autoclave, availability of adequate ventilation, electrical, and plumbing services should be assessed.

Ultraviolet Lights
Ultraviolet (UV) lights are often found in biological safety cabinets and other biomedical laboratory equipment. UV light may be used to kill certain types of microorganisms; however, its effectiveness depends on the radiation intensity and exposure time. UV light should not be considered a sterilizing agent except in certain exceptional circumstances. Typically, UV light is intended only to reduce the number of microorganisms on surfaces and in the air.

UV light presents hazards to laboratory workers. The eyes and skin should not be exposed to direct or strongly reflected UV radiation. Some individuals are more susceptible to UV injury than others. Overexposure of the eyes and/or skin will result in symptoms (redness, irritation) developing up to 9 or so hours following exposure. The symptoms usually disappear in a day or two. A hazard warning sign should be displayed prominently on the doors of laboratories with UV light installations. Adequate eye and skin protection must be worn when working in an irradiated area. Safety glasses with side shields or goggles with solid sidepieces must be worn. Skin protection can be afforded by face shields, caps, lab coats, gloves, aprons, and gowns.

Secondary Barriers

Facility Design
The design and construction of the facility contribute to laboratory workers' protection, provide a barrier to protect persons outside the laboratory, and protect persons or animals in the community from infectious agents that may be accidentally released from the laboratory. Laboratory management is responsible for providing facilities commensurate with the laboratory's function and the recommended biosafety level for the agents being manipulated.

The recommended secondary barrier(s) will depend on the risk of transmission of specific agents. For example, the exposure risks for most laboratory work in Biosafety Level 1 and 2 facilities (discussed earlier) will be direct

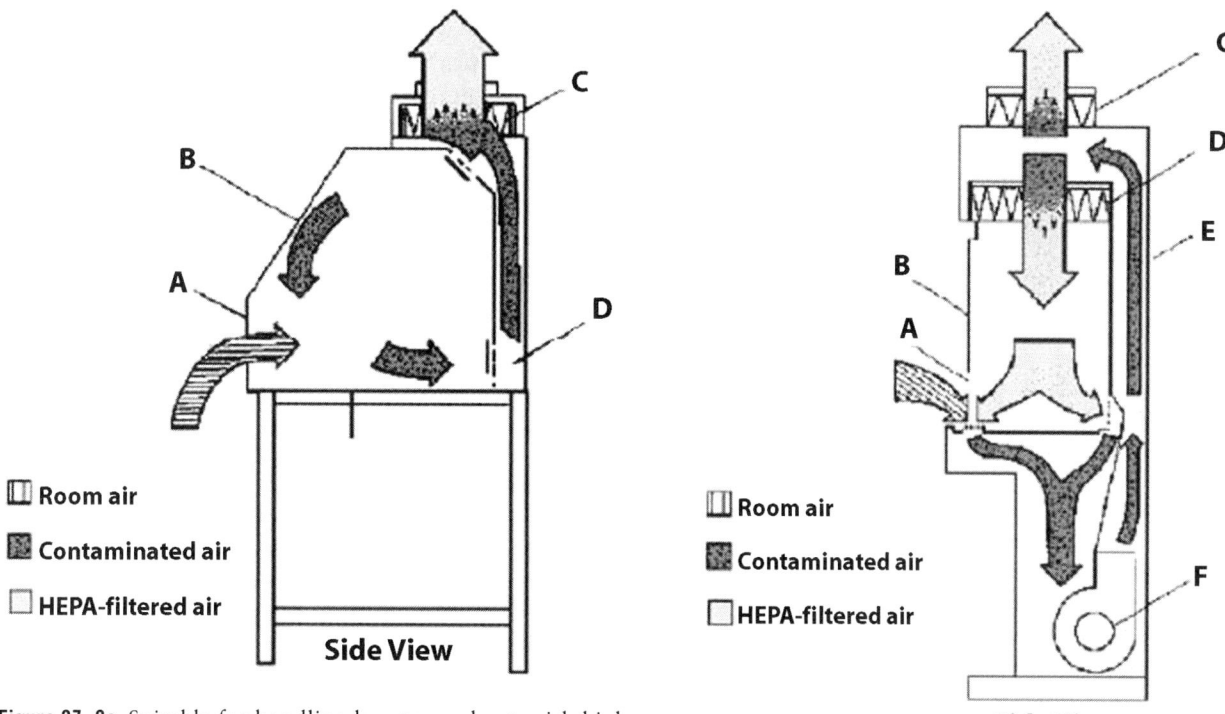

Figure 27–2a. Suitable for handling low-to-moderate-risk biohazardous aerosols when product protection is not essential, Class I cabinets provide partial personnel protection and no product protection. *(Courtesy CDC/National Institutes of Health. Biosafety in Microbiological and Biomedical Laboratories. 4th Edition. Atlanta, GA: U.S. Department of Health and Human Services, Public Health Service, CDC and NIH, 1999.)*

Figure 27–2b. Designed for the handling of low- and moderate-risk biohazards, Class II cabinets provide both product and partial personnel protection. Type B-2 cabinets do not recirculate air and are more suitable for handling carcinogens and other hazardous chemicals than either Type A or Type B-1 cabinets.

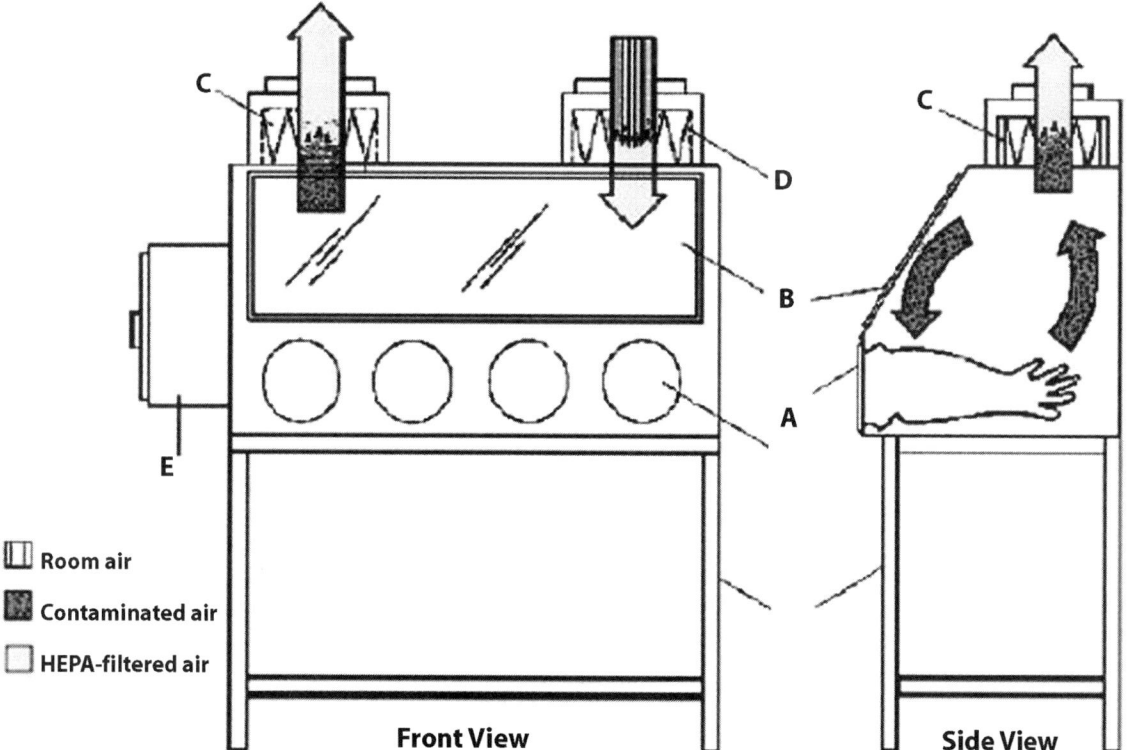

Figure 27–2c. Class III safety cabinets, or glove boxes, are closed-front, gas-tight boxes and provide the highest degree of personnel protection and a clean work environment.

contact with the agents or inadvertent contact exposures through contaminated work environments (Tables 27–C and 27–D). Secondary barriers in these laboratories may include separation of the laboratory work area from public access, availability of a decontamination facility (e.g., autoclave), and hand-washing facilities.

When the risk of infection by exposure to an infectious aerosol is present, higher levels of primary containment and multiple secondary barriers may become necessary to prevent infectious agents from escaping into the environment. Such design features include specialized ventilation systems to ensure directional airflow, air treatment systems to decontaminate or remove agents from exhaust air, controlled-access zones, airlocks as laboratory entrances, or separate buildings or modules to isolate the laboratory.

Bloodborne Pathogens

Biomedical laboratories, especially those in health care facilities, commonly work with human specimens for a variety of research, diagnostic, and clinical support services. Typically, the infectious nature of these clinical materials is unknown.

For this reason, human specimens can be handled safely at BSL-2, the recommended level for work with bloodborne pathogens such as hepatitis B, hepatitis C, and HIV.

BSL-2 recommendations and OSHA requirements focus on the prevention of percutaneous and mucous membrane exposures to clinical material. Primary barriers such as biological safety cabinets (Class I or II) should be used when performing procedures that might cause splashing, spraying, or splattering of droplets.

The segregation of clinical laboratory functions and limited access to such areas is the responsibility of the laboratory director. It is also the director's responsibility to establish standard, written procedures that address the potential hazards of clinical specimens and the required precautions to be implemented.

The Bloodborne Pathogens Standard—Infection Control Plan (29 CFR 1910.1030)

OSHA standard 29 CFR 1910.1030 limits occupational exposure for all employees who, as the result of performing their job duties, could be "reasonably anticipated" to face contact with blood and other potentially infectious materials. "Good Samaritan" acts such as assisting a coworker with a nosebleed would probably not be considered occupational exposure.

Other potentially infectious materials include semen, vaginal secretions, cerebrospinal fluid, synovial fluid, pleural fluid, pericardial fluid, peritoneal fluid, amniotic fluid, cell lines, saliva in dental procedures, any body fluid visibly contaminated with blood, and all body fluids in situations

TABLE 27–C. Summary of Recommended Biosafety Levels for Infectious Agents

BSL	Agents	Practices	Safety Equipment (Primary Barriers)	Facilities (Secondary Barriers)
1	Not known to consistently cause disease in healthy adults	Standard Microbiological Practices	None required	Open bench top sink required
2	Associated with human disease; hazard = percutaneous injury, ingestion, mucous membrane exposure	BSL-1 practice plus: Limited access Biohazard warning signs "Sharps" precautions Biosafety manual defining any needed waste decontamination or medical surveillance policies	Primary barriers = Class I or II BSCs or other physical containment devices used for all manipulations of agents that cause splashes or aerosols of infectious materials; PPEs: laboratory coats; gloves; face protection as needed	BSL-1 plus: Autoclave available
3	Indigenous or exotic agents with potential for aerosol transmission; disease may have serious or lethal consequences	BSL-2 practice plus: Controlled access Decontamination of all waste Decontamination of lab clothing before laundering Baseline serum	Primary barriers = Class I or II BCSs or other physical containment devices used for all open manipulations of agents; PPEs: protective lab clothing; gloves; respiratory protection as needed	BSL-2 plus: Physical separation from access corridors Self-closing, double-door access Exhausted air not recirculated Negative airflow into laboratory
4	Dangerous/exotic agents which pose high risk of life-threatening disease; aerosol-transmitted lab infections; or related agents with unknown risk of transmission	BSL-3 practices plus: Clothing change before entering Shower on exit All material decontaminated on exit from facility	Primary barriers = All procedures conducted in Class III BSCs or Class I or II BSCs *in combination with* full-body, air-supplied, positive pressure personnel suit	BSL-3 plus: Separate building or isolated zone Dedicated supply and exhaust, vacuum, and decon systems Other requirements outlined in the text

(Courtesy CDC/National Institutes of Health. *Biosafety in Microbiological and Biomedical Laboratories.* 4th Edition. Atlanta, GA: U.S. Department of Health and Human Services, Public Health Service, CDC and NIH, 1999.)

TABLE 27–D. Summary of Recommended Biosafety Levels for Activities in Which Experimentally or Naturally Infected Vertebrate Animals Are Used

BSL	Agents	Practices	Safety Equipment (Primary Barriers)	Facilities (Secondary Barriers)
1	Not known to consistently cause disease in healthy human adults.	Standard animal care and management practices, including appropriate medical surveillance programs	As required for normal care of each species.	Standard animal facility No recirculation of exhaust air Directional airflow recommended Hand-washing sink recommended
2	Associated with human disease. Hazard: percutaneous exposure, ingestion, mucous membrane exposure.	ABSL-1 practices plus: Limited access Biohazard warning signs Sharps precautions Biosafety manual Decontamination of all infectious wastes and of animal cages prior to washing	ABSL-1 equipment plus primary barriers: containment equipment appropriate for animal species; PPE: laboratory coats, gloves, face and respiratory protection as needed.	ABSL-1 facility plus: Autoclave available Hand-washing sink available in the animal room Mechanical cage washer used
3	Indigenous or exotic agents with potential for aerosol transmission; disease may have serious health effects.	ABSL-2 practices plus: Controlled access Decontamination of clothing before laundering Cages decontaminated before bedding removed Disinfectant foot bath as needed	ABSL-2 equipment plus: Containment equipment for housing animals and cage-dumping activities Class I or II BSCs available for manipulative procedures (inoculation, necropsy) that may create infectious aerosols. PPEs: appropriate respiratory protection	ABSL-2 facility plus: Physical separation from access corridors Self-closing, double-door access Sealed penetrations Sealed windows Autoclave available in facility
4	Dangerous/exotic agents that pose high risk of life-threatening disease; aerosol transmission, or related agents with unknown risk of transmission.	ABSL-3 practices plus: Entrance through change room where personal clothing is removed and laboratory clothing is put on; shower on exiting All wastes are decontaminated before removal from the facility	ABSL-3 equipment plus: Maximum containment equipment (i.e., Class III BSC or partial containment equipment in combination with full-body, air-supplied positive-pressure personnel suit) used for all procedures and activities	ABSL-3 facility plus: Separate building or isolated zone Dedicated supply and exhaust, vacuum, and decontamination systems Other requirements outlined in the text

(Courtesy CDC/National Institutes of Health. *Biosafety in Microbiological and Biomedical Laboratories.* 4th Edition. Atlanta, GA: U.S. Department of Health and Human Services, Public Health Service, CDC and NIH, 1999.)

in which it is difficult or impossible to differentiate between body fluids.

The standard requires employers to develop a written occupational exposure control plan that identifies tasks and procedures as well as job classifications in which the potential for occupational exposure to blood occurs (without regard to personal protective clothing and equipment). The plan must specify the procedure for evaluating circumstances surrounding exposure incidents and be accessible to employees and available to OSHA.

Training must be conducted initially upon assignment and annually thereafter. Employees who have received appropriate training within the past year need only receive additional training in items not previously covered. The standard requires that warning labels, including the orange or orange-red biohazard symbol, be affixed to containers of regulated waste, refrigerators and freezers, and other containers that are used to store or transport blood or other potentially infectious materials. Red bags or containers may be used instead of labeling. When a facility uses universal precautions in its handling of all specimens, labeling is not required within the facility. Likewise, when all laundry is handled with universal precautions, the laundry need not be labeled. Decontaminated regulated waste and blood that has been tested and found free of HIV or HBV and released for clinical use need not be labeled. Nevertheless, identification, labeling, and disposal of biomedical waste must be done according to the respective state regulations. Employers must provide, at no cost, and require employees to use appropriate personal protective equipment such as gloves, gowns, masks, mouthpieces, and resuscitation bags and must clean, repair, and replace these when necessary.

Hepatitis B vaccinations must be made available to all employees who have occupational exposure to blood within 10 working days of assignment. These vaccinations should be made available according to the latest recommendations of the U.S. Public Health Service. Employees must sign a declination form if they choose not to be vaccinated, although may later decide to receive the vaccine.

The standard specifies procedures after an employee has been exposed to materials potentially infected with bloodborne pathogens. Follow-up must include a confidential medical evaluation documenting the circumstances of exposure, identifying and testing the source individual if feasible, testing the exposed employee's blood if he or she consents, post-exposure prophylaxis, counseling, and evaluation of reported illnesses. This work will be conducted at no cost to the employee.

RADIATION SAFETY

Principles of Radiation

There are two basic kinds of radiation. One kind of radiation is tiny, fast-moving particles that have both energy and mass (weight), known as *particle radiation*. The other kind of radiation is pure energy with no weight. This kind of radiation is like vibrating or pulsating waves of electrical and magnetic energy. The radiation waves are called *electromagnetic waves* or *electromagnetic radiation*.

Radioactive materials and radiation-producing machines are commonly found in laboratories. The types of radiation used vary widely. Providing a safe and healthy workplace that uses radiation sources requires an understanding of the hazards and the regulations that apply.

In this section, the term *radiation* refers to the process in which energy is emitted by a source, transmitted through an intervening medium, and absorbed by another body. The energy transfer occurs in the form of subatomic particles (charged particles) or electromagnetic waves (photons). *Ionizing radiation* is any radiation consisting of particles or electromagnetic waves with sufficient energy to produce ions.

Ionization is the process of removing electrons from atoms, leaving two electrically charged particles (ions) behind. Ionization occurs when enough energy is transferred to displace an electron from an atom. Some forms of radiation, such as visible light, microwaves, and radio waves, do not have sufficient energy to remove electrons from atoms and, hence, are called *nonionizing radiation*. The negatively charged electrons and positively charged nuclei may cause changes in living tissue.

Ionization can be caused by alpha-particles, beta-particles, gamma-rays, neutrons, and x-rays. The first three types of radiation are emitted from the nucleus of a radioactive atom. Neutrons are created from nuclear fission and are not normally found in a laboratory setting. Radioactive materials are present in our natural environment, and they can also be artificially produced. For example, x-rays are produced when high-speed electrons are decelerated; some of the electrons' kinetic energy is converted into x-rays. Because x-rays are produced by high voltage, turning the x-ray–producing machine off stops the x-rays.

Alpha-, beta-, and gamma-radiation cannot be turned off like x-rays. The emitted radiation is produced from the decay of a radioactive atom. Radioactive isotopes decay with a specific *half-life*. This half-life is the length of time it takes for half of the original radioactive atoms to decay to another nuclear configuration. The duration of a half-life ranges from fractions of a second to billions of years. Each radioactive material has a unique half-life. After a radioactive material goes through seven half-lives, less than 1% of the radioactive material is left.

Number of Half-Lives Elapsed	Percentage of Radioactivity Remaining
0	100%
1	50%
2	25%
3	12.55%
4	6.25%
5	3.13%
6	1.56%
7	0.78%

Radiation Hazards

As radiation passes through a substance (such as air, water, tissue, or bone), a transfer of energy can result in ionization. In tissue or bone, this ionization can produce damage related to the amount of transferred energy. Unless high exposures occur, the affected individual will notice no immediate effects. The deposited or absorbed energy can produce damage, generally assumed to be proportional to the energy absorbed (dose). At high dose levels, this relationship can be shown. At low dose levels, the exact relationship of energy and damage is still being debated. As a conservative approach, the dose-response relationship is assumed to be constant (linear) regardless of dose or dose rate (Table 27–E).

Radioactive contamination is different from radiation. *Contamination* is the unwanted presence of radioactive material, which emits energy in the form of alpha-, beta-, and gamma-rays. When unwanted radioactive materials are in the body, the body is considered internally contaminated. There will always be some radioactivity in the body because of the natural radioactivity in our environment. When the radioactive contamination is on the skin of the body or the surface of an object, it is called *skin* or *surface contamination*.

TABLE 27–E. Effect versus Dose for a Full-Body Exposure Received in a Few Days or Less

Dose	Effect
1 Rem (0.01 Sv)	No detectable change
10 Rem (0.1 Sv)	Some blood changes may be detectable
100 Rem (1 Sv)	Some injury, no disability
200 Rem (2 Sv)	Injury and some disability
400 Rem (4 Sv)	50% mortality within 30 days*
600 Rem (6 Sv)	100% mortality within 30 days*
10,000 Rem (100 Sv)	50% mortality within 4 days*
Rem:	(Roentgen equivalent man) is a radiation dose unit that equals the dose in rads multiplied by the appropriate value of relative biological effect or Quality Factor for the particular radiation.

*Deaths at 4 days are due to intestinal damage; deaths at 30 days are due to blood cell damage.

Equal doses of different types of ionizing radiation are not equally damaging. For a given absorbed dose, alpha-particles produce greater harm than do beta-particles, gamma-rays, and x-rays. Radiation dose is expressed as equivalent dose in units of roentgen equivalent man (rem) or sievert (Sv), discussed shortly, to account for this difference.

Alpha-Radiation
Alpha-radiation is not an external hazard. Alpha-particles on the skin deposit most of their energy in the dead layer of epidermal tissue that covers the body, resulting in little damage or harm. But once inside the body, there is no dead layer of cells to protect the living tissues, and alpha-particles then deposit all of their energy within a few cell layers. Alpha-radiation potentially can cause greater damage to the cells in the immediate vicinity of the source than do other types of radiation. A few sheets of paper can stop alpha-particles.

Beta-Radiation
Beta-radiation is an external and internal hazard. A highly energetic beta-particle can penetrate a centimeter in tissue, about 500 times farther than an alpha-particle. This radiation can deposit energy in the base layers of the skin, but not the internal organs. The lens of the eye is particularly vulnerable to beta-radiation. Beta-particles deposited in the lens may cause cataracts. Though beta-particles are not as energetic as alpha-particles, the potential damage area for beta-radiation extends beyond the immediate vicinity to adjacent tissues, including nearby internal organs.

Gamma- and X-Ray Radiation
Gamma- and x-ray radiation are external and internal hazards. The properties of these two types of radiation are identical. Gamma- and x-ray radiation can pass completely through the human body, including the internal organs and bones. Because of their ability to penetrate tissue, gamma-rays and x-rays are whole-body radiation hazards. Gamma-radiation does not lose energy continuously as do alpha- and beta-radiation. As a result, gamma-rays are capable of causing less damage but inflict the damage over a wider area.

Radiation Protection Methods
The guiding principle or goal in radiation safety is to keep radiation exposures as low as reasonably achievable (ALARA). Exposure is the potential of a radiation field to deposit energy in an individual. The basic radiation protection methods use the concepts of time, distance, shielding, quantity, and signs and labels.

- *Time*—Reducing the length of time the individual is exposed to a radiation field reduces the individual's radiation exposure.
- *Distance*—If the distance to the source is doubled, the radiation exposure is reduced by a factor of 4. This is referred to as the *inverse-square law*.
- *Shielding*—Placing radiation-absorbing materials between the source and the potentially exposed individual will reduce radiation exposure. The effectiveness of the shielding will depend on type of material and thickness used, as well as the strength of the source and type of radiation present. Beta-radiation can be shielded by relatively thin pieces of material. Gamma- and x-rays are effectively shielded by materials with a higher atomic number (e.g., lead and concrete).
- *Quantity*—Limiting the amount of radioactive material will limit the potential for external and internal exposure (Table 27–F).
- *Signs and labels*—Proper identification and labeling of radioactive materials and radiation-producing devices alert persons to the presence of a radiation hazard.

TABLE 27–F. Exposure Limits for Workers in Radiation Areas (10 CFR 20,101)

Body Area	Rems/Calendar/Quarter
Whole body; head and trunk; active blood-forming organs, lens of eyes; or gonads	1 ¼
Hands and forearms; feet and ankles	18 ¾
Skin of whole body	7 ½

Source: 29 CFR 1910.1096 (b) (1), Table G-18.

Radiation Standards

Radiation standards provide dose limits to ionizing radiation. The occupational limit in the United States for whole-body radiation dose is 5 rem (0.05 Sv) per year. A more common unit of measure of dose is the millirem, or 0.001 rem.

The U.S. Nuclear Regulatory Commission (NRC) standard for units of radiation dose (10 CFR 20.1004) defines units as follows:

- The *gray* (Gy) is the SI unit of absorbed dose. One gray is equal to an absorbed dose of 1 joule/kilogram (100 rads).
- The *rad* is the special unit of absorbed dose. One rad is equal to an absorbed dose of 100 ergs/gram or 0.01 joule/kilogram (0.01 Gy).
- The *rem* is the special unit of any of the quantities expressed as dose equivalent. The dose equivalent in rems is equal to the absorbed dose in rads multiplied by the quality factor (1 rem = 0.01 sievert).
- The *sievert* is the SI unit of any of the quantities expressed as dose equivalent. The dose equivalent in sieverts is equal to the absorbed dose in grays multiplied by the quality factor (1 Sv = 100 rem).

The quality factors for converting absorbed dose to dose equivalent are shown in the following table.

Quality Factors and Absorbed Dose Equivalencies

Type of Radiation	Quality Factor (Q)	Absorbed Dose Equal to a Unit Dose Equivalent[a]
X-, gamma-, or beta-radiation	1	1
Alpha-particles, multiple-charged particles, fission fragments, and heavy particles of unknown charge	20	0.05
Neutrons of unknown energy	10	0.1
High-energy protons	10	0.1

[a]Absorbed dose in rads equal to 1 rem or the absorbed dose in grays equal to 1 sievert.

There are limits for different organs, including the extremities, the eyes, and the skin.

The basic standards for protection against ionizing radiation, such as those for the extremities and skin or for pregnant women, are written into NRC regulations. The NRC regulates the use of radioactive materials through 10 CFR 20, Standards for Protection against Radiation. Part 20 includes agency requirements for the following:

- dose limits for radiation workers and members of the public
- monitoring and labeling radioactive materials
- posting radiation areas
- reporting the theft or loss of radioactive material.

Although the NRC sets radiation protection standards and rules for radioactive materials, it does not govern x-rays, because x-rays are not produced by radioactive materials, or x-ray machine design or use. The Food and Drug Administration provides x-ray machine manufacturers with performance standards for x-ray machine design.

The NRC has agreements with most states that these states will regulate radiation protection at least as strictly as the NRC does. In addition to radioactive materials usage, most states (nonagreement states included) regulate the use of x-rays because this type of radiation is not covered by the NRC. Some states now have regulations that also cover a hazard known as the use of naturally occurring radioactive materials.

NONIONIZING RADIATION SAFETY

In terms of overall workplace exposures to the electromagnetic spectrum, nonionizing radiation sources have become extremely common. Examples of nonionizing radiation such as visible light, microwaves, and radio waves do not have sufficient energy to remove electrons from atoms.

Electric and magnetic fields are present whenever electricity is generated, transmitted, or used. Electric fields represent the forces that electric charges exert on charges at a distance, whereas magnetic fields are produced by the motion of the charge. These fields vary as a function of the frequency, or the number of times that the field oscillates per second. Power systems produce an electromagnetic field (EMF) that oscillates 50 to 60 times per second, or at 50 to 60 hertz (Hz). At frequencies between 3 and 3,000 Hz, the electric and magnetic fields are essentially independent of each other.

Health regulations applying to electric fields tend to fall into two categories: exposure criteria, which describe how much of an electrical field a person can be exposed to, and emission criteria, which describe how much can be released into or present in the environment near a source. The International Commission on Non-Ionizing Radiation Protection (ICNIRP) has issued general public and occupational exposure criteria based on induced current flow considerations. The American Conference of Governmental Industrial Hygienists (ACGIH) also established Threshold Limit Values (TLVs). A number of states have established emission criteria for transmission power lines by stating the maximum fields that can exist along the edges of the right-of-way occupied by a power transmission line.

The demand for guidelines to protect workers on the job has resulted in several ideas, two of which deserve mention. The first is prudent avoidance, which relies on reducing magnetic fields when and wherever possible. The second approach is a specific limit of 2 milligauss, which represents an averaged exposure level used to describe scenarios from epidemiological studies. However, this value does not define the field strengths that actually might cause harmful effects. Much work remains to be done in this area of worker protection and accident prevention from magnetic fields, and it is important to manage the risk from this potential hazard in settings where strong magnets are used.

Radiation Safety Program

Companies that use radiation sources are required to institute a radiation safety program to protect workers and to comply with safety standards and regulations. The basic objective of any effective radiation control program is to reduce unnecessary exposure to ionizing radiation. A radiation safety program should use some or all of the elements discussed next. The exact mix of these elements will depend on the size and number of the radiation sources and the exposure potentials during their use.

Radiation Safety Officer

A radiation safety officer (RSO) should administer the radiation safety program. RSOs should be well trained and educated to meet the radiation needs of the program as well as the managerial requirements. The RSO should provide advice and assistance to the users and the radiation safety committee, if a committee is needed. The RSO is the individual authorized by a company to officially communicate with state and federal regulatory agencies. This person should have the authority to act on behalf of the organization when addressing radiation-related matters.

The RSO is also the individual specifically named on a license to receive communications from nuclear regulatory agencies. An RSO should act as a resource to help users comply with applicable regulations.

Exposure Control

Limiting worker exposure to radiation is an essential element of a radiation program, whether the radiation comes from radioactive sources or x-ray–producing equipment. Reducing exposure supports the philosophy of keeping radiation exposures as low as reasonably achievable (ALARA), which is now a regulatory requirement. The control of radiation exposure includes minimizing both external and internal sources of radiation exposure (dose). Methods of controlling radiation exposures (doses) include posting signs marking radiation areas, restricting access to these areas or to radiation-emitting equipment, and training users in safe procedures and practices. Workers can also be separated from radiation hazards by distance or shielding with appropriate materials to reduce the radiation received to below the maximum permissible dose.

Contamination Control

Loose radioactive material is the primary source of internal contamination. Radioactive materials can enter the body via four basic paths: ingestion, inhalation, injection, and absorption. In most cases, the potential for ingestion and inhalation is higher than the potential for injection or absorption. Preventing radioactive materials from gaining access to these paths helps prevent internal contamination and subsequent exposure. Control methods and practices to protect individuals from internal radioactive contamination can include wearing gloves; prohibiting smoking, eating, or drinking in work areas; and providing respiratory protection and adequate ventilation.

Radiation Detection Instrumentation

There are two common types of portable survey equipment: gas-filled detectors and scintillation detectors. Gas-filled detectors are the most popular and versatile. Within this category are two types of meters: ionization chambers and Geiger-Mueller (G-M) detectors. Each type of meter has strengths and weaknesses. G-M detectors respond much faster and can be designed to measure radiation or contamination levels. Ionization chambers measure radiation dose independent of the radiation energy. Both instruments are useful and belong in a sound radiation program.

Scintillation detectors are generally more sensitive than gas-filled detectors. Scintillation detectors are normally designed to measure a single radiation type, such as alpha-particles. An alpha detector is designed to measure surface contamination levels. If the radioactive materials program uses alpha-emitting materials, an alpha detector helps locate contamination so it can be removed. However, the information obtained from such survey meters is only as good as the instrument calibration. Survey instrument calibrations should be based on standards set by the National Institute of Standards and Technology (NIST). The survey instruments should be calibrated to within 10% of the dose rate standard. These calibrations should be performed at least annually but can be required as often as once a quarter. Recalibration helps ensure the accuracy and consistency of radiation survey measurements. In addition, each time a portable survey meter is used, a pre-survey test should be done that includes checking the battery level, taking a background reading, and using a radioactive check source.

Dosimetry is used to measure the dose received by

individual workers. Commercial dosimetry processors use either film or thermoluminescent dosimeters as the radiation detector. These dosimetry processors have tested their dosimeters' accuracy through the National Voluntary Laboratory Accreditation Program (NVLAP). This program tests the dosimetry in eight radiation categories. Any dosimetry processor used by a radiation safety program should be accredited through the NVLAP process.

Bioassay involves determining the extent of a worker's internal contamination by radioactive materials. This process can be accomplished by direct measurement or analyzing the individual's urine or feces. The method of bioassay will depend greatly on the type of material suspected in the contamination. The bioassay process should be performed only by well-qualified experts who have the equipment and protocols to provide accurate results.

Other Program Elements

Safety training programs and course content will depend on a company's use of radiation machines or radioactive materials. As a minimum, users should be trained in radiation protection methods and regulatory requirements for relevant operations. Users should also be trained in the operations they will perform before starting work. The qualifications of principal users should be specified by a knowledgeable radiation safety officer or a radiation safety committee. Principal users will manage the use of the radiation source and employees who work with that source.

Transport of radioactive materials either as sources or as waste on a public road is controlled by the Department of Transportation (DOT). The DOT specifies the packaging, marking, and labeling of these and other hazardous materials offered for transport. The specifics for radioactive materials are found in 49 CFR 400–478. The NRC uses 10 CFR 71 to determine acceptability of transport packaging.

Since the enactment of the Low-Level Radioactive Waste Policy Act (LLRWPA) in 1980, the disposal of radioactive waste and materials has become an expensive but necessary part of doing business. Waste reduction methods, perceived as too costly 10 years ago, are now routinely used to reduce current volumes and disposal charges. Waste minimization and effective contamination control will help reduce the amount and thereby the cost of waste disposal. The amount of radioactive waste generated has been drastically reduced since the passage of the LLRWPA.

LASER SAFETY

The term *laser* is an acronym for *light amplification by stimulated emission of radiation*. Laser radiation is a form of electromagnetic radiation characterized by its wavelength. The wavelengths typically associated with laser radiation range from 180 nm to 1 mm. This range of wavelengths encompasses several regions of the electromagnetic spectrum: ultraviolet, visible, and infrared. Although laser radiation is often referred to as "light," many lasers produce radiation in the ultraviolet and infrared regions, which cannot be seen by the human eye.

Principles of Lasers

As a specific type of radiation, lasers are versatile and beneficial when used properly but potentially hazardous when used unwisely. Lasers produce a beam that is coherent and directional. Lasers are also monochromatic, meaning that the output of energy of a laser is a single wavelength.

Probably the most important quality of laser radiation is coherence—that is, all the photons act in unison. Coherent radiation is an efficient way of delivering energy because there is no destructive interference. A coherent beam produces a concentrated and powerful effect when striking a surface.

Lasers are also directional. Directional beams keep their shape for long distances and retain these properties even after reflecting off a mirror-like surface. This directionality allows the laser to be guided but can also present a hazardous condition at a distance from the laser itself.

Although lasers may take various forms, they are all constructed with an *optical cavity* and a source of excitation. An optical cavity contains the active medium—a solid, liquid, or gas—and two mirrors. The medium is excited by high-voltage electricity until it begins to emit radiation in all directions. Mirrors placed in parallel at each end of the medium reflect the laser radiation between them until standing electromagnetic waves form in the cavity. One type of mirror is totally reflective and the other is partially reflective to allow the release or emission of the laser radiation. Although the active medium generally dictates the type of laser wavelength emitted, the length of the optical cavity plays a substantial role as well. Figure 27-3 shows a schematic diagram of a simple laser system.

Hazards of Laser Radiation

Hazards associated with laser equipment can be divided into beam hazards and nonbeam hazards. Beam hazards, from direct or scattered laser radiation, affect primarily two organs: the eyes and the skin.

The skin is the less susceptible of the organs, because it can repair itself when injured. The main hazard to the skin is thermal burns produced from the extreme heat of a laser beam. Some lasers emitting UV laser radiation add

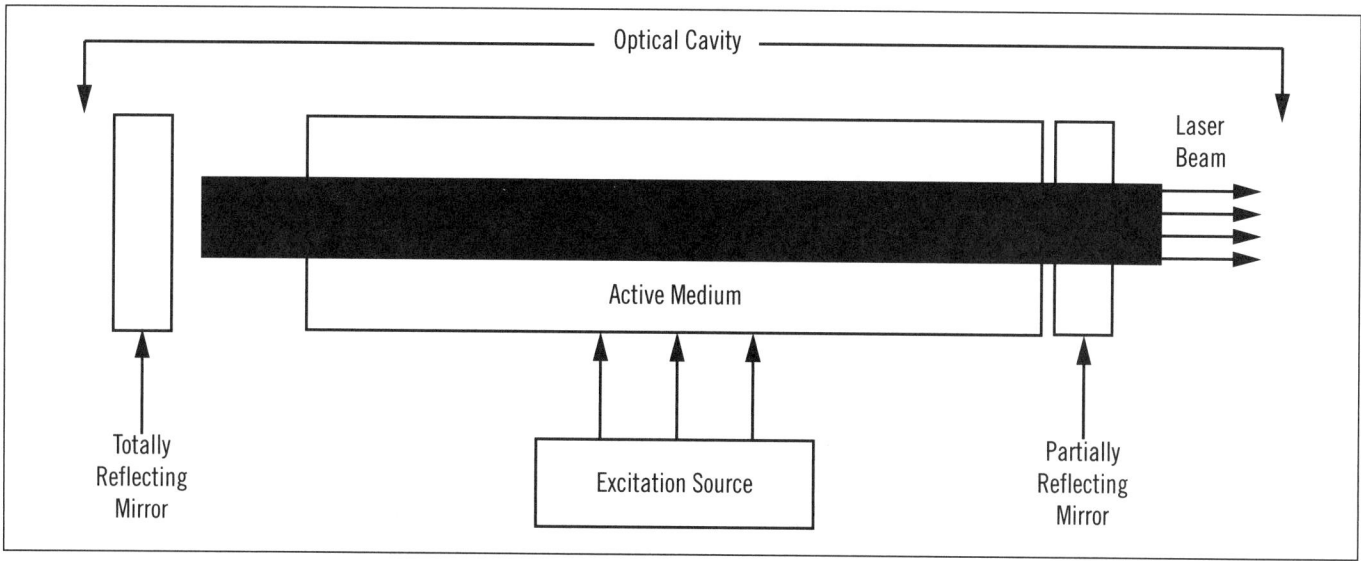

Figure 27-3. This is a schematic diagram of a simple laser system.

a photochemical effect, which produces the equivalent of sunburn on exposed skin. Workers burned by lasers should seek immediate medical attention.

Compared to the skin, eyes have less ability for repair and so are more vulnerable to injury. The eyes have three basic structures that can be affected by laser radiation: the cornea, the lens, and the retina. The cornea, which is the outer covering of the eye, helps focus visible-light radiation used in normal vision and absorbs the energy of deep UV and far infrared laser radiation. The hazard for the cornea is both thermal and photochemical laser burns.

The retina is the most sensitive structure in the eye. Visible and near-infrared light passes through the cornea and lens to the retina. The light receptors in this structure focus the laser beam and can intensify the thermal and photochemical damage to the retina. Lasers can also damage the retina through thermoacoustic effects. This occurs when the highly focused beam forms a steam bubble near the retina. When the bubble pops, it sends out shock waves that damage retinal tissue and blood vessels. The degree of permanent damage caused by thermoacoustic injury is determined by what part of the retina was struck and whether blood reached nerve tissue. In general, brief pulses of laser light lasting only millionths of a second can produce more damage to the eye than can longer pulses, because shorter bursts retain more heat.

Nonbeam hazards associated with lasers are numerous and potentially more dangerous than beam hazards. Lasers generally use high-voltage electrical power as a source of excitation for the active medium. As a result, electric shock is by far the greatest hazard associated with lasers. In addition, because of the inefficiency of many lasers, the power supply provides much more power than the laser can convert, and the remaining energy is given off as heat. This heat must be safely dissipated to minimize risk of personal injury, equipment damage, and property damage.

Chemical hazards are associated with the use of some lasers. Potential chemical hazards include asphyxiant gases such as carbon dioxide. Acid gases such as hydrogen chloride and potentially toxic by-products from the laser action on a surface are also chemical hazards of concern.

Other nonbeam hazards requiring evaluation in the laser workplace are noise, dust, hot surfaces, cryogenic gases, explosion of high-pressure flash lamps, toxic laser dyes, electrocution or shock, fire, and laser-generated airborne contaminants (LGACs). High-powered lasers also have the potential to ignite combustible materials. Workers should keep flammable materials out of areas where Class 4 and some focused Class 3 lasers are used.

Hazard Classification

Laser products are classified by the manufacturers according to the requirements of 21 CFR 1040.10, the U.S. federal performance standard. The classification scheme, which designates hazards as Classes 1–4, is based on the hazard level the laser presents (Table 27–G). These classifications are based on normal operation of the laser product. Maintenance and service procedures may expose the individual to levels of laser radiation far above those during normal operation. Before work begins, the company should evaluate the hazards that workers will encounter under actual working conditions.

TABLE 27-G. Laser Classification Scheme

Class 1:	Levels of laser radiation are not considered to be hazardous.
Class 2a:	Levels of laser radiation are not considered to be hazardous if viewed for any period of time less than or equal to 1,000 seconds but are considered to be a chronic viewing hazard for any period of time greater than 1,000 seconds.
Class 2:	Levels of laser radiation are considered to be a chronic viewing hazard.
Class 3a:	Levels of laser radiation are considered to be, depending upon the irradiance, either an acute intrabeam viewing hazard or chronic viewing hazard, and an acute viewing hazard if viewed directly with optical instruments.
Class 3b:	Levels of laser radiation are considered to be an acute hazard to the skin and eyes from direct radiation.
Class 4:	Levels of laser radiation are considered to be an acute hazard to the skin and eyes from direct and scattered radiation.

Hazard Controls

Engineering Controls

In general, engineering controls—such as protective housings, interlocked enclosures, ventilation systems, and beam enclosures—should be the primary method used to control exposure to laser hazards. Ideally, these controls require no action by the individual worker to be effective.

Interlocked laser beam enclosures include plastic panels on a framework and metal boxes with doors large enough to accommodate a person or vehicles. Laser beams can also be routed over or below walkways by the use of mirrors or elevated enclosures or tunnels. Interlocked laser beams either shut off the electrical power or drop a shutter into the beam, cutting it off. Interlocks can also be operated remotely and are used for laser systems that cover large areas.

For Class 4 lasers, viewing portals, viewing screens, and optical instruments must be connected to interlocks to reduce beam intensity to acceptable levels. Status lights, loudspeakers, and buzzers can be installed to protect employees working inside and outside the area of laser operations.

Although engineering controls are effective when lasers are installed, they are less useful when workers set up the device or carry out maintenance work. As a result, management should make sure that workers use special caution during these times, including the use of administrative controls and warning signs. For example, nearly 37% of laser accidents occur during the process of aligning the beam. Workers should use special protective equipment, such as laser alignment goggles, so that the laser beam will not go into the eyes. ANSI Z136.1 specifies the types of special warning signs that workers should use during setup.

Finally, workers and management should remove all unnecessary reflecting surfaces, such as chrome chairs, mirrors, or highly polished floors or furnishings. Even painted walls and stipple-polished tools can serve as highly reflective surfaces for gas laser beams.

The personal protective equipment used for lasers, such as safety eyewear, must meet the provisions of the latest edition of ANSI Z87.1. Workers should select laser eyewear that is tailored to the specific laser wavelength used and severity of exposure. This eyewear can be color-coded to indicate the specific use.

Administrative Controls

Maintenance and service operations may require more stringent safety controls than during operation of a laser. The Laser Institute of America's LIA/ANSI standard Z136.1 specifies calculating the nominal hazard zone (NHZ) (where people could be exposed to laser beam levels above the maximum permissible exposures [MPEs]) and blocking people from entering this zone by barrier signs, flashing lights, and warning signs. Specifications for these items can be found in the ANSI Z535 series of standards, which specify colors, warning words, standard symbols, and warning sign layout.

Because administrative controls require the individual to take some proactive action, they should not be used in place of engineering controls. Instead, they should serve as backup in case engineering controls fail. When engineering controls are determined ineffective or impractical, administrative controls can assume a primary role. Examples of administrative laser safety controls are adherence to written standard operating procedures, observing warning signs and labels, establishing a nominal hazard zone for a laser product, training laser users and maintenance personnel, and any other procedures that will enhance the laser safety process.

Personal Protective Equipment

The use of personal protective equipment (PPE) is the third type of control measure. PPE is used to minimize a worker's exposure to a hazard when engineering or administrative controls have not been fully implemented or are infeasible. The most recognizable type of laser PPE is laser goggles or glasses, which provide a defined amount of eye protection at specific wavelengths. However, companies must make sure that the correct laser PPE is specifically chosen for each laser. Eye protection is determined by reducing the exposure to less than the MPE. The required protection value is called the *optical density* (OD). The OD calculation can be found in the ANSI Z136.1 standard. The ODs for specific wavelengths must be provided on the laser glasses to help management select the proper eye protection (Table 27-H).

TABLE 27-H. Selecting Laser Safety Glass

Intensity, CW Maximum Power Density (watts/cm³)	Optical Density (OD)	Attenuation Factor
10^2	5	10^5
10^4	6	10^6
1	7	10^7
10	8	10^8

Source: 29 CFR 1926.102(b)(2)(I)

Recommendations for Laser Safety Programs

Laser safety programs developed by companies should be based on laser hazard classifications and hazard evaluations. Although no two laser safety programs will be identical, an effective laser safety program should follow the requirements of the applicable ANSI and FDA standards.

Safety Regulations

The main U.S. standard for laser safety is the Laser Institute of America's LIA/ANSI Z136.1, American National Standard for Safe Use of Lasers, revised in 2014. As the parent document of the Z136 series of laser safety standards, Z136.1 is the foundation of laser safety programs for industry, military, research and development (labs), and higher education (universities).

The Z136.1 standard provides guidance for individuals who work with high-power Class 3B and Class 4 lasers and laser systems, for example:
- industrial users who cut metals and ceramics (manufacturers of automobiles, computers, medical devices, etc.)
- industrial workers who weld using lasers
- those utilizing other laser material applications, including laser engraving, drilling, and ablation.

It provides a practical means for establishing a laser safety program to protect the employer, facility, and personnel from:

Beam Hazards:
- potential injury to the eye
- potential injury to the skin.

Nonbeam Hazards:
- electrocution
- fire
- exposure to air contaminants, hazardous gases, laser dyes and solvents.

Through engineering, administrative, and procedural controls, including:

- protective housing
- standard operating procedures (SOPs)
- personal protective equipment (PPE)
- warning sign and label requirements.

The Z136.1 standard defines:
- duties and responsibilities of the LSO
- laser and laser systems hazard classifications
- education and training requirements.

Other information includes:
- medical examinations (i.e., When are they necessary?)
- technical information provided on measurements, calculations, and biological effects
- comparison of national and international standards for classification.

The Laser Institute of America (LIA) is the international society for laser applications and safety information (laserinstitute.org). The Center for Devices and Radiological Health (CDRH) of the Food and Drug Administration has promulgated regulations for safety in the use of commercial laser devices (21 CFR 1040). These standards focus on the hazard control class of regulation, which requires classifying the severity of the hazard posed by a laser device. The two main measures of severity are wavelength and power or energy output; other measures include pulse duration and size or spread of reflection of the beam. The LIA/ANSI and CDRH standards set accessible emission limits (AELs) for each class of laser. Safety precautions become more stringent as the output of the laser increases from Class 1 to Class 4.

The LIA also specifies maximum permissible exposures (MPEs) for lasers, which are similar to OSHA PELs set for chemicals or the MPEs set for radiofrequency/microwave radiation by ANSI/IEEE C95.1–1991. The standard also gives protocols for evaluating the risks of exposure to pulsed lasers, although the protocols are complicated and difficult to apply. Managers and other employees working with pulsed lasers should read paragraph 8.2.2 of the LIA/ANSI standard carefully.

The Food and Drug Administration publishes regulations that contain the performance standards that laser manufacturers must meet to sell lasers in the United States. The laser product standard is found in 21 CFR 1040, Performance Standards for Light-Emitting Products. These standards provide the requirements for the construction, classification, and labeling of laser products. Laser manufacturers and importers must certify that their laser products comply with the performance standards according to the requirements in 21 CFR 1000–1010.

Currently, OSHA does not have an all-encompassing and comprehensive laser standard. In the construction industry there is an OSHA standard that covers the use of lasers (29 CFR 1926). However, OSHA has issued citations associated with laser use under the General Duty Clause of the Occupational Safety and Health Act of 1970. In these cases, employers were asked to revise their reportedly unsafe workplace using the recommendations and requirements of such industry consensus standards as the ANSI Z136.1 standard.

The Written Program

The program document should describe the overall purpose of the organization's laser safety program, with responsibilities and authorities assigned. The document should reference the ANSI Z136.1 or similar standard as the basis for the company's laser safety practices.

The appointment of a laser safety officer (LSO) is critical because his or her duty is to manage the laser safety program. Training for the LSO is very important. Because hazard recognition and control methods form the basis of a successful laser safety program, the LSO should be viewed as the company resource for information and regulatory requirements for laser use.

Medical surveillance monitors employees' health to determine whether they have the physical capabilities needed to perform a task and whether they have experienced symptoms that indicate overexposure to the hazard. In the case of lasers, the practice deals specifically with evaluation of the eyes or skin for laser damage. The examination (generally done by an ophthalmologist) helps identify any preexisting conditions and the extent of injury from exposures. The ANSI standard recommends that users of Class 3b and Class 4 lasers receive examinations before starting laser work and following any suspected laser injury.

The type and content of laser training depend on the particular hazards. Important ideas to communicate are the recognition of hazards and methods of protection from laser hazards. Service and maintenance personnel should be trained to protect themselves against the highest hazard level of accessible laser radiation because many laser products have lower hazard classifications based on enclosure of the beam. Repair and maintenance technicians should also be trained in high-voltage safety and cardiopulmonary resuscitation. Sometimes, these workers must remove or circumvent safety interlocks to perform a maintenance or service task. With safety interlocks defeated, the accessible laser radiation can be far in excess of the laser hazard indicated by the classification. Companies should prepare specific procedures and checklists to help protect employees properly during and after such service.

Several types of inspections should be included in an effective laser safety program. These would include installation, post-service, and annual inspections. The inspections should focus on engineering controls (e.g., enclosures, interlocks) and administrative controls (e.g., labels, signs, procedures). Personal protective equipment (PPE) use and applicability should be an important part of these inspections. Users should demonstrate the proper selection, care, and maintenance of their safety equipment.

PHYSICAL HAZARDS AND OTHER CONSIDERATIONS

Cryogenic Liquids

Cryogenic liquids are extremely cold refrigerated liquids (below –130°F [–90°C]) normally stored at low pressures in specially constructed, multi-walled, vacuum-insulated containers. Potential hazards include the following:

- exposure to extreme cold that can freeze human tissue on contact
- extreme pressure as a result of rapid vaporization of the refrigerated liquid when leakage occurs
- asphyxiation due to displacement of air by escaping liquid that changes to gas/vapor when released to the ambient surroundings.

Transfer operations involving open containers (dewars) must be conducted slowly to minimize boiling and splashing of the cryogenic liquid and must be conducted only in well-ventilated areas to prevent accumulation of inert gas, which can lead to asphyxiation. All personnel handling cryogenic liquids should be trained in the use of specialized equipment designed for the storage, transfer, and handling of these products. PPE that must be worn while handling these materials include heavy leather protective gloves, safety shoes, aprons, and eye protection.

Fiber-Optic Cables

Fiber-optic technology has revolutionized communications, improving fidelity, distance of transmission, and safety. The cables are made of fiberglass. Fiber-optic cable is basically a cylindrical mirror that reflects light internally. In fiber-optic cable, photons flow, not electrons, as in electrical transmission. These photons originate from a laser.

Laser energy is the principal hazard of fiber-optic devices. The signal is transmitted through the fiber by means of lasers, so anyone working with fiber-optic cables should also be familiar with basic laser safety. Fortunately, the energy spreads rapidly from the end of a fiber-optic cable, unlike many laser devices in which the energy remains concentrated over distance. But fiber-optic devices can still be hazardous when in close proximity. Fiber optics may need to be covered by a lockout/tagout-type program to protect individuals from inadvertent exposure to laser energy if the cable is connected to the laser source.

Further information on laser and fiber-optic safety can be found in ANSI standards Z136.1; Z136.2, Safe Use of Optical Fiber Communication Systems Utilizing Laser Diode and LED Sources; Z136.3, Safe Use of Lasers in Health Care Facilities; Z136.4, Recommended Practice for Laser Safety Measurements for Hazard Evaluation; Z136.5, Safe Use of Lasers in Educational Institutions; and Z136.6, Safe Use of Lasers Outdoors. In addition, there are ISO counterparts to these American standards.

Caution should be used when handling fiber-optic cables to prevent splinters. Also, scraps should be collected in a container for disposal to minimize the chance of exposure to splinters. Safety glasses should be worn when working with fiber optics.

ALTERED ENVIRONMENTS

Altered environments are special-purpose workplaces that require special conditions in order to carry out the work they host. These are often associated with laboratories and can include clean rooms, animal quarters, dark rooms, warm rooms, and cold rooms. Traditionally, safety issues surrounding altered environments have focused on exposure to physical hazards such as heat, cold, humidity, and altitude. Over the past 25 years, workers have found that they now are confronting new physical and psychological exposures and stresses. The solutions to the problems raised by these environments require specialized information and expertise from medical, safety, hygiene, ergonomics, engineering, and human resources personnel.

Clean Rooms

Clean rooms are designed and built to control dust and particulates in order to protect work materials from contamination by airborne particles. This is important in a number of settings, including electronic chip manufacturing. Dust control at this level requires a series of engineering controls that can also create potential hazards for the work force. For example, clean-room ventilation hoods typically push air away from the product and toward the worker, thereby increasing employees' potential exposure to chemical vapors from the work zone.

An additional problem that can increase worker exposure to air contaminants is the recycling of internal air within the clean-room environment. This means that any chemical vapors in the room are also rapidly redistributed into the work area. Safety professionals must be prepared to understand the mechanical engineering aspects of clean-room ventilation so that they can successfully address worker questions and problems.

The clean room is typically a semi-arid environment: hot (74°F), dry (35% relative humidity), and drafty because of the rapid number of air changes per hour. This environment creates a number of potential medical problems for workers, including dermatologic, allergic, and respiratory ailments. Chronic skin problems are one of the most frequent causes of medical transfer from the clean room.

Upper respiratory problems, such as sinus, throat, and lung complaints, have also been attributed to working in a clean-room environment. Because these problems are usually attributed to the environmental conditions required by the manufacturing process, it is often unlikely that anything short of medical transfer will be a satisfactory treatment. A similar, but less difficult, problem is eye irritation, which can usually be treated with simple over-the-counter drops or lubricating fluids.

An additional problem associated with protective garments is the development and aggravation of facial acne due to constant use of headgear or respiratory protection. This has been attributed to the combination of relatively hot ambient air temperatures and the constant rubbing and irritation that can be produced by full facial garments. Similarly, health problems, primarily dermatologic, have been associated with the polyvinyl chloride gloves that are usually used in the clean-room environment. Any individual experiencing health issues should be examined by an occupational health physician or health care practitioner experienced in dealing with occupational health exposures.

Animal Facilities

Animal facilities, which house laboratory animals, face many of the same health and safety challenges as described for clean rooms. An additional hazard in these workplaces is exposure to dander from experimental mice and rats. Some individuals are allergic to this dander, or develop allergies with time, which is likely to require them to cease working in this environment. Thus, animal care workers must be monitored and provided with careful oversight of this concern by occupational doctors, and good use of personal protective equipment is important in controlling this problem. Other potential issues include the sensitivity of animals to environmental conditions such as temperature, noise, and vibration.

Dark Rooms

Developing film in dark rooms can result in chemical exposures and hazardous wastes. Potential exposures should be addressed by engineering and administrative controls where possible using personal protective equipment where needed. Chemical wastes should be evaluated for proper disposal.

Warm and Cold Rooms

Many biological procedures require specific temperature and humidity conditions to proceed. These conditions are often provided by environmental rooms that are kept at specific temperature and humidity. The temperature control required for this is sometimes achieved by eliminating fresh air into the room; air is simply recirculated in the room on an ongoing basis. Therefore, if laboratory workers use hazardous chemicals or store liquid nitrogen or dry ice in such rooms, hazardous concentrations of the chemicals can build up or decreased levels of oxygen can result. Workers in these rooms must be educated as to potential hazards, and the rooms must have signs pointing this concern out to laboratory workers working in them.

Potential Psychological Problems

Given the conditions that are required in these specialized rooms, it should not be surprising that workers who work in them on a regular basis report a myriad of psychological problems. Large amounts of time are spent in a constant, unchanging environment. The use of protective equipment further accentuates the sense of isolation and dehumanization. Overall, these conditions can make the clean room a relatively harsh workplace environment. To date, no large-scale psychological studies of workers in this environment have been conducted. However, it would not be surprising to find somewhat increased rates of alienation, depression, or other mood disorders. Clearly, this environment presents both an opportunity and a challenge for specialists in occupational health, safety, hygiene, and human resources.

LABORATORY ERGONOMICS

CDC conducts a significant amount of biomedical laboratory research, which has led the agency to recognize that laboratory work presents significant ergonomic challenges to its workers. The information on this subject below is adapted from the CDC website.

Laboratory workers are at risk for repetitive-motion injuries during routine laboratory procedures such as pipetting, working at microscopes, operating microtomes, and using cell counters and video display terminals. Repetitive-motion injuries develop over time and occur when muscles and joints are stressed, tendons are inflamed, nerves are pinched, and the flow of blood is restricted. Standing and working in awkward positions in laboratory hoods and biological safety cabinets can also present ergonomic problems.

General solutions include providing fully adjustable seating and encouraging workers to take short breaks of 3 to 5 minutes for every 20 to 30 minutes of intense, awkward work. These breaks can be spent doing mild hand exercises or stretches. The following are further suggestions specific to certain kinds of equipment.

Biosafety Cabinets, Fume Hoods, and Laboratory Workbenches

Biosafety cabinets, fume hoods, and laboratory workbenches present similar ergonomic challenges, largely because of the lack of adjustability and leg room associated with use of the equipment. To manage this hazard, cabinets should be removed from under the workbenches when possible in order to provide leg room. Use a turntable to store equipment near the worker. This reduces excessive reaching and twisting, which places an increased load on the lower back. Position materials in the cabinet and on the bench top as close as possible to avoid extended reaching without compromising containment of the cabinet.

Computer Workstations

Many laboratory workers spend 50% or more of their day entering data with their keyboard and mouse resting on a lab bench. Because of the nature of laboratory furniture, many of these lab benches are too high and require users to elevate their arms and excessively stretch their wrists while inputting data. In addition to standard computer ergonomic considerations, laboratory personnel should not go from keyboarding to pipetting activities (or vice versa) without an adequate break to allow the hands to recover and to wash them.

Microscopy

The following practices are recommended for control of ergonomic hazards associated with the use of microscopes:
- Don't use a microscope for more than 5 hours per day. Spread the use out over the entire workday, avoiding long, uninterrupted periods of microscope work.
- Use a cut-out microscope table. This puts the worker close to the scope and gives an area for supporting forearms.
- Maintain a neutral spine posture. A neutral spine is neither rounded forward nor arched back too much.

Repetitive Pipetting

Pipetting is one of the most common tasks performed in the research laboratory. It involves several ergonomic stressors: thumb force, repetitive motions, and awkward postures, especially of the wrists, arms, and shoulders. The following are recommendations for ergonomics hazards associated with the use of pipettes:

- Use pipettes with newer trigger mechanisms requiring less force to activate; use the pointer finger to aspirate and the thumb to dispense.
- Use pipettes that fit comfortably in the user's hand.
- Use shorter pipettes. This decreases hand elevation and consequent awkward postures.
- Rotate pipetting activities between laboratory tasks, hands, and people.

Microtome and Cryostat Work

The following are recommended for control of ergonomic hazards associated with the use of a microtome or cryostat:
- Lower the workstation to keep the arms closer to the body.
- Apply padding to the front edge of the work surface to eliminate sharp edges and increase the amount of blood flow to the hands.
- Retrofit the existing handle with an adapter that will allow the operator to use the handwheel in a pistol grip position. This will help alleviate repetitive wrist flexion and extension.
- Consider the use of an automatic foot-operated cryostat when frequent cryosectioning is performed.

Glove Boxes

Working in glove boxes or anaerobic chambers requires extended static loading on the shoulders. Extending the arms for more than a couple of minutes can be exhausting. In addition to static loading and frequent side reaching, the thick gloves also make the user overcompensate on grip strength. Where possible, the following controls are recommended for ergonomic hazards associated with using a glove box:
- Move all needed materials for the experiment from the side chamber to the main chamber at one time to reduce the amount of side reaching.
- Use highly absorbent hand powder for glove comfort.
- Utilize job enlargement (periodically performing other activities) to avoid long, continuous use of glove boxes.

Centrifuges

Centrifuge rotors present a unique lifting hazard in the laboratory. Centrifuge rotors can weigh up to 35 lb and are awkward in shape. For this reason, if possible, use a second person to assist with lifting and removing the rotors. It is also a good idea to use a cart to transport rotors.

EMERGENCY PLANNING

Laboratory emergencies occur infrequently, but because of the variety of hazards found in laboratories, planning for effective response can be a challenge. Although emergency responders such as police, fire fighters, and emergency medical technicians are trained in how to assess the risk associated with entering the scene of an emergency, they are usually not familiar with the combinations of chemicals, biological agents, radiation sources, and other hazards found in the laboratory setting. Further, the training they receive in managing hazardous materials is typically built around industrial settings, where the nature of the hazard is very different from that found in laboratories.

This means that there is no standard way to plan for emergencies in laboratory settings. However, it is imprudent to wait until the emergency is occurring to address these issues. The laboratory safety program must proactively undertake an emergency-planning effort that involves the agencies that are expected to respond to fire, medical, or security emergencies. This usually involves touring the laboratory facilities and discussing the oversight programs in place to maintain a safe workplace.

The nature of the emergency plan developed through this interaction will depend on the resources available to both the laboratory organization and the local responders. In some settings, local response agencies may have specific requirements for laboratories. In other settings, they may be completely unfamiliar with laboratories and their special hazards. On the other hand, laboratory workers should have received specific training about how to effectively work with emergency responders in explaining the hazards associated with the laboratory. For the safety of all parties and to avoid undue disruption of laboratory work, an ongoing emergency-planning and preparedness effort should be maintained.

LABORATORY WASTE AND CLOSING PROCEDURES

As mentioned earlier, disposal of hazardous chemical waste is heavily regulated by federal and local agencies. This is also true of biological and radioactive wastes. Although the development of a complete laboratory waste management program is beyond the scope of this chapter, significant management advantages can be realized by connecting this waste program with the rest of the laboratory safety effort. A common element of these programs is that a risk assessment should be performed before work begins to ensure that proper practices are implemented and that the use of hazardous materials is minimized as much as possible.

In addition to routine waste disposal protocols, each organization should have a documented protocol in place for laboratory closures or moves. This protocol should be developed into a formalized procedure involving input

and participation from laboratory administration, facilities management, housekeeping, and the safety staff. The following issues should be addressed in this procedure:
- evaluation of chemicals and equipment for continued usefulness
- proper disposal of chemical, biohazardous, and medical waste
- clear labeling of all materials in the laboratory, whether to be disposed of or transferred to another laboratory
- use of appropriate cleaning solutions for cleaning and decontaminating equipment
- packaging unused chemicals for transfer to another location
- radiological safety decommissioning through contamination surveys
- removal and proper disposal of refrigerant and mercury switches in scientific equipment.

SUMMARY

- Laboratory safety management presents special challenges because of the wide variety of activities that occur in laboratories, which change frequently. Development of a proactive safety culture among laboratory workers is necessary to manage laboratory hazards successfully.
- Laboratory risks include chemical, biological, radiation, and physical hazards that vary widely in magnitude. In general, hazardous materials are small in quantity but can be high in hazards that must be managed carefully.
- An ongoing laboratory safety inspection program is necessary to provide adequate oversight of laboratory operations as conditions and processes change.
- A successful chemical safety program requires ongoing evaluation of the risks of the chemical processes being used; careful housekeeping, handling, storage, and disposal of hazardous chemicals; proper use and maintenance of equipment and PPE; and training and compliance appropriate to the hazards involved.
- A biological safety program involves training in and compliance with CDC biosafety guidelines, containment principles, and bloodborne pathogen issues. Risk assessments of biological practices in laboratories should be conducted to ensure that the correct biosafety-level practices are used for the work being performed.
- An effective radiation safety program includes the following elements: appointment of a radiation safety officer, exposure and contamination control, use of radiation detection instrumentation, inventory control, guidelines for transporting radioactive materials, and guidelines for complying with radioactive waste disposal.
- The basic radiation protection methods are limiting workers' time of exposure, keeping the source of radiation at a distance, shielding workers, careful labeling of all radiation sources, and limiting the amount of radioactive material used.
- Nonionizing radiation potentially presents some acute hazards, but the long-term health effects of this radiation are not well understood.
- Lasers produce high-energy beams that are coherent, directional, and monochromatic. Hazards associated with laser equipment are classified as Classes 1–4 (least hazardous to most hazardous) and involve beam and nonbeam hazards. Beam hazards affect primarily the skin and eyes, subjecting these organs to varying degrees of thermal and photochemical burns. Nonbeam hazards include electric shock, heat, toxic by-products, noise, dust, and explosion of high-pressure flash lamps.
- The FDA and OSHA have established specific guidelines and regulations for the use of lasers that require manufacturers and employers to comply with performance standards and safe practices.
- Altered environments, such as clean rooms, animal facilities, and temperature-controlled environments, also pose threats to human health. Hazards associated with clean rooms include arid environmental conditions, air pollution, health problems such as dermatitis, and psychological isolation. Animal facilities present exposures to allergenic animal dander. Temperature-controlled environments such as warm rooms and cold rooms generally do not have ventilation; therefore, hazardous atmospheres can develop in them.
- Laboratory workers are at risk for repetitive-motion injuries during routine laboratory procedures such as pipetting, working at microscopes, operating microtomes, and using cell counters and video display terminals. Careful planning of workstations to minimize these hazards and frequent breaks from repetitive tasks are key to minimizing this hazard.
- Emergency planning for laboratories is a significant challenge that requires ongoing, proactive attention to be successful.
- Waste disposal regulations require that laboratory wastes be minimized as much as possible. Extra attention should be paid to the transition of laboratories from one occupant to another to ensure that hazards are not passed from one to the other.

REFERENCES

Alaimo, R. J., ed. *Handbook of Chemical Health and Safety*. Oxford, England: Oxford University Press, 2001.

American Conference of Governmental Industrial Hygienists, 1330 Kemper Meadow Drive, Cincinnati, OH 45240.

Documentation of the Threshold Limit Values and Biological Exposure Indices. Updated annually.

American National Standards Institute, 11 West 42nd Street, New York, NY 10036.

Emergency Eyewashes and Shower Equipment, ANSI Z358.1-2007.

ASHRAE Standard 110-1995, "Method of Testing Performance of Laboratory Fume Hoods."

Balf, T., F. Churchill, G. Hall, Z. S. Graham, and R. Stuart. "Piloting an EMS-Based Regulation for Chemical Waste in Laboratories: A Lab XL Progress Report." *Chemical Health and Safety* 10, no. 3 (2003): 22–28.

Centers for Disease Control and Prevention/National Institutes of Health. *Biosafety in Microbiological and Biomedical Laboratories.* 5th ed. Atlanta, GA: U.S. Department of Health and Human Services, Public Health Service, 2009.

Centers for Disease Control and Prevention, Office of the Director/Administrator. "Laboratory Ergonomics." cdc.gov/od/ohs/Ergonomics/labergo.htm, October 2002.

Deming, W. E. *The New Economics for Industry, Government*, and Education. Boston, MA: MIT Press, 1993, p. 132.

Department of Homeland Security (DHS). Chemical Facility Anti-Terrorism Standards (6 CFR 27).

Department of Transportation (DOT). The Hazardous Materials Regulations (49 CFR 100–185).

Environmental Protection Agency (EPA). Resource Conservation and Recovery Act Regulations (RCRA) (40 CFR 239–299).

Furr, A. K. *CRC Handbook of Laboratory Safety.* 5th ed. Boca Raton, FL: CRC Press, 2000.

Gollnick, D. A. *Basic Radiation Protection Technology.* 4th ed. Altadena, CA: Pacific Radiation, 2000.

Laser Institute of America, 12424 Research Parkway, Suite 130, Orlando, FL 32826.

American National Standard for Recommended Practice for Laser Safety Measurements for Hazard Evaluation, ANSI Z136.4–2005.

American National Standard for Safe Use of Lasers, Z136.1–2014.

American National Standard for Safe Use of Lasers in Educational Institutions, ANSI Z136.5–2009.

American National Standard for Safe Use of Lasers in Health Care Facilities, ANSI Z136.3–2005.

American National Standard for Safe Use of Lasers Outdoors, ANSI Z136.6–2005.

American National Standard for Safe Use of Optical Fiber Communication Systems Using Laser Diode and LED Sources, Z136.2–2012.

American National Standard for Testing and Labeling of Laser Protective Equipment, ANSI Z136.7–2008.

National Fire Protection Association, 1 Batterymarch Park, Quincy, MA 02269-9101.

Standard for Portable Fire Extinguishers, NFPA 10.

National Research Council. *Prudent Practices in the Laboratory: Handling and Disposal of Chemicals.* Washington DC: National Academy Press, 1995.

NIOSH Publication No. 2007-107, *School Chemistry Laboratory Safety Guide.* Cincinnati, OH: National Institute for Occupational Safety and Health, 2007.

Occupational Safety and Health Administration.

Hazard Communication Standard (29 CFR 1910.1200).

Laboratory Standard (29 CFR 1910.1450).

OSHA Voluntary Protection Programs—www.osha.gov/dcsp/vpp/index.html.

Noz, M. E., and G. Q. Maguire, Jr. *Radiation Protection in the Health Sciences.* 2nd ed. Hackensack, NJ: World Scientific, 2007.

Sliney, D., and M. Wolbarsht. *Safety with Lasers and Other Optical Sources: A Comprehensive Handbook.* New York: Plenum Press, 1985.

Stuart, R. B., and C. Moore. *Safety and Health on the Internet.* 3rd ed. Rockville, MD: Government Institutes, 1999.

Urben, P. *Bretherick's Handbook of Reactive Chemical Hazards. 7th ed.* Academic Press, 2006.

Varanelli, A. G. "Electrical Hazards Associated with Lasers." *Journal of Laser Applications* 7 (1995): 62–64.

REVIEW QUESTIONS

1. How close should emergency showers and eyewashes be located to areas where hazardous chemicals are used?
2. What are the four stages of the management system approach to safety?
   a.
   b.
   c.
   d.
3. Name three key sources of chemical safety information.
   a.
   b.
   c.
4. Which of the following is a key piece of chemical safety information in assessing the risk of a particular chemical?
   a. molecular weight
   b. number of electron shells
   c. concentration of the chemical solution
   d. electronegativity
5. Distinguish between primary and secondary containment in biosafety.
6. List and describe the different levels of biosafety.
7. At what biosafety level is human blood appropriately handled?
8. The role of a radiation safety officer (RSO) includes
   a. performing medical evaluations of employees' eyes or skin for laser damage.
   b. managing the laser safety program.
   c. communicating with regulatory agencies on behalf of the company.
   d. all of the above
9. What are the four basic radiation protection methods?
   a.
   b.
   c.
   d.
10. The term *laser* is an acronym for what words?
11. Lasers use which of the following as the active medium?
    a. liquid
    b. gas
    c. solid
    d. all of the above
12. At what wavelengths of light do lasers operate?
    a. Visible
    b. Infrared
    c. Ultraviolet
    d. All of the above
13. List the general types of health problems that are associated with clean rooms.
14. Which of the following is not an important strategy for dealing with laboratory ergonomic issues?
    a. taking frequent, short breaks
    b. stretching exercises for limbs that are placed in awkward postures
    c. providing background music in the laboratory
    d. providing adjustable seating at workstations
15. Which emergency responders should be consulted in planning for laboratory emergencies?
    a. fire fighters
    b. emergency medical technicians
    c. hazard materials technicians
    d. police
    e. all of the above

Contractor and Customer Safety

28

Michael O'Berry, MEd
John Leyenberger, CSP, ARM, CPCU
John F. Montgomery, PhD, CSP, CHCM, CHMM

Safety and Contractor Oversight
Factors Influencing Contractor Safety ▶ Guidelines for Selecting Safe Contractors

Contractor Employees' Safety
Pre-Award Meetings ▶ Safety Requirements in Work Contract ▶ Safety Orientation ▶ Safety Record Keeping

Contractor Safety Responsibilities

Customer (Third-Party) Incident Prevention

Legal Side of Third-Party Injuries
Tort ▶ Negligence ▶ Degree of Care ▶ Invitee ▶ Licensee ▶ Limitations of Recovery ▶ Assumption of Risk ▶ Hold Harmless ▶ Attractive Nuisance ▶ Burden of Proof ▶ Third-Party Incident Trends

Problem Areas
Parking Lots ▶ Entrances to the Building ▶ Glazing ▶ Types of Ramps ▶ Walking Surfaces ▶ Falls on Floors ▶ Merchandise Displays ▶ Housekeeping

Escalators, Elevators, and Stairways
Escalators ▶ Elevators ▶ Stairways

Protection against Fire, Explosion, and Smoke
Fire Detection ▶ Response Methods ▶ High-Rise Building Fire and Evacuation Controls ▶ Crowd and Panic Control ▶ Self-Service Operations ▶ Evacuation of People with Physical Disabilities

Transportation
Courtesy Cars ▶ Company-Owned Vehicles

Protection of Attractive Nuisances

Summary

References

Review Questions

Good management is evident in the conduct of a business's routine operations and in areas that directly affect relations among managers, employees, outside contractors, their employees, and the firm's customers. Two of these areas, contractor and nonemployee safety, also affect company operations, employees, and the products or services they provide or sell. Incorporating safe work practices and controls into the business is the low-cost model of running a business. When unwanted incidents occur, operations are interrupted no matter who is involved. With the increase in consumer product safety legislation (see Chapter 20, Product Safety Management) and greater cost of liability claims, management must pay close attention to these areas of potential loss. To help management establish effective safety policies and work procedures with contractors, this chapter covers the following topics:

- safety issues and guidelines in hiring contractor workers
- how to manage contractor safety at all stages of the contractor–employer relationship
- basic safety responsibilities of contractors toward employees
- safety and legal issues in nonoccupational incidents and injuries
- work and other areas in companies that pose special liability issues
- loss control efforts regarding escalators, elevators, stairways, fire, transportation, and attractive nuisances.

A company must ensure that its loss control plans include preventing contractor and customer/guest (sometimes called *third-party*) incidents. These unwanted events can be minor or catastrophic. A firm can be liable for damages or for injuries sustained by someone (such as a customer or temporary office worker) from the minute that person enters the property (including the parking lot). In addition, a firm can be liable for the actions of its employees when they are off company property on business should their actions result in damage or personal injury to others.

SAFETY AND CONTRACTOR OVERSIGHT

Today, many companies hire outside contractors to perform a variety of tasks, such as maintenance and repair, clerical duties, construction, computer installation, and training. Based on the general-duty clause of the OSH Act and a 1993 OSHA regulation regarding contractor safety performances, employers have a responsibility for the safety of outside employees who either work on the employer's premises or are sent off site on the employer's business. In 1993, OSHA published a rule emphasizing the qualifications of contractors on pre-bid reviews and the employer's responsibility to obtain information about the contractor's safety record. Although the rule, Process Safety Management of Highly Hazardous Chemicals, Explosives (29 CFR 1910.119), applies to places where hazardous chemicals are used, subparagraph (h)(2)(i) states:

> The owner (employer) when selecting a contractor, shall obtain and evaluate information regarding the contractor employer's safety performance and programs.

More employers are accepting responsibility not only for the safety performance of their own workers but for the improved safety performance of outside contractors they hire. A comprehensive, systematic safety program can produce the following immediate and long-term benefits:

- reduced injuries and liability risks
- reduced potential regulatory action
- reduced potential for damage to the employer's facility and contractor's equipment
- increased productivity and lower overall costs.

With the publishing of OSHA's multi-employer citation policy, employers may find themselves citable for actions caused or created by another contractor. Employers with the best safety records understand the key factors that influence contractor safety and how to select contractors who have good safety records. These employers also develop and implement safety programs for outside workers as part of their overall company safety policy.

Factors Influencing Contractor Safety

Studies of West Coast companies that routinely hired contractors and subcontractors revealed that employers could strongly influence the safety performance of these outside workers (CII 1991). The studies highlighted five key factors in companies with excellent safety records:

1. *Strong employer management*—Employers clearly define the jobs contracted workers will do and keep tight control of schedules, responsibilities, training, and problem solving.
2. *Effective coordination of job tasks*—All in-house and outside employees know their roles and responsibilities and to whom they report. Management makes an effort to foster a spirit of teamwork and cooperation among all employees.
3. *Employer emphasis on safety*—Employers emphasize safety in their daily communications, reflecting top management's commitment to safe work practices and to safety in general.

4. *Strong interpersonal skills of supervisory personnel*—Recognition of the individuality of workers and respect for their experience, ideas, and feelings enhance their adherence to safety practices and procedures.
5. *Safe work environment in the employer's facility*—If outside employees see that safety is already a top priority in the company, they will take safety more seriously. Employers should make sure that their workplaces conform to all current safety regulations and that written materials, posters, Safety Data Sheets (SDSs; formerly Material Safety Data Sheets [MSDSs]), and other safety-related materials are posted in a prominent location.

Guidelines for Selecting Safe Contractors

One way to help manage the risk of hiring outside workers is to select contractors that have good safety records. The preamble to OSHA's new process safety management ruling (29 CFR 1910.119) explains that

> ... an employer should be fully informed about a contract employer's safety performance. Therefore, [OSHA] is requiring an evaluation of a contract employer's safety performance ... and safety programs.... The final rule ... does not require that employers refrain from using contractors with less than perfect safety records. However, the employer does have the duty to evaluate the contract employer's safety record....

Employers have three sources of objective information to evaluate the safety performance of outside contractors: contractor safety practices and policies, experience modification rates (EMRs) for workers' compensation insurance, and OSHA incidence rates for recordable injuries and illnesses.

Safety Practices and Policies

In general, companies that hold their management and workers accountable for safety have the best safety records. Employers can interview contractor management to determine their attitudes and commitment toward health/safety. They can do this by examining contractors' safety training materials, on-the-job training, hazard recognition program, substance abuse program, and other safety-related matters. Some key items to look for include written policies stating management's commitment to safety, thoroughness/accuracy of incident and illness data, and the frequency of safety inspections and scheduled safety meetings. Additionally, work practices and disciplinary measures that stress safe methods of working and protecting employees are another indicator of an integrated safety management system.

Experience Modification Rates for Workers' Compensation Insurance

The insurance industry uses experience modification rates (EMRs) as a means of determining equitable premiums to charge companies for workers' compensation insurance. These rating systems consider the average incident losses for a given firm's type of work and amount of payroll and predict the dollar amount of expected losses due to work-related injuries and illnesses that the employer will pay over a set period of time. In general, the lower a firm's rates, the better its safety record is likely to be. A company with a rating of less than 1.0 has a better loss record compared to similar companies (within the United States) doing the same type of work. A company with a 0.70 rating is 30% better than the average company and pays less for workers' compensation insurance.

Incidence Rates

Employers can also check a contractor's safety record by asking to see the incident and illness records. OSHA requires employers to record and report information on its annual Log of Occupational Injuries and Illnesses, OSHA Form 300. These records must be retained for 5 years. It is easy to calculate a contractor's or subcontractor's incidence rate by multiplying the number of incidents by the standard base number of 200,000 and dividing the total by the number of hours employees worked. The lower the incidence rate, the fewer incidents and illnesses the company sustained.

These three sources of information can help employers make good choices among the range of outside firms that are available. Employers should request from prospective contractors the EMRs and OSHA incidence rates for the three most recent years. Although the EMR is a more objective measure of a company's safety record than is the OSHA incidence rate, both indicate past safety performance.

CONTRACTOR EMPLOYEES' SAFETY

Research shows that concerned employers can do a great deal to improve the on-the-job safety performance of outside workers. The Stanford University Department of Civil Engineering sent questionnaires to owners and contractors to determine what measures they took to ensure safe working conditions (Business Roundtable 1989). The responses, arranged in order of importance, were as follows:

1. Outside workers were required to obtain permits certifying their skill and knowledge to perform potentially hazardous activities.
2. Employers required the contractor or subcontractor to designate a supervisor to be responsible for safety coordination on the job site.

3. Employers provided outside employees with company safety guidelines that had to be followed.
4. Employer/contractor safety meetings were regularly held.
5. Employers conducted regular safety audits of outside employees' work.
6. All incidents were reported to employer management immediately.
7. All incidents involving outside employees were investigated promptly.
8. Employers stressed safety as part of the pre-bid process.
9. Employers maintained statistics of outside workers' incidents.
10. Employers conducted periodic safety inspections.
11. Employers set goals for safety on the job.
12. Safety was considered as a major factor in prequalifying bids.
13. Where appropriate, employers set up a safety department or function specifically to monitor outside worker safety.
14. Safety guidelines were established in the body of the contract.
15. Employer orientation sessions informed outside workers of safety hazards on the job and how to control, avoid, or eliminate them.

Although none of the employers carried out all 15 measures, those with the best safety records implemented at least 8 to 10. However, several key activities, described in the following sections, were common to all employers.

Pre-Award Meetings

Before the contract or bid is awarded to outside firms, the employer should clearly specify the scope of the work to be done and emphasize the importance of safety. This information is best communicated in face-to-face meetings with contractor management. These meetings are particularly important when the work to be done is hazardous or will be carried out in a hazardous area.

Safety Requirements in Work Contract

Once the contract is awarded, the employer should specify in the contract the contractor's safety responsibilities and a measurement system to verify that the safety process is in place and effective. The contract language should go beyond the usual, standard statements (e.g., "The contractor must comply with all federal, state, and local safety regulations"). Employers should require a detailed safety plan that includes such information as safety inspections, enforcement, staffing, permits required, testing for substance abuse, basic safety training, and incident reporting and investigation. The contract should also specify regular reporting of incident and illness statistics to the employer.

Safety Orientation

Before work begins, the employer should meet with contractor management and safety staff to conduct a safety orientation and review session. The employer should address and document the following items:

- worksite safety requirements, including providing safety manuals and standards related to the proposed work
- a detailed outline of the safety responsibilities of the employer, outside management, and outside workers
- any special hazards that exist at the worksite, including a review of hazardous materials and relevant SDSs
- training requirements, including orientation, for outside personnel
- a schedule to review safety auditing, performance, and training programs after the first few days and weeks on the job.

Safety Record Keeping

Employers can require contractors to turn in monthly and, if appropriate, annual reports of incidence rates. Employers will need to total the number of incidents and contractor employee hours worked in order to analyze this data. To show that they are serious about the safety of outside workers, employers can establish a monitoring program such as the following:

- regular (weekly) safety progress reviews between employer and outside management
- regular safety progress reports that include such items as new contractor employees, documentation of training given, summary of injuries and lost workdays, safety meetings conducted, and number of attendees
- reports of safety inspections by contractor management and a summary of steps taken to correct safety problems
- safety inspections by employers of the work areas that are affected by the contractor's work activities and a review of the findings
- investigation of any injuries or illnesses that occur at the facility; employers should review the results of incident investigation reports with contractor management and keep detailed files for third-party litigation situations
- chemical screening programs administered by contractor management that are equal to or are at least as strict as the employer's
- annual evaluation by the employer of the safety performance of each contractor; the results of these evaluations should be placed in the employer's file
- careful auditing of each contractor firm for compliance with safety obligations and adherence to various safety

programs as specified in the contract and in discussions with the employer. Non-safety-related activities are not included in this audit.

Employers must require contractor management to maintain a safe work environment and ensure employee health and safety on the job.

CONTRACTOR SAFETY RESPONSIBILITIES

When requesting bids from contractors or subcontractors, employers should request that each firm submit a written safety program. The extent of the program will depend on the safety requirements of the particular job. The following items are some of the safety provisions that should be included.

1. The contractor should have a written safety program document containing:
   - a management policy statement
   - safety goals and objectives and a method of measuring the program's effectiveness
   - safety responsibilities for managers, supervisors, safety representatives, and employees
   - written procedures for safety activities such as new employee safety orientation, training, enforcement, inspections, personal protective equipment, medical services, and substance abuse program
   - written operating procedures, with methods on how to obtain permits.
2. The safety document should contain hazard communication procedures that conform to the OSHA hazard communication standard (29 CFR 1926.59 and/or 1910.1200):
   - a written hazard communication program and designated administrators
   - SDSs available for all materials used on the job
   - all containers properly labeled and showing hazard warnings
   - a documented employee hazard training program.
3. The contractor should have a safety handbook for employees, covering responsibilities of management and employees and important safety procedures, warning signs and symbols, etc.
4. Contractor management should provide orientation for all new employees, covering safety responsibilities, safety rules, first-aid facilities, incident reporting, safety meetings, PPE use, and reporting of unsafe conditions or actions.
5. Safety enforcement procedures should spell out the consequences of safety violations and the disciplinary actions that will be taken.
6. Job safety inspections should be conducted by contractor or subcontractor managers, safety personnel, and supervisors. Specific areas for inspection include facility operations, building and grounds conditions, housekeeping programs, electrical circuits and equipment, lighting, heating and ventilation, machinery, chemicals, fire prevention, maintenance, and personal protective equipment.
7. Incident investigation procedures should cover how to report OSHA-recordable injuries and illnesses to the employer in a timely manner, how to investigate the causes of the incident, and how to prevent a recurrence.
8. Contractors should issue work permits, such as safe work permit, hot work permit, excavation permit, confined-space entry permit, crane permit, and lockout/tagout.
9. Contractor management must provide specified personal protective equipment for hazardous conditions and employee training in their proper use and care.
10. First-aid and medical services must be provided to meet OSHA requirements. However, small contractor firms may elect to use the employer's services. If so, this decision should be stated in the contract.
11. Contractors' substance abuse programs should be as comprehensive as the employer's.
12. Emergency-response guidelines must be prepared in coordination with the employer's safety personnel.
13. Periodic (monthly) safety reports can be submitted to the employer, including a record of new contractor employees, documentation of training, summary of injuries and illnesses, and safety meetings conducted and number/names of employees who attend.

Safety on the job is the shared responsibility of the employer, contractor management, and the employees. A strong partnership between the two organizations can reduce or eliminate risks, injuries, and illnesses; help control health and insurance costs; and improve employee productivity and morale.

The second major area of risk for many companies involves safety issues related to nonemployees, such as customers, visitors, and others who enter the building or worksite of a company. The next sections describe the liability of companies and steps they can take to reduce that liability.

CUSTOMER (THIRD-PARTY) INCIDENT PREVENTION

In today's business environment, more people are involved in serving and selling to the public than are engaged in manufacturing the products that are sold. Service businesses include specialty shops, department stores, discount stores, shopping center complexes, restaurants, fast-food service operations, hotels/motels, automotive service facilities/dealerships, hardware or building supply stores, amusement parks, banks, office buildings, and franchise businesses. Providing good customer service is critical in sustaining this type of business, but it cannot be accomplished without providing a safe environment for the customer or guest. Injuries involving customers, guests, or visitors of these businesses make up the category of third-party injuries.

Service and sales operations must constantly monitor customer activities for potential loss exposures, even though they cannot control customers' personal activities or habits. For example, inebriated guests or patrons represent a serious risk when they improperly dispose of smoking materials or their instability causes them to slip, fall, or incur other injuries. Children can also be involved in incidents in the business environment. Whenever possible, design the original facility for predictable use and misuse.

LEGAL SIDE OF THIRD-PARTY INJURIES

Customer or product claim cases can end up in a courtroom. The legal terminology and aspects of the law that deal with incident claims include the following definitions and explanations. These are not intended to be comprehensive but rather to provide a quick review of some principal considerations.

Tort
A tort is a private or civil wrong resulting in an injury—a violation of a right not arising out of a contract. It may be (1) a direct invasion of some legal right of the individual, (2) an infraction of some public duty by which special damage accrues to the individual, or (3) a violation of some private obligation by which damage accrues to the individual. Torts result from negligence, accidents, trespass, assault, battery, seduction, deceit, conspiracy, malicious persecution, and many other wrongs or injuries.

Negligence
Negligence is the failure to exercise the degree of care that an ordinarily careful and prudent person ("reasonable individual") would exercise under similar circumstances. To establish a proper claim of negligence, however, there must be (1) a legal duty to use care, (2) a breach of that duty, and (3) injury or damage.

Degree of Care
The degree of legally required attention, caution, concern, diligence, discretion, prudence, or watchfulness depends on the circumstances. For example, a reasonable degree of care is required from people who invite others onto their premises by written, verbal, or implied invitation. All sales and service enterprises must exercise a reasonable degree of care for the safety of their patrons. As long as a business is open, it assumes a responsibility for the well-being of its customers.

Invitee
An invitee is one whose presence on the premises is at the invitation of another, such as a patron at a sports stadium or a person who visits an exhibition hall, even though no admission is charged.

Licensee
Licensees are neither invitees nor trespassers. They have not been specifically invited to enter the property, but they have a reasonable excuse (by permission or by operation of the law) for being there. These could be vendors, delivery personnel, people visiting executives, purchasing agents for business purposes, and the like. Police officers and fire fighters who enter property in the course of their duties have sometimes been held by the courts to be invitees (patrons) and sometimes licensees (nonpatrons).

Limitations of Recovery
American businesses are not automatically insurers of their clients or patrons. All states apply some formula to limit the ability of insured persons to recover monies from a company through a civil lawsuit. These legal doctrines fall into two broad categories. The older, called *contributory negligence*, bars any financial recovery by an injured party if he or she contributed to the original incident in any way.

A newer, and perhaps more humane, approach is found in the legal doctrine called *comparative negligence*. This doctrine, adopted by most states, requires a court to limit the recovery of an injured party based on how much his or her own action contributed to the original incident. Both doctrines illustrate the need for thorough and truthful investigation of every incident, which can strengthen a company's ability to reduce the overall cost of doing business and help prevent future lawsuits.

Assumption of Risk
Claimants cannot collect damages when they were aware of peril or danger yet were willing to proceed with their original intention and undertook the action. "That to which a person assents is not regarded by law as an injury." For example, a skier who falls while descending

a slope legally assumes the normal risks associated with this sport, unless some special negligence in the design or maintenance of the slope and its environment is a contributing factor. If so, the skier assumes no risk when the owner or operator is negligent. However, an injury caused by a mechanical defect in a chair lift or towrope could prove costly for the company running the operation or for the owners of the ski resort.

Hold Harmless

A clause in a contract agreeing that one party will assume all liabilities, losses, or expenses involved is a *hold-harmless agreement*. For example, a department store may have a hold-harmless agreement with a manufacturer who supplies a particular type of merchandise. If a consumer claims that an injury was caused by that product, the manufacturer will reimburse the store should it be held liable and will pay for legal and other expenses incurred by the store in defending itself. Even though the consumer purchased the item from the department store, a hold-harmless agreement may relieve the store of the financial effect of a tort.

Attractive Nuisance

This item refers to liability associated with a dangerous condition that is generally a threat to children. It excuses trespassing and penalizes an organization for failure to keep children away from the hazard or for failure to protect or eliminate a hazard that may reasonably be expected to attract children to the premises. A swimming pool is often considered an attractive nuisance because children are drawn to it, regardless of the protective features.

Burden of Proof

The burden of proof means that the injured party must prove injury or damage and its causal relation to the event or item that created the incident. A defendant is not liable if he or she is without fault. Proof must be established by facts, not opinion, suspicion, rumor, hearsay, gossip, or emotional reaction. Proof is a conclusion drawn from the evidence.

Honest and sincere witnesses often report widely different impressions about the same events. Thus, it is essential to assemble and preserve evidence quickly. Signed statements taken shortly after the incident or an all-important photograph can often make a big difference in a trial. Do not include the plaintiff in photos. In liability claims, the burden of proof rests upon the plaintiff (claimant).

This chapter describes the most common ways that a third party can incur an injury. Incident prevention techniques are broad and varied. Each safety professional must examine the company's operations in light of these guidelines. The following sections review some major third-party incident trends and discuss typical incidents that occur in modern business facilities.

Third-Party Incident Trends

When starting a nonemployee risk control process, the safety professional should identify incident trends by looking over the records of past third-party injuries. Next, choose general trend areas, such as slips and falls in entryways and those due to poor lighting, faulty interior design, and high traffic flow. Change facilities and/or operations so that both employees and customers are given a safe business and shopping environment.

PROBLEM AREAS

Parking Lots

Most shopping or business centers have parking lots for customer use, usually self-service. Bicycles, mopeds, and motorcycles as well as automobiles, vans, and trucks require parking facilities. If self-service parking lots are well designed and attractive, they can minimize the risk of damage to vehicles in the lot.

Parking lot loss exposures depend to some extent on how the lots are used. Self-service parking lots for industrial facility employees, for instance, usually differ from public lots at shopping centers, theaters, stadiums, stores, schools, restaurants, and motels. Usually, public lots have a more steady flow of traffic, more children present, and more obstacles such as shopping carts obstructing traffic ways. More important, the lots are used by persons with varied levels of driving skill who may be less familiar with the lot's layout. Some public parking lots at places such as theaters, sports arenas, and schools have the combined hazards of extreme traffic fluctuation and great variations in driver skill and alertness.

An easy-to-use layout, adequate signs, and conspicuous markings help make a parking lot safe and attractive. These steps represent the first way to reduce hazards. Consider enclosing the parking lot with a curb or fence so that cars cannot enter or leave traffic unexpectedly. This step can significantly reduce incidents in the area. If possible, parking lots should have separate, well-marked entrances and exits, placed so that they favor right-hand turns. Single-lane entrances should be at least 15 ft (4.5 m) wide; exits should be at least 10 ft (3 m) wide. Where entrance and exit must be combined, the double-lane drive should have at least 26 ft (8 m) of usable width. When the double-lane combination is necessary, it should have median curbs or strips to positively control the flow of traffic.

In general, entrances and exits should be:
- at least 50 ft (15 m) from intersections
- away from heavily traveled highways or streets
- well marked and well illuminated
- as few in number as possible.

Pedestrian traffic must be considered in parking lots. The most common injury occurs from hazards such as potholes, changes in elevation, and cracks in the walking surfaces due to uneven settling of the ground; wear of heavily loaded vehicles; and effects of cold weather. Maintenance staff should conduct periodic inspections to identify and rectify these hazards and to maintain a safe walking surface for customers. Stairways and ramps in parking garages should be constructed in accordance with ANSI standard A1264.1–1995, Safety Requirements for Workplace Floor and Wall Openings, Stairs, and Railing Systems. They should be well marked, well illuminated, and guarded with handrails. The needs of people who are physically disabled must also be considered in the design of the lot and in any walkway to the building. See Chapter 23, Workers with Disabilities, in this volume.

Parking aisles should be perpendicular to the buildings, so that pedestrians walk down the aisles rather than between parked cars. Where possible, marked lanes, islands, or raised sidewalks should be provided between parking rows, particularly where pedestrian traffic is especially heavy. These walks should be wide enough so that car bumpers overhanging them do not restrict pedestrians from walking to or from their cars.

Companies should minimize the use of curbs because they present a tripping and falling hazard.

Angle parking has both advantages and disadvantages. Although fewer cars may fit in the lot, angle parking is easier for customers and does not require as much room for sharp turns.

The area allowed per car in parking lots varies from 200 ft^2 to more than 300 ft^2 (18.5–28 m^2) if aisles are included. (See Chapter 2, Buildings and Facility Layout, in the *Engineering & Technology* volume.)

Parking Stalls
The design of parking stalls depends on (1) the size and shape of the lot, (2) the traffic pattern in the lot, (3) the type of drivers and pedestrians who use the lot, and (4) local and or state regulations/ordinances. (See the figures in Chapter 23, Workers with Disabilities, for stall dimensions for people who are physically disabled.)

Bumpers on stalls have the following advantages:
- They prevent drivers from driving forward through facing stalls and proceeding in the wrong direction in one-way aisles.
- They encourage drivers to pull forward against the bumper, thereby preventing rear overhang, which may reduce aisle width.
- They break up the huge expanse of an open lot, which may tempt drivers to cut across aisles and endanger pedestrians and other drivers.
- They block cars from rolling down inclines or running through walkway areas.

Stall bumpers can cause some problems, however:
- They may require maintenance.
- They may interfere with drainage or snow removal.
- They may cause pedestrians to trip and fall. (Painting them a bright yellow or other distinctive, contrasting color may help reduce such incidents.)
- They may reduce the flexibility of traffic flow.

Signs and Lighting
Traffic signs in parking lots should conform to recommended standards. They should be like those used as street and highway signs for easy recognition. For example, stop or yield signs should be installed at main crosswalks for pedestrians, wherever exits cross public sidewalks, and wherever exits enter main thoroughfares.

A well-lit parking lot reduces incidents and discourages crime. The amount of light recommended for parking at night usually ranges from 0.5 to 1.0 footcandles (decalux) per sq ft at a height of 36 in. (0.9 m). Lights are usually mounted 30 to 35 ft (9 to 10 m) high on poles whose bases are protected against impact from cars.

Design
Built-in bumps (speed bumps/humps) on lot surfaces have been used to discourage speeding. Disadvantages include possible damage to cars and creating a trip hazard, which must be considered. An alternative device is to keep straight lanes short and to provide sharp curves that require drivers to reduce their speed.

Parking lot operators should control unauthorized use of company lots. Local police or organizational security personnel can check parking areas after hours. If lots cannot be fenced or entrances cannot be protected against unauthorized entry, prominent warning signs stating that the lot is guarded or under video surveillance will help minimize the risks of unauthorized use and vandalism.

Those responsible for supervising lots should cooperate closely with local traffic authorities. It is also important to cooperate with police, sheriff, and fire departments. Fire trucks will need access to hydrants in a lot and to buildings served by the lot. Zoning boards are also interested in problems connected with parking lots because of the impact on traffic flow and congestion. Planted islands should be monitored to control mulch, rock, and other materials from getting onto the lot. Surface areas should be cleaned regularly.

Shopping Carts
Many companies want to keep shopping carts out of parking lots. To achieve this goal, companies can install barriers and have customers drive their cars to a loading point where an attendant puts their goods into the car or wheels the cart to the car, unloads it, and brings it back.

If shopping carts are permitted in the parking lot, they should be collected frequently and temporarily stored in a designated space. Do not allow carts to accumulate in stalls in heavily trafficked aisles. Cart-collecting areas should be well marked, well illuminated, and, preferably, separated from traffic by barriers in the pavement. Encourage customers to leave carts in those areas by installing signs and access lanes. People who supervise the lot should make sure that children do not play with the empty carts or use them as scooters or for racing.

In inclement or cold weather, shopping carts should be stored indoors or cleaned off before bringing them inside. The build-up of ice and snow on carts that are stored outdoors will melt when the carts are brought inside and can cause a slipping hazard on the floor. A well-drained floor surface should be provided in this area.

In addition to shopping carts, stores should also provide motorized carts to help transport customers with physical disabilities. These carts should be regularly inspected and serviced by qualified personnel. This provision will help stores comply with regulations in the Americans with Disabilities Act (ADA).

Entrances to the Building
Entrances to stores and office buildings must be adequate in number and size to meet local building codes. Maintenance should replace worn weather-stripping and keep sidewalks and driveways cleared and in good repair to prevent tripping and falling hazards. Entrance lighting should be a minimum of 10 footcandles. Revolving doors should have governors that limit their speed to 12 turns per minute.

Entrances represent a high-traffic area, and special precautions should be taken during wet weather to minimize the slip hazard.

Stores that serve the public also must be accessible to people with disabilities under the ADA. Some significant issues include width of floor aisles, accessibility of entrances and exits, bathroom access, and assistance in reaching shelves of different heights.

Glazing
Modern office building, store, and facility design often uses glass extensively in doors, show windows, panels, and enclosures. Such areas can result in a confusing pattern, especially to first-time visitors.

Unmarked glass panels and doors with untempered glass can cause injuries and cuts if employees or patrons walk or fall through them. Companies must follow recommended practices and standards to correct or eliminate this hazard. The safety professional or manager should understand regulatory standards on this issue.

Tempered glass panels, required for many uses, are given a special heat treatment during production so that if broken, the pieces of glass will fall as small, blunt beads instead of sheet glass plates.

Improper use of glazing materials is a continuing source of legal problems for companies dealing with the public. A complete check of all glazed items is important.

Some plastic glazing offers advantages similar to glass but without the risk of damage or injury. Unfortunately, some plastics have higher impact resistance and tend to scratch or be more easily abraded than glass. As a result, workers may find it harder to break plastic windows or doors than glass if they need to escape from an area.

A few precautions will allow companies to enjoy the beauty and functionality of glass while lessening the chances for injury:
- Areas around doorways should be properly illuminated.
- Make glass doors more visible to adults and children by placing decals or pressure-sensitive tape at their respective eye levels. Sandblasted or etched designs serve the same purpose.
- Decals or pressure tape will also prevent glass panels from appearing to be doorways. A tall, attractive plant, placed in the center of the panel, will let people know the panel is there.
- Installing safety bars reduces the size of the open glass areas and lessens the chance of glass breakage. The bars should be at the door-handle level on sliding doors and should be on both sides of a swinging door.
- Keep doorways and areas that are close to glass panels free of tripping hazards; common hazards include scatter rugs and toys. Indeed, this is a good rule for all occupied areas.

Types of Ramps
To provide passage from the sidewalk to the parking lot leading to the customer's car, it is common in shopping centers to construct a ramp. Indented ramps are the best type, ones that actually cut into the sidewalk with an easy slope. Rails on each side can prevent people from falling into one of the recessed sides if they are sufficiently deep to be a hazard. Although highlighting differences in elevation

is effective in drawing attention to ramps, do not paint the ramp surface itself because paint seals the concrete surface and can create a slipping hazard.

Entrances to buildings are of considerable importance from a safety viewpoint. Often entrance and exit doors are automatic and are activated by a photoelectric eye or when pedestrians step onto a touch-sensitive pad. With increased use of automatic entry/exit doors to provide customers with easy access to retail locations, these stores can have increased exposure to customer injury claims caused by malfunctioning automatic doors if they are not properly managed. Cameras built into the door frames can monitor proper operation and also document incidents to reduce claims. Companies must provide periodic maintenance and adjustment to ensure that the doors operate safely and are free of mechanical problems. If the vestibules are glass-enclosed, decals should be applied to alert people against walking into glass panels, as discussed earlier.

Walking Surfaces

Slips and falls are the most common type of customer injury. Although people can adjust their stride to walk across ice without falling, if they cannot see a wet spot on a floor, they may slip and fall. Footwear and physical characteristics such as balance and vision can also contribute to the cause of slip-and-fall incidents. These injuries can occur almost anywhere at any time. Few surfaces can be ignored, including asphalt roads; concrete walks; wooden, tiled, or rug-covered floors; special surfaces on stairs and conveyances (moving sidewalks, escalators, elevators); and bridges and catwalks.

The natural properties of any surface can change substantially when people track in mud, snow, dirt, and moisture. Moisture-absorbent mats, runners, or rugs can reduce such hazards. Patterned concrete surfaces are also effective in providing traction in wet conditions. Floor maintenance requires special attention to eliminate the hazard of torn or curled-up floor coverings. The company should implement administrative controls to ensure that these hazards are quickly removed. Liquid spills on floors require regular checks and cleanup procedures for employees to follow to minimize the time the condition exists on the floor. Another method that can provide a safer walking surface is to use slip-resistant flooring products to help prevent falls. Most slip-resistant materials can be installed in the same manner as other vinyl tile products. However, slip-resistant products still require periodic maintenance to preserve their appearance and to ensure that the floor surface remains hazard free.

Slip meters have been developed by testing agencies and insurance companies to measure the slipperiness of floors. One type of instrument (Figure 28–1), mounted on three leather "feet," is pulled across the floor by a motorized winch. The dial on top measures the intensity of the pull required to start moving it. This is converted into a *slip index*, or the *coefficient of friction* (0.5 to 0.6 is ideal). Although important, the coefficient of friction is not foolproof. Testing results can vary by the technique used by the person conducting the test, and floor slipperiness may change and usually increase because of moisture, oil, grease, foreign or waste materials, and incorrect cleaning.

Falls on Floors

Falls on floors occur in various ways and from various causes. Changes in elevation or an unnoticed change in the coefficient of friction (such as clear water on a tiled surface) can cause a person to slip or trip and possibly fall. A person may slip and thus lose traction or trip over an open drawer, a box in the aisle, or another object. The primary mechanical causes of falls on floors are unobserved, misplaced, or movable equipment, fixtures, or displays; poor housekeeping; and defective equipment. The condition of a person's shoes or type of footwear soles and heels are likely to be major contributing factors in slips and falls.

Inadequate illumination can also cause falls. Light values at floor level should be uniform with minimal glare or shadows. Also, there should be no major contrasts in light levels between floor areas, such as from bright sunlight outside the entrance to a dimly lit lounge or restaurant. Stair lighting in dimly lit areas is effective in drawing attention to the change in elevation.

Other factors causing falls include a patron's age, illness, emotional distractions, fatigue, lack of familiarity with the environment, and poor vision. Because a company cannot readily control many of these factors, it should make sure that the walking surface is properly illuminated and free from trip and fall hazards. For example, management should not place mirrors and other distracting decorations

Figure 28–1. Slip meters, ranging from the motorized type (shown here) to a simple spring scale and heavy block pulled by hand, can be used to gauge the slipperiness of floors. *(Courtesy Liberty Mutual Insurance Company.)*

in areas visible from steps or from approaches to steps or escalators.

Types of Floor Surfaces

A wide variety of floor surfaces are available. In office buildings, hotels, mercantile stores, and similar establishments, the use of masonry (terrazzo, cement, or quarry tile) floors is common at entrances, in lobbies, on stairways, and sometimes throughout the ground floor and in upper-floor corridors. Decorative materials such as terrazzo, marble, and ceramic tile are most often applied to interiors; concrete and granite are generally considered more practical for exterior use (Tables 28–A and 28–B).

In other public areas of these buildings, the base floor, usually of concrete or wood, is generally surfaced with one or more of the popular resilient floor-covering materials. Management generally uses carpeting in limited areas of various stores and hotels. Elsewhere, asphalt, linoleum, rubber, or plastic in either sheet or tile form will usually be found. Obviously, safety, initial investment, durability, and maintenance costs are some factors governing the choice of floor covering. Follow the recommendations of the floor material manufacturer for the product that best fits the application. Incidents can happen when managers attempt to improve the original appearance of a floor material at the expense of safety.

Most flooring materials, whether wood, masonry, or the resilient types, are reasonably slip resistant in their original, untreated condition. However, some masonry materials are exceptions to the rule. A highly polished marble, terrazzo, or ceramic tile, used for ornamental effect, can be slippery even when dry. When moisture is present, its slipperiness will increase. Improper surface-treating preparations and improper cleaning materials and methods also increase a floor's slipperiness. Unless a slip-resistant material is added to the compound during construction, the only preparation that should be used on such floors is a penetrating, slip-resistant sealer. Waxes are available that provide both a shine and good coefficient of traction. Ask your supplier to provide information on the performance of their products and the best floor treatment for your application. Refer to National Safety Council Data Sheet 12304-0495, Slips, Trips, and Falls on Floors, for more detail.

Floor Coverings and Mats

Reduce the possibility of slips and falls by using good carpeting, bound edging, and flush floor-level mats and runners (Figure 28–2). Over a given period of time, carpet threads tend to become loose, creating a tripping and falling hazard. This condition is more common in wall-to-wall carpeting because worn areas appear in high-traffic areas and it is hard to replace a small section of the carpet. Maintenance staff should inspect carpeting regularly for potential hazards. When loose threads are found, have the carpet restretched or replaced to eliminate the hazard.

Whenever possible, provide a contrasting color on car-

TABLE 28–A. Physical Properties of Floor Finishes

	Resistance to			Quality of			
Types of Finish	Abrasion	Impact	Indentation	Slipperiness	Warmth	Quietness	Ease of Cleaning
Portland cement concrete in situ	VG-P	G-P	VG	G-F	P	P	F
Portland cement concrete precast	VG-G	G-F	VG	G-F	P	P	F
High-alumina cement concrete in situ	VG-P	G-P	VG	G-F	P	P	F
Magnesite	G-F	G-F	G	F	F	F	G
Latex-cement	G-F	G-F	F	G	F	F	G-F
Resin emulsion cement	G-F	G-F	F	G	F	F	G-F
Bitumen emulsion cement	G-F	G-F	F-P	G	F	F	F
Pitch mastic	G-F	G-F	F-P	G-F	F	F	G
Wood block (hardwood)	VG-F	VG-F	F-P	G-F	F	F	G
Metal tiles	VG	VG	VG	F	P	P	G-F
Clay tiles and bricks	VG-G	VG-F	VG	G-F	P	P	VG
Epoxy resin compositions	VG	VG	VG	VG	F	F	VG

Code: VG—Very Good; G—Good; F—Fair; P—Poor; VP—Very Poor.

TABLE 28–B. Guide to Floor Materials and Surfacings

Floor Types*	Characteristics	Use of Abrasives	Dressing Materials
Asphalt tile	Composed of blended asphaltic and/or resinous thermoplastic binders, asbestos fibers, and/or other inert filler materials and pigments.	Abrasive materials of various types may be used to reduce slipperiness of floors. Colloidal silica** can be incorporated in wax and synthetic resin floor coatings.	Wax or wax-base products—For most purposes, wax has several advantages. This is especially true of Carnauba wax, an ingredient generally used in so-called wax products. This wax, a Brazilian palm tree product, dries in place with a very hard and glossy finish, but with a characteristically slippery surface. Because of its many good qualities, it is widely used as a base for floor surface preparations, both in paste and emulsion forms. Other waxes, notably petroleum wax and beeswax, have their place in floor dressing formulas; they are softer and less slippery than Carnauba, but are still slippery to a degree depending on the formulation.
Linoleum	Cork dust, wood flour, or both held together by binders consisting of linseed oil or resins and gum. Pigments are added for color.		
Rubber	Vulcanized, natural, synthetic, or combination rubber compound cured to a sufficient density to prevent creeping under heavy foot traffic.	Slip-resistant except when wet.	
Vinyls	Composed of inert, nonflammable, nontoxic resins compounded with other filler and stabilizing ingredients.	Adhesive fabric with ingrained abrasives can be used. They are patterned in strips, tiles, and cleats.	
Terrazzo	Consists of marble or granite chips mixed with a cement matrix.	Silicon carbide or aluminum oxide can be included in mix when floor is laid. Also an abrasive-reinforced plastic coating can be painted on.	Slip-resistant sealant will typically improve slip-resistant quality if renewed periodically.
Concrete	Made of Portland cement mixed with sand, gravel, and water and then poured.		
Mastic	Like asphalt tile in composition but is heated on the job and troweled onto the floor to form a seamless flooring. Such floors are often used over concrete to give a new durable, resilient surface.	(Same as asphalt tile)	Synthetic resins—These preparations, known as "synthetics," "resins," or "polishes," are intended to supply the desirable characteristics of wax without producing the same degree of surface slipperiness. They include soaps, oils, resins, gums, and other ingredients, compounded in various ways to produce the desired result.
Wood	May be either soft or hard, in a variety of thicknesses and designs.	Metallic particles and artificial abrasives in varnish or paint give good nonslip qualities to various floors.	
Cork tile	Made of molded and compressed ground cork bark with natural resins of the cork to bind the mass together when heat cured under pressure.	(Same as asphalt tile)	Other materials—Paint products (paint, enamel, shellac, varnish, plastic) are semipermanent finishes used principally on wood and concrete floors. They do not materially increase the slipperiness of the base.
Steel	Iron containing carbon in any amount up to about 1.7 percent as an alloying constituent, and malleable when used under suitable conditions.	Surface can be touched up with an arc welding electrode so the shape of raised places on the surface resembles angle worms. Also an abrasive reinforced plastic coating can be painted on to any desired thickness, dries hard as cement, and has a sandpaper-like finish. If a temporary nonskid surface is needed, two uses of mats can be employed: (1) flexible rubber mats made from old automobile tires; (2) rubber or vinyl runners.	
Clay and quarry tile	Kiln-dried clay products are similar to bricks and are extensively used in areas requiring wet cleaning.	Typically resistant to abrasives.	May be treated by etching. May be formulated as nonslip by adding carborundum or aluminum oxide when mixing the clay before kilning.

*Floors and stairways should be designed to have slip-resistant surfaces insofar as possible; adhesive carborundum strips may be used on stair treads or ramps and at critical concrete areas. Etching with mild hydrochloric (muriatic) acid solution will lessen slip problems.

**Colloidal silica is an opalescent, aqueous solution containing 30 percent amorphous silicon dioxide and a small amount of alkali as a stabilizer.

peted areas that meet and continue on treads and risers (or treads only) of stairways. If material such as an extruded metal runner is used to provide self-cleaning removal of snow, ice, or mud at entryways, it should be flush and not present a tripping hazard. Maintenance staff must pay careful attention to rubber mats, rug runners, and the like to prevent them from becoming tripping hazards. Often, their edges become rumpled, corners and ends are torn or do not

Figure 28–2. A heavyweight rubber or plastic mat with a nubby finish or raised design and beveled edges tends to lie flat and stay in place. Rotating the mat distributes the wear and minimizes "bald" spots in high-traffic locations.

lie flat, and excess wear causes tears. Management should replace mats and runners at the first sign of wear or other unsafe conditions.

Eliminate slips and falls on throw, oriental, and area rugs by using skid-resistant rug pads. When using rugs over carpet, attach a skid-resistant underlay to the rugs. These underlay pads can be purchased from carpet dealers. Floor mats, runners, and carpeting are used wherever water, oil, food, waste, and other material on the floor may make it slippery.

A company should establish clear procedures for placing, cleaning, removing, and storing mats. Those who put mats in place during inclement weather should have specific instructions about when and where mats should be put down and removed. If workers do not place the mats promptly and close enough to the door, entranceways may become slippery, and customers and others may track in water and dirt beyond the entranceway, creating a hazard and maintenance problem in other locations. Maintenance staff should follow definite procedures for inspecting and checking the condition of mats and for maintaining them in a safe condition. Consider a dry mop by the entryway in high-traffic areas for easy access to help keep floors dry and free of slipping hazards.

Stair rails, treads, and surfaces should be designed per standards, codes, and regulations; kept in good repair; and checked frequently for defects. Make sure nothing is stored on stairways and landings that could contribute to falls. Refer to National Safety Council Data Sheet 12304-0595, Floor Mats and Runners, for more detail.

Merchandise Displays

Workers must build displays so they are stable to prevent articles from falling and injuring or tripping customers. Repair and smooth sharp or broken displays and counters so they do not cut or catch passers-by.

Merchandising people sometimes try to stack merchandise high on a display table to catch customers' attention. However, if these displays are stacked too high, they become a hazard for the shopper. A customer reaching for an item can set off a cascade of boxes, cans, bottles, or other items that can strike or injure the person or create a tripping hazard.

When a product is stacked too high, the customer may need to step onto a lower shelf to reach it. Staff should stack shelves in a stable manner, placing heavy items on the lower shelves and lighter items on the top shelves (Figure 28–3, left).

In many stores, workers may hang nonfood accessories such as hardware items, notions, and kitchen equipment on pegboard panels (Figure 28–3, right). These panels or sections should be adequately recessed to accommodate the extended J-hooks. Remember, because shoppers will be bending over to reach lower items, the hooks above should not extend so far as to be a potential risk to a person's eyes or face. Specially safeguarded extenders can eliminate this problem (Figure 28–3, center).

Merchandise with sharp or cutting edges should not be in open displays unless the edges are covered or otherwise protected. Protective sleeves or plastic coatings or covers guard the edges from customer handling. Likewise, all electrical and mechanical display elements should have adequate protective features.

Construct display platforms using colors or lighting that contrasts with the floor or carpet, and do not allow them to obstruct aisles. Round or clip corners of platforms facing

Figure 28–3. Point-of-purchase display hangers must be located so that the human eye cannot contact them (such as shown at left, where both the upper and lower shelves project beyond the hangers). Shelf hangers must be safeguarded if they project into the aisle (center). Locating projections above eye level is another safe method (right).

customer traffic areas. Displays and mannequins should be at least 6 in. (15 cm) off the floor so that no one will tip them over accidentally. Make sure the display or the mannequin is fastened to its base.

Floor displays should be at least 3 ft (0.9 m) high so they are visible without becoming a tripping hazard. They should not be arranged at the ends of aisles so that shopping carts can dislodge them.

If a company hangs its displays from the ceiling, it should be sure that ceilings have adequate weight capacity to handle the load and that all code requirements have been met.

Housekeeping

In mercantile establishments, fixtures, displays, and other portable equipment are involved in many customer falls. Therefore, it is important that management provide safe equipment and that the risk control process particularly emphasize safe placement and use of that equipment. Also, workers should remove dress racks, pallets, and stock carts from the sales area and return them to the stockroom as soon as the items have been emptied.

Electrical wiring and extension cords connected to store machines, displays, special decorations, and the like should not lie exposed across floor aisles. They can present a significant tripping hazard. Where necessary, install wires or cords in low-profile channels.

Poor housekeeping accounts for one-third of all customer falls. Each employee needs to realize that it is part of his or her responsibility to maintain good housekeeping in the sales area and to promptly report unsafe floor conditions, such as tears in carpets and holes in the floors. Train workers to wipe up or barricade spills as soon as possible until the hazard can be removed. A special warning sign can be used. Spill-cleanup materials should be readily available throughout the establishment. Spills of potentially hazardous materials must be responded to by an individual trained in proper spill cleanup, and care must be taken to manage the spilled material per EPA or state regulations.

ESCALATORS, ELEVATORS, AND STAIRWAYS

Many business establishments move people between floors via escalators, elevators, stairways, and ramps. See also the discussion in Chapter 13, Hoisting and Conveying Equipment, in the *Engineering & Technology* volume.

Escalators

Escalators can be operated at speeds ranging from a low of 70 ft per min (fpm) to a speed of 125 fpm (0.36–0.64 m/s). The average recommended speed is 90 fpm (0.46 m/s). A 4-ft- (1.2-m-) wide escalator can move 4,000 to 8,000 passengers per hour. If escalators are operated faster or slower than people anticipate, passengers can be injured, especially children or elderly patrons.

Incidents can occur most often when passengers are entering or leaving escalators but can also happen while they are riding them. Common conditions and situations that can cause such incidents include the following:

- unsafe floor conditions and poor housekeeping at landings
- overshoes, particularly the thin plastic type; sneakers; or other objects catching in the combplate when pressed against a riser or catching at the side of the moving steps
- sales counters, mannequins, display bases, and similar units that hamper the movement of passengers
- lights or spotlights facing passengers as they step on or off at landings
- mirrors near escalator landings that cause passengers to misjudge their step and stumble
- merchandise signs or displays distracting riders, causing them to bump into one another or to fall
- failure of passengers to step on the center of a step tread, causing them to fall, possibly against others
- a passenger "riding" a hand on the handrail beyond the combplate and back into the handrail return, which can cause an injury if exposed fingers run into the handrail guard
- an emergency stop button that is unprotected against unintended contact, which can cause an incident if a passenger (or object) mistakenly presses it and halts the escalator unexpectedly
- a package, stroller, or other conveyance placed on the escalator and jamming or slipping from the grasp of the person trying to hold it
- parts of the body becoming involved with moving parts of the escalator, resulting in falls, lacerations, and amputations of the toes, fingers, or other parts of the body
- a passenger walking or running up or down a moving escalator
- children sitting on escalator steps
- a passenger who fails to hold on to a moving handrail and stumbles or falls.

Escalator Standards and Regulations

When installing or modifying escalators, check local and state ordinances. Also refer to the Elevator Escalator Safety Foundation's, *General Safety—Escalators*. The American National Standards Institute (American Society of Mechanical Engineers) Safety Code for Elevators and Escalators A17.1, is known as the Elevator Code. Some of its important points are as follows:

- Hand and finger guards are to be provided at the point where the handrail enters the balustrade.
- Prominent caution signs should be displayed (Figure 28–4).
- Combplates with broken teeth should be replaced immediately.
- Strollers, carts, and the like must be prohibited on escalators.

The Elevator Code also states that balustrades must have

Figure 28–4. Caution sign recommended for escalators by ANSI/ASME standard A17.1–1996. The sign should be 4 in. (102 mm) wide by 7 in. (197 mm) high.

handrails that move in the same direction and substantially at the same speed as the steps. The company should also follow city, state or provincial, and local code requirements.

Inspections

Examine all escalators from landing to landing every day. This includes riding them before the store opens to discover any defects.

Once a week, inspect the escalators as follows:
- Replace any broken treads or fingers in step treads and combplates.
- Examine handrails for damage.
- Check balustrades for loose or missing screws and for damaged or misaligned trim.

Semi-annually, inspect the following:
- step chain switches
- governor
- top and bottom oil pans
- skirt clearances
- step treads and risers
- steps
- machine brake
- skirt switch
- handrail brushes.

Start and Stop Controls

Escalators should have an emergency stop button or switch located at the top and the bottom landing. These must stop but not start the escalator. By placing a stop button at the base of a newel, with either a recessed design or a cover, the company can prevent a person or object from pressing the button unintentionally. All employees should be familiar with the location and use of emergency switches.

Elevators

Business establishments with elevators face other risk exposures. One of the most common is a customer being struck by the door. This is particularly common in self-operating elevators.

Inspections

A logbook containing the following information should be kept:
- day, month, year, and time of inspection
- observations by mechanics or inspectors
- all breakdowns, including causes and corrective action(s).

Entries should be initialed and dated.

A company must comply with all regulatory standards and should use the ANSI/ASME standard A17.1, the Elevator Code referred to under Escalators. At least every 3 years, all elevators should have a balance test and a contract load test. Make any required adjustments.

Once a year, hand-test the following controls:
- governors
- governor cable grip jaws
- gripping jaws of car and counterweight
- releasing carrier
- cutoff switches
- tail rope and trip rod drums
- safety rails.

Spot-check automatic elevators at the start of each day for level floor stops, brakes, and other mechanical operations. Test the alarm bell to be sure it rings and that its signal registers in the maintenance department (if available).

Stairways

Because stairways are so important in emergency exits from buildings, they are covered in detail in the National Fire Protection Association (NFPA) publication Life Safety Code, NFPA 101, specifically Chapter 5, Means of Egress. The chapter also covers ramps, exit passageways, smokeproof towers, outside stairs, fire escape stairs and ladders, illumination, exit marking, and escalators and moving walks. Only inside stairways are discussed in the following section.

The typical stair incident occurs when someone slips while descending a stairway. Most people who fall tend to look at stairway treads far less than do other people, and thus they are more prone to incidents. See the U.S. Department of Commerce study *Guidelines for Stair Safety* (listed in the References at the end of this chapter).

The following list represents major areas of concern, along with their corrective actions, that companies need to address.

- Because stairways are not level walking surfaces, they require different walking habits. Companies can minimize their use by posting signs directing people to the escalators and elevators. Also, never chain or otherwise lock designated exits to and from stairs so they cannot be used in an emergency.
- Adequate lighting is essential; NFPA specifies a minimum of 1 footcandle (decalux), measured at the floor. Arrange lights so they do not create glare surfaces or temporarily blind stair users and so the failure of one unit will not leave any area in darkness. If one side of the stairwell is open to adjoining space, it is a good idea to close off that view to prevent distractions that may cause people to fall. Emergency lighting is required in exit stairwells to provide adequate illumination in the event of a power failure.
- Color-contrast or highlight treads and handrails to help riders quickly and easily distinguish them from the riser and wall surfaces.
- The edge of the step or tread (nosing) should be easy to see. If possible, use contrasting carpeting to distinguish the approach to the steps from the stairs themselves. If not, special nosing may be installed to provide definition between the steps. Nosings must be securely fastened. Uncarpeted stairs should be edge marked.
- Stair treads need to be stable and provide good traction. Outdoor stairs need extra slip resistance and should have adequate water runoffs that lead away from other walking surfaces.
- Worn or defective treads or other parts should be replaced immediately.
- A continuous handrail must be provided; see details in Chapter 26, Office Safety, under Aisles and Stairs. The railing preferably should be of a lighter color because people seem to be more inclined to use a lighter-colored railing than a dark one that looks dirty and greasy. The railing should be kept clean. The handrail should extend to the top and bottom of the staircase so that it may be grasped before stepping on the first step or leaving the last step of the flight.
- Design handrails so that employees and patrons can grasp

them firmly and slide their hands along the rail without encountering obstructions. When designing handrails keep in mind that a simple, round profile, which permits the fingers to curl around the rail, in effect "locks" the hand in place. Two inches is the recommended maximum diameter for a typical handrail.
- Stairs should be kept clear of obstructions. Make sure sharp edges on handrail ends are removed or covered to prevent injury. Clearly mark and protect glass areas adjacent to stair landings or at either end of the stairway to prevent people from walking through them. Fixtures should not project into the stairway.

PROTECTION AGAINST FIRE, EXPLOSION, AND SMOKE

If fire breaks out or an explosion occurs, a quick, orderly evacuation of the premises will protect employees and visitors. Base the evacuation plans on the premise that most visitors will be on the property for the first time. They need to be directed to exits by numerous direction signs, even though exit locations may be well known to regular employees.

Businesses that generally attract large numbers of customers, guests, or patrons should provide well-marked exits, emergency lighting, and enough direction signs to guide people to safe exits. In a fire, dense, penetrating smoke can be as deadly as the heat or flame and can quickly sear people's lungs. Therefore, it is essential to evacuate the fire area immediately.

Enclosed stairwells provide the best fire escape routes. Doors to such stairwells must never be obstructed, locked, or propped open. A public address system, staffed by a qualified and trained person, can be used to direct the building's evacuation and issue lifesaving instructions.

Usually early detection of a fire and use of a good evacuation plan can prevent panic and personal injuries. More details are given later in this chapter and in Chapter 18, Emergency Preparedness.

Consider installing a fire detection and suppression system for early detection and control, which is required in many jurisdictions. Management should make sure that the fire plan is updated regularly and that fire drills are conducted at least semi-annually. Also consider asking your local fire department to help develop the plan so they are familiar with your facility. You can coordinate the arrival path of emergency fire-fighting equipment and your meeting locations outside the building to minimize congestion during an emergency. These steps are necessary so that employees are familiar with the plan and know where to meet outside the building should a fire break out.

Fire Detection

Properly engineered fire detection systems are sound investments, but the best installation is useless if no one responds to the alarm. Systems should have a direct connection to the local fire department or security service.

The Four Stages of Fire

Fire is a chemical combustion process created by the rapid combination of fuel, oxygen, and heat. A full discussion is found in Chapter 9, Fire Protection, in the *Engineering & Technology* volume. Most fires develop in four distinct stages: incipient, smoldering, flame, and heat. Detectors are available for each stage.
- *Incipient stage*—No visible smoke, flame, or significant heat develops, but a large number of combustion particles are generated over time. These particles, created by chemical decomposition, have weight and mass but are too small to be visible to the human eye. They behave according to gas laws and quickly rise to the ceiling. Ionization detectors respond to these particles.
- *Smoldering stage*—As the incipient stage continues, the combustion particles increase until they become visible—a condition called *smoke*. No flame or significant heat has developed. Photoelectric detectors "see" visible smoke.
- *Flame stage*—As the fire condition develops further, ignition occurs and flames start. The level of visible smoke decreases and the heat level increases. Infrared energy is given off that can be picked up by infrared detectors.
- *Heat stage*—At this point, large amounts of heat, flame, smoke, and toxic gases are produced. This stage develops very quickly, usually in seconds. Thermal detectors respond to heat energy.

Burning Plastics

Some fuels such as plastic waste receptacles can produce considerable toxic smoke when they burn. Therefore, the use of nontoxic and noncombustible materials (such as metal cans) is preferable. Managers and supervisors should be aware that other materials, such as building finishes, carpentry, furniture, and office fixtures, can also give off toxic smoke. For example, polyvinyl chloride (PVC) in a single foot of 1-in. (2.5-cm) PVC rigid nonmetallic conduit involved in a fire:
- can produce a sufficiently heavy, dense smoke to obscure 3,500 ft^3 (100 m^3) of space
- can generate enough hydrogen chloride to provide a lethal concentration of HCl in approximately 1,650 ft^3 (45 m^3) of space.

Retailers should verify that the fire protection system is designed to handle the fire load generated by merchandise made from plastic materials.

Engineering and Control Procedures

The best fire detection system is only as good as its weakest component. Management should use the services of a fire protection consultant in engineering the system and establishing the control procedures. Here are four steps to consider:

1. Select the proper detector(s) for the hazard areas. For example, a computer area may involve ionization or combination detectors. A warehouse may have infrared and ionization detectors. In low-risk areas, thermal detectors or combinations of detectors may be used.
2. Determine the spacing and locations of detectors to provide the earliest possible warning.
3. Select the best control system arrangement to provide fast identification of the exact source of alarm initiation.
4. Ensure notification of responsible authorities who can immediately respond to the alarm and can take appropriate action. Every detection system must have an alarm signal transmitted to a constantly supervised point. If this cannot be ensured on the premises, the signal must be transmitted to a central station, fire department, or other reporting source.

Sprinkler systems can also be viewed as detection systems as well as extinguishing systems. Routine maintenance must be provided for these systems on a regular basis.

Response Methods

Early warning systems are as important during hours of occupancy as they are when the premises are closed. When detection pinpoints a trouble spot, immediate response by a responsible trained company representative is important. Shaving seconds and minutes off the response time can mean the difference between lives lost or saved and between fire confined or allowed to spread out of control.

Some systems that companies use include the following.

Twenty-Four-Hour Supervisory Service

If an installation has 24-hour, 7-days-a-week supervision at some point in the building or building complex, then management should hook alarm, trouble, and zone signals into this location.

Less-Than-24-Hour Supervisory Service

For periods when an installation does not have responsible personnel to respond to the alarm, a backup system should be provided. The NFPA advises connecting to a central station supervisor's service or other service.

Such systems should include a way to initiate fire and trouble signals to the central station transmitter equipment. In the case of a trouble signal, a representative can be dispatched immediately to investigate the trouble and to notify proper representatives of the property under surveillance.

Central Station Monitoring Unavailable

In areas where no central station supervision is available, the local fire or police department may accept installation of a remote fire alarm panel at their headquarters or firehouse.

Central Station or Telephone Leased-Line Tie-In Unavailable

If none of the foregoing possibilities exists, then management should consider using qualified and licensed telephone answering services. Also, automatic dialing units connected to responsible officials are another alternative.

High-Rise Building Fire and Evacuation Controls

Just what is a high-rise building? Four basic criteria designate a high-rise structure. First, the size of the building makes personnel evacuation impossible or impractical. Second, part or most of the building is beyond the reach of fire department aerial equipment. Third, any fire within the building must be attacked from within because of building height. Fourth, the building has the potential for "stack effect."

High-rise structures all share one characteristic—they are intended to house people. As building heights go higher, the population density per square foot of ground area increases, posing a new set of risk exposures concerning the health, safety, and welfare of building occupants. Actually, each building is a sealed life-support system. Although engineering approaches have improved heating and cooling systems, these buildings are usually airtight. Stairways are designed with positive pressure to reduce smoke penetration, but occupants can be at greater risk of death or injury from smoke and toxic combustion by-products if the systems fail to operate correctly. This adds to the importance of proper design, monitoring the proper operation of emergency-response equipment, and testing up-to-date emergency evacuation procedures.

The ladders on most fire department aerial equipment are limited to approximately 85 ft (26 m). This means that a building higher than about 8 to 10 stories cannot be served by this equipment. Because of the height limitations of fire-fighting equipment and increased floor areas, fires must be fought from within the building. The principal fire-fighting equipment includes automatic sprinklers, hose standpipes, and portable extinguishers, as well as hose lines from the building exterior.

Evacuation Plan

When a building must be evacuated, management should have an orderly plan in place. An evacuation checklist is given in

Evacuation Preparedness Checklist

All questions in this checklist should be answered with "yes," "no," "NA" (not applicable), or "U" (undetermined). For all answers that are not "yes," or "NA," the persons responsible for the specific areas in question that need correction should be noted.

Floor Diagrams
- Are floor plans prominently posted on each floor?
- Is each plan legible?
- Does the plan indicate every emergency exit on the floor?
- Does a person looking at the plan see an "X" indicating "you are here"?
- Are room number identifications for the floor as well as compass directions given?
- Are directions to stairwells clearly indicated?
- Are local and familiar terms used on the diagram to define directions to emergency exit stairwells? For example, are particular areas identified, such as mail room, cafeteria, personnel department, wash rooms?

Exit Paths to Stairwells
- If color coding of pillars and doors, or stripes and markings on floors are used, are they properly explained?
- Is additional clarification needed?
- Are paths to exits relatively straight and clear of all obstructions?
- Are proper instructions posted at changes of direction en route to an emergency exit?
- Are overpressure systems and venting systems operative?

Elevators
- Are signs prominently posted at and on elevators warning of the possible dangers in using elevators during fire and emergency evacuation situations?
- Do these signs indicate the direction of emergency exit stairwells which are available for use?

Elderly and Physically Disabled
- Are there elderly or physically disabled persons who will need assistance during a fire and emergency evacuation of premises?
- What provision is made for their removal during an emergency?
- Who will assist? How will the disabled be moved?

Emergency Exit Doors
- Are all emergency exits properly identified?
- Are exit door location signs adequately and reliably illuminated?
- Do exit doors open easily and swing in the proper direction (open out)?
- Are any exit doors blocked, chained, locked, partially blocked, obstructed by cabinets, coat racks, umbrella stands, packages, etc.?
  Note: Blockage must be removed immediately and subsequently prohibited.
- Are all exit doors self-closing?
- Are there complete closures of each door?
- Are all exit doors kept closed, or are they occasionally propped open for convenience or to allow for ventilation?
  Note: This practice must be prohibited.

Emergency Stairwells
- Are stair treads and risers in good condition?
- Are stairwells free of mops, pails, brooms, rags, packages, barrels, or any other obstructing material?
- Are all stairwells equipped with proper handrails?
- Does each emergency stairwell go directly to the ground floor exit level without interruption?
- Does the stairwell terminate at some interim point in the building?
- If so, are there clear directions at that point which show the way to completion of exit?
- Is there provision for directing occupants to refuge areas out of and away from the building when they reach the ground floor?
- Are directions provided where evacuees can congregate for a "head count" during and after the evacuation has been completed?
- Is there adequate lighting in the stairwell?
- Are any bulbs and/or fixtures broken or missing?
- Where? Describe locations?
- Are exits properly identified?
- Are they illuminated for day, night, and power-loss situations?
- Are any confusing nonexits clearly marked for what they are?
- Are floor numbers displayed prominently on both sides of exit doors?

Emergency Lighting
- In the event of an electrical power failure or interruption of service in the building, is automatic or manually operated emergency lighting available?
- If not, what will be used?
- Where are stand-by lights kept?
- Who controls them?
- How would they be made available during an emergency?
- Is there an emergency generator in the building?
- Is it operable?
- Is it secured against sabotage?
- Is a "fail-safe" type of emergency lighting system available for the exit stairwells that will function automatically in event of total power failure?
- How long can it provide light?
- Is the emergency lighting tested on a regular monthly basis with results recorded? Who maintains such record?

Communications
- How should occupants of the building be notified that an emergency evacuation is necessary?
- Is one or more communication system available to each floor? (P.A. system, Muzak, stand-pipe phones, battery-operated "pagers," etc.)
- If messengers must be used, have they been properly instructed?
- Is the communications system(s) in good working condition?
- Under what emergency conditions is it used and who operates it?
- Can announcements be prerecorded by someone with a calm but authoritative voice?
- Is the communications system protected from sabotage?
- Do all occupants know how to contact building control to report a dangerous situation?
- Is the building's emergency communications system tested monthly? By whom and to what extent?

Figure 28–5. Reprinted from National Safety Council Occupational Safety and Health Data Sheet 656, Evacuation System for High-Rise Buildings (available from the NSC Library).

Figure 28–5. Several sources are available to help companies devise a plan. Title 29 CFR mandates in Section 1910.38:

(a) The emergency action plan shall be in writing. (The only exception to this regulation is for locations with ten or fewer employees.)
(a) (5) Training (ii) The employer shall review the plan with each employee covered by the plan at the following times:
  (A) Initially when the plan is developed,
  (B) Whenever the employee's responsibilities or designated actions under the plan change, and
  (C) Whenever the plan is changed.

Although OSHA is oriented to the employees, obviously such a plan should include delegation of evacuation leaders and a series of steps to evacuate both employees and guests/visitors in an emergency.

Stack Effect

Every building has its own peculiarities for creation of a "stack effect," or spread of fire, smoke, and toxic fumes from one area to another. Among the factors are structure configuration; height, number, and size of openings; wind velocities; temperature extremes; and number and location of mail chute openings, and elevator shafts—all of which create varying airflows that can accelerate and intensify an interior fire. Unprotected air-conditioning systems represent a major exposure to fire and smoke spread if not properly safeguarded. Without automatic smoke and heat detection, automatic fan shutdown, and automatic fire dampers, smoke and toxic fumes can be quickly drawn into the exhaust or return air duct system and promptly distributed to other floors and areas of the building served by the air-conditioning system.

The NFPA recognizes this potential. Its Standard 90A, Installation of Air Conditioning and Ventilating Systems, states that in systems of more than 15,000 cubic feet per minute (cfm) (7 m^3/s) capacity, smoke detectors should be installed in the main supply duct downstream of the filters. These detectors automatically shut down fans and close smoke dampers to stop the recirculation of the smoke, or they may incorporate automatic exhaust.

In planning evacuation, assume that children, elderly people, and people who are physically disabled will be involved. In addition, some people panic in a fire situation. The quantity and size of staircases may prohibit complete evacuation. Tests indicate that with an occupant load of 240 persons per floor, total evacuation of an 11-story building can take up to 6 min; an 18-story building can take up to 7 min. Exits in high-rise buildings are not designed to handle all occupants simultaneously.

Most codes do not consider elevators to be an exit component and prohibit their use during fire emergencies. But codes generally also require that one or more elevators be designated and equipped for fire fighters. Key operations can transfer automatic elevators to manual operation and bring the elevators to the street floor for use by the fire service. The elevators should be situated so the fire department can find and use them easily.

Some older elevators use capacitance-type call buttons, which may bring them to a stop on the fire-involved floor. Then they cannot move because smoke interrupts the light beam and keeps the doors open. Other possibilities include the inadvertent arrival of the elevator at the fire-involved floor by a passenger who does not know there is a fire, or a person who pushes the call button and then, in panic, uses the staircase for exit. With human nature the way it is, assume that complete evacuation is not going to happen in a timely manner.

Crowd and Panic Control

Any commercial establishment may be faced with an unruly crowd because of an emergency, panic, or even a planned demonstration. Management should plan for such events to protect the facility and its employees, along with patrons and bystanders. Different measures are required depending on whether the panic occurs in the building or outside it and who is involved. In shopping areas, companies should have directional signs displayed at many areas in the building. Exit signs are especially important. But these alone do not reduce the higher risk potential involved when a great number of people are gathered, such as at sports events, entertainments, and schools.

In almost every emergency, some panic can be anticipated. Employees need periodic drills and practice in handling emergency situations with customers, some of whom may be confused and others who may be physically disabled. Emergency-response plans should not overlook the threat of an unusual occurrence (such as a riot or bombing).

Although high-rise buildings pose special risk exposures, other public places such as theaters and amusement and recreation facilities must also provide well-planned emergency procedures. Panic and the press of frantic people have caused considerable loss of life in many emergencies. People tend to try to exit using the same way they entered the building, causing congestion (as in the Rhode Island nightclub fire), so well-marked alternate exits are important. Often, such losses could have been prevented by observing building and fire codes to eliminate hazards caused by improper design and lack of fire-fighting and disaster control measures.

Demonstrations

The following procedures concern controlling public demonstrations at a store, but they can easily be adapted to the needs of any establishment.

Demonstration Outside Store Building. Advise employees to call security and/or management. Security should telephone police, advise them of the situation, and follow their instructions.

Arrange for two or more key personnel to assume previously assigned positions at all store entrances and other key points; they should know security's telephone number to relay information and to receive instructions. They should never leave their assigned posts unless relieved or advised accordingly. Caution them to remain calm and not to interfere with the entrance or exit of customers or employees.

Those employed in portable, high-valued merchandise departments (diamonds, furs, etc.) should arrange to have such merchandise placed in an assigned secure area. Employees working in departments selling firearms, knives, axes, straight razors, bows and arrows, and even meat cutlery should have them removed from the selling floor to a secure area. Proceed with "business as usual" in all other departments.

Because rumors can create panic or problems among employees and customers present, all employees should be advised of two or three emergency interior telephone numbers to verify information and squelch rumors.

Demonstration Moving into Store. (People carrying signs, groups linking arms across aisles and taunting employees, fights between individuals.)
1. Advise all employees to avoid any comments, antagonism, or physical contact with demonstrators; refer all queries to a member of management; and above all keep calm.
2. In areas where demonstrations are taking place, have employees stop selling, lock their registers, remain in their areas keeping as calm as possible, and await further instructions from their supervisors.
3. In areas where "business as usual" is being maintained, arrange for frequent cash pickups.
4. Key personnel and employees should take their assigned places, as discussed earlier.
5. If such a demonstration turns into group looting or group "hit-and-run" stealing, employees should not attempt to make any apprehensions. Security personnel will follow previous orders for such conditions, as advised by management. Remember, personal safety is more important than property protection.

Self-Service Operations

The best-known self-service operation is found in gasoline stations. Here, customers put the gasoline in their vehicles without employee assistance. The employee merely sees that customers observe the safety rules and follow the prescribed procedure, usually posted on the pump housing or on a nearby sign. All states require that customers turn off their engines and refrain from smoking or using open flames. Emergency shut-off devices should be readily visible.

Evacuation of People with Physical Disabilities

The evacuation of people with physical disabilities from hotels, stores, and other facilities is an added consideration for both management and the safety professional. Clear communication is important in a successful evacuation. Special written instructions can be given to people with a hearing impairment; verbal instructions can be given to people with visual impairment. Workers should be assigned to assist people with disabilities during emergency evacuations.

More details on how to help people with physical disabilities are presented in Chapter 23, Workers with Disabilities.

TRANSPORTATION

Some businesses by their very nature of activity have a public roadway exposure within the fleet operations. Much of their risk control effort must be directed at protecting the passengers from harm.

Different types of transportation, such as commercial airlines, railroads, ships and boats, public transit, taxis, and school bus operations must focus their safety efforts on preventing passenger injuries. This involves not only maintenance and vehicle or unit operation, but also the safety of areas around vehicles, such as terminals, stations, and school bus loading areas.

Courtesy Cars

When a company provides transportation for customer courtesy and convenience, such as a hotel or motel courtesy car or bus, it should follow the same precautions used in commercial operations. These include providing a safe vehicle that is easy to enter and exit, maintaining it in proper working condition, and providing a professional driver who is trained and skilled in the vehicle's operation.

Company-Owned Vehicles

Another risk exposure arises out of the operation of company motor vehicles by employees. Because the odds are that one of every five drivers will have a collision in any given year, liability from such collisions is a constant exposure to loss. Management must be aware of the liability resulting from an employee's use of his or her own car on company business. Safety procedures should require all occupants to use seat belts and address the use of cell phones and smartphones and tablets during vehicle operation.

PROTECTION OF ATTRACTIVE NUISANCES

Some businesses have a risk exposure caused by the public's curiosity about their equipment or operations. Some examples follow.

Unattended vehicles or machines left in an operable condition are attractive to the young (and the so-called young at heart). Tamperproof locks discourage unauthorized use of vehicles or machines and, if necessary, security officers should be employed to watch over the equipment.

Partially finished road repairs and construction sites can present serious hazards for motor vehicles. Companies should barricade such hazards and warn motorists away from them. Refer to the U.S. Department of Transportation's *Manual on Uniform Traffic Control Devices for Streets and Highways*.

Many contractors or builders provide special, safe observation facilities for members of the public who like to watch construction projects. Local authorities and insurance loss control representatives can be consulted for regulations and control measures.

One of the more common attractive nuisances is a swimming pool, particularly one where the public can easily enter the swimming area. Many hotels, motels, and public areas provide swimming pools for their patrons (Figure 28–6). Pool incidents can result from inadequate protective barriers around pools, absence of lifeguards or qualified adult supervision, disregard for the rules of good pool conduct, and failure to teach youngsters drowning prevention techniques.

Pool precautions include the following:
- Screens, fences, or other enclosures should be erected to control admittance. A tamperproof lock and pool alarm may provide additional protection.
- Pool depths should be accurately marked in feet or meters, both on the pool deck and poolside.
- The pool should have no diving boards unless the pool has been constructed and staffed for diving.
- Adequate safety measures should be taken to prevent diving injuries. The depth of water in a pool is extremely difficult for the eye to measure. This presents a hazard for persons diving. Every year numerous injuries occur when people unknowingly dive into shallow water. To prevent such injuries, the following actions are recommended.
  - In shallow-water areas, post warning signs such as DANGER, NO DIVING ALLOWED, or SHALLOW WATER.
  - Designate specific diving areas where the water is sufficiently deep.
  - Use either pool markings or plastic or wood floats strung on rope to differentiate between the diving and no-diving areas.
- Do not store pool cleaning agents near flammable or reactive chemicals. Containers of pool cleaners should be disposed of in the proper manner.
- Make sure that after chemical treatment, pool water has the proper pH balance before allowing swimmers to use the pool.
- Provide basic lifesaving equipment, which should include a lightweight, strong pole with blunt ends at least 12 ft (3.7 m) long, or a ring buoy with long throwing rope.
- Consider posting a lifeguard on duty whenever the pool is in use and a sign listing pool rules and a statement that visitors swim at their own risk.
- Provide a telephone nearby, such as in the bathhouse or changing room. Emergency telephone numbers should be on hand for the nearest available physician, ambulance service, hospital, police, and fire/rescue unit.
- Keep the decks around the pool clear of debris, and do

Figure 28–6. Basic lifesaving equipment must be available at every pool. Shown here at the lifeguard platform is a ring buoy with a throwing rope attached. Note the fence enclosure. *(Courtesy National Spa and Pool Institute.)*

not allow breakable bottles in the area. Make sure all cups and dishes used at poolside are unbreakable. Provide litter baskets and replace defective matting.
- Prohibit games near the pool that could injure anyone.
- Ensure that electrical equipment conforms to local regulations and the latest National Electrical Code requirements (NFPA 70). Any electrical appliance used near the pool must be protected by a ground-fault circuit interrupter (GFCI).
- Do not allow swimming in the pool during a thunderstorm.
- Maintain all pool appliances and equipment properly. Make periodic safety checks.
- Establish, post, and enforce sensible pool rules.

SUMMARY

- Employers can set clear safety expectations for outside contract workers without directing work activities of those contractors. To help ensure a safe workplace and to protect contractor workers' health and safety, employers can establish criteria for an effective contractor safety program; develop procedures for selecting safe contractors; and insist on written, implemented safety programs developed by contractors.
- Companies must be familiar with their legal duties and liabilities regarding guests and customer safety. Two of the most common risk exposures are slips and falls and parking lots.
- Other areas that commonly present hazards and must be safeguarded include building entrances, walking surfaces, merchandise displays, escalators, elevators, and stairways. Companies should implement safety measures to reduce or eliminate tripping and falling hazards and post warning signs and other safety markings where guests/customers can easily see them.
- Businesses should install adequate fire detection equipment to sound the alarm at any of the four stages of a fire. Fire detection and fire response systems must be adequate for each building's requirements and should be tied into company and municipal fire-fighting departments.
- Management and employees must be trained in and practice evacuation procedures to handle crowds and control panic among guests/customers. Company employees should be able to help customers and visitors, including people with physical disabilities, quickly and efficiently to well-marked exits.
- Companies should also develop procedures to handle public demonstrations on their property. Employees should know how to communicate with security staff, summon police, handle demonstrators who enter the building, and guard their own safety.
- Courtesy cars and company-owned vehicles can be sources of nonemployee injuries. Companies must make sure their courtesy vehicles are in good repair and operated by a skilled driver.
- Firms must follow state and local ordinances and guidelines for preventing the public from gaining access to attractive nuisances.

REFERENCES

American National Standards Institute, 11 West 42nd Street, New York, NY 10036.
    Safety Code for Elevators and Escalators, ANSI/ASME A17.1–1996.
    Safety Requirements for Workplace Floor and Wall Openings, Stairs, and Railing Systems, ANSI A1264.1–1995.

Business Roundtable. *Improving Construction Safety Performance: A Construction Industry Cost Effectiveness Project Report.* Report A-3. New York: Business Roundtable, 1982, 1989.

Construction Industry Institute (CII). *Managing Subcontractor Safety.* Austin: University of Texas, 1991.

Elevator Escalator Safety Foundation, 356 Morgan Avenue, Mobile, AL 36606.
    General Safety—Escalators.

International Conference of Building Officials, 5360 South Workman Mill Road, Whittier, CA 90601.
    Uniform Building Code, section 3303(j).

National Fire Protection Association, 1 Batterymarch Park, Quincy, MA 02269-9101.
    Installation of Air Conditioning and Ventilating Systems, NFPA 90A, 1989.
    Life Safety Code, NFPA 101, 1990.
    National Electrical Code, NFPA 70, 1993.

National Safety Council, 1121 Spring Lake Drive, Itasca, IL 60143.
    Occupational Safety and Health Data Sheets:
        Floor Mats and Runners, 12304-0595, 1986.
        Slips, Trips, and Falls on Floors, 12304-0495, 1991.
    Safety and Health. Issued monthly.

U.S. Department of Commerce. *Guidelines for Stair Safety.* Washington DC: U.S. GPO, 1979.

U.S. Department of Labor, Occupational Safety and Health Administration (OSHA), 200 Constitution Avenue NW, Washington DC 20210.
    Code of Federal Regulations, Title 29. Section 1910.119, Process Safety Management of

Highly Hazardous Chemicals, Explosives. Code of Federal Regulations, Title 29. Section 1910.1200 and 1926.59, Hazard Communication Standard.

U.S. Department of Transportation, 400 Seventh Street SW, Washington DC 20590.
Manual on Uniform Traffic Control Devices for Streets and Highways, D6.1.

REVIEW QUESTIONS

1. How does 29 CFR 1910.119, Process Safety Management of Highly Hazardous Chemicals, Explosives, apply to contractor and guest/customer safety?
2. Studies of West Coast companies that routinely hired contractors and subcontractors highlighted five key factors in companies with excellent safety records. Cite three.
    a.
    b.
    c.
3. Regarding 29 CFR 1910.119, does the final rule require that employers refrain from using contractors with less-than-perfect safety records?
4. Why do insurance companies use EMRs (experience modification rates) when considering workers' compensation insurance?
5. What is indicative of lower rates?
6. When should the safety orientation between the employer and contractor management take place?
7. Identify three of the five recommended items to be addressed and documented by the employer.
    a.
    b.
    c.
8. What is an attractive nuisance? Describe one.
9. What is the most important feature of a parking lot?
10. How often should escalators be examined?
11. Where should emergency stop buttons be located for an escalator?
12. What three types of information should an elevator logbook contain?
    a.
    b.
    c.
13. How often should all elevators have a balance test and contract load test?
14. What are the four stages of a fire?
    a.
    b.
    c.
    d.
15. Identify two of the criteria that designate a high-rise building.
    a.
    b.
16. Briefly discuss the advantages of a protected air-conditioning system.
17. Your place of employment is suddenly the target of an outside demonstration. What are your first two duties?
    a.
    b.
18. The demonstration mentioned in #17 has moved inside. What five precautions should be taken?
    a.
    b.
    c.
    d.
    e.
19. What are two of the most common sources of guest/customer incidents on the premises?
    a.
    b.
20. Before allowing swimmers into the pool after a chemical treatment, what should be done?

PART 4

Program Implementation and Maintenance

The previous parts have explained the background and the systems that underlie successful safety, health, and environmental programs. This part focuses the safety professional's attention on training, incentives, awareness programs, and the media to present such programs. In addition, Part 4 opens with a chapter that addresses the growing concern of homeland security compliance in the workplace and the potential for intentional terrorist attacks on high-risk facilities such as chemical plants and refineries. The chapter identifies the federal agencies responsible for creating and negotiating protective measures and helps prepare operators for critical threats, should they occur.

Homeland Security Compliance in the Workplace

James T. O'Reilly, JD

Introduction

Federal Rules to Defend against Sabotage and Terrorism

What Sites Are Affected?

What Should Safety Managers Do?
Terrorism, Worksite Attacks, Preparation, and Response
- Dealing with Federal Risk Planners

Reporting Obligations

Local and State Involvement

Information Sharing

Conclusions

Summary

References

Review Questions

INTRODUCTION

This chapter discusses the potential for intentional attacks on chemical-using facilities by terrorists and the government's plans to reduce the risks of such attacks. The federal agency responsible for avoidance of attacks places chemical-using facilities into several categories and then negotiates plans for the protective measures needed at each high-risk site. Unlike other chapters that discuss accidents, this chapter covers what is hoped will never occur—an intentional attack. However, factory and refinery operators must be ready if this type of critical threat occurs.

Why should a book on preventing accidents discuss intentional attacks? The whole point of terrorism is to intentionally cause as much harm, and as much widespread fear, as possible so as to create fear that hurts the civilian population of the opponent. Homeland security has been a multibillion-dollar federal and state objective since September 11, 2001. But in the private sector, at the operating facility, the same management team that planned for the organization's response to chemical accidents will be called upon to deal with risks of sabotage, arson, explosives, and other external threats to the safety of workers and the community. The 2007 federal rules for protecting facilities against attacks should be studied and understood by the same team of well-informed, well-prepared planners who have already prepared for potential accidental events.

FEDERAL RULES TO DEFEND AGAINST SABOTAGE AND TERRORISM

The U.S. Department of Homeland Security (DHS) has been given the vital responsibility of protecting U.S. infrastructure, including industrial facilities, against attacks that may be terrorist acts or inspired by a desire to cripple U.S. productivity. DHS protective efforts in the context of industrial facilities are implemented through several means:
1. regulations on chemical-use facilities
2. state coordination of emergency responses to high-hazard-facility events
3. industry-sector coordinating committees to consider joint or mutual efforts by similar industries to avoid or mitigate damage from terrorism.

Of these efforts, the first is mandatory as an obligation for compliance by facility operators that are affected. The second is an intergovernmental cooperation effort, operating out of the limelight, that will interact with facility operations managers only after an incident occurs and emergency responders are called. And the third mechanism is information-based sharing of potential threats so that companies can be more aware of the direction from which certain types of threats might come.

The chemical facility security issues that DHS identified are in three main categories: chemical release, theft/diversion, and sabotage/contamination. Two categories have three subcategories each. The chemical release category has three subcategories described in DHS rules:
1. *release—toxic*: chemicals with the potential to create a toxic cloud that would affect populations within and beyond the facility, if intentionally released
2. *release—flammables*: chemicals with the potential to create a vapor cloud explosion that would affect populations within and beyond the facility, if intentionally released
3. *release—explosives*: chemicals with the potential to affect populations within and beyond the facility, if intentionally detonated.

The theft/diversion category also has three subcategories:
1. *theft/diversion—chemical weapons (CW)/chemical weapons precursors (CWP)*: chemicals that could be stolen or diverted and used as CW or easily converted into CW
2. *theft/diversion—weapons of mass effect (WME)*: chemicals that could be stolen or diverted and used directly as WME
3. *theft/diversion—explosives (EXP)/improvised explosive device precursors (IEDP)*: chemicals that could be stolen or diverted and used in explosives or improvised explosive devices (IEDs).

The third category, sabotage/contamination, refers to chemicals that, if mixed with other readily available materials, have the potential to create significant adverse consequences for human life or health.

A specialized approach for propane, chlorine, and ammonium nitrate was included in the final regulations. These can be found on the DHS website (dhs.gov).

WHAT SITES ARE AFFECTED?

The chemical facility protection rules do not apply everywhere; they apply only to facilities that have a certain quantity of certain high-hazard chemicals. DHS requires a "Top Screen" report from facilities. The DHS center receives many reports each year that go into the screening process; only a smaller number—perhaps 6,000—go to the more intensive level of scrutiny and are designated as high risk. The facility will be regulated if its chemicals of interest (CoIs) exceed a DHS-set amount called the *screening threshold quantity* (*STQ*). The amounts are published in 6 CFR 27, Chemical Facility Anti-Terrorism Standards.

There are four tiers of regulatory controls, based on risks. In the higher-risk tiers, security vulnerability assessments identifying security risks are required; a site security plan must be completed. For example, a facility with a substantial quantity of methyl bromide, hydrogen sulfide, or phosphorus could be attacked by a terrorist, with lethal consequences to those living downwind; a scenario similar to the Bhopal tragedy is possible.

Facilities excluded from the rule are those with extensive federal controls already in place; these include railroads, public water systems, and certain port activities. When the DHS rule conflicts with a state law, DHS prevails under the express congressional preemption authority. But when separate federal programs apply—such as the Clean Air Act requirements for planning for worst-case scenarios of large chemical release—each program continues and the filing of a DHS report does not excuse the firm from also filing an EPA report.

Beyond the precision of DHS rules is the practical recognition that the "product becomes a target" when a plant makes weapons. Drones used to bomb remote villages elsewhere on the planet are known to be made in certain U.S. factories. Munitions and rocket fuel used to power the deadly strikes are also made in certain U.S. factories. In addition, radio guidance tools for bombs and the satellite tracking equipment that pinpoints locations of the enemy are known to be made in certain U.S. factories. These are the targets that enemy agents (a cold war term, but still literally correct) will be considering when they send over a team to wreak revenge on the civilian producers of weapons and weapon support systems.

WHAT SHOULD SAFETY MANAGERS DO?

This chapter addresses personal and physical security from terrorist attacks and threats. It is important to note that the terrorism seen around the world has a purpose: interrupting the productivity and positive feelings that make the American workplace so successful. If threats of terrorism and violence at the workplace can frighten Americans, they will interrupt our economic strength, and terrorists will have won their psychological battle without actually having to attack any particular site. This is not a remote hypothetical; "going postal" entered the vocabulary because mentally ill persons with guns shot people at postal facilities. The point is not to arm each safety manager with a machine gun, but to make the preparation for disasters as robust as possible, whether a facility produces doughnuts or drone aircraft.

The safety manager must anticipate the needs for evacuation, which will occur if a bombing or shooting incident takes place. The evacuation of buildings is routinely practiced during fire drills; in the context of terrorism, the evacuation drill might need to be expanded so that the persons assigned to conduct the drill learn to direct the exiting crowd away from a certain area of the facility and toward an evacuation site. If threats have been received by site managers, the company's management—especially the safety manager—should begin to work with the police and the state homeland security coordinator to evaluate how best to protect the facility.

One of the important uses of technology for emergencies will be to communicate readily between the entry guard station, if any, and the first responders, such as fire, ambulance, and police. Backup radios to augment the facility phone system will be worth the investment. Testing that emergency communication and alert system is an important aspect of the safety manager's checklist of preparation steps for the facility.

Routes for evacuation must be planned. There must be a safe route identified to evacuate the site, preferably through more than one door or exit route. If there were an attack on a facility conducted by a shooter or an external terrorist group, the terrorists might come looking for the most likely means of entrance and exit, and this would suggest that the well-informed safety manager should prepare a site map and perform drills for safety-related team members and employees at least once a year.

Within the evacuation plan, in addition to having alternative routes of exit, there should be provisions for the step of counting which of the facility employees and other persons have escaped and which persons are still inside. To do this, there should be an area designated for those who have left the building to safely assemble in order to be accounted for by the members of the facility safety team.

Terrorism, Worksite Attacks, Preparation, and Response

Several practical events have caused more attention to be paid to the need for practice by safety professionals in preparations against violent acts. In September 2013, a highly secure Navy building was the scene of brutal carnage by a rogue contract employee who fired down on workers from an internal atrium balcony and killed 12 employees.

To reduce the risks of terrorist events, the company and its contractors should screen applicants for prior violent crime arrests, and effective background checking is an important protection at the very outset of the process. Longer term, applicants should be screened, informed, and required to sign a notice that the facility will not permit guns to be brought onto the property. Any applicants who refuse to accept this condition probably should not be considered.

As part of the workplace violence prevention effort, the safety manager should advocate for human resources officials to publicize a company's zero-tolerance policy for weapons and for physical violence on site. The individual who carries a weapon into the building is not intending to help the company, but might be intent upon killing or injuring co-workers, ex-spouses, or other persons, whose only crime was to show up for work on time.

The facility managers should readily accept a well-stated and clear company policy that guns and violence are not permitted or tolerated on the grounds of their facility. Showing the workers that management is behind this effort will increase the level of voluntary compliance among those workers who might otherwise have assumed that this was only one more boring paper policy. Gun violence and the threat of worker revenge against co-workers are risks that explicit policies should directly oppose.

Dealing with Federal Risk Planners

The principal challenge for facility management, once the federal Department of Homeland Security rules are applied, is the task of tying the rules in with accident prevention systems, such as the federal EPA risk management plans, that have been operational and are well understood. Those systems have been tested (hopefully) and work to reduce the risks of an accident.

Certain high-risk facilities must negotiate site security plans with DHS, and the level of detail needed exceeds what this text can cover. Expert consultant advice is essential. Safety managers should study the DHS final regulations, available at fdsys.gov under the Code of Federal Regulations (6 CFR 27). Then consulting firms should be carefully screened to determine the experience base from which they might draw in presenting and advocating for the private entity's desired terms in the final site security plan.

REPORTING OBLIGATIONS

Once the firm (or DHS local representatives) determines that the facility is high risk, detailed emergency plans must be developed and submitted to DHS. Inspections by DHS are likely to follow. The best advice of legal counsel and of skilled consultants should be obtained immediately because of the narrow window of time allowed for compliance.

LOCAL AND STATE INVOLVEMENT

State inspections and follow-up to meet the new federal requirements for chemical facilities will be important to facility managers. There will likely be a state planning organization that considers how environmental responders will mitigate the damage if a terrorist incident occurs. The state planners will probably integrate these defensive plans with the existing worst-case scenario plans for environmental spills that were adopted under section 112(r) of the Clean Air Act.

INFORMATION SHARING

The final of the three methods of interaction addressed in this chapter is the government's sharing with industry of incoming and specific threat data—for example, a telephone intercept of a terrorist cell planning to explode a truck bomb on the outside wall of the Jones Refinery in Jones, Texas. Federal officials will participate in a series of special organizations that allow their member entities to learn rapidly about targeted threats and the federal agencies' responses to those threats.

In this hypothetical case, the Jones Company would be a member of the Chemical Sector Information Sharing and Analysis Center (Chemical Sector ISAC). Its designated manager would receive word of the threat from DHS through the Protected Critical Infrastructure Information (PCII) reporting system. As an ISAC member, Jones would be expected to have contact persons in its management who will be ready to interact with the DHS specialists as they plan a defensive measure to thwart the possible attack. To the extent that the possible risk could be avoided (e.g., by closing a nearby street to trucks that could contain remotely triggered explosives), the facility and DHS would cooperate, and the threat would be removed or reduced.

CONCLUSIONS

"As if my plate weren't already full...." Some safety managers dislike having to deal with the paperwork of DHS requirements, and others lack the staffing to hold the meetings and training sessions that may be required. Of course, if terrorism ended tomorrow, there would still be plenty of work for site managers to perform on accident prevention. But prudent companies and their managers take steps to comply with federal "high-hazard" notices and keep their safety teams aware of the potential that DHS may someday alert the facility to a potential vulnerability or risk situation.

SUMMARY

- This chapter discusses the potential for attacks by terrorists on facilities that use chemicals and the government's plans to reduce such attacks.
- The Department of Homeland Security is responsible for negotiating plans, identifying high-risk facilities, and developing protective measures where threats might occur.
- Factory and refinery operators, in particular, must be ready for the critical threat of an intentional attack.

REFERENCES

6 CFR 27, Chemical Facility Anti-Terrorism Standards.
Federal Register, Vol. 72, pp. 17687 (April 9, 2007) and 65395 (November 20, 2007).
Public Law 109-295, section 550 (2006).

REVIEW QUESTIONS

1. What are the categories of "chemical releases" covered by federal rules?
2. Why are some sites considered "high risk"?
3. Why are some sites excluded from coverage under the chemical facility protection rule?
4. What plans will sites negotiate with the government?
5. What role does the ISAC play?

Motivation

John Montgomery, PhD, CSP, CHMM
J. B. Gregory, MEd

Achievement of Commitment and Involvement

Psychological Factors in Safety
Individual Differences ▸ Motivation ▸ Emotions ▸ Stress ▸ Attitudes and Behaviors ▸ Learning Processes

Individual Differences
Average-Person Fallacy ▸ Fitting the Person to the Job ▸ Methods of Measuring Characteristics

Motivation
Internal Motivation ▸ External Motivation ▸ Workplace Motivation Models

Selected Motivational Considerations
Directing Motivation ▸ Communication ▸ Awards, Incentives, and Recognition ▸ Employee Surveys ▸ Specifying Safety Objectives ▸ Behavior Reinforcement ▸ Measuring Job Satisfaction

Attitude, Behavior, and Change
Attitude ▸ Determination of Attitudes ▸ Attitude Change ▸ Behaviors versus Attitudes ▸ Behavioral Management ▸ Structural Change

Learning Processes
Motivational Requirements ▸ Principles of Learning ▸ Short-Term and Long-Term Memory

Summary

References

Suggested Readings

Review Questions

The industrial safety professional helps line management achieve maximum production by preventing or mitigating work-related accidents involving injury or death. The occupational environment is composed of interacting components, primarily the worker, materials, and equipment. A comprehensive safety program addresses all aspects of the work environment and recognizes that safe-behavior management stands at the center of the program. Consequently, a major responsibility of line management, with the safety professional's assistance, is to educate and motivate workers to follow safe practices and procedures. This chapter is designed to promote understanding of human behavior in the work environment and covers the following topics:
- motivation and learning
- the importance of gaining employee commitment and involvement in safety programs
- worker attitudes and behavior
- motivational techniques that work
- key psychological factors in safety issues
- methods used to motivate workers.

Motivating workers is an important facet of an effective safety and health program. Motivation involves moving people to action that supports or achieves desired goals. In occupational safety and health, motivation increases the awareness, interest, and willingness of employees to act in ways that increase safety and health for themselves and their co-workers and support the organization's stated goals and objectives. The ultimate success of a motivational model in changing employee attitudes and behavior depends on visible management leadership. In addition, the motivational techniques used should support the mainline safety and health management system, not take its place. Similarly, these techniques should be evaluated in terms of how well they achieve their support roles, such as maintaining employees' interest in their own safety, rather than by their effect on injury rates.

Many companies are moving away from traditional approaches to managing employee safety and health. Traditional approaches exhibited such characteristics as top-down communication, minimal employee participation, and dependence on discipline to influence safety behavior. The challenges of motivating employees, changing attitudes, and modifying behavior continue to resist uniform, simple solutions.

ACHIEVEMENT OF COMMITMENT AND INVOLVEMENT

Achievement of permanent performance improvement is a product of employee and management commitment and involvement. Employee acceptance must occur, and management leadership must be visible. When behaviors are repeated and reinforced, the foundation has been laid for personal acceptance. In other words, employees are willing to take responsibility for paying attention to the behavior or process changes necessary to achieve specified safety and health objectives until they become part of their habitual action patterns. A change in work habits is not attained overnight. Rather, it takes time to accomplish and must have continuous management support. Employees must perceive that management at all levels is behind the objectives of the safety and health program and supports the methods chosen to achieve them. If at all possible, the techniques used to motivate employees should blend with an organization's culture. They should not be viewed as special or outside the management mainstream, but rather as part and parcel of the motivational methods that are used to achieve high levels of production and quality. When this mesh is present, the likelihood that safety and health needs will be met within the normal business function is high, and therefore, the potential for permanent improvement is increased. When motivation and behavior change techniques do not fit in with a company's management style or are seen as nonmainstream activities, even though they receive considerable initial attention, they are unlikely to become permanent.

PSYCHOLOGICAL FACTORS IN SAFETY

The full role and responsibilities of the industrial psychologist are too broad to be covered in this chapter. However, the psychological factors that most directly influence safety program success—individual differences, motivation, emotions, stress, attitudes and behaviors, and learning processes—will be described.

Individual Differences
Individual differences among employees are an ongoing challenge to both managers and safety professionals. Although individual differences are not always obvious and constant, some factors are common to groups of people, a fact that managers and safety personnel need to understand when dealing with work groups.

Motivation
When someone has an internal drive to acquire something, it can be said that he or she is *motivated* to acquire it. A person can be equally motivated *not* to want something. For instance, using a guard to protect one's fingers from a saw may indicate motivation for following safe practices. However, the desire to ignore a safety device because it

may decrease production is also a motivator. Safety personnel and others need to consider the concept of conflicting motivators in any attempt to understand human actions.

Emotions
Although emotions can be constructive at times, they also can be counterproductive—working to the detriment of both the individual and the safety program. Emotions such as anger, fear, and excitement can interfere with the thought processes, resulting in behavior that conflicts with a rational approach to work tasks.

Stress
Stress—physical, chemical, or emotional—is always present within an individual and can be a factor in producing either positive or negative results. How employees choose to react to the stress on or off the job is very important, affecting decisions to use safe or unsafe behaviors.

Attitudes and Behaviors
Industry has recognized the effects that employee attitudes can have on production, workplace morale, turnover, absenteeism, workplace safety, and other work-related issues. As a result, management has spent considerable time and money trying to determine workers' attitudes. Although workers' attitudes toward safety and health procedures and policies can make a difference between an effective and ineffective safety program, accurately determining an individual's attitude can be difficult. Attitudes are internal and difficult to measure. Accordingly, it is necessary to observe and measure employee behaviors. Changing behaviors can help modify attitudes.

Learning Processes
Finally, management should be concerned with how people learn on the job. Motivation, attitudes, emotions, or even individual differences cannot be understood without considering the learning process involved in establishing each worker's perspective and personality.

Much of the success of a safety program depends on its acceptance by those to whom it is directed. The safety professional and line management must find factors common to the group that can be used to promote safe conduct by all employees. The basic question is, what factors associated with human behavior can be used to increase the effectiveness of safety programs?

INDIVIDUAL DIFFERENCES

When a chemist analyzes a chemical compound, its exact nature and composition can be accurately specified. Therefore, when this compound is used, its behavior can be predicted and duplicated; the action and reaction of any one sample are the same for all other samples of the compound.

When encountering human behavior, however, the safety professional is not dealing with the same degree of certainty that is afforded to a chemist. Due to the individuality and complexity of the human psyche, the safety professional is faced with the challenge of not understanding why a person is acting in a safe or unsafe manner. Therefore, the safety professional relies on what is being communicated by the individual. Furthermore, the individuality of the human psyche creates a state in which psychological equivalency cannot be relied upon for prediction purposes. In other words, what motivates person A to act in a safe or unsafe manner might or might not be the same for person B, C, or D. The variances that occur among persons are referred to as "personal equation" or "individual differences."

However, within the framework of individual differences, there are general psyche patterns that are common to all persons. For example, human behavior is typically motivated by some belief, need, or drive. Regardless of what the individual does, some purpose is driving the behavior. For example, the purpose may be to resolve or reduce some basic tension. The degree and nature of this tension, or dissonance, will depend in part on the individual's attitude, values, or perception of situational variables (e.g., job expectations, work conditions, physical ability). Whether an employee works safely depends on the following:

1. the present situation—is the employee rushed, stressed, fatigued, or in poor health?
2. past experiences—were accidents avoided in the past? What amount of training does the employee have?
3. workplace and methods design—were the job procedures and work setting designed to promote safe and healthy behaviors?

Average-Person Fallacy
Because the individual psyche differs from person to person, the safety professional oftentimes turns to making decisions based on generalizations. When these choices are made, it leads to what is known as the "average-person fallacy."

Figure 30–1 presents the distribution of scores in a normal curve. Many human characteristics (e.g., anthropometric characteristics or IQ) are assumed to be distributed according to the *normal distribution* (often referred to as the *normal curve*). Note that in such a distribution, half the population is below the average line (called the *mean*) and half the population is above it. Unfortunately, inappropriate use of the "average" as the main characteristic of a population has distorted statistical data and produced personal misconceptions. People, in reality, are highly unlikely to be average. For example, in body dimensions, which can be measured more accurately than emotional, behavioral, or intellectual characteristics, less than 4% of a test group had three common average dimensions. Less than 1% were average in five or more dimensions at the same time. Therefore, when an appeal is aimed at the average, it misses much of the population.

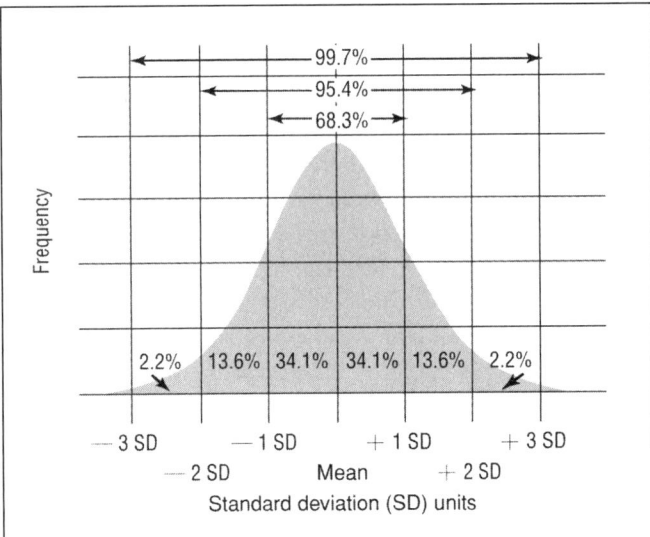

Figure 30–1. Many human characteristics are assumed to be distributed according to the normal curve shown here. Because the normal curve has a known shape, it is possible to state the percentage that lies between +1 and −1 standard deviation (SD) or between any other two points expressed in SD units. Thus, it is easy to describe a normal distribution of data using simple statistics—the mean (average) and the standard deviation (SD). For example, 68% of all the measured values lie within ±1 standard deviation about the mean; 95% are within ±2 SD, and so on. Given the mean and the standard deviation, therefore, the complete distribution in a normal distribution can be determined.

When designing the system, a better approach is to use percentiles. Instead of setting the standards to fit the average, design them to fit all but the upper 5% and the lower 5%. The system will then fit 90% of the population. The use of percentiles (e.g., between the 5th and 95th percentiles), rather than averages, has long been the technique for relating individual test scores to a group as a whole. This system can also serve as a guiding rule in the design of human–machine relations.

Fitting the Person to the Job

When replacing or relocating personnel, managers oftentimes makes decisions based on the ideal candidate possessing the most desirable characteristics deemed necessary, including technical knowledge, company loyalty, and process productivity. Unfortunately, this mind-set leads management to hire the theoretically "perfect person for the job" rather than seeking out the "right person for the job." Psychological factors are seldom expressed within a minimum and maximum (see Chapter 16, Ergonomics Yesterday, Today, and Tomorrow).

For example, a deaf person properly trained for a task in which appropriate noise levels cannot be controlled would experience fewer problems than a hearing person properly trained for the same task because of the issue of having to wear protective hearing devices.

Another example is the person who comes to work under any condition, no matter what physical or mental impairments are present. This person may be as unsuitable as someone who uses every excuse to take time off from work. A worker with a high fever, cold, flu, or sore back is a potential risk in the workplace, especially if he or she is taking prescription medication.

Physical Characteristics of the Individual

Reaction time, psychomotor skills (e.g., manual dexterity), and visual abilities have some bearing on safe performance of a job. Whether they are directly responsible for accidents is not clear, but a certain degree of physical competence seems to be required for successful, accident-free performance. Some types of jobs require significant physical demands; others may not.

Individual Characteristics and the Job

Human resources departments screen candidates on the basis of the specific characteristics required to perform a particular job. In many cases, both physical characteristics (such as size, visual acuity, and steadiness) and personality characteristics are important. As a result, many individuals are not considered for a job because they do not have all the necessary qualifications.

The same principle applies in other areas of matching the worker to the job. When designing equipment, for example, human engineering specialists consider physical limitations as well as other human characteristics in an effort to make human–machine interfaces as compatible as possible. When hazards cannot be eliminated, safeguarding provides protection against the potential failure of people to use equipment correctly. Both safe design and safeguarding minimize the effect of individual differences on accident frequency and severity.

Methods of Measuring Characteristics

Regardless of the technique used to screen, place, and motivate employees, management needs methods for measuring safety program effectiveness. Techniques used to obtain feedback include comparing cross-industry accident rates and the internal approach of safety sampling or critical-incident techniques (see also Chapter 11, Injury and Illness Record Keeping, Incidence Rates, and Analysis).

Measuring techniques can be evaluated by determining reliability and validity. *Reliability* refers to consistency of measurement. The reliability of a given measurement, such as a test or instrument, can be estimated in various ways. In general, however, the concept of reliability refers to how stable results remain. This is the case whether stability is assessed (1) across time or between settings, (2) using the same or a different group of individuals, or (3) for internal consistency or consistency between alternate forms of the same test or instrument.

Underlying the concept of reliability is measurement error. All gauges used to assess performance have, in differing degrees, some inherent measurement error. *Measurement error* refers to the estimated fluctuations likely to occur in individual or group responses because of personal factors. *Validity* refers to how well the test or instrument measures what it is supposed to measure. The validity of a test or instrument can be assessed in various ways: (1) the relevance or plausibility of items (or of the overall instrument) with regard to the given behavior (*face validity*); (2) whether all aspects within a given concept are adequately covered (*content validity*); (3) how well the test or instrument measures a specific construct or trait (*construct validity*); and (4) a demonstrable relationship between performance, as measured by the given test or instrument, and some other related behavior (such as IQ and school aptitude) (Anastasi 1988). A measurement can be reliable without being valid, but a valid measurement also must be reliable. To illustrate that a reliable measure is sometimes not valid, consider a yardstick. A yardstick is reliable—it gives a consistent measurement every time it is used. It is also valid for measuring the length of a table. But if a yardstick is used for weighing the table, it is no longer a valid measure.

In addition to reliability and validity, a measuring technique must be *practical*. A technique can possess high validity but be so cumbersome and intricate that it can be used only in special situations by highly skilled technicians. In spite of its statistical value, such a technique is virtually worthless in most workplaces. Two sampling techniques used for evaluating potential accident-producing behavior are the critical-incident technique (described in Chapter 6, Loss Control Programs) and behavior sampling.

- The critical-incident technique involves the following process. A random sample of employees is interviewed to collect accident information concerning near misses; difficulties in operations; and conditions that could have resulted in death, injury, or property loss. Those participating are asked to describe any incidents that have come to their attention. This technique can be useful in investigating worker–equipment relationships in past or existing systems, evaluating modifications to existing systems, or developing new systems.
- The behavior-sampling or activity-sampling technique involves observing worker behaviors at random intervals and classifying these behaviors according to whether they are safe or unsafe (place the worker at risk). Managers then calculate either the percentage of time the workers are involved in at-risk practices, the percentage of workers involved in at-risk practices during the observation period, or the percentage of unsafe versus safe behavior observed. Using this technique, management can apply various components of a safety program (such as safety lectures, posters, brief safety talks, safety inspections, training films, and supervisory training) and immediately note their influence on workers' unsafe behavior.

Studies have shown that feedback can be introduced into this process with positive effects on worker behavior and accident rates. Feedback charts are prepared that show the percentage of safe behavior observed during each sample. These charts are posted in the workplace and serve as positive reinforcement for safe behavior. Human resources personnel must be careful when screening employees or potential employees in regard to their accident potential. No systematic screening procedures have been developed that meet both reliability and validity criteria *and* are adequate for use in all industries or even in a specific industry. Although, in theory, such a screening procedure is possible, the state of scientific knowledge in occupational safety research is too limited to support the development of such screening measures. Therefore, the reduction in human-error potential by means of a good ergonomics program (see Chapter 16, Ergonomics Yesterday, Today, and Tomorrow) is still the best line of defense.

MOTIVATION

When a person has the internal drive to act, it can be said that he or she is motivated. The athlete is inspired to improve his or her performance, or a worker may be driven to increase productivity. However, even though these two individuals are motivated to act, the inner force driving them to achieve their goal and how the process is externally managed may differ. This section briefly examines theories surrounding the interpersonal or endogenous factors that affect personal motivation. Furthermore, it discusses work-based management or exogenous models, which are used to manage employee motivation with regards to creating a safer work environment.

Internal Motivation

Motivation centers on the psychological forces present in an individual that create a specific action. It asks the question, within the individual, what drives the internal desire to act? Either intrinsically (internal) or extrinsically (external) based motivation can be seen as the cause agent for employees' behaviors. Because motivation is so dependent on psychological and psychosocial factors, it is one of the most complex fields of study in the area of human behavior. Furthermore, there are countless theories that attempt to explain the motivational factors that account for human behavior. This section does not attempt to identify and cover all the psychological models for motivation found within the literature. Instead, it focuses on specific models that attempt to explain the role that fulfilling personal needs has on influencing personal perceptions.

A person's needs are dependent on the individual's psyche. There are multiple theories within this area of study, but the following are two well-known endogenous models that attempt to identify the underlying psychological need that drives an individual toward a particular action: Maslow's *hierarchy of needs* and Reiss's *16 basic desires theory*.

Maslow's hierarchy of needs was developed by Abraham Maslow in 1943. This theory proposes that, in a person's life, there are five specific stages, or needs, that serve to motivate or provoke an individual to act (Figure 30–2). Furthermore, even though the stages are hierarchical, they are not static, which means that once an individual has satisfied his or her most basic needs of sustaining food, clothing, and shelter (i.e., physiological stage), the need for personal safety becomes the dominant force in the person's decision making. Once safety is established, then love/belonging is sought out, followed by the desire for esteem, and then finally self-actualization. Here is an example of Maslow's hierarchy of needs: a line manager may feel the major need for acceptance and recognition by his line workers. Therefore, he may care less about encouraging or enforcing an unpopular safety rule in order to gain or maintain his popularity with the workers. Remember, the strength of certain needs will change depending on the individual's level of satisfaction. As an individual satisfies a need, it is no longer as strong of a motivator as an unsatisfied need.

Unlike Maslow's five needs, the second model that proposes to identify the internal forces that shape a person's personality and motivates him or her to action is Reiss's

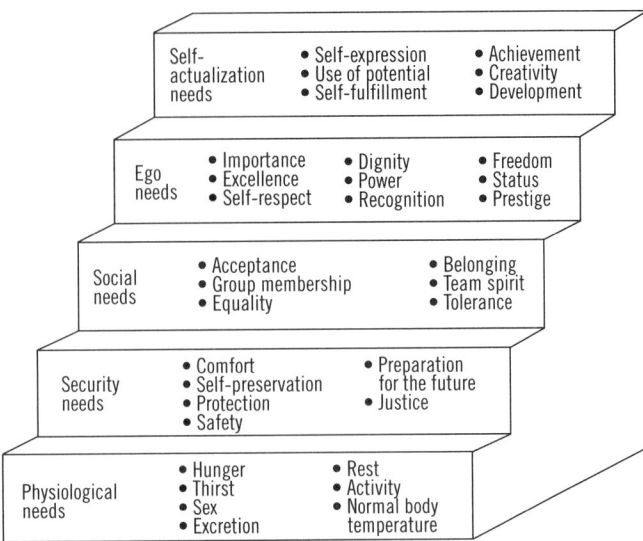

Figure 30–2. Abraham Maslow investigated people's various needs and arranged them into a hierarchy. Note that safety needs are near the base and more intellectually satisfying needs are near the top.

16 basic desires theory. Following are the 16 items that Reiss proposes:

1. acceptance
2. curiosity
3. eating
4. family
5. honor
6. idealism
7. independence
8. order
9. physical activity
10. power
11. romance
12. saving
13. social contact
14. social status
15. tranquillity
16. vengeance.

Furthermore, according to Reiss, there are three points that must be kept in mind when applying this theory to motivation: (1) each desire can be a motivator; (2) depending on the individual, desires vary in strengths and weaknesses; and (3) desires exist in combination with each other, never in isolation.

Safety professionals will find it easier to work effectively with employees when they realize that individuals seek to satisfy their own needs. Therefore, the safety professional must develop the ability to identify workers' needs in order to not only effectively communicate with workers, but also to help create a culture of safety. This awareness is based on cues in the individuals' behavior that the safety professional and manager can learn to recognize, provided he or she has made the effort to get to know workers.

Maslow's hierarchy of needs and Reiss's 16 basic desires theory are just two of the many models that can be used to help identify the needs of workers. However, a word of caution is in order. It should never be assumed that a worker's motivational needs remain static. For example, a situational change at work or home can create changes within the needs of the individual. Therefore, when dealing with people, the safety professional must be cautious not to assume that what served as a motivator today may serve as such next week.

External Motivation

Many theories in the psychological literature address human motivation. However, there are also specific theories that reference the role of management perception of employee motivation within organizational settings. Although only two such theories are presented here, other equally cogent points of view, such as Blake and Mouton's attitude-based Managerial Grid III or Hersey and Blanchard's behavior-based Situational Leadership, could be discussed. (For additional information, see the Suggested Readings at the end of this chapter.)

Because theories of human motivation lack sufficient data to support all their tenets, they may best be viewed as philosophies of management. They are important to the safety professional because, if accepted, they can influence the direction in which management seeks to develop and implement a safety program.

Theory X and Theory Y

In an attempt to analyze how management regards human motivation, McGregor (1985) discovered two basic ways in which managers view workers. Theory X, according to McGregor, assumes that workers are essentially uninterested and unmotivated to work. In order to change this attitude, management must motivate workers through various external motivators such as rewards (e.g., bonuses, prizes, etc.) or punishments (e.g., docked pay, fewer privileges, etc.). In effect, the worker is motivated to work by virtue of the external rewards and punishments offered. Thus, under Theory X, management uses both positive and negative outcomes as the means to motivate the worker.

Unlike Theory X, however, McGregor's Theory Y assumes that workers have the potential to be interested and motivated to work. In fact, work is assumed to be as natural and desirable as other forms of human activity, such as sleep and recreation. Under such circumstances, management must organize work so the worker's job coincides with the goals and objectives of the organization. Thus, a Theory Y manager views the task as constructive, using the worker's self-control and self-direction as the instruments through which work is accomplished. By emphasizing responsibility and goal orientation, management capitalizes on the motivation already present within the worker. If conflicts occur between the worker's goals and management's goals, they are resolved through mutual exploration and discussion. It is always assumed under Theory Y that the worker's inherent motivation is essential to accomplish the organization's goals.

Both Theory X and Theory Y proponents exist, and successful management systems operating under each of these theories can be found throughout American industry. The important point is for safety professionals to remember that safety programs can work under whichever system is operating within an organization. Although the technique of implementation may differ, Theory X and Theory Y approaches to human motivation can both be used to motivate workers to adopt safe behavior.

Job Enrichment Theory

Another analysis of human motivation used in occupational environments was developed by Fredrick Herzberg (1966). Although quite comparable to Theory X and Theory Y, Herzberg's two-factor theory is explicit in both detail and philosophy. His concept of job enrichment, in many ways an extension of Theory Y, has been a major force in the development of management and leadership strategies.

The classic approach to motivation concerns itself with changing the environment in which a person works. This includes the circumstances surrounding the individual on the job (good or poor lighting, an agreeable or offensive manager, etc.) and the incentives given in exchange for work (money, a pat on the back, inclusion in the company bulletin, etc.).

Herzberg's theory proposes that worker concerns about the work environment are important but are not sufficient alone for effective motivation. People are motivated optimally by using the work itself to satisfy their desires and needs.

Herzberg holds that there is no conflict between the classic (environmental) approach to motivation and his approach to motivation through work itself. He considers both important. The classic approach is called *hygiene*, whereas Herzberg's approach is called *job enrichment*. Figure 30–3 presents a contrast of the classic, hygiene approach and the job enrichment approach to motivation.

The hygiene approach may be understood using the following analogy: a person is provided with pure drinking water

Hygiene Approach (Classic)	Job Enrichment Approach
Company policies and administration	Achievement
Supervision	Recognition
Working conditions	Work itself
Interpersonal relations	Responsibility
Money, status, security	Professional growth

Figure 30–3. Contrast between the hygiene approach to motivation and the job enrichment approach.

and waste disposal; both are necessary to keep this person healthy, but neither makes him or her any healthier. By extension, good rapport with others may enhance an individual's job satisfaction, but job satisfaction alone will not necessarily motivate the individual to develop safe work habits.

Further, treating a person better does not enrich the job, although the individual can become unhappy if he or she is not treated well. Again, a salary increase can keep an employee from becoming dissatisfied for a period of time, but sooner or later another increase will be required to boost motivation again. Nor does worker protection enrich a job. Even in hazardous industries such as coal mining and bridge building, the worker may regard protection such as gas testing and lifelines as part of the job. In such cases, the protection is not a motivating force. On the other hand, the worker will be very unhappy if no protective effort is made.

Herzberg's idea that work itself can be a motivator represents an important behavioral science breakthrough. Traditionally, work has been regarded as an unpleasant necessity, but it has not been thought of as a potential motivator. Although automation is helping to phase out the nonstimulating aspect of many jobs, a job should provide an opportunity for personal satisfaction or growth. When it does, it becomes a powerful motivating force.

People, Herzberg further theorized, must be given the opportunity to do work they think is meaningful. Merely complimenting an individual who is doing a routine job or saying that the worker is doing something meaningful accomplishes little. The worker often does not feel that this is true. Job rotation is not the answer, either; it does not enrich a job, but only enlarges a worker's responsibilities and tasks.

Herzberg also observed that new technology can make workers feel obsolete. As Herzberg noted, "Resurrection is more difficult than giving birth." Obsolescence must therefore be eliminated by continued retraining—not just an occasional session. Jobs should be updated, and people doing the jobs must keep abreast of the latest equipment and techniques.

Even though a company provides the hygiene factors, it must also provide a task that has challenge, meaning, and significance. An unchallenged individual who does not quit but stays on is usually resigned to the work and has poor morale. That, says Herzberg, is the price a company pays for not motivating people.

In summary, the following are seven principles of job enrichment:
- Organize the job to give each worker a complete and natural unit of work.
- Provide new and more difficult tasks to each worker.
- Allow the worker to perform specialized tasks in order to provide a unique contribution.
- Increase the authority of the worker.
- Eliminate unneeded controls on the worker while maintaining accountability.
- Require increased accountability of the worker.
- Provide direct feedback through periodic reports to the worker.

Workplace Motivation Models

Among the many models that can be used to motivate employees to improve safety and health performance, two have had considerable acceptance. One, the organization behavior management (OBM) model, is tied directly to the use of reinforcement and feedback to modify behavior. The other, the total quality management (TQM) model, is based on attitude adjustment methods used to achieve quality improvement goals in industry.

As shown in Table 30–A, both management approaches respond to the motivational variables and supporting actions that have been described. The models vary, however, on the safety emphasis used to motivate employees. As a result, they differ in their efficiency in accomplishing the requirements of Kanfer's three motivational variables, which are discussed on the next page.

The theoretical orientation of the models differs with regard to the specification of safety and health objectives. As Table 30–A shows, OBM emphasizes external behavior change, while TQM emphasizes internal attitudinal changes as a prerequisite to behavior change. Accordingly, OBM objectives can probably be communicated in a more direct and simple fashion to employees, while TQM principles may be more complex.

Training provided to employees through OBM and TQM differs as well. OBM training is specific in its coverage of critical behavior to be changed, while TQM education includes behavioral as well as attitudinal skill development, such as team building and problem solving.

Reinforcement and feedback are provided in both OBM and TQM models, but the observational techniques used in

TABLE 30–A. Comparison of OBM vs. TQM Approaches to Employee Motivation

Motivational Variable	Supporting Action	Safety Emphasis OBM Model	Safety Emphasis TQM Model
Direction of behavior	Specify objectives	Behavior	Attitude/Behavior
	Provide training	Behavior training	Process education
Intensity of action	Give reinforcement	Behavior occurrence	Process improvement
	Maintain feedback	Behavior data	Operating indicators
Persistence of effort	Commit employee	Behavior change	Continuous improvement
	Commit management	Style change	Cultural change

OBM are more formal and specific to the objectives (i.e., specific behavior changes) than are those used in TQM. Also, the reinforcement/feedback schedules used in OBM are more structured than those found in TQM, thus facilitating behavior changes more rapidly.

Employee commitment to permanently improved safety performance as reflected in both attitude and behavior changes appears more directly attained by TQM than OBM because of TQM's emphasis on process or root cause improvements. System and process change are produced by TQM and result from employee and team involvement and empowerment. Employees are much more likely to support change if the objectives and the methods used to achieve them are based on their own recommendations rather than imposed by management.

OBM's emphasis on critical behaviors or special cause improvements includes some degree of employee participation in the selection of behaviors to be changed and, possibly, the incentives to be used. Control is maintained through a relatively rigid observation and reinforcement system. As a result, the permanence of the behavior change remains questionable unless the system is continued.

With regard to management commitment, TQM requires a definite change from the philosophy that employees should be "managed" to conform to existing systems to a viewpoint that processes and systems can be improved or even completely revamped and employees are in the best position to know how those processes work and should work. It also calls for the creation of a corporate culture that emphasizes flexibility and responsiveness to employee needs and trust between labor and management.

The OBM approach does not demand a shift from the traditional management philosophy that seeks to manage employee behavior. It does, however, call for acceptance of a new management style that focuses on a different and relatively exacting observation system that is not widely used and that requires a relatively substantial amount of time and resources to maintain.

SELECTED MOTIVATIONAL CONSIDERATIONS

Directing Motivation

According to Kanfer (1990), motivation includes three variables:

1. direction of actions that will accomplish defined objectives
2. intensity in the amount of personal thought given to the performance of goal-oriented actions
3. persistence of effort in which desired performance lasts over time.

Direction of behavior requires four supporting actions: (1) the goal to be achieved must be specified; (2) employees must clearly understand the steps required to achieve such goal; (3) the employee must accept, acknowledge, and desire to meet the new goal; and (4) the employee must persist to achieve such goal.

Safety and health objectives can be defined narrowly in terms of specific behaviors or broadly in terms of process improvements. Once objectives have been set, employees must attain the knowledge necessary and learn the skills to achieve them. Unless these two elements are present, motivational efforts are futile.

Intensity of action means that employees integrate safety and health objectives into their job assignments with the same degree of mental and emotional effort that they expend for other work objectives. It also implies that employees may have to be willing to spend extra time to incorporate safety and health practices into their routine work patterns and, if necessary, accept that additional work may be required as some new behaviors are acquired.

To ensure that employees are involved, the reinforcement of appropriate behaviors and/or performance levels should be as strong as possible. Also, communicating performance results to employees is critical to the change process in order to demonstrate achievement, emphasize its importance, and identify areas still needing improvement.

Persistence of effort relates directly to the nurturing and maturing of the attitudes or action tendencies that support improved safety and health performance throughout the organization. For this continuity of purpose to occur, both employees and management must be committed. Employees must be willing to modify personal behavior in accordance with company safety goals and objectives. Management must be visibly committed and active in its support for employee safety and health.

No matter how intrinsically effective a motivational model is in changing employee attitudes and behavior, its ultimate success depends on visible management leadership. This is a prerequisite for any program, whether it focuses on production, product or service quality, or employee safety and health. This is consistently true for all successful safety programs, despite their differences. As Cohen and Cleveland (1983), in their study of outstanding National Safety Council member companies, conclude, "No safety program was quite like any other. However, all had one major thing in common: safety in each instance was a real priority in corporate policy and action."

Besides this essential ingredient, many other productive motivational techniques can be used to promote employee safety and health or otherwise influence employees to protect themselves against workplace hazards. These have been reviewed by a number of authors, including Cohen, Smith, and Anger (1979); Cohen (1987); and Peters (1991).

The role of motivational efforts is to support the safety and health management system, not to take its place. Safety and health programs are more likely to succeed when they are energized through slogans, performance recognition, and discipline in combination with safe and healthful workplace conditions; well-designed tools, equipment, and workplace layout; effective maintenance; and appropriate training and supervision.

Similarly, evaluation of the worth of motivation techniques should not be measured by reduction in injuries, property damage, or workers' compensation costs. Rather, their effectiveness should be judged in terms of how well they achieve their support roles, such as maintaining employees' interest in their own safety, communicating management's interest in employee safety and health, generating employee involvement in safety activities, increasing morale, and reminding employees to take special precautions.

Cohen and colleagues (1979) point out that these approaches "seek to establish a generalized tendency to act in safe and healthful ways through appropriate attitude change, increased knowledge, or heightened awareness." The following discussion briefly highlights three of them.

Communication

Communications of various kinds are used to enhance the general effectiveness of any motivational effort. The communication process can be summed up thus: Who says what, in what way, to whom, and with what effect? Accordingly, communication programs usually involve a source, message, media, target, and objectives.

Communications vary in terms of their coverage and impact. Safety posters, banners, and other mass media are high in coverage and most effective in increasing general awareness about safety and health issues and in presenting on-the-spot directions or safety reminders. They can also be a useful vehicle for making employees aware of management's general interest in their welfare. As with mass media, their impact is lessened when used alone because they provide no opportunity for interaction, further information, or response to questions. However, the impact of their message can be increased when combined with opportunities for person-to-person or two-way communication, through either group discussions or individual contacts. Though low in coverage value, these methods can be high in impact and lead to changes in behavior.

Credibility of source is important in safety and health communications. Bauer (1965) describes two types of credibility:
- problem solving—based on competence and trust
- compliance—based on likability and prestige

In this model, the problem-solving source is more likely to prompt lasting behavior change than the compliance-based source. Cohen and colleagues (1979) note that in the workplace, supervisor competence (i.e., knowledge of the task and the hazards involved, ability to set a good example) is an important factor in enabling employees to regard supervisors as credible sources of safety and health information. With regard to communication content, the use of fear or scare tactics has been a topic of research and controversy for years. This strategy attempts to change attitudes about the risks involved in hazardous behaviors by instilling fear in a target audience and then reducing that fear by providing methods to prevent the danger or lower the risk. Workplace examples include personal protective equipment–use campaigns; non-occupational examples include antismoking and seat belt use programs. The main argument against using scare tactics is that receivers may block out or suppress the message and that they are not long-lasting in effect.

Based on his review of the research literature, Peters (1991) recommends the following when using fear messages:
- The message should attempt to evoke a high (versus low) level of fear.
- The suggested preventive actions should be relatively

detailed, specific, and presented immediately after the fear response is evoked.
- The preventive actions should be presented in a block, rather than interspersed with information designed to elicit fear.
- The suggested preventive actions must be perceived by the target audience to be effective in preventing danger.
- The source of the communication should have high credibility. Most research suggests the direct supervisor as being the best source for fear messages.
- Person-to-person, two-way communication is more effective than written, one-way communication.
- Increased effectiveness can be achieved by gaining the support of significant others, such as co-workers, supervisors, and family.
- Safety and health communications should consider the target groups at which messages are aimed. For example, research has shown that fear messages are more effective with new employees than with seasoned employees, who can use their experiences to discount the message. Additionally, fear messages have been found especially effective in influencing employees who are not under direct supervision and are expected to comply with safety regulations on their own.

As an aid to both defining targets and establishing objectives, employee surveys are recommended to assess current levels of safety and health knowledge, attitudes toward safety management programs and practices, and compliance with rules and procedures. Such measurements assist in pinpointing education and persuasion priorities and set a baseline for later evaluations of the effectiveness of communication efforts. Cohen (1987) provides these general guidelines for communications:
- Statistics or risk data have the greatest impact if they are specific to a particular workplace or location.
- Persons in the target group should be involved in planning communication content.
- Pilot testing of communication programs is advisable.

Finally, Planek and Fearn (1993) point out that because of the unknowns involved in the attitude/behavior change process, evaluations of the success of communications programs should be guided by the following principle: observed success in conveying knowledge or changing attitude is not an indicator of behavior change unless further observation of behavior is made.

Awards, Incentives, and Recognition

The use of awards, incentives, and recognition to motivate employees to perform safely is an accepted feature of both OBM and TQM models. In the OBM model, use of incentives to reinforce employee behavior is critical to program success. In TQM, rewards, promotions, and other incentives are used to recognize individuals for contributions to process improvement. Also, at the group, team, or company level, special days or other functions are used to celebrate achievement.

Broadly speaking, the use of incentives of any type—including salary increases and promotions—can be viewed as having a positive influence on employee attitudes and behavior. When safety and health are made part of the decision to reward employee performance, these factors take on added significance as important job-related requirements.

Cohen and colleagues (1979) comment that the use of incentive and award programs is controversial. Given the wide use of incentives and awards of some kind in all phases of social enterprise, some wonder why this controversy exists. The problems with such programs tend to arise when they are misused or abused. At the organization level, incentive or award programs for employees can be substituted for legitimate safety and health management programs. As has been indicated, motivational techniques support mainline programs but are never a substitute for them. At the employee level, abuse can result in failure to report an injury or incident for fear that either an individual or a workgroup will not receive an award.

Many companies use incentives and awards as a continuing part of their safety and health program. The OBM literature is filled with success stories supporting the use of incentives, and a great amount of anecdotal evidence cites the successful use of a wide variety of reward programs.

Deci (1975) also presents evidence that distinguishes between the effects of rewards that are perceived as "controlling," as can happen in the OBM approach to the reinforcement of safe behaviors, and those that are viewed as "informational," which is the emphasis in the TQM model. Studies that explore these differences (Kanfer 1990) have found that rewards that recognize personal competence, as in TQM, are stronger than those that simply provide positive performance feedback, as in OBM. One explanation is that employees perceive informational rewards, which recognize achievement and personal competence, to be under their own control rather than in the hands of another person who gives or withholds rewards based on the performance being observed. Accordingly, the focus for control of informational rewards is within the employee or intrinsic, as opposed to being outside the employee or extrinsic, as is the case for controlling rewards.

Awards that recognize achievement at the company or corporate level should also be evaluated. In the arena of quality, there are two national awards of note: the President's Quality Award and the Malcolm Baldrige

National Quality Award, which have both attracted the attention of top U.S. corporations. Similarly, the National Safety Council provides awards for outstanding records in occupational safety and health.

The purpose of these awards is also to recognize achievement and reinforce the actions that lead to successful corporate performance at all organizational levels. When these awards are applauded throughout recipient companies, they can be a powerful stimulus for continuous improvement.

Here again, however, problems of misuse can occur insidiously in organizations that receive national awards for excellence in safety and health. When management uses attainment of an annual award as the sole criterion of program achievement, two forms of misuse are possible. Management may have a self-satisfied attitude that causes it to ignore or be otherwise unresponsive to the safety and health system and performance inadequacies that have not caused lost workdays, injuries, or deaths. Or it may adopt a "win at all costs" posture that generates counterproductive stress and can lead to abuses in reporting or in the proper handling of employee injuries and illnesses. In either case, a national organizational award loses its effectiveness and becomes a hindrance to continuous safety and health improvement.

In summary, the appropriate use of awards, incentives, and recognition can play an important role for organizations that use them wisely. They can, as Cohen and colleagues (1979) conclude, "add interest to an established hazard control program, which could enhance self-protective actions on the part of the work force."

Employee Surveys

Employee surveys in the form of questionnaires and/or interviews are a widely used means of uncovering strengths and weaknesses of safety and health management systems. They play a critical role in TQM approaches to employee motivation as a source of information on safety and health needs as well as a measure of performance improvement.

Surveys also stimulate employee participation in safety and health. As Planek and Fearn (1993) point out, "It makes sense to include employees when designing program plans intended to increase their safety and health. When gathered in an objective and representative fashion, candid employee input can provide unique insights about the current status of safety and health and assist the selection of improvement priorities." DeJoy (1986) points out other positive effects of employee participation through surveys, such as heightened employee awareness of and interest in safety and the perception that management not only is interested in employee safety and health but also is open to employee input on the subject.

At a minimum, a basic employee survey should consider the following five factors:
1. *Management leadership.* Clearly communicates a safety and health vision supported in words and action.
2. *Supervision involvement.* Reinforces and communicates management's vision through open employee interactions, example, training, control, and recognition.
3. *Employee responsibility.* Personally supports company safety/health objectives and adopts the attitudes and actions necessary to achieve them.
4. *Safety support activities.* Generates employee cooperation and action to achieve safety and health objectives, such as a management/nonmanagement safety committee.
5. *Safety support climate.* Sustains employees' perception that management is completely committed to employee safety and health as a top priority.

As with the other techniques that are discussed in this chapter, the benefits that can be derived from employee surveys depend on how they are used. For example, if managers act upon survey results, they can greatly enhance employee morale and provide a strong indication that management is serious about employee safety and health. This image can be further enhanced if results are communicated to employees, at least in summary form. If, in contrast, management fails to follow up on findings, the survey activity can be detrimental to morale and damage management's image with regard to employee safety and health.

This situation is particularly critical in organizations whose management style has been traditionally nonparticipatory. An employee survey can be an effective way to initiate increased employee participation and to communicate this change in management style to employees. However, if there is no intention to change the prevailing management style, the use of a survey will be seen simply as "window dressing."

The primary reason for conducting an employee survey is to obtain as accurate a picture as possible of how the safety and health management system is operating. Whether or not an organization has open communications, it is frequently difficult to get an objective and balanced view of employee reactions to safety and health programs and activities. Communication barriers, whether personal (due to individual employees) or organizational in nature, can produce biased impressions about program strengths as well as weaknesses. For this reason, all employees must be given an equal opportunity to participate in a survey. This may be done by including all employees in the activity or by selecting a random sample of employees that truly represents the workplace population. If the former approach is taken, the activity takes on added morale-building value because all employees are given an opportunity to express their opinion. However, economic considerations may preclude this possibility, particularly in large organizations. In such cases, a random sample of respondents is appropriate if the proper selection techniques are used. It takes time, effort, and resources to conduct an employee survey. Accordingly, management must be clear about how it intends to use survey findings and resolved to do whatever it takes to obtain the most objective and, therefore, most accurate and reliable results possible.

Specifying Safety Objectives

Conard (1983) and Cohen (1987) have suggested several broad classes of behaviors that have worksite hazard control implications. Most focus on prevention, but some deal with the mitigation of unwanted effects during and after incident occurrence. In the area of prevention, critical behaviors include the following:

- proper use and operation of equipment to maximize safe performance
- adhering to work procedures that maximize safe performance
- avoiding actions that increase risk of injury or illness
- recognizing physical and process-related hazards
- observing good housekeeping, maintenance, and personal hygiene practices.

With respect to injury/illness/incident mitigation, critical behaviors include the following:

- proper and consistent use of personal protective equipment and other controls
- recognizing illness-related symptoms
- proper response to emergency situations.

In each of these cases, the specific behaviors to be identified and practiced or avoided vary from company to company. Company operating policies, rules, and procedures normally address general safety practices. OSHA and related laws specify selected areas of compliance. The company's way of doing things, bolstered by group norms and motivation, also tends to influence safety priorities.

Often, however, information from these sources is not sufficiently specific to cover the complete array of tasks that must be performed at the most basic operational levels. Furthermore, policies, rules, and regulatory standards seldom, by themselves, carry the motivational understanding of workplace safety needs and improvement opportunities that can come from injury/illness/incident investigations as well as through periodic inspections and audits. These procedures frequently pinpoint safety deficiencies in a concrete way and with an immediacy that directly influences employees to improve their performance.

To supplement investigations, inspections, and audits and add considerably to their potential for stimulating safe behavior, employee involvement in identifying safety needs is also advisable. Job safety analysis, which allows employees to participate in identifying hazards and means for their mitigation, lays an important foundation for behavior change. This type of participation should be reinforced and continued through an open communication system that welcomes safety discussions by employees and offers a means for action in response to suggestions and corrective action when problems or hazards are reported. In this regard, the use of employee safety program perception and attitude assessments can play a significant role both in specifying process safety priorities and in enhancing employee morale. The role of surveys as a motivational stimulus to improved performance is detailed later in this chapter.

Once safety improvement objectives are specified, employees must learn how to attain or achieve them. Training or retraining of employees is often necessary and lays the groundwork for immediate positive results. With the proper thrust and scope, training in hazard identification and other skills develops an employee's capacity to make knowledgeable contributions to improved job safety performance.

Behavior Reinforcement

Once safety objectives are defined and the behaviors necessary to attain them are known, they must become part of the job performance action pattern. To achieve this, employees must pay attention to job safety requirements and put what they have learned into practice. This means that undesirable, habitual, and comfortable behaviors must be eliminated or altered, while new behaviors are substituted.

If employees are involved in specifying safety and health improvement objectives and receive adequate training, the process of behavior change and reinforcement has already begun. Reinforcement in this context refers to anything that encourages the repetition of a behavior.

Two important characteristics influence the effectiveness of reinforcement. First, positive reinforcement—such as personal recognition or performance awards, which focus on increasing the occurrence of desired behaviors—is more efficient in achieving higher levels of safety performance than forms of disciplinary action focusing on eliminating unwanted behaviors. Second, the closer in time reinforcement is associated with a behavior, the stronger its effect.

Both of these principles flow from learning theories that were developed largely on the basis of animal behavior studies. Principles derived from these studies have been used to explain how learning occurs in humans. These studies have also led to the development of a variety of "need" and "drive" theories about why people behave as they do. Based on these theories, timely, positive reinforcement of specific behaviors and action patterns and general feedback about performance that has produced the desired results are considered essential features of the learning and motivation process.

Group feedback helps create a positive atmosphere for behavior change. When a work group accepts common safety objectives, its members tend to reinforce one another's behaviors. To make this happen, feedback at the group level should be developed. Whether or not positive change is achieved, feedback provides momentum for future accomplishments.

Group feedback requires objective evaluation or measurement of progress. The establishment of a measurement system begins with the definition of the specified safety and health objectives so that they can be observed. Observation can be direct when specific behaviors are involved and can take the form of counting the occurrences of unsafe behaviors. To observe progress toward achieving broadly defined process improvement goals, surveys or other performance indicators may be required. These examine such factors as the current level of employee involvement in safety and health activities and the effectiveness of safety and health management system processes.

Whatever measurements are used, feedback at the group level requires the collection of baseline data and a commitment to continue the measurement process. Feedback maintains individual interest in and attention to the desired safety objectives. It also helps make the newly elicited safety action patterns part of a work group's job performance norm.

One of the best ways to motivate employees is to form a joint labor–management safety committee. Such a committee can have extremely positive results by involving workers and management toward a common goal of safety and health.

Peer pressure, positive involvement, and recognition combined with using expertise from every company level to improve safety and health can be a strong motivational tool for the safety professional.

Measuring Job Satisfaction

In the interest of further understanding motivation, many studies have been conducted to determine what constitutes job satisfaction. Generally, these studies assess the job elements that workers claim contribute to their job satisfaction (or dissatisfaction). The results of these investigations suggest that satisfaction of psychosocial needs rather than physiological needs may be the major motivational aspect of job satisfaction.

Table 30–B presents the results of various surveys of job satisfaction conducted over the years. The numbers represent rankings of the factors considered in each study. Because different language and alternatives were used in each survey, the factors have been paraphrased to represent the various elements.

The results of these types of surveys suggest that compensation is not in itself a primary motivator. Although workers expect a just and equitable income, they appear to expect only what others would be paid for comparable work. Although workers may feel dissatisfied if they are underpaid, higher pay alone does not guarantee job satisfaction.

TABLE 30–B. Summary of Different Surveys on Job Satisfaction in Order of Importance of Significant Factors

	Women Factory Workers	Union Workers	Nonunion Workers	Men	Women	Employees of Five Factories
Steady work	1	1	1	1	3	1
Type of work				3	1	3
Opportunity for advancement	5	4	4	2	2	4
Good working companions	4			4	5	
High pay	6	2 ½	2	5 ½	8	2
Good boss	3	5 ½	5	5 ½	4	6
Comfortable working conditions	2	2 ½	3	8 ½	6	7
Benefits		5 ½	6	8 ½	9	5
Opportunity to learn a job	8					
Good hours	9	7 ½	7	7	7	
Opportunity to use one's ideas	7	7 ½	8			
Easy work	10					

On the other hand, steady work or job security does appear to be a primary job motivator. Workers want the security of knowing that if they perform their jobs well, they will have a job in the future. Job security, as a component of job satisfaction, can explain the willingness of a worker to remain in a lower-paying, stable job instead of accepting a higher-paying, less stable job.

Other factors involved in job satisfaction include type of work, opportunities for advancement, and good working companions. Note that all of these factors are related to the psychosocial needs of feeling important and belonging to a peer group. Likewise, comfortable working conditions (rated high by a number of employees) are probably associated with a desire to be humanely treated.

The results of these job satisfaction surveys are important when considering the safety program within the context of personnel policy. A safety program designed to ensure employees' well-being helps maintain employees' continued ability to do the work (which, in turn, promotes job security).

Likewise, the safety program represents management's interest in the co-workers and working conditions of the employee. All of these aspects of safety programming should be anticipated and incorporated into the approach that is taken with both supervisory staff and employees. Honest and sincere positioning of the safety program to enhance employee welfare makes practical sense in light of current knowledge about job satisfaction.

ATTITUDE, BEHAVIOR, AND CHANGE

The link between attitude and behavior, on one hand, and motivation, on the other, is still a matter of debate. Although many social and industrial psychologists believe that motivation is primarily about changing or molding worker attitudes and behaviors, others believe that there is still no clear evidence indicating that attitudes predict future behavior. This section discusses some of the current theories on how people's attitudes are formed, how they affect behavior, and the problems encountered when management attempts to influence or change worker attitudes and behavior.

Attitude

Many theories on attitude formation and change suggest that attitudes consist of three components: feeling, knowing, and acting. Such a theoretical approach distinguishes, yet integrates, such concepts as knowledge, emotions, and behavior or behavioral tendencies. Other theories regard attitudes as the expression of a person's values toward something or someone. Most of these theories agree, however, that attitudes (at least in part) reveal a person's tendencies to evaluate objects, persons, or situations favorably or unfavorably.

Katz and Stotland (1959) identified the following three attitude components:

- *Affective (feeling) components*. Affective, or feeling, components are the positive or negative emotions underlying attitudes. Although some attitudes can be quite irrational, they involve little more than feelings. People may not like doing something safely, but cannot explain why.
- *Cognitive (knowing) components*. Attitudes can differ because of the extent of knowledge involved in the attitudes. For example, some people work safely because they have worked through the problem and have decided that working safely is the best way. Others may not work safely because their attitude toward safety is based on incorrect information or knowledge. In fact, misinformation can be a source for many attitudes.
- *Action components*. As stated earlier, attitudes may bear little relationship to behavior. An individual may express a strong attitude about working safely, and yet fail to wear the personal protective equipment provided for the job. In such a case, the attitude may lack a significant action component.

A person's response or attitude depends partly on that individual's previous experiences. For example, if someone sees a person on the street who resembles a friend, the immediate response may be a smile, friendly gestures, and warm voice quality. When that person is perceived to be a stranger, there is an immediate change to another facial expression, gesture, voice quality, and so on. Such reactions can be triggered by many factors—a certain look from another person, a manner of speaking, a mustache or the lack of one, hair color, or style of handshake. All individuals have certain feelings about other people or situations and tend to act according to their attitudes, which have been formed over the years.

Some of these attitudes are *latent*, or hidden. Responses based on latent attitudes can lie dormant within the individual until triggered by some event. Given the appropriate stimulus, the attitude surfaces and the behaviors exhibited are consistent with the individual's feelings. A certain word, for example, elicits different responses in different people. The words *union, management, labor,* and even *safety* create certain connotations that touch off different attitudinal reactions in individuals, depending on the kinds of experiences they have had with the subject.

Because attitudes play such an important role in everyday relationships, the safety professional should understand how they are developed, their effects on individuals, and what can be done to affect them.

Determination of Attitudes

Direct personal experience is thought to be one way attitudes are established. The three components of attitude—feeling, knowing, and action—can be influenced through personal experiences. In particular, experiences individuals have had in the past determine attitudes—especially those involving strong emotions. Many people have had experiences that they learned to associate with fear, sorrow, pain, or happiness. All of these emotions tend to make people act the same way when a similar experience occurs.

For example, if a worker has been fired several times for what the individual believes are superficial reasons, the worker's attitude toward managers may be fearful and hostile. There may have been sound reasons for the dismissals or layoffs, but to admit this would be a threat to the person's pride. Thus, the individual puts the blame elsewhere. Because of the effects these dismissals have had on the employee and the family, such as causing financial stress, the employee may become angry. Management now is the enemy. The person's present hostile, fearful attitude is set and will be difficult for the next manager to change without concerted effort on both parties.

Attitude Change

Numerous factors are associated with attitude change. Commitment and responsibility are core concepts in the formation and changing of attitudes. Simply imitating a particular view will have little impact on an individual's attitude. If the person truly wants to develop a new attitude or change an existing attitude, he or she must learn it in a specific situation in which the attitude is tied to action. Attitudes that are acquired at only a verbal level cannot be expected to influence behavior.

For example, suppose the safety professional wants to bring about a change in the safety attitude of an individual or group. He or she must design the change to include an activity that requires participants to link the new attitude to the activity. Without providing individuals with the means or methods to experience change, the safety professional cannot expect simple slogans to change workers' attitudes.

In summarizing the large body of psychological literature on attitude formation and change, McGuire (1968) suggested three components of the process (Table 30–C): (1) types of attitude-change situations (column 1), (2) variables associated with the communication process (column 2), and (3) behaviors associated with attitude change (column 3). The main points of the process are as follows:
- *Relevant situational factors* in changing attitudes include the following:
    - suggestive situations (in which the desired attitude is repeatedly presented)
    - conformity (in which social or peer pressure is used to elicit the desired response)
    - group discussion, persuasive messages, intensive indoctrination.

Each situation is associated with varying degrees of attitude change. Also, the implications of each depend on the type of communication variables involved.
- *Source variables* refers to characteristics associated with the person or represented organization presenting the message. The effect of the message on subsequent attitudes can vary depending on the following:
    - the credibility, attractiveness, and power of the message
    - the order in which specific issues are presented
    - the differences between the sender and receiver of the message.
- *Channel factors* consist of ways the message is presented; for example, direct personal experience or mass media.
- *Receiver factors* include whether the audience is actively involved in the process and how much the audience can be influenced.
- *Destination factors* refers to the degree of postcommunication message decay across time and time latency associated with delayed action.

According to McGuire (1968), the receiver (or audience) must proceed through five steps for attitude change to occur: attention, comprehension, yielding, retention, and action. According to this model, each step depends on the preceding one.

To add to the complexity of this model, communication variables, especially in daily situations, can affect one another and interfere with the message. Also, the sequential-step process may have more intuitive than objective support. Nevertheless, this model should alert safety professionals and others to the fact that attitude change is not a simple process.

Organizational Development

Behavioral scientists, concerned about the amount and rate of change in a technological society, have studied the impact of such change on the industrial environment. These researchers have evolved an approach called *organizational development*, which attempts to assess a corporation's ability to adjust to such conditions as rapid growth, new technology, increasing diversity, and management system problems. Organizational development is designed to assess and improve the attitude and structure of organizations so they can better adapt to change.

Effective organizational development requires a strong organizational culture. The development of a strong culture

TABLE 30-C. Factors Associated with Persuasive Communication

Attitude Change: Situation Types	Communication Process	Attitude Change: Behavioral Step
1. Suggestive	1. Source	1. Attention
2. Conformity	2. Message factors	2. Comprehension
3. Group discussion	3. Channel factors	3. Yielding
4. Persuasive messages	4. Receiver factors	4. Retention
5. Intensive doctrination	5. Destination factors	5. Action

will aid a company's success by giving employees the means to identify and act on the values of the company (Deal and Kennedy 1982). The organization's culture provides workers with informal rules on how they are to act. Through the company's culture, the organization can either reinforce the safety efforts of the workers or create barriers to the development of an effective safety program.

Generally, an organizational development approach gathers information from employees and management to determine the climate and capacity of the organization to adapt its objectives to the new environment. Based on such feedback, an attempt is made to develop organic systems to replace mechanical systems within the organization. *Organic systems* are characterized by a preoccupation with people as they work together; *mechanical systems* are characterized by a preoccupation with the structure of a company. Figure 30-4 presents a summary of the differences between mechanical and organic systems.

To implement the ideas represented by organizational development, management can use a number of procedures to effect changes in the organization. Because the changes are typically people oriented, management should recognize the need to motivate employees to accept the changes within both management and work groups. Typically, people can be educated and trained to learn how to work with the new system. These techniques have merit in implementing change only as long as top management agrees to the changes and provides incentives within the organization for their adoption. Organizational development seeks to implement organic systems in place of mechanical systems.

Ultimately, organizational development seems to provide the necessary means for preparing an organization to plan for the future. Included in such an effort should be the recognition of how changes in an organization will affect the current safety program. Involving all levels of managers and workers in developing and managing safety programs is likely to ensure that the programs will be able to meet the future needs of modern organizations.

Behaviors versus Attitudes

Attitudes refers to internal predispositions to behavior; as such, they are difficult to observe and measure. *Behavior* refers to observable actions, which can be measured. This distinction is vitally important because measurement of behaviors lays the groundwork for effective safety management. Although attitudes are difficult to affect directly, changing behavior is not as difficult. A good example of this is people's attitudes toward using seat belts. As the behavior of wearing seat belts has changed over the years, so have attitudes toward wearing them.

Behavioral Management

Significant attention and recognition have been given to the importance of behavioral management in safety programs. *Behavioral management* refers to the systematic identification, measurement, and modification of safety-related behaviors. This section briefly reviews this approach.

Developing and planning for behavior change requires the safety professional or manager to understand the situational variables that influence the behavior of individuals. Why would a person have a strong desire to work safely, yet continually perform in an unsafe manner? Why are there

Mechanical Systems	Organic Systems
Emphasis upon individual performance	Emphasis on relationships in group
Chain of command concepts	Confidence and trust among everyone
Adherence to delegated responsibility	Adherence to shared responsibility
Division of labor and management	Participation in multimember teams
Centralized management control	Decentralized sharing of control
Resolution of conflict through grievance procedures	Resolution of conflict through problem solving

Figure 30-4. Organizational development seeks to implement organic systems in place of mechanical systems.

discrepancies between the expectations and results of the safety efforts in the work environment?

Often, when there is a discrepancy between results and expectations, managers may believe that there is a need for additional employee education or training. Ferdinand Fournies (1987) asked more than 4,000 managers, "Why don't subordinates do what they are supposed to do?" The responses included the following:
- They don't know what they are supposed to do.
- They don't know how to do it.
- They don't know why they should do it.
- There are obstacles beyond their control.
- They don't think it will work.
- They think their way is better.
- They're not motivated—poor attitude.
- They're personally incapable of doing it.
- There's not enough time for them to do it.
- They are working on the wrong priority items.
- They think they are doing it.
- There's poor management.
- They have personal problems.

Robert Mager and Peter Pipe (1984) developed a useful procedure that gives safety professionals and managers a way to systematically and accurately analyze performance discrepancies and create change. Mager and Pipe's first step in analyzing performance is to identify the nature of the discrepancy by asking, "What is the issue?" The safety professional must accurately describe the safety discrepancy. Examples would include driving a forklift with the forks elevated above a specified level, not wearing personal protective equipment, or conducting maintenance without locking out the equipment. The second step is to ask, "Is it important?" If the answer is yes, the process continues to the third step and question, "Is it a skill deficiency?"

If a person drives the forklift with the fork too high because of a lack of skill, the safety professional may need to provide formal training if the individual has not had prior experience on a forklift. A second factor could be that the person has not driven the forklift often and requires additional practice. The third factor could be that the person has driven the forklift often but requires appropriate feedback.

If the person's forklift-driving behavior is not due to a skill deficiency—"They could do it if they wanted to"—additional knowledge from education and training may not change the individual's performance. In these types of situations, the safety professional must determine whether:
- the appropriate performance will result in punishment
- there is no reward for the appropriate performance
- the appropriate performance may not matter to the individual
- there are obstacles to attaining the appropriate performance.

In these types of situations, it is important to remove the punishment, arrange positive consequences, or remove the obstacles. Does the facility layout require the forklift driver to constantly raise and lower the forks to clear congested areas, thus reinforcing the driver's decision to drive with the forks elevated? Can the work area be redesigned?

In some situations, such as not wearing personal protective equipment, the safety professional may not need to determine whether the discrepancy is related to a skill deficiency. By analyzing the situation, the safety professional or manager may be able to design a simpler way to do the job that does not require the use of the protective equipment. If a simpler way is identified, the job should be changed or on-the-job training should be provided.

In addition to redesigning the job, poor performance or nonperformance may be due to an individual's potential. The individual may be unable to learn the job, lack the physical or mental potential, or be overqualified for the job. If the safety professional or manager determines that the person lacks the potential, he or she must decide to either transfer the person to a different job or terminate the relationship. Mager and Pipe believe that transfer or termination options are used more than they should be. This could be due in part to the decision maker's lack of understanding regarding human performance and how to make improvements or an indication of the decision maker's inability to identify other options.

Because an individual's behavior can be the result of competing factors—created by either internal needs or situational variables—the safety professional or manager may identify a variety of solutions. Understanding the factors that can influence performance discrepancies can help management select and implement the best possible solution.

Evaluating critical behaviors in the workplace can assist in changing safety-related behaviors. The basic steps of the process are as follows:
- *Identify critical behaviors*. This means to write, in observable terms, what employees should do to properly perform their jobs. The safety professional can list a few critical behaviors or a complete inventory, depending on the scope and results desired.
- *Conduct measurement through observation*. Trained observers watch the workplace to determine whether the listed behaviors are performed safely or unsafely. The total number of observed behaviors is divided into the number of safe behaviors to obtain a percentage figure for safe behaviors.
- *Give performance feedback*. The percentage figure for

safe behaviors is shown on a graph displayed in the workplace. At regular intervals, behaviors are again observed and the new safe behavior figures added to the graph. Studies show that this critical feedback will improve safety behaviors. Praise and recognition from managers or peer pressure can be effective ways to encourage and reinforce safe behaviors.

Structural Change

In the previous discussion of attitude formation and change, no direct relationship between attitude change and behavioral change was assumed. In practice, however, line managers and safety personnel are primarily interested in ways of changing employees' attitudes only if their behavior also changes—for example, if workers not only change their attitude about wearing personal protective equipment, but also change their behavior by actually using it consistently. Although the research literature indicates that many techniques—training and counseling, for example—can change attitudes, there is much less support for a direct relationship between attitude changes and subsequent behavior changes (Ajzen and Fishbein 1977).

In effect, line managers and safety professionals must consider techniques other than simple attitude change when considering employee behavior. Evidence suggests that it is possible to change behavior by changing the structure of the work environment. Examples of structural changes involve changing situational variables, including changing job contents, modifying the physical arrangements of work, changing worker interaction patterns, and rearranging work procedures.

In each case, it is not necessary to expend the time and effort to change attitudes before changing behavior directly. Rather, the introduction of the structural change can modify behavior and, with it, possibly change employee attitudes.

Although the structural change model has not been applied extensively in the past to occupational safety, it warrants consideration by safety professionals. Human behavior can be changed by altering the circumstances under which the individual works. Unsafe practices, for example, cannot occur when the structure of the work and the workgroups prevents them from happening.

LEARNING PROCESSES

Not only is motivation a key component in a successful safety program, the role that training plays in sustaining motivation is equally important. Therefore, safety professionals must take into consideration the importance of employee learning.

Motivational Requirements

With the changing demographics of the U.S. work force, effective employee education and training programs are now more important than ever. Therefore, training materials must be designed to not only develop the workers' interest in the topic, but must also pertain to the practical situations that workers encounter every day.

The safety professional must recognize that if workers are going to learn safe procedures, they must be motivated to do so. Based on the previous sections on motivation, it is not wise to assume that because management sees the value of implementing safe work procedures, the workers will automatically share the same sentiment. Therefore, to have employees buy into a safety program, the safety professional must know what will motivate the work force. (Chapter 33, Safety Awareness Programs, discusses this in more detail.)

To help motivate workers toward following safe work procedures, the safety professional should emphasize to workers:
- the risks of improper work habits and their costs to both the company and the individual
- how improper work habits affect the product or service provided
- the long-term impact of improper work habits on the employee, the company, and the product
- the need for workers' help in addressing these problems in a functional manner for the benefit of all.

Principles of Learning

Some consideration of basic learning principles is valid whether the learning is to be done in a college classroom or in a work area. When training procedures are based on these principles, learning is more efficient and thorough.

Reinforcement

Through experimentation, psychologists have found that positive reinforcement can often facilitate learning. In practice, reinforcement can make work more efficient and more productive. When a worker's pay increases because he or she produces more units, the person has received a reward for learning. This type of reinforcement can have negative aspects as well—the worker who figures out a hazardous shortcut to produce more units may actually be rewarded for an unsafe practice.

Employees must be recognized and rewarded only for safe work methods. Higher productivity based on safe work methods satisfies the dual need for achievement and recognition. This fact not only reinforces employee learning, but also influences other employees who see or learn about it. A supervisor who recognizes the same needs in other workers can reinforce this learning through praise of greater productivity and telling workers how much improvement has been shown. Thus, the publicity given to a safety record, a bonus, a promotion, or anything else that satisfies individual needs would serve as a reward to reinforce whatever learning has brought about the desired behavior.

Reinforcement tends to be more effective if it is given soon after the desired behavior occurs. Praise, for example, should be given at the time or within a reasonably short time after the desired behavior happens or, if delayed, be associated verbally with the desired behavior. If reinforcement is delayed, the safety professional or supervisor should let the individual know why he or she is receiving the reward.

The principle of positive reinforcement also applies to employee participation in a safety program whether through suggestion systems, safety and health committees, discussion groups, or training sessions. In all these areas, an individual gains personal worth if his or her opinions are requested and positively reinforced.

A safety program that gathers ideas from all constituents satisfies the need people have for being "in the know." In this way, the safety professional can create a positive atmosphere for the program and instill in employees a sense of obligation and responsibility for its success. Research has demonstrated that when employees feel they helped to create a program, there is more chance for its success.

Research on the effectiveness of disciplinary action suggests that it can have mixed consequences. Discipline is thought to be less effective than reinforcement, perhaps because discipline provides indirect cues or information (what *not* to do), as opposed to positive reinforcement, which provides direct information about what employees *should* do (Church 1963).

Recognizing an individual's correct performance will lead to a more positive attitudinal response on the worker's part than will any form of discipline. A supervisor's praise for a task performed well may influence an employee's attitude toward other parts of the job, including training procedures, safety devices, and the safety program. Proper use of recognition can lead to worker acceptance of efficient and safe work methods.

Knowledge of Results

Another aspect of recognition is knowledge of results. People like and need to know how they are doing on the job. When workers know that a new procedure is helping to increase their production, they have received reinforcement for their learning and effort.

Knowledge that a job was done correctly and safely is rewarding to employees. However, the contrary is also true. If a job was not done correctly, then employees must know the correct rules and procedures that are a part of job performance.

One phenomenon that often occurs in the learning process is known as a *plateau*. At this point, a worker's progress levels off for some time before he or she experiences another increase in learning. Often, some individuals become discouraged during this time, and their learning may decline if it is not supported. The trainer needs to understand this phenomenon and communicate to the individual that such a leveling off is normal and that further progress will come later.

Practice

The safety professional is interested in developing safe working habits in employees that will become almost automatic on the job. As a result, merely putting a worker through training sessions is not enough. Despite any apparent mastery of safety methods, an employee may make one or more mistakes the next time he or she performs the task. To be sure habit patterns are firmly entrenched, workers must practice what they learn. Job instruction training (JIT) programs include one of the important aspects of training—follow-up by the manager. This helps ensure mastery of safe working methods. Within a reasonable time, depending on the job complexity, follow-up will no longer be needed.

In addition to the learning principles covered so far, whole versus part learning has been a challenge to industrial trainers for many years. Whether trainers should teach a work procedure as a whole or break it down and teach it part by part has advantages and disadvantages depending on job complexity, the trainee, the kind of job breakdown used, and the level of support provided by the manager. A combination of the two approaches—using the whole method, but with sufficient flexibility to emphasize meaningful parts of the task wherever necessary—may be most effective.

Meaningfulness

Studies in verbal learning have demonstrated how important it is that the material used be meaningful to learners. Meaningfulness to management and workers is understanding the value of producing a marketable product or service (which contributes to job satisfaction, job retention, profit, etc.) and the negative consequences of failing to engage in appropriate actions, leading to higher production costs, higher product rejection rates, accidents, and their associated losses.

Meaningfulness is essential in safety training because the worker needs to understand why a certain procedure is better than another. Adequate explanation of how a movement or change in position can eliminate hazards with no decrease in production gives meaningfulness to the procedure. With this understanding, workers are more likely to be motivated to learn the safe procedure. Without it, they will be inclined to use their own methods until they learn, perhaps by an accident, how inadequate unapproved work procedures are. The safety professional and line supervisor should reinforce the advantage of workers' understanding the reason for having protective clothing, safety devices, safety meetings, and discussions, as well as the need for full and complete accident reports.

Selective Learning

Out of each day's many experiences, people for a myriad of reasons select the events they will retain. This selection process is closely related to motivation, and for that reason motivation must be considered as part of a job-training program. Safety trainers must try to ensure that workers retain the most important facts. Relating subject matter to individual needs will help ensure proper selection and retention.

Frequency

People do best those things they practice most often. This principle is certainly important to the safety professional because it emphasizes the necessity of frequently applying safety practices in training programs as well as on the job. Giving the worker a copy of safety rules and regulations in hopes that they will be learned and used is not enough. Instructors must devise ways to bring these rules and regulations to workers' attention frequently and regularly. Trainers must insist not only on frequent practice of the correct procedures, but also on the trainee following through with such methods.

Recall

Closely related to the principle of frequency is that of recall. This theory holds that what is learned last usually can be most easily recalled. As just indicated, handing a worker a set of safety rules and regulations does not ensure learning. Those who received printed instructions or a few training demonstrations a long time ago may not be able to recall what the rules and regulations are. Safety professionals must devise ways for workers to come into frequent contact with these regulations through activities such as frequent review of safety regulations and committee work.

Primacy

The law of primacy must be taken into consideration in two aspects of the safety program. First, the worker's initial contact with safety procedures should be positive and of major importance. If this initial contact is of a negative nature—such as being tossed a book of rules and told, "Make sure you learn them"—the worker is left with the impression that safety is unimportant. From the first day of employment, the importance of individual safety rules as well as the whole safety program should be emphasized. Second, employees developing habit patterns using safe methods should assume primary importance. Therefore, the manager must be certain the worker does not learn how to work any way other than by the safe method. It is more difficult to establish good patterns if poor ones have been learned first.

Intensity

Those things made most vivid to the worker will be retained the longest. Safety programs already use this principle in safety publicity with eye-catching posters, slogans, and so forth. In this way, the training material will be retained longer. Therefore, to some degree, the trainer must position the safety program in the same way advertisers position their commercials to catch the public's attention.

Transfer of Training

Positive transfer of training occurs when the previous learning experience helps current learning or enhances current performance. Negative transfer of training occurs when previous learning makes the current learning experience more difficult or in some way inhibits current performance. Therefore, learning can be influenced by previous learning, a phenomenon known as *transfer of training*.

Generally, learning results in positive transfer if there is identical response to the new stimuli. For example, learning to drive in a new (or different) automobile is facilitated by the fact that identical responses (such as accelerating or steering) are required in the new automobile, similar to those in the old automobile. Positive transfer should be high if the new stimuli are similar to previous stimuli (e.g., the location of the controls, their texture, and their direction of movement). In such situations, positive transfer increases with a higher level of similarities between the two learning situations.

However, negative transfer can also occur. For example, if the controls of the cars appear identical but the "fifth" gear is in the opposite position, negative transfer can be expected for a time. If the fifth-gear position is located at the upper right position on the floor shifter in one car, the person will experience some difficulty learning to drive a car in which the fifth-gear position appears at the lower right. Or, if the off position of a toggle switch is the same as the on position of another, the potential for accidents is increased. Such negative transfer can cause errors and delayed reactions and create inefficient performance on the new task. In task situations, the amount of negative transfer increases the more similar the two learning situations seem to be.

With an understanding of transfer phenomena, it is possible to maximize, within the industrial environment, opportunities for positive transfer and minimize those for negative transfer. This requires careful planning of machine purchases and work procedures to ensure that new tasks required of a worker use (and do not conflict with) previous learning experiences.

Short-Term and Long-Term Memory

The topic of learning would not be complete without a discussion of individual memory. Learning and forgetting are not mutually exclusive. As a person learns, he or she also forgets some of what was previously learned. Learning curves indicate that most knowledge is lost immediately after it has been received unless the information received is put into immediate practice. Therefore, depending on the complexity of the job and the time frame in which the information learned will be utilized, a certain amount of learning will be lost after a training session. Even though the amount of information lost following formal training varies among individuals, there may be a greater need to hold frequent refresher sessions to reinforce what was learned originally.

SUMMARY

- Performance improvement is a product of employee and management commitment and involvement. Employees must be willing to take responsibility for accepting the behavior changes needed to achieve safety and health objectives.
- The psychological factors that influence safety behavior include individual differences, motivation, emotions, stress, attitudes, behaviors, and learning processes.
- Individual differences among employees present an ongoing challenge in occupational settings. Yet employers often make the mistake of gearing their safety efforts toward the nonexistent "average person" instead of matching individuals to a specific job.
- Motivation includes direction of behavior, intensity of action, and persistence of effort. To affect employee motivation, employers must establish specific safety objectives, reinforce desired behaviors, understand the needs theory of Maslow, recognize the motivational power of job satisfaction, understand the X and Y theories of management, and be familiar with job enrichment theory.
- Attitudes are composed of three elements: feeling, knowing, and action. Attitudes are formed in a variety of ways but tend to be persistent and difficult to influence. Commitment and responsibility are the key factors in forming and changing attitudes.
- Attitudes are important to a safe work environment, but safety-related behaviors are critical. Behavioral management's "action"-directed focus involves identifying, measuring, and controlling employees' safe work-related behaviors. Steps involved in changing safety-related behaviors can include (1) providing training, (2) arranging practice, (3) arranging positive consequences that reinforce correct performance, (4) removing obstacles that prevent satisfactory performance, and (5) finding a safer way to do the job.
- Two motivational models used in many companies are organization behavior management and total quality management. Selected motivational techniques that have proven successful include communication, awards, incentives, recognition, and employee surveys.
- By understanding the principles of how people learn, management can establish effective employee development and training programs. These programs can influence, change, and reinforce the concept of safe work-related behavior. Management must understand and apply learning principles such as reinforcement, feedback, practice, frequency, transfer of training, and supporting short-term and long-term memory.

REFERENCES

Ajzen, I., and M. Fishbein. "Attitude-Behavior Relations: A Theoretical Analysis and Review of Empirical Research." *Psychological Bulletin* 84 (1977): 888–918.

Anastasi, A. *Psychological Testing*. 6th ed. Toronto, Canada: Macmillan, 1988.

Bauer, R. A. "Communication as Transaction." In *The Obstinate Audience*, edited by D. E. Payne. Ann Arbor, MI: Foundation for Research on Human Behavior, 1965.

Church, R. M. "The Varied Effects of Punishment on Behavior." *Psychological Review* 70 (1963): 369–402.

Cohen, A. "Protective Behaviors in the Workplace." In *Taking Care: Understanding and Encouraging Self-Protective Behavior*, edited by N. Weinstein. Cambridge, UK: Cambridge University Press, 1987.

Cohen, A., and B. Cleveland. "Safety Program Practices in Record-Holding Plants." *Professional Safety* 28, no. 3 (1983): 26–33.

Cohen, A., M. Smith, and W. Anger. "Self-Protective Measures against Workplace Hazards." *Journal of Safety Research* 11 (1979): 121–31.

Conard, R. J. *Employee Work Practices.* NIOSH Contact Report 81-2905. Cincinnati, OH: National Institute for Occupational Safety and Health, 1983.

Deal, T., and A. Kennedy. *Corporate Cultures: The Rites and Rituals of Corporate Life.* Reading, MA: Addison-Wesley, 1982.

Deci, E. L. *Intrinsic Motivation.* New York: Plenum Press, 1975.

DeJoy, D. "A Behavioral-Diagnostic Model for Self-Protective Behavior in the Workplace." *Professional Safety* 31, no. 12 (1986): 26–30.

Fournies, F. *Coaching for Improved Work Performance.* Blue Ridge Summit, PA: Liberty House, 1987.

Herzberg, F. *Work and the Nature of Man.* Cleveland, OH: World, 1966.

Kanfer, R. "Motivation Theory and Industrial and Organizational Psychology." In *Handbook of Industrial and Organizational Psychology*, edited by M. D. Dunnette and L. M. Hough. Palo Alto, CA: Consulting Psychologists Press, 1990.

Katz, D., and E. Stotland. "A Preliminary Statement to a Theory of Attitude Structure and Change." In *Psychology: A Study of a Science*, edited by S. Kock, vol. 3, 423–75. New York: McGraw-Hill, 1959.

Mager, R. F., and P. Pipe. *Analyzing Performance Problems or You Really Oughta Wanna.* 2nd ed. Belmont, CA: Lake, 1984.

McGregor, D. *The Human Side of Enterprise.* New York: McGraw-Hill, 1985.

McGuire, W. J. "The Nature of Attitudes and Attitude Change." In *The Handbook of Social Psychology*, edited by G. Lindzey and E. Aronson, 2nd ed., vol. 3. Reading, MA: Addison-Wesley, 1968.

Peters, R. "Strategies for Encouraging Self-Protective Employee Behavior." *Journal of Safety Research* 22 (1991): 53–70.

Planek, T. W., and K. Fearn. "Reevaluating Occupational Safety Priorities." *Professional Safety* 10 (1993): 16–21.

SUGGESTED READINGS

"Accident Causation Theory: The Emerging Pluralism." *Occupational Hazards* 45 (October 1983): 10.

Bennis, W. G., and B. Nanus. *Leaders: The Strategies for Taking Charge.* New York: Harper & Row, 1986.

Blake, R. R., and J. S. Mouton. *The Managerial Grid III.* Houston, TX: Gulf, 1985.

Bolton, R. *Social Style/Management Style.* New York: American Management Association, 1984.

Brown, W. D. *Welcome Stress! It Can Help You Be Your Best.* Minneapolis, MN: CompCare, 1983.

Chhokar, J. S., and A. Wallin. "Improving Safety through Applied Behavior Analysis." *Journal of Safety Research* 15 (1984): 141–51.

Cooper, C. L., and M. J. Smith. *Job Stress and Blue Collar Work.* Chichester, UK: Wiley, 1985.

Dennis, L. E., and M. E. Onion. *Out in Front: Effective Supervision in the Workplace.* Chicago: National Safety Council, 1990.

Dipboye, R. L., and W. C. Howell. *Essentials of Industrial and Organizational Psychology.* Rev. ed. Homewood, IL: Dorsey Press, 1982.

Glasser, W. *Control Theory.* New York: Harper & Row, 1984.

Hersey, P., and K. Blanchard. *Management of Organizational Behavior: Utilizing Human Resources.* 5th ed. Englewood Cliffs, NJ: Prentice-Hall, 1988.

Herzberg, F., et al. *The Motivation to Work.* Cleveland, OH: World, 1993.

Hulse, S. H., J. Deese, and H. Egeth. *The Psychology of Learning.* 5th ed. New York: McGraw-Hill, 1980.

Machlowitz, M. *Workaholics.* New York: Addison-Wesley, 1980.

Marks, L. N. "OHN's Can Provide Listening Ear Dealing with Employee's Problems." *Occupational Health and Safety* 55 (May 1986): 5.

Maslow, A. H., et al. *Motivation and Personality.* 3rd ed. New York: Harper & Row, 1987.

"Open Your Ears and Learn to Listen." *Safety and Health* (July 1990): 44–47.

Patrick, P. *Health Care Worker Burnout: What Is It, What to Do about It.* New York: Inquiry Books, 1980.

Rader, C. M., and P. B. Haber. "A Psychological Profile of Industrial Injured Workers." *Occupational Health Nursing* 32 (November 1984): 11.

ReVelle, J. B., and L. Boulton. "Worker Attitudes and Perceptions of Safety." *Professional Safety* 26 (December 1981): 12.

Weisinger, H. *Dr. Weisinger's Anger Work-Out Book.* New York: Quill, 1985.

"What Does It Take to Change Unsafe Behavior?" *Safety and Health* (July 1990): 60–63.

Zimbardo, P. G. *Psychology and Life.* 12th ed. Glenview, IL: Scott, Foresman, 1988.

REVIEW QUESTIONS

1. Define *motivation*.
2. List five of the seven psychological factors that directly influence the success of a safety program.
   a.
   b.
   c.
   d.
   e.
3. What two characteristics influence the effectiveness of reinforcement of desired behaviors?
   a.
   b.
4. Whether an employee works safely depends on what three factors?
   a.
   b.
   c.
5. What was Abraham Maslow's contribution to the study of motivation?
6. According to various types of studies and surveys, satisfaction of which of the following is the major motivational aspect of job satisfaction?
   a. physiological needs
   b. high salary
   c. psychosocial needs
   d. all of the above
7. McGregor discovered two basic ways in which managers view workers. Give a brief description of each.
   a.
   b.
8. What did Herzberg theorize to be the most effective concept of motivation in the workplace?
   a. job enrichment
   b. environment approach
   c. hygiene approach
   d. new technology
9. What are the three components of attitude?
   a.
   b.
   c.
10. On what does a person's response or attitude usually depend?
11. List four of the seven principles of job enrichment.
    a.
    b.
    c.
    d.
12. Briefly explain the difference between behaviors and attitudes.
13. List 7 of the 10 principles of learning.
    a.
    b.
    c.
    d.
    e.
    f.
    g.

31 Safety and Health Training

John F. Montgomery, PhD, CSP, CHMM
Teddy G. Gil, BSME, MBA, DAL Level II VT, UT, RT, PT, MT, IR

Training as Part of the Safety and Health Program
Introduction ▸ Training Definition ▸ Benefits of Training ▸ Training and Nontraining Solutions ▸ Training Roles of the Safety Professional

Developing the Training Program
Analysis/Identification ▸ Design ▸ Training Methods ▸ Development

Delivering Training
Implementation ▸ Evaluation/Maintenance

Regulatory Obligations
New Employee Training and Orientation ▸ Additional Resources

Summary

References

Review Questions

TRAINING AS PART OF THE SAFETY AND HEALTH PROGRAM

Introduction

Often, a safety or health problem can be caused by lack of appropriate attitude, knowledge, skills, and/or awareness. One solution to mitigate this problem is to provide training. The goal for training specifically tailored for safety and health is for workers to learn and apply techniques that will keep them and others safe and healthy.

A key element in every successful safety and health program is training built to promote behavior, skills, knowledge, and/or awareness as they relate to accident prevention and the occupational safety and health program within the organization. Workers must have proper training specific to the organization to do their jobs safely and efficiently.

The responsibility for implementing and conducting employee training rests with management. Management must recognize that for the organization to achieve its objective of providing a safe and healthy workplace, employees must perform work at a certain competency level and meet certain minimum expectations and performance objectives. Therefore, management must establish business operations and strategies that include a safety and health program with defined goals. These goals need to be tailored for all levels of the organization from top management, supervisors/line managers, and safety professionals to front-line employees. These goals then need to be incorporated into the appropriate training programs for each level. Hence, everyone in the organization must be trained to understand his or her roles and responsibilities as they relate to the safety and health program. It should be noted that for a safety and health program to be successful, it will need more than just training. There needs to be a firm and unwavering management commitment along with committed employee involvement. An organization's safety and health program must include:

- training in the proper methods of leadership and supervision to hold personnel accountable to the safety and health requirements/policies/procedures
- safety professional training on how to detect, eliminate, and/or control hazards; investigate accidents; and handle emergency situations
- targeted training based on the position held within the organization.

The safety and health program must also be reviewed on a regular basis and reconciled with current training to ensure that it meets all regulatory training obligations and complements/supports the goals and objectives of the company (refer to Training Roles of the Safety Professional later in this chapter). The rest of this chapter provides more specifics relating to the definition of training, its benefits, choosing training as a solution, how the safety professional supports training objectives, and the characteristics of effective safety and health training programs. Effective training programs ensure that employees know how to do their jobs using safe methods.

Training Definition

Training is often confused with education. *Education* focuses on information that may or may not be used to improve a process or a specific task on the job. It can be present based, but it is more commonly focused on future applications, with theoretical emphasis rather than process-oriented emphasis. Merriam-Webster defines *training* as "a process by which someone is taught the skills that are needed for an art, profession, or job." In business organizations, training is primarily focused on enhancing performance in a work setting by targeting changes to an individual's behavior, skills, knowledge, aptitude, and/or awareness. Hence, the goal of effective training in a business is to provide learning that leads to improved on-the-job performance. Training is one specific solution to meet a safety or health need caused by inappropriate behavior or lack of needed skills, knowledge/aptitude, and/or awareness. It focuses on the present, providing information on the process necessary to accomplish task, objective, or goal. It provides the *why* only to the extent that people need to know the information (see also Chapter 30, Motivation).

Benefits of Training

Training benefits can be categorized into immediate, short-term, and long-term benefits. It should be noted that a well-rounded safety and health program will have benefits that have implications far beyond just safety and health. The benefits of safety and health training include, but are not limited to, the following:

Immediate:
- increased awareness
- increased knowledge
- increased skill level
- increased aptitude
- increased confidence
- reduced complacency.

Short term:
- increased employee satisfaction
- improved morale
- improved performance
- improved decision making.

Long term:
- reduced employee turnover
- reduced OSHA-recordable incident rates
- reduced total incident rate
- reduced lost time
- reduced lost workday rates
- decreased severity rate
- increased return on training investment
- improved productivity.

All these benefits will culminate into improved business performance directly affecting the organization's bottom-line financial performance. It should be noted that to achieve optimal benefits, training needs to be properly designed and developed, implemented, and evaluated. The method(s) used for the delivery of the training will also have a major impact on the total benefit to the organization (see also Chapter 32, Media).

Training and Nontraining Solutions

Sometimes, a training program is not needed in order to affect on-the-job performance. Millions, perhaps billions, of dollars are wasted every year because leadership takes the easy road by concluding that training will solve a problem that, in reality, requires an all-together-different solution. In many situations, the solution may be a nontraining solution or a combination of training and nontraining solutions. Nontraining solutions are actions taken to resolve safety and health problems that do not require training. The following are some hypothetical examples of nontraining solutions:

- A supervisor may offer an incentive to the employee with the best overall safety record.
- A construction company may require the use of cut-resistant gloves while working.
- A manufacturing facility may require the use of safety glasses in the assembly division or beyond a certain point where manufacturing is being completed.
- The engineering department may redesign the work flow of a manufacturing site to decrease the number of injuries caused by people walking near heavy machinery.
- A company may institute a program to stretch and flex at the beginning of the workday to reduce stress and strains.

It is important to analyze the problems and assess training needs before making decisions about solutions. The following are some analyses to consider using when assessing training needs, deficiencies, or requirements:
- safety and health deficiency/problem analysis (aka root cause analysis)
- organizational impact analysis
- performance analysis
- procedural analysis
- strengths, weaknesses, opportunities, and threats analysis (SWOT)
- root cause analysis
- training needs analysis
- course content delivery method and organizational analysis.

When facing safety and health challenges, management should consider all training and nontraining alternatives. Much of the lack of return on investment in training occurs because of inadequate analysis of the safety and health problem/deficiency leading up to the incident. One example to consider is the frequency of the task being performed. A thorough root cause analysis followed by a training needs analysis will provide information such as whether a worker failed to remember all the steps (or perhaps the task is too complex to perform from memory).

In situations such as provided in the preceding examples, nontraining solutions such as the following specified processes and procedures are more effective than continued training without any changes. When considering what actions to take to resolve a problem, it is important to consider the repetition or frequency of the task accomplishment; if there is sufficient follow-up after the initial completion of training to prevent the introduction of bad behavior; and whether the task is very repetitive, whereby complacency may have been a contributing factor. Therefore, for more complex and nonfrequent tasks, the overall imperative is to have established, written processes and procedures. Examples of processes and procedures include:
- Safety Data Sheets (SDSs), with sections on protection information, special precautions, spill/leak procedures
- flowcharts
- checklists
- diagrams
- troubleshooting guides
- decision tables
- reference manuals
- help desks or hotlines.

Training and nontraining problems or needs and their related solutions generally can be categorized as selection and assignment, information and practice, environment, and motivation/incentive. These four major categories of training and nontraining problems and solutions are defined as follows:

1. *Selection and assignment* refers to considerations and processes used to hire people and assign them specific

responsibilities and on-the-job tasks. This area includes improving the hiring selection process by identifying the best candidate for the task performance, thus minimizing related safety and health issues.

2. *Information and practice* is the act of providing all necessary information and developing the skills needed to use safe work practices. This area includes clearly communicating tasks, goals, and instructions and providing feedback, training, job aids, guided practice, and experience. This is the only area that can sometimes be improved through a training solution because effective training provides many opportunities for skill development.

3. *Environment* refers to all day-to-day influences on performance. These include, but are not limited to, equipment, floor plan, access to tools, working relationships with others, and working relationships with the safety professionals of the organization. For example, in work areas where people should wear safety goggles but may forget, a sign can be posted showing the use of the proper safety goggles. Or, if walkways are located near areas where heavy machinery is used, the engineering department may redesign the work flow of the site to decrease the potential for injuries.

4. *Motivation/incentive* refers to any link between safety and rewards for safe behavior, such as bonuses, pay increases, contest prizes, and team competition. For example, a supervisor may offer a bonus to the employee with the best overall safety record. In safety training, in particular, the reduced risk of injury and even death is a potential motivator. For detailed discussion of motivation, see Chapter 30, Motivation.

Training Roles of the Safety Professional

A safety professional has four primary functional areas, which include the following:

1. Anticipate, evaluate, and identify hazardous conditions and practices.
2. Develop hazard control designs, methods, procedures, and programs.
3. Implement, administer, and advise on hazard control programs.
4. Audit, evaluate, and measure the effectiveness of hazard control programs

The role of a safety professional is very broad and one of the drivers that has led to specializations within the functional areas. This chapter's focus is on the role of a safety professional as it relates to training. The role of training lies primarily within the third functional area: implement, administer, and advise on hazard control programs. This function involves directing and/or assisting in the design, development, implementation, and evaluation of training materials and/or courses. Additionally, the safety professional is responsible for overseeing and conducting or assisting with courses related to safety and health. The safety professional may be required to coordinate the function of training development with internal or external instructional designers, purchase materials, contract with a trainer, or coordinate the entire training function. No matter the role, it is important to have a clear idea of the training goals, based on the needs of the company or facility and the workers. As previously mentioned, because of the functional areas and because the role of the safety professional is so vast, it is important for the safety professional to utilize training analysis, design, and development subject matter experts as required to ensure best practices are implemented unless this is the safety professional's area of specialization.

DEVELOPING THE TRAINING PROGRAM

Unless training is the safety professional's area of specialization, the safety professional should work with internal or external training subject matter experts (SMEs) who utilize generally accepted instructional systems design (ISD). Instructional designers utilize ISD models such as the Dick and Carey Model and the ADDIE model. These two models differ in sequence and interaction but share the major categories described in the analysis, design, development, implementation, and evaluation (ADDIE) model as illustrated in Figure 31–1. This figure differs slightly in that it incorporates a continuous evaluation process to ensure the training is always being evaluated to the desired outcome(s). The development processes' major activities include the following training cycle primary categories:

- analysis/identification
- design
- development
- implementation
- evaluation/maintenance
- continuous improvement loop/performance evaluation.

Each of these major categories will have subcategories that will include tasks such as training needs analysis, written performance objectives analysis, outline of content, selection of training delivery method, duration and recurrent frequency (if any), testing/assessment, and evaluation, which would include the competency levels required to indicate whether the performance objectives were achieved (to name a few). All of the primary categories once completed are independently evaluated for completion and reanalyzed

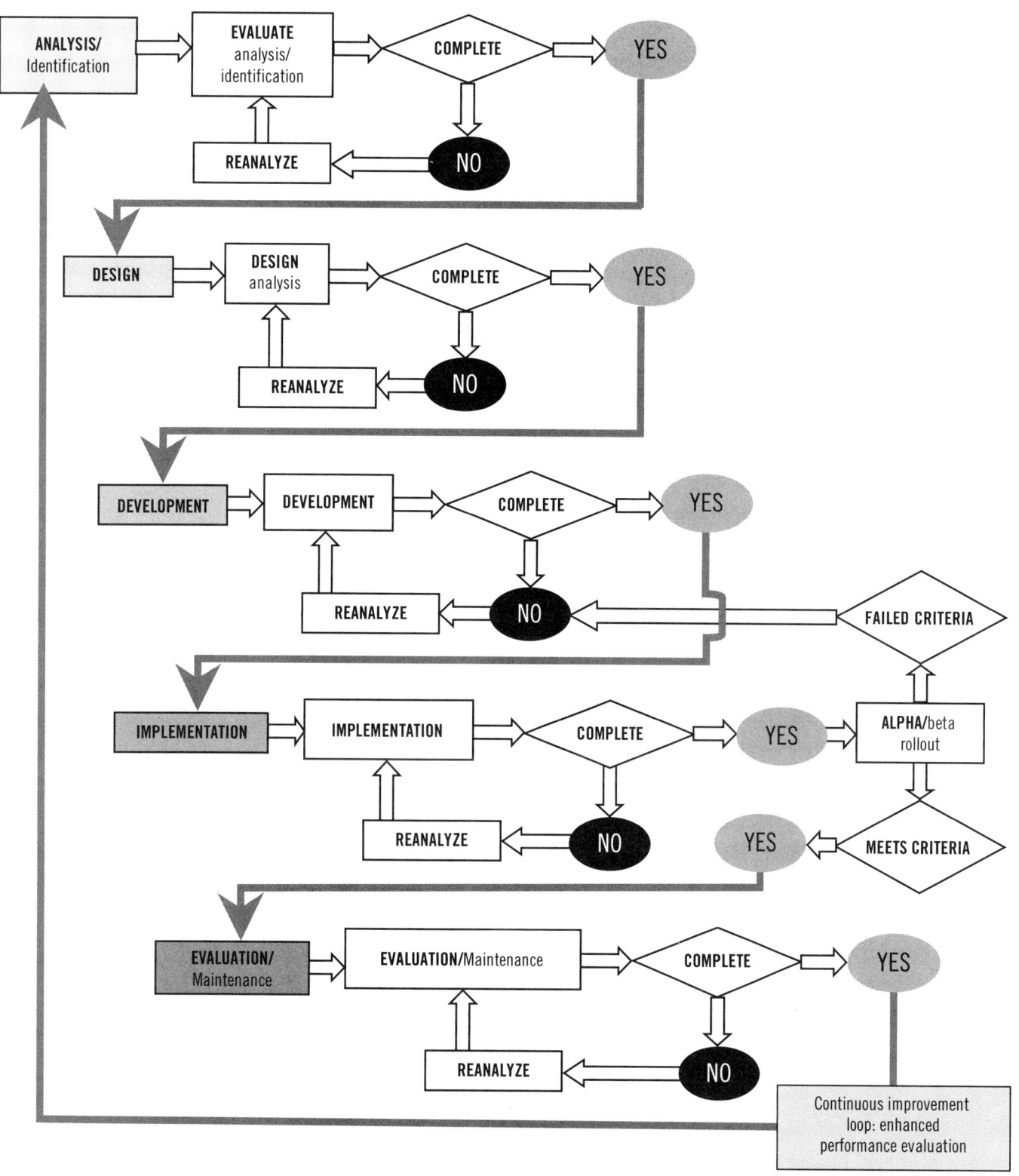

Figure 31–1. Training cycle.

as required to ensure all aspects of the category have been addressed. Following the flow of the training cycle (see Figure 31–1), an evaluation of the training cycle system is accomplished as a last step and on an ongoing basis to ensure that it meets or exceeds the desired results. This is sometimes referred to as the maintenance step, whereby evaluations are analyzed and improvements made based on the outcomes of the analysis.

This process is directly tied to the safety professional's four primary functional areas, specifically the first: anticipate, evaluate, and identify hazardous conditions and practices. Hence, a safety professional's overall success lies in the success of the training cycle. It should be noted that the training development cycle should meet the same criteria whether it is internally or externally sourced.

Analysis/Identification

In the analysis/identification category, the primary focus is on the foundational aspects of what will ultimately constitute the bulk of the decisions that will drive training content and delivery. In order to create an effective safety training program, the safety professional must first begin by analyzing and identifying the objective of the training to be designed. Is the training being developed in response to a regulatory requirement, because of increased risk associated with a task, as part of a performance improvement initiative, or as a specific training course as it relates to a safety and health program or to a specific job or task? The objective can be as simple as a new regulatory requirement requiring a certain number of hours of training to be provided to employees performing certain jobs or tasks. In these cases, much of the course curriculum/topics to be covered will likely be provided. More difficult cases require more in-depth investigation, such as cases where management has set new, more aggressive goals to reduce on-the-job injuries and lost workdays. In these cases, more detailed analysis is required to determine current worker performance and compare it to desired worker performance to identify training needs and goals. The curriculum will then need to be developed specifically for the intended purpose of improving current worker performance as it relates to the goal. For cases where the curriculum is not predefined, a broad range of analysis may be required as previously discussed under Training and Nontraining Solutions.

For the purposes of this chapter, analysis specifically related to training will be covered, but it is important to note that the safety professional has many other methods of analysis available that may be more relevant and may require incorporation into the analysis/identification category of the training development cycle in order for the training outcome to have maximum impact. In completing a training needs analysis, any additional analysis will be reviewed and incorporated. Even in the event that the required training is a regulatory requirement with all aspects of the training defined, the training needs analysis will provide the basis for the analysis/identification category from Figure 31–1.

Training Needs Analysis

In order to create an effective safety training program, the safety professional needs to assess current worker performance and compare it to desired worker performance to identify training needs and goals as they relate to the deficiencies found—hence, the need to have learning and performance objectives as previously discussed. A training needs analysis is important to an organization because it helps to:
- distinguish between training and nontraining requirements
- distinguish and understand the problem or need before designing a solution
- save time and money by ensuring that solutions address the problems that need to be solved
- identify factors that will affect the training before its development.

A training needs analysis is the process of determining the "who, what, when, where, and why" of performing the work process effectively and safely. Key points to be considered when completing the training needs analysis include the following:
- knowledge and skills
- capacity
- standards
- measurement
- feedback
- conditions
- incentives.

The analysis ensures that training addresses the performance gap as it relates to safety or health issues. Identifying specific training needs is the first phase of developing training that will ensure an effective and cost-efficient deployment. In analyzing training needs, the safety professional determines who may need training, what specifically they need to be trained to do, when they need to be trained, where they need to be trained, and why they need the training. This may be done by answering the following questions:
- Who does the job?
- What do they do in their jobs?
- When do they do the job?
- Where do they do the job?
- Why do they do the job?
- How do they do the job?

Primarily, training needs are analyzed by monitoring performance through observations, interviews, quality scores, audits, and questionnaires to determine the specifics of the problem or need that must be addressed through the required training. The training needs analysis may also show that other, nontraining solutions are necessary. When a regulatory agency is not involved, the safety professional must determine whether training is even needed before creating a complete picture of the training need, which may, in fact, only be perceived. This is one of the main reasons safety professionals must understand nontraining solutions. In the initial stages of an assessment, always consider the many solutions that could solve the problem. Training for the sake of training can and often does increase employee frustration, lower morale, and increase nonperformance and lack of compliance. Alternatively, mandated regulatory training is often seen as noncritical and nonessential to employees. In some cases, if a training needs analysis is not accomplished, training is provided to personnel who, under the regulation, do not require the training. This can only serve to diminish and negatively affect the overall reputation of the safety and health programs and the training programs and employees' perception of the organization.

In the training needs analysis phase, the safety professional should identify training goals—statements that describe how the training will satisfy the safety and health need or solve the problem, as well as state the general purpose of the training. Training goals provide direction for further analysis and help develop learning objectives. Training goals are also valuable tools for the safety professional to use in communicating the purpose of training to others. A training goal should clearly describe how the training will solve the safety or health problem. If the training needs analysis identified nontraining solutions, the goals must describe how the nontraining solution will solve the safety or health problem.

While the training goals and objectives are being identified, the safety professional/training developer should also identify the specific characteristics of the learners (workers) who will be involved in the training. This is the identification stage of the training development cycle analysis/identification category. These characteristics include general demographics; learners' preferences, attitudes, knowledge, skills, and language and reading comprehension levels; and their previous experiences. The results of the analysis/identification include the following:

- specific characteristics that will influence the design and delivery of the training (also called an audience identification/description)
- recommendations on how the design of the training will accommodate the specific characteristics of learners.

Key points to be considered when completing training needs analysis include the following:
- knowledge and skills
- capacity
- standards
- measurement
- feedback
- conditions
- incentives.

Training focused on the effectiveness in solving the problem and influencing behavior is also called *performance-based* training. Performance-based training is a learning experience (training) that is implemented to solve a specific on-the-job problem or to encourage a specific behavioral change. Performance-based training can be measured or evaluated by analyzing a worker's performance. This training is directly related to the job the worker is expected to perform, linked to the organization's safety and health goals, measured and evaluated by the worker's observable performance, and created using a systematic approach illustrated in Figure 31–1. In determining training requirements for performance, a performance analysis should be conducted. A *performance analysis* should be conducted by a safety and health professional for each specific job or task to determine what knowledge, skills, and attitudes people need to develop to meet the safety and health goals.

There are two other types of analyses conducted in the needs assessment phase: job analysis and setting analysis. *Job analysis* is the process used to determine the procedures, decisions, knowledge, and skills required for a worker to perform a job function. *Setting analysis* is the process of identifying specific characteristics of the training environment (where training will occur) that will affect training. Ideally, it is best to obtain setting information before designing the training. However, this is not always possible. The training course should be flexible enough to be delivered in a variety of settings. Chapter 32, Media, provides further information on the types and available training delivery methods. Once all the analysis has been completed, it can then be used to determine what it is the training is seeking to change, improve, or reinforce as it relates to the particular job or task, also known as the performance objectives.

Written Learning and Performance Objectives

Objectives are especially critical in the safety and health arena because there is too much risk (potential for injury, illness, or loss of life) to workers to allow a "hit-or-miss" approach. Additionally, a vast amount of time and money is potentially wasted if training objectives are not determined in advance and training is conducted simply to place a checkmark as having performed the training. With the completion of the analysis and determination of the general or specific areas in which the learner needs to become proficient, the learning and performance objectives can be developed. The importance of the learning and performance objectives cannot be overstated and are usually considered the most critical step in the instruction system design.

Learning and performance objectives provide the safety professional with a structure or framework for developing training that allows for everyone involved with the training—as well as those investing in the training—to know what needs to be taught, what is expected to be learned, the performance associated with the learning, and whether the training is worth the investment in time and money. Objectives are also important guides in the following areas:
- selection and development of course content
- selection and development of learning activities
- measurement of the learner's performance.

Unlike goals, objectives are measurable and observable. Objectives differ from learning activities in that they describe the results, not how to get results in the classroom. The four parts of an effective objective (sometimes referred to as the *ABCD method* of objective writing) are as follows:
- audience (learners)—always identify the audience
- behavior—identify what learners must do in order to demonstrate mastery
- condition—identify what learners will be given or not given in order to perform the behavior
- degree—specify how well the audience members (learners) must perform the behavior.

Thus, objectives are important to the workers being trained because they provide a target for performance (or behavior), they help learners identify the performance areas being evaluated, and they inform learners how they will be evaluated. Objectives have three primary parts, which include the job or task (observable action), the environment/conditions, and the standard to which the job or task under the conditions must be performed. Each of these parts can be defined as follows:
- *Job or task* (observable action) is defined as what the learner must be able to do.
- *Environment/conditions* is defined as the environment and/or conditions in which the job/task is expected to occur.
- *Standard* is defined as the quantity and/or quality that is considered acceptable.

Generally, most people will assume that 100% is the standard performance if the objective does not specify a level.

Performance can be measured in terms of quantity, quality, or both. The current philosophy behind loss control supports measuring quality and quantity. Because of human nature, safeguards should be implemented to provide for the highest achievable performance level on a continuous basis within the training program development as well as for the job or task that the training aims to provide, set, and/or change. Once this determination is made, then the level of proficiency required to satisfactorily complete the training program can be determined. For example, if a training program has been developed for the use of a radioactive material in the performance of radiographic tests, is 75% an acceptable standard or level of competency? In this case, specifically as it relates to the training of the actual exposure of the radioactive material outside its contained area, 100% proficiency is required to ensure the safety and health of everyone on the team and all those who may be in the general area. Would the same standard or level of competency be required for a training program on the safe handling of food? These are the questions that the safety professional needs to answer and reevaluate on a continuous basis as they relate to the performance outcome/goals that were originally set by the organization and the regulatory requirements for minimum levels of acceptable safety standards.

Design

In the design category, the primary focus is on the design of the course, which includes, but is not limited to, course delivery design, course content, major sections, and course sequence. Additionally, the safety and health professional or contracted ISD identifies instructional goals, chooses training media, and most importantly, chooses the sequence in which workers will learn new information and skills. This work is based on the analysis/identification stage (which includes training needs analysis results, an audience analysis, and specific performance-based objectives).

After selecting the course delivery method(s), the types of training materials required can be determined. A major carryover from the analysis/identification stage is the written learning and performance objectives, which are now expanded to include subcategories as required to meet the overall learning and performance objectives. The design category cannot be fully addressed without the analysis category having been properly and thoroughly completed because the design hinges on the outcomes of the analysis section. The design section should take advantage of all available resources, such as company policy and procedures manuals, regulatory brochures, and pamphlets in the design of the training.

The use of written policies and procedures manuals has been a traditional means of providing information to new employees. The company should exercise care in preparing the manuals to ensure that the information is complete and easy to understand and that all rules are enforced. Policies and procedures manuals should cover items such as first aid, personal protective equipment, electrical safety, and housekeeping. These manuals can be used as key references by the participants during training and should be used as the major source for references as the training is being designed and developed. References should always be included in all materials, thus providing the necessary recognition but also providing the training with credibility.

The design stage includes but is not limited to the following:
1. literacy level consideration
2. adult learning needs
3. training delivery method selection
4. evaluation design.

Literacy Levels

Many workers read, write, use math skills, comprehend information, and use problem-solving abilities at a level called low literacy or mid-literacy. *Low literacy* generally refers to workers whose skills and/or abilities are at or below a fourth-grade level. *Mid-literacy* generally refers to workers whose skills and/or abilities are at a fifth- to eighth-grade level.

Workers with low literacy or mid-literacy may have trouble understanding large amounts of information that is presented in complex textbooks, manuals, and lectures. This must be taken into consideration when designing and delivering safety and health training programs. For instance, training can be modified to be written at a fourth- or fifth-grade reading level or to rely less on printed material. Use of symbols and illustrations in safety and health training and promotion can minimize the risk of the message being misunderstood. Incorporating course assessments announced as part of the course objectives also helps provide feedback on comprehension and appropriate competency target levels. (See Evaluation/Maintenance later in this chapter.) Therefore, while the analysis phase determines the specific target audience as it relates to the job or task, the design phase must take into consideration and determine the appropriate literacy levels for the training design. As with literacy levels, the design phase needs to consider the audience. As a safety and health professional, the primary audience—barring a few exceptions—will normally be adults. Hence, proper consideration to adult learning needs to be given when designing training.

Adult Learning Needs

A major factor in the success of training is the extent to which adult learning needs are taken into consideration. Safety and health issues can be technical and difficult to learn. Also, people may see safety and health training as a disruption in their busy workday that does not really relate or add value to the work they are doing. To keep participants motivated and involved in safety and health training, the participants' learning needs must be considered. To accomplish this, adult learning principles must be used in the design and delivery of training. Different people learn in different ways, and using a combination method based on adult learning concepts such as hearing, seeing, and doing is key when designing and delivering training.

The following percentages demonstrate potential learning retention based on the type of sensory concepts designed into the training program:
- reading—10%
- hearing—20% (20% of humans are auditory learners)
- seeing—30% (30% of humans are visual learners)
- hearing and seeing—50%
- speaking—70%
- speaking, hearing, and touching or doing an activity—90%.

Additionally, research conducted by the National Training Laboratories in Bethel, Maine, found retention percentages as represented in Figure 31–2. These figures are a culmination of various methods of teaching and the average retention rates. Therefore, it is important to understand and apply all available adult learning concepts in the development and implementation of a training program.

More generally applicable are the four needs common to all adult learners. This information is adapted from Clay Carr, a widely respected training professional and author of the book *Smart Training: The Manager's Guide to Training for Improved Performance*. These principles are generally accepted by most adult learning experts:

1. **Need to know why (WIFM):** Adults need to know why they are learning a particular topic or skill because they need to apply learning to immediate, real-life challenges. This is also known as the WIFM statement, or "What's in it for me?"
2. **Need to apply experience:** Adults have experience that they apply to all new learning.
3. **Need to be in control:** Adults need to be in control of their learning.
4. **Need for success:** Adults want to learn things that will make them more effective and successful.

These principles should be applied to any safety and health training developed, contracted, or designed for electronic learning applications.

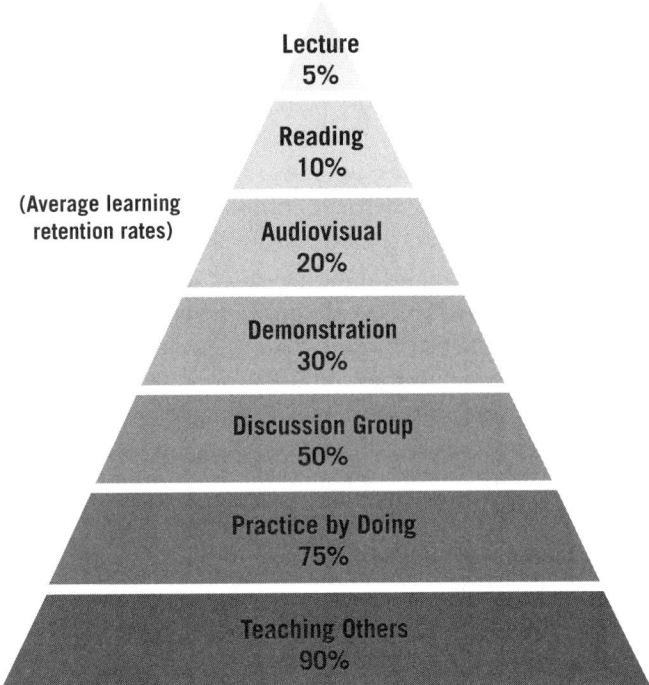

Figure 31-2. Learning pyramid.

Need to Know Why (WIFM). Adults need to know why they are learning a particular topic or skill because adults relate to the learning better once they understand why the training is advantageous. They also want to know how the training or learning can be applied immediately to real-life challenges. Therefore, they will learn best when:
- they see clear demonstrations of how the training directly applies to their jobs.
- they have opportunities to apply the new information or skill to solve problems during the training.
- they have opportunities to think about how the new information or skill can be used.

An example of meeting this need is a supervisor demonstrating how to safely operate machinery; providing real-life experiences and observations on the importance of proper use, including examples of accidents, illnesses, and injuries associated with improper use; and then providing feedback while watching workers operate the machinery.

Need to Apply Experience. Adults have experience that they apply to all new learning. Therefore, they will learn best when:
- they have opportunities to share their experiences during training, using comparison and contrast, relationships, and association of ideas.
- they have opportunities to think about how the new information or skill relates to their past, present, or future experiences.
- their experience, and opinions formed from their experience, is taken seriously by the facilitator.

For example, during training on personal protective equipment (PPE), provide time for participants to talk about their own PPE practices at their facility. They can also share their observations on the practices of their co-workers. Allow discussions on challenges, hardships, dislikes, concerns, and any other negative information and be prepared to provide positive solutions or negative action result examples that further support the need and requirement to wear PPE. These discussions are not only healthy, but also increase the credibility and awareness of the participants. Always work to resolve issues and concerns by answering honestly and following up as required on unanswered questions.

Need to Be in Control. Adults need to be in control of their learning. Therefore, they will learn best when:
- they choose, or at least influence, the training they receive.
- they are in a flexible learning environment.
- they are actively involved in the learning process rather than passive receivers of information. (The facilitator is a guide for the learning experience.)
- they have opportunities to voice concerns and see them addressed and can take part in activities such as group discussions, role-playing, and simulations.

For example, at a fluorescent light factory, workers are involved in a safety and health committee. Their responsibilities include assessing workers' needs for training, identifying appropriate training solutions, and evaluating the training once it has been delivered. Committee members are actively involved in the training and promote it as part of the committee and as engaged workers.

Need for Success. Adults want to learn things that will make them more effective and successful. Therefore, they will learn best when:
- they know why they are taking the training and how it will benefit them.
- they can ask and answer the questions, "What is in it for me (WIFM)?" "How will it help me get ahead?" "What hazards will it help me control?"

An example of meeting this need is a safety management techniques course that includes action planning for future personal and professional development. A clearly stated and articulated objective at the beginning of every course or section is paramount and helps set the stage for all the learners by answering the WIFM question. Hence, while the analysis/identification stage defines the learning and performance objectives, the design stage may further break down those defined overall learning and performance objectives into more detailed objectives that relate to specific topics needed to create the basis or foundation to achieve the overall learning and performance goals.

Training Delivery Method Selection
Once the training needs analysis has been completed and the performance objectives have been written, the safety professional should carefully consider the most effective delivery method based on the audience. In selecting the delivery method, the safety professional may consider analyzing the return on investment (ROI) to determine the most cost-effective training delivery approach. This analysis is not necessarily the best, but it will show the economic impact of the different options. The safety professional should consider the following when selecting the delivery method type:
- return on investment analysis
- delivery method yielding the highest performance results
- audience skill and literacy level.

This list is not inclusive and should be used as a general guide. Safety and health risks as well as regulatory requirements are also important factors.

There are three major types of performance-based training delivery methods:
1. *Instructor-led training*—This training can be presented in one of two ways. The first is presented in a classroom-like setting in which everyone follows the pace set by the instructor (also commonly called facilitator-led or trainer-led training). This type of training generally includes participant materials. The second is presented via an online platform such as Online Meeting or WebEx. This platform mimics the classroom-like training except that the training is all conducted via a computer through the use of presentation slides, video, webcams, and interactive use of computer software that all participants see and is controlled by the trainer/facilitator. The trainer has the ability to pass on the control and has additional functionality that is not available to the participants.
2. *Self-paced training*—This is usually completed on an individual basis, by working through a textbook and/or workbook. Self-paced training may also be completed through computer- or web-based training or computer-assisted instruction, in which the learner is guided through the course through the use of a computer or online software. The software provides the course content, guides learning activities, and administers tests. Computer- or web-based training, as part of a learning management system, can also provide the safety professional with a quick list of who has completed the training, and some programs can provide tests, scores, time to complete, number of attempts, and lowest and highest scores achieved. Computer software and the data available to the safety professional continue to evolve and expand, making analysis of the training much easier and more in depth.
3. *Structured on-the-job training*—This training, also called on-the-job training (OJT), is similar to instructor-led training in that it is conducted by an instructor, but it is conducted at the learner's workplace. The instructor is usually a trained supervisor or employee who acts as a coach or guide. This type of training may or may not include participant materials, but it does incorporate all resources normally used during the course of completing the task(s) to be performed.

Evaluation Design
Evaluations are a necessary tool in eliciting information that can then be used to determine required changes to the training program design, can help determine effectiveness, and can be referred to as a major source of information. Constructing a questionnaire requires care and diligence to ensure that the questions asked are pertinent and useful and provide enough detail to enable tabulation and additional analysis. There are four primary types of information that can be discerned using questionnaires, and they include:
1. behavior—what people do
2. attributes—personal and demographic characteristics
3. beliefs—what people think, attitudes, opinions, etc.
4. knowledge—what people know and understand.

Evaluation design should always be accomplished such that the ISD, safety and health professional, and management have direct input to maximize data collection with purposeful information gathering. The evaluation design itself must be tailored to the culture of the organization and must consider all aspects of the training design such as literacy, adult learning concepts, and the like.

Training Methods
Training methods fall under the design stage of the training cycle. Instruction system designers as well as safety and health professionals can select one of several different training methods when preparing a program. Each method has strengths and weaknesses. The technique selected should be determined by the objectives to be met, type of student

participation, time allocated, facility being used, and equipment available. Everyone learns at different speeds and through different methods. Trainers must have the training and teaching skills to address these elements of human behavior. The following are the three most common types of training techniques used in industry:

1. on-the-job training (OJT)
   a. job instruction training (JIT)
   b. coaching
   c. internship
   d. apprenticeship
   e. mentoring
2. group methods
   a. conference/meeting/tailgating
   b. brainstorming
   c. case studies
   d. incident process
   e. facilitated discussion
   f. role-playing
   g. lecture
   h. question-and-answer (Q&A) sessions
   i. simulation
3. individual methods
   a. drill
   b. demonstration
   c. testing
   d. video-based training
   e. computer-assisted training
   f. reading
   g. independent study
   h. seminars and short courses.

These methods are discussed in detail in the following sections.

On-the-Job Training

On-the-job training (OJT) is widely used because it allows for the workers to be productive during the training period. However, three considerations must be addressed when using OJT:

1. The trainer must possess proper training skills and have been trained on the OJT being provided.
2. A training program should be developed to ensure that all workers are trained in the same way to perform their tasks in the safest and most productive manner.
3. Adequate time must be allotted to the trainer and trainee to be sure the subject is well covered and thoroughly understood.

Advantages to on-the-job training include the fact that additional resources and/or tools are not needed because the training is accomplished with resources and equipment normally used during the course of performing the job. On-the-job training is normally conducted where the job is performed, which provides further trainee familiarity and increased comfort in completing the job function. On-the-job training also has inherent disadvantages because the trainer must be well trained and vigilant while performing the job tasks, especially for work involving safety-sensitive tasks.

Job Instruction Training. A variation of OJT is known as job instruction training (JIT), also referred to as the four-point method. Instruction is broken down into four simple steps:

1. preparation
2. presentation
3. performance
4. follow-up.

This method of OJT has been highly successful. Workers are taught each job skill from a formal schedule of training. The program is adjusted to each student so that workers learn one task before beginning the next. In all training programs, selection of the trainer is critical to program success. The one-on-one relationship between trainer and trainee allows for better consistency and communication. In the four-point method, this trainer–trainee relationship works in the following ways:

Step 1: *Preparation*. During the preparation stage, the trainer puts the workers at ease. The trainer explains the job and determines what the workers currently know about the subject. This stage also includes preparation of the proper learning/working environment.

Step 2: *Presentation*. In the second step, the trainer demonstrates the work process. The student watches the performance and asks questions. The trainer should present the steps in sequence and stress all key points.

Step 3: *Performance*. In the next step, the worker performs the task under close supervision (Figure 31–3). The trainer should identify any discrepancies in the work performance and note good performance. The worker should explain the steps being performed. This ensures that the worker not only can perform the task but understands how and why the task is done. This stage continues until the trainer is satisfied with the worker's competence at the job.

Step 4: *Follow-Up*. The final step is follow-up on performance. The trainer and/or the supervisor must monitor the worker's performance to be sure the job is being performed as instructed and to answer any questions the worker may have.

Of the various OJT methods, JIT is probably the most flexible and direct. By practicing, the trainee is expected

Figure 31–3. This supervisor is explaining an enclosed operation to one of his workers. The supervisor is instructing the worker in the correct and safe operation of the machine and will make frequent follow-ups on his progress until certain of his proficiency.

to develop and apply the learned skills in a typical work environment while under the guidance of a trained worker/trainer. The trainer must know the job thoroughly; be a safe worker; and have the patience, skill, and desire to train.

The advantages of training in the JIT method are as follows:
- The worker can be more easily motivated because the training/guidance is personal.
- The trainer can identify and correct deficiencies as they occur.
- Results of the training are immediately evaluated because the worker is performing the actual job on actual equipment. The work performed can be judged against reasonable standards.
- Training is practical, realistic, and demonstrated under actual conditions. Workers can easily ask questions.

Timing is an important element in this type of training. The trainee can receive help when needed, and the trainer can provide feedback as the training progresses. The trainer can also determine when the trainee is ready to move on to new levels of training. It may be helpful to prepare a chart of tasks or subjects for which workers must receive training. This chart will make it easier to keep track of workers' training progress and the levels of competence they achieve.

Coaching. Of the other methods of OJT that can be used, the most common is known as coaching or the buddy system. This system is considered effective in some situations, but the following challenges are associated with it:

- Trainers may be selected for their availability rather than for their training skills. A trainer who lacks basic teaching skills can undermine the entire orientation of a new worker.
- Each trainer may have his or her own way of performing the tasks being taught. This lack of continuity can make it difficult to control hazards in the workplace and can lead to many accidents.
- Key elements of orientation can be overlooked in the training program and may not be realized until an incident or accident occurs.
- Poor techniques or bad habits can be spread from one worker to another. Shortcuts or safety violations can be demonstrated to new workers as the "way we do it."
- Safety performance may not be emphasized during the training. Job performance should never be separated from safety standards in any training provided to workers.

Advantages and disadvantages of job instruction training mirror those previously covered under On-the-Job Training earlier in this chapter.

Internship. Internships are usually short-duration programs that last no more than 1 year in total. They are normally accomplished after the completion of the academic portion of the program, but sometimes they are integrated into the last portion of the program. These programs, unlike an apprenticeship program, are normally comprised of two different entities working together to provide a collated program of study for completion of a degree or certification. Successful completion of the internship is required to meet the requirements of the degree or certificate to enable the learner to advance into the workforce.

Apprenticeship. Apprenticeship training is one that combines both the on-the-job training discussed earlier along with additional training specific to the job or task being performed. Unlike internships, apprenticeship programs are much longer in duration, lasting anywhere from more than a year to upwards of 4 years. This type of training is sometimes referred to as dual training because of the training delivery combinations of both on-the-job (practical) and internal training instruction, which usually consists of in-class training but can also be online, computer-based, web-based, or self-paced training. Apprenticeships also differ from internships in that the worker is usually paid, with salary increases associated with satisfactory completion of each level within the program. The Office of Apprenticeship within the Department of Labor's Employment and Training Administration provides a number of registered apprenticeship programs that are available for special incentives. These apprenticeships are approved by the government and may be eligible to receive tax benefits and workforce development grants. Registered apprenticeship programs offer career training in areas such as carpentry, dentistry, and law enforcement, to name a few.

Mentoring. Mentoring is a one-on-one professional relationship in which an experienced person (mentor) provides direct and personal guidance to assist another person (mentored person) in developing specific skills and knowledge to enhance the less-experienced, mentored person in his or her professional and personal growth. While mentoring is, in many cases, an organizationally driven initiative, it is seldom that the mentor receives training on mentoring. Like trainers, mentoring is a skill that is obtained through training and experience. It is a long process and not one that should be used if immediate results are sought. Mentorships can last years as the mentored individual is mentored to the level of competency usually sought after for, say, higher-level leadership positions. For the scope of a safety and health professional, mentoring should be done frequently and with very defined and achievable goals/objectives.

Group Methods

Group techniques encourage participation from a selected audience. These methods allow trainees to share ideas and evaluate information and, depending on the organization, could include being actively involved in the planning and implementation of company policy. Several types of group training are used, but all require skilled facilitators to be successful.

Conference/Meeting/Tailgating. The conference method of training is widely used in business and industry for education-sharing purposes. This method is readily accepted because of the knowledge each participant brings to the group. In this process, sometimes called meetings, the trainer/facilitator controls the flow of the session as participants share their knowledge and experience. The skill of the facilitator can mean the success or failure of these sessions. Facilitators must use various techniques to draw information and opinions from members. The number of people involved should be limited to allow open discussion from all participants. The opinions of each member should be recorded, and a summary of the group's conclusions should be provided to those who were involved as well as to those who should be kept abreast of the information (see Figure 31–4).

The conference technique is also a valuable method of problem solving. There are several situations in which a safety professional can use the knowledge and expertise of members of the organization to address safety and health

Figure 31–4. In a participation class, the number of people in the conference group should be small enough so open discussion can occur.

issues. At the beginning of the conference, members should identify their goals and expectations for the session. For example, if a conference has been called to identify possible solutions to a safety and health concern, the group must understand that they are to make recommendations only, not to establish policy or procedures. By defining the actual role of the conference at the start, the group can avoid misconceptions and misdirection. When a group is asked to make recommendations, participants should be kept informed of the results of those recommendations.

If management fails to establish these ground rules and to provide follow-up information, the members may feel that their efforts are ineffective. On the other hand, proper control and guidance of a conference can ensure its success and make it a gratifying experience for the participants.

This method has several advantages, including participation of the entire group and the fact that small groups can reach consensus. Disadvantages include the fact that large groups may discourage everyone's participation, the group needs to have a purpose that everyone agrees on, and consensus may be difficult if not impossible to reach in large groups.

Brainstorming. Brainstorming is a technique of group interaction that encourages each participant to present ideas on a specific issue. The method is normally used to find new, innovative approaches to issues. There are four ground rules:
1. Ideas presented are not criticized.
2. Freewheeling creative thinking and building on ideas are positively reinforced.
3. As many ideas as possible should be presented quickly.
4. Combining several ideas or improving suggestions is encouraged.

A recorder should be selected to write down all of the ideas presented. The moderator must control the flow of suggestions, cut off negative comments, and solicit ideas from each member.

The advantage of brainstorming is that it allows ideas to be developed quickly, encourages creative thinking, and involves everyone in the process. The group can go beyond old stereotypes or the "way it's always done," but success in this method depends on the maturity level of the group. Additionally, the technique can lead to unfocused ideas and hence the need to limit the sessions to short durations to prevent straying from the known reality of ideas that have true potential of solving the problem.

Case Studies. Case studies are written descriptions of business decisions or problems that learners will use as a basis for demonstrating predetermined skills and knowledge. They have two distinct advantages. First, case studies provide an opportunity for the learners to use skills and knowledge acquired during the course. Second, case studies can serve as an evaluation tool for trainers to measure the degree of proficiency attained during a course or module. Here are some other key benefits:
- During a case study, students begin to internalize the critical principles being taught and retain the information for longer periods of time.
- Case studies emphasize practical or critical-thinking skills.
- The student's perspective is broadened through interaction with others.
- Case studies encourage reflection, application, and analysis.
- Case studies reinforce the value of discussion and interaction with others.

Planning, thinking about, and adhering to the instructional objectives are paramount in designing an effective case study. The key is to start at the end and work backward toward the beginning. Ask the following questions:
- What questions must be answered?
- What skills or knowledge should be exercised?
- What specific performance objectives should be measured?
- What learning objectives are to be addressed by the case study?

A case study may involve an actual situation or be fictitious. The goal of the activity is to develop group members' insight and problem-solving skills. The case study is normally presented by defining what happened in a particular incident and the events leading up to the incident. The group is then given the task of determining the actual causes or problems, the significance of each element, and/or acceptable solutions.

The advantage to the case study method is that it enables the trainee to develop analytical and problem-solving skills, investigate and solve complex issues, and apply learned knowledge and skills. Disadvantages include the trainees' potential inability to understand or picture the relevance to the problem because of immaturity or lack of information.

Incident Process. The incident process is a type of case study in which the group works with a written account of an incident. The group is allowed to ask questions about facts, clues, and details. The trainer provides the answers to the questions, and the group must assemble the information, determine what happened, and arrive at a decision. The facilitator must guide the group to prevent arguments and to prevent one or two members from dominating the discussion. This is a useful training method that encourages employee participation in the accident prevention program. Situations can be real cases in the company or potential hazards that exist.

The facilitator/moderator must be able to control the group process and progress and to prevent the group from missing the true or root causes. For example, suppose an employee was struck in the eye by a foreign body, and an investigation revealed that the employee was not wearing safety glasses while operating a bench grinder. The group must seek the root cause of the accident and not settle for the common conclusion that "the worker failed to follow procedures" or "the worker failed to wear eye protection." The group must specify why the supervisor or management failed to enforce the proper procedures (assuming there was an established procedure). Why did management allow this lax supervision?

Facilitated Discussion. Facilitated discussion or dialogue is the management of discussion about the course content so that the learning objectives are met, the discussion flows logically from topic to topic, and the applications to the learners' jobs are made clear. Facilitated discussion requires that the trainer/facilitator have the skills to accept all ideas and contributions as valid, show how they relate to the course objectives, and manage the time element and the flow of information to meet the course objectives.

The benefits of facilitated discussion include the following:
- It ensures that the learners are involved and challenged.
- It builds a bond between the trainer/facilitator and learners that encourages the free exchange of ideas and information.

Role-Playing. Role-playing is effective for evaluating human relations issues. Group members attempt to identify the ways people behave under various conditions. Although this technique is not an effective method of problem solving, it can uncover issues not previously considered. This method is particularly helpful in identifying and changing personnel issues such as poor morale or negative attitudes.

Advantages to role-playing include giving trainees an opportunity to assume the role of others, thereby obtaining better understanding and a different perspective on the position, task, or responsibility. This method of learning also offers an opportunity to practice skills and gain knowledge in a controlled environment. However, this method is difficult to apply effectively to large groups, and trainees may feel self-conscious and even threatened. A good facilitator is essential to making this method successful.

Lecture. Using the lecture method, a single person can impart information to a large group in a relatively short time. This method is normally used to communicate facts, give motivational speeches, or summarize events for trainees. There is little time or opportunity for interaction by the attendees, and follow-up for these sessions must be well planned in advance to be successful.

Advantages to this method of learning include the fact that the material is presented in a direct and logical manner. It enables the trainer to provide experiences that motivate and inspire the group, whether small or very large, effectively. The trainer must have excellent oral skills, and assessments are necessary to effectively gauge the learning.

Question-and-Answer (Q&A) Sessions. Normally, Q&A sessions follow training periods, after the trainer has summarized the material presented. Workers can use this method to clarify individual concerns or facts. However, workers often need time to prepare and organize their thoughts before they can ask questions. When workers must absorb a large amount of material, allow them time to reflect on or apply the knowledge and to formulate questions. The trainer can plan to have a follow-up session or allow workers to present their questions personally. Question-and-answer sessions are helpful in clarifying issues of policy or changes in schedules or events.

One advantage to this method is that it allows everyone to participate in an active process. The disadvantages include the fact that Q&A is not practical with large groups and can lead to a few trainees dominating the session. This type of session can also can be time consuming, get off track, and yield little participation from most of the trainees if the trainer/facilitator does not work the audience for best participation.

Simulation. When actual materials or machines cannot be used, trainers can use a simulation device. This method is used effectively in aircraft pilot training, railroad engineer training, and other applications in which external and internal inputs can have a varied number of possible outputs, all requiring the same outcome. Various methods are employed in management training programs as well, such as the "in-basket technique," "war games," and manufacturing-line process simulations. One simulation demonstrates the loss of eyesight to workers to encourage them to wear safety glasses on the job. The only limit to this technique is the trainer's creativity.

Simulation is most effective when workers can participate. Careful planning and attention to detail are required. The initial costs of these sessions can be high because of the equipment required and time involved in conducting training.

Individual Methods

Individual methods entail learning and/or training that is individualized to allow for each learner to go at his or her own pace, taking into consideration the differences in learner abilities. With individual learning, the trainer/facilitator must consider and cater to the individual needs of the learner.

Drill. Using the elements of practice and repetition, the drill method of instruction is valuable for developing worker skill in fundamental tasks and for performing under pressure. Workers required to perform in crisis situations should be trained under conditions that resemble the crisis as closely as possible. For example, when instructing workers in cardiopulmonary resuscitation (CPR), the trainer must try to instil the tension that workers will experience when they attempt to resuscitate a real victim. This method helps workers develop control and builds confidence; in real-life situations, they react competently and revert to the trained actions they have come to learn and know, rather than panicking or having knee-jerk reactions that result in the loss of critical time, energy, and potentially life.

Demonstration. As discussed in the section on JIT, the demonstration method allows the trainer to perform the actual task and then have the worker repeat the performance. Trainers must be sure the job is performed exactly as required to prevent workers from developing poor habits and performance deficiencies (and supervisors must see that employees follow the designated procedure). If the conditions used in demonstrations are not similar to the actual workstation or equipment on the job, this method will yield few if any desirable results.

Testing. Testing is normally used to determine whether workers understand the necessary information and can apply the knowledge when required. Developing good tests is a skill that requires constant review to ensure that training objectives are being met. Poor tests can reduce workers' morale, undermine training objectives, and yield poor training assessment data, preventing continual improvements of the training program. For additional information on testing, see the Evaluation/Maintenance section later in the chapter.

Video-Based Training/Podcast. An increasing number of training programs are designed as videotape presentations and, now more than ever, podcasts. These training formats are available on nearly any subject. This method of instruction is effective if properly applied. The use of videotapes/podcasts does not eliminate the need for professional instruction but can enhance a classroom presentation. Videos are available from the National Safety Council and numerous private companies, as are podcasts. Trainers should screen the training to make sure it fulfills the needs of the training program.

Production companies can produce training videos on budgets ranging from hundreds of dollars to millions of dollars. The first step before selecting a production company—or, for that matter, deciding to make a video for the organization—is to determine how the video will fit into the overall training program. The major factors are the course design, the purpose of the course, and how the objectives are met. The same consideration is taken when developing podcasts. Some distinct advantages to using a video in a training class are as follows:

- Video offers the learner an opportunity to see examples of tasks and processes being performed correctly.
- The difficulty of producing training videos has been diminished. Newer formats, such as Hi8 and S-VHS, digital media, small personal computer cameras, and now even smartphone technology, offer lighter cameras and greater ease of use.

If the video portion is the core of the training program, a "higher-end" type of production should be used. This usually involves hiring a production company (independent producer), professional scriptwriter, director, and editor. Expect postproduction costs (editing, graphics, animation, sound design, and music) to consume two-fifths of the overall budget. However, if the video portion of the training program is supplemental or ancillary, the safety professional can tape the material and hire a professional editor to assemble the footage.

This method provides an entertaining way to introduce content and usually keeps the trainees' attention. A disadvantage of this method includes raising too many issues to allow for follow-up, focused discussion. This can be remedied in the design phase by limiting the focus to help in retention.

Computer-Assisted Training. Interactive computer programs, also known as *computer-based training (CBT)* or *web-based training (WBT)*, are being developed for many areas of employee training. They enable workers to receive information by reading and/or watching a video presentation and then responding to situations and questions via a computer. For example, if a trainee is completing a computer training session and enters the correct response to a question, the computer will advance the program to the next section; if the wrong response is entered, the program will repeat the information and retest the trainee. The system is valuable for several reasons:

- Trainees can work at their own pace.
- Records can be automatically kept of all training. The amount and type of records maintained can be modified to meet company requirements.
- Correct answers are required before a worker can proceed to the next lesson, or remediation methods can be built into the program.
- Workers receive training as time allows, rather than having to meet training schedules.

With computer-assisted training programs, instructions can guide workers step-by-step through a curriculum designed to meet individual or company goals and regulatory obligations. The company can keep records of the amount of time each worker spends in training, the type of material or information presented, and the success of the training. This type of program works extremely well for organizations with small work forces or those that cannot remove large groups of workers from their jobs at any one time. (See also Chapter 32, Media.)

The differences between CBT and WBT have to do more with where the information for the training is housed than differences in the actual course content. Web-based training uses the Internet or offsite server to deliver content and allows for greater interactivity through live chats with an instructor and bulletin boards. Computer-based training provides multimedia instruction via CD-ROM on a desktop computer; it usually has no connection to the Internet and hence no interactivity.

Reading. Companies should provide employees with written safety materials such as monthly newsletters and safety magazines. In addition, organizations may establish a resource library or Internet access to allow employees to research information on subjects such as work procedures, safety, leadership, health care, and family or home safety. However, management must not assume everyone has the ability to read and comprehend all of the written material provided. Companies cannot replace instruction or training programs simply by handing an employee a training manual. Written material is meant to provide a supplement or reference for training, not replace it.

A major advantage of having a resource library for employees is that it enables motivated employees to continue learning and personal development when no other training is available. This type of training is usually also accomplished at the person's own pace and on his or her own time. The cost of implementing such a program varies, but considering advancements in technology, the availability of Internet and intranet services, and the affordability of website management software, it can be readily affordable and accessible to anyone from anywhere.

Independent Study. Home-study courses or correspondence courses are used by many companies. They can help employees advance within the organization or improve their knowledge of their jobs and the industry. A major advantage of this method of training is that the employee does not lose any time from work and can complete the course at his or her own pace. Another advantage is the low cost of implementing and maintaining home-study programs. Normally, they center on textbook assignments, followed by self-tests using multiple choice, true/false, fill-in, or essay questions. Although independent study sometimes includes computer-assisted training (discussed earlier), it more commonly involves textbooks. Several independent courses also provide laboratory or performance materials as part of the curriculum. Examples include television, radio, and computer repair programs that work on actual equipment. Some home-study programs come with media presentations for workers to view and follow.

The National Safety Council offers home-study programs for supervisor training—"Supervising for Safety" and "Protecting Workers' Lives." For more information visit: www.nsc.org.

Seminars and Short Courses. Seminars, short courses, and workshops for safety and leadership information and skills are offered by many colleges and universities as well as by insurance companies and private organizations. The National Safety Council offers onsite instruction for workers, supervisors, and management. Seminars, short courses, and workshops range from 1-hour to several-day sessions.

These type of courses, though traditionally more expensive, usually provide employees with specific, relevant, and in-depth information on the topics being discussed. These types of training programs are similar to facilitated discussions and lecture courses.

Development

In the development stage, the primary focus is on building the specific individual parts to the training such as presentation, handout materials, precourse work, course work, facilitator training guides, and props, to name a few. This is essentially the assembly phase of the training where all the design content is taken and organized into sections. The development stage must begin with a content outline. The content outline can be developed upon completion of the training needs analysis and the written performance analysis. The outline should be the basis for the development of the training curriculum and should contain the following:
- safety and health policy
- course objective
- statement addressing "What's in it for me?"
- introduction
- main performance objective training content
- review and summary
- assessment.

The organization of the content of the safety and health training program is achieved by outlining the content. To outline, the training developer selects an objective and identifies any actions the worker must take to achieve the objective and related topics the worker must know to accomplish the actions or change the behavior. The training developer then sequences the content in the appropriate order, preferably with input from the safety professional, other subject matter experts as required, and management as needed. With this information, the developer can move onto the types of materials or deliverables that will be used in the delivery of the training.

The materials or deliverables used in an effective safety and health training program should include (1) an introduction; (2) presentation of the information; (3) practice and feedback; and (4) a summary, evaluation, review, or transition phase. The introduction can be an overview, a rationale, a warm-up activity, questions, or a story or analogy. The information should be presented in a manner that helps the workers organize and remember the important facts. The following types of materials can be used to achieve this goal:
- table of contents
- text

- graphics
- examples
- job aids
- checklists
- graphs
- tables
- data
- reports
- relevant articles
- glossary.

Practice and feedback should be provided to workers as an opportunity to try out the new knowledge and skills and to receive information about their performance. Materials that help guide the process of practice and feedback include the following:
- activity directions
- activity time allowance
- activity worksheets
- question-and-answer worksheets
- business scenarios
- performance checklists
- sample solutions and expert answers.

The summary, evaluation, review, or transition phase should stress important training points, answer workers' questions, and make a connection to the next training topic or the job. Supporting materials for this phase include the following:
- text
- graphics
- question-and-answer sheets
- listing of next steps
- goal sheets for implementing new knowledge and skills on the job
- duration and recurrent frequency.

Depending on operational requirements, the timing of the content of the safety and health training program is determined either before or after the outline is created. The training developer sequences the content in the appropriate order (perhaps with input from the safety professional or management) and estimates the amount of time needed for each point. If more time is needed, the trainer can either request it or narrow the scope of the objectives. In addition, regulations may require that certain content be covered and that it be covered with a certain frequency. The safety professional is responsible for knowing and understanding these requirements and for providing this information for the ROI analysis and to the training developer as the training is designed.

DELIVERING TRAINING

Implementation

In the implementation category, the primary focus is on the implementation of the developed training. As the threshold between development and implementation is crossed, there must be an alpha and beta rollout of the training implemented to test the effectiveness and validity of the training and to assess the training development's ability to meet the design and criteria (see Figure 31–1). It should be noted that the trainer/facilitator (if applicable) must exhibit safety and health expertise, sound instructional skills, and flexibility. There are many cases where the trainer/facilitator does not have the necessary expertise or level of expertise, in which case it will negatively affect the learning and performance objectives. For training that will be conducted by a trainer/facilitator, it is especially critical for the alpha and beta testing to be conducted by an individual who has expertise in both safety and health as well as being a seasoned trainer/facilitator with experience delivering newly developed training. The latter is particularly important because all the participant feedback, trainer/facilitator comments, material discrepancies as they relate to all materials, and evaluations must be analyzed and compared with the original design criteria to ensure the developed training meets or exceeds the desired results.

Upon completion of the alpha and beta rollouts, and only after confirming that the developed training meets or exceeds the desired results, the implementation stage is released. Once the training has reached the implementation stage, it must be evaluated periodically as part of the continuous improvement cycle to ensure that it continues to meet the developed training needs or that it exceeds the desired results—that is, as maintenance. Thus, good training is never really fully developed because it should always be transitioning either as a result of development improvements and/or because of ongoing maintenance refinements to further ensure that the learning and performance objectives are meet and/or exceeded. Advantages of the alpha/beta rollout process include, but are not limited to, the following:
- verification of course and, more specifically, module/section timing and duration
- verification that objectives were covered
- verification of content order
- verification that material content is best suited for media method used.

The implementation phase, then, is much more than just rolling out the training; rather, it affords the opportunity to prove the design stage concept, verify the development stage, ensure the selected media are appropriate to the training being conducted, verify that the trainer/facilitator skill level is appropriate, and allow for continuous improvement of the system through feedback loops back into the development stage and the analysis/identification stage.

Evaluation/Maintenance

Evaluation Design

While the actual evaluation methodologies may be determined in the analysis/identification stage, it is in this stage that the methodology is expanded to incorporate best practices and potential benchmarking to enable further analysis after implementation. In their commitment to improving the knowledge, skills, and attitudes of workers, and as part of their roles and responsibilities, safety professionals must continually measure and evaluate whether workers are meeting or exceeding the learning and performance objectives. Performance testing is one way to measure whether learners have met the learning objective. Performance tests can be created only after performance objectives have been developed in the analysis/identification stage. There are three main types of performance tests: pretests, review tests, and posttests.

Pretests

A pre-test measures how well learners can perform objectives before training. Pretests are generally optional and are often provided as a means for learners to identify how much they already know before starting the training. Trainers also use the results to customize the program to meet audience needs. In other situations, pretest and posttest results are used to assess the learning progress made by an individual or class. Pretests can be a very useful tool as a means to benchmark the learner's competency before training, after training, and at some later point in time. Comparisons can be taken and reviewed against the learning and performance objectives to determine proper recurrent training intervals as well as to provide feedback to the organizational leadership on new, more challenging goals if they were to be adopted.

Review Tests

A review test measures how well learners can perform the objectives during training. Learners take review tests to determine when they are ready to move on to the next objective (topic or task). Review tests often take the form of activities or exercises. Participants' progress is informally assessed by the trainer/facilitator or by the participants themselves. While these tests are not normally collected and analyzed, they can provide good information as it relates to developing training subtopics as well as in evaluating the individual trainer/facilitator and/or the module.

Post-tests

A post-test measures how well the learner can perform the objectives after training. Posttests often test on the same information included in the pretest but may differ in complexity or format. When designing a training program, safety professionals must decide whether and when to use pretests, review tests, and posttests to measure learners' performance. In safety and health training, tests are highly recommended in the following situations:
- when the training involves a certification or qualification process
- when the organizational culture supports its use
- when the risks of not mastering the objectives may include significant financial loss, injury, or death
- when the effectiveness of training may be questioned
- when qualitative and quantitative data on training and/or safety and health issues are needed.

Posttests are an effective tool in assessing trainees' competency at completion of a training program and at 30, 90, and 180 days afterward, as an example. This type of follow-up, when coupled with a continuous improvement loop that revises the training based on post-test findings, helps identify training deficiencies that require additional emphasis. This helps facilitators and ISDs focus on the identified areas of deficiency, increase trainee retention and competency, and assess the training's effectiveness in meeting the original learning and performance goals.

Additionally, it is important to have determined in the analysis/identification stage what the minimum level of acceptable performance will be. This is sometimes the goal, while at other times, it is a minimum competency level. This is important because knowledge retention rates fall over time, especially if the task is not performed frequently. Initial retention rates were discussed previously in the design stage under Adult Learning Needs, but posttesting accomplished some period of time after training completion will provide a better picture for the safety and health professional, as well as for the ISD, of what needs more emphasis during training or if there is a nontraining solution that needs to be implemented. Indiana University performed a study in which it was found that game-based learning increased retention by more than 300% in immediate posttesting. It was further determined that the same training provided as much as 10 times the retention rate when subjects were tested 6 weeks later than traditional training did (see Figure 31–5).

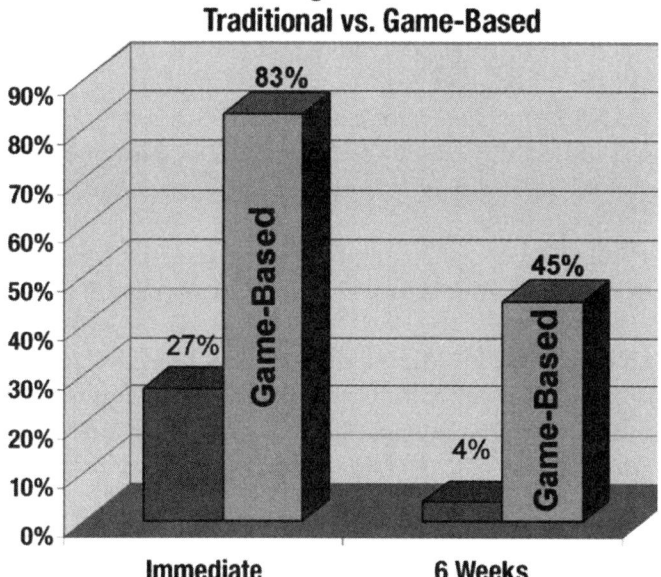

Figure 31–5. Traditional versus game-based knowledge retention rates.

As they relate to safety and health, the safety and health professional will need to analyze the specific training retention rates and determine whether or not they are acceptable based on the business's goals and objectives. In some cases, the learning and performance goals will be achieved, yet the goal set out by the leadership may not have been obtained. In such cases, it is critical to reevaluate the whole training cycle starting with the analysis/identification stage, specifically the reaccomplishment of the training needs analysis to ensure that all training and nontraining solutions have been adequately identified and implemented for the goals set.

REGULATORY OBLIGATIONS

New Employee Training and Orientation

Many companies regard orientation as an excellent opportunity to begin training workers in safety policies and safe work practices. In fact, regulatory agencies state that an employer not only should provide such training but is obligated to do so. Safety and health training can be delivered through an orientation program and through written policies and procedures manuals.

Safety and health training begins when a new worker is hired. The person is usually open to ideas and information about how the company operates. From the first day, the new employee formulates opinions about management, supervisors, workers, and the organization. Some personnel managers say they have never hired a worker with a bad attitude; this may not be entirely true, but many employees' negative attitudes are developed after they have been on the job.

Timeliness of instruction is a key issue in the orientation program. For example, although statistical data differ, studies generally agree that new employees are significantly more prone to work-related accidents. This fact is attributed to the inexperience of new workers, their unfamiliarity with procedures and facilities, and their zealousness to do the work. Also, a significant number of workers are injured during the 4- to 6-year range. This is generally attributed to either a change in work duties (new or transferred employee) or worker complacency. These issues can be covered in an ongoing training program, which will help reduce the number of accidents.

The following subjects are suggested as part of the orientation program:
- company orientation: history and goals
- policy statements
- benefit packages
- organized labor agreements (if applicable)
- safety and health policy statement (if separate)
- acceptable dress code (as required)
- personnel introductions
- housekeeping standards
- communications about hazards
- personal protective equipment
- emergency-response procedures: fire, spill, etc.
- incident reporting procedures
- near-miss incident reporting
- incident investigation (supervisors)
- lockout/tagout procedures
- machine guarding
- electrical safety awareness
- ladder use and storage (if applicable)
- confined-space entry (if applicable)
- medical facility support
- first aid/CPR
- hand tool safety
- ergonomic principles
- eyewash and shower locations
- fire prevention and protection
- access to exposure and medical records.

These subjects, among many others, cover essential information that is far too important to overlook or leave to casual learning. A formal program should be developed not only to provide the worker with this information, but to create a strong link between employees and the safety and health philosophy of the organization.

Many of these subjects are left to the manager, supervisor, or team leader to cover with employees when they arrive at the worksite. This person must have time to cover these points with the employees. Unfortunately, the importance of going over these subjects with the employees is not often recognized until an accident occurs.

Managers, supervisors, and team leaders should be familiar with current information presented to new workers to avoid the age-old problem of contradicting the training material. They should also become involved in the development of training programs to keep the information current and practical in relation to work demands. When they start disregarding or contradicting the training program, the entire program—and the company's image—loses credibility with the workers.

Each manager, supervisor, or team leader should reinforce the training program content by demonstrating how it will apply to the worker's specific job assignment. For example, the new employee may have been trained how to read the warning labels on chemical containers. The manager, supervisor, or team leader can then enhance this information by conducting a tour of the department, pointing out all hazardous materials, and identifying the protection provided through personal protective equipment and control measures.

Additional Resources

All regulatory agencies publish information about employers' responsibilities to provide training and retraining to their employees. The final interpretation may still present difficulties; the best solution is to communicate with the agency involved, if necessary, and the organization's legal counsel.

The continued importance of training is evidenced by the requirements of both the Occupational Safety and Health Administration (OSHA) and the Mine Safety and Health Administration (MSHA). The major parts of the OSHA regulations (Title 29—Labor, Code of Federal Regulations) covering training requirements include the following:

- Part 1910, Safety and Health Training Requirements for General Industry
- Parts 1915–1918, Safety and Health Training Requirements for Maritime Employment
- Part 1926, Safety and Health Training Requirements for Construction
- Part 1928, Occupational Safety and Health Requirements for Agriculture.

Table 31–A gives a convenient index, by types of hazards, of the locations in OSHA regulations requiring training to protect against the hazards. OSHA also provides additional guidance and documents for review. The following is a short list of additional documents and resources. Additional information may also be found on the OSHA website, www.osha.gov.

- ANSI: American National Standards Institute
- ANSI/ASSE Criteria for Accepted Practices in Safety, Health, and Environmental Training, ANSI/ASSE 2490.1–2009
- ASSE: American Society of Safety Engineers
- NIEHS: National Institute of Environmental Health Sciences
- NIEHS WETP Minimum Health and Safety Training Criteria Guidance for Hazardous Waste Operations and Emergency Response (HAZWOPER); HAZWOPER Supporting and All Hazards Prevention, Preparedness and Response.
- WETP: Worker Education and Training Program.

The requirements for training and retraining of miners are defined by MSHA in 30 CFR 48—Training and Retraining of Miners (see §§ 48.2 and 48.22 for definitions of terms used in Part 48). Remember that these regulatory obligations represent only the minimum requirements; therefore, a thorough evaluation as previously discussed is necessary to ensure a truly safe and health-promoting environment for employees.

SUMMARY

Training focuses mainly on behavior change: showing workers how to perform tasks properly and apply their knowledge on the job. In some cases, however, nontraining solutions are more appropriate. In developing a quality safety and health training program, it is important to use a system that allows for full understanding and continuous improvement, such as the training cycle listed in Figure 31–1. This method provides the system needed to develop the best training that will provide the necessary knowledge and skills, set the appropriate behaviors to protect workers' health and lives, and prevent work-related injury or illness. Effective training will help ensure that workers develop the appropriate skills and are educated and empowered to improve the working conditions in their places of employment.

One important factor that helps to ensure successful training implementation is the selection of a training facilitator who exhibits safety and health expertise, sound instructional skills, and flexibility. In developing effective training, the following are important considerations:

- Training and nontraining problems and solutions are cat-

TABLE 31–A. Index to OSHA Training Requirements

Hazard	Part	Subpart	Section	Hazard	Part	Subpart	Section
Blasting or Explosives	1910.109	H	(d)(3)(i)(iii)	Medical and First Aid	1910.94	G	(d)(9)(i) and (vi)
	1926.901	U	(c)		1910.151	K	(a) and (b)
	1926.902	U	(i)		1915.58	K	(a)
	1915.10	B	(a) thru (b)		1917.58	F	(a)
	1916.10	B	(a) thru (b)		1926.50	D	(c)
	1917.10	B	(a) thru (b)				
				Personal Protective Equipment	1910.94	G	(d)(11)(v)
Carcinogens					1910.134	I	(a)(3)
4-Nitrobiphenyl	1910.1003	Z	(e)(5)(i) thru (ii)		1910.134	I	(b)(1), (2) and (3)
alpha-Naphthylamine	1910.1004	Z	(e)(5)(i) thru (ii)		1910.134	I	(e)(2), (3) and (5)
4, 4'-Methylene bis					1910.134	I	(e)(5)(i)
(2-chloroaniline)	1910.1005	Z	(e)(5)(i) thru (ii)		1910.161	K	(a)(2)
Methyl chloromethyl ether	1910.1006	Z	(e)(5)(i) thru (ii)		1915.82	I	(a)(4)
3, 3'-Dichlorobenzidine					1915.82	I	(b)(4)
(and its salts)	1910.1007	Z	(e)(5)(i) thru (ii)		1916.57	F	(f)
bis-Chloromethyl ether	1910.1008	Z	(e)(5)(i) thru (ii)		1916.58	F	(a)
beta-Naphthylamine	1910.1009	Z	(e)(5)(i) thru (ii)		1916.82	I	(a)(4)
Benzidine	1910.1010	Z	(e)(5)(i) thru (ii)		1916.82	I	(b)(4)
4-Aminodiphenyl	1910.1011	Z	(e)(5)(i) thru (ii)		1917.57	F	(f)
Ethyleneimine	1910.1012	Z	(e)(5)(i) thru (ii)		1918.102	J	(a)(4)
beta-Propiolactone	1910.1013	Z	(e)(5)(i) thru (ii)		1926.21	C	(b)(2) thru (6)
2-Acetylaminofluorene	1910.1014	Z	(e)(5)(i) thru (ii)		1926.103	E	(c)(1)
4-Dimethylamino-azobenzene	1910.1015	Z	(e)(5)(i) thru (ii)		1920.800	S	(e)(xii)
N-Nitrosodimethylamine	1910.1016	Z	(e)(5)(i) thru (ii)	Pulpwood Logging	1910.266	R	(c)(5)(i) thru (xi)
Vinyl chloride	1910.1017	Z	(j)(1)(i) thru (ix)				
					1910.266	R	(c)(6)(i) thru (xxi)
Cranes and Derricks	1910.179	N	(m)(3)(ix)		1910.266	R	(c)(7)
	1910.180	N	(h)(3)(xii)		1910.266	R	(e)(2)(i) and (ii)
Decompression or Compression	1926.803	S	(a)(2)		1910.26	R	(e)(9)
	1926.803	S	(b)(10)(xii)		1910.266	R	(e)(1)(iii) thru (vii)
	1926.803	S	(e)(1)	Powder-Actuated Tools	1915.75	H	(b)(1) thru (6)
Employee Responsibility	1910.109	H	(g)(3)(iii)(a)		1916.75	H	(b)(1) thru (6)
	1926.609	U	(a)	Power Press	1910.217	O	(e)(3)
Equipment Operations	1910.217	O	(f)(2)	Power Trucks, Motor Vehicles, or Agricultural Tractors	1910.109	H	(d)(3)(iii)
	1926.20	C	(b)(4)				
	1926.53	D	(b)		1910.109	H	(g)(3)(iii)(a)
	1926.54	D	(a)		1910.178	N	(1)
	1910.252	Q	(c)(6)		1910.266	R	(e)(9)
Fire Protection	1916.32	D	(e)		1910.266	R	(e)(6)(viii)

Hazard	Part	Subpart	Section	Hazard	Part	Subpart	Section
	1917.32	D	(b)		1928.51	C	(d)
	1926.150	F	(a)(5)	Radioactive Material	1916.37	D	(b)
	1926.155	F	(e)	Signs—Danger, Warning, Instruction	1910.96	G	(f)(3)(viii)
	1926.351	J	(d)(1) thru (5)		1910.145	J	(c)(1)(ii)
	1926.901	U	(c)		1910.145	J	(c)(2)(ii)
Forging	1910.218	O	(a)(2)(i) thru (iv)		1910.145	J	(c)(3)
Gases, Fuel, Toxic Material, Explosives	1910.109	H	(d)(3)(i) and (iii)		1910.264	R	(d)(1)(v)
	1910.111	H	(b)(13)(ii)				
	1910.266	R	(c)(5)(i) thru (xi)	Tunnels and Shafts	1926.800	S	(e)(xiii)
	1910.106	H	(b)(5)(vi)(v)(3)				
	1916.35	D	(d)(1) thru (6)	Welding	1910.252	Q	(b)(1)(iii)
	1926.21	C	(a) and (b)(2) thru (6)		1910.252	Q	(c)(1)(iii)
	1926.350	J	(d)(1) thru (6)		1915.35	D	(d)(1) thru (6)
					1915.36	D	(d)(1) thru (4)
General	1926.21	C	(a)		1916.35	D	(d)(1) thru (6)
Hazardous Material	1915.57	F	(d)		1917.35	D	(d)(1) thru (6)
	1916.57	F	(d)				
	1917.57	F	(d)				

Table 31–A. Concluded.

egorized as selection and assignment, information and practice, environment, and motivation/incentive.
- Adult learners have special needs and requirements that trainers must recognize for programs to be effective. Performance-based instruction generally works well with adult learners.
- To design an effective training program, the safety professional must assess workers' needs, analyze learners' characteristics, develop specific objectives, develop materials and schedules, and design testing and evaluation methods by following the training cycle provided in Figure 31–1.
- Training begins with new employee orientation. Using written policies and procedures manuals is one way to meet new employee training needs and conform to regulatory standards.
- Training methods include on-the-job training, job instruction, group methods, and individual methods.
- Maintenance of the overall training cycle is essential for training that currently meets and/or exceeds expectations, and a training program that does not degrade becomes obsolete over time.

REFERENCES

The ADDIE Model, Copyright 2013. www.instructionaldesignexpert.com www.instructionaldesignexpert.com/implementation.html.

American Society of Safety Engineers. *Safety Scope and Function*. www.asse.org/about/scope_function.php. © Copyright 1996 American Society of Safety Engineers. All rights reserved.

Brothers, S. K. "Game-Based E-Learning: The Next Level of Staff Training." www.ltlmagazine.com/article/game-based-e-learning-next-level-staff-training, March 1, 2007.

Bureau of Industrial Relations, Department of Training Materials for Industry, University of Michigan, Graduate School of Business Administration, Ann Arbor, MI 48104. "The In-Basket Method."

Carr, C. *Smart Training: The Manager's Guide to Training for Improved Performance*. New York: McGraw-Hill, 1992.

Dennis, L. E., and M. E. Onion. *Out in Front: Effective*

Supervision in the Workplace. Chicago: National Safety Council, 1990.

Didactic Systems. PO Box 4, Cranford, NJ 07016. "A Catalog of Ideas for Action Oriented Training."

DuPont. Education and Applied Technology Division, Wilmington, DE 20017. "Library of Programmed Instruction Courses." Hendershot, C. "A Bibliography of Programs and Presentation Devices." 4114 Ridgewood Drive, Bay City, MI 48706.

Indiana University. Private study commissioned by QBInternational. Details available from QBInternational, 900 Larkspur Landing Circle, Suite 115, Larkspur, CA 94939. www.ltlmagazine.com/article/game-based-e-learning-next-level-staff-training.

Mager, R. F. *Preparing Instructional Objectives.* 2nd ed. Belmont, CA: Lake, 1984.

Mager, R. F., and K. M. Beach, Jr. *Developing Vocational Instruction.* Belmont, CA: Lake, 1984.

Mager, R. F., and P. Pipe. *Analyzing Performance Problems or You Really Oughta Wanna.* 2nd ed. Belmont, CA: Lake, 1984.

Merriam-Webster, www.merriam-webster.com/dictionary/training.

National Safety Council. *Supervisors' Safety Manual.* 9th ed. Itasca, IL: National Safety Council, 1997.

National Society for Programmed Instruction, PO Box 137, Cardinal Station, Washington DC 20017.

Powell, E. T. "Questionnaire Design: Asking Questions with a Purpose." University of Wisconsin-Extension, Cooperative Extension, G3658-2. http://learningstore.uwex.edu/assets/pdfs/g3658-2.pdf.

Science Research Associated. Department of Management Services, 155 North Wacker Drive, Chicago, IL 60606. "Simulation Series for Business and Industry."

U.S. Department of Labor. Training Requirements in OSHA Standards and Training Guidelines. OSHA No. 2254. Washington DC: U.S. GPO, 1998. Available at local OSHA regional offices.

U.S. Department of Labor, Occupational Safety and Health Administration. "Susan Harwood Training Grant Program Best Practices for the Development, Delivery, and Evaluation of Susan Harwood Training Grants." www.osha.gov/dte/sharwood/best-practices.html, September 2010.

———. "Training Requirements in OSHA Standards and Training Guidelines." OSHA 2254 1998 (Revised). www.osha.gov/Publications/osha2254.pdf.

Zoll, A. A. *The In-Basket Kit.* Reading, MA: Addison-Wesley, n.d.

REVIEW QUESTIONS

1. List four benefits of safety and health training.
   a.
   b.
   c.
   d.
2. Identify 7 of the 12 nontraining solutions that can be as effective as a training program.
   a.
   b.
   c.
   d.
   e.
   f.
   g.
3. Define *performance-based training.*
4. What are the five phases of the systematic approach of performance-based training?
   a.
   b.
   c.
   d.
   e.
5. Adults learn best through which of the following senses?
   a. hearing
   b. sight
   c. touch
   d. all of the above
6. Describe the four adult learning principles that should be applied to all safety and health training.
   a.
   b.
   c.
   d.

7. List the three most common types of training methods used in industry.
   a.
   b.
   c.
8. Group training encompasses which of the following techniques?
   a. brainstorming
   b. case study
   c. role-playing
   d. simulation
   e. all of the above
   f. b and c
9. Define *computer-assisted training*.
10. List the five steps in the ADDIE system or training cycle.
    a.
    b.
    c.
    d.
    e.

32 Media

Teddy G. Gil, BSME, MBA, DAL Level II VT, UT, RT, PT, MT, IR
Sherrie C. Wilson
John F. Montgomery, PhD, CSP, CHMM

Training and Media
Introduction ▸ Media Comprehension Levels and Formats ▸ Media Documentation

Selection and Design of Training Media
Establishing Objectives ▸ Audience Characteristics and Setting ▸ Role of the Trainer ▸ Use of Complementary Training Tools ▸ Organizing the Media ▸ Text Charts ▸ Graphic Formats ▸ Graphs ▸ Drawings and Sketches ▸ Use of Colors in Media ▸ Computer Presentation Software ▸ Projected Visuals

Video Media
Showing Prerecorded Videos ▸ Working with a Video Production Company ▸ Commercial versus In-House Production

Computer-Based Training
Benefits and Drawbacks ▸ Assessing Needs ▸ Computer Learning Formats ▸ Computer-Managed Training ▸ Interactive Training

Supplemental Materials
Posters ▸ Pictures ▸ Handouts ▸ Audiocassettes/CDs and Podcasting

Training Presentation and Aids
Tips for Presentations ▸ Seating ▸ Trainer/Speaker Aids ▸ Organizing Props and Presentation Aids

Evaluation
Creating an Evaluation Climate ▸ Evaluating Media Materials ▸ Evaluating Media Use ▸ Evaluating Audience Reaction

Cost Factors

Safety Considerations
Safe Background ▸ Safety Equipment ▸ Safe Practices ▸ Lighting Equipment ▸ Fall Protection ▸ Review of Photos

Summary

References

Review Questions

TRAINING AND MEDIA

Introduction

A *medium* is a channel of communication. It can be anything that carries information between a source and a receiver, such as television, films, diagrams, printed materials, computers, the Internet, and trainers. The purpose of media is to facilitate communication, and the purpose of this chapter is to facilitate effective use of media, whether a particular type of medium is being used for safety and health training, general instruction, or presentations in meetings. The term *trainer* is used most frequently in this chapter, but the information presented applies to speakers in all types of communication situations. This chapter discusses the following topics:
- selection and design of training media
- computer-based training and considerations
- use of supplemental materials
- presentation materials and preparation requirements
- how to evaluate the effectiveness of a training program.

Figure 32-1. The more concrete the medium of communication, the more effective it is.

(Inverted triangle, top to bottom — Scale of Sensory Experience, Concrete to Abstract)
- Actual, Direct Experience
- Simulations: "Hands-On" Devices
- Demonstrations: "Show and Tell"
- Field Trips: Familiarization
- Exhibits: Displays
- Live Television
- Motion Pictures
- Still Pictures: Photos and Slides
- Auditory Aids
- Graphics: Charts and Graphs
- Words: Spoken and Written

Media Comprehension Levels and Formats

Media used for any purpose take advantage of the two senses people use most when learning—sight and hearing (Figure 32-1). Learners retain 83% of what they experience through sight and 11% through hearing. Only 1.0% is learned through taste, 1.5% through touch, and 3.5% through smell. Further, extensive research shows that people retain 50% of what they both see and hear, but only 20% of what they learn through hearing alone and only 30% of what they learn through sight alone. (See Chapter 31, Safety and Health Training, for additional information.)

The full range of instructional or training media includes books and workbooks, chalkboards, whiteboards, tablets, bulletin boards, display cases, flannel boards, posters, flip charts, slides and other projected transparencies, videos, computer disks and tape, films, recordings, online videos and media, and models. Media equipment (hardware) includes projectors, simulators, tape and cassette players for sound and video, cameras for stills and video, monitors, television equipment, and equipment for computer-interactive training, such as personal computers, CD-ROM players, and web-based equipment that includes servers (Figure 32-2). Live demonstrations and experiments also are classified as audio and visual media events.

Media Documentation

Safety professionals who want to make full use of the power of media must first understand employee training needs and how media can help support training objectives. Regulatory

Figure 32-2. This portable multimedia projector allows the trainer to conduct presentations virtually anywhere and to enhance learning through effective visuals. *(Courtesy of 3M Visual Systems Division.)*

agencies and legislative bodies are mandating safety and health training in far more detail than ever before. For an in-depth discussion of safety and health training, see Chapter 31, Safety and Health Training. In general, employers no longer rely on class rosters as sufficient evidence to prove that workers received adequate training. Instead, they must train workers in specific competencies and document that such training has been effective. Existing regulations require that training time, even for veteran employees, be documented and kept available for inspection by regulatory agencies upon request (see Chapter 6, Safeguarding, in the *Engineering & Technology* volume). Therefore, if the medium is not facilitated or supervised, it must include a means of documenting the event, duration, content, and success rate. Systems such as learning management systems (LMSs) have become an increasingly important tool in documenting and storing completed training delivered in various media forms. These systems have also evolved from simple databases that stored only the learners' information to sophisticated platforms that can store different training media to be delivered on demand. Additionally, these systems can be linked to other sources to deliver, capture, and store delivered training information. Hence, LMSs are not in themselves a medium but are rather an effective way to deliver multiple types of media and to store all types of media meant for training, which allows the safety professional to store training duration, location delivered, and learner success rates, and to schedule training recurrence information and make assignments based on competency required for the job. The job requirements, level of worker education, and required performance mastery will influence the optimal media-delivery selection.

SELECTION AND DESIGN OF TRAINING MEDIA

This section introduces the factors involved in selecting and designing media for training, including advantages and limitations related to various media forms.

Establishing Objectives
For both management and employees to get the most out of training, managers must establish clear objectives early in the development of training programs. Once trainers know what they need to accomplish, they can begin to concentrate on the details of how to achieve the objectives, including which medium is best suited for the situation. A good example of a training objective might be the following: "After completing the program, the worker will respond to a simulated hazardous materials emergency by obtaining the proper SDS and handling the emergency according to the SDS directions with less than 10% error." An example of a poor training objective is: "After viewing this training tape, workers will feel better motivated to follow recommended procedures for wearing appropriate eye protection when working with lasers." *Feel, believe, support,* and similar words are not performance or behavior oriented because how well the audience must perform the behavior is not indicated. Teaching, reinforcement, and psychomotor skills practice are necessary if workers are to retain and assimilate information. Carefully chosen media help employees understand the information they receive, shorten the learning curve, reinforce concepts and skills, and create interest in the topic.

Audience Characteristics and Setting
A training presentation for a group of 200 in an auditorium setting will require a significantly different approach than a hands-on, "how-to" session given by a supervisor to a small group of employees. The larger, more formal session can be handled best by a lecture format, with appropriate media such as computer-generated slides, video, film, and handouts used to reinforce key points. The factory floor discussion, which may take place in a small conference or training room, probably can benefit from a discussion format. A flip chart, chalkboard, or large pad on which the supervisor can list ideas the group contributes would be an appropriate medium for this setting, although such environments are increasingly using computer-created presentations shown via projector or television/monitor. Training materials should be designed with the target audience in mind. Language levels should match the audience's comprehension or literacy level. For example, if an instructor uses too many technical terms without explanation, the talk will be difficult for a general audience to understand. On the other hand, too simple a presentation is not appropriate for a group of technical engineers. (See also Literacy Levels in Chapter 31, Safety and Health Training.)

Role of the Trainer
Each type of media training material available has an appropriate role to play in training. Major features and limitations of many widely used media tools are given in Table 32–A. Note that closed-circuit television, videocassettes, and computer-based interactive training can be used in conjunction with Internet-accessible materials with appropriate electronic connection. One factor sometimes neglected in the selection of media is the role a trainer expects to play in a particular training session and how he or she perceives training responsibilities. Good trainers continually ask themselves these questions: Is what I want to do appropriate for the training objective in this particular session? Does the medium I am considering enhance my training role as I see it? (See Table 32–B.) These questions are applicable to facilitated sessions and are considered instrumental in the design phase of training.

TABLE 32–A. Major Features and Limitations of Various Media

Type and Popular Size	Audience Size	Limitations	Strong Points	Comments
MOTION PICTURES 16mm	M/L	Camera and projector expensive; require trained operator, except for self-threading models. Film not easily changed or updated.	Effective for training and motivating. Uniform professional message. Optical sound unerasable. Sharper image than 8 mm for given projection size. Single-frame, stop-motion projectors are available.	Silent version less costly, but less effective. Many companies prefer video.
SLIDES 2 x 2 in. (35mm, 126, or 127) 2 ¼ x 2 ¼ in. (120 film) 3 ¼ x 4 in. (superslides) (theater projector)	S/L	Slides may get out of sequence, reversed, etc. Cardboard mounts not durable.	Effective for training and motivating. Less of a "canned" show since slides may be rearranged. Slides can be made and processed quickly. Color inexpensive.	Taped message or reading script easily added or changed. Remote control and multiple projection possible. Many companies prefer video.
FILMSTRIPS 35mm sound	S/L	Strips and records not easily updated. Message might not be effective or suitably paced for user's needs.	Effective for training and motivating. Message uniform. Sounds easily added to tape or disk.	Silent strips with scripts less expensive, but still effective. Seldom used anymore.
OVERHEAD PROJECTORS 10 x 10 in. 7 x 7 in.	S/L	Transparencies positioned by hand. Projector close to screen; it or user may block view unless screen is raised or set at an angle. Ready-made material not widely available.	Effective for training. User can write on transparency while facing audience. No need to darken room. Transparencies easily made and filed. Presentation informal and flexible.	Color transparencies or overlays easily made.
OPAQUE PROJECTORS 10 x 10 in. max.	S/M	Projectors require manual operation. Material in books may be difficult to store or ship. Room must be darkened. Copy may be too small.	Effective for training. No transparencies required; small objects, printed material, drawings, and photographs used "as is."	Copies or originals can be hinged or put on rolls to maintain sequence. Seldom used anymore.
CLOSED-CIRCUIT TELEVISION, VIDEOTAPE CASSETTE	S/M	Initial investment expensive. Requires adequate lighting. In color or black and white. Copies must be made one at a time unless duped by lab.	Instant replay. Excellent for training situation where trainee must "see himself or herself in action." Has relatively low operating cost. Can be shown in lighter room.	Small number of people can view screen. TV is a culturally natural transmission medium.
COMPUTER-BASED INTERACTIVE TRAINING	S	High initial expense. May be difficult to copy. Requires that trainees have access not only to computers, but videodiscs, videotapes, and so on.	Low operating costs. High retention of material presented, user paced. Can automatically record and verify training and test scores for individual employees. Presentation and message uniform, does not require presence of instructor. Effective if a machine or process must be shown in detail or simulated. Best method for simulating hazardous conditions or those where a mistake would have serious consequences in the "real world."	Computer-based training works best if delivered through computers that trainees use regularly.
FLANNEL, HOOK AND LOOP, MAGNETIC 12 x 36 in. to 48 x 24 in.	S/M	Presentation requires advance preparation. Few ready-made presentations available. Flannel board material may fall off if not applied correctly or if board too nearly vertical.	Effective for training. Message easily changed, yet can be filed and reused. Permits informal presentation with desirable audience contact. Dramatic, "slap-on" effect builds interest.	Boards suitable for heavier displays; cost slightly higher than cards or pads.
FLIP CHARTS, CARDS, AND PAPER SHEETS 38 x 48 in. 28 x 36 in. 18 x 24 in.	S	Limited to small groups. Good lighting necessary. Speaker must print legibly.	Effective for training, informing, or discussion. Good audience contact. Material easily prepared; can be added during talk and saved. Low-cost pads easily obtainable. Permits reference to other sheets both during the discussion and later for writing of minutes.	Ready-made letters, color, sketches, cut-outs easily added. Colored paper effective. Paper sheets used in place of chalkboards, no erasing.
CHALKBOARDS (portable and wall mounted) 36 x 48 in., large for wall mounted	S	Board must be erased before reuse and recall not possible. Good lighting necessary. Ordinary chalk marks hard to see. Dust from chalk and erasers annoying.	Effective for training or discussing a limited number of points. Presentation informal. Portable chalkboards also useful for holding cards or displays.	Colored or fluorescent chalk adds life to talk. Magnetic boards available.
POSTERS AND BANNERS 8 ½ x 11 ½ in., 17 x 23 in., and larger	S/L	Only one or two ideas can be presented at a time; considerable time needed for changing.	Effective for motivating; support training. Specific messages can be posted at points of hazard or to meet timely situations. Ample posters available.	Homemade posters supplement general posters.
WORKING MODELS, EXHIBITS, AND DEMONSTRATIONS	S/L	May require special training to use. Live action is subject to errors.	Action can closely simulate actual conditions. Permit group participation.	

TABLE 32–B. Some Safety Uses for Videotape

Videotape is a versatile medium. Here are a few ways it is currently being used to put across the industrial safety message:

Security surveillance—Mounted cameras observe distant gates, loading docks, and payroll departments.

Job review—Used in conjunction with job safety analysis, task safety analysis, and step safety analysis for observation and review of both good and bad procedures.

Management training and development procedures—First train the trainer, then have him or her train employees down the line. Playbacks ensure that procedures are practiced correctly by the instructor before others are taught to do likewise.

Incident investigation—When brought to the accident scene, a recording is made of the physical set-up, personnel present, time, weather, lighting, and other conditions pertinent to the accident.

Motivation and enforcement—Showing employees how their performance can lead to an accident is an effective way of gaining their cooperation. OSH Act violations can be spotted and corrected. Before-and-after scenes can prove dramatic.

Training in sophisticated equipment, complex procedures—Videotapes can show and repeat processes a step at a time, permitting interruptions for questions.

Informing distant audiences of a procedure or announcement affecting all units—Duplicates of a tape can be made and mailed anywhere so that all personnel are informed simultaneously.

Trainer as Expert

Trainers can choose the *expert* role, communicating information in a more formal presentation. The trainer answers questions from the audience but allows little opportunity for dialogue. Such a choice may be appropriate when trainers need to convey information to a large audience—for example, explaining new or revised regulations or policies to the work force. Appropriate media would include projected visuals, such as slides and overheads, and a take-home handout.

Trainer as Mentor

At times, trainers can assume the role of a *mentor* who teaches his or her skills to other workers. Supervisors or team leaders who have come up from assembly-line work or who have had experience on specific equipment often are called upon to pass on their skills and expertise to new employees or to promoted or transferred workers. Hands-on film demonstrations, either commercially available or company-produced, would be appropriate media. If the organization uses interactive training, a CD-ROM module—if one is available on the subject—is another choice. For example, such a module might show a worker how to use a full-face mask respirator: the respirator parts and their functions; how to put on and take off the respirator; and how to test, maintain, and store the equipment.

Trainer as Facilitator

A third role the trainer can play is that of a *facilitator*—that is, resource person or guide—who helps match individual student needs with appropriate materials. Such a role assumes that the trainer is flexible and has a broad knowledge of available teaching aids, including various media. Trainers who function successfully in this role create a training environment in which students are not afraid to ask questions or to admit their need for extra help.

Student workbooks or worksheets, which can be completed at the employee's own pace, can serve as the basis for individualized discussions. The trainer uses the material to help employees understand their shortcomings and to suggest remedies. The safety professional can also provide audiotapes, audio CDs, or video DVDs that employees can take home, enabling students to go over the material until they understand it. Interactive training using CD-ROMs or web-based training solutions also may be an option if the organization has such equipment. Workers can practice privately, without feeling pressured to keep up with faster learners in the class, while sophisticated online training or e-learning solutions can permit trainers to monitor the progress of students and adapt training as needed.

Use of Complementary Training Tools

In addition to materials that stress sight and hearing, training designers and trainers can also use training tools that engage the sense of touch. These tools help reinforce learning by allowing workers to practice the psychomotor skills of what they have just learned or to experience a real-world application of the lesson. (See Chapter 31, Safety and Health Training, for a thorough discussion of live presentations, including lectures, discussion, demonstrations, and performances.) For example, exhibits and models make effective 3-D displays for use in training. Cardiopulmonary resuscitation (CPR) training, for instance, requires a mannequin on which students can practice. The use of psychomotor skills increases learning retention and the possibility that the student will be able to perform the skill when necessary. A cutaway model showing layers of muscles or a skeleton showing vertebrae can be effective in demonstrating how injuries can occur and what body parts are affected (Figures 32–3a and 32–3b). Designers and safety professionals can design training such that it will use exhibits to demonstrate the safe operation of a machine or process. Exhibits can also feature examples of first-aid equipment, protective clothing, rescue equipment, respiratory protective equipment, and fire protection appliances. Companies at times may organize demonstrations of fire prevention and suppression with the assistance of the local fire department (Figures 32–4a and 32–4b). Designers and safety professionals also can arrange

Here Is How to Use It

There is a correct and an incorrect way to pick up a heavy object. The correct way is to keep the back straight, the knees bent and spread, and the load close to the body. The incorrect way is to reach way over and lift. (Twisting the back complicates the bad effect.)

To demonstrate this effectively, a special model can be used. If used with its block spine locked, the model simulates lifting with strong leg muscles; the ribbon on the spine remains limp, indicating very little tension of the back muscles. It demonstrates that the back cannot be kept straight without bending the knees.

To demonstrate improper lifting, the model is used as is shown in the sketch. The legs are bent only slightly (or held straight). One hand lifts the handle just ahead of the fulcrum at the hips in order to lift the weight in the hands. The back arches under the strain and visibly pulls each block apart.

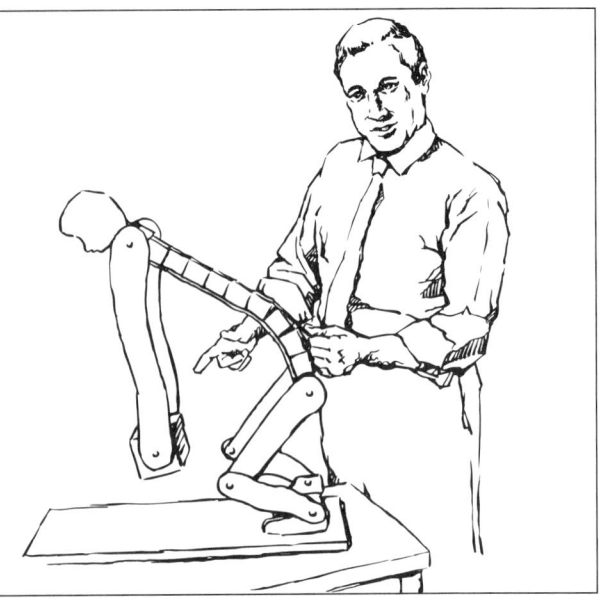

Figure 32–3a. A model is used to demonstrate lifting techniques.

Construction Specifications

The spine is made with 9 blocks, each 2 in. square and 1⅝ in. high. Each is drilled at the center to accommodate a standard screen door spring. The T-shaped head and shoulder piece is about 8 in. long, 2 in. thick, and supports the arms on shoulders about 5 in. apart. The hip block is 2 in. square, with sloping sides so the legs will spread open in front. Add a ⅜-in. spacer between the hip block and each leg.

Assemble the body by using wood screws to attach ends of spring to the head and hip pieces. Tack a piece of 2-in. wide belting to the *front* of the spine blocks to hold them in alignment.

The arms and legs can be shaped from ¼-in. plywood.

A block, approximately 5 in. square and 3½ in. high, represents the lifting weight. Elastic tape or multiple rubber bands are stretched from the shoulder to the hip, along the *back* of the spine blocks to represent the spine muscles.

A metal handle can be secured to the lower end of the hip block, as shown in the drawing.

The model can be mounted on a 1×12-in. board, 18 to 24 in. long.

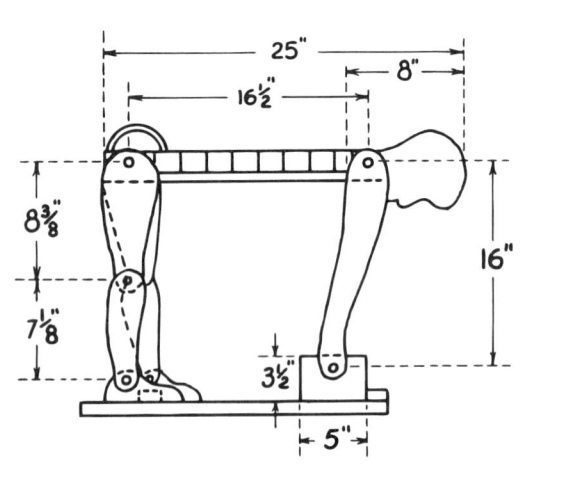

Figure 32–3b. Instructions for constructing the model.

Figure 32–4a. Railroad safety officers are receiving hands-on training in fire fighting. *(Courtesy of the Association of American Railroads.)*

Figure 32–4b. As part of hazardous material response training, these students extinguish a liquefied petroleum gas fire at the AAR Transportation Test Center. *(Courtesy of the Association of American Railroads.)*

demonstrations of good and bad lighting; impacts on safety hats, safety shoes, and eye protection devices; and first-aid treatment and transport of injured workers.

Organizing the Media

Designing effective media is easy to do if designers and trainers remember the basic purpose of media: to communicate a message. After creating a training objective, considerations must be given to how to enhance the training sessions with the media. Every presentation begins with an idea or concept and builds from the basics. Here are several ways to organize a session:
- Start with a question or problem and move toward an answer or solution.
- Present a problem, fill in the details, and end with a solution (a good way to use case histories effectively).
- Describe "What if?" possibilities; after discussion, reach a workable solution.
- Do a time sequence, developing visuals to illustrate chronological order.
- Present information, using a near-to-far or far-to-near sequence.
- List "for" and "against" points in support of, or opposed to, a main idea.
- Explain the features and benefits of an idea or problem.

To check the organization, develop a simple outline of the media to be used in the session. In advance, work through key training points that media can help explain. Media should clarify or dramatize all important points. Designers and trainers should ask themselves the following questions: Does the outline flow logically? Does it match the order in which the trainer presents ideas? Although the training will want to build from simple thoughts to more complex concepts, it should not forget to put at least one visual early in the presentation to communicate clearly what topics will be explained. Such a visual (called an *advance organizer*) focuses attention on succeeding material as this lets the audience know what is coming.

Text Charts

Text charts are one of the most commonly used media. They are easy to create, by either typing text into the computer, using a typewriter, lettering by hand, or using press-on letters available at art stores or office supply stores. Text charts are made in four styles:
- *Titles*—Title charts are good for introducing presentations. For maximum effectiveness, include no more than eight lines. As a general rule, use the top one-third of the chart for titles and the center of the chart for the message or announcement of the topic.
- *Lists*—Lists are a good way to help people organize their thoughts about a central idea. They can show sequential steps in a process, such as lockout/tagout training for a particular machine, or can present ideas that then serve as a springboard for discussion. In a simple list, each idea is emphasized equally. Items in the list can be highlighted by putting a bullet or icon in front of each one.

  A good way to stimulate discussion on each bulleted idea is to present them as single thoughts. The trainer can make the first visual using only the first line of the list, the second visual with the second line, the third visual with the third line, and so on, building the list as the talk continues. If the trainer is using projected visuals, another easy way to control what the audience sees, bit by bit, is to create a single visual with three or four ideas on the list. This technique is especially good for ideas that follow a logical sequence.
- *Organization charts*—An organization chart is a quick way to illustrate lines of authority and corresponding responsibilities. Use an organization chart, for instance, to indicate safety committees for each shift. Many available computer graphics programs automatically do the drawing once the trainer has entered the names and titles of the staff and indicated the reporting relationships.
- *Tables*—Tables are good for handouts but difficult to follow on projected visuals unless they are extremely simple. Resist the temptation to put too much information in a table. Viewers cannot absorb details that quickly. A variation of a table is a column chart, with related data displayed in two or three columns. "Days without Reportable Accidents" could be the overall title, with June, July, and August listed as column headers, and number of days under each month. Column charts are also useful for listing ideas or problems, with potential benefits in one column and drawbacks in another.

Graphic Formats

Media work best in training when used with easily understood charts and graphs that provide accurate information. For projected visuals, designers and trainers have a choice of formats. Although the most common types used in the past have been overhead projector transparencies and overlays or 35-mm transparencies, most presentations since the mid-1990s have been prepared using computer presentation software that permits the user to integrate words, pictures, video, audio, and other images with special effects such as wipes, dissolves, and other animations that move elements on and off the screen.

The value of using computer presentation software is that the program content can be tailored for each audience, using a standard presentation's contents as a starting point for creation of a completely new program. Thus, the title slide can be personalized to acknowledge the specific group, location, and date of presentation, even including a company logo or facility location photo. Presentation software also enables sharing of the program contents by automatically preparing handout pages for students, with space for notes for each slide image. The most popular and widely used presentation software is Microsoft PowerPoint, but there are many other alternatives, including several inexpensive or free programs available through web sources that may substitute for MS PowerPoint with similar features and effectiveness. For projected images, use the horizontal or *landscape* format. Like a landscape, such visuals are wider than they are long. Because fewer lines fit on a horizontal visual than on a vertical one, designers and trainers are forced to limit their content, resulting in a clearer visual. Long lists, forms, and detailed flowcharts generally should not be used in projected images.

Font Styles

As a rule, the clearer the posters, slides, and overheads are, the more memorable they will be because less is better than more. Thus, avoid selecting fonts that are too complicated, placed unattractively on a page or on the screen, or too small to be read comfortably. *Font* is a term for a family of letters with a particular design (also called a *typeface*). Some fonts have *serifs*, which are lines or curves that project from the ends of letters. Times New Roman, Bookman, and Bodoni are examples of *serif fonts*. Serif fonts are usually easier to read when used for the body of text charts because they are widely found in newspapers and magazines. Fonts without these lines are called *sans serif*. Helvetica, News Gothic, and Arial are examples of sans serif fonts. They look more modern, but some studies show that multiple lines of sans serif type may be harder to read, especially for people more accustomed to serif fonts. Most designers recommend restricting use of sans serif fonts to headlines and outlines rather than to the body text.

For big, bold lettering, Helvetica and Gothic Bold (both sans serif) have long proved popular. They make attractive headlines and display type to accompany body text set in Times New Roman or Bookman (both of which are serif fonts). *Script* fonts, which look like handwriting, are a third style of font, but are the hardest of all to read. If another, less common font is desired, designers and trainers should consult with a graphic designer or compositor (typesetter) to make sure fonts and graphic images do not clash, distracting from the importance of the safety or health message being conveyed.

Font Size

Any visual is ineffective when lettering is too small to be read comfortably. People who wear prescription lenses, especially bifocals, and older workers may need to have larger type. Design lettering so it can be read easily from the back row of the audience; the message will have a much better chance of getting across. In general, use 24-point or larger type for projected visuals to make them easy to read. Remember, larger is better. Keep at least a 2-point size difference between text and headline, or text and subheads, to make a visual distinction between the two. For example, if the text font is 24 point, set headlines in 26-point font. For projected visuals to be read comfortably, the designer should be able to read the smallest line of type on the visuals at arm's length without a magnifying glass. Use Figures 32–5 and 32–6 as guidelines for recommended heights and thicknesses of lines and symbols. Block letters show up better than handwritten copy. To be easily read at a distance of 50 ft (15 m), printed or projected letters should be at least 2 in. (5 cm) high and ¼ in. (8 mm) thick.

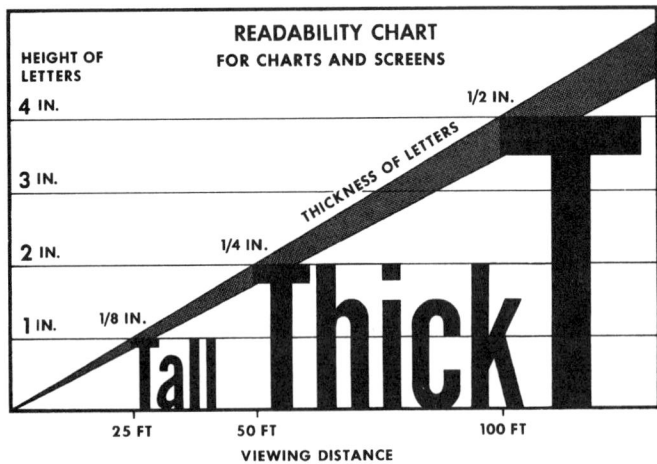

Figure 32–5. This chart shows what heights and thicknesses letters and symbols should be for easy viewing of projected and nonprojected visuals.

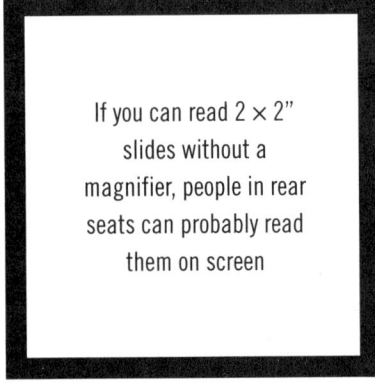

Figure 32–6. Rule of thumb for lettering to be shown on a 2 × 2-in. slide. *(Printed with permission from Eastman Kodak Company.)*

Aligning Text

Text set on the page is aligned in straight or ragged (uneven) margins. Text said to be *flush left* has a straight left margin and a *ragged*, or uneven, right margin. Text set *flush right* has a straight right margin and a ragged left margin. When text is *full justified*, it has both left and right straight margins. Text charts are generally easier to read, and easier to produce, if designers set the text flush left. This is because viewers are accustomed to looking first at the left-hand side of a page. For clarity, double-space between lines on a text chart and keep lines short. Try to put no more than eight lines on a single text chart, and the addition of bullets for each line is recommended.

Capitalization

Studies show that viewers find it easier to read text set in uppercase and lowercase than text set in all capital letters, or *all caps*. All caps works for short headlines, but not for body text. If designers and trainers want to emphasize certain words, *boldface* (heavy) fonts work better in headlines. Inside a paragraph, however, *italic* fonts work better than either boldface or all capital letters. In this modern computer era, underlining is rarely used to emphasize words or sentences; however, it is more of a personal preference than an effectiveness issue.

Plenty of white space in visuals makes them more effective. Double-space text for high readability. For projected visuals, use no more than six lines of text in a title chart, six words per line as a general rule. Computer graphics and desktop publishing software make it easy to create text charts and legends for graphs by letting users change and control the font. Traditionally, font size is measured vertically in points, and line lengths are measured in units called *picas*. As a rough rule, 12 points = 1 pica, and 6 picas = 1 in. (2.54 cm). The conversion from picas to inches is slightly different on a type gauge (a special, inexpensive ruler available in most art supply stores) because the computer software rounds off the conversion.

Computer instruction manuals and "how-to" books about popular software, available at most computer stores, are good sources of helpful information. They often suggest appropriate font sizes for varying usage. Some even specify a particular size as the default, unless the user deliberately overrides the program. Usually, the designer can preview how a chart will look and make any adjustments before printing/publishing the final product. Font specification sheets show different styles and sizes of fonts. If designers and trainers consider working with an outside media service bureau to have overheads, transparencies, or even posters made, the bureau's designers can give a quick overview of what will look best for that particular purpose.

Because each font has its own personality, the number of characters per line may vary even for fonts of the same size. Some fonts have a relatively high *character count* per line and look smaller than others. Commercial printers and service bureaus can help designers and trainers choose a font that will give visuals the clearest, cleanest look. Barring quotes and specific definitions, it is notably emphasized that any visual media should present only the main and needed words, omitting filler words wherever possible. At no time should the visual media be a written training book read to the trainee verbatim.

Graphs

Graphs are a visual summary of data. To make them easier to understand, always put a legend or short title alongside the graph to tell viewers what the information is about. Graphs work most effectively when they are matched to the information presented. Although there is no rule about which type of graph to use, designers and trainers should find the following suggestions helpful.

Circle Graphs (Pie Charts)

A *circle graph*, more commonly called a *pie chart* because of its circular form and pie-shaped wedges, is useful to show the relationship of each part to the whole. For example, a circle graph can show that out of 1,000 reportable accidents, 250 of them took place at only one facility. This would clearly demonstrate the need to improve safety at this facility. Circle graphs also help illustrate good safety records. Of those 1,000 reportable accidents, for example, perhaps only 16 took place in the shipping department. Remember: use a circle graph to show data from only one time period. Do not compare two years' accident records within a single circle. Instead, use two separate circles, one for each year.

Creating a circle graph is fairly easy. Computer presentation software will make a graph automatically if a user merely enters the data. If drawing the circle graph by hand, remember that the total circle contains 360 degrees. The angle needed for each component of the graph must correspond to the correct percentage of the circle. In the preceding example, the 250 accidents represent ¼ of the 1,000 accidents. Consequently, the circle "wedge" drawn must represent ¼ of the 360 degrees of the circle, or a 90-degree angle. All the components in a circle graph should add up to 100% of the total amount represented, just as the angles for each of the wedges should total 360 degrees. To break data down further, create a chart within a chart. For example, a wedge is only part of the whole circle, but can become its own series of data. Pull the wedge to one side of the visual for extra emphasis, and turn the information contained in that wedge into a column chart.

Assume that of those 250 hypothetical accidents, 50 occurred in the warehouse, 175 happened in the factory, and 25 took place in the facility office. For the column chart, which is basically a vertically stacked circle, 250 becomes the 100% total. Computer presentation software such as Microsoft PowerPoint or similar software will calculate the percentages and draw the column automatically. The column will look more balanced if the largest section (with the biggest percentage) is placed on the bottom. Emphasize that section's importance by giving it a dark color or heavy pattern. Link the circle and the column for an effective visual (Figure 32–7).

Because most people look at the right-hand side of a graph first, place the most important "wedge" of the circle at 3 o'clock on an imaginary clock face. Moving counter-clockwise (backward) around the clock, each wedge following the most important one should be a slightly lighter color. If the chart is in black and white, rather than color, give the most important segments black or dark gray colors. Make the least important elements light gray or white. Avoid putting too many wedges in a circle so as not to confuse the audience; six or seven wedges is about the limit an average viewer can comprehend. When there are too many wedges, it is hard to compare their sizes.

Computer presentation software allows designers and trainers to make proportional circle graphs or three-dimensional ones. Simply because they can, however, does not mean they should. Two circles on a single visual, equal in size but representing different amounts of data, are often difficult for viewers to interpret. For example, suppose that facility A has only half as many accidents as facility B. One circle could represent the number of accidents at facility A, and a circle twice as large could represent the number of accidents at facility B (Figure 32–8). Although the proportional circles shown in Figure 32–8 are statistically accurate, they may mislead the audience. Most people find it hard to judge circular proportions quickly because they tend to think in terms of line lengths. Three-dimensional circles (which show the "depth" of the graph) also can be misleading. If the top of the circle is made lighter to show perspective, that section may appear too large to represent its data correctly. It must be noted that most people will make a judgement within 1 to 3 seconds, which is hardly enough time for in-depth analysis, so avoid confusion and keep it simple and easily discernable at a glance.

Bar Charts

Bar charts are one of the simplest ways to display data to an audience. They are used to illustrate comparisons of volume over time. For example, a bar chart can show the dollar amounts a company has paid in workers' compensation over the past 5 years. Designers can set the bars horizontally or vertically, provided there are only 10 to 12 bars on a single visual. If bar labels are long, a horizontal bar chart is a good choice. When the data permit, arrange the bars in sequential order, either low to high, or high to low. Ranking lets viewers compare data more easily. Computer graphics software allows users to choose among several designs, such as overlapping bars, stacked bars, or paired bars. When choosing an overlap option, start the series with the smaller values first, so that the shorter bars in the overlapping groups come before larger ones.

Line Charts

Line charts are plots of points that trace changes in data over time. For example, "How many accidents did the company

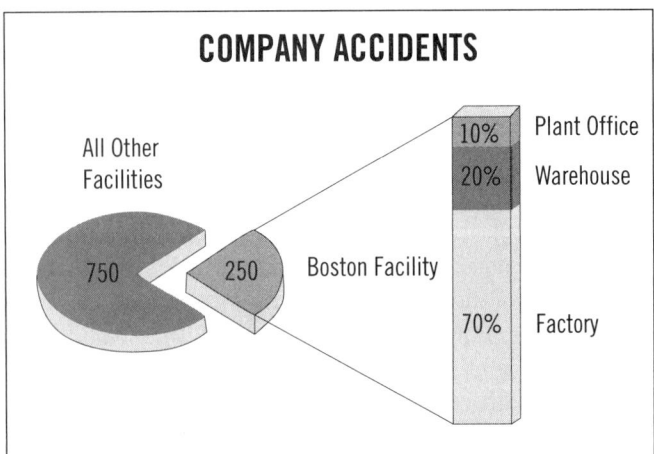

Figure 32–7. A pie chart can be created by a computer graphics software package to show accident (or other) data at a glance. *(Courtesy of Karen Zmrhal.)*

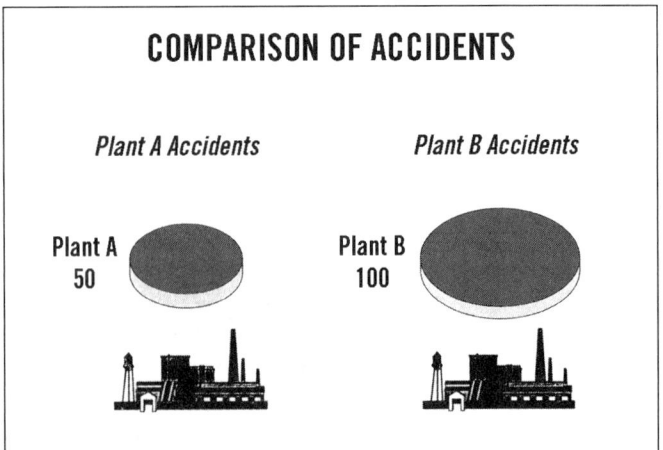

Figure 32–8. Two pies of different size may be clearer than one pie with several slices. In this computer-generated chart, the area of the pie on the left is 50% of the area of the pie on the right. *(Courtesy of Karen Zmrhal.)*

have in the machine shop from 1982 to 1991?" is a good question to answer with a line chart. The years, which are the independent variable, go on the *X*, or horizontal, axis, and the number of accidents per year go on the *Y*, or vertical, axis. Plot the points and draw a line connecting them, or use computer graphics software to draw the line chart.

Area Charts

Area charts are a good way to show volume, especially to emphasize changes in data. These charts use lines and patterns to separate different kinds of information. Designers and trainers can use area charts as an alternative to a line chart if they are working with a single series of data. For example, to show dramatically how accidents have declined at a factory, an area chart can make the information easy to remember. Designers can combine data series by constructing an area chart with several layers: e.g., the total number of accidents and the number of accidents in the office, the factory, and the warehouse. To make a multiple stacked chart, put the darkest pattern on the bottom and work from dark to light, with the lightest pattern on the top of the stack. Either shade the chart by hand or use computer graphics software to create the chart from data values that have been entered.

Drawings and Sketches

A simple drawing, done by hand or with a computer program, can often help make a process or procedure clearer. Keep the illustration as simple as possible, and always include a brief title or explanation next to the drawing. For consistency within a single presentation, try to keep elements of the visuals, if they are repeated, in approximately the same place for each of the slides or transparencies. If a company logo is used on each visual, always place it in the same location on each slide, transparency, or video media. Do the same on all drawings for titles or explanations.

Use of Colors in Media

Skillful use of colors will add greatly to the impact of visuals. On charts (including those that are all text), a too-light background makes information difficult to read. Overheads with lettering on a clear background project well; so do transparencies and slides that use *reverse lettering*, or white text on a darker background. Remember that approximately 10% of the population has difficulty distinguishing the colors red and green. If lettering or graph markings are in either color, such persons will see them only as varying shades of gray, and the impact of the message will be lost.

When using computer presentation software to generate text charts and graphs, limit color combinations to no more than four colors in a projected visual. Presentations are generally more readable in a darkened room if the background of the slide is a dark color such as blue or deep red and the words on the slide are in a lighter color such as white or yellow. Slides with dark type on a white background are difficult to read and should be avoided. Experiment with color schemes and test them for visibility and legibility. A good color concept on paper may not look good on screen, especially for viewers sitting in the back of the room. Make sure lettering and background offer sufficient contrast; green lettering on a gray background, for instance, blurs at a distance if the values of each color are too similar.

Computer Presentation Software

Presentation software such as Microsoft PowerPoint comes with *templates*, or predesigned slide styles that specify type size, justification, and color schemes. Users only have to enter data or text, and the program's automatic layout features ensure consistent, attractive results. Software like PowerPoint usually includes a predefined color palette, which can be modified. Some software can read data directly from existing spreadsheets and worksheets, including text, titles, labels, and graph settings, thus eliminating time-consuming double entry of data.

Projected Visuals

Projected visuals include slides, PowerPoint presentations, films, television, video, and podcasts, to name a few. This chapter does not cover projected visuals in great detail because excellent information is available from the manufacturers of films and software stores, libraries, schools, and publications. The National Safety Council's *Safety+Health* magazine, for instance, carries detailed articles on producing projected visuals, and data sheets are available through the National Safety Council Library (see the References at the end of this chapter).

VIDEO MEDIA

Technical training by videotape and DVD is used nearly everywhere because of its portability and other considerations.

- *Job site brought to the classroom.* A major advantage of video is its ability to show worksite situations. Students easily identify with on-screen visual representations of conditions or problems.
- *Familiar format.* Almost everyone is used to receiving information from a television screen, and most U.S. households have video recording devices, computers, and Internet connections that enable video streaming.
- *Instant playback.* Trainers can replay a sequence, pointing out specific techniques or incidents. Students viewing

the video can see material over and over again until they feel they have mastered the content.
- *Availability of easy-to-use equipment.* It is easy to transport and play videotapes or DVDs. Increasingly, video can be digitized and included in computer-based presentations or streamed through an Internet connection.

Showing Prerecorded Videos

Safety videos on a wide variety of subjects are available, either as single topics or as part of a series. Many videos, including a number of those offered by the National Safety Council, come with additional learning material designed for reinforcement: textbooks, instructor's guides, and program guides. These are easy to distribute, and the Internet is increasingly becoming a place for training departments to deliver training to dispersed work forces through online training programs.

Working with a Video Production Company

Because people are so familiar with television programs, videos, and DVDs, they are highly conscious of production quality. Unless a company staffs experienced professionals, the firm should consider using an outside training organization, the company's own media production department, or an outside video production company to achieve the visual quality and impact needed for effective training. However, the cost of video cameras and editing software has declined dramatically, and it may be cost effective for an organization to produce its own training videos using consumer-grade equipment. Most audiences are familiar with online video-sharing sites where consumer-generated media such as short video clips are made available, and there may even be an additional cachet of credibility associated with a program if the audience knows that it was created by their colleagues in the organization rather than by professional actors and videographers. In many organizations, the employees covet the ability to be considered for such opportunities, which, when utilized, raises morale and provides a greater credibility to the video.

Before trainers commission a custom-designed training video, they will need to answer the following questions:
- *Benefit/impact*—How will the project save the company money or improve productivity, safety, or efficiency? Can the trainer provide "hard data" on potential benefits?
- *Background*—Why does the safety professional believe that a video is necessary? Can he or she describe the situation or problem that needs to be communicated? What solutions has the company already tried? What solutions, other than video, is it considering?
- *Objectives*—Can the designer or trainer describe the session objectives in behavioral terms and list them in order of priority? Should the video primarily train (impart specific skills), inform (give background, ideas, facts), or motivate (difficult to measure effectiveness or results)? What should the audience be able to do—or stop doing—after they see the video?
- *Audience demographics*—What are workers' ages, years with the company, gender, education, and so on?
- *Audience interest/need*—Why should the audience be interested in the situation shown in the tape or in solving the work-related problem? How will they benefit from the video?
- *Audience knowledge/experience*—What does the audience know about the situation or problem? What is their past experience or involvement? (Before video production is started, the producer should poll the target audience for their input.)
- *Audience attitudes/prejudices*—What attitudes do the audience members already have about the situation or problem? What motives might the audience have for paying attention to, or ignoring, the video?
- *Company goals and strategies*—How does the intended video support the company goals and strategies?
- *Generic value*—Can the proposed video be used in more than one facility? Can it be written generically and still meet company objectives? Can alternative scenes be shot to make the program generic? Would other organizations be interested in the program?
- *Use*—How will the program be used? What will the viewing environment be? Will the video be used as a "stand-alone" program or as part of a course? Will any support materials (leader's guide, student study guides, handouts, booklets, etc.) be needed to go along with the program? Who will develop them?
- *Evaluation*—Will the video be evaluated formally, with specific questions, or informally through comments? Who will do the evaluation, and what is the evaluation strategy? Will the video be previewed with a sample from the target audience before production is finalized? Will the video be a part of a training program or a stand-alone video? If it is meant to be a stand-alone video, will it be tied to an LMS for viewing documentation?

Commercial versus In-House Production

Today's workers are far more experienced and sophisticated as viewers than employees were in the pretelevision era. According to the American Academy of Pediatrics, by the time the average person reaches age 70, he or she will have spent approximately 7 years watching television. Data from A.C. Nielsen, a market research organization, indicate that adolescents age 12 to 17 watch more than 23 hours of television a week, with this number rising. As a result, viewers expect that in-house-produced media will equal commercial productions in quality.

Companies with in-house production organizations can consult with training designers and the safety professional about their needs for slides, transparencies, posters, and audio or video materials. The media department can provide cost estimates, show trainers what has already been done for other company departments, and—if a custom-designed presentation is required—work with the trainers from conception to execution. If trainers want a second or third opinion (and they should always get more than one bid on any major film or video project), they can include their in-house organization as one possible alternative. They also should look outside the company for others, such as independent production houses or organizations that specialize in custom training materials.

State-of-the-art technology and equipment can develop complex programs at surprisingly affordable costs. Although the average training department's personnel may not have the skills and expertise of media professionals, they can produce training materials with sophisticated production values using computer software and inexpensive audio and video equipment. A variety of training materials can be made available to widely dispersed employees via the World Wide Web, reducing travel costs for training. The safety professional, training designer, and trainer are the experts on what should be communicated and on the required training methods. Media professionals are the experts on how to manipulate the media to get the best effects and quality. With the advancements in small portable video equipment currently entering the market, the best source for video-quality charts and recommended settings is often the video equipment's instruction manual.

The wide range of commercial products, however, should not exclude the use of homemade materials. Trainers can easily make inexpensive flip charts, paper-and-pencil charts, chalkboard presentations, digital photos and slide shows, and audio and even digital video interviews. Some software packages are fairly easy to use and will produce acceptable media materials.

The choice between professionally developed or homemade materials always depends on a number of factors that the trainer must weigh according to their relative importance for his or her organization. These factors include the number of showings, how many people will be delivering the training, the size and composition of the audience, the degree of customization (is this aid for general use, or is the information for a particular department or facility only?), and the importance of the message. Hence, all cost factors need to be considered and determination based on acceptable return on investment.

Choosing a Production Company

If management decides to produce a custom-designed video, trainers will want to get several bids before selecting one firm. The organization's media production company, if it exists, should be one of the candidates. As a basis for bid comparisons and to help manage client expectations, one major company has developed a program classification matrix describing four levels of video production. These levels range from a "talking head" video in which someone simply faces the camera and delivers the message (Class I) to a comprehensive, top-of-the-line program comparable to broadcast television (Class IV), using professional talent, studio sets, extensive editing, and custom music. Use the program classification matrix as a tool in selecting the right video production.

Designers and trainers may want an outside consultant to help them evaluate the bids, check references, and evaluate past productions of any company they are considering. Remember that production costs do not include distribution. Get bids on duplicating and distributing the video from several vendors. If a company has a culturally diverse work force, trainers should consider using a production firm specializing in translations to do voice-overs or narration for the video. For maximum effectiveness, use a competent, experienced organization that is sensitive to nuances of language and cultural differences.

Shooting In-House Videos

Photographing short onsite videos with a camcorder is not difficult to do and may be an option when budget restraints have priority over production quality. Often, camera stores run free or low-cost classes in "how-to" techniques. Local television studios may offer cable access classes in video production; high schools and community colleges frequently have training classes. Before investing time or money in making videos, be sure that the video can be played on company-owned equipment or that management can acquire appropriate, cost-efficient equipment.

Designers and safety professionals/trainers who decide to shoot their own videos should make them short and simple. They may want to have a colleague videotape their training sessions before giving presentations to students in order to review, analyze, and improve their performance. For onsite shooting, plan a script carefully. Using storyboard techniques will help to yield the best results for the message that is to be conveyed. Be sure to obtain appropriate releases from those who appear in the video.

COMPUTER-BASED TRAINING

Computer-based training, also called *computer-assisted training*, is a viable, high-tech media option that companies have used with good results. With direct connection to Internet training services in a wide variety of applications, computer-based training opportunities have greatly expanded in recent years.

Benefits and Drawbacks
The benefits of this media form include the following:
- *More efficient learner-centered training*—Students proceed at their own learning rates.
- *More timely training*—A training module can be given to a student whenever he or she needs it.
- *Increased student-to-instructor ratio*—One instructor can monitor more students who are using computers than is possible in a conventional classroom setting if enough computer workstations are available in one location.

Because concepts and "how-to" training are standardized, computer-based training can be more effective. All students receive the same information, regardless of their location or their teacher. Interactive computer-based training provides standardized feedback, giving the same responses each time a student answers a question. Also, because students have the opportunity to practice until they master the skills addressed in the training, they become more proficient (Figure 32–9). If individualized training is desired, interactive computer-based training can provide pretests and posttests for students, tracking and recording their performance as they complete various segments of the program. Some interactive computer-based training programs can be customized to adjust the work within a module, based on a student's performance.

Drawbacks to computer-based training include the cost and time necessary to develop or purchase high-quality training, based on specific training needs. Generic computer-based training packages are difficult to match with an individual company's objectives. Organizations that wish to develop their own programs (or work with an outside organization that will do so) must be prepared to invest time and money. Programs often need to be tested and refined several times during development. A second drawback to computer-based training lies in its standardization. Not all people learn in the same way or think alike. If computer-based training is used by itself, without an instructor to answer questions, students may learn to respond with the "correct" answers but may not understand those answers or why they are "correct." When faced with a situation on the job that is slightly different from the computer version, they may not know how to handle the problem. Because a computer-based training program is impersonal, keeping students "on target" and motivated can be difficult. Well-designed training materials can help, but the student's attention span may be shorter than would be the case in a more conventional training class that promotes interaction between participants and the trainer.

Assessing Needs
Deciding whether computer-based training is appropriate for an organization is easier if managers use a systematic approach. They will need to evaluate their training requirements carefully. Are they training for specific skills needed to operate certain equipment? If so, hands-on time with the equipment is necessary. If the problem is shared by many companies across a variety of industries, computer-based training software packages may have already been developed.

Programs in widespread use in business, such as word processing and spreadsheets, often come with self-paced on-screen tutorials. Training on how to use such software may be available on audiotape or video. On the other hand, if managers want to show new employees how to change machine tool setups by reprogramming a computer-controlled machine, the company will almost certainly have to develop its own computer-based training or find a different vehicle for training. If company training programs must meet regulations that require documented employee training, trainers may find that commercial packages already have been developed on these topics for the industry. Before investing in any packages and the necessary hardware to run them, however, designers and trainers should ask whether vendors will release names of companies that have used the computer-based training successfully or that can provide evidence that the training does the job it is designed to do. A substantial investment in computer-based training also may be hard to justify if new procedures, new equipment, or new regulations make the program outdated soon after its production and distribution. (For additional

Figure 32–9. Interactive, computer-based training is designed to allow the student to work independently, correct mistakes, and reinforce correct responses. *(Courtesy of the Eastman Kodak Company.)*

information on training needs analysis, see Chapter 31, Safety and Health Training.)

Computer Learning Formats
Computer learning is achieved by a question-and-answer or challenge-and-response format. These formats require the student to participate actively in the training. Two types of interactive materials are often stressed: conceptualization and simulation.

Conceptualization
Primarily used for training that depends on theory or other abstract ideas, *conceptualization* is a good choice when employees must learn "soft skills," such as human relations or management training. This type of training helps communicate psychological or philosophical viewpoints.

Simulation
The use of *simulations* that duplicate the look and feel of real experiences has proven to be a valuable learning tool. The most sophisticated simulators are probably those used by the aerospace industry to train pilots. To the extent possible, the simulated cockpit replicates visual, auditory, and tactile experiences of real flight. The primary advantage is to place the pilot in emergency situations under controlled conditions so that safety procedures can be practiced. Trainers can evaluate both the procedures and the pilot's performance.

Simulated electrical panels are used widely for electrical training. The electronic components for a machine or system are installed on a board so that trainers can create problems requiring students to practice troubleshooting skills in a classroom setting. Fire training institutes and some fire departments use simulations to create various fire-fighting and rescue situations. Trainees then practice the recommended techniques under controlled conditions. The National Aeronautics and Space Administration (NASA) uses special water-filled tanks and weighted suits for astronauts to simulate working in a weightless environment. Some mining companies have replicated underground conditions in surface training labs to enable miners to practice machine operations under controlled conditions.

Computer-Managed Training
Trainers can choose to use *computer-managed training* along with computer-based training. Under computer-managed training, computers monitor students on their learning time, attendance, and participation; the training materials students have used; tests taken; and the final results achieved. The computer can keep track of how a student performs and suggest additional materials or extra drills if the performance needs improvement. If some students complete the modules ahead of schedule and are ready for a higher level of training, the computer can generate a list of names for the instructor's review.

Interactive Training
Interactive training or *e-learning*, using computer-based programs, sometimes in stand-alone kiosks, allows a person to be trained systematically by simply touching the screen to choose from various options. The user does not have to know anything about computers nor key in the answers. Increasingly, organizations are moving away from stand-alone kiosk approaches to e-learning distributed via the Internet that, in many cases, is tied to LMSs. The system combines sight, sound, and animation in a single presentation, thus reinforcing learning. Because training is one-on-one and requires computer or touchscreen interaction by the worker before the module continues, employees stay motivated and involved in learning. Training modules can also include quizzes and interactive feedback. Interactive technology offers a number of advantages, including the following:

- *Reduced learning time*—In some studies, learning time can be reduced up to 50%. Immediate feedback provides constant reinforcement of concepts and content.
- *Reduced delivery costs per student*—In interactive training, the largest costs cover design and production, not duplicating or distributing the film. Delivery cost per student decreases as more students use the same program. In traditional training systems, which depend heavily on teachers, delivery cost remains constant or rises as numbers of students increase. Even for a custom interactive program, the cost-per-student breakeven point occurs with a relatively small number of students. For large organizations with many students who need to learn the same material, savings can increase dramatically.
- *Reduced risk*—Interactive systems can allow students to explore potentially dangerous subjects without compromising safety. For example, a course on basic electronics and maintenance lets students accidentally "touch" the wrong parts without risking electrocution, yet "see" the consequences on screen.

Because interactive systems do not allow students to move on to new material until they have demonstrated their mastery of fundamentals, trainees gain a strong foundation on which to build further skills. Yet because training is self-paced, students can repeat and review materials until they are confident they have mastered the information.

SUPPLEMENTAL MATERIALS

Supplemental materials can enhance and support a company's media training tools. Whether supplemental materials are stand-alone safety reminders or designed as part of a comprehensive presentation package, they can increase safety awareness and reinforce training messages.

Posters

The purpose of posters is to create safety awareness and educate or reinforce. A number of commercially available safety posters on various topics are offered by a variety of organizations, including the National Safety Council. State-of-the-art procedures, single-topic reminders, and compliance/regulatory warnings are good visuals. Displayed at various locations throughout an organization, posters can help employees avoid hazards and can be a useful part of a company's comprehensive safety program. If a designer, trainer, safety professional, or other employee is making posters, keep them simple and short for maximum impact. Each poster should convey only one main idea, with no more than two or three additional details for reinforcement. Eye-catching visuals can be used; for example, a drawing of wiggling toes sticking out of a cast that is decorated with signatures of fellow workers, and the message "Wearing Safety Shoes Might Have Prevented This," make the point graphically.

Ideally, posters should be large: 16 × 20 in. (41 × 51 cm) is a minimum size, although 22 × 28 in. (56 × 71 cm) is better. If posters are too small, people will pass by without noticing them. Vivid colors such as golds, yellows, hot orange, and brilliant blues grab attention immediately. For even more emphasis, highlight with a splash of contrasting color. Type should be as large as possible—38 to 72 points, or letters at least 1 in. (2.5 cm) high for headlines. Designers, trainers, and/or safety professionals can use markers, transfer letters, cut paper, or India ink for lettering. Keep in mind the issues of adults with color blindness previously discussed in this chapter when finalizing the color selection.

Pictures

A good safety photograph also attracts attention. Photos can show the safe way to perform an activity, such as lifting a package, putting on a respirator, or using a computer keyboard correctly to minimize the risk of carpal tunnel syndrome. Like posters, photographs should convey one main idea to be most effective. Pictures of people are always more interesting than those of equipment alone; close-ups are more attention-getting than long, "establishing" shots. If the photographer is shooting a series of photos, however, he or she may want a long "establishing" shot to set the location and orient the viewer, plus some medium shots and close-ups for variety. Someone skilled at focusing may want to use a tighter close-up by including only the important details, such as focusing on a worker's hand to illustrate proper tool use.

To ensure top-quality photos, be aware of the "color balance" of the light in the environment being photographed. Professional-quality digital cameras and even many consumer-level cameras can be adjusted to balance the light in the scene. To improve the chances of getting successful photographs, overshoot; that is, take substantially more pictures than needed. Taking four or five photos of a single setup to get one usable picture makes economic sense. Bracket exposures; adjust the camera meter and settings for what will be the ideal exposure. After snapping the picture, take at least two more, one f-stop under and one f-stop over the recommended setting. For trainers who want to brush up on their photography skills, camera stores, equipment manufacturers, high schools, and community colleges often offer free or low-cost classes on simple techniques. Digital cameras are more forgiving than film cameras, and underexposed or overexposed images can often be salvaged in photo-editing software.

On the other hand, if the project calls for a number of photographs, particularly in a factory or large office setting, or if the upcoming event is a major one, it is almost always more cost-efficient to use a professional photographer. Professional photographers have the lighting equipment and know-how to make a shoot run smoothly. Clear the way with the supervisor and with whatever authorities might be involved, including the facility's manager, and any governmental body that might have jurisdiction. Get releases from all persons photographed (including employees and managers) before including them in pictures. The company's legal department can draw up the correct form, which should assign to the company the right to use the photo in a wide variety of ways. It is also a good idea to negotiate with the photographer in advance to purchase the digital image files, often called a *digital reprint license*. In some cases this will culminate in paying a slight premium for the right to own the digital images. Photographers used to earn significant income from customers who ordered reprints of images or duplicate slides, but in the digital photography environment, those sources of revenue are less common, so it is appropriate to offer the photographer a premium payment in return for possession of the actual digital images on CD-ROM, DVD, or larger storage media or to have the images e-mailed to the safety professional. Such digital images can be used repeatedly, e-mailed, and infinitely copied.

Handouts

Handouts represent another supplemental way of reinforcing training concepts or of testing specific skills taught in the presentation. Employee safety booklets, guidebooks, manuals, compilations of case studies, and "hard copy" of overhead transparencies and slides are good learning aids. Many are available commercially from various sources, including trade associations and the National Safety Council. Service bureaus can turn overheads or slides into hard-copy handouts. However, the projected visuals will be summaries, with few words and lines on each overhead or slide. Handouts give the speaker a chance to expand on the information shown on screen, including more details or cases. When designers and trainers plan to use such handouts as part of a comprehensive program that includes other visual media, they should make sure not only to preview visuals, but to check the printed handouts carefully. Both media should give the same message, not confuse students.

Audiocassettes/CDs and Podcasting

Used as stand-alone training or as components in a comprehensive package, audiocassettes can promote safety effectively. Purchased cassettes or, increasingly, CDs—some available with response booklets and self-tests—are valuable for individualized training or refresher training. Some are available commercially with training in a choice of languages. Because students can replay audio programs, they can go over material until they have mastered it. Audio also can be an integrated part of classroom presentations. Since 2003, distribution of audio and video programs over the Internet has been facilitated by the use of RSS feed technology. RSS is a version of web page coding that enables a content producer to alert interested parties when new rich-media content such as audio or video has been made available. Its advantage is that the content producer does not need to transmit large digital audio or video files to the recipients, merely information about where the content is stored. The recipient determines how and when to retrieve the content material. This distribution channel is sometimes referred to as *podcasting*, a term that recognizes the Apple iPod digital media player as a common end-user choice for playing the content and the origins of the distribution concept in traditional broadcasting.

Designers and safety professionals who make their own recordings should use a microphone on a desk or floor stand to ensure a high-quality recording. Handheld microphones will record sounds from the manipulation of the microphone and should be avoided.

Digital audio recorders are widely available and offer substantial recording times, often 4 hours or more. Digital recordings can also be edited and mixed with music using sound-editing software. The resulting files can be integrated with computer slide presentations, transferred to audio CDs, e-mailed to recipients, stored on storage devices, and/or uploaded to a server for easy access via any Internet-ready computer or device.

Additionally, safety professionals and trainers can do brief, on-the-job interviews that may prove useful later for bulletins, newsletters, and future meetings. They also can dictate a running commentary while photographing safe or unsafe operating conditions, new processes, or equipment that requires subsequent study. They can record minutes in safety meetings or valuable discussions in training sessions for future use. In addition, safety professionals can record safe, efficient job procedures, provided they allow sufficient time for performing each task. The trainee then listens to the tape and follows the training while performing the work.

TRAINING PRESENTATION AND AIDS

Most modern meeting rooms and classrooms are well designed for media presentation. Occasionally, trainers may find themselves leading a training session in a less-than-ideal environment. The tips in the following sections will help in any situation.

Tips for Presentations

The following are some common tips for presenters whether providing a training session, facilitating an exercise, or presenting a business case.

- Make sure the room to be used is available. Know in advance what training activities are planned and where the participants can go if forced to switch rooms at the last minute. Will the planned equipment work in an alternative location? Can three-pronged adapters or extension cords be obtained fast, if necessary? Is a chalkboard on wheels available? Is there chalk and an eraser? Will the number of tables and chairs needed fit in the substitute room? Will someone move additional tables and/or chairs there, if needed?
- Collect everything trainers could possibly need for the presentation: if they are using a chalkboard, extra chalk and erasers; if they are using an audiocassette player, extra batteries, AC extension cords, and power strips. In many locations, rooms are retrofitted for use as training rooms but were originally designed with electrical outlets in locations inconvenient to the trainer's need to set up projection equipment. Power "squids," a kind of power strip with outlets on long, extending tentacles, can be useful in making difficult outlets more accessible. Trainers should have essential items, such as spare projector bulbs and fuses, and make sure trainers or staff members know how to replace bulbs and fuses.

- Be sure the person responsible knows how to operate all the equipment to be used. Have trainers coach and rehearse their talk with a substitute in case the first helper is unable to attend the meeting.
- Make an agenda of the meeting with approximate times listed for each part of the presentation. Trainers should give the agenda to their helpers well ahead of the session. Have an extra copy of the agenda on hand in case a last-minute substitute helper becomes necessary.
- Staff should know when the speaker wants the lights on and off or adjusted to a particular level. Determine who will adjust the lighting controls.
- Check the equipment a day ahead of time to be sure it is in good working order. If equipment is shared with other users, safety professionals and/or trainers should make sure they have signed up for the required equipment well in advance. Double-check the day before the presentation to make sure no one has preempted the reservation. Does all equipment "match"? That is, is the VCR/DVD player the correct model? Don't take anything for granted; assume that if something can go wrong, it will, and prepare accordingly.
- Make sure that prerecorded media (e.g., videotapes, DVDs, CDs, and audiotapes) are of acceptable quality for presentation. Preview everything well in advance, if possible, in case alternatives are needed. Time any prerecorded media in advance to be sure the presentation will run smoothly and on schedule.
- Also preview projected visuals made in-house. If possible, use the same room, equipment, and lighting chosen for the presentation. Slides that look good on a slide sorter may be under- or overexposed when seen on the actual screen to be used at the meeting. Leave slides or transparencies on screen and sit in the farthest chair. Can the visuals be read comfortably? Are they sharp and in focus?
- Before the presentation begins, preset the brightness and audio controls on projectors, audio equipment, and video monitors to avoid disrupting the program later to make adjustments.
- Make sure all the slides to be used in a presentation are in order and good condition. When the slides are in proper sequence, draw a line diagonally across the top of the pack. In this way, a slide that happens to be turned accidentally on its side or moved out of sequence will be noticeable at a glance.
- Use trip guards to tape down electric cords so they won't pose a fall hazard. (Most professional music stores carry special gaffer's tape specifically designed to tape down cables and power cords. Avoid using duct tape for this, as the adhesive is difficult to remove from cables and can often damage floors and carpets.)
- For effective communication, presenters should face the audience or keep their faces turned in profile while discussing slide shows, movies, and videos.
- If a loudspeaker is used, place it near the screen and slightly elevated for realistic-sounding audio reproduction. Presenters can achieve more freedom of motion around the room if they use a wireless lapel microphone rather than a fixed mike on a lectern.
- Test microphones ahead of time to be sure they actually work and that the volume is set at the correct level for the meeting room conditions.
- Make sure microphones are placed at convenient locations if audience participation is planned so that everyone can hear what is said. Wireless handheld mikes can be used to amplify audience questions.

Seating

Check the presentation room for safety features and for seating arrangement. Aisle space should be adequate, and exits should be available for emergency use. The seating arrangement should give every member of the audience an unobstructed view of the visual. White matte screens do better at an angle of 40 degrees or less; beaded screens are not as effective for viewing at an angle. To view a projected visual, the most desirable seating area is within a 30-degree angle from the projectors for a matte screen and within a 20-degree angle for a beaded screen.

The recommended minimum viewing distance is at least twice the screen width; the maximum should be no more than six times the screen width. A 4-ft- (1.2-m-) wide image, therefore, could be viewed at a maximum distance of about 24 ft (6.3 m); a 6-ft- (1.83-m-) wide image could be viewed comfortably at a maximum distance of about 48 ft (14.6 m). For meeting room or auditorium seating, allow 21 to 24 in. (53 to 61 cm) width for each chair and 36 in. (91 cm) for each row of chairs, or about 5 to 6 sq ft (0.46 to 0.56 m^2) per viewer. In classrooms or conference rooms, allow twice as much space.

Various seating arrangements work better for certain kinds of meetings. If a conference table is not available and the group consists of 5 persons or fewer, set up an overhead projector on a desk and project visuals onto a portable screen or against light-colored walls (see also Figure 32–2). Meetings with 6 to 12 people work well with a single center table. For groups of 20 or fewer, use a U-shaped table to encourage participation and discussion. Tables arranged in a herringbone pattern are good when the presentation is mostly a lecture format and the speaker wants to focus attention on talk and visuals. The tables provide work space, yet the participants can still see each other and interact. A more formal auditorium setting, or an amphitheater with elevated levels, probably provides little or no audience participation. In such settings, the speaker almost certainly needs a microphone and a screen and visuals large enough so the audience can see them clearly. In some extremely wide rooms, the seats cannot be arranged so the audience sits within the recommended viewing angle. For these situations, project slides or overheads simultaneously on two or three screens, separated by at least 25 ft (7 m).

Trainer/Speaker Aids

To be most effective in a training session, trainers should take full advantage of attention-getting devices, such as pointers, easels, charts, and teleprompters.

Pointers

A pointer, usually made of wood or aluminium, is a necessary item when a speaker is using a chalkboard or charts on an easel. Many computer presentations are controlled with wireless remote devices that allow the presenter to walk around the room while sequencing a presentation. Most of these handheld devices also incorporate a laser pointer that will display a red or green indicator dot on the screen. (*Caution:* Make sure no one in the audience looks directly at the source of the laser light to prevent injury to sight.) Using a pointer allows the speaker to face the audience while referring to the visual material. A speaker should not fidget with the pointer, however, because it distracts from the presentation.

Easels

Portable aluminium easels are helpful for holding flip charts and paper pads. Be sure the easel is sturdy enough to hold materials and will not collapse during a speech. Some easels fold for use on a table. To be effective, easels should be used close to the audience. However, the impact of the display is often lost when audiences are large or are more than 20 to 30 ft (6 to 9 m) away from the board.

Charts

Well-chosen and well-prepared charts and diagrams allow the audience to "see" what the speaker is saying. Charts tell a story faster and more clearly than can an oral presentation alone. Trainers who use charts (rather than projected visuals) should practice showing them before their talks to make sure the charts appear in the correct sequence and that the text is readable. Trainers who fumble with the charts or switch them haphazardly distract from their message, losing any impact the information may have had. If trainers discuss material other than topics illustrated by the charts, they should cover the charts temporarily to focus audience attention on what they are saying.

Teleprompters

Teleprompters or other cuing devices can be used for dramatic or formal presentations in which trainers/speakers are required to follow a prepared script. Teleprompters require technical help to set up. Trainers must know how to use the equipment so their presentations will appear natural.

Boards and Pads

Chalkboards, whiteboards (or dry erase boards), and paper pads lend themselves to small-group training and provide inexpensive visuals. Each is an excellent tool for stimulating discussion. Participants can comment on points the trainer previously listed, and the trainer can quickly record student responses or suggestions. Although chalkboards come in several colors, light green is considered standard. A dustless chalk should be used, preferably a strong, bright color for maximum visibility. Fluorescent chalks are particularly effective. Whiteboards have replaced many chalkboards. Colored felt-tip erasable markers are used to write or draw on the polished white surface. Marks are erased with an ordinary cloth with little or no dust. One advantage to these boards is that they serve a dual purpose as a projection screen and marker board for small audiences.

Large paper pads, available at art stores or office supply stores, are easy to use. Clamped on a lightweight or folding easel, such pads are portable and always ready for use. A speaker can keep training material written on the pad or discard it after the session. Feedback or questions from students, recorded and saved on the paper, can serve as a nucleus of ideas for a future training session. In brainstorming sessions, in which the leader's objective is to stimulate quickly as many ideas as possible without detailed analysis or criticism, pads are of real value. As each chart or page is filled, it can be tacked to a bulletin board or clipped to a wire running along one side of the conference room. Group members then have a continuous record of what has been discussed. Latecomers can use these sheets to catch on to the discussion, and the speaker has a good record of the meeting.

Inexpensive rolls of white paper can be used for charts; occasionally newsprint "ends" are available at small cost from a local publisher. Even brown wrapping paper will do if bright-colored, contrasting markers are used to write down the ideas. The speaker can write and read material while unrolling the chart like a scroll.

Flip charts, similar to paper pads but usually prepared in advance, often are used in more formal meetings. Blank pages can be provided for on-the-spot notes. Portable units are easy to make and transport to different locations. Magnetic boards also provide an inexpensive way to display materials. Trainers can make a magnetic board of either a spray-painted sheet metal plate or a steel-backed chalkboard. Magnetic boards are also available commercially. Small objects or cut-outs mounted on small magnets or on magnetic tape can be placed on the board and moved at will. Safety professionals often use this type of visual for training operators of vehicles such as forklift trucks. The mobility of the objects—toy vehicles or cut-outs of trucks, together with the cut-outs of aisles and loads—enables the safety professional and/or trainer to give a realistic demonstration of safe practices.

Organizing Props and Presentation Aids

As a first step in preparing for a training session or speech, go through the talk and the media materials well ahead of time. Make a duplicate copy of the outline, and mark items needed for the talk in the margin. Make a separate list of those items to ensure that nothing is left out. Use a simple notepad or a ruled accounting notebook page with multiple columns to keep track of all requirements. Head the columns "Have," "Need to Reserve," "Need to Buy," and "Need to Rent." Down the left side, list all required media props and presentation aids in the order in which they will be used during the talk. Don't forget to include items such as chalk, erasers, markers, and extra pencils or pens. Next, check the appropriate column for that particular presentation aid. In this way, a speaker can easily see what has to be acquired. Revise the list at least once before the presentation is scheduled, updating the status of each item. Use the far right column to indicate who will be responsible for seeing that the item is on hand. However, even if some tasks are delegated, the trainer is still responsible for double-checking that everything is ready and in the proper place well before the presentation.

Preparation and Rehearsals

Successful trainers and speakers never give presentations without preparing well in advance. The smooth, seamless talk that looks so effortless was undoubtedly rehearsed ahead of time, often more than once. By practicing their presentations, trainers/speakers are able to stay within time limits and to deliver a more effective talk. When a trainer or speaker is confident in what he or she is doing and handles accompanying media well, audiences are free to concentrate on the message being delivered. If a presentation includes several speakers, it is a good idea for all participants to show their materials to one another in advance. This is especially important if company executives or important visitors will be in the audience. This approach will help avoid duplicating information or media. Individuals may wish to revise their delivery or practice certain skills, depending on suggestions they receive from other trainers/speakers.

EVALUATION

Evaluation helps designers and trainers get the maximum benefit from the time and effort they have spent in creating and using media materials. They need to know how they are doing and what areas require improvement. Effective evaluation involves more than handing out forms asking audiences how they liked a speaker or presentation. A well-designed evaluation strategy can yield a great deal of useful information. Such a strategy, planned in advance, should consider the following:

- What will be evaluated? At times, trainers want their audiences to evaluate media quality. Other times, they ask for feedback on how well the media materials were used. A third question to consider is, "Did the materials and training session influence worker motivation or behavior?"
- Who will do the evaluation? The trainer? Audiences? The supervisor? Executives? Outside consultants? Establish who will evaluate the sessions.
- When will the evaluation be done? Trainers can lose thoughtful comments and feedback if evaluating becomes a routine, fill-out-the-form exercise after each training session. Instead, consider polling individuals in the audience to keep interest alive.
- How will the evaluation be implemented? Will formal rating sheets be used? Will the audience fill out 3 × 5 index cards and drop them in a suggestion box? Do trainers want to be ranked "by the numbers," or are they looking for descriptive comments? One way of self-evaluation is for the instructor to perform a validation on the students' written exams. The trainer could take an answer sheet and track all the questions missed by each student. This allows the trainer to see what is missing that would make a difference to the students moving forward.

No one strategy or form is "right" for a particular organization or situation. By working together with others in the facility, designers and trainers should be able to devise an effective evaluation strategy for their organization's needs. Also, they should evaluate the strategy itself, at least once a year, to see how it can be improved. (See Chapter 31, Safety and Health Training, for additional information on training development and evaluation.)

Creating an Evaluation Climate

For training with live presentations, designers, safety professionals, and trainers want honest comments, so they need to create a climate that encourages feedback and respect for each other's ideas. If trainers become defensive when criticized, or if workers believe that those who make suggestions will be penalized, trainers will not get the thoughtful appraisal they need. One useful approach that often works involves asking employees who say they don't like something to tell the trainer how they would make it better. By forcing them to focus on the positives, the trainer not only gets suggestions that might work, but reinforces the team feeling of "we're in this together." Just as trainers do in a brainstorming session, they should accept all comments without allowing negative discussion on any ideas presented. The purpose is to reinforce creative thinking and involvement, weaning the group from the "we've always done it this way" limitations.

Coaching students in how to provide feedback requires them to edit out the emotions. This is done by asking, "What is missing that could make a difference to you moving forward?" This shows them that the trainer is not only working toward positive change but is willing to accept input. In addition, it creates ownership for both parties. Another way to state this is to say to the student, "If you could have it perfectly laid out for you, how would it look, feel, and be presented?" This type of questioning sparks creativity and creates both ownership and excitement about the future. Most importantly, careful choices of words must create possibilities for the future of the organization, the trainers, and the employee/students.

Evaluating Media Materials

Criteria trainers can use as a starting point to judge the quality of media materials include the following and should be incorporated into the overall evaluation designed and created in Chapter 31, Safety and Health Training. Some questions to consider for alpha and beta rollouts of a good sampling of workers trained are as follows.

Appearance

Were diagrams and graphs appropriate for the information being displayed? Were easy-to-read fonts chosen? Was lettering large enough to be seen easily by the audience, no matter where a viewer was sitting or whether he or she wore glasses? If more than one color or pattern was used, were they distinct enough so viewers could distinguish information easily? Were the visuals focused properly? Were they appropriate for the message? Was the information they contained up-to-date? If "how-to" skills were shown, was the training paced slowly enough so that viewers could understand and follow the teaching?

Audio

If sound was part of the media presentation, was it clear and distinct? Was the volume level comfortable, but adequate? Was the sound consistent throughout the session? Was the sound appropriate for the visuals? Was the level of language appropriate for the audience and message? If technical terms were used, were they defined immediately? For live presentations, were microphones free of annoying electronic feedback or distortion? Were there enough microphones, correctly placed, for all who spoke? If a soundtrack was used in a second language, was the message accurately translated? Did the translation take into account any cultural differences or sensitivities?

Props

If chalkboards, flannel boards, or easels were used, were they placed so the audience could see them? Could each member of the audience see the material they contained easily?

Evaluating Media Use

When evaluating media use, it is important to take the time to fully understand what are the key points trying to be conveyed as part of the message. This relates back to the adult learning concepts and the WIFM ("What's in it for me?"). Survey questions to evaluate media need to be reviewed to ensure they provide the clear answers being sought since in many cases, it will depend on the questions' format and the audience providing the feedback. However, every attempt should be made to answer basic questions to ensure that presentations are effective and meet the objectives for which the presentations were developed and include:

- Did the media help make the presentation message clear?
- Did they reinforce what was being taught? Were the materials well suited to the presentation format, whether it was individual or small-group training, computer-based training or simulation, discussion, lecture, or a presentation by a panel?
- Were the materials integrated into the presentation appropriately—that is, at appropriate places and for an appropriate length of time?
- Had the trainer rehearsed enough with the materials to use them well and with confidence?
- Was the room comfortable?
- Was the seating arrangement satisfactory?
- Was lighting appropriate for the materials used?
- If projected visuals were used (overheads, slides, films, videos), did the person responsible for the projection handle the equipment properly and efficiently?

Evaluating Audience Reaction

Trainers can use the following questions as a checklist to evaluate audience reaction, adding their own questions to tailor the checklist for each presentation.

- Was the objective of the presentation motivational? If so, did the session stimulate or reinforce motives? Did the materials make a direct link to the core desire or identified need?
- Was the objective of the presentation skills improvement? If so, did the materials provide illustrations and examples of application? Did the presentation reinforce the message by also using learning activities?
- Was the objective of the presentation informational? If so, did the materials present facts and supporting material in a clear, logical structure? Was the pacing of the total presentation appropriate so that the audience could acquire new knowledge?

Many of these media questions are required for proper design of the overall training and hence will need to be evaluated early in the training cycle. These questions can then be removed for subsequent training deliveries, only to be reconsidered and/or reintroduced at varying time intervals as part of the ongoing training cycle review.

COST FACTORS

To determine whether the cost of a training medium is justified, management must consider what the expenditure will buy, how quickly the material will be outdated, how it will be used, and how many students will view it. Numerous options should be considered: selecting commercial training materials, modifying materials or having them custom designed, and designing and producing materials in-house.

A bottom-line cost for a single medium never tells the whole story; management must always rely on personal judgment to determine whether the return on investment (ROI) is appropriate for the organization and its needs. For example, interactive computer simulation of an airplane cockpit may seem expensive. However, by allowing a pilot to practice correct responses to emergency conditions, such simulation may prevent an airplane crash, with its potential loss of life and property damage. Because the up-front expense of interactive training is in design and production, costs per student for training are affordable if many students will use the material. One major company estimates that interactive training technology is saving $100 million in training costs.

Even in small companies, a training department can easily justify purchasing a computer graphics package to produce attractive slides and posters if the company conducts worker training frequently. Virtually every company using personal computers has access to some version of a presentation-creation software package installed on its computers, and cost-effective or free alternatives have also become available on the Web. (See also Computer Presentation Software earlier in this chapter.) The training department can make the case that effective training will help cut down on employee turnover as well as accidents, thus lowering company costs. Inexpensive teaching aids that also can be used for in-the-field training or mobile classrooms (e.g., portable chalkboards, flip charts, audiotape or CD players, and videotape or DVD players) make it possible for organizations with limited resources to present information with the added impact of sight and sound reinforcement. With the dramatic reduction in cost associated with the production of videos for use on the Web or in small social settings, a wide range of custom-produced videos can be employed—whether produced professionally under commission by a trade association or other industry group, or produced by a firm's own employees using inexpensive consumer video equipment and editing software—to give users greater access to the latest in technology and training materials.

SAFETY CONSIDERATIONS

Photographers—whether the trainers themselves, their company professionals, or outside photographers from a production company—should observe the following safety precautions when on location.

Safe Background

It is not enough simply to make sure that the subject of a photo is in complete compliance with all pertinent safety regulations. Persons in the background who might show up in the picture also need to be observing safety precautions and wearing all the personal protective equipment required for the job they are doing.

Note: Evidence of a safety violation or poor housekeeping in a photo that is supposed to be promoting safety and health is likely to attract a lot more attention than the main safety message. On the other hand, this is a great teaching tool to allow students to find what is missing in the picture or photo.

Safety Equipment

Photographers, their helpers, and everyone else who, for the purpose of taking safety photos, goes into an area where personal protective equipment is required must wear all the equipment specified, even though it may appear that they personally face little risk of injury. All persons promoting safety must follow all the rules for using personal protective equipment, as well as all safety rules for that facility or worksite.

Safe Practices

Photographers, crew members, and visitors must observe all safety rules, including no-smoking rules and rules that prohibit certain types of electrical equipment in areas where sparks, radio frequency (RF), or electromagnetic interference (EMI) could be hazards. Rules must be observed both for the protection of the visitors and to demonstrate respect for safety principles.

Lighting Equipment

Photo floodlights and other special equipment used on a photo shoot should be in good working condition and use wiring, plugs, and outlets that are automatically grounded. Electrical cables must be strung in a way so they will not cause a fall or some other accident. When appropriate, a facility electrician or safety personnel should be on hand to help a photographer avoid mistakes or accidents. All the equipment used on a safety assignment should be UL-listed or approved by another, equivalent agency.

Fall Protection

Photographers should use sturdy ladders and work platforms to climb on when going after elevated photo angles. They should not climb on machinery or tables. Fall-arrest harnesses and lifelines should be used when working from heights.

Review of Photos

Any photographs taken to depict good safety practices should be reviewed by the most senior safety official of the location or company to ensure that nothing in the photos inadvertently depicts a safety violation or poor safety practice. Attention to detail is essential. One safety director given the opportunity to review photos was able to prevent use of a training photo that depicted an employee who was killed in a job-related accident shortly after the photo was taken.

SUMMARY

- Because people learn and retain material better when they use both sight and hearing, media have become indispensable tools in occupational safety, health, and environmental training.
- The selection of media depends on the role the trainer must play, audience size, cost of the materials, whether in-house or outside personnel will make the materials, and whether the materials justify the time and cost.
- Management must establish measurable training objectives tailored to a specific audience and message.
- Media should be designed to reinforce a message. Graphics used in media presentations include circle graphs, bar charts, line charts, and area charts. Companies can use computer graphics software to design their own materials. These visuals should be easy to read from all angles of a meeting room and from the back of the seating area.
- Projected visuals, films and videos, and computer-based training are three of the most commonly used media in training. A company can make its own materials or hire a service bureau or production company to create them.
- The most common computer-based systems include computer-managed training and interactive training, which uses conceptualizations and simulations to teach workers. Simulations allow students to practice high-risk procedures and safety techniques without incurring injury.
- Supplemental materials can enhance and support media and make a trainer's/speaker's presentation more effective and memorable.
- To create effective media presentations, trainers should plan carefully to ensure that they have confirmed or checked such items as the meeting area, all supplies and materials, equipment, seating arrangements, and all safety concerns.
- The final step is evaluation of the training session. Trainers need to establish a climate that encourages open, honest feedback that includes "creating possibilities for the future" from participants. Media materials should be evaluated on the basis of their quality and their use.

REFERENCES

American Society for Training and Development, 1630 Duke Street, Alexandria, VA 22313.

Association of Audio-Visual Technicians, 2378 South Broadway, Denver, CO 80210. *Fast Forward.*

Association for Multi-Image International, Inc., 8019 North Himes Avenue, Suite 401, Tampa, FL 33614.

Bruccoli, M., ed. *Audio Visual Market Place.* New York: R.R. Bowker. Published annually.

Bunyan, J. A. *Why Video Works: New Applications for Management.* White Plains, NY: Knowledge Industry Publications, 1988.

Carlberg, S. *Corporate Video Survival.* New York: Macmillan, 1991.

Cole, M., and S. Odenwald. *Desktop Presentations.* New York: American Management Association, 1990.

Hansell, K. J. *The Teleconferencing Manager's Guide.* White Plains, NY: Knowledge Industry Publications, 1989.

International Association of Business Communicators, 1 Hallidie Plaza, Suite 600, San Francisco, CA 94102.

International Television Association, 6311 North O'Connor Road, LB 511, Irving, TX 75039.

Kerlow, I. V., and J. Rosebush. *Computer Graphics for Designers and Artists.* 2nd ed. New York: Van Nostrand Reinhold, 1993.

Media Horizons, 50 West 23rd Street, New York, NY 10010. *Audio-Visual Communications.*

National Safety Council, 1121 Spring Lake Drive, Itasca, IL 60143.
- *Safety+Health* magazine. www.safetyandhealthmagazine.com.
- Occupational Safety and Health Data Sheets (listing of other applicable data sheets is available from NSC Library):
  - Nonprojected Visual Aids, 12304-0564, 1993.
  - Photography for Safety, 12304-0619, 1991.
  - Posters, Bulletin Boards, and Safety Displays, 12304-0616, 1991.
  - Projected Still Pictures, 12304-0574, 1989.

PTN Publishing, 2210 Crossways Park Drive, Woodbury, NY 11797.
- *In-Plant Video/AV Communicator.*
- *Technical Photography.*

U.S. Office of Education, Bureau of Adult and Vocational Education, 400 Maryland Avenue SW, Washington DC 20036.

Wiese, M. *Film and Video Budgets.* Rev. ed. Studio City, CA: Michael Wiese Productions, 1990.

Women in Communications Inc., National Headquarters, 2101 Wilson Boulevard, Suite 417, Arlington, VA 22201.

(Other books and pamphlets are available from distributors and manufacturers of cameras, films, computer graphics software, and other visual media equipment. Trade journals in the fields of computers, visual media, photography, education, sales management, advertising, and training also contain excellent information.)

REVIEW QUESTIONS

1. Define *medium* and its purpose.
2. To be effective, media should target what human senses the most?
3. What are the benefits of computer-based training?
4. How do evaluations help trainers and their use of media?
5. What should trainers take advantage of to be most effective?
6. How does management justify the cost of training?
7. Name four of the five factors to consider in deciding between commercial and in-house training materials.
   a.
   b.
   c.
   d.
8. In computer learning formats, what two types of interactive materials are often stressed?
   a.
   b.
9. How many colors should be used in a visual?
10. What are some disadvantages to using slides in a presentation?
11. Should personal protective equipment be used by a photographer when working to create training media?

Safety Awareness Programs

Philip E. Hagan, JD, MBA, MPH, ARM, CIH, CET, CHMM, CHCM, CHSP, CEM
Patrick Moylan, CSP, MBA

Reasons for Maintaining Interest
Indications of Need for Improving a Program ▶ Program Objectives and Benefits

Selection of Program Activities
Basis for the Program ▶ Factors to Consider

Staff Functions
Role of the Safety and Health Professional ▶ Role of the Line Manager

Safety Committees and Observers

Quality Circles and Safety Circles

Meetings
Types of Meetings ▶ Planning Programs

Contests
Purpose and Principles ▶ Injury Rate Contests ▶ The "XYZ Company" Contest ▶ Interdepartmental Contests ▶ Intergroup Competitions ▶ Intraplant or Intradepartmental Contests ▶ Personalized Contests ▶ Overcoming Difficulties ▶ Noninjury Rate Contests ▶ Publicity ▶ Meaningful Awards ▶ Award Presentations

Posters
Purposes of Posters ▶ Effectiveness of Posters ▶ Types of Posters ▶ Changing and Mounting Posters ▶ Bulletin Boards

Displays and Exhibits

Other Promotional Methods
Campaigns ▶ Unique Safety Ideas ▶ Courses and Demonstrations ▶ Publications ▶ Public Address Systems ▶ Suggestion Systems ▶ Suggestion Awards

Campaigns

Public Relations

The Voice of Safety
Working within the Company ▶ Sixteen Ways to Make Safety News

Some Basics of Publicity
Select the Audience ▶ Use Humor and Human Interest ▶ Names Are News ▶ Friendly Rivalry ▶ Techniques ▶ Communications by the Safety Office ▶ Hints for TV Interviews ▶ Handling a Serious Incident Story

Working with Company Publications
Producing a Publication ▶ Getting Ideas

Summary

References

Review Questions

This chapter deals primarily with promoting and maintaining managers', supervisors', and employees' interest in safety. The chapter covers promotions (what an organization does to promote safe behavior, conditions, and practices within its workplace) and campaigns (what it does to publicize safe behavior and practices to workers' families and areas served by the organization).

To promote and maintain interest in safety, the safety and health professional must inform and train key executives and managers about the safety program's objectives and how to achieve them. However, management must also demonstrate its commitment and interest and actively support a solid safety program. Only with top management leadership, support, and commitment will activities to promote employees' interest be successful. A well-promoted program involves both management and employees in safety and health activities. (See Meetings of Executives, later in this chapter.)

If top management seems uninterested in such issues, the safety professional should not assume that executives are indifferent or opposed to safety. Often, they are simply unaware of the importance and basic benefits of an organized safety program. This chapter will discuss the following topics:

- the importance of promoting and maintaining interest in safety awareness
- objectives and benefits of safety awareness
- factors to consider when initiating a safety awareness program
- the role of safety and health professionals and line managers in promoting safety awareness
- the function of safety committees and observers in safety programs
- how to involve workers in quality and safety circles
- types of meetings and planning programs in safety promotion
- the uses of contests, awards, and competitions in promoting safety
- posters, displays, and other promotional methods for promoting safety
- the benefits of good public relations
- how organizations can make safety news
- how to plan and produce a safety publication.

REASONS FOR MAINTAINING INTEREST

Why is it necessary to maintain interest in safety (1) if the workplace has been designed for safety, (2) when work procedures have been made as safe as possible, and (3) after supervisors have trained their crews thoroughly and continue to enforce safe work procedures? The answer is simple: even with these optimum work conditions, incident prevention basically depends on the desire of people to work safely.

It is not possible to anticipate every hazardous conditions, unsafe practice, and loss control problem. Nor is it possible to develop written procedures or protocols for every imaginable situation; sometimes, an employee has to improvise. Employees frequently must use their own imagination, common sense, and self-discipline to protect themselves and others; they must think beyond the immediate work procedures to act safely in potentially hazardous situations when on their own. They must know how to assess risk and know when to ask for help in stopping work in a potentially hazardous situation.

Indications of Need for Improving a Program

Various yardsticks indicate supervisor and employee attitudes toward incident prevention. These measures also indicate problems that the safety professional needs to address when improving a program to stimulate and maintain employee interest in safety.

- An increased rate of injuries, incidents, and near misses may signal that a change in a program is needed. If such an increase cannot be explained through engineering problems, ineffective training, or supervision, then perhaps employees are forgetting or ignoring work rules, failing to stay alert, or taking unnecessary chances. A program to develop and maintain their interest in safety will help to reverse this trend.
- If housekeeping is deteriorating, protective equipment is not being used, and guards are not being replaced, it is time to improve supervisory interest in safety awareness and to promote renewed and increased management interest in safety.
- Incomplete or missing incident reports also indicate decreasing supervisory participation and, perhaps, even a failure to report minor incidents and injuries. Motivation to ensure better reporting is then in order.

Program Objectives and Benefits

A well-planned and executed program, although it cannot be expected to solve all problems, can help workers become more committed to safety. For example, the program can:

1. Help to develop safe work habits and attitudes—but it cannot compensate for unsafe conditions and unsafe procedures.
2. Focus attention on specific causes of incidents, although it cannot eliminate them.
3. Supplement safety training, yet it cannot be considered a substitute for a good training program.

4. Give employees a chance to participate in incident-prevention activities, such as suggesting safety improvements in job procedures.
5. Provide a channel for communication and cooperation between workers and management; incident prevention is certainly a common meeting ground.
6. Improve employee, customer, and community relations because it can be evidence of management's commitment to incident prevention.

The major objective of safety awareness programs is to improve interest and participation in safe work practices by involving management and employees. Usually, though, it is difficult to determine the degree of success such a program achieves. This is because companies with these activities usually have sound basic safety programs: working conditions are safe, employees are well trained and safety minded, and supervision is heavily involved in promoting safety measures.

However, there are documented successes showing that safety awareness programs can be successful in companies with strong safety programs. One company that already had a good basic safety program attributed a reduction in its work injury rate to stepped-up efforts to maintain an ongoing interest in safety. It was based on an idea submitted by an employee: each month, candy bars were distributed to injury-free employees. Wrapped with some of the candy bars were slips that could be traded for free pairs of safety shoes. Another company gave each employee who had an injury during the month a package of gum with the slogan "Something to chew on" along with a friendly safety message and wishes for an injury-free future.

One poultry-processing company was able to boost its safety program by giving away "big-ticket" items. Prizes ranged from appliances up to an all-expense-paid trip to Hawaii for eligible employees and their guests. An expenditure of $40,000 netted the organization a 29% reduction in lost-workday cases and a reduction of $450,000 in workers' compensation costs.

SELECTION OF PROGRAM ACTIVITIES

Modern advertising and marketing techniques have much in common with those used to "sell" safety. Just as steady and imaginative sales promotion is needed to sell most products and services, safety also requires constant, skillful promotion. This approach makes the basic elements of incident prevention easier for workers to understand and accept.

Many safety and health professionals make the common mistake of instituting promotional activities with no preliminary planning or objectives in mind. This haphazard approach should not be used to develop an entire safety program. It is important to develop clearly defined objectives and goals with performance indicators that can be measured so that the return on investment can be quantified and modified as necessary.

Basis for the Program

For a program to be effective in maintaining interest in safety, it must be based on employer and worker needs. Management should select activities not simply because they will be popular, but because they will yield results. To develop suitable activities and promotional materials, management must know the needs and wants of supervisors and employees.

To find out what employees really thought of their safety program and just how interested they were, one company inaugurated a safety inventory plan. Each year, after the regular stock inventory had been taken, safety inventory cards were distributed to all employees—salaried workers as well as hourly. The cards were distributed by supervisors, who asked employees to take stock of their jobs and environment with regard to safety. The following year's safety program was planned on the basis of the questionnaire returns, which ran better than 90%. Many suggestions for improving the safety program were received and subsequently put into practice.

Factors to Consider

When planning safety awareness activities, several major factors should be considered.

Company Policy and Experience

If a company ordinarily uses activities such as committees, mass meetings, award programs, and contests in areas other than safety, then these should be used for the safety program. Purely outcome-based promotions can be unwise if focused only on results (recordable injuries) and not on inputs and behaviors. It is generally unwise to spend much time on activities that are not in keeping with existing company policies and experiences, unless it is believed that a new "sales pitch" is justified.

On the other hand, if supervisors and employees are too involved in committee work and activities such as sales promotion, quality control, and tool damage programs, similar activities for safety may be lost or considered burdensome. Other approaches might prove more effective. However, if management believes safety is important, effective measures will be taken regardless of the burden.

Once a promotional program is under way, it should not impair other aspects of the incident-prevention program or other company activities. For example, safety meetings should not take much more time than meetings for quality control, sales, or industrial relations.

Budget and Facilities
Budget considerations always affect plans for a safety promotion program. At first, the program will require extra effort, time, and money. The safety and health professional can justify this expenditure, however, as an investment that will produce direct and indirect benefits from both a financial and an employee relations standpoint. If the program is to be successful, management must allocate sufficient funds to carry it out.

In selecting program activities, line management and the safety and health professional should consider the available facilities. In many companies, safety-oriented facilities, publications, or services may be obtained through the industrial or public relations departments. Public relations also may be a valuable source in helping to plan promotional activities.

Types of Operations
The nature and organization of company operations affect the choice of activities and materials for maintaining interest in safety. When operations are widely scattered and diversified, as in the construction, railroad, marine, motor transport, sales, and air transport industries, the job of selecting and disseminating safety information becomes more complicated.

In decentralized operations, the safety and health professional must help management choose materials, such as publications, videos, and other media, that can be used easily in the field. Local supervisors must be able to conduct meetings, present material, and handle posters.

The safety and health professional must also consider the needs of employees involved in widely different kinds of work at remote locations. For example, a poster program may be used as one means for maintaining interest in safety at each outlying location. If so, the supervisor can designate a trustworthy employee to receive posters and take care of their distribution and posting. This individual can also be sure that poster boards and display cases are kept clean, attractive, and free of extraneous paper.

The company website can facilitate this internal sharing of safety information through a password-protected blog, wiki, or other intranet function with which most employees, especially younger ones, are very familiar.

Another method is to equip a trailer with permanent displays and transport the company safety story to remote locations. Rear-screen projectors (35-mm) and videotapes are also well suited for use in scattered locations and in the field. A relatively inexpensive method of communication is the video camera. Videos can be used to communicate a variety of short subjects to distant worksites. This method will ensure that the same message is communicated in the same way to all employees.

For many organizations, the types of educational materials used for different groups of employees may vary considerably. For instance, movies may be suitable for the day shift when many employees can be taken off the job, but would not be appropriate for a small night shift when no one can leave the job site even for a short time.

It is important that night crews and maintenance employees are not overlooked. Their work is as vital as that of other employees to the overall incident-prevention effort. Programs for them will have to be planned to fit their specific requirements and time constraints.

Types of Employees
The types, backgrounds, and educational levels of employees must be considered when choosing safety promotion activities. For example, migrant workers frequently do not receive sufficient job training. Material for these employees should present basic safe practices for their jobs in brief, easily understood form and in the appropriate languages when their English skills are poor. Management can use local colleges or universities as a resource to assist in translating material into other languages. Be aware that different dialects, even in the same language, may result in the safety message being misunderstood. A similar approach should be used with temporary workers or those assigned from union halls. Employees who have difficulty understanding or reading English need visual material. Material in Spanish is available from many government websites. In addition, the National Safety Council provides safety literature in Spanish that can be used to bolster marketing efforts.

Basic Human Interest
If employees seem bored or uninterested in safety activities, an extra push is needed to involve them. One approach is to base promotional activities on employee interests, such as bowling or fishing, as an incentive.

The wise choice of these activities depends on an understanding of basic human needs and emotions, such as those listed in Figure 33–1. The safety activities suggested for employees should have general appeal. Delivering "safety content" embedded in "general-interest" content is an idea that has merit. For instance, use of free Internet sites that have ads on them and using sports stories, weather, news, and the like on breakroom screens, integrated with a few safety items from time to time, can be quite effective. Pure safety content streams will typically not be watched.

BASIC HUMAN INTERESTS AND CORRESPONDING ACTIVITIES

Basic Interest Factors

Fear of painful injury, death, loss of income, family hardship, group disapproval or ridicule, supervisory criticism.

Pride in safe workmanship, in good records, both individual and group.

Recognition: desire for approval of others in group and family, for praise from supervisors.

Participation: desire to be "one of the gang," "to get in the act."

Competition: desire to win over others, such as shown in sports.

Financial gain through increased departmental or company profits.

Ways to Use These Factors

Visual material: emotional or shocker posters, dramatic films, pictures, and reports of serious injuries on bulletin boards, in company papers.

Recognition for individual and group achievement; trophies, personal awards, letters of appreciation.

Publicity: photos and stories in company and community papers, on bulletin boards.

Group and individual activities: safety committees, suggestion plans, safety stunts, campaigns.

Contests with attractive awards.

Monetary awards through suggestion systems, profit-sharing plans, promotions, increased responsibility.

Figure 33–1. Ways to put six basic human-interest factors to effective use in promoting safety. Basic needs and desires that motivate people are shown on the left. The right column lists direct appeals safety promotional programs can make.

Other Considerations

The tasteful use of employee human-interest materials featuring children, animals, and cartoon figures, along with activities and contests fostering good-natured competition, play a big part in most companies' safety promotion programs. These elements can be as effective in safety promotion as they are in sales and marketing. Their use should not compromise company policy or offend anyone.

Ideas for maintaining interest often use humor. The "light touch" is essential and should be good natured. Ridicule should never be used; it only arouses resentment.

A positive, constructive approach is generally better than a negative approach. However, the latter sometimes is preferable if it is more dramatic. A picture showing the consequences of an incident, such as a fall on a slippery floor, may have a greater impact than one depicting the safe practice that could have prevented the incident.

Variety Is Essential

Often a simple change—such as a different type of contest, a bulletin board redesign, or revising the format of safety meetings—can renew workers' interest. The activity itself may not be more effective, but its new form stimulates thought and discussion. Although safe work practices should become routine, their presentation should not.

Activities that require employee participation generate more interest than do those that involve only seeing and listening. Many companies have worked with the National Safety Council to produce films, videos, and movies in their plants. These firms report an upsurge in safety interest when some employees are asked to act in a video or film emphasizing safety procedures or practices that may have seemed routine.

In some instances, management urges employees to submit suggestions for equipment guards or to help in the selection of personal protective equipment. Companies find that these workers are more inclined to use the guards and the personal equipment than they would be if they had not been asked for their opinions. For the same reasons, including employees in the process of drafting safety rules encourages compliance on the part of those workers who participate in the project. Companies have also found that workers who serve on a safety and health committee tend to develop increased awareness of safety responsibilities.

STAFF FUNCTIONS

Creating and maintaining employee interest in a line management responsibility falls to the safety and health professional. This individual is responsible for assisting line management in planning the safety program. Line managers and supervisors are responsible for planning, organizing, and executing the program.

Role of the Safety and Health Professional

Safety and health professionals coordinate the program; supply the ideas and inspiration; and enlist the support of management, supervision, and employees. Programs should be designed to involve both management and employees. Employees like to receive awards; managers like the public relations aspects of presenting them.

Safety and health professionals may work with local safety councils, chapters of the American Society of Safety Engineers, and other civic or technical groups interested in incident prevention. They can gain much by attending the National Safety Congress and Exposition and regional safety conferences. Here, they learn what other companies are doing and how those ideas can be translated into practical activities for their own organizations. After participating in roundtable discussions, listening to speakers, and meeting people with similar interests, safety and health professionals often return to their jobs with renewed enthusiasm. Internet websites, blogs, and list-servs are also popular sources of current information that can support safety program efforts.

Because they frequently are called upon to address groups, safety and health professionals should be able to present their ideas clearly, effectively, and convincingly. Polishing public speaking skills will also help when dealing with people on a one-to-one basis. Safety professionals should know when and how to use visual aids.

Showmanship tactics, however, should be used only with considerable discretion. If they backfire, they can hurt the credibility of the safety and health professional, the company, and the safety program in workers' eyes.

Safety and health professionals should help educate line management in two areas. First, managers should learn how to keep working conditions as safe as possible. Second, they must know how to motivate workers to follow safe procedures consistently, as a part of good job performance.

A vast amount of program material useful to the safety professional is available through National Safety Council publications. Safety and health professionals also are invited to submit interesting data, difficult problems, or "gimmicks" of any nature to the Council for help in solving safety problems and to share ideas and solutions with others. The Council has information on every phase of safety, gleaned from the experience of members in various industries. One company's solution to a problem may prove invaluable to other firms.

Role of the Line Manager

The line manager is the key person in any program designed to create and maintain interest in safety. This is because the manager is responsible for translating management's policies into action and for promoting safety activities directly among the employees. How well this responsibility is met will determine to a large extent how favorably employees will view safety activities.

Ranking managers must learn that under existing laws, they are directly accountable to their organization and society for their employees' safety. Management, with the safety professional's assistance, must see that supervisors receive adequate safety training.

The supervisor's attitude toward safety is a direct reflection of upper management's attitude toward safety and is a key factor in the success not only of specific promotional activities, but also of the entire safety program. Supervisors look to upper management and employees look to the supervisor for leadership. Line managers, including supervisors, who are sincere and enthusiastic about incident prevention can usually maintain employee interest. Conversely, if line managers give only lip service to the program or ridicule any part of it, their attitude can cancel any good that the safety and health professional may be able to accomplish.

Many supervisors are reluctant to change their modes of operation or to accept new safety engineering ideas, much less to enthusiastically support contests, safety stunts, committee projects, and other activities used to promote and maintain interest in safety. It is line management's task to sell these supervisors on the benefits of incident prevention. They must be convinced that promotional activities are not "frills" but important projects that can help to prevent injuries. Line management's wholehearted cooperation is essential to the success of the entire program. It can be pointed out that a successful program makes management's job easier and less time consuming.

One way line managers can support safety is by setting a good example and educating their workers in safe practices. They must follow safety rules and procedures and wear safety glasses and other personal protective equipment whenever they are required. Teaching safety is also an important function of supervisors. To be successful in this area, they cannot depend on safety posters, a few warning signs, or even general rules to do their teaching for them. Supervisors themselves must first be trained to teach if they are to be competent in this area. (See Chapter 31, Safety and Health Training, for details on training courses and techniques.)

When enforcing safety and health rules, line managers should not suddenly adopt a "get tough" approach after being lax. They should be consistently firm and fair all along. Otherwise, workers who have the impression the supervisor cannot recognize unsafe conditions and practices, or simply does not care, will not take the new, tougher line very seriously.

Supervisors should be encouraged to take every opportunity to exchange ideas on incident prevention with workers, to commend them for their efforts to do the job safely, and to invite them to submit safety suggestions. They can also be most effective in relaying personal reminders on safety to employees. Such reminders are particularly appropriate in the transportation and utility industries, where crews are on their own from terminal to terminal.

Line managers, including supervisors, should request and receive help from the safety department. Help may also come through correspondence, educational materials for distribution, and personal visits from safety personnel. Supervisors should also receive adequate recognition for independent and original activity.

SAFETY COMMITTEES AND OBSERVERS

Various types of safety and health committees have many different functions. However, the basic function of every safety committee is to create and maintain interest in safety and health and thereby help to reduce incidents.

Some organizations prefer other types of employee participation to formal safety committees. This is because they feel safety and health committees require a disproportionate amount of administrative time, generally tend to pass the buck, may stir up more trouble than they are worth, and may be a scapegoat for supervisors who want to unload their responsibilities.

The answer to these objections is not to abolish the committees but rather to reexamine their duties, responsibilities, and methods of operation. Such analysis often can lead to constructive changes enabling a committee to fulfill its original objective—that of stimulating and maintaining interest in safety.

Committee membership should be rotated periodically. This ensures a fresh perspective and also increases the number of employees who are trained to look at operations with safety in mind.

Involving employees in safety inspections, either alone, as observers, or as part of a formal safety and health committee, has the same basic objective: to get more employees actively involved and interested in the safety and health program. Planning, organizing, publicizing, and following definite procedures will streamline the work of both committees and observers and help ensure effective results.

QUALITY CIRCLES AND SAFETY CIRCLES

Hazard recognition and control is one area in which employee involvement produces substantial results. The following discussion is taken from the Indiana Labor and Management Council's booklet *Worker Involvement in Hazard Control* (see References), used with permission.

There are two reasons for the popularity of involving employees in hazard recognition and control. The first is management's desire to use all available resources to increase productivity and quality in the face of growing competition. The second is management's understanding that employees want to accept new challenges and to participate in activities that affect their work life.

Popular forms of worker involvement programs are quality and safety circles. A quality circle is a group of employees, performing similar work or sometimes varied work, who meet weekly. In the meetings, they learn about and apply basic techniques to identify problems within their area(s), analyze them, and recommend solutions to management. In some instances, circle members discover hazards during their analysis of other plant problems and make excellent recommendations for their control.

Safety circles are a type of quality circle used by some companies to reduce the number of unintentional injuries by keeping safety and all its important features foremost in the minds of the employees. This implies a change in the employee's role from passive to active, while management's role becomes less negative (fewer don'ts) and more positive (more dos).

In many situations, safety circles are established on a plantwide and departmental level. Safety circle meetings are held monthly for approximately 0.5 hour to 1 hour. Each circle meeting is usually preceded by a presentation explaining the successes and problems that the safety circle team experienced during the preceding month. The circle reviews a breakdown of all injuries, including first-aid cases, as a means of measuring safety progress and as a means of pinpointing trouble areas. Usually each member of the safety circle is assigned a specific responsibility to review. Companies employing the safety circle concept report an improvement in their incident and injury experience.

MEETINGS

Safety and health meetings may be conducted for managers, supervisors, employees, or other groups. In every case, the purpose of the meeting is to stimulate and maintain interest and commitment in safety and health issues. If meetings fail to achieve this, their format or content should be changed to make them effective or they should be discontinued and a new approach taken.

A formal agenda for all attendees should be distributed prior to the safety meeting. An agenda helps to focus attention on specific issues, and avoids wasting time on nonrelevant topics.

Types of Meetings

The following types of safety and health meetings commonly arouse and maintain interest in incident prevention (see also Chapter 31, Safety and Health Training):

- meetings of operating executives and supervisors to for-

mulate policies, initiate and maintain a safety and health program, or plan special activities
- departmental meetings to discuss special problems, plan campaigns, or analyze incidents
- small-group meetings to plan the day's or week's work so that it can be done safely, to discuss specific incidents, or to review safety instructions. These are sometimes referred to as "toolbox sessions."
- mass meetings of all employees, sometimes including families or even the entire community, to serve special purposes such as launching a major new program or contest and "selling" safety to everyone affected by unintentional injuries and illnesses

Meetings of Executives

When a safety and health program is inaugurated, it is especially important that the chief executive officer (CEO) of the company or top plant manager call a meeting to announce the general injury prevention plans and policies to all line managers and other operating executives. Under some labor contracts, this must be, or should be, a joint union–employee announcement. If these people meet at regular intervals to discuss operating problems, this announcement can be made at a regular gathering. Otherwise, the CEO or manager should call a special meeting. Periodic safety inspections by executives help to lend credence to management's interest in safety. Inclusion on the company intranet website of a message from the CEO about safety can be very helpful to setting a positive tone of top-down and bottom-up alignment about safety.

After this first meeting, the group may hold sessions periodically (usually monthly) to evaluate and coordinate the safety program, to check on the progress being made in injury prevention, to appraise proposed activities, to set policies, and to make decisions. It is also desirable for this group to review and/or investigate all fatalities, multiple amputations, and other serious injuries.

Departmental Meetings

Departmental meetings serve many safety and health purposes. They may be used to discuss the company safety program so that employees will better understand its policies and procedures, to provide information about causes and types of injuries and illnesses, or, in a purely inspirational manner, to create an awareness of hazards and a desire to prevent incidents. Departmental meetings also can be used to review and/or investigate all injuries involving lost workdays or restricted work.

Many departmental safety meetings are held monthly, and most are conducted by the supervisor. The safety department usually assists in planning and provides materials, such as visual aids.

The program for a departmental meeting may include:
- reports on injuries, illnesses, or other concerns in the department since the last meeting; a safety inspection in the department; and the department's standing in a contest (the total time spent on reports should be limited so that this part of the meeting does not become tiresome)
- discussion by the supervisor of specific safety practices or unsafe conditions that need to be improved
- talk, demonstration, or audiovisual presentation on an appropriate incident-prevention subject; the speaker may be the supervisor, a department employee, the company safety and health professional, an outside expert, or a company executive
- a housekeeping walk prior to the meeting will give visibility to the safety committee members and provide specific topics to discuss and correct.

Departmental meetings give the supervisor an opportunity to point out the dangers of particular unsafe practices or conditions. By condemning those practices, the supervisor sets a good example and lets workers know they are to follow the same rules and procedures. Most workers welcome an opportunity to share their safety ideas in these meetings.

At the conclusion of departmental meetings, the supervisor may prepare written reports for the plant (or company) safety committees and managers.

Small-Group Meetings

Supervisors also can hold small-group meetings with people doing similar work at or near the workplace. The supervisor may discuss the causes of a recent incident workers have witnessed or learned about. Employees should be encouraged to join in the discussion, with the goal of reaching some conclusions about how the incident might have been prevented.

The supervisor may present a problem that has developed because of new work or new equipment. Again, all should participate and offer their views.

At times, the supervisor may present a film or talk on a subject related to the work of the group members. Other audiovisuals such as charts, models, or exhibits may be used. Safety devices or pieces of equipment or material may be brought in and discussed.

"Production huddles" are instruction sessions about a specific job that include safety information. Such meetings are particularly useful with maintenance crews when an unusual job is about to start. The plans for doing the job safely and efficiently are discussed, and a procedure is agreed upon. Public utility line crews that use this type of meeting call it a "tailboard" or "tailgate" conference. Before starting a job, the crew gathers around the truck and discusses the job, laying out the tools and materials they will need and agreeing upon each person's tasks.

A particular advantage of small-group meetings is that they provide excellent opportunities for presenting all types of information, including safety information, directly to employees. They also stimulate an exchange of ideas that can benefit the incident-prevention program. To be successful, each safety and health meeting must have a tangible message, imaginative presentation, opportunity for audience participation, and a conclusion that spurs action toward an attainable goal.

Mass Meetings

Mass meetings are held for special purposes, such as launching a contest or an award program, presenting awards, introducing new equipment, explaining a change in company policy, or celebrating an exceptionally fine safety record with an event such as "safety day."

In companies with plants in different cities, a top executive may call a meeting of employees during a plant visit. The talk may cover safety as well as other subjects. One company president makes an annual round of plants with the safety director and speaks at a safety rally of all employees at each plant.

Under certain conditions, particularly in smaller communities, large meetings can be held in a local theater or public hall. These meetings require fairly elaborate arrangements and considerable publicity to ensure good attendance on employees' part.

A mass meeting in a public hall using the "family safety night" theme allows not only employees to attend but also their families and friends. In addition to a presentation or speech targeted at the program theme, management should provide for some other entertainment. Often, good talent can be found right in the plant or shop.

Mass meetings afford an excellent opportunity to use an outside speaker who can talk with authority and in a crowd-pleasing manner on general incident-prevention work. Such speakers can be found in nearby plants, insurance companies, city administrations, automobile clubs, or community safety councils.

At the first yearly mass meeting, always hand out the company Safety Mission Statement. This needs to be a yearly statement from the chief executive officer outlining his or her views on safety for the location. If management wants to include movies, videocassettes, or slide shows relating to incident prevention at the meeting, it can find a suitable selection from those available through the National Safety Council or regional film service organizations, whose locations may be obtained from the Council.

Planning Programs

Making the safety meeting interesting is of the utmost importance. Speakers should not complain about or scold workers on their job performance. Talks should be brief and start and end on time. The subject matter should be studied in advance to make sure it is pertinent and is not a repeat of other, recent talks.

Large occasional meetings need even more preparation and timing than do small meetings. Speakers, including company executives, should review their remarks with the meeting planner to ensure they will add to the desired purpose. Films, videos, and other visuals should also be reviewed in advance.

People responsible for employee meetings should analyze them periodically to determine whether they are accomplishing their purpose. It is all too easy for meetings to degenerate into dull routine. Only continual effort and planning will prevent this trend.

A plan of action to develop a successful safety and health meeting includes these points:

- Prepare in advance. The preliminary arrangement determines the results. Do not conduct a meeting without preparation.
- Select a major topic. Make it timely and practical—one that the group can discuss.
- Obtain facts and figures. Be sure they are correct and complete. Prepare a visual, such as a simple chart or table, whenever possible.
- Map the presentation. Decide on the best way to present the subject of the meeting. Try to anticipate the group's reaction and questions. Outline results you hope to accomplish.
- Set a timetable. Allow adequate time, but set a reasonable limit.
- Have an agenda distributed in advance.
- Be sincere. Managers' sincerity and interest in employees' welfare must be unmistakable.
- Introduce the topic. Tell in simple terms what the meeting is all about. Use a punch line or other short, to-the-point lead-in.
- Present facts, arouse interest. State pertinent facts in an interesting manner.
- Promote group discussion. Ask questions that cannot be answered "yes" or "no." Prompt members of the group to think individually and collectively. Let them talk.
- Agree on some action. Try for group agreement on methods of correction and improvement. Write these down.
- Always post minutes of all safety meetings to let all employees know what the safety and health committee is doing for them to improve their work environment.
- Summarize the meeting. Review briefly what has been discussed and decided. Follow up in the various departments in writing, if possible.

CONTESTS

Management and supervisors should always keep in mind that contests (such as housekeeping contests, interdepartment contests, etc.) are not substitutes for management interest, safe procedures, and "built-in" safety. Nevertheless, although a successful incident-prevention program depends on good management, good training, and efficient operations, some special effort may be needed to maintain worker interest in safety. Moreover, the interest value of contests has direct bonus values in good publicity and improved employee morale.

Therefore, a competition should be held only after management has taken the basic steps in a safety program: developed a policy statement, adopted a record-keeping system, safeguarded equipment, and installed a first-aid department. Such substantial demonstrations of management's interest, sincerity, and responsibility greatly help win the active participation of supervisors and workers in a contest. However, a contest serves only as part of a safety program, not as a substitute for one.

Purpose and Principles

Safety contests are operated purely for their motivational value. A contest that does not motivate workers to be more safety minded is worthless. In most contests, the competing groups are departments of the same plant or divisions of the same company. This type of contest is useful when investigations of incident injuries reveal that most of them are caused by employees' unsafe practices. If incidents are the result instead of unsafe conditions or a combination of unsafe practices and unsafe conditions, the contest should be reconsidered. It might well focus negative attention on one of the company's most conscientious, safety-minded employees injured through no fault of his or her own.

Generally, contests are based on incident experience and are operated over a stated period. A prize is offered to the group attaining the best record according to the contest rules. Contests have been important almost from the time of the first safety programs. Over the years, companies have developed a fairly well-established set of operating principles:
- A contest should be planned and conducted by a committee representing all competing groups.
- Competing groups should be natural units, not arbitrary divisions.
- Methods of grading must be simple and easily understood.
- The grading system must be fair to all groups.
- Awards must be worth winning and generate interest among employees.
- Good publicity and enthusiasm are important.

Contests may run for various periods—from a few months to a year. Those who recommend longer periods believe that if workers are kept on their toes for a longer time, safe working practices are more likely to become a habit. Some safety professionals, however, believe that greater interest can be aroused and maintained during a short period and, therefore, prefer short and more frequent competitions. Management should devise contests of different lengths to see which is most effective.

A safety contest stock certificate idea was developed by the Maxwell House Division of Kraft General Foods and ran for 1 year. For each week of the contest that a department worked without a disabling injury, employees received a stock certificate worth 50 cents. Dividends were paid on this stock at the rate of 10 cents for the first 1,000 consecutive safe hours worked; 25 cents for the first 10,000 consecutive safe hours worked; 50 cents for the first 50,000 consecutive safe hours worked; and $1 for the first 100,000 consecutive safe hours worked. This meant that if a department worked 100,000 consecutive work hours without a disabling injury, it received a total of $1.85 in dividends for each share of stock held.

There were penalties, however. If a disabling injury occurred in a department, it was penalized 1 month's stock earnings. This meant that during that particular month, the department could not be awarded any stock or dividends. It also meant that if an incident occurred in the latter few days of the month, the department would lose all dividends and stock certificates previously issued for that month.

Injury Rate Contests

In a contest based on injury rates, the typical measure of safety performance is the Occupational Safety and Health Administration (OSHA) incidence rate (described in detail in Chapter 11, Injury and Illness Record Keeping, Incidence Rates, and Analysis). These types of contests have generally fallen out of favor for a variety of reasons.

Injury rate contests carry with them an inherent risk of possible abuse and must be closely monitored to ensure honesty. All injuries must be reported. These contests tend to put peer pressure on employees not to report their injuries, to bring administrative pressure to bear on those doing the recording not to count certain injuries, and to make management focus attention on the contest rather than on the safety program. If these conditions arise, the entire safety program is discredited. Another example would be an injury to an employee from a third-party action that could result in a negative impact. A bee sting or spider bite could affect contest results in an unfair manner.

Also, contests should not be based on injury severity because it is difficult to evaluate severity data in a timely fashion. The classification is frequently a matter of chance and contributes little to devising incident-prevention measures. Using a combination of frequency and severity classifications is also not helpful, for the same reasons. Finally, contests should not be based on reducing the number of reported first-aid cases; people will simply stop reporting such injuries.

In-Plant Contests

To be effective, a contest must run for a specific length of time. The contest will be more successful if the campaign is launched with advance publicity and fanfare. Often, the president or other high official makes the original announcement, presents awards, and otherwise lends prestige to the contest.

Competition, if properly organized, can do much to develop teamwork in safety. Some employees who may give little thought to their own work habits can be influenced to cooperate with their fellow workers if they know that their actions can either discredit their department or "team" or result in a payoff for safe work.

Council Award Programs

The National Safety Council firmly believes in the value of award programs for maintaining interest in employee safety and health. The award programs are available to employer members, and some include participation of nonmembers through cosponsor agreements with trade associations.

The Occupational Safety/Health Award Program is a noncompetitive recognition program. Awards can be earned by organizations achieving a perfect safety record—no deaths or cases involving days away from work. Organizations with nonperfect records may also earn recognition if they achieve the criteria for a significant reduction of their incidence rates. There are four levels of awards:
- Award of Honor
- Award of Merit
- Award of Commendation
- President's Citation.

The Occupational Safety/Health Contest is a competitive award program. Participants are able to compete in one or more of 20 different industry contests. Each industry contest is subdivided into divisions according to the products being produced. Whenever possible, organizations of like size are grouped together.

Awards are given based on rank within each division. The site having the lowest incidence rate for cases involving days away from work is ranked first. First-, second-, and third-place awards can be earned.

The Council designated the Safe Worker Award Program to reward and encourage the incident-free performance of individual employees. Each employee is awarded a Safe Worker lapel pin or other incentive for each 12-month period during which no cases of days away from work are incurred. The employee continues to earn an incentive item for each additional year of safe work achieved, striving for the pinnacle of 45 years.

The National Fleet Safety Contest is a competitive award program comparing fleets from the same industries, having similar vehicles, or of the same size. The program is open to all National Safety Council members and members of cosponsoring associations whose fleets operate in the United States and Canada.

The National Safety Council Safe Driver Award is the recognized trademark of professional drivers and is designed to prevent incidents and promote safety in the professional driving industry. Member organizations of the National Safety Council enrolled in the Motor Fleet Safety Service or the School Transportation Safety Service are eligible to participate in this program.

Companies certify drivers who have driven a continuous number of years without a preventable incident. Through this certification, the driver receives a wallet-sized certificate denoting the number of years of safe driving service. Other incentive items, such as lapel pins and patches, may be purchased to signify this accomplishment.

The Safe Driver Award rules spell out what is expected of professional drivers in the way of safety performance—driving a motor vehicle without having a preventable incident. The award rules constitute a yardstick by which drivers can measure their own performance and by which supervisors can measure the performance of individual drivers.

The "XYZ Company" Contest

The "XYZ Company" is engaged in steel fabrication and erection. Rules for one of its safety contests are given in Figure 33–2.

Rule 1

Although contests can run for any period of time, the company chose a 1-year period. This was done for two reasons. First, the period allows development of safe working habits—the useful objective of the contest. Second, because some of the departments are relatively small, the 1-year period eliminates random (chance) factors from influencing an individual department's experience record.

Rule 2a

Hazards in steel fabrication and erection vary greatly; therefore, fabricating units compete in one division and erection units in another. Operations are similar enough to have a common basis for determining standings—see rule 4.

RULES FOR THE "XYZ COMPANY" SAFETY CONTEST

Rule 1. The contest will begin January 1, and end December 31, 20...

Rule 2. The contest will consist of two divisions.
  a. Fabricating units will participate in Division I.
  b. Field erection departments under the direction of a superintendent will participate as separate units in Division II, which will be divided on the basis of size into two groups: Group A will consist of the five erection units working the largest number of employee-hours, and Group B will consist of all other units. Units will be tentatively grouped by size during the first 3 months, and the final classification will be made on the basis of total employee-hours worked at the end of 4 months. No further changes in size groups will be made after April 30, 20...

Rule 3. Recognition awards will be:
  a. Trophies to the winners in Division I and Groups A and B of Division II.
  b. Engraved certificates to plants and erection units ranking second and third in Division I and in Groups A and B of Division II.

Rule 4.
  a. The winners will be the contestants having the lowest weighted incidence rates.
  b. In the event that two or more contestants in any classification have had no chargeable injuries during the contest period, the winner will be the contestant who has worked the largest number of employee-hours since the last chargeable injury.

Rule 5. All injuries and illnesses resulting in death or days away from work will be counted as defined by OSHA record-keeping requirements.

Rule 6. A sum of $50.00 will be presented to the units that have had no lost-workday cases during the first 6 months of the contest or have reduced their incidence rates for the first 6 months 50 percent in comparison with the rate for the preceding 6 month period. The award will be divided into prizes of $12, $10, $8, $6, $4, $2, and eight $1 prizes and raffled to employees. No employee may win more than one prize.

Rule 7. Standings will be compiled monthly and published in a bulletin that will be distributed to all managers, superintendents, and foremen.

Rule 8. All questions pertaining to the definitions of injuries and rules should be referred to the Contest Committee, whose decisions will be final.

Rule 9. Awards will be presented at an appropriate ceremony to be announced at the end of the contest.

Figure 33–2. A typical set of rules for a safety contest.

Other companies may find that hazards differ sharply from one plant or operation to another, and the similar units may be too few to group. Such plants or operations may compete on an equitable basis in several ways:
- The participant achieving the largest percentage reduction in its frequency rate in comparison with a base period, such as the previous year, may be the winner. Because each unit competes only against its past record, this method provides a fair basis for comparison of rates.
- The workers' compensation insurance rates for different types of units in the same state have been used to establish a handicap factor to compensate for differences in hazards. If the rates per $100 payroll are $3.00 for plant A, $2.00 for plant B, and $1.50 for plant C, factors for the units are 3, 2, and 1.5, respectively. The frequency rate of each plant is adjusted by dividing the rate by its factor. Plants are ranked from the lowest to the highest on the basis of the adjusted rates.
- The national average rates for different types of units may be used similarly for determining standings. If plant A achieved a frequency rate of 10.0 and the national average for units in this industry was 12.0, plant A's rate is 0.83 of the national average. Plant B's rate in comparison with its national average rate is 0.75. Therefore, plant B would rank nearer the top than plant A. Average rates for most industries are published annually in the National Safety Council's *Injury Facts*.

Rule 2b

Separation of large and small units is essential. Generally, a small unit finds it easier to go through an entire contest period without a disabling case than does a larger unit. If the number of units is sufficient, management can set up three size classifications. Unequal-size groups can compete if they include units that can attain no-injury records. For this reason, the erection departments of the company in this example were divided into group A—the five largest departments—and group B—the remaining units. Five contestants in a group are usually considered minimum.

Rule 3

Awards should be specified. "XYZ Company" follows the general practice of giving first-, second-, and third-place awards.

Rule 4a

The frequency rate is most often used as the basis for determining standings because it can be computed promptly and easily.

Rule 4b
There should be a satisfactory method of determining the winner between two or more units having perfect records to give the smallest contestant a fair chance. For example, selecting the winner on the basis of the largest number of employee-hours worked since the last lost-time case is fair to all units, regardless of size.

Rule 5
A standard method of counting cases is essential to avoid controversies and maintain confidence in standards. Use the incidence rate described in Chapter 11 of this volume.

First-aid and other minor injuries should be excluded for purposes of the contest. If they are counted, workers may fail to obtain treatment so that cases will not be recorded. The result could be an increase in infections.

Rule 6
Various "special rules" may be used in a contest that runs 6 or more months to maintain employee interest.

Rule 7
Frequent contest bulletins keep everyone informed about standings. Supervisors can post bulletins and discuss standings at safety meetings. Safety personnel can announce contest results in plant publications and in other ways to build and maintain interest.

Rule 8
Questions about rule interpretation will arise and must be settled fairly. This is an important function of the contest committee, which, in turn, may refer decisions about disputable cases to outside judges. A committee of the American National Standards Institute has been set up to interpret standard definitions.

Rule 9
See the section Meaningful Awards, later in this chapter for award ideas. It is important that competing plants or groups be informed of the latest findings. Figure 33–3 shows a typical contest bulletin.

Interdepartmental Contests

Interdepartmental competitions get "close to home" and place responsibility for a good showing on supervisors. Because workers have a greater personal interest in the standing of their own department than in the record of the entire plant or another unit, this type of competition has proved popular for creating interest in safety among both supervisors and workers.

An interplant contest plan often may be adapted to an interdepartmental competition. A company operating a number of similar facilities may take advantage of workers' interest in their departmental records by conducting a competition among the same departments in various organizations. Public utilities may conduct a contest among districts, and other nonmanufacturing companies may follow a similar plan. Because hazards in similar operating units are about the same, the basis of standings may be the frequency rate. Departments or other units may be grouped according to size.

Most departmental contests, however, are conducted among dissimilar departments in one plant; the departments often differ greatly in size. The difficulties due to variations in hazards and numbers of employees from one department to another may be overcome using the same techniques discussed earlier regarding differences among plants.

Always measuring the numbers of incidents of this year versus last year can have unwanted results. Too much emphasis on negative measurements can result in underreporting because participants do not want to look bad or lose the contests.

Another approach can be considered; that is, measuring the success of a safety program by using positive numbers. Here are a few examples:

1. Count the number of safety and health suggestions submitted by a department and count the number of positive ideas and corrections made because of these suggestions. This effort could be tied into production and quality suggestion programs.
2. Give credit for incident investigation reports that are submitted complete, on time, and with positive recommendations for preventing an incident from happening again (avoid accepting reports recommending, "I told Joe to be more careful next time!").
3. Encourage employees to report all incidents, including near incidents, first-aid cases, property damage without injuries, and incidents without injuries.
4. Give credit for safety meetings that are planned, outlined, and run by supervisors and managers.
5. Use safety and health inspections in a system that counts the number of corrections that were made because of these inspections.

Intergroup Competitions

Intergroup contests are particularly suitable for units that employ fewer than 400 people and have small departments in which hazards vary sharply. Employees are divided into teams of from 20 to 50 workers. To equalize factors of size and differences in hazards, each team has a proportionate number of employees from the most and least hazardous occupations. Each group is led

STANDINGS IN THE "XYZ COMPANY" SAFETY CONTEST
JANUARY–JUNE

Oakland leads at the halfway mark!
Tulsa moved into second place.
Chicago slipped from second to third place in June.
Corbin, in last place, had no lost-workday cases during June. Good work!
Tulsa still leads for the President's Award for largest improvement over the previous year's record.

Plant	Rank	January–June Incidence Rates*	% Increase + or Decrease – over Last Year
Oakland	1	2.8	−30%
Tulsa	2	2.9	−40%
Chicago	3	3.5	+5%
Cincinnati	4	4.4	−20%
Corbin	5	5.0	+22%

*Incidence rate is number of cases resulting in death or days away from work per 200,000 employee-hours.

Tips on How to Win, No. 6

One-sixth of our lost-workday cases occur in the use of cranes and hoists. Have foremen hold a safety meeting on safe practices in hitching loads and other crane operations for shop workers. We'll furnish a film. Use posters on the subject. Require safe methods. See the enclosed bulletin for suggestions on how to solve this major accident problem.

Figure 33–3. A simple monthly contest bulletin gives essential information about current standings and also includes a suggestion for improvement in "Tips on How to Win."

by a captain, whose principal duty is to contact members of the team and create interest in winning. Team members may be members of contest committees.

Interest is often promoted by naming teams after prominent baseball, football, or other outstanding sports teams, and the entire competition may be named after a league or other sports organization. Workers can draw their team names from a hat and show their membership by wearing colored pin-on buttons. Supervisors can place contest signs in competing departments.

Using colored buttons to identify workers helps supervisors or others to keep score more accurately. Because members of different teams often work together, the buttons make it easy to ensure that every case involving a competitor is charged against that group's record.

Intraplant or Intradepartmental Contests

With set standards of performance, an intraplant or intradepartmental safety contest helps to emphasize the supervisor's and employees' responsibilities for avoiding incidents. These standards often include no-injury records for varying periods, achievement of lower injury rates compared with a previous period, and improvement over the average injury rates of similar units or of the industry. In this type of contest, units of an organization do not compete with one another; rather, each unit attempts to match or surpass the established standards.

Personalized Contests

Contests that focus on individual skills and achievements are very effective motivators.

Safe Driver Award Program

This award program is the recognized trademark of the professional driving industry for those who have proven their skills as incident-free drivers. More than 7 million drivers have earned this prestigious award since 1930. Employees of these eligible organizations must be certified by their company in order to receive awards designating the number of years of safe driving service. It is also important to define an "incident" for your workers so they are clear on reporting procedures.

Safe Worker Award Program

Recognize your safe workers with distinctive incentive items that encourage continued safe performance. Recognition items indicating 1 or more years of incident-free performance are part of the program. All full-time employees of

Council member organizations and members of Council chapters are eligible to participate in this program.

Some plants single out for special recognition employees who have safe records. Managers give certificates to those who have worked 1, 2, 5, and 10 years without an injury. Holders of certificates have found them useful in obtaining a promotion and even in seeking employment with other organizations.

Periodic raffles of merchandise or cash, or the use of a new car for 3 or 4 months, regardless of the injury record of a department or plant, also have been used successfully to encourage and acknowledge the efforts of safety-conscious workers. Employees who are involved in disabling incidents (or drivers who have had a "preventable incident") during the period become ineligible for drawings.

Various sweepstakes plans have proved popular and effective in maintaining interest in a good record from month to month. One such plan was operated successfully by a branch plant of a well-known paper company. Here's how it worked.

On the payday prior to the beginning of each month, all hourly employees received cards with serial numbers at the pay office window. The workers wrote their names and departments on the cards, tore off the stubs and put them in a box, and retained their portions of the cards. Names of employees in departments having no disabling injuries during the month then were drawn for prizes ranging from $5 to $25. The company contributed a total of $75 per month.

Because eligibility for the drawing depended on a perfect departmental record, each worker had to be careful about all job-related actions. Workers frequently corrected others for unsafe practices that might spoil chances for prizes.

If no incidents occurred during a 3-month period, supervisors participated in a drawing for their own prizes ranging in prices from $5 to $15. This feature proved helpful in enlisting their cooperation.

Some companies give awards like first-aid kits or trading stamps to those in every department who have worked during a given period without an incident. This approach is effective only if going a month, for instance, without an incident is unusual.

Overcoming Difficulties

Although there are several difficulties in operating contests, they can be overcome rather easily. One of the most common difficulties is that some departments are inherently more hazardous than others. To overcome this problem, management can establish handicaps based on annual rates set by insurance companies or on average incident frequency for the different kinds of work.

OSHA can be a good resource. Contact the local OSHA office and request industry information. Using OSHA data will ensure workers that the numbers used are fair and impartial.

Another method is to base standings on improvement over past records. Thus, a department with a past average incident rate of 20 and a current rate of 15 would be acknowledged as achieving a 25% improvement. They would win over a department with a past frequency of 15 and a current frequency of 12—only a 20% improvement. Usually, in both methods, an average of rates over 3 to 5 years is used as the base.

Another problem is that a department may have so many incidents at the beginning of the contest period that it is out of the running and loses interest. Having shorter contest periods helps overcome this difficulty. Another remedy is to have different awards for different achievements. An award for the department having the longest run of injury-free work hours is one example.

Dividing responsibility for the cause of an incident may become a problem. For example, the unsafe practices of one supervisor's worker (or an unsafe condition in this supervisor's department) can result in an incident involving an employee from another department. This would affect another supervisor's record. These situations must be anticipated and the contest planned to deal with them fairly.

Noninjury Rate Contests

The use of noninjury rate contests is gaining favor among safety professionals to motivate program performance. They are used to target problem areas and are not subject to the same potential misuse as are injury rate contests.

For example, an employee contest can be based on the safe worker. The contest uses either a monetary award or a status award, such as a drawing for a television set or a private parking space. All employee names are included in the competition. During the contest period, each worker spotted performing an unsafe practice or not wearing required protective equipment has his or her name removed from the contest. To maintain credibility, those experiencing recordable injuries would also be eliminated.

Supervisory contests, such as a "Supervisor's Safety Club," generally are based on how well supervisors achieve program objectives. These objectives range from the percentage of a supervisor's employees receiving safety talks to the number of self-inspections conducted and deficiencies promptly corrected. Here again, to maintain credibility, the recordable injury of an employee would remove the supervisor from the contest. Awards range from steak dinners and sports events to a Supervisor's Safety Club jacket. Managers overseeing several supervisors or departments would have similar competitions and similar awards.

Contests at the plant level generally cover multiplant

performance in a corporation. In addition to how well corporate safety program objectives are met, injury experience should be included in the contest criteria. Such contests usually have a corporate trophy to be displayed at the winning plant rather than providing any monetary consideration.

Other noninjury rate contests include safety slogan, poster, housekeeping, and community contests. No matter what the contest, the objective is to get the maximum number of people talking, thinking, and participating in safety.

Slogan, Limerick, and Poster Contests

Safety slogan contests vary. One can be for the best safety slogan submitted by an employee. Another may ask the employees or their spouses if they can repeat the "slogan of the week" or the message on a certain safety poster.

Company magazines may conduct contests to "finish the limerick" or "write a rhyme" or "write 25 words on the best way to be safe." Often, these are open to both employees and family members.

The value of homemade posters is in their special application to a particular industry or company. If the posters are the result of an employee contest, their motivational value will be increased, possibly exceeding that of the industry-made variety of posters. An important ingredient of such a contest is to get employees and their families participating in the planning and judging stages as well.

Departments often conduct housekeeping contests. This type of competition is aimed at fundamental causes of incidents and usually tries to eliminate unsafe practices and conditions. Housekeeping contest plans differ from one company to another. The following plan has been used successfully by a metals firm.

Once a week, a committee of three management representatives inspects each department and reports unsafe conditions to the superintendent. A copy of the report is furnished to the works manager, and another is kept for the use of later inspection committees. A demerit for each unsatisfactory condition is charged to the department. If the condition is not rectified within 1 week, an additional demerit is added.

At the end of the month, the demerits are totaled, and departments are rated on the basis of the proportion of demerits to the total number of employees in the department. If department A employs 175 people and had 25 demerits, its rating would be 85.7. This figure is obtained by dividing 25 (number of demerits) by 175 (number of employees), multiplying by 100, and subtracting the product from 100. Standings are posted monthly on the bulletin boards in each department.

Awards are made at a mass meeting of the employees and guests. Names of employees in winning departments are placed in a box from which is drawn the name of the winner for the month. The winner's picture is posted on a special bulletin board, and a short talk on safety by a representative of management is broadcast throughout the plant. (A general rule prohibits an employee from winning more than one award during the contest.)

The name of the winning department is inscribed on a plaque each month. The head of the winning department receives the plaque from the previous month's winner at the mass meeting. At the end of the year, the department that has won the plaque the greatest number of times receives it permanently.

Community and Family Contests

Many companies have stimulated interest by sponsoring safety essay or poster contests for children of employees, local schoolchildren, or young art students. The publicity before and after such contests, plus the interest generated by the posters themselves and the judging, not only stimulates the interest of employees, but also promotes the company's community and public relations.

More than one company has launched a safety poster or essay contest for the children of employees. Management knows full well that employees will give their children considerable help and that there will be much favorable discussion about the contest in locker rooms, lunchrooms, and car pools.

To promote greater safety at railroad crossings, the railroads in Texas sponsored a contest at the Texas Press Association Convention. The journalist who made the closest guess of the total number of grade crossings in Texas won a prize. Because of the contest, safety at grade crossings received considerable attention in the press.

Miscellaneous Contests

There are an endless number of different types of contest possibilities. Often, they can be combined with injury reduction contests. Contests can be held for attending safety meetings, for wearing safety shoes, for reporting unsafe conditions or unsafe practices, or for off-the-job or public safety activities of individuals, departments, or branch plants.

Safety contests are popular with both management and employees. However, before recommending such a contest, the safety and health professional should always determine whether the contest will take valuable time and effort away from providing safer equipment or better training for supervisors and employees.

Publicity

Management should publicize all stages of a contest. It can announce the contest through intranet postings, e-mails, placards, and stories in employee newsletters. Special signs, banners, or posters should publish standings at frequent intervals. Supervisors can hand out bulletins to employees urging them to keep the record perfect. Trade journals like *Safety+Health* and National Safety Council newsletters are other publicity outlets. In a smaller community, outstanding safety performance by a well-known company deserves—and usually gets—excellent publicity in the local newspaper and on radio stations.

The publicity value of a successful contest is considerable, although difficult to estimate. It should be commensurate with the significance of the occasion.

The presentation of an award to an employee who has gone 25 years without a lost-time injury has human-interest value for both internal and external communications resources. Recognition of an exceptionally fine "no-injury" record by a plant or company or presentation of a National Safety Council award would also call for widespread publicity.

Some companies purchase radio or television time to announce the results of a contest. One company had large campaign-type buttons made, and photos of children wearing them appeared in the local press.

Publicity (including photographs) can be sent to the local newspaper. The information should be prepared as a press release, indicating the nature of the contest or award. (See Some Basics of Publicity later in this chapter for more details.)

Meaningful Awards

An award serves several purposes: an inducement, a goodwill builder, a continuing reminder, and a publicity tool. To serve these purposes, however, an award must be meaningful to those participating in the contest.

Employees sense when management is giving awards only for "sales" or publicity purposes and is providing little or no safety effort. If a number of awards are given, the sheer volume can detract from the value of the program.

The value of awards lies in their appeal to basic human-interest factors, such as pride, need for recognition, urge to compete, and desire for financial gain (refer back to Figure 33–1). Some organizations present monetary awards as a bonus for making an extra safety effort. The distribution of U.S. savings bonds for safe records gets away from the direct monetary nature of a financial award. Other organizations, however, prefer gift or plaque awards appealing to the other basic human-interest factors.

The originality or cleverness of an award or of its presentation is an important factor. Select awards that are worthy of good publicity, photograph well, and provoke employee conversation. Refreshments for all employees in a department after the completion of a set number of injury-free hours probably will create more favorable comment than would the presentation of a fancy plaque to the department supervisor. Drawing for a small cash prize or a grab-bag prize would attract more interest than a routine presentation of the same award. An award to an employee's spouse for completing a home safety checklist, for writing a safety slogan, or for contributing to the company paper also generates more interest than would the same award given on the job.

- One Council member reported an award idea that received an unusual amount of publicity, both within the plant and locally. A local automobile agency loaned the company a new car that an injury-free employee, whose name was drawn from a hat, drove for a week. The employee also had a special "reserved for John Doe" parking space in the company lot. The only expense for the employer was a few dollars to cover special insurance. Even a special parking space awarded on a rotating basis can be effective.
- Another way to gain interest is to let the employees participate in selecting the award, planning its presentation, and helping with publicity. Frequently, they will suggest a humorous or novel award or publicity approach that may attract more interest than one planned by management.
- Payment of bonuses as awards for good safety records evokes considerable differences of opinion. Some managements and safety people feel that this approach is unwarranted because all employees are paid to work safely. Others believe it can enhance an already successful program.
- Some companies raffle household or sports merchandise. Interest in safety among supervisors and workers often is developed by personal awards like wallets, knives, or key cases, often suitably inscribed.
- Many employees value attractive pins or engraved cards commending them for years of employment without an incident. One company places a safety record sticker on the employee's hard hat. Others provide special badges, pins, or shoulder patches to recognize safety achievement or service on a safety committee.

In addition to contest awards, recognition should be given to those who have saved lives, served on safety committees, submitted valuable safety suggestions, or made other significant contributions to incident prevention.

Award Presentations

To make an award presentation something special, one company rented an auditorium and invited civic and labor leaders, company executives, other dignitaries, and employees' families to a large celebration party. Another approach is for the president of the company, or some other high official, to present the awards at a general meeting, a picnic, or a dinner (or breakfast) that may even include entertainment. The reason for inviting VIPs is not only to add prestige to the presentation, but also to promote their interest in and commitment to safety.

Such presentations require planning and must be in keeping with the importance of the occasion. The location chosen for the event should be appropriate and comfortable, not noisy or crowded. An award to an individual might be made in an executive office; groups might receive awards in a conference room, private dining room, or company lounge or cafeteria (during a nonrush period).

Inform participants about the agenda. Familiarize those who make the presentation with the significance of the award, the achievement it recognizes, and the background of the individual(s) who earned it. Ask the press to cover the event. Managers can post press photographs after the actual presentation to take advantage of desirable backgrounds, such as the plant or company name, some prominent trademark, or another eye-catching effect. Feature the award itself prominently in the picture. (See the Campaigns section [and those that follow it] later in this chapter and the National Safety Council Data Sheet 619, Photography for the Safety Professional, available through the Council Library, for other ideas.)

For a group award—such as a company, plant, or department completing an injury-free year—management can provide free refreshments for a specified time ranging from one coffee break (per shift) to a full 24-hour period. Another, more elaborate award was given by a company employing about 300. At the end of a year in which there had been no disabling injuries, the president took the group to a Major League Baseball game. The following year, after the company maintained its no-injury record, the employees, plus their families, were invited to an all-day picnic and cruise.

Such activities help build better employee relations, as well as promote more interest in safety on the job and within communities.

POSTERS

Safety posters are one of the most visible evidences of incident-prevention work. Because of this, some companies have mistakenly assumed that posters alone can accomplish safety instruction and have neglected such essentials as real management support, guarding, and job instruction. In fact, hit-or-miss use of posters in a plant where no other safety work is done is likely to have a negative influence, making employees feel that the company is not sincere. (See the National Safety Council Data Sheet 616, Posters, Bulletin Boards, and Safety Displays, available in the Council Library.)

Posters can reach a large audience with brief, simple messages designed to accomplish one or more missions: to convey information, to change attitudes, and/or to change behavior. They are used to communicate with people who are going about their normal activities. As a result, the audience must notice, read, and take in the message in a very short time—often in a matter of seconds (Figure 33–4).

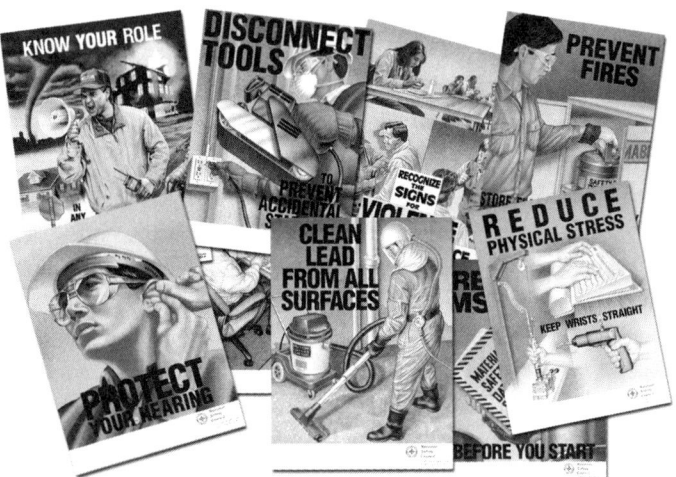

Figure 33–4. Posters convey safety messages forcefully and often humorously and help emphasize a safety message in a memorable way.

The most effective poster program has two major components:
1. The posters are changed frequently (preferably weekly) and never used twice.
2. The posters should use employees and equipment from the location.

Purposes of Posters

Posters properly used have great value in a safety program through their influence on attitudes and behavior. A look at the efforts that commercial advertisers make to acquire space in business areas or near factory gates can lead to an appreciation of the value of posters inside the workplace. When selecting posters, keep their specific purposes in mind. They are used to:
1. remind employees of common human traits that cause incidents
2. impress people with the value of working safely
3. suggest behavior patterns that help prevent incidents

4. inspire a friendly interest in the company's safety efforts
5. foster the attitude that incidents are mistakes and safety is a mark of skill
6. remind employees of specific hazards.

Posters also help to support special campaigns, for instance, using guards, wearing eye protection, maintaining good housekeeping, offering safety suggestions, or driving carefully. They promote traffic, home, and even pedestrian safety by encouraging safe habits both on and off the job.

Effectiveness of Posters

A number of studies have underscored the effectiveness of posters for training and motivating.
- One study, conducted by the British Iron and Steel Research Association, used three posters that reminded workers to hook cable slings. The posters were displayed in six plants over a period of 6 weeks. A seventh plant was used as a control. Tallies made in the six plants before and after display of the posters showed about an 8% increase in compliance with the rule. The seventh plant, in which the posters were not used, showed a very slight decrease in compliance.
- Although use of the posters merely supplemented previous training, the plants that originally had the lowest rates of compliance showed some of the best gains. Using the three test posters separately, on a biweekly basis, proved slightly more effective than simultaneous use of all three during the entire 6-week test.
- In a survey conducted by a prominent casualty insurance company, more than 200 employees were interviewed in depth on the effectiveness of safety posters, films, and leaflets. Results indicated that all the media were instrumental in bringing workers to a higher level of safety awareness and that all were effective in sustaining that awareness. Employees preferred posters to leaflets, although they acknowledged the value of leaflets for more detailed coverage of safety issues.
- A survey of Council members indicated that about three-fourths of the nearly 800 respondents use a variety of poster subjects, with one-third preferring cartoons and all-industry posters. Horror or shocker posters were least preferred.
  - Fifty-four percent of the respondents used posters to influence general attitudes; 27% to cover special operations; 18% to meet special or seasonal problems; and 14% to promote off-the-job safety.
  - Sixty-nine percent of the respondents in the Council's poster survey preferred posters of 8 × 11 in. and 17 × 23 in. sizes (21.5 × 28 cm and 43 × 58 cm). Of these two sizes, the smaller was preferred by a six-to-one margin over the larger.

Frequently, an unusually striking photo or overly elaborate (and expensive) artwork actually detracts from the safety message. This is not to say that "eye-popping" illustrations should never be used; they may serve to attract attention to the more conservative, serious messages on other posters. However, be sure photos are related to the message being presented.

Types of Posters

Industrial posters available from the National Safety Council, insurance companies, associations, and other sources fall into three broad categories: general, special industry, and special hazard.
- General posters are concerned with such subjects as risky behavior, disregarding safety rules, forgetting to replace guards, and other human failures.
- Special industry posters, as the term indicates, have application only in specific industries, such as mining or logging.
- Special hazard posters, for example, on lifting, ladders, or the storage or handling of flammable liquids, are useful in every industry where the particular hazards are encountered. In some cases, a hazard is so serious that a special poster is developed based on the severity and not the frequency of exposure.

Other Locations

Posters and stickers can be mounted on delivery trucks, buses, industrial trucks, and mail trucks and carts; in elevators; and even on doors. Pocket cards or plastic pocket protectors, such as those available from the National Safety Council, might be called "walking safety posters."

Other Materials

Safety messages need not be limited to printing and artwork. They can be very effective when used in illuminated or changeable signs. One company paints a safety message on a plywood welding screen.

Homemade Posters

A company can develop its own posters to deal with special hazards not covered by posters available from outside sources. Even the smallest company can make an occasional special poster inexpensively—using colored paper, crayons, or felt marking pens—to call attention to a special hazard, to commemorate the winning of a safety award, or to point up a problem not likely to be covered by a commercial poster. (See Chapter 32, Media, for details.)

Safety personnel or supervisors can make effective posters using photographs of local conditions or incidents, even if the situations must be posed. A common type of homemade poster is the "testimonial," featuring a photo of an employee and a close-up of damaged safety glasses or safety shoes. The poster carries a brief statement explaining how this equipment protected the wearer.

Homemade posters on new processes, new guards, or new rules personalize the safety program and augment even the best selection of commercial posters. An excellent source of this type of poster is an employee contest, which is easily administered at a nominal cost.

Changing and Mounting Posters

No specific rule can be given for how often posters should be changed because of varying definitions of the term "poster." Some types probably should be mounted permanently. For example, a poster on artificial respiration can be kept in the first-aid room, or one on the use of a certain kind of fire extinguisher can be posted near it.

Most companies change general-interest posters at definite intervals, usually weekly. They may rotate them from one area to another or file and reuse them after a year or so.

Management should vary the type of posters displayed. Consecutive posting of several health-related messages, for instance, or of new machinery posters is not desirable unless the company is conducting a special campaign on the topic. It is better to mix poster topics; for example, present an eye safety poster, then a machinery guarding poster, followed by a poster warning against wound infections, and so on.

For maximum effectiveness, posters not only must be selected carefully and changed on a definite schedule, but also displayed attractively in well-lighted locations where they will be seen by the greatest number of people. They should be placed on safety bulletin boards, near time clocks, in cafeterias, and at points of special hazard, such as paint storage rooms, rubbish cans, hazardous machinery, or dangerous intersections. Consider placing safety posters in unique places such as in the washroom above urinals, on the back of stall doors, on mirrors, or by paper towel dispensers.

Bulletin Boards

Bulletin boards should be large enough to allow convenient change of posters and should be placed where employees can see them during break times, such as near drinking fountains. They should be centered at eye level, about 63 in. (1.6 m) from the floor. Place them in a well-lighted location, or provide special lighting. Use a bulletin board size that is appropriate for the work space.

Boards should be painted attractively and glass-covered. Although a single board in the workplace is usually enough, several panels may be better in lunchrooms or locker rooms. Flashing lights or other eye-catching extras may be suitable for nonproduction areas but are likely to distract and annoy employees in workplaces.

A bulletin board should be used for only one display at a time, but need not carry safety posters exclusively. Any program of mutual interest to company and employees can be displayed on the bulletin boards. In fact, safety posters may have a stronger appeal if they appear along with other displays and announcements.

Bulletin boards in the same company may range from large, enclosed, illuminated boards with special sections for posters, safety bulletins, and other messages to a number of small frames or other inexpensive poster mounts installed at strategic points.

Poster frames to which clip-on literature racks can be added are available from the National Safety Council. This arrangement allows the company to conveniently distribute leaflets and other pickup literature that supports the safety message.

DISPLAYS AND EXHIBITS

The company can also use displays and exhibits to promote off-the-job safety as well as workplace safety. For example, many companies try to educate their employees to understand that a safe vacation is a happy one.

Personal protective devices, tools, and pieces of fire-fighting equipment can be used to make up displays or exhibits, with or without corresponding posters (Figures 33–5 and 33–6). Another good eye-catcher is a seasonal exhibit combining a Council poster and a safety display featuring such items as proper footwear for winter walking or skin and eye protection items in the summer to guard against sunburn and eyestrain. Management can purchase signs with changeable letters, electric tape messages, or unusual lighting for safety topics.

Companies and associations have devised many simple and attractive displays for presenting statistical data to workers. One is a safety clock, whose face is marked off to indicate the frequency of disabling injuries. Twin clocks or dials often are used, one recording the present injury rate and the other the rate for the corresponding period of the past month or year.

One company used large thermometer-like boards, placed at every gate and clock-house. Arrows indicated the present and previous month's records. The comparative standings of departments appeared below the symbols.

An auto race was the theme of another display. Each car represented a department, and the cars moved daily to denote progress being made. Airplanes can be used similarly. Another exhibit featured racehorses participating in

Figure 33–5. Personal protective equipment is the theme of this display at NRPC's Beech Grove, Indiana, maintenance shop. The display is lighted at night and is in Amtrak's impressive 15-foot display case.

Figure 33–6. In this scaled-down trailer-mounted house, children learn to escape from fires. They learn to use the proper procedures as they exit the bedroom when the smoke detector goes off (Operation EDITH: Exit Drills In The Home). *(Courtesy Springfield, IL, Fire Department.)*

a "Safety Derby"; the horses were named after items of personal protective equipment. A display of highway signs with photographs of incidents below them was placed near the plant entrance of another company.

OTHER PROMOTIONAL METHODS

Other methods that can be used effectively to spark and maintain interest in safety are campaigns, safety stunts, courses and demonstrations, publications, public address systems, and suggestion systems. These will include intranet postings, safety blogs, list-servs, and other e-mail-based communications.

Campaigns

Campaigns focus attention on one specific incident problem. They are additions to, and not substitutes for, continuous incident-prevention efforts the year round.

Campaigns may promote home safety, vacation safety, fire safety, or the use of safety equipment such as safety shoes. Companies can sponsor a "Clean-Up Week" or run a "Stop Incidents" campaign to promote the development of safe attitudes both on and off the job.

Many large corporations have conducted extensive campaigns to promote safety on and off the job. Much of their safety awareness material is aimed at employee families, local citizens, and even groups outside the United States.

Companies must plan suitable publicity for the campaign from kickoff to conclusion, similar to that discussed earlier in this chapter for safety contests. Signs, flags, desktop symbols, and other items can be used to dramatize the campaign. To wind it up, schedule a special event such as giving each employee an inexpensive novelty item, free coffee for a week, or a free breakfast or dinner.

Many of the same promotional ideas that help to maintain interest in contests also can be effective in special campaigns—for example, a first-aid drill or a demonstration of cardiopulmonary resuscitation (CPR). Some companies use safety parades, exhibits of unsafe and safe tools and equipment, pledge cards, and other such features.

Timeliness may be an important factor in the way employees respond to a campaign. Successful safety campaigns have been linked to elections, the World Series, the football season, Thanksgiving, and other special events.

Unique Safety Ideas

Unique safety ideas or "stunts" capitalize on all the effective aspects of showmanship. Companies can develop them as separate devices for maintaining interest or use them to supplement contests and campaigns. *Safety+Health* and other publications regularly give details on various stunts.

Most companies agree that constructive stunts help inspire employees to high standards while stunts that ridicule usually do more harm than good. Moreover, employees and supervisors who are the objects of ridicule may have just cause to blame management for not setting up safe procedures or providing safe facilities and equipment.

Safety stunts can involve an entire company, a department, a small group, or just the individual. A stunt may be humorous, novel, or dramatic, and occasionally even shocking.

A simple stunt is often the most effective. A pivoted hammer, mounted over a pair of safety glasses in a display case, can be operated by a string to demonstrate the impact resistance of the glasses. To dramatize the importance of eye protection, the "let's pretend" stunt can be used. Several volunteers are blindfolded and then asked to eat, write, and move around.

Stunts developed for the company safety program often serve equally well at company open houses or safety picnics and in community safety projects as well. Such stunts, when supported by visual aids, signs, and printed material, demonstrate the company's interest in incident prevention. They also give employees a chance to participate in programs that help create safer attitudes on the job.

Some firms use a somewhat unique poster or card as an attention getter. For example, safety cards handed out by employees can alert co-workers that they have been risking an incident by their behavior.

Courses and Demonstrations

Most safety and health professionals agree that courses in first aid, lifesaving, water safety, civil defense, and disaster control have bonus values that help prevent work injuries. The worker who has completed a course in first aid and has learned to do CPR will be more aware of the hazards of electric shock and more likely to help maintain electrical equipment in safe condition. Likewise, the employee who learns how to stop arterial bleeding better appreciates the serious consequences of using an unguarded saw, power press, or cutter.

Home-study and extension courses also serve to stimulate and maintain interest by giving the employee a better understanding of hazards involved in the job. Most safety training courses are designed specifically to improve the attitudes of both supervisors and employees. Using appropriate videotapes and other visual aids will enhance the effectiveness of the courses. (See Chapter 32, Media.)

Public participation in courses involving employees promotes community goodwill. Many industrial safety people are doing an excellent job of promoting safety and fire prevention by arranging or teaching courses on these subjects for the Boy Scouts and Explorer Scouts, the Girl Scouts, Camp Fire groups, Junior Achievement companies, and other youth or school groups (Figure 33–7).

The National Safety Council's "Defensive Driving Courses" provide an excellent means of promoting good employee and public relations. They stimulate safer attitudes both on the job as well as off.

Fire equipment and fire safety demonstrations have a practical value beyond that of teaching employees how to react in an emergency. The fact that the equipment is provided for their use reminds them of management's concern about their welfare. Moreover, the demonstrations make

Figure 33–7. Management, employees, and their families participate in a safety fair. Before the children can slide down this authentic fire fighters' pole, they are given a safety question to answer.

employees more aware of the dangers of fire and point up the need for obeying fire prevention rules.

Such demonstrations are easily arranged through local fire departments or fire equipment distributors. Many companies conduct their own demonstrations, using extinguishers that require recharging, or "not-in-service" extinguishers kept specifically for this purpose.

Publications

Reports

Safety program progress reports should be written in an interesting and concise manner for superiors and supervisors. Visuals can be used effectively where appropriate. (See Chapter 32, Media.)

The safety and health professional can present the cost of incidents and, perhaps, the cost of prevention in terms significant to management. These costs include medical and compensation costs, production losses, sales losses, increased maintenance costs, and the less tangible, but perhaps more important, hidden costs involved in administrative problems and in impaired public, customer, or employee relations. Reports need not be dull. Photographs, for example, can pinpoint a company's major sources of disabling work incidents.

A statement of incident losses and safety achievements may be included in the company's annual report or in a special annual or monthly safety report issued to top executives and supervisors. If departmental incident losses, like incidence rates, can be charged on an equitable basis, such as "per hundred thousand dollars of sales" or "per one thousand employee-days of production," the comparative standings of departments and improvement in departments or units are easy to evaluate.

The fact that the company records and publicizes such information is, in itself, an incentive to supervisors. It reminds everyone concerned that incident costs are just as much an integral part of profit and loss as production, sales, maintenance, distribution, and advertising.

Special charts, graphs, and statistical reports also can visually show facts about incidents. One chart can track the number of lost-time injuries, others the number of days lost, injury causes, incident causes, or type of injuries. It cannot be too strongly emphasized that unless such charts are kept up to date, they can do more harm than good. An outmoded chart indicates management is losing interest in maintaining a strong safety program.

Annual Reports

In recent years, companies have worked hard to make their annual reports to stockholders interesting and clearly understandable. In many cases, annual reports are also distributed to employees so that they can become better acquainted with the company's purposes and problems. A section on aims and accomplishments in incident prevention attracts employee interest and further serves to emphasize the interest of management in the safety of its employees.

Newsletters

Monthly or weekly Internet dissemination of newsletters or videos is especially important as a means of maintaining interest. This keeps employees and supervisors informed, particularly in decentralized or field operations where bulletin boards are not practical. Newsletters can give detailed information on standings in a safety contest and publicize unusual incidents or serious hazards. They can help explain safety rules, remind employees of safe work practices, and support the safety program in general. If workers can serve as "reporters" or help to produce such a newsletter, so much the better (see details later in this chapter). A case history of a particularly unusual or spectacular incident can sometimes be featured.

Booklets, leaflets, and personalized messages take many forms: safety rule booklets, NSC's "New Employee Booklets," special "one-shot" leaflets, and monthly publications such as the National Safety Council's *Today's Supervisor* newsletter. The content of an employee rule booklet, except for material involving company policy, may be developed with the help of safety committees or selected workers to stimulate interest in the topic and help to ensure compliance with the rules.

Larger companies may have their own editors and artists, even their own printing facilities. Smaller companies, however, also can issue attractive booklets, leaflets, and personalized messages, and at negligible expense, by using local consultants and printers or their desktop publishing systems.

The National Safety Council, trade associations, and professional organizations publish a wide variety of booklets and leaflets that are authoritative, attractive, and relatively inexpensive. They cover a wide range of subjects—material handling, first aid, housekeeping, fire prevention, vacation safety, safe driving, and the like. Such materials, carefully selected and regularly distributed, effectively supplement company-prepared publications.

Letters commending meritorious service, signed by the manager and addressed to individuals, also make an excellent impression on workers.

Safety calendars, together with holiday letters from the manager, have a direct appeal that reaches the workers' homes. Such mailings should include all employees. *Family Safety & Health*, a quarterly magazine, is sent to more than 2 million homes by organizations that are interested in the welfare of both employees and their families.

Buttons, blotters, book matches, pencils, and other small novelties all conveying a safety message also may be used. For example, packets of silicone tissues for cleaning glasses, imprinted with brief messages or safety rules, serve to remind employees of the rules and to encourage proper use of safety equipment.

Public Address Systems

Public address systems often are used to broadcast announcements and page employees. Many companies have taken advantage of these installations to broadcast safety information as well.

Such messages should be planned carefully. Employees might readily lose interest in long speeches or too-frequent safety reminders. When the public address system is used for broadcasting music, safety announcements can be made between songs.

Suggestion Systems

Because incident prevention is closely associated with efficient operation, many suggestion systems not only help to prevent incidents, but also lower production costs, improve manufacturing conditions and methods, and change the outlook of workers.

If a company does not have a general suggestion system, it is probably better not to establish one for safety suggestions alone. Setting safety apart from ordinary operating procedures may deemphasize its importance.

Near-miss reporting has gained traction in many industries. In some cases, near-miss reporting is a leading indicator, but is a lagging indicator in other cases. Reporting of near misses can be anonymous or tied to an incentive program.

Getting good suggestions is essential and must be encouraged by all levels of management. Company posters, contests, campaigns, merchandise incentives, direct mail, printed handouts, personal appeals, supervisory training, safety clubs, and press releases can motivate employees to submit safety suggestions.

Many effective suggestion systems offer money for ideas, and companies may pay considerable sums for employee suggestions. One firm awards more than $10 million annually but considers the money well spent on valuable suggestions.

To merit an award, a safety suggestion, like a production suggestion, should be substantial and practical and should offer a real solution to a problem. For example, suggestions worthy of awards include changing a method or material, guarding a hazard, or inventing a safety tool or device. Suggestions that would not merit awards include erecting a sign, cautioning workers, or publishing slogans.

Safety suggestions generally are regarded as a highly desirable way to avoid safety grievances. Improved employee interest and personal involvement are additional benefits.

Suggestion Awards

It is easy to measure the monetary value of suggestions that result in greater efficiency, lower material cost, decreased labor cost, or reduced waste. Usually, awards for suggestions in these categories are in proportion to the savings derived by the company. In other instances, safety suggestions may have a monetary value but be harder to evaluate. As a result, payments for suggestions that contribute to the welfare of the employees but result in no direct savings to the company are most often estimates or composite judgments. Some firms have developed guidelines that consider such factors as the degree of hazard, originality, extent of application, and so forth. One company has also developed an award guide based on disability cost experience.

If management downplays "safety awards" and gives more weight to other awards, employees will believe that the company regards safety as an unimportant sideline.

Payment for a safety suggestion must be equal to awards given for other suggestions—it should be based on its real worth if it can be determined. If a suggested safety device enables an operation to be run at a speed that would be dangerous without the device, a saving may be measured. If a number of incidents have occurred in an operation and a suggested device will eliminate them, the cost of those incidents can be projected and a saving calculated.

Awards should reflect the merit of the suggestions. Most companies award cash and/or bonds. Some award merchandise, all-expense-paid trips, company stock, certificates of merit, medals, gifts for the suggester's spouse and family, or attendance at a recognition luncheon.

Some companies exclude superintendents, supervisors, designers, methods and systems personnel, and other supervisory or technical personnel from receiving awards, so that the other workers will have someone to whom they can go for assistance. Some companies feel supervisory personnel should not be excluded; several firms have separate plans and award schedules for salaried, supervisory, technical, and management personnel.

Suggestion Committee

If necessary, a special subcommittee can be set up to determine the monetary value of safety suggestions so that employees will be rewarded for them exactly as they would be for other money-saving suggestions. However, some firms believe such special treatment sets safety ideas apart from other ideas.

Many established suggestion systems now in operation are producing excellent results in monetary and "people" savings. No company should consider a suggestion plan without first studying carefully the plans now in existence.

Boxes and Forms

Suggestion boxes should be attractive and well placed and stocked with special, blank submission forms. Commercial suggestion forms also are available. To increase the interest of the employees and establish a spirit of cooperation and importance, it is essential that management acknowledge and resolve all suggestions promptly.

CAMPAIGNS

The preceding sections covered internal publicity within a company or organization. The remainder of this chapter discusses how to influence the way a company looks to people on the outside. Favorable publicity is an unmistakable bonus to a good safety program. Yet many organizations often disregard it.

Any company likes to have someone—especially a prospective customer—say, "I like what I hear about this company. I understand that it really takes care of its employees. So I figure it must treat its customers right; therefore, I'll be treated right." One good way for a company to get a reputation for taking care of its employees is to be known as a safe place to work. Surprisingly, an amazing number of companies do little or nothing to let their public—customers, stockholders, the community—know that the safety and health of their employees are important to them. This chapter presents a simple and sensible formula for letting people know that "at my company, the health and safety of the workers are important."

Most companies have a professional public relations (PR) department to handle the public communication program. In smaller companies, the safety and health professional may have to handle public communication. In both cases, the information in the following pages should prove useful. The safety professional should be aware of overall company policies and programs and know when to turn to specialists and when to ask for creative help.

PUBLIC RELATIONS

Public relations is the "management function which evaluates public attitudes, identifies the policies and procedures of an individual or an organization with the public interest, and plans and executes a program of action to earn public understanding and acceptance," according to the magazine *Public Relations News*. Abraham Lincoln knew about PR. Speaking at Ottawa, Illinois, in 1858, Lincoln said: "Public sentiment is everything. With public sentiment nothing can fail; without it, nothing can succeed. He who molds public sentiment goes deeper than he who enacts statutes or pronounces decisions. He makes statutes or decisions possible or impossible to execute."

Lincoln's classic quotation on public sentiment can be traced to Jean-Jacques Rousseau, the 18th-century French philosopher who is generally credited with developing the term *public opinion* as we use it today. At its simplest, PR is the promotion of good public opinion about a person, a company, a government, or other entity. Thus, every employee, every activity, every facility of a company contributes in many ways to the overall opinion that people outside the company have about that company. This is true PR.

Anyone concerned with incident prevention—safety and health professional, supervisor, member of the plant safety and health committee, or officer of the school or community safety council—should realize that any time they communicate with someone outside the committee, department, or company, they are involved in PR. Even an event such as a company-sponsored family picnic can strengthen PR.

To be successful, a PR program must be backed by an effective management team and a quality organization. Public relations can then reflect the nature of that organization. A strong safety program merits good publicity, but false publicity or publicity based on inflated facts or specious statistics will be recognized for what it is—company propaganda—and can do more harm than good.

The basis of a successful PR program is successful management. Such a management team makes sure that staff and employees produce good products safely and efficiently, that they cooperate with each other and with the customers, and that all give the best and friendliest service humanly possible at all times.

The plain fact is that poor PR is costing individuals and organizations in this country millions of dollars each year. The remedy is simple: a better understanding and use of fundamental PR on the part of everyone—and a sincere effort to put it into practical use. For lack of good PR, many a worthy cause has failed to get the support it deserves, and many an organization has failed to gain the business it needs. A good PR program need not cost a great deal of money. But it is worth all the time, effort, and reasonable expense allocated to support its activities.

THE VOICE OF SAFETY

In any genuine, effective PR program, emphasizing safety can be a real help. In fact, it is hard to imagine a PR program where sincere and effective concern for the protection of employees from incidents is not a top priority.

If a company does not have a safety program, it misses vital opportunities for good PR and dramatically increases the chance for adverse public attitudes, not to mention the risk of an incident. A safety program, properly managed and communicated to internal and external audiences, can help offset or at least mitigate adverse news should incidents occur. In addition, the safety program is long range and sustained, whereas "news" is here today and replaced by other news tomorrow. The time to implement a safety PR program is not when trouble strikes. The program must be in place and functioning, and the public must know about it.

The safety and health professional should not only welcome publicity for safety efforts but should energetically seek it.

Working within the Company

In evaluating a company's PR function, begin by asking two questions:

1. Does the company have a PR department?
2. Is there an employee publication in the company?

If both answers are "yes," the safety professional should get in touch with both these units before doing anything about publicizing the safety program. This step is important. It not only ensures professional skill and consistency of efforts to publicize safety activities, but it will save confusion, avoid duplication, and possibly prevent conflict between safety personnel and the PR department.

Why does such an obvious step have to be mentioned? The reason is that, too often, there is little, if any, communication between a safety and health professional and the PR department and publications editor. Communication between the safety and health professional and the PR staff and editor is indispensable because these three must work together, or safety will not receive the attention it deserves and needs.

The safety and health professional should explain to the PR staff and the editor, if this has not been done before, that positive safety motivation and incident reduction at the company are priorities and that their communications help and advice are needed to reach these goals. They need to understand the necessity of employee and public acceptance of the safety program and how much the safety and health professional is depending on them to help earn employee acceptance.

Safety activities contain a wealth of real news. It may be that everyone in the safety business has assumed that safety activities and programs by their very nature must be on the dull side. In recent times, however, more writers and others have begun to recognize that safety can be made interesting. It takes the combined efforts of safety and health professionals, PR people, and editors to do the job.

Sixteen Ways to Make Safety News

Public relations information, to deliver an effective message, need not always be red-hot news. However, it must have an element of spot news, human interest, or self-help. These qualities will give it feature value for the media. Activity—real, honest, legitimate activity—makes news. Of course, dramatic circumstances such as a rescue or near miss make news. But until they turn up, the best stories are those that describe a company's constructive and creative efforts to promote and maintain a worker health and safety program.

It might be useful to list some of the items and activities that can make safety news in an organization and that editors and PR staff ought to know.

1. No-injury records for the entire company or for any one unit—in terms of either days or worker-hours.
2. Improved safety records for the company or for any one unit, even if no prolonged no-injury period is involved.
3. An interplant safety contest, or an intercompany contest—especially if anyone has dreamed up an unusual angle or prize.
4. Any unusual safety record for safety performance by an officer or employee of the company—either in length of time or character of the job done.
5. Innovations in safety programs that will prevent incidents. An invention, too, has special news value if the company has been plagued with incidents that the new gadget may prevent.
6. An unusual or highly valuable safety suggestion by an employee.
7. Safety conventions or meetings, either those held by the company or those held elsewhere, to which company representatives will go. A digest of such meetings should be publicized.
8. Other special safety events besides conventions—a safety banquet, a safety training course, fire and first-aid demonstration, a special meeting, or an award ceremony.
9. Some unusual event intended to get the employees to take their safety training home to their families, or something the company is doing directly with workers' families to promote around-the-clock safety. Open-house tours, local water safety shows, or public showings of safety films are examples.
10. Some pronouncement or statement by the president or other high officer of the company on some unusual or new safety device or company safety service, such as free inspection of employees' cars.
11. A speech by the head of the company or the safety and health professional at a local, regional, state, or national safety convention or conference. The editors and PR staff should have advance copies of it. The person making the speech should be sure to say something worthy of public attention.
12. The company's annual report. This report is the foundation for corporate communications. Stockholders do read these reports. A paragraph or two on the good safety record for the past year will go a long way in achieving sound publicity within the corporate family, as well as to inform the analysts, who recommend stocks, what the company is doing beyond its financial performance.
13. Any act of heroism by someone in the company. This is a sure-fire story for local papers as well as company publications. Maybe this type of news is not pure safety, but news media regard it as part of safety, and it can always be tied in with an indirect safety message.
14. A survey or study of some phase of incident preven-

tion in the company. If the investigators discover that married people who own their own homes are safer than their single counterparts, they have provided a ready-made story.

15. Election or appointment of a company officer or safety professional to an important post as a volunteer officer of the National Safety Council, American Society of Safety Engineers, Board of Certified Safety Professionals, local safety organization, or governmental agency.
16. A company or employee winning an award in a National Safety Council award program or contest. Winners, not losers, are publicized.

A good rule of thumb is to stress the positive, rather than the negative, side of an event. For example, instead of a story that says, "Single people are less safe," it could say that "A company study shows a need for special safety efforts by singles." Another way to stress the positive is to report, "Last year 290 employees worked without an occupational injury. Only 10 did not."

SOME BASICS OF PUBLICITY

If a company does not have a PR department, this fact need not prevent its getting information into local papers or on the air. Without PR personnel, the task may be more difficult, because PR people are more experienced and know how the media work better than the safety and health professional does. However, even in the absence of a company PR department, the safety and health professional can get publicity for safety activities by going directly to the newspapers, magazines, and radio and TV. It helps to solicit and secure PR advice from local safety organizations or even business or trade associations with such service. The safety and health professional should always keep the company management informed of what is going on at all times.

The safety and health professional should not pretend to be a PR expert. Instead, he or she should be sincere in approaching editors and TV/radio program managers with sound, worthwhile information and not worry about whether it is packaged perfectly for media acceptance. A good story is a good story. Most editors and reporters are glad to have the material and often will use it. News value and reader interest are what count the most.

Editors and program people are usually not difficult to approach, provided the safety and health professional is courteous and friendly and admits to a lack of specialized knowledge of PR techniques. Naturally, no editor or anyone else likes to have someone come charging in and claim to be doing a big favor by delivering the story the world has been waiting for.

In general, the common sense and salesmanship needed for success in the safety field are enough to enable the safety professional to present his or her case clearly and effectively to the media. Nevertheless, the person should be willing to accept advice from PR professionals on how best to tell the story. The safety and health professional needs to be careful about being quoted out of context.

Select the Audience

"Who must be reached with safety information, and is it of interest to them?" These seem like fundamental questions, but many companies never think to answer them. Because publicity cannot reach everyone, the company must choose its audience carefully, especially if the budget is tight or time is limited. In that case, it is logical to assume that—in addition to the in-plant (company) audience—the company would prefer to reach people who might buy the product or help the company in some other direct and profitable manner.

Use Humor and Human Interest

It is worthwhile to try to brighten safety, to make it positive, rather than ponderous and dreary. It is even possible to evoke a chuckle now and then.

Editors are familiar with the solemn pronouncement that "Safety is a serious subject, and must be taken seriously—safety is no laughing matter." No one can argue with that position. But does it follow, therefore, that no one can put into safety some of the same techniques, the same sales appeal, the same sparkle that is used so successfully to sell commercial products? If those techniques can sell shampoo or a personal care product or an automobile, is it unreasonable to expect they can also sell safety?

Or how about a cartoon treatment? This just may brighten what might otherwise be a slightly dull and drab presentation. The safety and health professional should not be too disturbed if someone points out that a cartoon has shown a negative aspect of a safety message—that is, what not to do. It may well have done just that. This is the very thing that gives a cartoon its punch. A pratfall cartoon will draw attention to the slippery, icy sidewalk in a way that cannot be shown by a person walking and not falling. However, care should be taken to avoid ridiculing or negatively portraying victims of incidents or the negative results of incidents.

It is even possible to get a child or baby into the act, or even a faithful, shaggy dog, in order to get the spark that lifts safety activities out of a dreary, impersonal rut and presents them in terms of human interest.

Names Are News

Remember that facts and figures about injuries and their frequency and severity are made more interesting by injecting human-interest elements into safety news. Because human interest means people, the safety and health professional should include more information on people and safety and less on things and safety. Mention which managers or employees accomplished safety goals, contributed ideas to the program, and performed other such activities. Whenever a person is identified in such an endeavor, ensure that the person is comfortable with the use of his or her name. Obtaining a signature on a release would be prudent in many cases.

Friendly Rivalry

Safety awards, safety records, safety contests, safety inventions, and gimmicks—these are only a few of the many things that make good safety news. For example, if the company has reached a new injury-free record in its industry, it is headed for headlines. The editor, the PR department, and senior management must be kept informed of safety progress all along the way. They will help arouse public interest in the performance and also stimulate greater interest and greater effort among the employees themselves.

An award loses its motivational and publicity value if kept a secret. The safety professional must publicize the event. Photos of award presentations are sometimes used, but they should be interesting and even unusual to attract special attention. Editors dislike the typical "grip and grin" award photos, so be inventive in trying to get a picture worthy of publication. For example, a representative from an organization conferring an award could be shown handing the award to a foreman or worker in the workplace setting.

In some instances, top safety awards have been accepted by some companies as if they were a "dime a dozen." On the other hand, companies that have made similar awards the occasion for a major bell-ringing celebration have seen their "safety stock" rise sharply among employees.

A public utility company in Michigan, for example, made so much of the award-winning safety records of its various units throughout the state that an outsider might have thought the company had won the World Series. At one celebration, more than 1,000 employees, from top management on down, jammed the closed-off street in front of company headquarters for presentations, followed by an all-company picnic. This celebration got coverage in the papers and on the air throughout the state of Michigan. Here was public relations that any organization would welcome. Safety had made news.

Techniques

These pointers are offered to the safety and health professional who wants to make the most of PR opportunities:

1. Be honest in what you say. Never exaggerate.
2. Deliver what you promise. If you say to the media that something is going to happen, make certain it happens—as you said it would when you said it would. This often calls for a "run-through" in advance.
3. If, for any reason, there is a change in plans from what you have announced, immediately notify the papers, radio, and TV stations in addition to social media networks if appropriate.
4. Be scrupulously accurate with all names, dates, places, and other facts. There is no such thing as being too careful in this respect. If an editor misspells names, the only thing to do is to resubmit the names correctly spelled and hope for the best.
5. Offer ideas but do not ask for specific space or time. Complaining about your company PR department, or complaining that the local paper or station has treated your company shabbily will not only be fruitless but will create a strained relationship between you and the press.
6. Do not alert your PR department or publications editor (or put out anything yourself) unless you have real news or features to offer. You must not issue material just to be issuing it. Be reasonable with the amount of material you send out. You can wear out your welcome.
7. Tip off your local or industry association, safety councils, and your PR department (or if you do not have one, the papers or radio and TV stations) to anything worthwhile you run across that might make an interesting news story. Do so even if it has no relation to you or your company; the media will appreciate it and be more inclined to view your own materials favorably.
8. Above all, do things that make news. Almost every routine safety item can, with a little extra effort by the safety and health professional and the editor, become a more readable, more constructive piece. News will be published only if something is being done for safety that makes news. News can always be heightened by intelligent, imaginative treatment, but the story must be worth telling. Advertising space can even be purchased for special items.
9. Use good sense and an honest approach if the news is bad. Prove to press representatives and the public that you can roll with the punches. Know your organization's policy and procedure to follow in all situations.

Communications by the Safety Office

If a safety and health professional must handle his or her own publicity with the local media, here are some tips. Many stories have died because someone failed to observe the following seven tips:

1. Editors and news directors are busy people. Unless a situation is an emergency, media people prefer to receive

possible story ideas in writing. Send them a brief release or a letter outlining what you have to offer. Include facts and, by all means, make your material interesting. Timing is vital: don't notify the media today of something you are planning for tomorrow. Never suggest that a reporter or news crew be sent out—this is the editor's or news director's province, not yours.

2. Generally, the person to write to is the city or business editor of the paper or the news director of the radio or TV station. Of course, if an item is specifically written for a certain columnist or commentator, it is better to make direct contact. If it applies only to a specialized area (e.g., finance or sports), it should be brought to the attention of that editor. Know the publication, whether it is general news or a trade journal. If you are working with TV or radio, don't waste their time on information or items they'll never use. (Example: Broadcast media seldom mention personnel changes.)

3. Write not for your boss, but for your reader or listener. Answer objectively the questions: who, what, when, where, how, and why? Do not load your releases with propaganda for the company. There is no surer way to kill your positive relations with the media.

4. Make your releases as professional in style, appearance, and general quality as you possibly can.

5. In writing a release, be brief and to the point. Printed space is limited and costly. Try to keep the piece to one page. Publications receive thousands of releases each month. These are skimmed, and only the best get into print. Likewise, remember that TV and radio broadcasts usually are measured in seconds, not minutes. Keep your material short and factual.

6. If you are sending a picture with the release, the caption should be typed on a piece of plain white paper and taped to the back or bottom of the photo. Never use paper clips, and never write with a pen or pencil on the back of the picture. Either of these will likely damage the photo and make it difficult to reproduce clearly. Think visuals when working with TV—what can they show that will help to tell or illustrate your story?

7. Leave script writing to the professionals, but check facts. If a radio or TV station requests material, send them the facts, figures, and whatever narrative is necessary. The people at the station will put them into the proper form.

Hints for TV Interviews

The safety and health professional must often be the spokesperson for his or her company, not only for newspaper coverage, but also for radio and television. Although getting the facts correct may be adequate for a newspaper interview, a television or radio interview reflects more of the company than merely what the facts show. It projects a company image through the company spokesperson. If the safety and health professional does not feel that he or she projects a good image over the radio or television, then someone who does should be picked from the safety department or from the PR department.

Here are four tips that will help the safety and health professional give a better radio or TV appearance:

1. Remember that you are being interviewed for your knowledge, not for your personality, entertainment value, or good looks. Be yourself. Don't put on a special voice or worry how the lavaliere microphone looks with your clothes. Don't wave your hands or touch your face or hair. Don't jingle coins or play with jewelry. Keep your hands down.

2. If your TV appearance is preplanned, dress conservatively. Avoid loud clothing or busy patterns that can affect the camera or overpower your message.

3. Review with the interviewer in advance what areas will be discussed. You can steer away from areas you cannot discuss, and you can get help on questions that you might not be able to answer. If you don't know the answer to a question, say so. But add that you will check and get back with the answer in a specific time period.

4. Be natural and cordial. Smile, if the situation is friendly. Be sure to maintain eye contact with the interviewer or camera. Do not memorize a statement and rattle it off. You will waste everyone's time. It is better to be well prepared and let the on-air material be a question and answer or discussion between you and the interviewer. Don't rush, and don't try to cram a lot of information into a TV slot. Keep it short and light. Avoid statistics.

Handling a Serious Incident Story

In any PR program, it is just as important to know what not to do as to know what to do. In fact, it can be even more important.

The foremost warning is this: do not cover up bad news. Good media relations are of utmost importance. It is at such a time that a sound PR program pays off.

Every safety and health professional hopes the day will never come when a serious incident damages a company's safety record, but it happens. In some instances, the repercussions of how the incident reporting and news coverage are handled have been even more unfortunate for the company than the incident itself.

Here is an example of how not to handle a press representative: in a midwestern city some time ago, two workers were killed by a crane. This company enjoyed a first-rate relationship with the newspapers and radio and TV people in that city. It had worked hard at safety and at PR. It was good to its employees and had a fine reputation for honest and open communication.

On this particular occasion, however, someone in the company's top management became anxious over the deaths. As a result, when a reporter came out to the plant to get what was to her paper a routine story of the incident, she ran into a maze of censorship.

First, the safety and health professional turned the reporter over to the personnel manager. This person switched her to the general manager, who gave her some "double talk" about the incident and told her the company doctor was the one to see. The doctor said the safety and health professional was the person to talk to.

By this time the reporter's anger and suspicions were rising. What had started out as a routine assignment now had become a challenge to discover what everyone in management appeared to be covering up.

It is important to remember that a reporter cannot lose in a contest like this. Because the workers had been killed, the coroner would have all the facts. If they had been badly hurt, one of the hospitals would have the information. If the workers had not been killed or hurt badly, then there would have been no story in the first place.

The reporter obtained the facts regarding the workers' deaths from the coroner's office. She then wrote a story that was as harsh toward the company as it could be without committing libel. The story was edited, headed, and set in type and laid out in the composing room, awaiting its turn to get into the paper.

Ordinarily, newspapers set more news in type than they can print. Each day dozens of items get left out—the "overset," as it is called in newspaper parlance. Under normal circumstances, the story of the incident might well have ended up as "overset" or, if printed, might not have received much attention. However, the cover-up and runaround the reporter had received at the plant changed all that. As a result, the story was marked "must" when sent to the composing room and ended up as front-page news.

When there is bad news to report, the safety and health professional must back up the PR department 100% to help get out the news as quickly as if the tidings were all in the company's favor. Unless directed to deal personally with the media, the safety professional should stay in the background and provide the proper information to management and public relations.

Along with the bad news, the safety professional's material should mention the good things. It may be that this is the first incident in months or years, that the company has a safety record far better than the national average for its type of operation, and that it has won a number of safety awards. Such counterbalancing information will help to take the sting out of the story. Reporters are usually willing to include these facts, too.

It is not only good policy but a wise precaution to be honest and fair to press representatives whenever there is news—whether pleasant or unpleasant. This principle is vital to a good PR program and will pay considerable dividends in terms of getting fair treatment from the media.

It would be wise for a safety and health professional to anticipate that someday, he or she may have to serve as a company spokesperson at an incident or disaster scene. It is imperative to follow company policy and procedure when acting as a company spokesperson. This needs to be planned ahead of time.

News media can actually help during a big emergency. Families, friends, and neighbors will be clamoring for news, and the media can get it to them fast.

WORKING WITH COMPANY PUBLICATIONS

If the safety and health professional thinks the company publication has neglected safety, it is time to correct the situation. He or she should ask the editor how more news value and human interest can be worked into safety stories. With some editorial assistance, the safety program can be just as newsworthy as other stories. The safety professional should offer to provide details and descriptions of events. The idea is to give the editor plenty of good, current information. The editor is just as eager as anyone to publish interesting news and features and will go more than halfway to think up ways to put news value and reader interest into safety doings.

Producing a Publication

Materials such as safety newsletters, instruction cards, bulletins, broadsides, booklets, and manuals for communicating safety rules, information, and ideas in print require careful planning and preparation. Among steps to be taken in planning both internal (to a company audience) and external (to the general public or other out-of-company groups) publications are as follows:

1. Clearly define the objectives of the publication. Consider the type of audience to be reached by those objectives.
2. Determine how general or how restricted the message is to be.
3. Decide what form of publication will best convey the message.
4. Estimate the cost of preparing and printing the publica-

tion in whatever forms, sizes, and quantities needed. An expenditure for a new publication must, of course, be provided for in the budget, whether or not the item is produced in-house.

If the objective is to give the worker specific rules to follow in doing a job safely and efficiently, an instruction card may be suitable. To stimulate general safety consciousness, a single sheet printed on one side may be effective. If a series of short reminders, for example, on fire prevention, is needed, posters may be the answer. To treat a topic of general interest, such as methods of materials handling, a leaflet can be the best choice. Here, posters or leaflets from an insurance company or other organization may be more effective and more economical than in-house materials. Manuals may be required for highly technical jobs or for more thorough coverage of a plant's safety policies and rules. Even a companywide (or plantwide) public address system, closed-circuit TV network, or videotapes would be appropriate.

When the safety professional and editor are deciding on the form of the publication, they should keep in mind that there is a direct relationship between the appearance of a printed piece and the degree of interest it arouses. Most readers will react unfavorably to a bulletin, newsletter, or booklet with text in small type, few or no illustrations, narrow margins, and long paragraphs.

Reasonably large type (12 point), selected to fit the size of the page and, of course, to accommodate the volume of material, will make the material more readable. In addition, judicious use of white space and variety in size and placement of illustrations help make a publication both pleasing to the eye and easy to read. In safety, as in other fields, ideas conveyed in print are best received and absorbed if they are well organized and attractively presented.

Illustrations break up the text and help to convey information to the reader. Photographs that show action described in the text add realism to instructional materials such as manuals. Human-interest photos are desirable in newsletters. Line drawings and sketches are valuable to clarify technical points on instruction cards, in manuals, and in other training materials. Awards can also be publicized.

If the printing process permits reproduction of photos and other illustrations, pictures of award winners, safety devices, and safe and unsafe practices can be used. To avoid embarrassing or ridiculing employees who have been injured or caught in an unsafe procedure, their features can be blocked out or the pictures can be posed (and so identified) to reproduce the event. Or cartoon illustrations can be made up depicting the unsafe procedures from the photograph.

Some state laws forbid publication of a person's photograph without his or her written permission. As a result, the editor should obtain a signed release from every person who appears in recognizable form in any picture. Often, having a new employee sign a photo release is part of the employment routine. Asking for a photo release is just good manners.

Preparation of Material

Once the objectives, scope, and form of a publication are determined, the person preparing it should make an outline of the subject or subjects to be covered. For most types of material, the outline need not be elaborate, but it should be logical and complete, showing how each topic is a part of the overall plan.

Before gathering material, the writer should study the audience for whom the message is intended in order to get some idea of the readers' knowledge and level of comprehension. In the interests of accuracy, completeness, and balance, material should be gathered from several sources—including articles, books, and especially supervisors, workers, and others in the company who have had experience in the matters to be treated. To ensure technical accuracy, it may be necessary to solicit help from specialists in specific areas.

No matter what the form of the publication, the writer should keep in mind certain basic rules of good writing. To get ideas across quickly and easily, use short sentences, simple words, and brief paragraphs.

In a piece of some length, such as a booklet, a system of headings, kept as informal as possible, will both arouse the reader's interest and guide his or her thinking. In a piece designed to instruct, numbered lists of job steps, for instance, will prove helpful. In any case, the writer should follow closely the line of logical thought developed in the outline.

Copy should be written in a positive, constructive style. When the nature of the material and the form of publication permit, a friendly—but never condescending—tone can be used effectively. Personal references and names, as in a newsletter, will increase readership. Humor tied to the message and pitched to the employees' sense of what is funny can add a great deal to some types of publications. For instance, cartoon illustrations and a light touch may be particularly effective in a rule booklet. The proposed publication's readability can be gauged by having a few people from the intended audience test-read the piece for understanding.

Production of Publications

For the technical details of printing, the advertising department or experts in the publishing field can be consulted. In terms of layout and typography, readability should be the first consideration.

How the piece is to be used will determine its size, paper, binding, cover, and similar details. If the materials are to be filed or if revised pages will be inserted, loose-leaf binders may be used. The in-company or outside editor or printer who will handle the job should be asked for technical advice.

Getting Ideas

Everyone in the PR business runs out of ideas now and then. Anyone suffering from this affliction should not hesitate to call on others for help. Employee publications do not compete with one another, so ideas can be borrowed freely from them. Some national agencies produce and supply safety materials that may serve as inspiration. (See Appendix 1, Sources of Help.)

SUMMARY

- Companies must work to maintain management and employee interest in safety and health programs because incident prevention depends on the desire of people to work safely. Thus, the major objective of a program designed to maintain safety awareness is to involve employees in incident prevention.
- Indications of the need for such a program include an increased rate of incidents and injuries, deteriorating housekeeping, and decreasing interest on the part of supervisors in enforcing safety and health regulations and procedures.
- A well-designed program to maintain interest in safety and health issues attempts to "sell" safety to employees. It must be based on management and worker needs and opinions. When developing the program and its related activities, the safety professional will need to consider company policy and experience, budget and facilities, types of operations, types of employees, basic human-interest factors for promotional activities, and other considerations. Activities and promotional efforts should emphasize a positive, constructive approach to safety and be changed often to catch employees' attention.
- Safety and health professionals are responsible for assisting line management in developing and coordinating the program and for encouraging the support of management, supervisors, and employees. Safety and health directors also work with local, community, state, and national groups interested in promoting incident prevention. They must be able to communicate their ideas clearly and effectively and to educate supervisors to maintain high levels of safety performance in their work areas.
- The ranking managers play a key role in any program designed to create and maintain interest in safety. Under existing laws, they are directly accountable to management and society for their employees' safety. Line managers, including supervisors, must (1) set a good example for their workers, (2) teach safe practices and procedures to employees, (3) enforce the rules fairly and consistently, and (4) keep abreast of the latest developments in safety and health issues.
- Programs designed to create and maintain interest in safety can include safety committees and observers along with quality and safety circles. Committees should consist of management and worker representatives, and membership should be rotated periodically. Quality and safety circles involve employees in hazard recognition and control. These groups meet regularly to discuss safety and health-related issues.
- Safety and health meetings can be conducted for managers, supervisors, employees, or other groups within a company. These gatherings can be of top executives, departmental meetings, small-group meetings, or mass meetings of all employees. The gatherings must be carefully planned to focus on specific problems and solutions, or their effectiveness will be diluted.
- Various promotional activities such as contests can help to stimulate worker interest in safety and health policies and procedures. Contests should be undertaken only after management has established a sound safety program in the company—they must not be viewed as a substitute for such a program.
- To be successful, contests must be planned and conducted by a committee representing all competing groups, competing groups must be composed of natural units, methods of grading must be simple and fair, awards must be worth winning, and the contest should be well publicized and promoted. Types of contests include injury rate contests, Council contests and awards, associate contests, intergroup competitions, personalized contests (raffles, sweepstakes), noninjury rate contests, community and family contests, and miscellaneous contests.
- The value of contest awards lies in their appeal to basic human-interest factors such as pride, need for recognition, urge to compete, and desire for financial gain. Awards must be meaningful to workers to be properly motivating. Many companies allow employees to select their own awards. Award presentations should be staged to provide maximum recognition for the recipients and maximum publicity for the company.
- Posters and displays can reach a large audience with brief, simple messages designed to convey information, to change attitudes, and to change behavior. Properly used, posters and displays have great value in safety programs.

They are used to remind employees to work safely, suggest behavior patterns that prevent incidents, inspire interest in safety efforts, foster attitudes that incident prevention is a skill, and remind employees of specific hazards.
- There are three types of posters: general posters concerned with broad safety topics, special industry posters that apply only to specific industries, and special hazard posters emphasizing particular hazards.
- Posters and displays should be set up where employees are most likely to see them. They should be prominently visible and changed often to maintain worker interest in their messages. Companies can use professionally designed posters or ask managers and employees to design their own. Posters can be mounted on bulletin boards or hung in their own frames.
- Other promotional methods include various campaigns to promote home safety, vacation safety, fire safety, and the like. These campaigns can be timed to coincide with elections, sports events, holidays, and so on. Unique safety ideas or stunts can be effective if they are not overdone. Courses and demonstrations on safety and health topics serve the dual purpose of stimulating interest in safety and teaching workers valuable skills and techniques. Finally, various publications such as reports, annual reports, newsletters, and booklets, leaflets, and personalized messages can help generate interest in safety among workers.
- Company public address systems and suggestion systems are also effective ways to maintain workers' interest. Public address messages should be brief and well timed, or employees tend to tune them out. Suggestion systems must be carefully planned and administered to reward good ideas and demonstrate the company's interest in safety and health policies.
- Many companies do little or nothing to publicize their internal safety and health programs and activities. Yet, the best way for a company to develop a reputation for taking care of employees is to be known as a safe place to work.
- Public relations is defined as the management function that evaluates public attitudes, identifies the firm's policies and activities with the public interest, and brings company news to the public's attention. Most organizations have either a professional public relations department or delegate public relations responsibilities to the safety professional.
- To be successful, a public relations program must be actively supported by top management, a strong safety program, and newsworthy stories supplied by the safety professional. Lack of a safety program and poor public relations efforts can cost companies millions of dollars each year in lost productivity and potential business.
- The safety professional can conduct publicity work within a company by coordinating efforts with the public relations department and with the employee publications team to publicize the safety program. The safety professional can point out how the PR director and publications editor can help to gain employee acceptance and support of the safety program.
- Safety activities contain a wealth of real news. Items such as no-incident records, interplant safety contests, safety suggestions by employees, safety award presentations, and the like can be the focus of a company media event.
- In companies without public relations departments, the safety and health professional can still obtain publicity for the firm's safety activities. Stories should be selected for a specific audience, contain human interest and appropriate humor, include graphics (cartoons, photographs, posters, etc.), and focus on people, not things.
- Safety professionals who want to make the most of public relations opportunities must (1) be honest in what they say; (2) deliver what they promise; (3) notify the media of any changes immediately; (4) supply accurate information; (5) offer ideas, not demand space or time; (6) feature only real news; (7) tip off local media to any worthwhile story; (8) give stories news value; and (9) be forthright about bad news (injuries, etc.).
- In working with local media, the safety and health professional should keep in mind the following tips: (1) keep press releases brief, timely, accurate, and complete; (2) be familiar with the publication or radio/TV station and know which editor or news director to contact; (3) tailor press releases for the audience; (4) include captions with photographs; and (5) leave script writing to the professionals but make sure all facts are accurate.
- When the safety and health professional must give a radio or TV interview, the individual should (1) be natural and refrain from distracting gestures or adopting an "official" tone, (2) dress conservatively, (3) review in advance what topics will be covered, and (4) respond naturally to the interviewer's questions without attempting to cram information into the time allotted.
- When a serious incident occurs at a plant or company, the public relations and safety professional must work together to provide accurate, honest information about the incident. Bad news about the incident should be balanced by pointing out the company's overall safety record or safety achievements.
- Company publications represent excellent opportunities to publicize safety programs, activities, and issues. In planning to start internal and/or external publications, the company should take the following steps: (1) clearly define the objectives and purpose of the publication, (2)

determine how general or restricted the message will be, (3) decide what form of publication is best, and (4) estimate costs of preparing and printing the publication.
- Eye-catching illustrations and concisely written, lively copy add interest to publications. The more audience-based the publication is, the more readers are likely to accept the messages it delivers. Companies should also seek the advice and expertise of professionals in the public relations field to help ensure the success of their publications.

REFERENCES

Alliance of American Insurers, 3025 Highland Parkway, Suite 800, Downers Grove, IL 60515.
   Fire Prevention and Control, A/V.
Culligan, M. J., and D. Crewe. *Getting Back to the Basics of Public Relations and Publicity.* New York: Crown Publishers, 1982.
Cutlip, S. M., et al. *Effective Public Relations.* 8th ed. New York: Prentice-Hall, 1999.
Heath, R. L. *Handbook of Public Relations.* Thousand Oaks, CA: Sage Publications, 2000.
Indiana Labor and Management Council Inc., 2780 Waterfront Parkway, Indianapolis, IN 46214. *Worker Involvement in Hazard Control* booklet. 1985.
Moore, H. F., and F. B. Kalupa. *Public Relations Principles, Cases and Problems.* Homewood, IL: Richard D. Irwin, 1985.
National Safety Council, 1121 Spring Lake Drive, Itasca, IL 60143.
   Catalog and Poster Directories.
   Family Safety & Health magazine.
   Focus (Industrial Division newsletter).
   Injury Facts™ (formerly *Accident Facts*). Annual.
   101 More Ideas That Worked.
   Safedriver magazine.
   Safety+Health magazine.
   Safeworker magazine.
   Today's Supervisor newsletter.
   Occupational Safety and Health Data Sheets (available in the Council Library):
   Nonprojected Visual Aids, 12304–0564, 1993.
   Photography for the Safety Professional, 12304–0619, 1991.
   Posters, Bulletin Boards, and Safety Displays, 12304–0616, 1991.
   Writing, Publishing, and Administering Employee Safety Regulations, 12304–0664, 1991.
Nolte, L. W., and D. L. Wilcox, eds. *Fundamentals of Public Relations: Professional Guidelines, Concepts, and Integrations.* 2nd ed. Elmsford, NY: Pergamon Press, 1979.
Parkhurst, W. *How to Get Publicity.* New York: Times Books, 1985.
Public Relations News, http://www.prnewsonline.com/.

REVIEW QUESTIONS

1. What is the main objective of a safety awareness program?
2. What are five factors a safety and health professional needs to consider when developing a safety awareness program?
    a.
    b.
    c.
    d.
    e.
3. List three indications of the need for a safety awareness program.
    a.
    b.
    c.
4. Which of the following can help to stimulate worker interest in safety and health policies and procedures?
    a. Contests
    b. Unique safety ideas
    c. Posters and displays
    d. Campaigns
    e. Publications
    f. All of the above
5. List five ways to make a contest successful.
    a.
    b.
    c.
    d.
    e.
6. Name the three broad categories of posters.
    a.
    b.
    c.

7. When working with local media, what five tips should the safety and health professional keep in mind?
   a.
   b.
   c.
   d.
   e.
8. What are the four steps the company should take in planning to start internal and/or external publications?
   a.
   b.
   c.
   d.
9. When a serious incident occurs at a facility or company, who must work together to provide information about the incident?
   a. Top management and line managers
   b. Line managers and supervisors
   c. Public relations and the safety professional
10. To be successful, a public relations program must be actively supported by which of the following?
    a. A strong safety program
    b. Newsworthy stories supplied by the safety professional
    c. Top management
    d. All of the above
    e. Only a and c

# APPENDIX 1: Sources of Help

Service Organizations

Government Information Sources

Canadian Government Agencies, Associations, and Boards

International Safety Organizations

Safety and health professionals sometimes need specialized information. This appendix offers resources to assist you.

SERVICE ORGANIZATIONS

National Safety Council
1121 Spring Lake Drive
Itasca, IL 60143-3201
www.nsc.org

The National Safety Council (NSC) is a nonprofit, nongovernmental public service organization dedicated to protecting life and promoting health. The NSC is a membership organization, founded in 1913 and chartered by the U.S. Congress in 1953. Members include more businesses, labor organizations, schools, public agencies, and private groups and individuals. The NSC saves lives by preventing injuries and deaths at work, in homes and communities, and on the roads through leadership, research, education and advocacy.

Publications
The following NSC publications have proven to be useful for industrial and off-the-job safety programs.
- *Safety+Health* (monthly)
- *Family Safety & Health* (quarterly)
- *Today's Supervisor* (monthly)
- *OSHA Up-to-Date* (monthly)
- *Traffic Safety* (monthly)
- *Journal of Safety Research* (five times per year)

In addition, the following statistical publication is also available from the NSC:
- *Injury Facts* (annually)

Technical Materials
The following is a partial list of technical manuals and publications. See www.nsc.org for a complete listing.
- *Accident Prevention Manual for Business & Industry* (four volumes)
  - Administration & Programs
  - Engineering & Technology
  - Environmental Management
  - Security Management
- *Aviation Ground Operations Safety Handbook*
- *Fundamentals of Industrial Hygiene*
- *Motor Fleet Safety Manual*
- *Occupational Health & Safety*
- *Power Press Safety Manual*
- *Safeguarding Concepts Illustrated*
- Data Sheets

Library
With a collection of more than 176,000 documents, the NSC Library is one of the most comprehensive safety and occupational health libraries in the world. It is recognized by government agencies, universities, and research organizations as a primary source of safety and health information. The staff of professional librarians can provide assistance with searching the library's catalog and other resources. The catalog is also accessible to NSC members via the nsc.org.

American Association of Occupational Health Nurses (AAOHN)
7794 Grow Drive
Pensacola, FL 32514
URL: www.aaohn.org

American Board of Industrial Hygiene (ABIH)
6015 West St. Joseph
Lansing, MI 48917
URL: www.abih.org

American Chemical Society (ACS)
1155 16th Street NW
Washington, DC 20036
URL: www.acs.org

American Chemistry Council (ACC)
700 Second St., NE
Washington, DC 20002
URL: www.americanchemistry.com

American College of Occupational and Environmental Medicine (ACOEM)
25 Northwest Point Boulevard
Elk Grove Village, IL 60007
URL: www.acoem.org

American Conference of Governmental Industrial Hygienists (ACGIH)
1330 Kemper Meadow Drive
Cincinnati, OH 45240
URL: www.acgih.org

American Foundry Society (AFS)
1695 North Penny Lane
Schaumburg, IL 60173
URL: www.afsinc.org

American Fuel & Petrochemical Manufacturers (AFPM)
1667 K Street NW
Washington, DC 20006
URL: www.afpm.org

American Gas Association (AGA)
400 North Capitol Street NW
Washington, DC 20001
URL: www.aga.org

American Industrial Hygiene Association
3141 Fairview Park Drive, Suite 777
Falls Church, VA 22042
URL: www.aiha.org

American Institute of Chemical Engineers (AICHE)
120 Wall Street
New York, NY 10005
URL: www.aiche.org

American Iron and Steel Institute (AISI)
25 Massachusetts Ave NW, Suite 800
Washington, DC 20001
URL: www.steel.org

American National Standards Institute (ANSI)
1899 L Street NW
Washington, DC 20036
URL: www.ansi.org

American Petroleum Institute (API)
1220 L Street NW
Washington, DC 20005
URL: www.api.org

American Road and Transportation Builders Association (ARTBA)
1219 28th Street NW
Washington, DC 20007
URL: www.artba.org

American Society of Heating, Refrigerating and Air Conditioning Engineers (ASHRAE)
1791 Tullie Circle NE
Atlanta, GA 30329
URL; www.ashrae.org

American Society of Mechanical Engineers (ASME)
3 Park Avenue
New York, NY 10016-5902
URL: www.asme.org

American Society of Safety Engineers (ASSE)
1800 East Oakton
Des Plaines, IL 60018
URL: www.asse.org

American Trucking Associations Inc. (ATA)
950 North Glebe Road
Arlington, VA 22203
URL: www.truckline.com

American Water Works Association (AWWA)
6666 West Quincy Avenue
Denver, CO 80235
URL: www.awwa.org

American Welding Society (AWS)
550 NW LeJeune Road
Miami, FL 33126
URL: www.aws.org

Associated General Contractors of America Inc. (AGC)
2300 Wilson Boulevard
Arlington, VA 22201
URL: www.agc.org

Association for Talent Development
1640 King Street
Alexandria, VA 22314
URL: www.td.org

Association of American Railroads (AAR)
425 3rd Street SW
Washington, DC 20024
URL: www.aar.org

ASTM International
100 Barr Harbor Drive
West Conshohocken, PA 19428
URL: www.astm.org

Board of Certified Safety Professionals
2301 W. Bradley Avenue
Champaign, IL 61821
URL: www.bcsp.org

Compressed Gas Association Inc. (CGA)
14501 George Carter Way
Chantilly, VA 20151
URL: www.cganet.com

Flight Safety Foundation (FSF)
801 N. Fairfax Street
Alexandria, VA 22314
URL: www.flightsafety.org

Forest Resources Association (FRA)
1901 Pennsylvania Avenue NW
Washington, DC 20006
URL: www.forestresources.org

Human Factors and Ergonomics Society (HFES)
PO Box 1369
Santa Monica, CA 90406
URL: www.hfes.org

Illuminating Engineering Society (IES)
120 Wall Street
New York, NY 10005-4001
URL: www.ies.org

International Safety Equipment Association Inc. (ISEA)
1901 North Moore Street
Arlington, VA 22209
URL: www.safetyequipment.org

International System Safety Society (SSS)
PO Box 70
Unionville, VA 22567
URL: www.system-safety.org

Laser Institute of America (LIA)
13501 Ingenuity Drive
Orlando, FL 32826
URL: www.laserinstitute.org

National Fire Protection Association (NFPA)
1 Batterymarch Park
Quincy, MA 02269
URL: www.nfpa.org

National Mining Association (NMA)
101 Constitution Avenue NW
Washington, DC 20001
URL: www.nma.org

National Propane Gas Association (NPGA)
1899 L Street NW
Washington, DC 20036
URL: www.npga.org

National Restaurant Association (NRA)
2055 L Street NW
Washington, DC 20036
URL: www.restaurant.org

Network of Employers for Traffic Safety (NETS)
344 Maple Avenue W
Vienna, VA 22180
URL: www.trafficsafety.org

Power Tool Institute Inc. (PTI)
1300 Sumner Avenue
Cleveland, OH 44115
URL: www.powertoolinstitute.com

Prevent Blindness America
211 West Wacker Drive
Chicago, IL 60606
URL: www.preventblindness.org

Risk & Insurance Management Society (RIMS)
5 Bryant Park
New York, NY 10018
URL: www.rims.org

Underwriters Laboratories Inc. (UL)
333 Pfingsten Road
Northbrook, IL 60062
URL: www.ul.com

Voluntary Protection Programs Participant Association (VPPPA)
7600 E Leesburg Pike
Falls Church, VA 22043
URL: www.vpppa.org

GOVERNMENT INFORMATION SOURCES

The U.S. government collects and distributes volumes of information concerning safety, health, and environmental issues. Some websites to access for safety, health, and environmental information include the following:

Occupational Safety and Health

- Bureau of Labor Statistics, Worker Safety and Health, www.stats.bls.gov/bls/safety.htm
- Mine Safety and Health Administration, www.msha.gov
- National Institute for Occupational Safety and Health, www.cdc.gov/niosh
- Occupational Safety and Health Administration, www.osha.gov

Environment

- Environmental Protection Agency, www.epa.gov

Traffic and Transportation Safety
- Federal Highway Administration, www.fhwa.dot.gov
- Federal Motor Carrier Safety Administration, www.fmcsa.dot.gov
- Federal Railroad Administration, www.fra.dot.gov
- National Highway Traffic Safety Administration, www.nhtsa.dot.gov

Health and Medicine
- Centers for Disease Control and Prevention, www.cdc.gov
- National Institutes of Health, www.nih.gov
- National Library of Medicine, www.nlm.nih.gov

CANADIAN GOVERNMENT AGENCIES, ASSOCIATIONS, AND BOARDS

Association of Workers Compensation Boards of Canada
40 University Avenue
Toronto, ON M5J 1T1
URL: www.awcbc.org

Canada Safety Council
1020 Thomas Spratt Place
Ottawa, ON K1G 5L5
URL: www.canadasafetycouncil.org

Canadian Centre for Occupational Health and Safety
135 Hunter Street East
Hamilton, ON L8N 1M5
URL: www.ccohs.ca

Transport Canada
330 Sparks Street
Ottawa, ON K1A 0N5
URL: www.tc.gc.ca

Workplace Safety and Prevention Services
5100 Creekbank Road
Mississauga, ON L4W 0A1
www.wsps.ca

INTERNATIONAL SAFETY ORGANIZATIONS

International Labor Organization (ILO)
International Labor Office
4 Route des Morillons, CH 1211
Geneva 22 Switzerland
URL: www.ilo.org

International Organization for Standardization (ISO)
Chemin de Blandonnet 8
CP 401
1214 Vernier
Geneva, Switzerland

Pan American Health Organization (PAHO)
525 23rd Street NW
Washington, DC 20037
URL: www.paho.org

Royal Society for the Prevention of Accidents (ROSPA)
RoSpa House
28 Calthorpe Road
Edgbaston, Birmingham
B15 1RP, United Kingdom
URL: www.rospa.com

World Health Organization (WHO)
20 Avenue Appia,
1211 Geneva 27, Switzerland
URL: www.who.int/en/

APPENDIX 2: Bibliography

- Safety and Health Periodicals
- General Principles
- Risk Assessment—Industrial Hygiene
- Sampling Methods
- Toxicology
- Medical/Health
- Ergonomics
- Chemical Hazards
- Pollution/Hazardous Waste
- Control
- Handbooks and Manuals
- NIOSH Publications
- Criteria Documents
- Safety Management
- Emergency Preparedness
- Safety Training
- Risk Assessment/Accident Investigation and Analysis
- Product Safety
- Fire
- Loss Control
- OSHA

The reference material cited in this bibliography was selected to provide safety and health professionals with sources of information that are likely to prove useful in coping with problems of worker health protection and hazard assessment. This compilation is not to be viewed as a comprehensive coverage of the abundant literature on this subject, nor is any endorsement implied. Also listed are safety and health periodicals that may be of interest. Again, the list is by no means comprehensive.

SAFETY AND HEALTH PERIODICALS

Accident Analysis & Prevention (monthly)
http://www.journals.elsevier.com/accident-analysis-and-prevention/
American Journal of Industrial Medicine (monthly)
http://onlinelibrary.wiley.com/journal/10.1002/(ISSN)1097-0274
American Journal of Public Health (monthly)
http://ajph.aphapublications.org/
The Annals of Occupational Hygiene (9 issues)
http://annhyg.oxfordjournals.org/
Applied Ergonomics (bimonthly)
http://www.journals.elsevier.com/applied-ergonomics/
EHS Today (monthly)
http://ehstoday.com/
Environmental Health Perspectives (monthly)
http://ehp.niehs.nih.gov/
Facility Safety Management (monthly)
http://www.fsmmag.com
Fire Engineering (monthly)
http://www.fireengineering.com
Fleet Solutions (bimonthly)
http://www.nafa.org/publications/fleetsolutions/#
Human Factors (8 issues)
http://hfs.sagepub.com/
Industrial Safety and Hygiene News (monthly)
http://www.ishn.com
Journal of Agromedicine (quarterly)
http://www.tandfonline.com/loi/wagr20#
Journal of Occupational and Environmental Hygiene (monthly)
https://www.aiha.org/publications-and-resources/JOEH/Pages/default.aspx
Journal of Occupational & Environmental Medicine (monthly)
http://journals.lww.com/joem/Pages/default.aspx
National Safety Magazine (bimonthly)
http://nsca.org.au/knowledge-centre/resources/national-safety-magazine/
NFPA Journal (bimonthly)
http://www.nfpa.org/newsandpublications/nfpa-journal/
Occupational Health & Safety (monthly)
http://www.ohsonline.com
Occupational Health & Safety Canada (bimonthly)
http://www.ohscanada.com
Professional Safety (monthly)
http://www.asse.org /professional-safety
The Synergist (monthly)
https://www.aiha.org/publications-and-resources/TheSynergist/Pages/default.aspx
Workplace Health & Safety—official journal of the American Association of Occupational Health Nurses, Inc. (formerly AAOHN Journal)
http://whs.sagepub.com/

GENERAL PRINCIPLES

Balge, M. Z., and G. R. Krieger, eds. *Occupational Health and Safety*. 3rd ed. Itasca, IL: National Safety Council, 2000.

Block, M. R., and I. R. Marash. *Integrating ISO 14001 Into a Quality Management System*. 2nd ed. Milwaukee, WI: ASQ Quality Press, 2002.

Finkelstein, E. A., and P. S. Corso. *The Incidence and Economic Burden of Injuries in the United States*. New York: Oxford University Press, 2006.

Janicak, C. A. *Safety Metrics: Tools and Techniques for Measuring Safety Performance*. 3rd ed. Rockville, MD: Government Institutes, 2015.

Manuele, F. A. *On the Practice of Safety*. 4th ed. Hoboken, NJ: Wiley InterScience, 2013.

Oxenburgh, M., P. Marlow, and A. Oxenburgh. *Increasing Productivity and Profit Through Health and Safety: The Financial Returns from a Safe Working Environment*. 2nd ed. Boca Raton, FL: CRC Press, 2004.

Protecting Workers' Lives: A Safety and Health Guide for Unions. 3rd ed. Itasca, IL: National Safety Council, 2007.

Tompkins, N. C. *Basics of Safety and Health*, rev. ed. Itasca, IL: National Safety Council, 2004.

RISK ASSESSMENT—INDUSTRIAL HYGIENE

Asfahl, C. R. *Industrial Safety and Health Management*. 6th ed. Upper Saddle River, NJ: Prentice-Hall, 2010

Cohen, B. S., and S. V. Hering, eds. *Air Sampling Instruments—For Evaluation of Atmospheric Contaminants*. 9th ed.

Cincinnati, OH: American Conference of Governmental Industrial Hygienists, 2001.

SAMPLING METHODS

Godish, T. *Air Quality*. 5th ed. Boca Raton, FL: Lewis, 2015.
Rose, V. E. *Patty's Industrial Hygiene*. 6th ed. New York: Wiley InterScience, 2011.
Industrial Ventilation: A Manual of Recommended Practice for Design. 28th ed. Cincinnati, OH: American Conference of Governmental Industrial Hygienists, 2013.

TOXICOLOGY

Derelanko, M. J., and C. S. Auletta. *Handbook of Toxicology*, 3rd ed. Boca Raton, FL: CRC Press, 2014.
Lioy, P., and C. Weisel. *Exposure Science: Basic Principles and Applications*. Waltham, MA: Academic Press, 2014.
Watkins, J. B., and C. D. Klaassen. *Casarett & Doull's Essentials of Toxicology*, 2nd ed. Cincinnati, OH: American Conference of Governmental Industrial Hygienists, 2010.
Wexler, P. *Encyclopedia of Toxicology*, 3rd ed. Waltham, MA: Academic Press, 2014.

MEDICAL/HEALTH

Guidotti, T. L., et al. *Occupational Health Services: A Practical Approach*. 2nd ed. Chicago, IL: American Medical Association, 2012.
Kahn, A. P. *The Encyclopedia of Work-Related Illnesses, Injuries, and Health Issues*. New York: Facts on File, 2004.
O'Donnell, M. P., and J. S. Harris, eds. *Health Promotion in the Workplace*. 2nd ed. Albany, NY: Delmar, 2002.
Root Cause Analysis in Health Care: Tools and Techniques. 4th ed. Oakbrook Terrace, IL: Joint Commission Resources, 2009.

ERGONOMICS

Chaffin, D. B., and J. B. Andersson. *Occupational Biomechanics*. 4th ed. Hoboken, NJ: Wiley InterScience, 2006.
Kodak's Ergonomic Design for People at Work. 2nd ed. New York: Wiley, 2003.
Karwowski, W., and W. S. Marras. *Occupational Ergonomics: Design and Management of Work Systems*. Boca Raton, FL: CRC Press, 2003.
———. *Occupational Ergonomics: Principles of Work Design*. Boca Raton, FL: CRC Press, 2003.
Marras, W. S., and W. Karwowski. *Fundamentals and Assessment Tools for Occupational Ergonomics*. Boca Raton, FL: CRC Press, 2006.
———*Interventions, Controls, and Applications in Occupational Ergonomics*. Boca Raton, FL: CRC Press, 2006.
Salvendy, G., ed. *Handbook of Human Factors and Ergonomics*. 4th ed. Hoboken, NJ: Wiley, 2012.
Stanton, N., and A. Hedge. *The Handbook of Human Factors and Ergonomics Methods*. Boca Raton, FL: CRC Press, 2004.

CHEMICAL HAZARDS

Compressed Gas Association. *Handbook of Compressed Gases*. 5th ed. Chantilly, VA: 2013.
Haynes, W. M. *CRC Handbook of Chemistry and Physics: A Ready-Reference Book of Chemical and Physical Data*. 95th ed. Boca Raton, FL: CRC Press, 2014
Lewis, R. J. Sr., ed. *Rapid Guide to Hazardous Chemicals in the Workplace*. 4th ed. New York: Wiley InterScience, 2000.
Lewis, R. J. Sr. *Sax's Dangerous Properties of Industrial Materials*. 12th ed. Hoboken, NJ: Wiley InterScience, 2012.
Patnaik, P. *A Comprehensive Guide to the Hazardous Properties of Chemical Substances*. 3rd ed. Hoboken, NJ: Wiley, 2007.
Proctor, N. H., J. P. Hughes, and G. J. Hathaway. *Chemical Hazards in the Workplace*. 5th ed. New York: Wiley, 2014.
Threshold Limit Values and Biological Exposure Indices. Cincinnati, OH: American Conference of Government Industrial Hygienists. Published annually.

POLLUTION/HAZARDOUS WASTE

Blackman, W. C. *Basic Hazardous Waste Management*. 3rd ed. Boca Raton, FL: CRC Press, 2001.
Emergency Response Guidebook: A Guidebook for First Responders During the Initial Phase of a Dangerous Goods/Hazardous Materials Incident. Washington

DC: Department of Transportation, 2012. http://phmsa.dot.gov/pv_obj_cache/pv_obj_id_7410989F4294AE44A2EBF6A80ADB640BCA8E4200/filename/ERG2012.pdf

Martin, W. F., and M. Gochfeld. *Protecting Personnel at Hazardous Waste Sites.* 3rd ed. Boston: Butterworth, 2000.

McDermott, H. J. *Air Monitoring for Toxic Exposures.* 2nd ed. Hoboken, NJ: Wiley, 2004.

Pohanish, R. P. *HazMat Data: For First Response, Transportation, Storage, and Security.* 2nd ed. Hoboken, NJ: Wiley InterScience, 2004.

Shah, K. L. *Basics of Solid and Hazardous Waste Management Technology.* Upper Saddle River, NJ: Prentice-Hall, 2000.

Woodside, G. *Hazardous Materials and Hazardous Waste Management.* 2nd ed. New York: Wiley, 1999.

CONTROL

Industrial Ventilation: A Manual of Recommended Practice for Design. 28th ed. Cincinnati, OH: American Conference of Governmental Industrial Hygienists, 2013.

McDermott, H. *Handbook of Ventilation for Contaminant Control.* 3rd ed. Cincinatti, OH: American Conference of Governmental Industrial Hygienists, 2001.

HANDBOOKS AND MANUALS

Burkart, M. J., M. McCann, and D. M. Paine. *Elevated Work Platforms and Scaffolding: Job Site Safety Manual.* New York: McGraw-Hill, 2004.

Fanning, F. *Basic Safety Administration: A Handbook for the New Safety Specialist.* Des Plaines, IL: American Society of Safety Engineers, 2003.

Harris, M. K. *Welding Health and Safety: A Field Guide for OEHS Professionals.* Fairfax, VA: American Industrial Hygiene Association, 2002.

Hopwood, D., and S. Thompson. *Workplace Safety: A Guide for Small and Midsized Companies.* Hoboken, NJ: Wiley, 2006.

Kovacic, T. M. *An Illustrated Guide to Electrical Safety.* 6th ed. Des Plaines, IL: American Society of Safety Engineers, 2011.

MacCollum, D. V. *Construction Safety Planning.* New York: Wiley, 1997.

NFPA 70E: Standard for Electrical Safety in the Workplace. Quincy, MA: National Fire Protection Association, 2015.

Plog, B.A. *Fundamentals of Industrial Hygiene.* 6th ed. Itasca, IL: National Safety Council, 2012.

Reese, C. D., and J. V. Eidson. *Handbook of OSHA Construction Safety and Health.* 2nd ed. Boca Raton, FL: CRC/Taylor & Francis, 2006.

Ridley, J., and J. Channing. *Safety at Work.* 6th ed. Boston: Butterworth-Heinemann, 2003. http://ohph.sbmu.ac.ir/uploads/355_1287_1376281720161_Safety_at_Work_6E.pdf

NIOSH PUBLICATIONS

The National Institute for Occupational Safety and Health (NIOSH) has published many useful publications in the field of industrial hygiene. Many NIOSH publications are available online by publication order or by topic at http://www.cdc.gov/niosh/pubs/all_date_desc_nopubnumbers.html or http://www.cdc.gov/niosh/pubs/type.html.

Copies may also be available from the Centers for Disease Control and Prevention at 1-800-CDC-INFO (1-800-232-4636) or http://www.cdc.gov/publications.

Many of these publications are also available from the Superintendent of Documents, U.S. Government Printing Office, 732 North Capitol Street, NW, Washington, DC 20401-0001 or http://www.gpo.gov.

Some NIOSH publications can also be obtained from the National Technical Information Services (NTIS), Alexandria, VA 22312, or http://www.ntis.gov.

CRITERIA DOCUMENTS

NIOSH is responsible for providing relevant data from which valid criteria for effective standards can be derived. Recommended standards for occupational exposure, which are the result of this work, are based on the health effects of exposure.

The single most comprehensive source of information on a particular material will probably be found in the NIOSH Criteria Document for that substance. The table of contents for a Criteria Document is as follows:

I. Recommendations for an Occupational Exposure Standard

Section 1—Environmental (workplace air)

Section 2—Medical

Section 3—Labeling and posting

Section 4—Personal protective equipment and clothing

SAFETY MANAGEMENT

Brauer, R. L. *Safety and Health for Engineers*. 2nd ed. Hoboken, NJ: Wiley, 2006.

Czerniak, J. A., and D. M. Ostrander. *Nine Elements of a Successful Safety and Health System*. Itasca, IL: National Safety Council, 2005.

Davies, J., A. Ross, and B. Wallace. *Safety Management: A Qualitative Systems Approach*. Boca Raton, CRC Press, 2003.

Della-Giustina, D. *Developing a Safety and Health Program*. 2nd ed. Boca Raton, FL: CRC Press, 2010.

Goetsch, D. L. *Occupational Safety and Health: For Technologists, Engineers, and Managers*. 6th ed. Upper Saddle River, NJ: Pearson Prentice-Hall, 2008.

Hammer, W., and D. Price. *Occupational Safety Management and Engineering*. 5th ed. Upper Saddle River, NJ: Pearson Prentice-Hall, 2001.

Krause, T. R. *Leading with Safety*. Hoboken, NJ: Wiley, 2005.

Lack, R. W. *Safety, Health, and Asset Protection: Management Essentials*. 2nd ed. Boca Raton, FL: Lewis, 2001.

McSween, T. E. *The Values-Based Safety Process: Improving Your Safety Culture with a Behavioral Approach*. 2nd ed. New York: Wiley, 2003.

Mroszczyk, J.W. *Safety Engineering*. 4th ed. Des Plaines, IL: American Society of Safety Engineers, 2013.

Onion, M. L., and M. F. O'Toole. *You've Just Been Made Supervisor . . . Now What? Bringing Safety to the Front Line*. Itasca, IL: National Safety Council, 2003.

Petersen, D. *Measurement of Safety Performance*. Des Plaines, IL: American Society of Safety Engineers, 2005.

———. *Techniques of Safety Management: A Systems Approach*. 4th ed. Des Plaines, IL: American Society of Safety Engineers, 2003.

Reese, C. D. *Occupational Health and Safety Management: A Practical Approach*. 2nd ed. Boca Raton, FL: CRC Press, 2008.

Rothstein, M. A. *Occupational Safety and Health Law*. Eagan, MN: Thompson/West, 2015.

Roughton, J. E., and J. J. Mercurio. *Developing an Effective Safety Culture: A Leadership Approach*. Boston: Butterworth-Heinemann, 2002. http://www.taksimdanismanlik.com/inc/uploads/katalog_images/katalogs-1075-Safety%20culture%20A%20leadership%20approach.pdf

Stephens, R. A. *System Safety for the 21st Century*. Hoboken, NJ: Wiley, 2004.

Stewart, J. M. *Managing for World Class Safety*. New York: Wiley, 2001.

Swartz, G. *Safety Culture and Effective Safety Management*. Itasca, IL: National Safety Council, 2000.

EMERGENCY PREPAREDNESS

Canton, L.G. *Emergency Management: Concepts and Strategies for Effective Programs*. Hoboken, NJ: Wiley, 2006.

Fagel, M. J. *Crisis Management and Emergency Planning: Preparing for Today's Challenges*. Boca Raton, FL: CRC Press, 2013.

———. *Principles of Emergency Management and Emergency Operations Centers (EOC)*. Boca Raton, FL: CRC Press, 2010.

McEntire, D. A. *Disaster Response and Recovery: Strategies and Tactics for Resilience*. 2nd ed. Hoboken, NJ: Wiley, 2014.

Mullen, S. *Emergency Planning Guide for Utilities*. 2nd ed. Boca Raton, FL: CRC Press, 2013.

Phillips, B. D., and D. M. Neal. *Introduction to Emergency Management*. Boca Raton, FL: CRC Press, 2011.

Veasey, D. A., and L. C. McCormick. *Confined Space Entry and Emergency Response*. Hoboken, NJ: Wiley, 2006.

SAFETY TRAINING

Roughton, J., and N. E. Whiting. *Safety Training Basics: A Handbook for Safety Training Program Development*. Rockville, MD: Government Institutes, 2000.

RISK ASSESSMENT/ACCIDENT INVESTIGATION AND ANALYSIS

Brown, J. F., Obenski KS, and Osborn TR. *Forensic Engineering Reconstruction of Accidents*. 2nd ed. Springfield, IL: Thomas, 2002.

Latino, R. J., and K. C. Latino. *Root Cause Analysis: Improving Performance for Bottom-line Results*, 4th ed. Boca Raton, FL: Taylor & Francis, 2011.

Main, B. W. *Risk Assessment: Basics and Benchmarks*. Ann Arbor, MI: Design Safety Engineering, 2004.

Oakley, J. S. *Accident Investigation Techniques*. 2nd

ed. Des Plaines, IL: American Society of Safety Engineers, 2012.

Swartz, G. *Job Hazard Analysis: A Guide to Identifying Risks in the Workplace.* Rockville, MD: Government Institutes, 2001.

Vincoli, J. W. *Basic Guide to Accident Investigation and Loss Control.* New York: Wiley, 1994.

PRODUCT SAFETY

Hammer, W. *Product Safety and Management Engineering.* 2nd ed. Des Plaines, IL: American Society of Safety Engineers, 1993.

FIRE

Cote, R. *Life Safety Code Handbook.* 15th ed. Quincy, MA: National Fire Protection Association, 2015.

Cote, R., J. R. Hall, and P. A. Powell, eds. *Fire Protection Handbook.* 2 vols. 20th ed. Quincy, MA: National Fire Protection Association, 2008.

LOSS CONTROL

Nolan, D. P. *Loss Prevention and Safety Control: Terms and Definitions.* Boca Raton, FL: CRC Press, 2011.

Dunlap, E. S. *Loss Control Auditing: A Guide for Conducting Fire, Safety, and Security Audits.* Boca Raton, FL: CRC Press, 2011.

OSHA

Kaletsky, R. *OSHA Inspections: Preparation and Response,* 2nd ed. Itasca, IL: National Safety Council, 2012.

Moran, M. M. *Construction Safety Handbook: A Practical Guide to OSHA Compliance and Injury Prevention.* 2nd ed. Rockville, MD: ABS Consulting, Government Institutes, 2003.

Friend, M. *Fundamentals of Occupational Safety and Health.* 5th ed. Lanham, MD: Government Institutes, 2010.

Reese, C. D. *Handbook of OSHA Construction Safety and Health.* 2nd ed. Boca Raton, FL: CRC/Taylor & Francis, 2006.

Index

A

ABCD objective writing method, 712
Abduction, 397
Accidental injury, 132
Accident Prevention Manual for Business & Industry: Environmental Management (National Safety Council), 330
Accident Prevention Manual for Business & Industry (National Safety Council), 10
Accident-prone employees, 20
Accidents. *See also* Injuries
    chain reaction to, 444, 446
    early 20th-century reporting of, 6
    emergency preparedness for, 444, 446
    office, 598–600
    off-the-job, 260–262, 290
    prevention as objective of workers' compensation, 186
    reasons to prevent, 4
    reporting requirements for, 58, 196
    source of injury vs. types of, 252, 253
    use of term, 132, 238
Accommodations, for individuals with disabilities, 20, 423, 526, 527, 532–533
Accounting. *See* Audits
Acetic acid, 378
Acids, storage of, 623–624
Action level (AL) (OSHA), 228
Active surveillance, 406–407
Activity-sampling technique, 685
Adduction, 397
Administrative law judge (ALJ)
    in contested cases, 64–65
    Mine Act and, 89
    Occupational Safety and Health Review Commission and, 51
Affirmative action, individuals with disabilities and, 528, 529
Aflatoxin B_1, 376–377
After-the-fact incident investigation, 132–133
Aggression. *See* Workplace violence
Agreement system, for workers' compensation claims, 192
Air filters, 377–378
Air pollutants. *See also* Indoor air quality (IAQ); Indoor air quality (IAQ) management
    approaches to, 379–380
    Clean Air Act and, 331
    housekeeping activities and, 360–361
    maintenance activities and, 361–364
    pathways for, 366–367
    regulatory requirements for, 380–382
    sources of indoor, 355, 357–366

Air quality. *See* Indoor air quality (IAQ)
Aisles
    office, 604
    standards for, 116
Alarm systems, 458–459
Alcohol abuse, 309–310
Alcohol/drug testing, 592–594
Allergic reactions, to molds, 376
Alpha-radiation, 635
Altered environments, 643–644
Ambiance, office, 601–602
American Board of Industrial Hygiene (ABIH), certification and, 324
American Chemical Council (ACC), 326
American Chemical Society, 326
American Conference of Governmental Industrial Hygienists (ACGIH), 70, 324, 326, 399
American Consulting Engineers Council, 200
American Engineering Standards Committee (AESC), 7
American Industrial Hygiene Association (AIHA)
    certification programs and, 11
    establishment of, 9, 326
American Institute of Chemical Engineers (AIChE), 326
American Medical Association, 307
American National Standards Institute (ANSI)
    certification, 342
    on confined-space requirements, 120
    on drinking cups and ladles, 504
    on elevator safety, 663
    establishment of, 8
    on incident investigation and analysis, 240
    Occupational Injury and Illness Classification System and, 245
    on purchasing agents, 155
    on recording work injuries, 9
    standards enacted by, 15
    on ventilation and indoor air quality, 382–383
    on washroom and locker room sanitation, 504
American Petroleum Institute (API), 326
American Red Cross, 464
American Society for Testing and Materials (ASTM), on auditing practices, 178
American Society of Heating, Refrigerating, and Air-Conditioning Engineers (ASHRAE)
    on air filtration, 377, 378
    on environmental tobacco smoke, 374
    on nitrogen oxides, 374
    on ventilation and indoor air quality, 365–366, 382–383
American Society of Safety Engineers (ASSE), 11
American Standards Association (ASA), background of, 7–8

Americans with Disabilities Act (ADA). *See also* Employment discrimination
    background of, 528–529
    compliance with, 534, 657
    definitions for disability in, 531
    discrimination under, 530–531
    ergonomics and, 423
    hiring and placement and, 553, 554
    physical examinations and, 539
    provisions of, 526–528
    workplace violence injuries and, 483
Analysis, statistical method of, 240
Analysis of Workers' Compensation Laws (U.S. Chamber of Commerce), 185
Analytical trees, 204, 206
Animal facilities, 643
Annual reports, 777
ANSI. *See* American National Standards Institute (ANSI)
Anthropometric Source Book Vols. I-III (National Aeronautics and Space Administration), 420
API. *See* American Petroleum Institute (API)
Applications Manual for the Revised NIOSH Lifting Equation (NIOSH), 412, 413
Apprenticeships, 717–718
Architectural and Transportation Barriers Board, 423
Area charts, 741
Arrestance, 377–378
Artifacts, 34
Asbestos
    exposure to, 79–81, 378, 379
    indoor air quality and, 378, 379
Asbestos-containing material (ACM), 80
ASHRAE. *See* American Society of Heating, Refrigerating, and Air-Conditioning Engineers (ASHRAE)
As low as reasonably achievable (ALARA), 637
Aspergillus flavus, 376
Aspergillus parasiticus, 376, 377
Aspergillus versicolor, 376
Association of Iron and Steel Electrical Engineers, 6, 7, 10
Assumption of risk, 6
Assumptions, culture and, 34–35
Assurance reports, 180
Asthma, molds and, 376
Attitudes
    behaviors vs., 697
    change in, 696–697
    components of, 695
    determination of, 696
    motivation and, 695–697
Attractive nuisances
    liability associated with, 655
    protection of, 670–671
Audiocassettes, 747
Auditors, 168, 178
Auditory displays, 418–419
Audit programs
    design of, 174–175
    growth in, 179
    in retail/service industries, 554, 556–558
Audit reports, 172–173, 179, 180

Audits. *See also* Safety, health, and environmental (SH&E) audits
    compliance, 118
    consultant, 177–178
    corporate, 177
    enterprise-related, 177–178
    environmental, 175–177, 345–346
    explanation of, 168
    field work for, 171–173
    follow-up to, 173
    future outlook for, 178–180
    government-related regulatory, 175–177
    internal, 177
    organization and staffing for, 175
    planning process for, 169–171
    protocol for, 173–174
    purposes of, 168
    scope and focus of, 168–169
    types of, 169
    uses of, 178
    working papers for, 173–174
Auster, Ellen, 334, 335
Autoclaves, 630
Automated external defibrillators (AEDs), 297
Average-person fallacy, 683–684
Awards. *See also* Contests
    meaningful, 771
    motivation and, 691–692
    National Safety Council on, 765
    presentations of, 771–772
    suggestion, 778
Awkward posture, 397

B

Back injuries, 400
Bar charts, 740
Barrier creams, 505
Battery charging, 590
Before-the-fact incident investigation, 133
Behavior
    attitudes and, 697
    motivation and, 697–699
Behavioral management, 697–699
Behavior-based safety (BBS), as change catalyst, 43–44
Behavior reinforcement, 689–690, 693–694
Behavior-sampling technique, 685
Benches, locker room, 506
Beta-radiation, 635
Beyond a reasonable doubt, 140
Bhopal (India) disaster, 19
Bioassay, 638
Biological hazards
    evaluation and control of, 321–322
    in laboratories, 618–619
    nature of, 144
Biological monitoring, 225
Biological safety (biosafety)
    bloodborne pathogens and, 632–634
    containment practices and, 628–629
    guidelines for, 627–628
    personal protective equipment for, 630

primary barriers to, 630, 631
secondary barriers to, 630, 632, 633
Biological safety cabinets (BSCs), 630
Biomarkers, 520–521
Biosafety cabinets, 644
Biosafety in Microbiological and Biomedical Laboratories (CDH-NIH), 627
Biosafety levels (BSL), 627–628
Black lung disease, workers' compensation and, 189
Blind performance tests, 593
Bloodborne pathogens
exposure to, 79, 321–322
OSHA standard for, 79, 632–634
precautions for handling, 627, 632
Board of Certified Hazard Control Management, 11
Board of Certified Safety Professionals (BCSP), 11
Boards, presentation, 749–750
Braille, 534
Brainstorming, 718–719
Brakes, 591
Breath testing, 593
Brownfields redevelopment, 332
BS-7750 standard (Great Britain), 333
Building Air Quality: A Guide for Building Owners and Facility Managers (BAQ) (Environmental Protection Agency), 354
Building Air Quality Action Plan (Environmental Protection Agency), 354
Buildings
air quality and occupants of, 360
entrances to, 657
fires in high-rise, 666–668
individuals with disabilities and access to, 540
Bulletin boards, 283, 774
Burden of proof, 655
Bureau of Labor Statistics (BLS)
function of, 49
on injury and illness incidence, 10, 52
on musculoskeletal disorders, 395
Occupational Injury and Illness Classification System and, 245, 280–281
on workplace deaths, 518
on workplace violence, 468
Burns, chemical, 590

C

Cafeterias. *See* Food-service facilities
Calculative culture, 37
California State Department of Industrial Relations, 598, 599
Campaigns, safety, 775, 778–779
Canada
driver selection standards in, 576
industrial safety legislation in, 16
Responsible Care Program in, 332
Canadian Standards Association (CSA), 155
Carbon dioxide, 339, 378
Carbon footprints, 339
Carbon monoxide, 227, 372–373
Carcinogens, formaldehyde as, 373
Cardiopulmonary resuscitation (CPR), 297, 461, 735

Carpal tunnel disease (CTD), 401, 417–418
Carr, Clay, 713
Carrying, 415
Carter, Jimmy, 82
Case studies, 719
Causation. *See* Incident causation
C charts, 287, 288
CDs, 747
Ceiling Threshold Limit Value (TLV-C), 227
Center for Devices and Radiological Health (CDRH), 641
Centers for Disease Control and Prevention (CDC)
biohazard guidelines of, 627
health and safety standards and, 51
safety evaluation checklist, 216
Centrifuges, 645
CERCLA. *See* Comprehensive Environmental Response, Compen-sation, and Liability Act (CERCLA) (Superfund)
Certification
of environmental management staff, 337
industrial hygienist, 324, 326
Certified Associate Industrial Hygienists (CAIHs), 324
Certified Employee Assistance Professional (CEAP), 431, 432
Certified Industrial Hygienists (CIHs), 11, 229, 324
Chain hoists, 591
Chairs
ergonomics and, 417
falls from, 600, 607
office, 604–605
Change
in attitudes, 696–697
cultural, 41–43
lifestyle, 308
structural, 699
Charts, training presentation, 739–741, 749
Chemical burns, 590
Chemical facilities, 676–677
Chemical Facility Anti-Terrorism Standards, 331
Chemical hygiene plan (CHP), 626
Chemical Manufacturers Association (CMA), 326
Chemicals. *See also* Hazardous chemicals
exposure to, 144
labeling of, 623
in laboratories, 623–627
laser use and, 639
in offices, 605
safety information on, 622–623
storage of, 623–624
violations related to, 335–336
Chemical safety regulations, 626–627
Chemical Sector Information Sharing and Analysis Center (Chemical Sector ISAC), 678
Chemical Transportation Emergency Center (CHEMTREC), 464
Chemical waste management, 624, 645–646
Chernobyl (Ukraine) disaster, 19, 36
Chief executive officer (CEO), 762
Child labor, in early 20th century, 5
Chlorine, 500–501
Circadian cycles, 310–311

Circle graphs, 739–740
Citations, for OSH Act violations, 63
Civil liability
    administrative penalties against organizations and, 138–139
    individual, 141
    of organizations and potential damages, 137–138
    statutory penalties and, 139–140
Civil Rights Act of 1964, Title VII, 483
Civil strife, 444
Civil tort remedies, 137
Clean Air Act Amendments (CAAA), 331, 381
Clean Air Act (CAA), 16, 331, 339, 381
Clean rooms, 643
Clean Water Act (CWA), 16, 177, 331
Climate change, 339–340. *See also* Greenhouse gases (GHGs)
Cloud-to-ground lightning, 450
Coaching, 717
Coal Mine Health and Safety Act (1969), 189
Code of Federal Regulations (CFR)
    accessibility standards in, 530
    DHS final regulations, 678
    explanation of, 92
    OSHA regulations in, 55–56, 98, 266
    system safety directives and, 200, 201
Codes of ethics
    for industrial hygienists, 325
    safety and health and, 200
Cognitive ergonomics, 394
Cold
    effects on worker of, 416–417
    emergency preparedness for extreme, 449
Cold rooms, 644
Collisions. *See also* Vehicles; Vehicle safety programs
    cost of, 566, 567
    frequency controls for, 575
    reporting procedures for, 570–574
Common law, favoring employers, 5–6
Communication. *See also* Public relations (PR); Training media; Training media presentations
    factors in persuasive, 696, 697
    on indoor air quality, 379–380
    in laboratories, 620
    motivation and, 690–691
    safety office, 782–783
Communities
    contests in, 770
    environmental risk in, 20
    off-the-job safety and, 161–162
    treatment resources in, 429
Company-owned vehicles, 669–670
Company picnics, 162
Competition, environmental management and, 346
Compliance audits, 118, 169
Compliance safety and health officer (CSHO)
    employee rights and, 55
    inspections and, 53, 58–61
    violations and, 61, 62
Comprehensive Environmental Response, Compensation, and Liability Act (CERCLA) (Superfund), 16, 178, 330, 332

Compressed-gas cylinders, 625–626
Computer-based training (CBT)
    benefits and drawbacks of, 744
    explanation of, 721, 743–744
    formats for, 745
    interactive, 745
    needs assessment for, 744–745
Computer glasses, 610, 612
Computer-managed training, 745
Computer presentation software, 737–738, 741
Computers, safety precautions for, 605
Computer workstation checklist, 612
Computer workstations
    ergonomics and, 408–413, 417–418, 644
    in laboratories, 644
Concealed weapons, workplace violence and, 484
Conference training method, 718
Confidentiality, in employee assistance programs, 429, 431, 434
Confined spaces
    entry process for, 119–120
    explanation of, 73, 118
    permit-required, 73, 74, 118–119
    training for work in, 74, 120
Confined-spaces standard, 73–74, 118–120
Construction Advancement Foundation SAFE Committee, 155
Construction Safety Act (1969), 16
Constructive confrontation, 429
Construct validity, 685
Consultants' audits, 177–178
Consultants/consulting
    on indoor air quality, 380–381
    industrial hygiene, 323–324
    measuring and testing inspections by, 229
    OSH Act and, 53–54
    professional liability of, 28–29
    SH&E professionals and, 28
Consumer Product Safety Act, 67, 201
Consumer Product Safety Commission (CPSC), 139, 379
Contact stress, 398
Containment practices, in laboratories, 628–629
Contaminants
    in drinking water, 498
    indoor air quality, 372–379
    inhalation of, 226–227
Contamination
    of drinking water, 496–501
    in food-service facilities, 509–511
    groundwater, 338–339
    radioactive, 448–449, 634–635, 637
Content validity, 685
Contested cases, OSHA actions and, 64–65
Contests
    awards for, 771–772
    community and family, 770
    example of, 765–768
    injury-rate, 764–765
    interdepartmental, 767
    intergroup, 767–768
    intraplant or intradepartmental, 768

noninjury-rate, 769–770
for off-the-job safety programs, 162
personalized, 768–769
problems in organizing, 769
publicity for, 770–771
purpose of, 764
slogan, limerick, and poster, 770
Continuous inspections, 212–213
Contractors
confined-space work and, 120
function of, 650
incidence rates of, 651
incident investigation and analysis by, 252–254
in indoor air quality management and, 360
inspections conducted by, 219
liability and government, 141
oversight of, 650–651
pre-award meeting with, 652
safety record keeping for, 652–653
safety responsibilities of, 653
selection of, 651
training in hazardous chemical handling for, 117
work contracts for, 652
Contractor safety
contractor responsibilities for, 653
factors that influence, 650–651
guidelines for, 651–652
Contract Workers and Safety Standards Act (Construction Safety Act), 49
Contributory negligence, 5
Control, of operations, 135
Control charts
explanation of, 287
method to draw, 288
types of, 287
use of, 288–290
Control of hazardous energy (lockout/tagout) standard, 74–76, 103–105
Controls (machine)
arrangement of, 419
design of, 419
evaluation of, 419–420
explanation of, 419
Corporate Audit Agreements, 177
Corporate culture. *See* Organizational culture
Corrective actions
after incidents, 233, 244–251
after inspections, 223–224
Cost-effectiveness method of hazard analysis, 204
Costs
estimation of incident-related, 254–261
insured, 254–255
off-the-job injury, 260–262
training media, 752
uninsured, 255–256
of vehicle collisions, 566, 567
of work-related musculoskeletal disorders, 394, 395
Courtesy cars, 669
Credibility, 690
Crime, in retail/service industries, 551–552
Criminal law, 140

Criminal liabilities, 140–141
Criteria for a Recommended Standard for Hand-Arm Vibration (NIOSH), 399
Critical-incident stress debriefing (CISD), 133, 435
Critical-incident technique, 685
Crowd control, 668–669
Cryogenic liquids, 642
Cryostat use, 645
CSHO. *See* Compliance safety and health officer (CSHO)
Cultural change
implementation of, 42–43
levels, goals and strategies for, 41–42
measurement systems and, 42
overview of, 41
Culture. *See also* Organizational culture; Safety culture
artifacts and, 34
basic assumptions and, 34–35
climate vs., 35
espoused values and, 34
explanation of, 34
group, 35, 41
levels of, 37–38
organizational, 35–36
Customers
incidents involving, 653–654
injuries to, 654–655
Customer safety
building entrances and, 657
escalators, elevators, and stairways and, 662–665
falls on floors and, 658–661
fire, explosion, and smoke and, 665–669
glazing and, 657
housekeeping and, 662
legal issues related to, 654–655
merchandise displays and, 661–662
observation facilities and, 670
parking lots and, 655–657
pools and, 655, 670–671
ramps and, 657–658
transportation and, 669–670
walking surfaces and, 658
Customer Store Incident Report, 556

D
Dark rooms, 643–644
Daubert test, 28
Death rates
off-the-job vs. on-the-job, 12–13
before OSH Act, 49
from unintentional injuries, 9
Decision trees, 204–206
Deductive method of hazard analysis, 204
Defensive Driving Program (National Safety Council), 162, 776
Degree of care, 654
De minimus violations, 63
Demonstrations, 669, 720
De Morbis Artificum Diatriba (Ramazzini), 393
Denison, D., 35
Department of Commerce, 664

Department of Health and Human Services (DHHS), 49, 592–593
Department of Homeland Security (DHS), 676–678
Department of Labor (DOL), 49, 468, 529, 717
Department of Transportation (DOT)
    alcohol and drug testing requirements, 592–594
    employee treatment and recovery monitoring and, 431, 435
    on hazardous material shipment, 619
    penalties and, 139
    on radioactive material transport, 638
    on uniform traffic control devices, 670
Design
    laboratory, 624–626, 630, 632
    parking lot, 656–657
    tool, 421
Direct payment system, for workers' compensation claims, 192
Disabilities. *See also* Individuals with disabilities
    accommodations for, 20, 423, 526, 527, 532–533
    definition of, 526
    exclusive remedy for work-related, 187–188
    permanent partial, 184, 194–195
    permanent total, 195
    safety issues and, 526
    temporary total and partial, 194
    terms used for, 529–531
    total and permanent, 184
    workers' compensation and, 187–188, 190, 191, 194–196
Disability insurance, 191
Disabled employee safety evaluation form, 535
Disabled veterans, 531
Disaster drills, 452, 453
Disaster services, 463
Discipline, employee, 477
Discrimination. *See* Employment discrimination
Diseases. *See also* Illnesses; Occupational diseases
    waterborne, 496
    workers' compensation and, 189
Displays, safety, 774–775
District of Columbia Workers' Compensation Act (1982), 186
Documentation, of incidents, 240–241
Domestic violence, 483–484. *See also* Workplace violence
Domestic violence prevention, 485–486
Doors, 601–602
Dosimetry, 637–638
Drills, 720
Drinking fountains, 504
Drinking water
    disinfection of, 500
    in-facility contamination of, 496–497
    plumbing and, 497
    private supplies of, 497–498
    purification of, 500–501
    quality of, 498
    storage of, 501
    wells and, 498–500
Driver Performance Measurement Test (Michigan State University), 579–581

Drivers
    awards for safe, 570, 575
    collision reports by, 570–573
    information gathering for prospective, 576–581
    post-collision interviews of, 570
    record maintenance for, 570, 575
    safety programs for, 567–570
    selection of, 575–576
    training of, 582, 583
Driving tests, 579
Droughts, 449
Drug testing, 592–594
Dryers, 505
Due professional care, 178
Dust, 377–378

E

EAP. *See* Employee assistance programs (EAP)
Earthquake emergency preparedness, 444–446
Earthquake-resistant construction, 444
Easels, 749
Eco-Management and Audit Scheme (EMAS) (European Union), 333, 345, 346
Economic issues, 26
Education. *See also* Safety and health training; Safety and health training programs; Training
    in ergonomics, 403
    on indoor air quality, 379–380
    NIOSH and, 51
    safety engineering, 9
    training vs., 706
Egregious policy, for OSH Act violations, 63–64
Egress, means of, 550–551
E-learning, 745. *See also* Computer-based training (CBT)
Electrical hazards, 110–111, 553, 602–603, 625
Electrical safety standards, 110–111, 553
Electromagnetic radiation, 634
Electromagnetic waves, 634
Elevator Escalator Safety Foundation, 663
Elevators, 664
Emergencies/emergency preparedness
    action plans for, 72–73
    chain of command in, 450–452
    civil strife and sabotage as, 444
    earthquakes, 444–446
    equipment for, 458
    fires and explosions as, 442
    floods as, 442–443
    hazardous materials causing, 447, 454–455, 457
    hurricanes and tornadoes as, 443–444
    individuals with disabilities and, 543
    for laboratories, 645
    obtaining outside help to assist with, 463–464
    for offices, 609
    radioactive materials causing, 447–449
    in retail/service industries, 560–561
    shutdowns as, 446–447
    training in, 73, 452–453
    types of, 441–450
    weather-related, 449–450
    work accidents and rumors causing, 444, 446

Emergency directors, 450–451
Emergency management plans (EMP)
    alarm systems and, 458–459
    chain of command and, 450–452
    command headquarters and, 455
    development of, 441
    emergency equipment and, 458
    emergency medical services and, 461–462
    fire brigades and, 459–461
    function of, 440, 447
    overview of, 440
    program considerations for, 450
    steps in, 440–441
    training and, 452–453
    transportation and, 462–463
    uniform incident command system and, 455–458
    warden service and evacuation and, 462
Emergency medical services (EMS), 297, 300–301, 461–462
Emergency medical technicians, 462
Emissions
    from carbon dioxide, 378
    from paints, 364
    standards for, 381
Employee Assistance Professionals Association (EAPA), 428
Employee assistance programs (EAP)
    community treatment resources and, 429
    confidentiality in, 429, 431, 434
    critical incident stress debriefing and, 435
    delivery systems for, 430
    employee training and, 436
    employer benefits of, 428, 429, 432
    external, 432–434
    function of, 428–429
    history of, 428
    internal, 431–432
    management consultation and organizational development and, 436
    method to choose, 432–434
    method to establish, 429–431
    obtaining services of, 429
    overview of, 428
    staffing and philosophy of, 429–431
    statistics related to, 428
    substance abuse and, 309–310, 435
    task force members of, 430
    treatment services and, 431–432, 434–435
    work/family services and, 435–436
Employee discipline, 477
Employee Injury Data Entry Report, 555
Employees. *See also* Individuals with disabilities
    access to medical records of, 68–69
    accident-prone, 20
    alcohol and drug testing of, 592–594
    duties under OSH Act, 53
    economic losses of, 184
    health records for, 305–307
    individual differences in, 682–685
    inspections conducted by, 218
    learning needs of, 713–715
    literacy levels of, 713
    loss-control programs and, 136, 154–155
    monitoring laboratory, 626–627
    older, 310
    orientation for new, 725–726
    relationship between inspectors and, 220–221
    rights under OSH Act, 54–55
    surveys of, 692–694
    termination procedures for, 477
    workplace violence prevention and, 470–471
Employers. *See also* Management; Supervisors/foremen
    duties under OSH Act, 53
    NIOSH technical services for, 51
    record-keeping requirements of, 57–58
    rights under OSH Act, 53, 64–65
Employment discrimination. *See also* Americans with Disabilities Act (ADA)
    individuals with disabilities and, 528, 530–531
    Supreme Court on, 526–527
EMS. *See* Environmental management systems (EMS)
Energy control programs, 75–76
Energy expenditure, 408, 410
Engineering controls
    ergonomics and, 421–422
    fire detection and, 666
    laser hazards and, 640
Enterprise-related audits, 177–178
Enterprise resource planning (ERP) systems, 277
Environmental aspects review, 343–344
Environmental assessments (EA), 330
Environmental audits
    by consultants, 177–178
    explanation of, 175–176
Environmental costing, 348–349
Environmental hazards
    evaluation and control of, 320–322
    SH&E professionals and, 29–30
Environmental impact statements (EIS), 330
Environmental management
    environmental costing and, 348–349
    environmental regulations and, 330–332
    global solutions for, 339–340
    historical background of, 341
    ISO 14000:2004 and, 340–347
    ISO 14031:2013 and, 347–348
    life-cycle assessment and, 349
    new approaches to, 332–333
    strategies for, 334
    trends in, 349–350
Environmental management programs
    developmental stages for, 334–335
    groundwater contamination and, 338–339
    organization of, 335–336
    staff skills and backgrounds for, 337
    waste minimization, 337–338
Environmental management systems (EMS)
    auditing and, 345–346
    certification and, 342
    competition and, 346
    disclosures and, 346
    environmental aspects review and, 343–344

function of, 333
gap analysis and, 343
implementation of, 343
integration and alignment of, 345
ISO 14031 guidelines for, 347–348
key elements of, 342–343
legal considerations related to, 344–345
market forces and, 346
mentoring and, 347
objectives and targets of, 345
suppliers and, 346–347
Environmental monitoring, 225
Environmental performance evaluation (EPE), 347
Environmental performance indicators (EPIs), 348
Environmental Protection Agency (EPA)
    on air pollutants, 354, 381
    audits and, 175–177
    CERCLA and, 330
    on greenhouse gases, 331, 339–340
    on lead exposure, 379
    mentoring programs and, 347
    on mold remediation, 377
    penalties issued by, 139
    on pesticides, 374
    on radon, 379
    redevelopment and, 332
Environmental regulations
    by-product of, 332
    indoor air quality and, 354, 381–383
    overview of, 330–332
Environmental tobacco smoke (ETS), 374–375
Environments
    altered, 643–644
    ergonomic issues and, 399–400, 415–417
    incident causation and, 144–145
    loss-control programs and, 136
Equal Employment Opportunity Commission (EEOC), disabilities and, 423, 526, 527
Equipment
    biosafety, 630
    emergency command headquarters, 458
    food-service facility, 508
    guidelines to condemn, 221–222
    in laboratories, 625
    loss-control programs and, 136
    office, 603–606
    OSHA standard on, 553
    personal protective, 76–77, 149, 322, 630
    in vehicle repair shops, 588
Ergonomic Design for People at Work, Vol. 1 (Eastman Kodak), 420
Ergonomic risk mitigation
    active surveillance and, 406–408
    passive surveillance and, 406
Ergonomics
    business of, 394–396
    commitment to and involvement in, 422
    computer workstations and, 408–411, 417–418, 611–613
    cumulative-trauma disorders and, 401
    definition and scope of, 392–393

engineering and administrative controls in, 421–422
environment and, 399–400, 415–417
evaluation and control of, 322
history of, 393–394
in laboratories, 644–645
machine displays and controls and, 418–420
mental challenges in, 400–401
muscle exertion and, 611
office, 609–613
physical challenges and, 396–399
physical demands and, 412–415
physiological demands and, 408, 410
posture and reach and, 610–611
purpose of, 394
repetitive motion and, 611
in retail/service industries, 560
risk factors and, 396–401, 405–406, 421–422
standards in, 401–402
static work and, 410–412
training in, 403, 404, 422
visual demands and, 609–610
workplace characteristics and, 420–421
Ergonomics programs
    accommodating disabilities and, 423
    active surveillance and, 406–407
    checklists for, 407, 409–411
    components of, 403–405
    focus of, 403
    medical management, 422
    passive surveillance and, 406
    risk factor management and, 421–422
    systemic evaluation and, 408
Ergonomic work wheel, 403
Ergonomists, 392
Escalators, 662–664
Espoused values, 34
Ethical issues
    biomarkers and, 521
    SH&E professionals and, 26–27
    system safety as, 200
European Union (EU)
    chemical regulation by, 333
    Eco-Management and Audit Scheme, 333, 345
    regulations on products and goods, 21
    standardization and, 90
Evacuation
    following terrorist attacks, 677
    of high-rise buildings, 666–668
    of individuals with disabilities, 669
    routes for, 677
    wardens in, 462
Evacuation preparedness checklist, 667
Evaluation design, 724
Excavations standard, 115
Exclusive remedy provision, for work-related disabilities, 187–188
Exhaust gases, 591
Exhibits, safety, 774–775
Exits, office, 601
Experience modification rates (EMRs), for workers' compensation, 651

Expert witnesses, consultants as, 28
Explosions, emergency preparedness for, 442
Exposure
    to asbestos, 79–81, 378, 379
    to bloodborne pathogens, 79, 321–322
    to carbon monoxide, 227, 372–373
    to formaldehyde, 81
    to hazardous chemicals, 70–72, 309
    lead, 82, 379
    measuring and evaluating for, 226–228
    to molds, 375–376
    to radiation, 635–638
    records access for, 68–69
    screening requirements for, 516
    toxicity and, 226
External motivation, 687–688
Eye, first aid for burns of, 590
Eyewashes, 625

F

Face validity, 685
Facilitated discussion, 719
Facilities. *See also* Retail/service industries
    identifying hazards in, 551–552
    individuals with disabilities and access to, 539–543
    inspection of, 58–61 (*See also* Inspections)
Factor analysis, 38
Factories, safety and health movement and, 5–6
Factory inspectors, background of, 6
Factory Mutual Handbook of Industrial Loss Prevention Data (Factory Mutual System), 155
Factory Mutual System, 155
Factory system, 5
Failure mode and effect analysis (FMEA), 204, 205
Fall-protection programs, 101, 115
Fall-protection standard, 101
Falls
    chair, 600, 607
    customer, 658–662
    in offices, 599, 600
Family contests, 770
Family night programs, 162
Fault trees, 204
Federal agency directory, 91–92
Federal Aviation Administration (FAA), safety directives and, 200–201
Federal Black Lung Act, 189
Federal Coal Mine Safety and Health Act (1969), 49
Federal emergency assistance, 464
Federal Emergency Management Agency (FEMA), 464
Federal Employees' Compensation Act (FECA), 185, 186
Federal Employers' Liability Act (FELA), 187
Federal Hazardous Substances Act, 201
Federal Insecticide, Fungicide, and Rodenticide Act (FIFRA), 139
Federal Metal and Nonmetallic Mine Safety Act (1966), 49
Federal Motor Carrier Safety Administration (FMCSA), 592
Federal Railroad Administration (FRA), 266
Federal Register
    accessibility standards in, 530
    explanation of, 92
    filing of variance requests and rulings in, 57
    injury statistics in, 215
    posting of standards in, 85
Fellow servant rule, 5
Fiber-optic cables, 642–643
Field Operations Manual (OSHA), 62
Field work, for audits, 171–173, 179
Finance, in incident command system, 458
Fire brigades, 459–460
Fire extinguishers
    fire brigades and, 460–461
    in laboratories, 626
    in offices, 608–609
    in vehicle repair shops, 589
Fire fighters, 463
Fire-prevention plan, 72–73
Fire pump team, 461
Fires
    computers and, 605
    crowd and panic control during, 668–669
    detection of, 665–666
    emergency preparedness for, 442
    evacuation during, 666–668
    in high-rise buildings, 666, 668–669
    individuals with disabilities and, 669
    in mercantile establishments, 669
    in offices, 608–609
    radioactive materials and, 447
    response methods for, 666–667
    stages of, 665
    in vehicle repair shops, 589
First aid
    for burns, 590
    emergency, 297
    explanation of, 125
    prompt attention, 297
    supplies for, 297
First Aid Institute (National Safety Council), 461
First-aid kits, 299–300
First-aid programs
    components of, 297–298
    stretchers for, 300–301
    supplies for, 297, 299–300
    training and, 298, 299
First-aid reports, 268, 269
First Cooperative Safety Congress, 7
First responders, hazardous materials accidents and, 454–455
Fit test, for respirators, 107
Flammable Fabrics Act, 201
Flash floods, 442
Fleet safety programs. *See also* Vehicle safety programs
    collision frequency and, 575
    in retail/service industries, 559–560
    safe-driving programs and, 559–560
Flexible culture, 38
Flip boards, 749–750
Floods, emergency preparedness for, 442–443
Floors
    coverings and mats on, 659–660

customer falls on, 658–659
loads for, 116
office, 603–604
standards for, 506
surface types for, 659, 660
Folding machines, 606
Follow-up inspections, 61. *See also* Inspections
Fonts, 738
Food and Drug Administration (FDA)
on laser performance standards, 641
Model Food Code, 509
penalties and, 139–140
Food-borne illness, 509–510
Food-service facilities
contamination control in, 509–511
eating areas in, 508–509
equipment installation and maintenance in, 508
kitchens in, 509
nutrition and, 508
types of, 508
Force
ergonomics and, 397–398, 405
repetitive work and, 415
Ford Motor Company, 200
Foremen. *See* Supervisors/foremen
Forklifts, 111–112, 558, 559
Formaldehyde, 81, 373
Fourteenth Amendment, 6
Fraud, 141
Fume hoods, 624–625, 644
Fundamentals of Industrial Hygiene (National Safety Council), 4, 19, 225, 227, 365, 501, 605

G

Gap analysis, 343
Gary, Elbert, 6
Gas-filled detectors, 637
Gasoline safety, 590–591
Gathering and stitching machines, 606
Gender, office injuries and, 600
General-duty clause (Occupational Safety and Health Act), 100, 366, 550, 650
General inspections, 214
Generative culture, 37–38
Gilbreth, Frank, 393
Gilbreth, Lillian, 393
Glazing, 657
Global Harmonization System (GHS), 622
Globalization, 27
Glove boxes, 645
Glucans, 377
Goodwill, 11
Government contractors, 141. *See also* Contractors
Graphs, 739–741
Great Britain, 5
Greenfields, 332
Greenhouse Gas Endangerment Findings (Environmental Protection Agency), 339
Greenhouse gases (GHGs), 331, 339–340
Greenhouse gas inventories, 339
Greyfields, 332

Ground-fault circuit interrupters (GFCI), 110–111
Groundwater contamination, 338–339
Groundwater flow, 338
Groups
culture of, 35
training in, 718–720
Guardrails, 115, 116
Guide for Identifying Causal Factors and Corrective Actions, 244–251
Guidelines for Stair Safety (Department of Commerce), 664
Guns, workplace violence and, 469, 484

H

Half-life of radioactive material, 634
Hand-arm vibration (HAV), 399
Handouts, 746–747
Hawthorne study, 36
Hazard analysis. *See also* Job safety analysis (JSA); System safety
factors considered in, 206–207
formal methods of, 204–206
function of, 146, 203–204
participants in, 206
philosophy of, 203
Hazard communication program
employee training and, 71
function of, 69–70, 101–102
hazard evaluation and, 70–71
hazardous chemical list and, 102
labels and, 71, 102, 103
in retail/service industries, 551
safety data sheets and, 71, 102
written, 70, 102
Hazard communication standard (HCS), 55, 69–71, 101–103, 360, 551, 622
Hazard control inspection inventory, 214, 219
Hazard-control programs. *See also* Loss-control programs
asbestos, 80–81
corrective action in, 223–224
incident investigation in, 229–233
measures for, 148–149
workplace violence and, 474–477
Hazard-control specialists, 203
Hazardous chemicals. *See also* Toxic substances
evaluation and control of, 320
females and exposure to, 309
hazard communication program and, 70–71
highly, 77–78
labeling of, 71, 102, 103, 623
in laboratories, 71–72, 618, 623–627
process safety management for, 77–78, 116–118, 650
safety data sheets for, 69–71, 102, 103, 622
toxicity and, 226
Hazardous energy, 74–76, 103–105
Hazardous materials. *See also* Radioactive materials
loading and unloading of, 584–585
safeguards for handling of, 447
site supervisors and, 454
site workers and, 454
treatment, storage, and disposal facility workers and, 454

Hazardous materials emergency responders, 454–455
Hazardous materials specialists, 455
Hazardous materials/spills emergencies (HAZWOPER), 454–455
Hazardous materials technicians, 455
Hazardous waste, 335–336
Hazards
  control measures for, 134–135
  effects on work process, 134
  electrical, 110–111
  explanation of, 134, 200, 226
  guidelines to record, 221
  identification and evaluation of, 145–146, 238
  identification of, 210–211
  information sources on, 144–146
  inhalation, 226–227
  office, 601–609
  ranked by risk, 146–147
  solubility of chemical, 227
  toxicity vs., 322–323
Health and Safety at Work Act (United Kingdom, 1974), 16
Health care workers, 481–482
Health history, 305
Health Insurance Portability and Accountability Act (HIPAA), 306
Health promotion, 308
Health promotion programs, 307–308
Health records, 305–307
Healthy People 2000: National Health Promotion and Disease Prevention Objectives (U.S. Department of Health and Human Services), 307
Hearing loss, 189, 306
Hearing tests, 306
Heat
  effects on worker, 415–416
  emergency preparedness for extreme, 449
Heating, ventilation, and air conditioning (HVAC) systems. See HVAC systems
Heating devices, in laboratories, 625
Heinrich's dominoes, 201
Hepatitis B (HEV), 321, 627
Hepatitis B vaccination, 633
Hepatitis C (HCV), 321
Herzberg, Fredrick, 687, 688
Hierarchy of needs (Maslow), 686
High-efficiency particulate air (HEPA) filters, 630
Highly hazardous chemicals (HHCs), 77–78
Hiring
  in retail/service industries, 553–554
  screening for potential violence and, 477
  workers' compensation programs and, 196
HIV. See Human immunodeficiency virus (HIV)
Hold-harmless agreement, 655
Homeland security compliance
  affected sites and, 676–677
  federal rules and, 676
  information sharing and, 678
  local and state involvement and, 678
  reporting obligations and, 678
  safety manager's role in, 677–678

Homicides, 468, 484
Horizons, 338
Hot-work permits, 118
Housekeeping
  hazardous chemicals and, 623
  indoor air quality and, 360–361
  loss-control programs and, 152–154
  in mercantile establishments, 662
  to prevent industrial accidents, 116
Human factors, in incident causation, 144
Human Factors and Ergonomics Society (HFES), 392
Human-factors engineering. See Ergonomics
Human immunodeficiency virus (HIV), 321, 627
Human interest, 781
Humidity, 366
Humor, 781
Hunt, Christopher, 334, 335
Hurricanes, emergency preparedness for, 443–444
HVAC Checklist—Short Form, 362–364
HVAC systems. See also Ventilation
  carbon dioxide and, 378
  indoor air quality and, 361–366
  in laboratories, 624, 625
  maintenance of, 379
Hydraulic interconnects, 338
Hydrogen fluoride, 226–227
Hygiene approach, 687, 688
Hypersensitivity pneumonitis, 376

I
IAQ. See Indoor air quality (IAQ)
IESNA Lighting Handbook, 602
Illnesses. See also Occupational diseases
  assessment and handling of, 126
  classification of, 125, 126
  data sources for, 519
  food-borne, 509–510
  injury vs., 9
  mold exposure and, 375–376
  shiftwork and, 311, 313
  waterborne, 496
  while traveling, 124
  work environment and, 19
  workers' compensation and, 189
  work-related, 123–124
Illness reporting
  exemptions for, 121–122
  for office employees, 614
  OSHA and, 121, 122
  record keeping and, 121
  work-relatedness determination and, 123–124
Incentives, 691–692. See also Awards
Incident causation
  environmental factors as, 144–145
  examination of, 132–133
  factors related to, 135
  identification of, 210–211, 244–251
  in negligence action, 137
  situational factors as, 143–145
  unsafe practices or procedures as, 142–143
Incident commanders (ICs), 455–457

Incident command system (ICS), 455–456
Incident investigation/analysis
    causation in, 244, 246–251
    classifying incident data for, 245–251
    corrective actions following, 233, 244–251
    cost estimation in, 254–261
    costs of off-the-job disabling injuries, 260–262
    in hazard-control programs, 229–233
    hazardous chemical, 118
    interviews in, 232
    involving outside employees and contractors, 252, 254
    key concepts in, 240–242
    minimum data required for, 242–244
    overview of, 238–239
    personnel involved in, 240
    process for, 245, 251–253
    purpose of, 242
    report preparation for, 232–233
    terminology in, 238
    timing of, 231
    types of, 239–240
Incident Investigation Form, 274–276
Incident Investigation Reports
    examples of, 242, 270–271
    function of, 268
    minimum data set for, 242, 244
    preparation of, 232–233
Incident investigations
    function of, 229–230
    individuals that conduct, 231–232
    questions for, 232
    reasons for, 230
Incident (No-Injury Accident) Report Form, 238, 239
Incident prevention
    cooperation in, 10
    early breakthroughs in, 8
    factors associated with, 238
    investigations and, 238–239
    philosophy of, 4
Incident process, 719
Incident rate formula, 285–287
Incident rates. *See also* Record keeping
    calculation of, 287–290
    contractors and, 651
    data classification and analysis and, 245–251
    hazard information and, 146
    noninjury/illness, 286–287
    third-party, 655
Incident records. *See also* Record keeping
    monthly, 281–282
    for office employees, 614
    requirements to maintain, 57–58, 121, 266, 284
    types of data for, 269, 278
    uses of, 266
Incident reports
    annual summary, 283–284
    first-aid, 267, 268
    function of, 266, 268, 282
    hazard information and, 146
    incident investigation form, 274–276
    periodic and on-demand, 281–284

    record-keeping tools and, 276–279
    supervisor's, 268, 270–271
    supplementary record of occupational injuries and illnesses, 272–273
    workplace violence and, 479–480
Incidents
    documentation of, 240–241
    estimating cost of, 254–261
    list of typical, 133
    loss control and, 134–137
    near-miss, 238
    profit motive and, 200
    public relations for serious, 783–784
    use of term, 238, 266
Income replacement, workers' compensation and, 186, 189–190
Increased-risk test, 188
Indemnity costs, 395
Independent study, 722
Individuals with disabilities. *See also* Disabilities
    accommodations for, 20, 423, 526, 527, 532–533
    discrimination against, 530–531
    emergency procedures for, 543, 669
    evacuation procedures for, 669
    facility access and, 539–543
    insurance considerations for, 530
    job analysis and, 537–539
    job placement for, 534–537
    legislation related to, 526–530
    safety and health professional's responsibilities for, 533–534
    safety evaluation form for, 534, 535
    safety issues for, 527
    support groups for, 537
    terms used for, 529–531
    types of, 531
    veterans as, 531
    workers' compensation and, 187–188, 190, 191, 194–196
Indoor Air Quality Building Education and Assessment Model (I-BEAM) software (EPA), 354
Indoor air quality (IAQ). *See also* Air pollutants
    asbestos and, 378–379
    building-related illness and, 355
    building systems affecting, 365–366
    carbon dioxide and, 378
    carbon monoxide and, 372–373
    common pollutant sources and, 355, 357–366
    dust and, 377–378
    elements related to, 355
    environmental tobacco smoke and, 374–375
    formaldehyde and, 373
    housekeeping and, 360–361
    HVAC systems and, 361–366
    investigations related to, 371–372, 380–381
    lead and, 379
    legislation on, 382
    microbial contaminants and, 375
    miscellaneous pollutants and, 378
    molds and, 375–377
    nitrogen oxides and, 373–374

non-HVAC building components and, 366
occupant activities and, 360–364
outside pollutant sources and, 365
overview of, 354
painting and, 364
pesticides and, 374
pollutant pathways and, 366–367
radon and, 378–379
remodeling and renovation and, 364
sick-building syndrome and, 355, 356
thermal comfort and, 365–366
vocabulary for, 383–388
volatile organic compounds and, 373
Indoor air quality (IAQ) management. *See also* Air pollutants
checklist for, 368–370
consultants and, 380–381
control strategies for, 355
function of, 367, 370
hypothesis development and testing and, 372
IAQ profile in, 371
maintenance activities and, 361–364
problem resolution and, 379–380
record keeping in, 371
regulatory requirements and, 354, 381–383
role of manager in, 371
SH&E professionals and, 29
strategies to address in, 370–371
training and, 360, 361
Indoor air quality (IAQ) managers, 371
Indoor air quality (IAQ) profiles, 371
Indoor Air Quality Management Checklist, 368–370
Indoor environmental quality (IEQ), 354
Inductive method of hazard analysis, 204
Industrial accidents. *See* Accidents
Industrial hygiene
environmental hazards and, 320–322
explanation of, 318
practice of, 319–322
toxicity and, 322–323
Industrial hygienists
certification and licensure of, 324, 326
code of ethics of, 325
as consultants, 323–324
in-house, 319
measurement and testing by, 225–228
on occupational health and safety team, 318–319
professional organizations for, 326
role of, 318
Industrial Revolution, 4–5
Industrial Ventilation—A Manual of Recommended Practice (American Conference of Governmental Industrial Hygienists), 326
Inflation, cost estimates and, 255
Inhalation hazards
measurement of, 226–227
of toxic substances, 323
Injuries. *See also* Accidents
assessment and handling of, 126
body parts and nature of, 252
classification of, 125, 126

covered by workers' compensation, 188–189
ergonomic-related, 400–401
financial costs of, 12
illness vs., 9
medical treatment and first-aid definitions and, 125
office, 598–600
off-the-job, 260–262, 290
before OSH Act, 49
statistics related to, 12
third-party, 654–655
travel-related, 124
trends in nonfatal, 9–10
type of accident and, 252, 253
unintentional, 9, 132, 238
use of term, 266
work-related, 123–124, 393–394
Injury Cost Capturing Tool (National Safety Council), 256
Injury data
analysis of, 252, 254
classification of, 245–251
Injury Facts (National Safety Council), 15
Injury/illness data analysis
control charts and, 287–290
implementation and follow-up to, 290
incidence rates and, 284–285
noninjury/illness incidence rates and, 286–287
Pareto diagram and, 286
severity measures and, 285–286
statistical measures in, 284
tests for significance of change in rates and, 286–287
Injury-rate contests, 764–765
Injury reporting
exemptions for, 121–122
OSHA and, 121, 122
record keeping and, 121
work-relatedness determination and, 123–124
Insect contamination, 503
Inspection reports, 222–223
Inspections
closing conference for, 61
conditions examined in, 215–217
continuous, 212–213
corrective action following, 223–224
energy control program, 76
of equipment, 221–222
escalator and elevator, 663, 664
follow-up to, 61, 223–224
frequency of, 217
general procedures for, 59–61
hazard-control, 214
informal conferences following, 61
interval, 213–214
items included in, 214–215
measurement and testing, 225–229
mine, 87
opening conference for, 59–60
OSHA and, 58–59
planning for, 214–219
preparation for, 219–220
priorities for, 59

purpose of, 212
recommended penalties following, 139
recording hazards found during, 221
regulatory compliance, 224–225
relationships of personnel involved in, 220–221
responsibility for, 217–219
in retail/service industries, 554, 556–558
during shutdowns, 447
tools for, 220
views of, 212
workplaces, 58–61
Inspectors
background of, 218–219
company compliance, 221
hazards recorded by, 221
independent consultants as, 229
qualifications of, 217–218
relationship between employees and, 220–221
relationship between supervisors and, 220
Instructional media. *See* Training media
Insurance incentives, of workers' compensation, 195–196
Insured costs, 254–255
Integrated pest management (IPM), 503
Intentional torts, 138
Interdepartmental contests, 767
Intergroup contests, 767–768
Intermittent inspections, 213–214
Internal audits, 177
Internal motivation, 686–697
International Agency for Research on Cancer (IARC), 70
International Ergonomics Association (IEA), 392
International issues, 339–340
International Organization for Standardization (ISO)
ISO 9000, 17–18, 90, 346
ISO 9001, 18
ISO 9002, 18
ISO 9004, 18
ISO 14000, 18–19, 90, 333, 340–347
ISO 14001, 333, 350
ISO 14031, 347–348
Technical Committee, 207, 340
on vibration, 399
water management and, 339
International standards
Framework Directive and, 17
goals of, 89–90
for health and safety regulations, 17
ISO 9000 series and, 17–18, 90, 346
ISO 14000 series, 18–19, 90, 333, 340–347
Internships, 717
Interval inspections, 213–214
Interviews
following collisions, 570
during incident investigations, 232
television, 783
of vehicle driver applicants, 577–578
Intradepartmental contests, 768
Intraplant contests, 768
Investigations. *See* Incident investigation/analysis
Investigator's Cost Data Sheet and Summary Report, 256–259

Invitees, 654
Ionizing radiation, 634. *See also* Radiation; Radiation safety
ISO Continual Improvement Model, 548
ISO standards. *See* International Organization for Standardization (ISO)

J
Jacks, 591
Janitorial services, 508
Jastrzebowski, Woyciech, 393
Job analysis
ergonomic, 406–408
individuals with disabilities and, 537–539
Job description, 538–539
Job enrichment theory, 687–688
Job placement
for individuals with disabilities, 534–537
in retail/service industries, 553–554
Job safety analysis (JSA)
benefits of, 208
effective use of, 211–212
explanation of, 207
identifying hazards for, 210–211
job breakdown for, 209–210
job selection for, 208–209
solution development in, 211
Job safety analysis (JSA) forms, 207, 208
Job safety analysis (JSA) worksheets, 209
Job satisfaction
measurement of, 694–695
safety culture and, 38, 40
surveys of, 694
Joint safety and health committees, 6, 15, 218–219
Jones Act, 187
Judicial review, of mine safety standards, 85
Just culture, 38

K
Kitchens, 509
Knowledge, of safety professionals, 11–12

L
Labeling
hazardous chemicals, 71, 102, 103, 623
respirators, 105, 106
Labor and Management Council, 761
Laboratories
biological hazards in, 618–619, 627–634
chemical hazards in, 618, 623–627
closing procedures for, 645–646
design of, 624–626, 630, 632
ergonomic issues in, 644–645
lasers in, 638–642
management of, 619–621
physical hazards in, 619, 642–643
radiation hazards in, 619, 634–642
safety culture and, 621
waste disposal in, 624, 645–646
Laboratory Chemical Safety Summaries (LCSSs), 622
Laboratory safety
altered environments and, 643–644

biological, 627–634
chemical, 622–627
communication and leadership skills and, 620
cryogenic liquids and, 642
emergency planning and, 645
ergonomics and, 644–645
fiber-optic cables and, 642–643
model for, 620–621
nature of, 619
radiation, 634–642
regulatory information and, 619
scientific information and, 619
tools for, 620
Laboratory tests, employee, 306
Laboratory workbenches, 644
Labor Division (National Safety Council), 13
Labor unions, 13–15
Ladders, 113–114, 604
Laser Institute of America, 641
Laser radiation
administrative controls for, 640
classification of, 639, 640
engineering controls for, 640
hazards of, 638–639
personal protective equipment use for, 640, 641
safety recommendations for, 641–642
Lasers, 638, 641
Laser safety officer (USO), 642
Law enforcement occupations, 482
Layout, office, 601–602
Lead, indoor air quality and, 379
Lead exposure, 82, 379
Learning
memory and, 702
motivation and, 699
principles of, 699–702
selective, 701
Learning culture, 38
Learning management systems (LMSs), 733
Learning pyramid, 714
Lecture training method, 720
Legal issues
biomarkers and, 521
environmental management systems and, 344–345
overview of, 98
system safety and, 200
of third-party injuries, 654–655
Legionella, 375, 376
Legislation. *See also* Environmental regulations; Occupational Safety and Health Act (OSH Act) (1970); Regulations; *specific legislation*
individuals with disabilities, 526–530
indoor air quality, 382
industrial safety, 16–17
occupational safety and health, 49
overview of safety-related, 16–17
product safety, 67
workers' compensation, 184–187
Liability
civil, 137–140
criminal, 140–141

government contractors and, 141
individual civil, 141
organizational, 138
prevention strategies for, 142
SH&E professionals and, 28–29
strict, 141
Liability issues
audit and inspection programs as, 554, 556–558
emergency preparedness as, 560–561
ergonomics as, 560
fleet safety programs as, 559–560
hiring and placement as, 553–554
materials handling as, 556, 558, 559
risk control as, 554–556
violence and crime in workplace as, 561–562
Licensees, 654
Licensure, industrial hygienist, 324, 326
Life-cycle assessment (LCA), 339
Life Safety Code (National Fire Protection Association), 550, 601, 602, 664
Lifestyle change programs, 308
Lifting
equation for, 412, 414
guidelines for, 412, 413, 604
in offices, 604, 608
Light duty, 305
Lighting
office, 602, 603, 610
parking lot, 656
shiftwork and, 312
stairway, 664
ultraviolet, 630
Lightning, 450
Limerick contests, 770
Limited duty, 305
Line charts, 740–741
Literacy, 713
Loans, small-business, 65
Local emergency planning committees (LEPCs), 464
Locker rooms, 504–506
Lockers, 506
Lockout/tagout
hazardous energy control and, 74–75, 103–105
OSHA standard on, 552–553
Logistics, in incident command system, 458
Log of Occupational Injuries and Illnesses (Occupational Safety and Health Administration), 651
Long-term memory, 702
Loss control, 145, 184
Loss-control programs
background of, 132
contractors and, 650
establishing objectives for, 150
establishing organizational policy for, 150–151
evaluation of, 149–150
examination of incident causation and, 132–134
housekeeping and, 152–154
incident causes and controls and, 134–137, 142–145
liability and, 137–142
off-the-job safety and, 159–163
oversight and omission management and, 132

preventive and corrective measures in, 152–154
principles of, 145
processes of, 145–150
professionals in, 156
purchasing agents and, 155–156
purchasing and, 156–159
purchasing-safety liaison and, 156–157
responsibility for, 151–156
safety and health committees and, 159
safety considerations and, 157
Loss of wage-earning capacity theory, 194, 195
Low-Level Radioactive Waste Policy Act (LLRWPA) (1980), 638
Lubrication operations, 589
Lunchrooms. *See* Food-service facilities

M
Machine-guarding standard, 108–110, 553
Machines
  displays and controls for, 418–420
  employee training in use of, 110
  requiring point-of-operation guarding, 108–109
  safety measures in use of, 109, 553
  servicing and maintenance of, 109–110
Mager, Robert, 698
Maintenance
  indoor air quality and, 361–364
  loss-control programs and, 152–154
Maintenance personnel, inspections conducted by, 218
Malcolm Baldrige National Quality Award, 691–692
Management
  environmental, 337
  explanation of, 201
  hazard identification by, 147–148
  health and safety concerns of, 16–17
  incident investigations by, 231
  inspections conducted by, 218
  joint safety and health committees with unions, 15
  leadership in health and safety, 548–549
  loss-control programs and, 132, 135, 151–152
  presenting SH&E benefits to, 27–28
  record-keeping role of, 268
  safety awareness program role of, 760–761
  safety culture role of, 36
  workplace violence prevention and, 470–471
Management consultation, 436
Manual Materials Handling Tables (Liberty Mutual), 412
Manual on Uniform Traffic Control Devices for Streets and Highways (Department of Transportation), 670
Marshall v. Barlow's Inc., 58
Maslow, Abraham, 686
Massachusetts vs. the Environmental Protection Agency, 339
Materials handling
  carrying and, 415
  lifting and, 412–414
  pushing and pulling and, 414–415
  in retail/service industries, 556, 558, 559
Maximum permissible exposure (MPE), 641
McGregor, D., 687

Meaningfulness, 700–701
Measurement and testing inspections
  for exposure, 226–228
  function of, 225
  independent consultants conducting, 229
  for influence of solubility, 227
  for inhalation hazards, 226–227
  for permissible exposure limits, 227–228
  phases of, 228–229
  for Threshold Limit Values, 227
  for toxicity, 226
  types of, 225–226
Measurement error, 686
Mechanical engineers, inspections conducted by, 218
Mechanical systems, 697
Media. *See* Training media; Training media presentations
Media audits, 169
Medical care
  for asbestos exposure, 81
  in emergencies, 73
  explanation of, 125
  for exposure to bloodborne pathogens, 79
  for hazardous substance exposure, 72
  workers' compensation and, 190–191
Medic alert tags, 306
Medical examinations
  to determine respirator use suitability, 107
  for driver applicants, 579
  exit, 306
  function and scope of, 304
  individuals with disabilities and, 539
  laboratory tests and, 306
  periodic, 304–305
  preplacement, 304
  return-to-work, 305
  vision and hearing tests and, 306
Medical management program, for ergonomics, 422
Medical monitoring, 225–226
Medical records, access to employee, 68–69
Medical rehabilitation programs, workers' compensation and, 192–193
Medical screening
  biomarkers and, 520–521
  ethical and legal issues in, 521
  function of, 519–520
  prevention and, 520
  programs in, 521
Medical surveillance
  explanation of, 308, 516
  exposure effects and, 516, 517
  types of, 518–519
  working populations and, 516, 518
Medium, 732
Meetings, safety awareness program, 761–763
Memory, short- and long-term, 702
Mental demands, ergonomics and, 400
Mentoring, 718
  ISO 14001 and, 347, 350
Merchandise displays, 661–662
Merit Partnership (Environmental Protection Agency), 347
Merit rating, 195

Microbial volatile organic compounds (mVOCs), 377
Microbiological techniques, 629
Microorganisms, indoor air quality and, 375
Microscopes, 644
Microsoft PowerPoint, 741
Microtome use, 645
Mine operators
  citations for, 88
  contested cases and, 89
  penalties for, 88–89
  requirements for, 85
Miners, 84, 85
Mines
  health and safety standards in, 85–89
  inspection of, 87
Mine Safety and Health Act (1977)
  advisory committees and, 84
  background of, 82–83
  contested cases and, 89
  coverage of, 84
  mine operators requirements and, 85
  miners' rights and, 84
  miner training and, 85
  provisions of, 52
Mine Safety and Health Administration (MSHA), 726
  directory of, 92
  function of, 83
  record-keeping requirements of, 266
  services of, 51
Mine Safety and Health Review Commission, 83–84
Mishap models, 201
Mishaps, 201
Mishra, A., 35
Model Domestic Violence Policy (New York State), 485–486
Molds
  explanation of, 375
  glucans or fungal cell wall components of, 377
  health effects of exposure to, 375–376
  indoor air quality and, 375
  microbial volatile organic compounds produced by, 377
  toxic, 376–377
Mold spores, 376
Monitoring
  biological, 225
  environmental, 225
  ergonomics and, 417
  of hazards, 149
  medical, 225–226
  personal, 225
Moreell, Ben, 8
Motivation
  attitude and, 695–697
  behavior and, 697–699
  external, 687–688
  individual differences in, 683–686
  internal, 686–687
  job satisfaction, 694–695
  learning processes and, 699–702
  methods to direct, 689–695
  overview of, 682

psychological factors in safety and, 682–683
structural change and, 699
variables that affect, 689
workplace models of, 688–689
Motivation strategies
  awards, incentives, and recognition as, 691–692
  behavior reinforcement as, 693–694
  communication and, 690–691
  employee surveys and, 692–694
  function of, 689–690
  specifying safety objectives as, 693
Motor vehicles. *See* Vehicles
Motor vehicle safety programs. *See* Vehicle safety programs
Muscle exertion, 611
Musculoskeletal disorders (MSD)
  computer workstations and, 417
  ergonomics and, 393–398, 401
  posture and, 396, 397
  repetitiveness and, 398
  risk factors for, 405, 406
  symptoms of, 401
  trends in, 394–396
  workplace analysis to identify areas with potential for, 406–408
Mutual-aid plans, 463

N

Naps, 312
National Academy of Social Insurance, 395
National Agricultural Chemical Association, 326
National Air Filtration Association (NAFA), 377
National Ambient Air Quality Standards (NAAQS), 373
National Bureau of Standards, 7
National Council for Industrial Safety, 7
National Electrical Code (ANSI/NFPA), 602, 603
National Emission Standards for Hazardous Air Pollutants (NESHAP), 381
National Environmental Policy Act (NEPA), 330
National Fire Protection Association (NFPA)
  on alarm systems, 458
  on fire hazard management, 215
  Life Safety Code, 550, 601, 602, 664
National Fleet Safety Contest (National Safety Council), 575
National Foundation on Arts and Humanities Act (1965), 49
National Highway Traffic Safety Administration (NHTSA), 340, 593
National Institute for Occupational Safety and Health (NIOSH)
  on energy expenditure, 408, 411
  functions of, 51–52
  on heat exposure, 416
  on manual material handling, 412–414
  on safety practices, 36
  on vibration, 399
  on workplace violence, 468
National Institutes of Health (NIH), 627
National Oceanic and Atmospheric Administration (NOAA), 464

National Safety Council (NSC)
    on accident prevention, 132
    award program of, 692, 765
    establishment of, 7, 10
    evacuation preparedness checklist, 667
    on first aid, 298, 300
    First Aid Institute, 461
    Fundamentals of Industrial Hygiene, 4, 19, 225, 227, 365, 501, 605
    Injury Cost Capturing Tool, 256
    Labor Division of, 13
    on labor–management incident prevention, 13
    mission of, 4
    National Fleet Safety Contest, 575
    on shiftwork, 313
    Supervisor Incident Investigation Report, 277–279
    on Threshold Limit Values, 227
    training courses offered by, 10
    Work Injury and Illness Rates, 503
    World War II and, 9
National Safety Management Society, 11
National Toxicology Program (NTP), 70
National Training Laboratories, 713
National Voluntary Laboratory Accreditation Program (NVLAP), 638
National Weather Service, 442, 443, 464
Near-miss incidents, 238
Needs analysis, 710–711
Needs assessment, for employee assistance programs, 430–431
Negligence
    in civil context, 137–138, 141
    contributory, 6
    criminal prosecution for, 140–141
    explanation of, 654
Neutral posture, 396–397, 420
Newsletters, 777
New York Central Railroad v. White, 6
NIOSH. *See* National Institute for Occupational Safety and Health (NIOSH)
Nitrogen dioxide, 373–374
Nitrogen oxides, 373–374
Nixon, Richard M., 48
Noise, evaluation and control of, 320–321
Nongovernmental organizations (NGOs), 30, 492
Noninjury/illness incidence rate, 286–287
Noninjury-rate contests, 769–770
Nonionizing radiation. *See also* Radiation; Radiation safety
    explanation of, 634, 636
    safety guidelines for, 636–638
Normal distribution, 683, 684
NSC. *See* National Safety Council (NSC)
Nuclear Regulatory Commission (NRC)
    penalties and, 139–140
    radiation standard, 636
Nuisance bird control, 503
Nurse Line Programs, 303–304
Nurse Practice Act, 298
Nurses, occupational health, 297, 298, 301–302
Nutrition, in food-service facilities, 508

O

Occupant Diary, 356
Occupational diseases. *See also* Diseases; Illnesses
    black lung disease as, 189
    reasons to prevent, 4
    workers' compensation and, 189
Occupational Exposure to Hazardous Chemicals in Laboratories standard, 71–72
Occupational Health and Safety (National Safety Council), 4
Occupational health clinics, 297
Occupational health nurses, 297, 298, 301–302
Occupational Health Nursing Newsletter (National Safety Council), 308
Occupational health physicians, 302–303
Occupational health professionals
    responsibilities of, 297, 301
    types of, 297, 298, 301–304
Occupational health programs
    components of, 296, 319
    emergency medical planning in, 306
    employee health records in, 306–307
    first-aid provisions of, 297–301
    first-aid training for, 298
    health care for women aspect of, 308–309
    health promotion and wellness, 307–308
    location of, 297
    medic alert tags in, 307
    medical surveillance, 308
    objectives of, 296–297
    older-worker, 310
    overview of, 301
    personnel in, 301–304
    physical examinations in, 304–306
    shiftwork and, 310–313
    substance abuse and employee health, 309–310
Occupational Health & Safety (National Safety Council), 313
Occupational Injury and Illness Classification System (OIICS), 245, 281
Occupational Safety and Health Act (OSH Act) (1970)
    access to employee exposure and medical records and, 68–69
    administration of, 49–52
    background of, 9, 16, 48–49
    citations for violations of, 63
    contested cases and, 64–65
    coverage of, 52
    egregious policy and, 63
    emphasis of, 132
    employee rights in, 54–55
    employer and employee duties in, 53
    employer rights in, 53
    federal-state relationships and, 65–67
    general-duty clause of, 100, 366, 550, 650
    implications of, 67
    legislative history of, 49
    onsite consultation and, 53–54
    OSHA poster and, 55, 58, 100
    OSHA standards and, 55–56
    penalties and, 63, 139

posting requirements under, 550
private-sector input and, 56
on record-keeping requirements, 57–58, 121, 266, 283–284
regulations governing workplace and, 100
reporting requirements and, 58
scope of, 98, 100
small-business loans and, 65
state plans and, 65–67
variances from standards and, 56–57
violations of, 61–63
on worker participation in safety and health activities, 15
workplace inspection and, 58–61
Occupational Safety and Health Administration (OSHA). *See also* Standards
area offices of, 50
authority of, 50
on bloodborne pathogens, 79, 627, 632–634
compliance issues related to, 17, 30
on contractor safety performance, 650
directory of, 91–92
duties of, 50
on ergonomics, 401–402
on exposure limits, 227–228, 373, 516
on facility hazards, 551–552
on first aid, 461
General Industry Standards worksheet, 216
Hazard Communication Standard, 55, 69–71, 101–103, 360, 551
on high-risk occupations, 481, 482
on indoor air quality, 366
on industrial exposure standards, 381–382
on industry-specific training, 297
on injury and illness reporting, 121–126
on inspections, 58–61
on laboratories, 618, 623, 626
on lasers, 642
on lockout/tagout, 552–553
Log of Occupational Injuries and Illnesses, 651
on medical records access, 306
on medical screening, 516, 517
on medical services and first aid, 551
on musculoskeletal disorders, 395, 396
on permissible exposure limits, 22, 81, 227–229, 373
on powered equipment, 553
purchasing checklist, 408–410
on radioactive materials, 448
on record-keeping requirements, 9, 57–58, 121, 177, 266, 268
respiratory protection standard, 72, 78–79, 105–108, 537
on service industry, 549–550
staff of, 50
on training requirements, 726
on variances from standards, 56–57
on workplace emergencies, 72–73
on workplace violence, 468, 469, 482
Occupational safety and health programs. *See also* Loss-control programs
establishing organizational policy for, 150–151
function of, 150
management of, 548–549
responsibility for, 151–156
Occupational Safety and Health Review Commission (OSHRC), 50–51, 64
Occupational Safety/Health Contest (National Safety Council), 765
Occupational sentinel health events, 518
Occupations
at high risk for workplace violence, 468, 478, 481–482
most affected by ergonomics-type injuries, 400–401
Odors, formaldehyde, 373
Office hazard prevention
aisles and stairs and, 604
electrical equipment and, 602–603
fires and, 608–609
floors and, 603–604
layout and ambiance and, 601–602
lighting and, 602
office equipment and, 603–606
outdoor areas and, 604
procedures for, 606–608
storage and, 604
ventilation and, 602
Office injuries
background of, 598–599
gender and, 600
new or young employees involved in, 599–600
in new surroundings, 599
types of, 600
Office safety
emergency plan for, 609
ergonomics and, 609–613
hazard prevention in, 601–608
incident and illness records on, 614
for individuals with disabilities, 540–543
training in, 613–614
Offset presses, 606
Off-the-job (OTJ) injuries
cost of disabling, 260–262
explanation of, 260
record keeping of, 290
World War II and, 9
Off-the-job (OTJ) safety programs
background of, 159–160
benefits of, 160
development of, 160–161
effects of, 162
family and community involvement in, 161–162
methods for, 162–163
promotion of, 9, 160
safety policy and, 160
Oil Pollution Act (1990), 331
Omnibus Transportation Employee Testing Act (1991), 435, 592
On-the-job training (OJT), 716–718
Operations hazard analysis (OHA), 204, 205
Organic systems, 697
Organizational culture. *See also* Culture
environmental management programs and, 335
explanation of, 35–36

implementing change in, 41–43
safety culture and, 36–38 (*See also* Safety culture)
Organizational development (OD)
   components of, 696–697
   employee assistance programs and, 436
Organizational liability, 138
Organization behavior management (OBM) model, 688–689, 691
Organizations
   administration penalties against, 138–139
   cooperation between, 21
   criminal remedies and, 140
   nongovernmental, 30
   safety rules of, 155
OSHA. *See* Occupational Safety and Health Administration (OSHA)
OSH Act. *See* Occupational Safety and Health Act (OSH Act) (1970)
OSHA poster, 55, 58
OSHRC. *See* Occupational Safety and Health Review Commission (OSHRC)
Other-than-serious violations, 63
Overexertion, 600
Oxygen-deficiency levels, 119
Ozone, 378

P

Paint emissions, 364
Panic control, 668–669
Pareto diagram, 286
Parking lots, 655–657
Parking spaces, for individuals with disabilities, 540, 541
Parking stalls, 656
Particle radiation, 634
Particulates, 226
Passive surveillance, 406
Path analysis, 40
Pathological culture, 37
Peculiar-risk test, 188
Penalties
   federal, 139–140
   for OSHA violations, 63, 139
   statutory, 139–140
Performance objectives, 711–712
Performance reports, 180
Periodic inspections, 213
Periodic physical examinations, 304–306
Permanent partial disability, 194–195
Permanent total disability, 195
Permissible exposure limits (PELS)
   background of, 227–228
   formaldehyde, 81, 373
   individuals with disabilities and, 537
   OSHA and, 22, 516
   Threshold Limit Values and, 228, 229
   tobacco smoke and, 375
Permit-Required Confined Space standard, 73, 74, 118–119
Personalized contests, 768–769
Personal monitoring, 225
Personal protective equipment (PPE)
   function of, 76–77
   as hazard control measure, 149, 322
   in laboratories, 630
   for laser protection, 640, 641
Personal protective equipment (PPE) standard, 76–77
Personal service facilities
   drinking fountains and, 504
   function of, 503
   janitorial services and, 508
   showers and, 506–507
   toilets and, 507
   washrooms and locker rooms and, 504–506
Pesticides, 374
Petition for Modification of Abatement (PMA), 63
Petition request, for mine safety alternatives, 86
Photographs, 748
Physical capacities form, 536
Physical demands
   carrying and, 415
   explanation of, 412
   lifting and, 412–414
   pushing and pulling and, 414–415
   repetition and, 415
Physical demands form, 536
Physical examinations. *See* Medical examinations
Physical exercise, 312
Physical hazards
   evaluation and control of, 320–321
   in laboratories, 619
Physicians
   incident investigations by, 231
   occupational health, 302–303
   selection of, 197
Picnics, company, 162
Pie charts, 739–740
Pinch grip, 397–398
Pipe, Peter, 698
Pipetting, 644–645
Pittsburgh Survey (1906), 6
Planning, for SH&E audits, 169–171
Plastics, burning, 665
Plateau, 700
Plea bargain, 140
Plumbing systems, contaminated drinking water and, 497
Podcasts, 721, 747
Pointers, 749
Police, emergencies and, 463
Politics, regulations and, 21
Pollutant and Source Inventory, 357–359
Pollutants. *See* Air pollutants
Polyvinyl chloride (PVC), 665
Pools, 655, 670–671
Population stereotypes, 419
Positional-risk test, 188
Positive pressure, 367
Positive reinforcement, 699–700
Positive trees, 204
Poster contests, 770
Posters
   on bulletin boards, 774
   changing and mounting, 774

effectiveness of, 773
function of, 746, 772–773
OSHA, 55, 58, 100
types of, 773–774
Post-tests, 724–725
Posture
awkward, 397
at computer workstations, 417
ergonomics and, 610–611
neutral, 396–397, 420
repetitive work and, 415
Powered industrial trucks standard, 111–113
Power grip, 398
PowerPoint (Microsoft), 741
Pregnancy, 309
Pregnancy Discrimination Act, 309
Preplacement physical examinations, 304
Presentations. *See* Training media presentations
Presentation software, 737–738, 741. *See also* Training media presentations
President's Quality Award, 691–692
Presumed asbestos-containing material (PACM), 80
Pretests, 724
Prevention of Significant Deterioration (PSD) permit (Title V permit), 339
Preventive maintenance
loss-control programs and, 152–154
motor vehicle, 585–588
Price considerations, purchasing and, 157–158
Primacy, 701
Primary prevention, 308, 520
Printing services, 606
Private rights of action, 140
Proactive culture, 37
Problem reports, 180
Process Hazard Analysis (PHA), 117
Process safety management
compliance elements for, 77–78
function of, 77
for new facilities and modified worksites, 78
Process Safety Management of Highly Hazardous Chemicals standard, 77–78, 116–118, 650
Production methods, 4–5
Productivity, 36, 135
Product safety, 67, 491
Product safety management programs
elements of, 490–492
evaluation of, 492
function of, 490
leadership commitment to, 490
relationship building in, 492
Professionalism, 11
Professional organizations, 326
Profit margins, injury cost recovery and, 260, 261
Profit motive, system safety and, 200
Projected visuals, 741
Pronation, 397
Protected Critical Infrastructure Information (PCII) system, 678
Protective clothing, for cold environments, 417
Prudent Practices in the Laboratory (National Research Council), 624
Psychometric models, 38
Public address systems, 777
Publications
company, 784–786
promoting safety, 776–777
Publicity. *See* Public relations (PR)
Public relations (PR)
audience selection and, 781
contests and, 770–771
friendly rivalry and, 782
function of, 779–781
humor and human-interest focus of, 781
safety news development and, 780–781
safety office communications and, 782–783
for serious incidents, 783–784
techniques for, 782
television interviews and, 783
Pulling, 414–415
Punitive damages, 138
Purchasing
codes and standards in, 158
loss-control programs and, 155–159
office equipment, 605–606
price considerations in, 157–158
safety considerations in, 157
specifications in, 158–159
Purchasing Checklist (OSHA), 409–410
Pushing, 414–415

Q

Quality circles, 761
Question-and-answer (Q&A) sessions, 720

R

Radial deviation, 397
Radiation
ionizing and nonionizing, 320
in laboratories, 619, 634–638
types of, 634–635
Radiation safety
contamination control and, 637
detection instruments and, 637–638
exposure control and, 637
hazards and, 634–635
nonionized, 620, 636–638
principles of, 634
programs for, 637
protection methods for, 635
standards for, 636
training in, 638
Radiation safety officer (RSO), 637
Radiators, 591
Radioactive materials, 447–449
Radium, 379
Radon, 378
Railroad industry
accidents resulting in death in, 6, 7
safety programs in, 6
workers' compensation laws and, 186–187
Ramazzini, Bernardino, 393

Ramps, 657–658
Rea, Mark, 602
REACH (European Union), 333
Reactive culture, 37
Reasonable accommodations, 532–533
Recall, 701
Recognition, 691–692. *See also* Awards
Recommended weight limit (RWL), 412, 414
Recording and Reporting Occupational Injuries and Illness Standards, 121–126, 266
Record keeping. *See also* Incident records; Incident reports
    contractors and, 652–653
    environmental management and, 336
    on indoor air quality, 371
    injury and illness, 121, 122
    process of, 266, 267
    requirements for, 57–58, 121, 266, 284, 550
    tools for, 276–281
    types of data for, 269, 276
    of workplace violence, 481
Recordkeeping Handbook (Occupational Safety and Health Administration), 268, 284
Recovery, 654
Recovery time, 415
Recreational programs, 163
Recycling
    hazards related to, 502
    in waste minimization programs, 337–338
Refuse collection, 503
Regulations. *See also* Standards
    air pollutant, 380–382
    alcohol and drug testing, 435
    chemical safety, 626–627
    coverage of OSHA by state, 98, 99
    environmental, 330–332, 335
    future outlook for, 22
    hazard identification and preparedness, 446
    laboratory, 619
    pollutant-related, 380–382
    SH&E professionals and, 30
    ventilation, 382–383
    waste disposal, 501
    workplace violence prevention, 482–483
Regulatory compliance inspections, 224–225
Regulatory directives, for system safety, 200–201
Rehabilitation
    medical, 192–193, 305–306, 308
    vocational, 193–194
    workers' compensation and, 192–194, 197
Rehabilitation Act. *See* Vocational Rehabilitation Act of 1973
Rehabilitation Act Amendments of 1974, 529–530
Reinforcement
    behavior, 689–690, 693–694
    positive, 699–700
Reinspections, 59. *See also* Inspections
Reliability, 685
Remodeling, indoor air quality and, 364
Renovation, indoor air quality and, 364
Repair shop safety practices. *See* Vehicle repair shop safety practices

Repeated violations, 62
Repetitive work
    ergonomics and, 398, 415
    in laboratories, 644–645
    in offices, 611
Reporting culture, 38
Reports, safety, 776–777
Reproductive health, 309
Research, NIOSH and, 51
Research and Statistical Services group, 16
Reservoirs, 501
Resource Conservation and Recovery Act (RCRA), 16, 330, 335, 337, 619
Resource libraries, 721–722
Respiratory protection program
    employee training and, 107
    file testing and, 107
    function of, 78–79, 105, 107
    labels and, 105, 106
    maintenance, storage, and replacement and, 107–108
    medical evaluation and, 107
Respiratory protection standard (OSHA), 72, 78–79, 105–108, 537
Respondeat superior, 138
Responsible Care Program, 332
Restricted work activity, 125
Retail/service industries
    audit and inspection program in, 554, 556
    emergency preparedness in, 560–561
    ergonomic issues in, 560
    fleet safety program in, 559–560
    health and safety management program and, 548–549
    hiring and placement in, 553–554
    liability issues in, 553–562
    materials handling in, 556, 558, 559
    OSHA regulations and, 549–553
    risk control in, 554–558
Retail workers, 481
Return-to-work examinations, 305
Review tests, 724
Right to Privacy Act, 576
Risk
    assumption of, 6, 654–655
    explanation of, 201
    hazard ranking by, 146, 147
    in retail/service industries, 554
    SH&E professionals and, 29
Risk assessment, 147
Risk assessment code (RAC), 147
Risk factors
    ergonomic, 396–401, 405–406, 421–422
    for musculoskeletal disorders, 405, 406
    workplace violence, 469–470
Risk management development tree, 202
Risk mitigation. *See* Ergonomic risk mitigation
Rodent contamination, 503
Role-playing, 720
Roosevelt, Theodore, 6, 185
Routine job function, 125
Rumors, 444
Russell Sage Foundation, 6

S

Sabotage, 444
Safe Drinking Water Act (SDWA), 331
Safe Driver Award (National Safety Council), 768
Safety
    behavior-based, 43–44
    explanation of, 201
    during incident investigations, 231–232
    individuals with disabilities and, 526
    legislation addressing industrial, 16–17
    psychological factors in, 682–683
    training media preparation, 752–753
Safety and health committees
    background of, 13
    joint union/management, 15
    labor unions and, 14–15
    responsibilities of, 159, 614
    safety awareness program role of, 761
    scope of, 14
Safety and health training. *See also* Education; Training; Training media
    benefits of, 706–707
    in bloodborne pathogen handling, 633
    confined-space work, 74, 120
    driver, 582, 583
    for emergencies, 73, 452–453
    employee assistance programs and, 436
    energy control, 76
    for environmentally sensitive substances handling, 336–337
    ergonomic, 403, 404, 422
    fire-fighter, 461
    for formaldehyde exposure reduction, 82
    function of, 706
    hazardous chemical exposure, 71, 72, 103
    hazardous chemical handling, 117
    in indoor air quality management, 360, 361
    machine use, 110
    National Safety Council courses, 10
    new employee, 725–726
    NIOSH and, 51
    nontraining solutions and, 707–708
    powered industrial truck operation, 112–113
    radiation safety, 638
    resources for, 726
    respirator use, 107
    for vehicle repair personnel, 591–592
    for workplace violence prevention, 478
Safety and health training programs
    analysis/identification aspect of, 710–712
    components of, 708–710
    design of, 712–715
    development of, 722–723
    evaluation and maintenance of, 724–725
    implementation of, 723–724
    methods for, 716–722
    models of, 708
    resources for, 726
Safety awareness programs
    activity selection for, 757–759
    campaigns and, 775, 778–779
    company publications and, 784–786
    contests and, 764–772
    courses and demonstrations and, 776
    displays and exhibits and, 774–775
    importance of, 756–757
    meetings and, 761–763
    posters and, 772–774
    public address systems and, 777
    publications and, 776–777
    publicity and, 779–781, 786
    quality circles and safety circles and, 761
    safety committees and observers and, 761
    staff functions in, 759–761
    suggestion awards and, 778
    suggestion systems and, 777–778
    unique safety ideas and, 775–776
Safety circles, 761
Safety climate
    job satisfaction and, 38, 40
    safety culture and, 35 (*See also* Safety culture)
    survey of, 38, 41
Safety climate perception surveys, 38
Safety Codes Correlating Committee, 8
Safety culture
    background of, 36
    behavior-based safety and, 43–44
    benefits of, 549
    elements of, 34
    group behavior and, 41
    levels of positive, 37–38
    measurement of, 38–41
    organizational change and, 41–43
    trust and, 40–41
Safety culture perception surveys, 38
Safety data sheets (SDSs)
    availability of, 297, 320
    for hazardous chemicals, 69–71, 102, 103, 622
    regulations related to, 360
Safety devices, for vehicles, 582, 584–587
Safety engineering programs, in colleges and universities, 9
Safety, health, and environmental (SH&E) audits
    current practices in, 169
    development of, 168–169
    enterprise-related, 177–178
    field work step in, 171–173
    follow-up in, 173
    future outlook for, 178–180
    government-related regulatory, 175–177
    planning process for, 169–171
    process for, 169
    program design for, 174–175
    program organization and staffing for, 175
    protocols for, 173
    standards for, 178–179
    working papers for, 173–174
Safety, health, and environmental professionals (SH&Es). *See also* Safety professionals
    benefits to management of, 27–28
    consulting and expert witness roles of, 28
    economic issues and, 26

ethical issues and, 26–27
future opportunities for, 29–30
globalization and, 27
personal liability issues and, 28
professional liability and, 28–29
role of, 26
training for, 28
Safety incentives, workers' compensation and, 196
Safety management
elements of, 548
evaluation and corrective action in, 549
implementation and operation of, 548–549
potential terrorist attacks and, 677
safety culture and, 549
Safety movement
achievements of, 12–22
cooperation in, 10
current issues in, 20–21
evaluation of, 9–10
future outlook for, 21–22
Safety professionals. *See also* Safety, health, and environmental professionals (SH&Es)
in incident investigation, 240
individuals with disabilities and role of, 533–534
inspections conducted by, 218
knowledge of, 11–12
public interest and, 21
record-keeping role of, 268
responsibilities of, 533–534
safety awareness program role of, 759–760
training and, 12, 708
Safety programs. *See also specific safety programs*
off-the-job, 9, 159–163
record-keeping requirements for, 57–58, 121, 266, 284
vehicle, 566–594
in World War II period, 8–9
Safety resources
cooperation and, 10
goodwill and, 11
knowledge and experience and, 10
professionalism and, 11
scientific advancements and, 11–12
Safety sampling, 133–134
Safety/Sanitation Checklist, 557–558
Safe Worker Award Program (National Safety Council), 765, 768–769
Salvation Army, 464
Santayana, George, 233
Satellite accumulation area (SAA), 624
"Save Manpower for Warpower" (National Safety Council), 9
Scaffolding standard, 114–115
Schein, E. H., 34
Scintillation detectors, 637
Screening threshold quantity (STQ), 677
Sealed sources of radiation, 448
Seating, for media presentations, 748–749
Secondary prevention, 308, 520
Securities and Exchange Commission (SEC), 345, 346
Segmental vibration, 399
Self-audits, 177

Self-Inspection Security Checklist, 475–476
Seminars, 722
Sentinel health events, 518
Septic tanks, 502
Serious violations, 61–62
Service Contract Act (1965), 49
Service industries. *See* Retail/service industries
Severity measures, 285–286
Sewers, 501
SH&Es. *See* Safety, health, and environmental professionals (SH&Es)
Shiftwork
effects of, 310–311
medical issues related to, 313
rescheduling efforts for, 312–313
solutions to, 312
trends in, 310
Shipping, safety issues related to, 158–159
Shopping carts, 657
Short-term exposure limit (TLV-STEL)
explanation of, 227
formaldehyde, 81
Short-term memory, 702
Showers
in employee facilities, 506–507
in laboratories, 625
Shutdowns, emergency preparedness for, 446–447
Sick-building syndrome (SBS), 355
Simulation, 720
Sitting neutral posture, 496
Situational factors, incident causation and, 143–145
16 basic desires theory, 686, 687
Skin contamination, 634
Skinner, B. F., 43
Slip meters, 658
Slogan contests, 770
Small Business Administration (SBA), 65
Small businesses
loans for, 65
safety and health problems in, 13
workplace violence and, 484
SMART system, 42
Smart Training: The Manager's Guide to Training for Improved Performance (Carr), 713
Smoking. *See* Tobacco smoke exposure
Soaps, 505
Social Security, 191
Social service workers, 481–482
Solicitor of Labor, 50
Solid waste disposal, 502–503
Solubility, of chemical compounds, 227
Spills, reporting plan for, 336
Spores, 376
Stachybotrys atra, 376
Stachybotrys chartarum, 376
Stack effect, 367, 668
Staff Accounting Bulletin No. 92 (SAB 92), 345, 346
Stairways, 601, 604, 662–663
Standards
access to employee exposure and medical records, 68–69

asbestos exposure, 79–81
bloodborne pathogens, 79
confined-spaces, 73–74, 118–120
control of hazardous energy (lockout/tagout), 74–76, 103–105, 552–553
electrical safety, 110–111
ergonomics, 401–402
excavations, 115
fall-protection, 101
formaldehyde exposure, 81
hazard communication, 55, 69–71, 101–103
hazardous chemicals in laboratories exposure, 71–72
individuals with disabilities and, 536–537
injury and illness recording and reporting, 121–126
international, 17–19
ISO 9000 Series, 17–18, 90, 346
ladder, 113–114
lead exposure in construction, 82
machine-guarding, 108–110
OSHA, 55–57
OSHA and, 55–56, 100–101
personal protective equipment, 76–77
powered industrial trucks, 111–113
private sector and, 56
process safety management of highly hazardous chemicals, 77–78, 116–118
respiratory protection, 72, 78–79, 105–108, 537
scaffolding, 114–115
for service industry, 549–550
variances from, 56–57
variances from OSHA, 56–57
violations of, 61–63
walking/working surfaces, 116
workplace emergencies, 72–73
Standing neutral posture, 397
States
disability accommodations by, 530, 540
OSHA regulations and, 65–67, 82, 98, 99
Static work, 410–412
Statistical method of analysis, 240
Statutory penalties, 139–140
Steel industry, safety programs in, 6
Steiger, William, 48
Storage
of chemicals, 623–624
of drinking water, 501
hazardous material, 454
office, 604, 607–608
Stress, individual reactions to, 683
Stretchers, 300–301
Strict liability, 141
Student Trip Report, 583
Substance abuse
Department of Transportation testing requirements for, 592–594
employees assistance programs and, 309–310, 435
shiftwork and, 311
Substance abuse professionals (SAPs), 435
Suggestion awards, 778
Suggestion boxes, 778
Suggestion committee, 778

Suggestion systems, 777–778
Superfund. *See* Comprehensive Environmental Response, Compensation, and Liability Act (CERCLA) (Superfund)
Superfund Amendments and Reauthorization Act (SARA or Community Right-to-Know), 16, 177, 178, 330
Supervisor Incident Investigation Report (National Safety Council), 277–279
Supervisors/foremen
employment assistance programs and, 429, 431
incident reports for, 282–283
inspections conducted by, 218
record-keeping role of, 268, 270–271
relationship between inspectors and, 220
workplace violence prevention and, 470–471
Supination, 397
Supplementary Record of Occupational Injuries and Illnesses Form, 268, 272–273
Suppliers, 346–347
Support groups, for individuals with disabilities, 537
Supreme Court, U.S.
on employment discrimination, 526–527
on fetal protection policies, 309
on greenhouse gas regulation, 339
on workers' compensation, 6, 185
Surface radiation, 634
Surveillance. *See also* Medical surveillance
active, 406–407
medical, 308
passive, 406
Surveys
employee, 692–694
to identify potential for workplace violence risk, 472, 473
Swimming pools, 655, 670–671
System, 201
System safety. *See also* Hazard analysis
design of, 201
ethical motives in, 200
legal vulnerability in, 200
mishap models and, 201–203
overview of, 200
profit motive in, 200
regulatory directives in, 200–201
terminology in, 201
System safety development tree, 202
System Safety Society, 11

T

Tabletop drills. *See* Disaster drills
Tailgaiting, 718
Taxi drivers, 482
Taylor, Frederick W., 393
Technological advances
public interest in, 21
record-keeping tools and, 277
safety issues and, 11–12
SH&E professionals and, 29
Teleprompters, 749
Television interviews, 783

Temperature, indoor, 365–366
Temperature extremes
    effect on workers, 415–417
    evaluation and control of, 321
Temporary partial disability, 194
Temporary total disability, 194
Terrorism, 469, 676–678
Tertiary prevention, 308, 520
Testing
    alcohol and drug, 592–594
    in training programs, 721, 724–725
Textile industry, 5, 6
Theory X, 687
Theory Y, 687
Thermal comfort requirements, 365–366
Third-party injuries, 654–655
Threat assessment teams, 472
Three Mile Island incident, 19
Threshold Limit Values (TLVs)
    categories of, 227–228
    for vibration, 399
Threshold Limit Values (TLVs) *and Biological Exposure Indices* (BEIs) (American Conference of Governmental Industrial Hygienists), 326, 399
Time-weighted average (TLV-TWA)
    for asbestos exposure, 80
    explanation of, 227
    for formaldehyde exposure, 81
    tests to define, 228, 229
Tires, 588–589
Title VII (Civil Rights Act of 1964), 483
Tobacco smoke exposure, 374–375, 536
Toilets, 507
Tool checklist, 407
Tool design, 421
Top-down audits, 177
Tornadoes, emergency preparedness for, 443–444
Torts, 138, 654
Total quality management (TQM)
    explanation of, 90
    incentives and, 691
    motivation and, 688–689
    surveys and, 692
Toxicity
    explanation of, 226, 322
    hazard vs., 322–323
    measurement for, 226
    molds and, 376–377
Toxic substances. *See also* Hazardous chemicals
    explanation of, 68
    ingestion of, 323
    inhalation of, 323
    injection of, 323
    pesticides as, 374
    routes of entry into body of, 323
    skin absorption of, 323
Toxic Substances Control Act (TSCA), 16, 139, 331
Toxoplasma spp., 627
Toynbee, A., 5
Toynbee, Arnold J. (nephew), 5
Trade secrets
    disclosure of chemical identity in, 68–69, 103
    process safety management and, 118
Traffic safety
    off-the-job safety programs and, 162
    in vehicle repair shops, 591
Traffic signs, 656
Trainers, 732, 733, 735
Training. *See also* Education; Safety and health training; Safety and health training programs
    benefits of, 706–707
    in bloodborne pathogen handling, 633
    computer-assisted, 721
    computer-based, 743–745
    confined-space work, 74, 120
    driver, 582, 583
    for emergencies, 73, 452–453
    employee assistance programs and, 436
    energy control, 76
    energy control procedures, 105
    for environmentally sensitive substances handling, 336–337
    ergonomics, 403, 404, 422
    explanation of, 706
    fire-fighter, 461
    first-aid, 298, 299
    formaldehyde exposure reduction, 82
    hazardous chemical exposure, 71, 72, 103
    hazardous chemical handling, 117
    hazardous materials responders, 455
    in indoor air quality management, 360, 361
    machine use, 110
    National Safety Council courses, 10
    need for safety professionals, 12
    needs analysis for, 710–711
    NIOSH and, 51
    nontraining solutions and, 707–708
    on-the-job, 716–718
    podcast, 721
    powered industrial truck operation, 112–113
    radiation safety, 638
    resources for, 726
    respirator use, 107
    transfer of, 701–702
    for vehicle repair personnel, 591–592
    video-based, 721
    for workplace violence prevention, 478
Training cycle system, 709, 710
Training media
    audience characteristics and, 733
    color use in, 741
    comprehension levels and formats for, 732
    computer-based, 743–745
    cost of, 752
    drawings and sketches in, 741
    features of, 734
    function of, 732–733
    graphic formats for, 737–739
    graphs in, 739–741
    objectives for, 733
    organization of, 737
    safety considerations for, 752–753

supplemental materials to enhance, 746–747
text charts as, 737
trainer's role and, 733, 735
types of, 742
video, 741–743
Training media presentations
evaluation of, 750–752
guidelines for, 747–748
models' use during, 735–737
props for, 749–750
rehearsals for, 750
seating arrangements for, 748–749
software for, 737–738, 741
Transfer of training, 701–702
Transportation
company-owned vehicles for, 669–670
courtesy cars for, 669
in emergencies, 462–463
of radioactive materials, 638
regulations related to, 335
Transportation safety programs. See also Vehicle safety programs
alcohol and drug testing and, 592–594
overview of, 566
Travel, illness or injury during, 124, 188–189
Treatment, storage, and disposal (TSD) facility workers, 454
Triangle Fire (1911), 6
Trichoderma, 376
Trust, safety culture and, 40–41
Two-factor theory (Herzberg), 687

U

U charts, 287, 289
Ulnar deviation, 397
Ultraviolet (UV) lights, 630
Underground Storage Tank (UST) regulations, 330
Undue hardship, 532
Unified Soil Classification System, 338
Uninsured costs, 255–256
Unintentional injuries
prevention of, 9
use of term, 132, 238
Unions. See Labor unions
Unsafe practices/procedures, incidents and, 142–143
Unsealed sources of radiation, 448
Urea-formaldehyde foam insulation (UFFI), 373
Urea-formaldehyde (UF) resins, 373
U.S. Army Corps of Engineers, 442, 464
U.S. Chamber of Commerce, 185
User seal check, 107
U.S. Geological Survey, 442
U.S. National Response Center, 336
U.S. Public Health Service, 464
Utility Air Regulatory Group v. EPA, 339

V

Vacation/holiday programs, 163
Validity, 685
Values, espoused, 34
ValuJet, 200

Variances
from mine safety standards, 86
from OSHA standards, 56–57
permanent, 57
Vehicle repair shop safety practices
battery charging and, 590
equipment maintenance and, 591
fire protection and, 589
gasoline safety and, 590–591
lubrication and service operations and, 589
personnel training in, 591–592
tire operations and, 588–589
wash rack operations and, 589–590
Vehicles
battery charging for, 590
collision costs involving, 566, 567
fire protection for, 589
gasoline safety for, 590–591
lubrication and service operations for, 589
safety devices for, 582, 584–587
servicing and maintaining, 588
tire operations and, 588–589
wash rack operations for, 589–590
Vehicle safety programs
collision-reporting procedures in, 570–574
driver interviews in, 570
driver record cards for, 570, 575
driver safety and, 567–570
driver selection in, 575–578
driver training in, 582, 583
drug and alcohol testing and, 592–594
fleet collision frequency controls in, 575
function of, 566–567
information gathering for, 576–582
overview of, 566
preventive maintenance in, 585–588
repair shop and wash rack personnel training in, 591–592
responsibility in, 567
in retail/service industries, 559–560
Ventilation. See also HVAC systems
indoor air quality and, 365, 366, 379
in laboratories, 624
office, 602
in personal service facilities, 507
pesticides and, 374
regulation of, 382–383
Veterans with disabilities, 531
Vibration
ergonomics and, 398–399
evaluation and control of, 321
Victoria Occupational Health and Safety Act (Australia, 1985), 16
Video-based training, 721
Video media
characteristics of, 741–742
prerecorded, 742
production of, 742–743
Vietnam Era Veterans' Readjustment Assistance Act of 1974, 531
Violations
chemical and hazardous waste, 335–336

citations for, 63
egregious policy for, 63
of OSHA standards, 61–63
penalties for, 63
Violence. *See* Domestic violence; Workplace violence; Workplace violence prevention
Virginia Workers' Compensation Act, 189
Vision tests, 306
Visual displays, 418
Vocational Rehabilitation Act of 1973, 529–532
Vocational rehabilitation programs, workers' compensation and, 193–194
Volatile organic compounds (VOCs), air quality and, 366, 373
Voluntary Protection Programs (VPPs), 54

W
Wage loss theory, 194, 195
Wainwright Law (1910), 6
Walking/working surfaces standards, 116, 551–552
Walsh-Healy Act, 16
Warden service, 462
Warm rooms, 644
Washing facilities, 504–505
Wash rack operations, 589–590
Washrooms, 504–506
Waste disposal
    building drains and sewers and, 501
    insect, rodent, and nuisance bird infestations and, 503
    in laboratories, 624, 645–646
    refuse collection and, 503
    regulation of, 501
    solid-waste, 502–503
    wastewater, 502
Waste management, chemical, 624, 645–646
Waste minimization programs, 337–338
Wastewater, 442, 502
Water. *See* Drinking water
Water purification, 500–501. *See also* Drinking water
Weapons, workplace violence and, 469, 484
Weather-related emergencies, 449–450
Wellness programs, 307–308
Wells, 498–500
Wet-bulb globe temperature (WBGT), 416
Wheelchair accommodations, 539–543
Whitney, Eli, 4
Whole-body vibration (WBV), 399
Whole person theory, 194, 195
Willful actions, 140–141
Willful violations, 61
Williams, Harrison A., 48
Williams-Steiger Act. *See* Occupational Safety and Health Act (OSH Act) (1970)
Windchill factors, 449
Withdrawal orders, mine safety and, 87–88
Women, 308–309
Work
    energy cost of, 408, 411
    restricted, 125
    static, 410–412

Worker Involvement in Hazard Control (Labor and Management Council), 761
Workers' compensation
    accident prevention and reduction as objective of, 186
    administration of, 191–192
    background of, 6, 184–185
    benefits of, 189–191
    black lung disease and, 189
    civil tort liability lawsuits and, 138
    coverage of, 186–187
    degree of disability and, 194–196
    economic losses of workers and, 184
    ergonomics and, 392, 394–395
    exclusive remedy for work-related disabilities and, 187–188
    experience modification rates for, 651
    hearing loss and, 189
    income replacement and, 186, 189–190
    for individuals with disabilities, 190, 191
    injuries covered by, 188–189
    insurance incentives of, 195–196
    legislation addressing, 184–187
    management of, 196–197
    medical benefits and, 190–191
    objectives of, 185–186
    occupational disease and, 189
    rehabilitation and, 192–194, 197
    subrogation of, 138
    Supreme Court on, 6, 185
Workers' Compensation Act (Wisconsin), 6, 185
Work/family services, 435–436
Work force, 308–309
Working papers, audit, 173–174
Work Injury and Illness Rates (National Safety Council), 503
Workplace
    ergonomics and, 420–421 (*See also* Ergonomics; Ergonomics programs)
    gender and, 308–309
    lighting in, 312
    music in, 312
    security analysis of, 472–474
Workplace Hazardous Management Information System (WHMIS) (Canada), 16
Workplace inspections. *See* Inspections
Workplace violence
    assessment of threats of, 478–479
    categories of, 469
    discipline for, 477
    domestic violence and, 483–484
    government policies and, 468
    high-risk occupations for, 468, 481–482
    incident reporting system for, 479–480
    nature of, 468–469
    record keeping and, 488
    regulatory requirements and, 482–483
    in retail/service industries, 551–552
    risk factors for, 469–470
    in small businesses, 484
    statistics on, 468

termination for, 477
weapons and, 469, 484
Workplace violence prevention
    domestic violence prevention policy and, 485–486
    factors in, 470
    hazard prevention and control and, 474, 476–477
    policy statements for, 471–472
    procedures for, 477
    program evaluation for, 480
    in retail/service industries, 561–562
    threat assessment team for, 472
    training needs for, 477–478
    worker and manager commitment to, 470–471
    workplace analysis and, 472–476
Workstations
    ergonomics and, 408–411, 417–418, 420, 644
    laboratory, 644
World War II
    human-factors studies and, 393, 394
    safety programs and policies during, 8–9

Y

Youth activities, 162–163